FOR STUDENTS

- Career Opportunities
- Career Fitness Program
- Becoming an Electronics Technician
- Free On-Line Study Guides (companion web sites)

FOR FACULTY*

- Supplements
- On-Line Product Catalog
- Electronics Technology Journal
- Prentice Hall Book Advisor

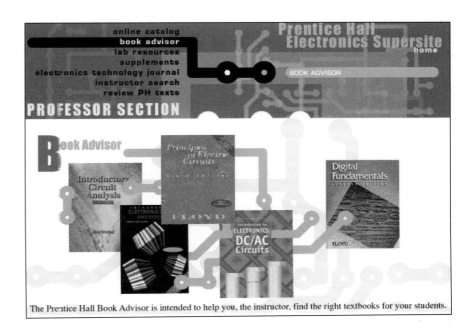

The Prentice Hall Book Advisor is intended to help you, the instructor, find the right textbooks for your students.

Please contact your Prentice Hall representative for passcode

Introductory DC/AC Electronics

Introductory DC/AC Electronics

Fifth Edition

Nigel P. Cook

Prentice
Hall

Upper Saddle River, New Jersey
Columbus, Ohio

Library of Congress Cataloging-in-Publication Data
Cook, Nigel P.
   Introductory DC/AC electronics / Nigel P. Cook.—5th ed.
     p.   cm.
   Includes index.
   ISBN 0-13-031085-9
   1. Electronics.   I. Title.

TK7816.C653   2002
621.381—dc21

                                                        00-053746

Editor in Chief: Stephen Helba
Acquistions Editor: Scott J. Sambucci
Associate Editor: Kate Linsner
Production Editor: Rex Davidson
Design Coordinator: Karrie Converse-Jones
Text Designer: Rebecca Bobb
Cover Designer: Rod Harris
Cover Photo: PhotoDisc, Inc.
Illustrations: Rolin Graphics
Production Manager: Pat Tonneman
Project Management: Holly Henjum, Clarinda Publication Services

This book was set in Times Roman by The Clarinda Company and was printed and bound by R. R. Donnelley & Sons Company. The cover was printed by Phoenix Color Corp.

Electronics Workbench™ and MultiSim™ are trademarks of Electronics Workbench.

10 9 8 7 6 5 4 3 2 1
ISBN: 0–13–031085–9

To Dawn, Candy, and Jon

Books by Nigel P. Cook

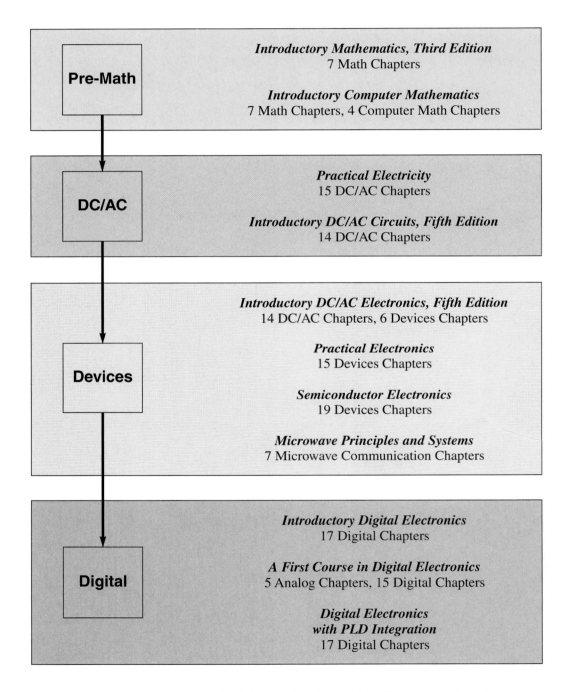

Pre-Math

Introductory Mathematics, Third Edition
7 Math Chapters

Introductory Computer Mathematics
7 Math Chapters, 4 Computer Math Chapters

DC/AC

Practical Electricity
15 DC/AC Chapters

Introductory DC/AC Circuits, Fifth Edition
14 DC/AC Chapters

Devices

Introductory DC/AC Electronics, Fifth Edition
14 DC/AC Chapters, 6 Devices Chapters

Practical Electronics
15 Devices Chapters

Semiconductor Electronics
19 Devices Chapters

Microwave Principles and Systems
7 Microwave Communication Chapters

Digital

Introductory Digital Electronics
17 Digital Chapters

A First Course in Digital Electronics
5 Analog Chapters, 15 Digital Chapters

*Digital Electronics
with PLD Integration*
17 Digital Chapters

For more information on any of the other textbooks by Nigel Cook, see his web page at www.prenhall.com/cook or ask your local Prentice Hall representative.

Preface

TO THE STUDENT

The early pioneers in electronics were intrigued by the mystery and wonder of a newly discovered science, whereas people today are attracted by its ability to lend its hand to any application and accomplish almost anything imaginable. If you analyze exactly how you feel at this stage, you will probably discover that you have mixed emotions about the journey ahead. On one hand, imagination, curiosity, and excitement are driving you on, while apprehension and reservations may be slowing you down. Your enthusiasm will overcome any indecision you have once you become actively involved in electronics and realize that it is as exciting as you ever expected it to be.

ORGANIZATION OF THE TEXTBOOK

This textbook has been divided into four basic parts. Chapters 1 through 3 introduce you to the world of electronics and the fundamentals of electricity. Chapters 4 through 7 cover direct current, or dc, circuits; Chapters 8 through 14 cover alternating current, or ac, circuits; and Chapters 15 through 20 cover semiconductor principles, devices, and circuits.

The material covered in this book has been logically divided and sequenced to provide a gradual progression from the known to the unknown, and from the simple to the complex.

ANCILLARIES ACCOMPANYING THIS TEXT

The following ancillaries accompanying this text provide extensive opportunity for further study and support:

- ■ *Electronics Workbench/MultiSim CD-ROM.* Packaged with each copy of this text, the CD-ROM contains over 100 circuits from the text, created in both Electronics Workbench Version 5 and Electronics Workbench MultiSim Version 6. Forty (40) of these circuits can be simulated in the free demonstration version of MultiSim. The remaining circuits require that the user have the Electronics Workbench software, as this software is not included. Electronics Workbench software can be obtained by contacting your local bookstore, or by visiting http://www.electronicsworkbench.com.

- ■ *Laboratory Manual.* Co-authored by Nigel Cook and Gary Lancaster, the lab manual offers numerous experiments designed to translate all of the textbook's theory into practical experimentation.

- ■ *Practical Circuit Applications in DC/AC Electronics, with Study Wizard CD-ROM,* by Bradley Thompson. This important study guide lends additional support, including a math review and many practical circuit applications.

- ■ *Companion Website,* located at http://www.prenhall.com/cook. Numerous interactive study questions are provided on this site to reinforce the concepts covered in the book.

To complete the ancillary package, the following supplements are essential elements for any instructor using this text for a course:

- ■ *Instructor's Solutions Manual*
- ■ *Solutions Manual to Accompany Laboratory Manual*
- ■ *Instructor's Answer Key to Practical Circuit Applications in DC/AC Electronics*
- ■ *PowerPoint™ Transparencies.* This CD-ROM includes a full set of **lecture presentations** as well as transparencies for all schematics appearing in the text.
- ■ *Test Item File*
- ■ *PH Custom Test Manager*

CIRCUIT SIMULATION CD-ROM USING EWB VERSION 5 AND EWB MULTISIM VERSION 6

In the back of this book is a CD-ROM containing the circuit simulation software *Electronics Workbench*® *(EWB)*. Using the demo on this CD, you can simulate forty circuits taken directly from this text. The EWB icon shown here in the margin indicates which circuits in this text have been prebuilt and stored on the CD, ready for simulation.

DEVELOPMENT, CLASS TESTING, AND REVIEWING

The first phase of development for this manuscript was conducted in the classroom with students and instructors as critics. Each topic was class-tested by videotaping each lesson, and the results were then evaluated and implemented. This invaluable feedback enabled me to fine-tune my presentation of topics and instill understanding and confidence in the students.

The second phase of development was to forward a copy of the revised manuscript to several instructors at schools throughout the country. Their technical and topical critiques helped to mold the text into a more accurate form.

The third and final phase was to class-test the final revised manuscript and then commission the last technical review in the final stages of production.

ACKNOWLEDGMENTS

My appreciation and thanks are extended to the following instructors who have reviewed and contributed greatly to the development of this textbook: Venkata Anadu, Southwest Texas State University; Don Barrett, Jr., DeVry Institute of Technology; Lynnette Garetz, Heald College; Joe Gryniuk, Lake Washington Technical College; Jerry M. Manno, DeVry Institute of Technology; George Sweiss, ITT Technical Institute; and Bradley J. Thompson, State University of New York College of Technology at Alfred.

Nigel P. Cook

Timeline Photo Credits

P. 2, Charles Steinmetz, General Electric Company; p. 10, James Maxwell, American Institute of Physics/Emilio Segre Visual Archives; p. 23, André Ampère, American Institute of Physics/Emilio Segre Visual Archives; p. 26, Alessandro Volta, American Institute of Physics/Emilio Segre Visual Archives; p. 29, Benjamin Franklin, EMG Education Management Group; p. 50, Seymour Cray, Cray Inc.; p. 64, Georg Ohm, Library of Congress; p. 80, James Joule, Library of Congress; p. 82, James Watt, Library of Congress; p. 98, Steve Jobs and Steve Wozniak, Apple Computer, Inc.; p. 144, Grace Murray Hopper, Navy Visual News Service; p. 174, John Napier, Steven S. Nau/Pearson Education/PH College; p. 216, Isaac Newton, Library of Congress; p. 250, Sir Charles Wheatstone, Corbis; p. 306, Theodore Maiman, Archive Photos; p. 329, Guglielmo Marconi, Hulton Getty/Archive Photos; p. 361, Vladymir Zworykin, Keystone View Company/Hulton Getty/Archive Photos; p. 382, Charles Babbage, Library of Congress; p. 385, Robert Moog and Jon Weiss, courtesy of Bob Moog, Big Briar, Inc., Asheville, NC; p. 430, David Packard, UPI/Corbis; p. 470, Michael Faraday, Library of Congress; p. 506, Thomas Edison, Library of Congress; p. 510, Joseph Henry, Michael A. Gallitelli/Metroland Photo, Inc./Pearson Education/PH College; p. 552, Kenneth Olson, Digital Equipment Corporation; p. 586, William Shockley, courtesy of Lucent Technologies Bell Labs Innovations; p. 634, Alan Turing, Princeton University Archives, Department of Rare Books and Special Collections, Princeton University Library; p. 636, Konrad Zuse, Karsten Thielker/AP/Wide World News; p. 639, Sir Joseph Thomson, Library of Congress; p. 640, Dr. John Mauchly and Dr. J. Presper Eckert, Unisys Corporation; p. 662, René Descartes, Library of Congress; p. 728, Karl Friedrich Gauss, Steven S. Nau/Pearson Education/PH College; p. 790, Nikola Tesla, Nikola Tesla Museum, Belgrade, Yugoslavia; p. 840, Jack Kilby, Corbis/Sygma; p. 872, Gottfried Wilhelm von Liebniz, Library of Congress; p. 875, John Baird, Hulton Getty/Archive Photos; p. 879, Ted Hoff, Intel Corporation Museum Archives and Collections.

Photos in Introduction to Electronics, pp. xxv–xxxii, courtesy of Hewlett-Packard Company.

NEW! The accompanying **CD-ROM** includes over 100 circuits from the text, created in both Electronics Workbench Version 5 and Electronics Workbench MultiSim Version 6. Forty of these circuits can be simulated in the free demo version of MultiSim.

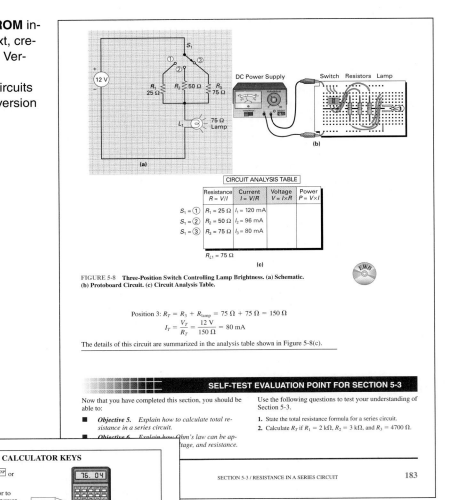

CIRCUIT ANALYSIS TABLE

	Resistance $R = V/I$	Current $I = V/R$	Voltage $V = I \times R$	Power $P = V \times I$
$S_1 = ①$	$R_1 = 25\ \Omega$	$I_1 = 120\ mA$		
$S_1 = ②$	$R_2 = 50\ \Omega$	$I_2 = 96\ mA$		
$S_1 = ③$	$R_3 = 75\ \Omega$	$I_3 = 80\ mA$		
	$R_{L1} = 75\ \Omega$			

(c)

FIGURE 5-8 **Three-Position Switch Controlling Lamp Brightness. (a) Schematic. (b) Protoboard Circuit. (c) Circuit Analysis Table.**

$$\text{Position 3: } R_T = R_3 + R_{lamp} = 75\ \Omega + 75\ \Omega = 150\ \Omega$$
$$I_T = \frac{V_T}{R_T} = \frac{12\ V}{150\ \Omega} = 80\ mA$$

The details of this circuit are summarized in the analysis table shown in Figure 5-8(c).

SELF-TEST EVALUATION POINT FOR SECTION 5-3

Now that you have completed this section, you should be able to:

■ **Objective 5.** Explain how to calculate total resistance in a series circuit.

■ **Objective 6.** Explain how Ohm's law can be applied to voltage, and resistance.

Use the following questions to test your understanding of Section 5-3.

1. State the total resistance formula for a series circuit.
2. Calculate R_T if $R_1 = 2\ k\Omega$, $R_2 = 3\ k\Omega$, and $R_3 = 4700\ \Omega$.

CALCULATOR KEYS

Name: Exponent entry key $\boxed{\text{EXP}}$ or $\boxed{\text{EE}}$

Function: Prepares calculator to accept next digits entered as a power-of-ten exponent. The sign of the exponent can be changed by using the change-sign key $(+/-)$.

Example: $76\ \boxed{\text{EXP}}4$

Press keys: $\boxed{7}\boxed{6}\boxed{\text{EXP}}\boxed{4}$

Display shows: $\boxed{76.\ 04}$

Example: 85×10^{-5}

Press keys: $\boxed{8}\boxed{5}\boxed{\text{EXP}}\boxed{5}\boxed{+/-}$

Display shows: $\boxed{85.\ -05}$

EXP   +/−

Scientific Notation
A widely used floating-point system in which numbers are expressed as products consisting of a number between 1 and 10 multiplied by an appropriate power of 10.

Engineering Notation
A widely used floating-point system in which numbers are expressed as products consisting of a number that is greater than 1 multiplied by a power of 10 that is some multiple of 3.

TIME LINE
As a small boy, James C. Maxwell (1831-1879) was persistently inquisitive. He built many scientific toys before he was 8. At the age of 14 he wrote a paper on how to construct oval curves, and at 18 two of his papers were published. The supreme achievement of this Scottish physicist, however, was to translate Michael Faraday's experiments into scientific notation. This set of mathematical equations, known as Maxwell's equations, shows the relationship between electricity and magnetism.

Scientific and Engineering Notation

As mentioned previously, powers of ten are used in science and technology as a shorthand due to the large number of zeros in many values. There are basically two systems or notations used, involving values that have exponents that are a power of ten. They are called **scientific notation** and **engineering notation.**

A number in *scientific notation* is expressed as a base number between 1 and 10 multiplied by a power of ten. In the following examples, the values on the left have been converted to scientific notation.

■ **EXAMPLE A:**

$$32{,}000 = 3.2000_. = 3.2 \times 10^4$$

Scientific notation

Decimal point is moved to a position that results in a base number between 1 and 10. If decimal point is moved left, exponent is positive. If decimal point is moved right, exponent is negative.

■ **EXAMPLE B:**

$$0.0019 = 0.001\,9 = 1.9 \times 10^{-3}$$

Scientific notation

■ **EXAMPLE C:**

$$114{,}300{,}000 = 1.14300000_. = 1.143 \times 10^8$$

Scientific notation

NEW! Extensive **First-Section Mini-Math Reviews** are included in their appropriate positions to overview and test the students on the math concepts needed for electronics concepts that follow.

NEW! An **Introduction to Electronics** section before Chapter 1 overviews the electronics industry. In addition, it details the Product Development Process and the responsibilities and requirements of different technician job types.

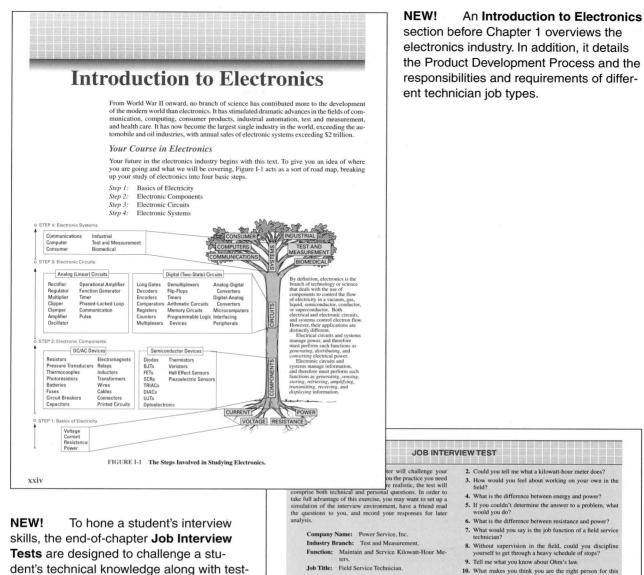

Introduction to Electronics

From World War II onward, no branch of science has contributed more to the development of the modern world than electronics. It has stimulated dramatic advances in the fields of communication, computing, consumer products, industrial automation, test and measurement, and health care. It has now become the largest single industry in the world, exceeding the automobile and oil industries, with annual sales of electronic systems exceeding $2 trillion.

Your Course in Electronics

Your future in the electronics industry begins with this text. To give you an idea of where you are going and what we will be covering, Figure I-1 acts as a sort of road map, breaking up your study of electronics into four basic steps.

Step 1: Basics of Electricity
Step 2: Electronic Components
Step 3: Electronic Circuits
Step 4: Electronic Systems

STEP 4: Electronic Systems

Communications	Industrial
Computer	Test and Measurement
Consumer	Biomedical

STEP 3: Electronic Circuits

Analog (Linear) Circuits

Rectifier	Operational Amplifier
Regulator	Function Generator
Multiplier	Timer
Clipper	Phased-Locked Loop
Clamper	Communication
Amplifier	Pulse
Oscillator	

Digital (Two-State) Circuits

Long Gates	Demultiplexers	Analog-Digital
Decoders	Flip-Flops	Converters
Encoders	Timers	Digital-Analog
Comparators	Arithmetic Curcuits	Converters
Registers	Memory Circuits	Microcomputers
Counters	Programmable Logic	Interfacing
Multiplexers	Devices	Peripherals

STEP 2: Electronic Components

DC/AC Devices

Resistors	Electromagnets
Pressure Transducers	Relays
Thermocouples	Inductors
Photoresistors	Transformers
Batteries	Wires
Fuses	Cables
Circuit Breakers	Connectors
Capacitors	Printed Circuits

Semiconductor Devices

Diodes	Thermistors
BJTs	Varistors
FETs	Hall Effect Sensors
SCRs	Piezoelectric Sensors
TRIACs	
DIACs	
UJTs	
Optoelectronic	

STEP 1: Basics of Electricity

Voltage
Current
Resistance
Power

By definition, electronics is the branch of technology or science that deals with the use of components to control the flow of electricity in a vacuum, gas, liquid, semiconductor, conductor, or superconductor. Both electrical and electronic circuits, and systems control electron flow. However, their applications are distinctly different.

Electrical circuits and systems manage power, and therefore must perform such functions as *generating, distributing,* and *converting* electrical power.

Electronic circuits and systems manage information, and therefore must perform such functions as *generating, sensing, storing, retrieving, amplifying, transmitting, receiving,* and *displaying* information.

FIGURE I-1   The Steps Involved in Studying Electronics.

xxiv

NEW! To hone a student's interview skills, the end-of-chapter **Job Interview Tests** are designed to challenge a student's technical knowledge along with testing their character traits with personal questions.

JOB INTERVIEW TEST

...ter will challenge your ...ou the practice you need ...re realistic, the test will comprise both technical and personal questions. In order to take full advantage of this exercise, you may want to set up a simulation of the interview environment, have a friend read the questions to you, and record your responses for later analysis.

Company Name: Power Service, Inc.
Industry Branch: Test and Measurement.
Function: Maintain and Service Kilowatt-Hour Meters.
Job Title: Field Service Technician.

1. What do you know about us?

Answers

1. Visit the company's web site before the interview to get an overall understanding of the company's ownership, production line, service, and support.
2. Section 2-5.
3. Discuss how you have worked alone on project assignments and other related work history.
4. Section 2-5.
5. Say that you would report the problem to your supervisor and ask for help.

2. Could you tell me what a kilowatt-hour meter does?
3. How would you feel about working on your own in the field?
4. What is the difference between energy and power?
5. If you couldn't determine the answer to a problem, what would you do?
6. What is the difference between resistance and power?
7. What would you say is the job function of a field service technician?
8. Without supervision in the field, could you discipline yourself to get through a heavy schedule of stops?
9. Tell me what you know about Ohm's law.
10. What makes you think you are the right person for this job?

6. Sections 2-3 and 2-5.
7. See introduction, Field Service Technician.
8. Discuss how you have juggled school, studying, and a job to get to the position you currently hold, and other related work experience.
9. Section 2-3.
10. Describe why you were first attracted to the company's advertisement, why it felt compatible with your career goals, and how the company's positive attributes listed in its web site matched your expecations.

NEW! Over 35 **Margin Timeline Photographs** highlight key historical milestones by industry entrepreneurs, connecting their discoveries to appropriate text topics.

Resistance and Power

Genius of Chippewa Falls

In 1960, Seymour R. Cray, a young vice-president of engineering for Control Data Corporation, informed president William Norris that in order to build the world's most powerful computer he would need a small research lab built near his home. Norris would have shown any other employee the door, but Cray was his greatest asset, so in 1962 Cray moved into his lab, staffed by 34 people and nestled in the woods near his home overlooking the Chippewa River in Wisconsin. Eighteen months later the press was invited to view the 14- by 6-foot 6600 supercomputer that could execute 3 million instructions per second and contained 80 miles of circuitry and 350,000 transistors, which were so densely packed that a refrigeration cooling unit was needed due to the lack of airflow.

Cray left Control Data in 1972 and founded his own company, Cray Research. Four years later the $8.8 million Cray-1 scientific supercomputer outstripped the competition. It included some revolutionary design features, one of which is that since electronic signals cannot travel faster than the speed of light (1 foot per billionth of a second) the wire length should be kept as short as possible, because the longer the wire the longer it takes for a message to travel from one end to the other. With this in mind, Cray made sure that none of the supercomputer's conducting wires exceeded 4 feet in length.

In the summer of 1985, the Cray-2, Seymour Cray's latest design, was installed at Lawrence Livermore Laboratory. The Cray-2 was 12 times faster than the Cray-1, and its densely packed circuits are encased in clear Plexiglas and submerged in a bath of liquid coolant. The 60-year-old genius has moved on from his latest triumph, nicknamed "Bubbles," and is working on another revolution in the supercomputer field, because for Seymour Cray a triumph is merely a point of departure.

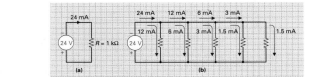

FIGURE 7-39 *R–2R* **Circuit with Data Inserted.**

SELF-TEST EVALUATION POINT FOR SECTION 7-7

Now that you have completed this section, you should be able to:

■ **Objective 4.** *Explain what loading effect a piece of equipment will have when connected to a voltage divider.*

■ **Objective 5.** *Identify and describe the Wheatstone bridge circuit in both the balanced and unbalanced condition.*

■ **Objective 6.** *Describe the R–2R ladder circuit used for digital-to-analog conversion.*

Use the following questions to test your understanding of Section 7-7.

1. What is meant by loading of a voltage-divider circuit?
2. Sketch a Wheatstone bridge circuit and list an application of this circuit.
3. What value will the total resistance of an *R–2R* ladder circuit always equal?
4. For what application could the *R–2R* ladder be used?

NEW! The section review breaks are now **Self-Test Evaluation Points,** with a list of objectives that should have been met up to that point, and a set of questions designed to test the student's level of comprehension.

7-8 TROUBLESHOOTING SERIES–PARALLEL CIRCUITS

Troubleshooting is defined as the process of locating and diagnosing malfunctions or breakdowns in equipment by means of systematic checking or analysis. As discussed in previous resistive-circuit troubleshooting procedures, there are basically only three problems that can occur:

1. A component will open. This usually occurs if a resistor burns out or a wire or switch contact breaks.
2. A component will short. This usually occurs if a conductor, such as solder, wire, or some other conducting material, is dropped or left in the equipment, making or connecting two points that should not be connected.
3. There is a variation in a component's value. This occurs with age in resistors over a long period of time and can eventually cause a malfunction of the equipment.

Using the example circuit in Figure 7-40, we will step through a few problems, beginning with an open component. Throughout the troubleshooting, we will use the voltmeter whenever possible, as it can measure voltage by just connecting the leads across the component, rather than the ammeter, which has to be placed in the circuit, in which case the circuit path has to be opened. In some situations, using an ammeter can be difficult.

To begin, let's calculate the voltage drops and branch current obtained when the circuit is operating normally.

New Technology Topics have been added, such as how to solder and desolder surface mount components (SMCs), how to use the handheld scopemeter, and so on.

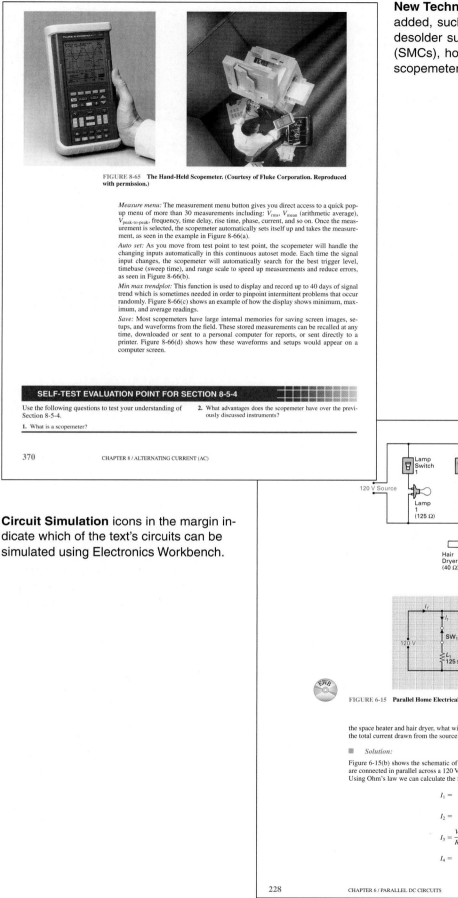

FIGURE 8-65   The Hand-Held Scopemeter. (Courtesy of Fluke Corporation. Reproduced with permission.)

Measure menu: The measurement menu button gives you direct access to a quick pop-up menu of more than 30 measurements including: V_{rms}, V_{mean} (arithmetic average), $V_{peak-to-peak}$, frequency, time delay, rise time, phase, current, and so on. Once the measurement is selected, the scopemeter automatically sets itself up and takes the measurement, as seen in the example in Figure 8-66(a).

Auto set: As you move from test point to test point, the scopemeter will handle the changing inputs automatically in this continuous autoset mode. Each time the signal input changes, the scopemeter will automatically search for the best trigger level, timebase (sweep time), and range scale to speed up measurements and reduce errors, as seen in Figure 8-66(b).

Min max trendplot: This function is used to display and record up to 40 days of signal trend which is sometimes needed in order to pinpoint intermittent problems that occur randomly. Figure 8-66(c) shows an example of how the display shows minimum, maximum, and average readings.

Save: Most scopemeters have large internal memories for saving screen images, setups, and waveforms from the field. These stored measurements can be recalled at any time, downloaded or sent to a personal computer for reports, or sent directly to a printer. Figure 8-66(d) shows how these waveforms and setups would appear on a computer screen.

SELF-TEST EVALUATION POINT FOR SECTION 8-5-4

Use the following questions to test your understanding of Section 8-5-4.

1. What is a scopemeter?

2. What advantages does the scopemeter have over the previously discussed instruments?

Circuit Simulation icons in the margin indicate which of the text's circuits can be simulated using Electronics Workbench.

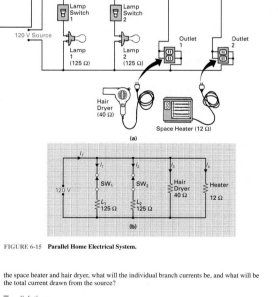

FIGURE 6-15   Parallel Home Electrical System.

the space heater and hair dryer, what will the individual branch currents be, and what will be the total current drawn from the source?

■   *Solution:*

Figure 6-15(b) shows the schematic of the pictorial in Figure 6-15(a). Since all resistances are connected in parallel across a 120 V source, the voltage across all devices will be 120 V. Using Ohm's law we can calculate the four branch currents:

$$I_1 = \frac{V_{lamp1}}{R_{lamp1}} = \frac{120\ V}{125\ \Omega} = 960\ mA$$

$$I_2 = \frac{V_{lamp2}}{R_{lamp2}} = \frac{120\ V}{125\ \Omega} = 960\ mA$$

$$I_3 = \frac{V_{hairdryer}}{R_{hairdryer}} = \frac{120\ V}{40\ \Omega} = 3\ A$$

$$I_4 = \frac{V_{heater}}{R_{heater}} = \frac{120\ V}{12\ \Omega} = 10\ A$$

Circuit Analysis Tables train the student to collect circuit facts in an easy-to-read table that enables the student to clearly see the total and individual current, voltage, resistance, and power values and relationships.

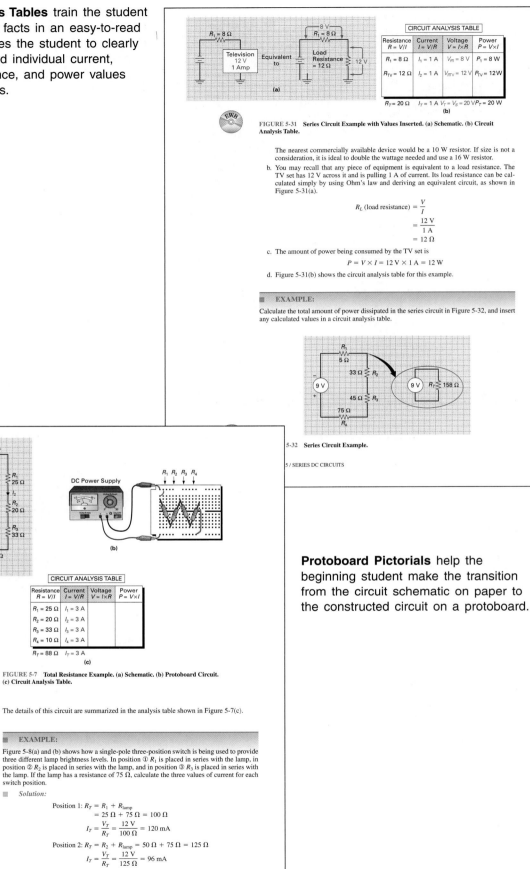

FIGURE 5-31 Series Circuit Example with Values Inserted. (a) Schematic. (b) Circuit Analysis Table.

The nearest commercially available device would be a 10 W resistor. If size is not a consideration, it is ideal to double the wattage needed and use a 16 W resistor.

b. You may recall that any piece of equipment is equivalent to a load resistance. The TV set has 12 V across it and is pulling 1 A of current. Its load resistance can be calculated simply by using Ohm's law and deriving an equivalent circuit, as shown in Figure 5-31(a).

$$R_L \text{ (load resistance)} = \frac{V}{I}$$
$$= \frac{12 \text{ V}}{1 \text{ A}}$$
$$= 12 \text{ Ω}$$

c. The amount of power being consumed by the TV set is

$$P = V \times I = 12 \text{ V} \times 1 \text{ A} = 12 \text{ W}$$

d. Figure 5-31(b) shows the circuit analysis table for this example.

■ **EXAMPLE:**

Calculate the total amount of power dissipated in the series circuit in Figure 5-32, and insert any calculated values in a circuit analysis table.

5-32 Series Circuit Example.

5 / SERIES DC CIRCUITS

FIGURE 5-7 Total Resistance Example. (a) Schematic. (b) Protoboard Circuit. (c) Circuit Analysis Table.

The details of this circuit are summarized in the analysis table shown in Figure 5-7(c).

■ **EXAMPLE:**

Figure 5-8(a) and (b) shows how a single-pole three-position switch is being used to provide three different lamp brightness levels. In position ① R_1 is placed in series with the lamp, in position ② R_2 is placed in series with the lamp, and in position ③ R_3 is placed in series with the lamp. If the lamp has a resistance of 75 Ω, calculate the three values of current for each switch position.

■ *Solution:*

Position 1: $R_T = R_1 + R_{\text{lamp}}$
$$= 25 \text{ Ω} + 75 \text{ Ω} = 100 \text{ Ω}$$
$$I_T = \frac{V_T}{R_T} = \frac{12 \text{ V}}{100 \text{ Ω}} = 120 \text{ mA}$$

Position 2: $R_T = R_2 + R_{\text{lamp}} = 50 \text{ Ω} + 75 \text{ Ω} = 125 \text{ Ω}$
$$I_T = \frac{V_T}{R_T} = \frac{12 \text{ V}}{125 \text{ Ω}} = 96 \text{ mA}$$

Protoboard Pictorials help the beginning student make the transition from the circuit schematic on paper to the constructed circuit on a protoboard.

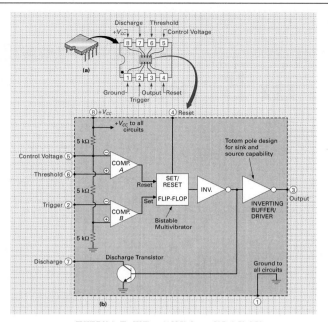

FIGURE 20-4   The 555 Timer. (a) IC Pin Layout. (b) Basic Block Diagram.

appear as a LOW at the output pin 3. If the output of comparator *B* were to go high, the set/reset flip-flop would be set HIGH to +5 V. This HIGH output would be inverted by the INVERTER to a LOW, and then inverted and buffered by the final INVERTER to appear as a HIGH at the output pin 3. When the output of the set/reset flip-flop is LOW (reset), the input at the base of the discharge transistor will be HIGH. The transistor will therefore turn ON, and its low emitter-to-collector resistance will ground pin 7. When the output of the set/reset flip-flop is HIGH (set), the input at the base of the discharge transistor will be LOW. The transistor will therefore turn OFF and its high emitter-to-collector resistance will cause pin 7 to ground.

The 555 Timer as an Astable Multivibrator

Figure 20-5(a) shows how the 555 timer can be connected to operate as an astable or free-running multivibrator. The waveforms in Figure 20-5(b) show how the externally

CHAPTER 20 / TIMERS, THYRISTORS, AND TRANSDUCERS

Contents in Brief

Contents

11

Electromagnetism and Electromagnetic Induction   470

12

Inductance and Inductors   506

13

Transformers   552

Appendixes 917

Index 989

Introduction to Electronics

From World War II onward, no branch of science has contributed more to the development of the modern world than electronics. It has stimulated dramatic advances in the fields of communication, computing, consumer products, industrial automation, test and measurement, and health care. It has now become the largest single industry in the world, exceeding the automobile and oil industries, with annual sales of electronic systems exceeding \$2 trillion.

Your Course in Electronics

Your future in the electronics industry begins with this text. To give you an idea of where you are going and what we will be covering, Figure I-1 acts as a sort of road map, breaking up your study of electronics into four basic steps.

Step 1:   Basics of Electricity
Step 2:   Electronic Components
Step 3:   Electronic Circuits
Step 4:   Electronic Systems

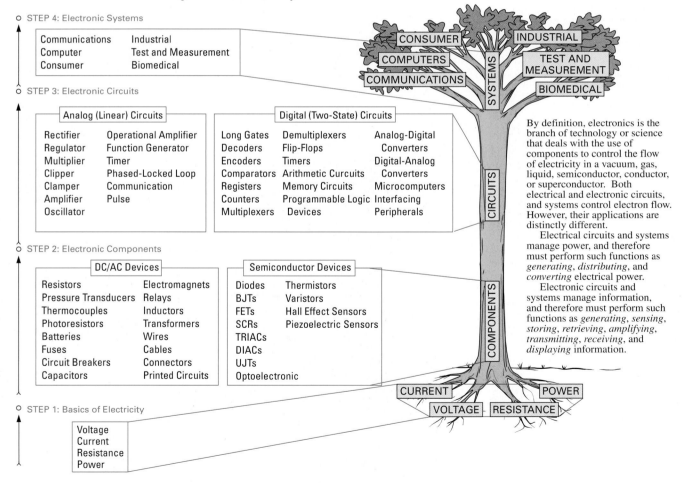

STEP 4: Electronic Systems

Communications	Industrial
Computer	Test and Measurement
Consumer	Biomedical

STEP 3: Electronic Circuits

Analog (Linear) Circuits

Rectifier	Operational Amplifier
Regulator	Function Generator
Multiplier	Timer
Clipper	Phased-Locked Loop
Clamper	Communication
Amplifier	Pulse
Oscillator	

Digital (Two-State) Circuits

Long Gates	Demultiplexers	Analog-Digital
Decoders	Flip-Flops	Converters
Encoders	Timers	Digital-Analog
Comparators	Arithmetic Circuits	Converters
Registers	Memory Circuits	Microcomputers
Counters	Programmable Logic	Interfacing
Multiplexers	Devices	Peripherals

STEP 2: Electronic Components

DC/AC Devices

Resistors	Electromagnets
Pressure Transducers	Relays
Thermocouples	Inductors
Photoresistors	Transformers
Batteries	Wires
Fuses	Cables
Circuit Breakers	Connectors
Capacitors	Printed Circuits

Semiconductor Devices

Diodes	Thermistors
BJTs	Varistors
FETs	Hall Effect Sensors
SCRs	Piezoelectric Sensors
TRIACs	
DIACs	
UJTs	
Optoelectronic	

By definition, electronics is the branch of technology or science that deals with the use of components to control the flow of electricity in a vacuum, gas, liquid, semiconductor, conductor, or superconductor. Both electrical and electronic circuits, and systems control electron flow. However, their applications are distinctly different.

Electrical circuits and systems manage power, and therefore must perform such functions as *generating*, *distributing*, and *converting* electrical power.

Electronic circuits and systems manage information, and therefore must perform such functions as *generating*, *sensing*, *storing*, *retrieving*, *amplifying*, *transmitting*, *receiving*, and *displaying* information.

STEP 1: Basics of Electricity

| Voltage |
| Current |
| Resistance |
| Power |

FIGURE I-1   **The Steps Involved in Studying Electronics.**

The main purpose of this introduction is not only to introduce you to the terms of the industry, but also to show you why the first two chapters in this text begin at the very beginning with "voltage and current," and then "resistance and power." **Components,** which are the basic electronic building blocks, were developed to control these four roots or properties, and when these devices are combined they form **circuits.** Moving up the tree to the six different branches of electronics, you will notice that just as components are the building blocks for circuits, circuits are in turn the building blocks for **systems.** Take a glance at the following pages, which will list many of the different types of components, circuits, and systems in the electronics industry.

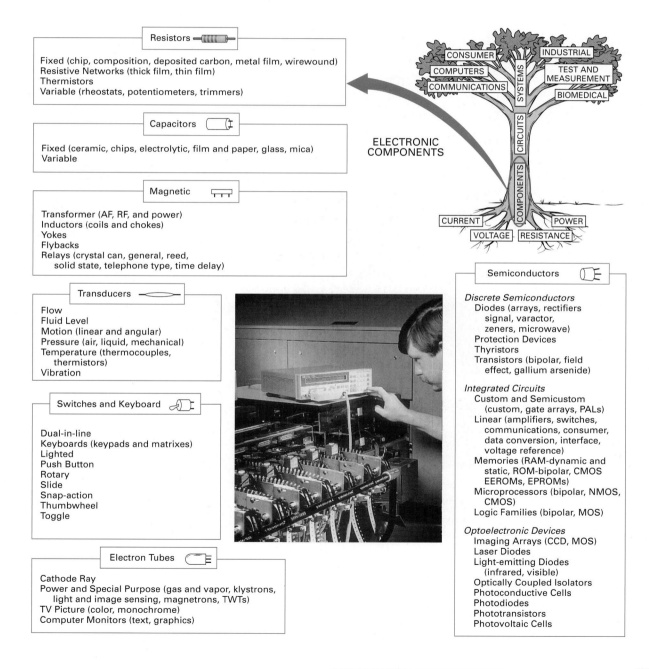

Resistors

Fixed (chip, composition, deposited carbon, metal film, wirewound)
Resistive Networks (thick film, thin film)
Thermistors
Variable (rheostats, potentiometers, trimmers)

Capacitors

Fixed (ceramic, chips, electrolytic, film and paper, glass, mica)
Variable

Magnetic

Transformer (AF, RF, and power)
Inductors (coils and chokes)
Yokes
Flybacks
Relays (crystal can, general, reed,
    solid state, telephone type, time delay)

Transducers

Flow
Fluid Level
Motion (linear and angular)
Pressure (air, liquid, mechanical)
Temperature (thermocouples,
    thermistors)
Vibration

Switches and Keyboard

Dual-in-line
Keyboards (keypads and matrixes)
Lighted
Push Button
Rotary
Slide
Snap-action
Thumbwheel
Toggle

Electron Tubes

Cathode Ray
Power and Special Purpose (gas and vapor, klystrons,
    light and image sensing, magnetrons, TWTs)
TV Picture (color, monochrome)
Computer Monitors (text, graphics)

ELECTRONIC
COMPONENTS

CONSUMER
COMPUTERS
COMMUNICATIONS
INDUSTRIAL
TEST AND
MEASUREMENT
BIOMEDICAL
SYSTEMS
CIRCUITS
COMPONENTS
CURRENT
POWER
VOLTAGE
RESISTANCE

Semiconductors

Discrete Semiconductors
    Diodes (arrays, rectifiers
        signal, varactor,
        zeners, microwave)
    Protection Devices
    Thyristors
    Transistors (bipolar, field
        effect, gallium arsenide)

Integrated Circuits
    Custom and Semicustom
        (custom, gate arrays, PALs)
    Linear (amplifiers, switches,
        communications, consumer,
        data conversion, interface,
        voltage reference)
    Memories (RAM-dynamic and
        static, ROM-bipolar, CMOS
        EEROMs, EPROMs)
    Microprocessors (bipolar, NMOS,
        CMOS)
    Logic Families (bipolar, MOS)

Optoelectronic Devices
    Imaging Arrays (CCD, MOS)
    Laser Diodes
    Light-emitting Diodes
        (infrared, visible)
    Optically Coupled Isolators
    Photoconductive Cells
    Photodiodes
    Phototransistors
    Photovoltaic Cells

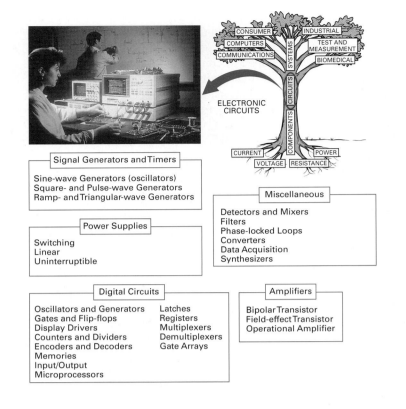

ELECTRONIC CIRCUITS

Signal Generators and Timers

Sine-wave Generators (oscillators)
Square- and Pulse-wave Generators
Ramp- and Triangular-wave Generators

Power Supplies

Switching
Linear
Uninterruptible

Miscellaneous

Detectors and Mixers
Filters
Phase-locked Loops
Converters
Data Acquisition
Synthesizers

Digital Circuits

Oscillators and Generators   Latches
Gates and Flip-flops   Registers
Display Drivers   Multiplexers
Counters and Dividers   Demultiplexers
Encoders and Decoders   Gate Arrays
Memories
Input/Output
Microprocessors

Amplifiers

Bipolar Transistor
Field-effect Transistor
Operational Amplifier

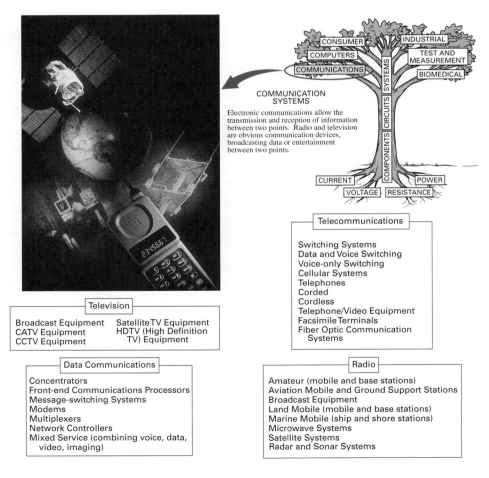

COMMUNICATION SYSTEMS

Electronic communications allow the
transmission and reception of information
between two points. Radio and television
are obvious communication devices,
broadcasting data or entertainment
between two points.

Telecommunications

Switching Systems
Data and Voice Switching
Voice-only Switching
Cellular Systems
Telephones
Corded
Cordless
Telephone/Video Equipment
Facsimile Terminals
Fiber Optic Communication
  Systems

Television

Broadcast Equipment   Satellite TV Equipment
CATV Equipment   HDTV (High Definition
CCTV Equipment   TV) Equipment

Data Communications

Concentrators
Front-end Communications Processors
Message-switching Systems
Modems
Multiplexers
Network Controllers
Mixed Service (combining voice, data,
  video, imaging)

Radio

Amateur (mobile and base stations)
Aviation Mobile and Ground Support Stations
Broadcast Equipment
Land Mobile (mobile and base stations)
Marine Mobile (ship and shore stations)
Microwave Systems
Satellite Systems
Radar and Sonar Systems

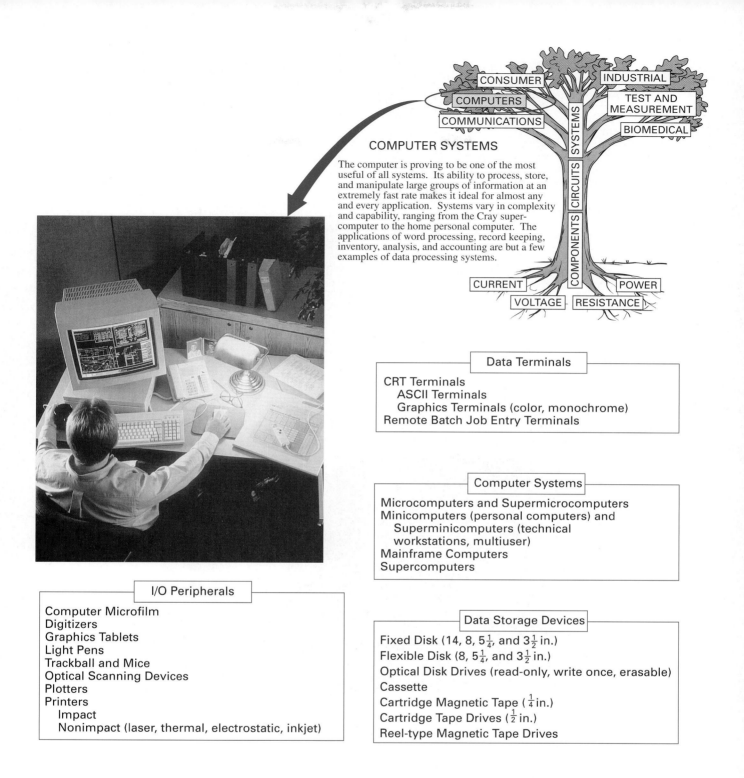

CONSUMER
COMPUTERS
COMMUNICATIONS
INDUSTRIAL
TEST AND MEASUREMENT
BIOMEDICAL

SYSTEMS

CIRCUITS

COMPONENTS

CURRENT — VOLTAGE — RESISTANCE — POWER

COMPUTER SYSTEMS

The computer is proving to be one of the most useful of all systems. Its ability to process, store, and manipulate large groups of information at an extremely fast rate makes it ideal for almost any and every application. Systems vary in complexity and capability, ranging from the Cray supercomputer to the home personal computer. The applications of word processing, record keeping, inventory, analysis, and accounting are but a few examples of data processing systems.

Data Terminals

CRT Terminals
  ASCII Terminals
    Graphics Terminals (color, monochrome)
Remote Batch Job Entry Terminals

Computer Systems

Microcomputers and Supermicrocomputers
Minicomputers (personal computers) and
    Superminicomputers (technical
    workstations, multiuser)
Mainframe Computers
Supercomputers

I/O Peripherals

Computer Microfilm
Digitizers
Graphics Tablets
Light Pens
Trackball and Mice
Optical Scanning Devices
Plotters
Printers
    Impact
    Nonimpact (laser, thermal, electrostatic, inkjet)

Data Storage Devices

Fixed Disk (14, 8, $5\frac{1}{4}$, and $3\frac{1}{2}$ in.)
Flexible Disk (8, $5\frac{1}{4}$, and $3\frac{1}{2}$ in.)
Optical Disk Drives (read-only, write once, erasable)
Cassette
Cartridge Magnetic Tape ($\frac{1}{4}$ in.)
Cartridge Tape Drives ($\frac{1}{2}$ in.)
Reel-type Magnetic Tape Drives

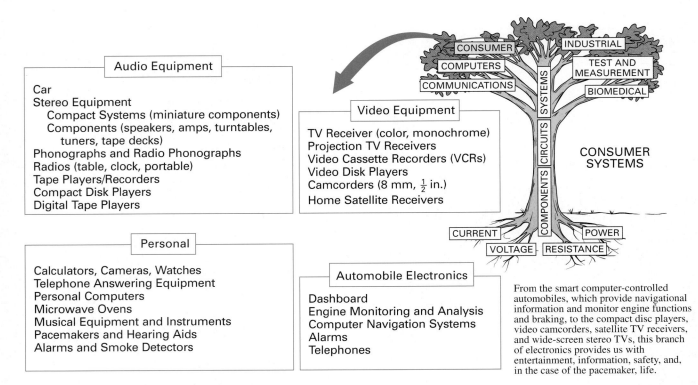

Audio Equipment

Car
Stereo Equipment
   Compact Systems (miniature components)
   Components (speakers, amps, turntables,
     tuners, tape decks)
Phonographs and Radio Phonographs
Radios (table, clock, portable)
Tape Players/Recorders
Compact Disk Players
Digital Tape Players

Video Equipment

TV Receiver (color, monochrome)
Projection TV Receivers
Video Cassette Recorders (VCRs)
Video Disk Players
Camcorders (8 mm, $\frac{1}{2}$ in.)
Home Satellite Receivers

CONSUMER SYSTEMS

Personal

Calculators, Cameras, Watches
Telephone Answering Equipment
Personal Computers
Microwave Ovens
Musical Equipment and Instruments
Pacemakers and Hearing Aids
Alarms and Smoke Detectors

Automobile Electronics

Dashboard
Engine Monitoring and Analysis
Computer Navigation Systems
Alarms
Telephones

From the smart computer-controlled automobiles, which provide navigational information and monitor engine functions and braking, to the compact disc players, video camcorders, satellite TV receivers, and wide-screen stereo TVs, this branch of electronics provides us with entertainment, information, safety, and, in the case of the pacemaker, life.

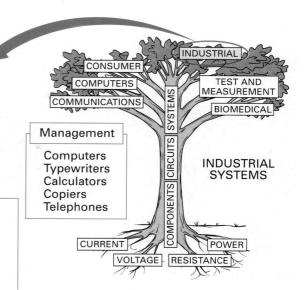

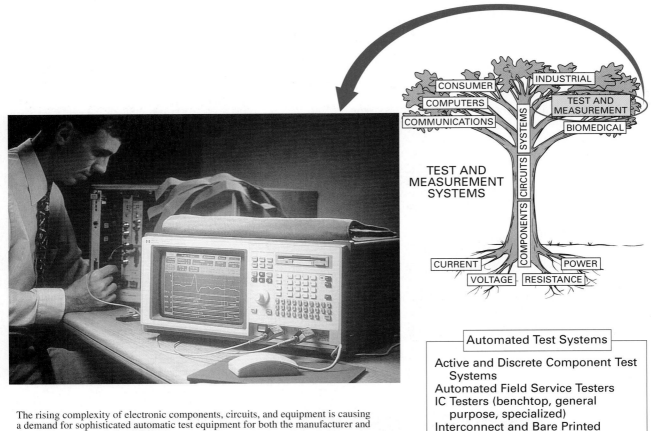

The rising complexity of electronic components, circuits, and equipment is causing a demand for sophisticated automatic test equipment for both the manufacturer and the customer to test their products.

Automated Test Systems

Active and Discrete Component Test
   Systems
Automated Field Service Testers
IC Testers (benchtop, general
   purpose, specialized)
Interconnect and Bare Printed
   Circuit-Board Testers
Loaded Printed Circuit-Board Testers
   (in-circuit, functional, combined)

General Test and Measurement Equipment

Amplifiers (lab)
Arbitrary Waveform Generators
Analog Voltmeters, Ammeters, and Multimeters
Audio Oscillators
Audio Waveform Analyzers and Distortion Meters
Calibrators and Standards
Dedicated IEEE–488 Bus Controllers
Digital Multimeters
Electronic Counters (RF, Microwave, Universal)
Frequency Synthesizers
Function Generators
Pulse/Timing Generators
Signal Generators (RF, Microwave)

Logic Analyzers
Microprocessor Development Systems
Modulation Analyzers
Noise-measuring Equipment
Oscilloscopes (Analog, Digital)
Panel Meters
Personal Computer (PC) Based Instruments
Recorders and Plotters
RF/Microwave Network Analyzers
RF/Microwave Power-measuring Equipment
Spectrum Analyzers
Stand-alone In-circuit Emulators
Temperature-measuring Instruments

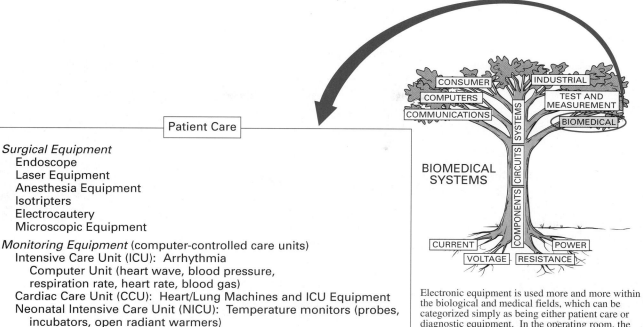

Patient Care

Surgical Equipment
  Endoscope
  Laser Equipment
  Anesthesia Equipment
  Isotripters
  Electrocautery
  Microscopic Equipment

Monitoring Equipment (computer-controlled care units)
  Intensive Care Unit (ICU): Arrhythmia
    Computer Unit (heart wave, blood pressure,
    respiration rate, heart rate, blood gas)
  Cardiac Care Unit (CCU): Heart/Lung Machines and ICU Equipment
  Neonatal Intensive Care Unit (NICU): Temperature monitors (probes,
    incubators, open radiant warmers)

Electronic equipment is used more and more within the biological and medical fields, which can be categorized simply as being either patient care or diagnostic equipment. In the operating room, the endoscope, which is an instrument used to examine the interior of a canal or hollow organ, and the laser, which is used to coagulate, cut, or vaporize tissue with extremely intense light, both reduce the use of invasive surgery. A large amount of monitoring equipment is used both in and out of operating rooms, and the equipment consists of generally large computer-controlled systems that can have a variety of modules inserted (based on the application) to monitor, on a continuous basis, body temperature, blood pressure, pulse rate, and so on. In the diagnostic group of equipment, the clinical laboratory test results are used as diagnostic tools. With the advances in automation and computerized information systems, multiple tests can be carried out at increased speeds. Diagnostic imaging, in which a computer constructs an image of a cross-sectional plane of the body, is probably one of the most interesting equipment areas.

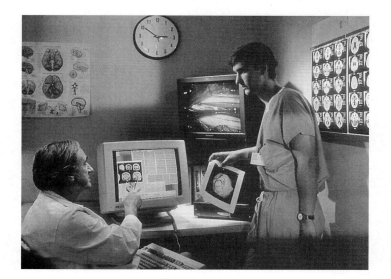

Diagnostic

Diagnostic Equipment
  X-Ray (computed tomography)
  Magnetic Resonance Imaging (MRI)
  Diagnostic Sounder
  Electrocardiograph (EKG)
  Electromyograph
  Electroencephalograph (EEG)
  Coagulograph
  Ultrasound (computed sonography)
  Nuclear Medicine (isotopes, spectroscopy)

Clinical Laboratory
  Automated Clinical Analyzers
  Centrifuge Incubators
  Cell Counters

Development of an Electronic Product

The flow chart below expands on the basic block diagram on the previous page, showing the order, from top to bottom, in which an electronic product is developed from conception to shipping.

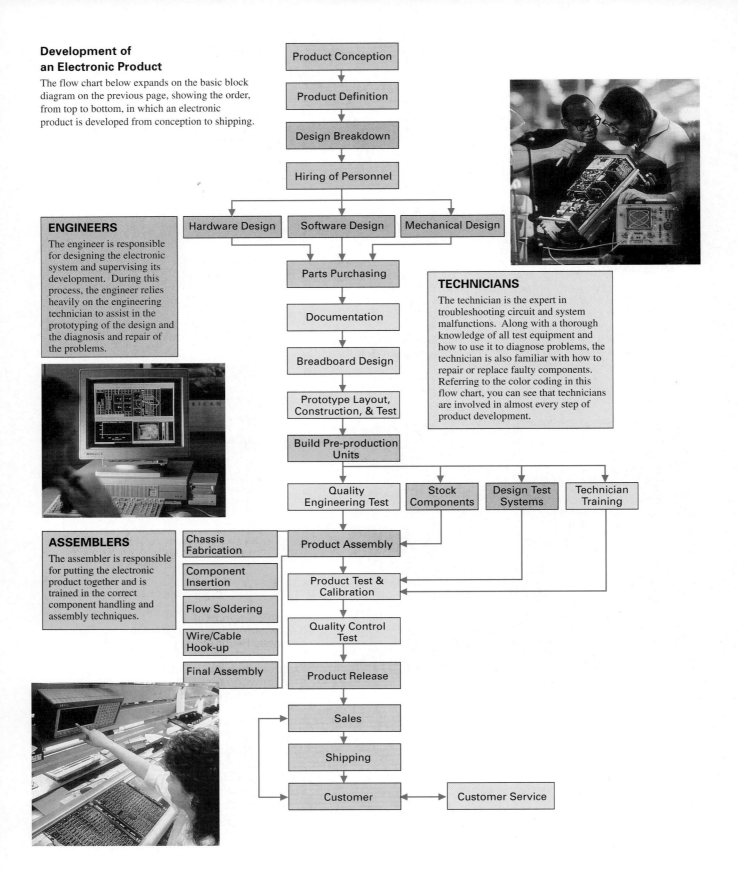

Product Conception

Product Definition

Design Breakdown

Hiring of Personnel

Hardware Design | **Software Design** | **Mechanical Design**

Parts Purchasing

Documentation

Breadboard Design

Prototype Layout, Construction, & Test

Build Pre-production Units

Quality Engineering Test | **Stock Components** | **Design Test Systems** | **Technician Training**

Chassis Fabrication

Component Insertion

Flow Soldering

Wire/Cable Hook-up

Final Assembly

Product Assembly

Product Test & Calibration

Quality Control Test

Product Release

Sales

Shipping

Customer | **Customer Service**

ENGINEERS

The engineer is responsible for designing the electronic system and supervising its development. During this process, the engineer relies heavily on the engineering technician to assist in the prototyping of the design and the diagnosis and repair of the problems.

TECHNICIANS

The technician is the expert in troubleshooting circuit and system malfunctions. Along with a thorough knowledge of all test equipment and how to use it to diagnose problems, the technician is also familiar with how to repair or replace faulty components. Referring to the color coding in this flow chart, you can see that technicians are involved in almost every step of product development.

ASSEMBLERS

The assembler is responsible for putting the electronic product together and is trained in the correct component handling and assembly techniques.

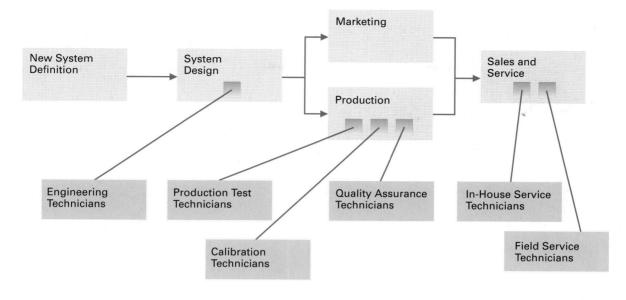

New System Definition → System Design

Marketing

Production

Sales and Service

Engineering Technicians

Production Test Technicians

Quality Assurance Technicians

In-House Service Technicians

Calibration Technicians

Field Service Technicians

ENGINEERING TECHNICIAN

Here you can see an engineering technician breadboarding the design. From sketches supplied by the engineers, a breadboard model of the design is constructed. The breadboard model is an experimental arrangement of a circuit in which the components are temporarily attached to a flat board. In this arrangement, the components can be tested to prove the feasability of the circuit. A breadboard facilitates making easy changes when they are necessary.

Working under close supervision, the engineering technician performs all work assignments as given by all levels of engineers.

RESPONSIBILITIES:

• Breadboard electronic circuits from schematics.

• Test, evaluate, and document circuits and system performance under the engineer's direction.

• Check out, evaluate, and take data for the engineering proto-types including mechanical assembly of prototype circuits.

• Help to generate and maintain preliminary engineering documentations. Assist engineers and senior Engineering Technicians in ERN documentation.

• Maintain the working station equipment and tools in orderly fashion.

• Support the engineers in all aspects of the development of new products.

REQUIREMENTS:

AS/AAS Degree in Electronics or equivalent plus 1–2 years technician experience. Ability to read color code, to solder properly, and to bond wires where the skill is required to complete breadboards and prototypes. Ability to use common machinery required to build prototype circuits. Working knowledge of common electronic components: TTL logic circuits, op-amps, capacitors, resistors, inductors, semiconductor devices.

Once the system is fully operational, it is calibrated by a calibration technician. The more complex problems are handled by the production test technicians seen in this photograph.

TECHNICIAN I

Working under close supervision, the production test technician performs all work assignments, and exercises limited decision-making.

RESPONSIBILITIES:

• Perform routine, simple operational tests and fault isolation on simple components, circuits, and systems for verification of product performance to well-defined specifications.

• May perform standard assembly operations and simple alignment of electronic components and assemblies.

• May set up simple test equipment to test performance of products to specifications.

REQUIREMENTS:

AS/AAS Degree in Electronic Technology or equivalent work experience.

TECHNICIAN II

Working under moderate supervision, the production test technician exercises general decision-making involving simple cause and effect relationships to identify trends and common problems.

RESPONSIBILITIES:

• Perform moderately complex operational tests and fault isolation on components, circuits, and systems for verification of product performance to well-defined specifications.

• Perform simple mathematical calculations to verify test measurements and product performance to well-defined specifications.

• Set up general test stations, utilizing varied test equipment, including some sophisticated equipment.

• Perform standard assembly operations.

REQUIREMENTS:

AS/AAS Degree in Electronic Technology or equivalent plus 2–4 years directly related experience. Demonstrated experience working mathematical formulas and equations. Working knowledge of counters, scopes, spectrum analyzers, and related industry standard test equipment.

CALIBRATION TECHNICIAN

The calibration technician shown here is undertaking a complete evaluation of the newly constructed breadboard's mechanical and electrical form, design, and performance.

Working under general supervision, the calibration technician interfaces with test equipment in a system environment requiring limited decision-making.

RESPONSIBILITIES:

• Work in an interactive mode with test station. Test, align, and calibrate products to defined specifications.

• May set up own test stations and those of other operators.

• Perform multiple alignments to get products to meet specifications.

• May perform other manufacturing-related tasks as required.

REQUIREMENTS:

1–2 years experience with test and measurement equipment, experience with multiple alignment and calibration of assemblies including test station setups. Able to follow written instructions and write clearly.

QUALITY ASSURANCE TECHNICIANS

The quality assurance (QA) technician takes one of the pre-production units through an extensive series of tests to determine whether it meets the standards listed. This technician is evaluating the new product as it is put through an extensive series of tests.

Working under direct supervision, the quality assurance technician performs functional tests on completed instruments.

RESPONSIBILITIES:

• Using established Acceptance Test Procedures, perform operational tests of all completed systems to ensure that all functional and electrical parameters are within specified limits.

• Perform visual inspection of all completed systems for cleanliness and absence of cosmetic defects.

• Reject all systems that do not meet specifications and/or established parameters of function and appearance.

• Make appropriate notations on the system history sheet.

• Maintain the QA Acceptance Log in accordance with current instructions.

• Refer questionable characteristics to supervisor.

REQUIREMENTS:

AS/AAS Degree in Electronics or equivalent including the use of test equipment. Must know color code and be able to distinguish between colors. Must have a working knowledge of related test equipment. Must know how to read and interpret drawings.

IN-HOUSE SERVICE TECHNICIAN

This photograph shows some in-house service technicians troubleshooting problems on returned units. Once the customer has received the electronic equipment, customer service provides assistance in maintenance and repair of the unit through direct in-house service or at service centers throughout the world.

Working under moderate supervision, the in-house service technician performs all work assignments given by lead tech or direct supervisor. Work necessary overtime as assigned by supervisors.

RESPONSIBILITIES:

- Utilizing all appropriate tools, troubleshoot and repair customer systems in a timely, quality manner, to the general component level.

- Working with basic test equipment, perform timely quality calibration of customer systems to specifications.

- Timely repair of QA rejects.

- Solder and desolder components where appropriate, meeting company standards.

- Aid marketing in solving customer problems via the telephone.

- When appropriate, instruct customers in the proper methods of calibration and repair of products.

REQUIREMENTS:

AS/AAS Degree in Electronics or equivalent plus 2–3 years experience troubleshooting analog and/or digital systems, at least 6–12 months of which should be in a service environment. Must be able to read and understand flow charts, block diagrams, schematics, and truth tables. Must be able to operate and utilize test equipment such as oscilloscopes, counters, voltmeters, and analyzers.

Ability to effectively communicate and work with customers.

FIELD SERVICE TECHNICIAN

The field service technician seen here has been requested by the customer to make a service call on a malfunctioning unit that currently is under test.

Working under moderate supervision, the field service technician performs all work assignments given by the lead tech or direct supervisor. Work necessary overtime as assigned by supervisors.

RESPONSIBILITIES:

- Utilizing all appropriate tools, troubleshoot and repair customer systems in a timely, quality manner, to the general component level.

- Working with basic test equipment, perform timely quality calibration of customer systems to specifications.

- When appropriate, instruct others in proper soldering techniques meeting company standards.

- Timely repair of QA rejects.

- Aid marketing in solving customer problems via the telephone or at the customers' facility at marketings' discretion.

- When appropriate, aid marketing with sales applications.

- Help QA, Production, and Engineering in solving field problems.

- Evaluate manuals and other customer documents for errors or omissions.

REQUIREMENTS:

AS/AAS Degree in Electronics or equivalent plus 4–5 years experience troubleshooting analog and/or digital systems, at least 6–12 months of which should be in a service environment. Must be able to read and understand flow charts, block diagrams, schematics, and truth tables. A demonstrated ability to effectively communicate and work with customers, and suggest alternative applications for product utilization is also required.

Introductory
DC/AC Electronics

Voltage and Current

Problem-Solver

Charles Proteus Steinmetz (1865–1923) was an outstanding electrical genius who specialized in mathematics, electrical engineering, and chemistry. His three greatest electrical contributions were his investigation and discovery of the law of hysteresis, his investigations in lightning, which resulted in his theory on traveling waves, and his discovery that complex numbers could be used to solve ac circuit problems. Solving problems was in fact his specialty, and on one occasion he was commissioned to troubleshoot a failure on a large company system that no one else had been able to repair. After studying the symptoms and schematics for a short time, he chalked an X on one of the metal cabinets, saying that this was where they would find the problem, and left. He was right, and the problem was remedied to the relief of the company executives; however, they were not pleased when they received a bill for $1000. When they demanded that Steinmetz itemize the charges, he replied—$1 for making the mark and $999 for knowing where to make the mark.

The strong message this vignette conveys is that you will get $1 for physical labor and $999 for mental labor, and this is a good example as to why you should continue in your pursuit of education.

Outline and Objectives

Introduction

Before we begin, let me try to put you in the right frame of mind. As you proceed through this chapter and the succeeding chapters, it is imperative that you study every section, example, self-test evaluation point, and end-of-chapter question. If you cannot understand a particular section or example, go back and review the material that led up to the problem and make sure that you fully understand all the basics before you continue. Since each chapter builds on previous chapters, you may find that you need to return to an earlier chapter to refresh your understanding before moving on with the current chapter. This process of moving forward and then backtracking to refresh your understanding is very necessary and helps to engrave the material in your mind. Try never to skip a section or chapter because you feel that you already have a good understanding of the subject matter. If it is a basic topic that you have no problem with, read it anyway to refresh your understanding about the steps involved and the terminology because these may be used in a more complex operation in a later chapter.

If the introduction were described as the big picture, Chapter 1 would be called the small picture. In this chapter we will examine the smallest and most significant part of electricity and electronics—the **electron.** By having a good understanding of the electron and the atom, we will be able to obtain a clearer understanding of our four basic electrical quantities: **voltage, current, resistance,** and **power.** In this chapter we will be discussing the basic building blocks of matter, the electrical quantities of voltage and current, and the difference between a conductor and insulator. To begin with, though, let's review the math skills you will need for this chapter's material.

Electron

Smallest subatomic particle of negative charge that orbits the nucleus of the atom.

Voltage (*V* or *E*)

Term used to designate electrical pressure or the force that causes current to flow.

Current (*I*)

Measured in amperes or amps, it is the flow of electrons through a conductor.

Resistance

Symbolized R and measured in ohms (Ω), it is the opposition to current flow with the dissipation of energy in the form of heat.

Power

Amount of energy converted by a component or circuit in a unit of time, normally seconds. It is measured in units of watts (joules/second).

1-1 MINI-MATH REVIEW—EXPONENTS AND METRIC PREFIXES

This first "Mini-Math Review" is included to overview the mathematical details you need for the electronic concepts covered in this chapter. In this review, we will be examining exponents and metric prefixes.

Like many terms used in mathematics, the word **exponent** sounds like it will be complicated. However, once you find out that exponents are simply a sort of "math shorthand," the topic loses its intimidation. Many of the values used in science and technology contain numbers that have exponents. An exponent is a number in a smaller type size that appears to the right of and slightly higher than another number, for example:

Exponent
A symbol written above and to the right of a mathematical expression to indicate the operation of raising to a power.

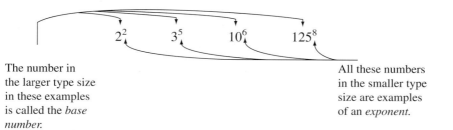

The number in the larger type size in these examples is called the *base number.*

All these numbers in the smaller type size are examples of an *exponent.*

However, what does a term like 2^3 mean? It means that the base 2 is to be used as a factor 3 times; therefore,

$$2^3 = 2 \times 2 \times 2 = 8$$

Similarly, 3^5 means that the base 3 is to be used as a factor 5 times; therefore,

$$3^5 = 3 \times 3 \times 3 \times 3 \times 3 = 243$$

As you can see, it is much easier to write 3^5 than to write $3 \times 3 \times 3 \times 3 \times 3$, and since both mean the same thing and equal the same amount (which is 243), exponents are a quick and easy math shorthand.

In this chapter we will examine the details relating to squares, roots, exponents, scientific and engineering notation, and prefixes.

1-1-1 *Raising a Base Number to a Higher Power*

A base number's exponent indicates how many times the base number must be multiplied by itself. This is called *raising a number to a higher power*. For example, $5 \times 5 \times 5 \times 5$ can be written as 5^4, which indicates that the base number 5 is raised to the fourth power by the exponent 4. As another example, 3×3 can be written as 3^2, which indicates that the base number 3 is raised to the second power by the exponent 2. The second power is also called the **square** of the base number, and therefore 3^2 can be called "three squared" or "three to the second power."

Square
The product of a number multiplied by itself.

■ **EXAMPLE**

What does 10^3 mean?

■ *Solution:*

10^3 indicates that the base number 10 is raised to the third power by the exponent 3. Described another way, it means that the base 10 is to be used as a factor 3 times; therefore,

$$10^3 = 10 \times 10 \times 10 = 1000$$

Therefore, instead of writing $10 \times 10 \times 10$, or 1000, you could simply write 10^3 (pronounced "ten to the three," "ten to the third power," or "ten cubed").

Because the square of a base number is used very frequently, let us begin by discussing raising a base number to the second power.

Square of a Number

The square of a base number means that the base number is to be multiplied by itself. For example, 4^2, which is pronounced "four squared" or "four to the second power," means 4×4. The squares of the first ten base numbers are used very frequently in numerical problems and are as follows:

$$0^2 = 0 \times 0 = 0$$
$$1^2 = 1 \times 1 = 1$$
$$2^2 = 2 \times 2 = 4$$
$$3^2 = 3 \times 3 = 9$$
$$4^2 = 4 \times 4 = 16$$
$$5^2 = 5 \times 5 = 25$$
$$6^2 = 6 \times 6 = 36$$
$$7^2 = 7 \times 7 = 49$$
$$8^2 = 8 \times 8 = 64$$
$$9^2 = 9 \times 9 = 81$$
$$10^2 = 10 \times 10 = 100$$

Many people get confused with the first three of the squares. Be careful to remember that $0^2 = 0$ because nothing $\times$ nothing equals nothing; $1^2 = 1$ because one times one is still one; and 2^2 means 2×2, not $2 + 2$, even though the answer works out both ways to be 4.

◫ EXAMPLE:

Give the square of the following.

   a. 12^2 b. 7^2

◼ *Solution:*

   a. $12^2 = 12 \times 12 = 144$
   b. $7^2 = 7 \times 7 = 49$

Finding the square of a base number is done so frequently in science and technology that most calculators have a special key just for that purpose. It is called the *square key* and operates as follows.

CALCULATOR KEYS

Name: Square key

Function: Calculates the square of the number in the display.

Example: $16^2 = ?$

Press keys: $\boxed{1}\,\boxed{6}\,\boxed{x^2}$
Display shows: 256

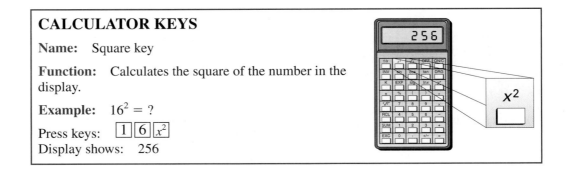

Most calculators also have a key for raising a base number of any value to any power. This is called the *y to the x power key,* and it operates as follows.

CALCULATOR KEYS
Name: y to the x power key
Function: Raises the displayed value y to the xth power.
Example: $5^4 = ?$
Press keys: $\boxed{5}\;\boxed{y^x}\;\boxed{4}\;\boxed{=}$
Display shows: 625

Root of a Number

What would we do if we had the result 64 and we didn't know the value of the number that was multiplied by itself to get 64? In other words, we wanted to *find the source or root number that was squared to give us the result.* This process, called **square root,** uses a special symbol called a *radical sign* and a smaller number called the *index.*

$$\sqrt[2]{64} = ?$$

Smaller number is called the *index.* It indicates how many times a number was multiplied by itself to get the value shown inside the radical sign. A 2 index is called the *square root.*

Radical sign ($\sqrt{\ }$) indicates that the value inside is the result of a multiplication of a number two or more times.

Square Root

A factor of a number that when squared gives the number.

In this example the index is 2, which indicates that the number we are trying to find was multiplied by itself two times. Of course, in this example we already know that the answer is 8 because $8 \times 8 = 64$.

If squaring a number takes us forward ($8^2 = 8 \times 8 = 64$), taking the square root of a number must take us backward ($\sqrt[2]{64} = 8$). Almost nobody extracts the squares from square root problems by hand because calculators make this process more efficient. Most calculators have a special key just for determining the square root of a number. Called the square root key, it operates as follows.

CALCULATOR KEYS
Name: Square root key
Function: Calculates the square root of the number in the display.
Example: $\sqrt{81} = ?$
Press keys: $\boxed{8}\;\boxed{1}\;\boxed{\sqrt{x}}$
Display shows: 9

Some of the more frequently used *square root* values are as follows:

$$\sqrt{1} = 1$$
$$\sqrt{2} = 1.414$$
$$\sqrt{3} = 1.732$$
$$\sqrt{4} = 2$$
$$\sqrt{9} = 3$$
$$\sqrt{16} = 4$$
$$\sqrt{25} = 5$$
$$\sqrt{36} = 6$$
$$\sqrt{49} = 7$$
$$\sqrt{64} = 8$$
$$\sqrt{81} = 9$$
$$\sqrt{100} = 10$$

Check each of these answers using the $\boxed{\sqrt{x}}$ key on your calculator.

In some cases the index may be 3, meaning that some number was multiplied by itself three times to get the value within the radical sign; for example,

$$\sqrt[3]{125} = ?$$

Cube Root

A number that when multiplied by itself three times gives the number.

Using the index 3 instead of the index 2 is called taking the **cube root.** The answer to this problem is 5 because 5 multiplied by itself three times equals 125.

$$\sqrt[3]{125} = 5$$

Most calculators also have a key for calculating the root of any number with any index. It operates as follows.

CALCULATOR KEYS

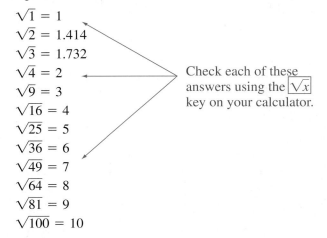

Name: xth root of y key

Function: Calculate the xth root of the displayed value y.

Example: $\sqrt[3]{512} = ?$

Press keys: $\boxed{5}\boxed{1}\boxed{2}\boxed{\sqrt[x]{y}}\boxed{3}\boxed{=}$

Display shows: 8

In most cases a radical sign will not have an index, in which case you can assume that the index is 2, or square root; for example,

$$\sqrt{16} = 4$$
Square root of sixteen = four

1-1-2 *Powers of Ten*

Many of the sciences deal with numbers that contain a large number of zeros, for example:

$$14,000$$
$$0.000032$$

By using exponents, we can eliminate the large number of zeros to obtain a shorthand version of the same number. This method is called *powers of ten.*

As an example, let us remove all the zeros from the number 14,000 until we are left with simply 14. However, this number (14) is not equal to the original number (14,000) and therefore simply removing the zeros is not an accurate shorthand. Another number needs to be written with the 14 to indicate what has been taken away—this is called a *multiplier*. The multiplier must indicate what you have to multiply 14 by to get back to 14,000; therefore,

$$14,000 = 14 \times 1000$$

As you know from our discussion on exponents, we can replace the 1000 with 10^3 because $1000 = 10 \times 10 \times 10$. Therefore, the powers-of-ten notation for 14,000 is 14×10^3. To convert 14×10^3 back to its original form (14,000), simply remember that each time a number is multiplied by 10, the decimal place is moved one position to the right. In this example 14 is multiplied by 10 three times (10^3, or $10 \times 10 \times 10$), and therefore the decimal point will have to be moved three positions to the right.

$$14 \times 10^3 = 14 \times 10 \times 10 \times 10 = 14{\overset{\frown}{.}}0{\overset{\frown}{0}}0{\overset{\frown}{0}}0. = 14,000$$

As another example, what is the powers-of-ten notation for the number 0.000032? If we once again remove all the zeros to obtain the number 32, we will again have to include a multiplier with 32 to indicate what 32 has to be multiplied by to return it to its original form. In this case 32 will have to be multiplied by 1/1,000,000 (one millionth) to return it to 0.000032.

$$32 \times \frac{1}{1,000,000} = 0.000032$$

This can be verified because when you divide any number by 10, you move the decimal point one position to the left. Therefore, to divide any number by 1,000,000, you simply move the decimal point six positions to the left.

$$32 \times \frac{1}{1,000,000} = \frac{32}{1,000,000} = \frac{32}{10 \times 10 \times 10 \times 10 \times 10 \times 10}$$
$$= 0.0{\overset{\frown}{0}}0{\overset{\frown}{0}}0{\overset{\frown}{0}}3{\overset{\frown}{.}}2$$

Once again an exponential expression can be used in place of the 1/1,000,000 multiplier, namely,

$$\frac{1}{1,000,000} = \frac{1}{10^6} = 0.000001 = 10^{-6}$$

Whenever you divide a number into 1, you get the *reciprocal* of that number. In this example, when you divide 1,000,000 into 1, you get 0.000001, which is equal to power-of-ten notation with a negative exponent of 10^{-6}. The multiplier 10^{-6} indicates that the decimal point must be moved back (to the left) by six places, and therefore

$$32 \times \frac{1}{1,000,000} = 32 \times 0.000001 = 32 \times 10^{-6} = 0.0{\overset{\frown}{0}}0{\overset{\frown}{0}}0{\overset{\frown}{0}}3{\overset{\frown}{.}}2$$

Now that you know exactly what a multiplier is, you have only to remember these simple rules:

1. A *negative exponent* tells you how many places *to the left* to move the decimal point.
2. A *positive exponent* tells you how many places *to the right* to move the decimal point.

Remember one important point: a negative exponent does not indicate a negative number; it simply indicates a fraction. For example, 4×10^{-3} meter means that 1 meter has been broken up into 1000 parts and we have 4 of those pieces, or 4/1000.

Most calculators have a key specifically for entering powers of ten. It is called the *exponent key* and operates as follows.

CALCULATOR KEYS

Name: Exponent entry key $\boxed{\text{EXP}}$ or $\boxed{\text{EE}}$

Function: Prepares calculator to accept next digits entered as a power-of-ten exponent. The sign of the exponent can be changed by using the change-sign key $(+/-)$.

Example: $76 \boxed{\text{EXP}} 4$

Press keys: $\boxed{7}\boxed{6}\boxed{\text{EXP}}\boxed{4}$

Display shows: $\boxed{76.04}$

Example: 85×10^{-5}

Press keys: $\boxed{8}\boxed{5}\boxed{\text{EXP}}\boxed{5}\boxed{+/-}$

Display shows: $\boxed{85.\,-05}$

EXP

+/−

Scientific Notation

A widely used floating-point system in which numbers are expressed as products consisting of a number between 1 and 10 multiplied by an appropriate power of 10.

Engineering Notation

A widely used floating-point system in which numbers are expressed as products consisting of a number that is greater than 1 multiplied by a power of 10 that is some multiple of 3.

TIME LINE

As a small boy, James C. Maxwell (1831-1879) was persistently inquisitive. He built many scientific toys before he was 8. At the age of 14 he wrote a paper on how to construct oval curves, and at 18 two of his papers were published. The supreme achievement of this Scottish physicist, however, was to translate Michael Faraday's experiments into scientific notation. This set of mathematical equations, known as Maxwell's equations, shows the relationship between electricity and magnetism.

Scientific and Engineering Notation

As mentioned previously, powers of ten are used in science and technology as a shorthand due to the large number of zeros in many values. There are basically two systems or notations used, involving values that have exponents that are a power of ten. They are called **scientific notation** and **engineering notation.**

A number in *scientific notation* is expressed as a base number between 1 and 10 multiplied by a power of ten. In the following examples, the values on the left have been converted to scientific notation.

EXAMPLE A:

$$32{,}000 = 3.2000_\circ = 3.2 \times 10^4$$

Decimal point is moved to a position that results in a base number between 1 and 10. If decimal point is moved left, exponent is positive. If decimal point is moved right, exponent is negative.

— Scientific notation

EXAMPLE B:

$$0.0019 = 0_\circ 001.9 = 1.9 \times 10^{-3}$$

— Scientific notation

EXAMPLE C:

$$114{,}300{,}000 = 1.14300000. = 1.143 \times 10^8$$

— Scientific notation

EXAMPLE D:

$$0.26 = 0.2\overset{\frown}{6} = 2.6 \times 10^{-1}$$

└ Scientific notation

As you can see from the preceding examples, the decimal point floats backward and forward, which explains why scientific notation is called a **floating-point number system.** Although scientific notation is used in science and technology, the engineering notation system, discussed next, is used more frequently.

In *engineering notation* a number is represented as a base number that is greater than 1 multiplied by a power of ten that is some multiple of 3. In the following examples, the values on the left have been converted to engineering notation.

Floating-point Number System

A system in which numbers are expressed as products consisting of a number and a power-of-10 multiplier.

EXAMPLE A:

$$32,000 = 32.\overset{\frown}{0\,0\,0} = 32 \times 10^3$$

Decimal point is moved to a position that results in a base number that is greater than 1, and a power-of-ten exponent that is some multiple of 3.

EXAMPLE B:

$$0.0019 = 0.0\,0\,0\,\overset{\frown}{1.9} = 1.9 \times 10^{-3}$$

EXAMPLE C:

$$114,300,000 = 114.\overset{\frown}{3\,0\,0\,0\,0\,0} = 114.3 \times 10^6$$

EXAMPLE D:

$$0.26 = 0.\overset{\frown}{2\,6\,0}. = 260 \times 10^{-3}$$

1-1-3 *Metric Prefixes*

The **metric system** *is a decimal system of weights and measures.* This system was developed to make working with values easier. Comparing two examples, you will see that working with the metric system, which uses multiples of 10 (decimal), is much easier than working with the *U.S. customary system of units.*

Metric System

A decimal system of weights and measures based on the meter and the kilogram.

EXAMPLE: THE METRIC SYSTEM

We are using a system like the metric system with our money. It is a decimal system, which means that it is based on multiples of 10. For instance, there are 100 cents in 1 dollar ($100 \times 1¢ = \$1$), or 10 dimes in 1 dollar ($10 \times 10¢ = \$1$).

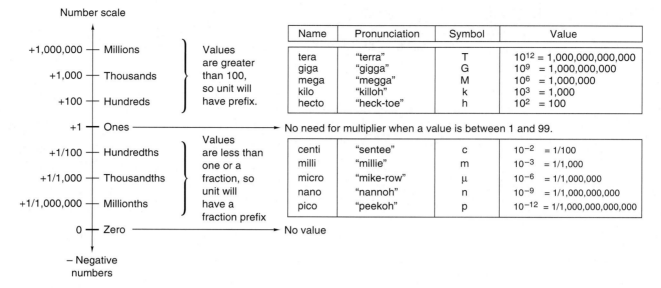

FIGURE 1-1 Metric Prefixes

The number scale (left) shows:

Number scale		
+1,000,000	Millions	Values are greater than 100, so unit will have prefix.
+1,000	Thousands	
+100	Hundreds	
+1	Ones	No need for multiplier when a value is between 1 and 99.
+1/100	Hundredths	Values are less than one or a fraction, so unit will have a fraction prefix
+1/1,000	Thousandths	
+1/1,000,000	Millionths	
0	Zero	No value
−	Negative numbers	

Name	Pronunciation	Symbol	Value
tera	"terra"	T	10^{12} = 1,000,000,000,000
giga	"gigga"	G	10^{9} = 1,000,000,000
mega	"megga"	M	10^{6} = 1,000,000
kilo	"killoh"	k	10^{3} = 1,000
hecto	"heck-toe"	h	10^{2} = 100
centi	"sentee"	c	10^{-2} = 1/100
milli	"millie"	m	10^{-3} = 1/1,000
micro	"mike-row"	μ	10^{-6} = 1/1,000,000
nano	"nannoh"	n	10^{-9} = 1/1,000,000,000
pico	"peekoh"	p	10^{-12} = 1/1,000,000,000,000

▢ EXAMPLE: THE U.S. CUSTOMARY SYSTEM

We are still not using the metric system of measurements. For instance, there are 12 inches in 1 foot (12 in. = 1 ft), and 3 feet in 1 yard (3 ft = 1 yd). None of these values are multiples of 10. To add even more to the confusion of this system, inches are divided up into strange fractions such as fourths, sixteenths, thirty-seconds, and sixty-fourths.

From these two examples you can see that it is easier to count, represent, and perform mathematical operations on a system based on multiples of 10.

To help understand what we mean by units and prefixes, let us begin by examining how we measure length using the metric system. The standard **unit** of length in the metric system is the *meter* (abbreviated m). You have probably not seen the unit "meter" used a lot on its own. More frequently you have heard and seen the terms *centimeter, millimeter,* and *kilometer.* All these words have two parts: a *prefix name* and *unit.* For example, with the name *centimeter, centi* is the prefix and *meter* is the unit. Similarly, with the names *millimeter* and *kilometer, milli* and *kilo* are the prefixes and *meter* is the unit. The next question, therefore, is: What are these prefixes? A **prefix** *is simply a power of ten or multiplier that precedes the unit.* Figure 1-1 shows the names, symbols, and values of the most frequently used metric prefixes.

Unit

A determinate quantity adopted as a standard of measurement.

Prefix

An affix attached to the beginning of a word, base, or phrase.

SELF-TEST EVALUATION POINT FOR SECTION 1-1

Now that you have completed this section, you should be able to:

■ *Objective 1.* Define the term exponent.

■ *Objective 2.* Describe what is meant by raising a number to a higher power.

■ *Objective 3.* Explain how to find the square and root of a number.

■ *Objective 4.* Explain the powers-of-ten method and how to convert to powers of ten.

■ *Objective 5.* Describe the two following floating-point number systems:
a. Scientific notation
b. Engineering notation

■ *Objective 6.* Define and explain the purpose of the metric system.

■ *Objective 7.* List the metric prefixes and describe the purpose of each.

Use the following questions to test your understanding of Section 1-1:

1. Raise the following base numbers to the power indicated by the exponent, and give the answer.
   a. $16^4 = ?$
   b. $32^3 = ?$
   c. $112^2 = ?$
   d. $15^6 = ?$
   e. $2^3 = ?$
   f. $3^{12} = ?$

2. Give the following roots.
   a. $\sqrt[2]{144} = ?$
   b. $\sqrt[3]{3375} = ?$
   c. $\sqrt{20} = ?$
   d. $\sqrt[3]{9} = ?$

3. Convert the following to powers of ten.
   a. $100 = ?$
   b. $1 = ?$
   c. $10 = ?$
   d. $1,000,000 = ?$
   e. $\dfrac{1}{1,000} = ?$
   f. $\dfrac{1}{1,000,000} = ?$

4. Convert the following to common numbers without exponents.
   a. $6.3 \times 10^3 = ?$
   b. $114,000 \times 10^{-3} = ?$
   c. $7,114,632 \times 10^{-6} = ?$
   d. $6624 \times 10^6 = ?$

5. Perform the indicated operation on the following:
   a. $\sqrt{3} \times 10^6 = ?$
   b. $(2.6 \times 10^{-6}) - (9.7 \times 10^{-9}) = ?$
   c. $\dfrac{(4.7 \times 10^3)^2}{3.6 \times 10^6} = ?$

6. Convert the following common numbers to engineering notation.
   a. $47,000 = ?$
   b. $0.00000025 = ?$
   c. $250,000,000 = ?$
   d. $0.0042 = ?$

7. Give the power-of-ten value for the following prefixes.
   a. kilo
   b. centi
   c. milli
   d. mega
   e. micro

1-2 THE STRUCTURE OF MATTER

All of the matter on the earth and in the air surrounding the earth can be classified as being either a solid, liquid, or gas. A total of approximately 107 different natural elements exist in, on, and around the earth. An **element,** by definition, is a substance consisting of only one type of atom; in other words, every element has its own distinctive atom, which makes it different from all the other elements. This **atom** is the smallest particle into which an element can be divided without losing its identity, and a group of identical atoms is called an element, shown in Figure 1-2.

For the sake of discussion, let us take a small amount of either a solid, a liquid, or a gas and divide it into two pieces. Then we divide a resulting piece into two pieces, and keep repeating the process until we finally end up with a tiny remaining part. Viewing the part under

Element

There are 107 different natural chemical substances or elements that exist on the earth and can be categorized as either being a gas, solid, or liquid.

Atom

Smallest particle of an element.

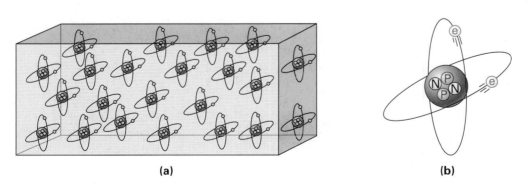

(a)   (b)

FIGURE 1-2 (a) Element: Many Similar Atoms. (b) Atom: Smallest Unit.

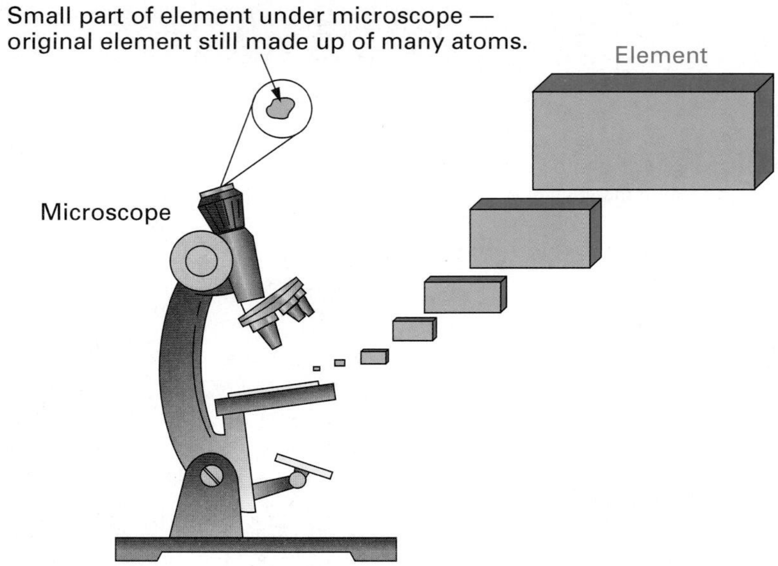

Small part of element under microscope —
original element still made up of many atoms.

Element

Microscope

FIGURE 1-3   **Elements under the Microscope.**

the microscope, as shown in Figure 1-3, the substance can still be identified as the original element as it is still made up of many of the original solid, liquid or gas atoms. A small amount of gold, for example, the size of a pinpoint, will still contain several billion atoms. If the element subdivision is continued, however, a point will be reached at which a single atom will remain. Let us now analyze the atom in more detail.

1-2-1   *The Atom*

The word atom is a Greek word meaning a particle that is too small to be subdivided. At present, we cannot clearly see the atom; however, physicists and researchers do have the ability to record a picture as small as 12 billionths of an inch (about the diameter of one atom), and this image displays the atom as a white fuzzy ball.

In 1913, a Danish physicist, Neils Bohr, put forward a theory about the atom, and his basic model outlining the **subatomic** particles that make up the atom is still in use today and is illustrated in Figure 1-4. Bohr actually combined the ideas of Lord Rutherford's (1871–1937) nuclear atom with Max Planck's (1858–1947) and Albert Einstein's (1879–1955) quantum theory of radiation.

The three important particles of the atom are the *proton,* which has a positive charge, the *neutron,* which is neutral or has no charge, and the *electron,* which has a negative charge. Referring to Figure 1-4, you can see that the atom consists of a positively charged central mass called the *nucleus,* which is made up of protons and neutrons surrounded by a quantity of negatively charged orbiting electrons.

Subatomic

Particles such as electrons, protons, and neutrons that are smaller than atoms.

TIME LINE
The electron was first discovered by Jean Baptiste Perrin (1870-1942), a French physicist who was awarded the Nobel prize for physics.

FIGURE 1-4   **The Atom.**

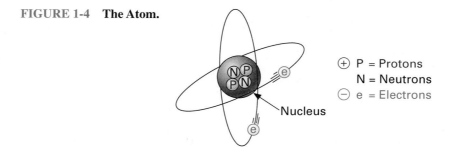

$\oplus$ P = Protons
    N = Neutrons
$\ominus$ e = Electrons

Nucleus

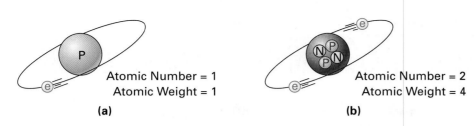

Atomic Number = 1
Atomic Weight = 1

(a)

Atomic Number = 2
Atomic Weight = 4

(b)

FIGURE 1-5   **(a) Hydrogen Atom. (b) Helium Atom.**

Table 1-1 lists the periodic table of the elements, in order of their atomic number. The **atomic number** of an atom describes the number of protons that exist within the nucleus.

The proton and the neutron are almost 2000 times heavier than the very small electron, so if we ignore the weight of the electron, we can use the fourth column in Table 1-1 (weight of an atom) to give us a clearer picture of the protons and neutrons within the atom's nucleus. For example, a hydrogen atom, shown in Figure 1-5(a), is the smallest of all atoms and has an atomic number of 1, which means that hydrogen has a one-proton nucleus. Helium, however [Figure 1-5(b)], is second on the table and has an atomic number of 2, indicating that two protons are within the nucleus. The **atomic weight** of helium, however, is 4, meaning that two protons and two neutrons make up the atom's nucleus.

The number of neutrons within an atom's nucleus can therefore be calculated by subtracting the atomic number (protons) from the atomic weight (protons and neutrons). For example, Figure 1-6(a) illustrates a beryllium atom which has the following atomic number and weight.

Atomic Number

Number of positive charges or protons in the nucleus of an atom.

Atomic Weight

The relative weight of a neutral atom of an element, based on a neutral oxygen atom having an atomic weight of 16.

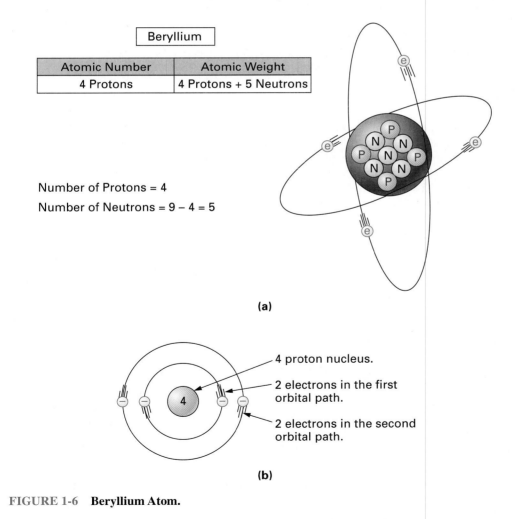

Beryllium	
Atomic Number	**Atomic Weight**
4 Protons	4 Protons + 5 Neutrons

Number of Protons = 4

Number of Neutrons = 9 − 4 = 5

(a)

4 proton nucleus.

2 electrons in the first orbital path.

2 electrons in the second orbital path.

(b)

FIGURE 1-6   **Beryllium Atom.**

TABLE 1-1   Periodic Table of the Elements

ATOMIC NUMBER	ELEMENT NAME	SYMBOL	ATOMIC WEIGHT	ELECTRONS/SHELL K L M N O P Q							DISCOVERED	COMMENT
				K	L	M	N	O	P	Q		
1	Hydrogen	H	1.007	1							1766	Active gas
2	Helium	He	4.002	2							1895	Inert gas
3	Lithium	Li	6.941	2	1						1817	Solid
4	Beryllium	Be	9.01218	2	2						1798	Solid
5	Boron	B	10.81	2	3						1808	Solid
6	Carbon	C	12.011	2	4						Ancient	Semiconductor
7	Nitrogen	N	14.0067	2	5						1772	Gas
8	Oxygen	O	15.9994	2	6						1774	Gas
9	Fluorine	F	18.998403	2	7						1771	Active gas
10	Neon	Ne	20.179	2	8						1898	Inert gas
11	Sodium	Na	22.98977	2	8	1					1807	Solid
12	Magnesium	Mg	24.305	2	8	2					1755	Solid
13	Aluminum	Al	26.98154	2	8	3					1825	Metal conductor
14	Silicon	Si	28.0855	2	8	4					1823	Semiconductor
15	Phosphorus	P	30.97376	2	8	5					1669	Solid
16	Sulfur	S	32.06	2	8	6					Ancient	Solid
17	Chlorine	Cl	35.453	2	8	7					1774	Active gas
18	Argon	Ar	39.948	2	8	8					1894	Inert gas
19	Potassium	K	39.0983	2	8	8	1				1807	Solid
20	Calcium	Ca	40.08	2	8	8	2				1808	Solid
21	Scandium	Sc	44.9559	2	8	9	2				1879	Solid
22	Titanium	Ti	47.90	2	8	10	2				1791	Solid
23	Vanadium	V	50.9415	2	8	11	2				1831	Solid
24	Chromium	Cr	51.996	2	8	13	1				1798	Solid
25	Manganese	Mn	54.9380	2	8	13	2				1774	Solid
26	Iron	Fe	55.847	2	8	14	2				Ancient	Solid (magnetic)
27	Cobalt	Co	58.9332	2	8	15	2				1735	Solid
28	Nickel	Ni	58.70	2	8	16	2				1751	Solid
29	Copper	Cu	63.546	2	8	18	1				Ancient	Metal conductor
30	Zinc	Zn	65.38	2	8	18	3				1746	Solid
31	Gallium	Ga	69.72	2	8	18	4				1875	Liquid
32	Germanium	Ge	72.59	2	8	18	4				1886	Semiconductor
33	Arsenic	As	74.9216	2	8	18	5				1649	Solid
34	Selenium	Se	78.96	2	8	18	6				1818	Photosensitive
35	Bromine	Br	79.904	2	8	18	7				1898	Liquid
36	Krypton	Kr	83.80	2	8	18	8				1898	Inert gas
37	Rubidium	Rb	85.4678	2	8	18	8	1			1861	Solid
38	Strontium	Sr	87.62	2	8	18	8	2			1790	Solid

Beryllium
Atomic number: 4 (protons)
Atomic weight: 9 (protons and neutrons)

Neutral Atom

An atom in which the number of positive charges in the nucleus (protons) is equal to the number of negative charges (electrons) that surround the nucleus.

Subtracting the beryllium atom's weight from the beryllium atomic number, we can determine the number of neutrons in the beryllium atom's nucleus, as shown in Figure 1-6(a).

In most instances, atoms like the beryllium atom will not be drawn in the three-dimensional way shown in Figure 1-6(a). Figure 1-6(b) shows how a beryllium atom could be more easily drawn as a two-dimensional figure.

A **neutral atom** or *balanced atom* is one that has an equal number of protons and orbiting electrons, so the net positive proton charge is equal but opposite to the net negative

TABLE 2-1 *(continued)*

ATOMIC NUMBER[a]	ELEMENT NAME	SYMBOL	ATOMIC WEIGHT	ELECTRONS/SHELL K L M N O P Q							DISCOVERED	COMMENT
39	Yttrium	Y	88.9059	2	8	18	9	2			1843	Solid
40	Zirconium	Zr	91.22	2	8	18	10	2			1789	Solid
41	Niobium	Nb	92.9064	2	8	18	12	1			1801	Solid
42	Molybdenum	Mo	95.94	2	8	18	13	1			1781	Solid
43	Technetium	Tc	98.0	2	8	18	14	1			1937	Solid
44	Ruthenium	Ru	101.07	2	8	18	15	1			1844	Solid
45	Rhodium	Rh	102.9055	2	8	18	16	1			1803	Solid
46	Palladium	Pd	106.4	2	8	18	18				1803	Solid
47	Silver	Ag	107.868	2	8	18	18	1			Ancient	Metal conductor
48	Cadmium	Cd	112.41	2	8	18	18	2			1803	Solid
49	Indium	In	114.82	2	8	18	18	3			1863	Solid
50	Tin	Sn	118.69	2	8	18	18	4			Ancient	Solid
51	Antimony	Sb	121.75	2	8	18	18	5			Ancient	Solid
52	Tellurium	Te	127.60	2	8	18	18	6			1783	Solid
53	Iodine	I	126.9045	2	8	18	18	7			1811	Solid
54	Xenon	Xe	131.30	2	8	18	18	8			1898	Inert gas
55	Cesium	Cs	132.9054	2	8	18	18	8	1		1803	Liquid
56	Barium	Ba	137.33	2	8	18	18	8	2		1808	Solid
57	Lanthanum	La	138.9055	2	8	18	18	9	2		1839	Solid
72	Hafnium	Hf	178.49	2	8	18	32	10	2		1923	Solid
73	Tantalum	Ta	180.9479	2	8	18	32	11	2		1802	Solid
74	Tungsten	W	183.85	2	8	18	32	12	2		1783	Solid
75	Rhenium	Re	186.207	2	8	18	32	13	2		1925	Solid
76	Osmium	Os	190.2	2	8	18	32	14	2		1804	Solid
77	Iridium	Ir	192.22	2	8	18	32	15	2		1804	Solid
78	Platinum	Pt	195.09	2	8	18	32	16	2		1735	Solid
79	Gold	Au	196.9665	2	8	18	32	18	1		Ancient	Solid
80	Mercury	Hg	200.59	2	8	18	32	18	2		Ancient	Liquid
81	Thallium	Tl	204.37	2	8	18	32	18	3		1861	Solid
82	Lead	Pb	207.2	2	8	18	32	18	4		Ancient	Solid
83	Bismuth	Bi	208.9804	2	8	18	32	18	5		1753	Solid
84	Polonium	Po	209.0	2	8	18	32	18	6		1898	Solid
85	Astatine	At	210.0	2	8	18	32	18	7		1945	Solid
86	Radon	Rn	222.0	2	8	18	32	18	8		1900	Inert gas
87	Francium	Fr	223.0	2	8	18	32	18	8	1	1945	Liquid
88	Radium	Ra	226.0254	2	8	18	32	18	8	2	1898	Solid
89	Actinium	Ac	227.0278	2	8	18	32	18	9	2	1899	Solid

[a]Rare earth series 58–71 and 90–107 have been omitted

electron charge, resulting in a balanced or neutral state. For example, Figure 1-7 illustrates a copper atom, which is the most commonly used metal in the field of electronics. It has an atomic number of 29, meaning that 29 protons and 29 electrons exist within the atom when it is in its neutral state.

Orbiting electrons travel around the nucleus at varying distances from the nucleus, and these orbital paths are known as **shells** or **bands.** The orbital shell nearest the nucleus is referred to as the first or K shell. The second is known as the L, the third is M, the fourth is N, the fifth is O, the sixth is P, and the seventh is referred to as the Q shell. There are seven

Shells or Bands

An orbital path containing a group of electrons that have a common energy level.

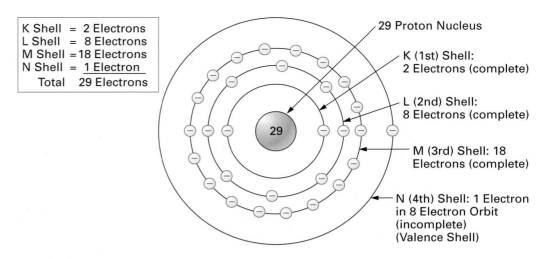

```
K Shell  =  2 Electrons
L Shell  =  8 Electrons
M Shell = 18 Electrons
N Shell  =  1 Electron
   Total   29 Electrons
```

29 Proton Nucleus

K (1st) Shell:
2 Electrons (complete)

L (2nd) Shell:
8 Electrons (complete)

M (3rd) Shell: 18
Electrons (complete)

N (4th) Shell: 1 Electron
in 8 Electron Orbit
(incomplete)
(Valence Shell)

FIGURE 1-7   Copper Atom.

shells available for electrons (K, L, M, N, O, P, and Q) around the nucleus, and each of these seven shells can only hold a certain number of electrons, as shown in Figure 1-8.

An atom's outermost electron-occupied shell is referred to as the **valence shell** or **ring,** and electrons in this shell are termed *valence electrons.* In the case of the copper atom, a single valence electron exists in the valence N shell.

All matter exists in one of three states: solid, liquid, or gas. The atoms of a solid are fixed in relation to one another but vibrate in a back-and-forth motion, unlike liquid atoms, which can flow over each other. The atoms of a gas move rapidly in all directions and collide with one another. The far-right column of Table 1-1 indicates whether the element is a gas, a solid, or a liquid.

Valence Shell or Ring
Outermost shell formed by electrons.

1-2-2   *Laws of Attraction and Repulsion*

For the sake of discussion and understanding, let us theoretically imagine that we are able to separate some positive and negative subatomic particles. Using these separated protons and electrons, let us carry out a few experiments, the results of which are illustrated in Figure 1-9. Studying Figure 1-9, you will notice that:

1. *Like charges* (positive and positive or negative and negative) repel one another.
2. *Unlike charges* (positive and negative or negative and positive) attract one another.

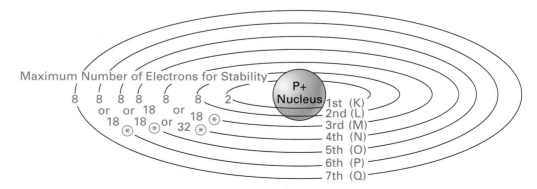

Maximum Number of Electrons for Stability

P+ Nucleus

1st (K)
2nd (L)
3rd (M)
4th (N)
5th (O)
6th (P)
7th (Q)

*The maximum number of electrons in these shells is dependent on the element's place in the periodic table.

FIGURE 1-8   Electrons and Shells.

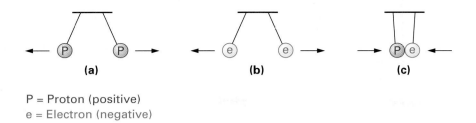

(a)   (b)   (c)

P = Proton (positive)
e = Electron (negative)

FIGURE 1-9   Attraction and Repulsion. (a) Positive Repels Positive. (b) Negative Repels Negative. (c) Unlike Charges Attract.

Orbiting negative electrons are therefore attracted toward the positive nucleus, which leads us to the question of why the electrons do not fly into the atom's nucleus. The answer is that the orbiting electrons remain in their stable orbit due to two equal but opposite forces. The centrifugal outward force exerted on the electrons due to the orbit counteracts the attractive inward force trying to pull the electrons toward the nucleus due to the unlike charges.

Due to their distance from the nucleus, valence electrons are described as being loosely bound to the atom. These electrons can easily be dislodged from their outer orbital shell by any external force, to become a **free electron.**

1-2-3   *The Molecule*

An atom is the smallest unit of a natural element, or an element is a substance consisting of a large number of the same atom. Combinations of elements are known as **compounds,** and the smallest unit of a compound is called a **molecule,** just as the smallest unit of an element is an atom. Figure 1-10 summarizes how elements are made up of atoms and compounds are made up of molecules.

Free Electron

An electron that is not in any orbit around a nucleus.

Compound

A material composed of united separate elements.

Molecule

Smallest particle of a compound that still retains its chemical characteristics.

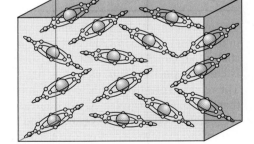

Element   Compound

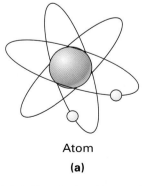   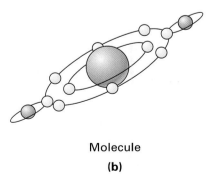

Atom   Molecule

(a)   (b)

FIGURE 1-10   (a) An Element is Made Up of Many Atoms. (b) A Compound Is Made Up of Many Molecules.

Water is an example of a liquid compound in which the molecule (H_2O) is a combination of an explosive gas (hydrogen) and a very vital gas (oxygen). Table salt is another example of a compound; here the molecule is made up of a highly poisonous gas atom (chlorine) and a potentially explosive solid atom (sodium). These examples of compounds each contain atoms that, when alone, are both poisonous and explosive, yet when combined the resulting substance is as ordinary and basic as water and salt.

SELF-TEST EVALUATION POINT FOR SECTION 1-2

Now that you have completed this section, you should be able to:

■ **Objective 8.** *Explain the atom's subatomic particles.*

■ **Objective 9.** *State the difference between an atomic number and atomic weight.*

■ **Objective 10.** *Understand the term* natural element.

■ **Objective 11.** *Describe what is meant by and state how many shells or bonds exist around an atom.*

■ **Objective 12.** *Explain the difference between:*
  a. *An atom and a molecule*
  b. *An element and a compound*
  c. *A proton and an electron*

■ **Objective 13.** *Describe the terms:*
  a. *Neutral atom*
  b. *Negative ion*
  c. *Positive ion*

Use the following questions to test your understanding of Section 1-2:

1. Define the difference between an element and a compound.
2. Name the three subatomic particles that make up an atom.
3. What is the most commonly used metal in the field of electronics?
4. State the laws of attraction and repulsion.

1-3 CURRENT

The movement of electrons from one point to another is known as *electrical current*. Energy in the form of heat or light can cause an outer shell electron to be released from the valence shell of an atom. Once an electron is released, the atom is no longer electrically neutral and is called a **positive ion,** as it now has a net positive charge (more protons than electrons). The released electron tends to jump into a nearby atom, which will then have more electrons than protons and is referred to as a **negative ion.**

Let us now take an example and see how electrons move from one point to another. Figure 1-11 shows a broken metal conductor between two charged objects. The metal conductor could be either gold, silver, or copper, but whichever it is, one common trait can be noted: The valence electrons in the outermost shell are very loosely bound and can easily be pulled from their parent atom.

Positive Ion

Atom that has lost one or more of its electrons and therefore has more protons than electrons, resulting in a net positive charge.

Negative Ion

Atom that has more than the normal neutral amount of electrons.

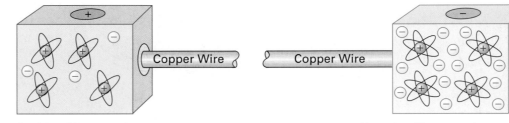

Positive Charge or Ions—
More Protons than Electrons
(absence of free electrons).

Negative Charge or Ions—
More Electrons than Protons
(abundance of free electrons).

FIGURE 1-11 **Positive and Negative Charges.**

In Figure 1-12, the conductor between the two charges has been joined so that a path now exists for current flow. The negative ions on the right in Figure 1-12 have more electrons than protons, while the positive ions on the left in Figure 1-12 have fewer electrons than protons and so display a **positive charge.** The metal joining the two charges has its own atoms, which begin in the neutral condition.

Let us now concentrate on one of the negative ions. In Figure 1-12(a), the extra electrons in the outer shells of the negative ions on the right side will feel the attraction of the positive ions on the left side and the repulsion of the outer negative ions, or **negative charge.** This will cause an electron in a negative ion to jump away from its parent atom's orbit and land in an adjacent atom to the left within the metal wire conductor, as shown in Figure 1-12(b). This adjacent atom now has an extra electron and is called a negative ion, while the initial parent negative ion becomes a neutral atom, which will now receive an electron from one of the other negative ions, because their electrons are also feeling the attraction of the positive ions on the left side and the repulsion of the surrounding negative ions.

The electrons of the negative ion within the metal conductor feel the attraction of the positive ions, and eventually one of its electrons jumps to the left and into the adjacent atom, as shown in Figure 1-12(c). This continual movement to the left will produce a stream of electrons flowing from right to left. Millions upon millions of atoms within the conductor

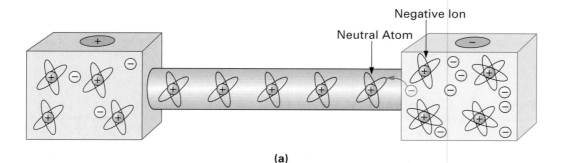

(a)

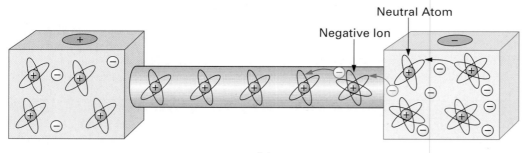

(b)

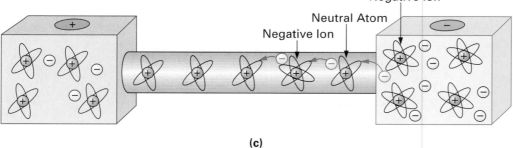

(c)

FIGURE 1-12   Electron Migration Due to Forces of Positive Attraction and Negative Repulsion on Electrons.

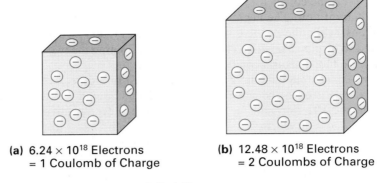

(a) 6.24×10^{18} Electrons
= 1 Coulomb of Charge

(b) 12.48×10^{18} Electrons
= 2 Coulombs of Charge

FIGURE 1-13 **(a) 1 C of Charge. (b) 2 C of Charge.**

Electric Current (I)

Measured in amperes or amps, it is the flow of electrons through a conductor.

pass a continuous movement of billions upon billions of electrons from right to left. This electron flow is known as **electric current.**

To summarize, we could say that as long as a force or pressure, produced by the positive charge and negative charge, exists it will cause electrons to flow from the negative to the positive terminal. The positive side has a deficiency of electrons and the negative side has an abundance, and so a continuous flow or migration of electrons takes place between the negative and positive terminal through our metal conducting wire. This electric current or electron flow is a measurable quantity, as will now be explained.

1-3-1 *Coulombs per Second*

Coulomb of Charge

Unit of electric charge. One coulomb equals 6.24×10^{18} electrons.

There are 6.24×10^{18} electrons in 1 **coulomb of charge,** as illustrated in Figure 1-13. To calculate coulombs of charge (designated Q), we can use the formula

$$\text{charge, } Q = \frac{\text{total number of electrons } (n)}{6.24 \times 10^{18}}$$

where Q is the electric charge in coulombs.

EXAMPLE:

If a total of 3.75×10^{19} free electrons exist within a piece of metal conductor, how many coulombs (C) of charge would be within this conductor?

Solution:

By using the charge formula (Q) we can calculate the number of coulombs (C) in the conductor.

$$Q = \frac{n}{6.24 \times 10^{18}}$$
$$= \frac{3.75 \times 10^{19}}{6.24 \times 10^{18}}$$
$$= 6 \text{ C}$$

A total of 6 C of charge exists within the conductor.

In the calculator sequence you will see how the exponent key (E, EE, or EXP) on your calculator can be used.

CALCULATOR SEQUENCE

Step	Keypad Entry	Display Response
1.	③ . ⑦ ⑤ Ⓔ (Exponent) ① ⑨	3.75E19
2.	÷	
3.	⑥ . ② ④ Ⓔ ① ⑧	6.24E18
4.	=	6.0096

1-3-2   *The Ampere*

A coulomb is a **static** amount of electric charge. In electronics, we are more interested in electrons in motion. Coulombs and time are therefore combined to describe the number of electrons and the rate at which they flow. This relationship is called *current (I)* flow and has the unit of **amperes (A).** By definition, 1 ampere of current is said to be flowing if 6.24×10^{18} electrons (1 C) are drifting past a specific point on a conductor in 1 second of time. Stated as a formula:

$$\text{current } (I) = \frac{\text{coulombs } (Q)}{\text{time } (t)}$$

$$1 \text{ ampere} = 1 \text{ coulomb per 1 second}$$

$$1 \text{ A} = \frac{1 \text{ C}}{1 \text{ s}}$$

In summary, 1 ampere equals a flow rate of 1 coulomb per second, and current is measured in amperes.

TIME LINE

The unit of electrical current is the ampere, named in honor of André Ampère (1775-1835), a French physicist who pioneered in the study of electromagnetism. After hearing of Hans Oerstad's discoveries, he conducted further experiments and discovered that two current-carrying conductors would attract and repel one another, just like two magnets.

EXAMPLE:

If 5×10^{19} electrons pass a point in a conductor in 4 s, what is the amount of current flow in amperes?

Static

Crackling noise heard on radio receivers caused by electric storms or electric devices in the vicinity.

Solution:

Current (I) is equal to Q/t. We must first convert electrons to coulombs.

Ampere (A)

Unit of electric current.

$$Q = \frac{n}{6.24 \times 10^{18}}$$

$$= \frac{5 \times 10^{19}}{6.24 \times 10^{18}}$$

$$= 8 \text{ C}$$

Now, to calculate the amount of current, we use the formula

$$I = \frac{Q}{t}$$

$$= \frac{8 \text{ C}}{4 \text{ s}}$$

$$= 2 \text{ A}$$

This means that 2 A or 1.248×10^{19} electrons (2 C) are passing a specific point in the conductor every second.

CALCULATOR SEQUENCE

Step	Keypad Entry	Display Response
1.	[5] [E] (exponent) [1] [9]	5E19
2.	[÷]	
3.	[6] [.] [2] [4] [E] [1] [8]	6.24E18
4.	[=]	8.012
5.	[÷]	
6.	[4]	2.003
7.	[=]	2.003

1-3-3   *Units of Current*

Current within electronic equipment is normally a value in milliamperes or microamperes and very rarely exceeds 1 ampere. Table 1-2 lists all the prefixes related to current. For example, 1 milliampere is one-thousandth of an ampere, which means that if 1 ampere were divided into 1000 parts, 1 part of the 1000 parts would be flowing through the circuit.

TABLE 1-2 Current Units

NAME	SYMBOL	VALUE
Picoampere	pA	$10^{-12} = \dfrac{1}{1{,}000{,}000{,}000{,}000}$
Nanoampere	nA	$10^{-9} = \dfrac{1}{1{,}000{,}000{,}000}$
Microampere	μA	$10^{-6} = \dfrac{1}{1{,}000{,}000}$
Milliampere	mA	$10^{-3} = \dfrac{1}{1000}$
Ampere	A	$10^{0} = 1$
Kiloampere	kA	$10^{3} = 1000$
Megaampere	MA	$10^{6} = 1{,}000{,}000$
Gigaampere	GA	$10^{9} = 1{,}000{,}000{,}000$
Teraampere	TA	$10^{12} = 1{,}000{,}000{,}000{,}000$

EXAMPLE:

Convert the following:

    a. 0.003 A = _____ mA (milliamperes)

    b. 0.07 mA = _____ μA (microamperes)

    c. 7333 mA = _____ A (amperes)

    d. 1275 μA = _____ mA (milliamperes)

Solution:

    a. 0.003A = _____ mA. In this example, 0.003 A has to be converted so that it is represented in milliamperes (10^{-3} or $\frac{1}{1000}$ of an ampere). The basic algebraic rule to be remembered is that both expressions on either side of the equals must be equal.

LEFT		RIGHT	
Base	Multiplier	Base	Multiplier
0.003×10^{0}		= _____	$\times 10^{-3}$

The multiplier on the right in this example is going to be decreased 1000 times (10^{0} to 10^{-3}), so for the statement to balance the number on the right will have to be increased 1000 times; that is, the decimal point will have to be moved to the right three places (0.003 or 3). Therefore,

$$0.003 \times 10^{0} = 3 \times 10^{-3}$$

or

$$0.003 \text{ A} = 3 \times 10^{-3} \text{ A or 3 mA}$$

b. 0.07 mA = _____ μA. In this example the unit is going from milliamperes to microamperes (10^{-3} to 10^{-6}) or 1000 times smaller, so the number must be made 1000 times greater.

$$0.0\overset{\frown}{7}\overset{\frown}{0} \text{ or } 70.0$$

Therefore, 0.07 mA = 70 μA.

c. 7333 mA = _____ A. The unit is going from milliamperes to amperes, increasing 1000 times, so the number must decrease 1000 times.

$$7\overset{\frown}{3}\overset{\frown}{3}\overset{\frown}{3}. \text{ or } 7.333$$

Therefore, 7333 mA = 7.333 A.

d. 1275 μA = _____ mA. The unit is changing from microamperes to milliamperes, an increase of 1000 times, so the number must decrease by the same factor.

$$1\overset{\frown}{2}\overset{\frown}{7}\overset{\frown}{5}.0 \text{ or } 1.275$$

Therefore, 1275 μA = 1.275 mA.

1-3-4   *The Speed of Current Flow*

Electrons will in fact move very slowly as they jump from parent to adjacent atom; however, the chain reaction occurs at the speed of light, which is 186,000 miles per second or 300,000,000 meters (m) per second (3×10^8 m/s). This chain reaction is best understood by using an analogy where we compare free electrons within a conductor to a string of ping-pong balls within a tube.

If one extra ball is inserted in one end, a ball will appear out of the other end almost instantly. Although each ball within the tube has only moved a small distance, the effect has traveled almost instantly toward the end of the tube, as seen in Figure 1-14(a).

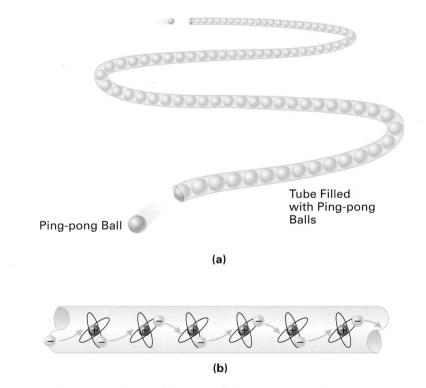

Tube Filled with Ping-pong Balls

Ping-pong Ball

(a)

(b)

FIGURE 1-14   **Chain Reaction. (a) Ping-Pong Ball Analogy. (b) Electrons.**

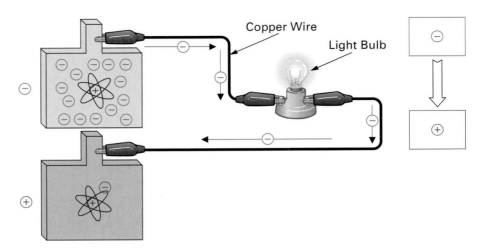

Copper Wire

Light Bulb

FIGURE 1-15 Electron Current Flow.

TIME LINE

The unit of voltage, the volt, was named in honor of Alessandro Volta (1745-1827), an Italian physicist who is famous for his invention of the electric battery. In 1801, he was called to Paris by Napoleon to show his experiment on the generation of electric current.

Electron Flow

A current produced by the movement of free electrons toward a positive terminal.

Conventional Current Flow

A current produced by the movement of positive charges toward a negative terminal.

As far as electrons go, the first electron jump from parent to adjacent atom causes the next to jump, and the next, and next, and so on, as shown in Figure 1-14(b). The first electron therefore causes a chain reaction with the others, and even though the actual speed of the electrons is only a fraction of an inch per second, the effective velocity is at the speed of light.

1-3-5 *Conventional versus Electron Flow*

Electrons drift from a negative to a positive charge, as illustrated in Figure 1-15. As already stated, this current is known as **electron flow.**

In the eighteenth and nineteenth centuries, when very little was known about the atom, researchers believed that current was a flow of positive charges. Although this has now been proved incorrect, many texts still use **conventional current flow,** which is shown in Figure 1-16.

Whether conventional flow or electron flow is used, the same answers to problems, measurements, and designs are obtained. The key point to remember is that direction is not important, but the amount of current flow is.

Throughout this book we will be using electron flow so that we can relate back to the atom when necessary. If you wish to use conventional flow, just reverse the direction of the arrows. To avoid confusion, be consistent with your choice of flow.

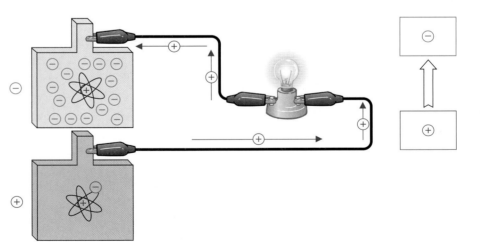

TIME LINE

The galvanometer, which is used to measure electrical current, was named after Luigi Galvani (1737-1798), who conducted many experiments with electrical current, or, as it was known at the time, "galvanism."

FIGURE 1-16 Conventional Current Flow.

1-3-6 How Is Current Measured?

Ammeters (ampere meters) are used to measure the current flow within a circuit. Stepping through the sequence detailed in Figure 1-17, you will see how an ammeter is used to measure the value of current within a circuit.

In the simple circuit shown in Figure 1-17, an ON/OFF switch is being used to turn ON or OFF a small light bulb. One of the key points to remember with ammeters is that if you wish to measure the value of current flowing within a wire, the current path must be opened and the ammeter placed in the path so that it can sense and display the value.

Ammeter

Meter placed in the path of current flow to measure the amount.

SELF-TEST EVALUATION POINT FOR SECTION 1-3

Now that you have completed this section, you should be able to:

■ **Objective 14.** *Define electrical current.*

■ **Objective 15.** *Describe the ampere in relation to coulombs per second.*

■ **Objective 16.** *List the different currents and values.*

■ **Objective 17.** *Explain why the effect of current flow travels at the speed of light.*

■ **Objective 18.** *Describe the difference between conventional current flow and electron flow.*

■ **Objective 19.** *List the three rules to apply when measuring current.*

Use the following questions to test your understanding of Section 1-3:

1. What is the unit of current?
2. Define current in relation to coulombs and time.
3. What is the difference between conventional and electron current flow?
4. What test instrument is used to measure current?

1-4 VOLTAGE

Voltage is the force or pressure exerted on electrons. Referring to Figure 1-18(a) and (b), you will notice two situations. Figure 1-18(a) shows highly concentrated positive and negative charges or potentials connected to one another by a copper wire. In this situation, a large potential difference or voltage is being applied across the copper atom's electrons. This force or voltage causes a large amount of copper atom electrons to move from right to left. On the other hand, Figure 1-18(b) illustrates a low concentration of positive and negative potentials, so a small voltage or pressure is being applied across the conductor, causing a small amount of force, and therefore current, to move from right to left.

In summary, we could say that a highly concentrated charge produces a high voltage, whereas a low concentrated charge produces a low voltage. Voltage is also appropriately known as the "electron moving force" or **electromotive force (emf),** and since two opposite potentials exist (one negative and one positive), the strength of the voltage can also be referred to as the amount of **potential difference (PD)** applied across the circuit. To compare, we can say that a large voltage, electromotive force, or potential difference exists across the copper conductor in Figure 1-18(a), while a small voltage, potential difference, or electromotive force is exerted across the conductor in Figure 1-18(b).

Voltage is the force, pressure, potential difference (PD), or electromotive force (emf) that causes electron flow or current and is symbolized by italic uppercase V. The unit for voltage is the volt, symbolized by roman uppercase V. This can become a bit confusing. For example, when the voltage applied to a circuit equals 5 volts, the circuit notation would appear as

$$V = 5 \text{ V}$$

You know the first V represents "voltage," not "volt," because 1 volt cannot equal 5 volts. To avoid confusion, some texts and circuits use E, symbolizing electromotive force, to represent voltage; for example,

$$E = 5 \text{ V}$$

In this text, we will maintain the original designation for voltage (V).

Electromotive Force (emf)

Force that causes the motion of electrons due to a potential difference between two points.

Potential Difference (PD)

Voltage difference between two points, which will cause current to flow in a closed circuit.

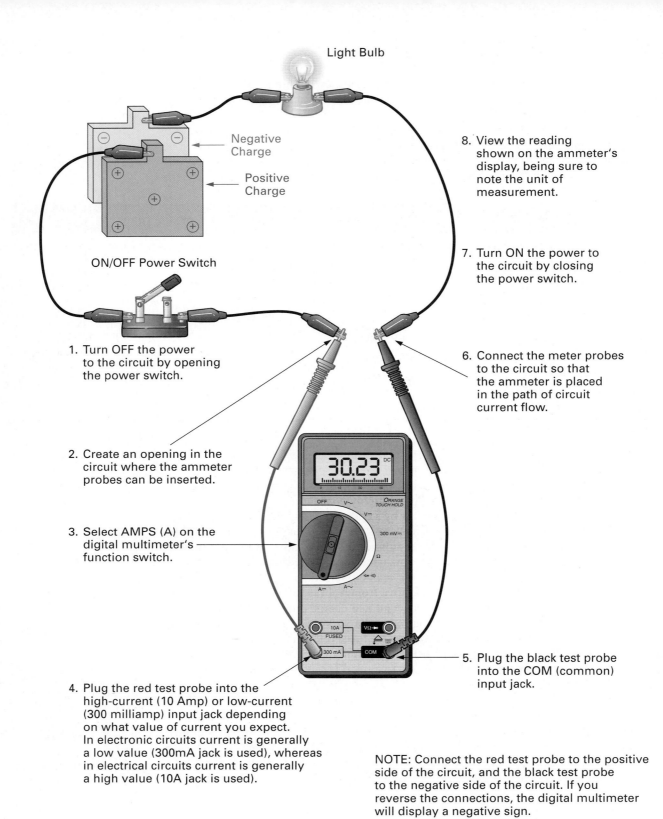

Light Bulb

Negative Charge

Positive Charge

ON/OFF Power Switch

8. View the reading shown on the ammeter's display, being sure to note the unit of measurement.

7. Turn ON the power to the circuit by closing the power switch.

6. Connect the meter probes to the circuit so that the ammeter is placed in the path of circuit current flow.

1. Turn OFF the power to the circuit by opening the power switch.

2. Create an opening in the circuit where the ammeter probes can be inserted.

3. Select AMPS (A) on the digital multimeter's function switch.

5. Plug the black test probe into the COM (common) input jack.

4. Plug the red test probe into the high-current (10 Amp) or low-current (300 milliamp) input jack depending on what value of current you expect. In electronic circuits current is generally a low value (300mA jack is used), whereas in electrical circuits current is generally a high value (10A jack is used).

NOTE: Connect the red test probe to the positive side of the circuit, and the black test probe to the negative side of the circuit. If you reverse the connections, the digital multimeter will display a negative sign.

FIGURE 1-17   Measuring Current with the Ammeter.

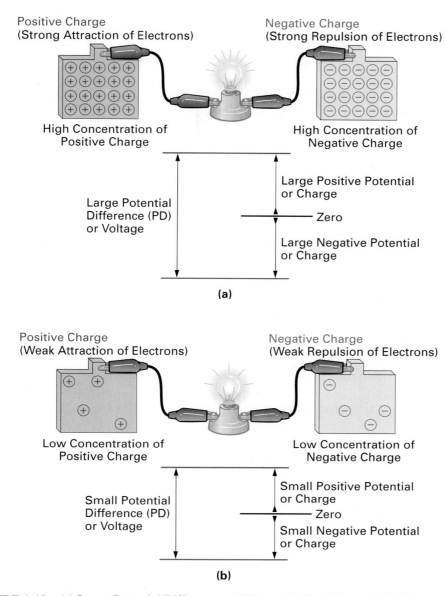

Positive Charge
(Strong Attraction of Electrons)

Negative Charge
(Strong Repulsion of Electrons)

High Concentration of
Positive Charge

High Concentration of
Negative Charge

Large Positive Potential
or Charge

Zero

Large Potential
Difference (PD)
or Voltage

Large Negative Potential
or Charge

(a)

Positive Charge
(Weak Attraction of Electrons)

Negative Charge
(Weak Repulsion of Electrons)

Low Concentration of
Positive Charge

Low Concentration of
Negative Charge

Small Positive Potential
or Charge

Zero

Small Potential
Difference (PD)
or Voltage

Small Negative Potential
or Charge

(b)

FIGURE 1-18   (a) Large Potential Difference or Voltage. (b) Small Potential Difference or Voltage.

1-4-1   *Symbols*

A **battery,** like the one shown in Figure 1-19(a), converts chemical energy into electrical energy. At the positive terminal of the battery, positive charges or ions (atoms with more protons than electrons) are present, and at the negative terminal, negative charges or ions (atoms with more electrons than protons) are available to supply electrons for current flow within a circuit. A battery, therefore, chemically generates negative and positive ions at its respective terminals. The symbol for the battery is shown in Figure 1-19(b).

In Figure 1-20(a) you can see the schematic symbols for many of the devices discussed so far, along with their physical appearance.

In the circuit shown in Figure 1-20(b), a 9 V battery chemically generates positive and negative ions. The negative ions at the negative terminal force away the negative electrons, which are attracted by the positive charge or absence of electrons at the positive terminal. As the electrons proceed through the copper conductor wire, jumping from one atom to the next, they eventually reach the bulb. As they pass through the bulb, they cause it to glow.

Battery

DC voltage source containing two or more cells that converts chemical energy into electrical energy.

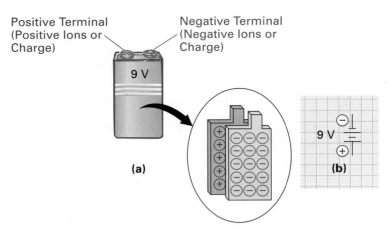

FIGURE 1-19　The Battery—A Source of Voltage. (a) Physical Appearance. (b) Schematic Symbol.

When emerging from the light bulb, the electrons travel through another connector cable and finally reach the positive terminal of the battery.

Studying Figure 1-20(b), you will notice two reasons why the circuit is drawn using symbols rather than illustrating the physical appearance:

1. A circuit with symbols can be drawn faster and more easily.
2. A circuit with symbols has less detail and clutter, and is therefore more easily comprehended since it has fewer distracting elements.

1-4-2　*Units of Voltage*

The unit for voltage is the volt (V). Voltage within electronic equipment is normally measured in volts, whereas heavy-duty industrial equipment normally requires high voltages that are generally measured in kilovolts (kV). Table 1-3 lists all the prefixes and values related to volts.

TABLE 1-3　**Voltage Units**

NAME	SYMBOL	VALUE
Picovolts	pV	$10^{-12} = \dfrac{1}{1,000,000,000,000}$
Nanovolts	nV	$10^{-9} = \dfrac{1}{1,000,000,000}$
Microvolts	μV	$10^{-6} = \dfrac{1}{1,000,000}$
Millivolts	mV	$10^{-3} = \dfrac{1}{1000}$
Volts	V	$10^{0} = 1$
Kilovolts	kV	$10^{3} = 1000$
Megavolts	MV	$10^{6} = 1,000,000$
Gigavolts	GV	$10^{9} = 1,000,000,000$
Teravolts	TV	$10^{12} = 1,000,000,000,000$

Component	Symbol	Name	Description
		Incandescent lamp	Incandescence: Release of visible radiation (light) by a heated object
		Connecting wire with end alligator clips	Used to connect different components
9 V		Battery	Source of voltage and current
	OPEN CLOSED	Switch	A device used to open or close a current path
30.23	(AM)	Ammeter	Used to measure the current flow within a circuit

(a)

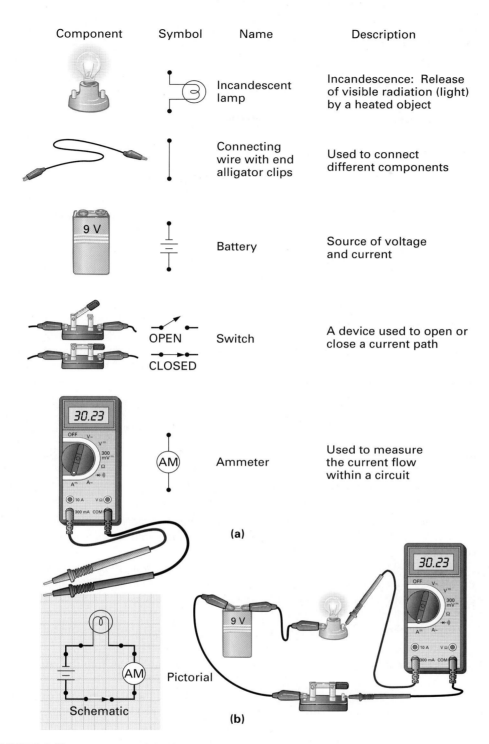

Schematic

Pictorial

(b)

FIGURE 1-20   **(a) Components. (b) Example Circuit.**

EXAMPLE:

Convert the following:

    a. 3000 V = _____ kV (kilovolts)

    b. 0.14 V = _____ mV (millivolts)

    c. 1500 kV = _____ MV (megavolts)

■ *Solution:*

    a. 3000 V = 3 kV or 3×10^3 volts (multiplier ↑ 1000, number ↓ 1000)

    b. 0.14 V = 140 mV or 140×10^{-3} volt (multiplier ↓ 1000, number ↑ 1000)

    c. 1500 kV = 1.5 MV or 1.5×10^6 volts (multiplier ↑ 1000, number ↓ 1000)

1-4-3   *How Is Voltage Measured?*

Voltmeter

Instrument designed to measure the voltage or potential difference. Its scale can be graduated in kilovolts, volts, or millivolts.

Voltmeters (voltage meters) are used to measure electrical pressure or voltage. Stepping through the sequence detailed in Figure 1-21, you will see how a voltmeter can be used to measure a battery's voltage.

    In many instances, the voltmeter is used to measure the potential difference or voltage drop across a device, as shown in Figure 1-22(a) and (b). Figure 1-22(a) shows how to measure the voltage across light bulb 1 (L1), and Figure 1-22(b) shows how to measure the voltage across light bulb 2 (L2).

    In some applications, fixed-range voltmeters are used, like the one shown in Figure 1-23. In this example, a voltage indicator on a car's dashboard shows the condition of the car's battery.

1-4-4   *Fluid Analogy of Current and Voltage*

An analogy is a comparison that describes the similarities between two otherwise different things. In this section, we will use a fluid analogy to reinforce your understanding of current and voltage.

    In Figure 1-24(a), a system using a pump, pipes, and a waterwheel is being used to convert electrical energy into mechanical energy. When electrical energy is applied to the pump, it will operate and cause water to flow. The pump generates:

1. A high pressure at the outlet port, which pushes the water molecules out and into the system
2. A low pressure at the inlet port, which pulls the water molecules into the pump

    The water current flow is in the direction indicated, and the high pressure or potential within the piping will be used to drive the waterwheel around, producing mechanical energy. The remaining water is attracted into the pump due to the suction or low pressure existing at the inlet port. In fact, the amount of water entering the inlet port is the same as the amount of water leaving the outlet port. It can therefore be said that the water flow rate is the same throughout the circuit. The only changing element is the pressure felt at different points throughout the system.

    In Figure 1-24(b), an electric circuit containing a battery, conductors, and a bulb is being used to convert electrical energy into light energy. The battery generates a voltage just as the pump generates pressure. This voltage causes electrons to move through conductors, just as pressure causes water molecules to move through the piping. The amount of water flow is dependent on the pump's pressure, and the amount of current or electron flow is de-

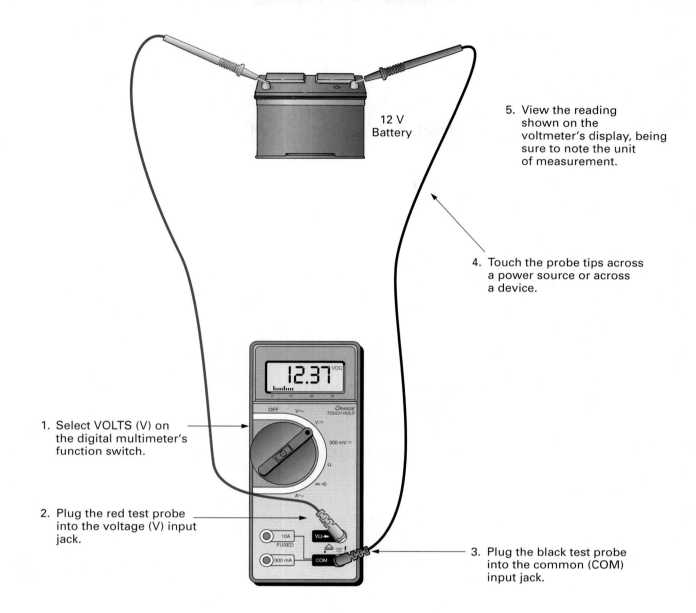

5. View the reading shown on the voltmeter's display, being sure to note the unit of measurement.

4. Touch the probe tips across a power source or across a device.

12 V Battery

12.37 VDC

1. Select VOLTS (V) on the digital multimeter's function switch.

2. Plug the red test probe into the voltage (V) input jack.

3. Plug the black test probe into the common (COM) input jack.

NOTE: If test leads are reversed, a negative sign will show in the display.

(a)

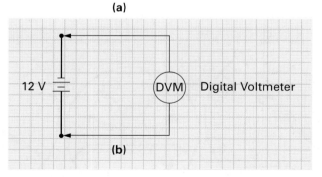

12 V   (DVM)   Digital Voltmeter

(b)

FIGURE 1-21 **Using the Voltmeter to Measure Voltage.**

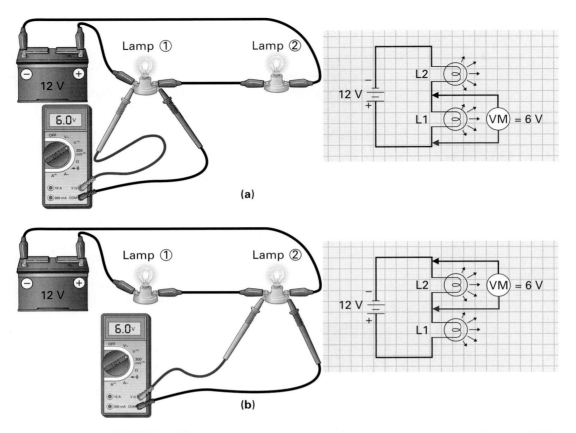

FIGURE 1-22 Measuring the Voltage Drop across Components (a) Lamp 1 (b) Lamp 2.

pendent on the battery's voltage. Water flow through the wheel can be compared to current flow through the bulb. The high pressure is lost in turning the wheel and producing mechanical energy, just as voltage is lost in producing light energy out of the bulb. We cannot say that pressure or voltage flows: Pressure and voltage are applied and cause water or current to flow, and it is this flow that is converted to mechanical energy in our fluid system and light energy in our electrical system. Voltage is the force of repulsion and attraction needed to cause current to flow through a circuit, and without this potential difference or pressure there cannot be current.

FIGURE 1-23 Fixed-Range Digital Voltmeter.

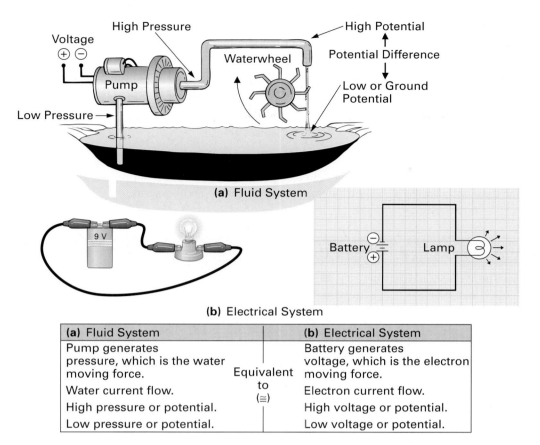

(a) Fluid System

(b) Electrical System

(a) Fluid System		(b) Electrical System
Pump generates pressure, which is the water moving force.	Equivalent to ($\cong$)	Battery generates voltage, which is the electron moving force.
Water current flow.		Electron current flow.
High pressure or potential.		High voltage or potential.
Low pressure or potential.		Low voltage or potential.

FIGURE 1-24 **Comparison between a Fluid System (a) and an Electrical System (b).**

1-4-5 *Current Is Directly Proportional to Voltage*

Referring to the fluid system and then the electrical circuit in Figure 1-24, you can easily see that the flow is proportional to the pressure, which means: If the pump were to generate a greater pressure, a larger amount of water would flow through the system.

$$\text{pressure} \uparrow \qquad \text{water flow} \uparrow$$

Similarly, if a larger voltage were applied to the electrical circuit, this larger electron moving force (emf) would cause more electrons or current to flow through the circuit, as shown in Figure 1-25.

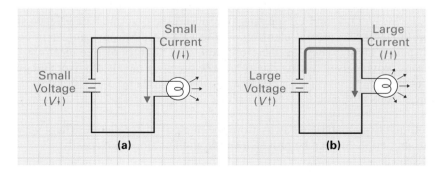

FIGURE 1-25 **Current Is Directly Proportional to Voltage. (a) Small Voltage Produces a Small Current. (b) Large Voltage Produces a Large Current.**

Current (I) is therefore said to be *directly proportional (∝) to voltage (V),* as a voltage decrease *(V ↓)* results in a current decrease *(I ↓),* and similarly, a voltage increase *(V ↑)* causes a current increase *(I ↑).*

> Current is directly proportional to voltage ($I \propto V$)

$$V \downarrow \text{ causes a } I \downarrow$$
$$V \uparrow \text{ causes a } I \uparrow$$

Proportional

A term used to describe the relationship between two quantities that have the same ratio.

As you can see from this example, directly **proportional** (∝) is a phrase which means that one term will change in proportion, or in size, relative to another term.

SELF-TEST EVALUATION POINT FOR SECTION 1-4

Now that you have read this section, you should be able to:

■ **Objective 20.** *Define electrical voltage.*

■ **Objective 21.** *List the various voltage units and values.*

■ **Objective 22.** *List the three rules to apply when measuring voltage.*

■ **Objective 23.** *Relate the electrical system to a fluid system.*

■ **Objective 24.** *Describe why current is directly proportional to voltage.*

Use the following questions to test your understanding of Section 1-4:

1. What is the unit of voltage?
2. Convert 3 MV to kilovolts.
3. Which meter is used to measure voltage?
4. What is the relationship between current and voltage?

1-5   CONDUCTORS

A lightning bolt that sets fire to a tree and the operation of your calculator are both electrical results achieved by the flow of electrons. The only difference is that your calculator's circuits control the flow of electrons, while the lightning bolt is the uncontrolled flow of electrons. In electronics, a **conductor** is used to channel or control the path in which electrons flow.

Any material that passes current easily is called a conductor. These materials are said to have a "low resistance," which means that they will offer very little opposition to current flow. This characteristic can be explained by examining conductor atoms. As mentioned previously, the atom has a maximum of seven orbital paths known as shells, which are named K, L, M, N, O, P, and Q, stepping out toward the outermost or valence shell. Conductors are materials or natural elements whose valence electrons can easily be removed from their parent atoms. They are therefore said to be sources of free electrons, and these free electrons provide us with circuit current. The precious metals of silver and gold are the best conductors.

Conductor

Length of wire whose properties are such that it will carry an electric current.

More specifically, a better conductor has:

1. Electrons in shells farthest away from the nucleus; these electrons feel very little nucleus attraction and can be broken away from their atom quite easily.
2. More electrons per atom.
3. An incomplete valence shell. This means that the valence shell does not have in it the maximum possible number of electrons. If the atom had its valence ring complete (full), there would be no holes (absence of an electron) in that shell, so no encourage-

ment for adjacent atom electrons to jump from their parent atom into the next atom would exist, preventing the chain reaction known as current.

Economy must be considered when choosing a conductor. Large quantities of conductors using precious metals are obviously going to send the cost of equipment beyond reach. The conductor must also satisfy some physical requirements, in that we must be able to shape it into wires of different sizes and easily bend it to allow us to connect one circuit to the next.

Copper is the most commonly used conductor, as it meets the following three requirements:

1. It is a good source of electrons.
2. It is inexpensive.
3. It is physically pliable.

Aluminum is also a very popular conductor, and although it does not possess as many free electrons as copper, it has the two advantages of being less expensive and lighter than copper.

1-5-1   *Conductance*

Conductance is the measure of how good a conductor is at carrying current. Conductance (symbolized G) is equal to the reciprocal of resistance (or opposition) and is measured in the unit siemens (S):

$$
\text{conductance } (G) = \frac{1}{\text{resistance } (R)}
$$

Conductance (G) values are measured in siemens (S)
and resistance (R) in ohms (Ω).

This means that conductance is inversely proportional to resistance. For example, if the opposition to current flow (resistance) is low, the conductance is high and the material is said to have a good conductance.

$$
\text{high conductance } G \uparrow = \frac{1}{R \downarrow \text{ (low resistance)}}
$$

On the other hand, if the resistance of a conducting wire is high ($R \uparrow$), its conductance value is low ($G \downarrow$) and it is called a poor conductor. A good conductor therefore has a high conductance value and a very small resistance to current flow.

◼ EXAMPLE:

A household electric blanket offers 25 ohms of resistance to current flow. Calculate the conductance of the electric blanket's heating element.

◼ *Solution:*

$$
\text{conductance} = \frac{1}{\text{resistance}}
$$

$$
G = \frac{1}{R} = \frac{1}{25 \text{ ohms}} = 0.04 \text{ siemens}
$$

or 40 mS (millisiemens)

TIME LINE
Stephen Gray (1693-1736), an Englishman, discovered that certain substances would conduct electricity.

Conductance (*G*)
Measure of how well a circuit or path conducts or passes current. It is equal to the reciprocal of resistance.

TIME LINE
Stephen Gray's lead was picked up in 1730 by Charles du Fay, a French experimenter, who believed that there were two types of electricity, which he called *vitreous* and *resinous* electricity.

◼ **EXAMPLE:**

To further reinforce your understanding of a conductor, compare a good conductor (copper) to a poor conductor (carbon) using the data given in the atomic periodic table (Table 1-1).

◼ *Solution:*

Carbon

1. The valence ring is the L or second shell, which feels a strong nucleus attractive force, discouraging the release of free electrons and thus of circuit current.
2. Only six electrons orbit the carbon atom, making it a poor source of electrons.
3. The valence shell (L, maximum eight electrons) is already half complete (four electrons), so few holes (only four) exist, discouraging the jumping of electrons from parent atoms into adjacent atom holes.

Copper

1. The valence ring is the N or fourth shell, which feels only a weak nucleus attractive force, encouraging the release of free electrons and thus of circuit current.
2. A total of 29 electrons orbit the nucleus, making it a good source of electrons.
3. The valence shell (N, maximum 32 electrons) is incomplete, as only one electron occupies it when neutral, so 31 holes exist, encouraging the jumping of electrons from parent atoms into adjacent atom holes.

To express how good or poor a conductor is, we must have some sort of reference point. The reference point we use is the best conductor, silver, which is assigned a relative conductivity value of 1.0. Table 1-4 lists other conductors and their relative conductivity values with respect to the best, silver. The formula for calculating relative conductivity is

$$\text{Relative Conductivity} = \frac{\text{Conductor's Relative Conductivity}}{\text{Reference Conductor's Relative Conductivity}}$$

◼ **EXAMPLE:**

What is the relative conductivity of tungsten if copper is used as the reference conductor?

TABLE 1-4   Relative Conductivity of Conductors

MATERIAL	CONDUCTIVITY (RELATIVE)
Silver	1.000
Copper	0.945
Gold	0.652
Aluminum	0.575
Tungsten	0.297
Nichrome	0.015

Solution:

$$\text{tungsten} = 0.297$$
$$\text{copper} = 0.945$$

$$\text{relative conductivity} = \frac{\text{conductivity of conductor}}{\text{conductivity of reference}}$$

$$= \frac{0.297}{0.945}$$

$$= 0.314$$

CALCULATOR SEQUENCE

Step	Keypad Entry	Display Response
1.	0 . 2 9 7	0.297
2.	÷	
3.	0 . 9 4 5	0.945
4.	=	0.314

EXAMPLE:

What is the relative conductivity of silver if copper is used as the reference?

Solution:

$$\text{relative conductivity} = \frac{\text{silver}}{\text{copper}}$$

$$= \frac{1.000}{0.945} = 1.058$$

SELF-TEST EVALUATION POINT FOR SECTION 1-5

Now that you have read this section, you should be able to:

■ **Objective 26.** *Describe conductance.*

Use the following questions to test your understanding of Section 1-5.

1. True or false: A conductor is a material used to block the flow of current.

2. List the three atomic properties that make a better conductor.

3. What is the most commonly used conductor in the field of electronics?

4. Calculate the conductance of a 35 ohm heater element.

1-6 INSULATORS

Any material that offers a high resistance or opposition to current flow is called an **insulator.** Conductors permit the easy flow of current and so have good conductivity, while insulators allow small to almost no amount of free electrons to flow. Insulators can, with sufficient pressure or voltage applied across them, "break down" and conduct current. This **breakdown voltage** must be great enough to dislodge the electrons from their close orbital shells (K, L shells) and release them as free electrons.

Insulators are also called "dielectrics," and the best insulator or dielectric should have the maximum possible resistance and conduct no current at all. The **dielectric strength** of an insulator indicates how good or bad an insulator is by indicating the voltage that will cause the insulating material to break down and conduct a large current. Table 1-5 lists some of the more popular insulators and the value of kilovolts that will cause a centimeter of insulator to break down. As an example, if 1 centimeter of paper is connected to a variable voltage source, a voltage of 500 kilovolts is needed to break down the paper and cause current to flow.

Insulator
A material that has few electrons per atom and those electrons are close to the nucleus and cannot be easily removed.

Breakdown Voltage
The voltage at which breakdown of an insulator occurs.

Dielectric Strength
The maximum potential a material can withstand without rupture.

TABLE 1-5 Breakdown Voltages of Certain Insulators

MATERIAL	BREAKDOWN STRENGTH (kV/cm)
Mica	2000
Glass	900
Teflon	600
Paper	500
Rubber	275
Bakelite	151
Oil	145
Porcelain	70
Air	30

The following formula can be used to calculate the dielectric thickness needed to withstand a certain voltage.

$$\text{dielectric thickness} = \frac{\text{voltage to insulate}}{\text{insulator's breakdown voltage}}$$

■ **EXAMPLE:**

What thickness of mica would be needed to withstand 16,000 V?

■ *Solution:*

CALCULATOR SEQUENCE

Step	Keypad Entry	Display Response
1.	1 6 E (exponent) 3	16E3
2.	÷	
3.	2 E 6 (or 2000 E3)	2E6
4.	=	0.008

$$\text{mica strength} = 2000 \text{ kV/cm}$$

$$\text{dielectric thickness} = \frac{16{,}000 \text{ V}}{2000 \text{ kV/cm}} = 0.008 \text{ cm}$$

■ **EXAMPLE:**

What maximum voltage could 1 mm of air withstand?

■ *Solution:*

There are 10 mm in 1 cm. If air can withstand 30,000 V/cm, it can withstand 3000 V/mm.

Now that you have completed this section, you should be able to:

■ **Objective 25.**   *Explain the difference between:*
  a. A conductor
  b. An insulator
■ **Objective 27.**   *Explain what makes a good:*
  a. Conductor
  b. Insulator

Use the following questions to test your understanding of Section 1-6:

1. True or false: An insulator is a material used to block the flow of current.
2. What is considered to be the best insulator material?
3. Define *breakdown voltage.*
4. Would the conductance figure of a good insulator be large or small?

1-7   THE OPEN, CLOSED, AND SHORT CIRCUIT

In this section we will examine some of the terms used in association with circuits.

1-7-1   *Open Circuit (Open Switch)*

Figure 1-26 shows how an opened switch can produce an "open" in a circuit. The open prevents current flow as the extra electrons in the negative terminal of the battery cannot feel the attraction of the positive terminal due to the break in the path. The opened switch therefore has produced an **open circuit.**

Open Circuit
Break in the path of current flow.

1-7-2   *Open Circuit (Open Component)*

In Figure 1-27, the switch is now closed so as to make a complete path in the circuit. An open circuit, however, could still exist due to the failure of one of the components in the circuit. In the example in Figure 1-27, the light bulb filament has burned out creating an open in the circuit, which prevents current flow. An open component therefore can also produce an open circuit.

1-7-3   *Closed Circuit (Closed Switch)*

If all of the devices in a circuit are operating properly and connected correctly, a closed switch produces a **closed circuit,** as shown in Figure 1-28. A closed circuit provides a

Closed Circuit
Circuit having a complete path for current to flow.

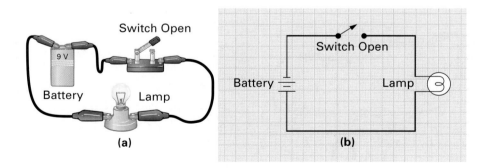

FIGURE 1-26   An Open Switch Causing an Open Circuit. (a) Pictorial. (b) Schematic.

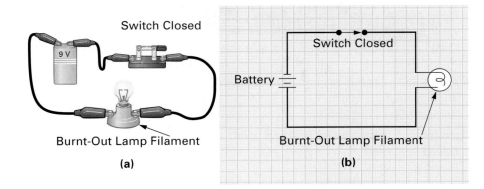

FIGURE 1-27 **An Open Lamp Filament Causing an Open Circuit. (a) Pictorial. (b) Schematic.**

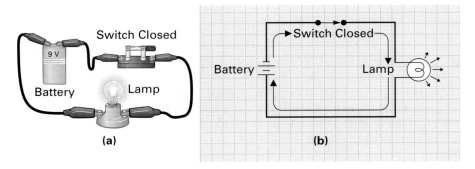

FIGURE 1-28 **Closed Switch Causing a Closed Circuit. (a) Pictorial. (b) Schematic.**

complete path from the negative terminal of the battery to the positive, and so current will flow through the circuit.

1-7-4   *Short Circuit (Shorted Component)*

A **short circuit** normally occurs when one point in a circuit is accidentally connected to another. Figure 1-29(a) illustrates how a short circuit could occur. In this example, a set of pliers was accidentally laid across the two contacts of the light bulb. Figure 1-29(b) shows the circuit's schematic diagram, with the effect of the short across the light bulb drawn in. In this

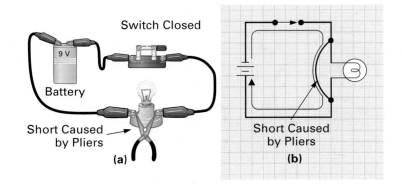

FIGURE 1-29 **Short Circuit Due to Pliers across a Lamp. (a) Pictorial. (b) Schematic.**

instance, nearly all the current will flow through the metal of the pliers, which offers no resistance or opposition to the current flow. Since the light bulb does have some resistance or opposition, very little current will flow through the bulb, and therefore no light will be produced.

SELF-TEST EVALUATION POINT FOR SECTION 1-7

Now that you have completed this section, you should be able to:

■ **Objective 28.** *Explain the terms:*
  a. *Open circuit*
  b. *Closed circuit*
  c. *Short circuit*

Use the following questions to test your understanding of Section 1-7:

1. Does current flow through an open circuit?
2. Does current flow through a short circuit?
3. Describe the difference between a closed circuit and a short circuit.
4. Describe the difference between an open switch and a closed switch.

SUMMARY

1. An *element,* by definition, is a substance consisting of only one type of atom.

2. The *atom* is the smallest particle into which an element can be divided without losing its identity, and a group of identical atoms is called an element.

3. The word atom is a Greek word meaning a particle that is too small to be subdivided.

4. In 1913, a Danish physicist, Neils Bohr, put forward a theory about the atom, and his basic model outlining the *subatomic* particles that make up the atom is still in use today.

5. The three important particles of the atom are the *proton,* which has a positive charge, the *neutron,* which is neutral or has no charge, and the *electron,* which has a negative charge.

6. The atom consists of a positively charged central mass called the *nucleus,* which is made up of protons and neutrons surrounded by a quantity of negatively charged orbiting electrons.

7. The *atomic number* of an atom describes the number of protons that exist within the nucleus.

8. The number of neutrons within an atom's nucleus can be calculated by subtracting the atomic number (protons) from the atomic weight (protons and neutrons).

9. A *neutral atom* or *balanced atom* is one that has an equal number of protons and orbiting electrons, so the net positive proton charge is equal but opposite to the net negative electron charge, resulting in a balanced or neutral state.

10. Orbiting electrons travel around the nucleus at varying distances from the nucleus, and these orbital paths are known as *shells* or *bands.* The orbital shell nearest the nucleus is referred to as the first or K shell. The second is known as the L, the third is M, the fourth is N, the fifth is O, the sixth

is P, and the seventh is referred to as the Q shell. There are seven shells available for electrons (K, L, M, N, O, P, and Q) around the nucleus, and each of these seven shells can only hold a certain number of electrons. The outermost electron-occupied shell is referred to as the *valence shell or ring,* and its electrons are termed *valence electrons.*

11. All matter exists in one of three states: solids, liquids, and gases. The atoms of a solid are fixed in relation to one another but vibrate in a back-and-forth motion, unlike liquid atoms, which can flow over each other. The atoms of a gas move rapidly in all directions and collide with one another.

12. *Like charges* (positive and positive or negative and negative) repel one another. *Unlike charges* (positive and negative or negative and positive) attract one another.

13. Due to their distance from the nucleus, valence electrons are described as being loosely bound to the atom. The electrons can, therefore, easily be dislodged from their outer orbital shell by any external force, to become free electrons.

14. Combinations of elements are known as *compounds,* and the smallest unit of a compound is called a *molecule,* just as the smallest unit of an element is an atom.

15. Many of the values inserted into formulas contain numbers that have *exponents.* An exponent is a smaller number that appears to the right and slightly higher than another number.

16. *Exponents allow us to save space,* as you can see in this example:

$$5 \times 5 \times 5 \times 5 = 5^4$$

Here the exponent (^{4}) is used to indicate how many times its base (5) is used as a *factor.* You can also see from this example that the other advantage is that *exponents are a sort of math shorthand.* It is much easier to write 5^4 than to write $5 \times 5 \times 5 \times 5$. The "$y^x$" key on your calculator allows you to enter in any base number (y) and any exponent (x).

17. Many of the sciences, such as electronics, deal with numbers that contain a large number of zeros. By using exponents we can eliminate the large number of zeros to obtain a shorthand version of the same number. This method is called *scientific notation* or *powers of 10*.

18. A *negative exponent* tells you how many places *to the left* to move the decimal point. A *positive exponent* tells you how many places *to the right* to move the decimal point. Remember one important point: that a negative exponent does not indicate a negative number; it simply indicates a fraction.

19. *Scientific notation* is expressed as a base number between 1 and 10 with an exponent that is a power of 10. *Engineering notation,* however, is expressed as a base number that is greater than 1 with an exponent that is some multiple of three.

Current (Figure 1-30)

20. Once an electron is released, the atom is no longer electrically neutral and is called a *positive ion,* as it now has a net positive charge (more protons than electrons). The released electron tends to jump into a nearby atom, which will then have more electrons than protons and is referred to as a *negative ion.*

21. Millions upon millions of atoms within a conductor pass a continuous movement of billions upon billions of electrons from negative to positive. This electron flow is known as *electric current.*

22. A force or pressure produced by a positive charge and negative charge will cause electrons to flow from the negative to the positive terminal. The positive side has a deficiency of electrons and the negative side has an abundance, and so a continuous flow or migration of electrons takes place through our metal conducting wire. This electric current or electron flow is a measurable quantity.

23. There are 6.24×10^{18} electrons in 1 *coulomb of charge.*

24. Coulombs and time are therefore combined to describe the number of electrons and the rate at which they flow. This relationship is called *current (I)* flow and has the unit of *amperes* (A). If 6.24×10^{18} electrons (1 C) were to drift past a specific point on a conductor in 1 second of time, 1 ampere of current is said to be flowing.

25. Current within electronic equipment is normally a value in milliamperes or microamperes and very rarely exceeds 1 ampere.

26. Electrons drift from a negative to a positive charge. This current is known as *electron flow.* In the eighteenth and nineteenth centuries when very little was known about the atom, researchers believed that current was a flow of positive charges. Although this has now been proved incorrect many texts still use *conventional current flow.* Whether conventional current flow or electron flow is used, the same answers to problems, measurements, or designs are obtained. The key point to remember is that direction is not important, but the amount of current flow is.

27. *Ammeters* (ampere meters) are used to measure the current flow within a circuit.

Voltage (Figure 1-31)

28. *Voltage* is the force or pressure exerted on electrons.

29. A highly concentrated charge produces a high voltage, whereas a low concentrated charge produces a low voltage. Voltage is also appropriately known as the electron moving force or *electromotive force (emf),* and since two opposite potentials exist, one negative and one positive, the strength of the voltage can also be referred to as the amount of *potential difference (PD)* applied across the circuit.

30. Voltage is the force, pressure, potential difference (PD), or electromotive force (emf) that causes electron flow or current and is symbolized by italic uppercase *V*. The unit for voltage is the volt, symbolized by roman uppercase V.

31. A *battery* converts chemical energy into electrical energy. Positive charges or ions (atoms with more protons than electrons) are present at the positive terminal of a battery, while negative charges or ions (atoms with more electrons than protons) are present at the negative terminal of a battery.

32. Voltage within electronic equipment is normally measured in volts, whereas heavy-duty industrial equipment normally requires high voltages that are generally measured in kilovolts (kV).

33. *Voltmeters* (voltage meters) are used to measure electrical pressure or voltage. To measure the voltage drop across a device, a voltmeter is connected across the device with the two leads of the voltmeter touching either side of the component.

34. If a larger voltage were applied to the electrical circuit, this larger electron moving force (emf) would cause more electrons or current to flow through the circuit. *Current (I)* is therefore said to be *directly proportional* ($\propto$) to *voltage (V),* as a voltage decrease ($V\downarrow$) results in a current decrease ($I\downarrow$), while a voltage increase ($V\uparrow$) causes a current increase ($I\uparrow$).

35. Directly proportional ($\propto$) is a phrase which means that one term will change in proportion, or in size, relative to another term.

Conductors and Insulators

36. Materials that pass current easily are called *conductors.* Conductors are materials or natural elements whose valence electrons can easily be removed from their parent atoms. The precious metals of silver and gold are the best conductors.

37. Copper is the most commonly used conductor, as it meets the following three requirements:
   a. It is a good source of electrons.
   b. It is inexpensive.
   c. It is physically pliable.

38. *Conductance* is the measure of how good a conductor is at carrying current. Conductance (symbolized *G*) is equal to the reciprocal of resistance and is measured in the unit siemens (S).

Figure 1-30   Current.

Electric Charge (Q) in coulombs (C) = $\dfrac{\text{total number of electrons } (n)}{6.24 \times 10^{18}}$

Electric Current (I) in amperes (A) = $\dfrac{\text{coulombs } (Q)}{\text{time } (t)}$

Current Units

NAME	SYMBOL	VALUE
Picoampere	pA	$10^{-12} = \dfrac{1}{1{,}000{,}000{,}000{,}000}$
Nanoampere	nA	$10^{-9} = \dfrac{1}{1{,}000{,}000{,}000}$
Microampere	µA	$10^{-6} = \dfrac{1}{1{,}000{,}000}$
Milliampere	mA	$10^{-3} = \dfrac{1}{1000}$
Ampere	A	$10^{0} = 1$
Kiloampere	kA	$10^{3} = 1000$
Megaampere	MA	$10^{6} = 1{,}000{,}000$
Gigaampere	GA	$10^{9} = 1{,}000{,}000{,}000$
Teraampere	TA	$10^{12} = 1{,}000{,}000{,}000{,}000$

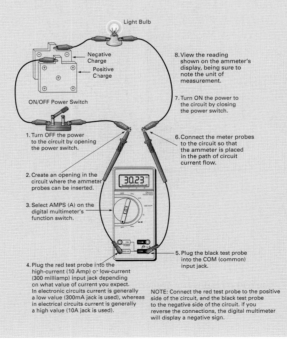

Light Bulb

Negative Charge
Positive Charge

ON/OFF Power Switch

1. Turn OFF the power to the circuit by opening the power switch.

2. Create an opening in the circuit where the ammeter probes can be inserted.

3. Select AMPS (A) on the digital multimeter's function switch.

4. Plug the red test probe into the high-current (10 Amp) or low-current (300 milliamp) input jack depending on what value of current you expect. In electronic circuits current is generally a low value (300mA jack is used), whereas in electrical circuits current is generally a high value (10A jack is used).

5. Plug the black test probe into the COM (common) input jack.

6. Connect the meter probes to the circuit so that the ammeter is placed in the path of circuit current flow.

7. Turn ON the power to the circuit by closing the power switch.

8. View the reading shown on the ammeter's display, being sure to note the unit of measurement.

NOTE: Connect the red test probe to the positive side of the circuit, and the black test probe to the negative side of the circuit. If you reverse the connections, the digital multimeter will display a negative sign.

Figure 1-31   Voltage.

Voltage Units

NAME	SYMBOL	VALUE
Picovolts	pV	$10^{-12} = \dfrac{1}{1,000,000,000,000}$
Nanovolts	nV	$10^{-9} = \dfrac{1}{1,000,000,000}$
Microvolts	µV	$10^{-6} = \dfrac{1}{1,000,000}$
Millivolts	mV	$10^{-3} = \dfrac{1}{1000}$
Volts	V	$10^{0} = 1$
Kilovolts	kV	$10^{3} = 1000$
Megavolts	MV	$10^{6} = 1,000,000$
Gigavolts	GV	$10^{9} = 1,000,000,000$
Teravolts	TV	$10^{12} = 1,000,000,000,000$

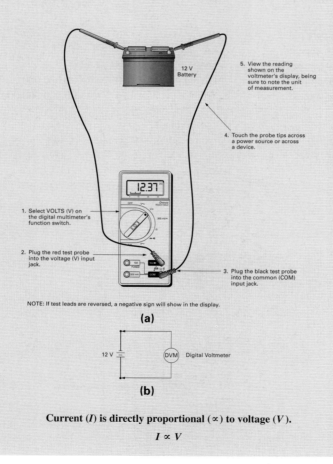

5. View the reading shown on the voltmeter's display, being sure to note the unit of measurement.

4. Touch the probe tips across a power source or across a device.

1. Select VOLTS (V) on the digital multimeter's function switch.

2. Plug the red test probe into the voltage (V) input jack.

3. Plug the black test probe into the common (COM) input jack.

NOTE: If test leads are reversed, a negative sign will show in the display.

(a)

12 V ⎓ (DVM) Digital Voltmeter

(b)

Current (I) is directly proportional ($\propto$) to voltage (V).

$$I \propto V$$

39. Relative conductivity describes how good or poor a conductor is by comparing it to a reference point. The reference point we use is the best conductor, silver, which has a conductivity value of 1.0.

40. Materials that are used to block current are called *insulators*. Insulators can, with sufficient pressure or voltage applied across them, "break down" and conduct current. The dielectric strength describes how good or poor an insulator is by listing what voltage will cause a small section to break down and conduct a large current.

41. An opened switch produces an "open" in the circuit, preventing circuit current. If the switch is now closed so as to make a complete path in the circuit, an open could still exist due to the failure of one of the components. An open component can also produce an open circuit.

42. A closed switch produces a *closed circuit*. If all devices are operating correctly and all connections are correct, current will be present in a closed circuit.

43. A *short circuit* normally occurs when one point is accidentally connected to another.

REVIEW QUESTIONS

Multiple-Choice Questions

1. If $x^2 = 16$, then $\sqrt[2]{16} = x$. What is the value of x?
a. 4 **b.** 2 **c.** 16 **d.** 8

2. Calculate $\sqrt[5]{32}$.
a. 2 **b.** 1.3 **c.** 4 **d.** 6.4

3. Raise the following number to the power of ten indicated by the exponent, and give the answer: 3.6×10^2.
a. 3600 **b.** 36 **c.** 0.036 **d.** 360

4. Convert the following number to powers of ten: 0.00029.
a. 0.29×10^{-3} **c.** 29×10^{-5}
b. 2.9×10^{-4} **d.** All the above

5. What is the power-of-ten value for the metric prefix *milli*?
a. 10^{-3} **c.** 10^3
b. 10^{-2} **d.** All the above

6. Hydrogen is a:
a. Gas **b.** Solid **c.** Liquid **d.** Other

7. Which of the following is a liquid?
a. Calcium **b.** Magnesium **c.** Mercury **d.** Helium

8. The atomic number of an atom describes:
a. The number of neutrons **c.** The number of nuclei
b. The number of electrons **d.** The number of protons

9. The most commonly used metal in the field of electronics is:
a. Silver **b.** Copper **c.** Mica **d.** Gold

10. The smallest unit of an element is:
a. A compound **c.** A molecule
b. An atom **d.** A proton

11. The smallest unit of a compound is:
a. An element **c.** An electron
b. A neutron **d.** A molecule

12. A water molecule is made up of:
a. 2 parts hydrogen and 1 part oxygen
b. Chlorine and sodium
c. 1 part oxygen and 2 parts sodium
d. 3 parts chlorine and 1 part hydrogen

13. A negative ion has:
a. More protons than electrons
b. More electrons than protons
c. More neutrons than protons
d. More neutrons than electrons

14. A positive ion has:
a. Lost some of its electrons **c.** Lost neutrons
b. Gained extra protons **d.** Gained more electrons

15. One coulomb of charge is equal to:
a. 6.24×10^{18} electrons **c.** 6.24×10^8 electrons
b. $10^{18} \times 10^{12}$ electrons **d.** 6.24×10^{81} electrons

16. If 14 C of charge passes by a point in 7 seconds, the current flow is said to be:
a. 2 A **b.** 98 A **c.** 21 A **d.** 7 A

17. How many electrons are there within 16 C of charge?
a. 9.98×10^{19} **b.** 14 **c.** 16 **d.** 10.73×10^{19}

18. Current is measured in:
a. Volts **b.** Coulombs/second **c.** Ohms **d.** Siemens

19. Voltage is measured in units of:
a. Amperes **b.** Ohms **c.** Siemens **d.** Volts

20. Another word used to describe voltage is:
a. Potential difference **c.** Electromotive force (emf)
b. Pressure **d.** All of the above

21. Conductors offer a _____ resistance to current flow.
a. High **b.** Low **c.** Medium **d.** Maximum

22. Conductors have:
a. Electrons in shells farthest away from the nucleus
b. Relatively more electrons per atom
c. An incomplete valence shell
d. All of the above
e. None of the above

23. Conductance is the measure of how good a conductor is at passing current, and is measured in:
a. Siemens **b.** Volts **c.** Current **d.** Ohms

24. Insulators have:
a. Electrons close to the nucleus
b. Relatively few electrons per atom
c. An almost complete valence shell
d. All of the above
e. None of the above

25. An open circuit will cause:
a. No current flow
b. Maximum current flow
c. A break in the circuit
d. Both (a) and (c)

Communication Skill Questions

26. What is an exponent? (1–1)

27. Define the following terms: (1–1)
  a. Square of a number
  b. Square root of a number

28. What is a power-of-ten exponent? (1–1)

29. Describe the scientific and engineering notation systems. (1–1)

30. List the metric prefixes and their value. (1–1)

31. Describe the three factors that make a good conductor. (2-4)

32. Why is current directly proportional to voltage? (1-4-5)

33. List the four most commonly used current units and their values in terms of the basic unit. (1-3-3)

34. What is the speed of light in (a) miles and (b) meters per second? (1-3-4)

35. List the four most commonly used voltage units and their values in terms of the basic unit. (1-4-2)

36. Describe what is meant by: (1-7)
  a. An open circuit
  b. A closed circuit
  c. A short circuit

37. What is the difference between conventional and electron current flow? (1-3-5)

38. Give the unit and symbol for the following:
  a. Voltage (V) is measured in _____ (__). (1-4).
  b. Current (I) is measured in _____ (__). (1-3-2).
  c. Conductance (G) is measured in _____ (__). (1-5-1)
  d. Resistance (R) is measured in _____ (__). (1-5-1)

39. In relation to the structure of matter and the atom, describe:
  a. The atom's subatomic particles (1-2)
  b. An element (1-2)   **e.** A neutral atom (1-2-1)
  c. A compound (1-2-3)   **f.** A positive ion (1-3)
  d. A molecule (1-2-3)   **g.** A negative ion (1-3)

40. Describe how to use a voltmeter to measure voltage and how to use an ammeter to measure current.

Practice Problems

41. Determine the square of the following values.
  a. 9^2 **b.** 6^2 **c.** 2^2 **d.** 0^2 **e.** 1^2 **f.** 12^2

42. Determine the square root of the following values.
  a. $\sqrt{81}$ **d.** $\sqrt{36}$
  b. $\sqrt{4}$ **e.** $\sqrt{144}$
  c. $\sqrt{0}$ **f.** $\sqrt{1}$

43. Raise the following base numbers to the power indicated by the exponent, and give the answer.
  a. 9^3 **c.** 4^6
  b. 10^4 **d.** 2.5^3

44. Determine the roots of the following values.
  a. $\sqrt[3]{343} = ?$ **c.** $\sqrt[4]{760}$
  b. $\sqrt{1296} = ?$ **d.** $\sqrt[6]{46,656}$

45. Convert the following values to powers of ten.
  a. $\dfrac{1}{100}$ **c.** $\dfrac{1}{1000}$
  b. 1,000,000,000   **d.** 1000

46. Convert the following to proper fractions and decimal fractions.
  a. 10^{-4} **c.** 10^{-6}
  b. 10^{-2} **d.** 10^{-3}

47. Convert the following values to powers of ten in both scientific and engineering notation.
  a. 475   **c.** 0.07
  b. 8200   **d.** 0.00045

48. Convert the following to whole-number values with a metric prefix.
  a. 0.005 A
  b. 8000 m
  c. 15,000,000 Ω
  d. 0.000016 s

49. What is the value of conductance in siemens for a 100 ohm resistor?

50. Calculate the total number of electrons in 6.5 C of charge.

51. Calculate the amount of current in amperes passing through a conductor if 3 C of charge passes by a point in 4 s.

52. Convert the following:
  a. 0.014 A = _____ mA
  b. 1374 A = _____ kA
  c. 0.776 μA = _____ nA
  d. 0.91 mA = _____ μA

53. Convert the following:
  a. 1473 mV = _____ V
  b. 7143 V = _____ kV
  c. 0.139 kV = _____ V
  d. 0.390 MV = _____ kV

54. What is the relative conductivity of copper if silver is used as the reference conductor?

55. What minimum thickness of porcelain will withstand 24,000 V?

56. To insulate a circuit from 10 V, what insulator thickness would be needed if the insulator is rated at 750 kV/cm?

57. Convert the following:
  a. 2000 kV/cm to _____ meters
  b. 250 kV/cm to _____ mm

58. What maximum voltage may be placed across 35 mm of mica without it breaking down?

Web Site Questions

Go to the web site http://www.prenhall.com/cook, select the textbook *Introductory DC/AC Electronics* or *Introductory DC/AC Circuits*, this chapter, and then follow the instructions when answering the multiple-choice practice problems.

These tests at the end of each chapter will challenge your knowledge up to this point, and give you the practice you need for a job interview. To make this more realistic, the test will be comprised of both technical and personal questions. In order to take full advantage of this exercise, you may want to set up a simulation of the interview environment, have a friend read the questions to you, and record your responses for later analysis.

Company Name: Multi-Meters, Inc.

Industry Branch: Test and Measurement.

Function: Design and Manufacture Multimeters.

Job Title: QA Technician.

1. What do you know about our company?
2. What steps would you follow to use a multimeter to measure current and voltage?

3. Have you had much experience working in a team environment?
4. Could you define, simply, the difference between current and voltage?
5. What is your understanding of the responsibilities of a QA technician?
6. If your meter showed a reading of 10 milliamps, what does the "milli" mean?
7. Could you use the multimeter to measure potential difference?
8. What would you say are your faults?
9. If you discovered an error in a circuit design, would you report it to your supervisor, or go and find the engineer responsible?
10. Why would you want to work for our company?

Answers

1. Visit the company's web site before the interview to get an overall understanding of the company's ownership, product line, service and support.
2. Sections 1-3-6 and 1-4-3.
3. Discuss how you have worked with a partner in the lab on experiments, and any other related work history.
4. Chapter 1 margin definitions.

5. See introduction, QA Technician.
6. One-thousandth.
7. Yes, pd is voltage.
8. Make sure your faults are positive. For example, you always like to finish what you start, and so on.
9. Your supervisor.
10. Refer back to the company's positive attributes listed in its web site.

Resistance and Power

Genius of Chippewa Falls

In 1960, Seymour R. Cray, a young vice-president of engineering for Control Data Corporation, informed president William Norris that in order to build the world's most powerful computer he would need a small research lab built near his home. Norris would have shown any other employee the door, but Cray was his greatest asset, so in 1962 Cray moved into his lab, staffed by 34 people and nestled in the woods near his home overlooking the Chippewa River in Wisconsin. Eighteen months later the press was invited to view the 14- by 6-foot 6600 supercomputer that could execute 3 million instructions per second and contained 80 miles of circuitry and 350,000 transistors, which were so densely packed that a refrigeration cooling unit was needed due to the lack of airflow.

Cray left Control Data in 1972 and founded his own company, Cray Research. Four years later the $8.8 million Cray-1 scientific supercomputer outstripped the competition. It included some revolutionary design features, one of which is that since electronic signals cannot travel faster than the speed of light (1 foot per billionth of a second) the wire length should be kept as short as possible, because the longer the wire the longer it takes for a message to travel from one end to the other. With this in mind, Cray made sure that none of the supercomputer's conducting wires exceeded 4 feet in length.

In the summer of 1985, the Cray-2, Seymour Cray's latest design, was installed at Lawrence Livermore Laboratory. The Cray-2 was 12 times faster than the Cray-1, and its densely packed circuits are encased in clear Plexiglas and submerged in a bath of liquid coolant. The 60-year-old genius has moved on from his latest triumph, nicknamed "Bubbles," and is working on another revolution in the supercomputer field, because for Seymour Cray a triumph is merely a point of departure.

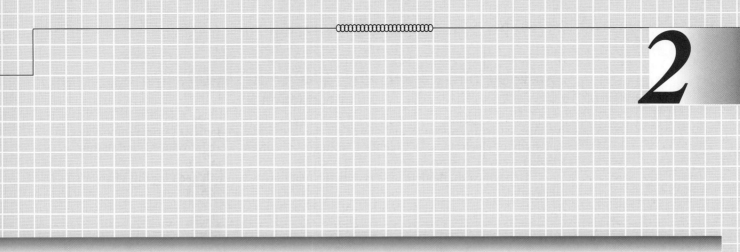

Outline and Objectives

Introduction

Voltage, current, resistance, and power are the four basic properties of prime importance in our study of electronics. In Chapter 1, voltage and current were introduced, and in this chapter we will examine resistance and power.

To begin with, though, let's review the math skills you will need for this chapter's material.

2-1 MINI-MATH REVIEW—ALGEBRA, EQUATIONS AND FORMULAS

This first section, "Mini-Math Review," is included to overview the mathematical details you need for the electronic concepts covered in this chapter. In this review, we will be examining algebra, equations, and formulas.

I contemplated calling this chapter "using letters in mathematics" because the word *algebra* seems to make many people back away. If you look up *algebra* in the dictionary, it states that it is "a branch of mathematics in which letters representing numbers are combined according to the rules of mathematics." In this chapter you will discover how easy algebra is to understand and then see how we can put it to some practical use, such as rearranging formulas. As you study this chapter you will find that algebra is quite useful, and because it is used in conjunction with formulas, an understanding is essential for anyone entering a technical field.

2-1-1 *The Basics of Algebra*

Algebra

A generalization of arithmetic in which letters representing numbers are combined according to the rules of arithmetic.

Formula

A general fact, rule, or principle expressed usually in mathematical symbols.

Equation

A formal statement of the equality or equivalence of mathematical or logical expressions.

Literal Number

A number expressed as a letter.

Algebra by definition is a branch of mathematics in which letters representing numbers are combined according to the rules of mathematics. The purpose of using letters instead of values is to develop a general statement or **formula** that can be used for any values. For example, distance (d) equals velocity (v) multiplied by time (t), or

$$d = v \times t$$

where d = distance in miles
v = velocity or speed in miles per hour
t = time in hours

Using this formula, we can calculate how much distance was traveled if we know the speed or velocity at which we are traveling and the time for which we traveled at that speed. Therefore, if I were to travel at a speed of 20 miles per hour (20 mph) for 2 hours, how far, or how much distance, would I travel? Replacing the letters in the formula with values converts the problem from a formula to an **equation,** as shown:

$$d = v \times t \ \leftarrow \text{ (Formula)}$$
$$d = 20 \text{ mph} \times 2 \text{ h} \ \leftarrow \text{ (Equation)}$$
$$d = 40 \text{ mi} \ \leftarrow \text{ (Answer)}$$

Letters such as d, v, t and a, b, c are called **literal numbers** (letter numbers), and are used, as we have just seen, in general statements showing the relationship between quantities. They are also used in equations to signify an *unknown quantity,* as will be discussed in the following section.

The Equality on Both Sides of the Equals Sign

All equations or formulas can basically be divided into two sections that exist on either side of an equals sign, as shown:

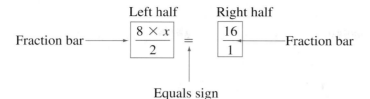

Everything in the left half of the equation is equal to everything in the right half of the equation. This means that 8 times x (which is an unknown value) divided by 2 equals 16 divided by 1. Generally, it is not necessary to put 1 under the 16 in the right section because any number divided by 1 equals the same number ($^{16}/_1 = 16 \div 1 = 16$). However, it was included in this introduction to show that each section has both a top and a bottom. If the equation is written without the fraction bar or 1 in the denominator position, it appears as follows:

$$\frac{8 \times x}{2} = 16$$

Although not visible, you should always assume that a number on its own on either side of the equals sign is above the fraction bar, as shown:

$$\frac{8 \times x}{2} = \frac{16}{}$$

Now that we understand the basics of an equation, let us see how we can manipulate it yet keep both sides equal to one another.

Treating Both Sides Equally

If you do exactly the same thing to both sides of an equation or formula, the two halves remain exactly equal, or in balance. This means that as long as you add, subtract, multiply, or divide both sides of the equation by the same number, the equality of the equation is preserved. For example, let us try adding, subtracting, multiplying, and dividing both sides of the following equation by 4 and see if both sides of the equation are still equal.

$$\boxed{2 \times 4 = 8} \quad \leftarrow \quad \text{Original equation}$$

1. Add 4 to both sides of the equation:

$$2 \times 4 = 8 \quad \leftarrow \text{(Original equation)}$$
$$(2 \times 4) \boxed{+ 4} = 8 \boxed{+ 4} \quad \leftarrow \text{(Add 4 to both sides.)}$$
$$8 + 4 = 8 + 4$$
$$12 = 12$$

Both sides of the equation remain equal.

2. Subtract 4 from both sides of the equation:

$$2 \times 4 = 8 \quad \leftarrow \text{(Original equation)}$$
$$(2 \times 4) \boxed{- 4} = 8 \boxed{- 4} \quad \leftarrow \text{(Subtract 4 from both sides.)}$$
$$8 - 4 = 8 - 4$$
$$4 = 4$$

Both sides of the equation remain equal.

3. Multiply both sides of the equation by 4:

$$2 \times 4 = 8 \qquad \leftarrow \text{(Original equation)}$$
$$(2 \times 4)\,\boxed{\times 4} = 8\,\boxed{\times 4} \qquad \leftarrow \text{(Multiply both sides by 4.)}$$
$$8 \times 4 = 8 \times 4$$
$$32 = 32$$

Both sides of the equation remain equal.

4. Divide both sides of the equation by 4:

$$2 \times 4 = 8 \qquad \leftarrow \text{(Original equation)}$$
$$(2 \times 4)\,\boxed{\div 4} = 8\,\boxed{\div 4} \qquad \leftarrow \text{(Divide both sides by 4.)}$$
$$\frac{2 \times 4}{4} = \frac{8}{4}$$
$$\frac{8}{4} = \frac{8}{4}$$
$$2 = 2$$

Both sides of the equation remain equal.

As you can see from the four preceding procedures, if you add, subtract, multiply, or divide both halves of an equation by the same number, the equality of the equation is preserved. In the next section we will see how these operations can serve some practical purpose.

2-1-2 *Transposition*

It is important to know how to transpose, or rearrange, equations and formulas so that you can determine the unknown quantity. This process of rearranging, called **transposition,** is discussed in this section.

Transposing Equations

As an example, let us return to the original problem introduced at the beginning of this chapter and try to determine the value of the unknown quantity x.

$$\frac{8 \times x}{2} = 16$$

To transpose the equation we must follow two steps:

Step 1. Move the unknown quantity so that it is above the fraction bar on one side of the equals sign.

Step 2. Isolate the unknown quantity so that it stands by itself on one side of the equals sign.

Looking at the first step, let us see if our unknown quantity is above the fraction bar on either side of the equals sign.

$$\text{The unknown quantity} \qquad \frac{8 \times \boxed{x}}{2} = 16$$
$$x \text{ is above the fraction bar.}$$

Since step 1 is done, we can move on to step 2. Looking at the equation, you can see that x does not stand by itself on one side of the equals sign. To satisfy this step, we must somehow move the 8 above the fraction bar, and the 2 below the fraction bar away from the left side of the equation, so that x is on its own.

Let us begin by removing the 2. To remove a letter or number from one side of a formula or equation, simply remember this rule: *To move a quantity, simply apply to both sides the arithmetic opposite of that quantity.* Multiplication is the opposite of division, so to remove a "divide by 2" ($\div 2$), simply multiply both sides by 2 ($\times 2$), as follows:

$$\frac{8 \times x}{2} = 16 \quad \leftarrow \text{(Original equation)}$$

$$\frac{8 \times x}{2} \underbrace{\times 2} = 16 \underbrace{\times 2} \leftarrow \text{(Multiply both sides by 2.)}$$

$$\frac{8 \times x}{\cancel{2}} \times \frac{\cancel{2}}{1} = 16 \times 2 \leftarrow \begin{array}{l}\text{(Because } \tfrac{2}{2} = 1, \text{ the two 2s on the} \\ \text{left side of the equation cancel.)}\end{array}$$

$$(8 \times x) \times 1 = 16 \times 2 \quad \begin{array}{l}\text{(Because anything multiplied} \\ \text{by 1 equals the same}\end{array}$$

$$8 \times x = 16 \times 2 \quad \begin{array}{l}\text{number, the 1 on the left} \\ \text{side of the equation can be} \\ \text{removed: } 8x \times 1 = 8x.)\end{array}$$

Looking at the result so far, you can see that by multiplying both sides by 2, we effectively moved the 2 that was under the fraction bar on the left side to the right side of the equation above the fraction bar.

$$\frac{8 \times x}{2} = 16 \quad \leftarrow \text{(Original equation)}$$

$$8 \times x = 16 \times 2 \leftarrow \begin{array}{l}\text{(Result after both sides of} \\ \text{equation were multiplied by 2)}\end{array}$$

However, we have still not completed step 2, which was to isolate x on one side of the equals sign. To achieve this we need to remove the 8 from the left side of the equation. Once again we will do the opposite: Because the opposite of multiply is divide, to remove a "multiply by 8" we must divide both sides by 8 ($\div 8$), as follows:

$$\frac{8 \times x}{2} = 16 \quad \leftarrow \text{(Original equation)}$$

$$8 \times x = 16 \times 2 \leftarrow \begin{array}{l}\text{(Equation after both sides} \\ \text{were multiplied by 2)}\end{array}$$

$$\frac{8 \times x}{8} = \frac{16 \times x}{8} \leftarrow \text{(Divide both sides by 8.)}$$

$$\frac{\cancel{8} \times x}{\cancel{8}} = \frac{16 \times 2}{8} \quad \begin{array}{l}\text{(Because } \tfrac{8}{8} = 1, \text{ the two} \\ \text{8s on the left side of} \\ \text{the equation cancel.)}\end{array}$$

$$1 \times x = \frac{16 \times 2}{8} \quad \begin{array}{l}\text{(Anything multiplied by 1 equals} \\ \text{the same number, so the 1 on the} \\ \text{left side of the equation can be} \\ \text{removed: } 1 \times x = x.)\end{array}$$

$$x = \frac{16 \times 2}{8}$$

Now that we have completed step 2, which was to isolate the unknown quantity so that it stands by itself on one side of the equation, we can calculate the value of the unknown x by performing the arithmetic operations indicated on the right side of the equation.

$$x = \frac{16 \times 2}{8} \quad (16 \times 2 = 32)$$

$$x = \frac{32}{8} \quad (32 \div 8 = 4)$$

$$x = 4$$

To double-check this answer, let us insert this value into the original equation to see if it works.

$$\frac{8 \times x}{2} = 16 \qquad \text{(Replace } x \text{ with 4, or substitute 4 for } x.\text{)}$$

$$\frac{8 \times 4}{2} = 16 \qquad (8 \times 4 = 32)$$

$$\frac{32}{2} = 16 \qquad (32 \div 2 = 16)$$

$$16 = 16 \qquad \text{(Answer checks out because } 16 = 16.\text{)}$$

Transposing Formulas

As mentioned previously, *a formula is a general statement using letters and sometimes numbers that enables us to calculate the value of an unknown quantity.* Some of the simplest relationships are formulas. For example, to calculate the distance traveled, you can use the following formula, which was discussed earlier.

$$d = v \times t$$
where $\quad d$ = distance in miles
$\quad\quad\quad v$ = velocity in miles per hour
$\quad\quad\quad t$ = time in hours

Using this formula we can calculate distance if we know the speed and time. If we need only to calculate distance, this formula is fine. But what if we want to know what speed to go ($v = ?$) to travel a certain distance in a certain amount of time, or what if we want to calculate how long it will take for a trip ($t = ?$) of a certain distance at a certain speed? The answer is: Transpose or rearrange the formula so that we can solve for any of the quantities in the formula.

▢ EXAMPLE:

As an example, calculate how long it will take to travel 250 miles when traveling at a speed of 50 miles per hour (mph). To solve this problem, let us place the values in their appropriate positions in the formula.

$$d = v + t$$
$$250 \,\text{mi} = 50 \,\text{mph} \times t$$

The next step is to transpose just as we did before with equations to determine the value of the unknown quantity t.

	Steps:
$250 = 50 \times t$	To isolate t, divide both sides by 50.
$\dfrac{250}{50} = \dfrac{\cancel{50} \times t}{\cancel{50}}$	$50 \div 50 = 1, 1 \times t = t$
$\dfrac{250}{50} = t$	$250 \div 50 = 5$
$t = 5$	

It will therefore take 5 hours to travel 250 miles at 50 miles per hour ($250 \,\text{mi} = 50 \,\text{mph} \times 5 \,\text{h}$).

All we have to do, therefore, is transpose the original formula so that we can obtain formulas for calculating any of the three quantities (distance, speed, or time) if two values are known. This time, however, we will transpose the formula instead of the inserted values.

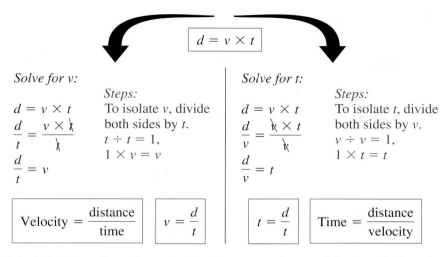

To test if all three formulas are correct, we can replace each letter with the values from the preceding example (also called "plugging in the values"), as follows:

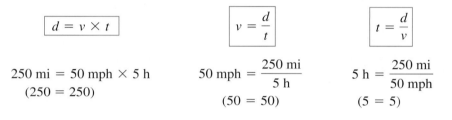

Directly Proportional and Inversely Proportional

The word **proportional** is used to describe a relationship between two quantities. For example, reconsider the formula

$$\text{Time } (t) = \frac{\text{distance } (d)}{\text{velocity } (v)}$$

With this formula we can say that the time (t) it takes for a trip is directly proportional (symbolized by $\propto$) to the amount of distance (d) that needs to be traveled. In symbols, the relationship appears as follows:

$$t \propto d \qquad \text{(Time is directly proportional to distance.)}$$

The term **directly proportional** can therefore be used to describe the relationship between time and distance because an increase in distance will cause a corresponding increase in time ($d \uparrow$ causes $t \uparrow$), and similarly, a decrease in distance will cause a corresponding decrease in time ($d \downarrow$ causes $t \downarrow$).

Proportional
The relation of one part to another or to the whole with respect to magnitude, quantity, or degree.

Directly Proportional
The relation of one part to one or more other parts in which a change in one causes a similar change in the other.

◼ **EXAMPLE:**

Using the previous example,

$$5 \text{ h } (t) = \frac{250 \text{ mi } (d)}{50 \text{ mph } (v)}$$

let us try doubling the distance and then halving the distance to see what effect it has on time. If time and distance are directly proportional to each other, they should change in proportion.

$$\text{Original example} \rightarrow \text{Time} = \frac{\text{distance}}{\text{velocity}} = \frac{250 \text{ mi}}{50 \text{ mph}} = 5 \text{ h}$$

Doubling the distance from 250 mi to 500 mi should double the time it takes for the trip, from 5 h to 10 h.

$$t = \frac{d}{v} = \frac{\boxed{500 \text{ mi}}}{50 \text{ mph}} = \boxed{10 \text{ h}}$$

Doubling the distance did double from 5 h to 10 h. the time ($d \uparrow, t \uparrow$).

Halving the distance from 250 mi to 125 mi should halve the time it takes for the trip, from 5 h to 2½ h.

$$t = \frac{d}{v} = \frac{\boxed{125 \text{ mi}}}{50 \text{ mph}} = \boxed{2.5 \text{ h}}$$

Halving the distance did halve the time ($d \downarrow, t \downarrow$).

As you can see from this exercise, time is directly proportional to distance ($t \propto d$). Therefore, *formula quantities are always directly proportional to one another when both quantities are above the fraction bars on opposite sides of the equals sign.*

On the other hand, **inversely proportional** means that the two quantities compared are opposite in effect. For example, consider again our time formula:

$$t = \frac{d}{v}$$

Observing the relationship between these quantities again, we can say that the time (t) it takes for a trip is *inversely proportional* (symbolized by ½) to the speed or velocity (v) we travel. In symbols, the relationship appears as follows:

$$t \frac{1}{\propto} v \quad \text{(Time is inversely proportional to velocity.)}$$

The term *inversely proportional* can therefore be used to describe the relationship between time and velocity because an increase in velocity will cause a corresponding decrease in time ($v \uparrow$ causes $t \downarrow$), and similarly, a decrease in velocity will cause a corresponding increase in time ($v \downarrow$ causes $t \uparrow$).

EXAMPLE:

Using the previous example,

$$5 \text{ h } (t) = \frac{250 \text{ mi } (d)}{50 \text{ mph } (v)}$$

let us try doubling the speed we travel, or velocity, and then halving the velocity to see what effect it has on time. If time and velocity are inversely proportional to each other, a velocity change should have the opposite effect on time.

$$\text{Original example} \rightarrow \text{Time} = \frac{\text{distance}}{\text{velocity}} \quad \frac{250 \text{ mi}}{50 \text{ mph}} = 5 \text{ h}$$

CHAPTER 2 / RESISTANCE AND POWER

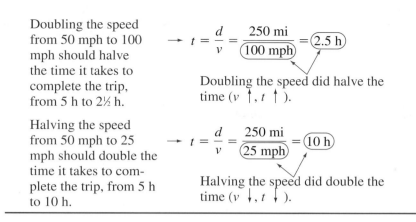

| Doubling the speed from 50 mph to 100 mph should halve the time it takes to complete the trip, from 5 h to 2½ h. | $\rightarrow t = \dfrac{d}{v} = \dfrac{250 \text{ mi}}{\boxed{100 \text{ mph}}} = \boxed{2.5 \text{ h}}$ |

Doubling the speed did halve the time ($v \uparrow, t \uparrow$).

| Halving the speed from 50 mph to 25 mph should double the time it takes to complete the trip, from 5 h to 10 h. | $\rightarrow t = \dfrac{d}{v} = \dfrac{250 \text{ mi}}{\boxed{25 \text{ mph}}} = \boxed{10 \text{ h}}$ |

Halving the speed did double the time ($v \downarrow, t \downarrow$).

As you can see from this exercise, time is inversely proportional to velocity ($t \frac{1}{x} v$). There-fore, *formula quantities are always inversely proportional to one another when one of the quantities is above the fraction bar and the other is below the fraction bar on opposite sides of the equals sign.*

2-1-3 *Substitution*

Substitution is a mathematical process used to develop alternative formulas by replacing or substituting one mathematical term with an equivalent mathematical term. To explain this process, let us try an example.

A **circle** is a perfectly round figure in which every point is at an equal distance from the center.

Some of the elements of a circle are as follows.

a. The **circumference** (C) of a circle is the distance around the perimeter of the circle.

b. The **radius** (r) of a circle is a straight line drawn from the center of the circle to any point of the circumference.

c. The **diameter** (d) of a circle is a straight line drawn through the center of the circle to opposite sides of the circle. The diameter of a circle is therefore equal to twice the cir-cle's radius ($d = 2 \times r$).

To review, *pi* (symbolized by π) is the name given to the ratio, or comparison, of a cir-cle's circumference to its diameter. This constant will always be equal to approximately 3.14 because the circumference of any circle will always be about 3 times larger than the same circle's diameter. Stated mathematically:

$$\pi = \frac{\text{circumference } (C)}{\text{diameter } (d)} = 3.14$$

Substitution

The replacement of one mathematical entity by another of equal value.

Circle

A closed plane curve, every point of which is equidistant from a fixed point within the curve.

Circumference

The perimeter of a circle.

Radius

A line segment extending from the center of a circle or sphere to the circumference.

Diameter

The length of a straight line passing through the center of an object.

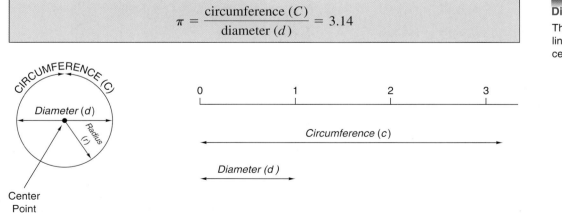

■ **EXAMPLE:**

Determine the circumference of a bicycle wheel that is 26 inches in diameter.

■ *Solution:*

To calculate the wheel's circumference, we will have to transpose the following formula to isolate C.

$$\pi = \frac{C}{d} \qquad \text{Multiply both sides by } d.$$

$$\pi \times d = \frac{C}{d} \times d \qquad d \div d = 1, \text{ and } C \times 1 = C.$$

$$\pi \times d = C$$

The circumference of the bicycle wheel is therefore:

$$C = \pi \times d$$
$$C = 3.14 \times 26 \text{ in.} = 81.64 \text{ in.}$$

In some examples, only the circle's radius will be known. In situations like this, we can substitute the expression $\pi = C/d$ to arrive at an equivalent expression comparing a circle's circumference to its radius.

$$\pi = \frac{C}{d} \qquad \text{Because } d = 2 \times r, \text{ we can substitute } 2 \times r \text{ for } d.$$

$$\pi = \frac{C}{d} \qquad \text{Therefore, this is equivalent to } \pi = \frac{C}{2 \times r}.$$

■ **EXAMPLE:**

How much decorative edging will you need to encircle a garden that measures 6 feet from the center to the edge?

■ *Solution:*

In this example *r* is known, so to calculate the garden's circumference, we will have to transpose the following formula to isolate *C*:

$$\pi = \frac{C}{2 \times r} \qquad \text{Multiply both sides by } 2 \times r.$$

$$\pi \times (2 \times r) = \frac{C}{2 \times r} \times (2 \times r) \qquad (2 \times r) \div (2 \times r) = 1, \text{ and } C \times 1 = C.$$

$$\pi \times (2 \times r) = C$$

The edging needed for the circumference of the garden will therefore be:

$$C = \pi \times (2 \times r)$$
$$C = 3.14 \times (2 \times 6 \text{ ft}) = 37.68 \text{ ft}$$

To continue our example of the circle, how could we obtain a formula for calculating the area within a circle? The answer is best explained with the diagram shown in Figure 2-1. In Figure 2-1(a) a circle has been divided into 8 triangles, and then these triangles have been separated as shown in Figure 2-1(b). Figure 2-1(c) shows how each of these triangles can be thought of as having an A and a B section that, if separated and rearranged, will form a square. The area of a square is easy to calculate because it is simply the product of the

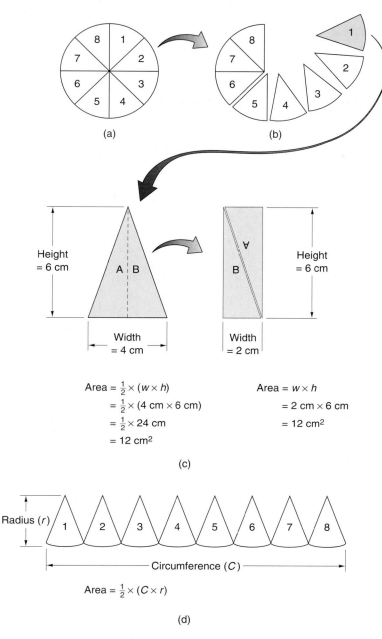

FIGURE 2-1   **Calculating the Area of a Circle.**

square's width and height (area $= w \times h$). The area of the triangle shown in Figure 2-1(c), therefore, will be

$$\text{Area} = \frac{1}{2} \times (w \times h)$$

Referring to Figure 2-1(d), you can see that the width of all eight triangles within a circle is equal to the circle's circumference ($w = C$), and the height of the triangles is equal to the circle's radius ($h = r$). The area of the circle, therefore, can be calculated with the following formula:

$$\text{Area of a circle} = \frac{1}{2} \times (C \times r)$$

Using both substitution and transposition, let us now try to reduce this formula with $C = \pi \times (2 \times r)$.

$$\text{Area of a circle} = \frac{1}{2} \times (C \times r) \qquad \longleftarrow \text{Because } C = \pi \times (2 \times r), \text{ we can replace } C \text{ with } \pi \times (2 \times r).$$

$$= \frac{1}{2} \times [\pi \times (2 \times r) \times r] \qquad \longleftarrow \text{We can now use transposition to reduce.}$$

$$= \frac{1 \times \pi \times 2 \times r \times r}{2} \qquad \longleftarrow 2 \div 2 = 1$$

$$= 1 \times \pi \times r \times r \qquad \longleftarrow r \times r = r^2$$

$$= 1 \times \pi \times r^2 \qquad \longleftarrow 1 \times (\pi \times r^2) = \pi \times r^2$$

$$\boxed{\text{Area of a circle} = \pi \times r^2}$$

EXAMPLE:

Calculate the area of a circle that has a radius of 14 inches.

Solution:

$$\text{Area of a circle} = \pi \times r^2$$
$$= 3.14 \times (14 \text{ in.})^2$$
$$= 3.14 \times 196 \text{ in.}^2$$
$$= 615.4 \text{ in.}^2$$

EXAMPLE:

Calculate the radius of a circle that has an area of 624 square centimeters (cm^2).

Solution:

To determine the radius of a circle when its area is known, we have to transpose the formula area of a circle $= \pi \times r^2$ to isolate r.

$$\text{Area of a circle} = \pi \times r^2$$
$$A = \pi \times r^2 \qquad \longleftarrow \text{Divide both sides by } \pi.$$
$$\frac{A}{\pi} = \frac{\pi \times r^2}{\pi} \qquad \longleftarrow \pi \div \pi = 1, \ 1 \times r^2 = r^2$$
$$\frac{A}{\pi} = r^2 \qquad \longleftarrow \text{Take the square root of both sides.}$$
$$\sqrt{\frac{A}{\pi}} = \sqrt{r^2} \qquad \longleftarrow \sqrt{r^2} = r$$
$$\sqrt{\frac{A}{\pi}} = r$$

Now we can calculate the radius of a circle that has an area of 624 cm^2.

$$r = \sqrt{\frac{A}{\pi}} = \sqrt{\frac{624}{3.14}} = \sqrt{198.7} = 14.1 \text{ cm}$$

Now that you have completed this section, you should be able to:

- **Objective 1.** *Define the terms* algebra, equation, formula, *and* literal numbers.
- **Objective 2.** *Describe the rules regarding the equality on both sides of the equals sign.*
- **Objective 3.** *Demonstrate how to remove a quantity by performing the same arithmetic operation on both sides of an equation or formula.*
- **Objective 4.** *Explain the steps involved in transposing in equation to determine the unknown.*
- **Objective 5.** *Demonstrate the process of transposing a formula.*
- **Objective 6.** *Describe how the terms* directly proportional *and* inversely proportional *are used to show the relationship between quantitites in a formula.*
- **Objective 7.** *Demonstrate how to use substitution to develop alternative fomulas.*

Use the following questions to test your understanding of Section 2-1:

1. Is the following equation true?

$$\frac{\frac{56}{14}}{2} = 2 \quad \text{or} \quad \frac{56 \div 14}{2} = 2$$

2. Fill in the missing values.

$$\frac{144}{12} \times \boxed{} = \frac{36}{6} \times 2 \times \boxed{}$$
$$60 = 60$$

3. Is the following equation balanced?

$$3.2 \text{ k}\Omega = 3200 \text{ } \Omega$$

4. Transpose the following equations to determine the unknown quantity.
   a. $x + 14 = 30$
   b. $8 \times x = \dfrac{80 - 40}{10} \times 12$
   c. $y - 4 = 8$
   d. $(x \times 3) - 2 = \dfrac{26}{2}$
   e. $x^2 + 5 = 14$
   f. $2(3 + 4x) = (2x + 13)$

5. Transpose the following formulas.
   a. $x + y = z, y = ?$
   b. $Q = C \times V, C = ?$
   c. $x_L = 2 \times \pi \times f \times L, L = ?$
   d. $V = I \times R, R = ?$

6. Determine the unknown in the following equations by using transposition.
   a. $I^2 = 9$ **b.** $\sqrt{z} = 8$

7. Use substitution to calculate the value of the unknown.

$$x = y \times z \quad \text{and} \quad a = x \times y$$

Calculate a if $y = 14$ and $z = 5$.

2-2   WHAT IS RESISTANCE?

By definition, resistance is the opposition to current flow accompanied by the dissipation of heat. To help explain the concept of resistance, Figure 2-2 compares fluid opposition to electrical opposition.

In the fluid circuit in Figure 2-2(a), a valve has been opened almost completely, so a very small opposition to the water flow exists within the pipe. This small or low resistance within the pipe will not offer much opposition to water flow, so a large amount of water will flow through the pipe and gush from the outlet. In the electrical circuit in Figure 2-2(b), a small value **resistor** (which is symbolized by the zigzag line) has been placed in the circuit. The resistor is a device that is included within electrical and electronic circuits to oppose current flow by introducing a certain value of circuit **resistance.** In this circuit, a small value resistor has been included to provide very little resistance to the passage of current flow. This low resistance or small opposition will allow a large amount of current to flow through the conductor, as illustrated by the thick current line.

Resistor

Device constructed of a material that opposes the flow of current by having some value of resistance.

Resistance

Symbolized R and measured in ohms (Ω), it is the opposition to current flow with the dissipation of energy in the form of heat.

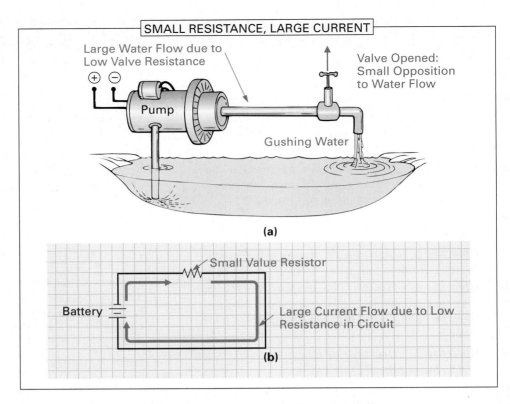

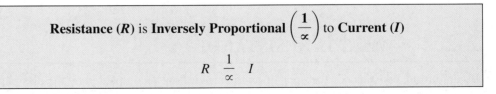

FIGURE 2-2 **The Opposite Relationship between Resistance and Current.**

In the fluid circuit in Figure 2-2(c), the valve is almost completely closed, resulting in a high resistance or opposition to water flow, so only a trickle of water passes through the pipe and out from the outlet. In the electrical circuit in Figure 2-2(d), a large value resistor has been placed in the circuit, causing a large resistance to the passage of current. This high resistance allows only a small amount of current to flow through the conductor, as illustrated by the thin current line.

In the previous examples of low resistance and high resistance shown in Figure 2-2, you may have noticed that resistance and current are inversely proportional ($1/\infty$) to one another. This means that if resistance is high the current is low, and if resistance is low, the current is high.

Resistance (R) is Inversely Proportional $\left(\dfrac{1}{\infty}\right)$ to Current (I)
$R \quad \dfrac{1}{\infty} \quad I$

This relationship between resistance and current means that if resistance is increased by some value, current will be decreased by the same value (assuming a constant electrical pressure or voltage). For example, if resistance is doubled current is halved and similarly, if resistance is halved current is doubled.

$$R\uparrow\frac{1}{\infty}I\downarrow \qquad R\downarrow\frac{1}{\infty}I\uparrow$$

In Figure 2-2, the fluid analogy was used alongside the electrical circuit to help you understand the idea of low resistance and high resistance. In the following section, we will examine resistance in more detail, and define clearly exactly how much resistance exists within a circuit.

TIME LINE

Ohm's law, the best known law in electrical circuits, was formulated by Georg S. Ohm (1787-1854), a German physicist. His law was so coldly received that his feelings were hurt and he resigned his teaching post. When his law was finally recognized, he was reinstated. In honor of his accomplishments, the unit of resistance is called the ohm.

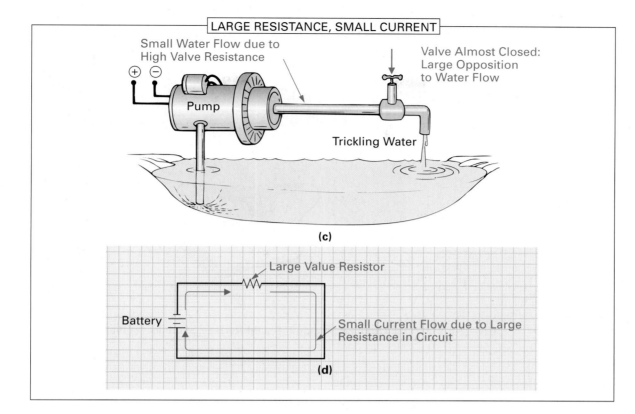

LARGE RESISTANCE, SMALL CURRENT

Small Water Flow due to High Valve Resistance

Valve Almost Closed: Large Opposition to Water Flow

Pump

Trickling Water

(c)

Large Value Resistor

Battery

Small Current Flow due to Large Resistance in Circuit

(d)

FIGURE 2-2 (continued)   **The Opposite Relationship between Resistance and Current.**

SELF-TEST EVALUATION POINT FOR SECTION 2-2

Use the following questions to test your understanding of Section 2-2:

1. What is resistance?

2. What is the difference between low and high resistance?

3. What is the relationship between current and resistance?

4. Would a very high resistance have a small or a large conductance figure?

2-3   THE OHM

As discussed in the previous chapter, current is measured in amperes and voltage is measured in volts. Resistance is measured in **ohms,** in honor of Georg Simon Ohm, who was the first to formulate the relationship between current, voltage, and resistance.

The larger the resistance, the larger the value of ohms and the more the resistor will oppose current flow. The ohm is given the symbol Ω, which is the Greek capital letter omega. By definition, 1 ohm is the value of resistance that will allow 1 ampere of current to flow through a circuit when a voltage of 1 volt is applied, as shown in Figure 2-3(a), where the resistor is drawn as a zigzag. In some schematics or circuit diagrams the resistor is drawn as a rectangular block, as shown in Figure 2-3(b). The pictorial of this circuit is shown in Figure 2-3(c).

The circuit in Figure 2-4 reinforces our understanding of the ohm. In this circuit, a 1 V battery is connected across a resistor, whose resistance can be either increased or decreased. As the resistance in the circuit is increased the current will decrease and, conversely, as the resistance of the resistor is decreased the circuit current will increase. So how much is 1 ohm

Ohm

Unit of resistance, symbolized by the Greek capital letter omega (Ω).

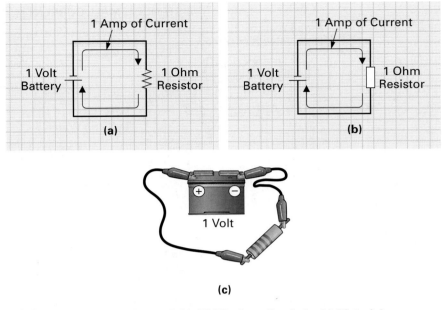

FIGURE 2-3   One Ohm. (a) New and (b) Old Resistor Symbols. (c) Pictorial.

of resistance? The answer to this can be explained by adjusting the resistance of the variable resistor. If the resistor is adjusted until exactly 1 amp of current is flowing around the circuit, the value of resistance offered by the resistor is referred to as 1 ohm (Ω).

2-3-1   *Ohm's Law*

Current flows in a circuit due to the electrical force or voltage applied. The amount of current flow in a circuit, however, is limited by the amount of resistance in the circuit. It can be said, therefore, that the amount of current flow around a circuit is dependent on both voltage and resistance. This relationship between the three electrical properties of current, voltage, and resistance was discovered by Georg Simon Ohm, a German physicist, in 1827. Published originally in 1826, **Ohm's law** states that the current flow in a circuit is directly proportional ($\propto$) to the source voltage applied and inversely proportional ($1/\propto$) to the resistance of the circuit.

Stated in mathematical form, Ohm arrived at this formula:

$$\text{current } (I) = \frac{\text{voltage } (V)}{\text{resistance } (R)}$$

$$\text{current } (I) \propto \text{voltage } (V)$$

$$\text{current } (I) \frac{1}{\propto} \text{resistance } (R)$$

The question you may have now is, how can you calculate voltage ($V = ?$) if I and R are known, and how do you calculate resistance ($R = ?$) if V and I are known? The answer is, transpose the original formula, as follows:

$$\text{current } (I) = \frac{\text{voltage } (V)}{\text{resistance } (R)}$$

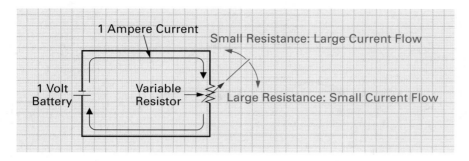

FIGURE 2-4   1 Ohm Allows 1 Amp to Flow When 1 Volt Is Applied.

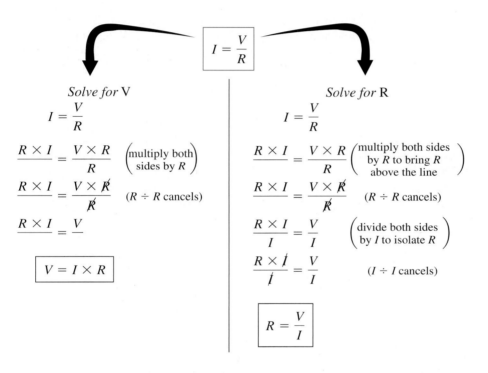

To test if all these three formulas are correct, we can insert into each the values of an example, as follows:

$$I = \frac{V}{R} \qquad\qquad V = I \times R \qquad\qquad R = \frac{V}{I}$$

$$2\,A = \frac{6\,V}{3\,\Omega} \qquad 6\,V = 2\,A \times 3\,\Omega \qquad 3\,\Omega = \frac{6\,V}{2\,A}$$

Since the values on either side of the equals sign are equal in all three equations, the transposition has been successful.

2-3-2   *The Ohm's Law Triangle*

Figure 2-5 shows how the three Ohm's law formulas can be placed within a triangle for easy memory recall.

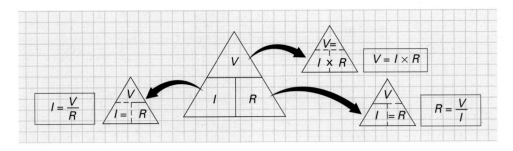

FIGURE 2-5   Ohm's Law Triangle.

2-3-3   *Current Is Proportional to Voltage* (I ∝ V)

Ohm's law states that the current flow in a circuit is proportional (∝) to the circuit's applied voltage, and inversely proportional (1/∝) to the circuit's resistance. In this section we will again use the fluid analogy to help you clearly see the first relationship described in Ohm's law.

Referring to the fluid system in Figure 2-6(a), you can see that an increase in the pump pressure will result in an increase in water flow. Similarly, in the electrical circuit in Fig-

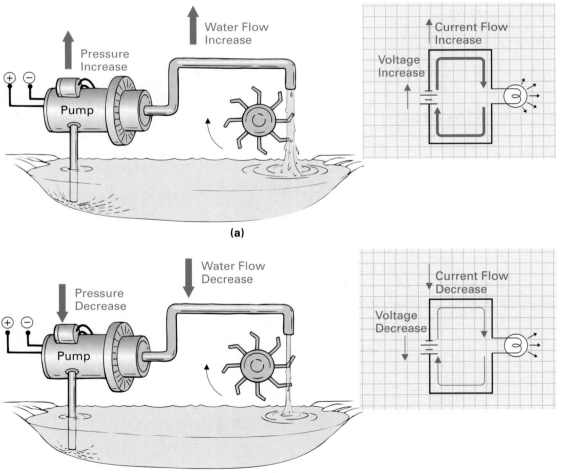

(a)

(b)

FIGURE 2–6   Current Flow Is Proportional to Voltage. (a) Pressure or Voltage Increase Causes a Water or Current Flow Increase (V↑, I↑). (b) Pressure or Voltage Decrease Results in a Water or Current Flow Decrease (V↓, I↓).

ure 2-6(a), an increase in voltage (electron moving force) will exert a greater pressure on the circuit electrons and cause an increase in current flow. This point is reinforced with the Ohm's law formula:

$$I\uparrow = \frac{V\uparrow}{R}$$ Both properties are above the line and are therefore proportional.

In the two following examples, you will see how an increase in the voltage applied to a circuit results in a proportional increase in the circuit current.

EXAMPLE:

If the resistance in the circuit in Figure 2-7(a) remains constant at 1 Ω and the applied voltage equals 2 V, what is the value of current flowing through the circuit?

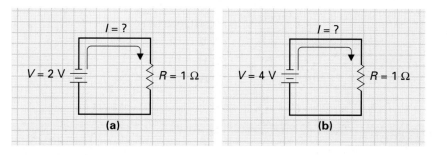

(a)   **(b)**

FIGURE 2-7

Solution:

$$\text{current } (I) = \frac{\text{voltage } (V)}{\text{resistance } (R)}$$
$$= \frac{2\text{ V}}{1\text{ }\Omega}$$
$$= 2\text{ A}$$

CALCULATOR SEQUENCE

Step	Keypad Entry	Display Response
1.	[2]	2
2.	[÷]	
3.	[1]	1
4.	[=]	2

EXAMPLE:

If the voltage from the previous example is now doubled to 4 V as shown in Figure 2-7(b), what would be the change in current?

Solution:

$$\text{current } (I) = \frac{\text{voltage } (V)}{\text{resistance } (R)}$$
$$= \frac{4\text{ V}}{1\text{ }\Omega}$$
$$= 4\text{ A}$$

On the other hand, the fluid system shown in Figure 2-6(b) shows how a decrease in pump pressure will result in a decrease in water flow. Similarly, the electrical circuit in Figure 2-6(b) shows how a decrease in the circuit's applied voltage will cause a decrease in circuit current. This is again reinforced by Ohm's law.

$$I\downarrow = \frac{V\downarrow}{R}$$

In the following two examples, you will see how a decrease in the applied voltage will result in a proportional decrease in circuit current.

■ **EXAMPLE:**

If a circuit has a resistance of 2 Ω and an applied voltage of 8 V, what would be the circuit current?

■ *Solution:*

$$\text{current } (I) = \frac{\text{voltage } (V)}{\text{resistance } (R)}$$
$$= \frac{8 \text{ V}}{2 \text{ }\Omega}$$
$$= 4 \text{ A}$$

■ **EXAMPLE:**

If the circuit described in the previous example were to have its applied voltage halved to 4 V, what would be the change in circuit current?

■ *Solution:*

$$\text{current } (I) = \frac{\text{voltage } (V)}{\text{resistance } (R)}$$
$$= \frac{4 \text{ V}}{2 \text{ }\Omega}$$
$$= 2 \text{ A}$$

To summarize the relationship between circuit voltage and current, we can say that if the voltage were to double, the current within the circuit would also double (assuming the circuit's resistance were to remain at the same value). Similarly, if the voltage were halved, the current would also halve, making the two properties directly proportional to one another.

2-3-4 *The Current Is Inversely Proportional to Resistance* $\left(I \underset{\propto}{\frac{1}{}} R \right)$

The second relationship in Ohm's law states that the circuit current is inversely proportional to the circuit's resistance. The fluid analogy is again used in Figure 2-8 to help you clearly see this second Ohm's law relationship.

In the fluid system in Figure 2-8(a), an open valve offers very little opposition to the water flow, and so a large water flow exists in a system with a low resistance. Similarly, in the electrical circuit in Figure 2-8(a), a small resistance ($R\downarrow$) allows a large current ($I\uparrow$) flow around the circuit.

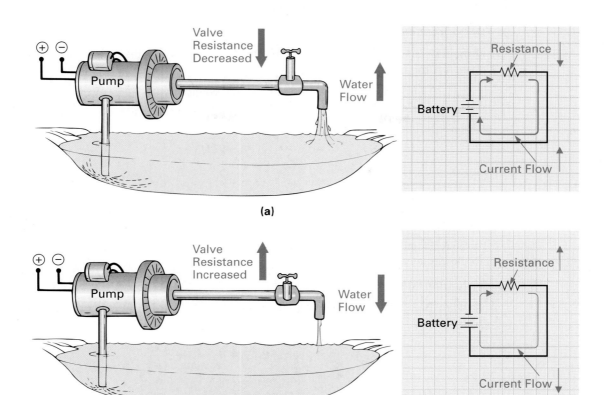

(a)

(b)

FIGURE 2-8   Current Is Inversely Proportional to Resistance. (a) Resistance Decrease Causes a Water or Current Flow Increase ($R\downarrow, I\uparrow$). (b) Resistance Increase Causes a Water or Current Flow Decrease ($R\uparrow, I\downarrow$).

This point is reinforced with the Ohm's law formula:

$$R\downarrow = \frac{V}{I\uparrow}$$ One property is above the line, the other is below the line, and therefore the two are inversely proportional.

On the other hand, Figure 2-8(b) shows how a large resistance ($R\uparrow$) will always result in a small current ($I\downarrow$).

$$R\uparrow = \frac{V}{I\downarrow}$$

To help reinforce this relationship, let us look at a couple of examples.

■   EXAMPLE:

Figure 2-9 shows a circuit that has a constant 8 volts applied. If the resistance in this circuit were doubled from 2 ohms to 4 ohms, what would happen to the circuit current?

FIGURE 2-9

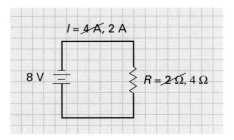

■ *Solution:*

If the resistance in Figure 2-9 were doubled from 2 Ω to 4 Ω, the circuit current would halve from 4 A to:

$$\text{current } (I) = \frac{\text{voltage } (V)}{\text{resistance } (R)}$$

$$= \frac{8 \text{ V}}{4 \text{ Ω}}$$

$$= 2 \text{ A}$$

■ **EXAMPLE:**

If the resistance in Figure 2-9 were returned to its original value of 2 Ω, what would happen to the circuit current?

■ *Solution:*

If the resistance in Figure 2-9 were halved from 4 Ω to 2 Ω, the circuit current would double from 2 A to:

$$\text{current } (I) = \frac{\text{voltage } (V)}{\text{resistance } (R)}$$

$$= \frac{8 \text{ V}}{2 \text{ Ω}}$$

$$= 4 \text{ A}$$

To summarize the relationship between circuit current and resistance, we can say that if the resistance were to double, the current within the circuit would be halved (assuming the circuit's voltage were to remain at the same value). Similarly, if the circuit resistance were halved, the circuit current would double, confirming that current is inversely proportional to resistance.

SELF-TEST EVALUATION POINT FOR SECTION 2-3

Now that you have completed this section, you should be able to:

■ *Objective 8.* *Define* resistance *and* ohm.

■ *Objective 9.* *Explain Ohm's law and its application.*

■ *Objective 10.* *Describe why:*
  a. Current is proportional to voltage.
  b. Current is inversely proportional to resistance.

Use the following questions to test your understanding of Section 2-3:

1. Define 1 ohm in relation to current and voltage.
2. Calculate I if $V = 24$ V and $R = 6$ Ω.
3. What is the Ohm's law triangle?
4. What is the relationship between: (a) current and voltage (b) current and resistance?
5. Calculate V if $I = 25$ mA and $R = 1$ kΩ.
6. Calculate R if $V = 12$ V and $I = 100$ μA.

2-4   CONDUCTORS AND THEIR RESISTANCE

If the current within a circuit needs to be reduced, a resistor is placed in the current path. Resistors are constructed using materials that are known to oppose current flow. Conductors, on the other hand, are not meant to offer any resistance or opposition to current flow. This, however, is not always the case since some are good conductors having a low resistance, while others are poor conductors having a high resistance. Conductance is the measure of how good a conductor is, and even the best conductors have some value of resistance. Up until this time, we have determined a conductor's resistance based on the circuit's electrical characteristics,

$$R = \frac{V}{I}$$

With this formula we could determine the conductor's resistance based on the voltage applied and circuit current. Another way to determine the resistance of a conductor is to examine the physical factors of the conductor.

The total resistance of a typical conductor, like the one shown in Figure 2-10, is determined by four main physical factors:

1. The type of conducting material used
2. The conductor's cross-sectional area
3. The total length of the conductor
4. The temperature of the conductor

Let's now address each of these topics separately and discuss the reason why each will vary a conductor's resistance.

2-4-1   *Conducting Material*

In a previous comparison, we said that copper is a better conducting material than carbon because it has electrons in outer shells, has more electrons per atom, and has an incomplete valence shell. The resistance of a conductor is therefore dependent on the type of material used. A good conducting material, like silver or copper, will have a high conductance figure and therefore a low resistance value.

2-4-2   *Conductor's Cross-Sectional Area*

The resistance of a conductor is inversely proportional to the conductor's cross-sectional area. This means that a conductor with a large cross-sectional area will have a small resistance as seen in Figure 2-11(a), while a conductor with a small cross-sectional area will have a larger resistance.

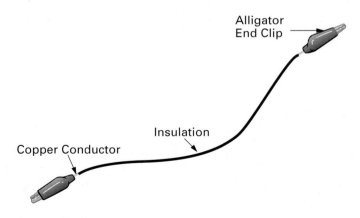

FIGURE 2-10   **Copper Conductor.**

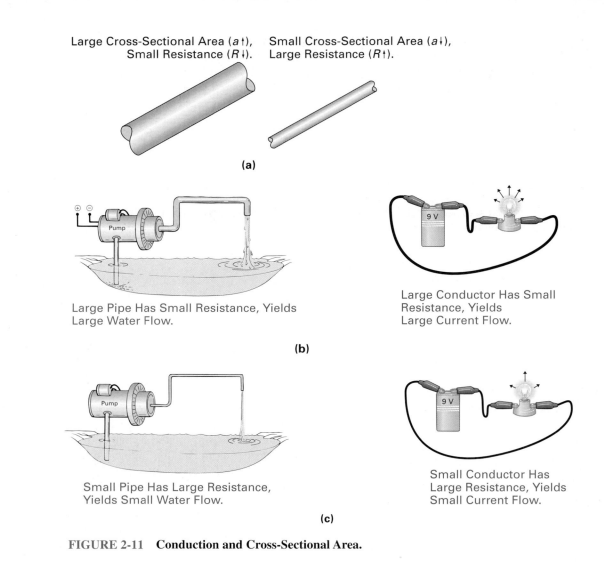

Large Cross-Sectional Area (a↑),
Small Resistance (R↓).

Small Cross-Sectional Area (a↓),
Large Resistance (R↑).

(a)

Large Pipe Has Small Resistance, Yields
Large Water Flow.

Large Conductor Has Small
Resistance, Yields
Large Current Flow.

(b)

Small Pipe Has Large Resistance,
Yields Small Water Flow.

Small Conductor Has
Large Resistance, Yields
Small Current Flow.

(c)

FIGURE 2-11   Conduction and Cross-Sectional Area.

In the fluid analogy in Figure 2-11(b), a larger pipe or conductor results in a larger water flow or current due to the lower resistance. On the other hand, Figure 2-11(c) shows that the smaller the pipe or conductor, the less the amount of water or electron flow, due to the larger resistance.

A unit of measure is needed to compare one thickness of wire conductor to another. Figure 2-12 illustrates a conductor with a diameter of 0.001 inch (¹⁄₁₀₀₀ of an inch) or 1 **mil.**

Mil

One thousandth of an inch
(0.001 in.).

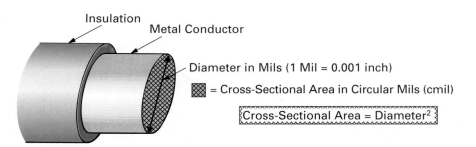

Insulation
Metal Conductor

Diameter in Mils (1 Mil = 0.001 inch)

= Cross-Sectional Area in Circular Mils (cmil)

Cross-Sectional Area = Diameter²

FIGURE 2-12   Measuring a Conductor's Cross-Sectional Area.

The diameters of all conductors are measured in mils. As nearly all conductors are circular in cross section, the area of a conductor is normally given in **circular mils** (cmil), which can be calculated by squaring the diameter.

Circular Mil (cmil)
A unit of area equal to the area of a circle whose diameter is 1 mil.

$$\text{circular mils (cmil)} = \text{diameter}^2$$

■ EXAMPLE:

If a conductor has a diameter of 71.96 mils, what is the circular mil area?

■ *Solution:*

$$
\begin{aligned}
\text{cmil} &= \text{diameter}^2 \\
&= 71.96^2 \\
&= 5178 \text{ cmil}
\end{aligned}
$$

CALCULATOR SEQUENCE

Step	Keypad Entry	Display Response
1.	① ⑤	71.96
2.	⊞	5178.24

2-4-3 *Conductor Length*

Increasing the length of the conductor used increases the amount of resistance within the circuit. For example, if a conductor has a resistance of 1 ohm for every 10 feet of conductor, then 30 feet of conductor are going to triple the conductor's resistance within the circuit to 3 ohms.

2-4-4 *Physical Resistance Formula*

By combining all of the physical factors of a conductor, we can arrive at a formula for its resistance.

$$R = \frac{\rho \times l}{a}$$

where R = resistance of conductor, in ohms
ρ = resistivity of conducting material
l = length of conductor, in feet
a = area of conductor, in circular mils

Resistivity, by definition, is the resistance (in ohms) that a certain length of conductive material (in feet) will offer to the flow of current. Table 2-1 lists the resistivity of the more commonly used conductors.

Resistivity
Measure of a material's resistance to current flow.

The resistance of a conductor, therefore, can be calculated based either on its physical factors or on electrical performance.

$$\frac{\rho \times l}{a} = R = \frac{V}{I}$$

Physical ⟺ Electrical

TABLE 2-1 Material Resistivity

MATERIAL	RESISTIVITY (cmil/ft) IN OHMS
Silver	9.9
Copper	10.7
Gold	16.7
Aluminum	17.0
Tungsten	33.2
Zinc	37.4
Brass	42.0
Nickel	47.0
Platinum	60.2
Iron	70.0

■ EXAMPLE:

Calculate the resistance of 333 ft of copper conductor with a conductor area of 3257 cmil.

■ *Solution:*

$$R = \frac{\rho \times l}{a}$$

$$= \frac{10.7 \times 333}{3257}$$

$$= 1.09 \, \Omega$$

CALCULATOR SEQUENCE

Step	Keypad Entry	Display Response
1.	[1] [0] [.] [7]	10.7
2.	[×]	
3.	[3] [3] [3]	333
4.	[÷]	3563.1
5.	[3] [2] [5] [7]	3257
6.	[=]	1.09

■ EXAMPLE:

Calculate the resistance of 1274 ft of aluminum conductor with a diameter of 86.3 mils.

■ *Solution:*

$$R = \frac{\rho \times l}{a}$$

If the diameter is equal to 86.3 mils, the circular mil area equals d^2:

$$86.3^2 = 7447.7 \text{ cmil}$$

Referring to Table 2-1 you can see that the resistivity of aluminum is 17.0, and therefore

$$R = \frac{17 \times 1274}{7447.7}$$

$$= 2.9 \, \Omega$$

2-4-5   *Temperature Effects on Conductors*

When heat is applied to a conductor, the atoms within the conductor convert this thermal energy into mechanical energy or movement. These random moving atoms cause collisions between the directed electrons (current flow) and the adjacent atoms, resulting in an opposition to the current flow (resistance).

CHAPTER 2 / RESISTANCE AND POWER

Metallic conductors are said to have a **positive temperature coefficient of resistance** (+ Temp. Coe. of R). This means that the greater the heat applied to the conductor, the greater the atom movement, causing more collisions of atoms to occur, and consequently the greater the conductor's resistance.

heat ↑   resistance ↑

2-4-6 *Maximum Conductor Current*

Any time that current flows through any conductor, a certain resistance or opposition is inherent in that conductor. This resistance will convert current to heat, and the heat further increases the conductor's resistance (+ Temp. Coe. of R), causing more heat to be generated due to the opposition. As a result, a conductor must be chosen carefully for each application so that it can carry the current without developing excessive heat. This is achieved by selecting a conductor with a greater cross-sectional area to decrease its resistance. The National Fire Protection Association has developed a set of standards known as the **American Wire Gauge** for all copper conductors, which lists their diameter, resistance, and maximum safe current in amperes. This table is given in Table 2-2. A rough guide for measuring wire size is shown in Figure 2-13.

Conducting wires are normally covered with a plastic or rubber type of material, known as insulation, as shown in Figure 2-14. This insulation is used to protect the users and technicians from electrical shock and also to keep the conductor from physically contacting other conductors within the equipment. If the current through the conductor is too high, this insulation will burn due to the heat and may cause a fire hazard. This is why a conductor's maximum safe current value should not be exceeded.

2-4-7 *Superconductivity*

Conductors have a positive temperature coefficient, which means that if temperature increases, so does resistance. But what happens if the temperature is decreased? In 1911, a Dutch physicist, Heike Onnes, discovered that mercury (a liquid conductor) lost its resistance to electrical current when the temperature was decreased to −459.7 Fahrenheit (0 on the Kelvin temperature scale). Mercury actually became a **superconductor,** allowing a supercurrent to flow and not encounter any resistance and therefore not generate any heat. (Heat

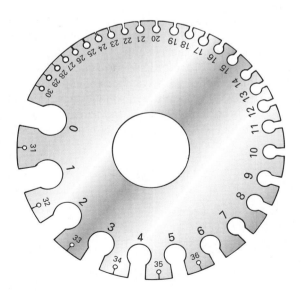

FIGURE 2-13   Wire Gauge Size.

TABLE 2-2 The American Wire Gauge (AWG) for Copper Conductor

AWG NUMBER[a]	DIAMETER (mils)	MAXIMUM CURRENT (A)	Ω/1000 ft
0000	460.0	230	0.0490
000	409.6	200	0.0618
00	364.8	175	0.0780
0	324.9	150	0.0983
1	289.3	130	0.1240
2	257.6	115	0.1563
3	229.4	100	0.1970
4	204.3	85	0.2485
5	181.9	75	0.3133
6	162.0	65	0.3951
7	144.3	55	0.4982
8	128.5	45	0.6282
9	114.4	40	0.7921
10	101.9	30	0.9981
11	90.74	25	1.260
12	80.81	20	1.588
13	71.96	17	2.003
14	64.08	15	2.525
15	57.07		3.184
16	50.82	6	4.016
17	45.26	Wires	5.064
18	40.30	of this 3	6.385
19	35.89	size	8.051
20	31.96	have	10.15
22	25.35	current	16.14
26	15.94	measured	40.81
30	10.03	in mA	103.21
40	3.145		1049.0

[a]The larger the AWG number, the smaller the size of the conductor.

dissipated by any resistance can be calculated by the power formula $P = I^2 \times R$. If $R = 0$, the power dissipated by the conductor is zero watts.)

In the spring of 1986, two IBM scientists discovered that a conductor compound made up of barium, lanthanum, copper, and oxygen would superconduct (have no resistance to current flow) at −406°F. Since then other scientists have increased the temperature to a point that now conductor compounds can be made to superconduct at −300°F. Some scientific projects that need to make use of the advantages of superconductivity immerse these conductor compounds in a bath of liquid nitrogen so that they can achieve superconductivity. Liquid nitrogen, however, is difficult to handle, so the real scientific achievement will be to find a conductor compound that will superconduct at room temperature.

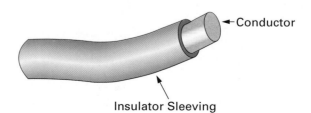

FIGURE 2-14   Conductor with Insulator.

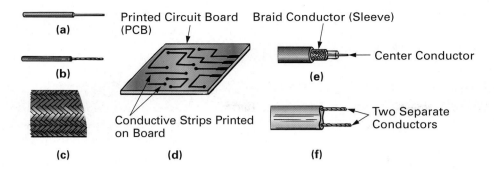

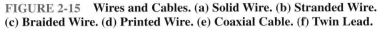

FIGURE 2-15 **Wires and Cables. (a) Solid Wire. (b) Stranded Wire.
(c) Braided Wire. (d) Printed Wire. (e) Coaxial Cable. (f) Twin Lead.**

When discussing the AWG table (Table 2-2), a maximum value of current was stated for a given thickness of wire. This is because a large current will generate a large amount of heat if the resistance is too large, so to decrease the resistance and therefore heat generated ($P\uparrow = I^2R\uparrow$), a thicker conductor with a lower resistance is used. Superconductivity allows a standard 1000 A conductor, which would be approximately 2 in. thick, to be replaced by a superconductor about the thickness of a human hair. Not only will conductors be reduced dramatically in size, but the increased efficiency (as no power is being wasted in the form of heat) will result in high energy savings.

2-4-8   *Conductor and Connector Types*

A cable is made up of two or more wires. Figure 2-15 illustrates some of the different types of **wires** and **cables.** In Figure 2-15(a), (b), and (c), only one conductor exists within the insulation, so they are classified as wires, whereas the cables seen in Figure 2-15(e) and (f) have two conductors. The coaxial and twin-lead cables are most commonly used to connect TV signals into television sets. Figure 2-15(d) illustrates how conducting copper, silver, or gold strips are printed on a plastic insulating board and are used to connect components such as resistors that are mounted on the other side of the board.

Wire
Single solid or stranded group of conductors having a low resistance to current flow.

Cable
Group of two or more insulated wires.

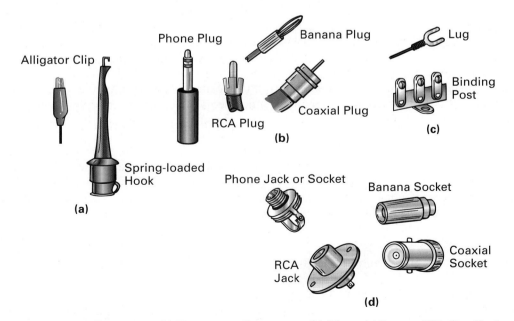

FIGURE 2-16 **Connectors. (a) Temporary Connectors. (b) Plugs. (c) Lug and Binding Post.
(d) Sockets.**

Wires and cables have to connect from one point to another. Some are soldered directly, whereas others are attached to plugs that plug into sockets. A sample of connectors is shown in Figure 2-16.

Now that you have completed this section, you should be able to:

■ **Objective 11.** *Explain what factors affect the resistance of conductors.*

■ **Objective 12.** *Explain superconductivity and its advantages.*

■ **Objective 13.** *List some of the different types of conductors and their connectors.*

Use the following questions to test your understanding of Section 2-4:

1. List the four factors that determine the total resistance of a conductor.

2. What is the relationship between the resistance of a conductor and its cross-sectional area, length, and resistivity?

3. True or false: Conductors are said to have a positive temperature coefficient.

4. True or false: The smaller the AWG number, the larger the size of the conductors.

Energy

Capacity to do work.

Work

Work is done anytime energy is transformed from one type to another, and the amount of work done is dependent on the amount of energy transformed.

TIME LINE

James P. Joule (1818-1889), an English physicist and self-taught scientist, conducted extensive research into the relationships between electrical, chemical, and mechanical effects, which led him to the discovery that one energy form can be converted into another. For this achievement, his name was given to the unit of energy, the joule.

2-5 ENERGY, WORK, AND POWER

The sun provides us with a consistent supply of energy in the form of light. Coal and oil are fossilized vegetation that grew, among other things, due to the sun, and are examples of **energy** that the earth has stored for millions of years. It can be said, then, that all energy begins from the sun. On the earth, energy is not created or destroyed; it is merely transformed from one form to another. The transforming of energy from one form to another is called **work.** The greater the energy transformed, the more work that is done.

The six basic forms of energy are light, heat, magnetic, chemical, electrical, and mechanical energy. The unit for energy is the **joule** (J). "Potential" (position) and "kinetic" (motion) are two terms used when describing energy. A cart on top of a hill has **potential** (position) **energy,** while a cart rolling down a hill has **kinetic** (motion) **energy.** Potential and kinetic energy are best described by looking at the example of a swinging pendulum, as seen in Figure 2-17. When the pendulum is in its upmost position shown in Figure 2-17(a), it has potential energy due to its position relative to the resting position, yet it has no motion and so no kinetic energy. When the pendulum is in the position shown in Figure 2-17(b), it has no potential energy, yet it has motion or kinetic energy. In between these two points, the pendulum possesses a combination of both potential and kinetic energy, as shown in Figure 2-17(c).

Looking at Figure 2-18, let us try to summarize our discussion so far on energy and work. Chemical energy within the battery is converted to electrical energy when the electrons are attracted and repelled and set in motion. To use the two terms, we could say that the battery has the potential energy needed to set electrons in motion, and these moving electrons are said to possess kinetic energy. This electrical energy drives two devices:

1. The light bulb, which converts electrical energy into light and heat energy
2. The pump, which converts electrical energy into mechanical energy

The pump has the potential energy to cause water flow or kinetic energy, just as the battery has the potential energy to cause kinetic energy in the form of electron flow. The kinetic energy within the water flow is finally used to turn the waterwheel, just as the kinetic energy within the electron flow is used to generate light from the bulb.

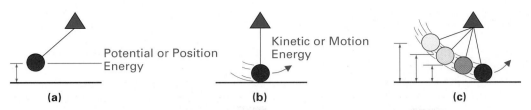

FIGURE 2-17 **Pendulum. (a) Potential Energy. (b) Kinetic Energy. (c) Potential and Kinetic Energy.**

Work is being done everytime one form of energy is transformed to another. A device that converts one form of energy to another is called a **transducer.** In Figure 2-18, the battery, light bulb, and pump were all examples of transducers doing work at each stage of conversion.

The amount of work done is equal to the amount of energy transformed, and in both cases an equal amount of energy was transformed, so an equal amount of work was done. As a result, energy and work have the same symbol (W), the same formula, and the same unit (the joule). Energy is merely the capacity, potential, or ability to do work, and work is done when a transformation of the potential, capacity, or ability takes place.

EXAMPLE:

One person walks around a track and takes 5 minutes, while another person runs around the track and takes 50 seconds. Both were full of energy before they walked or ran around the track, and during their travels around the track they converted the chemical energy within their bodies into the mechanical energy of movement.

a. Who exerted the most energy?
b. Who did the most work?

Solution:

Both exerted the same amount of energy. The runner exerted all his energy (for example, 100 J) in the short time of 50 seconds, while the walker spaced his energy (100 J) over 5 minutes. Since they both did the same amount of work, the only difference between the runner and the walker is time, or the rate at which their energy was transformed.

Joule

The unit of work and energy.

Potential Energy

Energy that has the potential to do work because of its position relative to others.

Kinetic Energy

Energy associated with motion.

Transducer

Any device that converts energy from one form to another.

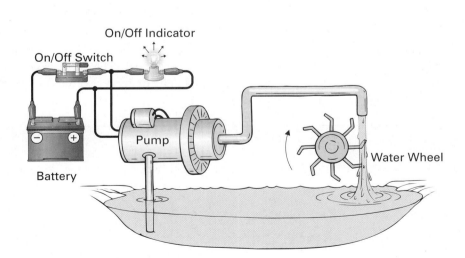

FIGURE 2-18 **Energy Transfer.**

2-5-1   *Power*

Power

Amount of energy converted by a component or circuit in a unit of time, normally seconds. It is measured in units of watts (joules/second).

TIME LINE

The unit of power, the watt, was named after Scottish engineer and inventor James Watt (1736-1819) in honor of his advances in the field of science.

Watt (W)

Unit of electric power required to do work at a rate of 1 joule/second. One watt of power is expended when 1 ampere of direct current flows through a resistance of 1 ohm.

Power (*P*) is the rate at which work is performed and is given the unit of **watt** (W), which is joules per second (J/s).

Returning to the example involving the two persons walking and running around the track, we could say that the number of joules of energy exerted in 1 second by the runner was far greater than the number of joules of energy exerted in 1 second by the walker, although the total energy exerted by both persons around the entire track was equal and therefore the same amount of work was done. Even though the same amount of energy was used, and therefore the same amount of work was done by the runner and the walker, the output power of each was different. The runner exerted a large value of joules/second or watts (high power output) in a short space of time, while the walker exerted only a small value of joules/second or watts (low power output) over a longer period of time.

Whether discussing a runner, walker, electric motor, heater, refrigerator, light bulb, or compact disk player—power is power. The output power, or power ratings, of electrical, electronic, or mechanical devices can be expressed in watts and describes the number of joules of energy converted every second. The output power of rotating machines is given in the unit *horsepower* (hp), the output power of heaters is given in the unit British thermal unit per hour (Btu/h), and the output power of cooling units is given in the unit *ton of refrigeration*. Despite the different names, they can all be expressed quite simply in the unit of watts. The conversions are as follows:

$$1 \text{ horsepower (hp)} = 746 \text{ W}$$
$$1 \text{ British thermal unit per hour (Btu/h)} = 0.293 \text{ W}$$
$$1 \text{ ton of refrigeration} = 3.52 \text{ kW (3520 W)}$$

Now we have an understanding of power, work, and energy. Let's reinforce our knowledge by introducing the energy formula and try some examples relating to electronics.

2-5-2   *Calculating Energy*

The amount of energy stored (*W*) is dependent on the coulombs of charge stored (*Q*) and the voltage (*V*).

$$W = Q \times V$$

where *W* = energy stored, in joules (J)
Q = coulombs of charge (1 coulomb = 6.24×10^{18} electrons)
V = voltage, in volts (V)

If you consider a battery as an example, you can probably better understand this formula. The battery's energy stored is dependent on how many coulombs of electrons it holds (current) and how much electrical pressure it is able to apply to these electrons (voltage).

■   **EXAMPLE:**

If a 1 V battery can store 6.24×10^{18} electrons, how much energy is the battery said to have?

■   *Solution:*

$$W = Q \times V$$
$$= 1 \text{ C} \times 1 \text{ V}$$
$$= 1 \text{ J of energy}$$

◼ EXAMPLE:

How many coulombs of electrons would a 9 V battery have to store to have 63 J of energy?

◼ *Solution:*

If $W = Q \times V$, then by transposition:

$$Q = \frac{W}{V}$$

coulombs of electrons $(Q) = \dfrac{\text{energy in joules } (W)}{\text{battery voltage } (V)}$

$$= \frac{63 \text{ J}}{9 \text{ V}}$$

$$= 7 \text{ C of electrons}$$

or

$$7 \times 6.24 \times 10^{18} = 4.36 \times 10^{19} \text{ electrons}$$

CALCULATOR SEQUENCE

Step	Keypad Entry	Display Response
1.	6 3	63
2.	÷	
3.	9	9
4.	=	7
5.	×	7
6.	6 . 2 4 E 1 8	6.24E18
7.	=	4.36E19

2-5-3 *Calculating Power*

Power, in connection with electricity and electronics, is the rate (t) at which electric energy (W) is converted into some other form. A power formula can therefore be derived as follows:

$$P = \frac{W}{t}$$

where P = power, in watts
W = energy, in joules
t = time, in seconds

Since $W = Q \times V$, we can replace W in the formula above with $Q \times V$ to obtain

$$P = \frac{W}{t} = \frac{Q \times V}{t}$$

where Q = coulombs of charge
V = voltage, in volts

Since coulombs of charge $Q = I \times t$, we can replace Q in the formula above with $I \times t$ to obtain

$$P = \frac{Q \times V}{t} = \frac{(I \times t) \times V}{t}$$

By canceling t, we arrive at a final formula for power:

$$P = I \times V$$

where P = power, in watts (W)
I = current, in amperes (A)
V = voltage, in volts (V)

This formula states that the amount of power delivered to a device is dependent on the electrical pressure or voltage applied across the device and the current flowing through the device.

EXAMPLE:

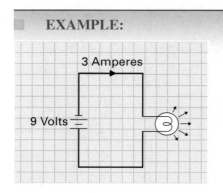

FIGURE 2-19 Calculating Power.

In regards to electrical and electronic circuits, power is the rate at which electric energy is converted into some other form. In the example in Figure 2–19, electric energy will be transformed into light and heat energy by the light bulb. Power has the unit of watts, which is the number of joules of energy transformed per second (J/s). If 27 J of electric energy is being transformed into light and heat per second, how many watts of power does the light bulb convert?

Solution:

$$\text{power} = \frac{\text{joules}}{\text{second}}$$

$$= \frac{27 \text{ J}}{1 \text{ s}}$$

$$= 27 \text{ W}$$

The power output in the previous example could have easily been calculated by merely multiplying current by voltage to arrive at the same result.

$$\text{power} = I \times V = 3 \text{ A} \times 9 \text{ V}$$
$$= 27 \text{ W}$$

We could say, therefore, that the light bulb dissipates 27 watts of power, or 27 joules of energy per second.

Studying the power formula $P = I \times V$, we can also say that 1 watt of power is expended when 1 ampere of current flows through a circuit that has 1 volt applied.

Like Ohm's law, we can transpose the power formula as follows:

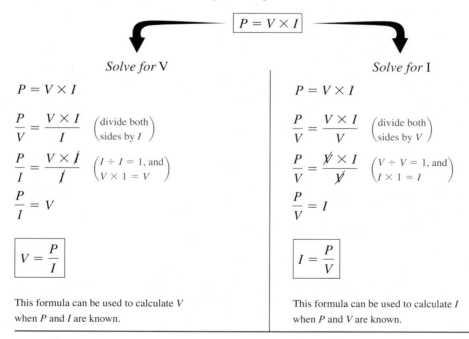

Solve for V	Solve for I
$P = V \times I$	$P = V \times I$
$\dfrac{P}{V} = \dfrac{V \times I}{I}$ $\left(\begin{array}{l}\text{divide both}\\\text{sides by } I\end{array}\right)$	$\dfrac{P}{V} = \dfrac{V \times I}{V}$ $\left(\begin{array}{l}\text{divide both}\\\text{sides by } V\end{array}\right)$
$\dfrac{P}{I} = \dfrac{V \times \cancel{I}}{\cancel{I}}$ $\left(\begin{array}{l}I \div I = 1\text{, and}\\ V \times 1 = V\end{array}\right)$	$\dfrac{P}{V} = \dfrac{\cancel{V} \times I}{\cancel{V}}$ $\left(\begin{array}{l}V \div V = 1\text{, and}\\ I \times 1 = I\end{array}\right)$
$\dfrac{P}{I} = V$	$\dfrac{P}{V} = I$

$$\boxed{V = \frac{P}{I}}$$

This formula can be used to calculate V when P and I are known.

$$\boxed{I = \frac{P}{V}}$$

This formula can be used to calculate I when P and V are known.

In the previous section we stated that $P = V \times I$, and by transposition we arrived at $V = P/I$ and $I = P/V$. When trying to calculate power we may not always have the values of V and I available. For example, we may know only I and R, or V and R. Substitution enables us to obtain alternative power formulas by replacing or substituting one mathematical term with an equivalent mathematical term. For example, since $I = V/R$, we could substitute I in any formula with V/R. Similarly, since $V = I \times R$, we could substitute V in any formula with $I \times R$. The following shows how we can substitute terms in the $P = V \times I$ formula to arrive at alternative power formulas for wattage calculations.

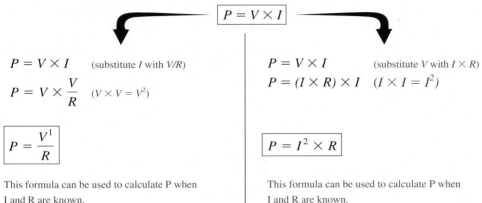

$$P = V \times I \quad \text{(substitute } I \text{ with } V/R)$$
$$P = V \times \frac{V}{R} \quad (V \times V = V^2)$$

$$\boxed{P = \frac{V^1}{R}}$$

This formula can be used to calculate P when I and R are known.

$$P = V \times I \quad \text{(substitute } V \text{ with } I \times R)$$
$$P = (I \times R) \times I \quad (I \times I = I^2)$$

$$\boxed{P = I^2 \times R}$$

This formula can be used to calculate P when I and R are known.

EXAMPLE:

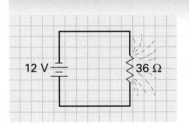

In Figure 2-20 a 12 V battery is connected across a 36 Ω resistor. How much power does the resistor dissipate?

FIGURE 2-20

Solution:

Since V and R are known, the V2/R power formula should be used.

$$\text{power} = \frac{\text{voltage}^2}{\text{resistance}}$$
$$= \frac{12\ \text{V}^2}{36\ \Omega} = \frac{144}{36}$$
$$= 4\text{W}$$

Four joules of heat energy are being dissipated every second.

CALCULATOR SEQUENCE

Step	Keypad Entry	Display Response
1.	☐1 ☐2	12
2.	X² (square key)	144
3.	÷	
4.	☐3 ☐6	36
5.	=	4

2-5-4 *Measuring Power*

The multimeter can be used to measure power by following the procedure described in Figure 2-21. In this example, the power consumed by a car's music system is being determined. After the current measurement and voltage measurement have been taken, the product of the two values will have to be calculated, since power = current × voltage ($P = I \times V$).

■ **EXAMPLE:**

Calculate the power consumed by the car music system shown in Figure 2-21.

■ *Solution:*

Current drawn from battery by music system = 970 mA. Voltage applied to music system power input = 13.6 V.

$$P = I \times V = 970 \text{ mA} \times 13.6 \text{ V} = 13.2 \text{ watts}$$

By connecting a special current probe, as shown in Figure 2-22, some multimeters are able to make power measurements. In this configuration, the current probe senses current, the standard meter probes measure voltage, and the multimeter performs the multiplication process and displays the power reading.

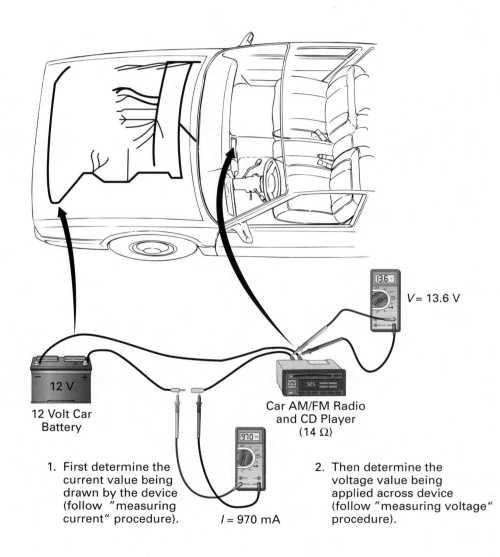

V = 13.6 V

12 Volt Car Battery

Car AM/FM Radio and CD Player (14 Ω)

1. First determine the current value being drawn by the device (follow "measuring current" procedure).

I = 970 mA

2. Then determine the voltage value being applied across device (follow "measuring voltage" procedure).

3. Multiply voltage value by current value to obtain power value ($P = V \times I$).

FIGURE 2-21 **Measuring Power with a Multimeter.**

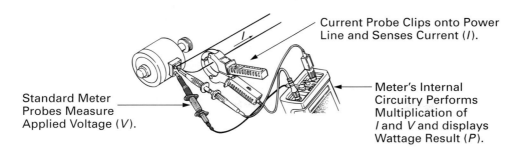

Current Probe Clips onto Power Line and Senses Current (*I*).

Standard Meter Probes Measure Applied Voltage (*V*).

Meter's Internal Circuitry Performs Multiplication of *I* and *V* and displays Wattage Result (*P*).

FIGURE 2-22 **Configuring the Multimeter to Measure Power.**

2-5-5   *The Kilowatt-Hour*

You and I pay for our electric energy in a unit called the **kilowatt-hour** (kWh). The **kilowatt-hour meter,** shown in Figure 2-23, measures how many kilowatt-hours are consumed, and the electric company then charges accordingly.

So what is a kilowatt-hour? Well, since power is the rate at which energy is used, if we multiply power and time, we can calculate how much energy has been consumed.

$$\text{energy consumed } (W) = \text{power } (P) \times \text{time } (t)$$

This formula uses the product of power (in watts) and time (in seconds or hours) and so we can use one of three units: the watt-second (Ws), watt-hour (Wh), or kilowatt-hour (kWh). The kilowatt-hour is most commonly used by electric companies, and by definition, a kilowatt-hour of energy is consumed when you use 1000 watts of power in 1 hour (1 kW in 1 h).

$$\text{energy consumed (kWh)} = \text{power (kW)} \times \text{time (h)}$$

Kilowatt-Hour
1000 watts for 1 hour.

Kilowatt-Hour Meter
A meter used by electric companies to measure a customer's electric power use in kilowatt-hours.

FIGURE 2-23 **Digital Kilowatt-Hour Meter.**

To see how this would apply, let us look at a couple of examples.

▢ EXAMPLE:

If a 100 W light bulb is left on for 10 hours, how many kilowatt-hours will we be charged for?

▢ *Solution:*

CALCULATOR SEQUENCE

Step	Keypad Entry	Display Response
1.	⓪ ⓒ ① Ⓔ ③	0.1E3
2.	⊗	
3.	① ⓪	10
4.	⊜	1E3

power consumed (kWh) = power (kW) × time (hours)
= 0.1 kW × 10 hours
(100 W = 0.1 kW)
= 1 kWh

▢ EXAMPLE:

Figure 2-24 illustrates a typical household electric heater and an equivalent electrical circuit. The heater has a resistance of 7 Ω and the electric company is charging 12 cents/kWh. Calculate:

    a. The energy consumed by the heater

    b. The cost of running the heater for 7 hours

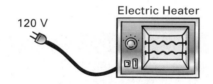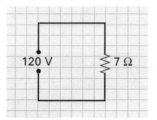

FIGURE 2-24   An Electric Heater

▢ *Solution:*

    a. Power $(P) = \dfrac{V^2}{R} = \dfrac{120^2}{7}$

$$= \frac{14400\text{ V}}{7} = \frac{14.4\text{ kV}}{7}$$

$$= 2057\text{ W (approximately 2 kilowatts or 2 kW)}$$

    b. Energy consumed = power (kW) × time (hours)

$$= 2.057 \times 7\text{ h}$$

$$= 14.4\text{ kWh}$$

    Cost = kWh × rate = 14.4 × 12 cents

$$= \$1.73$$

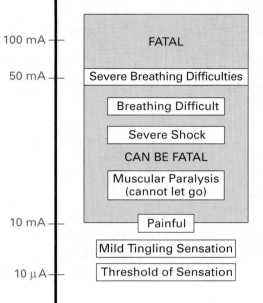

FIGURE 2-25   **Physiological Effects of Electric Current.**

2-5-6   *Body Resistance*

Safety precautions should always be your first priority when working on electronic equipment, as there is always the possibility of receiving an electric shock. An electric **shock** is a sudden, uncontrollable reaction as current passes through your body and causes your muscles to contract and deliver a certain amount of pain.

Figure 2-25 lists the physiological effects of different amounts of current. As you can see, even a current as small as 10 mA can be fatal. Any shock is dangerous, since even the mildest could surprise you and cause an involuntary action that could injure you or someone else. For example, a muscular spasm could throw you against a sharp object or move your arm to a point of higher voltage.

As you know, the value of current in a circuit is determined by the voltage applied and the circuit's resistance. When you come in contact with a high voltage point, therefore, the value of current passing through your body will be dependent on the value of voltage at that point and your body's resistance. The human body has a resistance of about 10,000 to 50,000 Ω, depending on how good a contact you make with the **"live"** (power present) conductor. Skin resistance is generally quite high and will subsequently oppose current flow; however, your resistance is lowered if your skin is wet due to perspiration or if your skin has a cut or an abrasion.

Shock
The sudden pain, convulsion, unconsciousness, or death produced by the passage of electric current through the body.

Body Resistance
The resistance of the human body.

Live
Term used to describe a circuit or piece of equipment that is ON and has current flow within it.

EXAMPLE:

Calculate the current through a body resistance of 10 kΩ if the body came in contact with 100 V. How much power is the body dissipating?

Solution:

$$I = \frac{V}{R} = \frac{100 \text{ V}}{10 \text{ k}\Omega} = 10 \text{ mA}$$

$$P = V \times 100 \text{ V} \times 10 \text{ mA} = 1 \text{ watt}$$

This value of current will be painful and possibly fatal.

Power is a point to consider. For example, if an infrared laser's 1 mW power supply generates a 50,000 V output, would this be dangerous if a person was to come in contact with it? The answer is probably not, since even though the power supply generates an intimidating voltage, it cannot deliver a dangerous value of current:

$$I = \frac{P}{V} = \frac{1 \text{ mW}}{50 \text{ kV}}$$

$$= \frac{1 \times 10^{-3} \text{ W}}{50 \times 10^3 \text{ V}}$$

$$= 0.02 \text{ } \mu\text{A}$$

This also works in the other way. For example, a car's battery may only be 12 V, but it can supply a very deadly value of current.

SELF-TEST EVALUATION POINT FOR SECTION 2-5

Now that you have completed this section, you should be able to:

- **Objective 14.** *Describe and define the terms* energy, work, *and* power.
- **Objective 15.** *Describe how to measure power.*
- **Objective 16.** *Explain what is meant by the term* kilowatt-hour.

Use the following questions to test your understanding of Section 2-5:

1. List the six basic forms of energy.

2. What is the difference between energy, work, and power?

3. List the formulas for calculating energy and power.

4. What is 1 kilowatt-hour of energy?

Rules of Algebra—A Summary

Monomial (One Term): $2a$
Binomial (Two Terms): $2a + 3b$
Trinomial (Three Terms): $2a + 3b + c$
Polynomial (Any Number of Terms).

<u>Order of Operations</u>: Parentheses, Exponents, Multiplication, Division, Addition, Subtraction.
 (Memory Aid—Please, Excuse, My, Dear, Aunt, Sally)

<u>Basic Rules</u>: $a + a = 2a$ $a - a = 0$ $a \cdot a = a^2$ $a \div a = 1$ $\sqrt{a^2} = a$
              $a + 1 = a + 1$ $a - 1 = a - 1$ $a \cdot 1 = a$ $a \div 1 = a$
              $a + 0 = a$ $a - 0 = a$ $a \cdot 0 = 0$ $a \div 0 = $ Impossible $0 \div a = 0$

<u>Parentheses First</u>: $(3 + 6) \cdot 2 = 9 \cdot 2 = 18$ $9 - (2 \cdot 3) = 9 - 6 = 3$ $6 + (9 - 7) = 6 + 2 = 8$

<u>No Parentheses</u>—Step 1:   Multiplications and Divisions, left to right. $3 + 6 \cdot 4 = 3 + 24 = 27$
                Step 2:   Additions and Subtractions, left to right. $4 \cdot 2 + 3 - 4 \div 2 = 8 + 3 - 2 = 9$

<u>Combine Only Like Terms</u>: $2x + x + y = 3x + y$ $3a + 4b - a = 2a + 4b$

<u>Commutative Property of Addition</u>—order of terms makes no difference: $a + b = b + a$

<u>Associative Property of Addition</u>—grouping of terms makes no difference: $(a + b) + c = a + (b + c)$

<u>Coefficients</u>: $x + x + x = 3x$ $x + y + x + y = 2x + 2y$

<u>Exponents</u>: $x \cdot x \cdot x = x^3$ $x \cdot y \cdot x \cdot y = x^2 \cdot y^2 = x^2y^2$

<u>Commutative Property of Multiplication</u>—order of factors makes no difference: $a \cdot b = b \cdot a$

<u>Associative Property of Multiplication</u>—grouping of factors makes no difference: $(ab)c = a(bc)$

<u>Distributive Property of Multiplication</u>—$a(b + c) = ab + ac$ $2(a - 3b) = 2 \cdot a - 2 \cdot 3b = 2a - 6b$

<u>Transposition</u>—Step 1: Ensure unknown is above fraction bar. $x \cdot 8 = 16$ (divide both sides by 8, $\div 8$)
              Step 2: Isolate unknown. $(x \cdot 8) \div 8 = (16) \div 8,$ $x = 2$

<u>Unknown Above Fraction Bar</u>	<u>Unknown Below Fraction Bar</u>	<u>Unknown on Both Sides</u>	<u>Combining Unknowns</u>
$(2 \cdot a) + 5 = 23$ (-5)	$\dfrac{72}{x} = 12$ $(\cdot x)$	$(6 \cdot y) + 7 = y + 27$ (-7)	$x + 2x + 4(x + 2x) = 100$
$2 \cdot a = 18$ $(\div 2)$		$6y = y + 20$ $(-y)$	$x + 2x + 4x + 8x = 100$
$a = 9$	$72 = 12 \cdot x$ $(\div 12)$	$5y = 20$ $(\div 5)$	$15x = 100$
	$6 = x$	$y = 4$	$x = 6.67$

<u>Positive and Negative Number Rules</u>
$\oplus$ $(+) + (+) = (+), (-) + (-) = (-)$
      $(+) + (-)$ or $(-) + (+) = $ Difference

<u>Positive and Negative Numbers</u>
$\oplus$ $2x + 5x = 7x$ $-6a + (+3a) = -3a$

$\ominus$ Change sign of second number, and then add.

$\ominus$ $-3y^2 - (-2y^2) = -3y^2 + (+2y^2) = -y^2$

$\otimes$ $(+) \cdot (+) = (+), (-) \cdot (-) = (+)$
      $(+) \cdot (-)$ or $(-) \cdot (+) = (-)$

$\otimes$ $-3(6a) = -3 \cdot (+6a) = -18a$

$\odot$ $(+) \div (+) = (+), (-) \div (-) = (+)$
      $(+) \div (-)$ or $(-) \div (+) = (-)$

$\odot$ $14a \div (-7) = -2a$

Adding Polynomials: $(4a^2 + 6b^2 + x - 2) + (7a^2 - b^2 + 3)$

$$
\begin{array}{l}
\;\; 4a^2 + 6b^2 + x - 2 \\
+ \;\; 7a^2 - b^2 + 3 \\
\hline
\, 11a^2 + 5b^2 + x + 1
\end{array}
$$

Subtracting Polynomials: $(6x^2 + y^2 - 4) - (-4x^2 + 2y^2 + 6)$

$$
\begin{array}{l}
\;\; 6x^2 + y^2 - 4 \\
- \;\; -4x^2 + 2y^2 + 6
\end{array}
\qquad
\boxed{\begin{array}{c}\text{ADD THE}\\\text{OPPOSITE}\end{array}}
\qquad
\begin{array}{l}
\;\; 6x^2 + y^2 - 4 \\
+ \;\; 4x^2 - 2y^2 - 6 \\
\hline
\, 10x^2 - y^2 - 10
\end{array}
$$

Multiplying Monomials, Binomials

$x^2 \cdot x^3 \cdot x = x \cdot x \cdot x \cdot x \cdot x \cdot x = x^{2+3+1} = x^6$

$(x^3)^2 = (x \cdot x \cdot x) \cdot (x \cdot x \cdot x) = x^{3 \cdot 2} = x^6$

$2(a + 2b) = 2a + 4b$

$4x^2 \cdot x^3 = 6 \cdot x \cdot x \cdot x \cdot x \cdot x = 4x^{2+3} = 4x^5$

$(-6y)^2 \cdot y^3 = -6^2 \cdot y^2 \cdot y^3 = 36y^5$

$-2(2a - 3b) = (-2 \cdot 2a) - (-2 \cdot 3b) = -4a - (-6b) = -4a + 6b$

Multiplying Polynomials

$$
\begin{array}{r}
x^2 + 2x - 1 \\
\times \qquad x - 3 \\
\hline
-3x^2 - 6x + 3 \\
x^3 + 2x^2 - x \\
\hline
x^3 - x^2 - 7x + 3
\end{array}
$$
   $(x^2 + 2x - 1) \cdot (x - 3) = ?$

   $\leftarrow$ $-3(x^3 + 2x - 1)$
   $\leftarrow$ $x(x^2 + 2x - 1)$

$$
\begin{array}{r}
a - 3 \\
\times \quad a - 3 \\
\hline
-3a + 9 \\
a^2 - 3a \\
\hline
a^2 - 6a + 9
\end{array}
$$
   $(a - 3)^2 = ?$

   $\leftarrow$ $-3(a - 3)$
   $\leftarrow$ $a(a - 3)$

Dividing Polynomials

$$
\frac{a^4}{a} = \frac{\overset{1}{\cancel{a}} \cdot a \cdot a \cdot a}{\underset{1}{\cancel{a}}} = a^3
$$

$$
\frac{-6ab^3}{2ab} = \frac{\overset{-3}{\cancel{-6}} \cdot \cancel{a} \cdot \cancel{b} \cdot b \cdot b}{\underset{1}{\cancel{2} \cdot \cancel{a} \cdot \cancel{b}}} = -3b^2
$$

$$
\frac{2ab - 4a^2b^2}{2a} = \frac{2ab}{2a} - \frac{4a^2b^2}{2a} = b - 2ab^2
$$

$$
\frac{2a^2 + 6a + 8ab}{2a} = \frac{2a^2}{2a} + \frac{6a}{2a} + \frac{8ab}{2a} = a + 3 + 4b
$$

SUMMARY

Resistance (Figure 2–26)

1. Resistance is the opposition to current flow accompanied by the dissipation of heat.

2. Resistance and current are inversely proportional ($1/\infty$) to one another. For example, if resistance is doubled, current is halved (assuming a constant voltage).

3. Current is measured in amperes, voltage is measured in volts, and resistance is measured in ohms, in honor of Georg Simon Ohm and his work with current, voltage, and resistance.

4. The larger the resistance, the larger the value of ohms and the more the resistor will oppose current flow.

5. The ohm is given the symbol Ω, which is the Greek capital letter omega. By definition, 1 ohm is the value of resistance that will allow 1 ampere of current to flow through a circuit when a voltage of 1 volt is applied.

6. Published originally in 1826, *Ohm's law* states that the current flow in a circuit is directly proportional (∞) to the source voltage applied and inversely proportional ($1/\infty$) to the resistance of the circuit.

7. Like all formulas, the Ohm's law formula provides a relationship between quantities, or values. It is important to know how to rearrange or transpose a formula so that you can solve for any of the formula's quantities—this process is known as transposition.

8. To transpose a formula, follow two steps: *First step:* Move the unknown quantity so that it is above the line. *Second step:* Move the unknown quantity so that it stands by itself on either side of the equals sign. *If you do exactly the same thing to both sides of the equation or formula, nothing is changed.*

9. If resistance were to remain constant and the voltage were to double, the current within the circuit would also double. Similarly, if the voltage were halved, the current would also halve, proving that current and voltage are directly proportional to one another.

10. If voltage were to remain constant and the resistance were to double, the current within the circuit would be halved. On the other hand, if the circuit resistance were halved, the circuit current would double, confirming that current is inversely proportional to resistance.

Conductors and Their Resistance

11. Resistors are normally made out of materials that cause an opposition to current flow. Conductance is the measure of how good a conductor is, and even the best conductors have some value of resistance.

12. The resistance of a conductor is inversely proportional to the conductor's cross-sectional area.

13. As nearly all conductors are circular in cross section, the area of a conductor is measured in *circular mils* (cmil).

FIGURE 2-26   Resistance.

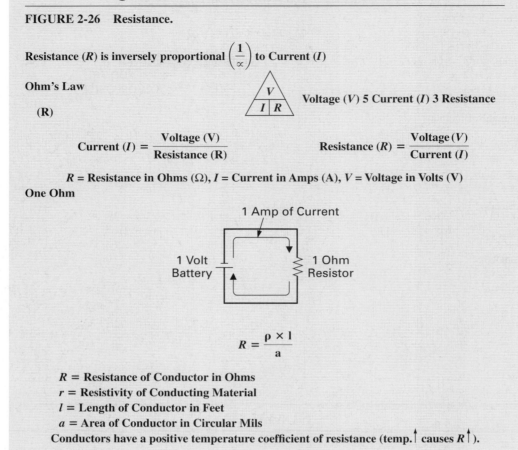

Resistance (*R*) is inversely proportional $\left(\dfrac{1}{\propto}\right)$ to Current (*I*)

Ohm's Law

  (R)

Voltage (*V*) 5 Current (*I*) 3 Resistance

$$\text{Current } (I) = \frac{\text{Voltage (V)}}{\text{Resistance (R)}}$$

$$\text{Resistance } (R) = \frac{\text{Voltage}\,(V)}{\text{Current }(I)}$$

R = Resistance in Ohms (Ω), *I* = Current in Amps (A), *V* = Voltage in Volts (V)

One Ohm

1 Amp of Current

1 Volt
Battery

1 Ohm
Resistor

$$R = \frac{\rho \times l}{a}$$

R = **Resistance of Conductor in Ohms**
r = **Resistivity of Conducting Material**
l = **Length of Conductor in Feet**
a = **Area of Conductor in Circular Mils**
Conductors have a positive temperature coefficient of resistance (temp.↑ causes *R*↑).

14. Increasing the length of the conductor used increases the amount of resistance within the circuit.

15. By combining all of a conductor's physical factors, we can arrive at a formula for resistance.

16. *Resistivity,* by definition, is the resistance (in ohms) that a certain length of material (in centimeters) will offer to the flow of current.

17. Metallic conductors are said to have a *positive temperature coefficient of resistance* (+ Temp. Coe. of R), because the greater the heat applied to the conductor, the greater the atom movement, causing more collisions of atoms to occur, and consequently the greater the conductor's resistance.

18. Any time that current flows through a conductor, a certain resistance or opposition is inherent in that conductor. This resistance will convert current to heat, and the heat further increases the conductor's resistance, causing more heat to be generated due to the opposition. Consequently, a conductor must be chosen carefully for each application so that it can carry the current without developing excessive heat.

19. Conducting wires are normally covered with a plastic or rubber type of material, known as *insulation*. This insulation is used to protect the users and technicians from electrical shock and also to keep the conductor from physically contacting other conductors within the equipment. If the current through the conductor is too high, this insulation will burn due to the heat and may cause a fire hazard. The National Fire Protection Association has developed a set of standards known as the *American Wire Gauge* for all copper conductors, which lists their diameter, resistance, and maximum safe current in amperes.

20. In 1911, a Dutch physicist, Heike Onnes, discovered that mercury (a liquid conductor) lost its resistance to electrical current when the temperature was decreased to −459.7 Fahrenheit (0 on the Kelvin temperature scale). Mercury actually became a *superconductor,* allowing a supercurrent to flow and not encounter any resistance and therefore not generate any heat.

21. In the spring of 1986, two IBM scientists discovered that a conductor compound made up of barium, lanthanum, copper, and oxygen would superconduct (have no resistance to current flow) at −406°F. The real scientific achievement will be to find a conductor compound that will superconduct at room temperature.

22. A cable is made up of two or more wires.

23. Wires and cables have to connect from one point to another. Some are soldered directly, whereas others are attached to plugs that plug into sockets.

Power (Figure 2-27)

24. The six basic forms of energy are light, heat, magnetic, chemical, electrical, and mechanical energy. The unit for energy is the *joule* (J).

25. Work is being done every time one form of energy is transformed to another. A device that converts one form of energy to another is called a *transducer.*

26. Energy and work have the same symbol *(W)*, the same formula, and the same unit (the joule). Energy is merely the capacity, potential, or ability to do work, and work is done when a transformation of the potential, capacity, or ability takes place.

27. *Power (P)* is the rate at which work is performed and is given the unit of watt *(W)*, which is joules per second.

28. Whether discussing a runner, walker, electric motor, heater, refrigerator, light bulb, or compact disk player—power is power. The output power, or power ratings, of electrical, electronic, or mechanical devices can be expressed in watts and describes the number of joules of energy converted every second.

   The output power of rotating machines is given in the unit *horsepower* (hp), the output power of heaters is given in the unit *British thermal units per hour* (Btu/h), and the output power of cooling units is given in the unit *ton of refrigeration.* Despite the different names, they can all be expressed in the unit of watts.

29. The amount of energy stored *(W)* is dependent on the coulombs of charge stored *(Q)* and the voltage *(V)*. The battery's energy stored is dependent on how many coulombs of electrons it holds (current) and how much electrical pressure it is able to apply to these electrons (voltage).

30. Power, in relation to electricity and electronics, is the rate *(t)* at which electric energy *(W)* is converted into some other form.

31. The $P = I \times V$ formula states that the amount of power delivered to a device is dependent on the electrical pressure or voltage applied across the device and the current flowing through the device. Looking at the power formula $P = I \times V$, we can say that 1 watt of power is expended when 1 ampere of current flows through a circuit that has 1 volt applied.

32. *Substitution* enables us to obtain alternative power formulas by replacing or substituting one mathematical term with an equivalent mathematical term.

33. The multimeter can be used to measure power by first making a current measurement, then a voltage measurement, and then calculating the product of the measured current and voltage values.

34. The *kilowatt-hour* meter measures how many kilowatt-hours are consumed, and the electric company then charges accordingly. Since power is the rate at which energy is used, if we multiply power and time, we can calculate how much energy has been consumed.

35. The kilowatt-hour is most commonly used by electric companies. A *kilowatt-hour* of energy is consumed when you use 1000 watts (1 kW) of power in 1 hour.

REVIEW QUESTIONS

Multiple-Choice Questions

1. What is the value of the unknown, *s,* in the following equation: $\sqrt{s^2} + 14 = 26$?
   a. 12   **b.** 26   **c.** 4   **d.** 5

2. What is the correct transposition for the following formula:
$a = \dfrac{b - c}{x}$, $x = ?$
   a. $x = \dfrac{b - a}{c}$ **c.** $x = \dfrac{b - c}{a}$
   b. $x = (b - a) \times c$ **d.** $x = x(b - c)$

3. If $V = I \times R$ and $P = V \times I$, develop a formula for *P* when *I* and *R* are known.
   a. $P = \dfrac{I}{R}$
   b. $P = I^2 \times R^2$
   c. $P = I^2 \times R$
   d. $P = \sqrt{\dfrac{I}{R}}$

4. If $E = M \times C^2$, M = ? and C = ?
   a. $\sqrt{\dfrac{E}{C}}, \dfrac{E^2}{M}$
   b. $\dfrac{E}{C^2}, \sqrt{\dfrac{E}{M}}$
   c. $E \times C^2, \sqrt{C^2 \times E}$
   d. $\dfrac{C^2}{E}, \sqrt{\dfrac{M}{E}}$

5. If $P = \dfrac{V^2}{R}$, transpose to obtain a formula to solve for V.
   a. $V = \sqrt{P \times R}$
   b. $V = \dfrac{P^2}{R}$
   c. $V = P \times R^2$
   d. $V = \sqrt{P} \times R$

6. Resistance is measured in:
   a. Ohms   **c.** Amperes
   b. Volts   **d.** Siemens

FIGURE 2-27   Power.

$$W = Q \times V$$

W = Energy Stored in Joules (J)

Q = Coulombs of Charge

V = Voltage in Volts (V)

$$P = \frac{W}{t}$$

P = Power in Watts

W = Energy in Joules

t = Time in Seconds

$$P = I \times V$$

P = Power in Watts (W), I = Current in Amps (A), V = Voltage in Volts (V)

$$V = \frac{P}{I}$$

$$I = \frac{P}{V}$$

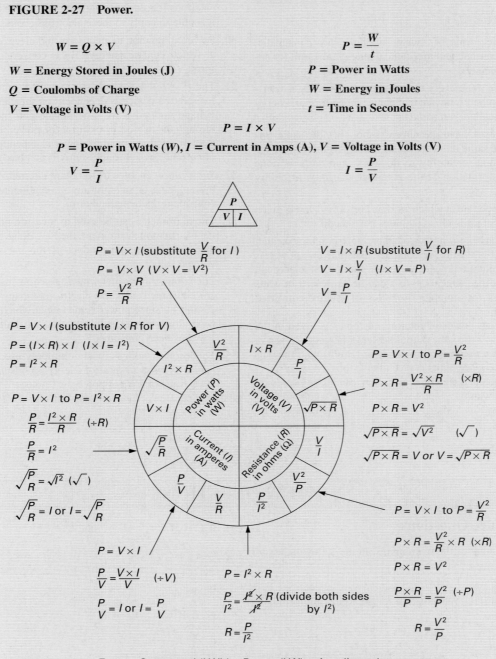

$P = V \times I$ (substitute $\frac{V}{R}$ for I)

$P = V \times V$ ($V \times V = V^2$)

$P = \frac{V^2}{R}$

$V = I \times R$ (substitute $\frac{V}{I}$ for R)

$V = I \times \frac{V}{I}$ ($I \times V = P$)

$V = \frac{P}{I}$

$P = V \times I$ (substitute $I \times R$ for V)

$P = (I \times R) \times I$ ($I \times I = I^2$)

$P = I^2 \times R$

$P = V \times I$ to $P = I^2 \times R$

$\frac{P}{R} = \frac{I^2 \times R}{R}$ ($\div R$)

$\frac{P}{R} = I^2$

$\sqrt{\frac{P}{R}} = \sqrt{I^2}$ ($\sqrt{\ }$)

$\sqrt{\frac{P}{R}} = I$ or $I = \sqrt{\frac{P}{R}}$

$P = V \times I$ to $P = \frac{V^2}{R}$

$P \times R = \frac{V^2 \times R}{R}$ ($\times R$)

$P \times R = V^2$

$\sqrt{P \times R} = \sqrt{V^2}$ ($\sqrt{\ }$)

$\sqrt{P \times R} = V$ or $V = \sqrt{P \times R}$

$P = V \times I$

$\frac{P}{V} = \frac{V \times I}{V}$ ($\div V$)

$\frac{P}{V} = I$ or $I = \frac{P}{V}$

$P = I^2 \times R$

$\frac{P}{I^2} = \frac{I^2 \times R}{I^2}$ (divide both sides by I^2)

$R = \frac{P}{I^2}$

$P = V \times I$ to $P = \frac{V^2}{R}$

$P \times R = \frac{V^2}{R} \times R$ ($\times R$)

$P \times R = V^2$

$\frac{P \times R}{P} = \frac{V^2}{P}$ ($\div P$)

$R = \frac{V^2}{P}$

Energy Consumed (kWh) = Power (kW) × time (hours)

7. Current is proportional to:
   a. Resistance c. Both (a) and (b)
   b. Voltage d. None of the above

8. Current is inversely proportional to:
   a. Resistance c. Both (a) and (b)
   b. Voltage d. None of the above

9. If the applied voltage is equal to 15 V and the circuit resistance equals 5 Ω, the total circuit current would be equal to:
   a. 4 A b. 5 A c. 3 A d. 75 A

10. Calculate the applied voltage if 3 mA flows through a circuit resistance of 25 kΩ.
    a. 63 mV b. 25 V c. 77 μV d. 75 V

11. Energy is measured in:
    a. Volts b. Joules c. Amperes d. Watts

12. Chemical energy within a battery is converted into:
    a. Electrical energy c. Magnetic energy
    b. Mechanical energy d. Heat energy

13. The water pump has the potential to cause water flow, just as the battery has the potential to cause:
    a. Voltage c. Current
    b. Electron flow d. Both (b) and (c)

14. Work is measured in:
    a. Joules c. Amperes
    b. Volts d. Watts

15. The device that converts one energy form to another is called a:
    a. Transformer c. Transit
    b. Transducer d. Transistor

16. Power is the rate at which energy is transformed and is measured in:
    a. Joules c. Volts
    b. Watts d. Amperes

17. Power is measured by using a (an):
    a. Ammeter c. Ohmmeter
    b. Voltmeter d. Wattmeter

18. A good conductor has a:
    a. Large conductance figure c. Both (a) and (b)
    b. Small resistance figure d. None of the above

19. The resistance of a conductor is:
    a. Proportional to the length of the conductor
    b. Inversely proportional to the area of the conductor
    c. Both (a) and (b)
    d. None of the above

20. AWG is an abbreviation for:
    a. Alternate Wire Gauge c. American Wave Guide
    b. Alternating Wire Gauge d. American Wire Gauge

Communication Skill Questions

21. Define the following terms: (2-1)
    a. Algebra c. Equation
    b. Formula d. Literal number

22. Why is it important to treat both sides of an equation equally? (2-1)

23. Describe how the following three factors can be combined in a formula: distance (d), velocity (v), and time (t). (2-1-2)

24. What two steps must be followed to transpose an equation? (2-1-2)

25. Define the terms *proportional* and *inversely proportional*. (2-1-2)

26. Explain how current (I), voltage (V), and resistance (R) can be combined in a formula. (2-1-2)

27. Transpose the formula from question 22 to solve for I, V, and R. (2-1-2)

28. If $P = V \times I$, describe how you would solve for V and I. (2-1-2)

29. If $\pi = \%$, and area $= \frac{1}{2} \times (C \times r)$, describe how substitution and transposition can be employed to obtain a formula for area when only the radius of a circle is known. (2-1-3)

30. Why is formula manipulation important? (2-1)

31. What is resistance? (2-2)

32. Briefly describe why:
    a. Current is proportional to voltage. (2-3-3)
    b. Current is inversely proportional to resistance. (2-3-4)

33. State Ohm's law. (2-3-1)

34. List the three forms of Ohm's law. (2-3-2)

35. List the six basic forms of energy. (2-5)

36. Briefly describe the following terms:
    a. Potential energy (2-5)
    b. Kinetic energy (2-5)

37. What is a transducer? (2-5)

38. Define power. (2-5-1)

39. Give three formulas for electric power. (2-5-3)

40. Give the units for each of the following:
    a. Energy e. Work
    b. Power f. Current
    c. Voltage g. Charge
    d. Resistance h. Conductance

41. What is the difference between work and power? (2-5)

42. How can you measure electrical power? (2-5-4)

43. State the formula used by electric companies to determine the amount of power consumed. (2-5-5)

44. What is 1 kilowatt-hour? (2-5-5)

45. List the four factors that determine a conductor's resistance. (2-4)

46. What is a circular mil? (2-4-2)

47. Define the resistivity of a conducting material. (2-4-4)

48. Describe why conductors have a positive temperature coefficient of resistance. (2-4-5)

49. What is the purpose(s) of placing an insulating sheath over conducting wires? (2-4-6)

50. What is the American Wire Gauge? (2-4-6)

51. What is a superconductor? (2-4-7)

52. List some of the advantages of superconductivity. (2-4-7)

53. What is the difference between a wire and a cable? (2-4-8)

54. Give some examples of different wires and cables. (2-4-8)

55. List examples of different conductor connectors. (2-4-8)

Practice Problems

56. Calculate the result of the following arithmetic operations involving literal numbers.
   a. $x + x + x = ?$ **g.** $4y \times 3y^2 = ?$
   b. $5x + 2x = ?$ **h.** $a \times a = ?$
   c. $y - y = ?$ **i.** $2y \div y = ?$
   d. $2x - x = ?$ **j.** $6b \div 3b = ?$
   e. $y \times y = ?$ **k.** $\sqrt{a^2} = ?$
   f. $2x \times 4x = ?$ **l.** $(x^3)^2 = ?$

57. Transpose the following equations to determine the unknown value.
   a. $4x = 11$
   b. $6a + 4a = 70$
   c. $5b - 4b = \dfrac{7.5}{1.25}$
   d. $\dfrac{2z \times 3z}{4.5} = 2z$

58. Apply transposition and substitution to Newton's second law of motion.
   a. $F = ma$ [Force (F) = mass (m) $\times$ acceleration (a)]
      If $F = m \times a$, $m = ?$ and $a = ?$
   b. If acceleration (a) equals change in speed (ΔV) divided by time (t), how would the $F = ma$ formula appear if acceleration (a) was substituted for $\dfrac{\Delta V}{t}$?

59. a. Using the formula power (P) = voltage (V) $\times$ current (I), calculate the value of electric current through a hairdryer that is rated as follows: power = 1500 watts and voltage = 120 volts.

$$\boxed{P = V \times I}$$

   where P = power in watts
      V = voltage in volts
      I = current in amperes
   b. Using the Ohm's formula voltage (V) = current (I) $\times$ resistance (R), calculate the resistance or opposition to current flow offered by the hairdryer in Question 14a.

60. An electric heater with a resistance of 6 Ω is connected across a 120 V wall outlet.
   a. Calculate the current flow. (2-3)
   b. Draw the schematic diagram.

61. What source voltage would be needed to produce a current flow of 8 mA through a 16 kΩ resistor? (2-3)

62. If an electric toaster draws 10 A when connected to a power outlet of 120 V, what is its resistance? (2-3)

63. Calculate the power used in Problems 41, 42, and 43. (2-5-3)

64. Calculate the current flowing through the following light bulbs when they are connected across 120 V: (2-5-3)
   a. 300 W   **b.** 100 W   **c.** 60 W   **d.** 25 W

65. If an electric company charges 9 cents/kWh, calculate the cost for each light bulb in Problem 45 if on for 10 hours. (2-5-5)

66. Indicate which of the following unit pairs is larger:
   a. Millivolts or volts
   b. Microamperes or milliamperes
   c. Kilowatts or watts
   d. Kilohms or megohms

67. Calculate the resistance of 200 ft of copper having a diameter of 80 mils. (2-4-4)

68. What AWG size wire should be used to safely carry just over 15 A? (2-4-6)

69. Calculate the voltage dropped across 1000 ft of No. 4 copper conductor when a current of 7.5 A is flowing through it. (2-4-6)

70. Calculate the unknown resistance in a circuit when an ammeter indicates that a current of 12 mA is flowing and a voltmeter indicates 12 V. (2-3-4)

71. What battery voltage would use 1000 J of energy to move 40 C of charge through a circuit? (2-5-2)

72. Calculate the resistance of a light bulb that passes 500 mA of current when 120 V is applied. What is the bulb's wattage? (2-5-3)

73. Which of the following circuits has the largest resistance and which has the smallest? (2-5)
   a. $V = 120\text{ V}, I = 20\text{ mA}$ **c.** $V = 9\text{ V}, I = 100\ \mu\text{A}$
   b. $V = 12\text{ V}, I = 2\text{ A}$ **d.** $V = 1.5\text{ V}, I = 4\text{ mA}$

74. Calculate the power dissipated in each circuit in Problem 54. (2-5-3)

75. How many watts are dissipated if 5000 J of energy is consumed in 25 s? (2-5-3)

76. Convert the following:
   a. 1000 W = _____ kW
   b. 0.345 W = _____ mW
   c. 1250×10^3 W = _____ MW
   d. 0.00125 W = _____ μW

77. What is the value of the resistor when a current of 4 A is causing 100 W to be dissipated? (2-5-3)

78. How many kilowatt-hours of energy are consumed in each of the following: (2-5-5)
   a. 7500 W in 1 hour   **c.** 127,000 W for half an hour
   b. 25 W for 6 hours

79. What is the maximum output power of a 12 V, 300 mA power supply? (2-5-3)

Web Site Questions

Go to the Web site http://www.prenhall.com/cook, select the textbook *Introductory DC/AC Electronics* or *Introductory DC/AC Circuits*, this chapter, and then follow the instructions when answering the multiple-choice practice problems.

These tests at the end of each chapter will challenge your knowledge up to this point, and give you the practice you need for a job interview. To make this more realistic, the test will comprise both technical and personal questions. In order to take full advantage of this exercise, you may want to set up a simulation of the interview environment, have a friend read the questions to you, and record your responses for later analysis.

Company Name: Power Service, Inc.

Industry Branch: Test and Measurement.

Function: Maintain and Service Kilowatt-Hour Meters.

Job Title: Field Service Technician.

1. What do you know about us?

2. Could you tell me what a kilowatt-hour meter does?

3. How would you feel about working on your own in the field?

4. What is the difference between energy and power?

5. If you couldn't determine the answer to a problem, what would you do?

6. What is the difference between resistance and power?

7. What would you say is the job function of a field service technician?

8. Without supervision in the field, could you discipline yourself to get through a heavy schedule of stops?

9. Tell me what you know about Ohm's law.

10. What makes you think you are the right person for this job?

Answers

1. Visit the company's web site before the interview to get an overall understanding of the company's ownership, production line, service, and support.

2. Section 2-5.

3. Discuss how you have worked alone on project assignments and other related work history.

4. Section 2-5.

5. Say that you would report the problem to your supervisor and ask for help.

6. Sections 2-3 and 2-5.

7. See introduction, Field Service Technician.

8. Discuss how you have juggled school, studying, and a job to get to the position you currently hold, and other related work experience.

9. Section 2-3.

10. Describe why you were first attracted to the company's advertisement, why it felt compatible with your career goals, and how the company's positive attributes listed in its web site matched your expecations.

Resistors

At the Core of Apple

Steven Jobs (left), and Stephen Wozniak

Stephen Wozniak and Steven Jobs met at their Los Altos, California, high school, and became friends due to their common interest in electronics.

Wozniak was a conservative youth whose obsession with technology left little room for social relationships or studying; in fact, after one year at the University of Colorado, he dropped out with his academic record littered with F's. In contrast to this serious nature, Wozniak was renowned for his high-tech pranks, for which, on one occasion, he spent a night in juvenile hall for wiring up a fake bomb in a friend's locker. In another incident, he devised a way to place a free telephone call to the Pope at the Vatican, identifying himself as then Secretary of State Henry Kissinger.

Jobs, on the other hand, had interests outside electronics. He searched for intellectual, emotional, and spiritual stimulation. After one semester at Reed College, he dropped out to pursue an interest in Eastern religions, which led him to temples in India, searching for the meaning of life.

The first enterprise built by Wozniak and sold by Jobs was an illegal device called a blue box that could crash the telephone system. It generated a set of tones that fooled the computerized telephone switching systems and opened up free, long-distance circuits; these systems allowed "phone phreaks" to take long and illegal joy rides throughout the world's telephone networks.

By selling a Volkswagen van and a programmable calculator, Wozniak and Jobs raised the initial capital of $1300 to start a business in April 1976 called Apple Computer, since Jobs had a passion for the Beatles, who recorded under the Apple record label.

In just five years, Apple Computer grew faster than any other company in history, from a two-man assetless partnership building computers in a family home to a publicly traded corporation earning its youthful entrepreneurs, Jobs (27) and Wozniak (24), fortunes of nearly $200 million each.

Outline and Objectives

3-6 FILAMENT RESISTOR

Objective 13: State the purpose of the filament and ballast resistors.

3-7 TESTING RESISTORS

Objective 14: Describe some of the more common resistor problems.

Introduction

Resistance would seem to be an undesirable effect, as it reduces current flow and wastes energy as it dissipates heat. Resistors, however, are probably used more than any other component in electronics, and in this chapter we discuss all the different types.

To begin with though, let's review the math skills you will need for this chapter's material.

3-1 MINI-MATH REVIEW—MATHEMATICAL OPERATIONS AND GRAPHS

This first section, "Mini-Math Review," is included to overview the mathematical details you need for the electronic concepts covered in this chapter. In this review, we will be examining certain mathematical operations and graphs.

3-1-1 *Ratios*

Ratio

The relationship in quantity, amount, or size between two or more things.

A **ratio** is a comparison of one number to another number. For example, a ratio such as 7:4 ("seven to four") is comparing the number 7 to the number 4. This ratio could be written in one of the following three ways:

$$\frac{7}{4} \leftarrow \text{Ratio is written as a fraction.}$$
$$7\!:\!4 \leftarrow \text{Ratio is written using the ratio sign (:).}$$
$$1.75 \text{ to } 1 \leftarrow \text{Ratio is written as a decimal.}$$

To express the ratio as a decimal, the number 7 was divided by the number 4, giving a result of 1.75 ($7 \div 4 = 1.75$). This result indicates that the number 7 is 1.75 times (or one and three-quarter times) larger than the number 4.

■ EXAMPLE:

If one store has 360 items and another store has 100 of the same items, express the ratio of 360 to 100:

   a. Using the divide sign

   b. Using the ratio sign

   c. As a decimal

■ *Solution:*

   a. $\dfrac{360}{100} = \dfrac{360 \div 20}{100 \div 20} = \dfrac{18}{5}$ ← 360 to 100 and 18 to 5 are equivalent ratios.

b. $360 : 100 = 18 : 5$

c. 3.6 ← The first store has 3.6 times as many items as the other store ($18 \div 5 = 3.6$).

■ **EXAMPLE:**

Calculate the ratio of the following two like quantities. If necessary, reduce the ratio to its lowest terms.

a. $0.8 \div 0.2 = ?$

b. $0.0008 \div 0.0002 = ?$

■ *Solution:*

a. How many 0.2s are in 0.8? The answer is the same as saying, How many 2s are in 8?

$$0.8 \div 0.2 = 4 \qquad \frac{8}{2} \; \frac{\div \; 2}{\div \; 2} = \frac{4}{1} \qquad \text{(a ratio of 4 to 1)}$$

b. How many 0.0002s are in 0.0008?

$$0.0008 \div 0.0002 = 4 : 1$$

The ratio in example (a) is the same as (b). Another way to describe this is that, because the size of each decimal fraction compared to the other decimal fraction (or the *ratio*) remained the same, the answer or quotient remained the same.

8 is four times larger than 2, or 2 is four times smaller than 8.
0.8 is four times larger than 0.2, or 0.2 is four times smaller than 0.8.
0.0008 is four times larger than 0.0002, or 0.0002 is four times smaller than 0.0008.
Therefore, all have the same 4 : 1 ratio.

3-1-2   *Rounding Off*

In many cases we will **round off** *decimal numbers because we do not need the accuracy indicated by a large number of decimal digits.* For example, it is usually unnecessary to have so many decimal digits in a value such as 74.139896428. If we were to round off this value to the nearest hundredths place, we would include two digits after the decimal point, which is 74.13. This is not accurate, however, since the digit following 74.13 was a 9 and therefore one count away from causing a reset and carry action into the hundredths column. To take into account the digit that is to be dropped when rounding off, therefore, we follow this basic rule: *When the first digit to be dropped is 6 or more, increase the previous digit by 1. On the other hand, when the first digit to be dropped is 4 or less, do not change the previous digit.* Therefore, the value 74.139896428, rounded off to the nearest hundredths place, would equal:

┌─ Hundredths place

$$\boxed{74.13}\,9896428 \;=\; 74.14$$

First digit to be dropped is greater than 6, so previous digit should be increased by 1.

Rounding Off
An operation in which a value is abbreviated by applying the following rule: When the first digit to be dropped is a 6 or more, or a 5 followed by a digit that is more than zero, increase the previous digit by 1. When the first digit to be dropped is a 4 or less, or a 5 followed by a zero, do not change the previous digit.

■ **EXAMPLE:**

Round off the value 74.139896428 to the nearest ten, whole number, tenth, hundredth, thousandth, and ten-thousandth.

$\boxed{7}$4.139896428 rounded off to the nearest ten = 70

⌐ First digit to be dropped is 4 or less; therefore, do not change previous
digit.

$\boxed{74.}$139896428 rounded off to the nearest whole number = 74

⌐ First digit to be dropped is 4 or less; therefore, do not change previous
digit.

$\boxed{74.1}$39896428 rounded off to the nearest tenth = 74.1

⌐ First digit to be dropped is 4 or less; therefore, do not change previous
digit.

$\boxed{74.13}$9896428 rounded off to the nearest hundredth = 74.14

⌐ First digit to be dropped is 6 or greater; therefore, increase previous
digit by 1 (3 to 4).

$\boxed{74.139}$896428 rounded off to the nearest thousandth = 74.140

⌐ First digit to be dropped is 6 or greater; therefore, increase previous
digit by 1. Since previous digit is 9, allow reset and carry action to
occur. The steps are:
74.1398 ← 8 carries 1 into thousandths column.
74.130 ← 9 resets to 0, and carries 1 into hundredths column.
74.140 ← Hundredths digit is increased by 1.

$\boxed{74.1398}$96428 rounded off to the nearest ten-thousandth = 74.1399

⌐ First number to be dropped is 6 or greater; therefore, increase previous number by 1.

In the preceding example we discovered how to round off if a digit was over halfway up the decimal scale (greater than 6), or if a digit was below the halfway point on the decimal scale (less than 4). The next question, therefore, is: How do we round off when the digit to be dropped is a 5 and therefore midway in the decimal scale of 10? In this instance, we follow this rule: When the 5 is followed by a digit that is more than zero, increase the previous digit by 1. On the other hand, when the 5 is followed by zero, do not change the previous digit. Let us apply this to an example:

$\boxed{79.5}$3 rounded off to the nearest whole number = 80

Let us examine why this is correct. Because the digit following the 5 is more than zero, the decimal fraction (0.53) is in this case more than one-half or 0.5 (it is actually 5 tenths or ½ plus 3 hundredths), and therefore the 9 should be changed to 0 and the 7 increased to 8.

■ EXAMPLE:

Round off the value 31,520.565 to the nearest ten thousand, thousand, hundred, whole number, tenth, and hundredth.

■ *Solution:*

$\boxed{3}$1,520.565 rounded off to the nearest ten thousand = 30,000

⌐First digit to be dropped is 4 or less; therefore, do not change previous digit.

$\boxed{31,}$520.565 rounded off to the nearest thousand = 32,000

⌐ The 5 is followed by a digit that is more than zero; therefore, increase previous
digit by 1.

$\boxed{31,5}$20.565 rounded off to the nearest hundred = 31,500
└First digit to be dropped is 4 or less; therefore, do not change previous digit.

$\boxed{31,520}$.565 rounded off to the nearest whole number = 31,521
└The 5 is followed by a digit that is more than zero; therefore, increase previous digit by 1.

$\boxed{31,520.5}$65 rounded off to the nearest tenth = 31,520.6
└Digit to be dropped is 6 or more; therefore, increase previous digit by 1.

$\boxed{31,520.56}$5 rounded off to the nearest hundredth = 31,520.56
└Because no digit follows 5, it is assumed to be a zero; therefore, do not change previous digit.

3-1-3   *Significant Places*

The number of significant places describes how many digits are in the value and how many digits in the value are accurate after rounding off. For example, a number such as 347.63 is a five-significant-place (or significant-figure) value because it has five digits in five columns. If we were to round off this value to a whole number, we would get 348.00. This value would still be a five-significant-place number; however, it would now be accurate to only three significant places. Let us examine a few problems to practice using the terms *significant places* and *accurate to significant places*.

EXAMPLE:

On a calculator, the value of π will come up as 3.141592654.

   a. Write the number π to six significant places.
   b. Give π to five significant places, and also round off π to ten-thousandths.
   c. 3.14159000 is the value of π to _____ significant places; however, it is accurate to only _____ significant places.

Solution:

   a. The value π to six significant places or figures is $\boxed{3.14159}$ 27

3.14159
└Uses six digits or columns.

   b. The value π to five significant places.
      $\boxed{3.1415}$92 rounded off to ten-thousandths = 3.1416

   c. 3.14159000 is the value of π to nine significant places (has nine digits); however, it is accurate to only six significant places (because the zeros to the right are just extra, and the value 3.14159 has only six digits).

3-1-4   *Percentages*

The **percent** sign (%) means hundredths, which as a proper fraction is $\frac{1}{100}$ or as a decimal fraction is 0.01. For example, 50% means 50 hundredths ($\frac{50}{100}$ $\frac{\div 50}{\div 50}$ = $\frac{1}{2}$), which is one-half.

Percent

In the hundred; of each hundred.

In decimal, 50% means 50 hundredths (50 × 0.01 = 0.5), which is also one-half. The following shows how some of the more frequently used percentages can be expressed in decimal.

$$1\% = 1 \times 0.01 = 0.01$$ (For example, 1% is expressed mathematically in decimal as 0.01)

$$5\% = 5 \times 0.01 = 0.05$$
$$10\% = 10 \times 0.01 = 0.10$$
$$25\% = 25 \times 0.01 = 0.25$$
$$50\% = 50 \times 0.01 = 0.50$$
$$75\% = 75 \times 0.01 = 0.75$$

Most calculators have a percent key that automatically makes this conversion to decimal hundredths. It operates as follows.

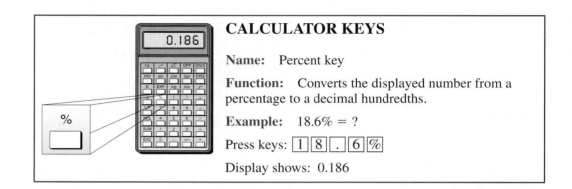

CALCULATOR KEYS

Name: Percent key

Function: Converts the displayed number from a percentage to a decimal hundredths.

Example: 18.6% = ?

Press keys: 1 8 . 6 %

Display shows: 0.186

Now that we understand that the percent sign stands for hundredths, what does the following question actually mean: What is 50% of 12? We now know that 50% is 50 hundredths ($^{50}\!/_{00}$), which is one-half. This question is actually asking: What is half of 12? Expressed mathematically as both proper fractions and decimal fractions, it would appear as follows:

Proper Fractions	Decimal Fractions
50% of 12 = ?	50% of 12 = ?
$= \dfrac{50}{100} \times 12$	$= (50 \times 0.01) \times 12$
$= \dfrac{1}{2} \times 12$	$= 0.5 \times 12$
$= 6$	$= 6$

Explaining this another way, we could say that percentages are fractions in hundredths. In the example above, therefore, the question is actually asking us to divide 12 into 100 parts and then to determine what value we have if we had 50 of those 100 parts. Therefore, if we were to split 12 into 100 parts, each part would have a value of 0.12 (12 ÷ 100 = 0.12). Having 50 of these 100 parts would give us a total of 50 × 0.12 = 6.

■   **EXAMPLE:**

If sales tax is 8% and the merchandise price is $256.00, what is the tax on the value, and what will be the total price?

■ *Solution:*

First, determine the amount of tax to be paid.

$$8\% \text{ of } \$256 =$$
$$0.08 \times 256 = \$20.48$$

Calculator sequence: $\boxed{8}\,\boxed{\%}\,\boxed{\times}\,\boxed{2}\,\boxed{5}\,\boxed{6}\,\boxed{=}$
Answer: 20.48
The total price paid will therefore equal

$$\$256.00 + \$20.48 = \$276.48$$

■ **EXAMPLE:**

Which is the better buy—a $599.95 refrigerator at 25% off or a $499.95 refrigerator with a $50 manufacturer's rebate?

■ *Solution:*

$$25\% \text{ of } \$599.95 = \$150$$
$$\text{Refrigerator A}: \$599.95 - \$150.00 = \$449.95$$
$$\text{Refrigerator B}: \$499.95 - \$50.00 = \$449.95$$

The refrigerators are the same price.

3-1-5 *Averages*

A mean **average** *is a value that summarizes a set of unequal values.* This value is equal to the sum of all the values divided by the number of values. For example, if a team has scores of 5, 10, and 15, what is their average score? The answer is obtained by adding all the scores (5 + 10 + 15 = 30) and then dividing the result by the number of scores (30 ÷ 3 = 10).

Average

A single value that summarizes or represents the general significance of a set of unequal values.

■ **EXAMPLE:**

The voltage at the wall outlet in your home was measured at different times in the day and equaled 108.6, 110.4, 115.5, and 123.6 volts (V). Find the average voltage.

■ *Solution:*

Add all the voltages.

$$108.6 + 110.4 + 115.51 + 123.6 = 458.1$$

Divide the result by the number of readings.

$$\frac{458.1}{4} = 114.525 \text{ V}$$

The average voltage present at your home was therefore 114.525 volts.

Statistics

A branch of mathematics dealing with the collection, analysis, interpretation, and presentation of masses of numerical data.

Statistics are generally averages and can misrepresent the actual conditions. For example, if five people earn an average salary of $204,600 a year, one would believe that all five are very well paid. Studying the individual figures, however, we find out that their yearly earnings are $8000, $6000, $4000, $20,000, and $985,000 per year. Therefore,

$$8000 + 6000 + 4000 + 20,000 + 985,000 = 1,023,000$$

$$\frac{1,023,000}{5} = 204,600$$

If one value is very different from the others in a group, this value will heavily influence the result, creating an average that misrepresents the actual situation.

▢ EXAMPLE:

A football player's "average number of yards gained per carry" ranking is calculated by dividing the total number of yards gained by the total number of times the player carried the football. Which of the following players had the higher average number of yards gained per carry?

Player	Total Yards	Total Carries
Payton	16,726	3,838
Brown	12,312	2,359

■ *Solution:*

$$\text{Payton: } \frac{16,726}{3,838} = 4.4 \text{ yards per carry average}$$

$$\text{Brown: } \frac{12,312}{2,359} = 5.2 \text{ yards per carry average}$$

On average, Brown gained a higher number of yards per carry.

3-1-6   *Graphs*

A **graph** is a diagram showing the relationship between two or more factors. As an example, the graph in Figure 3-1 shows how new technologies, such as electric light, radio, the telephone and television, have been adopted in homes in the United States. This type of graph is called a **line graph** because the data are represented by points joined by line segments. The vertical scale in this graph indicates U.S. households in percent (0% to 100%), while the horizontal scale indicates the year (1900 to 1980). The four plotted lines in this graph show how quickly these new technologies found their way into the home. The data used to create the four plotted lines was obtained by recording what percentage of homes in the United States had electric light, radio, telephone, and television every year from 1900 to 1980.

Reading Graphs

If the yearly facts for the graph shown in Figure 3-1 were listed in columns instead of illustrated graphically, it would be difficult to see the characteristics of each individual line graph, or to compare the characteristics of one technology with those of another. To explain this point, let us try an example.

▢ EXAMPLE:

From the graph shown in Figure 3-1, determine the following:

a. Approximately what percentage of U.S. households had radio in 1930?

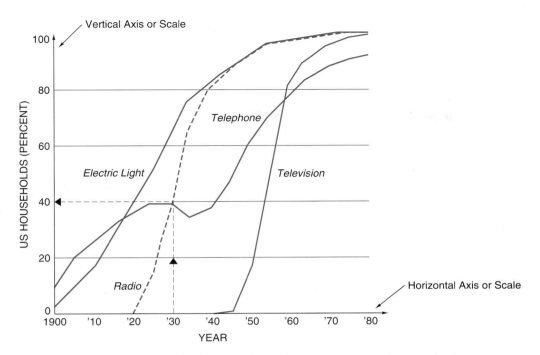

FIGURE 3-1   A Graph Showing How New Technologies Have Been Adopted in the United States.

b. Approximately what percentage of U.S. households had telephones in 1960?

c. Which technology was the slowest to be adopted into the home, and which was the fastest?

■ *Solution:*

a. To find this point on the graph, move along the horizontal scale until you reach 1930, and then proceed directly up until you connect with the plotted radio line graph (shown as a dashed line in Figure 3-1). Comparing this point with the vertical scale, you can see that it corresponds to about 40%. In 1930, therefore, approximately 40% of U.S. households had radio.

b. To find this point on the graph, move along the horizontal scale until you reach 1960, and then proceed directly up until you connect with the plotted telephone line graph. Comparing this point with the vertical scale, you can see that it corresponds to about 80%. In 1960, therefore, approximately 80% of U.S. households had a telephone.

c. Comparing the four line graphs shown in Figure 3-1, you can see that the telephone seemed to be the slowest technology to be adopted, while the television was the fastest.

Questions (a) and (b) in the previous example show how a graph can be used to analyze the characteristics of each individual line graph. Question (c) showed how a graph can be used to compare the characteristics of several line graphs.

Types of Graphs

There are three types of graphs: line graphs, bar graphs, and circle graphs. Line graphs, like the examples shown in Figure 3-1, are ideal for displaying historical data in which a factor is constantly changing.

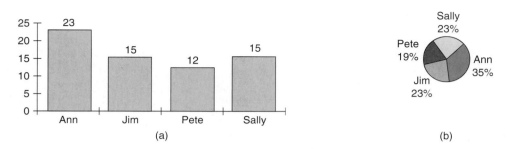

(a)   (b)

FIGURE 3-2   **Bar Graphs and Circle Graphs.**

Bar Graph or Bar Chart

A graphic means of quantitative comparison by rectangles with lengths proportional to the measure of the data being compared.

Bar graphs use parallel bars of different sizes to compare one variable with others. For example, Figure 3-2(a) shows how a bar graph can be used to compare the sales records of four computer salespersons for the month of May.

■ **EXAMPLE:**

From the graph shown in Figure 3-2(a), determine the following:

  a. How many sales representatives are there?

  b. Which sales rep sold the most computers in the month of May, and which sales rep sold the least?

  c. How many computers were sold in May?

  d. Calculate the number of computers Ann sold as a percentage of the total sold.

■ *Solution:*

  a. There are four bars in this bar graph, one for each of the four sales reps.

  b. Ann has the largest bar in the graph because she sold 23 computers, while Pete has the smallest bar because he sold only 12 computers.

  c. Ann sold 23, Jim sold 15, Pete sold 12, and Sally sold 15. A total of 65 computers were sold, therefore, in May.

  d. To calculate the number of computers Ann sold as a percentage of the total sold, use the following formula;

$$\text{Percentage} = \frac{\text{units sold by rep}}{\text{total units sold}} \times 100$$

$$\text{Percentage} = \frac{23}{65} \times 100 \qquad (23 \div 65 = 0.35)$$

$$= 0.35 \times 100 = 35\%$$

Pie Chart or Circle Graph

A circular chart cut by radii into segments illustrating relative magnitudes.

Figure 3-2(b) shows an example of a **circle graph,** which uses different size segments within a circle to compare one variable with others. These circle graphs are often called **pie charts** because the circle resembles a pie and the segments look like different slices in the pie. The size of the segments within a circle graph is usually given as a percentage, as shown in the example in Figure 3-2(b), with the whole circle equal to 100%. Like the bar graph in Figure 3-2(a), the circle graph in Figure 3-2(b) compares the sales records of four computer salespersons for the month of May. Because the data represented in these two graphs are the same, the only difference between these two graphs is the presentation format.

EXAMPLE:

From the graph shown in Figure 3-2(b), determine the following:

a. How many sales representatives are there?

b. Which sales rep sold the most computers in the month of May, and which sales rep sold the least?

c. If Ann sold 23 units, Jim sold 15 units, Pete sold 12 units, and Sally sold 15 units, calculate each sales rep's percentage of total sales.

d. If a total of 87 computers were sold in the month of June and Jim again sold 23%, how many did he sell?

Solution:

a. There are four segments in this circle graph, one for each of the four sales reps.

b. The largest segment is assigned to Ann because she sold 23 computers, and the smallest segment is assigned to Pete, who sold only 12 computers.

c. Each sales rep's percentage was calculated as follows:

$$\text{Ann's percentage} = \frac{\text{units sold by Ann}}{\text{total units sold}} \times 100 = \frac{23}{65} \times 100 = 0.35 \times 100 = 35\%$$

$$\text{Jim's percentage} = \frac{\text{units sold by Jim}}{\text{total units sold}} \times 100 = \frac{15}{65} \times 100 = 0.23 \times 100 = 23\%$$

$$\text{Pete's percentage} = \frac{\text{units sold by Pete}}{\text{total units sold}} \times 100 = \frac{12}{65} \times 100 = 0.19 \times 100 = 19\%$$

$$\text{Sally's percentage} = \frac{\text{units sold by Sally}}{\text{total units sold}} \times 100 = \frac{15}{65} \times 100 = 0.23 \times 100 = 23\%$$

d. $23\% \times 87 = 0.23 \times 87 = 20$ Jim sold 20 computers in June.

SELF-TEST EVALUATION POINT FOR SECTION 3-1

Now that you have completed this section, you should be able to:

■ **Objective 1.** *Explain the following mathematical operations:*

  a. Ratios

  b. Rounding off

  c. Significant places

  d. Percentages

  e. Averages

■ **Objective 2.** *Describe how to interpret the data displayed in a graph.*

■ **Objective 3.** *Explain the differences between the:*

  a. Line graph

  b. Bar graph

  c. Circle graph or pie chart

Use the following questions to test your understanding of Section 3-1:

1. Express a 176 meter to 8 meter ratio as a fraction, using the ratio sign and as a decimal.

2. Round off the following values to the nearest hundredth.

  a. 86.43760

  b. 12,263,415.00510

  c. 0.176600

3. Referring to the values in Question 2, describe:

  a. Their number of significant places

  b. To how many significant places they are accurate

4. Calculate the following.

  a. 15% of 0.5 = ?

  b. 22% of 1000 = ?

  c. 2.35% of 10 = ?

  d. 96% of 20 = ?

5. Calculate the average of the following values.

    a. 20, 30, 40, and 50

    b. 4000, 4010, 4008, and 3998

6. Plot a line graph to display the following data:

THE NUMBER OF INDOOR AND OUTDOOR MOVIE SCREENS

YEAR	INDOOR	OUTDOOR
1975	12,000	4000
1980	14,000	4000
1985	18,000	3000
1990	23,000	2000

7. Draw a circle graph and bar graph to indicate the following data: Each 24 h day Tom works for 8 h, studies for 3 h, showers and dresses for 2 h, eats for 1.5 h, sleeps for 8 h, and exercises for 1.5 h.

3-2   RESISTOR TYPES

Conductors are used to connect one device to another, and although they offer a small amount of resistance, this resistance is not normally enough. In electronic circuits, additional resistance is normally needed to control the amount of current flow, and the component used to supply this additional resistance is called a **resistor.**

    There are two basic types of resistors: fixed value and variable value. Fixed-value resistors, like the examples seen in Figure 3-3, have a resistance value that cannot be changed. On the other hand, variable-value resistors, like the examples seen in Figure 3-4, have a range of values that can be selected by adjusting a control.

3-2-1   *Fixed-Value Resistors*

In this section, we will examine some of the **fixed-value resistor** types.

Resistor

Component made of a material that opposes the flow of current and therefore has some value of resistance.

Fixed-Value Resistor

A resistor whose value cannot be changed.

Carbon Composition

Carbon Film

Metal Film

Wirewound

Metal Oxide

Thick Film

 SIP

 DIP

 Chip

FIGURE 3-3   **Fixed-Value Resistors.**

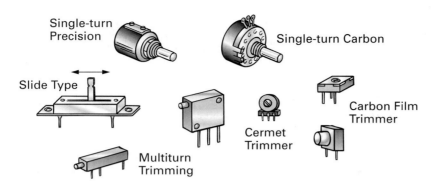

Single-turn
Precision

Single-turn Carbon

Slide Type

Carbon Film
Trimmer

Cermet
Trimmer

Multiturn
Trimming

FIGURE 3-4   Variable-Value Resistors.

Carbon Composition Resistors

This is the most common and least expensive type of fixed-value resistor, the appearance of which can be seen in Figure 3-5(a). It is constructed by placing a piece of resistive material, with embedded conductors at each end, within an insulating cylindrical molded case, as illustrated in Figure 3-5(b).

This resistive material is called carbon composition because powdered carbon and powdered insulator are bonded into a compound and used as the resistive material. By changing the ratio of powdered insulator to carbon, the value of resistance can be changed within the same area. For example, Figure 3-6 illustrates four fixed-value resistors, ranging from 2 Ω to 10 MΩ.

The 2 Ω resistor is exactly the same size as the 10 MΩ (10 million Ω) resistor. This is achieved by having more powdered carbon and less powdered insulator in the 2 W and less powdered carbon and more powdered insulator in the 10 MΩ. The color-coded rings or bands on the resistors in Figure 3-6 are a means of determining the value of resistance. This and other coding systems will be discussed later.

The physical size of the resistors lets the user know how much power in the form of heat can be **dissipated,** as shown in Figure 3-7. As you already know, resistance is the opposition to current flow, and this opposition causes heat to be generated whenever current is passing through a resistor. The amount of heat dissipated each second is measured in watts and each resistor has its own **wattage rating.** For example, a 2 watt size resistor can dissipate up to 2 joules of heat per second, whereas a ⅛ W size resistor can only dissipate up to ⅛ joule of heat per second. The key point to remember is that resistors in high current circuits

Carbon Composition Resistor
Fixed resistor consisting of carbon particles mixed with a binder, which is molded and then baked. Also called a composition resistor.

Dissipation
Release of electrical energy in the form of heat.

Wattage Rating
Maximum power a device can safely handle continuously.

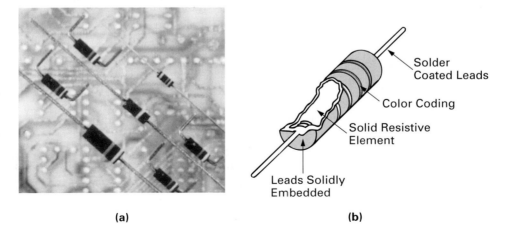

(a)

Solder
Coated Leads

Color Coding

Solid Resistive
Element

Leads Solidly
Embedded

(b)

FIGURE 3-5   Carbon Composition Resistors. (a) Appearance. (b) Construction. (Courtesy of Stackpole Electronics, Inc.)

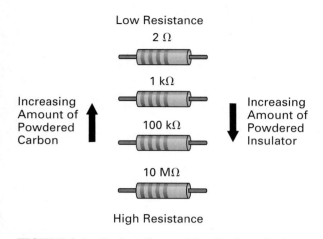

FIGURE 3-6   Carbon Composition Resistor Ratios.

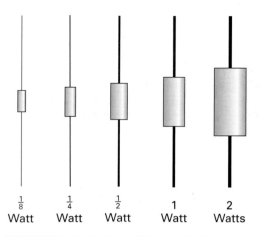

FIGURE 3-7   Resistor Wattage Rating Guide. All Resistor Silhouettes Are Drawn to Scale.

should have a surface area that is large enough to dissipate away the heat faster than it is being generated. If the current passing through a resistor generates heat faster than the resistor can dissipate it, the resistor will burn up and no longer perform its function.

Another factor to consider when discussing resistors is their **tolerance.** Tolerance is the amount of deviation or error from the specified value. For example, a 1000 Ω (1 kΩ) resistor with a ±10% (plus and minus 10%) tolerance when manufactured could have a resistance anywhere between 900 and 1100 Ω.

Tolerance

Permissible deviation from a specified value, normally expressed as a percentage.

$$\pm 10\% \text{ of } 1000 = 100$$

$$10\% - \boxed{1000} + 10\%$$
$$\downarrow \qquad\qquad \downarrow$$
$$900 \qquad\qquad 1100$$

This means that two identically marked resistors when measured could be from 900 to 1100 Ω, a difference of 200 Ω. In some applications, this may be acceptable. In other applications, where high precision is required, this deviation could be too large and so a more expensive, smaller tolerance resistor would have to be used.

◼ EXAMPLE:

Calculate the amount of deviation of the following resistors:

    a. 2.2 kΩ ± 10%

    b. 5 MΩ ± 2%

    c. 3 Ω ± 1%

CALCULATOR SEQUENCE

Step	Keypad Entry	Display Response
1.	[1] [0]	10
2.	[%]	10
3.	[×]	0.10
4.	[2] [.] [2] [E] [3]	2.2E3
5.	[=]	220

◼ *Solution:*

    a. 10% of 2.2 kΩ = 220 Ω. For + 10%, the value is

$$2200 + 220 \ \Omega = 2420 \ \Omega$$
$$= 2.42 \text{ k}\Omega$$

For −10%, the value is

$$2200 - 220 \ \Omega = 1980 \ \Omega$$
$$= 1.98 \text{ k}\Omega$$

The resistor will measure anywhere from 1.98 kΩ to 2.42 kΩ.

b. 2% of 5 MΩ = 100 kΩ

$$5 \text{ M}\Omega + 100 \text{ k}\Omega = 5.1 \text{ M}\Omega$$

$$5 \text{ M}\Omega - 100 \text{ k}\Omega = 4.9 \text{ M}\Omega$$

Deviation = 4.9 MΩ to 5.1 MΩ

c. 1% of 3 Ω = 0.03 Ω or 30 milliohms (mΩ).

$$3 \text{ }\Omega + 0.03 \text{ }\Omega = 3.03 \text{ }\Omega$$

$$3 \text{ }\Omega - 0.03 \text{ }\Omega = 2.97 \text{ }\Omega$$

Deviation = 2.97 Ω to 3.03 Ω

Carbon Film Resistors

Figure 3-8(a) illustrates the physical appearance of **carbon film resistors.** This resistor type is constructed, as shown in Figure 3-8(b), by first depositing a thin layer or film of resistive material (a blend of carbon and insulator) on a ceramic (insulating) **substrate.** The film is then cut to form a helix or spiral. A greater ratio of carbon to insulator will achieve a low-resistance helix. On the other hand, a greater ratio of insulator to carbon will create a higher-resistance helix. Carbon film resistors have smaller tolerance figures (±5% to ±2%) and have a better temperature stability than carbon composition resistors. The **temperature stability** of a resistor is the ability of the resistor to maintain its value of resistance despite changes in temperature. As discussed previously, there is a tendency when heat is increased for atoms to move, which in turn causes collisions resulting in an increase in resistance, which results in a further increase in temperature and resistance, and so on. This deviation from the desired resistance value can result in a circuit malfunction. Carbon film resistors also have less internally generated noise (random small bursts of voltage) than do carbon composition resistors. These voltage bursts are caused by small current surges within the resistor, as current overcomes the lack of conductivity within the resistor.

Carbon Film Resistor
Thin carbon film deposited on a ceramic form to create resistance.

Substrate
The mechanical insulating support on which a device is fabricated.

Temperature Stability
The ability of a resistor to maintain its value of resistance despite changes in temperature.

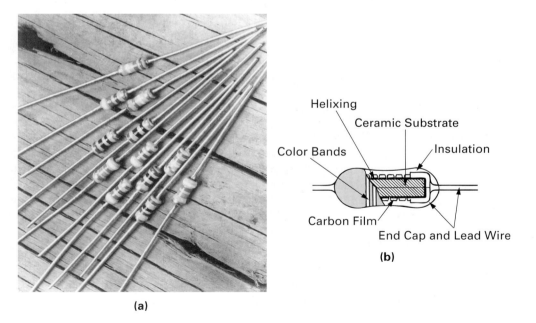

FIGURE 3-8   Carbon Film Resistors. (a) Appearance. (b) Construction. (Courtesy of Stackpole Electronics, Inc.)

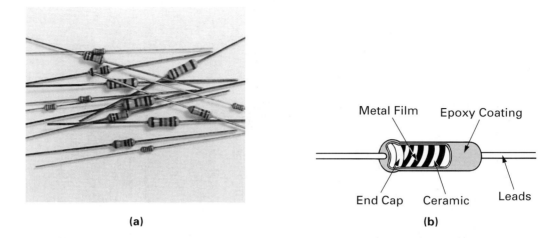

Metal Film   Epoxy Coating

End Cap   Ceramic   Leads

(a)   **(b)**

FIGURE 3-9   **Metal Film Resistors. (a) Appearance. (b) Construction. (Courtesy of Stackpole Electronics, Inc.)**

Metal Film Resistors

Figure 3-9(a) illustrates the physical appearance of some typical **metal film resistors.** This resistor type is constructed by spraying a thin film of metal on a ceramic cylinder (or substrate) and then cutting the film, as shown in Figure 3-9(b). Metal film resistors have possibly the best tolerances commercially available of $\pm 1\%$ to $\pm 0.1\%$. They also maintain a very stable resistance over a wide range of temperatures (good stability) and generate very little internal noise compared to any carbon resistor.

Wirewound Resistors

Figure 3-10(a) and (c) illustrate the appearance of different types of **wirewound resistors.** This resistor type is constructed, as can be seen in Figure 3-10(b), by wrapping a length of wire uniformly around a ceramic insulating core, with terminals making the connections at each end. The length and thickness of the wire are varied to change the resistance, which ranges from 1 Ω to 150 kΩ.

Since current flow is opposed by resistors and this opposition generates heat, the larger the physical size of the resistor, the greater the amount of heat that can be dissipated away and so the greater the current that can be passed through the resistor. Wirewound resistors are generally used in applications requiring low resistance values, which means that the current and therefore power dissipated are high ($R\downarrow = V/I\uparrow$, $P\uparrow = I^2\uparrow R$). As a result, these resistors are designed to have large surface areas so that they can safely dissipate away the heat. The amount of heat that can be dissipated in normal conditions is measured in watts and is indicated on the resistor. If the resistor's heat dissipation is assisted, it can handle a higher current. For example, a 10 W resistor can be used to dissipate 20 W if air is blown across its surface by a cooling fan or if it is immersed in a coolant. Wirewound resistors typically have good tolerances of $\pm 1\%$. Their large physical size and difficult manufacturing process, however, make them very expensive.

Metal Oxide Resistors

Figure 3-11(a) illustrates the physical appearance of some **metal oxide resistors.** This resistor type is constructed, as can be seen in Figure 3-11(b), by depositing an oxide of a metal such as tin onto an insulating substrate. The ratio of oxide (insulator) to tin (conductor) will determine the resistor's resistance.

Metal oxide resistors have excellent temperature stability, but are more expensive than carbon composition, carbon film, and metal film resistors.

Metal Film Resistor

A resistor in which a film of a metal, metal oxide, or alloy is deposited on an insulating substrate.

Wirewound Resistor

Resistor in which the resistive element is a length of high-resistance wire or ribbon, usually nichrome, wound onto an insulating form.

Metal Oxide Resistor

A metal film resistor in which an oxide of a metal (such as tin) is deposited as a film onto the substrate.

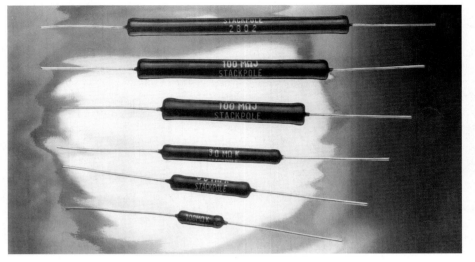

(a)

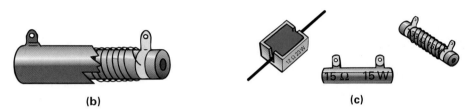

(b)   **(c)**

FIGURE 3-10   **Wirewound Resistors. (a) Appearance. (b) Construction. (c) Other Types. (Courtesy of Stackpole Electronics, Inc.)**

Thick-Film Resistors

Figure 3-12 illustrates examples of **thick-film resistors.** Figure 3-12(a) and (b) show the two different types of resistor networks, called SIPs and DIPs. The **single in-line package** is so called because all its lead connections are in a single line, whereas the **dual in-line package** has two lines of connecting pins. The chip resistors shown in Figure 3-12(c) are small thick-film resistors that are approximately the size of a pencil lead.

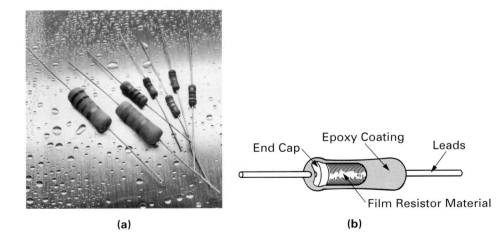

(a)   **(b)**

FIGURE 3-11   **Metal Oxide Resistors. (a) Appearance. (b) Construction. (Courtesy of Stackpole Electronics, Inc.)**

Thick-Film Resistors

Fixed-value resistor consisting of a thick-film resistive element made from metal particles and glass powder.

Single In-Line Package (SIP)

Package containing several electronic components (generally resistors) with a single row of external connecting pins.

Dual In-Line Package (DIP)

Package that has two (dual) sets or lines of connecting pins.

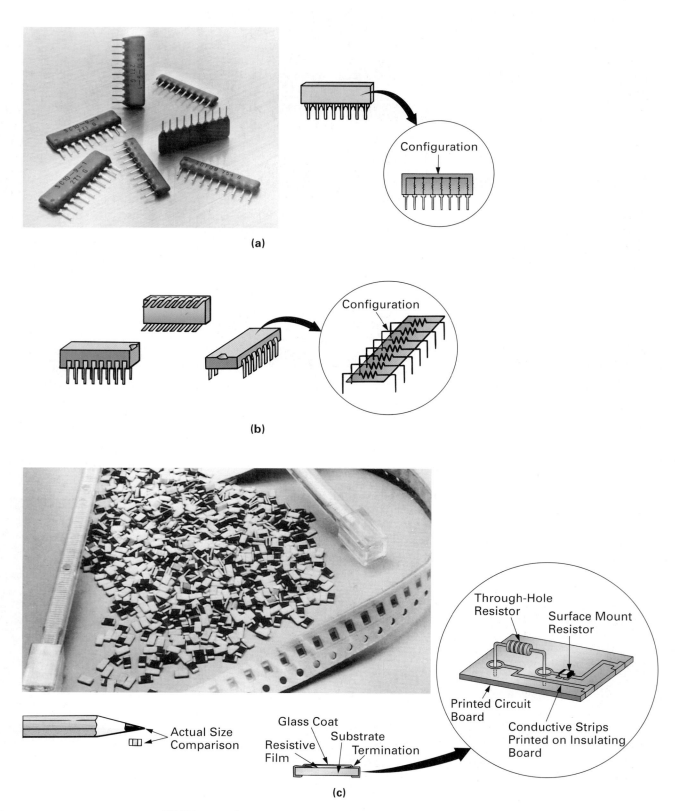

(a)

(b)

Configuration

Configuration

Through-Hole
Resistor

Surface Mount
Resistor

Printed Circuit
Board

Conductive Strips
Printed on Insulating
Board

Actual Size
Comparison

Glass Coat

Resistive
Film

Substrate
Termination

(c)

FIGURE 3-12   Thick-Film Resistors. (a) SIPs: Single In-Line Packages. (b) DIPs: Dual In-Line Packages. (c) Chips. (Courtesy of Stackpole Electronics, Inc.)

The SIP and DIP resistor networks are constructed by first screening on the internal conducting strip (silver) that connects the external pins to the resistive material, and then screening on the thick film of resistive paste (bismuth/ruthenate), the blend of which will determine resistance value. The chip resistor uses the same resistive film paste material, which is deposited onto an insulating substrate with two conductive end terminations and protected by a glass coat.

The SIP and DIP resistor networks, once constructed, are trimmed by lasers to obtain close tolerances of typically ±2%. Resistance values ranging from 22 Ω to 2.2 MΩ are available, with a power rating of ½ W.

The chip resistor shown in Figure 3-12(c) is commercially available with resistance values from 10 Ω to 3.3 MΩ, a ±2% tolerance, and a ⅛ W heat dissipation capability. They are ideally suited for applications requiring physically small sized resistors, as explained in the inset in Figure 3-12(c). The chip resistor is called a **surface mount technology** (SMT) device. The key advantage of SMT devices over "through-hole" devices is that a through-hole device needs both a hole in the printed circuit board (PCB) and a connecting pad around the hole. With the SMT device, no holes are needed since the package is soldered directly onto the surface of the PCB. Pads can be placed closer together. This results in a considerable space saving, as you can see in the inset in Figure 3-12(c) by comparing the through-hole resistor to the surface mount chip resistor.

3-2-2   *Variable-Value Resistors*

A **variable resistor** can have its resistance varied or changed while it is connected in a circuit. In certain applications, the ability to adjust the resistance of a resistor is needed. For example, the volume control on your television set makes use of a variable resistor to vary the amount of current passing to the speakers, and so change the volume of the sound.

Mechanically (User) Adjustable Variable Resistors

In this section we will discuss two variable resistor types, which will cause a change in resistance when a shaft is physically rotated.

Rheostat (two terminals: A and B).   Figure 3-13(a) shows the physical appearance of different rheostats, while Figure 3-13(b) shows the rheostat's schematic symbols. As can be seen in the construction of a circular rheostat in Figure 3-13(c), one terminal is connected to one side of a resistive track and the other terminal of this two-terminal device is connected to a movable wiper. As the wiper is moved away from the end of the track with the terminal, the resistance between the stationary end terminal and the mobile wiper terminal increases. This is summarized in Figure 3-13(d), where the wiper has been moved down by a clockwise rotation of the shaft. Current would have to flow through a large resistance as it travels from one terminal to the other. On the other hand, as the wiper is moved closer to the end of the track connected to the terminal, the resistance decreases. This is summarized in Figure 3-13(e), which shows that as the wiper is moved up, as a result of turning the shaft counterclockwise, current will see only a small resistance between the two terminals.

Rheostats come in many shapes and sizes, as can be seen in Figure 3-13(a). Some employ a straight-line motion to vary resistance, while others are classified as circular-motion rheostats. The resistive elements also vary; wirewound and carbon tracks are very popular. Cermet rheostats mix the ratio of ceramic (insulator) and metal (conductor) to produce different values of resistive tracks. A trimming rheostat is a miniature device used to change resistance by a small amount. Other circular-motion rheostats are available that require between two to ten turns to cover the full resistance range.

Potentiometer (three terminals: A, B, and C).   Figure 3-14(a) illustrates the physical appearance of a variety of potentiometers, also called pots (slang), while Figure 3-14(b) shows the potentiometer's schematic symbol. You will probably notice that the difference between a rheostat and potentiometer is the number of terminals; the rheostat has two

Surface Mount Technology

A method of installing tiny electronic components on the same side of a circuit board as the printed wiring pattern that interconnects them.

Variable Resistor

A resistor whose value can be changed.

Rheostat

Two-terminal variable resistor that, through mechanical turning of a shaft, can be used to vary its resistance and therefore its value of terminal to terminal current.

Potentiometer

Three-lead variable resistor that through mechanical turning of a shaft can be used to produce a variable voltage or potential.

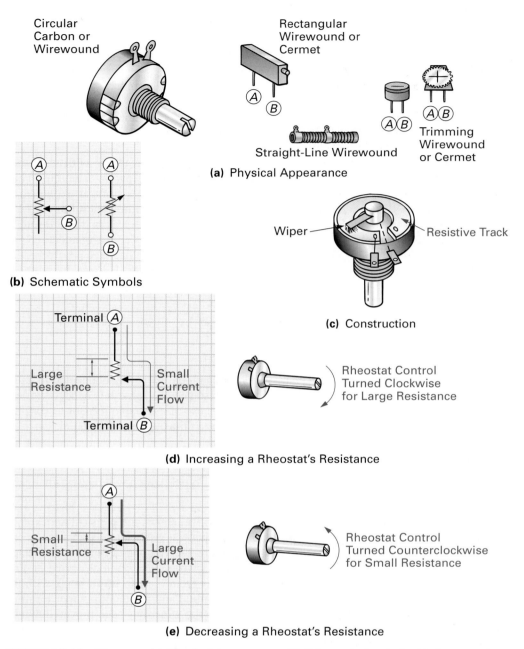

Circular Carbon or Wirewound

Rectangular Wirewound or Cermet

Straight-Line Wirewound

Trimming Wirewound or Cermet

(a) Physical Appearance

(b) Schematic Symbols

(c) Construction

Wiper — Resistive Track

Terminal (A)

Large Resistance

Small Current Flow

Terminal (B)

Rheostat Control Turned Clockwise for Large Resistance

(d) Increasing a Rheostat's Resistance

(A)

Small Resistance

Large Current Flow

(B)

Rheostat Control Turned Counterclockwise for Small Resistance

(e) Decreasing a Rheostat's Resistance

FIGURE 3-13 **Rheostat. (a) Physical Appearance. (b) Schematic Symbols. (c) Construction. (d) Increasing a Rheostat's Resistance. (e) Decreasing a Rheostat's Resistance.**

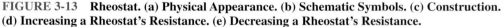

terminals while the potentiometer has three. With the rheostat, there were only two terminals and the resistance between the wiper and terminal varied as the wiper was adjusted. Referring to the potentiometer shown in Figure 3-14(c), you can see that resistance can actually be measured across three separate combinations: between *A* and *B* (*X*), between *B* and *C* (*Y*), and between *C* and *A* (*Z*).

The only difference between the rheostat and the potentiometer in construction is the connection of a third terminal to the other end of the resistive track, as can be seen in Figure 3-14(d), which shows the single-turn potentiometer. Also illustrated in this section of the figure is the construction of a multiturn potentiometer in which a contact arm slides along a shaft and the resistive track is formed into a helix of 2 to 10 coils.

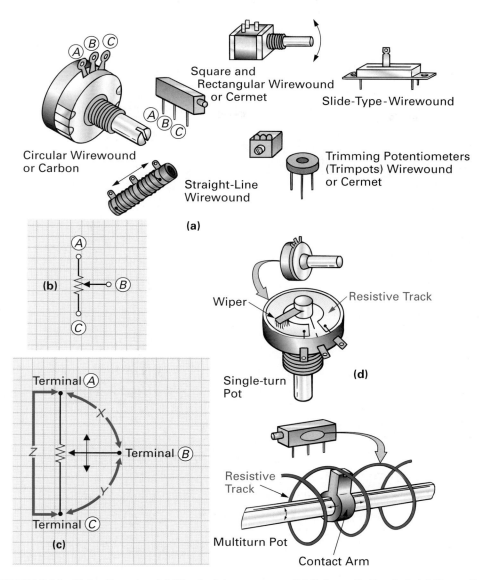

FIGURE 3-14 **Potentiometer. (a) Physical Appearance. (b) Schematic Symbol. (c) Operation. (d) Construction.**

To explain the operation of a potentiometer, Figure 3-15(a) shows a 10 kΩ potentiometer. Looking at Figure 3-15(b), you can see that the resistance measured between terminals A and C will always be the same (10 kΩ) no matter where we put the wiper, because current will still have to travel through the complete resistive track between A and C.

The resistance between A and B (X) and B and C (Y) will vary as the wiper's position is moved, as illustrated in Figure 3-15(c). If the user physically turns the shaft in a clockwise direction, the resistance between A and B increases, while the resistance between B and C decreases. Similarly, if the user mechanically turns the shaft counterclockwise, a decrease occurs between A and B and there is a resulting increase in resistance between B and C. This point is summarized in Figure 3-15(d).

In some applications, you may see the symbol shown in Figure 3-16, where B is connected to C and only two terminals are being used. In this situation, the potentiometer is being used as a rheostat.

Like the rheostat, the potentiometer comes in many different shapes and sizes. Wirewound, carbon, and cermet resistive tracks, circular or straight-line motion, 2 to 10 multiturn, and other variations are available to meet different application requirements.

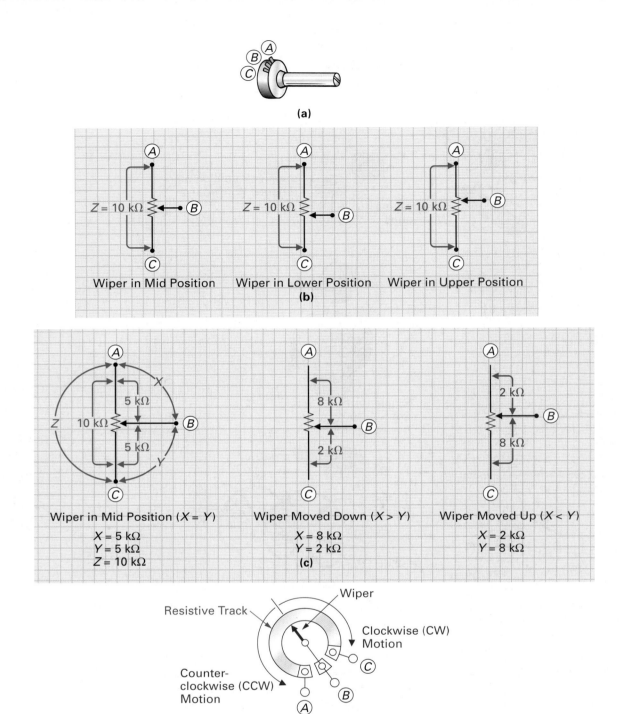

(a)

(b)

(c)

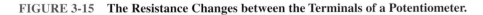

CW Motion: A to B Ω↑, B to C Ω↓.

CCW Motion: A to B Ω↓, B to C Ω↑.

(d)

FIGURE 3-15   The Resistance Changes between the Terminals of a Potentiometer.

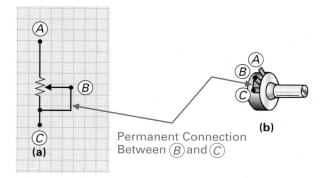

FIGURE 3-16 The Potentiometer as a Rheostat. (a) Schematic Symbol. (b) Physical Appearance.

Whether rheostat or potentiometer, the resistive track can be classified as having either a **linear** or a **tapered** (nonlinear) resistance. In Figure 3-17(a) we have taken a 1 kΩ rheostat and illustrated the resistance value changes between *A* and *B* for a linear and a tapered one-turn rheostat. The definition of linear is having an output that varies in direct proportion to the input. The input in this case is the user turning the shaft, and the output is the linearly increasing resistance between *A* and *B*.

With a tapered rheostat or potentiometer, the resistance varies nonuniformly along its resistor element, sometimes being greater or less for equal shaft movement at various points along the resistance element, as shown in the table in Figure 3-17(a). Figure 3-17(b) plots the position of the variable resistor's wiper against the resistance between the two output terminals, showing the difference between a linear increase and a nonlinear or tapered increase.

Thermally (Heat) Adjustable Variable Resistors

When first discussing variable-value resistors, we talked about the rheostat and potentiometer, both of which need a mechanical input (the user turning the shaft) to produce a change in resistance. A **bolometer** is a device that, instead of changing its resistance when mechanical energy is applied, changes its resistance when heat energy is applied. Before discussing the three basic types of bolometers, let's first consider the four units of temperature measurement detailed in Table 3-1.

The measurement of temperature **(thermometry)** is probably the most common type of measurement used in industry today. Commercially, temperature is normally expressed in degrees Celsius (°C) or degrees Fahrenheit (°F); however, although not commonly known, kelvin (K) and degrees Rankine (°R) are often used in industry, the kelvin being the international unit of temperature.

Linear
Relationship between input and output in which the output varies in direct proportion to the input.

Tapered
Nonuniform distribution of resistance per unit length throughout the element.

Bolometer
Device whose resistance changes when heated.

Thermometry
Relating to the measurement of temperature.

■ EXAMPLE:

Convert the following:

a. 74°F = _____ °C

b. 45°C = _____ °F

c. 25°C = _____ K

d. 10°F = _____ °R

	Fully CCW	$\frac{1}{4}$ CW	$\frac{1}{2}$ CW	$\frac{3}{4}$ CW	Fully CW
1 kΩ Rheostat					
Linear	0 Ω	250 Ω $\left(\begin{array}{c}\frac{1}{4}\text{ of}\\1000\text{ }\Omega\end{array}\right)$	500 Ω $\left(\begin{array}{c}\frac{1}{2}\text{ of}\\1000\text{ }\Omega\end{array}\right)$	750 Ω $\left(\begin{array}{c}\frac{3}{4}\text{ of}\\1000\text{ }\Omega\end{array}\right)$	1000 Ω
Tapered	0 Ω	350 Ω	625 Ω	900 Ω	1000 Ω

These values have been arbitrarily chosen to illustrate a nonlinear change.

(a)

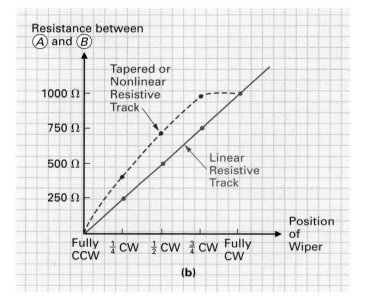

(b)

FIGURE 3-17   Linear versus Tapered Resistive Track.

TABLE 3-1   Four Temperature Scales and Conversion Formulas

	FAHRENHEIT	CELSIUS	KELVIN	RANKINE	
Absolute zero	−459.69°F	−273.16°C	0 K	0°R	$F = (^9/_5 \times C) + 32$
Melting point of ice (x)	32°F	0°C	273.16 K	491.69°R	$C = ^5/_9 \times (F - 32)$
(Division between x and y)	↓ (180°F) ↓	↓ (100°C) ↓	↓ (100 K) ↓	↓ (180°R) ↓	$R = F + 460$ $F = R - 460$
Boiling point of water (y)	212°F	100°C	373.16 K	671.69°R	$K = C + 273$ $C = K - 273$

■ **Solution:**

a. $C = \% \times (F - 32)$

    $= \% \times (74 - 32)$

    $= 0.555 \times 42 = 23.3°C$

b. $F = (\% \times C) + 32$

    $= (\% \times 45) + 32$

    $= (1.8 \times 45) + 32 = 113°F$

c. $K = C + 273$

    $= 25 + 273 = 298 \text{ K}$

d. $R = F + 460$

    $= 10 + 460 = 470°R$

Step	Keypad Entry	Display Response
1.	7 4	74
2.	−	
3.	3 2	32
4.	=	42
5.	STO (store result in memory)	
6.	C/CE (cancel display)	0
7.	5	5
8.	÷	
9.	9	9
10.	=	0.55555
11.	×	
12.	RCL (Recall value from memory)	42
13.	=	23.3

Bolometers are normally used as temperature sensors or detectors. The **resistive temperature detector** (RTD) and **thin-film detector** (TFD) shown in Figure 3-18(a) and (b) are both temperature sensors that make use of a copper, nickel, or platinum conducting element, and therefore have a positive temperature coefficient (resistance increases as temperature increases). Referring to the construction of the RTD in Figure 3-18(a), you can see that the sensing element consists of a coil of fine wire generally made of platinum, which gives a relatively linear increase in resistance as temperature increases, as indicated in the table in Figure 3-18(a).

The thin-film detector (TFD) shown in Figure 3-18(b) is used for very precise temperature readings, and is constructed by placing a thin layer of platinum on a ceramic substrate. Because of its small size, the TFD responds rapidly to temperature change and is ideally suited for surface temperature sensing.

Unlike the RTD and TFD, the **thermistor** contains a semiconductor material that has a negative temperature coefficient, which means that its resistance decreases as temperature increases. Semiconductor materials have characteristics that are midway between those of conductors and insulators. In addition to thermistors, semiconductor materials are also used to construct other electronic devices such as diodes and transistors, which will be discussed later.

A variety of thermistors can be seen in Figure 3-18(c). The thermistor is the most common type of temperature sensor and produces rapid and extremely large changes in resistance for very small changes in temperature, as shown in the associated table.

An application of the thermistor could be as a temperature sensor inside an oven. As the oven heats up, the thermistor's resistance decreases, and at a certain decreased resistance the thermistor turns off the oven. Once the temperature starts to drop, the thermistor's resistance increases, and this increase of resistance turns the oven back on.

    temperature ↑   thermistor's R ↓   oven OFF

    temperature ↓   thermistor's R ↑   oven ON

Optically (Light) Adjustable Variable Resistors

Photo means illumination, and the **photoresistor** is a resistor that is photoconductive. This means that as the material is exposed to light, it will become more conductive and less resistive. Figure 3-19 shows the construction, schematic symbol, and physical appearance of the photoresistor, which is also called a light-dependent resistor (LDR).

The photoresistor is made up of a thin slice of photoconductive material whose resistance decreases as light is applied. The light energy is absorbed by the atoms within the pho-

Resistive Temperature Detector (RTD)

A temperature detector consisting of a fine coil of conducting wire (such as platinum) that will produce a relatively linear increase in resistance as temperature increases.

Thin-Film Detector (TFD)

A temperature detector containing a thin layer of platinum, and used for very precise temperature readings.

Thermistor

Temperature-sensitive semiconductor that has a negative temperature coefficient of resistance (as temperature increases, resistance decreases).

Photoresistor

Also known as a photoconductive cell or light-dependent resistor, it is a device whose resistance varies with the illumination of the cell.

Platinum Winding			
°C	Ohms	°C	Ohms
−200	18.53	+200	175.84
−150	39.65	+250	194.08
−100	60.20	+300	212.03
−50	80.25	+350	229.69
±0	100.0	+400	247.06
+50	119.40	+450	264.14
+100	138.50	+500	280.93
+150	157.32	+550	297.16

(Conductors have a positive temperature coefficient of resistance—temp.↑, R↑)

°C	Ohms
−50	100,000
0	7,500
+50	7,400
+100	100
+150	50
+200	27
+250	10
+300	7.5

(Semiconductors have a negative temperature coefficient of resistance—T↑, R↓)

FIGURE 3-18 **Temperature Sensors. (a) Resistive Temperature Detector (RTD). (b) Thin-Film Detector (TFD). (c) Thermistor.**

toconductive material, causing these atoms to release their valence electrons. This results in an increase in electrons, and therefore current passing through the photoresistor, so its resistance will have decreased. To summarize, we can say:

light energy ↑   conduction ↑   resistance ↓

As an application, photoresistors can be used to turn on and off outdoor home security lights. During the day, the natural sunlight decreases the resistance of the photoresistor, and this low resistance is used to keep the lights off. At dusk, the sunlight is almost gone, and the photoresistor's resistance increases. This increased resistance is used to turn on the security lights.

sun up ↑   photoresistance ↓   security lights OFF
sun down ↓   photoresistance ↑   security lights ON

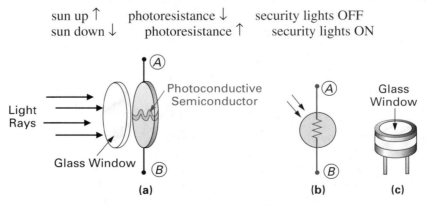

FIGURE 3-19 **Photoresistor. (a) Construction. (b) Schematic Symbol. (c) Physical Appearance.**

Now that you have completed this section, you should be able to:

■ **Objective 4.** *Describe the difference between a fixed- and a variable-value resistor.*

■ **Objective 5.** *Explain the difference between the six basic types of fixed-value resistors: carbon composition, carbon film, metal film, wirewound, metal oxide, and thick film.*

■ **Objective 6.** *Identify the different resistor wattage ratings, and their value and tolerance labeling methods.*

■ **Objective 7.** *Describe the SIP, DIP, and chip thick-film resistor packages.*

■ **Objective 8.** *Explain the following types of variable resistors:*

a. Mechanically adjustable: rheostat and potentiometer
b. Thermally adjustable: RTD, TFD, and thermistor
c. Optically adjustable: photoresistor

Use the following questions to test your understanding of Section 3-2:

1. List the six types of fixed-value resistors.

2. What is the difference between SIPs and DIPs?

3. Name the two types of mechanically adjustable variable resistors and state the difference between the two.

4. Describe a linear and a tapered potentiometer.

5. True or false: A thermistor has a negative temperature coefficient of resistance.

6. What name is given to the optically adjustable resistor?

3-3 HOW IS RESISTANCE MEASURED?

Resistance is measured with an **ohmmeter.** Stepping through the procedure described in Figure 3-20, you will see how the ohmmeter is used to measure the resistance of a resistor.

Ohmmeter

Measurement device used to measure electric resistance.

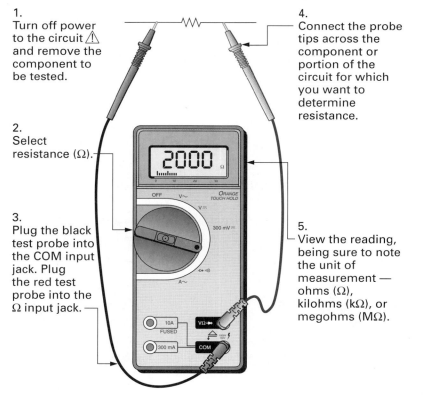

1.
Turn off power to the circuit ⚠ and remove the component to be tested.

2.
Select resistance (Ω).

3.
Plug the black test probe into the COM input jack. Plug the red test probe into the Ω input jack.

4.
Connect the probe tips across the component or portion of the circuit for which you want to determine resistance.

5.
View the reading, being sure to note the unit of measurement — ohms (Ω), kilohms (kΩ), or megohms (MΩ).

FIGURE 3-20   Measuring Resistance with the Ohmmeter.

Now that you have completed this section, you should be able to:

■ **Objective 9.** *List the three rules to remember when measuring resistance with an ohmmeter.*

Use the following questions to test your understanding of Section 3-3:

1. Name the instrument used to measure resistance.
2. Describe the five-step procedure that should be used to measure the resistance.

3-4   RESISTOR CODING

Manufacturers indicate the value and tolerance of resistors on the body of the component using either a color code (colored rings or bands) or printed alphanumerics (alphabet and numerals).

Resistors that are smaller in size tend to have the value and tolerance of the resistor encoded using colored rings, as shown in Figure 3-21(a). To determine the specifications of a color-coded resistor, therefore, you will have to decode the rings.

Referring to Figure 3-21(b), you can see that when the value, tolerance, and wattage of a resistor are printed on the body, no further explanation is needed.

3-4-1   *General-Purpose and Precision Color Code*

There are basically two different types of fixed-value resistors: general purpose and precision. Resistors with tolerances of ±2% or less are classified as *precision resistors* and have five bands. Resistors with tolerances of ±5% or greater have four bands and are referred to as *general-purpose resistors*. The color code and differences between precision and general-purpose resistors are explained in Figure 3-22.

When you pick up a resistor, look for the bands that are nearer to one end. This end should be held in your left hand. If there are four bands on the resistor, follow the general-purpose resistor code. If five bands are present, follow the precision resistor code.

General-Purpose Resistor Code

1. The first band on either a general-purpose or precision resistor can never be black, and it is the first digit of the number.
2. The second band indicates the second digit of the number.
3. The third band specifies the multiplier to be applied to the number, which ranges from × $\frac{1}{100}$ to 10,000,000.
4. The fourth band describes the tolerance or deviation from the specified resistance, which is ±5% or greater.

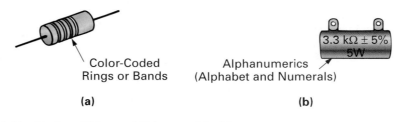

Color-Coded Rings or Bands

Alphanumerics (Alphabet and Numerals)

3.3 kΩ ± 5% 5W

(a)

(b)

FIGURE 3-21   Resistor Value and Tolerance Markings.

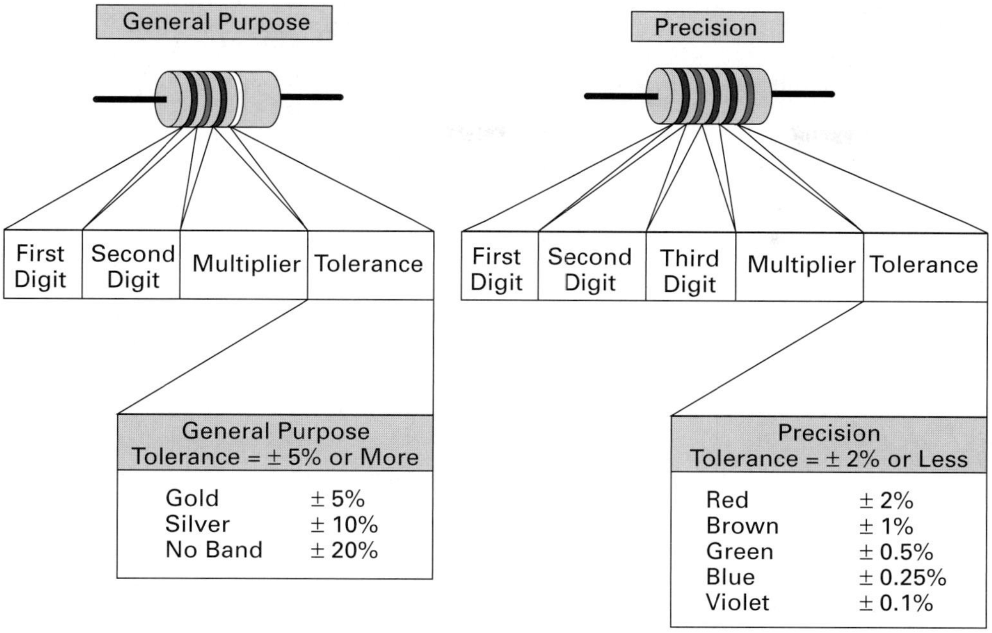

FIGURE 3-22 **General-Purpose and Precision Resistor Color Code.**

	Color	Digit Value	Multiplier	
Big	Black	0	1 One	1
Beautiful	Brown	1	10 One Zero	10
Roses	Red	2	100 Two Zeros	100
Occupy	Orange	3	1000 Three Zeros	1 k
Your	Yellow	4	10000 Four Zeros	10 k
Garden	Green	5	100000 Five Zeros	100 k
But	Blue	6	1000000 Six Zeros	1 M
Violets	Violet	7	10000000 Seven Zeros	10 M
Grow	Gray	8	–	
Wild	White	9	–	
So	Silver	–	10^{-2} or 0.01	1/100
Get some	Gold	–	10^{-1} or 0.1	1/10
Now	None	–		

Precision Resistor Code

1. The first band, like the general-purpose resistor, is never black and is the first digit of the three-digit number.

2. The second band provides the second digit.

3. The third band indicates the third and final digit of the number.

4. The fourth band specifies the multiplier to be applied to the number.

5. The fifth and final band indicates the tolerance figure of the precision resistor, which is always less than $\pm2\%$, which is why precision resistors are more expensive than general-purpose resistors.

Band 1 = Green
Band 2 = Blue
Band 3 = Brown
Band 4 = Gold

FIGURE 3-23

■ **EXAMPLE:**

Figure 3-23 illustrates a ½ W resistor.

    a. Is this a general-purpose or a precision resistor?

    b. What is the resistor's value of resistance?

    c. What tolerance does this resistor have, and what deviation plus and minus could occur to this value?

■ *Solution:*

    a. General purpose (four bands)

    b. green   blue × brown

        5   6 × 10   = 560 Ω

    c. Tolerance band is gold, which is ±5%.

$$\text{deviation} = 5\% \text{ of } 560 = 28$$
$$560 + 28 = 588 \text{ } \Omega$$
$$560 - 28 = 532 \text{ } \Omega$$

The resistor could, when measured, be anywhere from 532 to 588 Ω.

■ **EXAMPLE:**

State the resistor's value and tolerance and whether it is general purpose or precision for the examples shown in Figure 3-24.

Orange/Green/Black/Silver

(a)

Green/Blue/Red/None

(b)

Red/Red/Green/Gold/Blue

(c)

FIGURE 3-24

■ *Solution:*

    a. orange   green black silver   (four bands = general purpose)

        3   5 × 1   10%   = 35 Ω ± 10%

    b. green blue   red none   (four bands = general purpose)

        5   6 × 100 20%   = 5.6 kΩ ± 20%

    c. red red green gold blue   (five bands = precision)

        2   2   5 × 0.1 0.25% = 22.5 Ω ± 0.25%

3-4-2   *Zero-Ohms Resistor*

Figure 3-25(a) illustrates the only resistor color code that you will probably have trouble with, the zero-ohms resistor, which has one black band. Zero ohms is equivalent to a straight piece of wire. So why do they manufacture a resistor of zero ohms since it really is not a resistor at all?

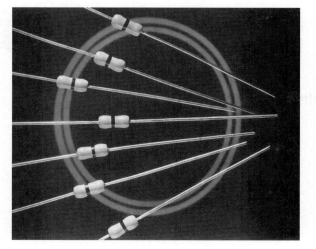

(a)

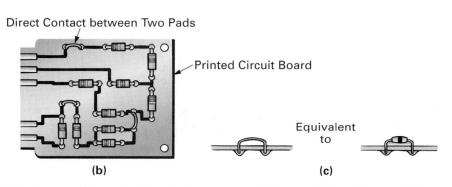

Direct Contact between Two Pads

Printed Circuit Board

Equivalent to

(b)   (c)

FIGURE 3-25   Zero-Ohms Resistor. (a) Appearance. (b) Printed Circuit Board. (c) Zero-Ohms Resistor as Jumper.

Printed circuit boards like the one shown in Figure 3-25(b) no longer have all their resistor and other components inserted by hand. *Automatic insertion equipment* replaces the human assembler and places the correct component in the correct position in a matter of seconds rather than hours. In some applications, a direct contact between two points is needed, in which case a piece of wire needs to be inserted, as shown in Figure 3-25(b). However, the resistor automatic insertion equipment will only handle resistors, not wire (called jumpers). In the past, manufacturers had to install jumpers manually after all the components were inserted, causing enormous delays. The zero-ohms resistor solves this problem, since it can be installed automatically during the resistor insertion process and have exactly the same effect, as shown in Figure 3-25(c).

3-4-3   *Other Resistor Identification Methods*

If a resistor color code is not found, some other form of marking will always be made on the resistor. The larger wirewound resistors and nearly all variable resistors normally have the resistance value, tolerance, and wattage printed on the resistor, as shown in Figure 3-26(a). SIP and DIP packages are normally coded with both letters and numerals, all of which have particular meaning, as shown in Figure 3-26(b). Chip resistors tend to be too small to have any intelligible form of marking and so will have to be measured with an ohmmeter. They are packaged, as previously shown in Figure 3-12(c), in either polythene bags, paper tape, or plastic magazines, and in all these cases the value, tolerance, and wattage will be printed on the packaging.

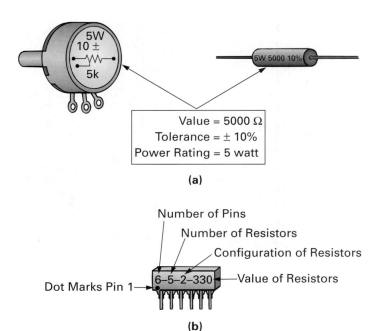

Value = 5000 Ω
Tolerance = ± 10%
Power Rating = 5 watt

(a)

Number of Pins
Number of Resistors
Configuration of Resistors
Dot Marks Pin 1 — 6–5–2–330 — Value of Resistors

(b)

FIGURE 3-26   **Typographical Value Indication. (a) Standard Variable and Fixed. (b) SIP and DIP.**

SELF-TEST EVALUATION POINT FOR SECTION 3-4

Now that you have completed this section, you should be able to:

■ **Objective 10.**   *Explain how a resistor's value and tolerance are printed on the body of the component either:*
a. *By color coding, or*
b. *Typographically*

■ **Objective 11.**   *Describe the difference between a general-purpose and a precision resistor.*

Use the following questions to test your understanding of Section 3-4:

1. What tolerance differences occur between a general-purpose and a precision resistor?

2. A red/red/red/gold/blue resistor has what value and tolerance figure?

3. What color code would appear on a 4.7 kΩ, ±10% tolerance resistor?

4. Would a three-band resistor be considered a general-purpose type?

3-5   POWER DISSIPATION DUE TO RESISTANCE

The atoms of a resistor obstruct the flow of moving electrons (current), and this friction between the two causes heat to be generated. Current flow through any resistance therefore will always be accompanied by heat. In some instances, such as a heater or oven, the device has been designed specifically to generate heat. In other applications it may be necessary to include a resistor to reduce current flow, and the heat generated will be an unwanted side effect.

Any one of the three power formulas can be used to calculate the power dissipated by a resistance.

$$P = V \times I \qquad P = I^2 \times R \qquad P = \frac{V^2}{R}$$

However, since power is determined by the friction between current (I) and the circuit's resistance (R), the $P = I^2 \times R$ formula is more commonly used to calculate the heat generated.

EXAMPLE:

What wattage rating should a 33 Ω resistor have if the value of current is 100 mA?

Solution:

$$P = I^2 \times R$$
$$= (100 \text{ mA}^2) \times 33 \text{ Ω}$$
$$= 0.33 \text{ W}$$

It would be safe therefore to use a ½ W (0.5 W) resistor.

EXAMPLE:

A coffee cup warming plate has a power rating of 120 V, 23 W. How much current is flowing through the heating element, and what is the heating element's resistance?

Solution:

$$P = V \times I \qquad I = \frac{P}{V}$$

$$I = \frac{23 \text{ W}}{120 \text{ V}} = 191.7 \text{ mA}$$

$$R = \frac{V}{I} = \frac{120 \text{ V}}{191.7 \text{ mA}} = 626 \text{ Ω}$$

EXAMPLE:

To illustrate a point, let us work through a hypothetical example. Calculate the current drawn by a 100 W light bulb if it is first used in the home (and therefore connected to a 120 V source) and then used in the car (and therefore connected across a 12 V source).

Solution:

$$P = V \times I \qquad \text{therefore,} \qquad I = \frac{P}{V}$$

In the home,

$$I = \frac{P}{V} = \frac{100 \text{ W}}{120 \text{ V}} = 0.83 \text{ A}$$

In the car,

$$I = \frac{P}{V} = \frac{100 \text{ W}}{12 \text{ V}} = 8.33 \text{ A}$$

This example brings out a very important point in relation to voltage, current, and power. In the home, only a small current needs to be supplied when the applied voltage is large ($P = V \uparrow \times I \downarrow$). However, when the source voltage is small, as in the case of the car, a large current must be supplied in order to deliver the same amount of power ($P = V \downarrow \times I \uparrow$).

Filament Resistor

The resistor in a light bulb or electron tube.

Figure 3-27(a) illustrates the **filament resistor** within a glass bulb. This component is more commonly known as the household light bulb. The filament resistor is just a coil of wire that glows white hot when current is passed through it and, in so doing, dissipates both heat and light energy.

Incandescent Lamp

An electric lamp that generates light when an electric current is passed through its filament of resistance, causing it to heat to incandescence.

This **incandescent lamp** is an electric lamp in which electric current flowing through a filament of resistive material heats the filament until it glows and emits light. Figure 3-27(b) shows a test circuit for varying current through a small incandescent lamp. Potentiometer R_1 is used to vary the current flow through the lamp. The lamp is rated at 6 V, 60 mA. Using Ohm's law we can calculate the lamp's filament resistance at this rated voltage and current:

$$\text{filament resistance, } R = \frac{V}{I} = \frac{6 \text{ V}}{60 \text{ mA}} = 100 \text{ }\Omega$$

The filament material (tungsten, for example) is like all other conductors in that it has a positive temperature coefficient of resistance. Therefore, as the current through the filament increases, so does the temperature and so does the filament's resistance ($I\uparrow$, temperature $\uparrow$, $R\uparrow$). Consequently, when R_1's wiper is moved to the right so that it produces a high resistance ($R_1\uparrow$), the circuit current will be small ($I\downarrow$) and the lamp will glow dimly. Since the circuit current is small, the filament temperature will be small and so will the lamp's resistance ($I\downarrow$, temperature $\downarrow$, lamp resistance $\downarrow$). This small value of resistance is called the lamp's **cold resistance.** On the other hand, when R_1's wiper is moved to the left so that it produces a small resistance ($R\downarrow$), the circuit current will be large ($I\uparrow$), and the lamp will glow brightly. With the circuit current high, the filament temperature will be high and so will the lamp's resistance ($I\uparrow$, temperature $\uparrow$, lamp resistance $\uparrow$). This large value of resistance is called the lamp's **hot resistance.**

Cold Resistance

The resistance of a device when cold.

Hot Resistance

The resistance of a device when hot due to the generation of heat by electric current.

Figure 3-27(c) plots the filament voltage, which is being measured by the voltmeter, against the filament current, which is being measured by the ammeter. As you can see, an increase in current will cause a corresponding increase in filament resistance. Studying this graph, you may have also noticed that the lamp has been operated beyond its rated value of 6 V, 60 mA. Although the lamp can be operated beyond its rated value (for example, 10 V, 80 mA), its life expectancy will be decreased dramatically from several hundred hours to only a few hours.

Ballast Resistor

A resistor that increases in resistance when current increases. It can therefore maintain a constant current despite variations in line voltage.

A coil of wire similar to the one found in a light bulb, called a **ballast resistor,** is used to maintain a constant current despite variation in voltage. Since the coil of wire is a conductor and conductors have a positive temperature coefficient, an increase in voltage will cause a corresponding increase in current and therefore in the heat generated, which will result in an increase in the wire's resistance. This increase in resistance will decrease the initial current rise. Similarly, a decrease in voltage and therefore current ($I \propto V$) will result in a decrease in heat and wire resistance. This decrease in resistance will permit an increase in current to counteract the original decrease. Current is therefore regulated or maintained constant by the ballast resistor, despite variations in voltage.

SELF-TEST EVALUATION POINT FOR SECTIONS 3-5 AND 3-6

Now that you have completed this section, you should be able to:

■ *Objective 12.* *Explain why current through any resistance is always accompanied by heat.*

■ *Objective 13.* *State the purpose of the filament and ballast resistors.*

Use the following questions to test your understanding of Sections 3-5 and 3-6:

1. Which formula is most commonly used to calculate the heat generated by a resistor?

2. An incandescent lamp has a _____ (large/small) cold resistance and a _____ (large/small) hot resistance.

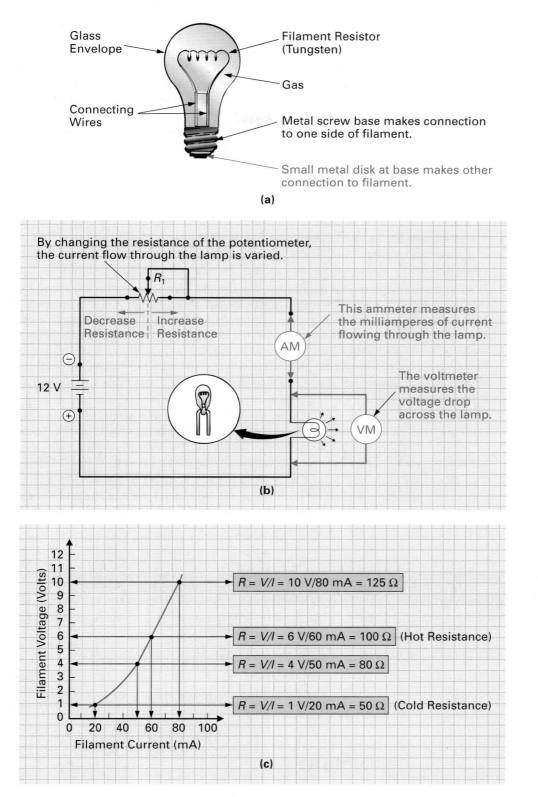

(a)

By changing the resistance of the potentiometer, the current flow through the lamp is varied.

R_1

Decrease Resistance | Increase Resistance

This ammeter measures the milliamperes of current flowing through the lamp.

AM

The voltmeter measures the voltage drop across the lamp.

12 V

VM

(b)

$R = V/I = 10 \text{ V}/80 \text{ mA} = 125\ \Omega$

$R = V/I = 6 \text{ V}/60 \text{ mA} = 100\ \Omega$ (Hot Resistance)

$R = V/I = 4 \text{ V}/50 \text{ mA} = 80\ \Omega$

$R = V/I = 1 \text{ V}/20 \text{ mA} = 50\ \Omega$ (Cold Resistance)

Filament Voltage (Volts)

Filament Current (mA)

(c)

FIGURE 3-27   **Filament Resistor.**

3-7 TESTING RESISTORS

It is virtually impossible for fixed-value resistors to internally short-circuit (no resistance or zero ohms). Generally, the resistor's internal elements will begin to develop a higher resistance than its specified value (due to a partial internal open) or in some cases go completely open circuit (maximum resistance or infinite ohms).

Variable-value resistors have problems with the wiper making contact with the resistive track at all points. Faulty variable-value resistors in sound systems can normally be detected because they generate a scratchy noise whenever you adjust the volume control.

The ohmmeter is the ideal instrument for verifying whether or not a resistor is functioning correctly. When checking a resistor's resistance, here are some points that should be remembered.

1. The ohmmeter has its own internal power source (a battery), so always turn off the circuit power and remove from the circuit the resistor to be measured. If this is not done, you will not only obtain inaccurate readings, but you can damage the ohmmeter.

2. With an autoranging digital multimeter, the suitable ohms range is automatically selected by the ohmmeter. The range scales on a non-autoranging digital multimeter have to be selected, and often confuse people. For example, if a 100 kΩ range is chosen the highest reading that can be measured is 100 kΩ. If a 1 MΩ resistor is measured, the meter will indicate an infinite-ohms reading and you may be misled into believing that you have found the problem (an open resistor). To overcome this problem, always start on the highest range and then work down to a lower range for a more accurate reading.

3. Another point to keep in mind is tolerance. For example, if a suspected faulty 1 kΩ (1000 Ω) resistor is measured with an ohmmeter and reads 1.2 kΩ (1200 Ω), it could be within tolerance if no tolerance band is present on the resistor's body (±20%). A 1 kΩ resistor with ±20% tolerance could measure anywhere between 800 and 1200 Ω.

4. The ohmmeter's internal battery voltage is really too small to deliver an electrical shock. You should, however, avoid touching the bare metal parts of the probes or resistor leads, as your body resistance of approximately 50 kΩ will affect your meter reading.

SELF-TEST EVALUATION POINT FOR SECTION 3-7

Now that you have completed this section, you should be able to:

■ *Objective 14. Describe some of the more common resistor problems.*

Use the following questions to test your understanding of Section 3-7:

1. What problem normally occurs with resistors? (open/shorts)

2. What instrument is used to verify a resistor's reistance?

SUMMARY

1. There are two basic types of resistors: fixed value and variable value. The fixed-value resistor has a value of resistance that cannot be changed, while the variable-value resistor has a range of values that can be selected generally by mechanically adjusting a control.

Fixed-Value Resistors (Figure 3-28)

2. Carbon composition resistors are the most common and least expensive type of fixed-value resistor. By changing the ratio of powdered insulator to carbon, the value of resistance can be changed within the same area.

FIGURE 3-28   Fixed-Value Resistors.

Type	Cost	Resistive Material	Advantages/Disadvantages
Carbon Composition 	$	Powdered carbon and carbon insulator.	Low cost but inherently noisy.
Carbon Film 	$$	A thin film of carbon and insulator.	Smaller tolerances ($\pm 5\%$ to $\pm 2\%$) and better temperature stability than carbon composition.
Metal Film	$$$	Thin metal film spiral on substrate.	Best tolerances ($\pm 1\%$ to $\pm 0.1\%$) and temperature stability; however, high cost.
Wirewound 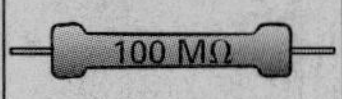	$$$$	Length of wire wrapped around ceramic core.	High power rating and good tolerance ($\pm 1\%$); however, very high cost.
Metal Oxide 	$$$	Oxide of a metal on an insulating substrate.	Has the best temperature stability; however, high cost.
Thick Film Networks Chips 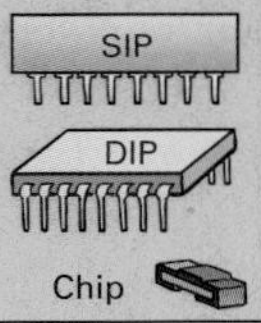	$$$$ $$	Thick film of resistive paste on insulating substrate.	Small size makes them ideal for high density circuits. Resistor networks (SIPs and DIPs) and chip resistors for SMT.

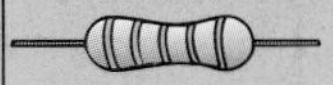

3. The physical size of the resistors lets the user know how much power in the form of heat can be *dissipated*. The amount of heat dissipated per unit of time is measured in watts and each resistor has its own *wattage rating*. The bigger the resistor, the more heat that can be dissipated.

4. Another factor to consider when discussing resistors is their *tolerance*. Tolerance is the amount of deviation or error from the specified value. In some applications where high precision is required, smaller tolerance resistors are used.

5. Carbon film resistors have smaller tolerance figures ($\pm 5\%$ to $\pm 2\%$), have good temperature stability (maintain same resistance value over a wide range of temperatures), and have less internally generated noise (random small bursts of voltage) than do carbon composition resistors.

6. Metal film resistors have possibly the best tolerances commercially available of $\pm 1\%$ to $\pm 0.1\%$. They also maintain a very stable resistance over a wide range of temperatures (good stability) and generate very little internal noise compared to any carbon resistor.

7. Wirewound resistors are constructed by wrapping a length of wire uniformly around a ceramic insulating core, with terminals making the connections at each end. The length and thickness of the wire are varied to change the resistance, which ranges from $1\ \Omega$ to $150\ k\Omega$. Wirewound resistors are generally used in applications requiring low resistance values, which means that the current and therefore power dissipated are high.

8. Metal oxide resistors have excellent temperature stability, which is the ability of the resistor to maintain its value of

resistance without change even when temperature is changed.

9. The two different types of *thick-film resistor* networks are called SIPs and DIPs. The *single in-line package* is so called because all its lead connections are in a single line, whereas the *dual in-line package* has two lines of connecting pins. The chip resistors are small thick-film resistors that are approximately the size of a pencil lead.

10. The SIP and DIP resistor networks, once constructed, are trimmed by lasers to obtain close tolerances of typically ±2%. Resistance values ranging from 22 Ω to 2.2 MΩ are available, with a power rating of ½ W.

11. The chip resistor is commercially available with resistance values from 10 Ω to 3.3 MΩ, a ±2% tolerance, and a ⅛ W heat dissipation capability.

12. The *rheostat* is a two-terminal variable resistor that, through mechanical turning of a shaft, can be used to vary its resistance and therefore its value of terminal to terminal current.

13. The difference between a rheostat and potentiometer is the number of terminals; the rheostat has two terminals and the potentiometer has three.

14. Like the rheostat, the potentiometer comes in many different shapes and sizes. Wirewound, carbon, and cermet resistive tracks, circular or straight-line motion, 2 to 10 multi-turn, and other variations are available for different applications.

15. Whether rheostat or potentiometer, the resistive track can be classified as having either a *linear* or a *tapered* (nonlinear) resistance. The definition of linear is having an output that varies in direct proportion to the input. The input, in this case, is the user turning the shaft, and the output is the linearly increasing resistance between *A* and *B*. With a ta-

pered rheostat or potentiometer, the resistance varies nonuniformly along its resistor element, sometimes being greater or less for equal shaft movement at various points along the resistance element.

16. A *bolometer* is a device that changes its resistance when heat energy is applied.

17. Commercially, temperature is normally expressed in degrees Celsius (°C) or degrees Fahrenheit (°F). Kelvin (K) and degrees Rankine (°R) are often used in industry, with the kelvin being the international unit of temperature.

18. There are basically three different types of temperature detectors. The *resistive temperature detector* (RTD) and *thin-film detector* (TFD) are both temperature sensors that contain a conducting material such as copper, nickel, or platinum and consequently have a positive temperature coefficient.

19. The *thermistor* contains a semiconductor material that has a negative temperature coefficient. The thermistor is the most common type of temperature sensor and produces rapid and extremely large changes in resistance for very small changes in temperature.

20. *Photo* means illumination, and the *photoresistor* is a resistor that is photoconductive. This means that as the material is exposed to light, it will become more conductive and less resistive.

Measuring Resistance (Figure 3-29)

21. To use the ohmmeter to measure resistance, follow the procedure described in Figure 3-29.

22. When measuring resistance with an ohmmeter, turn off the circuit power and remove the component (if practicable) to be measured from the circuit to protect the ohmmeter.

QUICK REFERENCE SUMMARY SHEET

FIGURE 3-29   Measuring Resistance.

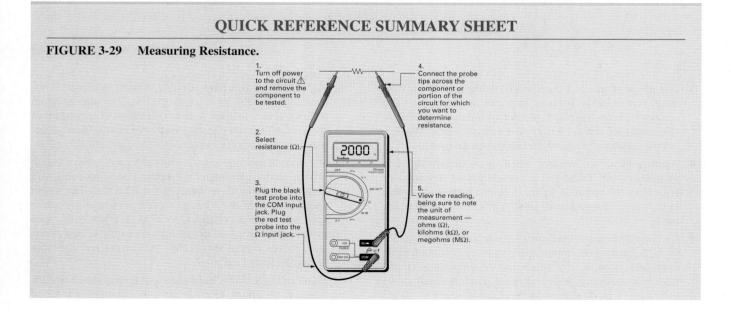

FIGURE 3-30   Resistor Color Code.

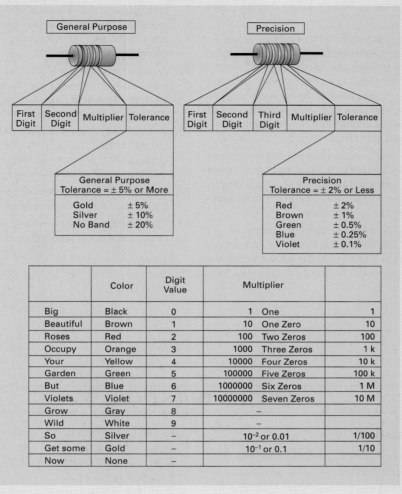

	Color	Digit Value	Multiplier	
Big	Black	0	1 One	1
Beautiful	Brown	1	10 One Zero	10
Roses	Red	2	100 Two Zeros	100
Occupy	Orange	3	1000 Three Zeros	1 k
Your	Yellow	4	10000 Four Zeros	10 k
Garden	Green	5	100000 Five Zeros	100 k
But	Blue	6	1000000 Six Zeros	1 M
Violets	Violet	7	10000000 Seven Zeros	10 M
Grow	Gray	8	–	
Wild	White	9	–	
So	Silver	–	10^{-2} or 0.01	1/100
Get some	Gold	–	10^{-1} or 0.1	1/10
Now	None	–		

Resistor Color Code (Figure 3-30)

23. There are basically two different types of fixed resistors: general purpose and precision. Resistors with tolerances of $\pm 2\%$ or less are *precision resistors* and have five bands. Resistors with tolerances of $\pm 5\%$ or greater have four bands and are referred to as *general-purpose resistors.*

24. For the general-purpose resistor code:
  a. The first band on either a general-purpose or precision resistor can never be black, and it is the first digit of the number.
  b. The second band indicates the second digit of the number.
  c. The third band specifies the multiplier to be applied to the number, which ranges from $\times \frac{1}{100}$ to 10,000,000.
  d. The fourth band describes the tolerance or deviation from the specified resistance, which is $\pm 5\%$ or greater.

25. For the precision resistor code:
  a. The first band, like the general-purpose resistor, is never black and is the first digit of the three-digit number.
  b. The second band provides the second digit.
  c. The third band indicates the third and final digit of the number.
  d. The fourth band specifies the multiplier to be applied to the number.
  e. The fifth and final band indicates the tolerance figure of the precision resistor, which is always less than $\pm 2\%$, which is why precision resistors are more expensive than general-purpose resistors.

26. If a resistor color code is not found, some other form of marking will always be made on the resistor. The larger wirewound resistors and nearly all variable resistors normally have the resistance value, tolerance, and wattage printed on the resistor. SIP and DIP packages are normally coded with both letters and numerals, all of which have particular meaning. Chip resistors will have to be measured with an ohmmeter. They are packaged in either polythene bags, paper tape, or plastic magazines, and in all these cases the packaging can be used to specify value, tolerance, and wattage.

27. The atoms of a resistor obstruct the flow of moving electrons (current), and this friction between the two causes heat to be generated. Current flow through any resistance therefore will always be accompanied by heat. In some instances, such as a heater or oven, the device has been designed specifically to generate heat. In other applications it may be necessary to include a resistor to reduce current flow, and the heat generated will be an unwanted side effect.

28. Any one of the three power formulas can be used to calculate the power dissipated by a resistance. However, since power is determined by the friction between current (I) and the circuit's resistance (R), the $P = I^2 \times R$ formula is more commonly used to calculate the heat generated.

29. The *filament resistor* is more commonly known as the household light bulb. The filament resistor is just a coil of wire that glows white hot when current is passed through it and, in so doing, dissipates both heat and light energy.

30. This *incandescent lamp* is an electric lamp in which electric current flowing through a filament of resistive material heats the filament until it glows and emits light.

31. The filament material (tungsten, for example) is like all other conductors in that it has a positive temperature coefficient of resistance. Therefore, as the current through the filament increases, so does the temperature and so does the filament's resistance ($I\uparrow$, temperature $\uparrow$, $R\uparrow$).

32. A coil of wire similar to the one found in a light bulb, called a *ballast resistor,* is used to maintain a constant current despite variation in voltage. Current is regulated or maintained constant by the ballast resistor, despite variations in voltage.

33. It is virtually impossible for fixed-value resistors to short-circuit internally. Generally, the resistor's internal resistive elements will begin to develop a higher resistance than its specified value or in some cases go completely open circuit.

34. Variable-value resistors have problems with the wiper making contact with the resistive track at all times.

35. The ohmmeter is the ideal instrument for verifying whether or not a resistor is functioning correctly.

REVIEW QUESTIONS

Multiple-Choice Questions

1. The Greek letter pi represents a constant value that describes how much bigger a circle's _____ is compared with its _____, and it is frequently used in many formulas involved with the analysis of circular motion.
   a. Circumference, radius
   b. Circumference, diameter
   c. Diameter, radius
   d. Diameter, circumference

2. Is π a ratio?
   a. Yes   b. No

3. Round off the following number to the nearest hundredth: 74.8552.
   a. 74.85   c. 74.84
   b. 74.86   d. 74.855

4. The number 27.0003 is a _____-significant-place number and is accurate to _____ significant places.
   a. 6, 2   c. 6, 6
   b. 2, 6   d. 4, 6

5. Express 29.63% mathematically as a decimal.
   a. 29.63   c. 0.02963
   b. 2.963   d. 0.2963

6. The most common type of fixed-value resistor is the:
   a. Wirewound   c. Film
   b. Carbon composition   d. Metal oxide

7. A SIP package has:
   a. A single line of connectors   c. No connectors
   b. A double line of connectors   d. None of the above

8. The most commonly used mechanically adjustable variable-value resistor is the:
   a. Rheostat   c. Potentiometer
   b. Thermistor   d. Photoresistor

9. A thermistor has a negative temperature coefficient, which means that:
   a. As temperature increases, resistance increases.
   b. As temperature increases, conductance decreases.
   c. As temperature increases, resistance decreases.
   d. All of the above are true.

10. A _____ band resistor is called a general-purpose resistor, while a _____ band resistor is known as a precision resistor.
    a. Four, five   c. One, four
    b. Three, four   d. Two, seven

11. What would be the power dissipated by a 2 kΩ carbon composition resistor when a current of 20 mA is flowing through it?
    a. 40 W   c. 1.25 W
    b. 0.8 W   d. None of the above

12. Which of the following is the most common type of temperature detector?
    a. RTD   c. Thermistor
    b. TFD   d. Barretter

13. A rheostat is a _____ terminal device, while the potentiometer is a _____ terminal device.
    a. 2, 3   c. 2, 4
    b. 1, 2   d. 3, 2

14. Which of the following is the unit of temperature in the International System of Units?
    a. Celsius   c. Kelvin
    b. Fahrenheit   d. Rankine

15. Which of the following temperature detectors has a positive temperature coefficient?
    a. RTD   d. Both (a) and (b)
    b. Thermistor   e. Both (a) and (c)
    c. TFD

16. Photoresistors are also called:
    a. TFDs   c. RTDs
    b. LDRs   d. TGFs

17. Photoresistors have:
    a. A positive temperature coefficient
    b. A negative temperature coefficient
    c. Neither (a) nor (b)

18. Resistance is measured with a (an):
    a. Wattmeter   c. Ohmmeter
    b. Milliammeter   d. A 100 meter

19. Which of the following is *not* true?
    a. Precision resistors have a tolerance of > 5%.
    b. General-purpose resistors can be recognized because they have either three or four bands.
    c. The fifth band indicates the tolerance of a precision resistor.
    d. The third band of a general-purpose resistor specifies the multiplier.

20. The first band on either a general-purpose or a precision resistor can never be:
    a. Brown   c. Black
    b. Red or black stripe   d. Red

21. What is the name of the resistor used to maintain a constant current despite variations in voltage by changing its resistance?
    a. RTD   c. Ballast resistor
    b. Thermistor   d. LDR

22. The term *infinite ohms* describes:
    a. A small finite resistance
    b. A resistance so large that a value cannot be placed on it
    c. A resistance between maximum and minimum
    d. None of the above

23. Resistance is always measured when:
    a. Circuit power is ON.
    b. Circuit power is OFF.
    c. The ohmmeter selector switch is in the A position.
    d. The ohmmeter is connected in the path of current.

24. The typical problem with resistors is that they will:
    a. Develop a partial internal open
    b. Completely open circuit
    c. Develop a higher resistance than specified
    d. All of the above
    e. Both (a) and (c)

25. A 10% tolerance, 2.7 MΩ carbon composition resistor measures 2.99 MΩ when checked with an ohmmeter. It is:
    a. Within tolerance   d. Both (a) and (c)
    b. Outside tolerance   e. Both (a) and (b)
    c. Faulty

Communication Skill Questions

26. What is a ratio, and how is it expressed? (3-1)

27. Briefly describe the rules for rounding off. (3-1)

28. What is a percentage? (3-1)

29. How is an average calculated? (3-1)

30. What is the approximate value of the constant pi (π), and what does it describe? (3-1)

31. Describe the difference between a fixed- and a variable-value resistor. (3-2)

32. List the six basic categories of fixed-value resistors. (3-2-1)

33. Explain why a resistor's size determines its wattage rating. (3-2-1)

34. Why are resistors given a tolerance figure, and which is best, a small or a large tolerance? (3-2-1)

35. Briefly describe the construction of:
    a. Carbon composition fixed-value resistors (3-2-1)
    b. Rheostats (3-2-2)
    c. Potentiometers (3-2-2)
    d. Multiturn precision potentiometers (3-2-2)

36. Draw the schematic symbol for a rheostat and potentiometer and describe the difference. Also show how a potentiometer can be used as a rheostat. (3-2-2)

37. Define linear and tapered resistive tracks. (3-2-2)

38. List the rules that should be applied when using an ohmmeter to measure resistance. (3-3)

39. Describe the construction, operation, and temperature coefficient of the following: (3-2-2)
    a. RTD   c. Thermistor
    b. Photoresistor   d. TFD

40. List the four temperature scales in use today. (3-2-2)

41. Give the color code for the following resistor values: (3-4)
    a. 1.2 MΩ, ± 10%   c. 27 kΩ, ± 20%
    b. 10 Ω, ± 5%   d. 273 kΩ, ±0.5%

42. Give the resistor values for the following color codes: (3-4)
  a. Orange, orange, black
  b. Red, red, green, red, red
  c. Brown, black, orange, gold
  d. White, brown, brown, silver

43. For what application would the zero-ohms resistor be used? (3-4-2)

44. Describe the difference between a SIP and a DIP resistor package. (3-2-1)

45. What are the common problems encountered with fixed- and variable-value resistors? (3-7)

Practice Problems

46. Express the ratio of the following in their lowest terms. (Remember to compare like quantities.)
  a. The ratio of 20 feet to 5 feet
  b. The ratio of 2½ minutes to 30 seconds

47. Round off as indicated.
  a. 48.36 to the nearest whole number
  b. 156.3625 to the nearest tenth
  c. 0.9254 to the nearest hundredth

48. Light travels in a vacuum at a speed of 186,282.3970 miles per second. This _____-significant-place value is accurate to _____ significant places.

49. Calculate the following percentages.
  a. 23% of 50 = ?
  b. 78% of 10 = ?
  c. 3% of 1.5 = ?
  d. 20% of 3300 = ?
  e. 5% of 10,000 = ?

50. Calculate the average value of the following groups of values (include units).
  a. 4, 5, 6, 7, and 8
  b. 15 seconds, 20 seconds, 18 seconds, and 16 seconds
  c. 110 volts, 115 volts, 121 volts, 117 volts, 128 volts, and 105 volts
  d. 33 ohms, 17 ohms, 1000 ohms, and 973 ohms
  e. 150 meters, 160 meters, and 155 meters

51. Refer to Figure 3-31(a), and answer the following questions:
  a. Which music type has remained the most popular?
  b. Which music type is the least popular?
  c. Which music type became less popular as country music had an increase in popularity in 1992 and 1993?
  d. What was the approximate market share for each of the music types in 1995?

52. Refer to Figure 3-31(b), and answer the following questions:
  a. What is the main reason for the average person to connect to the Internet?
  b. In order, what are the top five reasons why people want to go online?

53. If a 5.6 kΩ resistor has a tolerance of ±10%, what would be the allowable deviation in resistance above and below 5.6 kΩ? (3-4)

54. (a) If a current of 50 mA is flowing through a 10 kΩ, 25 W resistor, how much power is the resistor dissipating? (b) Can the current be increased, and if so by how much? (3-4)

55. What minimum wattage size could be used if a 12 Ω wire-wound resistor were connected across a 12 V supply? (3-4)

56. Calculate the resistance deviation for the tolerances of the resistors listed in Problems 41 and 42. (3-2-1)

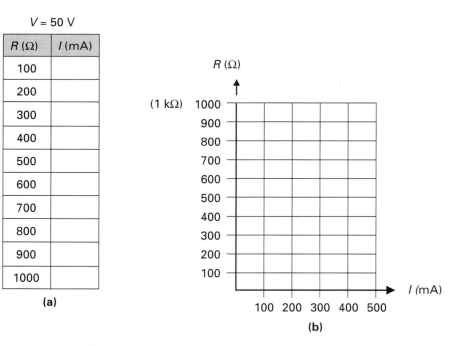

FIGURE 3-31 **Plotting a Rheostat's Current against Resistance.**

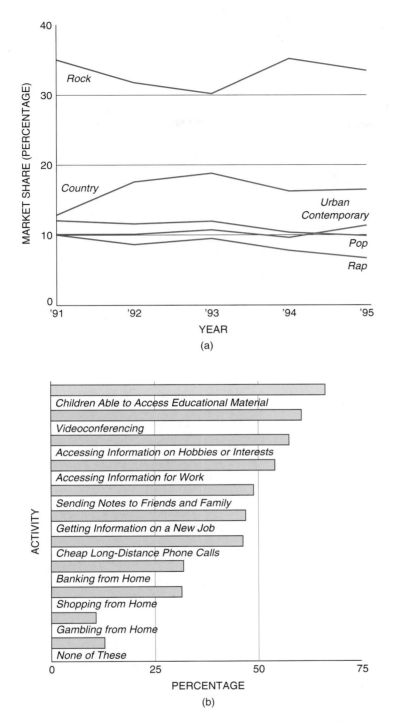

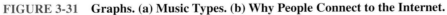

FIGURE 3-31   Graphs. (a) Music Types. (b) Why People Connect to the Internet.

57. If a 1 kΩ rheostat has 50 V across it, calculate the different values of current if the rheostat is varied in 100 Ω steps. Insert the values into the table in Figure 3-32(a), and then plot these values in the graph in Figure 3-32(b).

Troubleshooting Questions

58. Briefly describe some of the problems that can occur with fixed and variable resistors. (3-7)

59. Describe why the tolerance of a resistor can make you think that it has a problem when in fact it does not. (3-7)

60. If a 20 kΩ, 1 W, ±10% tolerance resistor measures 20.39 kΩ, is it in or out of tolerance? (3-2-1)

61. If 33 mA is flowing through a carbon film, 4.7 kΩ, ¼ W, ±5% resistor, will it burn up? (3-2-1)

62. What points should be remembered when using the ohmmeter to verify a resistor's value? (3-7)

Web Site Questions

Go to the web site http://www.prenhall. com/cook, select the textbook *Introductory DC/AC Electronics* or *Introductory DC/AC Circuits,* this chapter, and then follow the instructions when answering the multiple-choice practice problems.

These tests at the end of each chapter will challenge your knowledge up to this point, and give you the practice you need for a job interview. To make this more realistic, the test will be comprised of both technical and personal questions. In order to take full advantage of this exercise, you may want to set up a simulation of the interview environment, have a friend read the questions to you, and record your responses for later analysis.

Company Name: TRI, Inc.

Industry Branch: Industrial.

Function: Design and Manufacture Robotic Systems.

Job Title: Engineering Technician.

1. Would you say that your electronics education was complete?

2. What do you know about surface mount technology?

3. What is a tapered pot?

4. What is your attitude towards documentation?

5. What is a thermistor, and do you know what the international unit of temperature is?

6. Can you tell me the difference between hot resistance and cold resistance?

7. What would you say is the job function of an engineering technician?

8. Do you know anything about the Altera or Xilinx PLD application software?

9. Our systems use a wide variety of tranducers. What do you know about LDRs and RTDs?

10. Can you list the resistor color code in order?

Answers

1. The answer must be yes, but go on to explain that you see your education as an ongoing process, and like industry, you will always be in a learning mode.
2. Section 3-2-1.
3. Section 3-2-2.
4. Documenting the engineering process is vitally important. As an engineering technician you will be called on to document design successess, and failures, in explicit detail. They are testing here to see if you have a talent for technical writing, since this is a must in this position.
5. Section 3-2-2.
6. Section 3-6.
7. See introduction, Field Service Technician.
8. Always be prepared for the interviewer to test your limits. Be honest and say no if you have no idea, but show an interest to understand and learn. A good response would be, "No, are they electronic circuit simulation programs?"
9. Section 3-2-2.
10. Section 3-4. If you are not completely proficient at this, take a moment to quickly jot down the first letters of the memory aid, so that when you deliver your answer it will be smooth and accurate. Remember—less haste, more speed.

Direct Current (DC)

The First Computer Bug

Mathematician Grace Murray Hopper, an extremely independent U.S. naval officer, was assigned to the Bureau of Ordnance Computation Project at Harvard during World War II. As Hopper recalled, "We were not programmers in those days, the word had not yet come over from England. We were 'coders'," and with her colleagues she was assigned to compute ballistic firing tables on the Harvard Mark 1 computer. In carrying out this task, Hopper developed programming method fundamentals that are still in use.

Hopper is also credited, on a less important note, with creating a term frequently used today falling under the category of computer jargon. During the hot summer of 1945, the computer developed a mysterious problem. Upon investigation, Hopper discovered that a moth had somehow strayed into the computer and prevented the operation of one of the thousands of electromechanical relay switches. In her usual meticulous manner, Hopper removed the remains and taped and entered it into the logbook. In her own words, "From then on, when an officer came in to ask if we were accomplishing anything, we told him we were 'debugging' the computer," a term that is still used to describe the process of finding problems in a computer program.

4

Outline and Objectives

Introduction

Direct current, abbreviated dc, is a flow of continuous current in only one direction and it is what we have been dealing with up to this point. A dc current is produced when a dc voltage source, such as a battery, is connected across a closed circuit. The requirement for a dc voltage source is that its output voltage polarity remain constant. It is the fixed polarity of a dc voltage source that will produce a unidirectional (one direction) current through a circuit. In this chapter, we will see how a dc voltage can be generated using several different methods. We will also examine fuses, circuit breakers, switches, and the protoboard.

4-1   SOURCE AND LOAD

Source

Device that supplies the signal power or electric energy to a load.

Load

A component, circuit, or piece of equipment connected to the source. The load resistance will determine the load current.

Load Resistance

The resistance of the load.

Load Current

The current that is present in the load.

Figure 4-1(a) illustrates a dc circuit with a battery connected across a light bulb. In this circuit, the potential energy of the battery (voltage) produces kinetic energy (current) that is used to produce light energy from the bulb. The battery is the **source** in this circuit and the bulb is the **load.** By definition, a load is a device that absorbs the energy being supplied and converts it into the desired form. The **load resistance** will determine how hard the voltage source has to work. For example, the bulb filament has a resistance of 300 Ω. The connecting wires have a combined resistance of 0.02 Ω (20 mΩ), and since this value is so small compared to the filament's resistance, it will be ignored, as it will have no noticeable effect. Consequently, a 300 Ω load resistance will permit 10 mA of **load current** to flow when a 3 V source is connected across the dc circuit ($I = V/R = 3$ V/300 Ω = 10 mA). If you were to change the filament resistance by using a different bulb, the circuit current would be changed. To be specific, if the load resistance is increased, the load current decreases ($I \downarrow = V/R \uparrow$). Conversely, if the load resistance is decreased, the load current increases ($I \uparrow = V/R \downarrow$), assuming of course that the voltage source remains constant.

Figure 4-2 illustrates a heater, light bulb, motor, computer, robot, television, and microwave oven. All these and all other devices or pieces of equipment are connected to some source, such as a battery or dc power supply. This supply or source does not see all the circuitry and internal workings of the equipment; it simply sees the whole device or piece of

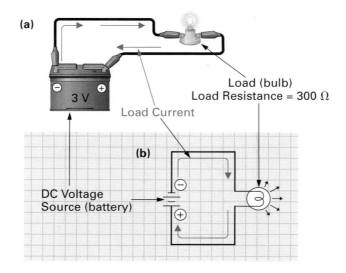

FIGURE 4-1   **Load on a Voltage Source. (a) Pictorial. (b) Schematic.**

    CHAPTER 4 / DIRECT CURRENT (DC)

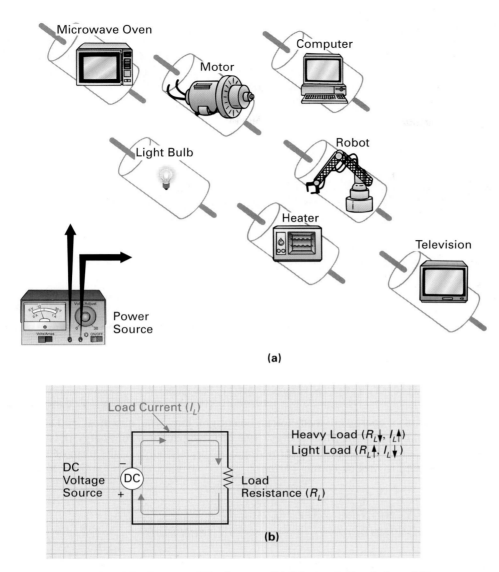

FIGURE 4-2 **(a) Load Resistance of Equipment. (b) Schematic Symbols and Terms.**

equipment as having some value of resistance. The resistance of the equipment is referred to as the device's *load resistance,* and it is this value of load resistance and the value of *source voltage* that determines the value of *load current.*

▦ EXAMPLE:

A small portable radio offers a 390 Ω load resistance to a 9 V battery source. What current will flow in this circuit?

▦ *Solution:*

$$\text{current } (I) = \frac{\text{voltage } (V)}{\text{resistance } (R)}$$

$$= \frac{9 \text{ V}}{390 \text{ }\Omega}$$

$$= 23 \text{ mA}$$

In summary, the phrase *load resistance* describes the device or equipment's circuit resistance, whereas the phrase *load current* describes the amount of current drawn by the device or equipment. A device that causes a large load current to flow (due to its small load resistance) is called a large or heavy load because it is heavily loading down or working the supply or source. On the other hand, a device that causes a small load current to flow (due to its large load resistance) is referred to as a small or light load because it is only lightly loading down or working the supply or source. It can be said, therefore, that load current and load resistance are inversely proportional to one another.

SELF-TEST EVALUATION POINT FOR SECTION 4-1

Use the following questions to test your understanding of Section 4-1:

1. Define *load resistance, load current, dc voltage source,* and *dc current.*

2. True or false: A large load current will flow when a large load resistance is connected across the supply.

3. What is the relationship between load resistance and load current?

4. Describe a heavy load resistance.

4-2   DIRECT-CURRENT SOURCES

The six basic forms of energy are mechanical, heat, light, magnetic, chemical, and electrical. To produce direct current (electrical energy), one of the other forms of energy must be used as a source and then converted or transformed into electrical energy in the form of a dc voltage that can be used to produce direct current.

4-2-1   *Mechanically Generated DC: Mechanical ————→ Electrical*

Pressure

The application of force to something by something else in direct contact with it.

Piezoelectric Effect

The generation of a voltage between the opposite sides of a piezoelectric crystal as a result of pressure or twisting. Also, the reverse effect in which the application of voltage to opposite sides causes deformation to occur at the frequency of the applied voltage.

Mechanical energy in the form of **pressure** can be used to generate electrical energy. Quartz and Rochelle salts are solid compounds that will produce electricity when a pressure is applied, as shown in Figure 4-3. Quartz is a natural or artificially grown crystal composed of silicon dioxide, from which thin slabs or plates are carefully cut and ground. Rochelle salt is a crystal of sodium potassium tartrate. They are both piezoelectric crystals. **Piezoelectric effect** is the generation of a voltage between the opposite faces of a crystal as a result of pressure being applied. This effect is made use of in most modern cigarette lighters. A voltage is produced between the opposite sides of a crystal (usually a crystalline lead compound) when it is either struck, pressured, or twisted. The resulting voltage is discharged across a gap to ignite the gas, which is released simultaneously by the operating button. The flame then burns for as long as the button is depressed. Piezoelectric effect also describes the reverse of what has just been discussed, as it was discovered that crystals would not only generate a voltage when mechanical force was applied, but would generate mechanical force when a voltage was applied.

Figure 4-3(e) illustrates the physical appearance and operation of a piezoresistive pressure sensor. Piezoresistance is described as a change in resistance due to a change in the applied pressure. Unlike quartz, which generates a voltage, the piezoresistive sensor changes its resistance based on the applied pressure. Piezoresistive sensors need an excitation voltage applied, and it is this value that will determine the output voltage range. As an application, this sensor could be used to measure either the cooling system, hydraulic transmission, or fuel injection pressure in an automobile, which all vary in the 0 to 100 psi (pounds per

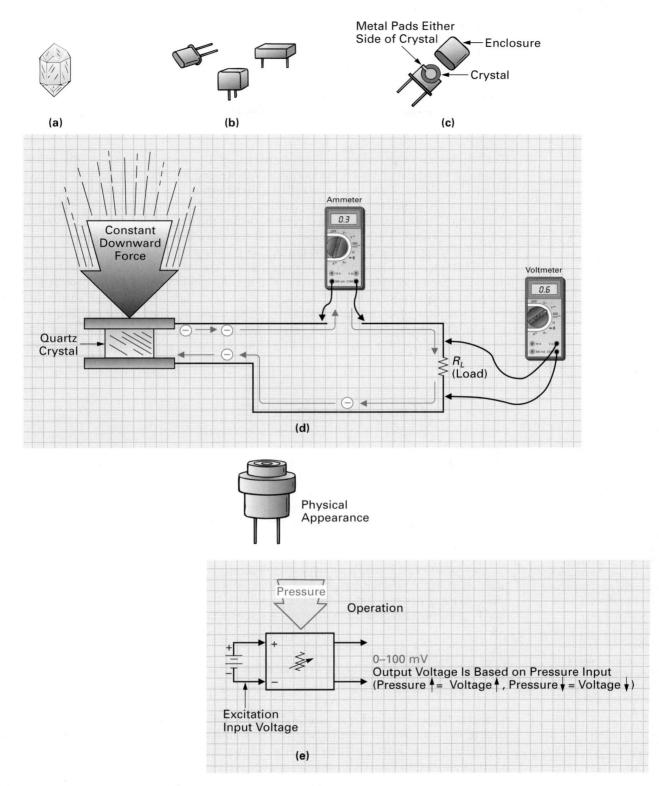

**FIGURE 4-3   Direct Current from Pressure Using Crystals. (a) Quartz Crystal.
(b) Physical Appearance. (c) Construction. (d) Operation. (e) Piezoresistive Sensor.**

square inch) range. The piezoresistive transducer would receive its excitation voltage from the car's battery (12 V) and produce a 0 to 100 mV output voltage, which would be sent to the car's computer, where the data would be analyzed and, if necessary, acted upon.

4-2-2 Thermally Generated DC: Heat ⟶ Electrical

Heat can be used to generate an electric charge, as illustrated in Figure 4-4(a). When two dissimilar metals, such as copper and iron, are welded together and heated, an electrical charge is produced. This junction is called a **thermocouple,** and the heat causes the release of electrons from their parent atoms, resulting in a meter deflection indicating the generation of a charge. The size of the charge is proportional to the temperature difference between the two metals. Figure 4-4(b) lists some typical thermocouples and their voltages generated when heat is applied, and shows the physical appearance of two types.

Thermocouples like the one in Figure 4-4(c) are normally used to indicate temperature on a display calibrated in degrees. The thermocouple seen in Figure 4-4(d) is being heated by the gas pilot light, and the electrical energy generated will allow a valve to open and let gas through to the water heater. If, by accident, the pilot is extinguished by a gust of wind, the thermocouple will not receive any more heat and therefore will not generate electrical energy, which will consequently close the electrically operated valve and prevent a large explosion.

4-2-3 Optically Generated DC: Light ⟶ Electrical

A **photovoltaic cell,** shown in Figure 4-5(a), also makes use of two dissimilar materials in its transformation of light to electrical energy. Figure 4-5(b) illustrates the construction and operation of a photovoltaic cell, which is also called a photoelectric cell or a **solar cell.**

A light-sensitive metal or semiconductor is placed behind a transparent piece of dissimilar metal. When the light illuminates the light-sensitive material, a charge is generated, causing current to flow and the meter to indicate current flow. This phenomenon is known as **photovoltaic action.**

To obtain a reasonable amount of power, solar cells are interconnected so that they work collectively on solar panels. These solar panels are used to power calculators, emergency freeway phones, and satellites, like the one shown in Figure 4-5(c). The power obtained on average is about 100 milliwatts per square centimeter.

4-2-4 Magnetically Generated DC: Magnetic ⟶ Electrical

A magnet like the one in Figure 4-6(a) can be used to generate electrical current by movement of the magnet past a conductor, as shown in Figure 4-6(b). When the magnet is stationary, no voltage is induced into the wire and therefore the light bulb is OFF. If the magnet is moved so that the invisible magnetic lines of force cut through the wire conductor, a voltage is induced in the conductor, and this voltage produces a current in the circuit causing the light bulb to emit light. An energy conversion from magnetic to electric can be achieved by moving either the magnet past the wire or the wire past the magnet.

To produce a continuous supply of electric current, the magnet or wire must be constantly in motion. The device that achieves this is the dc electric generator, illustrated in Figure 4-6(c).

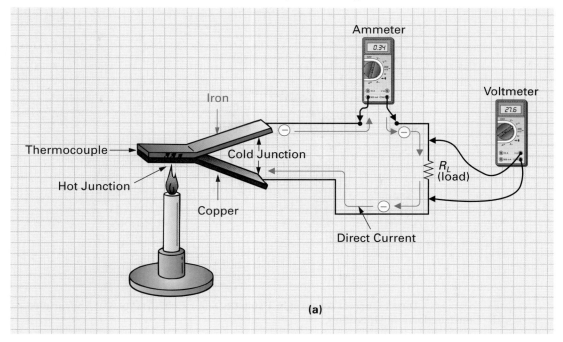

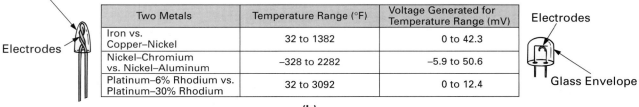

Two Metals	Temperature Range (°F)	Voltage Generated for Temperature Range (mV)
Iron vs. Copper–Nickel	32 to 1382	0 to 42.3
Nickel–Chromium vs. Nickel–Aluminum	–328 to 2282	–5.9 to 50.6
Platinum–6% Rhodium vs. Platinum–30% Rhodium	32 to 3092	0 to 12.4

(b)

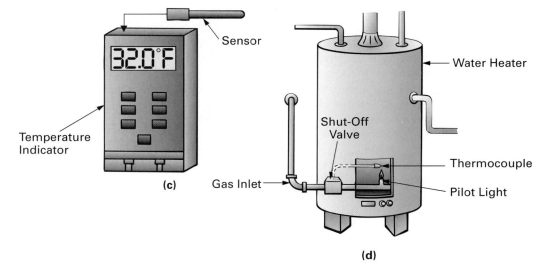

FIGURE 4-4   Direct Current from Heat Using the Thermocouple. (a) Operation.
(b) Typical Thermocouples. (c) Indicating Temperature. (d) Application: Water Heater.

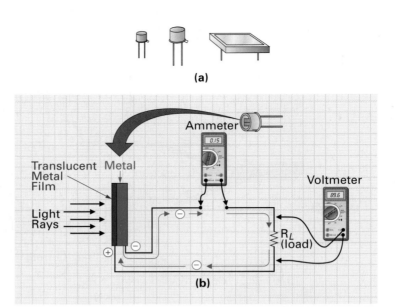

(a)

(b)

FIGURE 4-5 **Direct Current from Light Using the Solar Cell. (a) Physical Appearance. (b) Construction and Operation.**

4-2-5   *Chemically Generated DC:*
Chemical ──────────→ Electrical

The battery is a chemical voltage source that contains one or more voltaic cells. The **voltaic cell** was discovered by Alessandro Volta, an Italian physicist, in 1800. Figure 4-7 shows the three basic components within a battery's cell, which are a negative plate, a positive plate, and an electrolyte.

Cell Operation

Two dissimilar, separated metal plates such as copper and zinc are placed within a container that is filled with a substance known as the **electrolyte,** usually an acid. A chemical reaction causes electrons to pass through the electrolyte as they are repelled from one plate and attracted to the other. This action causes a large number of electrons to collect on one plate (negative plate), and an absence or deficiency of electrons to exist on the opposite plate (positive plate). The electrolyte therefore acts on the two plates and transforms chemical energy into electrical energy, which can be taken from the cell at its two output terminals as an electrical current flow.

If nothing is connected across the battery, as shown in Figure 4-8(a), a chemical reaction between the electrolyte and the negative electrode produces free electrons that travel from atom to atom but are held in the negative electrode. On the other hand, if a light bulb is connected between the negative and positive electrodes, as shown in Figure 4-8(b), the mutual repulsion of the free electrons at the negative electrode combined with the attraction of the positive electrode will cause a migration of free electrons (current flow) through the light bulb, causing its filament to produce light.

Primary Cells

The chemical reaction within the cell will eventually dissolve the negative plate. This discharging process also results in hydrogen gas bubbles forming around the positive plate, causing a resistance between the two plates, known as the battery's internal resistance, to increase (called *polarization*). To counteract this problem, all dry cells have a chemical within them known as a depolarizer agent, which reduces the buildup of these gas bubbles. With

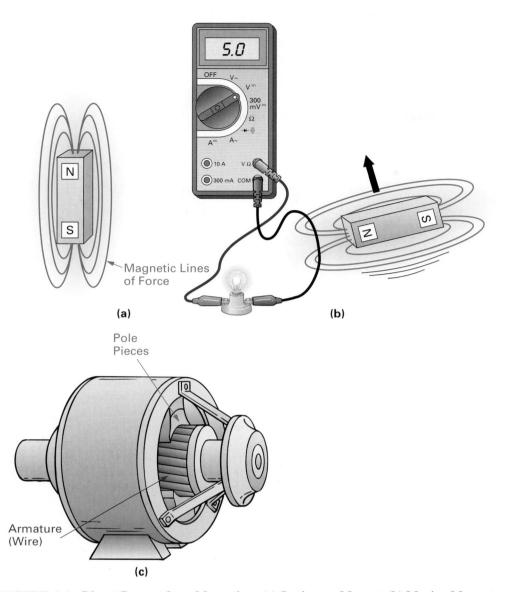

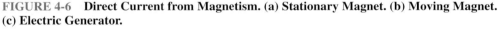

FIGURE 4-6   Direct Current from Magnetism. (a) Stationary Magnet. (b) Moving Magnet. (c) Electric Generator.

time, however, the depolarizer's effectiveness will be reduced, and the battery's internal resistance will increase as the battery reaches a completely discharged condition. These types of batteries are known as **primary cells** (first time, last time) because they discharge once and must then be discarded. Almost all primary cells have their electrolyte in paste form, which is why they are also referred to as **dry cells.** Wet cells have an electrolyte that is in liquid form.

     Primary cell types.   The shelf life of a battery is the length of time a battery can remain on the shelf (in storage) and still retain its usability. Most primary cells will deteriorate in a three-year period to approximately 80% of their original capacity. Of all the types of primary cells, five dominate the market as far as applications and sales are concerned. These are the carbon–zinc, alkaline–manganese, mercury, silver oxide, and lithium types, all of which are described and illustrated in Table 4-1. The cell voltage in Table 4-1 describes the voltage produced by one *cell* (one set of plates). If two or more sets of plates are installed in one package, the component is called a *battery.*

Primary Cell

Cell that produces electrical energy through an internal electrochemical action; once discharged, it cannot be reused.

Dry Cell

DC voltage-generating chemical cell using a nonliquid (paste) type of electrolyte.

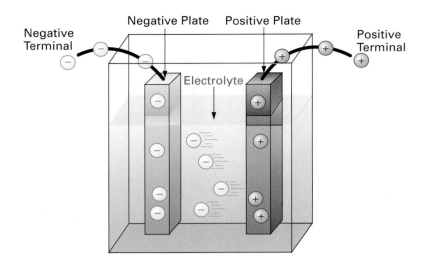

FIGURE 4-7 **A Battery's Cell.**

Secondary Cells

Secondary cells operate on the same basic principle as that of primary cells. In this case, however, the plates are not eaten away or dissolved, they only undergo a chemical change during discharge. Once discharged, the secondary cell can have the chemical change that occurred during discharge reversed by recharging, resulting once more in a fully charged cell. Restoring a secondary cell to the charged condition is achieved by sending a current through the cell in the direction opposite to that of the discharge current, as illustrated in Figure 4-9.

In Figure 4-9(a), we have first connected a light bulb across the battery, and free electrons are being supplied by the negative electrode and moving through the light bulb or load and back to the positive electrode. After a certain amount of time, any secondary cell will run down or will have discharged to such an extent that no usable value of current can be produced. The surfaces of the plates are changed, and the electrolyte lacks the necessary chemicals.

In Figure 4-9(b), during the recharging process, we use a battery charger, which reverses the chemical process of discharge by forcing electrons back into the cell and restoring the battery to its charged condition. The battery charger voltage is normally set to about 115% of the battery voltage. To use an example, a battery charger connected across a 12 V

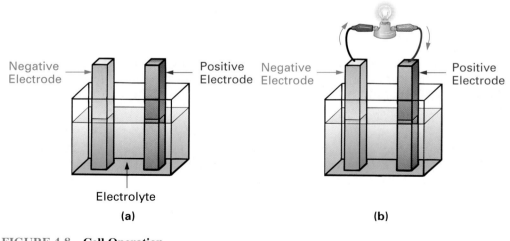

FIGURE 4-8 **Cell Operation.**

TABLE 4-1 Primary Cell Types

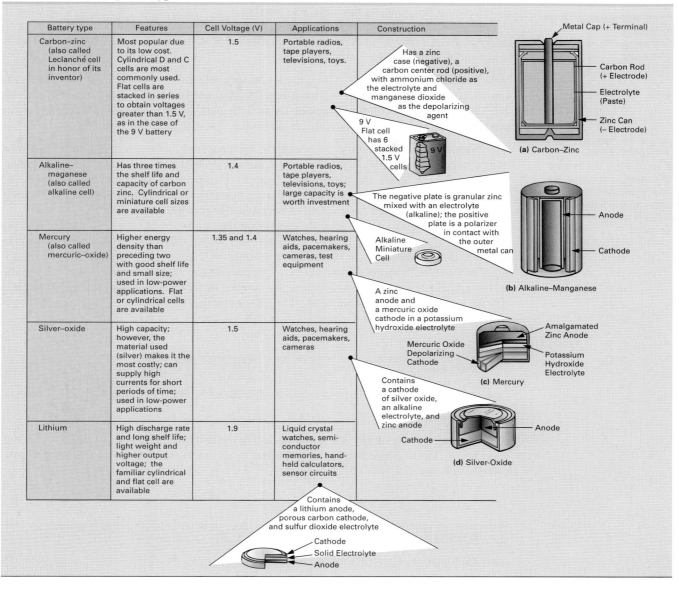

Battery type	Features	Cell Voltage (V)	Applications	Construction
Carbon–zinc (also called Leclanché cell in honor of its inventor)	Most popular due to its low cost. Cylindrical D and C cells are most commonly used. Flat cells are stacked in series to obtain voltages greater than 1.5 V, as in the case of the 9 V battery	1.5	Portable radios, tape players, televisions, toys.	
Alkaline–maganese (also called alkaline cell)	Has three times the shelf life and capacity of carbon zinc. Cylindrical or miniature cell sizes are available	1.4	Portable radios, tape players, televisions, toys; large capacity is worth investment	
Mercury (also called mercuric–oxide)	Higher energy density than preceding two with good shelf life and small size; used in low-power applications. Flat or cylindrical cells are available	1.35 and 1.4	Watches, hearing aids, pacemakers, cameras, test equipment	
Silver–oxide	High capacity; however, the material used (silver) makes it the most costly; can supply high currents for short periods of time; used in low-power applications	1.5	Watches, hearing aids, pacemakers, cameras	
Lithium	High discharge rate and long shelf life; light weight and higher output voltage; the familiar cylindrical and flat cell are available	1.9	Liquid crystal watches, semi-conductor memories, hand-held calculators, sensor circuits	

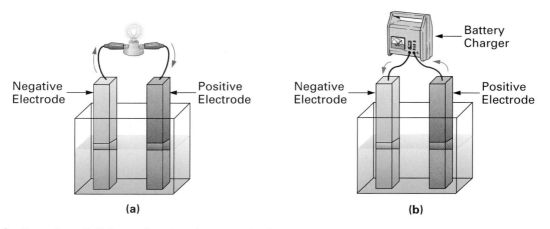

FIGURE 4-9 Secondary Cell Operation. (a) Discharge. (b) Charge.

battery would be set to 115% of 12 V, or 13.8 V. If this voltage is set too high, an excessive current can result, causing the battery to overheat.

Secondary cells are often referred to as **wet cells** because the electrolyte is not normally in paste form (dry) as in the primary cell, but is in liquid (wet) form and is free to move and flow within the container.

Secondary cell types.   Of all the secondary cells on the market today, the lead–acid and nickel–cadmium are the two most popular, and are illustrated and described in Table 4-2. Other secondary cell types include the nickel–metal hydride and lithium ion, which are used in cellular telephones, camcorders, and other portable electronic systems.

Ampere-Hours

If you multiply the amount of current, in amperes, that can be supplied for a given time frame in hours, a value of ampere-hours is obtained.

Secondary cell capacity.   Capacity (C) is measured by the amount of **ampere-hours** (Ah) a battery can supply during discharge. For example, if a battery has a discharge capacity of 10 ampere-hours, the battery could supply 1 ampere for 10 hours, 10 amperes for 1 hour, 5 amperes for 2 hours, and so on, during discharge. Automobile batteries (12 V) will typically have an ampere-hour rating of between 100 and 300 Ah.

Ampere-hour units are actually specifying the coulombs of charge in the battery. For example, if a lead–acid car battery was rated at 150 Ah, this value would have to be converted to ampere-seconds to determine coulombs of charge, as 1 ampere-second (As) is equal to 1 coulomb. Ampere-hours are easily converted to ampere-seconds simply by multiplying ampere-hours by 3600, which is the number of seconds in 1 hour. In our example, a 150 Ah battery will have a charge of 54×10^4 coulombs.

EXAMPLE:

How many coulombs of charge does a 3.8 Ah lead-acid gel battery store?

■   *Solution:*

A 3.8 Ah battery can supply 3.8 A in 1 hour. This would be equivalent to 13,680 A in a second (3.8×3600). Since current is measured in coulombs/second, this battery when fully charged can store 13.68×10^3 C.

EXAMPLE:

A Ni-Cd battery is rated at 300 Ah and is fully discharged.

a. How many coulombs of charge must the battery charger put into the battery to restore it to full charge?

b. If the charger is supplying a charging current of 3 A, how long will it take the battery to fully charge?

c. Once fully charged, the battery is connected across a load that is pulling a current of 30 A. How long will it take until the battery is fully discharged?

■   *Solution:*

a. The same amount that was taken out: $300 \text{ Ah} \times 3600 = 1080 \times 10^3$ C

b. $\dfrac{300 \text{ Ah}}{3 \text{ A}} = 100$ hours until charged

c. $\dfrac{300 \text{ Ah}}{30 \text{ A}} = 10$ hours until discharged

---

Table 4-2 Secondary Cell Types.

Battery type	Features	Applications	Construction
Lead–acid (lead cell)	High cycle (discharge–charge) life and high current capacity; 2.1 V per cell; must be operated upright if a wet lead–acid. Lead–acid gel cells can be used in any position	*Gel Cell Batteries* televisions, recorders, robots, alarms, and tools *Wet Cell Batteries* Starting source for automobiles and power for robotics equipment	The electrodes are lead oxide immersed in an electrolyte of dilute sulfuric acid Wet-type lead–acid battery with 6 internal 2.1 V cells making 12.6 V car battery Positive Plate / Insulating Separators (porous rubber) / Negative Plate Car Battery
Nickel–cadmium (Ni–Cd)	High capacity and high cost; three times discharge current of lead–acid for same amp-hour rating; can be sealed and operated in any position, so ideal for drills and other such portable equipment; 1.2 V per cell. Nickel cadmium batteries suffer from "memory effect." Short discharge and charge cycles cause the batteries' capacity to drop, because the cell appears to "remember" the lower capacity.	*Ni–Cd Gel Cell* televisions, radios, toys, recorders, shavers, toothbrushes *Ni–Cd Wet Cell Batteries* Starter source for jet engines (resembles lead–acid car battery in appearance). Robotics equipment	The positive plate is made of cadmium, the negative plate of nickel hydroxide, and the electrolyte used is potassium hydroxide Negative Sintered Plate / Separator / Positive Sintered (applied in powered form) Plate / Nickel-plated Steel Jacket
Nickel–metal Hydride (Ni–MH) and Lithium Ion	Discharge characteristics are very similar to those of the nickel–cadmium cell. The cell's nominal voltage is 1.2 V, and it does not suffer from memory effect.	Biggest applications include cellular phones, portable computers, camcorders, and other portable electronic systems	(+) Positive Terminal / Positive Electrode / Separator / Negative Electrode / (−) Negative Terminal

Batteries in Series and Parallel

Batteries are often connected in combination to gain a higher voltage or current than can be obtained from one cell.

Referring to Figure 4-10, you can see that when batteries are connected in series, the negative terminal of A is connected to the positive terminal of B. The total voltage across the bulb in this configuration will be the sum of the two cell voltages, which in this case will be 18 V.

Referring to Figure 4-11, you can see that when batteries are connected in parallel, the negative terminal connects to the other negative terminal and the positive terminal connects to the positive. The voltage in this configuration remains the same as for one cell. However, the current demand is now shared, and so the combined parallel batteries have the ability to supply a higher value of current if the load demands it.

4-2-6 *Electrically Generated DC*

Up until now we have only discussed electrical energy in the form of direct current (dc). However, as we will discover in Chapter 8, electrical energy is more readily available in the form of alternating current (ac). In fact, that is what arrives at the wall receptacle at your home and workplace. We can use a dc power supply, like the one seen in Figure 4-12 to convert the ac electrical energy at the wall outlet to dc electrical energy for experiments.

Power in the laboratory can be obtained from a battery, but since ac power is so accessible, dc power supplies are used. The power supply has advantages over the battery as a source of dc voltage in that it can quickly provide an accurate voltage that can easily be varied by a control on the front panel, and it never runs down.

During your electronic studies, you will use a power supply in the lab to provide different voltages for various experiments. In Figure 4-12 you can see how a dc power supply is being used to supply 12 V to a 12 kΩ resistor.

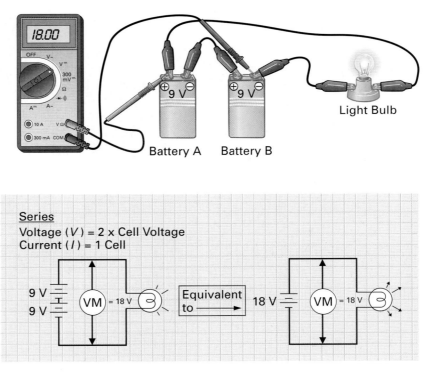

FIGURE 4-10 **Batteries in Series.**

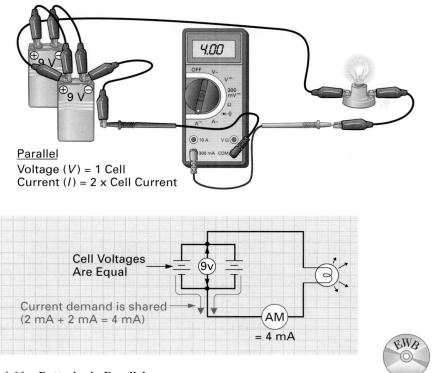

Parallel
Voltage (V) = 1 Cell
Current (I) = 2 x Cell Current

Cell Voltages
Are Equal

Current demand is shared
(2 mA + 2 mA = 4 mA)

9v

AM

= 4 mA

EWB

FIGURE 4-11   Batteries in Parallel.

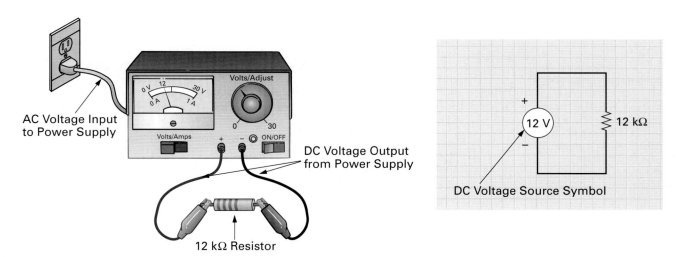

AC Voltage Input
to Power Supply

Volts/Adjust

0 V   12   30 V
0 A   1 A

Volts/Amps   +   −   ON/OFF

DC Voltage Output
from Power Supply

12 kΩ Resistor

12 V

+

−

12 kΩ

DC Voltage Source Symbol

FIGURE 4-12   Direct Current from Alternating Current.

Now that you have completed this section, you should be able to:

■ **Objective 1.** *Describe how direct current can be generated:*
  a. *Mechanically with pressure*
  b. *Thermally with a thermocouple*
  c. *Optically with a photocell*
  d. *Magnetically with a generator*
  e. *Chemically with a battery*
  f. *Electrically with a power supply*

■ **Objective 2.** *Describe the operation of both a primary and a secondary battery.*

■ **Objective 3.** *State the difference between a primary and a secondary cell.*

■ **Objective 4.** *Describe the features, applications, and construction of all the popular primary and secondary cells.*

■ **Objective 5.** *Explain the advantages of connecting batteries in series and parallel.*

Use the following questions to test your understanding of Section 4-2:

1. What is the difference between a piezoresistive device and a piezoelectric device?

2. What is a thermocouple?

3. How is light converted into a direct current?

4. What device is used to convert magnetic energy into electrical energy?

5. List the three basic components in a battery.

6. True or false: If two 1.5 V cells were connected in series, the total source voltage would be 1.5 V.

4-3   EQUIPMENT PROTECTION

Current must be monitored and not be allowed to exceed a safe level so as to protect users from shock, to protect the equipment from damage, and to prevent fire hazards. There are two basic types of protective devices: fuses and circuit breakers. Let's now discuss each of these in more detail.

4-3-1   *Fuses*

Fuse

This circuit- or equipment-protecting device consists of a short, thin piece of wire that melts and breaks the current path if the current exceeds a rated damaging level.

A **fuse** is an equipment protection device that consists of a thin wire link within a casing. Figure 4-13(a) shows the physical appearance of a fuse and Figure 4-13(b) shows the fuse's schematic symbol.

To protect a system, a fuse needs to disconnect power from the unit the moment current exceeds a safe value. To explain how this happens, Figure 4-13(c) shows how a fuse is mounted in the rear of a dc power supply. All fuses have a current rating, and this rating indicates what value of current will generate enough heat to melt the thin wire link within the fuse. In the example in Figure 4-13(c), if the current is less than the fuse rating of 2 amps, the metal link of the fuse will remain intact. If, on the other hand, the current through the thin metal element exceeds the current rating of 2 A, the excessive current will create enough heat to melt the element and "open" or "blow" the fuse, thus disconnecting the dc power supply from the 120 volt source and protecting it from damage.

One important point to remember is that if the current increases to a damaging level, the fuse will open and protect the dc power supply. This implies that something went wrong with the 120 V source and it started to supply too much current. This is almost never the case. What actually happens is that, as with all equipment, eventually something internally breaks down. This can cause the overall load resistance of the piece of equipment to increase or decrease. An increase in load resistance ($R_L\uparrow$) means that the source sees a higher resistance in the current path and therefore a small current would flow from the source ($I\downarrow$), and the user would be aware of the problem due to the nonoperation of the equipment. If an internal equipment breakdown causes the equipment's load resistance to decrease ($R_L\downarrow$), the

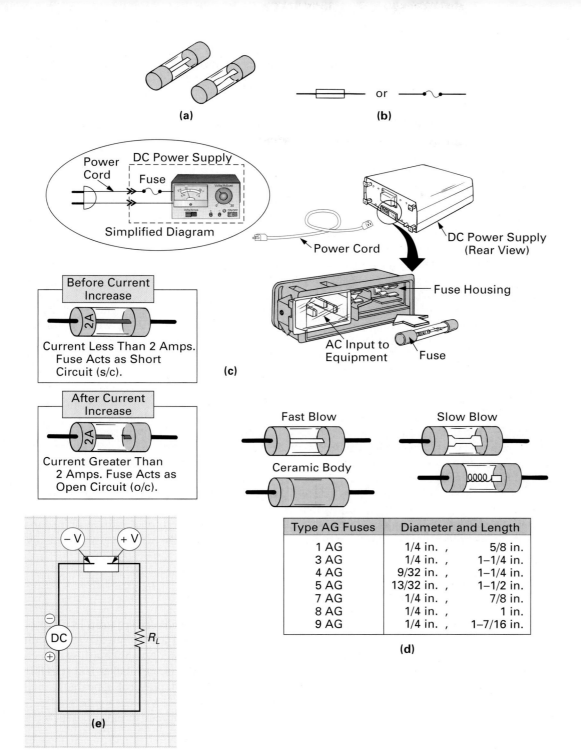

FIGURE 4-13 Fuses.

source will see a smaller circuit resistance and will supply a heavier circuit current ($I\uparrow$), which could severely damage the equipment before the user had time to turn it off. Fuse protection is needed to disconnect the current automatically in this situation, to protect the equipment from damage.

Fuse elements come in various shapes and sizes so as to produce either quick heating and then melting (fast blow) or delayed heating and then melting (slow blow), as illustrated in Figure 4-13(d). The reason for the variety is based in the application differences. When turned on, some pieces of equipment will be of such low resistance that a short momentary current surge, sometimes in the region of four times the fuse's current rating, will result for the first couple of seconds. If a fast-blow fuse were placed in the circuit, it would blow at the instant the equipment was turned on, even though the equipment did not need to be protected. The system merely needs a large amount of current initially to start up. A slow-blow fuse would be ideal in this application, as it would permit the initial heavy current and would begin to heat up and yet not blow due to the delay. Once the surge had ended, the current would decrease, and the fuse would still be intact (some slow-blow fuses will allow a 400% overload current to flow for a few seconds).

On the other hand, some equipment cannot take any increase in current without being damaged. The slow blow would not be good in this application, as it would allow an increase of current to pass to the equipment for too long a period of time. The fast blow would be ideal in this application, since it would disconnect the current instantly if the current rating of the fuse were exceeded, and so prevent equipment damage. The automobile was the first application for a fuse in a glass holder, and so a size standard named AG (automobile glass) was established. This physical size standard is still used today, and is listed in Figure 4-13(d).

Fuses also have a voltage rating that indicates the maximum circuit voltage that can be applied across the fuse by the circuit in which the fuse resides. This rating, which is important after the fuse has blown, prevents arcing across the blown fuse contacts, as illustrated in Figure 4-13(e). Once the fuse has blown, the circuit's positive and negative voltages are now connected across the fuse contacts. If the voltage is too great, an arc can jump across the gap, causing a sudden surge of current, damaging the equipment connected.

Fuses are mounted within fuse holders and normally placed at the back of the equipment for easy access, as was shown in Figure 4-13(c). When replacing blown fuses, you must be sure that the equipment is turned OFF and disconnected from the source. If you do not remove power you could easily get a shock, since the entire source voltage appears across the fuse as was shown in Figure 4-13(e). Also, make sure that a fuse with the correct current and voltage rating is used.

A good fuse should have a resistance of 0 Ω when it is checked with an ohmmeter. In circuit and with power ON, the fuse should have no voltage drop across it because it is just a piece of wire. A blown or burned-out fuse will, when removed, read infinite ohms, and when in its holder and power ON, will have the full applied voltage across its two terminals.

4-3-2   *Circuit Breaker*

Circuit Breaker

Reusable fuse. This device will open a current-carrying path without damaging itself once the current value exceeds its maximum current rating.

In many appliances and in a home electrical system, **circuit breakers** are used in place of fuses to prevent damaging current. A circuit breaker can open a current-carrying circuit without damaging itself, then be manually reset and used repeatedly, unlike a fuse, which must be replaced when it blows. We can say, therefore, that a circuit breaker is a reusable fuse. Figure 4-14 illustrates a circuit breaker's schematic symbol, physical appearance, and a typical application in which it is used to protect a television.

In this section we will examine the three basic types of circuit breakers, which are thermal type, magnetic type, and thermomagnetic type.

Thermal Circuit Breakers

The operation of this type of circuit breaker depends on temperature expansion due to electrical heating. Figure 4-15 illustrates the construction of a thermal circuit breaker. A U-shaped bimetallic (two-metal) strip is attached to the housing of the circuit breaker. This strip is composed of a layer of brass on one side and a layer of steel on the other. The current arrives at terminal A, enters the right side of the bimetallic U-shaped strip, leaves the left side, and travels to the upper contact, which is engaging the bottom contact, where it leaves and exits the circuit breaker from terminal B.

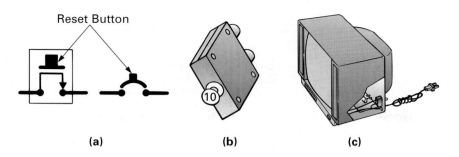

(a)   **(b)**   **(c)**

FIGURE 4-14   **Circuit Breaker. (a) Symbol. (b) Appearance. (c) Typical Application.**

If current were to begin to increase beyond the circuit breaker's current rating, heating of the bimetallic strip would occur. As with all metals, heat will cause the strip to expand. Some metals expand more than others, and this measurement is called the *thermal expansion coefficient*. In this situation, brass will expand more than steel, resulting in the lower end of the bimetallic strip bending to the right. This action allows a catch to release a pivoted arm. The arm's right side will be pulled down by the tension of a spring, lifting the left side of the arm and the attached top contact. The result is to separate the top contact surface from the bottom contact surface, thereby opening the current path and protecting the circuit from the excessive current. This action is also known as *tripping* the breaker. To reset the breaker, the reset button must be pressed to close the contacts. If the problem in the system still exists, however, the breaker will just trip once more, due to the excessive current.

Magnetic Circuit Breakers

Earlier in this chapter, we discussed how a generator converts a magnetic energy input into an electrical energy output. The magnetic type of circuit breaker makes use of a device that performs the opposite function. An electromagnet has a coil of insulated wire wrapped around a soft iron core, and it will generate a magnetic field whenever a current flows through its coil. The magnetic-type circuit breaker depends on the response of an electromagnet to break the circuit for protection.

The construction of a magnetic circuit breaker is shown in Figure 4-16, and it will operate as follows. A small level of current flowing through the coil of the electromagnet will

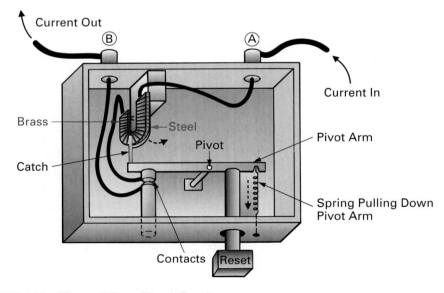

FIGURE 4-15   **Thermal-Type Circuit Breaker.**

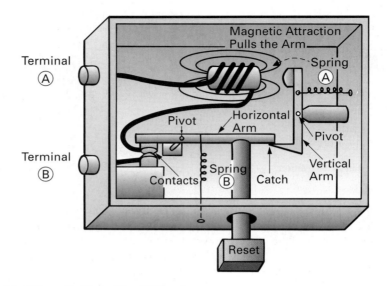

FIGURE 4-16   **Magnetic-Type Circuit Breaker.**

provide only a small amount of magnetic pull to the left on the iron arm. This magnetic force cannot overcome the pull to the right being generated by spring *A*. This safe value of current will therefore be allowed to pass from the *A* terminal, through the coil to the top contact, out of the bottom contact, and then exit the breaker at terminal *B*.

   If the current exceeds the current rating of the circuit breaker, an increase in current through the coil of the electromagnet generates a greater magnetic force on the vertical arm, which pulls the top half of the vertical arm to the left and the lower half below the pivot to the right. This action releases the catch holding the horizontal arm, allowing spring *B* to pull the right side of the lower arm down, opening the contact and disconnecting or tripping the breaker. The reset button must now be pressed to close the contacts. Once again, however, if the problem still exists, the breaker will continue to trip, as the excessive circuit current still exists.

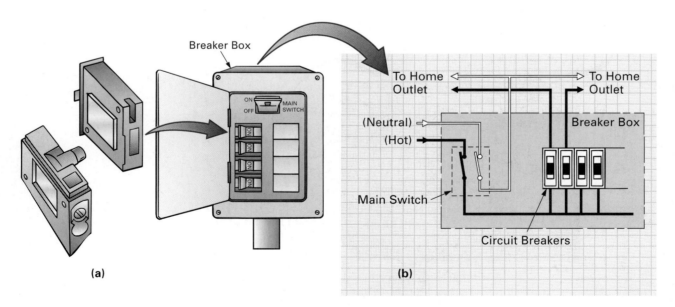

FIGURE 4-17   **Thermomagnetic-Type Circuit Breaker.**

Thermomagnetic Circuit Breakers

Figure 4-17(a) illustrates the typical residential-type thermomagnetic circuit breaker, and Figure 4-17(b) shows the physical appearance and schematic of a typical home breaker box. The thermal circuit breaker is similar to a slow-blow fuse in that it is ideal for passing momentary surges without tripping because of the delay caused by heating of the bimetallic strip. The magnetic circuit breaker, on the other hand, is most similar to a fast-blow fuse in that it is tripped immediately when an increase of current occurs.

The thermomagnetic circuit breaker combines the advantages of both previously mentioned circuit breakers by incorporating both a bimetallic strip and an electromagnet in its actuating mechanism. For currents at and below the current rating, the circuit breaker is a short circuit and connects the current source to the load. For momentary small overcurrent surges, the electromagnet is activated, but does not have enough force to trip the breaker. If overcurrent continues, however, the bimetallic strip will have heated and the combined forces of the bimetallic strip and electromagnet will trip the breaker and protect the load by disconnecting the source. Momentary large overcurrent surges will supply enough current to the electromagnet to cause it to trip the breaker independently of the bimetallic strip and so disconnect the source from the load.

4-4   SWITCHES

As we have seen previously, a *switch* is a device that completes (closes) or breaks (opens) the path of current, as seen in Figure 4-18. All mechanical switches can be classified into one of the eight categories shown in Figure 4-19. Many different variations of these eight different classifications exist.

SELF-TEST EVALUATION POINT FOR SECTION 4-3 AND 4-4

Now that you have completed this section, you should be able to:

■ **Objective 6.** *Describe the operation and use of various types of fuses, circuit breakers, and switches.*

Use the following questions to test your understanding of Sections 4-3 and 4-4:

1. What do the current and voltage ratings of a fuse indicate?

2. What is the difference between a slow-blow and a fast-blow fuse?

3. List the three types of circuit breakers.

4. List the eight basic types of switches.

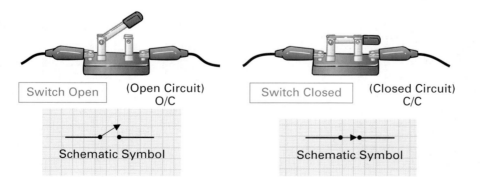

FIGURE 4-18   Open and Closed Switches.

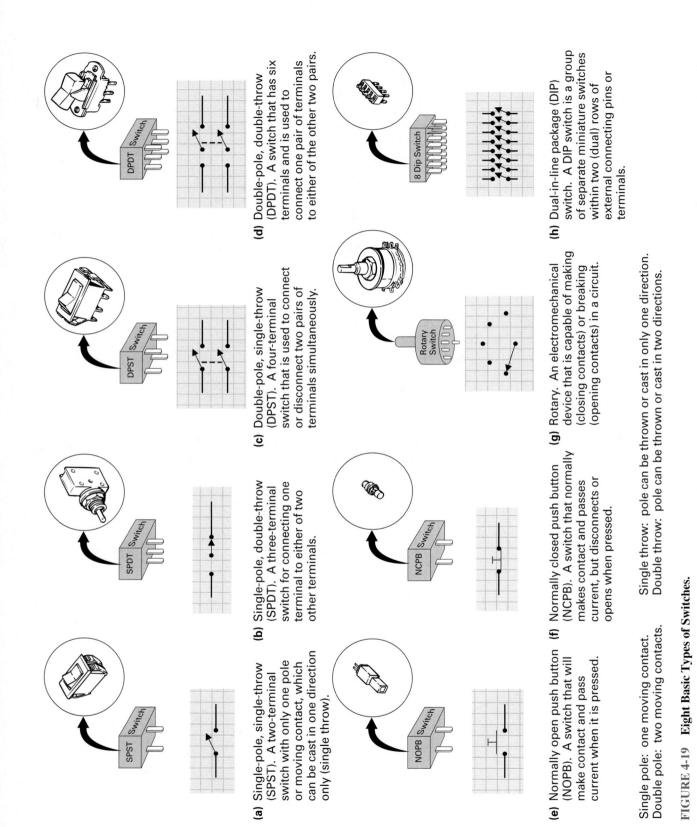

(a) Single-pole, single-throw (SPST). A two-terminal switch with only one pole or moving contact, which can be cast in one direction only (single throw).

(b) Single-pole, double-throw (SPDT). A three-terminal switch for connecting one terminal to either of two other terminals.

(c) Double-pole, single-throw (DPST). A four-terminal switch that is used to connect or disconnect two pairs of terminals simultaneously.

(d) Double-pole, double-throw (DPDT). A switch that has six terminals and is used to connect one pair of terminals to either of the other two pairs.

(e) Normally open push button (NOPB). A switch that will make contact and pass current when it is pressed.

(f) Normally closed push button (NCPB). A switch that normally makes contact and passes current, but disconnects or opens when pressed.

(g) Rotary. An electromechanical device that is capable of making (closing contacts) or breaking (opening contacts) in a circuit.

(h) Dual-in-line package (DIP) switch. A DIP switch is a group of separate miniature switches within two (dual) rows of external connecting pins or terminals.

Single pole: one moving contact.   Single throw: pole can be thrown or cast in only one direction.
Double pole: two moving contacts.   Double throw: pole can be thrown or cast in two directions.

FIGURE 4-19   Eight Basic Types of Switches.

4-5   PROTOBOARDS

The solderless prototyping board (**protoboard**) or breadboard is designed to accommodate the many experiments described in this textbook's associated lab manual. This protoboard will hold and interconnect resistors, capacitors, inductors, and many other components, as well as provide electrical power. Figure 4-20(a) shows an experimental circuit wired up on a protoboard.

Figure 4-20(b) shows the top view of a basic protoboard. As you can see by the cross section on the right side, electrical connector strips are within the protoboard. These conductive strips make a connection between the five hole groups. The bus strips have an electrical connector strip running from end to end, as shown in the cross section. They are usually connected to a power supply as seen in the example circuit in Figure 4-21(a). In this circuit, you can see that the positive supply voltage is connected to the upper bus strip, while the negative supply (ground) is connected to the lower bus strip. These power supply "rails" can then be connected to a circuit formed on the protoboard with hookup wire, as shown in Figure 4-21(a). In this example, three resistors are connected end-to-end, as shown in the schematic diagram in Figure 4-21(b).

Figure 4-22 illustrates how the multimeter could be used to make voltage, current, and resistance measurements of a circuit constructed on the protoboard.

Other protoboards may vary slightly as far as layout, but you should be able to determine the pattern of conductive strips by making a few resistance checks with an ohmmeter.

> **Protoboard**
> An experimental arrangement of a circuit on a board. Also called breadboard.

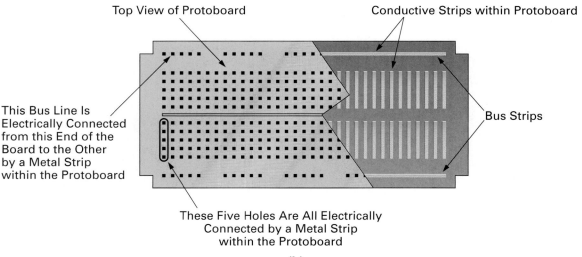

(a)

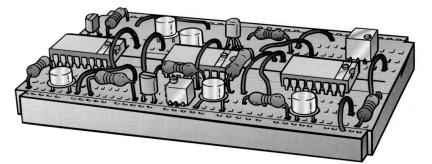

Top View of Protoboard

Conductive Strips within Protoboard

This Bus Line Is Electrically Connected from this End of the Board to the Other by a Metal Strip within the Protoboard

Bus Strips

These Five Holes Are All Electrically Connected by a Metal Strip within the Protoboard

(b)

FIGURE 4-20   Experimenting with the Protoboard.

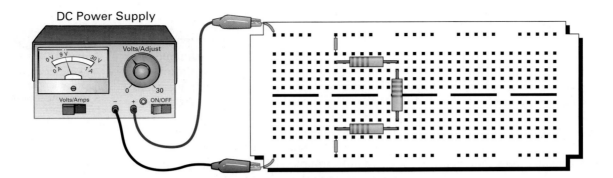

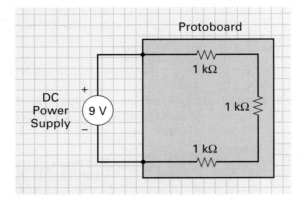

FIGURE 4-21 **Constructing Circuits on the Protoboard.**

SELF-TEST EVALUATION POINT FOR SECTION 4-5

Use the following questions to test your understanding of Section 4-5:

1. What is the purpose of the protoboard?

2. What is generally connected to the bus strips?

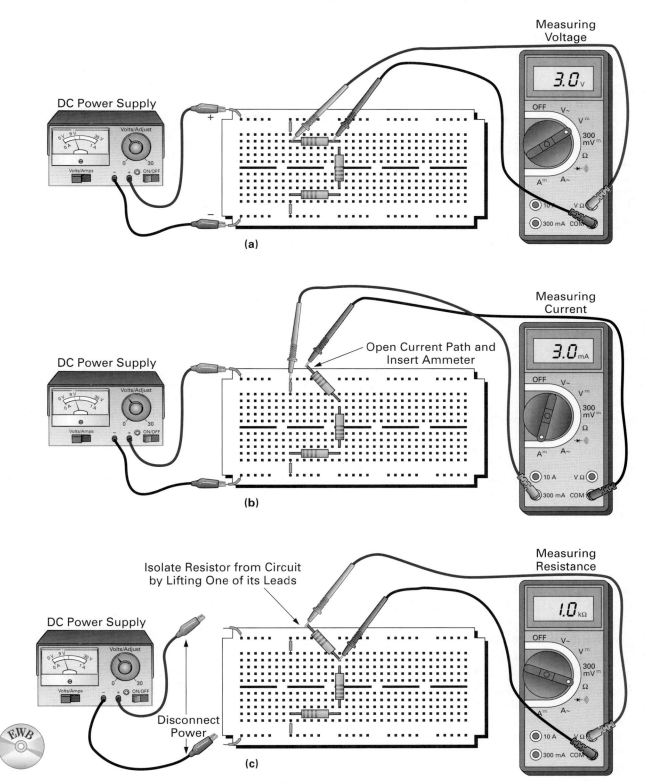

FIGURE 4-22 **Making Measurements of a Circuit on the Protoboard.**

SUMMARY

Source and Load

1. The phrase *load resistance* describes the device or equipment's circuit resistance, whereas the phrase *load current* describes the amount of current drawn by the device or equipment.

2. A device that causes a large load current to flow, due to its small load resistance, is called a large or heavy load because it is heavily loading down or working the supply or battery. On the other hand, a device that causes a small load current to flow, due to its large load resistance, is referred to as a small or light load because it is only lightly loading down or working the supply or battery.

3. Load current and load resistance are inversely proportional to one another.

4. Mechanical energy in the form of pressure can also be used to generate electrical energy.

5. Quartz is a natural or artificially grown crystal composed of silicon dioxide, from which thin slabs or plates are carefully cut and ground. Rochelle salt is a crystal of sodium potassium tartrate. They are both piezoelectric crystals. Piezoelectric effect is the generation of a voltage between the opposite faces of a crystal as a result of pressure being applied.

6. Piezoresistance is a change in resistance due to a change in the applied strain.

7. Unlike quartz, which generates a voltage, the piezoresistive sensor needs an excitation voltage applied, and its value will determine the value of output voltage.

8. Heat can be used to generate an electric charge.

9. When two dissimilar metals, such as copper and iron, are welded together and heated, an electrical charge is produced. This junction is called a thermocouple, and the heat causes the release of electrons from their parent atoms.

10. The size of the charge is proportional to the temperature difference between the two metals.

11. Thermocouples are normally used to indicate temperature on a display calibrated in degrees.

12. A photoelectric cell is also called a solar cell.

13. When the light illuminates the light-sensitive material, a charge is generated, causing current to flow.

14. This phenomenon, known as photovoltaic action, finds application as a light meter for photographic purposes and as an electrical power source for use by the satellite.

15. Electric power companies make use of solar cells to convert sunlight into electrical energy for the consumer.

16. A magnet can be used to generate electrical current by movement of the magnet past a conductor. When the magnet is stationary, no current flows through the circuit, so the current meter indicates this zero current. If the magnet is moved so that the magnetic lines of force cross the wire conductor, a voltage is induced in the conductor.

17. To achieve an energy conversion from magnetic to electric, you can either move the magnet past the wire or move the wire past the magnet.

18. To produce a continuous supply of electric current, the magnet or wire must be constantly in motion. The device that achieves this is the dc electric generator.

19. The chemical voltage source (the battery) is called the voltaic cell. Its principle of operation was discovered by Alessandro Volta, an Italian physicist, in 1800.

20. A battery has three basic components: a negative plate, a positive plate, and an electrolyte.

21. A chemical reaction causes electrons to be repelled from one plate and attracted to the other, passing through the electrolyte. A large number of electrons collect on one plate (negative plate), while an absence or deficiency of electrons exists on the opposite, positive plate.

22. The electrolyte acts on the two plates and transforms chemical energy into electrical energy, which can be taken from the cell at its two output terminals as an electrical current flow.

23. The mutual repulsion of the free electrons at the negative electrode combined with the attraction of the positive electrode results in a migration of free electrons (current flow) through the light bulb, causing its filament to produce light.

24. The chemical reaction within the cell will actually dissolve the negative plate until eventually it will be eaten away completely.

25. Primary (first time, last time) cells discharge once and must then be discarded. Almost all primary cells have their electrolyte in paste form and are therefore referred to as dry cells, as opposed to wet cells, whose electrolyte is in liquid form.

26. The shelf life of a battery is the length of time a battery can remain on the shelf (in storage) and still retain its usability.

27. The cell voltage describes the voltage produced by one cell (one set of plates). If two or more sets of plates are installed in one package, the component is called a battery.

28. Secondary cells operate on the same basic principle as that of primary cells. However, in this case the plates are not eaten away or dissolved. They only undergo a chemical change during discharge. Once discharged, the secondary cell can have the chemical change that occurred during discharge reversed by recharging, resulting once more in a fully charged cell.

29. The battery charger voltage is normally set to about 115% of the battery voltage.

30. Capacity (*C*) is measured by the amount of ampere-hours (Ah) a battery can supply during discharge.

31. Ampere-hour units are actually specifying the coulombs of charge in the battery.

32. Batteries are often connected together to gain a higher voltage or current than can be obtained from one cell.

33. When batteries are connected in series, the negative terminal of battery A is connected to the positive terminal of battery B. The total voltage across the load will be the sum of the two cell voltages.

34. When batteries are connected in parallel, the negative terminal connects to the other negative terminal and the positive terminal connects to the positive. The voltage remains the same as for one cell; however, the current demand can now be shared.

35. We can use ac electrical energy to generate dc electrical energy by use of a piece of equipment called a dc power supply.

36. The power supply has advantages over the battery as a source of voltage in that it can quickly provide an accurate voltage that can easily be varied by a control on the front panel, and it never runs down.

37. Current must be monitored and not be allowed to exceed a safe level so as to protect users from shock, to protect the equipment from damage, and to prevent fire hazards.

38. There are two basic types of protective devices: fuses and circuit breakers.

39. A fuse consists of a wire link or element of low melting point inside a casing. Fuses have a current rating that indicates the maximum amount of current they will allow to pass through the fuse to the equipment.

40. When current passes through the fuse, some of the electrical energy is transformed into heat. If the current through the thin metal element exceeds the current rating of the fuse, the excessive current will create enough heat to melt the element and "open" or "blow" the fuse, thus disconnecting the equipment and protecting it from damage.

41. An increase in load resistance ($R_L\uparrow$) means that the battery sees a higher resistance in the current path and therefore a small current would flow from the battery ($I\downarrow$), and the user would be aware of the problem due to the nonoperation of the equipment.

42. If an internal equipment breakdown causes the equipment's load resistance to decrease ($R_L\downarrow$), the battery will see a smaller circuit resistance and will supply a heavier circuit current, ($I\uparrow$), which could severely damage the equipment before the user had time to turn it off. Fuse protection is needed to disconnect the current automatically in this situation, to protect the equipment from damage.

43. Fuse elements come in various shapes and sizes so as to produce either quick heating and then melting (fast blow) or delayed heating and then melting (slow blow).

44. Fuses also have a voltage rating that indicates the maximum circuit voltage that can be applied across the fuse by the circuit in which the fuse resides. This rating, which is important after the fuse has blown, prevents arcing across the blown fuse contacts.

45. Fuses are mounted within fuse holders and normally placed at the back of the equipment for easy access.

46. When replacing fuses, you must be sure that the power is off, because a potential is across it, and that a fuse with the correct current and voltage rating is used.

47. In many appliances and in a home electrical system, circuit breakers are used in place of fuses to prevent damaging current.

48. A circuit breaker can open a current-carrying circuit without damaging itself, then be manually reset and used repeatedly, unlike a fuse, which must be replaced when it blows. There are three types of circuit breakers: thermal type, magnetic type, and thermomagnetic type.

49. The operation of thermal circuit breakers depends on temperature expansion due to electrical heating.

50. Magnetic circuit breakers depend on the response of an electromagnet to break the circuit for protection.

51. The thermal circuit breaker is similar to a slow-blow fuse in that it is ideal for passing momentary surges without tripping because of the delay caused by heating of the bimetallic strip.

52. The magnetic circuit breaker is most similar to a fast-blow fuse in that it is tripped immediately when an increase of current occurs.

53. The thermomagnetic circuit breaker combines both a bimetallic strip and an electromagnet as an actuating mechanism. For currents at and below the current rating, the circuit breaker is a short circuit and connects the current source to the load. For small momentary overcurrent surges, the electromagnet is activated, but does not have enough force to trip the breaker. However, if overcurrent continues, the bimetallic strip will have heated and the combined forces of the bimetallic strip and electromagnet will trip the breaker and disconnect the current source from the load. For a very large surge of current, the electromagnet receives enough current to trip the breaker independently of the bimetallic strip and disconnect the current source from the load.

54. A *switch* is a device that completes (closes) or breaks (opens) the path of current. All mechanical switches can be classified into one of eight categories, although many different variations of these eight different classifications exist.

REVIEW QUESTIONS

Multiple-Choice Questions

1. Direct current is:
   a. A reversing of current continually in a circuit
   b. A flow of current in only one direction
   c. Produced by an ac voltage
   d. None of the above

2. If load resistance were doubled, load current will _____.
   a. Halve
   b. Double
   c. Triple
   d. Remain the same

3. Quartz is a solid compound that will produce electricity when _____ is applied.
   a. Friction   c. A magnetic field
   b. An electrolyte   d. Pressure

4. Direct current is generated thermally by use of a:
   a. Crystal   c. Thermistor
   b. Thermocouple   d. None of the above

5. An application of a photovoltaic cell would be:
   a. A satellite power source
   b. To turn on and off security lights
   c. Both (a) and (b)
   d. None of the above

6. Which type of battery can be used for only one discharge (and cannot be rejuvenated)?
   a. A lead–acid cell   c. A nickel–cadmium cell
   b. A secondary cell   d. A primary cell

7. A battery with a capacity of 12 ampere-hours could supply:
   a. 12 A for 1 hour   c. 6 A for 2 hours
   b. 3 A for 4 hours   d. All of the above

8. Batteries are normally connected in series to obtain a higher total:
   a. Voltage   c. Resistance
   b. Current   d. None of the above

9. The dc power supply converts:
   a. ac to dc   c. High dc to a low dc
   b. dc to ac   d. Low ac to a high ac

10. A fuse's current rating states the:
    a. Maximum amount of current allowed to pass
    b. Minimum amount of current allowed to pass
    c. Maximum permissible circuit voltage
    d. None of the above

11. A thermal-type circuit breaker is equivalent to a _____ fuse, whereas a magnetic-type circuit breaker is equivalent to a _____ fuse.
    a. Slow-blow, slow-blow   c. Slow-blow, fast-blow
    b. Fast-blow, slow-blow   d. Glass case, ceramic case

12. The type of circuit breaker found in the home is the:
    a. Thermal type   c. Magnetic type
    b. Thermomagnetic type   d. Carbon-zinc type

13. A single-pole, double-throw switch would have _____ terminals.
    a. Two   b. Three   c. Four   d. Five

14. A switch will very simply _____ (open) or _____ (close) a path of current.
    a. Break, make   c. s/c, o/c
    b. make, break   d. o/c, break

15. With switches, the word *pole* describes a:
    a. Stationary contact   c. Path of current
    b. Moving contact   d. None of the above

Communication Skill Questions

16. Describe briefly how direct current can be generated:
    a. Mechanically with pressure (4-2-1)
    b. Thermally (4-2-2)
    c. Optically (4-2-3)

d. Magnetically (4-2-4)
e. Chemically (4-2-5)
f. Electrically (4-2-6)

17. What is the difference between a cell and a battery? (4-2-5)

18. Briefly describe the operation of a battery. (4-2-5)

19. Simply state the difference between a primary and a secondary cell. (4-2-5)

20. Describe what can be gained by connecting batteries in series and parallel. (4-2-5)

21. Describe a fuse's: (4-3-1)
    a. Current rating   b. Voltage rating

22. Briefly describe the operation of the following types of circuit breakers: (4-3-2)
    a. Thermal type   b. Magnetic type

23. List the eight different switch classifications. (4-4)

24. List the four popular primary cells. (4-2-5)

25. List the two popular secondary cells. (4-2-5)

Practice Problems

26. If a 50 W light bulb is connected across a 120 V source and the following fuses are available, which should be used to protect the circuit? (4-3-1)
    a. 0.5 A/120 V (fast blow)   c. 1 A/12 V (slow blow)
    b. ¾ A/120 V (fast blow)   d. 0.5 A/120 V (slow blow)

27. Draw a diagram and indicate the total source voltage of: (4-2-5)
    a. Six 1.5 V cells connected in series with one another
    b. Six 1.5 V cells connected in parallel with one another

28. Draw a diagram and indicate the polarities of a 12 V lead–acid battery being charged by a 15 V battery charger. If this battery is rated at 150 Ah: (4-2-5)
    a. How many coulombs of charge will be stored in the fully charged condition?
    b. How long will it take the battery to charge if a charging current of 5 A is flowing?

29. Show how to connect two 12 V batteries to increase (a) the voltage and (b) the current. (4-2-5)

30. If a 6 V nickel–cadmium battery discharges at a rate of 3.5 A in 4 hours: (4-2-5)
    a. Calculate its ampere-hour and ampere-second rating.
    b. How long will it take the battery to discharge if a current of 2 A is drawn?

Web Site Questions

Go to the web site http://www.prenhall.com/cook, select the textbook *Introductory DC/AC Electronics* or *Introductory DC/AC Circuits,* this chapter, and then follow the instructions when answering the multiple-choice practice problems.

These tests at the end of each chapter will challenge your knowledge up to this point, and give you the practice you need for a job interview. To make this more realistic, the test will be comprised of both technical and personal questions. In order to take full advantage of this exercise, you may want to set up a simulation of the interview environment, have a friend read the questions to you vocally, and record your responses for later analysis.

Company Name: LINK, Inc.

Industry Branch: Computers.

Function: Set up local area networks (LANs).

Job Title: Field Service Techician.

1. Is your interest in computers both personal and professional?

2. Could you define for me the difference between a source and a load?

3. Tell me what you know about nickel-metal hydride batteries for notebook computers.

4. What application software are you familiar with?

5. What is the difference between a photoconductive cell and a photovoltaic cell?

6. What is Ohm's Law?

7. Would you say that you are personable?

8. What is the difference between a fuse and a circuit breaker?

9. What does the prefix "mega" mean, in, for example, a 500 megahertz computer system?

10. What does the abbreviation ISP stand for?

Answers

1. Answer must be yes. Go on to explain what computer system you have at home and school, and in what capacity you have used them personally and professionally.
2. Section 4-1.
3. Table 4-2.
4. Discuss EWB's Multisym and/or Circuit Maker, and all other application software you are familiar with.
5. Sections 3-1-2 and 4-2-3.
6. Chapter 2.
7. They're asking if you can get along well with customers and colleagues. Mention any work experience in which you had direct contact with customers, and discuss how you worked with fellow students in your class.
8. Section 4-3.
9. Section 2-1.
10. Internet Service Provider.

Series DC Circuits

The First Pocket Calculator

During the seventeenth century, European thinkers were obsessed with any device that could help them with mathematical calculation. Scottish mathematician John Napier decided to meet this need, and in 1614 he published his new discovery of logarithms. In this book, consisting mostly of tediously computed tables, Napier stated that a logarithm is the exponent of a base number. For example:

The common logarithm (base 10) of 100 is 2 ($100 = 10^2$).

The common logarithm of 10 is 1 ($10 = 10^1$).

The common logarithm of 27 is 1.43136 ($27 = 10^{1.43136}$).

The common logarithm of 6 is 0.77815 ($6 = 10^{0.77815}$).

Any number, no matter how large or small, can be represented by or converted to a logarithm. Napier also outlined how the multiplication of two numbers could be achieved by simply adding the numbers' logarithms. For example, if the logarithm of 2 (which is 0.30103) is added to the logarithm of 4 (which is 0.60206), the result will be 0.90309, which is the logarithm of the number 8 ($0.30103 + 0.60206 = 0.90309$, $2 \times 4 = 8$). Therefore, the multiplication of two large numbers can be achieved by looking up the logarithms of the two numbers in a log table, adding them together, and then finding the number that corresponds to the sum in an antilog (reverse log) table. In this example, the antilog of 0.90309 is 8.

Napier's table of logarithms was used by William Oughtred, who developed, just 10 years after Napier's death in 1617, a handy mechanical device that could be used for rapid calculation. This device, considered the first pocket calculator, was the slide rule.

Photo Credit: From *Portraits of Eminent Mathematicians,* by David Eugene Smith, published by Pictorial Mathematics, New York, 1936.

Outline and Objectives

Introduction

A series circuit, by definition, is the connecting of components end to end in a circuit to provide a single path for the current. This is true not only for resistors, but also for other components that can be connected in series. In all cases, however, the components are connected in succession or strung together one after another so that only one path for current exists between the negative ($-$) and positive ($+$) terminals of the supply.

5-1   COMPONENTS IN SERIES

Figure 5-1 illustrates five examples of **series** resistive **circuits.** In all five examples, you will notice that the resistors are connected "in-line" with one another so that the current through the first resistor must pass through the second resistor, and the current through the second resistor must pass through the third, and so on.

☐   EXAMPLE:

In Figure 5-2(a), seven resistors are laid out on a table top. Using a protoboard, connect all the resistors in series, starting at R_1, and proceeding in numerical order through the resistors until reaching R_7. After completing the circuit, connect the series circuit to a dc power supply.

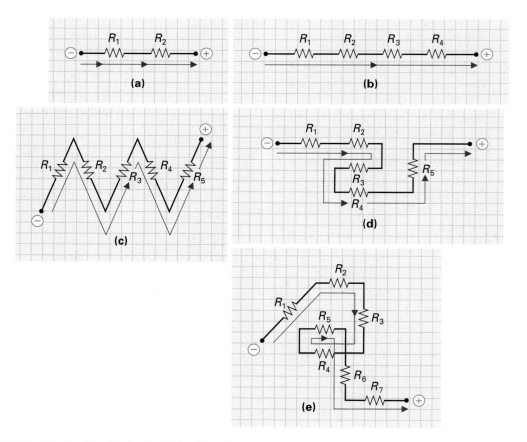

FIGURE 5-1   Five Series Resistive Circuits.

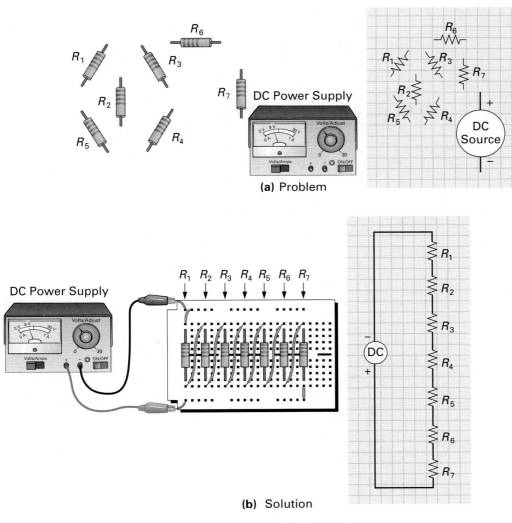

FIGURE 5-2 Connecting Resistors in Series. (a) Problem. (b) Solution.

■ *Solution:*

In Figure 5-2(b), you can see that all the resistors are now connected in series (end-to-end), and the current has only one path to follow from negative to positive.

■ **EXAMPLE:**

Figure 5-3(a) shows four 1.5 V cells and three lamps. Using wires, connect all of the cells in series to create a 6 V battery source. Then connect all of the three lamps in series with one another, and finally, connect the 6 V battery source across the three-series-connected-lamp load.

■ *Solution:*

In Figure 5-3(b) you can see the final circuit containing a source, made up of four series-connected 1.5 V cells, and a load, consisting of three series-connected lamps. As explained in Chapter 4, when cells are connected in series, the total voltage (V_T) will be equal to the sum of all the cell voltages:

$$V_T = V_1 + V_2 + V_3 + V_4 = 1.5\,\text{V} + 1.5\,\text{V} + 1.5\,\text{V} + 1.5\,\text{V} = 6\,\text{V}$$

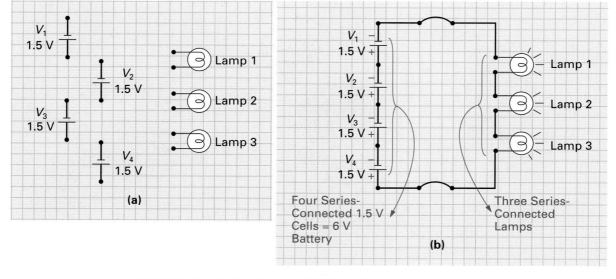

FIGURE 5-3   Series-Connected Cells and Lamps.

5-2   CURRENT IN A SERIES CIRCUIT

The current in a series circuit has only one path to follow and cannot divert in any other direction. The current through a series circuit, therefore, is the same throughout that circuit.

Returning once again to the water analogy, you can see in Figure 5-4(a) that if 2 gallons of water per second are being supplied by the pump, 2 gallons per second must be pulled into the pump. If the rate at which water is leaving and arriving at the pump is the

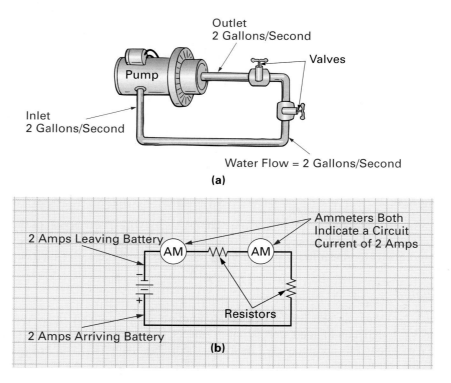

FIGURE 5-4   Series Circuit Current. (a) Fluid System. (b) Electric System.

same, 2 gallons of water per second must be flowing throughout the circuit. It can be said, therefore, that the same value of water flow exists throughout a series-connected fluid system. This rule will always remain true, for if the valves were adjusted to double the opposition to flow, then half the flow, or 1 gallon of water per second, would be leaving the pump and flowing throughout the system.

Similarly, with the electronic series circuit shown in Figure 5-4(b), there is a total of 2 A leaving and 2 A arriving at the battery, and so the same value of current exists throughout the series-connected electronic circuit. If the circuit's resistance were changed, a new value of series circuit current would be present throughout the circuit. For example, if the resistance of the circuit was doubled, then half the current, or 1 A, will leave the battery, but that same value of 1 A will flow throughout the entire circuit. This series circuit current characteristic can be stated mathematically as

$$I_T = I_1 = I_2 = I_3 = \cdots$$

Total current = current through R_1 = current through R_2 = current through R_3, and so on.

FIGURE 5-5   Total Current Example.
(a) Schematic. (b) Protoboard Circuit.
(c) Circuit Analysis Table.

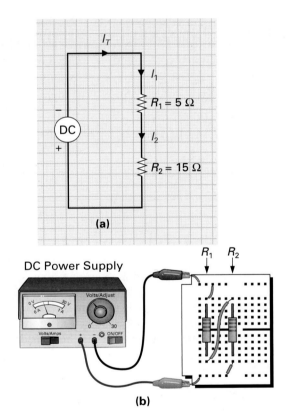

(a)

DC Power Supply

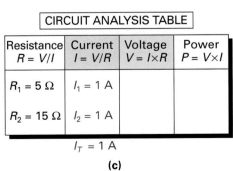

(b)

CIRCUIT ANALYSIS TABLE			
Resistance $R = V/I$	Current $I = V/R$	Voltage $V = I \times R$	Power $P = V \times I$
$R_1 = 5\ \Omega$	$I_1 = 1\ \text{A}$		
$R_2 = 15\ \Omega$	$I_2 = 1\ \text{A}$		

$I_T = 1\ \text{A}$

(c)

In Figure 5-5(a) and (b) on the previous page, a total current (I_T) of 1 A is flowing out of the negative terminal of the dc power supply, through two end-to-end resistors R_1 and R_2, and returning to the positive terminal of the dc power supply. Calculate:

    a. The current through R_1 (I_1)

    b. The current through R_2 (I_2)

■   *Solution:*

Since R_1 and R_2 are connected in series, the current through both will be the same as the circuit current, which is equal to 1 A.

$$I_T = I_1 \quad = I_2$$
$$1\,A = 1\,A = 1\,A$$

The details of this circuit are summarized in the analysis table shown in Figure 5-5(c).

SELF-TEST EVALUATION POINT FOR SECTION 5-1 AND 5-2

Now that you have completed these sections, you should be able to:

■ **Objective 1.** *Describe a series circuit.*

■ **Objective 2.** *Identify series circuits.*

■ **Objective 3.** *Connect components so that they are in series with one another.*

■ **Objective 4.** *Describe why current remains the same throughout a series circuit.*

Use the following questions to test your understanding of Sections 5-1 and 5-2.

1. What is a series circuit?

2. What is the current flow through each of eight series-connected 8 Ω resistors if 8 A total current is flowing out of a battery?

5-3   RESISTANCE IN A SERIES CIRCUIT

Resistance is the opposition to current flow, and in a series circuit every resistor in series offers opposition to the current flow. In the water analogy of Figure 5-4, the total resistance or opposition to water flow is the sum of the two individual valve opposition values. Like the battery, the pump senses the total opposition in the circuit offered by all the valves or resistors, and the amount of current that flows is dependent on this resistance or opposition.

    The total resistance in a series-connected electronic resistive circuit is thus equal to the sum of all the individual resistances, as shown in Figure 5-6(a) through (d). An equivalent circuit can be drawn for each of the circuits in Figure 5-6(b), (c), and (d) with one resistor of a value equal to the sum of all the series resistance values.

    No matter how many resistors are connected in series, the total resistance or opposition to current flow is always equal to the sum of all the resistor values. This formula can be stated mathematically as

$$R_T = R_1 + R_2 + R_3 + \cdots$$

Total resistance = value of R_1 + value of R_2 + value of R_3, and so on.

Equivalent Resistance (R_{eq})

Total resistance of all the individual resistances in a circuit.

    Total resistance (R_T) is the only opposition a source can sense. It does not see the individual separate resistors, but one **equivalent resistance.** Based on the source's voltage and the circuit's total resistance, a value of current will be produced to flow through the circuit (Ohm's law, $I = V/R$).

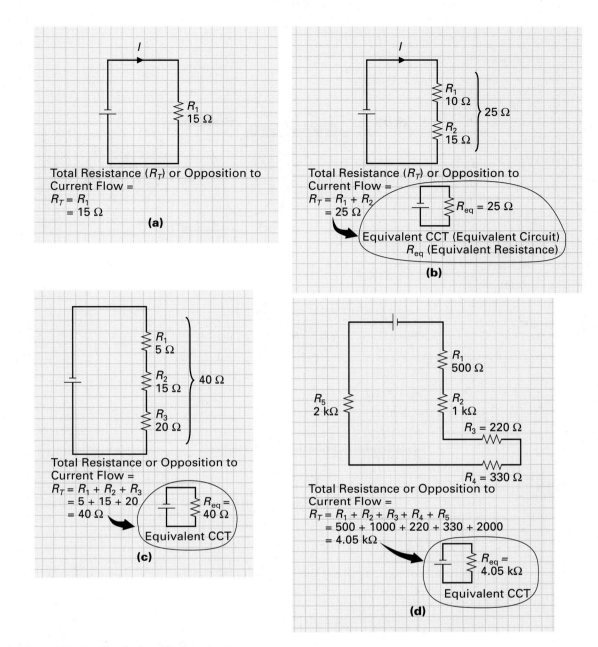

FIGURE 5-6   Total or Equivalent Resistance.

EXAMPLE:

Referring to Figure 5-7(a) and (b), calculate

a. The circuit's total resistance

b. The current flowing through R_2

Solution:

a. $R_T = R_1 + R_2 + R_3 + R_4$

   $= 25\ \Omega + 20\ \Omega + 33\ \Omega + 10\ \Omega$

   $= 88\ \Omega$

b. $I_T = I_1 = I_2 = I_3 = I_4$. Therefore, $I_2 = I_T = 3$ A.

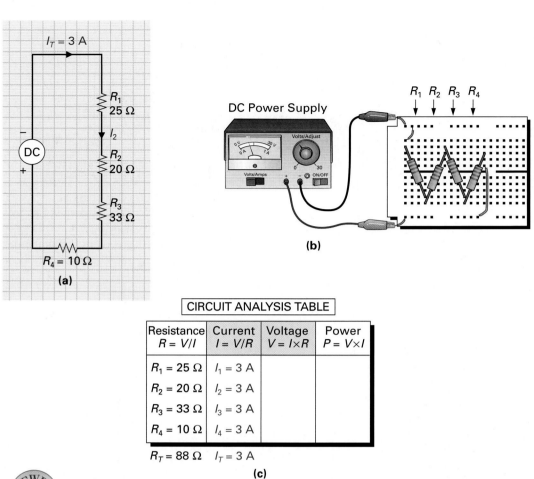

CIRCUIT ANALYSIS TABLE			
Resistance $R = V/I$	Current $I = V/R$	Voltage $V = I \times R$	Power $P = V \times I$
$R_1 = 25\,\Omega$	$I_1 = 3\,A$		
$R_2 = 20\,\Omega$	$I_2 = 3\,A$		
$R_3 = 33\,\Omega$	$I_3 = 3\,A$		
$R_4 = 10\,\Omega$	$I_4 = 3\,A$		
$R_T = 88\,\Omega$	$I_T = 3\,A$		

(c)

FIGURE 5-7 **Total Resistance Example. (a) Schematic. (b) Protoboard Circuit. (c) Circuit Analysis Table.**

The details of this circuit are summarized in the analysis table shown in Figure 5-7(c).

▢ EXAMPLE:

Figure 5-8(a) and (b) shows how a single-pole three-position switch is being used to provide three different lamp brightness levels. In position ① R_1 is placed in series with the lamp, in position ② R_2 is placed in series with the lamp, and in position ③ R_3 is placed in series with the lamp. If the lamp has a resistance of 75 Ω, calculate the three values of current for each switch position.

▨ *Solution:*

$$\text{Position 1: } R_T = R_1 + R_{\text{lamp}}$$
$$= 25\,\Omega + 75\,\Omega = 100\,\Omega$$
$$I_T = \frac{V_T}{R_T} = \frac{12\,V}{100\,\Omega} = 120\,mA$$

$$\text{Position 2: } R_T = R_2 + R_{\text{lamp}} = 50\,\Omega + 75\,\Omega = 125\,\Omega$$
$$I_T = \frac{V_T}{R_T} = \frac{12\,V}{125\,\Omega} = 96\,mA$$

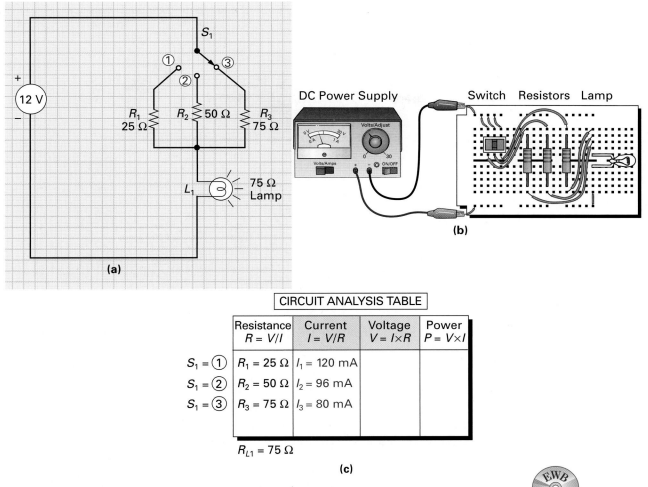

CIRCUIT ANALYSIS TABLE

	Resistance $R = V/I$	Current $I = V/R$	Voltage $V = I \times R$	Power $P = V \times I$
$S_1 = ①$	$R_1 = 25\ \Omega$	$I_1 = 120\ \text{mA}$		
$S_1 = ②$	$R_2 = 50\ \Omega$	$I_2 = 96\ \text{mA}$		
$S_1 = ③$	$R_3 = 75\ \Omega$	$I_3 = 80\ \text{mA}$		

$R_{L1} = 75\ \Omega$

(c)

FIGURE 5-8   Three-Position Switch Controlling Lamp Brightness. (a) Schematic. (b) Protoboard Circuit. (c) Circuit Analysis Table.

$$\text{Position 3: } R_T = R_3 + R_{\text{lamp}} = 75\ \Omega + 75\ \Omega = 150\ \Omega$$

$$I_T = \frac{V_T}{R_T} = \frac{12\ \text{V}}{150\ \Omega} = 80\ \text{mA}$$

The details of this circuit are summarized in the analysis table shown in Figure 5-8(c).

SELF-TEST EVALUATION POINT FOR SECTION 5-3

Now that you have completed this section, you should be able to:

■ **Objective 5.**   *Explain how to calculate total resistance in a series circuit.*

■ **Objective 6.**   *Explain how Ohm's law can be applied to calculate current, voltage, and resistance.*

Use the following questions to test your understanding of Section 5-3.

1. State the total resistance formula for a series circuit.
2. Calculate R_T if $R_1 = 2\ \text{k}\Omega$, $R_2 = 3\ \text{k}\Omega$, and $R_3 = 4700\ \Omega$.

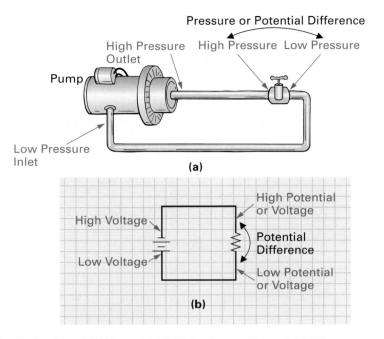

FIGURE 5-9 Series Circuit Voltage. (a) Fluid Analogy of Potential Difference. (b) Electrical Potential Difference.

5-4 VOLTAGE IN A SERIES CIRCUIT

A potential difference or voltage drop will occur across each resistor in a series circuit when current is flowing. The amount of voltage drop is dependent on the value of the resistor and the amount of current flow. This idea of potential difference or voltage drop is best explained by returning to the water analogy. In Figure 5-9(a), you can see that the high pressure from the pump's outlet is present on the left side of the valve. On the right side of the valve, however, the high pressure is no longer present. The high potential that exists on the left of the valve is not present on the right, so a potential or pressure difference is said to exist across the valve.

Similarly, with the electronic circuit shown in Figure 5-9(b), the battery produces a high voltage or potential that is present at the top of the resistor. The high voltage that exists at the top of the resistor, however, is not present at the bottom. Therefore, a potential difference or voltage drop is said to occur across the resistor. This voltage drop that exists across resistors can be found by utilizing Ohm's law: $V = I \times R$.

▢ **EXAMPLE:**

Referring to Figure 5-10, calculate:

    a. Total resistance (R_T)

    b. Amount of series current flowing throughout the circuit (I_T)

    c. Voltage drop across R_1

    d. Voltage drop across R_2

    e. Voltage drop across R_3

▢ *Solution:*

    a. Total resistance (R_T) $= R_1 + R_2 + R_3$
$$= 20\ \Omega + 30\ \Omega + 50\ \Omega$$
$$= 100\ \Omega$$

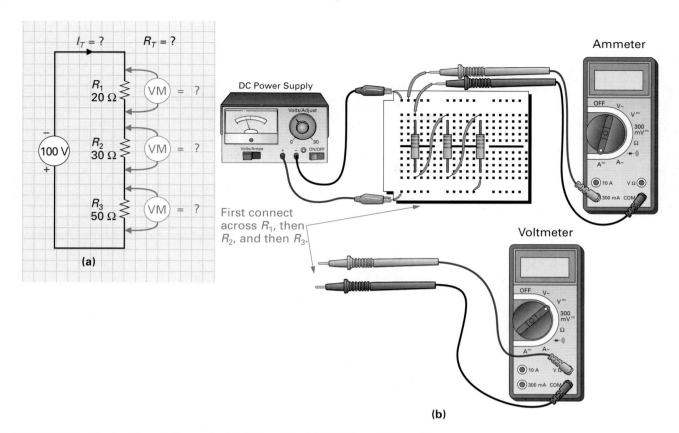

FIGURE 5-10 Series Circuit Example. (a) Schematic. (b) Protoboard Circuit.

b. Total current $(I_T) = \dfrac{V_{\text{source}}}{R_T}$

$= \dfrac{100\text{ V}}{100\ \Omega}$

$= 1\text{ A}$

The same current will flow through the complete series circuit, so the current through R_1 will equal 1 A, the current through R_2 will equal 1 A, and the current through R_3 will equal 1 A.

c. Voltage across R_1 $(V_{R1}) = I_1 \times R_1$

$= 1\text{ A} \times 20\ \Omega$

$= 20\text{ V}$

d. Voltage across R_2 $(V_{R2}) = I_2 \times R_2$

$= 1\text{ A} \times 30\ \Omega$

$= 30\text{ V}$

e. Voltage across R_3 $(V_{R3}) = I_3 \times R_3$

$= 1\text{ A} \times 50\ \Omega$

$= 50\text{ V}$

Figure 5-11(a) shows the schematic and Figure 5-11(b) shows the analysis table for this example, with all of the calculated data inserted. As you can see, the 20 Ω resistor drops 20 V, the 30 Ω resistor has 30 V across it, and the 50 Ω resistor has dropped 50 V. From this example, you will notice that the larger the resistor value, the larger the voltage drop. Resistance and voltage drops are consequently proportional to one another.

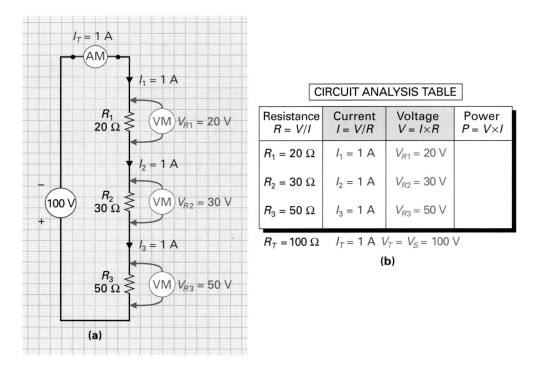

CIRCUIT ANALYSIS TABLE			
Resistance $R = V/I$	Current $I = V/R$	Voltage $V = I \times R$	Power $P = V \times I$
$R_1 = 20\ \Omega$	$I_1 = 1\ A$	$V_{R1} = 20\ V$	
$R_2 = 30\ \Omega$	$I_2 = 1\ A$	$V_{R2} = 30\ V$	
$R_3 = 50\ \Omega$	$I_3 = 1\ A$	$V_{R3} = 50\ V$	

$R_T = 100\ \Omega$ $I_T = 1\ A$ $V_T = V_S = 100\ V$

(b)

FIGURE 5-11   **Voltage Drop and Resistance. (a) Schematic. (b) Circuit Analysis Table.**

The Voltage Drop across a Device is $\propto$ to Its Resistance

---

$$V_{\text{drop}} \uparrow = I(\text{constant}) \times R \uparrow$$
$$V_{\text{drop}} \downarrow = I(\text{constant}) \times R \downarrow$$

Another interesting point you may have noticed from Figure 5-11 is that, if you were to add up all the voltage drops around a series circuit, they would equal the source (V_S) applied:

---

total voltage applied (V_S or V_T) $= V_{R1} + V_{R2} + V_{R3} + \cdots$

In the example in Figure 5-11, you can see that this is true, since

$$100\ V = 20\ V + 30\ V + 50\ V$$
$$100\ V = 100\ V$$

The series circuit has in fact divided up the applied voltage, and it appears proportionally across all the individual resistors. This characteristic was first observed by Gustav Kirchhoff in 1847. In honor of his discovery, this effect is known as **Kirchhoff's voltage law,** which states: The sum of the voltage drops in a series circuit is equal to the total voltage applied.

To summarize the effects of current, resistance, and voltage in a series circuit so far, we can say that:

1. The current in a series circuit has only one path to follow.
2. The value of current in a series circuit is the same throughout the entire circuit.
3. The total resistance in a series circuit is equal to the sum of all the resistances.
4. Resistance and voltage drops in a series circuit are proportional to one another, so a large resistance will have a large voltage drop and a small resistance will have a small voltage drop.
5. The sum of the voltage drops in a series circuit is equal to the total voltage applied.

TIME LINE

German physicist Gustav Robert Kirchhoff (1824-1879) extended Ohm's law and developed two important laws of his own, known as Kirchhoff's voltage law and Kirchhoff's current law.

Kirchhoff's Voltage Law

The algebraic sum of the voltage drops in a closed path circuit is equal to the algebraic sum of the source voltage applied.

EXAMPLE:

First, calculate the voltage drop across the resistor R_1 in the circuit in Figure 5-12(a) for a resistance of 4 Ω. Then, change the 4 Ω resistor to a 2 Ω resistor and recalculate the voltage drop across the new resistance value. Use a constant source of 4 V.

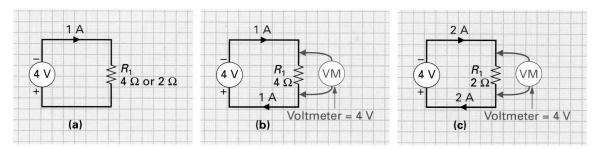

FIGURE 5-12   Single Resistor Circuit.

Solution:

Referring to Figure 5-12(b), you can see that when $R_1 = 4$ Ω, the voltage across R_1 can be calculated by using Ohm's law and is equal to

$$V_{R1} = I_1 \times R_1$$
$$V_{R1} = 1 \text{ A} \times 4 \text{ Ω}$$
$$= 4 \text{ V}$$

If the resistance is now changed to 2 Ω as shown in Figure 5-12(c), the current flow within the circuit will be equal to

$$I = \frac{V_S}{R} = \frac{4 \text{ V}}{2 \text{ Ω}} = 2 \text{ A}$$

The voltage dropped across the 2 Ω resistor will still be equal to

$$V_{R1} = I_1 \times R_1$$
$$= 2 \text{ A} \times 2 \text{ Ω}$$
$$= 4 \text{ V}$$

As you can see from this example, if only one resistor is connected in a series circuit, the entire applied voltage appears across this resistor. The value of this single resistor will determine the amount of current flow through the circuit and this value of circuit current will remain the same throughout.

EXAMPLE:

Referring to Figure 5-13(a), calculate the following, and then draw the circuit schematic again with all of the new values inserted.

a. Total circuit resistance

b. Value of circuit current (I_T)

c. Voltage drop across each resistor

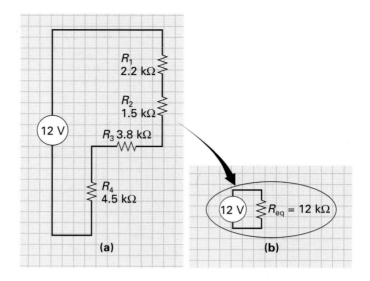

FIGURE 5-13   Series Circuit Example.

■   *Solution:*

a. $R_T = R_1 + R_2 + R_3 + R_4$
    $= 2.2\ k\Omega + 1.5\ k\Omega + 3.8\ k\Omega + 4.5\ k\Omega$
    $= (2.2 \times 10^3) + (1.5 \times 10^3) + (3.8 \times 10^3) + (4.5 \times 10^3)$
    $= 12\ k\Omega$ [Figure 5-13(b)]

b. $I_T = \dfrac{V_S}{R_T} = \dfrac{12\ V}{12\ k\Omega} = 1\ mA$

c. Voltage drop across each resistor:

$$V_{R1} = I_T \times R_1$$
$$= 1\ mA \times 2.2\ k\Omega$$
$$= 2.2\ V$$
$$V_{R2} = I_T \times R_2$$
$$= 1\ mA \times 1.5\ k\Omega$$
$$= 1.5\ V$$
$$V_{R3} = I_T \times R_3$$
$$= 1\ mA \times 3.8\ k\Omega$$
$$= 3.8\ V$$
$$V_{R4} = I_T \times R_4$$
$$= 1\ mA \times 4.5\ k\Omega$$
$$= 4.5\ V$$

Figure 5-14(a) shows the schematic diagram for this example with all of the values inserted, and Figure 5-14(b) shows this circuit's analysis table.

5-4-1   *A Voltage Source's Internal Resistance*

Figure 5-15(a) illustrates a battery on the left and its symbol on the right. When a circuit or system is connected across a battery, as shown in Figure 5-15(a), the circuit or system can be represented by its equivalent value of resistance, called the *load resistance* (R_L). In Figure 5-15(a) the switch is open, so the battery is not loaded and no load current flows. In Figure 5-15(b) the switch has been closed and so the battery has a completed current path. As a re-

FIGURE 5-14 Series Circuit Example with All Values Inserted.
(a) Schematic.
(b) Circuit Analysis Table.

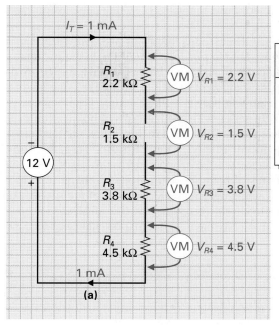

(a)

CIRCUIT ANALYSIS TABLE

Resistance $R = V/I$	Current $I = V/R$	Voltage $V = I \times R$	Power $P = V \times I$
$R_1 = 2.2\ k\Omega$	$I_1 = 1\ mA$	$V_{R1} = 2.2\ V$	
$R_2 = 1.5\ k\Omega$	$I_2 = 1\ mA$	$V_{R2} = 1.5\ V$	
$R_3 = 3.8\ k\Omega$	$I_3 = 1\ mA$	$V_{R3} = 3.8\ V$	
$R_4 = 4.5\ k\Omega$	$I_4 = 1\ mA$	$V_{R4} = 4.5\ V$	

$R_T = 12\ k\Omega$ $I_T = 1\ mA$ $V_T = V_S = 12\ V$

(b)

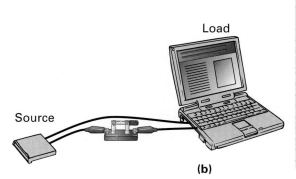

(a)

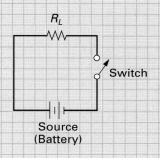

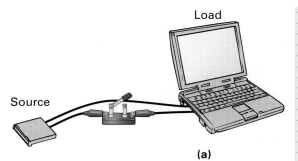

(b)

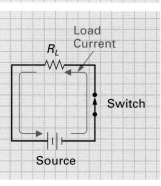

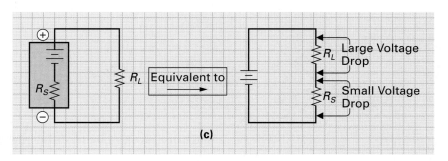

(c)

FIGURE 5-15 Loading a Battery.

sult, a load current will flow, the value of which depends on the load resistance and battery voltage of 12 V.

The battery just discussed is known as an *ideal voltage source* because its voltage did not change from a *no-load* to *full-load* condition. In reality, there is no such thing as an ideal voltage source. Batteries or any other type of voltage source are not 100% efficient. If a voltage source were 100% efficient, it would be able to generate electrical energy at its output without the accompanying heat. A voltage source's inefficiency is represented by an internal resistor connected in series with the battery symbol as seen in Figure 5-15(c). As you can see in the schematic circuit illustrated in this figure, the load resistance R_L and the source resistance R_S (pronounced "R sub L", and "R sub S") together form the total resistive load.

Very little voltage will be dropped across R_S, as it is normally small compared to R_L. Internal inefficiencies of batteries and all other voltage sources must always be kept small because a large R_S would drop a greater amount of the supply voltage, resulting in less output voltage to the load and therefore a waste of electrical power. To give an example, a lead–acid cell will typically have an internal resistance of 0.01 Ω (10 mΩ).

▣ EXAMPLE:

Figure 5-16 shows a 12 V battery that has an internal source resistance (R_S or R_{int}) of 0.5 Ω (500 mΩ). If the battery was to supply its maximum safe current, which in this example is 2.5 A, what would be the output terminal voltage of the battery?

▣ *Solution:*

If the battery was supplying its maximum safe current of 2.5 A, the output voltage (V_{out}) would equal the source voltage ($V_S = 12$ V) minus the voltage drop across the internal battery resistance (V_{Rint}). First let us calculate V_{Rint}.

$$V_{Rint} = I \times R_{int}$$
$$= 2.5 \text{ A} \times 0.5 \text{ Ω} = 1.25 \text{ V}$$

The output voltage under full load (maximum safe current) will therefore be

$$V_{out} = V_S - V_{Rint} = 12 \text{ V} - 1.25 \text{ V} = 10.75 \text{ V}$$

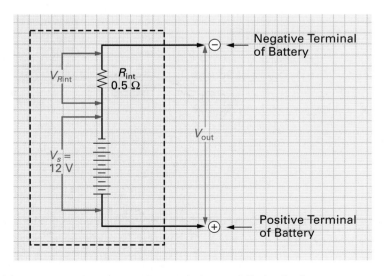

FIGURE 5-16   **Voltage Drop across a Battery's Internal Series Resistance.**

5-4-2   *Fixed Voltage Divider*

A series-connected circuit is often referred to as a *voltage-divider circuit,* because the total voltage applied (V_T) or source voltage (V_S) is divided and dropped proportionally across all the resistors in a series circuit. The amount of voltage dropped across a resistor is proportional to the value of resistance, and so a larger resistance will have a larger voltage drop, while a smaller resistance will have a smaller voltage drop.

The voltage dropped across a resistor is normally a factor that needs to be calculated. The voltage-divider formula allows you to calculate the voltage drop across any resistor without having to first calculate the value of circuit current. This formula is stated as

$$V_X = \frac{R_X}{R_T} \times V_S$$

where V_X = voltage dropped across selected resistor
R_X = selected resistor's value
R_T = total series circuit resistance
V_S = source or applied voltage

Figure 5-17 illustrates a circuit from a previous example. To calculate the voltage drop across each resistor, we would normally have to:

1. Calculate the total resistance by adding up all the resistance values.

2. Once we have the total resistance and source voltage (V_S), we could then calculate current.

3. Having calculated the current flowing through each resistor, we could then use the current value to calculate the voltage dropped across any one of the four resistors merely by multiplying current by the individual resistance value.

The voltage-divider formula allows us to bypass the last two steps in this procedure. If we know total resistance, supply voltage, and the individual resistance values, we can calculate the voltage drop across the resistor without having to calculate steps 2 and 3. For

FIGURE 5-17   Series Circuit Example.

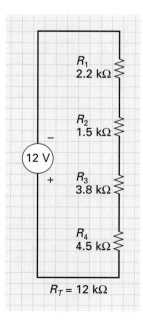

FIGURE 5-18   **Series Circuit Example.**

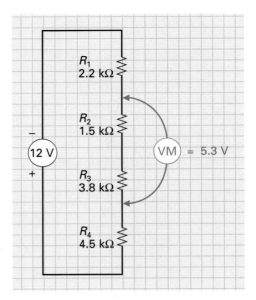

example, what would be the voltage dropped across R_2 and R_4 for the circuit shown in Figure 5-17? The voltage dropped across R_2 is

$$V_{R2} = \frac{R_2}{R_T} \times V_S$$

$$= \frac{1.5 \text{ k}\Omega}{12 \text{ k}\Omega} \times 12 \text{ V}$$

$$= 1.5 \text{ V}$$

The voltage dropped across R_4 is

$$V_{R4} = \frac{R_4}{R_T} \times V_S$$

$$= \frac{4.5 \text{ k}\Omega}{12 \text{ k}\Omega} \times 12 \text{ V}$$

$$= 4.5 \text{ V}$$

CALCULATOR SEQUENCE

Step	Keypad Entry	Display Response
1.	1 . 5 E 3	1.5E3
2.	+	
3.	1 2 E 3	12E3
4.	×	0.125
5.	1 2	12
6.	=	1.5

The voltage-divider formula could also be used to find the voltage drop across two or more series-connected resistors. For example, referring again to the example circuit shown in Figure 5-17, what would be the voltage dropped across R_2 and R_3 combined? The voltage across $R_2 + R_3$ can be calculated using the voltage-divider formula as follows:

$$V_{R2} \text{ and } V_{R3} = \frac{R_2 + R_3}{R_T} \times V_S$$

$$= \frac{5.3 \text{ k}\Omega}{12 \text{ k}\Omega} \times 12 \text{ V}$$

$$= 5.3 \text{ V}$$

As can be seen in Figure 5-18, the voltage drop across R_2 and R_3 is 5.3 V.

▢   **EXAMPLE:**

Referring to Figure 5-19(a), calculate the voltage drop across:

   a. R_1, R_2, and R_3 separately

   b. R_2 and R_3 combined

   c. R_1, R_2, and R_3 combined

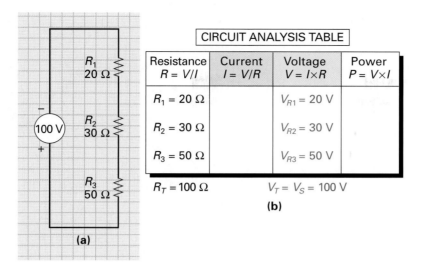

CIRCUIT ANALYSIS TABLE

Resistance $R = V/I$	Current $I = V/R$	Voltage $V = I \times R$	Power $P = V \times I$
$R_1 = 20\ \Omega$		$V_{R1} = 20$ V	
$R_2 = 30\ \Omega$		$V_{R2} = 30$ V	
$R_3 = 50\ \Omega$		$V_{R3} = 50$ V	
$R_T = 100\ \Omega$		$V_T = V_S = 100$ V	

(b)

(a)

FIGURE 5-19   **Series Circuit Example. (a) Schematic. (b) Circuit Analysis Table.**

■ *Solution:*

a. The voltage drop across a resistor is proportional to the resistance value. The total resistance (R_T) in this circuit is 100 Ω or 100% of R_T. R_1 is 20% of the total resistance, so 20% of the source voltage will appear across R_1. R_2 is 30% of the total resistance, so 30% of the source voltage will appear across R_2. R_3 is 50% of the total resistance, so 50% of the source voltage will appear across R_3. This was a very simple problem in which the figures worked out very neatly. The voltage-divider formula achieves the very same thing by calculating the ratio of the resistance value to the total resistance. This percentage is then multiplied by the source voltage in order to find the desired resistor's voltage drop:

$$V_{R1} = \frac{R_1}{R_T} \times V_S$$
$$= \frac{20\ \Omega}{100\ \Omega} \times V_S$$
$$= 0.2 \times 100\text{ V} \qquad (20\% \text{ of } 100\text{ V})$$
$$= 20\text{ V}$$

$$V_{R2} = \frac{R_2}{R_T} \times V_S$$
$$= \frac{30\ \Omega}{100\ \Omega} \times V_S$$
$$= 0.3 \times 100\text{ V} \qquad (30\% \text{ of } 100\text{ V})$$
$$= 30\text{ V}$$

$$V_{R3} = \frac{R_3}{R_T} \times V_S$$
$$= \frac{50\ \Omega}{100\ \Omega} \times V_S$$
$$= 0.5 \times 100\text{ V} \qquad (50\% \text{ of } 100\text{ V})$$
$$= 50\text{ V}$$

b. Voltage dropped across R_2 and R_3 = 30 + 50 = 80 V.

c. Voltage dropped across R_1, R_2, and R_3 = 20 + 30 + 50 = 100 V.

The details of this circuit are summarized in the analysis table shown in Figure 5-19(b).

FIGURE 5-20   **Series Circuit Example.**

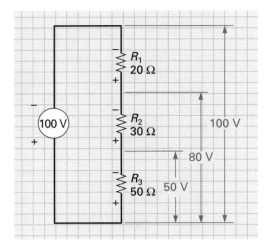

To summarize the voltage-divider formula, we can say: The voltage drop across a resistor or group of resistors in a series circuit is equal to the ratio of that resistance (R_X) to the total resistance (R_T), multiplied by the source voltage (V_S).

To show an application of a voltage divider, let us imagine that three voltages of 50, 80, and 100 V were required by an electronic system in order to make it operate. To meet this need, we could use three individual power sources, which would be very expensive, or use one 100 V voltage source connected across three resistors, as shown in Figure 5-20, to divide up the 100 V.

■ EXAMPLE:

Figure 5-21(a) shows a simplified circuit of an oscilloscope's cathode ray tube (CRT) or picture tube. In this example, a three-resistor series circuit and a -600 V supply voltage are being used to produce the needed three supply voltages for the CRT's three electrodes, called the focusing anode, the control grid, and the cathode. The heated cathode emits electrons that are collected and concentrated into a beam by the combined electrostatic effect of the control grid and focusing anode. This beam of electrons passes through the apertures of the control grid and focusing anode and strikes the inner surface of the CRT screen. This inner surface is coated with a phosphorescent material that emits light when it is struck by electrons. The voltage between the cathode (K) and grid (G) of the CRT (V_{KG}) determines the intensity of the electron beam and therefore the brightness of the trace seen on the screen. The voltage between the grid (G) and anode (A) of the CRT (V_{GA}) determines the sharpness of the electron beam and therefore the focus of the trace seen on the screen. For the resistance values given, calculate the voltages on each of the three CRT electrodes.

■ *Solution:*

Referring to the illustration and calculations in Figure 5-21(b), you can see how the voltage-divider formula can be used to calculate the voltage drop across R_1 ($V_{R1} = 300$ V), R_2 ($V_{R2} = 255$ V), and R_3 ($V_{R3} = 45$ V). Since the grid of the CRT is connected directly to the -600 V supply, the grid voltage (V_G) will be

$$V_G = V_T = -600 \text{ V}$$

The cathode voltage (V_K) will be equal to the total supply voltage (V_T) minus the voltage drop across R_3 (V_{R3}), so

$$V_K = V_T - V_{R3} = (-600) - (-45) = -555 \text{ V}$$

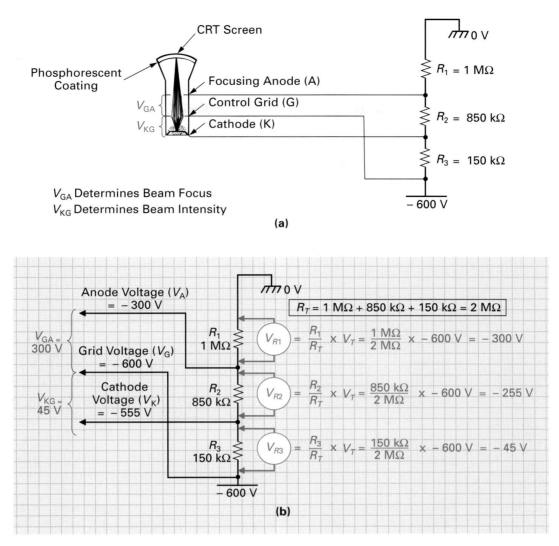

(a)

(b)

FIGURE 5-21   Fixed Voltage-Divider Circuit for Supplying Voltages to the Electrodes of a CRT.

The anode voltage (V_A) will be equal to the total supply voltage (V_T) minus the voltage drops across R_3 (V_{R3}) and R_2 (V_{R2}), and therefore

$$V_A = V_T - (V_{R3} + V_{R2}) = (-600) - (-45 + -225)$$
$$= -600 - 300 \text{ V} = -300 \text{ V}$$

The V_A, V_K, and V_G voltages are all negative with respect to 0 V. The voltages V_{GA} and V_{KG}, shown on the left of Figure 5-21(b), will be the potential difference between the two electrodes. Therefore, V_{KG} is equal to the difference between V_K and V_G (the difference between -600 V and -555 V), and is equal to 45 V ($V_{KG} = 45$ V $= V_{R3}$). The voltage between the CRT's grid and anode (V_{GA}) is equal to the difference between V_G and V_A (the difference between -600 V and -300 V), which will be 300 V ($V_{GA} = 300$ V $= V_{R1}$).

EXAMPLE:

Figure 5-22(a) shows a 24 V voltage source driving a 10 Ω resistor that is located 1000 ft from the battery. If two 1000 ft lengths of AWG No. 13 wire are used to connect the source to the load, what will be the voltage applied across the load?

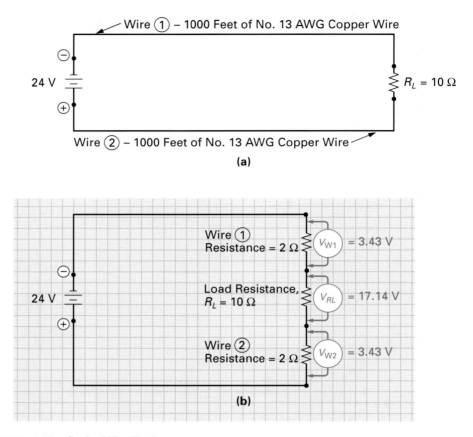

(a)

(b)

FIGURE 5-22   **Series Wire Resistance.**

■   *Solution:*

Referring back to Table 2-2, you can see that AWG No. 13 copper cable has a resistance of 2.003 Ω for every 1000 ft. To be more accurate, this means that our circuit should be re-drawn as shown in Figure 5-22(b) to show the series resistances of wire 1 and wire 2. Using the voltage-divider formula, we can calculate the voltage drop across wire 1 and wire 2.

$$V_{W1} = \frac{R_{W1}}{R_T} \times V_T$$

$$= \frac{2\ \Omega}{14\ \Omega} \times 24\ V = 3.43\ V$$

Since the voltage drop across wire 2 will also be 3.43 V, the total voltage drop across both wires will be 6.86 V. The remainder, 17.14 V (24 V − 6.86 V = 17.14 V), will appear across the load resistor, R_L.

5-4-3   *Variable Voltage Divider*

When discussing variable-value resistors in Chapter 3, we talked about a potentiometer, or variable voltage divider, which consists of a fixed value of resistance between two terminals and a wiper that can be adjusted to vary resistance between its terminal and one of the other two.

To review, Figure 5-23(a) shows the potentiometer's schematic symbol and physical appearance. If the wiper of the potentiometer is moved down, as seen in Figure 5-23(b), the resistance between terminals *A* and *B* will increase, while the resistance between *B* and *C* will decrease. On the other hand, if the wiper is moved up, as seen in Figure 5-23(c), the re-

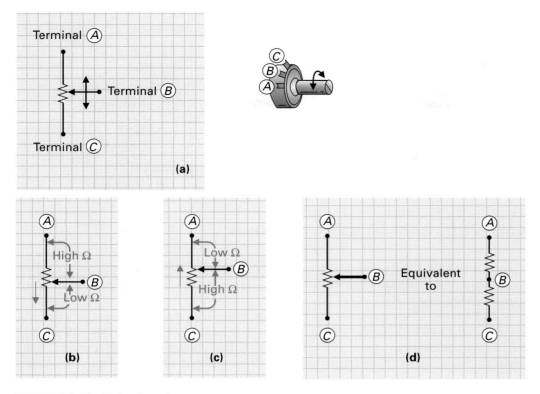

FIGURE 5-23   Potentiometer.

sistance between terminals *A* and *B* will decrease, while the resistance between *B* and *C* will increase. The resistances between *A* and *B* and *B* and *C* are therefore inversely proportional to one another in that if one increases the other decreases, and vice versa. The resistance values between terminals *A* and *B* and between *B* and *C* can be thought of as two separate resistors, as seen in Figure 5-23(d). No matter what position the wiper is put in, the total resistance between *A* and *B* (R_{AB}) and *B* and *C* (R_{BC}) will always be equal to the rated value of the potentiometer, and equal to the resistance between *A* and *C* (R_{AC}).

$$R_{AB} + R_{BC} = R_{AC}$$

As an example, Figure 5-24(a) illustrates a 10 kΩ potentiometer that has been hooked up across a 10 V dc source with a voltmeter between terminals *B* and *C*. If the wiper terminal is positioned midway between *A* and *C*, the voltmeter should read 5 V, and the potentiometer will be equivalent to two 5 kΩ resistors in series, as shown in Figure 5-24(b).

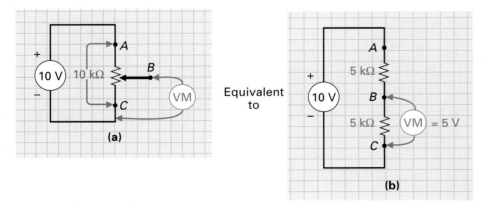

FIGURE 5-24   Potentiometer Wiper in Mid-Position.

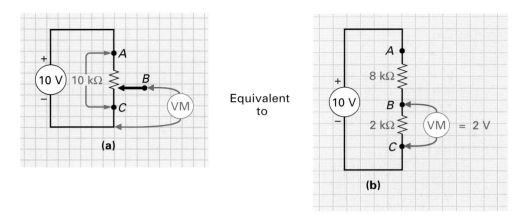

FIGURE 5-25   Potentiometer Wiper in Lower Position.

Kirchhoff's voltage law states that the entire source voltage will be dropped across the resistances in the circuit, and since the resistance values are equal, each will drop half of the source voltage, that is, 5 V.

In Figure 5-25(a), the wiper has been moved down so that the resistance between A and B is equal to 8 kΩ, and the resistance between B and C equals 2 kΩ. This will produce 2 V on the voltmeter, as shown in Figure 5-25(b). The amount of voltage drop is proportional to the resistance, so a larger voltage will be dropped across the larger resistance. Using the voltage-divider formula, you can calculate that the 8 kΩ resistance is 80% of the total resistance and therefore will drop 80% of the voltage:

$$V_{AB} = \frac{R_{AB}}{R_{\text{total}}} \times V_S = \frac{8 \text{ k}\Omega}{10 \text{ k}\Omega} \times 10 \text{ V} = 8 \text{ V}$$

The 2 kΩ resistance between B and C is 20% of the total resistance and consequently will drop 20% of the total voltage:

$$V_{BC} = \frac{R_{BC}}{R_{\text{total}}} \times V_S = \frac{2 \text{ k}\Omega}{10 \text{ k}\Omega} \times 10 \text{ V} = 2 \text{ V}$$

In Figure 5-26(a), the wiper has been moved up and now 2 kΩ exists between A and B, and 8 kΩ is present between B and C. In this situation, 2 V will be dropped across the 2 kΩ between A and B, and 8 V will be dropped across the 8 kΩ between B and C, as shown in Figure 5-26(b).

From this discussion, you can see that the potentiometer can be adjusted to supply different voltages on the wiper. This voltage can be decreased by moving the wiper down to

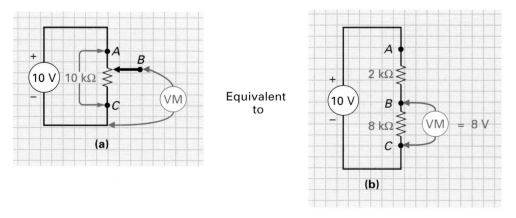

FIGURE 5-26   Potentiometer Wiper in Upper Position.

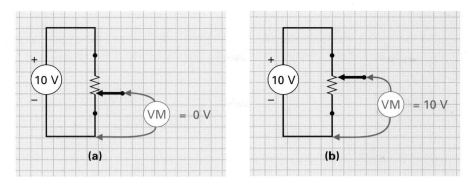

FIGURE 5-27 **Minimum and Maximum Settings of a Potentiometer.**

supply a minimum of 0 V as show in Figure 5-27(a), or the wiper can be moved up to supply a maximum of 10 V as shown in Figure 5-27(b). By adjusting the wiper position, the potentiometer can be made to deliver any voltage between its maximum and minimum value, which is why the potentiometer is known as a variable voltage divider.

EXAMPLE:

Figure 5-28 illustrates how a potentiometer can be used to control the output volume of an amplifier that is being driven by a compact disk (CD) player. The preamplifier is producing an output of 2 V, which is developed across a 50 kΩ potentiometer. If the wiper of the potentiometer is in its upper position, the full 2 V from the preamp will be applied into the input of the power amplifier. The power amplifier has a fixed voltage gain (A_V) of 12, and therefore the power amplifier's output is always 12 times larger than the input voltage. An input of 2 V (V_{in}) will therefore produce an output voltage (V_{out}) of 24 V $(V_{out} = V_{in} \times A_V = 2 \text{ V} \times 12 = 24 \text{ V})$.

As the wiper is moved down, less of the 2 V from the preamp will be applied to the power amplifier, and therefore the output of the power amplifier and volume of the music heard will decrease. If the wiper of the potentiometer is adjusted so that a resistance of 20 kΩ exists between the wiper and the lower end of the potentiometer, what will be the input voltage to the power amplifier and output voltage to the speaker?

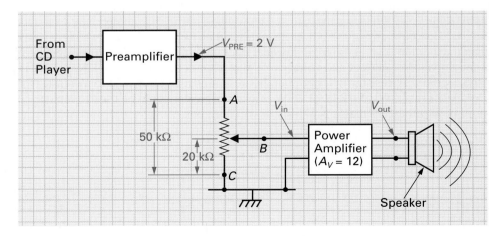

FIGURE 5-28 **The Potentiometer as a Volume Control.**

■ *Solution:*

By using the voltage-divider formula, we can determine the voltage developed across the potentiometer with 20 kΩ of resistance between the wiper (*B*) and lower end (*C*).

$$V_{in} = \frac{R_{AB}}{R_{AC}} \times V_{Pre}$$

$$= \frac{20 \text{ k}\Omega}{50 \text{ k}\Omega} \times 2 \text{ V} = 0.8 \text{ V (800 mV)}$$

The voltage at the input of the power amplifier (V_{in}), is applied to the input of the power amplifier, which has an amplification factor or gain of 12, and therefore the output voltage will be

$$V_{out} = V_{in} \times A_V$$

$$= 0.8 \text{ V} \times 12 = 9.6 \text{ V}$$

SELF-TEST EVALUATION POINT FOR SECTION 5-4

Now that you have completed this section, you should be able to:

■ *Objective 7.* *Describe why the series circuit is known as a voltage divider.*

■ *Objective 8.* *Describe a fixed and a variable voltage divider.*

Use the following questions to test your understanding of Section 5-4.

1. True or false: A series circuit is also known as a voltage-divider circuit.

2. True or false: The voltage drop across a series resistor is proportional to the value of the resistor.

3. If 6 Ω and 12 Ω resistors are connected across a dc 18 V supply, calculate I_T and the voltage drop across each.

4. State the voltage-divider formula.

5. Which component can be used as a variable voltage divider?

6. Could a rheostat be used in place of a potentiometer?

5-5   POWER IN A SERIES CIRCUIT

As discussed earlier in Chapter 3, power is the rate at which work is done, and work is said to have been done when energy is converted from one energy form to another. Resistors convert electrical energy into heat energy, and the rate at which they dissipate energy is called *power* and is measured in *watts* (joules per second). Resistors all have a resistive value, a tolerance, and a wattage rating. The wattage of a resistor is the amount of heat energy a resistor can safely dissipate per second, and this wattage is directly proportional to the resistor's size. Figure 5-29 reviews the size versus wattage rating of commercially available resistors. As you know, any of the power formulas can be used to calculate wattage. Resistors are manufactured in several different physical sizes, and if, for example, it is calculated that for a certain value of current and voltage a 5 W resistor is needed, and a ½ W resistor is put in its place, the ½ W resistor will burn out, because it is generating heat (5 W) faster than it can dissipate heat (½ W). A 10 W or 25 W resistor or greater could be used to replace a 5 W resistor but anything less than a 5 W resistor will burn out.

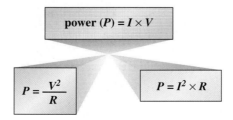

FIGURE 5-29   **Resistor Wattage Ratings.**

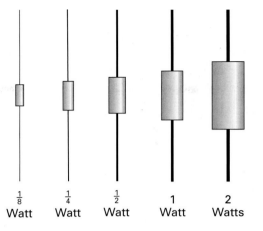

$\frac{1}{8}$ Watt $\frac{1}{4}$ Watt $\frac{1}{2}$ Watt   1 Watt   2 Watts

It is, therefore, necessary that we have some way of calculating which wattage rating is necessary for each specific application.

A question you may be asking is why not just use large-wattage resistors everywhere. The two disadvantages with this are:

1. The larger the wattage, the greater the cost.
2. The larger the wattage, the greater the size and area the resistor occupies within the equipment.

EXAMPLE:

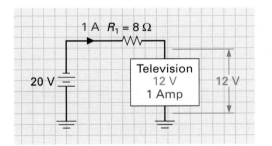

FIGURE 5-30   **Series Circuit Example.**

Figure 5-30 shows a 20 V battery driving a 12 V/1 A television set. R_1 is in series with the TV set and is being used to drop 8 V of the 20 V supply, so 12 V will be developed across the television.

a. What is the wattage rating for R_1?

b. What is the series load resistance of the TV set?

c. What is the amount of power being consumed by the TV set?

d. Compile a circuit analysis table for this circuit.

Solution:

a. Everything is known about R_1. Its resistance is 8 Ω, it has 1 A of current flowing through it, and 8 V is being dropped across it. As a result, any one of the three power formulas can be used to calculate the wattage rating of R_1.

$$\text{power } (P) = I \times V = 1\,\text{A} \times 8\,\text{V} = 8\,\text{W}$$

or

$$P = I^2 \times R = 1^2 \times 8 = 8\,\text{W}$$

or

$$P = \frac{V^2}{R} = \frac{8^2}{8} = 8\,\text{W}$$

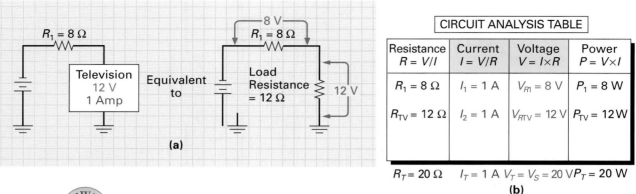

CIRCUIT ANALYSIS TABLE			
Resistance $R = V/I$	Current $I = V/R$	Voltage $V = I \times R$	Power $P = V \times I$
$R_1 = 8\ \Omega$	$I_1 = 1\ A$	$V_{R1} = 8\ V$	$P_1 = 8\ W$
$R_{TV} = 12\ \Omega$	$I_2 = 1\ A$	$V_{RTV} = 12\ V$	$P_{TV} = 12\ W$

$R_T = 20\ \Omega$ $I_T = 1\ A$ $V_T = V_S = 20\ V$ $P_T = 20\ W$

(b)

FIGURE 5-31 **Series Circuit Example with Values Inserted. (a) Schematic. (b) Circuit Analysis Table.**

      The nearest commercially available device would be a 10 W resistor. If size is not a consideration, it is ideal to double the wattage needed and use a 16 W resistor.

b. You may recall that any piece of equipment is equivalent to a load resistance. The TV set has 12 V across it and is pulling 1 A of current. Its load resistance can be calculated simply by using Ohm's law and deriving an equivalent circuit, as shown in Figure 5-31(a).

$$R_L \text{ (load resistance)} = \frac{V}{I}$$
$$= \frac{12\ V}{1\ A}$$
$$= 12\ \Omega$$

c. The amount of power being consumed by the TV set is

$$P = V \times I = 12\ V \times 1\ A = 12\ W$$

d. Figure 5-31(b) shows the circuit analysis table for this example.

   ▢ **EXAMPLE:**

Calculate the total amount of power dissipated in the series circuit in Figure 5-32, and insert any calculated values in a circuit analysis table.

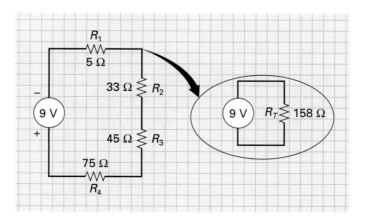

FIGURE 5-32 **Series Circuit Example.**

     CHAPTER 5 / SERIES DC CIRCUITS

Solution:

The total power dissipated in a series circuit is equal to the sum of all the power dissipated by all the resistors. The easiest way to calculate the total power is to simplify the circuit to one resistance, as shown in Figure 5-32.

$$R_T = R_1 + R_2 + R_3 + R_4$$
$$= 5\ \Omega + 33\ \Omega + 45\ \Omega + 75\ \Omega$$
$$= 158\ \Omega$$

We now have total resistance and total voltage, so we can calculate the total power:

$$P_T = \frac{V_S^2}{R_T} = \frac{9\ V^2}{158\ \Omega} = \frac{81}{158} = 512.7 \text{ milliwatts (mW)}$$

The longer method would have been to first calculate the current through the series current:

$$I = \frac{V_S}{R_T} = \frac{9\ V}{158\ \Omega} = 56.96 \text{ mA} \quad \text{or} \quad 57 \text{ mA}$$

We could then calculate the power dissipated by each separate resistor and add up all the individual values to gain a total power figure. This is illustrated in Figure 5-33(a).

$$P_T = P_1 + P_2 + P_3 + P_4 + \cdot \cdot \cdot$$

total power = addition of all the individual power losses

$$P_T = 16 \text{ mW} + 107 \text{ mW} + 146 \text{ mW} + 243 \text{ mW}$$
$$= 512 \text{ mW}$$

The calculated values for this example are shown in the circuit analysis table in Figure 5-33(b).

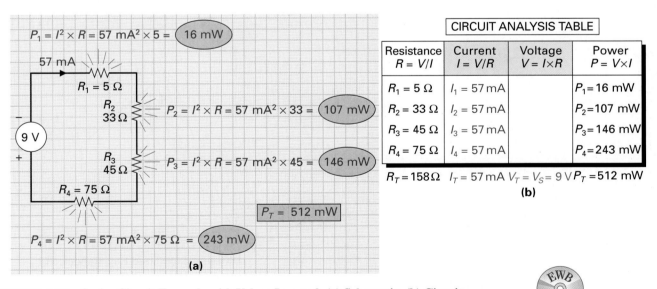

FIGURE 5-33 **Series Circuit Example with Values Inserted. (a) Schematic. (b) Circuit Analysis Table.**

5-5-1 *Maximum Power Transfer*

The **maximum power transfer theorem** states that maximum power will be delivered to a load when the resistance of the load (R_L) is equal to the resistance of the source (R_S). The best way to see if this theorem is correct is to apply it to a series of examples and then make a comparison.

EXAMPLE:

Figure 5-34(a) illustrates a 10 V battery with a 5 Ω internal resistance connected across a load. The load in this case is a light bulb, which has a load resistance of 1 Ω. Calculate the power delivered to this light bulb or load when $R_L = 1$ Ω.

Solution:

$$\text{circuit current, } I = \frac{V}{R}$$

$$= \frac{10 \text{ V}}{R_S + R_L}$$

$$= \frac{10 \text{ V}}{6 \text{ Ω}}$$

$$= 1.66 \text{ A}$$

The power supplied to the load is $P = I^2 \times R = 1.66^2 \times 1 \text{ Ω} = 2.8 \text{ W}$.

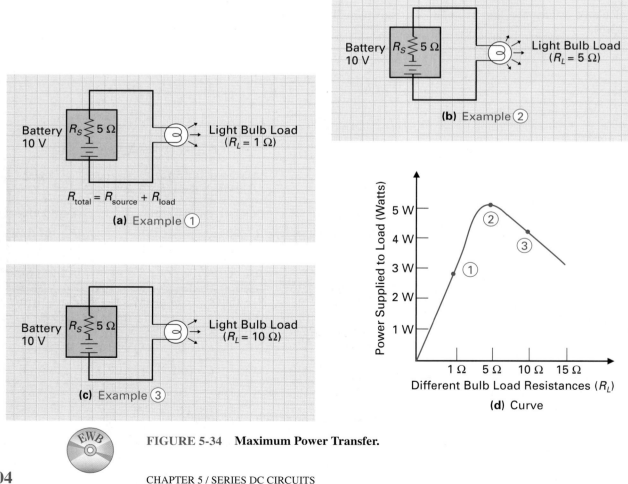

FIGURE 5-34   Maximum Power Transfer.

EXAMPLE:

Figure 5-34(b) illustrates the same battery and R_S, but in this case connected across a 5 Ω light bulb. Calculate the power delivered to this light bulb now that $R_L = 5$ Ω.

■ *Solution:*

$$
\begin{aligned}
I &= \frac{V}{R} \\
&= \frac{10 \text{ V}}{R_S + R_L} \\
&= \frac{10 \text{ V}}{10 \text{ Ω}} \\
&= 1 \text{ A}
\end{aligned}
$$

Power supplied is $P = I^2 \times R = 1^2 \times 5$ Ω $= 5$ W.

EXAMPLE:

Figure 5-34(c) illustrates the same battery again, but in this case a 10 Ω light bulb is connected in the circuit. Calculate the power delivered to this light bulb now that $R_L = 10$ Ω.

■ *Solution:*

$$
\begin{aligned}
I &= \frac{V}{R} \\
&= \frac{10 \text{ V}}{R_S + R_L} \\
&= \frac{10 \text{ V}}{15 \text{ Ω}} \\
&= 0.67 \text{ A}
\end{aligned}
$$

Thus $P = I^2 \times R = 0.67^2 \times 10$ Ω $= 4.5$ W.

As can be seen by the graph in Figure 5-34(d), which plots the power supplied to the load against load resistance, maximum power is delivered to the load (5 W) when the load resistance is equal to the source resistance.

The maximum power transfer condition is only used in special cases such as the automobile starter, where the load resistance remains constant and maximum power is needed. In most other cases, where load resistance can vary over a range of values, circuits are designed for a load resistance that will cause the best amount of power to be delivered. This is known as **optimum power transfer.**

Optimum Power Transfer

Since the ideal maximum power transfer conditions cannot always be achieved, most designers try to achieve optimum power transfer and have the source resistance and load resistance as close in value as possible.

SELF-TEST EVALUATION POINT FOR SECTION 5-5

Now that you have completed this section, you should be able to:

■ *Objective 9.* *Explain how to calculate power in a series circuit.*

■ *Objective 10.* *Explain the maximum power transfer theorem.*

Use the following questions to test your understanding of Section 5-5.

1. State the power formula.
2. Calculate the power dissipated by a 12 Ω resistor connected across a 12 V supply.
3. What fixed resistor type should probably be used for Question 2, and what would be a safe wattage rating?
4. What would be the total power dissipated if R_1 dissipates 25 W and R_2 dissipates 3800 mW?

A resistor will usually burn out and cause an open between its two leads when an excessive current flow occurs. This can normally, but not always, be noticed by a visual check of the resistor, which will appear charred due to the excessive heat. In some cases you will need to use your multimeter (combined ammeter, voltmeter, and ohmmeter) to check the circuit components to determine where a problem exists.

The two basic problems that normally exist in a series circuit are opens and shorts. In most instances a problem is not always as drastic as a short or an open, but may be a variation in a component's value over a long period of time, which will eventually cause a problem.

To summarize, then, we can say that one of three problems can occur to components in a series circuit:

1. A component will open (infinite resistance).
2. A component's value will change over a period of time.
3. A component will short (zero resistance).

The voltmeter is the most useful tool for checking series circuits as it can be used to measure voltage drops by connecting the meter leads across the component or resistor. Let's now analyze a circuit problem and see if we can solve it by logically **troubleshooting** the circuit and isolating the faulty component. To begin with, let us take a look at the effects of an open component.

Troubleshooting
The process of locating and diagnosing malfunctions or breakdowns in equipment by means of systematic checking or analysis.

5-6-1   *Open Component in a Series Circuit*

A component is open when its resistance is the maximum possible (infinity).

■ EXAMPLE:

Figure 5-35(a) illustrates a TV set with a load resistance of 3 Ω. The TV set is off because R_2 has burned out and become an open circuit. How would you determine that the problem is R_2?

■ *Solution:*

If an open circuit ever occurs in a series circuit, due in this case to R_2 having burned out, there can be no current flow, because series circuits have only one path for current to flow and that path has been broken ($I = 0$ A). Using the voltmeter to check the amount of voltage drop across each resistor, two results will be obtained:

1. The voltage drop across a good resistor will be zero volts.
2. The voltage drop across an open resistor will be equal to the source voltage V_S.

No voltage will be dropped across a good resistor because current is zero, and if $I = 0$, the voltage drop, which is the product of I and R, must be zero ($V = I \times R = 0 \times R = 0$ V). If no voltage is being dropped across the good resistor R_1 and the TV set resistance of 3 Ω, the entire source voltage will appear across the open resistor, R_2, in order that this series circuit comply with Kirchhoff's voltage law: V_S (9 V) $= V_{R1}$ (0 V) $+ V_{R2}$ (9 V) $+ V_L$ (0 V).

To explain this point further, refer to the fluid analogy in Figure 5-35(b). Like R_2, valve 2 has completely blocked any form of flow (water flow = 0). Looking at the pressure differences across all three valves, you can see that no pressure difference occurs across valves 1 and 3, but the entire pump pressure is appearing across valve 2, which is the component that has opened the circuit.

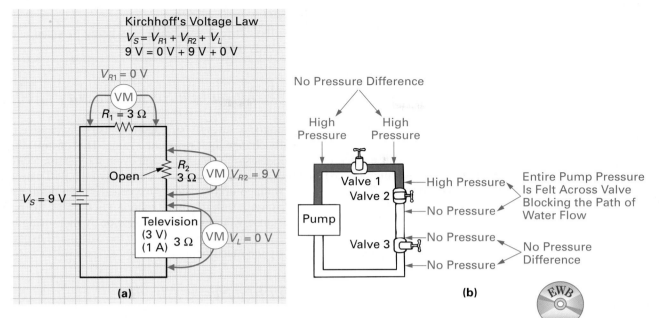

FIGURE 5-35   Troubleshooting an Open in a Series Circuit.

☐   **EXAMPLE:**

Figure 5-36 illustrates a set of three lights connected across a 9 V battery. Bulb 3 is open and therefore there is no current flow. With no current flow, all three bulbs are off, and we need to isolate which of the three is faulty.

▨   *Solution:*

Using the voltmeter, there will be:

1. Zero volts dropped across bulb 1 (9 V appears on both sides of bulb 1, so the potential difference or voltage drop across bulb 1 is zero), and so bulb 1 is OK.

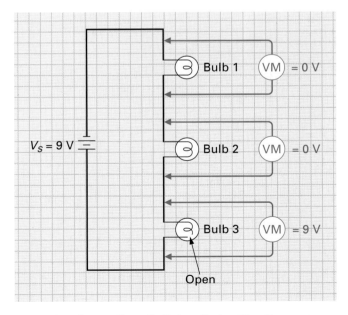

FIGURE 5-36   Troubleshooting an Open Bulb in a Series Circuit.

2. Zero volts dropped across bulb 2, and so bulb 2 is OK.

3. Nine volts dropped across bulb 3 (the entire source voltage is being dropped across bulb 3), and so bulb 3 is open and needs to be replaced.

5-6-2 Component Value Variation in a Series Circuit

Resistors will rarely go completely open unless severely stressed because of excessive current flow. With age, resistors will normally change their resistance value. This occurs slowly and will generally cause a decrease in the resistor's resistance and eventually cause a circuit problem. This lowering of resistance will cause an increase of current, which will cause an increase in the power dissipated. If the wattage of a resistor is exceeded, it can burn out. If they do not burn out but merely blow the circuit fuse due to an increase in current, the problem can be found by measuring the resistance values of each resistor or by measuring how much voltage is dropped across each resistor and comparing these to the calculated expected voltage, based on the parts list supplied by the manufacturer.

5-6-3 Shorted Component in a Series Circuit

A component has gone short when its resistance is 0 W. Let's work out a problem using the same example of the three bulbs across the 9 V battery, as shown in Figure 5-37.

EXAMPLE:

Bulbs 1 and 3 in Figure 5-37 are on and bulb 2 is off. A piece of wire exists between the two terminals of bulb 2, and the current is taking the lowest resistance path through the wire rather than through the filament resistor of the bulb. Bulb 2 therefore has no current flow through it, and light cannot be generated without current. How would you determine what the problem is?

Solution:

If bulb 2 was open (burned out), there could be no current flow in the circuit, so bulbs 1 and 2 would be off. Bulb 2 must therefore be shorted. Using the voltmeter to investigate this fur-

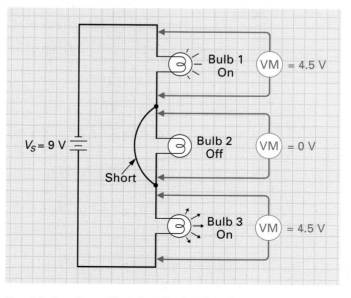

FIGURE 5-37   Troubleshooting a Short in a Series Circuit.

ther, you find that bulb 2 drops 0 V across it because it has no resistance except the very small wire resistance of the bypass, which means that 4.5 V must be dropped across each working bulb (1 and 3) in order to comply with Kirchhoff's voltage law. The loss of bulb 2's resistance causes the overall circuit resistance offered by bulbs 1 and 3 to decrease, which will cause an increase in current and so bulbs 1 and 3 should glow more brightly.

To summarize opens and shorts in series circuits, we can say that:

1. The supply voltage appears across an open component.
2. Zero volts appears across a shorted component.

SELF-TEST EVALUATION POINT FOR SECTION 5-6

Now that you have completed this section, you should be able to:

■ **Objective 11.** *Describe how to troubleshoot and recognize:*
  a. An open component
  b. A component value variation
  c. A short circuit in a series circuit

Use the following questions to test your understanding of Section 5-6.

1. List the three basic problems that can occur with components in a series circuit.
2. How can an open component be detected in a series circuit?
3. True or false: If a series-connected resistor's value were to decrease, the voltage drop across that same resistor would increase.
4. How could a shorted component be detected in a series circuit?

SUMMARY

Series Circuits (Figure 5-38)

1. A series circuit is the connecting of components end to end in a circuit to provide a single path for the current. In all cases the components are connected in succession or strung together one after another so that only one path for current exists between the negative (−) and positive (+) terminals of the supply.

2. The current in a series circuit has only one path to follow and cannot divert in any other direction. The current through a series circuit is the same throughout that circuit.

3. Resistance is the opposition to current flow, and in a series circuit every resistor in series offers opposition to the current flow.

4. The total resistance in a series-connected electronic resistive circuit is thus equal to the sum of all the individual resistances.

5. Total resistance (R_T) is the only opposition the voltage source can sense. It does not see the individual separate resistors, but one equivalent resistance. Based on its voltage and this total resistance, a value of current will be produced to flow through the circuit (Ohm's law, $I = V/R$).

6. A potential difference or voltage drop will occur across each resistor in a series circuit when current is flowing. The amount of voltage drop is dependent on the value of the resistor and the amount of current flow.

7. The voltage drop across resistors can be found by utilizing Ohm's law: $V = I \times R$.

8. The series circuit divides up the applied voltage, and it appears proportionally across all the individual resistors. This characteristic was first observed by Gustav Kirchhoff in 1847; in honor of his discovery, this effect is known as Kirchhoff's voltage law, which states: The sum of the voltage drops in a series circuit is equal to the total voltage applied.

9. If only one resistor is connected in a series circuit, the entire applied voltage appears across this resistor. The amount of current flow is determined by the value of this single resistor and remains the same throughout the circuit.

10. A series-connected circuit is often referred to as a *voltage-divider circuit*. The total voltage applied (V_T) or source voltage (V_S) is divided and dropped across the resistors in the series circuit. The amount of voltage dropped across a resistor is proportional to the value of resistance, and so a larger resistance causes a larger voltage drop across that resistor, while a smaller resistance causes a smaller voltage drop.

11. The voltage-divider formula allows you to calculate the voltage drop across any resistor without having to work out current.

12. The voltage-divider formula can be summarized by saying that: The voltage drop across a resistor or multiple resistors in a series circuit is equal to the ratio of that resistance (R_X) to the total resistance (R_T) multiplied by the source voltage (V_S).

13. A potentiometer, or variable voltage divider, consists of a fixed value of resistance between two terminals and a wiper that can be adjusted to vary resistance between its terminal and one of the other two. The resistances between terminals A and B and between B and C are inversely proportional to one another.

14. The potentiometer can be adjusted to supply different voltages on the wiper.

15. Power is the rate at which work is done. Work is said to have been done when energy is converted, in this case from electrical energy to heat energy.

16. Resistors dissipate heat energy, and the rate at which they dissipate energy is called *power* and is measured in *watts* (joules per second).

17. Resistors all have a resistive value, a tolerance, and a wattage rating. The wattage of a resistor is the amount of heat energy a resistor will dissipate per second. This wattage is directly proportional to the resistor's size and a larger resistor will be able to dissipate more heat than a smaller resistor.

18. The larger the wattage, the greater the cost and the greater the size and area the resistor occupies within the equipment.

19. The maximum power transfer theorem states that maximum power will be delivered to the load when the load resistance (R_L) is equal to the source resistance (R_S).

20. A resistor will usually burn out and result in an open between its two leads when an excessive current flow occurs. This can normally, but not always, be noticed by a visual check of the resistor, which will appear charred due to the excessive heat. In some cases you will need to use your multimeter to check the circuit components to determine where a problem exists.

21. The voltmeter is the most useful tool when checking series circuits as it can be used to measure voltage drops by connecting the meter leads across the component or resistors.

22. Using the voltmeter to check the amount of voltage drop across each resistor, two results will be obtained.
   a. The voltage drop across a good resistor will be zero volts.
   b. The voltage drop across an open resistor will be equal to the source voltage V_S.

23. With age, resistors will normally change their resistance value. This occurs slowly and will generally always cause a decrease in the resistor's resistance and eventually cause a circuit problem. This lowering of resistance will cause an increase of current, which will cause an increase in the power dissipated; if the wattage of the resistor is exceeded, it can burn out. If they do not burn out but merely blow the circuit fuse due to this increase in current, the problem can be found by measuring the resistance values of each resistor or by measuring how much voltage is dropped across each resistor and comparing these to the calculated voltage, based on the parts list supplied by the manufacturer.

24. To summarize opens and shorts in series circuits, we can say that:
   a. The supply voltage appears across an open component.
   b. Zero volts appears across a shorted component.

REVIEW QUESTIONS

Multiple-Choice Questions

1. A series circuit:
   a. Is the connecting of components end to end
   b. Provides a single path for current
   c. Functions as a voltage divider
   d. All of the above

2. The total current in a series circuit is equal to:
   a. $I_1 + I_2 + I_3 + \ldots$ **c.** $I_1 = I_2 = I_3 = \ldots$
   b. $I_1 - I_2$ **d.** All of the above

3. If R_1 and R_2 are connected in series with a total current of 2 A, what will be the current flowing through R_1 and R_2, respectively?
   a. 1 A, 1 A
   b. 2 A, 1 A
   c. 2 A, 2 A
   d. All of the above could be true on some occasions.

4. The total resistance in a series circuit is equal to:
   a. The total voltage divided by the total current
   b. The sum of all the individual resistor values
   c. $R_1 + R_2 + R_3 + \ldots$
   d. All of the above
   e. None of the above are even remotely true.

5. Which of Kirchhoff's laws applies to series circuits:
   a. His voltage law
   b. His current law
   c. His power law
   d. None of them apply to series circuits, only parallel.

6. The amount of voltage dropped across a resistor is proportional to:
   a. The value of the resistor
   b. The current flow in the circuit
   c. Both (a) and (b)
   d. None of the above

7. If three resistors of 6 kΩ, 4.7 kΩ, and 330 Ω are connected in series with one another, what total resistance will the battery sense?
   a. 11.03 MΩ **c.** 6 kΩ
   b. 11.03 Ω **d.** 11.03 kΩ

FIGURE 5-38   Series Circuits.

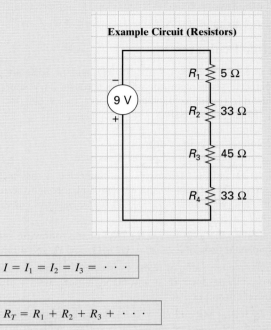

Example Circuit (Resistors)

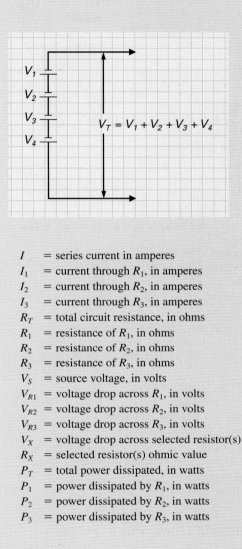

$$I = I_1 = I_2 = I_3 = \cdots$$

$$R_T = R_1 + R_2 + R_3 + \cdots$$

$$V_S = V_{R1} + V_{R2} + V_{R3} + \cdots$$

$$\frac{V_S}{I_T \mid R_T} \qquad I_T = \frac{V_S}{R_T}$$

$$V_{R1} = I_1 \times R_1,$$
$$V_{R2} = I_2 \times R_2,$$
$$V_{R3} = I_3 \times R_3, \text{ and so on}$$

Voltage-divider formula

$$V_X = \left(\frac{R_X}{R_T}\right) \times V_S$$

$$P_T = P_1 + P_2 + P_3 + \cdots$$

$$P_T = \frac{V_S{}^2}{R_T}$$
$$P_T = I^2 \times R_T$$
$$P_T = V_S \times I$$

I	= series current in amperes
I_1	= current through R_1, in amperes
I_2	= current through R_2, in amperes
I_3	= current through R_3, in amperes
R_T	= total circuit resistance, in ohms
R_1	= resistance of R_1, in ohms
R_2	= resistance of R_2, in ohms
R_3	= resistance of R_3, in ohms
V_S	= source voltage, in volts
V_{R1}	= voltage drop across R_1, in volts
V_{R2}	= voltage drop across R_2, in volts
V_{R3}	= voltage drop across R_3, in volts
V_X	= voltage drop across selected resistor(s)
R_X	= selected resistor(s) ohmic value
P_T	= total power dissipated, in watts
P_1	= power dissipated by R_1, in watts
P_2	= power dissipated by R_2, in watts
P_3	= power dissipated by R_3, in watts

8. The voltage-divider formula states that the voltage drop across a resistor or multiple resistors in a series circuit is equal to the ratio of that _____ to the _____ multiplied by the _____.
  a. Resistance, source voltage, total resistance
  b. Resistance, total resistance, source voltage
  c. Total current, resistance, total voltage
  d. Total voltage, total current, resistance

9. The _____ can be used as a variable voltage divider.
  a. Potentiometer c. SPDT switch
  b. Fixed resistor d. None of the above

10. A resistor of larger physical size will be able to dissipate _____ heat than a small resistor.
  a. More c. About the same
  b. Less d. None of the above

11. The _____ is the most useful tool when checking series circuits.
  a. Ammeter c. Voltmeter
  b. Wattmeter d. Both (a) and (b)

12. When an open component occurs in a series circuit, it can be noticed because:
  a. Zero volts appears across it
  b. The supply voltage appears across it
  c. 1.3 V appears across it
  d. None of the above

13. Power can be calculated by:
  a. The addition of all the individual power figures
  b. The product of the total current and the total voltage
  c. The square of the total voltage divided by the total resistance
  d. All of the above

14. A series circuit is known as a:
  a. Current divider c. Current subtractor
  b. Voltage divider d. All of the above.

15. In a series circuit only _____ path(s) exists for current flow, while the voltage applied is distributed across all the individual resistors.
  a. Three c. Four
  b. Several d. One

Communication Skill Questions

16. Describe a series-connected circuit. (5-1)

17. Describe and state mathematically what happens to current flow in a series circuit. (5-2)

18. Describe how total resistance can be calculated in a series circuit. (5-3)

19. Describe why voltage is dropped across a series circuit and how each voltage drop can be calculated. (5-4)

20. Briefly describe why resistance and voltage drops are proportional to one another. (5-4)

21. Describe a fixed and a variable voltage divider. (5-4-2 and 5-4-3)

22. How can individual and total power be calculated in a series circuit? (5-5)

23. How can you recognize shorts and opens in a series circuit when troubleshooting with a voltmeter? (5-6)

24. List the three problems that can occur in a series circuit. (5-6)

25. State Kirchhoff's voltage (series circuit) law. (5-4)

Practice Problems

26. If three resistors of 1.7 kΩ, 3.3 kΩ, and 14.4 kΩ are connected in series with one another across a 24 V source as shown in Figure 5-39, calculate:
  a. Total resistance (R_T)
  b. Circuit current
  c. Individual voltage drops
  d. Individual and total power dissipated

27. If 40 Ω and 35 Ω resistors are connected across a 24 V source, what would be the current flow through the resistors, and what resistance would cause half the current to flow?

28. Calculate the total resistance (R_T) of the following series-connected resistors: 2.7 kΩ, 3.4 MΩ, 370 Ω, and 4.6 MΩ.

29. Calculate the value of resistors needed to divide up a 90 V source to produce 45 V and 60 V outputs, with a divider circuit current of 1 A.

30. If R_1 = 4.7 kΩ and R_2 = 6.4 kΩ and both are connected across a 9 V source, how much voltage will be dropped across R_2?

31. What current would flow through R_1 if it were one-third the ohmic value of R_2 and R_3, and all were connected in series with a total current of 6.5 mA flowing out of V_S?

FIGURE 5-39

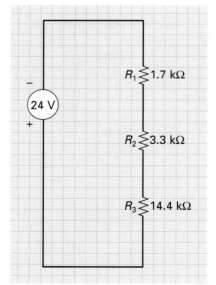

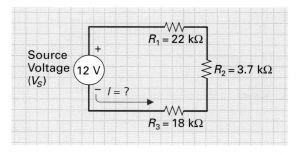

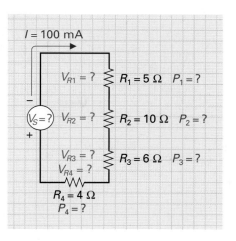

FIGURE 5-40

FIGURE 5-42

32. Draw a circuit showing $R_1 = 2.7$ kΩ, $R_2 = 3.3$ kΩ, and $R_3 = 0.027$ MΩ in series with one another across a 20 V source. Calculate:

   a. I_T **d.** P_2 **g.** V_{R2}
   b. P_T **e.** P_3 **h.** V_{R3}
   c. P_1 **f.** V_{R1} **i.** I_1

33. Calculate the current flowing through three light bulbs that are dissipating 120 W, 60 W, and 200 W when they are connected in series across a 120 V source. How is the voltage divided around the series circuit?

34. If three equal-value resistors are series connected across a dc power supply adjusted to supply 10 V, what percentage of the source voltage will appear across R_1?

35. Refer to the following figures and calculate:

   a. I (Figure 5-40)
   b. R_T and P_T (Figure 5-41)
   c. V_S, V_{R1}, V_{R2}, V_{R3}, V_{R4}, P_1, P_2, P_3, and P_4 (Figure 5-42)
   d. P_T, I, R_1, R_2, R_3, and R_4 (Figure 5-43)

Troubleshooting Questions

36. If three bulbs are connected across a 9 V battery in series, and the filament in one of the bulbs burned out, causing an open in the bulb, would the other lamps be on? Explain why.

37. Using a voltmeter, how would a short be recognized in a series circuit?

38. If one of three series-connected bulbs is shorted, will the other two bulbs be on? Explain why.

39. When one resistor in a series string is open, explain what would happen to the circuit's:

   a. Current
   b. Resistance
   c. Voltage across the open component
   d. Voltage across the other components

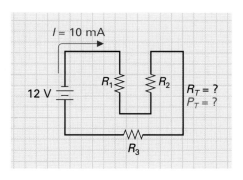

FIGURE 5-41

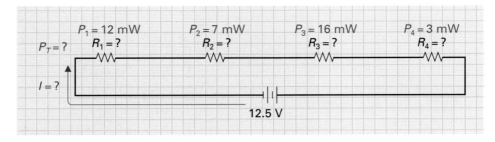

FIGURE 5-43

40. When one resistor in a series string is shorted, explain what would happen to the circuit's:
  a. Current
  b. Resistance
  c. Voltage across the shorted component
  d. Voltage across the other components

Web Site Questions

Go to the Web site http://www.prenhall.com/cook, select the textbook *Introductory DC/AC Electronics* or *Introductory DC/AC Circuits*, this chapter, and then follow the instructions when answering the multiple-choice practice problems.

These tests at the end of each chapter will challenge your knowledge up to this point, and give you the practice you need for a job interview. To make this more realistic, the test will be comprised of both technical and personal questions. In order to take full advantage of this exercise, you may want to set up a simulation of the interview environment, have a friend read the questions to you, and record your responses for later analysis.

Company Name: **NPC, Inc.**

Industry Branch: **Consumer Products.**

Function: **Audio/Video, Sales/Service.**

Job Title: **Customer Service Technician.**

1. What do you know about our product line?
2. What is a series circuit?
3. How do you use a multimeter to measure resistance?

4. Would you object to attending a six week out-of-state training course?
5. What originally attracted you to the electronics industry?
6. What is meant by the expression "loading a battery"?
7. Would you say you're a self-starter?
8. What audio/video equipment do you own?
9. What would you say are the duties of a customer service technician?
10. Are you someone who enjoys interaction on a daily basis?

Answers

1. Visit the company's web site before the interview to get an overall understanding of the company's ownership, product line, service, and support.
2. Section 5-1.
3. Chapter 2.
4. Additional training only serves to increase your skills and marketability. If you are unsure, however, don't object to anything at the time. Later, if you decide that it is something you do not want to do, you can call and decline the offer.
5. Describe early interest and reasons for pursuing a career in electronics.

6. Section 5-4-1.
7. Discuss how you have disciplined yourself to manage school, studying, and a job, and how well you performed in all. Also, describe the lab environment, which was more than likely self-paced.
8. Discuss systems, and mention any modifications or installations you have performed (such as installing your car's music system, etc.).
9. Quote intro. section in text and job description listed in newspaper.
10. They're asking if you can get along well with customers and colleagues. Mention any work experience in which you had direct contact with customers, and discuss how you worked with fellow students in your class.

Parallel DC Circuits

An Apple a Day

One of the early pioneers who laid the foundations of many branches of science was Isaac Newton. He was born in a small farmhouse near Woolsthorpe in Lincolnshire, England, on Christmas Day in 1642. He was an extremely small, premature baby, which worried the midwives who went off to get medicine and didn't expect to find him alive when they came back. Luckily for science, however, he did survive.

Newton's father was an illiterate farmer who died three months before he was born. His mother married the local vicar soon after Newton's birth and left him in the care of his grandmother. This parental absence while he was growing up had a traumatic effect on him and throughout his life affected his relationships with people.

At school in the nearby town of Grantham, Newton showed no interest in classical studies but rather in making working models and studying the world around him. When he was in his early teens, his stepfather died and Newton had to return to the farm to help his mother. Newton proved to be a hopeless farmer; in fact, on one occasion when he was tending sheep he became so engrossed with a stream that he followed it for miles and was missing for hours. Luckily, a schoolteacher recognized Newton's single-minded powers of concentration and convinced his mother to let him return to school, where he performed better and later went off to Cambridge University.

In 1665, in his graduation year, Newton left Cambridge to return home to escape an epidemic of bubonic plague that had spread throughout London. During this time, Newton reflected on his years of seclusion at his mother's cottage and called them the most significant time in his life. It was here on a warm summer's day that Newton saw the apple fall to the ground, leading him to develop his laws of motion and gravitation. It was here that he wondered about the nature of light and later built a prism and proved that white light contains all the colors in a rainbow.

Later, when Newton returned to Cambridge, he demonstrated many of his discoveries but was reluctant to publish the details and did so finally only at the insistence of others. Newton went on to build the first working reflecting astronomical telescope and wrote a paper on optics that was fiercely challenged by the physicist Robert Hooke. Hooke quarreled bitterly with Newton over the years, and there were also heated debates about whether Newton or the German mathematician Gottfried Leibniz invented calculus.

The truth is that many of Newton's discoveries roamed around with him in the English countryside, and even though many of these would, before long, have been put forward by others, it was Newton's genius and skill (and long walks) that tied together all the loose ends.

Outline and Objectives

By tracing the path of current, we can determine whether a circuit has series-connected or parallel-connected components. In a series circuit, there is only one path for current, whereas in parallel circuits the current has two or more paths. These paths are known as *branches*. A parallel circuit, by definition, is when two or more components are connected to the same voltage source so that the current can branch out over two or more paths. In a parallel resistive circuit, two or more resistors are connected to the same source, so current splits to travel through each separate branch resistance.

6-1 COMPONENTS IN PARALLEL

Parallel Circuit

Also called shunt; circuit having two or more paths for current flow.

Many components, other than resistors, can be connected in parallel, and a **parallel circuit** can easily be identified because current is split into two or more paths. Being able to identify a parallel connection requires some practice, because they can come in many different shapes and sizes. The means for recognizing series circuits is that if you can place your pencil at the negative terminal of the voltage source (battery) and follow the wire connections through components to the positive side of the battery and only have one path to follow, the circuit is connected in series. If, however, you can place your pencil at the negative terminal of the voltage source and follow the wire and at some point have a choice of two or more routes, the circuit is connected with two or more parallel branches. The number of routes determines the number of parallel branches. Figure 6-1 illustrates five examples of parallel resistive circuits.

EXAMPLE:

Figure 6-2(a) illustrates four resistors laid out on a table top.

a. With wire leads, connect all four resistors in parallel on a protoboard, and then connect the circuit to a dc power supply.
b. Draw the schematic diagram of the parallel-connected circuit.

■ *Solution:*

Figure 6-2(b) shows how to connect the resistors in parallel and the circuit's schematic.

EXAMPLE:

Figure 6-3(a) shows four 1.5 V cells and three lamps. Using wires, connect all of the cells in parallel to create a 1.5 V source. Then connect all of the three lamps in parallel with one another and finally connect the 1.5 V source across the three parallel-connected lamp load.

■ *Solution:*

In Figure 6-3(b) you can see the final circuit containing a source consisting of four parallel-connected 1.5 V cells, and a load consisting of three parallel-connected lamps. As explained in Chapter 4, when cells are connected in parallel, the total voltage remains the same as for one cell, but the current demand can now be shared.

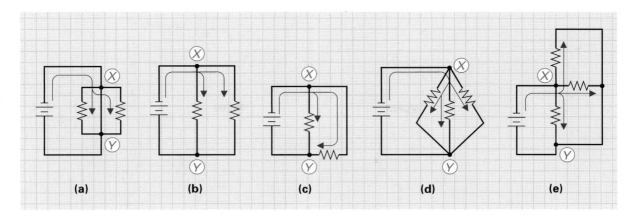

FIGURE 6-1   Parallel Circuits.

6-2   VOLTAGE IN A PARALLEL CIRCUIT

Figure 6-4(a) shows a simple circuit with four resistors connected in parallel across the voltage source of a 9 V battery. The current from the negative side of the battery will split between the four different paths or branches, yet the voltage drop across each branch of a parallel circuit is equal to the voltage drop across all the other branches in parallel. This means that if the voltmeter were to measure the voltage across *A* and *B* or *C* and *D* or *E* and *F* or *G* and *H,* they would all be the same or, in this example, would all drop 9 V.

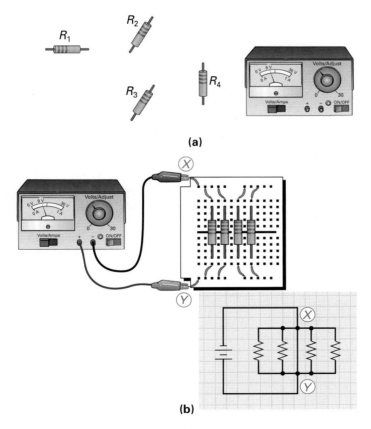

FIGURE 6-2   Connecting Resistors in Parallel. (a) Problem. (b) Solution.

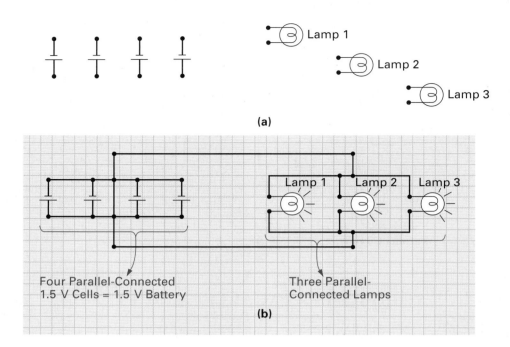

FIGURE 6-3　Parallel-Connected Cells and Lamps.

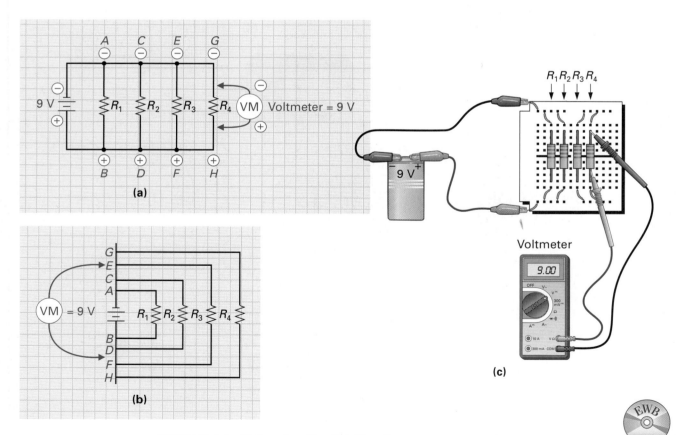

FIGURE 6-4　Voltage in a Parallel Circuit.

　　　　CHAPTER 6 / PARALLEL DC CIRCUITS

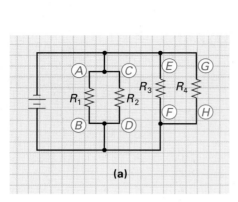
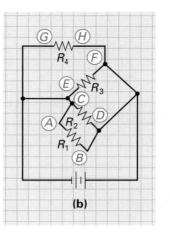

FIGURE 6-5   Parallel-Circuit Voltage Drop.

It is quite easy to imagine why there will be the same voltage drop across all the resistors, seeing that points *A, C, E,* and *G* are all one connection and points *B, D, F,* and *H* are all one connection. Measuring the voltage drop with the voltmeter across any of the resistors is the same as measuring the voltage across the battery, as shown in Figure 6-4(b). As long as the voltage source remains constant, the voltage drop will always be common (9 V) across the parallel resistors, no matter what value or how many resistors are connected in parallel. The voltmeter is therefore measuring the voltage between two common points that are directly connected to the battery, and the voltage dropped across all these parallel resistors will be equal to the source voltage.

In Figure 6-5(a) and (b), the same circuit is shown in two different ways, so you can see how the same circuit can look completely different. In both examples, the voltage drop across any of the resistors will always be the same and, as long as the voltage source is not heavily loaded, equal to the source voltage. Just as you can trace the positive side of the battery to all four resistors, you can also trace the negative side to all four resistors.

Mathematically stated, we can say that in a parallel circuit

$$V_{R1} = V_{R2} = V_{R3} = V_{R4} = V_S$$

voltage drop across R_1 = voltage drop across R_2 = voltage drop
across R_3 (etc.) = source voltage

To reinforce the concept, let us compare this parallel circuit characteristic to a water analogy, as seen in Figure 6-6. The pressure across valves *A* and *B* will always be the same,

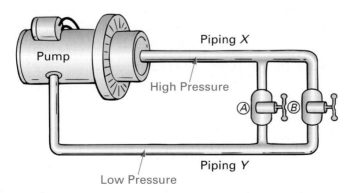

FIGURE 6-6   Fluid Analogy of Parallel-Circuit Pressure.

even if one offers more opposition than the other. This is because the pressure measured across either valve will be the same as checking the pressure difference between piping X and Y. Since the piping at point X and Y run directly back to the pump, the pressure across A and B is the same as the pressure difference across the pump.

■ EXAMPLE:

Refer to Figure 6-7 and calculate:

    a. Voltage drop across R_1

    b. Voltage drop across R_2

    c. Voltage drop across R_3

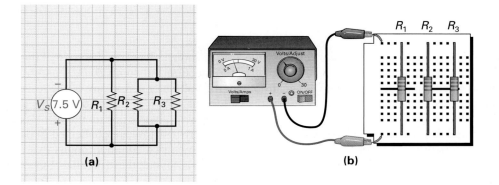

FIGURE 6-7   **A Parallel Circuit Example (a) Schematic. (b) Protoboard Circuit.**

■ *Solution:*

Since all these resistors are connected in parallel, the voltage across every branch will be the same and equal to the source voltage applied. Therefore,

$$V_{R1} = V_{R2} = V_{R3} = V_S$$
$$7.5 \text{ V} = 7.5 \text{ V} = 7.5 \text{ V} = 7.5 \text{ V}$$

SELF-TEST EVALUATION POINT FOR SECTIONS 6-1 AND 6-2

Now that you have completed these sections, you should be able to:

■ ***Objective 1.*** *Describe the difference between a series and a parallel circuit.*

■ ***Objective 2.*** *Be able to recognize and determine whether circuit components are connected in series or parallel.*

■ ***Objective 3.*** *Explain why voltage measures the same across parallel-connected components.*

Use the following questions to test your understanding of Sections 6-1 and 6-2.

1. Describe a parallel circuit.

2. True or false: A parallel circuit is also known as a voltage-divider circuit.

3. What would be the voltage drop across R_1 if $V_S = 12$ V and R_1 and R_2 are both in parallel with one another and equal to 24 Ω each?

4. Can Kirchhoff's voltage law be applied to parallel circuits?

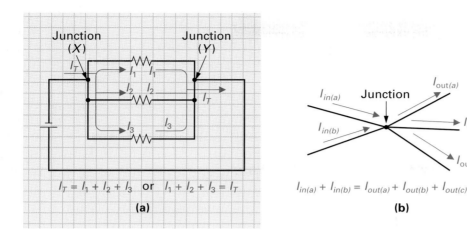

FIGURE 6-8   Kirchhoff's Current Law.

6-3   CURRENT IN A PARALLEL CIRCUIT

In addition to providing the voltage law for series circuits, Gustav Kirchhoff (in 1847) was the first to observe and prove that the sum of all the **branch currents** in a parallel circuit ($I_1 + I_2 + I_3$, etc.) was equal to the total current (I_T). In honor of his second discovery, this phenomenon is known as **Kirchhoff's current law,** which states that the sum of all the currents entering a junction is equal to the sum of all the currents leaving that same junction.

Figure 6-8(a) and (b) illustrate two examples of how this law applies. In both examples, the sum of the currents entering a junction is equal to the sum of the currents leaving that same junction. In Figure 6-8(a) the total current arrives at a junction X and splits to produce three branch currents, I_1, I_2, and I_3, which cumulatively equal the total current (I_T) that arrived at the junction X. The same three branch currents combine at junction Y, and the total current (I_T) leaving that junction is equal to the sum of the three branch currents arriving at junction Y.

$$I_T = I_1 + I_2 + I_3 + I_4 = \dots$$

As another example, in Figure 6-8(b) you can see that there are two branch currents entering a junction [$I_{in(a)}$ and $I_{in(b)}$] and three branch currents leaving that same junction [$I_{out(a)}$, $I_{out(b)}$, and $I_{out(c)}$]. As stated below the illustration, the sum of the input currents will equal the sum of the output currents: $I_{in(a)} + I_{in(b)} = I_{out(a)} + I_{out(b)} + I_{out(c)}$.

> **Branch Current**
>
> A portion of the total current that is present in one path of a parallel circuit.

> **Kirchhoff's Current Law**
>
> The sum of the currents flowing into a point in a circuit is equal to the sum of the currents flowing out of that same point.

☐ **EXAMPLE:**

Refer to Figure 6-9 and calculate the value of I_T.

■ *Solution:*

By Kirchhoff's current law,

$$I_T = I_1 + I_2 + I_3 + I_4$$
$$? = 2 \text{ mA} + 17 \text{ mA} + 7 \text{ mA} + 37 \text{ mA}$$
$$I_T = 63 \text{ mA}$$

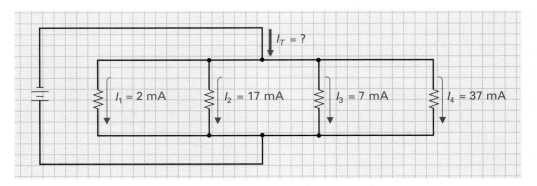

FIGURE 6-9   Calculating Total Current.

■ EXAMPLE:

Refer to Figure 6-10 and calculate the value of I_1.

FIGURE 6-10   Calculating Branch Current.

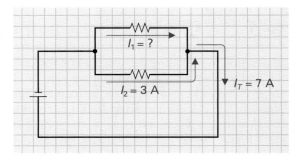

■ *Solution:*

By transposing Kirchhoff's current law, we can determine the unknown value (I_1):

$$I_T = I_1 + I_2 \qquad\qquad I_T = I_1 + I_2 \qquad (-I_2)$$
$$7\,\text{A} = ? + 3\,\text{A} \quad \text{or} \quad I_T - I_2 = I_1$$
$$I_1 = 4\,\text{A} \qquad\qquad I_1 = I_T - I_2 = 7\,\text{A} - 3\,\text{A} = 4\,\text{A}$$

As with series circuits, to find out how much current will flow through a parallel circuit, we need to find out how much opposition or resistance is being connected across the voltage source.

$$I_T = \frac{V_S}{R_T}$$

Total current equals source voltage divided by total resistance

When we connect resistors in parallel, the total resistance in the circuit will actually decrease. In fact, the total resistance in a parallel circuit will always be less than the value of the smallest resistor in the circuit.

To prove this point, Figure 6-11 shows how two sets of identical resistors (R_1, R_2, and R_3) were used to build both a series and a parallel circuit. The total current flow in the parallel circuit would be larger than the total current in the series circuit, because the parallel circuit has two or more paths for current to flow, while the series circuit only has one.

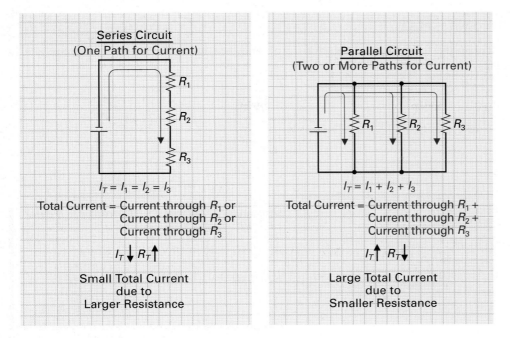

FIGURE 6-11   Series Circuit and Parallel Circuit Current Comparison.

To explain why the total current will be larger in a parallel circuit, let us take the analogy of a freeway with only one path for traffic to flow. A single-lane freeway is equivalent to a series circuit, and only a small amount of traffic is allowed to flow along this freeway. If the freeway is expanded to accommodate two lanes, a greater amount of traffic can flow along the freeway in the same amount of time. Having more lanes permits a greater total amount of traffic flow. In parallel circuits, more branches will allow a greater total amount of current flow because there is less resistance in more paths than there is with only one path. This concept is summarized in Figure 6-11.

Just as a series circuit is often referred to as a voltage-divider circuit, a parallel circuit is often referred to as a **current-divider** circuit, because the total current arriving at a junction will divide or split into branch currents (Kirchhoff's law), as shown in Figure 6-12.

The current division is inversely proportional to the resistance in the branch, assuming that the voltage across both resistors is constant and equal to the source voltage (V_S). This means that a large branch resistance will cause a small branch current ($I\downarrow = V/R\uparrow$), and a small branch resistance will cause a large branch current ($I\uparrow = V/R\downarrow$).

Current Divider

A parallel network designed to proportionally divide the circuit's total current.

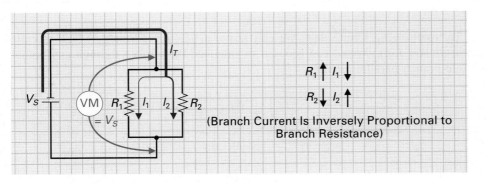

FIGURE 6-12   The Parallel Circuit Current Divider.

Calculate the following for Figure 6-13(a), and then insert the values in a circuit analysis table.

   a. I_1

   b. I_2

   c. I_T

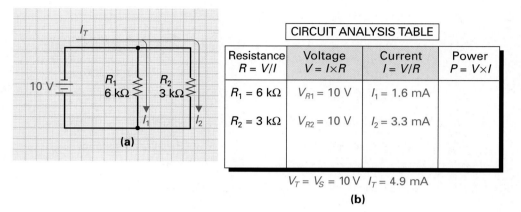

	CIRCUIT ANALYSIS TABLE		
Resistance $R = V/I$	Voltage $V = I \times R$	Current $I = V/R$	Power $P = V \times I$
$R_1 = 6\ k\Omega$	$V_{R1} = 10\ V$	$I_1 = 1.6\ mA$	
$R_2 = 3\ k\Omega$	$V_{R2} = 10\ V$	$I_2 = 3.3\ mA$	

$$V_T = V_S = 10\ V \quad I_T = 4.9\ mA$$

(b)

FIGURE 6-13   Parallel Circuit Example. (a) Schematic. (b) Circuit Analysis Table.

■   *Solution:*

Since R_1 and R_2 are connected in parallel across the 10 V source, the voltage across both resistors will be 10 V.

   a. $I_1 = \dfrac{V_{R1}}{R_1} = \dfrac{10\ V}{6\ k\Omega} = 1.6\ mA$ (smaller branch current through larger branch resistance)

   b. $I_2 = \dfrac{V_{R2}}{R_2} = \dfrac{10\ V}{3\ k\Omega} = 3.3\ mA$ (larger branch current through smaller branch resistance)

   c. By Kirchhoff's current law,

$$I_T = I_1 + I_2$$
$$= 1.6\ mA + 3.3\ mA$$
$$= 4.9\ mA$$

The circuit analysis table for this example is shown in Figure 6-13(b).

By rearranging Ohm's law, we can arrive at another formula, which is called the *current-divider formula* and can be used to calculate the current through any branch of a multiple-branch parallel circuit.

$$I_x = \frac{R_T}{R_x} \times I_T$$

where I_x = branch current desired
R_T = total resistance
R_x = resistance in branch
I_T = total current

☐ EXAMPLE:

Refer to Figure 6-14 and calculate the following if the total circuit resistance (R_T) is equal to 1 kΩ:

a. $I_1 =$

b. $I_2 =$

c. $I_3 =$

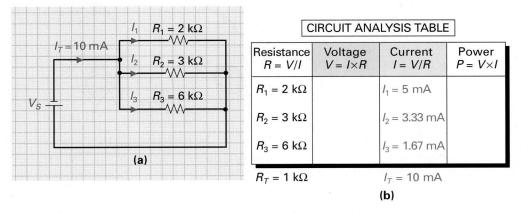

Resistance $R = V/I$	Voltage $V = I \times R$	Current $I = V/R$	Power $P = V \times I$
$R_1 = 2\ k\Omega$		$I_1 = 5\ mA$	
$R_2 = 3\ k\Omega$		$I_2 = 3.33\ mA$	
$R_3 = 6\ k\Omega$		$I_3 = 1.67\ mA$	
$R_T = 1\ k\Omega$		$I_T = 10\ mA$	

(b)

FIGURE 6-14   Parallel Circuit Example. (a) Schematic. (b) Circuit Analysis Table.

☐ *Solution:*

Since the source and therefore the voltage across each branch resistor are not known, we will use the current-divider formula to calculate I_1, I_2, and I_3.

a. $I_1 = \dfrac{R_T}{R_1} \times I_T = \dfrac{1\ k\Omega}{2\ k\Omega} \times 10\ mA = 5\ mA$

   (smallest branch resistance has largest branch current)

b. $I_2 = \dfrac{R_T}{R_2} \times I_T = \dfrac{1\ k\Omega}{3\ k\Omega} \times 10\ mA$

   $= 3.33\ mA$

c. $I_3 = \dfrac{R_T}{R_3} \times I_T = \dfrac{1\ k\Omega}{6\ k\Omega} \times 10\ mA$

   $= 1.67\ mA$

   (largest branch resistance has smallest branch current)

To double-check that the values for I_1, I_2, and I_3 are correct, you can apply Kirchhoff's current law, which is

$$I_T = I_1 + I_2 + I_3$$
$$10\ mA = 5\ mA + 3.33\ mA + 1.67\ mA$$
$$= 10\ mA$$

CALCULATOR SEQUENCE

Step	Keypad Entry	Display Response
1.	[1] [E] [3]	1. 3
2.	[÷]	
3.	[2] [E] [3]	2. 3
4.	[×]	0.5
5.	[1] [0] [E] [3] [+/−]	10 −3
6.	[=]	5.−03

☐ EXAMPLE:

A common use of parallel circuits is in the residential electrical system. All of the household lights and appliances are wired in parallel, as seen in the typical room wiring circuit in Figure 6-15(a). If it is a cold winter morning, and lamps 1 and 2 are switched ON, together with

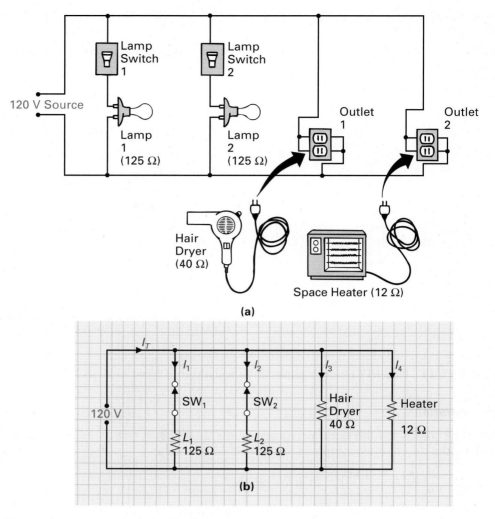

(a)

(b)

FIGURE 6-15   Parallel Home Electrical System.

the space heater and hair dryer, what will the individual branch currents be, and what will be the total current drawn from the source?

■ *Solution:*

Figure 6-15(b) shows the schematic of the pictorial in Figure 6-15(a). Since all resistances are connected in parallel across a 120 V source, the voltage across all devices will be 120 V. Using Ohm's law we can calculate the four branch currents:

$$I_1 = \frac{V_{\text{lamp1}}}{R_{\text{lamp1}}} = \frac{120 \text{ V}}{125 \text{ }\Omega} = 960 \text{ mA}$$

$$I_2 = \frac{V_{\text{lamp2}}}{R_{\text{lamp2}}} = \frac{120 \text{ V}}{125 \text{ }\Omega} = 960 \text{ mA}$$

$$I_3 = \frac{V_{\text{hairdryer}}}{R_{\text{hairdryer}}} = \frac{120 \text{ V}}{40 \text{ }\Omega} = 3 \text{ A}$$

$$I_4 = \frac{V_{\text{heater}}}{R_{\text{heater}}} = \frac{120 \text{ V}}{12 \text{ }\Omega} = 10 \text{ A}$$

By Kirchhoff's current law,

$$I_T = I_1 + I_2 + I_3 + I_4$$
$$= 960 \text{ mA} + 960 \text{ mA} + 3 \text{ A} + 10 \text{ A}$$
$$= 14.92 \text{ A}$$

SELF-TEST EVALUATION POINT FOR SECTION 6-3

Now that you have completed this section, you should be able to:

■ *Objective 4.* *State Kirchhoff's current law.*

■ *Objective 5.* *Describe why branch current and resistance are inversely proportional to one another.*

Use the following questions to test your understanding of Section 6-3.

1. State Kirchhoff's current law.
2. If $I_T = 4$ A and $I_1 = 2.7$ A in a two-resistor parallel circuit, what would be the value of I_2?
3. State the current-divider formula.
4. Calculate I_1 if $R_T = 1$ kΩ, $R_1 = 2$ kΩ, and $V_T = 12$ V.

6-4 RESISTANCE IN A PARALLEL CIRCUIT

We now know that parallel circuits will have a larger current flow than a series circuit containing the same resistors due to the smaller total resistance. To calculate exactly how much total current will flow, we need to be able to calculate the total resistance that the parallel circuit presents to the source.

The ability of a circuit to conduct current is a measure of that circuit's conductance, and you will remember from Chapter 1 that conductance (G) is equal to the reciprocal of resistance and is measured in siemens.

$$G = \frac{1}{R} \quad \text{(siemens)}$$

Every resistor in a parallel circuit will have a conductance figure that is equal to the reciprocal of its resistance, and the total conductance (G_T) of the circuit will be equal to the sum of all the individual resistor conductances. Therefore,

$$G_T = G_{R1} + G_{R2} + G_{R3} + \cdots$$

Total conductance is equal to the conductance of R_1 + the conductance of R_2 + the conductance of R_3 + $\cdots$.

Once you have calculated total conductance, the reciprocal of this figure will give you total resistance. If, for example, we have two resistors in parallel, as shown in Figure 6-16, the conductance for R_1 will equal

$$G_{R1} = \frac{1}{R_1} = \frac{1}{20 \ \Omega} = 0.05 \text{ S}$$

The conductance for R_2 will equal

$$G_{R2} = \frac{1}{R_2} = \frac{1}{40 \ \Omega} = 0.025 \text{ S}$$

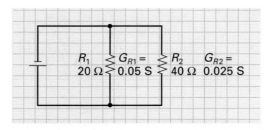

FIGURE 6-16 **Parallel Circuit Conductance and Resistance.**

The total conductance will therefore equal

$$G_{\text{total}} = G_{R1} + G_{R2}$$
$$= 0.05 + 0.025$$
$$= 0.075 \text{ S}$$

Since total resistance is equal to the reciprocal of total conductance, total resistance for the parallel circuit in Figure 6-16 will be

$$R_{\text{total}} = \frac{1}{G_{\text{total}}} = \frac{1}{0.075 \text{ S}} = 13.3 \ \Omega$$

Combining these three steps (first calculate individual conductances, total conductance and then total resistance) we can arrive at the following *reciprocal formula:*

$$R_{\text{total}} = \frac{1}{(1/R_1) + (1/R_2)}$$

this formula states that the conductance of R_1 (G_{R1})
+ conductance of R_2 (G_{R2})
= total conductance (G_T),
and the reciprocal of total conductance is equal to total resistance

In the example for Figure 6-16, this combined general formula for total resistance can be verified by plugging in the example values.

$$R_T = \frac{1}{(1/R_1) + (1/R_2)}$$
$$= \frac{1}{(1/20) + (1/40)}$$
$$= \frac{1}{0.05 + 0.025}$$
$$= \frac{1}{0.075}$$
$$= 13.3 \ \Omega$$

CALCULATOR SEQUENCE

Step	Keypad Entry	Display Response
1.	(Clear Memory)	
2.	[2] [0]	20.
3.	[1/x]	5.E-2
4.	[+]	5.E-2
5.	[4] [0]	40
6.	[1/x]	2.5E-2
7.	[=]	7.5E-2
8.	[1/x]	13.33333

The *reciprocal formula* for calculating total parallel circuit resistance for any number of resistors is

$$R_T = \frac{1}{(1/R_1) + (1/R_2) + (1/R_3) + (1/R_4) + \cdots}$$

EXAMPLE:

Referring to Figure 6-17(a), calculate:

a. Total resistance

b. Voltage drop across R_2

c. Voltage drop across R_3

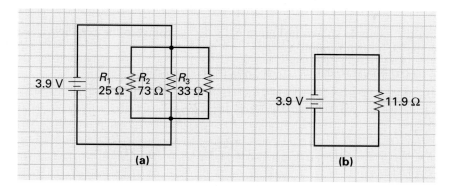

FIGURE 6-17 **Parallel Circuit Example. (a) Schematic. (b) Equivalent Circuit.**

Solution:

Total resistance can be calculated using the reciprocal formula.

$$R_T = \frac{1}{(1/R_1) + (1/R_2) + (1/R_3)}$$

$$= \frac{1}{(1/25\ \Omega) + (1/73\ \Omega) + (1/33\ \Omega)}$$

$$= \frac{1}{0.04 + 0.014 + 0.03}$$

$$= 11.9\ \Omega$$

With parallel resistance circuits, the total resistance is always smaller than the smallest branch resistance. In this example the total opposition of this circuit is equivalent to 11.9 Ω, as shown in Figure 6-17(b).

With parallel resistive circuits, the voltage drop across any branch is equal to the voltage drop across each of the other branches and is equal to the source voltage, in this example 3.9 V.

6-4-1 *Two Resistors in Parallel*

If only two resistors are connected in parallel, a quick and easy formula called the *product-over-sum formula* can be used to calculate total resistance.

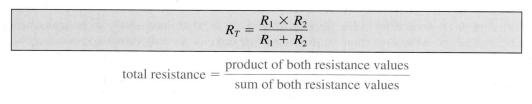

$$R_T = \frac{R_1 \times R_2}{R_1 + R_2}$$

$$\text{total resistance} = \frac{\text{product of both resistance values}}{\text{sum of both resistance values}}$$

FIGURE 6-18   Two Resistors in Parallel.

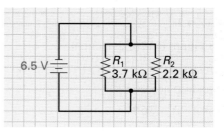

Using the example shown in Figure 6-18, let us compare the *product-over-sum* formula with the *reciprocal* formula.

(a) PRODUCT-OVER-SUM FORMULA

$$R_T = \frac{R_1 \times R_2}{R_1 + R_2}$$

$$= \frac{3.7 \text{ k}\Omega \times 2.2 \text{ k}\Omega}{3.7 \text{ k}\Omega + 2.2 \text{ k}\Omega}$$

$$= \frac{8.14 \text{ k}\Omega}{5.9 \text{ k}\Omega}$$

$$= 1.38 \text{ k}\Omega$$

(b) RECIPROCAL FORMULA

$$R_T = \frac{1}{(1/R_1) + (1/R_2)}$$

$$= \frac{1}{(1/3.7 \text{ k}\Omega) + (1/2.2 \text{ k}\Omega)}$$

$$= \frac{1}{(270.2 \times 10^{-6}) + (454.5 \times 10^{-6})}$$

$$= \frac{1}{724.7 \times 10^{-6}}$$

$$= 1.38 \text{ k}\Omega$$

As you can see from this example, the advantage of the product-over-sum parallel resistance formula (a) is its ease of use. Its disadvantage is that it can only be used for two resistors in parallel. The rule to adopt, therefore, is that if a circuit has two resistors in parallel use the *product-over-sum* formula, and in circuits containing more than two resistors, use the *reciprocal* formula.

6-4-2   *Equal-Value Resistors in Parallel*

If resistors of equal value are connected in parallel, a special case *equal-value formula* can be used to calculate the total resistance.

$$R_T = \frac{\text{value of one resistor } (R)}{\text{number of parallel resistors } (n)}$$

■   EXAMPLE:

Figure 6-19(a) shows how a stereo music amplifier is connected to drive two 8 Ω speakers, which are connected in parallel with one another. What is the total resistance connected across the amplifier's output terminals?

■   *Solution:*

Referring to the schematic in Figure 6-19(b), you can see that since both parallel-connected speakers have the same resistance, the total resistance is most easily calculated by using the equal-value formula:

$$R_T = \frac{R}{n} = \frac{8 \text{ }\Omega}{2} = 4 \text{ }\Omega$$

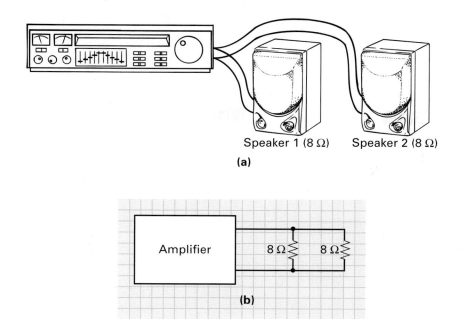

FIGURE 6-19   **Parallel-Connected Speakers.**

EXAMPLE:

Refer to Figure 6-20 and calculate:

a. Total resistance in part (a)

b. Total resistance in part (b)

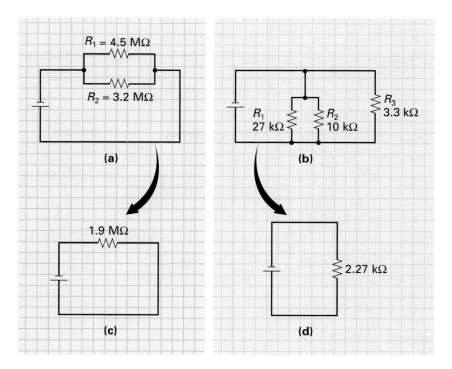

FIGURE 6-20   **Parallel Circuit Examples.**

■ *Solution:*

CALCULATOR SEQUENCE FOR (A)

Step	Keypad Entry	Display Response
1.	[4] [.] [5] [E] [6]	4.5E6
2.	[+]	
3.	[3] [.] [2] [E] [6]	3.2E6
4.	[=]	7.7E6
5.	[STO] (store in memory)	
6.	[C/CE]	0.
7.	[4] [.] [5] [E] [6]	4.5E6
8.	[×]	
9.	[3] [.] [2] [E] [6]	3.2E6
10.	[=]	1.44E13
11.	[÷]	
12.	[RM] (Recall memory)	7.7E6
13.	[=]	1.87E6

a. Figure 6-20(a) has only two resistors in parallel and therefore the product-over-sum resistor formula can be used.

$$R_T = \frac{R_1 \times R_2}{R_1 + R_2}$$

$$= \frac{4.5 \text{ M}\Omega \times 3.2 \text{ M}\Omega}{4.5 \text{ M}\Omega + 3.2 \text{ M}\Omega}$$

$$= \frac{14.4 \text{ M}\Omega}{7.7 \text{ M}\Omega}$$

$$= 1.9 \text{ M}\Omega$$

The equivalent circuit is seen in Figure 6-20(c).

b. Figure 6-20(b) has more than two resistors in parallel, and therefore the reciprocal formula must be used.

$$R_T = \frac{1}{(1/R_1) + (1/R_2) + (1/R_3)}$$

$$= \frac{1}{(1/27 \text{ k}\Omega + (1/10 \text{ k}\Omega + (1/3.3 \text{ k}\Omega)}$$

$$= \frac{1}{440.0 \times 10^{-6}}$$

$$= 2.27 \text{ k}\Omega$$

The equivalent circuit can be seen in Figure 6-20(d).

■ **EXAMPLE:**

Find the total resistance of the parallel circuit in Figure 6-21.

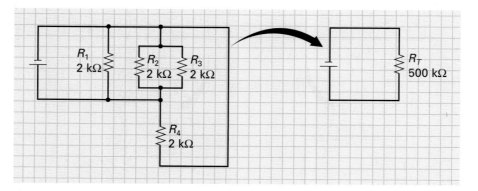

FIGURE 6-21   **Parallel Circuit Example**

■ *Solution:*

Since all four resistors are connected in parallel and are all of the same value, the equal-value resistors in parallel formula can be used to calculate total resistance.

$$R_T = \frac{R}{n} = \frac{2 \text{ k}\Omega}{4} = 500 \ \Omega$$

To summarize what we have learned so far about parallel circuits, we can say that:

1. Components are said to be connected in parallel when the current has to travel two or more paths between the negative and positive sides of the voltage source.
2. The voltage across all the components in parallel is always the same.
3. The total current from the source is equal to the sum of all the branch currents (Kirchhoff's current law).
4. The amount of current flowing through each branch is inversely proportional to the resistance value in that branch.
5. The total resistance of a parallel circuit is always less than the value of the smallest branch resistor.

6-5 POWER IN A PARALLEL CIRCUIT

As with series circuits, the total power in a parallel resistive circuit is equal to the sum of all the power losses for each of the resistors in parallel.

$$P_T = P_1 + P_2 + P_3 + P_4 + \cdots$$

total power = addition of all the power losses

The formulas for calculating the amount of power dissipated are

$$P = I \times V$$
$$P = \frac{V^2}{R}$$
$$P = I^2 \times R$$

EXAMPLE:

Calculate the total amount of power dissipated in Figure 6-22.

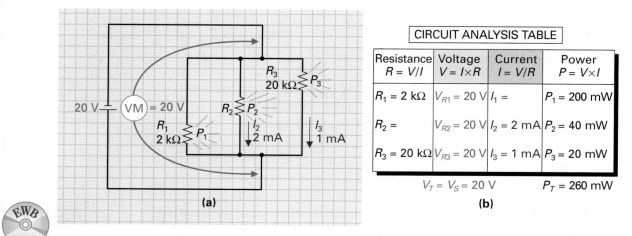

CIRCUIT ANALYSIS TABLE			
Resistance $R = V/I$	Voltage $V = I \times R$	Current $I = V/R$	Power $P = V \times I$
$R_1 = 2\ k\Omega$	$V_{R1} = 20\ V$	$I_1 =$	$P_1 = 200\ mW$
$R_2 =$	$V_{R2} = 20\ V$	$I_2 = 2\ mA$	$P_2 = 40\ mW$
$R_3 = 20\ k\Omega$	$V_{R3} = 20\ V$	$I_3 = 1\ mA$	$P_3 = 20\ mW$
	$V_T = V_S = 20\ V$		$P_T = 260\ mW$

FIGURE 6-22 Parallel Circuit Example. (a) Schematic. (b) Circuit Analysis Table.

■ *Solution:*

CALCULATOR SEQUENCE

Step	Keypad Entry	Display Response
1.	[2][0]	20.
2.	[x²]	400.
3.	[÷]	
4.	[2][E][3]	2.E3
5.	[=]	0.2

CALCULATOR SEQUENCE

Step	Keypad Entry	Display Response
1.	[2][E][3][+/−]	2.E-3
2.	[×]	
3.	[2][0]	20.
4.	[=]	40E-3

CALCULATOR SEQUENCE

Step	Keypad Entry	Display Response
1.	[1][E][3][+/−][x²]	1.E-6
2.	[×]	1.E-6
3.	[2][0][E][3]	20E3
4.	[=]	20.E-3

The total power dissipated in a parallel circuit is equal to the sum of all the power dissipated by all the resistors. With P_1, we only know voltage and resistance and therefore we can use the formula

$$P_1 = \frac{V_{R1}^2}{R_1} = \frac{20^2}{2\ k\Omega} = 2.0\ W \quad or \quad 200\ mW$$

With P_2, we only know current and voltage, and therefore we can use the formula

$$P_2 = I_2 \times V_{R2} = 2\ mA \times 20\ V = 40\ mW$$

With P_3, we know V, I, and R; however, we will use the third power formula:

$$P_3 = I_3^2 \times R_3 = 1\ mA^2 \times 20\ k\Omega = 20\ mW$$

Total power (P_T) equals the sum of all the power or wattage losses for each resistor:

$$\begin{aligned} P_T &= P_1 + P_2 + P_3 \\ &= 200\ mW + 40\ mW + 20\ mW \\ &= 260\ mW \end{aligned}$$

■ **EXAMPLE:**

In Figure 6-23, there are two ½ W (0.5 W) resistors connected in parallel. Should the wattage rating for each of these resistors be increased or decreased, or can they remain the same?

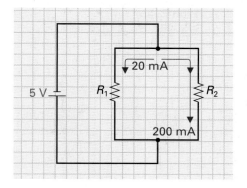

FIGURE 6-23   **Parallel Circuit Example.**

■ *Solution:*

Since current and voltage are known for both branches, the power formula used in both cases can be $P = I \times V$.

$$P_1 = I \times V \qquad\qquad P_2 = I \times V$$
$$ = 20 \text{ mA} \times 5 \text{ V} \qquad = 200 \text{ mA} \times 5 \text{ V}$$
$$ = 0.1 \text{ W} \qquad\qquad = 1 \text{ W}$$

R_1 is dissipating 0.1 W and so a 0.5 W resistor will be fine in this application. On the other hand, R_2 is dissipating 1 W and is only designed to dissipate 0.5 W. R_2 will therefore overheat unless it is replaced with a resistor of the same ohmic value, but with a 1 W or greater rating.

■ **EXAMPLE:**

Figure 6-24(a) shows a simplified diagram of an automobile external light system. A 12 V lead–acid battery is used as a source and is connected across eight parallel-connected lamps. The left and right brake lights are controlled by the brake switch, which is attached to the brake pedal. When the light switch is turned on, both the rear taillights and the low-beam headlights are brought into circuit and turned on. The high-beam set of headlights is activated only if the high-beam switch is closed. For the lamp resistances given, calculate the output power of each lamp, when in use.

■ *Solution:*

Figure 6-24(b) shows the schematic diagram of the pictorial in Figure 6-24(a). Since both V and R are known, we can use the V^2/R power formula.

$$\textit{Brake lights: } \text{Left lamp wattage } (P) \ = \frac{V^2}{R} = \frac{12\ V^2}{4\Omega} = 36 \text{ W}$$

Right lamp wattage is the same as left.

$$\textit{Taillights: } \quad \text{Left lamp wattage } (P) \ = \frac{V^2}{R} = \frac{12\ V^2}{6\Omega} = 24$$

Right lamp wattage is the same as left.

$$\textit{Low-beam headlights: } (P) \ = \frac{V^2}{R} = \frac{12\ V^2}{3\Omega} = 48 \text{ W}$$

Each high-beam headlight is a 48 W lamp.

$$\textit{High-beam headlights: } (P) \ = \frac{V^2}{R} = \frac{12\ V^2}{3\Omega} = 48 \text{ W}$$

Each high-beam headlight is a 48 W lamp.

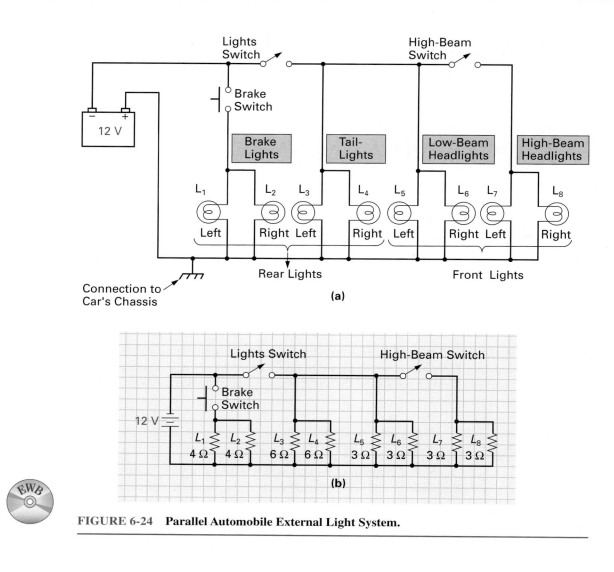

FIGURE 6-24   **Parallel Automobile External Light System.**

SELF-TEST EVALUATION POINT FOR SECTION 6-5

Now that you have completed this section, you should be able to:

■ **Objective 7.** *Describe and be able to apply all formulas associated with the calculation of voltage, current, resistance, and power in a parallel circuit.*

Use the following questions to test your understanding of Section 6-5.

1. True or false: Total power in a parallel circuit can be obtained by using the same total power formula as for series circuits.

2. If $I_1 = 2$ mA and $V = 24$ V, calculate P_1.

3. If $P_1 = 22$ mW and $P_2 = 6400$ μW, $PT = ?$

4. Is it important to observe the correct wattage ratings of resistors when they are connected in parallel?

6-6 TROUBLESHOOTING A PARALLEL CIRCUIT

In the troubleshooting discussion on series circuits, we mentioned that one of three problems can occur, and these three also apply to parallel circuits:

1. A component will open.
2. A component will short.
3. A component's value will change over a period of time.

As a technician, it is important that you know how to recognize these circuit malfunctions and know how to isolate the cause and make the repair. Since you could be encountering opens, shorts, and component value changes in the parallel circuits you construct in lab, let us step through the troubleshooting procedure you should use.

As an example, Figure 6-25(a) indicates the normal readings that should be obtained from a typical parallel circuit if it is operating correctly. Keep these normal readings in mind, since they will change as circuit problems are introduced in the following section.

6-6-1   *Open Component in a Parallel Circuit*

In Figure 6-25(b), R_1 has opened, so there can be no current flow through the R_1 branch. With one fewer branches for current, the parallel circuit's resistance will increase, causing the total current flow to decrease. The total current flow, which was 7 mA, will actually decrease by the amount that was flowing through the now open branch (1 mA) to a total current flow of 6 mA. If you had constructed this circuit in lab, you would probably not have noticed that there was anything wrong with the circuit until you started to make some multimeter measurements. A voltmeter measurement across the parallel circuit, as seen in Figure 6-25(b), would not have revealed any problem since the voltage measured across each resistor will always be the same and equal to the source voltage. A total current measurement, on the other hand, would indicate that there is something wrong since the measured reading does not equal the expected calculated value. Using your ammeter, you could isolate the branch fault by one of the following two methods:

1. If you measure each branch current, you will isolate the problem of R_1, because there will be no current flow through R_1. However, this could take three checks.
2. If you just measure total current (one check), you will notice that total current has decreased by 1 mA, and after making a few calculations you could determine that the only path that had a branch current of 1 mA was through R_1, so R_1 must have gone open.

We can summarize how method 2 could be used to isolate an open in this example circuit:

If R_1 opens, total current decreases by 1 mA to 6 mA ($I_T = I_1 + I_2 + I_3 = 0$ mA $+$ 4 mA $+$ 2 mA $=$ 6 mA).

If R_2 opens, total current decreases by 4 mA to 3 mA ($I_T = I_1 + I_2 + I_3 = 1$ mA $+$ 0 mA $+$ 2 mA $=$ 3 mA).

If R_3 opens, total current decreases by 2 mA to 5 mA ($I_T = I_1 + I_2 + I_3 = 1$ mA $+$ 4 mA $+$ 0 mA $=$ 5 mA).

Method 2 can be used as long as the resistors are unequal. If they are equal, method 1 may be used to locate the 0 current branch.

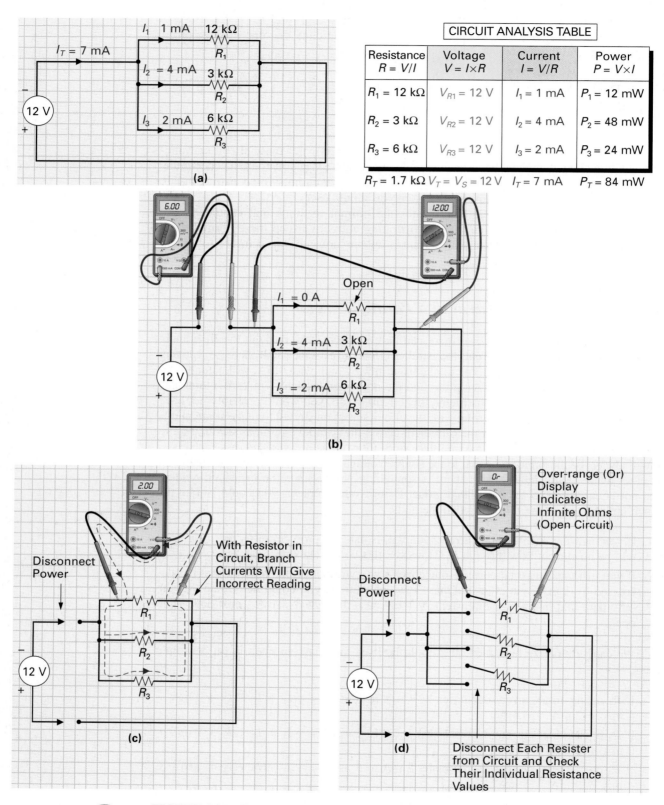

Resistance R = V/I	Voltage V = I×R	Current I = V/R	Power P = V×I
$R_1 = 12$ kΩ	$V_{R1} = 12$ V	$I_1 = 1$ mA	$P_1 = 12$ mW
$R_2 = 3$ kΩ	$V_{R2} = 12$ V	$I_2 = 4$ mA	$P_2 = 48$ mW
$R_3 = 6$ kΩ	$V_{R3} = 12$ V	$I_3 = 2$ mA	$P_3 = 24$ mW
$R_T = 1.7$ kΩ	$V_T = V_S = 12$ V	$I_T = 7$ mA	$P_T = 84$ mW

CIRCUIT ANALYSIS TABLE

(a)

(b)

(c)

With Resistor in Circuit, Branch Currents Will Give Incorrect Reading

Disconnect Power

(d)

Over-range (Or) Display Indicates Infinite Ohms (Open Circuit)

Disconnect Power

Disconnect Each Resister from Circuit and Check Their Individual Resistance Values

FIGURE 6-25   Troubleshooting an Open in a Parallel Circuit.

In most cases, the ohmmeter is ideal for locating open circuits. To make a resistance measurement, you should disconnect power from the circuit, isolate the device from the circuit, and then connect the ohmmeter across the device. Figure 6-25(c) shows why it is necessary for you to isolate the device to be tested from the circuit. With R_1 in circuit, the ohmmeter will be measuring the parallel resistance of R_2 and R_3, even though the ohmmeter is connected across R_1. This will lead to a false reading and confuse the troubleshooting process. Disconnecting the device to be tested from the circuit, as shown in Figure 6-25(d), will give an accurate reading of R_1, then R_2, and then R_3's resistance and lead you to the cause of the problem.

▢ EXAMPLE:

Figure 6-26 illustrates a set of three light bulbs connected in parallel across a 9 V battery. The filament in bulb 2 has burned out, causing an open, so there is no current flow through that path. There will, however, be current flow through the other two, so bulbs 1 and 3 will be on. How would the faulty bulb be located?

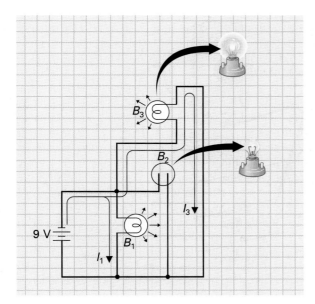

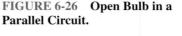

FIGURE 6-26   Open Bulb in a Parallel Circuit.

▢ *Solution:*

As always, a visual inspection would be your first approach. Since current is flowing through both B_1 and B_3, these two bulbs will be on. On the other hand, since B_2 is off, it would be easy for you to determine that bulb B_2 has blown.

Opens in parallel paths are generally quite easy to isolate since the device in the parallel path will not operate because its power source has been disconnected.

6-6-2   *Shorted Component in a Parallel Circuit*

In Figure 6-27(a), R_2, which has a resistance of 3 kΩ, has been shorted out. The only resistance in this center branch is the resistance of the wire, which would typically be a fraction of an ohm. In this example, let us assume that the resistance in the branch is 1 Ω, and therefore the current in this branch will attempt to increase to:

$$I_2 = \frac{12V}{1\ \Omega} = 12\ A$$

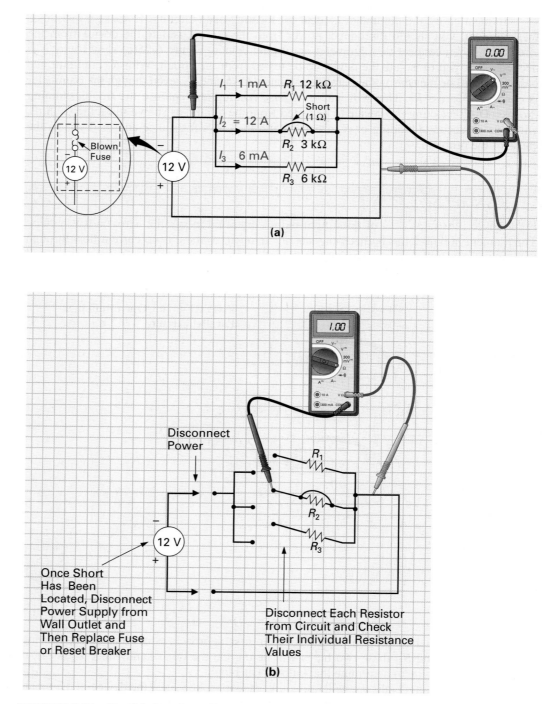

FIGURE 6-27   **Troubleshooting a Short in a Parallel Circuit.**

Before the current even gets close to 12 A, the fuse in the dc power supply will blow, as shown in the inset in Figure 6-27(a), and disconnect power from the circuit. This is typically the effect you will see from a circuit short. Replacing the fuse will be a waste of time and money since the fuse will simply blow again once power is connected across the circuit. As before, the tool to use to isolate the cause is the ohmmeter, as shown in Figure 6-27(b). To check the resistance of each branch, disconnect power from the circuit, disconnect the devices to be tested from the circuit to prevent false readings, and then test each branch until the shorted branch is located. Once you have corrected the fault, you will need to change the power supply fuse or reset its circuit breaker.

6-6-3 *Component Value Variation in a Parallel Circuit*

The resistive values of resistors will change with age. This increase or decrease in resistance will cause a corresponding decrease or increase in branch current and therefore total current. This deviation from the desired value can also be checked with the ohmmeter. Be careful to take into account the resistor's tolerance because this can make you believe that the resistor's value has changed, when it is, in fact, within the tolerance rating.

6-6-4 *Summary of Parallel Circuit Troubleshooting*

1. An open component will cause no current flow within that branch. The total current will decrease, and the voltage across the component will be the same as the source voltage.

2. A shorted component will cause maximum current through that branch. The total current will increase and the voltage source fuse will normally blow.

3. A change in resistance value will cause a corresponding opposite change in branch current and total current.

SELF-TEST EVALUATION POINT FOR SECTION 6-6

Now that you have completed this section, you should be able to:

■ **Objective 8.** *Describe how a short, open, or component variation will affect a parallel circuit's operation and how it can be recognized.*

Use the following questions to test your understanding of Section 6-6.

How would problems 1 through 3 be recognized in a parallel circuit?

1. An open component
2. A shorted component
3. A component's value variation
4. True or false: An ammeter is typically used to troubleshoot series circuits, whereas a voltmeter is typically used to troubleshoot parallel circuits.

SUMMARY

Parallel Circuits (Figure 6-28)

1. In a series circuit, there is only one path for current, whereas in parallel circuits the current has two or more paths. These paths are known as *branches*.

2. A parallel circuit is when two or more components are connected to the same voltage source so that the current can branch out over two or more paths.

3. In a parallel resistive circuit, two or more resistors are connected to the same source, so current splits to travel through each separate branch resistance.

4. Measuring the voltage drop with the voltmeter across any of the resistors in a parallel circuit is the same as measuring the voltage across the voltage source.

5. Gustav Kirchhoff was the first to observe and prove that the sum of all the branch currents ($I_1 + I_2 + I_3$, etc.) was equal to the total current (I_T).

6. Kirchhoff's current law states that the sum of all the currents entering a junction is equal to the sum of all the currents leaving that same junction.

7. When we connect resistors in parallel, the total resistance in the circuit will decrease. In fact, the total resistance in a

FIGURE 6-28   Parallel Circuits.

Example Circuit (Resistors)

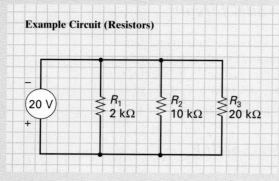

Example Circuit (Batteries)

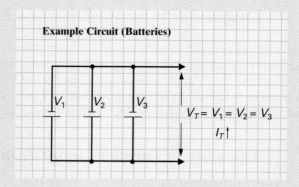

$$V_S = V_{R1} = V_{R2} = V_{R3} = \cdots$$

$$R_T = \frac{V_S}{I_T}$$

$$I_T = I_1 + I_2 + I_3 + \cdots$$

$$I_x = \frac{R_T}{R_x} \times I_T$$

$$R_T = \frac{1}{(1/R_1) + (1/R_2) + (1/R_3) + \cdots}$$

$$R_T = \frac{R_1 \times R_2}{R_1 + R_2}$$

$$R_T = \frac{R}{n}$$

$$P_T = P_1 + P_2 + P_3 + \cdots$$

$$P_T = \frac{V_S{}^2}{R_T} = I_T{}^2 \times R_T = V_S \times I_T$$

V_S = source voltage, in volts
V_{R1} = voltage across R_1, in volts
V_{R2} = voltage across R_2, in volts
V_{R3} = voltage across R_3, in volts
I_T = total current, in amperes
R_T = total resistance, in ohms
I_1 = current through R_1, in amperes
I_2 = current through R_2, in amperes
I_3 = current through R_3, in amperes
I_x = branch current desired, in amperes
R_x = resistance in branch
R_1 = resistance of R_1, in ohms
R_2 = resistance of R_2, in ohms
R_3 = resistance of R_3, in ohms
R_{EV} = resistance of one of the equivalent-value resistors
n = number of parallel resistors
P_T = total power dissipated, in watts
P_1 = power dissipated by R_1, in watts
P_2 = power dissipated by R_2, in watts
P_3 = power dissipated by R_3, in watts

parallel circuit will always be less than the value of the smallest resistor in the circuit.

8. Just as a series circuit is often referred to as a voltage-divider circuit, a parallel circuit is often referred to as a current-divider circuit, because the total current arriving at a junction will divide or split into branch currents (Kirchhoff's current law).

9. The current division is inversely proportional to the resistance in the branch seeing that the voltage across both resistors is constant and equal to the source voltage (V_S). This means that a large branch resistance will cause a small branch current ($I\downarrow = V/R\uparrow$), and a small branch resistance will cause a large branch current ($I\uparrow = V/R\downarrow$).

10. By rearranging Ohm's law, we can arrive at another formula, which is called the *current-divider formula* and can be used to calculate the current through any branch of a multiple-branch parallel circuit.

11. Parallel circuits will have a larger current flow than a series circuit containing the same resistors due to the smaller total resistance.

12. To calculate exactly how much total current will flow, we need to calculate the total resistance that the parallel circuit develops or contains.

13. The reciprocal formula allows us to calculate total parallel circuit resistance for any number of resistors.

14. With parallel resistance circuits, the total resistance is always smaller than the smallest branch resistance.

15. With parallel resistive circuits, the voltage drop across any branch is equal to the voltage drop across each of the other branches and is equal to the source voltage.

16. If only two resistors are connected in parallel, a quick and easy formula called the *product-over-sum formula* can be used to calculate total resistance.

17. If resistors of equal value are connected in parallel, a special case *equal-value formula* can be used to calculate the total resistance.

18. As with series circuits, the total power in a parallel resistive circuit is equal to the sum of all the power losses for each of the resistors in parallel.

19. An open component will cause no current flow within that branch. The total current will decrease, and the voltage across the component will be the same as the source voltage.

20. A shorted component will cause maximum current through that branch. The total current will increase and the voltage source fuse will normally blow.

21. A change in resistance value will cause a corresponding opposite change in branch current and total current.

REVIEW QUESTIONS

Multiple-Choice Questions

1. A parallel circuit has _____ path(s) for current to flow.
   a. One   c. Only three
   b. Two or more d. None of the above

2. If a source voltage of 12 V is applied across four resistors of equal value in parallel, the voltage drops across each resistor would be equal to:
   a. 12 V   c. 4 V
   b. 3 V   d. 48 V

3. What would be the voltage drop across two 25 Ω resistors in parallel if the source voltage were equal to 9 V?
   a. 50 V   c. 12 V
   b. 25 V   d. None of the above

4. If a four-branch parallel circuit has 15 mA flowing through each branch, the total current into the parallel circuit will be equal to:
   a. 15 mA   c. 30 mA
   b. 60 mA   d. 45 mA

5. If the total three-branch parallel circuit current is equal to 500 mA, and 207 mA is flowing through one branch and 153 mA through another, what would be the current flow through the third branch?
   a. 707 mA   c. 140 mA
   b. 653 mA   d. None of the above

6. A large branch resistance will cause a _____ branch current.
   a. Large   c. Medium
   b. Small   d. None of the above are true

7. What would be the conductance of a 1 kΩ resistor?
   a. 10 mS   c. 2 kΩ
   b. 1 mS   d. All of the above

8. If only two resistors are connected in parallel, the total resistance equals:
   a. The sum of the resistance values
   b. Three times the value of one resistor
   c. The product over the sum
   d. All of the above

9. If resistors of equal value are connected in parallel, the total resistance can be calculated by:
   a. One resistor value divided by the number of parallel resistors
   b. The sum of the resistor values
   c. The number of parallel resistors divided by one resistor value
   d. All of the above could be true

10. The total power in a parallel circuit is equal to the:
    a. Product of total current and total voltage
    b. Reciprocal of the individual power losses
    c. Sum of the individual power losses
    d. Both (a) and (b)
    e. Both (a) and (c)

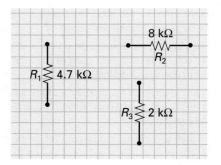

FIGURE 6-29 Connect in Parallel across a 12 Volt Source.

Communication Skill Questions

11. Describe the difference between a series and a parallel circuit. (6-1)

12. Explain and state mathematically the situation regarding voltage in a parallel circuit. (6-2)

13. State Kirchhoff's current law for parallel circuits. (6-3)

14. What is the current-divider formula? (6-3)

15. List the formulas for calculating the following total resistances: (6-4)
  a. Two resistors of different values
  b. More than two resistors of different values
  c. Equal-value resistors

16. Describe the relationship between branch current and branch resistance. (6-3)

17. Briefly describe total and individual power in a parallel resistive circuit. (6-5)

18. Discuss troubleshooting parallel circuits as applied to: (6-6)
  a. A shorted component **b.** An open component

19. Explain why parallel circuits have a smaller total resistance and larger total current than series circuits. (6-3)

20. Briefly describe why Kirchhoff's voltage law applies to series circuits and why Kirchhoff's current law relates to parallel circuits. (6-3)

Practice Problems

21. Calculate the total resistance of four 30 kΩ resistors in parallel.

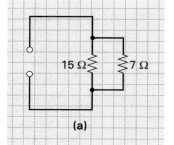

(a)

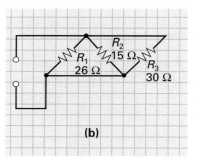

(b)

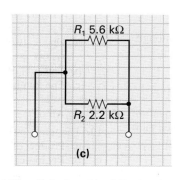

(c)

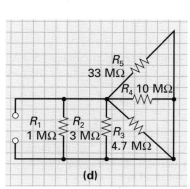

(d)

FIGURE 6-30 Calculate Total Resistance.

22. Find the total resistance for each of the following parallel circuits:
    a. 330 Ω and 560 Ω   **c.** 2.2 MΩ, 3 kΩ, and 220 Ω
    b. 47 kΩ, 33 kΩ, and 22 kΩ

23. If 10 V is connected across three 25 Ω resistors in parallel, what will be the total and individual branch currents?

24. If a four-branch parallel circuit has branch currents equal to 25 mA, 37 mA, 220 mA, and 0.2 A, what is the total circuit current?

25. If three resistors of equal value are connected across a 14 V supply and the total resistance is equal to 700 Ω, what is the value of each branch current?

26. If three 75 W light bulbs are connected in parallel across a 110 V supply, what is the value of each branch current? What is the branch current through the other two light bulbs if one burns out?

27. If 33 kΩ and 22 kΩ resistors are connected in parallel across a 20 V source, calculate:
    a. Total resistance   **d.** Total power dissipated
    b. Total current   **e.** Individual resistor power
    c. Branch currents   dissipated

28. If four parallel-connected resistors are each dissipating 75 mW, what is the total power being dissipated?

29. Calculate the branch currents through the following parallel resistor circuits when they are connected across a 10 V supply:
    a. 22 kΩ and 33 kΩ   **b.** 220 Ω, 330 Ω, and 470 Ω

30. If 30 Ω and 40 Ω resistors are connected in parallel, which resistor will generate the greatest amount of heat?

31. Calculate the total conductance and resistance of the following parallel circuits:
    a. Three 5 Ω resistors   **c.** 1 MΩ, 500 MΩ, 3.3 MΩ
    b. Two 200 Ω resistors   **d.** 5 Ω, 3 Ω, 2 Ω

32. Connect the three resistors in Figure 6-29 in parallel across a 12 V battery and then calculate the following:
    a. V_{R1}, V_{R1}, V_{R1} **d.** P_T
    b. I_1, I_2, I_3 **e.** P_1, P_2, P_3
    c. I_T **f.** G_{R1}, G_{R2}, G_{R3}

33. Calculate R_T in Figure 6-30 (a), (b), (c), and (d).

34. Calculate the branch currents through four 60 W bulbs connected in parallel across 110 V. How much is the total current, and what would happen to the total current if one of the bulbs were to burn out? What change would occur in the remaining branch currents?

35. Calculate the following in Figure 6-31:
    a. I_2 **c.** V_S, I_1, I_2
    b. I_T **d.** R_2, I_1, I_2, P_T

Troubleshooting Questions

36. An open component in a parallel circuit will cause _____ current flow within that branch, which will cause the total current to _____.
    a. Maximum, increase   **c.** Maximum, decrease
    b. Zero, decrease   **d.** Zero, increase

37. A shorted component in a parallel circuit will cause _____ current through a branch, and consequently the total current will _____.
    a. Maximum, increase   **c.** Maximum, decrease
    b. Zero, decrease   **d.** Zero, increase

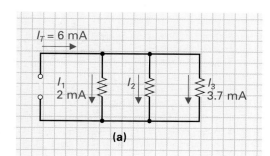

(a)

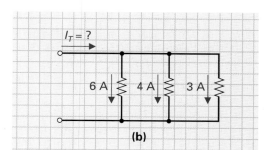

(b)

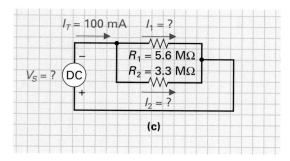

(c)

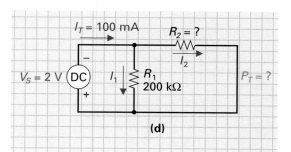

(d)

FIGURE 6-31   **Calculate the Unknown.**

38. If a 10 kΩ and two 20 kΩ resistors are connected in parallel across a 20 V supply, and the total current measured is 2 mA, determine whether a problem exists in the circuit and, if it does, isolate the problem.

39. What situation would occur and how would we recognize the problem if one of the 20 kΩ resistors in Question 38 were to short?

40. With age, the resistance of a resistor will _____, resulting in a corresponding but opposite change in _____.
   a. Increase, branch current
   b. Change, source voltage
   c. Decrease, source resistance
   d. Change, branch current

Web Site Questions

Go to the web site http://www.prenhall.com/cook, select the textbook *Introductory DC/AC Electronics* or *Introductory DC/AC Circuits,* this chapter, and then follow the instructions when answering the multiple-choice practice problems.

These tests at the end of each chapter will challenge your knowledge up to this point, and give you the practice you need for a job interview. To make this more realistic, the test will be comprised of both technical and personal questions. In order to take full advantage of this exercise, you may want to set up a simulation of the interview environment, have a friend read the questions to you, and record your responses for later analysis.

Company Name: Pony, Inc.

Industry Branch: Medical Electronics

Function: Manufacture ICU Equipment

Job Applying For: Production Test Technician

1. Who was your last employer?
2. What are Kirchhoff's laws?

Answers

1. Discuss previous employer and give a copy of letter of recommendation. If this job was not technical, refer back to how the lab work during training was structured like a work environment, and discuss other related topics.
2. Sections 5-4 and 6-3.
3. Yes, of course.
4. Describe the steps detailed in the Introduction section.
5. Chapter 6 introduction.
6. Up to this time, only the multimeter.

3. Would you report a fellow worker who was stealing from the company?
4. What do you know about the product development process?
5. What is the difference between a series circuit and a parallel circuit?
6. What test equipment are you familiar with?
7. Would you say that you would need close, moderate, or no supervision if you were to get this job?
8. What is the job function of a production test technician?
9. This job would require that you work under the watchful eye of an experienced technician for at least 6 months. Would that be a problem for you?
10. All of our techs work a 10-hour day, 4 days a week. Would this be a problem for you?

7. Say that you expect that in the beginning you would have many questions, but as you became more familiar with company procedures and the product line, you would need less and less supervision.
8. Quote intro. section in text and job description listed in newspaper.
9. This would be an opportunity, since experienced technicians have a wealth of knowledge they can impart.
10. If you are unsure, don't object to anything at the time. Later, if you decide that it is something you do not want to do, you can call and decline the offer.

Series–Parallel DC Circuits

The Christie Bridge Circuit

Charles Wheatstone

Who invented the Wheatstone bridge circuit? It was obviously Sir Charles Wheatstone. Or was it?

The Wheatstone bridge was actually invented by S.H. Christie of the Royal Military Academy at Woolwich, England. He described the circuit in detail in the *Philosophical Transactions* paper dated February 28, 1833. Christie's name, however, was unknown and his invention was ignored.

Ten years later, Sir Charles Wheatstone called attention to Christie's circuit. Sir Charles was very well known, and from that point on, and even to this day, the circuit is known as a Wheatstone bridge. Later, Werner Siemens would modify Christie's circuit and invent the variable-resistance arm bridge circuit, which would also be called a Wheatstone bridge.

No one has given full credit to the real inventors of these bridge circuits, until now!

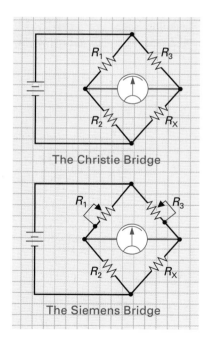

The Christie Bridge

The Siemens Bridge

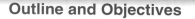

Outline and Objectives

Introduction

Very rarely are we lucky enough to run across straightforward series or parallel circuits. In general, all electronic equipment is composed of many components that are interconnected to form a combination of series and parallel circuits. In this chapter, we will be combining our knowledge of the series and parallel circuits discussed in the previous two chapters.

7-1   SERIES- AND PARALLEL-CONNECTED COMPONENTS

Series–Parallel Circuit

Network or circuit that contains components that are connected in both series and parallel.

Figure 7-1(a) through (f) shows six examples of **series–parallel resistive circuits.** The most important point to learn is how to distinguish between the resistors that are connected in series and the resistors that are connected in parallel, which will take a little practice.

One thing that you may not have noticed when examining Figure 7-1 is that:

Circuit 7-1(a) is equivalent to 7-1(b)

Circuit 7-1(c) is equivalent to 7-1(d)

Circuit 7-1(e) is equivalent to 7-1(f)

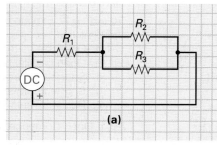

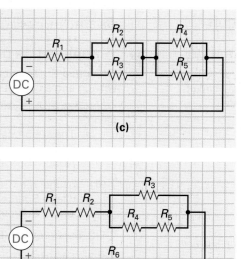

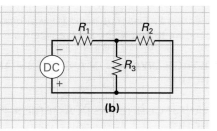

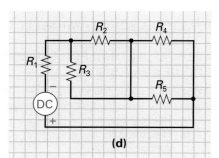

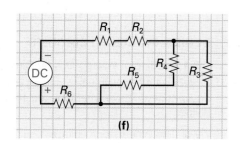

FIGURE 7-1   **Series–Parallel Resistive Circuits.**

When analyzing these series–parallel circuits, always remember that current flow determines whether the resistor is connected in series or parallel. Begin at the negative side of the battery and apply these two rules:

1. If the total current has only one path to follow through a component, that component is connected in series.

2. If the total current has two or more paths to follow through two or more components, those components are connected in parallel.

Referring again to Figure 7-1, you can see that series or parallel resistor networks are easier to identify in parts (a), (c), and (e) than in parts (b), (d), and (f). Redrawing the circuit so that the components are arranged from left to right or from top to bottom is your first line of attack in your quest to identify series– and parallel–connected components.

■ EXAMPLE:

Refer to Figure 7-2 and identify which resistors are connected in series and which are in parallel.

■ *Solution:*

First, let's redraw the circuit so that the components are aligned either from left to right as shown in Figure 7-3(a), or from top to bottom as shown in Figure 7-3(b). Placing your pencil at the negative terminal of the battery on whichever figure you prefer, either Figure 7-3(a) or (b), trace the current paths through the circuit toward the positive side of the battery, as illustrated in Figure 7-4.

The total current arrives first at R_1. There is only one path for current to flow, which is through R_1, and therefore R_1 is connected in series. The total current proceeds on past R_1 and arrives at a junction where current divides and travels through two branches, R_2 and R_3. Since current had to split into two paths, R_2 and R_3 are therefore connected in parallel. After the parallel connection of R_2 and R_3, total current combines and travels to the positive side of the battery.

In this example, therefore, R_1 is in series with the parallel combination of R_2 and R_3.

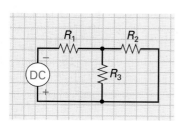

FIGURE 7-2   Series–Parallel Circuit Example.

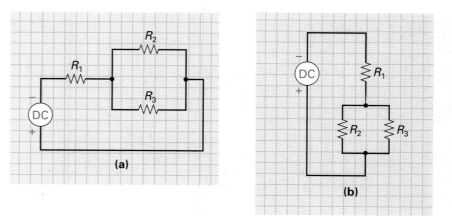

FIGURE 7-3   Redrawn Series–Parallel Circuit. (a) Left to Right. (b) Top to Bottom.

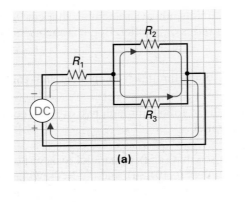

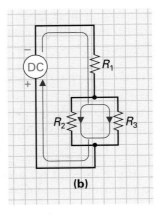

(a)

(b)

FIGURE 7-4   Tracing Current through a Series–Parallel Circuit.

■   **EXAMPLE:**

Refer to Figure 7-5 and identify which resistors are connected in series and which are connected in parallel.

■   *Solution:*

Figure 7-6 illustrates the simplified, redrawn schematic of Figure 7-5. Total current leaves the negative terminal of the battery, and all of this current has to travel through R_1, which is therefore a series-connected resistor. Total current will split at junction A, and consequently, R_3 and R_4 with R_2 make up a parallel combination. The current that flows through R_3 (I_2) will also flow through R_4 and therefore R_3 is in series with R_4. I_1 and I_2 branch currents combine at junction B to produce total current, which has only one path to follow through the series resistor R_5, and finally, to the positive side of the battery.

In this example, therefore, R_3 and R_4 are in series with one another and both are in parallel with R_2, and this combination is in series with R_1 and R_5.

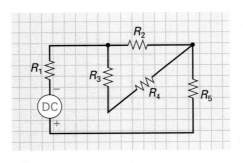

FIGURE 7-5   Series–Parallel Circuit Example.

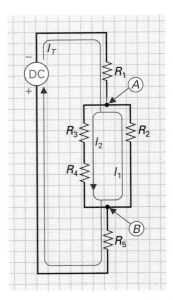

FIGURE 7-6   Redrawn Series–Parallel Circuit Example.

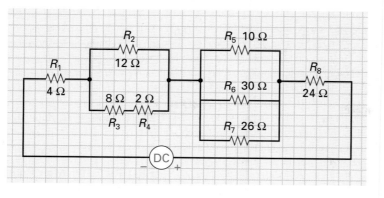

FIGURE 7-7   **Total Series–Parallel Circuit Resistance.**

7-2   TOTAL RESISTANCE IN A SERIES–PARALLEL CIRCUIT

No matter how complex or involved the series–parallel circuit, there is a simple three-step method to simplify the circuit to a single equivalent total resistance. Figure 7-7 illustrates an example of a series–parallel circuit. Once you have analyzed and determined the series–parallel relationship, we can proceed to solve for total resistance.

The three-step method is:

Step A: Determine the equivalent resistances of all branch series-connected resistors.

Step B: Determine the equivalent resistances of all parallel-connected combinations.

Step C: Determine the equivalent resistance of the remaining series-connected resistances.

Let's now put this procedure to work with the example circuit in Figure 7-7.

STEP A   Solve for all branch series-connected resistors. In our example, this applies only to R_3 and R_4, and since this is a series connection, we have to use the series resistance formula.

$$R_{3,4} = R_3 + R_4 = 8 + 2 = 10\ \Omega \qquad \text{(series resistance formula)}$$

With R_3 and R_4 solved, the circuit now appears as indicated in Figure 7-8.

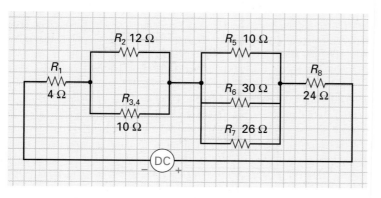

FIGURE 7-8   **After Completing Step A.**

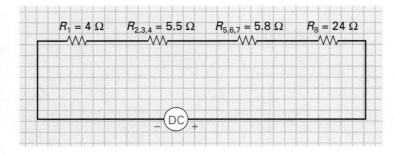

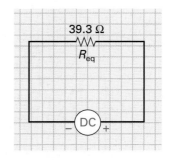

FIGURE 7-9   **After Completing Step B.**   FIGURE 7-10   **After Completing Step C.**

STEP B   Solve for all parallel combinations. In this example, they are the two parallel combinations of (a) R_2 and $R_{3,4}$ and (b) R_5 and R_6 and R_7. Since these are parallel connections, use the parallel resistance formulas.

$$R_{2,3,4} = \frac{R_2 \times R_{3,4}}{R_2 + R_{3,4}} = \frac{12 \times 10}{12 + 10} = 5.5\ \Omega \qquad \text{(product-over-sum formula)}$$

$$R_{5,6,7} = \frac{1}{(1/R_5) + (1/R_6) + (1/R_7)} = 5.8\ \Omega \qquad \text{(reciprocal formula)}$$

With $R_{2,3,4}$ and $R_{5,6,7}$ solved, the circuit now appears as illustrated in Figure 7-9.

STEP C   Solve for the remaining series resistances. There are now four remaining series resistances, which can be reduced to one equivalent resistance (R_{eq}) or total resistance (R_T). As seen in Figure 7-10, by using the series resistance formula, the total equivalent resistance for this example circuit will be

$$R_{eq} = R_1 + R_{2,3,4} + R_{5,6,7} + R_8$$
$$= 4\ \Omega + 5.5\ \Omega + 5.8\ \Omega + 24\ \Omega$$
$$= 39.3\ \Omega$$

☐   **EXAMPLE:**

Find the total resistance of the circuit in Figure 7-11.

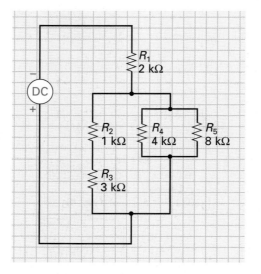

FIGURE 7-11   **Calculate Total Resistance.**

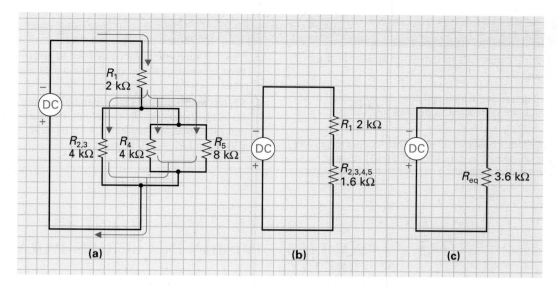

FIGURE 7-12 **Calculating Total Resistance. (a) Step A. (b) Step B. (c) Step C.**

■ *Solution:*

STEP A   Solve for all branch series-connected resistors. This applies to R_2 and R_3 (series connection):

$$R_{2,3} = R_2 + R_3 = 1 \text{ k}\Omega + 3 \text{ k}\Omega = 4 \text{ k}\Omega$$

The resulting circuit, after completing step A, is illustrated in Figure 7-12(a).

STEP B   Solve for all parallel combinations. Looking at Figure 7-12(a), which shows the circuit resulting from step A, you can see that current branches into three paths, so the parallel reciprocal formula must be used for this step.

$$R_{2,3,4,5} = \frac{1}{(1/R_{2,3}) + (1/R_4) + (1/R_5)}$$

$$= \frac{1}{(1/4 \text{ k}\Omega) + (1/4 \text{ k}\Omega) + (1/8 \text{ k}\Omega)} = 1.6 \text{ k}\Omega$$

The resulting circuit, after completing step B, is illustrated in Figure 7-12(b).

STEP C   Solve for the remaining series resistances. Looking at Figure 7-12(b), which shows the circuit resulting from step B, you can see that there are two remaining series resistances. The equivalent resistance (R_{eq}) is therefore equal to

$$R_{eq} = R_1 + R_{2,3,4,5} = 2 \text{ k}\Omega + 1.6 \text{ k}\Omega = 3.6 \text{ k}\Omega$$

The total equivalent resistance, after completing all three steps, is illustrated in Figure 7-12(c).

SELF-TEST EVALUATION POINT FOR SECTIONS 7-1 AND 7-2

Now that you have completed these sections, you should be able to:

■ *Objective 1.*   *Identify the difference between a series, a parallel, and a series-parallel circuit.*

■ *Objective 2.*   *Describe how to use a three-step procedure to determine total resistance.*

Continued

Use the following questions to test your understanding of Sections 7-1 and 7-2.

1. How can we determine which resistors are connected in series and which are connected in parallel in a series–parallel circuit?

2. Calculate the total resistance if two series-connected 12 kΩ resistors are connected in parallel with a 6 kΩ resistor.

3. State the three-step procedure used to determine total resistance in a circuit made up of both series and parallel resistors.

4. Sketch the following series–parallel resistor network made up of three resistors. R_1 and R_2 are in series with each other and are connected in parallel with R_3. If $R_1 = 470\ \Omega$, $R_2 = 330\ \Omega$, and $R_3 = 270\ \Omega$, what is R_T?

7-3 VOLTAGE DIVISION IN A SERIES–PARALLEL CIRCUIT

There is a simple three-step procedure for finding the voltage drop across each part of the series–parallel circuit. Figure 7-13 illustrates an example of a series–parallel circuit to which we will apply the three-step method for determining voltage drop.

STEP 1 Determine the circuit's total resistance. This is achieved by following the three-step method used previously for calculating total resistance.

Step A: $$R_{3,4} = 4 + 8 = 12\ \Omega$$

Step B: $$R_{2,3,4} = \frac{1}{(1/R_2) + (1/R_{3,4})} = 6\ \Omega$$

$$R_{5,6,7} = \frac{1}{(1/R_5) + (1/R_6) + (1/R_7)} = 12\ \Omega$$

Figure 7-14 illustrates the equivalent circuit up to this point. We end up with one series resistor (R_1) and two series equivalent resistors ($R_{2,3,4}$ and $R_{5,6,7}$). R_T is therefore equal to 28 Ω.

STEP 2 Determine the circuit's total current. This step is achieved simply by utilizing Ohm's law.

$$I_T = \frac{V_T}{R_T} = \frac{84\ \text{V}}{28\ \Omega} = 3\ \text{A}$$

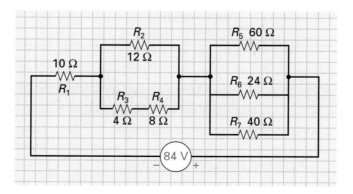

FIGURE 7-13 **Series–Parallel Circuit Example.**

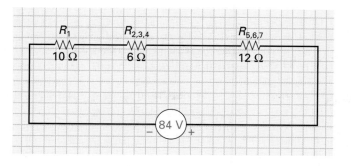

FIGURE 7-14   **After Completing Step 1.**

STEP 3   Determine the voltage across each series resistor and each parallel combination (series equivalent resistor) in Figure 7-14. Since these are all in series, the same current (I_T) will flow through all three.

$$V_{R1} = I_T \times R_1 = 3\,\text{A} \times 10\,\Omega = 30\,\text{V}$$
$$V_{R2,3,4} = I_T \times R_{2,3,4} = 3\,\text{A} \times 6\,\Omega = 18\,\text{V}$$
$$V_{R5,6,7} = I_T \times R_{5,6,7} = 3\,\text{A} \times 12\,\Omega = 36\,\text{V}$$

The voltage drops across the series resistor (R_1) and series equivalent resistors ($R_{2,3,4}$ and $R_{5,6,7}$) are illustrated in Figure 7-15.

Kirchhoff's voltage law states that the sum of all the voltage drops is equal to the source voltage applied. This law can be used to confirm that our calculations are all correct:

$$V_T = V_{R1} + V_{R2,3,4} + V_{R5,6,7}$$
$$= 30\,\text{V} + 18\,\text{V} + 36\,\text{V}$$
$$= 84\,\text{V}$$

To summarize, refer to Figure 7-16, which shows these voltage drops inserted into our original circuit. As you can see from this illustration:

30 V is dropped across R_1.

18 V is dropped across R_2.

18 V is dropped across both R_3 and R_4.

36 V is dropped across R_5.

36 V is dropped across R_6.

36 V is dropped across R_7.

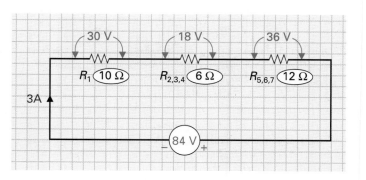

FIGURE 7-15   **After Completing Steps 2 and 3.**

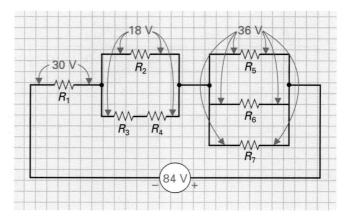

FIGURE 7-16   Detail of Step 3.

SELF-TEST EVALUATION POINT FOR SECTION 7-3

Use the following questions to test your understanding of Section 7-3.

1. State the three-step procedure used to calculate the voltage drop across each part of a series–parallel circuit.

2. Referring to Figure 7-13, double the values of all the resistors. Would the voltage drops calculated previously change, and if so, what would they be?

7-4 BRANCH CURRENTS IN A SERIES–PARALLEL CIRCUIT

In the preceding example, step 2 calculated the total current flowing in a series–parallel circuit. The next step is to find out exactly how much current is flowing through each parallel branch. This will be called step 4. Figure 7-17 shows the previously calculated data inserted in the appropriate places in our example circuit.

   STEP 4   Total current (I_T) will exist at points A, B, C, and D. Between A and B, current has only one path to flow, which is through R_1. R_1 is therefore a series resistor, so $I_1 = I_T = 3$ A. Between points B and C, current has two paths: through R_2 (12 Ω) and through R_3 and R_4 (12 Ω).

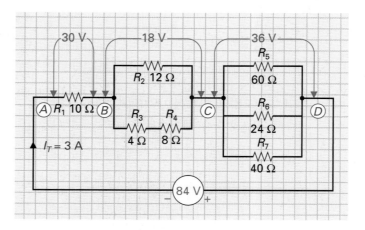

FIGURE 7-17   Series–Parallel Circuit Example with Previously Calculated Data.

$$I_2 = \frac{V_{R2}}{R_2} = \frac{18 \text{ V}}{12 \text{ }\Omega} = 1.5 \text{ A}$$

$$I_{3,4} = \frac{V_{R3,4}}{R_{3,4}} = \frac{18 \text{ V}}{12 \text{ }\Omega} = 1.5 \text{ A}$$

Not surprisingly, the total current of 3 A is split equally due to both branches having equal resistance.

The two 1.5 A branch currents will combine at point C to produce once again the total current of 3 A. Between points C and D, current has three paths to flow through, R_5, R_6, and R_7.

$$I_5 = \frac{V_{R5}}{R_5} = \frac{36 \text{ V}}{60 \text{ }\Omega} = 0.6 \text{ A}$$

$$I_6 = \frac{V_{R6}}{R_6} = \frac{36 \text{ V}}{24 \text{ }\Omega} = 1.5 \text{ A}$$

$$I_7 = \frac{V_{R7}}{R_7} = \frac{36 \text{ V}}{40 \text{ }\Omega} = 0.9 \text{ A}$$

All three branch currents will combine at point D to produce the total current of 3 A ($I_T = I_5 + I_6 + I_7 = 0.6 + 1.5 + 0.9 = 3$ A), proving Kirchhoff's current law.

7-5   POWER IN A SERIES–PARALLEL CIRCUIT

Whether resistors are connected in series or in parallel, the total power in a series–parallel circuit is equal to the sum of all the individual power losses.

$$P_T = P_1 + P_2 + P_3 + P_4 + \cdots$$

total power = addition of all power losses

The formulas for calculating the amount of power lost by each resistor are

$$P = \frac{V^2}{R} \qquad\qquad P = I \times V \qquad\qquad P = I^2 \times R$$

Let us calculate the power dissipated by each resistor. This final calculation will be called step 5.

STEP 5   Since resistance, voltage, and current are known, either of the three formulas for power can be used to determine power.

$$P_1 = \frac{V_{R1}^2}{R_1} = \frac{30 \text{ V}^2}{10 \text{ }\Omega} = 90 \text{ W}$$

$$P_2 = \frac{V_{R2}^2}{R_2} = \frac{18 \text{ V}^2}{12 \text{ }\Omega} = 27 \text{ W}$$

$$P_3 = I_{R3}^2 \times R_3 = 1.5 \text{ A}^2 \times 4 \text{ }\Omega = 9 \text{ W}$$

$$P_4 = I_{R4}^2 \times R_4 = 1.5 \text{ A}^2 \times 8 \text{ }\Omega = 18 \text{ W}$$

$$P_5 = \frac{V_{R5}^2}{R_5} = \frac{36 \text{ V}^2}{60 \text{ }\Omega} = 21.6 \text{ W}$$

$$P_6 = \frac{V_{R6}^2}{R_6} = \frac{36 \text{ V}^2}{24 \text{ }\Omega} = 54 \text{ W}$$

$$P_7 = \frac{V_{R7}^2}{R_7} = \frac{36 \text{ V}^2}{40 \text{ }\Omega} = 32.4 \text{ W}$$

$$P_T = P_1 + P_2 + P_3 + P_4 + P_5 + P_6 + P_7$$
$$= 90 + 27 + 9 + 18 + 21.6 + 54 + 32.4$$
$$= 252 \text{ W}$$

or

$$P_T = \frac{V_T{}^2}{R_T} = \frac{84 \text{ V}^2}{28 \text{ } \Omega}$$
$$= 252 \text{ W}$$

The total power dissipated in this example circuit is 252 W. All the information can now be inserted in a final diagram for the example, as shown in Figure 7-18.

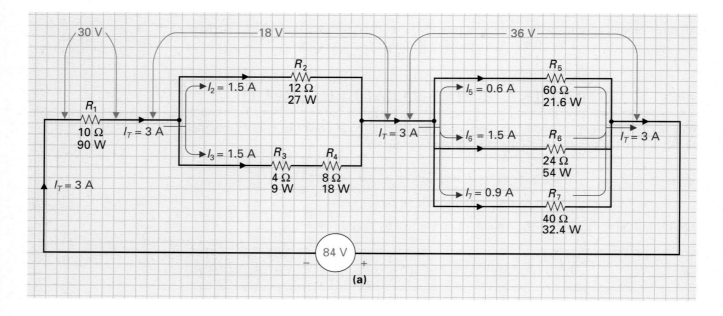

(a)

CIRCUIT ANALYSIS TABLE			
Resistance $R = V/I$	Voltage $V = I \times R$	Current $I = V/R$	Power $P = V \times I$
$R_1 = 10 \text{ } \Omega$	$V_{R1} = 30 \text{ V}$	$I_1 = 3 \text{ A}$	$P_1 = 90 \text{ W}$
$R_2 = 12 \text{ } \Omega$	$V_{R2} = 18 \text{ V}$	$I_2 = 1.5 \text{ A}$	$P_2 = 27 \text{ W}$
$R_3 = 4 \text{ } \Omega$	$V_{R3} = 6 \text{ V}$	$I_3 = 1.5 \text{ A}$	$P_3 = 9 \text{ W}$
$R_4 = 8 \text{ } \Omega$	$V_{R4} = 12 \text{ V}$	$I_4 = 1.5 \text{ A}$	$P_4 = 18 \text{ W}$
$R_5 = 60 \text{ } \Omega$	$V_{R5} = 36 \text{ V}$	$I_5 = 0.6 \text{ A}$	$P_5 = 21.6 \text{ W}$
$R_6 = 24 \text{ } \Omega$	$V_{R6} = 36 \text{ V}$	$I_6 = 1.5 \text{ A}$	$P_6 = 54 \text{ W}$
$R_7 = 40 \text{ } \Omega$	$V_{R7} = 36 \text{ V}$	$I_7 = 0.9 \text{ A}$	$P_7 = 32.4 \text{ W}$
$R_T = 28 \text{ } \Omega$	$V_T = V_S = 84 \text{ V}$	$I_T = 3 \text{ A}$	$P_T = 252 \text{ W}$

(b)

FIGURE 7-18   Series–Parallel Circuit Example with All Information Inserted. (a) Schematic. (b) Circuit Analysis Table.

FIVE-STEP METHOD FOR SERIES–PARALLEL CIRCUIT ANALYSIS

Let's now combine and summarize all the steps for calculating resistance, voltage, current, and power in a series–parallel circuit by solving another problem. Before we begin, however, let us review the five-step procedure.

Solving for Resistance, Voltage, Current, and Power in a Series–Parallel Circuit

STEP 1 Determine the circuit's total resistance.

    Step A   Solve for series-connected resistors in all parallel combinations.

    Step B   Solve for all parallel combinations.

    Step C   Solve for remaining series resistances.

STEP 2   Determine the circuit's total current.

STEP 3   Determine the voltage across each series resistor and each parallel combination (series equivalent resistor).

STEP 4   Determine the value of current through each parallel resistor in every parallel combination.

STEP 5   Determine the total and individual power dissipated by the circuit.

EXAMPLE:

Referring to Figure 7-19, calculate:

a. Total resistance

b. Total current

c. Voltage drop across all resistors

d. Current through each resistor

e. Total power dissipated by the circuit

FIGURE 7-19 Apply the Five-Step Procedure to This Series–Parallel Circuit Example.

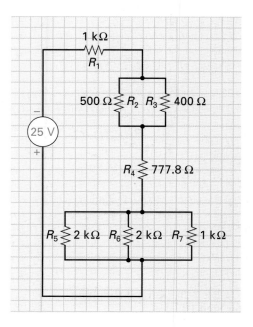

■ *Solution:*

This problem has asked us to calculate everything about the series–parallel circuit shown in Figure 7-19, and is an ideal application for our five-step series–parallel circuit analysis procedure.

STEP 1 Determine the circuit's total resistance.

Step A: There are no series resistors within parallel combinations.

Step B: There are two-resistor (R_2, R_3) and three-resistor (R_5, R_6, R_7) parallel combinations in this circuit.

$$R_{2,3} = \frac{1}{(1/R_2) + (1/R_3)} = 222.2 \ \Omega$$

$$R_{5,6,7} = \frac{1}{(1/R_5) + (1/R_6) + (1/R_7)} = 500 \ \Omega$$

Figure 7-20 illustrates the circuit resulting after step B.

Step C: Solve for the remaining four resistances to gain the circuit's total resistance (R_T) or equivalent resistance (R_{eq}).

$$R_{eq} = R_1 + R_{2,3} + R_4 + R_{5,6,7}$$
$$= 1000 \ \Omega + 222.2 \ \Omega + 777.8 \ \Omega + 500 \ \Omega$$
$$= 2500 \ \Omega \quad \text{or} \quad 2.5 \ k\Omega$$

Figure 7-21 illustrates the circuit resulting after step C.

STEP 2 Determine the circuit's total current.

$$I_T = \frac{V_S}{R_T} = \frac{25 \ \text{V}}{2.5 \ k\Omega} = 10 \ \text{mA}$$

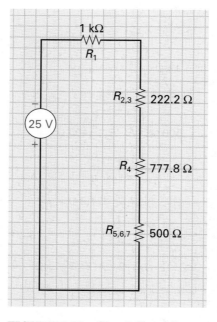

FIGURE 7-20 Circuit Resulting after Step 1B.

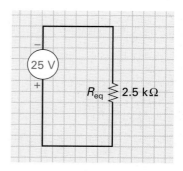

FIGURE 7-21 Circuit Resulting after Step 1C.

STEP 3 Determine the voltage across each series resistor and each series equivalent resistor. To achieve this, we utilize the diagram obtained after completing step B (Figure 7-20):

$$V_{R1} = I_T \times R_1 = 10 \text{ mA} \times 1 \text{ k}\Omega = 10 \text{ V}$$
$$V_{R2,3} = I_T \times R_{2,3} = 10 \text{ mA} \times 222.2 \text{ }\Omega = 2.222 \text{ V}$$
$$V_{R4} = I_T \times R_4 = 10 \text{ mA} \times 777.8 \text{ }\Omega = 7.778 \text{ V}$$
$$V_{R5,6,7} = I_T \times R_{5,6,7} = 10 \text{ mA} \times 500 \text{ }\Omega = 5 \text{ V}$$

Figure 7-22 illustrates the results after step 3.

STEP 4 Determine the value of current through each parallel resistor (Figure 7-23). R_1 and R_4 are series-connected resistors, and therefore their current will equal 10 mA.

$$I_1 = 10 \text{ mA}$$
$$I_4 = 10 \text{ mA}$$

The current through the parallel resistors is calculated by Ohm's law.

$$I_2 = \frac{V_{R2}}{R_2} = \frac{2.222 \text{ V}}{500 \text{ }\Omega} = 4.4 \text{ mA}$$

$$I_3 = \frac{V_{R3}}{R_3} = \frac{2.222 \text{ V}}{400 \text{ }\Omega} = 5.6 \text{ mA}$$

$$\left.\begin{array}{l} I_T = I_2 + I_3 \\ 10 \text{ mA} = 4.4 \text{ mA} + 5.6 \text{ mA} \end{array}\right\} \text{Kirchhoff's current law}$$

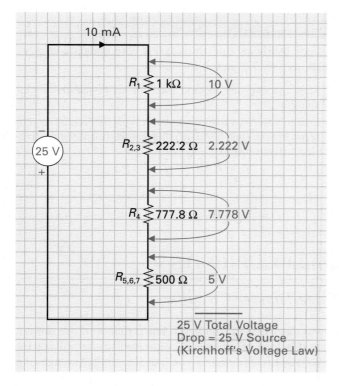

FIGURE 7-22 Circuit Resulting after Step 3.

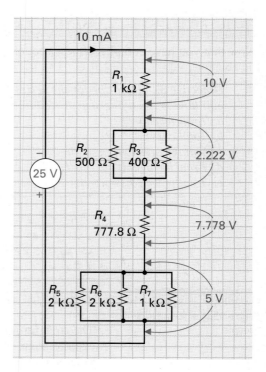

FIGURE 7-23 Series–Parallel Circuit Example with Step 1, 2, and 3 Data Inserted.

$$I_5 = \frac{V_{R5}}{R_5} = \frac{5 \text{ V}}{2 \text{ k}\Omega} = 2.5 \text{ mA}$$

$$I_6 = \frac{V_{R6}}{R_6} = \frac{5 \text{ V}}{2 \text{ k}\Omega} = 2.5 \text{ mA}$$

$$I_7 = \frac{V_{R7}}{R_7} = \frac{5 \text{ V}}{1 \text{ k}\Omega} = 5 \text{ mA}$$

$$\left. \begin{array}{l} I_T = I_5 + I_6 + I_7 \\ 10 \text{ mA} = 2.5 \text{ mA} + 2.5 \text{ mA} + 5 \text{ mA} \end{array} \right\} \text{Kirchhoff's current law}$$

STEP 5 Determine the total power dissipated by the circuit.

$$P_T = P_1 + P_2 + P_3 + P_4 + P_5 + P_6 + P_7$$

or

$$P_T = \frac{V_T{}^2}{R_T}$$

Each resistor's power figure can be calculated and the sum would be the total power dissipated by the circuit. Since the problem does not ask for the power dissipated by each individual resistor, but for the total power dissipated, it will be easier to use the formula:

$$\begin{aligned} P_T &= \frac{V_T{}^2}{R_T} \\ &= \frac{25 \text{ V}^2}{2.5 \text{ k}\Omega} \\ &= 0.25 \text{ W} \end{aligned}$$

SELF-TEST EVALUATION POINT FOR 7-4, 7-5, AND 7-6

Now that you have completed these sections, you should be able to:

■ **Objective 3.** *Describe for the series-parallel circuit how to use a five-step procedure to calculate:*
  a. Total resistance.
  b. Total current.
  c. Voltage division.
  d. Branch current.
  e. Total power dissipated.

Use the following questions to test your understanding of Sections 7-4, 7-5, and 7-6.

1. State the five-step method used for series–parallel circuit analysis.
2. Design your own five-resistor series–parallel circuit, assign resistor values and a source voltage, and then apply the five-step analysis method.

7-7   SERIES–PARALLEL CIRCUITS

7-7-1   *Loading of Voltage-Divider Circuits*

Loading
The adding of a load to a source.

The straightforward voltage divider was discussed in Chapter 5, but at that point we did not explore some changes that will occur if a load resistance is connected to the voltage divider's output. Figure 7-24 shows a voltage divider, and as you can see, the advantage of a voltage-divider circuit is that it can be used to produce several different voltages from one main voltage source by the use of a few chosen resistor values.

FIGURE 7-24   Voltage-Divider Circuit.

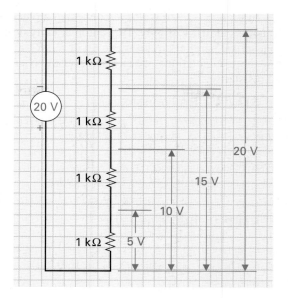

In our discussion on load resistance, we discussed how every circuit or piece of equipment offers a certain amount of resistance, and this resistance represents how much a circuit or piece of equipment will load down the source supply.

Figure 7-25 shows an example voltage-divider circuit that is being used to develop a 10 V source from a 20 V dc supply. Figure 7-25(a) illustrates this circuit in the unloaded condition, and by making a few calculations you can analyze this circuit condition.

STEP 1 $R_T = R_1 + R_2 = 1\ \text{k}\Omega + 1\ \text{k}\Omega = 2\ \text{k}\Omega$

STEP 2 $I_T = \dfrac{V_T}{R_T} = \dfrac{20\ \text{V}}{2\ \text{k}\Omega} = 10\ \text{mA}$

The current that flows through a voltage divider, without a load connected, is called the **bleeder current.** In this example the bleeder current is equal to 10 mA. It is called the bleeder current because it is continually drawing or bleeding this current from the voltage source.

STEP 3 $V_{R1} = V_{R2}$ (as resistors are the same value)
          $V_{R1} = 10\ \text{V}$
          $V_{R2} = 10\ \text{V}$

Bleeder Current

Current drawn continuously from a voltage source. A bleeder resistor is generally added to lessen the effect of load changes or provide a voltage drop across a resistor.

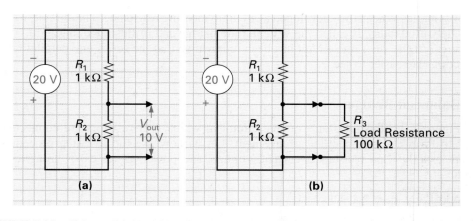

FIGURE 7-25   Voltage-Divider Circuit. (a) Unloaded Output Voltage. (b) Loaded Output Voltage.

In Figure 7-25(b) we have connected a piece of equipment represented as a resistance (R_3) across the 10 V supply. This automatically turns the previous series circuit of R_1 and R_2 into a series–parallel circuit made up of R_1, R_2, and the 100 kΩ load resistance. By making a few more calculations, we can discover the changes that have occurred by connecting this load resistance.

STEP 1 Total resistance (R_T)

Step B: $$R_{2,3} = \frac{R_2 \times R_3}{R_2 + R_3} = \frac{1 \text{ k}\Omega \times 100 \text{ k}\Omega}{1 \text{ k}\Omega + 100 \text{ k}\Omega} = 990.1 \text{ }\Omega$$

Step C:
$$R_{1,2,3} = R_1 + R_{2,3}$$
$$= 1 \text{ k}\Omega + 990.1$$
$$= 1.99 \text{ k}\Omega$$

STEP 2 Total current (I_T)

$$I_T = \frac{20 \text{ V}}{1.99 \text{ k}\Omega} = 10.05 \text{ mA}$$

STEP 3
$$V_{R1} = I_T \times R_1 = 10.05 \text{ mA} \times 1 \text{ k}\Omega = 10.05 \text{ V}$$
$$V_{R2,3} = I_T \times R_{2,3} = 10.05 \text{ mA} \times 990.1 \text{ }\Omega = 9.95 \text{ V}$$

STEP 4 $I_1 = I_T = 10.05$ mA

$$I_2 = \frac{V_{R2}}{R_2} = \frac{9.95 \text{ V}}{1 \text{ k}\Omega} = 9.95 \text{ mA}$$

$$I_3 = \frac{V_{R3}}{R_3} = \frac{9.95 \text{ V}}{100 \text{ k}\Omega} = 99.5 \text{ }\mu\text{A}$$

$$\left.\begin{array}{l} I_2 + I_3 = I_T \\ 9.95 \text{ mA} + 99.5 \text{ }\mu\text{A} = 10.05 \text{ mA} \end{array}\right\} \text{ Kirchhoff's current law}$$

As you can see, the load resistance is pulling 99.5 μA from the source, and this pulls the voltage down to 9.95 V from the required 10 V that was desired and is normally present in the unloaded condition.

When designing a voltage divider, design engineers need to calculate how much current a particular load will pull and then alter the voltage-divider resistor values to offset the loading effect when the load is connected.

7-7-2 *The Wheatstone Bridge*

In 1850, Charles Wheatstone developed a circuit to measure resistance. This circuit, which is still widely used today, is called the **Wheatstone bridge** and is illustrated in Figure 7-26(a). In Figure 7-26(b), the same circuit has been redrawn so that the series and parallel resistor connections are easier to see.

Balanced Bridge

Figure 7-27 illustrates an example circuit in which four resistors are connected together to form a series–parallel arrangement. Let us now use the five-step procedure to find out exactly what resistance, current, voltage, and power values exist throughout the circuit.

STEP 1 Total resistance (R_T)

Step A:
$$R_{1,3} = R_1 + R_3 = 10 + 20 = 30$$
$$R_{2,4} = R_2 + R_4 = 10 + 20 = 30$$

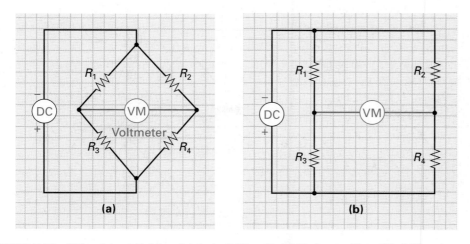

FIGURE 7-26 Wheatstone Bridge. (a) Actual Circuit. (b) Redrawn Simplified Circuit.

Step B:
$$R_T: (R_{1,2,3,4}) = \frac{R_{1,3} \times R_{2,4}}{R_{1,3} + R_{2,4}}$$
$$= \frac{30 \times 30}{30 + 30} = 15 \ \Omega$$
Total resistance $= 15 \ \Omega$

STEP 2 Total current (I_T)
$$I_T = \frac{V_T}{R_T} = \frac{30 \text{ V}}{15 \ \Omega} = 2 \text{ A}$$

STEP 3 Since $R_{1,3}$ is in parallel with $R_{2,4}$, 30 V will appear across both $R_{1,3}$ and $R_{2,4}$.
$$V_T = V_{R1,3} = V_{R2,4} = 30 \text{ V}$$

voltage-divider formula
$$V_{R1} = \frac{R_1}{R_{1,3}} \times V_T$$
$$= \frac{10}{30} \times 30 = 10 \text{ V}$$
$$V_{R3} = \frac{R_3}{R_{1,3}} \times V_T = 20 \text{ V}$$
$$V_{R2} = \frac{R_2}{R_{2,4}} \times V_T = 10 \text{ V}$$
$$V_{R4} = \frac{R_4}{R_{2,4}} \times V_T = 20 \text{ V}$$

STEP 4 $\quad I_{1,3} = \dfrac{V_{R1,3}}{R_{1,3}} = \dfrac{30 \text{ V}}{30 \ \Omega} = 1 \text{ A}$

$\qquad\qquad I_{2,4} = \dfrac{V_{R2,4}}{R_{2,4}} = \dfrac{30 \text{ V}}{30 \ \Omega} = 1 \text{ A}$

$\qquad\qquad \left.\begin{array}{l} I_{1,3} + I_{2,4} = I_T \\ 1 \text{ A} + 1 \text{ A} = 2 \text{ A} \end{array}\right\}$ Kirchhoff's current law

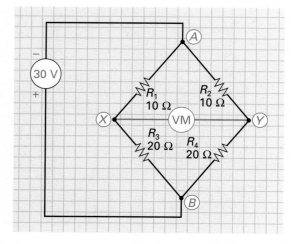

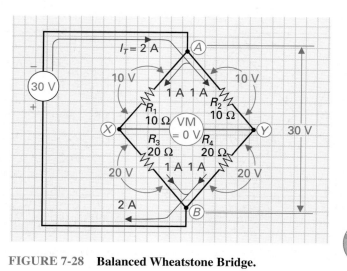

FIGURE 7-27   **Wheatstone Bridge Circuit Example.**

FIGURE 7-28   **Balanced Wheatstone Bridge.**

STEP 5   Total power dissipated (P_T)

$$P_T = I_T^2 \times R_T$$
$$= 2\,A^2 \times 15\,\Omega$$
$$= 4\,A^2 \times 15\,\Omega$$
$$= 60\,W$$

Figure 7-28 shows all of the step results inserted in the Wheatstone bridge example schematic. The Wheatstone bridge is said to be in the balanced condition when the voltage at point X equals the voltage at point Y ($V_{R3} = V_{R4}$, 20 V = 20 V). This same voltage exists across R_3 and R_4, so the voltmeter, which is measuring the voltage difference between X and Y, will indicate 0 V potential difference, and the circuit is said to be a *balanced bridge*.

Unbalanced Bridge

In Figure 7-29 we have replaced R_3 with a variable resistor and set it to 10 Ω. The R_2 and R_4 resistor combination will not change its voltage drop. However, R_1 and R_3, which are now equal, will each split the 30 V supply, producing 15 V across R_3. The voltmeter will indicate the difference in potential (5 V) from the voltage across R_3 at point X (15 V) and across R_4 at point Y (20 V). The voltmeter is actually measuring the imbalance in the circuit, which is why this circuit in this condition is known as an *unbalanced bridge*.

FIGURE 7-29   **Unbalanced Wheatstone Bridge.**

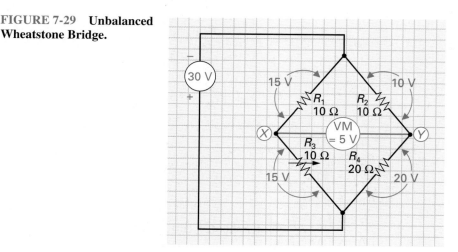

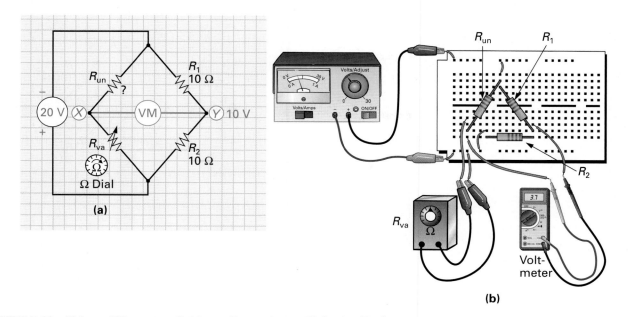

FIGURE 7-30 Using a Wheatstone Bridge to Determine an Unknown Resistance.
(a) Schematic. (b) Pictorial.

Determining Unknown Resistance

Figure 7-30 shows how a Wheatstone bridge circuit can be used to find the value of an unknown resistor (R_{un}). The variable-value resistor (R_{va}) is a calibrated resistor, which means that its resistance has been checked against a known, accurate resistance and its value can be adjusted and read from a calibrated dial.

The procedure to follow to find the value of the unknown resistor is as follows:

1. Adjust the variable-value resistor until the voltmeter indicates that the Wheatstone bridge is balanced (voltmeter indicates 0 V).
2. Read the value of the variable-value resistor. As long as $R_1 = R_2$, the variable resistance value will be the same as the unknown resistance value.

$$R_{va} = R_{un}$$

Since R_1 and R_2 are equal to one another, the voltage will be split across the two resistors, producing 10 V at point Y. The variable-value resistor must therefore be adjusted so that it equals the unknown resistance, and therefore the same situation will occur, in that the 20 V source will be split, producing 10 V at point X, indicating a balanced zero-volt condition on the voltmeter. For example, if the unknown resistance is equal to 5 Ω, then only when the variable-value resistor is adjusted and equal to 5 Ω would 10 V appear at point X and allow the circuit meter to read zero volts, indicating a balance. The variable-value resistor resistance could be read (5 Ω) and the unknown resistor resistance would be known (5 Ω).

EXAMPLE:

What is the unknown resistance in Figure 7-31?

Solution:

The bridge is in a balanced condition as the voltmeter is reading a 0 V difference between points X and Y. In the previous section we discovered that if $R_1 = R_2$, then:

$$R_{va} = R_{un}$$

FIGURE 7-31 **Wheatstone Bridge Circuit Example.**

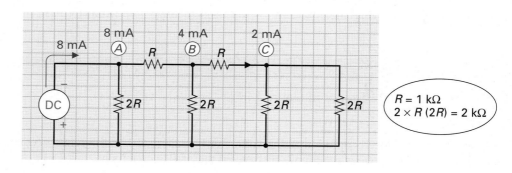

In this case, R_1 does not equal R_2, so a variation in the formula must be applied to take into account the ratio of R_1 and R_2.

$$R_{un} = R_{va} \times \frac{R_1}{R_2}$$

$$= 75 \ \Omega \times \frac{100}{30}$$

$$= 75 \ \Omega \times 3.33 = 250 \ \Omega$$

Since R_1 is 3.33 times greater than R_2, then R_{un} must be 3.33 times greater than R_{va} if the Wheatstone bridge is in the balanced condition.

CALCULATOR SEQUENCE

Step	Keypad Entry	Display Response
1.	1 0 0	100
2.	÷	
3.	3 0	30
4.	=	3.333
5.	×	
6.	7 5	75
7.	=	250

7-7-3   *The R–2R Ladder Circuit*

R–2R Ladder Circuit

A network or circuit composed of a sequence of *L* networks connected in tandem. This *R–2R* circuit is used in digital-to-analog converters.

Figure 7-32 illustrates an **R–2R ladder circuit,** which is a series–parallel circuit used within computer systems to convert digital information to analog information. To fully understand this circuit, our first step should be to find out exactly which branches will have which values of current. This can be obtained by finding out what value of resistance the current sees when it arrives at the three junctions A, B, and C. Let us begin with point C first and simplify the circuit. This is illustrated in Figure 7-33(a). No specific resistance value has been chosen, but in all cases 2R resistors (2 × R) are twice the resistance of an R resistor.

In Figure 7-33(a), if 2 mA of current arrives at point C, it sees 2R of resistance in parallel with a 2R resistance, and so the 2 mA of current splits and 1 mA flows through each branch. Two 2R resistors in parallel with one another would consequently be equivalent to one R, as seen in Figure 7-33(b).

FIGURE 7-32 *R–2R* **Ladder Circuit.**

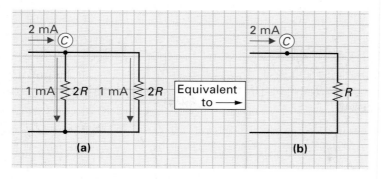

FIGURE 7-33 *R–2R* **Equivalent Circuit at Junction C.**

In Figure 7-34(a), if 4 mA of current arrives at point *B,* it sees two series resistances to the right, which is equivalent to 2*R* as seen in Figure 7-34(b), and one resistance down of 2*R*. The 4 mA therefore splits, causing 2 mA down one path and 2 mA down the other. The two 2*R* resistors in parallel, as seen in Figure 7-34(b), are equivalent to one *R,* as shown in Figure 7-34(c).

In Figure 7-35(a), if 8 mA of current arrives at point *A,* it sees two series resistors to the right, which is equivalent to 2*R* as seen in Figure 7-35(b), and one resistance down of 2*R*. The 8 mA of current therefore splits equally, causing 4 mA down one path and 4 mA down the other. The two 2*R* resistors in parallel, as seen in Figure 7-35(b), are equivalent to one resistance *R,* as shown in Figure 7-35(c).

Figure 7-36(a) through (f) summarizes the step-by-step simplification of this circuit.

The question you may have at this point is: What is the primary application of this circuit? The answer is as a current divider, as seen in Figure 7-37(a). The 8 mA of reference current is repeatedly divided by 2 as it moves from left to right, producing currents of 4 mA, 2 mA, and 1 mA. The result of this *R–2R* current division can be used in a circuit known as a *digital-to-analog converter* (DAC), which is illustrated in Figure 7-37(b).

Digital data or information exists within a computer, and this system expresses numbers and letters in two discrete steps; for example, on–off, high–low, open–closed, or 0–1. Only two conditions exist within the computer, and all information is represented by this two-state system.

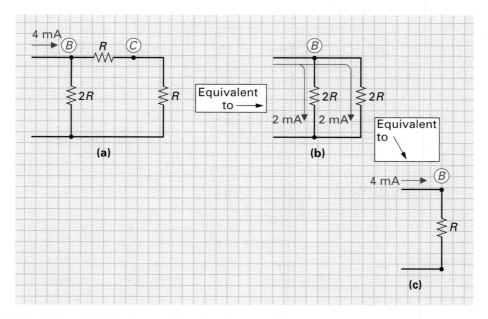

FIGURE 7-34 *R–2R* **Equivalent Circuit at Junction B.**

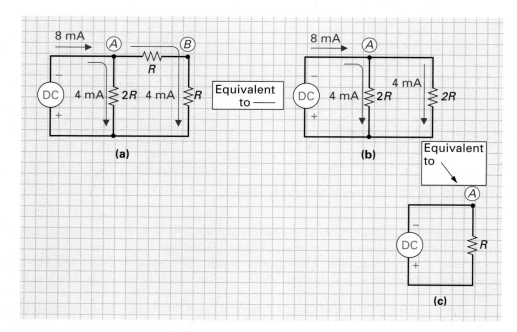

FIGURE 7-35 *R–2R* Equivalent Circuit at Junction A.

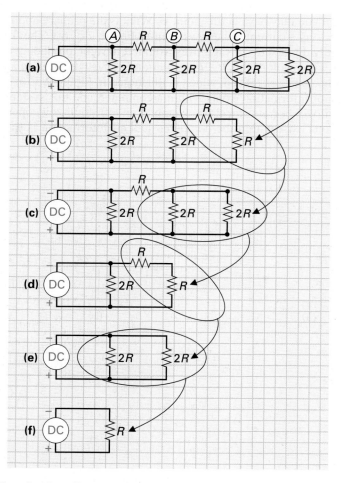

FIGURE 7-36 Step-by-Step Simplification of an *R–2R* Ladder Circuit.

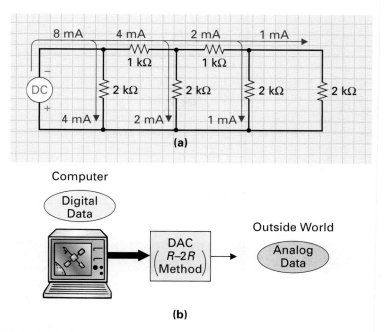

(a)

Computer

Digital
Data

DAC
$\left(\begin{array}{c} R-2R \\ \text{Method} \end{array}\right)$

Outside World

Analog
Data

(b)

FIGURE 7-37   *R–2R* **Ladder in a Digital-to-Analog Converter.**

Analog data or information exists outside a computer in our environment, and in this system, data are expressed in many different levels, as opposed to just two with digital information. Numbers are expressed in one of ten different levels (0–9), as opposed to only two (0–1) in digital.

Due to these differences, a device is needed that will interface (convert or link two different elements) the digital information within a computer to the analog information that you and I understand. This device, called a digital-to-analog converter, uses the *R–2R* ladder circuit that we just discussed.

EXAMPLE:

Determine the reference current and branch currents for the circuit in Figure 7-38.

Solution:

In our simplification of the ladder circuit, we discussed previously that any *R–2R* ladder circuit can be simplified to one resistor equal to *R,* as seen in Figure 7-39(a). The reference or total current supplied will therefore equal

$$I_T = \frac{V_T}{R_T} = \frac{24 \text{ V}}{1 \text{ k}\Omega} = 24 \text{ mA}$$

and the current will split through each branch, as shown in Figure 7-39(b).

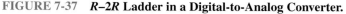

FIGURE 7-38   **Calculate *R–2R* Circuit Reference Current and Branch Currents.**

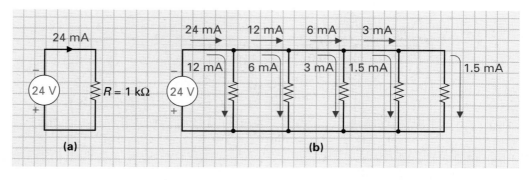

FIGURE 7-39   *R–2R* Circuit with Data Inserted.

Now that you have completed this section, you should be able to:

- **Objective 4.**   *Explain what loading effect a piece of equipment will have when connected to a voltage divider.*

- **Objective 5.**   *Identify and describe the Wheatstone bridge circuit in both the balanced and unbalanced condition.*

- **Objective 6.**   *Describe the R–2R ladder circuit used for digital-to-analog conversion.*

Use the following questions to test your understanding of Section 7-7.

1. What is meant by loading of a voltage-divider circuit?
2. Sketch a Wheatstone bridge circuit and list an application of this circuit.
3. What value will the total resistance of an *R–2R* ladder circuit always equal?
4. For what application could the *R–2R* ladder be used?

7-8   TROUBLESHOOTING SERIES–PARALLEL CIRCUITS

Troubleshooting is defined as the process of locating and diagnosing malfunctions or breakdowns in equipment by means of systematic checking or analysis. As discussed in previous resistive-circuit troubleshooting procedures, there are basically only three problems that can occur:

1. A component will open. This usually occurs if a resistor burns out or a wire or switch contact breaks.
2. A component will short. This usually occurs if a conductor, such as solder, wire, or some other conducting material, is dropped or left in the equipment, making or connecting two points that should not be connected.
3. There is a variation in a component's value. This occurs with age in resistors over a long period of time and can eventually cause a malfunction of the equipment.

Using the example circuit in Figure 7-40, we will step through a few problems, beginning with an open component. Throughout the troubleshooting, we will use the voltmeter whenever possible, as it can measure voltage by just connecting the leads across the component, rather than the ammeter, which has to be placed in the circuit, in which case the circuit path has to be opened. In some situations, using an ammeter can be difficult.

To begin, let's calculate the voltage drops and branch current obtained when the circuit is operating normally.

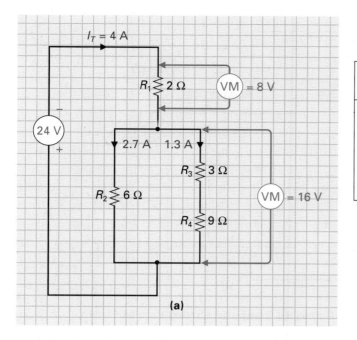

CIRCUIT ANALYSIS TABLE			
Resistance $R = V/I$	Voltage $V = I \times R$	Current $I = V/R$	Power $P = V \times I$
$R_1 = 2\ \Omega$	$V_{R1} = 8\ V$	$I_1 = 4\ A$	$P_1 = 32\ W$
$R_2 = 6\ \Omega$	$V_{R2} = 16\ V$	$I_2 = 2.7\ A$	$P_2 = 43.2\ W$
$R_3 = 3\ \Omega$	$V_{R3} = 4\ V$	$I_3 = 1.3\ A$	$P_3 = 5.2\ W$
$R_4 = 9\ \Omega$	$V_{R4} = 12\ V$	$I_4 = 1.3\ A$	$P_4 = 15.6\ W$

$R_T = 6\ \Omega \quad V_T = V_S = 24V \quad I_T = 4\ A \quad P_T = 96\ W$

(b)

FIGURE 7-40   Series–Parallel Circuit Example. (a) Schematic. (b) Circuit Analysis Table.

STEP 1   (A) $R_{3,4} = R_3 + R_4 = 3\ \Omega + 9\ \Omega = 12\ \Omega$

   (B) $R_{2,3,4} = \dfrac{R_2 \times R_{3,4}}{R_2 + R_{3,4}} = \dfrac{6\ \Omega \times 12\ \Omega}{6\Omega + 12\ \Omega} = 4\ \Omega$

   (C) $R_{1,2,3,4} = R_T = R_1 + R_{2,3,4}$
      $= 2\ \Omega + 4\ \Omega = 6\ \Omega$

STEP 2 $I_T = \dfrac{V_T}{R_T} = \dfrac{24\ V}{6\ \Omega} = 4\ A$

STEP 3 $V_{R1} = I_{R1} \times R_1 = 4\ A \times 2\ \Omega = 8\ V$
   $V_{R2,3,4} = I_{R2,3,4} \times R_{2,3,4} = 4\ A \times 4\ \Omega = 16\ V$
                  (Kirchhoff's voltage law)

STEP 4 $I_1 = 4\ A$ (series resistor)

   $I_2 = \dfrac{V_{R2}}{R_2} = \dfrac{16\ V}{6\ \Omega} = 2.7\ A$

   $I_{3,4} = \dfrac{V_{R3,4}}{R_{3,4}} = \dfrac{16\ V}{12\ \Omega} = 1.3\ A$

All these results have been inserted in the schematic in Figure 7-40(a) and in the circuit analysis table in Figure 7-40(b).

7-8-1   *Open Component*

R_1 Open (Figure 7-41)

With R_1 open, there cannot be any current flow through the circuit as there is not a path from one side of the power supply to the other. This fault can be recognized easily because approximately all of the applied 24 V will be measured across the open resistor (R_1), and 0 V will appear across all the other resistors.

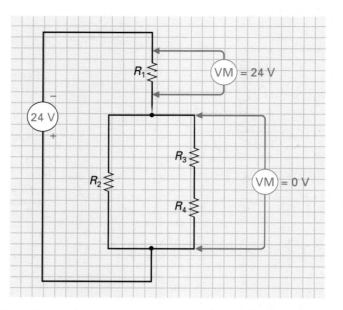

FIGURE 7-41 Open Series-Connected Resistor in a Series–Parallel Circuit.

R_3 Open (Figure 7-42)

With R_3 open, there will be no current through the branch made up of R_3 and R_4. The current path will be through R_1 and R_2, and therefore the total resistance will now increase ($R_T\uparrow$) from 6 Ω to

$$R_T = R_1 + R_2$$
$$= 2\ \Omega + 6\ \Omega = 8\ \Omega$$

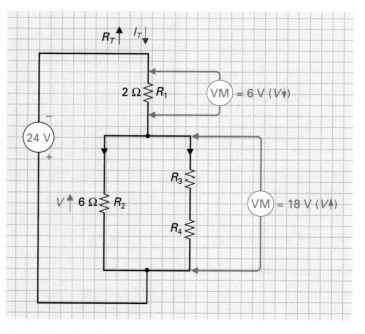

FIGURE 7-42 Open Parallel-Connected Resistor in a Series–Parallel Circuit.

This 8 Ω is an increase in circuit resistance from the normal resistance, which was 6 Ω, which implies that an open has occurred to increase resistance. The total current will decrease ($I_T\downarrow$) from 4 A to

$$I_T = \frac{V_T}{R_T} = \frac{24\text{ V}}{8\ \Omega} = 3\text{ A}$$

The voltage drop across the resistors will be

$$V_{R1} = I_T \times R_1 = 3\text{ A} \times 2\ \Omega = 6\text{ V}$$
$$V_{R2} = I_T \times R_2 = 3\text{ A} \times 6\ \Omega = 18\text{ V}$$

If one of the parallel branches is opened, the overall circuit resistance will always increase. This increase in the total resistance will cause an increase in the voltage dropped across the parallel branch (the greater the resistance, the greater the voltage drop), which enables the technician to localize the fault area and also to determine that the fault is an open.

The voltage measured with a voltmeter will be

$$V_{R1} = 6\text{ V}$$
$$V_{R2} = 18\text{ V}$$
$$V_{R3} = 18\text{ V (open)}$$
$$V_{R4} = 0\text{ V}$$

This identifies the problem as R_3 being open, as it drops the entire parallel circuit voltage (18 V) across itself, whereas normally the voltage would be dropped proportionally across R_3 and R_4, which are in series with one another.

7-8-2   *Shorted Component*

R_1 Shorted (Figure 7-43)

With R_1 shorted, the total circuit resistance will decrease ($R_T\downarrow$), causing an increase in circuit current ($I_T\uparrow$). This increase in current will cause an increase in the voltage dropped across the parallel branch. However, the fault can be located once you measure the voltage

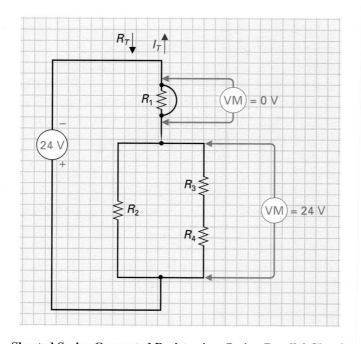

FIGURE 7-43 **Shorted Series-Connected Resistor in a Series–Parallel Circuit.**

across R_1, which will read 0 V, indicating that this resistor has almost no resistance, as it has no voltage drop across it.

R_3 Shorted (Figure 7-44)

With R_3 shorted, there will be a decrease in the circuit's total resistance from 6 Ω to

$$R_{2,3,4} = \frac{R_2 \times R_{3,4}}{R_2 + R_{3,4}} = \frac{6 \times 9}{6 + 9} = 3.6$$

$$\begin{aligned} R_{1,2,3,4} &= R_1 + R_{2,3,4} \\ &= 2\ \Omega + 3.6\ \Omega \\ &= 5.6\ \Omega \end{aligned}$$

This decrease in total resistance ($R_T\downarrow$) will cause an increase in total current ($I_T\uparrow$), which implies that a short has occurred to decrease resistance. The total current will now increase from 4 A to

$$I_T = \frac{V_T}{R_T} = \frac{24\text{ V}}{5.6\ \Omega} = 4.3\text{ A}$$

The voltage drops across the resistors will be

$$V_{R1} = I_T \times R_1 = 4.3\text{ A} \times 2\ \Omega = 8.6\text{ V}$$
$$V_{R2,3,4} = I_T \times R_{2,3,4} = 4.3\text{ A} \times 3.6\ \Omega = 15.4\text{ V}$$

When one of the parallel branch resistors is shorted, the overall circuit resistance will always decrease. This decrease in total resistance will cause a decrease in the voltage dropped across the parallel branch (the smaller the resistance, the smaller the voltage drop), and this enables the technician to localize the faulty area and also to determine that the fault is a short.

The voltage measured with the voltmeter (VM) will be

$$V_{R1} = 8.6\text{ V}$$
$$V_{R2} = 15.4\text{ V}$$
$$V_{R3} = 0\text{ V (short)}$$
$$V_{R4} = 15.4\text{ V}$$

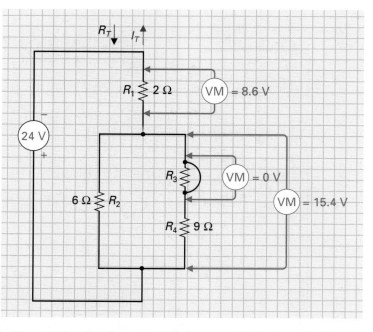

FIGURE 7-44   Shorted Parallel-Connected Resistor in a Series–Parallel Circuit.

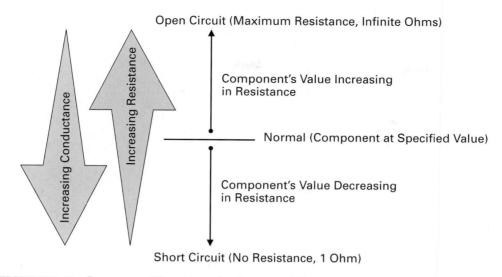

Open Circuit (Maximum Resistance, Infinite Ohms)

Component's Value Increasing in Resistance

Normal (Component at Specified Value)

Component's Value Decreasing in Resistance

Short Circuit (No Resistance, 1 Ohm)

FIGURE 7-45   **Summary of Symptoms for Opens and Shorts.**

This identifies the problem as R_3 being a short, as 0 V is being dropped across it.

In summary, you may have noticed that (Figure 7-45):

1. An open component causes total resistance to increase ($R_T\uparrow$) and, therefore, total current to decrease ($I_T\downarrow$), and the open component, if in series, has the supply voltage across it, and if in a parallel branch, has the parallel branch voltage across it.
2. A shorted component causes total resistance to decrease ($R_T\downarrow$) and, therefore, total current to increase ($I_T\uparrow$), and the shorted component, if in series or parallel branches, will have 0 V across it.

7-8-3   *Resistor Value Variation*

R_2 Resistance Decreases (Figure 7-46)

If the resistance of R_2 decreases, the total circuit resistance (R_T) will decrease and the total circuit current (I_T) will increase. The result of this problem and the way in which the fault can be located is that when the voltage drop across R_1 and the parallel branch is tested, there will be an increased voltage drop across R_1 due to the increased current flow, and a decrease in the voltage drop across the parallel branch due to a decrease in the parallel branch resistance.

With open and shorted components, the large voltage (open) or small voltage (short) drop across a component enables the technician to identify the faulty component. A variation in a component's value, however, will vary the circuit's behavior. With this example, the symptoms could have been caused by a combination of variations. So once the area of the problem has been localized, the next troubleshooting step is to disconnect power and then remove each of the resistors in the suspected faulty area and verify that their resistance values are correct by measuring their resistances with an ohmmeter. In this problem we had an increase in voltage across R_1 and a decrease in voltage across the parallel branch when measuring with a voltmeter, which was caused by R_2 decreasing. The same swing in voltage readings could also have been obtained by an increase in the resistance of R_1.

R_2 Resistance Increases (Figure 7-47)

If the resistance of R_2 increases, the total circuit resistance (R_T) will increase and the total circuit current (I_T) will decrease. The voltage drop across R_1 will decrease and the voltage across the branch will increase in value. Once again, these measured voltage changes could be caused by the resistance of R_2 increasing or the resistance of R_1 decreasing.

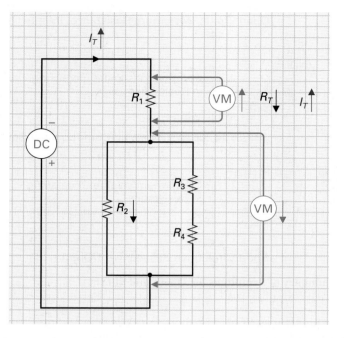

FIGURE 7-46 **Parallel-Connected Resistor Value Decrease in Series–Parallel Circuit.**

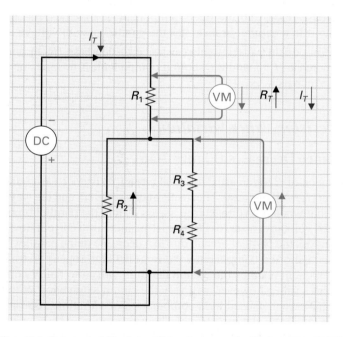

FIGURE 7-47 **Parallel-Connected Resistor Value Increase in Series–Parallel Circuit.**

To reinforce your understanding let's work out a few examples of troubleshooting other series–parallel circuits.

■ **EXAMPLE:**

If bulb 1 in Figure 7-48 goes open, what effect will it have, and how will the fault be recognized?

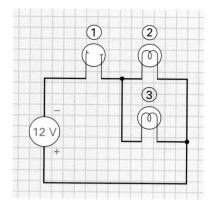

FIGURE 7-48   Series-Connected Open Bulb in Series–Parallel Circuit.

Solution:

A visual inspection of the circuit shows all bulbs off, as there is no current path from one side of the battery to the other, since bulb 1 is connected in series and has opened the only path. Since all bulbs are off, the faulty bulb cannot be visually isolated. However, you can easily localize the faulty bulb by using one of two methods:

1. Use the voltmeter and check the voltage across each bulb. Bulb 1 would have 12 V across it, while 2 and 3 would have 0 V across them. This isolates the faulty component to bulb 1, since all the supply voltage is being measured across it.

2. By analyzing the circuit diagram, you can see that only one bulb can open and cause all the bulbs to go out, and that is bulb 1. If bulb 2 opens, 1 and 3 would still be on, and if bulb 3 opens, 1 and 2 would still remain on. With power off, the ohmmeter could verify this open.

EXAMPLE:

One resistor in Figure 7-49 has shorted. From the voltmeter reading shown, determine which one.

Solution:

If the supply voltage is being measured across the parallel branch of R_2 and R_3, there cannot be any other resistance in circuit, so R_1 must have shorted. The next step would be to locate the component, R_1, and determine what has caused it to short. If we were not told that a resistor had shorted, the same symptom could have been caused if R_2 and R_3 were both open, and therefore the open parallel branch would allow no current to flow and maximum supply voltage would appear across it. The individual component resistance, when checked, will isolate the problem.

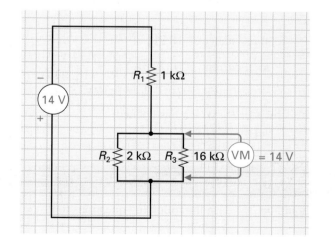

FIGURE 7-49   Find the Shorted Resistor.

EXAMPLE:

Determine if there is an open or short in Figure 7-50. If so, isolate it by the two voltage readings that are shown in the circuit diagram.

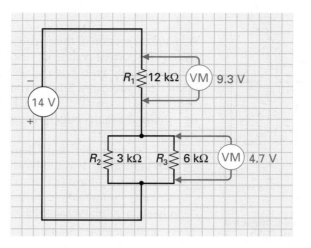

FIGURE 7-50 **Does a Problem Exist?**

■ *Solution:*

Performing a few calculations, you should come up with a normal total circuit resistance of 14 kΩ and a total circuit current of 1 mA. This should cause 12 V across R_1 and 2 V across the parallel branch under no-fault conditions. The decrease in the voltage drop across the series resistor R_1 leads you to believe that there has been a decrease in total circuit current, which must have been caused by a total resistance increase, which points to an open component (assuming that only an open or short can occur and not a component value variation).

If R_1 was open, all the 14 V would have been measured across R_1, which did not occur. If R_3 opened,

$$\text{total resistance} = 15 \text{ k}\Omega$$

$$\text{total current} = 0.93 \text{ mA} \left(\frac{V_T}{R_T} = \frac{14 \text{ V}}{15 \text{ k}\Omega} \right)$$

$$V_{R1} = I_T \times R_1 = 11.16 \text{ V}$$

Since the voltage dropped across R_1 was 9.3 V, R_3 is not the open. If R_2 opened,

$$\text{total resistance} = 18 \text{ k}\Omega$$

$$\text{total current} = 0.78 \text{ mA} \left(\frac{V}{R} = \frac{14 \text{ V}}{18 \text{ k}\Omega} \right)$$

$$V_{R1} = I_T \times R_1 = 9.36 \text{ V}$$

This circuit's problem is resistor R_2, which has opened.

SELF-TEST EVALUATION POINT FOR SECTION 7-8

Now that you have completed this section, you should be able to:

■ *Objective 7.* *Explain how to identify the following problems in a series-parallel circuit:*
a. Open series resistor.
b. Open parallel resistor.
c. Shorted series resistor.
d. Shorted parallel resistor.
e. Resistor value variation.

Use the following questions to test your understanding of Section 7-8.

1. An open component
2. A shorted component
3. A resistor value variation

Series–parallel circuits can become very complex in some applications, and the more help you have in simplifying and analyzing these networks, the better. The following theorems can be used as powerful analytical tools for evaluating circuits. To begin, let's discuss the differences between voltage and current sources.

7-9-1   *Voltage and Current Sources*

The easiest way to understand a current source is to compare its features to a voltage source, so let's begin by discussing voltage sources.

Voltage Source

The battery is an example of a **voltage source** that in the ideal condition will produce a fixed output voltage regardless of what load resistance is connected across its terminals. This means that even if a large load current is drawn from the battery (heavy load, due to a small load resistance) or if a small load current results (light load, due to a large load resistance), the battery will always produce a constant output voltage, as seen in Figure 7-51.

In reality, every voltage source, whether a battery, power supply, or generator, will have some level of inefficiency and not only generate an output electrical dc voltage but also generate heat. This inefficiency is represented as an internal resistance, as seen in Figure 7-52(a), and in most cases this internal source resistance (R_{int}) is very low (several ohms) compared to the load resistance (R_L). In Figure 7-52(a), no load has been connected, so the output or open-circuit voltage will be equal to the source voltage, V_S. When a load is connected across the battery, as shown in Figure 7-52(b), R_{int} and R_L form a series circuit, and some of the source voltage appears across R_{int}. As a result, the output or load voltage is

Voltage Source

The circuit or device that supplies voltage to a load circuit.

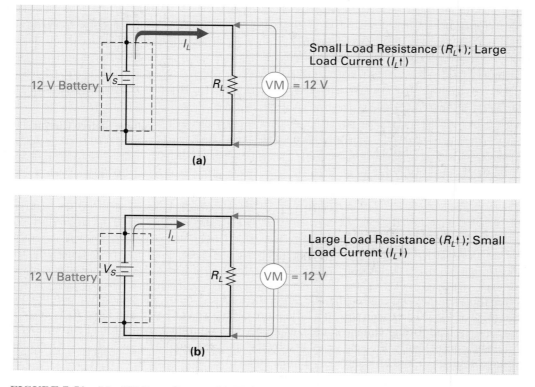

FIGURE 7-51   Ideal Voltage Source. (a) Heavy Load. (b) Light Load.

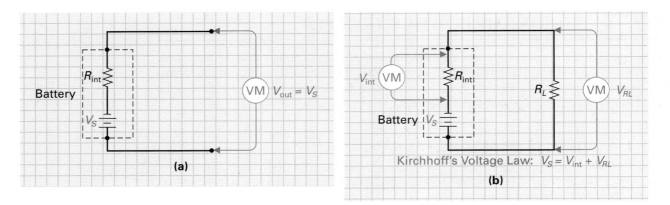

FIGURE 7-52 **Realistic Voltage Source. (a) Unloaded. (b) Loaded.**

always less than V_S. Since R_{int} is normally quite small compared to R_L, the voltage source approaches ideal, as almost all the source voltage (V_S) appears across R_L.

In conclusion, a voltage source should have the smallest possible internal resistance, so that the output voltage (V_{out}) will remain constant and approximately equal to V_S independent of whether a light load (large R_L, small I_L) or heavy load (small R_L, large I_L) is connected across its output terminals.

☐ **EXAMPLE:**

Calculate the output voltage in Figure 7-53 if R_L is equal to:

    a. 100 Ω

    b. 1 kΩ

    c. 100 kΩ

▨ *Solution:*

(Voltage-divider formula)

    a. $R_L = 100\ \Omega$:

$$V_{out} = \frac{R_L}{R_T} \times V_S$$

$$= \frac{100\ \Omega}{110\ \Omega} \times 100\ \text{V} = 90.9\ \text{V}$$

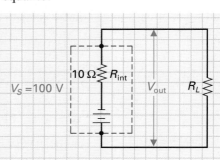

FIGURE 7-53 **Voltage Source Circuit Example.**

    b. $R_L = 1\ \text{k}\Omega\ (1000\ \Omega)$

$$V_{out} = \frac{1000\ \Omega}{1010\ \Omega} \times 100\ \text{V} = 99.0\ \text{V}$$

    c. $R_L = 100\ \text{k}\Omega\ (100,000\ \Omega)$

$$V_{out} = \frac{100,000\ \Omega}{100,010\ \Omega} \times 100\ \text{V} = 99.99\ \text{V}$$

From this example you can see that the larger the load resistance, the greater the output voltage (V_{out} or V_{RL}). To explain this in a little more detail, we can say that a large R_L is considered a light load for the voltage source, as it only has to produce a small load current ($R_L\!\uparrow$, $I_L\!\downarrow$), and consequently, the heat generated by the source is small ($P_{R_{int}}\!\downarrow = I_2\!\downarrow \times R$) and the voltage source is more efficient (V_{out} almost equals V_S), approaching ideal ($V_{out} = V_S$). The voltage source in this example, however, produced an almost constant output voltage (within 10% of V_S) despite the very large changes in R_L.

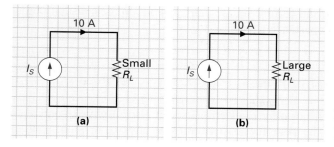

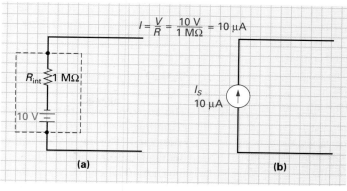

FIGURE 7-54   The Current Source Maintains a
Constant Output Current Whether the Load is (a) Heavy
(Small R_L) or (b) Light (Large R_L).

FIGURE 7-55   The Large Internal Resistance of the Current
Source.

Current Source

Just as a voltage source has a voltage rating, a **current source** has a current rating. An
ideal voltage source should deliver a constant output voltage, and similarly an ideal current
source should deliver its constant rated current, regardless of what value of load resistance is
connected across its output terminals, as seen in Figure 7-54.

A current source can be thought of as a voltage source with an extremely large internal
resistance, as seen in Figure 7-55(a) and symbolized in Figure 7-55(b).

The current source should have a large internal resistance so that, whatever the load resistance connected across the output, it will have very little effect on the total resistance and
the load current will remain constant. The symbol for a constant current source has an arrow
within a circle and this arrow points in the direction of current flow.

Current Source

The circuit or device that
supplies current to a load
circuit.

EXAMPLE:

Calculate the load current supplied by the current source in Figure 7-56 if the following values of R_L are connected across the output terminals:

a. 100 Ω

b. 1 kΩ

c. 100 kΩ

Solution:

a. $R_L = 100\ \Omega$ $(R_T = R_{int} + R_L = 1,000,000\ \Omega$
$+ 100\ \Omega = 1,000,100\ \Omega)$

$$I_L = \frac{V_S}{R_T}$$

$$= \frac{10\text{ V}}{1,000,100\ \Omega} = 9.999\ \mu A$$

b. $R_L = 1\ k\Omega$ $(R_T = R_{int} + R_L = 1,001,000\ \Omega)$

$$I_L = \frac{10\text{ V}}{1,001,000\ \Omega} = 9.99\ \mu A$$

c. $R_L = 100\ k\Omega$ $(R_T = R_{int} + R_L = 1,100,000\ \Omega)$

$$I_L = \frac{10\text{ V}}{1,100,000\ \Omega} = 9.09\ \mu A$$

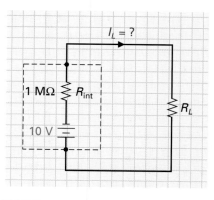

FIGURE 7-56   Current Source
Circuit Example.

From this example you can see that the current source delivered an almost constant output current regardless of the large load resistance change.

7-9-2   *Superposition Theorem*

Superposition Theorem

In a network or circuit containing two or more voltage sources, the current at any point is equal to the algebraic sum of the individual source currents produced by each source acting separately.

The **superposition theorem** is used not only in electronics, but also in physics and even economics. It is used to determine the net effect in a circuit that has two or more sources connected. The basic idea behind this theorem is that if two voltage sources are both producing a current through the same circuit, the net current can be determined by first finding the individual currents and then adding them together. Stated formally: *In a network containing two or more voltage sources, the current at any point is equal to the algebraic sum of the individual source currents produced by each source acting separately.*

The best way to fully understand the theorem is to apply it to a few examples. Figure 7-57(a) illustrates a simple series circuit with two resistors and two voltage sources. The 12 V source (V_1) is trying to produce a current in a clockwise direction, while the 24 V source (V_2) is trying to force current in a counterclockwise direction. What will be the resulting net current in this circuit?

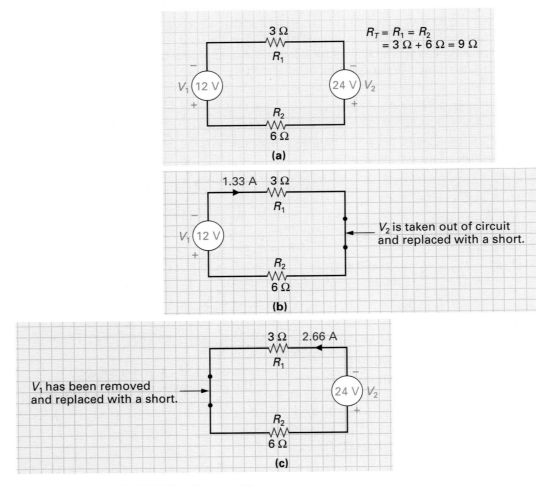

FIGURE 7-57   Superposition.

STEP 1   To begin, let's consider what current would be produced in this circuit if only V_1 were connected, as shown in Figure 7-57(b).

$$I_1 = \frac{V_1}{R_T} = \frac{12 \text{ V}}{9 \text{ }\Omega} = 1.33 \text{ A}$$

STEP 2   The next step is to determine how much current V_2 would produce if V_1 were not connected in the circuit, as shown in Figure 7-57(c).

$$I_2 = \frac{V_2}{R_T} = \frac{12 \text{ V}}{9 \text{ }\Omega} = 2.66 \text{ A}$$

V_1 is attempting to produce 1.33 A in the clockwise direction, while V_2 is trying to produce 2.66 A in the counterclockwise direction. The net current will consequently be 1.33 A in the counterclockwise direction.

EXAMPLE:

Calculate the current through R_2 in Figure 7-58 using the superposition theorem.

Solution:

The first step is to calculate the current through R_2 due only to the voltage source V_1. This is shown in Figure 7-59(a). R_1 is a series-connected resistor, while R_2 and R_3 are connected in parallel with one another. So:

$$R_{2,3} = \frac{R_{EV}}{n} = \frac{20 \text{ }\Omega}{2} = 10 \text{ }\Omega \qquad \text{(equal-value parallel resistor formula)}$$

$$R_T = R_1 + R_{2,3} = 20 \text{ }\Omega + 10 \text{ }\Omega = 30 \text{ }\Omega \qquad \text{(total resistance)}$$

$$I_T = \frac{V_1}{R_T} = \frac{12 \text{ V}}{30 \text{ }\Omega} = 400 \text{ mA} \qquad \text{(total current)}$$

$$I_2 = \frac{R_{2,3}}{R_2} \times I_T = \frac{10 \text{ }\Omega}{20 \text{ }\Omega} \times 400 \text{ mA} = 200 \text{ mA} \qquad \text{(current-divider formula)}$$

This 200 mA of current is flowing down through R_2.

The next step is to find the current flow through R_2 due only to the voltage source V_2. This is shown in Figure 7-59(b). In this instance, R_3 is a series-connected resistor, and R_1 and R_2 make up a parallel circuit. So:

$$R_{1,2} = 10 \text{ }\Omega$$

$$R_T = R_{1,2} + R_3 = 10 \text{ }\Omega + 20 \text{ }\Omega = 30 \text{ }\Omega$$

$$I_T = \frac{V_2}{R_T} = \frac{30 \text{ V}}{30 \text{ }\Omega} = 1 \text{ A} \qquad \text{(1000 mA)}$$

$$I_2 = \frac{R_{1,2}}{R_2} \times I_T = \frac{10 \text{ }\Omega}{20 \text{ }\Omega} \times 1000 \text{ mA} = 500 \text{ mA}$$

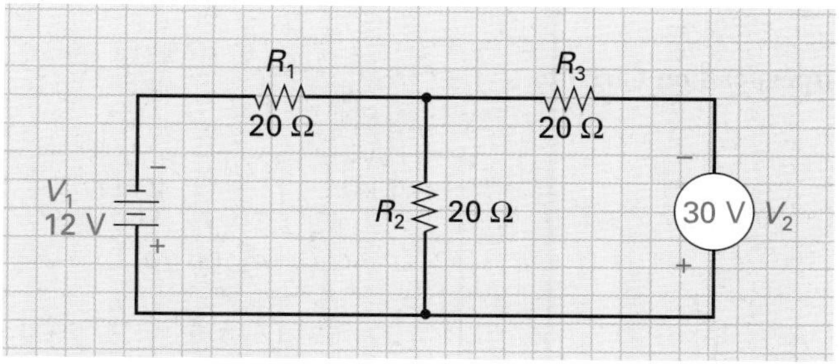

FIGURE 7-58   **Superposition Circuit Example.**

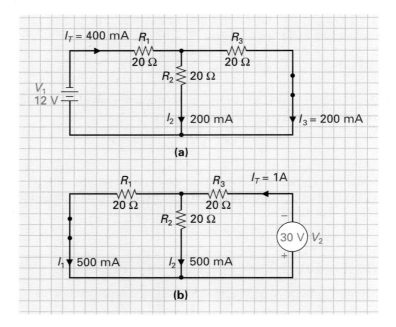

FIGURE 7-59   Superposition Circuit Solution.

This 500 mA of current will flow down through R_2.

Both V_1 and V_2 produce a current flow down through R_2, so I_{R2} and I_{R3} have the same algebraic sign, and the total current through R_2 is equal to the sum of the two currents produced by V_1 and V_2.

$$I_2 \text{ (total)} = I_2 \text{ due to } V_1 + I_2 \text{ due to } V_2$$
$$= 200 \text{ mA} + 500 \text{ mA}$$
$$= 700 \text{ mA}$$

☐   **EXAMPLE:**

Calculate the current flow through R_1 in Figure 7-60.

▧   *Solution:*

With the superposition theorem, current sources are treated differently from voltage sources in that each *current source is removed from the circuit and replaced with an open,* as illustrated in the first step of the solution shown in Figure 7-61(a). In this instance you can see that current has only one path (series circuit), so the current through R_1 is counterclockwise and equal to the I_{S1} source current, 100 μA.

FIGURE 7-60   **Superposition Circuit Example.**

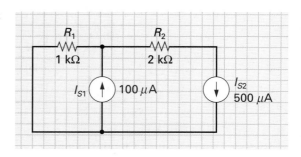

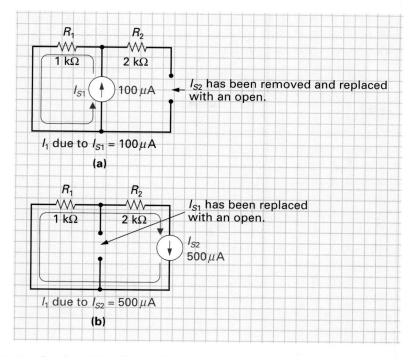

FIGURE 7-61 Superposition Circuit Solution.

In Figure 7-61(b), the current source I_{S1} has been removed and replaced with an open. R_1 and R_2 form a series circuit, so the total source current from I_{S2} flows through R_1 ($I_1 = 500$ μA) in a clockwise direction.

Since I_{S1} is producing 100 μA of current through R_1 in a counterclockwise direction and I_{S2} is producing 500 μA of current in a clockwise direction, the resulting current through R_1 will equal

$$I_1 = I_{S2} \text{ (cw)} - I_{S1} \text{ (ccw)}$$
$$= 500 \text{ μA} - 100 \text{ μA} = 400 \text{ μA clockwise}$$

7-9-3 *Thévenin's Theorem*

Thévenin's theorem allows us to replace the complex networks in Figure 7-62(a) with an equivalent circuit containing just one source voltage (V_{TH}) and one series-connected resistance (R_{TH}), as in Figure 7-62(b). Stated formally: *Any network of voltage sources and resistors can be replaced by a single equivalent voltage source (V_{TH}) in series with a single equivalent resistance (R_{TH}).*

Figure 7-63(a) illustrates an example circuit. As with any theorem, a few rules must be followed to obtain an equivalent V_{TH} and R_{TH}.

STEP 1 The first step is to disconnect the load (R_L) and calculate the voltage that will appear across points A and B, as in Figure 7-63(b). This open-circuit voltage will be the same as the voltage drop across R_2 (V_{R2}) and is called the *Thévenin equivalent voltage (V_{TH})*. First, let's calculate current:

$$I_T = \frac{V_S}{R_T} = \frac{12 \text{ V}}{9 \text{ Ω}} = 1.333 \text{ A}$$

Therefore, V_{R2} or V_{TH} will equal

$$V_{R2} = I_T \times R_2 = 1.333 \times 6 \text{ Ω} = 8 \text{ V}$$

so $V_{TH} = 8$ V.

> **Thévenin's Theorem**
> Any network of voltage sources and resistors can be replaced by a single equivalent voltage source (V_{TH}) in series with a single equivalent resistance (R_{TH}).

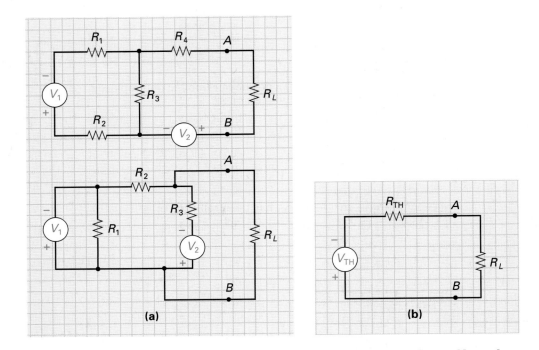

FIGURE 7-62 Thévenin's Theorem. (a) Complex Multiple Resistors and Source Networks Are Replaced by (b) One Source Voltage (V_{TH}) and One Series-Connected Resistance (R_{TH}).

STEP 2 Now that the Thévenin equivalent voltage has been calculated, the next step is to calculate the Thévenin equivalent resistance. In this step, the source voltage is removed and replaced with a short, as seen in Figure 7-63(c), and the Thévenin equivalent resistance is equal to whatever resistance exists between points A and B. In this example, R_1 and R_2 form a parallel circuit, the total resistance of which is equal to

$$ R_T = \frac{R_1 \times R_2}{R_1 + R_2} = \frac{3 \times 6}{3 + 6} = \frac{18}{9} = 2 \ \Omega $$

so $R_{TH} = 2 \ \Omega$. The circuit to be Thévenized in Figure 7-63(a) can be represented by the Thévenin equivalent circuit shown in Figure 7-63(d).

The question you may have at this time is why we would need to simplify such a basic circuit, when Ohm's law could have been used just as easily to analyze the network. Thévenin's theorem has the following advantages:

1. If you had to calculate load current and load voltage (I_{RL} and V_{RL}) for 20 different values of R_L, it would be far easier to use the Thévenin equivalent circuit with the series-connected resistors R_{TH} and R_L, rather than applying Ohm's law to the series–parallel circuit made up of R_1, R_2, and R_L.

2. Thévenin's theorem permits you to solve complex circuits that could not easily be analyzed using Ohm's law.

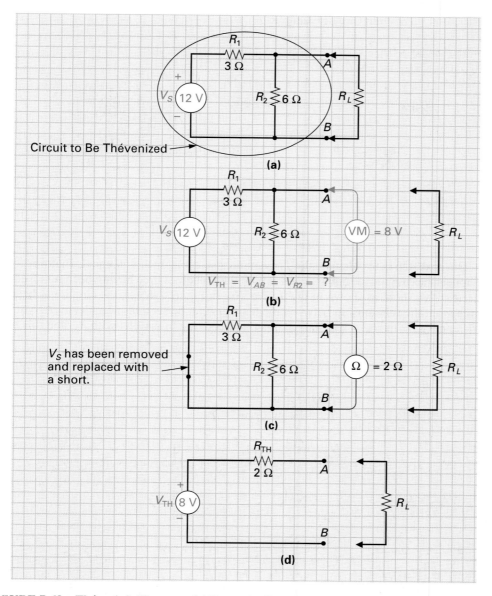

FIGURE 7-63 Thévenin's Theorem. **(a)** Example Circuit. **(b)** Obtaining Thévenin Voltage (V_{TH}). **(c)** Obtaining Thévenin Resistance. **(d)** Thévenin Equivalent Circuit.

EXAMPLE:

Determine V_{TH} and R_{TH} for the circuit in Figure 7-64.

Solution:

The first step is to remove the load resistor R_L and calculate what voltage will appear between points A and B, as shown in Figure 7-65(a). Removing R_L will open the path for current to flow through R_4, which will consequently have no voltage drop across it. The voltage between points A and B therefore will be equal to the voltage dropped across R_3, and since R_1, R_2, and R_3 form a series circuit, the voltage-divider formula can be used:

$$V_{R3} = \frac{R_3}{R_T} \times V_S$$

$$= \frac{10\ \Omega}{40\ \Omega} \times 10\ \text{V} = 2.5\ \text{V}$$

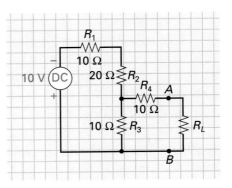

FIGURE 7-64   Thévenin Circuit Example.

Therefore,

$$V_{AB} = V_{R3} = V_{TH} = 2.5 \text{ V}$$

The next step is to calculate the Thévenin resistance, which will equal whatever resistance appears across the terminals *A* and *B* with the voltage source having been removed and replaced with a short, as shown in Figure 7-65(b). In Figure 7-65(c), the circuit has been redrawn so that the relationship between the resistors can be seen in more detail. As you can see, R_1 and R_2 are in series with one another and both are in parallel with R_3, and this com-

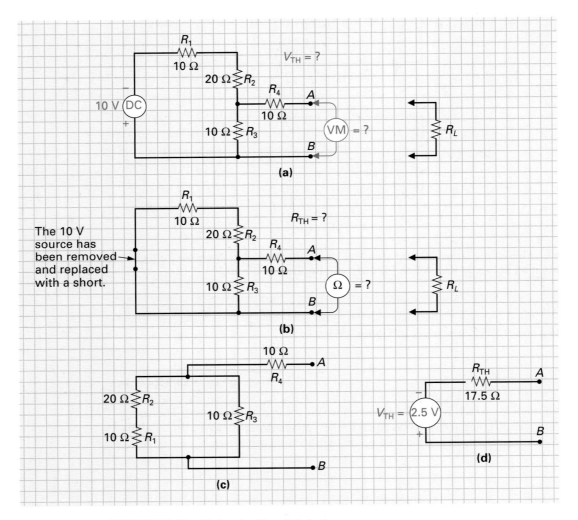

FIGURE 7-65   Thévenin Circuit Solution.

bination is in series with R_4. Using our three-step procedure for calculating total resistance in a series–parallel circuit, the following results are obtained:

1. $R_{1,2} = R_1 + R_2 = 10\ \Omega + 20\ \Omega = 30\ \Omega$

2. $R_{1,2,3} = \dfrac{R_{1,2} \times R_3}{R_{1,2} + R_3} = \dfrac{30\ \Omega \times 10\ \Omega}{30\ \Omega + 10\ \Omega} = \dfrac{300\ \Omega}{40\ \Omega} = 7.5\ \Omega$

3. $R_T = R_{1,2,3} + R_4 = 7.5\ \Omega + 10\ \Omega = 17.5\ \Omega$

Figure 7-65(d) on the previous page illustrates the Thévenin equivalent circuit.

7-9-4 *Norton's Theorem*

Norton's Theorem

Any network of voltage sources and resistors can be replaced by a single equivalent current source (I_N) in parallel with a single equivalent resistance (R_N).

Norton's theorem, like Thévenin's theorem, is a tool for simplifying a complex circuit into a more manageable one. Figure 7-66 illustrates the difference between a Thévenin equivalent and Norton equivalent circuit. Thévenin's theorem simplifies a complex network and uses an equivalent voltage source (V_{TH}) and an equivalent series resistance (R_{TH}). Norton's theorem, on the other hand, simplifies a complex circuit and represents it with an equivalent current source (I_N) in parallel with an equivalent Norton resistance (R_N), as shown in Figure 7-66.

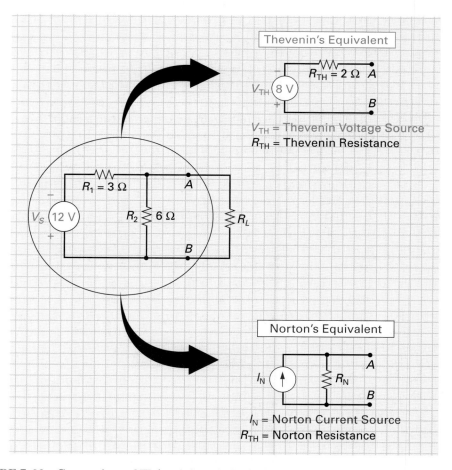

FIGURE 7-66   **Comparison of Thévenin's and Norton's Circuits.**

As with any theorem, a set of steps has to be carried out to arrive at an equivalent circuit. The example we will use is shown in Figure 7-67(a) and is the same example used in Figure 7-66.

STEP 1   Calculate the Norton equivalent current source, which will be equal to the current that would flow between terminals A and B if the load resistor were removed and replaced with a short, as seen in Figure 7-67(b). Placing a short between terminals A and B will short out the resistor R_2, so the only resistance in the circuit will be R_1. The Norton equivalent current source in this example will therefore be equal to

$$I_N = \frac{V_S}{R_T} = \frac{12 \text{ V}}{3 \text{ } \Omega} = 4 \text{ A}$$

STEP 2   The next step is to determine the value of the Norton equivalent resistance that will be placed in parallel with the current source, unlike Thévenin's equivalent resistance, which was placed in series. Like Thévenin's theorem, though, Norton's equivalent re-

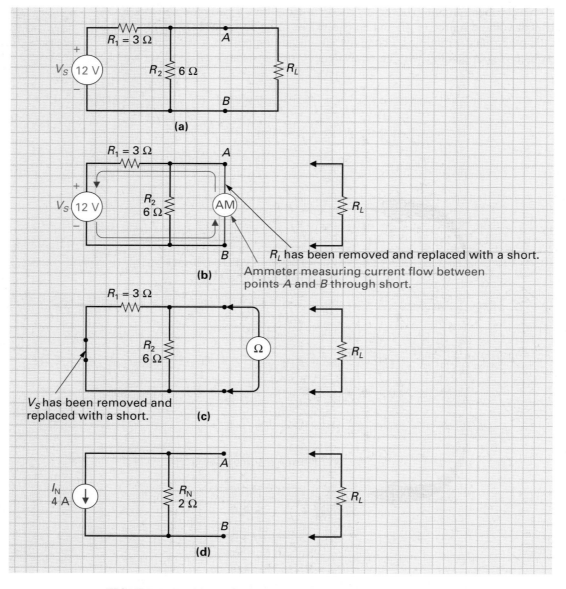

FIGURE 7-67   **Norton's Theorem. (a) Example Circuit. (b) Obtaining Norton Current. (c) Obtaining Norton Resistance (R_N). (d) Norton Equivalent Circuit.**

sistance (R_N) is equal to the resistance between terminals A and B when the voltage source is removed and replaced with a short, as shown in Figure 7-67(c). Since R_1 and R_2 form a parallel circuit, the Norton equivalent resistance will be equal to

$$R_N = \frac{R_1 \times R_2}{R_1 + R_2} = \frac{3\,\Omega \times 6\,\Omega}{3\,\Omega + 6\,\Omega} = \frac{18\,\Omega}{9\,\Omega} = 2\,\Omega$$

The Norton equivalent circuit has been determined simply by carrying out these two steps and is illustrated in Figure 7-67(d).

EXAMPLE:

Determine I_N and R_N for the circuit in Figure 7-68.

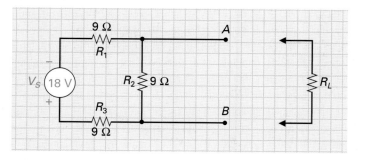

FIGURE 7-68 **Norton Circuit Example.**

Solution:

If a short is placed between terminals A and B, R_2 will be shorted out, so the current between points A and B, and therefore the Norton equivalent current, will be limited by only R_1 and R_3 and will equal [Figure 7-69(a)]

$$I_N = \frac{V_S}{R_T} = \frac{V_S}{R_1 + R_3} = \frac{18\text{ V}}{9\,\Omega + 9\,\Omega} = 1\text{ A}$$

Replacing V_S with a short, you can see that our Norton equivalent resistance between terminals A and B is made up of R_1 and R_3 in series with one another, and both are in parallel with R_2. R_N will therefore be equal to [Figure 7-69(b)]

$$R_{1,3} = R_1 + R_3 = 9\,\Omega + 9\,\Omega = 18\,\Omega$$
$$R_T = \frac{R_{1,3} \times R_2}{R_{1,3} + R_2} = \frac{18\,\Omega \times 9\,\Omega}{18\,\Omega + 9\,\Omega} = 6\,\Omega$$

The Norton equivalent circuit is shown in Figure 7-69(c).

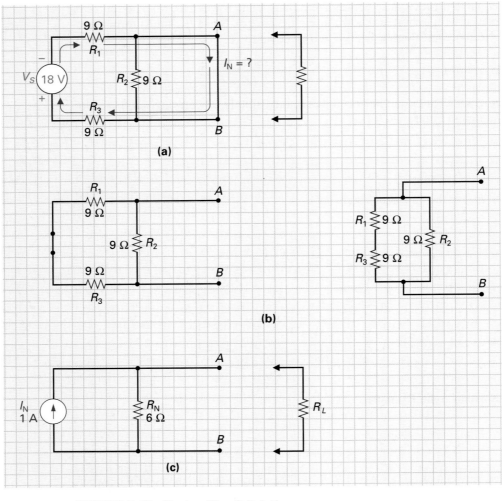

FIGURE 7-69   Norton Circuit Solution.

SELF-TEST EVALUATION POINT FOR SECTION 7-9

Now that you have completed this section, you should be able to:

■ **Objective 8.**   *Describe the differences between a voltage and a current source.*

■ **Objective 9.**   *Analyze series-parallel networks using:*
  a. The superposition theorem.
  b. Thévenin's theorem.
  c. Norton's theorem.

Use the following questions to test your understanding of Section 7-9.

1. A constant voltage source will have a _____ internal resistance, while a constant current source will have a _____ internal resistance.   (large or small)

2. The superposition theorem is a logical way of analyzing networks with more than one _____.

3. Thévenin's theorem represents a complex two-terminal network as a single _____ source with a series-connected single _____.

4. Norton's theorem also allows you to analyze complex two-terminal networks as a single _____ source in parallel with a single resistor.

SUMMARY

Series–Parallel Circuits (Figure 7-70)

1. If current has only one path to follow through a component, that component is connected in series.

2. If the total current has two or more paths to follow, these components are connected in parallel.

3. All electronic equipment is composed of many components interconnected to form a combination of series and parallel (series–parallel) circuits.

4. The R–2R ladder circuit is a series–parallel circuit used for digital-to-analog conversion.

5. Troubleshooting is the process of locating and diagnosing malfunctions or breakdowns in equipment by means of systematic checking or analysis.

6. **a.** If a series-connected resistor in a series–parallel circuit opens, there cannot be any current flow, and the source voltage will appear across the open and 0 V will appear across all the others.

   b. If a parallel branch in a series–parallel circuit opens, the overall circuit resistance will increase and the total current will decrease. A greater parallel resistance will cause a greater voltage drop across that parallel circuit.

QUICK REFERENCE SUMMARY SHEET

FIGURE 7-70   Series-Parallel Circuits.

Example Circuit

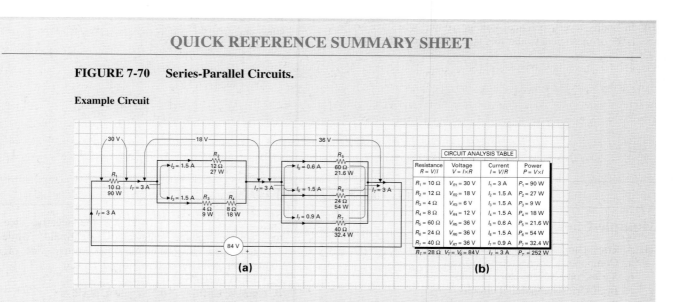

(a)   **(b)**

Solving for Resistance, Voltage, Current, and Power in a Series–Parallel Circuit

STEP 1   Determine the circuit's total resistance.

    Step A   Solve for series-connected resistors in all parallel combinations.

    Step B   Solve for all parallel combinations.

    Step C   Solve for remaining series resistances.

STEP 2   Determine the circuit's total current.

STEP 3   Determine the voltage across each series resistor and each parallel combination (series equivalent resistor).

STEP 4   Determine the value of current through each parallel resistor in every parallel combination.

STEP 5   Determine the total and individual power dissipated by the circuit.

Wheatstone bridge

$$R_{\text{variable}} = R_{\text{unknown}}$$

Branch resistances are equal.

$$R_{\text{unknown}} = R_{\text{variable}} \times \left(\frac{R_1}{R_2} \right)$$

Branch resistors, R_1 and R_2, are not equal.

7. **a.** If a series-connected resistor in a series–parallel circuit shorts, the total circuit resistance will decrease, resulting in an increase in total current. The fault can be located by measuring the voltage across the shorted resistor, which will be 0 V.

   b. If a parallel branch in a series–parallel circuit shorts, the total resistance will decrease, resulting in a total current increase. This smaller parallel circuit resistance will result in a smaller voltage drop across the parallel circuit.

8. A resistor's resistance will typically change with age, resulting in a total resistance change and therefore a total current change. The faulty component can be located due to its abnormal voltage drop.

9. An ideal voltage source will have an internal resistance of 0 Ω and provide a constant output voltage regardless of what load resistance is connected across its output.

10. In reality, every practical voltage source will have some level of inefficiency, and this is represented as an internal resistance, and the output voltage will remain relatively constant despite variations in load resistance.

11. An ideal current source will have an infinite internal resistance and provide a constant output current, regardless of what load resistance is connected across its output.

12. In reality, every practical current source has a relatively high internal resistance, and the output current will remain relatively constant despite variations in load resistance.

13. The superposition theorem is a useful tool when analyzing circuits with more than one voltage source.

14. Thévenin's theorem is another handy tool that can be used to represent complex series–parallel networks as a single voltage source (V_{TH}) in series with a single resistor (R_{TH}).

15. Norton's theorem can also simplify a complex series–parallel network to an equivalent form consisting of a single current source (I_N) in parallel with a single resistor (R_N).

REVIEW QUESTIONS

Multiple-Choice Questions

1. A series–parallel circuit is a combination of:
   a. Components connected end to end
   b. Series (one current path) circuits
   c. Both series and parallel circuits
   d. Parallel (two or more current paths) circuits

2. Total resistance in a series–parallel circuit is calculated by applying the _____ resistance formula to series-connected resistors and the _____ resistance formula to resistors connected in parallel.
   a. Series, parallel **c.** Series, series
   b. Parallel, series **d.** Parallel, parallel

3. Total current in a series–parallel circuit is determined by dividing the total _____ by the total _____.
   a. Power, current **c.** Current, resistance
   b. Voltage, resistance **d.** Voltage, power

4. Branch current within series–parallel circuits can be calculated by:
   a. Ohm's law
   b. The current-divider formula
   c. Kirchhoff's current law
   d. All of the above

5. All voltages on a circuit diagram are with respect to _____ unless otherwise stated.
   a. The other side of the component **c.** Ground
   b. The high-voltage source **d.** All of the above

6. A _____ ground has the negative side of the source voltage connected to ground, while a _____ ground has the positive side of the source voltage connected to ground.
   a. Positive, negative **c.** Positive, earth
   b. Chassis, earth **d.** Negative, positive

7. The output voltage will always _____ when a load or voltmeter is connected across a voltage divider.
   a. Decrease
   b. Remain the same
   c. Increase
   d. All of the above could be considered true.

8. A Wheatstone bridge was originally designed to measure:
   a. An unknown voltage **c.** An unknown power
   b. An unknown current **d.** An unknown resistance

9. A balanced bridge has an output voltage:
   a. Equal to the supply voltage **c.** Of 0 V
   b. Equal to half the supply voltage **d.** Of 5 V

10. The total resistance of a ladder circuit is best found by starting at the point _____ the source.
   a. Nearest to **c.** Midway between
   b. Farthest from **d.** All of the above

11. The R–2R ladder circuit finds its main application as a(an):
   a. Analog-to-digital converter
   b. Digital-to-analog converter
   c. Device to determine unknown resistance values
   d. All of the above

12. The Norton equivalent resistance (R_N) is always in _____ with the Norton equivalent current source (I_N).
   a. Proportion   **c.** Parallel
   b. Series   **d.** Either series or parallel

13. An ideal current source has _____ internal resistance, while an ideal voltage source has _____ internal resistance.
   a. An infinite, 0 Ω of   **c.** 0 Ω of, an infinite
   b. No, a large amount   **d.** No, an infinite

14. The Thévenin equivalent resistance (R_{TH}) is always in _____ with the Thévenin equivalent voltage (V_{TH}).
   a. Proportion   **c.** Parallel
   b. Series   **d.** Either series or parallel

15. The superposition theorem is useful for analyzing circuits with:
   a. Two or more voltage sources
   b. A single voltage source
   c. Only two voltage sources
   d. A single current source

16. A resistor, when it burns out, will generally:
   a. Decrease slightly in value   **c.** Short
   b. Increase slightly in value   **d.** Open

17. In a series–parallel resistive circuit, an open series-connected resistor will cause _____ current, whereas an open parallel-connected resistor will result in a total current _____.
   a. An increase in, decrease   **c.** Zero, decrease
   b. A decrease in, increase   **d.** None of the above

18. In a series–parallel resistive circuit, a shorted series-connected resistor will cause _____ current, whereas a shorted parallel-connected resistor will result in a total current _____.
   a. An increase in, increase   **c.** An increase in, decrease
   b. A decrease in, decrease   **d.** A decrease in, increase

19. A resistor's resistance will typically _____ with age, resulting in a total circuit current _____.
   a. Decrease, decrease   **c.** Decrease, increase
   b. Increase, increase   **d.** All of the above

20. The maximum power transfer theorem states that maximum power is transferred from source to a load when load resistance is equal to source resistance.
   a. True   **b.** False

Communication Skill Questions

21. State the five-step method for determining a series–parallel circuit's resistance, voltage, current, and power values. (7-6)

22. Illustrate the following series–parallel circuits:
   a. R_1 in series with a parallel combination R_2, R_3, and R_4.
   b. R_1 in series with a two-branch parallel combination consisting of R_2 and R_3 in series and R_4 in parallel.
   c. R_1 in parallel with R_2, which is in series with a three-resistor parallel combination, R_3, R_4, and R_5.

23. Using the example in Question 22(c), apply values of your choice and apply the five-step procedure. (7-6)

24. Describe what is meant by "loading of a voltage-divider circuit." (7-7-1)

25. Illustrate and describe the Wheatstone bridge in the: (7-7-2)
   a. Balanced condition
   b. Unbalanced condition
   c. Application of measuring unknown resistances

26. Describe how the ladder circuit acts as a current divider. (7-7-3)

27. Briefly describe the difference between a voltage source and a current source. (7-9-1)

28. Briefly describe the following theorems: (7-9)
   a. Superposition   **c.** Norton's
   b. Thévenin's   **d.** Maximum power transfer

29. What would be the advantages of Thévenin's and Norton's theorems to obtain an equivalent circuit? (7-9)

30. Draw the components that would exist in a: (7-9)
   a. Thévenin equivalent circuit
   b. Norton equivalent circuit

31. Describe the steps involved in obtaining a: (7-9)
   a. Thévenin equivalent circuit
   b. Norton equivalent circuit

32. When troubleshooting series–parallel circuits, describe what effect (7-8-1)
   a. An open series-connected
   b. An open parallel-connected
   resistor would have on total current and resistance, and how the opened resistor could be isolated.

33. When troubleshooting series–parallel circuits, describe what effect (7-8-2)
   a. A shorted series-connected
   b. A shorted parallel-connected
   resistor would have on total current and resistance, and how the shorted resistors could be isolated.

34. Describe what effect a resistor's value variation would have and how it could be recognized. (7-8-3)

35. Give the divider formula, Ohm's law, and Kirchhoff's laws used to determine: (7-6)
   a. Branch currents
   b. Voltage drops in a series–parallel circuit
   (List all six.)

Practice Problems

36. R_3 and R_4 are in series with one another and are both in parallel with R_5. This parallel combination is in series with two series-connected resistors, R_1 and R_2. $R_1 = 2.5$ kΩ, $R_2 = 10$ kΩ, $R_3 = 7.5$ kΩ, $R_4 = 2.5$ kΩ, $R_5 = 2.5$ MΩ, and $V_S = 100$ V. For these values, calculate:
   a. Total resistance
   b. Total current
   c. Voltage across series resistors and parallel combinations
   d. Current through each resistor
   e. Total and individual power figures

37. Referring to the example in Question 36, calculate the voltage at every point of the circuit with respect to ground.

38. A 10 V source is connected across a series–parallel circuit made up of R_1 in parallel with a branch made up of R_2 in series with a parallel combination of R_3 and R_4. $R_1 = 100\ \Omega$, $R_2 = 100\ \Omega$, $R_3 = 200\ \Omega$, and $R_4 = 300\ \Omega$. For these values, apply the five-step procedure, and also determine the voltage at every point of the circuit with respect to ground.

39. Calculate the output voltage (V_{RL}) in Figure 7-71 if R_L is equal to:
   a. 25 Ω   **b.** 2.5 kΩ   **c.** 2.5 MΩ

40. What load current will be supplied by the current source in Figure 7-72 if R_L is equal to:
   a. 25 Ω   **b.** 2.5 kΩ   **c.** 2.5 MΩ

41. Use the superposition theorem to calculate total current:
   a. Through R_2 in Figure 7-73(a)
   b. Through R_3 in Figure 7-73(b)

42. Convert the following voltage sources to equivalent current sources:
   a. $V_S = 10\ \text{V}, R_{int} = 15\ \Omega$
   b. $V_S = 36\ \text{V}, R_{int} = 18\ \Omega$
   c. $V_S = 110\ \text{V}, R_{int} = 7\ \Omega$

43. Use Thévenin's theorem to calculate the current through R_L in Figure 7-74. Sketch the Thévenin and Norton equivalent circuits for Figure 7-74.

44. Sketch the Thévenin and Norton equivalent circuits for the networks in Figure 7-75.

45. Convert the following current sources to equivalent voltage sources:
   a. $I_S = 5\ \text{mA}, R_{int} = 5\ \text{M}\Omega$
   b. $I_S = 10\ \text{A}, R_{int} = 10\ \text{k}\Omega$
   c. $I_S = 0.0001\ \text{A}, R_{int} = 2.5\ \text{k}\Omega$

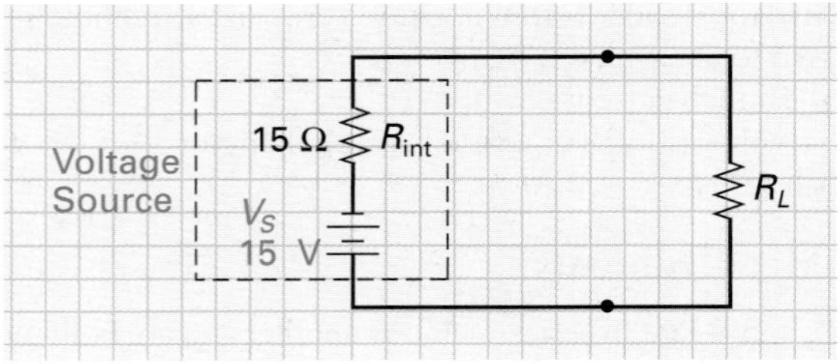

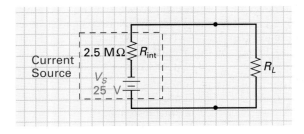

FIGURE 7-72

Troubleshooting Questions

46. Referring to the example circuit in Figure 7-49, describe the effects you would get if a resistor were to short, and if a resistor were to open.

47. Design a simple five-resistor series–parallel circuit and insert a source voltage and resistance values. Apply the five-step series–parallel circuit procedure, and then theoretically open and short all the resistors and calculate what effect would occur and how you would recognize the problem.

48. Carbon composition resistors tend to increase in resistance with age, while most other types generally decrease in resistance. What effects would resistance changes have on their respective voltage drops?

FIGURE 7-71

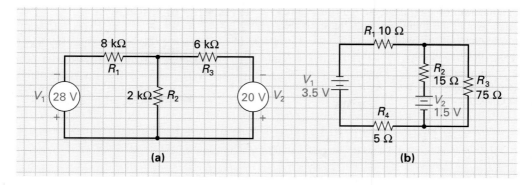

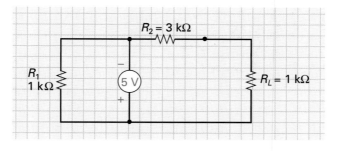

FIGURE 7-73

FIGURE 7-74

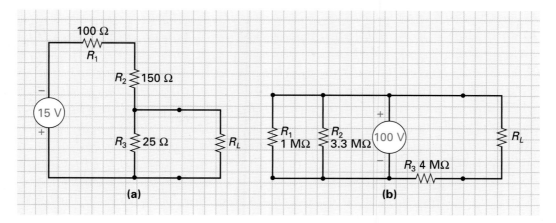

FIGURE 7-75

49. Use Thévenin's theorem to simplify the circuit in Figure 7-76. What effect would the following faults have on the Thévenin equivalent circuit?

    a. R_2 is shorted.   **b.** R_2 is open.

50. Calculate the Norton equivalent for Figure 7-76 and describe what circuit differences will occur for the same faults listed in Question 49.

Web Site Questions

Go to the Web site http://www.prenhall. com/cook, select the textbook *Introductory DC/AC Electronics* or *Introductory DC/AC Circuits*, this chapter, and then follow the instructions when answering the multiple-choice practice problems.

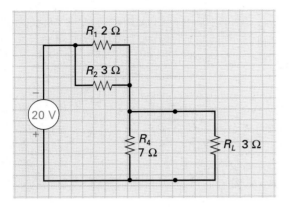

FIGURE 7-76

These tests at the end of each chapter will challenge your knowledge up to this point, and give you the practice you need for a job interview. To make this more realistic, the test will comprise both technical and personal questions. In order to take full advantage of this exercise, you may want to set up a simulation of the interview environment, have a friend read the questions to you, and record your responses for later analysis.

Company Name: **DVD, Inc.**

Industry Branch: **Consumer Electronics.**

Function: **Assisting engineers with R&D.**

Job Title: **Engineering Technician.**

1. How did you hear of the job opening?
2. What is a Wheatstone bridge circuit?

3. What quality advantages does a DVD have over a VHS tape?
4. What sort of responsibilities do you envision this job to have?
5. What could you tell me about Thévenin's theorem?
6. What symptoms would you expect from a series circuit open?
7. What symptoms would you expect from a parallel circuit short?
8. Would you be prepared to work a large number of extra hours for a short period of time if a project is behind schedule?
9. What is a series-parallel circuit?
10. Could you handle an emergency calmly and efficiently?

Answers

1. Describe source.
2. Section 7-7-2.
3. Knowing this company specializes in the design and development of DVD systems, you should have visited your local consumer electronics retailer prior to this interview and probed the sales staff for information on model types, advantages and features, etc.
4. Quote intro. section in text and job description listed in newspaper.
5. Section 7-9-3.

6. Chapter 5.
7. Chapter 6.
8. To be more competitive, most companies are bringing products to market in a much shorter space of time, and so this pattern has become more and more prevalent. If you are unsure, don't object to anything at the time. Later, if you decide that it is something you do not want to do, you can call and decline the offer.
9. Chapter 7.
10. The answer must be yes. Qualify this by discussing lab and job related safety practices and procedures.

Alternating Current (AC)

The Laser

Theodore Maiman

In 1898, H. G. Wells's famous book, *The War of the Worlds,* had Martian invaders with laserlike death rays blasting bricks, incinerating trees, and piercing iron as if it were paper. In 1917, Albert Einstein stated that, under certain conditions, atoms or molecules could absorb light and then be stimulated to release this borrowed energy. In 1954, Charles H. Townes, a professor at Columbia University, conceived and constructed with his students the first "maser" (acronym for "microwave amplification by stimulated emission of radiation"). In 1958, Townes and Arthur L. Shawlow wrote a paper showing how stimulated emission could be used to amplify light waves as well as microwaves, and the race was on to develop the first "laser." In 1960, Theodore H. Maiman, a scientist at Hughes Aircraft Company, directed a beam of light from a flash lamp into a rod of synthetic crystal, which responded with a burst of crimson light so bright that it outshone the sun.

An avalanche of new lasers emerged, some as large as football fields, while others were no bigger than a pinhead. They can be made to produce invisible infrared or ultraviolet light or any visible color in the rainbow, and the high-power lasers can vaporize any material a million times faster and more intensely than a nuclear blast, while the low-power lasers are safe to use in children's toys.

At present, the laser is being used by the FBI to detect fingerprints that are 40 years old, in defense programs, in compact disk players, in underground fiber optic communication to transmit hundreds of telephone conversations, to weld car bodies, to drill holes in baby-bottle nipples, to create three-dimensional images called holograms, and as a surgeon's scalpel in the operating room. Not a bad beginning for a device that when first developed was called "a solution looking for a problem."

Outline and Objectives

Introduction

In this chapter you will be introduced to alternating current (ac). While direct current (dc) can be used in many instances to deliver power or represent information, there are certain instances in which ac is preferred. For example, ac is easier to generate and is more efficiently transmitted as a source of electrical power for the home or business. Audio (speech and music) and video (picture) information are generally always represented in electronic equipment as an alternating current or alternating voltage signal.

In this chapter, we will begin by describing the difference between dc and ac, and then examine where ac is used. Following this we will discuss all the characteristics of ac waveform shapes, and finally, the differences between electricity and electronics.

To begin with, though, let's review the math skills you will need for this chapter's material.

8-1 MINI-MATH REVIEW—TRIGONOMETRY

This first "Mini-Math Review" is included to overview the mathematical details you need for the electronic concepts covered in this chapter. In this review, we will be examining trigonometry.

Electronic and electrical technicians and engineers, mechanical engineers, carpenters, architects, navigators, and many other people in technical trades make use of **trigonometry** on a daily basis. Although volumes of theory are available about this subject, most of the material is not that useful to us for everyday applications. In this chapter we will concentrate on the more practical part of *trigonometry—the right-angle triangle* and its applications.

Trigonometry

The study of the properties of triangles, trigonometric functions, and their applications.

8-1-1   *The 3, 4, 5 Right-Angle Triangle*

Right-Angle Triangle

A triangle having a 90° angle.

Like all triangles, the **right-angle triangle,** or *right triangle,* has three sides and three corners. Its distinguishing feature, however, is that two of the sides of this triangle are at right angles (at 90°) to each other, as shown in Figure 8-1(a). The small square box within the triangle is placed in the corner to show that sides *A* and *B* are square or at right angles to each other.

If you study this right triangle you may notice two interesting facts about the relative lengths of sides *A, B,* and *C*. These observations are as follows.

1. Side *C* is always longer than side *A* or side *B*.

2. The total length of side *A* and side *B* is always longer than side *C*.

The right triangle in Figure 8-1(b) has been drawn to scale to demonstrate another interesting fact about this triangle. If side *A* were to equal 3 cm and side *B* were to equal 4 cm,

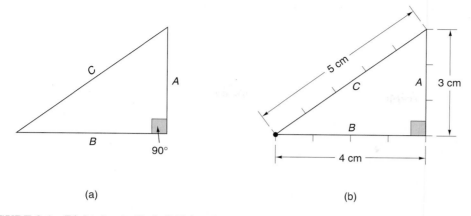

(a)   (b)

FIGURE 8-1   **Right-Angle (3, 4, 5) Triangle.**

side C would equal 5 cm. This demonstrates a basic relationship among the three sides and accounts for why this right triangle is sometimes referred to as a *3, 4, 5 triangle.* As long as the relative lengths remain the same, it makes no difference whether the sides are 3 cm, 4 cm, and 5 cm or 30 km, 40 km, and 50 km. Unfortunately, in most applications our lengths of A, B, and C will not work out as easily as 3, 4, 5.

When a relationship exists among three quantities, we can develop a formula to calculate an unknown when two of the quantities are known. It was Pythagoras who first developed this basic formula or equation (known as the **Pythagorean theorem**), which states that *the square of the length of the hypotenuse (side C) of a right triangle equals the sum of the squares of the lengths of the other two sides:*

$$\boxed{C^2 = A^2 + B^2}$$ $$\left(\begin{matrix} 5^2 = 3^2 + 4^2 \\ 25 = 9 + 16 \end{matrix} \right)$$

By using the rules of algebra, we can transpose this formula to derive formulas for sides C, B, and A.

Pythagorean Theorem
A theorem in geometry: The square of the length of the hypotenuse of a right triangle equals the sum of the squares of the lengths of the other two sides.

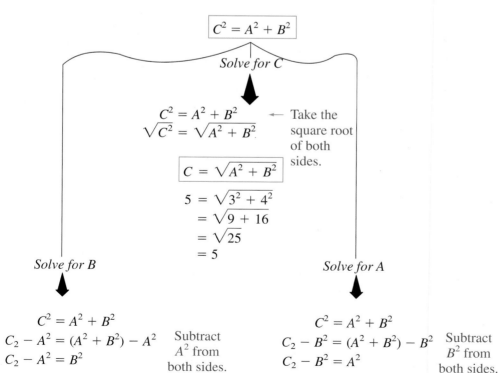

$$\sqrt{C^2-A^2} = \sqrt{B^2}$$
$$\sqrt{C^2-A^2} = B$$

Take the square root of both sides.

$$\boxed{B = \sqrt{C^2 - A^2}}$$

$$4 = \sqrt{5^2 - 3^2}$$
$$= \sqrt{25 - 9}$$
$$= \sqrt{16}$$
$$= 4$$

$$\sqrt{C^2-B^2} = \sqrt{A^2}$$
$$\sqrt{C^2-B^2} = A$$

Take the square root of both sides.

$$\boxed{A = \sqrt{C^2 - B^2}}$$

$$3 = \sqrt{5^2 - 4^2}$$
$$= \sqrt{25 - 16}$$
$$= \sqrt{9}$$
$$= 3$$

Let us now test these formulas with a few examples.

EXAMPLE:

In Figure 8-2(a), a 4-foot ladder has been placed in a position 2 feet from a wall. How far up the wall will the ladder reach?

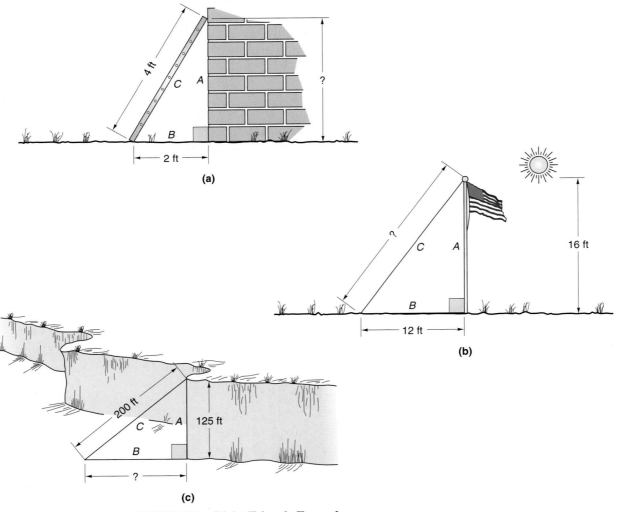

(a)

(b)

(c)

FIGURE 8-2 Right-Triangle Examples.

Solution:

In this example, A is unknown, and therefore

$$
\begin{aligned}
A &= \sqrt{C^2 - B^2} \\
&= \sqrt{4^2 - 2^2} \\
&= \sqrt{16 - 4} \\
&= \sqrt{12} \\
&= 3.46 \quad \text{or} \quad \approx 3.5 \text{ ft}
\end{aligned}
$$

EXAMPLE:

In Figure 8-2(b), a 16-foot flagpole is casting a 12-foot shadow. What is the distance from the end of the shadow to the top of the flagpole?

Solution:

In this example, C is unknown, and therefore

$$
\begin{aligned}
C &= \sqrt{A^2 + B^2} \\
&= \sqrt{16^2 + 12^2} \\
&= \sqrt{256 + 144} \\
&= \sqrt{400} \\
&= 20 \text{ ft}
\end{aligned}
$$

EXAMPLE:

In Figure 8-2(c), a 200-foot piece of string is stretched from the top of a 125-foot cliff to a point on the beach. What is the distance from this point to the cliff?

Solution:

In this example, B is unknown, and therefore

$$
\begin{aligned}
B &= \sqrt{C^2 - A^2} \\
&= \sqrt{200^2 - 125^2} \\
&= \sqrt{40,000 - 15,625} \\
&= \sqrt{24,375} \\
&= 156 \text{ ft}
\end{aligned}
$$

Vectors and Vector Diagrams

A **vector** or **phasor** is an arrow used to represent the magnitude and direction of a quantity. Vectors are generally used to represent a physical quantity that has two properties. For example, Figure 8-3(a) shows a motorboat heading north at 12 miles per hour. Figure 8-3(b) shows how vector **A** could be used to represent the vessel's direction and speed. The size of the vector represents the speed of 12 mph by being 12 cm long, and because we have made the top of the page north, vector **A** should point straight up so that it represents the vessel's direction. Referring to Figure 8-3(a) you can see that there is another factor that also needs to be considered, a 12 mile per hour easterly tide. This tide is represented in our vector diagram in Figure 8-3(b) by vector **B,** which is 12 centimeters long and pointing east. Since the motorboat is pushing north at 12 miles per hour and the tide is pushing east at 12 miles per hour, the resultant course will be northeast, as indicated by vector **C** in Figure 8-3(b). Vector **C** was determined by **vector addition** (as seen by the dashed lines) of vectors **A** and **B**. The result, vector **C,** is called the **resultant vector.** This resultant vector indicates

Vector or Phasor

A quantity that has magnitude and direction and that is commonly represented by a directed line segment whose length represents the magnitude and whose orientation in space represents the direction.

Vector Addition

Determination of the sum of two out-of-phase vectors using the Pythagorean theorem.

Resultant Vector

A vector derived from or resulting from two or more other vectors.

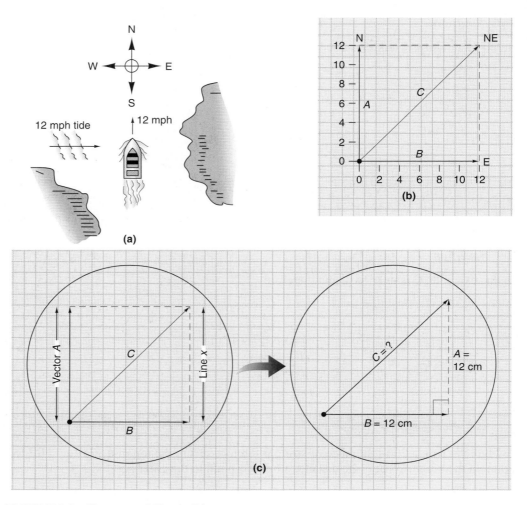

FIGURE 8-3   Vectors and Vector Diagrams.

the motorboat's course and speed. The direction or course, we can see, is northeast because vector **C** points in a direction midway between north and east. The speed of the motorboat, however, is indicated by the length or magnitude of vector **C.** This length, and therefore the motorboat's speed, can be calculated by using the Pythagorean theorem. Probably your next question is: How is the **vector diagram** in Figure 8-3(b) similar to a right-angle triangle? The answer is to redraw Figure 8-3(b), as shown in Figure 8-3(c). Because the dashed line x is equal in length to vector **A,** vector **A** can be put in the position of line x to form a right-angle triangle with vector **B** and vector **C.** Now that we have a right-angle triangle with two known lengths and one unknown length, we can calculate the unknown vector's length and therefore the motorboat's speed.

Vector Diagram

A graphic drawing using vectors that shows arrangement and relations.

$$C = \sqrt{A^2 + B^2}$$
$$= \sqrt{12^2 + 12^2}$$
$$= \sqrt{144 + 144}$$
$$= \sqrt{288} = 16.97$$

The length of vector **C** is 16.97 cm, and because each 1 mile per hour of the motorboat was represented by 1 centimeter, the motorboat will travel at a speed of 16.97 miles per hour in a northeasterly direction.

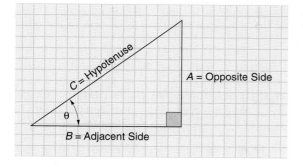

FIGURE 8-4 **Names Given to the Sides of a Right Triangle.**

8-1-2 *Opposite, Adjacent, Hypotenuse, and Theta*

Until this point we have called the three sides of our right triangle *A*, *B*, and *C*. Figure 8-4 shows the more common names given to the three sides of a right-angle triangle. Side *C*, called the **hypotenuse**, is always the longest of the three sides. Side *B*, called the **adjacent side**, always extends between the hypotenuse and the vertical side. An angle called **theta** (θ, a Greek letter) is formed between the hypotenuse and the adjacent side. The side that is always opposite the angle θ is called the **opposite side.**

 Angles are always measured in degrees because all angles are part of a circle, like the one shown in Figure 8-5(a). A circle is divided into 360 small sections called **degrees.** In Figure 8-5(b) our right triangle has been placed within the circle. In this example the triangle occupies a 45° section of the circle and therefore theta equals 45 degrees ($\theta = 45°$). Moving from left to right in Figure 8-5(c) you will notice that angle θ increases from 5° to 85°. The length of the hypotenuse (*H*) in all these examples remains the same; however, the adjacent side's length decreases (*A* $\downarrow$) and the opposite side's length increases (*O* $\uparrow$) as angle θ is increased (θ $\uparrow$). This relationship between the relative length of a triangle's sides and

Hypotenuse
The side of a right-angled triangle that is opposite the right angle.

Adjacent
The side of a right-angled triangle that has a common endpoint, in that it extends between the hypotenuse and the vertical side.

Theta
The eighth letter of the Greek alphabet, used to represent an angle.

Opposite
The side of a right-angle triangle that is opposite the angle theta.

Degree
A unit of measure for angles equal to an angle with its vertex at the center of a circle and its sides cutting off $\frac{1}{360}$ of the circumference.

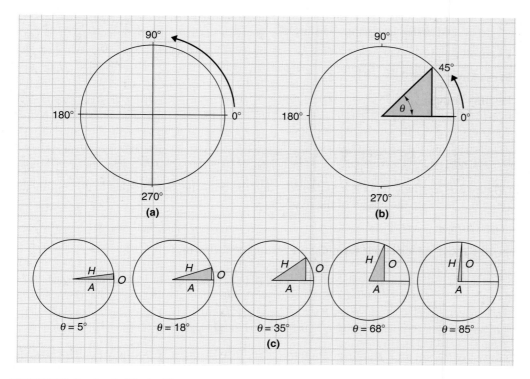

FIGURE 8-5 **Angle Theta (θ).**

theta means that we do not have to know the length of two sides to calculate a third. If we have the value of just one side and the angle theta, we can calculate the length of the other two sides.

Sine of Theta (Sin θ)

In the preceding section we discovered that a relationship exists between the relative length of a triangle's sides and the angle of theta. **Sine** is a comparison between the length of the opposite side and the length of the hypotenuse. Expressed mathematically,

$$\text{Sine of the theta (sin } \theta) = \frac{\text{opposite side } (O)}{\text{hypotenuse } (H)}$$

Because the hypotenuse is always larger than the opposite side, the result will always be less than 1 (a decimal fraction). Let us use this formula in a few examples to see how it works.

CALCULATOR KEYS

Name: Sine key

Function: Instructs the calculator to find the sine of the displayed value (angle $\rightarrow$ value).

Example: sin 37° = ?

Press keys: [3] [7] [SIN]

Display shows: 0.601815

Name: Arcsine ($\sin^{-1}$) or inverse sine sequence.

Function: Calculates the smallest angle whose sine is in the display (value $\rightarrow$ angle).

Example: invsin 0.666 = ?

Press keys: [.] [6] [6] [6] [INV] [SIN]

Display shows: 41.759°

☐ EXAMPLE:

In Figure 8-6(a), angle theta is equal to 41°, and the opposite side is equal to 20 centimeters in length. Calculate the length of the hypotenuse.

■ *Solution:*

Inserting these values in our formula, we obtain the following:

$$\sin \theta = \frac{O}{H}$$

$$\sin 41° = \frac{20 \text{ cm}}{H}$$

By looking up 41° in a sine trigonometry table or by using a scientific calculator that has all the trigonometry tables stored permanently in its memory, you will find that the sine of 41° is 0.656. This value describes the fact that when $\theta = 41°$, the opposite side will be 0.656, or 65.6%, as long as the hypotenuse side. By inserting this value into our formula

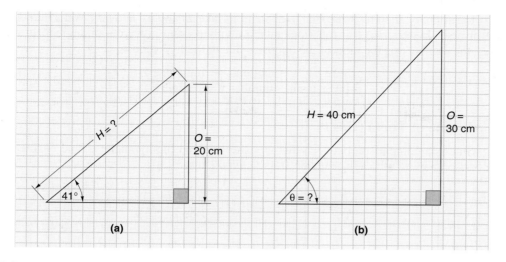

FIGURE 8-6 Sine of Theta.

and transposing the formula according to the rules of algebra, we can determine the length of the hypotenuse.

$$\sin 41° = \frac{20 \text{ cm}}{H}$$ *Calculator sequence:* $\boxed{4}\,\boxed{1}\,\boxed{\text{SIN}}$

$$0.656 = \frac{20 \text{ cm}}{H}$$

$$0.656 \times H = \frac{20 \text{ cm } \cancel{H}}{\cancel{H}}$$ Multiply both sides by H.

$$\frac{\cancel{0.656} \times H}{\cancel{0.656}} = \frac{20 \text{ cm}}{0.656}$$ Divide both sides by 0.656.

$$H = \frac{20 \text{ cm}}{0.656} = 30.5 \text{ cm}$$

EXAMPLE:

Figure 8-6(b) illustrates another example; however, in this case the lengths of sides H and O are known but θ is not.

Solution:

$$\sin \theta = \frac{O}{H}$$

$$\sin \theta = \frac{30 \text{ cm}}{40 \text{ cm}}$$

$$= 0.75$$

The ratio of side O to side H is 0.75, or 75%, which means that the opposite side is 75%, or 0.75, as long as the hypotenuse. To calculate angle θ we must isolate it on one side of the equation. To achieve this we must multiply both sides of the equation by **arcsine**, or **inverse sine** (invsin), which does the opposite of sine.

$$\sin \theta = 0.75$$

$$\cancel{\text{invsin}} \; (\cancel{\sin} \; \theta) = \text{invsin } 0.75$$ Take the inverse sine of 0.75.

$$\theta = \text{invsin } 0.75$$ *Calculator sequence:* $\boxed{.}\,\boxed{7}\,\boxed{5}\,\boxed{\text{INV}}\,\boxed{\text{SIN}}$

$$= 48.6°$$

Arcsine or Inverse Sine
The inverse function to the sine. (If y is the sine of θ, then θ is the arcsine of y.)

In summary, therefore, the sine trig functions take an angle θ and give you a number x. The inverse sine (arcsin) trig functions take a number x and give you an angle θ. In both cases, the number x is the ratio of the opposite side to the hypotenuse.

$$\text{Sine:} \qquad \text{angle } \theta \longrightarrow \text{ number } x$$
$$\text{Inverse sine:} \quad \text{number } x \longrightarrow \text{ angle } \theta$$

Cosine of Theta (Cos θ)

Sine is a comparison between the opposite side and the hypotenuse, and **cosine** is a comparison between the adjacent side and the hypotenuse.

$$\text{Cosine of theta (cos } \theta) = \frac{\text{adjacent } (A)}{\text{hypotenuse } (H)}$$

CALCULATOR KEYS

Name: Cosine key

Function: Instructs the calculator to find the cosine of the displayed value (angle $\longrightarrow$ value).

Example: $\cos 26° = ?$

Press keys: 2 6 COS

Display shows: 0.89879

Name: Arccosine ($\cos^{-1}$) or inverse cosine sequence.

Function: Calculates the smallest angle whose cosine is in the display (value $\longrightarrow$ angle).

Example: invcos $0.234 = ?$

Press keys: . 2 3 4 INV COS

Display shows: 76.467

EXAMPLE:

Figure 8-7(a) illustrates a right triangle in which the angle θ and the length of the hypotenuse are known. From this information, calculate the length of the adjacent side.

Solution:

$$\cos \theta = \frac{A}{H}$$

$$\cos 30° = \frac{A}{40 \text{ cm}} \qquad \textit{Calculator sequence:} \quad 3\ 0\ \text{COS}$$

$$0.866 = \frac{A}{40 \text{ cm}}$$

Looking up the cosine of 30° you will obtain the fraction 0.866. This value states that when $\theta = 30°$, the adjacent side will always be 0.866, or 86.6%, as long as the hypotenuse. By transposing this equation we can calculate the length of the adjacent side:

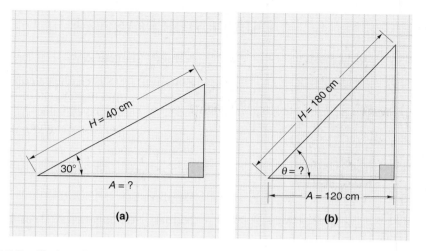

FIGURE 8-7 Cosine of Theta.

$$0.866 = \frac{A}{40 \text{ cm}}$$

$$40 \times 0.866 = \frac{A}{40 \text{ cm}} \times 40 \qquad \text{Multiply both sides by 40.}$$

$$A = 0.866 \times 40$$
$$= 34.64 \text{ cm}$$

▨ **EXAMPLE:**

Calculate the angle θ in Figure 8-7(b) if $A = 120$ centimeters and $H = 180$ centimeters.

▨ *Solution:*

$$\cos \theta = \frac{A}{H}$$

$$= \frac{120 \text{ cm}}{180 \text{ cm}}$$

$$= 0.667 \qquad A \text{ is } 66.7\% \text{ as long as } H.$$

$$\overline{\text{invcos}} \, \overline{(\cos \theta)} = \text{invcos } 0.667 \qquad \text{Multiply both sides by invcos.}$$

$$\theta = \text{invcos } 0.667 \qquad \textit{Calculator sequence:} \; \boxed{0}\boxed{.}\boxed{6}\boxed{6}\boxed{7}\;\boxed{\text{INV}}\boxed{\text{COS}}$$

$$= 48.2°$$

The inverse cosine trig function does the reverse operation of the cosine function:

Cosine: angle $\theta \;\rightarrow\;$ number x
Inverse cosine: number $x \;\rightarrow\;$ angle θ

Tangent of Theta (Tan θ)

Tangent is a comparison between the opposite side of a right triangle and the adjacent side.

$$\text{tangent of the theta (tan } \theta) = \frac{\text{opposite } (O)}{\text{adjacent } (A)}$$

Tangent

The trigonometric function that for an acute angle is the ratio between the leg opposite the angle when it is considered part of a right triangle, and the leg adjacent.

CALCULATOR KEYS

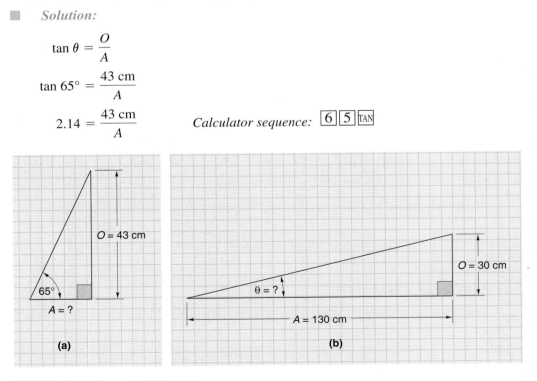

Name: Tangent key

Function: Instructs the calculator to find the tangent of the displayed value (angle ⟶ value).

Example: sin 73° = ?

Press keys: ⌷7⌷ ⌷3⌷ ⌷TAN⌷

Display shows: 3.2709

Name: Arctangent ($\tan^{-1}$) or inverse tangent sequence.

Function: Calculates the smallest angle whose tangent is in the display (value ⟶ angle).

Example: invtan 0.95 = ?

Press keys: ⌷.⌷ ⌷9⌷ ⌷5⌷ ⌷INV⌷ ⌷SIN⌷

Display shows: 43.5312

■ **EXAMPLE:**

Figure 8-8(a) illustrates a right triangle in which $\theta = 65°$ and the opposite side is 43 centimeters. Calculate the length of the adjacent side.

■ *Solution:*

$$\tan \theta = \frac{O}{A}$$

$$\tan 65° = \frac{43 \text{ cm}}{A}$$

$$2.14 = \frac{43 \text{ cm}}{A}$$

Calculator sequence: ⌷6⌷ ⌷5⌷ ⌷TAN⌷

FIGURE 8-8 **Tangent of Theta.**

$$2.14 \times A = \frac{43 \text{ cm}}{A} \times A$$ Whenever $\theta = 65°$, the opposite side will be 2.14 times longer than the adjacent side.

$$\frac{2.14 \times A}{2.14} = \frac{43 \text{ cm}}{2.14}$$ Multiply both sides by A.
Divide both sides by 2.14.

$$A = \frac{43 \text{ cm}}{2.14}$$

$$= 20.1 \text{ cm}$$

EXAMPLE:

Calculate the angle θ in Figure 8-8(b) if $O = 30$ centimeters and $A = 130$ centimeters.

Solution:

$$\tan \theta = \frac{O}{A}$$

$$= \frac{30 \text{ cm}}{130 \text{ cm}}$$

$= 0.231$ O is 0.231 or 23.1% as long as A.

$\text{invtan} (\tan \theta) = \text{invtan } 0.231$ Multiply both sides by invtan.

$\theta = \text{invtan } 0.231$ *Calculator sequence:* $\boxed{0}\boxed{.}\boxed{2}\boxed{3}\boxed{1}\boxed{\text{INV}}\boxed{\text{TAN}}$

$= 12.99$ or $13°$

Summary

As you have seen, trigonometry involves the study of the relationships among the three sides (O, H, A) of a right triangle and also the relationships among the sides of the right triangle and the number of degrees contained in the angle theta (θ).

If the lengths of two sides of a right triangle are known, and the length of the third side is needed, remember that

$$H^2 = O^2 + A^2$$

If the angle θ is known along with the length of one side, or if angle θ is needed and the lengths of the two sides are known, one of the three formulas can be chosen based on which variables are known and what is needed.

$$\sin \theta = \frac{O}{H} \quad \cos \theta = \frac{A}{H} \quad \tan \theta = \frac{O}{H}$$

When I was introduced to trigonometry, my mathematics professor spent 15 minutes having the whole class practice what he described as an old Asian war cry that went like this: "SOH CAH TOA." After he explained that it wasn't a war cry but in fact a memory aid to help us remember that SOH was in fact $\sin \theta = O/H$, CAH was $\cos \theta = A/H$, and TOA was $\tan \theta = O/A$, we understood the method in his madness.

Now that you have completed this section, you should be able to:

■ **Objective 1.** *Explain how the Pythagorean theorem relates to right-angle triangles.*

■ **Objective 2.** *Determine the length of one side of a right-angle triangle when the lengths of the other two sides are known.*

■ **Objective 3.** *Describe how vectors are used to represent the magnitude and direction of physical quantities and how they are arranged in a vector diagram.*

■ **Objective 4.** *Explain how the Pythagorean theorem can be applied to a vector diagram to calculate the magnitude of a resultant vector.*

■ **Objective 5.** *Define the trigonometric terms:*
   a. *Opposite*
   b. *Adjacent*
   c. *Hypotenuse*
   d. *Theta*
   e. *Sine*
   f. *Cosine*
   g. *Tangent*

■ **Objective 6.** *Demonstrate how the sine, cosine, and tangent trigonometric functions can be used to calculate:*
   a. *The length of an unknown side of a right-angle triangle if the length of another side and the angle theta are known.*
   b. *The angle theta of a right-angle triangle if the lengths of two sides of the triangles are known.*

Use the following questions to test your understanding of Section 8-1.

1. Assume that the two shorter sides of a right-angle triangle are 15 meters and 22 meters. Calculate the length of the other side.

2. Calculate the lengths of the unknown sides in the following triangles.

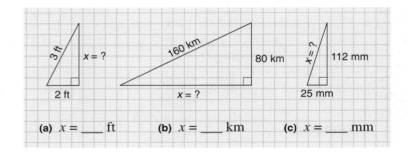

(a) *x* = ____ ft   (b) *x* = ____ km   (c) *x* = ____ mm

3. Calculate the length of the unknown side in the following triangles.

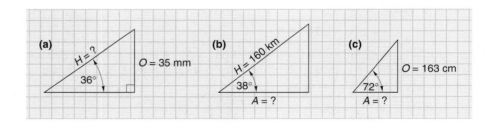

4. Calculate the angle θ for the following right-angle triangles.

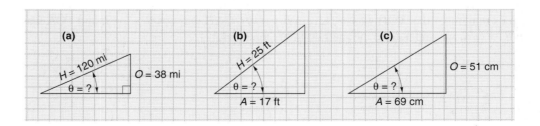

5. Calculate the magnitude of the resultant vectors in the following vector diagrams.

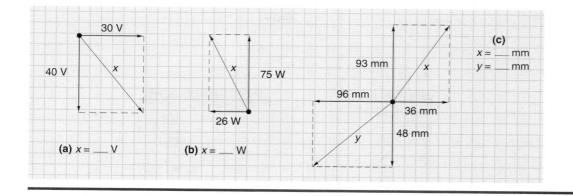

8-2   **THE DIFFERENCE BETWEEN DC AND AC**

One of the best ways to describe anything new is to begin by redescribing something known and then discuss the unknown. The known topic in this case is **direct current.** Direct current (dc) is the flow of electrons in one DIRECTion and one direction only. Dc voltage is non-varying and normally obtained from a battery or power supply unit, as seen in Figure 8-9(a). The only variation in voltage from a battery occurs due to the battery's discharge, but even then, the current will still flow in only one direction, as seen in Figure 8-9(b). A dc voltage of 9 or 6 V could be illustrated graphically as shown in Figure 8-9(c). Whether 9 or 6 V, the voltage can be seen to be constant or the same at any time.

Some power supplies supply a form of dc known as pulsating dc, which varies periodically from zero to a maximum, back to zero, and then repeats. Figure 8-10(a) illustrates the physical appearance and schematic diagram of a battery charger that is connected across two series resistors. The battery charger is generating a waveform, as shown in Figure 8-10(b), known as pulsating dc. At time 1 [Figure 8-10(c)], the power supply is generating 9 V and direct current is flowing from negative to positive. At time 2 [Figure 8-10(d)], the power supply is producing 0 V, and therefore no current is being produced. In between time 1 and time 2, the voltage out of the power supply will decrease from 9 V to 0 V. No matter what the voltage, whether 8, 7, 6, 5, 4, 3, 2, or 1 V, current will be flowing in only one direction (unidirectional) and is therefore referred to as dc.

Pulsating dc is normally supplied by a battery charger and is used to charge secondary batteries. It is also used to operate motors that convert the pulsating dc electrical energy into a mechanical rotation output. Whether steady or pulsating, direct current is current in only one DIRECTion.

Alternating current (ac) flows first in one direction and then in the opposite direction. This reversing current is produced by an alternating voltage source, as shown in Figure 8-11(a) on p. 324, which reaches a maximum in one direction (positive), decreases to zero, and then reverses itself and reaches a maximum in the opposite direction (negative). This is graphically illustrated in Figure 8-11(b). During the time of the positive voltage alternation, the polarity of the voltage will be as shown in Figure 8-11(c), so current will flow from negative to positive in a counterclockwise direction. During the time of the negative voltage alternation, the polarity of the voltage will reverse, as shown in Figure 8-11(d), causing current to flow once again from negative to positive, but in this case, in the opposite clockwise direction.

Direct Current

Current flow in only one direction.

Alternating Current

Electric current that rises from zero to a maximum in one direction, falls to zero, and then rises to a maximum in the opposite direction, and then repeats another cycle, the positive and negative alternations being equal.

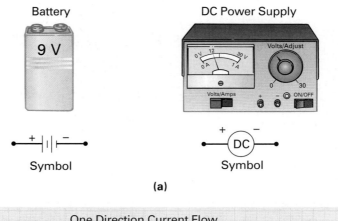

Battery

DC Power Supply

9 V

Volts/Adjust

Symbol

Symbol

(a)

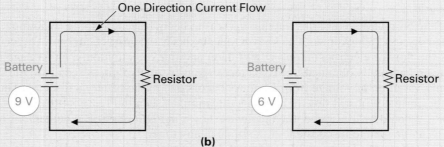

One Direction Current Flow

Battery

Resistor

9 V

Battery

Resistor

6 V

(b)

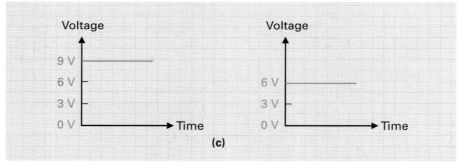

Voltage

Voltage

9 V

6 V

3 V

0 V

Time

6 V

3 V

0 V

Time

(c)

FIGURE 8-9   Direct Current. (a) DC Sources. (b) DC Flow. (c) Graphical Representation of DC.

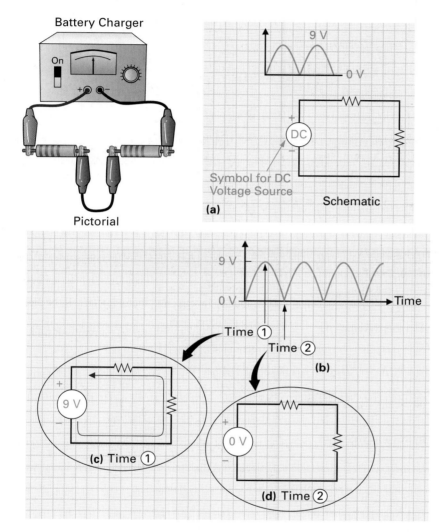

FIGURE 8-10   **Pulsating DC.**

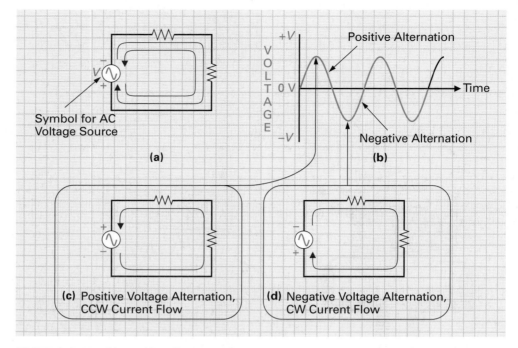

FIGURE 8-11 Alternating Current.

Now that you have completed this section, you should be able to:

■ **Objective 7.** *Explain the difference between alternating current and direct current.*

Use the following questions to test your understanding of Section 8-2.

1. Give the full names of the following abbreviations: (a) ac; (b) dc.

2. Is the pulsating waveform generated by a battery charger considered to be ac or dc? Why?

3. The polarity of a/an _____ voltage source will continually reverse, and therefore so will the circuit current.

4. The polarity of a/an _____ voltage source will remain constant, and therefore current will flow in only one direction.

5. List the two main applications of ac.

6. State briefly the difference between ac and dc.

8-3 WHY ALTERNATING CURRENT?

The question that may be troubling you at this point is: If we have been managing fine for the past chapters with dc, why do we need ac?

There are two main applications for ac:

1. *Power transfer:* to supply electrical power for lighting, heating, cooling, appliances, and machinery in both home and industry

2. *Information transfer:* to communicate or carry information, such as radio music and television pictures, between two points

To begin with, let us discuss the first of these applications, power transfer.

8-3-1 *Power Transfer*

There are three advantages that ac has over dc from a power point of view, and these are:

1. Flashlights, radios, and portable televisions all use batteries (dc) as a source of power. In these applications where a small current is required, batteries will last a good length of time before there is a need to recharge or replace them. Many appliances and most industrial equipment need a large supply of current, and in this situation a generator would have to be used to generate this large amount of current. Generators operate in the opposite way to motors, in that a **generator** converts a mechanical rotation input into an electrical output. Generators can be used to generate either dc or ac, but ac generators can be larger, less complex internally, and cheaper to operate, and this is the first reason why we use ac instead of dc for supplying power.

2. From a power point of view, ac is always used by electric companies when transporting power over long distances to supply both the home and industry with electrical energy. Recalling the power formula, you will remember that power is proportional to either current or voltage squared ($P \propto I^2$ or $P \propto V^2$), which means that to supply power to the home or industry, we would supply either a large current or voltage. As you can see in Figure 8-12, between the electric power plant and home or industry are power lines carrying the power. The amount of power lost (heat) in these power lines can be calculated by using the formula $P = I^2 \times R$, where I is the current flowing through the line and R is the resistance of the power lines. This means that the larger the current, the greater the amount of power lost in the lines in the form of heat and therefore the less the amount of power supplied to the home or industry. For this reason, power companies transport electric energy at a very high voltage between 200,000 and 600,000 V. Since the voltage is high, the current can be low and provide the same amount of power to the consumer ($P = V\uparrow \times I\downarrow$). Yet, by keeping the current low, the amount of heat loss generated in the power lines is minimal.

    Now that we have discovered why it is more efficient over a long distance to transport high voltages than high current, what does this have to do with ac? An ac voltage can easily and efficiently be transformed up or down to a higher or lower voltage by utilizing a device known as a **transformer,** and even though dc voltages can be stepped up and down, the method is inefficient and more complex.

3. Nearly all electronic circuits and equipment are powered by dc voltages, which means that once the ac power arrives at the home or industry, in most cases it will have to be converted into dc power to operate electronic equipment. It is a relatively simple process to convert ac to dc, but conversion from dc to ac is a complex and comparatively inefficient process.

Figure 8-12 illustrates ac power distribution from the electric power plant to the home and industry. The ac power distribution system begins at the electric power plant, which has the powerful large generators driven by turbines to generate large ac voltages. The turbines can be driven by either falling water (hydroelectric), or from steam, which is produced with intense heat by burning either coal, gas, or oil or from a nuclear reactor (thermoelectric). The turbine supplies the mechanical energy to the generator, to be transformed into ac electrical energy.

The generator generates an ac voltage of approximately 22,000 V, which is stepped up by transformers to approximately 500,000 V. This voltage is applied to the long-distance transmission lines, which connect the power plant to the city or town. At each city or town, the voltage is tapped off the long-distance transmission lines and stepped down to approximately 66,000 V and is distributed to large-scale industrial customers. The 66,000 V is

Generator
Device used to convert a mechanical energy input into an electrical energy output.

Transformer
Device consisting of two or more coils that are used to couple electric energy from one circuit to another, yet maintain electrical isolation between the two.

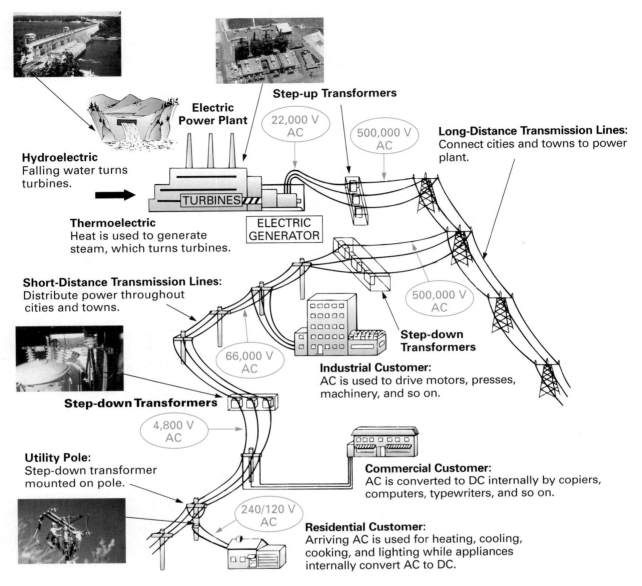

FIGURE 8-12   AC Power Distribution.

stepped down again to approximately 4800 V and distributed throughout the city or town by short-distance transmission lines. This 4800 V is used by small-scale industrial customers and residential customers who receive the ac power via step-down transformers on utility poles, which step down the 4800 V to 240 V and 120 V.

A large amount of equipment and devices within industry and the home will run directly from the ac power, such as heating, lighting, and cooling. Some equipment that runs on dc, such as televisions and computers, will accept the 120 V ac and internally convert it to the dc voltages required. Figure 8-13 illustrates a TV set that is operating from the 120 V ac from the wall outlet and internally converting it to dc so that it can power the circuits necessary to produce both audio (sound) and video (picture) information. This unit that converts ac to dc is called a **rectifier.**

Rectifier

Device that achieves rectification.

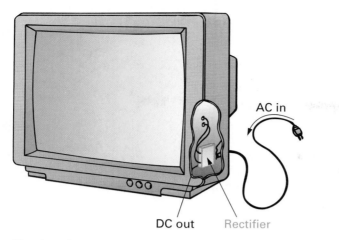

FIGURE 8-13 **Rectification (Converting AC to DC) by Appliances.**

Now that you have completed this section, you should be able to:

■ *Objective 8.* *Describe how ac is used to deliver power and to represent information.*

■ *Objective 9.* *Give the three advantages that ac has over dc from a power point of view.*

■ *Objective 10.* *Describe basically the ac power distribution system from the electric power plant to the home or industry.*

Use the following questions to test your understanding of Section 8-3-1.

1. In relation to power transfer, what three advantages does ac have over dc?

2. True or false: A generator converts an electrical input into a mechanical output.

3. What formula is used to calculate the amount of power lost in a transmission line?

4. What is a transformer?

5. What voltage is provided to the wall outlet in the home?

6. Most appliances internally convert the _____ input voltage into a _____ voltage.

8-3-2   *Information Transfer*

Information, by definition, is the property of a signal or message that conveys something meaningful to the recipient. **Communication,** which is the transfer of information between two points, began with speech and progressed to handwritten words in letters and printed words in newspapers and books. To achieve greater distances of communication, face-to-face communications evolved into telephone and radio communications.

A simple communication system can be seen in Figure 8-14(a).

The voice information or sound wave produced by the sender is a variation in air pressure, and travels at the speed of sound, as detailed in Figure 8-14. Sound waves or sounds are normally generated by a vibrating reed or plucked string in the case of musical instruments. In this example the sender's vocal cords vibrate backward and forward, producing a rarefaction or decreased air pressure, where few air molecules exist, and a compression or increased air pressure, where many air molecules exist. Like the ripples produced by a stone falling in a pond, the sound waves produced by the sender are constantly expanding and traveling outward.

The microphone is in fact a **transducer** (energy converter), because it converts the sound wave (which is a form of mechanical energy) into electrical energy in the form of voltage and current, which varies in the same manner as the sound wave and therefore contains the sender's message or information.

Communication

Transmission of information between two points.

Transducer

Any device that converts energy from one form to another.

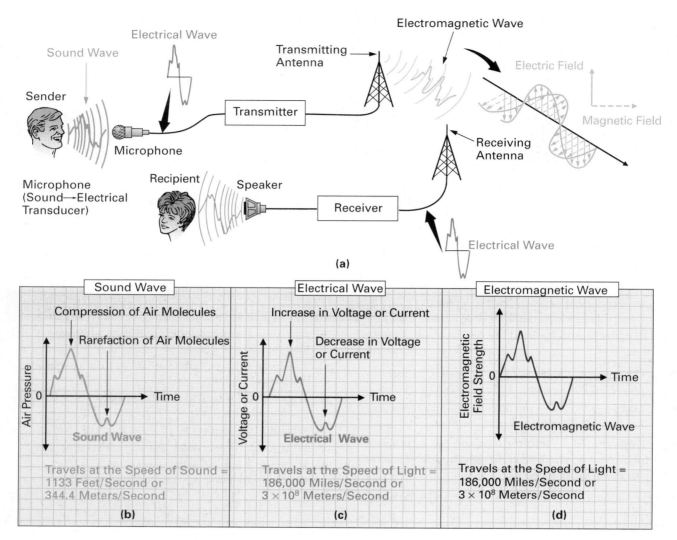

FIGURE 8-14   Information Transfer.

Electrical Wave

Traveling wave propagated in a conductive medium that is a variation in voltage or current and travels at slightly less than the speed of light.

The **electrical wave,** shown in Figure 8-14(c), is a variation in voltage or current and can only exist in a wire conductor or circuit. This electrical signal travels at the speed of light.

The speaker, like the microphone, is also an electroacoustical transducer that converts the electrical energy input into a mechanical sound-wave output. These sound waves strike the outer eardrum, causing the ear diaphragm to vibrate, and these mechanical vibrations actuate nerve endings in the ear, which convert the mechanical vibrations into electrochemical impulses that are sent to the brain. The brain decodes this information by comparing these impulses with a library of previous sounds and so provides the sensation of hearing.

To communicate between two distant points, a wire must be connected between the microphone and speaker. However, if an electrical wave is applied to an antenna, the electrical wave is converted into a radio or electromagnetic wave, as shown in the inset in Figure 8-14(a), and communication is established without the need of a connecting wire—hence the term **wireless communication.** Antennas are designed to radiate and receive electromagnetic waves, which vary in field strength, as shown in Figure 8-14(d), and can exist in either air or space. These radio waves, as they are also known, travel at the speed of light and allow us to achieve great distances of communication.

More specifically, radio waves are composed of two basic components. The electrical voltage applied to the antenna is converted into an electric field and the electrical current

Wireless Communication

Term describing radio communication that requires no wires between the two communicating points.

into a magnetic field. This **electromagnetic** (electric–magnetic) **wave** is used to carry a variety of information, such as speech, radio broadcasts, television signals, and so on.

In summary, the sound wave is a variation in air pressure, the electrical wave is a variation of voltage or current, and the electromagnetic wave is a variation of electric and magnetic field strength.

■ EXAMPLE:

How long will it take the sound wave produced by a rifle shot to travel 9630.5 feet?

■ *Solution:*

This problem makes use of the following formula:

$$\text{Distance} = \text{velocity} \times \text{time}$$

or

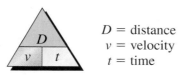

D = distance
v = velocity
t = time

If someone travels at 20 mph for 2 hours, the person will travel 40 miles ($D = v \times t = 20$ mph $\times$ 2 hours = 40 miles). In this problem, the distance (9630.5) and the sound wave's velocity (1130 ft/sec) are known, and so by rearranging the formula, we can find time:

$$\text{time} = \frac{\text{distance}}{\text{velocity}} = \frac{9630.5 \text{ ft}}{1130 \text{ ft/s}} = 8.5 \text{ s}$$

■ EXAMPLE:

How long will it take an electromagnetic (radio) wave to reach a receiving antenna that is 2000 miles away from the transmitting antenna?

■ *Solution:*

In this problem, both distance (2000 miles) and velocity (186,000 miles/s) are known, and time has to be calculated:

$$\text{time} = \frac{\text{distance}}{\text{velocity}} = \frac{2000 \text{ miles}}{186,000 \text{ miles/s}}$$
$$= 1.075 \times 10^{-2} \quad \text{or} \quad 10.8 \text{ ms}$$

8-3-3   *Electrical Equipment or Electronic Equipment?*

In the beginning of this chapter, it was stated that ac is basically used in two applications: (1) power transfer and (2) information transfer. These two uses for ac help define the difference between electricity and electronics. Electronic equipment manages the flow of information, while electrical equipment manages the flow of power. In summary:

EQUIPMENT	MANAGES
Electrical	Power (large values of V and I)
Electronic	Information (small values of V and I)

To use an example, we can say that a dc power supply is a piece of electrical equipment since it is designed to manage the flow of power. A TV set, however, is an electronic system since its electronic circuits are designed to manage the flow of audio (sound) and video (picture) information.

Since most electronic systems include a dc power supply, we can also say that the electrical circuits manage the flow of power, and this power supply enables the electronic circuits to manage the flow of information.

SELF-TEST EVALUATION POINT FOR SECTIONS 8-3-2 AND 8-3-3

Now that you have completed these sections, you should be able to:

■ **Objective 11.** *Describe the three waves used to carry infomation between two points.*

■ **Objective 12.** *Explain how the three basic information carriers are used to carry many forms of information on different frequencies within the frequency spectrum.*

Use the following questions to test your understanding of Section 8-3-2.

1. Define *information* and *communication*.
2. The _____ wave is a variation in air pressure, the _____ wave is a variation in field strength, and the _____ wave is a variation of voltage or current.
3. Which of the three waves described in Question 2:
   a. Can only exist in the air?
   b. Can exist in either air or a vacuum?
   c. Exists in a wire conductor?

4. Sound waves travel at the speed of sound, which is _____, while electrical and electromagnetic waves travel at the speed of light, which is _____.
5. A human ear is designed to receive _____ waves, an antenna is designed to transmit or receive _____ waves, and an electronic circuit is designed to pass only _____ waves.
6. Give the names of the following energy converters or transducers:
   a. Sound wave (mechanical energy) to electrical wave
   b. Electrical wave to sound wave
   c. Electrical wave to electromagnetic wave
   d. Sound wave to electrochemical impulses
7. _____ equipment manages the flow of information, and these ac waveforms normally have small values of current and voltage.
8. _____ equipment manages the flow of power, and these ac wave-forms normally have large values of current and voltage.

8-4   AC WAVE SHAPES

In all fields of electronics, whether medical, industrial, consumer, or data processing, different types of information are being conveyed between two points, and electronic equipment is managing the flow of this information.

Let's now discuss the basic types of ac wave shapes. The way in which a wave varies in magnitude with respect to time describes its wave shape. All ac waves can be classified into one of six groups, and these are illustrated in Figure 8-15.

8-4-1   *The Sine Wave*

Sine Wave

Wave whose amplitude is the sine of a linear function of time. It is drawn on a graph that plots amplitude against time or radial degrees relative to the angular rotation of an alternator.

The **sine wave** is the most common type of waveform. It is the natural output of a generator that converts a mechanical input, in the form of a rotating shaft, into an electrical output in the form of a sine wave. In fact, for one cycle of the input shaft, the generator will produce one sinusoidal ac voltage waveform, as shown in Figure 8-16. When the input shaft of the generator is at 0°, the ac output is 0 V. As the shaft is rotated through 360°, the ac output voltage will rise to a maximum positive voltage at 90°, fall back to 0 V at 180°, and then

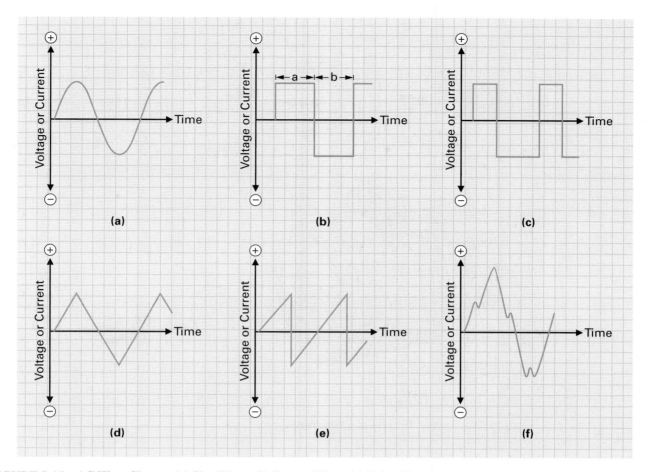

FIGURE 8-15   AC Wave Shapes. (a) Sine Wave. (b) Square Wave. (c) Pulse Wave. (d) Triangular Wave. (e) Sawtooth Wave. (f) Irregular Wave.

reach a maximum negative voltage at 270°, and finally return to 0 V at 360°. If this ac voltage is applied across a closed circuit, it produces a current that continually reverses or alternates in each direction.

Figure 8-17 illustrates the sine wave, with all the characteristic information inserted, which at first glance looks a bit ominous. Let's analyze and discuss each piece of information individually, beginning with the sine wave's amplitude.

Amplitude

Figure 8-18 plots direction and amplitude against time. The **amplitude** or magnitude of a wave is often represented by a **vector** arrow, also illustrated in Figure 8-18. The vector's length indicates the magnitude of the current or voltage, while the arrow's point is used to show the direction, or polarity.

Peak Value

The peak of an ac wave occurs on both the positive and negative alternation, but is only at the peak (maximum) for an instant. Figure 8-19(a) on p. 334 illustrates an ac current waveform rising to a positive peak of 10 A, falling to zero, and then reaching a negative peak of 10 A in the reverse direction. Figure 8-19(b) shows an ac voltage waveform reaching positive and negative peaks of 9 V.

Amplitude

Magnitude or size an alternation varies from zero.

Vector

Quantity that has both magnitude and direction. They are normally represented as a line, the length of which indicates magnitude and the orientation of which, due to the arrowhead on one end, indicates direction.

Peak Value

Maximum or highest-amplitude level.

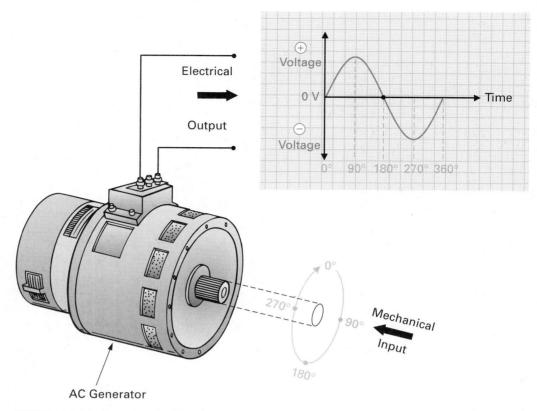

Electrical

Output

AC Generator

FIGURE 8-16 **Degrees of a Sine Wave.**

Peak-to-Peak Value

The **peak-to-peak value** of a sine wave is the value of voltage or current between the positive and negative maximum values of a waveform. For example, the peak-to-peak value of the current waveform in Figure 8-19(a) is equal to $I_{p-p} = 2 \times I_p = 20$ A. In Figure 8-19(b), it would be equal to $V_{p-p} = 2 \times V_p = 18$ V.

$$\text{p-p} = 2 \times \text{peak}$$

RMS or Effective Value

Both the positive and negative alternation of a sine wave can accomplish the same amount of work, but the ac waveform is only at its maximum value for an instant in time, spending most of its time between peak currents. Our examples in Figure 8-19(a) and (b), therefore, cannot supply the same amount of power as a dc value of 10 A or 9 V.

The effective value of a sine wave is equal to 0.707 of the peak value. Let's now see how this value was obtained. Power is equal to either $P = I^2 \times R$ or $P = V^2/R$. Said another way, power is proportional to the voltage or current squared. If every instantaneous value of either the positive or negative half-cycle of any voltage or current sinusoidal waveform is squared, as shown in Figure 8-20, and then averaged out to obtain the mean value, the square root of this mean value would be equal to 0.707 of the peak.

For example, if the process is carried out on the 10 A current waveform considered previously, the result would equal 7.07 A (0.707 × 10 A = 7.07 A), which is 0.707 of the peak value of 10 A.

$$\text{rms} = 0.707 \times \text{peak}$$

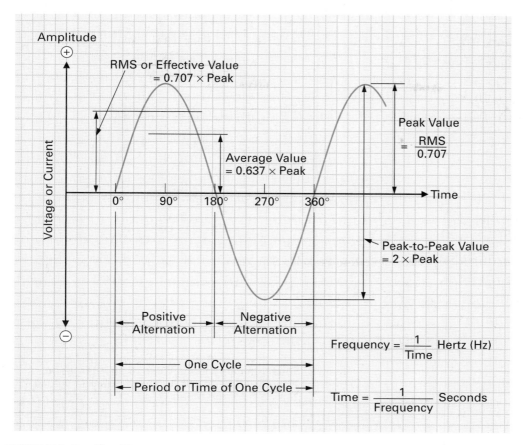

FIGURE 8-17   Sine Wave.

This **root-mean-square (rms) value** of 0.707 can always be used to tell us how effective an ac sine wave will be. For example, a 10 A dc source would be 10 A effective because it is continually at its peak value and always delivering power to the circuit to which it is connected, while a 10 A ac source would only be 7.07 A effective, as seen in Figure 8-21, because it is at 10 A for only a short period of time. As another example, a 10 V ac sine-wave alternation would be as effective or supply the same amount of power to a circuit as a 7.07 V dc source.

RMS Value

RMS value of an ac voltage, current, or power waveform is equal to 0.707 times the peak value. The rms value is the effective or dc value equivalent of the ac wave.

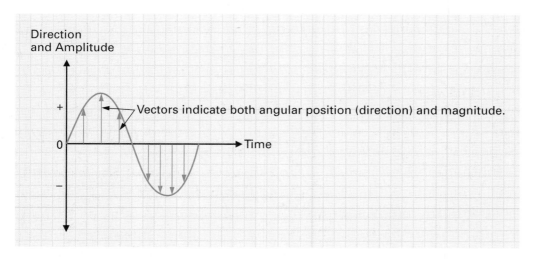

FIGURE 8-18   Sine-Wave Amplitude.

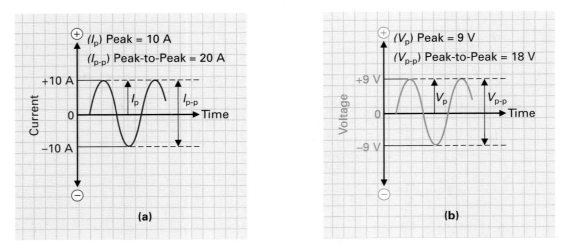

FIGURE 8-19   Peak and Peak-to-Peak of a Sine Wave.

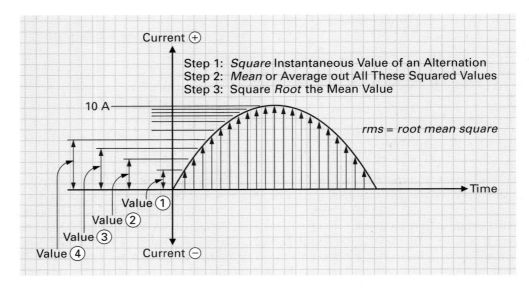

FIGURE 8-20   Obtaining the RMS Value of 0.707 for a Sine Wave.

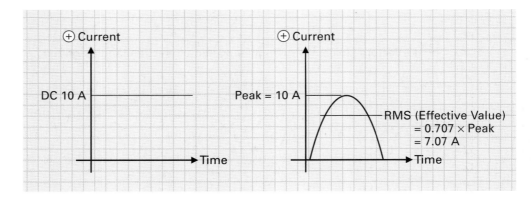

FIGURE 8-21   Effective Equivalent.

Unless otherwise stated, ac values of voltage or current are always given in rms. The peak value can be calculated by transposing the original rms formula of rms = peak × 0.707, and ending up with

$$\text{peak} = \frac{\text{rms}}{0.707}$$

Since 1/0.707 = 1.414, the peak can also be calculated by

$$\text{peak} = \text{rms} \times 1.414$$

Average Value

The **average value** of the positive or negative alternation is found by taking either the positive or negative alternation, and listing the amplitude or vector length of current or voltages at 1° intervals, as shown in Figure 8-22(a). The sum of all these values is then divided by the total number of values (averaging), which for all sine waves will calculate out to be

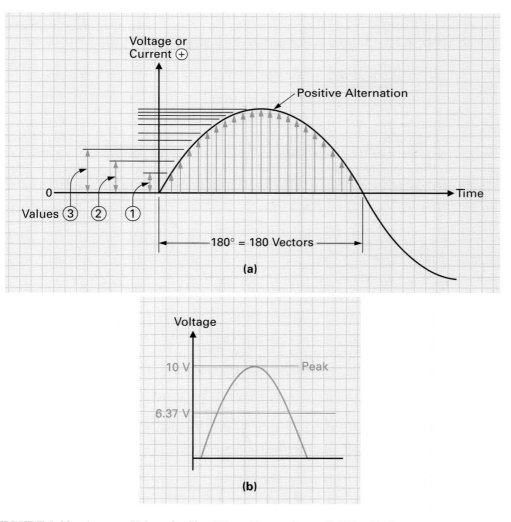

FIGURE 8-22   **Average Value of a Sine-Wave Alternation = 0.637 × Peak.**

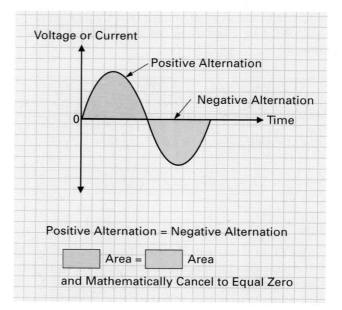

FIGURE 8-23   **Average Value of a Complete Sine-Wave Cycle = 0.**

0.637 of the peak voltage or current. For example, the average value of a sine-wave alternation with a peak of 10 V, as seen in Figure 8-22(b), is equal to

$$\text{average} = 0.637 \times \text{peak}$$

$$= 0.637 \times 10 \text{ V}$$
$$= 6.37 \text{ V}$$

The average of a positive or negative alternation (half-cycle) is equal to 0.637 × peak. However, the average of the complete cycle, including both the positive and negative half-cycles, is mathematically zero as the amount of voltage or current above the zero line is equal but opposite to the amount of voltage or current below the zero line, as shown in Figure 8-23.

▨   **EXAMPLE:**

Calculate V_p, $V_{p\text{-}p}$, V_{rms}, and V_{avg} of a 16 V peak sine wave.

▨   *Solution:*

$$V_p = 16 \text{ V}$$
$$V_{p\text{-}p} = 2 \times V_p = 2 \times 16 \text{ V} = 32 \text{ V}$$
$$V_{rms} = 0.707 \times V_p = 0.707 \times 16 \text{ V} = 11.3 \text{ V}$$
$$V_{avg} = 0.637 \times V_p = 0.637 \times 16 \text{ V} = 10.2 \text{ V}$$

▨   **EXAMPLE:**

Calculate V_p, $V_{p\text{-}p}$, and V_{avg} of a 120 V (rms) ac main supply.

▨   *Solution:*

$$V_p = \text{rms} \times 1.414 = 120 \text{ V} \times 1.414 = 169.68 \text{ V}$$
$$V_{p\text{-}p} = 2 \times V_p = 2 \times 169.68 \text{ V} = 339.36 \text{ V}$$
$$V_{avg} = 0.637 \times V_p = 0.637 \times 169.68 \text{ V} = 108.09 \text{ V}$$

The 120 V (rms) that is delivered to every home and business has a peak of 169.68 V. This ac value will deliver the same power as 120 V dc.

Frequency and Period

As shown in Figure 8-24, the **period** (t) is the time required for one complete cycle (positive and negative alternation) of the sinusoidal current or voltage waveform. A *cycle,* by definition, is the change of an alternating wave from zero to a positive peak, to zero, then to a negative peak, and finally, back to zero.

Frequency is the number of repetitions of a periodic wave in a unit of time. It is symbolized by f and is given the unit hertz (cycles per second), in honor of a German physicist, Heinrich Hertz.

Sinusoidal waves can take a long or a short amount of time to complete one cycle. This time is related to frequency in that period and is equal to the reciprocal of frequency, and vice versa.

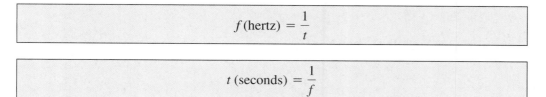

$$f \text{ (hertz)} = \frac{1}{t}$$

$$t \text{ (seconds)} = \frac{1}{f}$$

where t = period, f = frequency

For example, the ac voltage of 120 V (rms) arrives at the household electrical outlet alternating at a frequency of 60 hertz (Hz). This means that 60 cycles arrive at the household electrical outlet in 1 second. If 60 cycles occur in 1 second, as seen in Figure 8-25(a), it is actually taking ⅟₆₀ of a second for one of the 60 cycles to complete its cycle, which calculates out to be

$$\text{⅟₆₀ of 1 second} = \frac{1}{60 \text{ cycles}} \times 1 \text{ second} = 16.67 \text{ milliseconds (ms)}$$

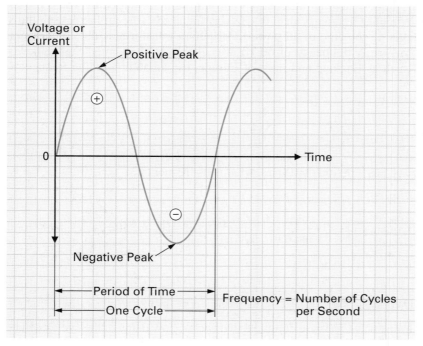

FIGURE 8-24   **Frequency and Period.**

Period
Time taken to complete one complete cycle of a periodic or repeating waveform.

Frequency
Rate of recurrences of a periodic wave normally within a unit of one second, measured in hertz (cycles/second).

TIME LINE
Heinrich R. Hertz (1857-1894), a German physicist, was the first to demonstrate the production and reception of electromagnetic (radio) waves. In honor of his work in this field, the unit of frequency is called the hertz.

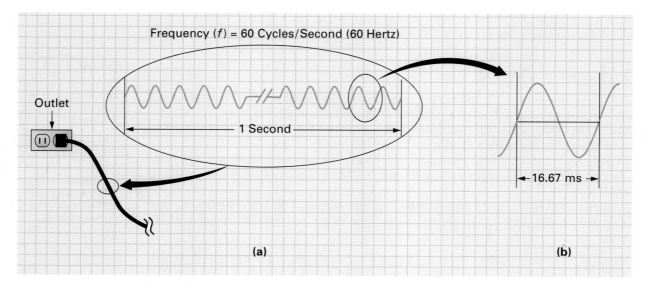

Figure 8-25 120 V, 60 Hz AC Supply.

So the time or period of one cycle can be calculated by using the formula period (t) = $1/f$ = 1/60 Hz = 16.67 ms, as shown in Figure 8-25(b).

If the period or time of a cycle is known, the frequency can be calculated. For example:

$$\text{frequency } (f) = \frac{1}{\text{period}} = 1/16.67 \text{ ms} = 60 \text{ Hz}$$

As illustrated in Figure 8-12, all homes in the United States receive at their wall outlets an ac voltage of 120 V rms at a frequency of 60 Hz. This frequency was chosen for convenience, as a lower frequency would require larger transformers, and if the frequency were too low, the slow switching (alternating) current through light bulbs would cause the lights to flicker. A higher frequency than 60 Hz was found to cause an increase in the amount of heat generated in the core of all power distribution transformers due to eddy currents and hysteresis losses. A frequency of 60 Hz was chosen in the United States; however, other countries, such as England and most of Europe, use an ac power line frequency of 50 Hz (240 V).

EXAMPLE:

If a sine wave has a period of 400 μs, what is its frequency?

Solution:

$$\text{frequency } (f) = \frac{1}{\text{time } (t)} = \frac{1}{400 \text{ μs}} = 2.5 \text{ kHz or 2500 cycles per second}$$

EXAMPLE:

If it takes a sine wave 25 ms to complete two cycles, how many of the cycles will be received in 1 s?

Solution:

If the period of two cycles is 25 ms, one cycle period will equal 12.5 ms. The number of cycles per second or frequency will equal

$$f = \frac{1}{t} = \frac{1}{12.5 \text{ ms}} = 80 \text{ Hz} \quad \text{or} \quad 80 \text{ cycles/second}$$

EXAMPLE:

Calculate the period of the following:

a. 100 MHz

b. 40 cycles every 5 seconds

c. 4.2 kilocycles/second

d. 500 kHz

Solution:

$$f = \frac{1}{t} \qquad \text{therefore, } t = \frac{1}{f}$$

a. $t = \dfrac{1}{100 \text{ MHz}} = 10$ nanoseconds (ns)

b. 40 cycles/5 s = 8 cycles/second (8 Hz)

$t = \dfrac{1}{8 \text{ Hz}} = 125$ ms

c. $t = \dfrac{1}{4.2 \text{ kHz}} = 238 \ \mu s$

d. $t = \dfrac{1}{500 \text{ kHz}} = 2 \ \mu s$

Wavelength

Wavelength, as its name states, is the physical length of one complete cycle and is generally measured in meters. The wavelength (λ, lambda) of a complete cycle is dependent on the frequency and velocity of the transmission:

$$\lambda = \frac{\text{velocity}}{\text{frequency}}$$

Wavelength
Distance between two points of corresponding phase and is equal to waveform velocity or speed divided by frequency.

Electromagnetic waves. Radio waves travel at the speed of light in air or a vacuum, which is 3×10^8 meters/second or 3×10^{10} cm/second.

$$\lambda \ (m) = \frac{3 \times 10^8 \text{ m/s}}{f \ (\text{Hz})} \qquad \text{or} \qquad \lambda \ (cm) = \frac{3 \times 10^{10} \text{ cm/s}}{\text{frequency (Hz)}}$$

(There are 100 centimeters [cm] in 1 meter [m], therefore cm = 10^{-2}, and m = 10^0 or 1.) Subsequently, the higher the frequency, the shorter the wavelength, which is why a shortwave radio receiver is designed to receive high frequencies ($\lambda \downarrow = 3 \times 10^8/f \uparrow$).

Calculate the wavelength of the electromagnetic waves illustrated in Figure 8-26.

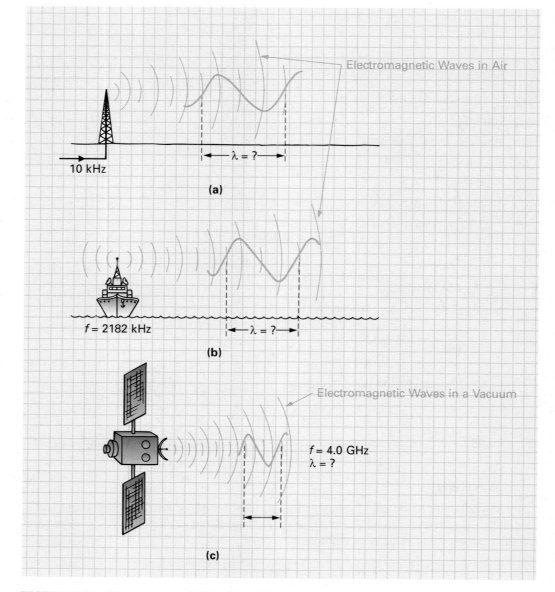

FIGURE 8-26 **Electromagnetic Wavelength Examples.**

Solution:

a. $\lambda = \dfrac{3 \times 10^8}{f\,(\text{Hz})}\ \text{m/s} = \dfrac{3 \times 10^8}{10\ \text{kHz}} = 30{,}000\ \text{m}$ or   30 km

b. $\lambda = \dfrac{3 \times 10^{10}}{f\,(\text{Hz})}\ \text{cm/s} = \dfrac{3 \times 10^{10}}{2182\ \text{kHz}} = 13{,}748.9\ \text{cm}$ or   137.489 m

c. $\lambda = \dfrac{3 \times 10^{10}}{f\,(\text{Hz})}\ \text{cm/s} = \dfrac{3 \times 10^{10}}{4.0\ \text{GHz}} = \dfrac{3 \times 10^{10}}{4 \times 10^9} = 7.5\ \text{cm}$ or   0.075 m

Sound waves. Sound waves travel at a slower speed than electromagnetic waves, as their mechanical vibrations depend on air molecules, which offer resistance to the traveling wave. For sound waves, the wavelength formula will be equal to

$$\lambda \text{ (m)} = \frac{344.4 \text{ m/s}}{f \text{ (Hz)}}$$

■ **EXAMPLE:**

Calculate the wavelength of the sound waves illustrated in Figure 8-27.

■ *Solution:*

a. $\lambda \text{ (m)} = \dfrac{344.4 \text{ m/s}}{f \text{ (Hz)}} = \dfrac{344.4}{35 \text{ kHz}} = 9.8 \times 10^{-3} \text{ m} = 9.8$

b. 300 Hz: $\lambda \text{ (m)} = \dfrac{344.4 \text{ m/s}}{300 \text{ Hz}} = 1.15 \text{ m}$

   3000 Hz: $\lambda \text{ (m)} = \dfrac{344.4 \text{ m/s}}{3000 \text{ Hz}} = 0.115 \text{ m}$ or   11.5 cm

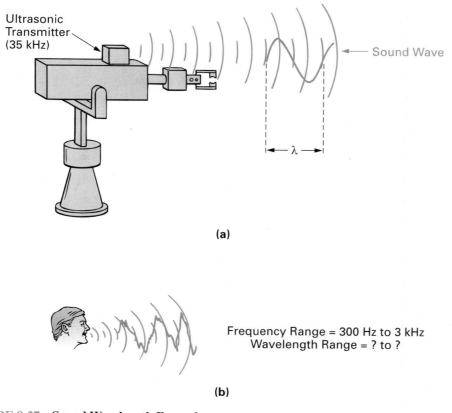

(a)

(b)

FIGURE 8-27 **Sound Wavelength Examples.**

Phase Relationships

The **phase** of a sine wave is always relative to another sine wave of the same frequency. Figure 8-28(a) illustrates two sine waves that are in phase with one another, while

Phase

Angular relationship between two waves, normally between current and voltage in an ac circuit.

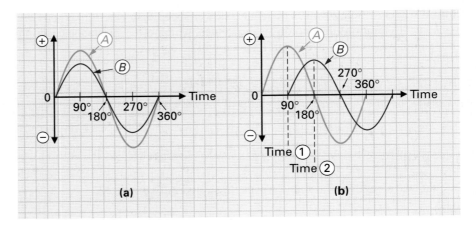

FIGURE 8-28   **Phase Relationship. (a) In Phase. (b) Out of Phase.**

Figure 8-28(b) shows two sine waves that are out of phase with one another. Sine wave *A* is our reference, since the positive-going zero crossing is at 0°, its positive peak is at 90°, its negative-going zero crossing is at 180°, its negative peak is at 270°, and the cycle completes at 360°. In Figure 8-28(a), sine wave *B* is in phase with *A* since its peaks and zero crossings occur at the same time as sine wave *A*'s.

In Figure 8-28(b), sine wave *B* has been shifted to the right by 90° with respect to the reference sine wave *A*. This **phase shift** or **phase angle** of 90° means that sine wave *A leads B* by 90°, or *B lags A* by 90°. Sine wave *A* is said to lead *B* as its positive peak, for example, occurs first at time 1, while the positive peak of *B* occurs later at time 2.

Phase Shift or Angle

Change in phase of a waveform between two points, given in degrees of lead or lag. Phase difference between two waves, normally expressed in degrees.

■   EXAMPLE:

What are the phase relationships between the two waveforms illustrated in Figure 8-29(a) and (b)?

■   *Solution:*

    a. The phase shift or angle is 90°. Sine wave *B* leads sine wave *A* by 90°, or *A* lags *B* by 90°.

    b. The phase shift or angle is 45°. Sine wave *A* leads sine wave *B* by 45°, or *B* lags *A* by 45°.

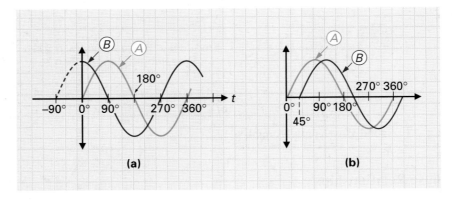

FIGURE 8-29   **Phase Relationship Examples.**

CHAPTER 8 / ALTERNATING CURRENT (AC)

The Meaning of Sine

The square, rectangular, triangular, and sawtooth waveform shapes are given their names because of their waveform shapes. The sine wave is the most common type of waveform shape, and why the name sine was given to this wave needs to be explained further.

Figure 8-30(a) shows the correlation between the 360° of a circle and the 360° of a sine wave. Within the circle a triangle has been drawn to represent the right-angle triangle formed at 30°. The hypotenuse side will always remain at the same length throughout 360°, and will equal the peak of the sine wave. The opposite side of the triangle is equal to the amplitude vector of the sine wave at 30°. To calculate the amplitude of the opposite side, and therefore the amplitude of the sine wave at 30°, we use the sine of theta formula discussed previously in the mini-math review on trigonometry.

$$\sin \theta = \frac{\text{opposite}}{\text{hypotenuse}}$$

$$\sin 30° = \frac{O}{H} \quad \left(\begin{array}{c}\textit{calculator sequence:} \\ \boxed{3}\,\boxed{0}\,\boxed{\text{SIN}}\end{array}\right)$$

$$0.500 = \frac{O}{H}$$

This tells us that at 30°, the opposite side is 0.500 or half the size of the hypotenuse. At 30°, therefore, the amplitude of the sine wave will be 0.5 (50%) of the peak value.

Figure 8-30(b) lists the sine values at 15° increments. Figure 8-30(c) shows an example of a 10-V-peak sine wave. At 15°, a sine wave will always be at 0.259 (sine of 15°) of the peak value, which for a 10 V sine wave will be 2.59 V (0.259 × 10 V = 2.59 V). At 30°, a sine wave will have increased to 0.500 (sine of 30°) of the peak value. At 45°, a sine wave will be at 0.707 of the peak, and so on. The sine wave is called a sine wave, therefore, because it changes in amplitude at the same rate as the sine trigonometric function.

8-4-2 *The Square Wave*

The **square wave** is a periodic (repeating) wave that alternates from a positive peak value to a negative peak value, and vice versa, for equal lengths of time.

In Figure 8-31 you can see an example of a square wave that is at a frequency of 1 kHz and has a peak of 10 V. If the frequency of a wave is known, its period or time of one cycle can be calculated by using the formula $t = 1/f = 1/1 \text{ kHz} = 1$ ms or $\frac{1}{1000}$ of a second. One complete cycle will take 1 ms to complete, so the positive and negative alternations will each last for 0.5 ms.

If the peak of the square wave is equal to 10 V, the peak-to-peak value of this square wave will equal $V_{\text{p-p}} = 2 \times V_{\text{p}} = 20$ V.

To summarize this example, the square wave alternates from a positive peak value of +10 V to a negative peak value of −10 V for equal time lengths (half-cycles) of 0.5 ms.

Duty Cycle

Duty cycle is an important relationship, which has to be considered when discussing square waveforms. The **duty cycle** is the ratio of a pulse width (positive or negative pulse or cycle) to the overall period or time of the wave and is normally given as a percentage.

$$\text{duty cycle (\%)} = \frac{\text{pulse width } (P_w)}{\text{period } (t)} \times 100\%$$

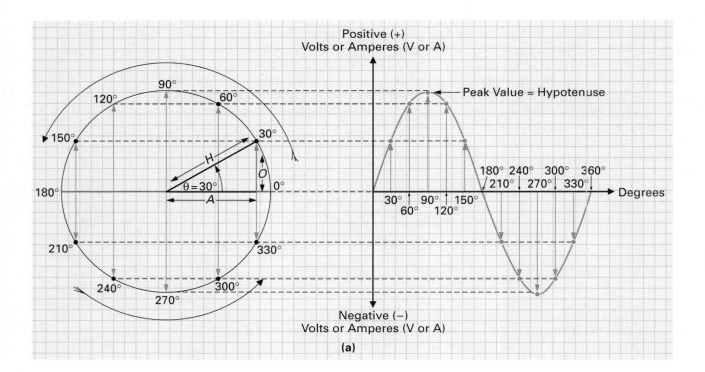

(a)

Angle	Sine
0	0.000
15	0.259
30	0.500
45	0.707
60	0.866
75	0.966
90	1.000

(b)

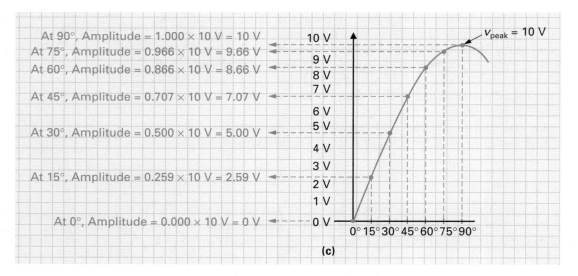

At 90°, Amplitude = 1.000 × 10 V = 10 V
At 75°, Amplitude = 0.966 × 10 V = 9.66 V
At 60°, Amplitude = 0.866 × 10 V = 8.66 V
At 45°, Amplitude = 0.707 × 10 V = 7.07 V
At 30°, Amplitude = 0.500 × 10 V = 5.00 V
At 15°, Amplitude = 0.259 × 10 V = 2.59 V
At 0°, Amplitude = 0.000 × 10 V = 0 V

(c)

FIGURE 8-30   Meaning of Sine.

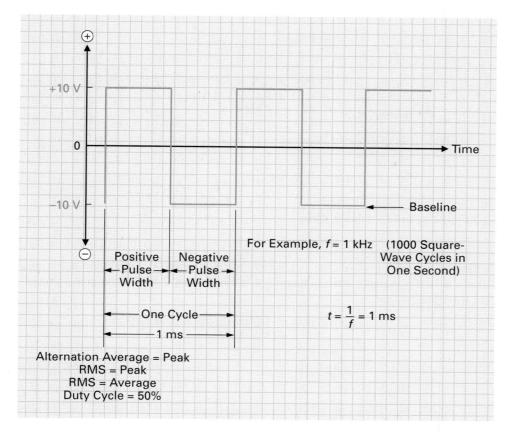

FIGURE 8-31 **Square Wave.**

The duty cycle of the example square wave in Figure 8-31 will equal

$$\text{duty cycle (\%)} = \frac{\text{pulse width } (P_w)}{\text{period } (t)} \times 100\%$$

$$= \frac{0.5 \text{ ms}}{1 \text{ ms}} \times 100\%$$

$$= 50\%$$

Since a square wave always has a positive and a negative alternation that are equal in time, the duty cycle of all square waves is equal to 50%, which actually means that the positive cycle lasts for 50% of the time of one cycle.

Average

The average or mean value of a square wave can be calculated by using the formula

$$V \text{ or } I \text{ average} = \text{baseline} + (\text{duty cycle} \times \text{peak to peak})$$

The average of the complete square-wave cycle in Figure 8-31 should calculate out to be zero, as the amount above the line equals the amount below. If we apply the formula to this example, you can see that

$$V_{avg} = \text{baseline} + (\text{duty cycle} \times \text{peak to peak})$$

$$= -10 \text{ V} + (0.5 \times 20 \text{ V})$$

$$= -10 \text{ V} + 10 \text{ V}$$

$$= 0 \text{ V}$$

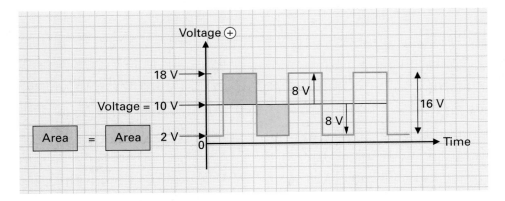

FIGURE 8-32   **2 to 18 V Square Wave.**

However, a square wave does not always alternate about 0. For example, Figure 8-32 illustrates a 16 $V_{\text{p-p}}$ square wave that rests on a baseline of 2 V. The average value of this square wave is equal to

$$V_{\text{avg}} = \text{baseline} + (\text{duty cycle} \times \text{peak to peak})$$
$$= 2\,\text{V} \times (0.5 \times 16\,\text{V})$$
$$= (+2) + (+8\,\text{V})$$
$$= 10\,\text{V}$$

◼ **EXAMPLE:**

Calculate the duty cycle and V_{avg} of a square wave of 0 to 5 V.

◼ *Solution:*

The duty cycle of a square wave is always 0.5 or 50%. The average is:

$$V_{\text{avg}} = \text{baseline} + (\text{duty cycle} \times V_{\text{p-p}})$$
$$= 0\,\text{V} + (0.5 \times 5\,\text{V})$$
$$= 0\,\text{V} + 2.5\,\text{V} = 2.5\,\text{V}$$

Up to this point, we have seen the ideal square wave, which has instantaneous transition from the negative to the positive values, and vice versa, as shown in Figure 8-33(a). In fact, the transitions from negative to positive (positive or leading edge) and from positive to negative (negative or trailing edge) are not as ideal as shown here. It takes a small amount of time for the wave to increase to its positive value (the **rise time**), and an equal amount of time for a wave to decrease to its negative value (the **fall time**). Rise time (T_R), by definition, is the time it takes for an edge to rise from 10% to 90% of its full amplitude, while fall time (T_F) is the time it takes for an edge to fall from 90% to 10% of its full amplitude, as shown in Figure 8-33(b).

With a waveform such as that in Figure 8-33(b), it is difficult, unless a standard is used, to know exactly what points to use when measuring the width of either the positive or negative alternation. The standard width is always measured between the two 50% amplitude points, as shown in Figure 8-34.

Frequency-Domain Analysis

A *periodic wave* is a wave that repeats the same wave shape from one cycle to the next. Figure 8-35(a) is a **time-domain** representation of a periodic sine wave, which is the same way it would appear on an oscilloscope display as it plots the sine wave's amplitude against

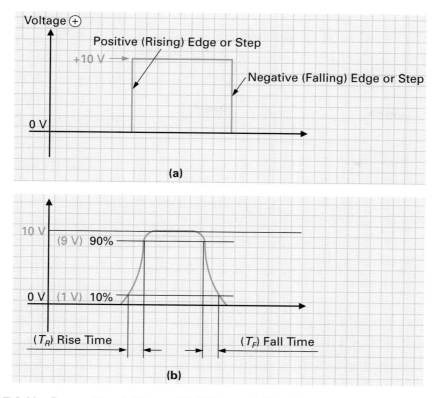

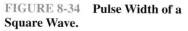

FIGURE 8-33 **Square Wave's Rise and Fall Times. (a) Ideal. (b) Actual.**

time. Figure 8-35(b) is a **frequency-domain** representation of the same periodic sine wave, and this graph, which shows the wave as it would appear on a spectrum analyzer, plots the sine wave's amplitude against frequency instead of time. This graph shows all the frequency components contained within a wave, and since, in this example, the sine wave has a period of 1 ms and therefore a frequency of 1 kHz, there is one bar at the 1 kHz point of the graph and its size represents the sine wave's amplitude. Pure sine waves have no other frequency components.

Other periodic wave shapes, such as square, pulse, triangular, sawtooth, or irregular, are actually made up of a number of sine waves having a particular frequency, amplitude, and phase. To produce a square wave, for instance, you would start with a sine wave, as shown in Figure 8-36(a), whose frequency is equal to the resulting square-wave frequency. This sine wave is called the **fundamental frequency,** and all the other sine waves that will be added to this fundamental are called **harmonics** and will always be lower in amplitude and higher in frequency. These harmonics or multiples are harmonically related to the

Frequency-Domain Analysis

A method of representing a waveform by plotting its amplitude versus frequency.

Fundamental Frequency

This sine wave is always the lowest frequency and largest amplitude component of any waveform shape and is used as a reference.

Harmonic Frequency

Sine wave that is smaller in amplitude and is some multiple of the fundamental frequency.

FIGURE 8-34 **Pulse Width of a Square Wave.**

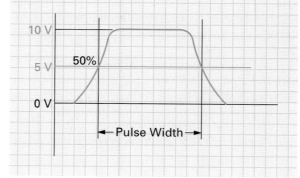

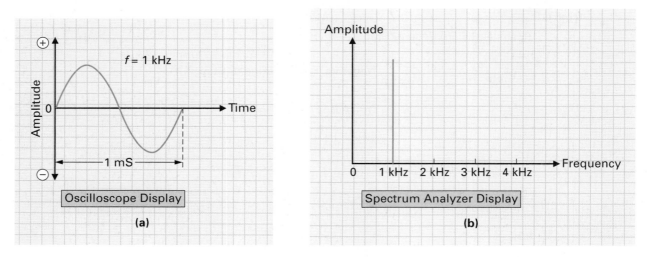

FIGURE 8-35   Analysis of a 1 kHz Sine Wave. (a) Time Domain. (b) Frequency Domain.

fundamental, in that the second harmonic is twice the fundamental frequency, the third harmonic is three times the fundamental frequency, and so on.

Square waves are composed of a fundamental frequency and an infinite number of odd harmonics (third, fifth, seventh, and so on). If you look at the progression in Figure 8-36(a) through (d), you see that by continually adding these odd harmonics the waveform comes closer to a perfect square wave, as shown in Figure 8-36(e).

Figure 8-37 plots the frequency domain of a square wave, with the bars representing the odd harmonics of decreasing amplitude.

8-4-3   *The Rectangular or Pulse Wave*

The **rectangular wave** is similar to the square wave in many respects, in that it is a periodic wave that alternately switches between one of two fixed values. The difference in the rectangular wave is that it does not remain at the two peak values for equal lengths of time, as shown in the examples in Figure 8-38(a) and (b).

In Figure 8-38(a), the rectangular wave remains at its negative level for a longer period than its positive, while the rectangular wave in Figure 8-38(b) stays at its positive value for the longer period of time and is only momentarily at its negative value.

PRF, PRT, and Pulse Length

When discussing a rectangular wave, a few terms change. Instead of stating the cycles per second as frequency, it is called **pulse repetition frequency** (PRF), which is far more descriptive. The reciprocal of frequency is time, and with rectangular pulse waveforms the reciprocal of the PRF is **pulse repetition time** (PRT), as summarized in Table 8-1. With rectangular or pulse waves, therefore, frequency is equivalent to PRF, time to PRT, and the only difference is the name.

Let us look at the example in Figure 8-39 of a 5 V rectangular wave at a frequency of 1 kHz and a pulse width of 1 μs, and practice with these new terms. With a pulse repetition frequency of 1 kHz, the time between the leading edges of pulses (PRT) will be 1/1 kHz = 1 ms. **Pulse width** (P_w), **pulse duration** (P_d), or **pulse length** (P_l) are all terms that describe the length of time for which the pulse lasts, and in this example it is equal to 1 μs, which means that 999 μs exists between the end of one pulse and the beginning of the next.

Rectangular (Pulse) Wave

Also known as a pulse wave; it is a repeating wave that only alternates between two levels or values and remains at one of these values for a small amount of time relative to the other.

Pulse Repetition Frequency

The number of times per second that a pulse is transmitted.

Pulse Repetition Time

The time interval between the start of two consecutive pulses.

Pulse Width, Pulse Length, or Pulse Duration

The time interval between the leading edge and trailing edge of a pulse at which the amplitude reaches 50% of the peak pulse amplitude.

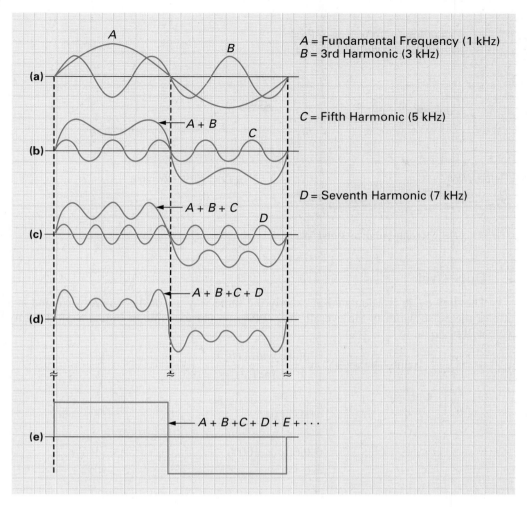

FIGURE 8-36   **Square-Wave Composition.**

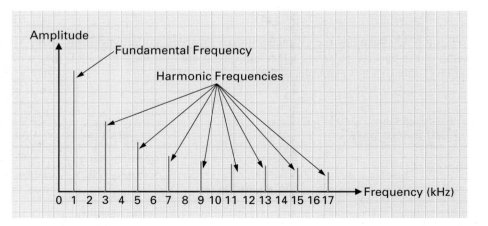

FIGURE 8-37   **Frequency-Domain Analysis of a Square Wave.**

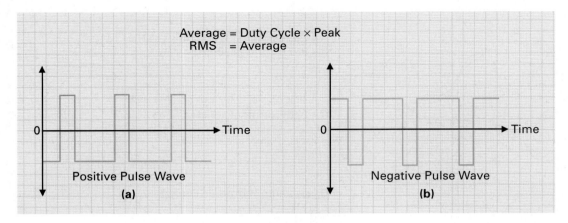

FIGURE 8-38　**Rectangular or Pulse Wave.**

Duty Cycle

The duty cycle is calculated in exactly the same way as for the square wave and is a ratio of the pulse width to the overall time (PRT). In our example in Figure 8-39 the duty cycle will be equal to

$$\text{duty cycle } (\%) = \frac{\text{pulse width } (P_w)}{\text{PRT}} \times 100\%$$

$$= \frac{1\ \mu s}{1000\ \mu s} \times 100\%$$

$$= \text{duty cycle figure of } 0.001 \times 100\%$$

$$= 0.001 \times 100\%$$

$$= 0.1\%$$

The result tells us that the positive pulse lasts for 0.1% of the total time (PRT).

Average

The average or mean value of this waveform is calculated by using the same square-wave formula. The average of the pulse wave in Figure 8-39 will be

$$V \text{ or } I \text{ average} = \text{baseline} + (\text{duty cycle} \times \text{peak to peak})$$
$$V_{\text{avg}} = 0\ V + (0.001 \times 5\ V)$$
$$= 0\ V + (5\ mV)$$
$$= 5\ mV$$

Figure 8-40 illustrates the average value of this rectangular waveform. If the voltage and width of the positive pulse are taken and spread out over the entire PRT, they will have a mean level equal, in this example, to 5 mV.

TABLE 8-1

SQUARE AND SINE WAVE		RECTANGULAR WAVE
$\text{Frequency } (f) = \dfrac{1}{\text{time } (t)}$		$\dfrac{\text{pulse repetition}}{\text{frequency (PRF)}} = \dfrac{1}{\dfrac{\text{pulse repetition}}{\text{time (PRT)}}}$
	Equivalent to	
$\text{Time } (t) = \dfrac{1}{\text{frequency } (f)}$		$\dfrac{\text{pulse repetition}}{\text{time (PRF)}} = \dfrac{1}{\dfrac{\text{pulse repetition}}{\text{frequency (PRF)}}}$

　CHAPTER 8 / ALTERNATING CURRENT (AC)

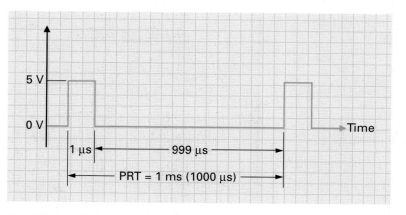

FIGURE 8-39   PRF and PRT of a Pulse Wave.

EXAMPLE:

Calculate the duty cycle and average voltage of the following radar pulse waveform:

$$\text{peak voltage, } V_p = 20 \text{ kV}$$
$$\text{pulse length, } P_l = 1 \text{ μs}$$
$$\text{baseline voltage} = 0 \text{ V}$$
$$\text{PRF} = 3300 \text{ pulses per second (pps)}$$

Solution:

$$\text{duty cycle} = \frac{\text{pulse length } (P_l)}{\text{PRT}} \times 100\%$$

$$= \frac{1 \text{ μs}}{303 \text{ μs}} \times 100\% \left(\text{PRT} = \frac{1}{\text{PRF}} = \frac{1}{3300} = 303 \text{ μs} \right)$$

$$= (3.3 \times 10^{-3}) \times 100\%$$

$$= 0.33\%$$

$$V_{avg} = \text{baseline} + (\text{duty cycle} \times V_{p\text{-}p})$$
$$= 0 \text{ V} + [(3.3 \times 10^{-3}) \times 20 \times 10^{3}]$$
$$= 66 \text{ V}$$

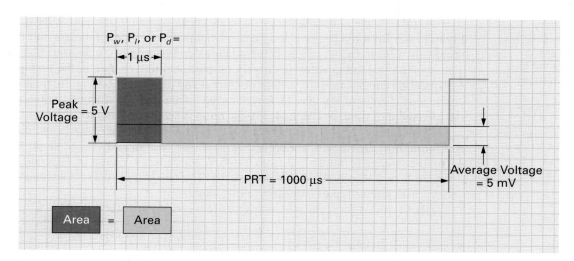

FIGURE 8-40   Average of a Pulse Wave.

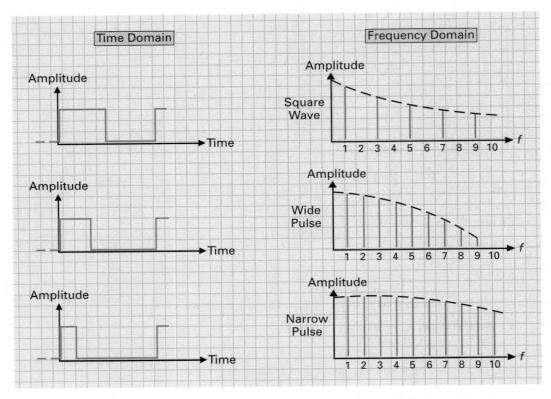

FIGURE 8-41   Time- and Frequency-Domain Analysis of a Pulse Waveform.

Frequency-Domain Analysis

The pulse or rectangular wave is closely related to the square wave, as shown in Figure 8-41, but some changes occur in its harmonic content. One is that even-number harmonics are present and their amplitudes do not fall off as quickly as do those of the square wave. The amplitude and phase of these sine-wave harmonics are determined by the pulse width and pulse repetition frequency, and the narrower the pulse, the greater the number of harmonics present.

8-4-4   *The Triangular Wave*

Triangular Wave

A repeating wave that has equal positive and negative ramps that have linear rates of change with time.

Linear

Relationship between input and output in which the output varies in direct proportion to the input.

A **triangular wave** consists of a positive and negative ramp of equal values, as shown in Figure 8-42. Both the positive and negative ramps have a linear increase and decrease, respectively. Linear, by definition, is the relationship between two quantities that exists when a change in a second quantity is directly proportional to a change in the first quantity. The two quantities in this case are voltage or current and time. As shown in Figure 8-42, if the increment of change of voltage ΔV (pronounced "delta vee") is changing at the same rate as the increment of time, Δt ("delta tee"), the ramp is said to be **linear.**

With Figure 8-43(a), the voltage has risen 1 V in 1 second (time 1) and maintains that rise through to time 2 and, consequently, is known as a linearly rising slope. In Figure 8-43(b), the voltage is falling first from 6 to 5 V, which is a 1 V drop in 1 second, and in time 2 from 6 V to 2 V, which is a 4 V drop in 4 s. The rate of fall still remains the same, so the waveform is referred to as a linearly falling slope.

This formula for slope will also apply to a current waveform, and is

$$\text{slope (A/s)} = \frac{\Delta I}{\Delta t}$$

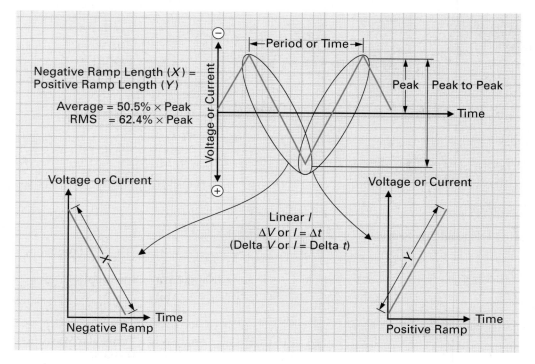

FIGURE 8-42　Triangular Wave.

where ΔI = increment of current change, Δt = increment of time change

With triangular waves, frequency and time apply as usual, as seen in Figure 8-44, with

$$\text{frequency} = \frac{1}{\text{time}} \quad (\text{Hz})$$

$$\text{time} = \frac{1}{\text{frequency}} \quad (\text{s})$$

Frequency-Domain Analysis

The time domain of a triangular wave is shown in Figure 8-45(a), while the frequency domain of a triangular wave is shown in Figure 8-45(b). Frequency domain analysis is often used to test electronic circuits as it tends to highlight problems that would normally not show up when a sine-wave signal is applied.

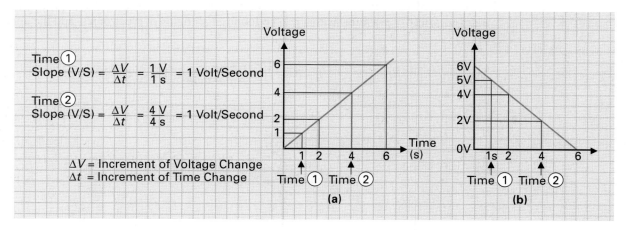

FIGURE 8-43　Linear Triangular Wave Rise and Fall.

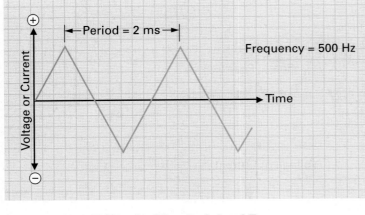

FIGURE 8-44   **Triangular Wave Period and Frequency.**

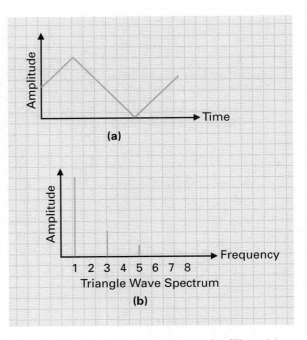

FIGURE 8-45   **Analysis of a Triangular Wave. (a) Time Domain. (b) Frequency Domain.**

8-4-5   *The Sawtooth Wave*

Sawtooth Wave

Repeating waveform that rises from zero to a maximum value linearly and then falls to zero and repeats.

On an oscilloscope display (time-domain presentation), the **sawtooth wave** is very similar to a triangular wave, in that a sawtooth wave has a linear ramp. However, unlike the triangular wave, which reverses and has an equal but opposite ramp back to its starting level, the sawtooth "flies" back to its starting point immediately and then repeats the previous ramp, as seen in Figure 8-46, which shows both a positive and a negative ramp sawtooth.

Figure 8-47 shows the odd and even harmonics contained in the frequency domain analysis of a negative-going ramp.

8-4-6   *Other Waveforms*

The waveforms discussed so far are some of the more common types; however, since every waveform shape (except a pure sine wave) is composed of a large number of sine waves combined in an infinite number of ways, any waveform shape is possible. Figure 8-48 illustrates a variety of waveforms that can be found in all fields of electronics, and the appendix

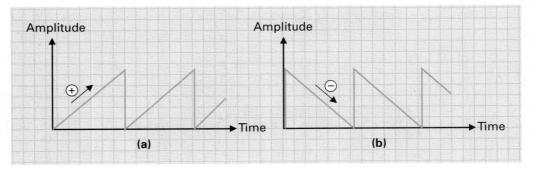

FIGURE 8-46   **Sawtooth Wave. (a) Positive Ramp. (b) Negative Ramp.**

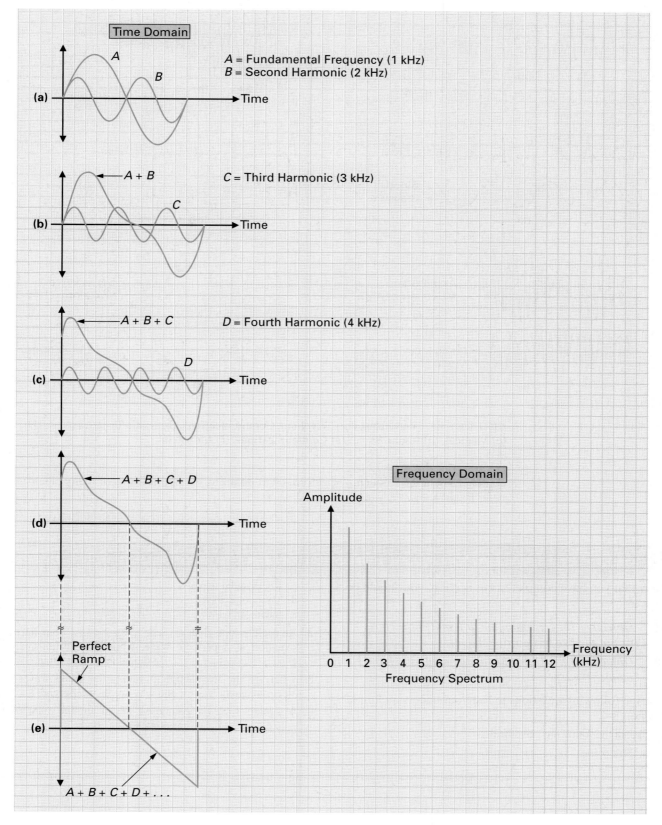

FIGURE 8-47 **Analysis of a Sawtooth Wave.**

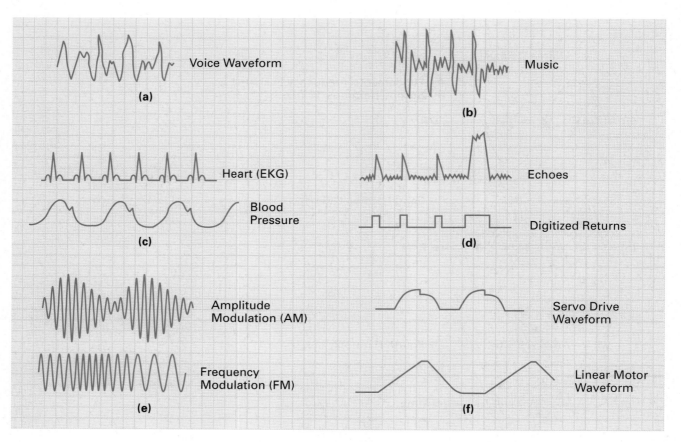

FIGURE 8-48 Waveforms. (a) Telephone Communications. (b) Radio Broadcast. (c) Medical. (d) Radar/Sonar. (e) Communications. (f) Industrial.

contains a detailed frequency spectrum chart detailing the applications of many electromagnetic frequencies.

SELF-TEST EVALUATION POINT FOR SECTION 8-4

Now that you have completed this section, you should be able to:

■ **Objective 13.** *Explain the different characteristics of the five basic wave shapes.*

■ **Objective 14.** *Describe fundamental and harmonic frequencies.*

Use the following questions to test your understanding of Section 8-4.

1. Sketch the following waveforms:

 a. Sine wave
 b. Square wave

 c. Rectangular wave
 d. Triangular wave
 e. Sawtooth wave

2. An oscilloscope gives a _____-domain representation of a periodic wave, while a spectrum analyzer gives a _____-domain representation.

3. What are the wavelength formulas for sound waves and electromagnetic waves, and why are they different?

4. A square wave is composed of an infinite number of _____ harmonics.

8-5 MEASURING AND GENERATING AC SIGNALS

As a technician or engineer, you are going to be required to diagnose and repair a problem in the shortest amount of time possible. To aid in the efficiency of this fault finding and repair process, you can make use of certain pieces of test equipment. Humans have five kinds of sensory systems: touch, taste, sight, sound, and smell. Four can be used for electronic troubleshooting: sight, sound, touch, and smell.

Electronic test equipment can be used either to *sense* a circuit's condition or to *generate* a signal to see the response of the component or circuit to that signal.

8-5-1 *The AC Meter*

Up to this point, we have seen how a multimeter, like the one shown in Figure 8-49, can be used to measure direct current and voltage. Most multimeters can be used to measure either dc or ac. When the technician or engineer wishes to measure ac, the ac current or voltage is converted to dc internally by a circuit known as a **rectifier,** as shown in Figure 8-50, before passing on to the meter.

The dc produced by the rectification process is in fact pulsating, as shown in Figure 8-51, so the current through the meter is a series of pulses rising from zero to maximum (peak) and from maximum back to zero. Frequencies below 10 Hz (lower frequency limit) will cause the digits on a digital multimeter to continually increase in value and then decrease in value as the meter follows the pulsating dc. This makes it difficult to read the meter. This effect will also occur with an analog multimeter which, from 10 Hz to approximately 2 kHz, will not be able to follow the fluctuation, causing the needle to remain in a position equal to the average value of the pulsating dc from the rectifier (0.637 of peak). Most meters are normally calibrated internally to indicate rms values (0.707 of peak) rather than average values, because this effective value is most commonly used when expressing ac voltage or current. The upper frequency limit of the ac meter is approximately 2 to 8 kHz. Beyond this limit the meter becomes progressively inaccurate due to the reactance of the capacitance in the rectifier. This reactance, which will be discussed later, will result in inaccurate indications due to the change in opposition at different ac input frequencies.

Rectifier
A device that converts alternating current into a unidirectional or dc current.

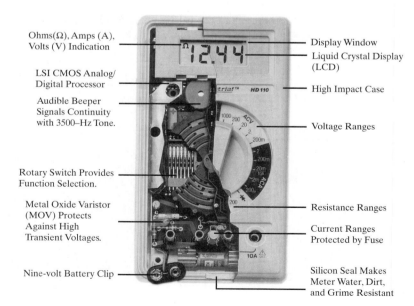

Ohms(Ω), Amps (A), Volts (V) Indication

LSI CMOS Analog/ Digital Processor

Audible Beeper Signals Continuity with 3500–Hz Tone.

Rotary Switch Provides Function Selection.

Metal Oxide Varistor (MOV) Protects Against High Transient Voltages.

Nine-volt Battery Clip

Display Window

Liquid Crystal Display (LCD)

High Impact Case

Voltage Ranges

Resistance Ranges

Current Ranges Protected by Fuse

Silicon Seal Makes Meter Water, Dirt, and Grime Resistant

FIGURE 8-49 **The Digital Multimeter.**

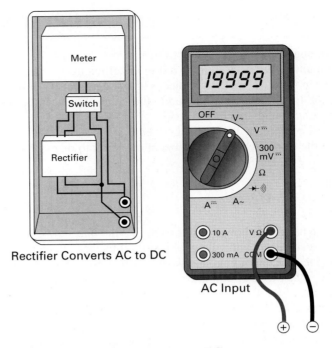

Rectifier Converts AC to DC

AC Input

FIGURE 8-50   **AC Meter Uses a Rectifier to Produce DC.**

Current Clamp

A device used in
conjunction with an ac
ammeter, containing a
magnetic core in the form
of hinged jaws that can be
snapped around the
current-carrying wire.

Current Clamps

The voltmeter is probably the most frequently used setting on the multimeter. A meter reading can be obtained by just connecting the probes across the component or source to be measured, unlike the ammeter, which requires that the circuit current path be opened, and the ammeter inserted in the path of current flow. If a current measurement is required, a clamp can be used, as seen in Figure 8-52, which allows us to sense the amount of current flow through the conductor without opening the current path.

The alternating current flowing through the conductor produces an expanding and collapsing magnetic field, which cuts across the coil of wire wound around the core of the clamp, and induces an alternating voltage in the coil (1 mA induces 1 mV). The induced alternating voltage causes an alternating current to flow, which is converted to dc by the rectifier and used to operate the meter. The larger the current flowing in the conductor, the larger the magnetic field surrounding the conductor, which results in a greater induced volt-

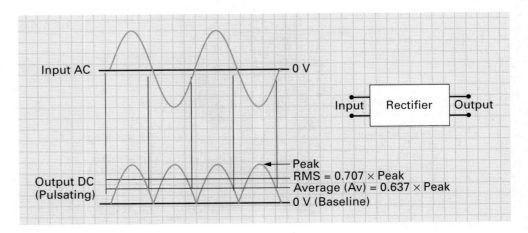

Input AC ———————————————— 0 V

Input   Rectifier   Output

Peak
RMS = 0.707 × Peak
Output DC   Average (Av) = 0.637 × Peak
(Pulsating)   0 V (Baseline)

FIGURE 8-51   **Rectifier.**

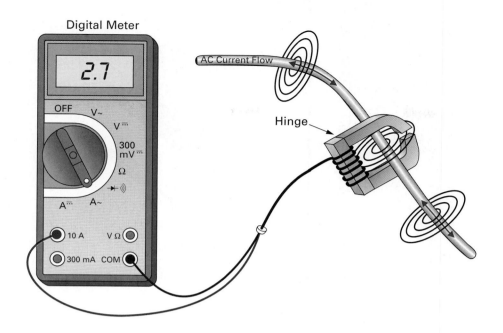

Digital Meter

2.7

AC Current Flow

Hinge

(a)

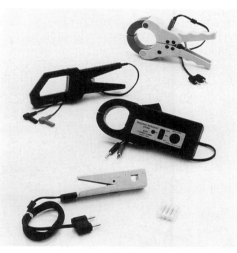

(b)

FIGURE 8-52 **Current Clamp.**

age, current, and consequent meter movement. These clamps are generally ineffective at measuring smaller currents (microamperes), because the magnetic field produced by the current is too weak.

Radio-Frequency Probe

As mentioned previously, the meter has a high range limit of approximately 2 kHz. If higher-frequency (radio frequencies) electrical waves are to be measured, a **radio-frequency (RF) probe,** as shown in Figure 8-53, can be used. The probe picks up the high-frequency ac voltage from a conducting point on the circuit and passes or couples it to a capacitor, which blocks any dc that could be present at the test point, as we only want to measure the high-frequency ac. The rectifier within the probe will convert this ac input into a dc output, which will be displayed as an rms value on the meter display.

Radio-Frequency Probe

A probe used in conjunction with an ac meter to measure high-frequency RF signals.

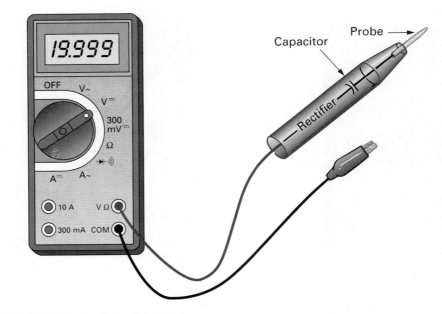

FIGURE 8-53   **Radio-Frequency Probe.**

High-Voltage Probe

The typical multimeter can handle voltages up to approximately 1000 V. If you wish to measure voltages higher than this, another component, known as a **high-voltage** (HV) **probe,** such as the one shown in Figure 8-54, can be used. The high-voltage probe has additional multiplier resistors to drop the extra voltage. Most high-voltage probes are designed so that $\frac{1}{100}$ of voltage at the probe tip from the test point will appear out of the probe and be applied to the meter. For example, if 10 kV is being measured, 100 V will appear out of the probe and be applied to the meter ($\frac{1}{100} \times 10{,}000\ V = 100\ V$). This probe would be called a

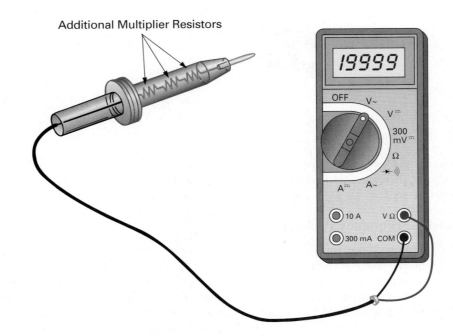

FIGURE 8-54   **High-Voltage Probe.**

$\times$ 100 probe because the 100 V shown on the meter display would now have to be multiplied by 100 for the operator to determine the voltage (100 V $\times$ 100 = 10 kV). The high-voltage probe is especially well insulated to protect its user, who should apply all safety precautions and exercise extreme caution.

EXAMPLE:

A DMM indicates 3.9 V on its display when a test point is probed by a $\times$ 1000 high-voltage probe. What is the voltage at this test point?

Solution:

A $\times$ 1000 probe will divide the voltage by 1000, so the displayed voltage must be multiplied by 1000 to obtain the correct value:

$$3.9 \text{ V} \times 1000 = 3.9 \text{ kV}$$

SELF-TEST EVALUATION POINT FOR SECTION 8-5-1

Use the following questions to test your understanding of Section 8-5-1.

1. True or false: The current clamp does not require the circuit current path to be broken in order to measure current. Can it be used to measure dc amps?

2. What is the difference between an RF and an HV probe?

8-5-2   *The Oscilloscope*

Figure 8-55 illustrates a typical **oscilloscope** (sometimes abbreviated to *scope*), which is used primarily to display the shape and spacing of electrical signals. The oscilloscope displays the actual sine, square, rectangular, triangular, or sawtooth wave shape that is occurring at any point in a circuit. This display is made on a cathode-ray tube (CRT), which is also used in television sets and computers to display video information. From the display on the CRT, we can measure or calculate time, frequency, and amplitude characteristics such as rms, average, peak, and peak-to-peak.

The oscilloscope allows us to see what is happening at every point through a circuit. In Figure 8-56 you can see the different waveforms at different points on the circuit. There are also voltage test points that can be tested with a voltmeter if a scope is not available. A voltmeter, as you can well imagine, does not supply the technician or engineer with as much information as the oscilloscope.

Controls

Oscilloscopes come with a wide variety of features and functions, but the basic operational features are almost identical. Figure 8-57 on p. 364 illustrates the front panel of a typical oscilloscope. Some of these controls are difficult to understand without practice and experience, so practical experimentation is essential if you hope to gain a clear understanding of how to operate an oscilloscope.

General controls (see Figure 8-57)
Intensity control: Controls the brightness of the trace, which is the pattern produced on the screen of a CRT.
Focus control: Used to focus the trace.

Oscilloscope

Instrument used to view signal amplitude, frequency, and shape at different points throughout a circuit.

TIME LINE

The father of television, Vladymir Zworykin (1889-1982), developed the first television picture tube, called the kinescope, in 1920.

(a)

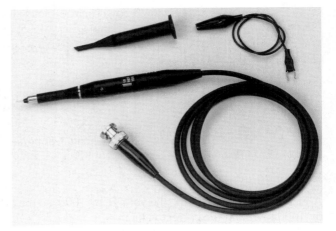

(b)

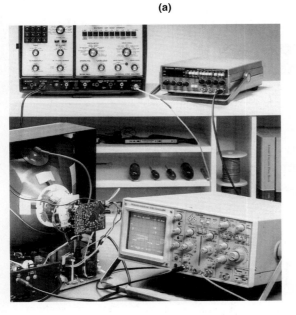

(c)

FIGURE 8-55   Typical Oscilloscope. (a) Oscilloscope. (b) Oscilloscope Probe. (c) Localizing the Faulty Block or Board.

Power OFF/ON: Switch will turn on oscilloscope while indicator shows when oscilloscope is turned on.

Some oscilloscopes have the ability to display more than one pattern or trace on the CRT screen, as seen by the examples in Figure 8-58. A dual-trace oscilloscope can produce two traces or patterns on the CRT screen at the same time, whereas a single-trace oscilloscope can trace out only one pattern on the screen. The dual-trace oscilloscope is very useful, as it allows us to make comparisons between the phase, amplitude, shape, and timing of two signals from two separate test points. One signal or waveform is applied to the channel *A* input of the oscilloscope, while the other waveform is applied to the channel *B* input.

Channel selection (see Figure 8-57)
Mode switch: This switch allows us to select which channel input should be displayed on the CRT screen.

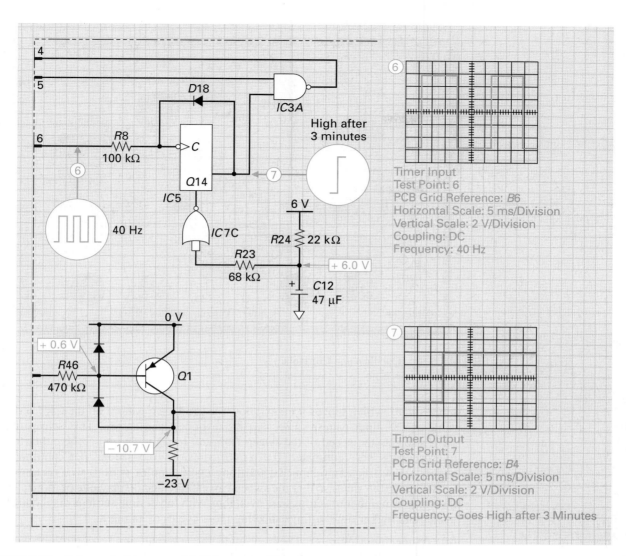

FIGURE 8-56 Schematic Diagram with Voltage and Waveform Test Points.

CHA: The input arriving at channel *A*'s jack is displayed on the screen as a single trace.

CHB: The input arriving at channel *B*'s jack is displayed on the screen as a single trace.

Dual: The inputs arriving at jacks *A* and *B* are both displayed on the screen, as a dual trace.

(a) **Calibration** Output   This output connection provides a point where a fixed 1 V peak-to-peak square-wave signal can be obtained at a frequency of 1 kHz. This signal is normally fed into either channel A or B's input to test probes and the oscilloscope operation.

(b) Channel *A* and *B* Horizontal Controls

↔ *Position control:* This control will move the position of the one (single-trace) or two (dual-trace) waveforms horizontally (left or right) on the CRT screen.

Sweep time/cm switch: The oscilloscope contains circuits that produce a beam of light that is swept continually from the left to the right of the CRT screen. When no input signal is applied, this sweep will produce a straight horizontal line in the center of the screen. When an input signal is present, this horizontal sweep is influenced by the

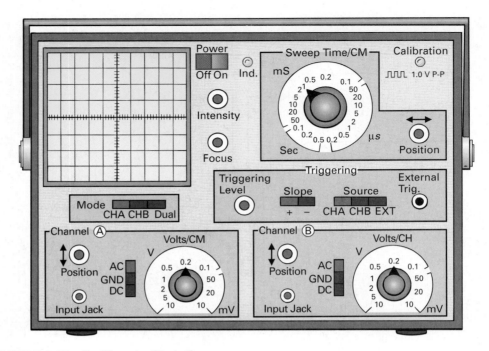

FIGURE 8-57   Oscilloscope Controls.

input signal, which moves it up and down to produce a pattern on the CRT screen the same as the input pattern (sine, square, sawtooth, and so on). This sweep time/cm switch selects the speed of the sweep from left to right, and it can be either fast (0.2 microseconds per centimeter; 0.2 μs/cm) or slow (0.5 second per centimeter; 0.5 s/cm). A low-frequency input signal (long cycle time or period) will require a long time setting (0.5 s/cm) so that the sweep can capture and display one or more cycles of the input. A number of settings are available, with lower time settings displaying fewer cycles and higher time settings showing more cycles of an input.

(c) Triggering Controls   These provide the internal timing control between the sweep across the screen and the input waveform.

Triggering level control: This determines the point where the sweep starts.

Slope switch (+): Sweep is triggered on positive-going slope.
            (−): Sweep is triggered on negative-going slope.

Source switch, CHA: The input arriving into channel *A* jack triggers the sweep.
            CHB: The input arriving into channel *B* jack triggers the sweep.
            EXT: The signal arriving at the external trigger jack is used to trigger the sweep.

(d) Channel *A* and *B* Vertical Controls   The *A* and *B* channel controls are identical.

Volts/cm switch: This switch sets the number of volts to be displayed by each major division on the vertical scale of the screen.

↕*Position control:* Moves the trace up or down for easy measurement or viewing.

AC-DC-GND *switch:* In the AC position, a capacitor on the input will pass the ac component entering the input jack, but block any dc components.
            In the GND position, the input is grounded (0 V) so that the operator can establish a reference.
            In the DC position, both ac and dc components are allowed to pass on to and be displayed on the screen.

Triggering

Initiation of an action in a circuit which then functions for a predetermined time, for example, the duration of one sweep in a cathode-ray tube.

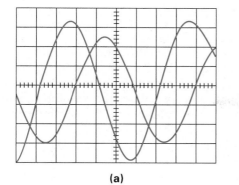
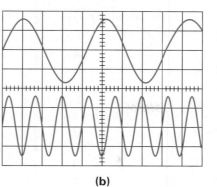

(a) (b)

(c)

FIGURE 8-58   Sample of Dual-Trace Oscilloscope Displays for Comparison.

Measurements

The oscilloscope is probably the most versatile of test equipment, as it can be used to test:

DC voltage
AC voltage
Waveform duration

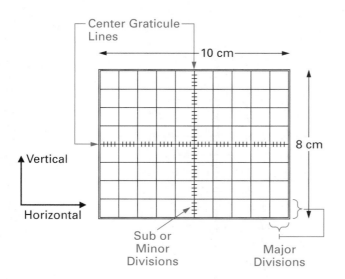

FIGURE 8-59   Oscilloscope Grid.

Waveform frequency
Waveform shape

(a) Voltage Measurement   The screen is divided into eight vertical and ten horizontal divisions, as shown in Figure 8-59 on previous page. This 8 × 10 cm grid is called the *graticule.* Every vertical division has a value depending on the setting of the volts/cm control. For example, if the volts/cm control is set to 5 V/div or 5 V/cm, the waveform shown in Figure 8-60(a), which rises up four major divisions, will have a peak positive alternation value of 20 V (4 div × 5 V/div = 20 V).

As another example, look at the positive alternation in Figure 8-60(b). The positive alternation rises up three major divisions and then extends another three subdivisions, which are each equal to 1 V because five subdivisions exist within one major division, and one major division is, in this example, equal to 5 V. The positive alternation shown in Figure 8-60(b) therefore has a peak of three major divisions (3 × 5 V/cm = 15 V), plus three subdivisions (3 × 1 V = 3 V), which equals 18 volts peak.

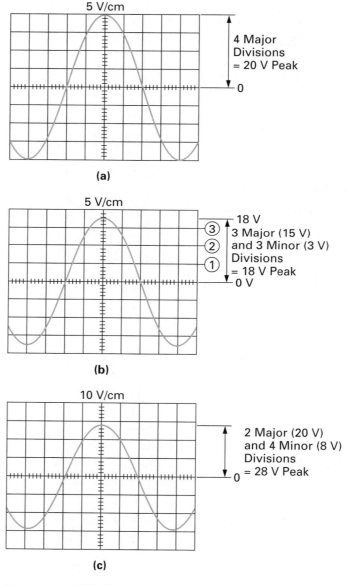

(a)

(b)

(c)

FIGURE 8-60   **Measuring AC Peak Voltage.**

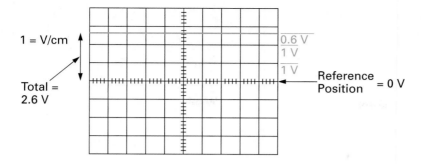

FIGURE 8-61   **Measuring DC Voltage.**

In Figure 8-60(c), we have selected the 10 volt/cm position, which means that each major division is equal to 10 V and each subdivision is equal to 2 V. In this example, the waveform peak will be equal to two major divisions ($2 \times 10 = 20$ V), plus four subdivisions (4×2 V $= 8$ V), which is equal to 28 V. Once the peak value of a sine wave is known, the peak to peak, average, and rms can be calculated mathematically.

When measuring a dc voltage with the oscilloscope, the volts/cm is applied in the same way, as shown in Figure 8-61. A positive dc voltage in this situation will cause deflection toward the top of the screen, whereas a negative voltage will cause deflection toward the bottom of the screen.

To determine the dc voltage, count the number of major divisions and to this add the number of minor divisions. In Figure 8-61, a major division equals 1 V/cm and, therefore, a minor division equals 0.2 V/cm, so the dc voltage being measured is interpreted as $+2.6$ V.

(b) Time and Frequency Measurement   The frequency of an alternating wave, such as that seen in Figure 8-62(a), is inversely proportional to the amount of time it takes to complete one cycle ($f = 1/t$). Consequently, if time can be measured, frequency can be determined.

The time/cm control relates to the horizontal line on the oscilloscope graticule and is used to determine the period of a cycle so that frequency can be calculated. For example, in Figure 8-62(b), a cycle lasts five major horizontal divisions, and since the 20 μs/division setting has been selected, the period of the cycle will equal 5×20 μs/division $= 100$ μs. If the period is equal to 100 μs, the frequency of the waveform will be equal to $f = 1/t = 1/100$ μs $= 10$ kHz.

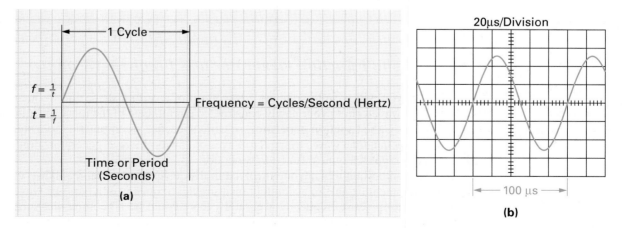

FIGURE 8-62   **Time and Frequency Measurement.**

■ **EXAMPLE:**

A complete sine-wave cycle occupies four horizontal divisions and four vertical divisions from peak to peak. If the oscilloscope is set on the 20 ms/cm and 500 mV/cm, calculate:

    a. V_{p-p}

    b. t

    c. f

    d. V_p

    e. V_{avg} of peak

    f. V_{rms}

■ *Solution:*

$$4 \text{ horizontal divisions} \times 20 \text{ ms/div} = 80 \text{ ms}$$
$$4 \text{ vertical divisions} \times 500 \text{ mV/div} = 2 \text{ V}$$

    a. $V_{p-p} = 2 \text{ V}$

    b. $t = 80 \text{ ms}$

    c. $f = \dfrac{1}{t} = 12.5 \text{ Hz}$

    d. $V_p = 0.5 \times V_{p-p} = 1 \text{ V}$

    e. $V_{avg} = 0.637 \times V_p = 0.637 \text{ V}$

    f. $V_{rms} = 0.707 \times V_p = 0.707 \text{ V}$

SELF-TEST EVALUATION POINT FOR SECTION 8-5-2

Use the following questions to test your understanding of Section 8-5-2.

1. The _____ is used primarily to display the shape and spacing of electrical signals.

2. Name the device within the oscilloscope on which the wave-forms can be seen.

3. On the 20 μs/div time setting, a full cycle occupies four major divisions. What is the waveform's frequency and period?

4. On the 2 V/cm voltage setting, the waveform swings up and down two major divisions (a total of 4 cm of vertical swing). Calculate the waveform's peak and peak-to-peak voltage.

5. What is the advantage(s) of the dual-trace oscilloscope?

8-5-3   *The Function Generator and Frequency Counter*

Function Generator

Signal generator that can function as a sine, square, rectangular, triangular, or sawtooth waveform generator.

Frequency Counter

Meter used to measure the frequency or cycles per second of a periodic wave.

In many cases you will wish to generate a waveform of a certain shape and frequency, and then apply this waveform to your newly constructed circuit. One of the most versatile wave-form generators is the **function generator,** so called because it can function as a sine wave, square wave, rectangular wave, triangular wave, or sawtooth wave generator. Figure 8-63 shows a photograph of a typical function generator.

It is very important that anyone involved in the design, manufacture, and servicing of electronic equipment be able to accurately measure the frequency of a periodic wave. Without the ability to measure frequency accurately, there could be no communications, home entertainment, or a great number of other systems. A **frequency counter,** like the one shown in Figure 8-64, can be used to analyze the frequency of any periodic wave applied to its input jack and provide a readout on the display of its frequency.

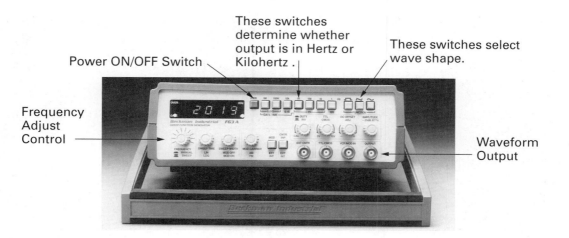

Power ON/OFF Switch

These switches
determine whether
output is in Hertz or
Kilohertz .

These switches select
wave shape.

Frequency
Adjust
Control

Waveform
Output

FIGURE 8-63   Function Generator.

SELF-TEST EVALUATION POINT FOR SECTION 8-5-3

Use the following questions to test your understanding of Section 8-5-3.

1. List some of the waveform shapes typically generated by a function generator.

2. What is the function of a frequency counter?

8-5-4   *Hand-Held Scopemeters*

The hand-held **scopemeter,** shown in Figure 8-65, combines a multimeter, oscilloscope, frequency counter, and signal generator in one easy-to-carry battery-operated unit. For a technician, this test instrument is portable, easy to set up, and easy to use since it will automatically change its settings for the best operating mode and continue to adjust itself as the input changes. To explain this test instrument in more detail, let us examine some of its key functions.

Scopemeter

A hand-held, battery-operated instrument that combines a multimeter, oscilloscope, frequency counter, and signal generator in one.

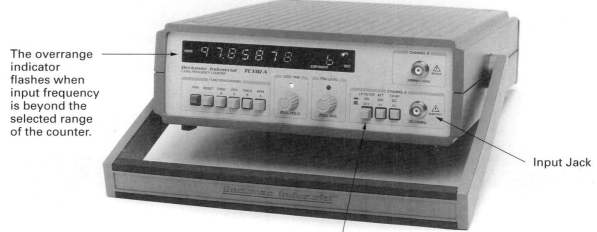

The overrange indicator flashes when input frequency is beyond the selected range of the counter.

Input Jack

Power ON/OFF Switch

FIGURE 8-64   Frequency Counter.

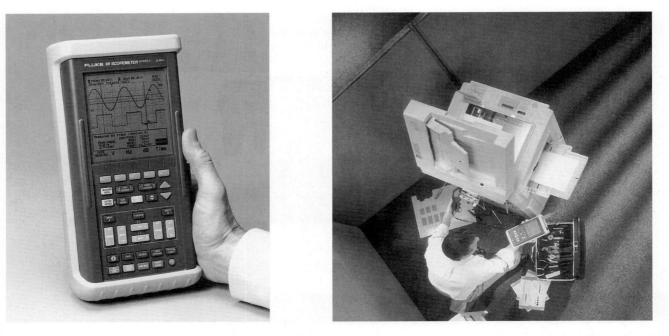

FIGURE 8-65 **The Hand-Held Scopemeter. (Courtesy of Fluke Corporation. Reproduced with permission.)**

Measure menu: The measurement menu button gives you direct access to a quick pop-up menu of more than 30 measurements including: V_{rms}, V_{mean} (arithmetic average), $V_{peak-to-peak}$, frequency, time delay, rise time, phase, current, and so on. Once the measurement is selected, the scopemeter automatically sets itself up and takes the measurement, as seen in the example in Figure 8-66(a).

Auto set: As you move from test point to test point, the scopemeter will handle the changing inputs automatically in this continuous autoset mode. Each time the signal input changes, the scopemeter will automatically search for the best trigger level, timebase (sweep time), and range scale to speed up measurements and reduce errors, as seen in Figure 8-66(b).

Min max trendplot: This function is used to display and record up to 40 days of signal trend which is sometimes needed in order to pinpoint intermittent problems that occur randomly. Figure 8-66(c) shows an example of how the display shows minimum, maximum, and average readings.

Save: Most scopemeters have large internal memories for saving screen images, setups, and waveforms from the field. These stored measurements can be recalled at any time, downloaded or sent to a personal computer for reports, or sent directly to a printer. Figure 8-66(d) shows how these waveforms and setups would appear on a computer screen.

SELF-TEST EVALUATION POINT FOR SECTION 8-5-4

Use the following questions to test your understanding of Section 8-5-4.

1. What is a scopemeter?

2. What advantages does the scopemeter have over the previously discussed instruments?

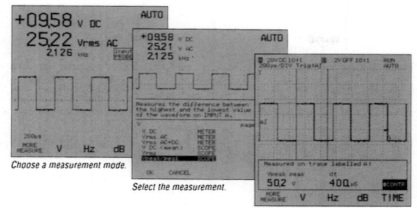

Choose a measurement mode.

Select the measurement.

The measurement is taken.

(a)

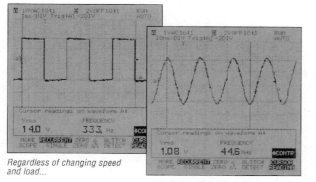

Regardless of changing speed and load...

...continuous Autoset gives a properly scaled signal display.

(b)

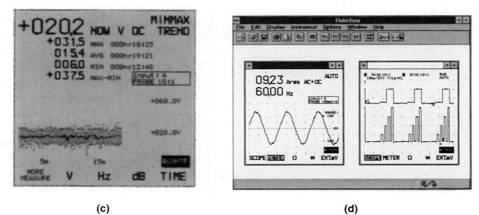

(c)

(d)

FIGURE 8-66   Scopemeter Operation. (Courtesy of Fluke Corporation. Reproduced with permission.)

SUMMARY

1. While direct current (dc) can be used in many instances to deliver power or represent information, there are certain instances in which ac is preferred. For example, ac is easier to generate and is more efficiently transmitted as a source of electrical power for the home or business. Audio (speech and music) and video (picture) information are generally always represented in electronic equipment as an alternating current or alternating voltage signal.

2. Direct current (dc) is the flow of electrons in one DIRECTion and one direction only. DC voltage is nonvarying and normally obtained from a battery or power supply unit. The only variation in voltage from a battery occurs due to the battery's discharge, but even then, the current will still flow in only one direction.

3. Some power supplies supply a form of dc known as pulsating dc, which varies periodically from zero to a maximum, back to zero, and then repeats. Pulsating dc is normally supplied by a battery charger and is used to charge secondary batteries. It is also used to operate motors that convert the pulsating dc electrical energy into a mechanical rotation output. Whether steady or pulsating, direct current is current in only one DIRECTion.

4. Alternating current (ac) flows first in one direction and then in the opposite direction. This reversing current is produced by an alternating voltage source, which reaches a maximum in one direction (positive), decreases to zero, and then reverses itself and reaches a maximum in the opposite direction (negative). During the time of the positive voltage alternation, the polarity of the voltage will cause current to flow from negative to positive in one direction. During the time of the negative voltage alternation, the polarity of the voltage will reverse, causing current to flow once again from negative to positive, but in this case, in the opposite direction.

5. There are two main applications for ac:
   a. *Power transfer:* to supply electrical power for lighting, heating, cooling, appliances, and machinery in both home and industry
   b. *Information transfer:* to communicate or carry information, such as radio music and television pictures, between two points.

6. There are three advantages that ac has over dc from a power point of view.
   a. Generators can be used to generate either dc or ac, but ac generators can be larger, less complex internally, and cheaper to operate, which is the first reason why we use ac instead of dc for supplying power.
   b. Power companies transport electric energy at a very high voltage between 200,000 and 600,000 V. Since the voltage is high, the current can be low and provide the same amount of power to the consumer. An ac voltage can easily and efficiently be transformed up or down to a higher or lower voltage by utilizing a device known as a transformer, and even though dc voltages can be stepped up and down, the method is inefficient and more complex.

   c. It is a relatively simple process to convert ac to dc, but conversion from dc to ac is a complex and comparatively inefficient process.

7. The ac power distribution system begins at the electric power plant, which has the powerful large generators driven by turbines to generate large ac voltages.

8. The turbines can be driven by either falling water (hydroelectric), or from steam, which is produced with intense heat by burning either coal, gas, or oil or from a nuclear reactor (thermoelectric).

9. The turbine supplies the mechanical energy to the generator, to be transformed into ac electrical energy.

10. The generator generates an ac voltage of approximately 22,000 V, which is stepped up by transformers to approximately 500,000 V.

11. At each city or town, the voltage is tapped off the long-distance transmission lines and stepped down to approximately 66,000 V and is distributed to large-scale industrial customers.

12. The 66,000 V is stepped down again to approximately 4800 V and distributed throughout the city or town by short-distance transmission lines. This 4800 V is used by small-scale industrial customers and residential customers who receive the ac power via step-down transformers on utility poles, which step down the 4800 V to 240 V and 120 V.

13. A large amount of equipment and devices within industry and the home will run directly from the ac power, such as heating, lighting, and cooling.

14. Some equipment that runs on dc, such as televisions and computers, will accept the 120 V ac and internally convert it to the dc voltages required.

15. The unit that converts ac to dc is called a rectifier.

16. Information, by definition, is the property of a signal or message that conveys something meaningful to the recipient.

17. Communication, which is the transfer of information between two points, began with speech and progressed to handwritten words in letters and printed words in newspapers and books. To achieve greater distances of communication, face-to-face communications evolved into telephone and radio communications.

18. Sound waves or sounds are normally generated by a vibrating reed or plucked string or a person's vocal cords.

19. Like the ripples produced by a stone falling in a pond, the sound waves are constantly expanding and traveling outward.

20. The microphone is in fact a transducer (energy converter), because it converts the sound wave (which is a form of mechanical energy) into electrical energy in the form of voltage and current, which varies in the same manner as the sound wave and therefore contains the sender's message or information.

21. The electrical wave is a variation in voltage or current and can only exist in a wire conductor or circuit. This electrical signal travels at the speed of light.

22. The speaker, like the microphone, is also an electroacoustical transducer that converts the electrical energy input into a mechanical sound-wave output.

23. To communicate between two distant points, a wire must be connected between the microphone and speaker. However, if an electrical wave is applied to an antenna, the electrical wave is converted into a radio or electromagnetic wave.

24. Antennas are designed to radiate and receive electromagnetic waves, which vary in field strength, and can exist in either air or space.

25. Radio waves are composed of two basic components. The electrical voltage applied to the antenna is converted into an electric field and the electrical current into a magnetic field.

AC Wave Shapes (Figure 8-67)

26. If the length of two sides of a right triangle are known, and the length of the third side is needed, remember that

$$C^2 = A^2 + B^2$$

27. If the angle θ is known along with the length of one side, or if angle θ is needed and the lengths of the two sides are known, one of the three formulas can be chosen based on what variables are known, and what is needed.

$$\sin \theta = \frac{O}{H} \quad \cos \theta = \frac{A}{H} \quad \tan \theta = \frac{O}{A}$$

28. The sine wave is the most common type of waveform. It is the natural output of a generator. If this ac voltage is applied across a closed circuit, it produces a current that continually reverses or alternates in each direction.

29. The amplitude or magnitude of a wave is often represented by a vector arrow. The vector's length indicates the magnitude of the current or voltage, while the arrow's point is used to show the direction, or polarity.

30. The peak of an ac wave occurs on both the positive and negative alternation, but is only at the peak (maximum) for an instant.

31. The peak-to-peak value of a sine wave is the value of voltage or current between the positive and negative maximum values of a waveform.

32. Both the positive and negative alternation of a sine wave can accomplish the same amount of work, but the ac waveform is only at its maximum value for an instant in time, spending most of its time between peak currents. The effective value of a sine wave is equal to 0.707 of the peak value.

33. This root-mean-square (rms) result of 0.707 can always be used to tell us how effective an ac sine wave will be.

34. Unless otherwise stated, ac values of voltage or current are always given in rms.

35. The average of a positive or negative alternation (half-cycle) is equal to 0.637 × peak.

36. The period (t) is the time required for one complete cycle (positive and negative alternation) of the sinusoidal current or voltage waveform. A *cycle,* by definition, is the change of an alternating wave from zero to a positive peak, to zero, then to a negative peak, and finally, back to zero.

37. Frequency is the number of repetitions of a periodic wave in a unit of time. It is symbolized by f and is given the unit hertz (cycles per second), in honor of a German physicist, Heinrich Hertz.

38. Sinusoidal waves can take a long or a short amount of time to complete one cycle. This time is related to frequency in that period is equal to the reciprocal of frequency.

39. All homes in the United States receive at their wall outlets an ac voltage of 120 V rms at a frequency of 60 Hz. This frequency was chosen for convenience, as a lower frequency would require larger transformers, and if the frequency were too low, the slow switching (alternating) current through the light bulb would cause it to flicker.

40. A higher frequency than 60 Hz was found to cause an increase in the amount of heat generated in the core of all power distribution transformers due to eddy currents and hysteresis losses.

41. Wavelength, as its name states, is the physical length of one complete cycle and is generally measured in meters. The wavelength (λ, lambda) of a complete cycle is dependent on the frequency and velocity of the transmission.

42. Electromagnetic waves or radio waves travel at the speed of light in air or a vacuum, which is 3×10^8 meters/second or 3×10^{10} cm/second.

43. Sound waves travel at a slower speed than electromagnetic waves, as their mechanical vibrations depend on air molecules, which offer resistance to the traveling wave.

44. The phase of a sine wave is always relative to another sine wave of the same frequency.

45. The sine wave is called a sine wave because it changes in amplitude at the same rate as the sine trigonometric function.

46. The square wave is a periodic (repeating) wave that alternates from a positive peak value to a negative peak value, and vice versa, for equal lengths of time.

47. The duty cycle is the ratio of a pulse width (positive or negative pulse or cycle) to the overall period or time of the wave and is normally given as a percentage.

48. Since a square wave always has a positive and a negative alternation that are equal in time, the duty cycle of all square waves is equal to 50%, which actually means that the positive cycle lasts for 50% of the time of one cycle.

49. The average of the complete square wave cycle should calculate out to be zero, as the amount above the line equals the amount below.

50. It takes a small amount of time for the wave to increase to its positive value (the rise time) and consequently an equal amount of time for a wave to decrease to its negative value

FIGURE 8-67   AC Wave Shapes.

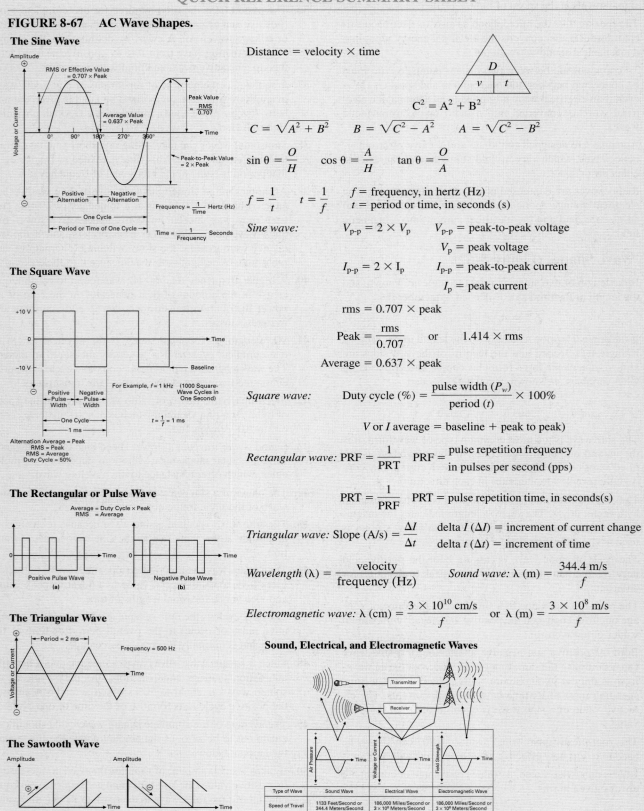

The Sine Wave

The Square Wave

The Rectangular or Pulse Wave

The Triangular Wave

The Sawtooth Wave

Distance = velocity × time

$$C^2 = A^2 + B^2$$

$$C = \sqrt{A^2 + B^2} \qquad B = \sqrt{C^2 - A^2} \qquad A = \sqrt{C^2 - B^2}$$

$$\sin \theta = \frac{O}{H} \qquad \cos \theta = \frac{A}{H} \qquad \tan \theta = \frac{O}{A}$$

$$f = \frac{1}{t} \qquad t = \frac{1}{f} \qquad \begin{aligned} &f = \text{frequency, in hertz (Hz)} \\ &t = \text{period or time, in seconds (s)} \end{aligned}$$

Sine wave: $\qquad V_{p\text{-}p} = 2 \times V_p \qquad V_{p\text{-}p} = \text{peak-to-peak voltage}$

$$V_p = \text{peak voltage}$$

$$I_{p\text{-}p} = 2 \times I_p \qquad I_{p\text{-}p} = \text{peak-to-peak current}$$

$$I_p = \text{peak current}$$

$$\text{rms} = 0.707 \times \text{peak}$$

$$\text{Peak} = \frac{\text{rms}}{0.707} \qquad \text{or} \qquad 1.414 \times \text{rms}$$

$$\text{Average} = 0.637 \times \text{peak}$$

Square wave: $\qquad \text{Duty cycle (\%)} = \dfrac{\text{pulse width } (P_w)}{\text{period } (t)} \times 100\%$

$$V \text{ or } I \text{ average} = \text{baseline} + \text{peak to peak)}$$

Rectangular wave: $\text{PRF} = \dfrac{1}{\text{PRT}} \qquad \begin{aligned} \text{PRF} = &\text{ pulse repetition frequency} \\ &\text{in pulses per second (pps)} \end{aligned}$

$$\text{PRT} = \frac{1}{\text{PRF}} \qquad \text{PRT} = \text{pulse repetition time, in seconds(s)}$$

Triangular wave: $\text{Slope (A/s)} = \dfrac{\Delta I}{\Delta t} \qquad \begin{aligned} &\text{delta } I \ (\Delta I) = \text{increment of current change} \\ &\text{delta } t \ (\Delta t) = \text{increment of time} \end{aligned}$

Wavelength $(\lambda) = \dfrac{\text{velocity}}{\text{frequency (Hz)}} \qquad$ *Sound wave:* $\lambda \text{ (m)} = \dfrac{344.4 \text{ m/s}}{f}$

Electromagnetic wave: $\lambda \text{ (cm)} = \dfrac{3 \times 10^{10} \text{ cm/s}}{f} \qquad \text{or} \qquad \lambda \text{ (m)} = \dfrac{3 \times 10^8 \text{ m/s}}{f}$

Sound, Electrical, and Electromagnetic Waves

Type of Wave	Sound Wave	Electrical Wave	Electromagnetic Wave
Speed of Travel	1133 Feet/Second or 344.4 Meters/Second	186,000 Miles/Second or 3 × 10⁸ Meters/Second	186,000 Miles/Second or 3 × 10⁸ Meters/Second

(the fall time). Rise time (T_R), by definition, is the time it takes for an edge to rise from 10% to 90% of its full amplitude, while fall time (T_F) is the time it takes for an edge to fall from 90% to 10% of its full amplitude.

51. A *periodic wave* is a wave that repeats the same wave shape from one cycle to the next.

52. A time-domain representation of a periodic sine wave, which is the same way it would appear on an oscilloscope display, plots the sine wave's amplitude against time.

53. A frequency-domain representation of the same periodic sine wave, which shows the wave as it would appear on a spectrum analyzer, plots the sine wave's amplitude against frequency instead of time. Pure sine waves have no other frequency components.

54. Other periodic wave shapes, such as square, pulse, triangular, sawtooth, or irregular, are actually made up of a number of sine waves having a particular frequency, amplitude, and phase.

55. The fundamental frequency sine wave is always the lowest frequency and largest amplitude component of any waveform shape and is used as a reference.

56. Sine wave harmonic frequencies are smaller in amplitude and are some multiple of the fundamental frequency.

57. Square waves are composed of a fundamental frequency and an infinite number of odd harmonics (third, fifth, seventh, and so on).

58. The rectangular wave is similar to the square wave in many respects, in that it is a periodic wave that alternately switches between one of two fixed values. The difference in the rectangular wave is that it does not remain at the two peak values for equal lengths of time.

59. Pulse repetition frequency is the number of times per second that a pulse is transmitted.

60. Pulse repetition time is the time interval between the start of two consecutive pulses.

61. Pulse width, pulse length, or pulse duration is the time interval between the leading edge and trailing edge of a pulse at which the amplitude reaches 50% of the peak pulse amplitude.

62. The triangular wave is a repeating wave that has equal positive and negative ramps that have linear rates of change with time.

63. The term linear describes a relationship between input and output in which the output varies in direct proportion to the input.

64. The sawtooth wave is a repeating waveform that rises from zero to a maximum value linearly and then falls to zero and repeats.

65. AC is basically used in two applications: (1) power transfer and (2) information transfer.

66. Electronic equipment manages the flow of information, while electrical equipment manages the flow of power.

67. A dc power supply is a piece of electrical equipment since it is designed to manage the flow of power. A TV set, however, is an electronic system since its electronic circuits are designed to manage the flow of audio (sound) and video (picture) information.

Measuring and Generating AC Signals (Figure 8-68)

68. Test equipment can be used to either sense a circuit's condition or generate a signal to see the response to the component or circuit to that signal.

69. Most multimeters can be used to measure either ac or dc.

70. When multimeters are selected to measure ac, an internal rectifier is used to convert the ac input into a dc voltage.

71. Multimeters are normally calibrated to indicate rms values of the ac being measured.

72. A current clamp allows the technician or engineer to measure ac current without opening the current path.

73. The RF probe can be used with the multimeter to more accurately measure higher frequencies above 2 kHz.

74. The high-voltage probe can be used to measure voltages in the kilovolt range by connecting additional multiplier resistors.

75. The analog multimeter unlike the DMM can be used to measure low frequency ac.

76. The digital multimeter is generally more popular because of its easy-to-read display and accuracy.

77. The oscilloscope displays the shape and spacing of electrical signals, and can therefore be used to measure dc voltage, ac voltage, waveform duration, waveform frequency, and waveform shape.

78. The oscilloscope can be used to display waveform shapes, and from this presentation we can calculate the waveform's time, frequency, and amplitude characteristics.

79. A dual-trace oscillocope can produce two traces or waveforms on the screen, which allows the technician or engineer to make comparisons between phase, amplitude, shape, or timing.

80. The function generator can produce a sine wave, square wave, rectangular wave, trangular wave, or sawtooth wave.

81. The frequency counter measures and displays the number of cycles per second (hertz) on a digital display.

FIGURE 8-68 **Making Measurements with the Digital Multimeter (DMM).**

Measuring Current

How to Make Current Measurements

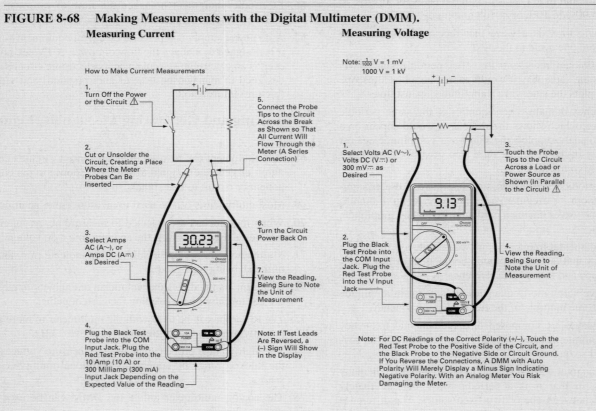

1.
Turn Off the Power
or the Circuit ⚠

2.
Cut or Unsolder the
Circuit, Creating a Place
Where the Meter
Probes Can Be
Inserted

3.
Select Amps
AC (A∼), or
Amps DC (A⎓)
as Desired

4.
Plug the Black Test
Probe into the COM
Input Jack. Plug the
Red Test Probe into the
10 Amp (10 A) or
300 Milliamp (300 mA)
Input Jack Depending on the
Expected Value of the Reading

5.
Connect the Probe
Tips to the Circuit
Across the Break
as Shown so That
All Current Will
Flow Through the
Meter (A Series
Connection)

6.
Turn the Circuit
Power Back On

7.
View the Reading,
Being Sure to Note
the Unit of
Measurement

Note: If Test Leads
Are Reversed, a
(–) Sign Will Show
in the Display

Measuring Voltage

Note: $\frac{1}{1000}$ V = 1 mV
1000 V = 1 kV

1.
Select Volts AC (V∼),
Volts DC (V⎓) or
300 mV⎓ as
Desired

2.
Plug the Black
Test Probe into
the COM Input
Jack. Plug the
Red Test Probe
into the V Input
Jack

3.
Touch the Probe
Tips to the Circuit
Across a Load or
Power Source as
Shown (In Parallel
to the Circuit) ⚠

4.
View the Reading,
Being Sure to
Note the Unit of
Measurement

Note: For DC Readings of the Correct Polarity (+/–), Touch the
Red Test Probe to the Positive Side of the Circuit, and
the Black Probe to the Negative Side or Circuit Ground.
If You Reverse the Connections, A DMM with Auto
Polarity Will Merely Display a Minus Sign Indicating
Negative Polarity. With an Analog Meter You Risk
Damaging the Meter.

Measuring Resistance

Note: 1,000 Ω = 1 kΩ
1,000,000 Ω = 1 MΩ

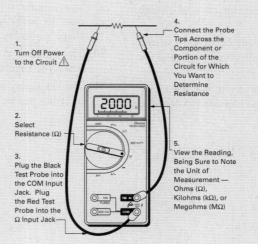

1.
Turn Off Power
to the Circuit ⚠

2.
Select
Resistance (Ω)

3.
Plug the Black
Test Probe into
the COM Input
Jack. Plug
the Red Test
Probe into the
Ω Input Jack

4.
Connect the Probe
Tips Across the
Component or
Portion of the
Circuit for Which
You Want to
Determine
Resistance

5.
View the Reading,
Being Sure to Note
the Unit of
Measurement —
Ohms (Ω),
Kilohms (kΩ), or
Megohms (MΩ)

Multiple-Choice Questions

1. Trigonometry is the study of:
   a. Triangles
   b. Angles
   c. Trigonometric functions
   d. All of the above

2. The Pythagorean theorem states that the _____ of the length of the hypotenuse of a right-angle triangle equals the sum of the _____ of the lengths of the other two sides.
   a. Square, square roots
   b. Square root, squares
   c. Square, squares
   d. Square root, square roots

3. A vector is an arrow whose length is used to represent the _____ of a quantity and whose point is used to indicate the same quantity's _____ .
   a. Magnitude, size
   b. Magnitude, direction
   c. Direction, size
   d. Direction, magnitude

4. If two vectors in a vector diagram are not working together, or are out of phase with one another, they must be _____ to obtain a _____ vector.
   a. Added, resultant
   b. Multiplied by one another, parallel
   c. Vectorially added, resultant
   d. Both (a) and (b)

5. The _____ of a right-angle triangle is always the longest side.
   a. Opposite side
   b. Adjacent side
   c. Hypotenuse
   d. Angle theta

6. The _____ of a right-angle triangle is always across from the angle theta.
   a. Opposite side
   b. Adjacent side
   c. Hypotenuse
   d. Angle theta

7. The _____ of a right-angle triangle is always formed between the hypotenuse and the adjacent side.
   a. Opposite side
   b. Adjacent side
   c. Hypotenuse
   d. Angle theta

8. Which trigonometric function can be used to determine the length of the hypotenuse if the angle theta and the length of the opposite side are known?
   a. Tangent
   b. Sine
   c. Cosine

9. Which trigonometric function can be used to determine the angle theta when the lengths of the opposite and adjacent sides are known?
   a. Tangent
   b. Sine
   c. Cosine

10. Which trigonometric function can be used to determine the length of the adjacent side when the angle theta and the length of the hypotenuse are known?
    a. Tangent
    b. Sine
    c. Cosine

11. A current that rises from zero to maximum positive, returns to zero, and then repeats is known as:
    a. Alternating current   c. Pulsating dc
    b. AC   d. Steady dc

12. A current that rises from zero to maximum positive, decreases to zero, and then reverses to reach a maximum in the opposite direction (negative) is known as:
    a. Alternating current   c. Steady direct current
    b. Pulsating direct current   d. All of the above

13. The advantage(s) of ac over dc from a power distribution point of view is/are:
    a. Generators can supply more power than batteries
    b. AC can be transformed to a high or low voltage easily, minimizing power loss
    c. AC can easily be converted into dc
    d. All of the above
    e. Only (a) and (c)

14. The approximate voltage appearing on long-distance transmission lines in the ac distribution system is:
    a. 250 V   c. 500,000 V
    b. 2500 V   d. 250,000 V

15. The most common type of alternating wave shape is the:
    a. Square wave   c. Rectangular wave
    b. Sine wave   d. Triangular wave

16. _____ equipment manages the flow of information.
    a. Electronic   c. Discrete
    b. Electrical   d. Integrated

17. _____ equipment manages the flow of power.
    a. Electronic   c. Discrete
    b. Electrical   d. Integrated

18. The peak-to-peak value of a sine wave is equal to:
   a. Twice the rms value
   b. 0.707 times the rms value
   c. Twice the peak value
   d. 1.14 × the average value

19. The rms value of a sine wave is also known as the:
   a. Effective value   c. Peak value
   b. Average value   d. All of the above

20. The peak value of a 115 V (rms) sine wave is:
   a. 115 V   c. 162.7 V
   b. 230 V   d. Two of the above could be true.

21. The mathematical average value of a sine wave cycle is:
   a. 0.637 × peak   c. 1.414 × rms
   b. 0.707 × peak   d. Zero

22. The frequency of a sine wave is equal to the reciprocal of _____.
   a. The period   d. Both (a) and (b)
   b. One cycle time   e. None of the above
   c. One alternation

23. What is the period of a 1 MHz sine wave?
   a. 1 ms   c. 10 ms
   b. One millionth of a second   d. 100 μs

24. The sine of 90° is:
   a. 0   c. 1
   b. 0.5   d. Any of the above

25. What is the frequency of a sine wave that has a cycle time of 1 ms?
   a. 1 MHz   c. 200 m
   b. 1 kHz   d. 10 kHz

26. The pulse width (P_w) is the time between the _____ points on the positive and negative edges of a pulse.
   a. 10%   c. 50%
   b. 90%   d. All of the above

27. The duty cycle is the ratio of _____ to period.
   a. Peak   c. Pulse length
   b. Average power   d. Both (a) and (c)

28. With a pulse waveform, PRF can be calculated by taking the reciprocal of:
   a. The duty cycle   c. P_d
   b. PRT   d. P_l

29. The sound wave exists in _____ and travels at approximately _____.
   a. Space, 1130 ft/s   c. Air, 3 × 106 m/s
   b. Wires, 186,282.397 miles/s   d. None of the above

30. The electrical and electromagnetic waves travel at a speed of:
   a. 186,000 miles/s   c. 162,000 nautical miles/s
   b. 3 × 10⁸ meters/s   d. All of the above

Communication Skill Questions

31. What is trigonometry?   (Introduction)

32. Describe the Pythagorean theorem. (8-1)

33. Describe how $c^2 = a^2 + b^2$ can be transposed to solve for a, b, and c. (8-1)

34. What is a vector, and what does it represent? (8-1)

35. Sketch a right triangle and indicate the adjacent and opposite sides and the hypotenuse. (8-2)

36. State and explain the purpose of the sine of theta formula. (8-2)

37. How would you transpose the sine of theta formula to determine the angle theta if the lengths of the two sides are known?   (8-2)

38. What is the difference between the sine and the inverse sine function? (8-2)

39. State the cosine of theta formula, and explain its function. (8-2)

40. State the tangent of theta formula, and explain its function. (8-2)

41. Describe the three advantages that ac has over dc from a power point of view. (8-3-1)

42. Describe briefly the ac power distribution system. (8-3-1)

43. What is the difference between electricity and electronics? (8-5)

44. What are the six basic ac information wave shapes? (8-4)

45. Describe briefly the following terms as they relate to the sine wave: (8-4-1)
   a. RMS   d. Average   g. Period
   b. Peak   e. The name sine   h. Wavelength
   c. Peak to peak   f. Frequency   i. Phase

46. Describe briefly the following terms as they relate to the square wave: (8-4-2)
   a. Duty cycle   b. Average

47. Describe briefly the following terms as they relate to the rectangular wave: (8-4-3)
   a. PRT   c. Duty cycle
   b. PRF   d. Average

48. Briefly describe the meaning of the terms *fundamental frequency* and *harmonics*. (8-4-2)

49. List and describe all the pertinent information relating to the following information carriers: (8-3-2)
   a. Sound wave   c. Electromagnetic wave
   b. Electrical wave

50. Describe the difference between frequency- and time-domain analysis. (8-4-2)

Practice Problems

51. Calculate the length of the hypotenuse (C side) of the following right triangles.
   a. $A = 20$ mi, $B = 53$ mi
   b. $A = 2$ km, $B = 3$ km
   c. $A = 4$ in., $B = 3$ in.
   d. $A = 12$ mm, $B = 12$ mm

52. Calculate the length or magnitude of the resultant vectors in the vector diagrams at the top of the next page.

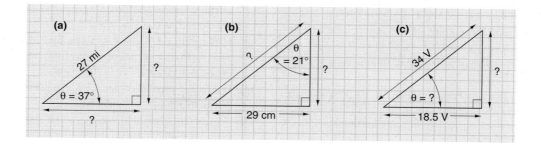

53. Calculate the value of the following trigonometric functions.

  a. sin 0°   **i.** cos 60°
  b. sin 30°   **j.** cos 90°
  c. sin 45°   **k.** tan 0°
  d. sin 60°   **l.** tan 30°
  e. sin 90°   **m.** tan 45°
  f. cos 0°   **n.** tan 60°
  g. cos 30°   **o.** tan 90°
  h. cos 45°

54. Calculate angle θ from the function given.

  a. $\sin \theta = 0.707$, $\theta = ?$
  b. $\sin \theta = 0.233$, $\theta = ?$
  c. $\cos \theta = 0.707$, $\theta = ?$
  d. $\cos \theta = 0.839$, $\theta = ?$
  e. $\tan \theta = 1.25$, $\theta = ?$
  f. $\tan \theta = 0.866$, $\theta = ?$

55. In the figure at the bottom of the page, name each of the triangle's sides, and calculate the unknown values.

56. Calculate the periods of the following sine-wave frequencies:

  a. 27 kHZ   **d.** 365 Hz
  b. 3.4 MHz   **e.** 60 Hz
  c. 25 Hz   **f.** 200 kHz

57. Calculate the frequency for each of the following values of time:

  a. 16 ms   **d.** 0.05 s
  b. 1 s   **e.** 200 μs
  c. 15 μs   **f.** 350 ms

58. A 22 V peak sine wave will have the following values:
  a. Rms voltage =   **c.** Peak-to-peak voltage =
  b. Average voltage =

59. A 40 mA rms sine wave will have the following values:
  a. Peak current =   **c.** Average current =
  b. Peak-to-peak current =

60. How long would it take an electromagnetic wave to travel 60 miles?

61. An 11 kHz rectangular pulse, with a pulse width of 10 μs, will have a duty cycle of _____%.

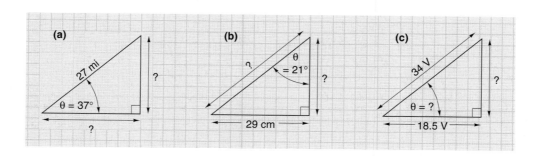

62. Calculate the PRT of a 400 kHz pulse waveform.

63. Calculate the average current of the pulse waveform in Question 61 if its peak current is equal to 15 A.

64. What is the duty cycle of a 10 V peak square wave at a frequency of 1 kHz?

65. Considering a fundamental frequency of 1 kHz, calculate the frequency of its:
   a. Third harmonic   **c.** Seventh harmonic
   b. Second harmonic

66. If one cycle of a sine wave occupies 4 cm on the oscilloscope horizontal grid and 5 cm from peak to peak on the vertical grid, calculate frequency, period, rms, average, and peak for the following control settings:
   a. 0.5 V/cm, 20 μs/cm   **c.** 50 mV/cm, 0.2 μs/cm
   b. 10 V/cm, 10 ms/cm

67. Assuming the same graticule and switch settings of the oscilloscope in Figure 8-9, what would be the lowest setting of the volts/cm and time/division switches to fully view a 6 V rms, 350 kHz sine wave?

68. If the volts/cm switch is positioned to 10 V/cm and the waveform extends 3.5 divisions from peak to peak, what is the peak-to-peak value of this wave?

69. If a square wave occupies 5.5 horizontal cm on the 1 μs/cm position, what is its frequency?

70. Which settings would you use on an autoset scopemeter to fully view an 8 V rms, 20 kHz triangular wave?

Web Site Questions

Go to the Web site http://www.prenhall.com/cook, select the textbook *Introductory DC/AC Electronics* or *Introductory DC/AC Circuits*, this chapter, and then follow the instructions when answering the multiple-choice practice problems.

These tests at the end of each chapter will challenge your knowledge up to this point, and give you the practice you need for a job interview. To make this more realistic, the test will comprise both technical and personal questions. In order to take full advantage of this exercise, you may like to set up a simulation of the interview environment, have a friend read the questions to you, and record your responses for later analysis.

Company Name: Free-2-Roam, Inc.

Industry Branch: Communications

Function: Quality testing wireless products

Job Appling For: QA Technician

1. Would you say that you have good written and oral communcation skills?
2. What is ac?

3. What is the difference between the ac waves we receive at the wall outlet, and the ac waves we receive with our car radio?
4. How are your team work skills?
5. What is wireless communication?
6. Can you tell me the differences between a sound wave and an electromagnetic wave?
7. How easily do you feel you could adapt to new technologies?
8. What is a sine wave?
9. What is the duty cycle of a pulse wave?
10. What test instrument would you use to measure the frequency domain of a waveform?

Answers

1. Your resume and cover letter were heavily scrutinized in order to evaluate your written communication skills, and they must have passed the test, otherwise you would not have been granted an interview. Your oral communication skills are now being tested, and it is important that all of your answers are clear and concise. You more than likely practiced communication skills during your training, and so quote these examples and any previous job related experience.
2. Section 8-2.
3. Section 8-3.
4. The reason for this question lies in industry's drive for a high-skill, high-quality, high-performance organization, and reduced cost. In order to achieve this goal, the work environment has changed, resulting in the integration of electronic system design, manufacture, and service. There is a need, therefore, for a technician to have better teamwork skills. Teamwork is defined in the dictionary as "combined effort or organized co-operation." To operate in a team environment, therefore, a person will need good communication and interpersonal skills. Emphasize your lab-team training and any previous job-related experience.
5. Section 8-3-2.
6. Section 8-3-2.
7. Technology, especially electronics technology, seems to advance in leaps and bounds, and so there will always be a need for you to be able to independently learn and adapt to the new changes. Your math skills, logic and reasoning skills, learning skills, and relationship skills develop intelligence, which in turn develops better adaptability, organizational and problem solving skills. Success in these areas will ensure career advancement. Discuss your training in each of these areas, and if you feel the same way I do, let them know that your interest in electronics is both personal and professional, and so there is a strong desire on your part to understand and embrace the advances.
8. Section 8-4-1.
9. Section 8-4-3.
10. Section 8-4-2.

Capacitance and Capacitors

BACK TO THE FUTURE

Born in England in 1791, Charles Babbage became very well known for both his mathematical genius and eccentric personality. Babbage's ultimate pursuit was that of mathematical accuracy. He delighted in spotting errors in everything from log tables (used by astronomers, mathematicians, and navigators) to poetry. In fact, he once wrote to poet Alfred Lord Tennyson, pointing out an inaccuracy in his line "Every moment dies a man—every moment one is born." Babbage explained to Tennyson that since the world population was actually increasing and not, as he indicated, remaining constant, the line should be rewritten to read "Every moment dies a man—every moment one and one-sixteenth is born."

In 1822, Babbage described in a paper and built a model of what he called "a difference engine," which could be used to calculate mathematical tables. The Royal Society of Scientists described his machine as "highly deserving of public encouragement," and a year later the government awarded Babbage £1500 for his project. Babbage originally estimated that the project should take 3 years; however, the design had its complications, and after 10 years of frustrating labor, in which the government grants increased to £17,000, Babbage was still no closer to completion. Finally, the money stopped and Babbage reluctantly decided to let his brainchild go.

In 1833, Babbage developed an idea for a much more practical machine, which he named "the analytical engine." It was to be a more general machine that could be used to solve a variety of problems, depending on instructions supplied by the operator. It would include two units, called a "mill" and a "store," both of which would be made of cogs and wheels. The store, which was equivalent to a modern-day computer memory, could hold up to 100 forty-digit numbers. The mill, which was equivalent to a modern computer's arithmetic and logic unit (ALU), could perform both arithmetic and logic operations on variables or numbers retrieved from the store, and the result could be stored in the store and then acted upon again or printed out. The program of instructions directing these operations would be fed into the analytical engine in the form of punched cards.

The analytical engine was never built. All that remains are the volumes of descriptions and drawings, and a section of the mill and printer built by Babbage's son, who also had to concede defeat. It was, unfortunately for Charles Babbage, a lifetime of frustration to have conceived the basic building blocks of the modern computer a century before the technology existed to build it.

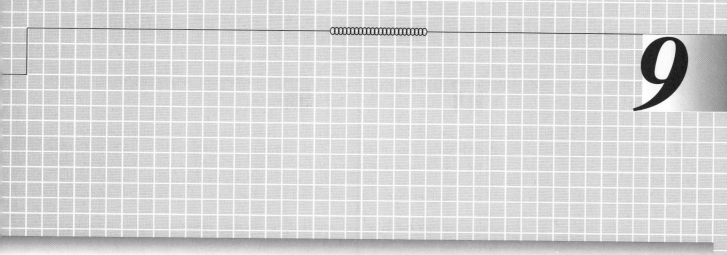

Outline and Objectives

Introduction

Up to this point we have concentrated on circuits containing only resistance, which opposes the flow of current and then converts or dissipates power in the form of heat. Capacitance and inductance are two circuit properties that act differently from resistance in that they will charge or store the supplied energy and then return almost all the stored energy back to the circuit, rather than lose it in wasted heat. Inductance will be discussed in a following chapter.

Capacitance is the ability of a circuit or device to store electrical charge. A device or component specifically designed to have this capacity or capacitance is called a *capacitor.* A capacitor stores an electrical charge similar to a bucket holding water. Using the analogy throughout this chapter that a capacitor is similar to a bucket will help you gain a clear understanding of the capacitor's operation. For example, the capacitor holds charge in the same way that a bucket holds water. A larger capacitor will hold more charge and will take longer to charge, just as a larger bucket will hold more water and take longer to fill. A larger circuit resistance means a smaller circuit current, and therefore a longer capacitor charge time. Similarly, a smaller hose will have a greater water resistance producing a smaller water flow, and therefore the bucket will take a longer time to fill. Capacitors store electrons, and basically, the amount of electrons stored is a measure of the capacitor's capacitance.

Capacitor

Device that stores electric energy in the form of an electric field that exists within a dielectric (insulator) between two conducting plates each of which is connected to a lead. This device was originally called a condensor.

9-1   CAPACITOR CONSTRUCTION

Several years ago, capacitors were referred to as condensers, but that term is very rarely used today. Figure 9-1 illustrates the main parts and schematic symbol of the **capacitor.** Two leads are connected to two parallel metal conductive plates, which are separated by an insulating material known as a *dielectric.* The conductive plates are normally made of metal foil, while the dielectric can be paper, air, glass, ceramic, mica, or some other form of insulator.

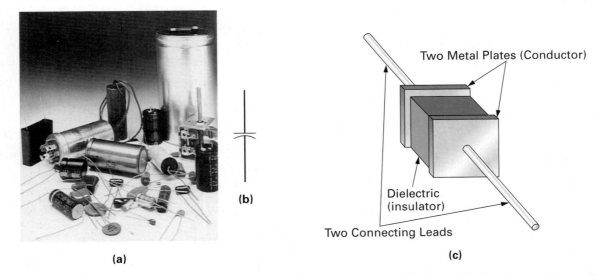

(a)

(b)

Two Metal Plates (Conductor)

Dielectric (insulator)

Two Connecting Leads

(c)

FIGURE 9-1   Capacitor. (a) Physical Appearance. (Courtesy of Sencore) (b) Schematic Symbol. (c) Basic Construction.

Now that you have completed this section, you should be able to:

■ **Objective 1.** *Define the term* capacitance.

■ **Objective 2.** *Describe the basic capacitor construction.*

Use the following questions to test your understanding of Section 9-1.

1. What are the principal parts of a capacitor?
2. What is the name given to the insulating material between the two conductive plates?

9-2   CHARGING AND DISCHARGING A CAPACITOR

Like a secondary battery, a capacitor can be made to charge or discharge, and when discharging it will return all of the energy it consumed during charge.

9-2-1   *Charging a Capacitor*

Capacitance is the ability of a capacitor to store an electrical charge. Figure 9-2 illustrates how the capacitor stores an electric charge. This capacitor is shown as two plates with air acting as the dielectric.

In Figure 9-2(a), the switch is open, so no circuit current results. An equal number of electrons exist on both plates, so the voltmeter (VM) indicates zero, which means no potential difference exists across the capacitor.

In Figure 9-2(b), the switch is now closed and electrons travel to the positive side of the battery, away from the right-hand plate of the capacitor. This creates a positive right-hand capacitor plate, which results in an attraction of free electrons from the negative side of the battery to the left-hand plate of the capacitor. In fact, for every electron that leaves the right-hand capacitor plate and is attracted into the positive battery terminal, another electron leaves the negative side of the battery and travels to the left-hand plate of the capacitor. Current appears to be flowing from the negative side of the battery, around to the positive, through the capacitor. This is not really the case, because no electrons can flow through the insulator or gap between the plates, and although there appears to be one current flowing throughout the circuit, there are in fact two separate currents—one from the battery to the capacitor and the other from the capacitor to the battery.

A voltmeter across the capacitor will indicate an increase in the potential difference between the plates, and the capacitor is said to be charging toward, in this example, 5 V. This potential difference builds up across the two plates until the voltage across the capacitor is equal to the voltage of the battery. In this example, when the capacitor reaches a charge of 5 V, the capacitor will be equivalent to a 5 V battery, as seen in Figure 9-2(c). Once charged, there will be no potential difference between the battery and capacitor, and so circuit current flow will be zero when the capacitor is fully charged.

9-2-2   *Discharging a Capacitor*

If the capacitor is now disconnected from the circuit by opening switch 1, as shown in Figure 9-3(a), it will remain in its charged condition. If switch 2 is now closed, a path exists across the charged capacitor, as shown in Figure 9-3(b), and the excess of electrons on the left plate will flow through the conducting wire to the positive plate on the right side. The capacitor is now said to be *discharging.* When an equal number of electrons exists on both

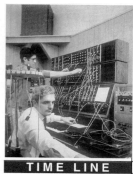

TIME LINE

Robert Moog (1934-   ), a self-employed engineer of Trumansburg, New York, built an analog music synthesizer that allowed the operator to control many tones simultaneously and rapidly alter tones during performances.

Synthesizers, devices used to generate a number of frequencies, had been around for a number of years, but electronic music did not find wide acceptance with the popular music audience until a young musician named Walter Carlos used a Moog instrument to produce a totally synthesized hit record. The album *Switched-On Bach,* released in 1968, was an interpretation of music by eighteenth-century composer Johann Sebastian Bach, and both the album and instrument were such breakthroughs in electronic music that in no time at all composers were writing a wide variety of original music specifically for Moog synthesizers.

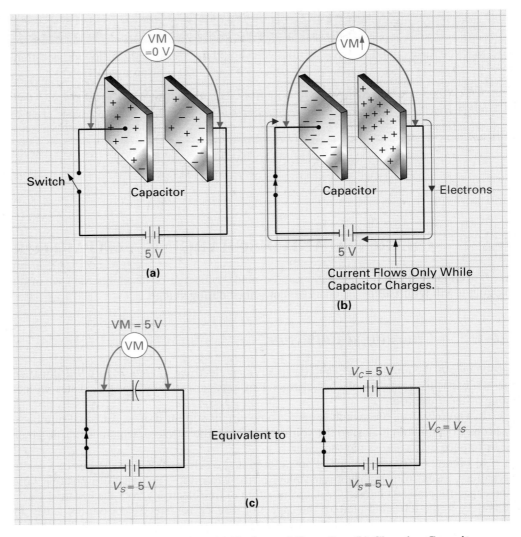

FIGURE 9-2 Charging a Capacitor. (a) Uncharged Capacitor. (b) Charging Capacitor. (c) Charged Capacitor.

sides, the capacitor is said to be *discharged* since both plates have an equal charge and the potential difference across the capacitor will be zero, as shown in Figure 9-3(c).

Figure 9-4(a) and (b) illustrate the capacitor's charge and discharge currents, which flow in opposite directions. In both cases the current flow is always from one plate to the other and never exists through the dielectric insulator.

SELF-TEST EVALUATION POINT FOR SECTION 9-2

Now that you have completed this section, you should be able to:

■ **Objective 3.** *Explain the charging and discharging process and its relationship to electrostatics.*

Use the following questions to test your understanding of Section 9-2.

1. True or false: When both plates of a capacitor have an equal charge, the capacitor is said to be charged.

2. If 2 μA of current flows into one plate of a capacitor and 2 μA flows out of the other plate, how much current is flowing through the dielectric of this working capacitor?

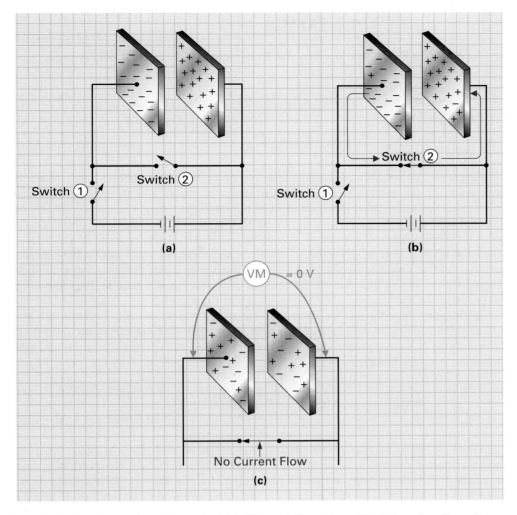

FIGURE 9-3 **Discharging a Capacitor. (a) Charged Capacitor. (b) Discharging Capacitor. (c) Discharged Capacitor.**

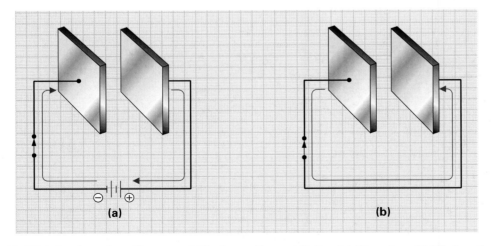

FIGURE 9-4 **Capacitor Charge and Discharge Current Paths. (a) Charging. (b) Discharging.**

9-3 ELECTROSTATICS

Electrostatic or Electric Field

Force field produced by static electrical charges. Also called a voltage field, it is a field or force that exists in the space between two different potentials or voltages.

Just as a magnetic field is produced by the flow of current, an **electric field** is produced by voltage.

Current generates a magnetic field

Voltage generates an electric field

Figure 9-5 illustrates an example capacitor circuit with the capacitor charged and the switch open. In this condition, the capacitor retains its charge and an invisible electric or electrostatic (voltage) field will be produced by nonmoving or static electrical (electrostatic) charges of different polarities. You will remember that like charges repel and unlike charges attract. Invisible electrostatic lines of flux or force can be illustrated to show this electrostatic force of attraction or repulsion. These lines are polarized away from a positive electrostatic (stationary electrical) charge and toward the negative electrostatic charge, as shown in Figure 9-6(a). If two like charges are in close proximity to one another, the electrostatic lines organize themselves into a pattern, as shown in Figure 9-6(b).

A charged capacitor has an electric or electrostatic field existing between the positively charged and negatively charged plates. The **strength of the electrostatic field** is proportional to the charge or potential difference on the plates and inversely proportional to the distance between the plates.

Field Strength

The strength of an electric, magnetic, or electromagnetic field at a given point.

$$\text{field strength (V/m)} = \frac{\text{charge difference } (V), \text{ volts}}{\text{distance between plates } (d), \text{ meters}}$$

The dielectric or insulator between the plates, like any other material, has its own individual atoms, and although the dielectric electrons are more tightly bound to their atoms than conductor electrons, stresses are placed on the atoms within the dielectric, as seen in Figure 9-7. The electrons in orbit around the dielectric atoms are displaced or distorted by the electric field existing between the positive and negative plate. If the charge potential across the capacitor is high enough and the distance between the plates is small enough, the attraction and repulsion exerted on the dielectric atom can be large enough to free the

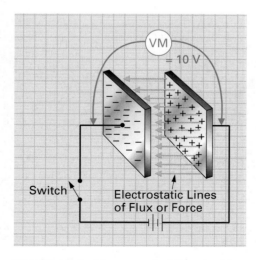

FIGURE 9-5 Electrostatic (Electric) Field between the Plates of a Charged Capacitor.

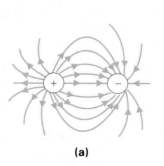

(a)

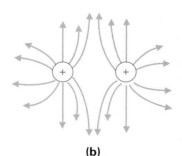

(b)

FIGURE 9-6 Electrostatic Field. (a) Electrostatic Lines of Attraction. (b) Electrostatic Lines of Repulsion.

dielectric atom's electrons. The material then becomes ionized, and a chain reaction of electrons jumping from one atom to the next in a right-to-left movement occurs. If this occurs, a large number of electrons will flow from the negative to the positive plate, and the dielectric is said to have broken down. This situation occurs if the capacitor is placed in a circuit where the voltages within the circuit exceed the voltage rating of the capacitor.

If the voltage rating of the capacitor is not exceeded, an electrostatic or electric field still exists between the two plates and causes this pulling of the atom's electrons within the dielectric toward the positive plate. This displacement, known as **electric polarization,** is similar to the pulling-back effect on a bow, as shown in Figure 9-7(b). When the capacitor is given a path for discharge, as shown in Figure 9-7(c), the electric field in the dielectric, which is causing the distortion, is the force field that drives the electrons, like the bow drives the arrow.

To summarize electrostatics and capacitors, the charges on the plates of a capacitor produce an electric field, the electric field causes the distortion of the atoms known as *electric polarization,* and this pulling back or distortion, which is held there by the electric field, is the electron moving force (emf) that drives the electrons when a discharge path is provided. The energy in a capacitor is actually stored in the electric or electrostatic field within the dielectric.

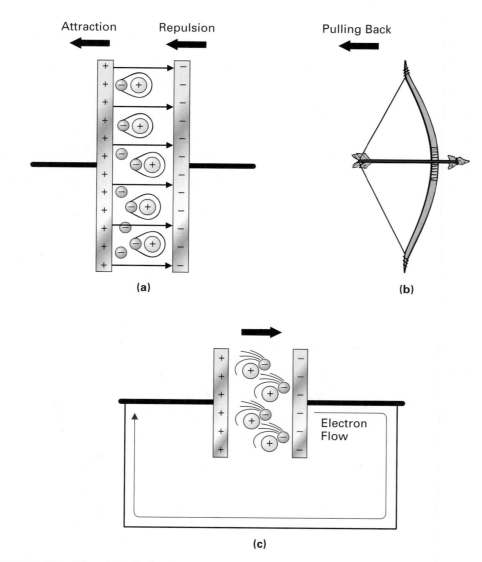

FIGURE 9-7   Electric Polarization.

Dielectric

Insulating material between two (di) plates in which the electric field exists.

Now that we understand these points, we can see where the word **dielectric** comes from. The dielectric is the insulating material that exists between two (di) plates and undergoes electric polarization when an electric field exists within it (dielectric).

Use the following questions to test your understanding of Section 9-3.

1. A/an _____ field is generated by the flow of current, while a/an _____ field is produced by voltage.

   a. Electric, magnetic

   b. Magnetic, electric

2. State the formula for calculating field strength.
3. What is electric polarization?
4. Describe why the term *dielectric* is used to indicate the insulating material between the plates of a capacitor.

9-4 THE UNIT OF CAPACITANCE

Farad

Unit of capacitance.

Capacitance is the ability of a capacitor to store an electrical charge, and the unit of capacitance is the **farad** (F), named in honor of Michael Faraday's work in 1831 in the field of capacitance. A capacitor with the capacity of 1 farad (1 F) can store 1 coulomb of electrical charge (6.24×10^{18} electrons) if 1 volt is applied across the capacitor's plates, as seen in Figure 9-8.

A 1 F capacitor is a very large value and not frequently found in electronic equipment. Most values of capacitance found in electronic equipment are in the units between the microfarad ($\mu F = 10^{-6}$) and picofarad ($pF = 10^{-12}$). A microfarad is 1 millionth of a farad (10^{-6}). So if a 1 F capacitor can store 6.24×10^{18} electrons with 1 V applied, a 1 μF capacitor, which has 1 millionth the capacity of a 1 F capacitor, can store only 1 millionth of a coulomb, or $(6.24 \times 10^{18}) \times (1 \times 10^{-6}) = 6.24 \times 10^{12}$ electrons when 1 V is applied, as shown in Figure 9-9.

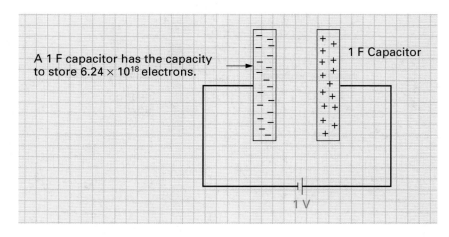

A 1 F capacitor has the capacity to store 6.24×10^{18} electrons.

1 F Capacitor

1 V

FIGURE 9-8   One Farad of Capacitance.

A 1 µF capacitor has the capacity to store one millionth of a coulomb, or 6.24×10^{12} electrons, at one volt.

1 µF Capacitor

1 V

FIGURE 9-9   One-Millionth of a Farad.

EXAMPLE:

Convert the following to either microfarads or picofarads (whichever is more appropriate):

   a. 0.00002 F
   b. 0.00000076 F
   c. 0.00047×10^{-7} F

Solution:

   a. 20 µF
   b. 0.76 µF
   c. 47 pF

Since there is a direct relationship between capacitance, charge, and voltage, there must be a way of expressing this relationship in a formula.

$$\text{capacitance, } C \text{ (farads)} = \frac{\text{charge, } Q \text{ (coulombs)}}{\text{voltage, } V \text{ (volts)}}$$

where C = capacitance, in farads
Q = charge, in coulombs
V = voltage, in volts

By transposition of the formula, we arrive at the following combinations for the same formula:

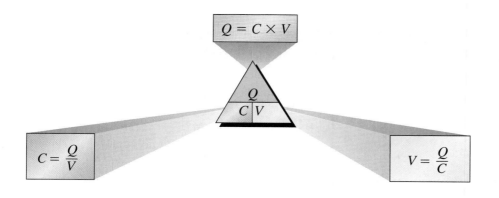

$$Q = C \times V$$

$$\frac{Q}{C \mid V}$$

$$C = \frac{Q}{V}$$

$$V = \frac{Q}{C}$$

EXAMPLE:

If a capacitor has the capacity to hold 36 C ($36 \times 6.24 \times 10^{18} = 2.25 \times 10^{20}$ electrons) when 12 V is applied across its plates, what is the capacitance of the capacitor?

Solution:

$$C = \frac{Q}{V}$$
$$= \frac{36 \text{ C}}{12 \text{ V}}$$
$$= 3 \text{ F}$$

EXAMPLE:

How many electrons could a 3 µF capacitor store when 5 V is applied across it?

Solution:

$$Q = C \times V$$
$$= 3 \text{ µF} \times 5 \text{ V}$$
$$= 15 \text{ µC}$$

(15 microcoulombs is 15 millionths of a coulomb.) Since $1 \text{ C} = 6.24 \times 10^{18}$ electrons, $15 \text{ µC} = (15 \times 10^{-6}) \times 6.24 \times 10^{18} = 9.36 \times 10^{13}$ electrons.

EXAMPLE:

If a capacitor of 2 F has stored 42 C of charge (2.63×10^{20} electrons), what is the voltage across the capacitor?

Solution:

$$V = \frac{Q}{C}$$
$$= \frac{42 \text{ C}}{2 \text{ F}}$$
$$= 21 \text{ V}$$

SELF-TEST EVALUATION POINT FOR SECTION 9-4

Now that you have completed this section, you should be able to:

■ **Objective 4.** *State the unit of capacitance and explain how it relates to charge and voltage.*

Use the following questions to test your understanding of Section 9-4.

1. What is the unit of capacitance?
2. State the formula for capacitance in relation to charge and voltage.
3. Convert 30,000 µF to farads.
4. If a capacitor holds 17.5 C of charge when 9 V is applied, what is the capacitance of the capacitor?

9-5　FACTORS DETERMINING CAPACITANCE

The capacitance of a capacitor is determined by three factors:

1. The plate area of the capacitor
2. The distance between the plates
3. The type of dielectric used

Let's now discuss these three factors in more detail, beginning with the plate area.

9-5-1　*Plate Area (A)*

The capacitance of a capacitor is directly proportional to the plate area. This area in square centimeters is the area of only one plate and is calculated by multiplying length by width. This is illustrated in Figure 9-10(a) and (b). In these two examples, the (b) capacitor plate is twice as large as the (a) capacitor plate, and since capacitance is proportional to plate area ($C \propto A$), the capacitor in example (b) will have double the capacity or capacitance of the capacitor in example (a). Since the energy of a charged capacitor is in the electric field between the plates and the plates of the capacitor (b) are double those of (a), there is twice as much area for the electric field to exist, and this doubles the capacitor's capacitance.

9-5-2　*Distance between the Plates (d)*

The distance or separation between the plates is dependent on the thickness of the dielectric used. The capacitance of a capacitor is inversely proportional to this distance between the plates, in that an increase in the distance ($d\uparrow$) causes a decrease in the capacitor's capacitance ($C\downarrow$). In Figure 9-11(a), a large distance between the capacitor plates results in a small capacitance, whereas in Figure 9-11(b) the dielectric thickness and the plate separation are half that of capacitor (a). This illustrates also how the capacitance of a capacitor can

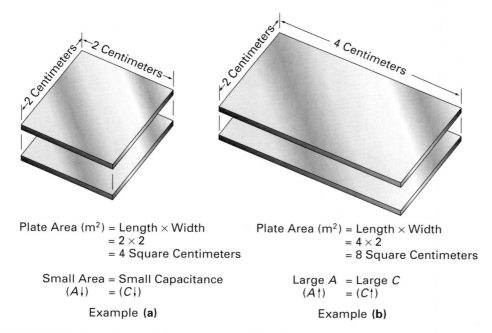

Plate Area (m²) = Length × Width
　　　　　　= 2 × 2
　　　　　　= 4 Square Centimeters

Small Area = Small Capacitance
　($A\downarrow$)　　= ($C\downarrow$)

Example **(a)**

Plate Area (m²) = Length × Width
　　　　　　= 4 × 2
　　　　　　= 8 Square Centimeters

Large A　= Large C
　($A\uparrow$)　= ($C\uparrow$)

Example **(b)**

FIGURE 9-10　Capacitance Is Proportional to Plate Area.

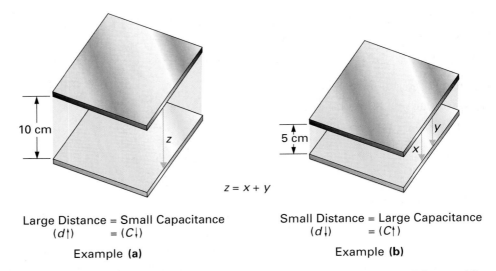

Large Distance = Small Capacitance
$(d\uparrow)$ $\quad = (C\downarrow)$

Example (a)

$z = x + y$

Small Distance = Large Capacitance
$(d\downarrow)$ $\quad = (C\uparrow)$

Example (b)

FIGURE 9-11 **Capacitance Is Inversely Proportional to Plate Separation or Distance (d).**

be doubled, in this case by halving the space between the plates. The gap across which the electric lines of force exist is halved in capacitor (b), and this doubles the strength of the electric field, which consequently doubles capacitance. Simply stated, an electric line of force (Z) in Figure 9-11(a) can be used to produce two electric lines of force (X and Y) in Figure 9-11(b) if the distance is half.

9-5-3 *Dielectric Constant*

The insulating dielectric of a capacitor concentrates the electric lines of force between the two plates. As a result, different dielectric materials can change the capacitance of a capacitor by being able to concentrate or establish an electric field with greater ease than other dielectric insulating materials. The **dielectric constant (K)** is the ease with which an insulating material can establish an electrostatic (electric) field. A vacuum is the least effective dielectric and has a dielectric constant of 1, as seen in Table 9-1. All the other in-

Dielectric Constant (K)
The property of a material that determines how much electrostatic energy can be stored per unit volume when unit voltage is applied. Also called permittivity.

TABLE 9-1 **Dielectric Constants**

MATERIAL	DIELECTRIC CONSTANT (K)[a]
Vacuum	1.0
Air	1.0006
Teflon	2.0
Wax	2.25
Paper	2.5
Amber	2.65
Rubber	3.0
Oil	4.0
Mica	5.0
Ceramic (low)	6.0
Bakelite	7.0
Glass	7.5
Water	78.0
Ceramic (high)	8000.

[a]The different material compositions can cause different values of K.

sulators listed in this table will support electrostatic lines of force more easily than a vacuum. The vacuum is used as a reference. All the other materials have dielectric constant values that are relative to the vacuum dielectric constant of 1. For example, mica has a dielectric constant of 5.0, which means that mica can cause an electric field five times the intensity of a vacuum; and, since capacitance is proportional to the dielectric constant ($C \propto K$), the mica capacitor will have five times the capacity of the same-size vacuum dielectric capacitor. In another example, we can see that the capacitance of a capacitor can be increased by a factor of almost 8000 by merely using ceramic rather than air as a dielectric between the two plates.

9-5-4 *The Capacitance Formula*

Thus plate area, separation, and the dielectric used are the three factors that change the capacitance of a capacitor. The formula that combines these three factors is

$$C = \frac{(8.85 \times 10^{-12}) \times K \times A}{d}$$

where C = capacitance, in farads (F)

8.85×10^{-12} is a constant

K = dielectric constant

A = plate area, in square meters (m²)

d = distance between the plates, in meters (m)

This formula summarizes what has been said in relation to capacitance. The capacitance of a capacitor is directly proportional to the dielectric constant (K) and the plates' area (A) and is inversely proportional to the dielectric thickness or distance between the plates (d).

EXAMPLE:

What is the capacitance of a ceramic capacitor with a 0.3 m² plate area and a dielectric thickness of 0.0003 m?

Solution:

CALCULATOR SEQUENCE

$$C = \frac{(8.85 \times 10^{-12}) \times K \times A}{d}$$

$$= \frac{(8.85 \times 10^{-12}) \times 6 \times 0.3 \text{ m}^2}{0.0003 \text{ m}}$$

$$= 5.31 \times 10^{-8}$$

$$= 0.0531 \ \mu\text{F}$$

Step	Keypad Entry	Display Response
1.	[8] [.] [8] [5] [E] [1] [2] [+/−]	8.85E−12
2.	[×]	
3.	[6]	6
4.	[×]	5.31E−11
5.	[0] [.] [3]	
6.	[÷]	1.59E−11
7.	[0] [.] [0] [0] [0] [3]	
8.	[=]	5.31E−8

Now that you have completed this section, you should be able to:

■ **Objective 5.** *List and explain the factors determining capacitance.*

Use the following questions to test your understanding of Section 9-5.

1. List the three variable factors that determine the capacitance of a capacitor.
2. State the formula for capacitance.
3. If a capacitor's plate area is doubled, the capacitance will _____.
4. If a capacitor's dielectric thickness is halved, the capacitance will _____.

9-6 DIELECTRIC BREAKDOWN AND LEAKAGE

Capacitors store a charge just as a container or tank stores water. The amount of charge stored by a capacitor is proportional to the capacitor's capacitance and the voltage applied across the capacitor ($Q = C \times V$). The charge stored by a fixed-value capacitor (C is fixed) therefore can be increased by increasing the voltage across the plates, as shown in Figure 9-12(a). If the voltage across the capacitor is increased further, the charge held by the capacitor will increase until the dielectric between the two plates of the capacitor breaks down and a spark jumps or arcs between the plates.

Using the water analogy shown in Figure 9-12(b), if the pressure of the water being pumped in is increased, the amount of water stored in the tank will also increase until a time is reached when the tank's seams at the bottom of the tank cannot contain the large amount of pressure and break down under strain. The amount of water stored in the tank is proportional to the pressure applied, just as the amount of charge stored in a capacitor is proportional to the amount of voltage applied.

Breakdown Voltage

The voltage at which breakdown occurs in a dielectric or insulation.

The **breakdown voltage** of a capacitor is determined by the strength of the dielectric used. Table 9-2 illustrates some of the different strengths of many of the common dielectrics. As an example, let's consider a capacitor that uses 1 mm of air as a dielectric between its two plates. This particular capacitor can withstand any voltage up to 787 V. If the voltage is increased further, the dielectric will break down and current will flow between the plates, destroying the capacitor (air capacitors, however, can recover from ionization).

The ideal or perfect insulator should have a resistance equal to infinite ohms. Insulators or the dielectric used to isolate the two plates of a capacitor are not perfect and therefore

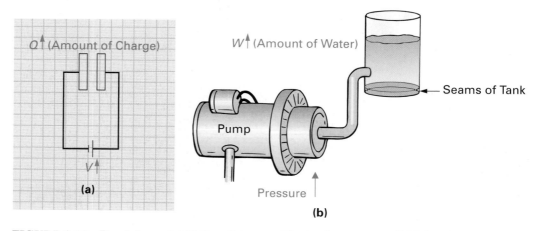

FIGURE 9-12 Breakdown. (a) Voltage Increase, Charge Increases until Dielectric Breakdown. (b) Pressure Increase, Water Increases until Seams Break Down.

TABLE 9-2 Dielectric Strengths

MATERIAL	DIELECTRIC STRENGTH (V/MM)
Air	787
Oil	12,764
Ceramic	39,370
Paper	49,213
Teflon	59,055
Mica	59,055
Glass	78,740

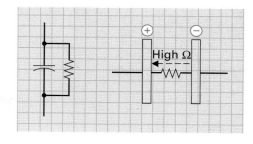

FIGURE 9-13   Capacitor Leakage.

have some very high values of resistance. This means that some value of resistance exists between the two plates, as shown in Figure 9-13; although this value of resistance is very large, it will still allow a small amount of current to flow between the two plates (in most applications a few nanoamperes or picoamperes). This small current, referred to as **leakage current,** causes any charge in a capacitor to slowly, over a long period of time, discharge between the two plates. A capacitor should have a large leakage resistance to ensure the smallest possible leakage current.

Leakage Current
Small, undesirable flow of current through an insulator or dielectric.

SELF-TEST EVALUATION POINT FOR SECTION 9-6

Now that you have completed this section, you should be able to:

■ **Objective 6.** *Describe capacitance breakdown and capacitor leakage.*

Use the following questions to test your understanding of Section 9-6.

1. Which dielectric material has the best breakdown voltage figure?

2. True or false: Current flows through the dielectric at and below the breakdown voltage of the material.

3. A capacitor should have a _____ leakage resistance to ensure a _____ leakage current.

4. Leakage current is normally in:
   a. Milliamperes   **c.** Microamperes
   b. Nanoamperes   **d.** Kiloamperes

9-7   CAPACITORS IN COMBINATION

Like resistors, capacitors can be connected in either series or parallel. As you will see in this section, the rules for determining total capacitance for parallel- and series-connected capacitors are opposite to series- and parallel-connected resistors.

9-7-1   *Capacitors in Parallel*

In Figure 9-14(a), you can see a 2 μF and 4 μF capacitor connected in parallel with one another. As the top plate of capacitor A is connected to the top plate of capacitor B with a wire, and a similar situation occurs with the bottom plates, you can see that this is the same as if the top and bottom plates were touching one another, as shown in Figure 9-14(b). When drawn so that the respective plates are touching, the dielectric constant and plate separation is the same as shown in Figure 9-14(a), but now we can easily see that the plate area is actually increased. Consequently, if capacitors are connected in parallel, the effective plate area is increased; and since capacitance is proportional to plate area [$C\uparrow = (8.85 \times 10^{-12}) \times K \times A\uparrow/d$], the capacitance will also increase. Total capacitance is actually calculated by

FIGURE 9-14   **Capacitors in Parallel.**

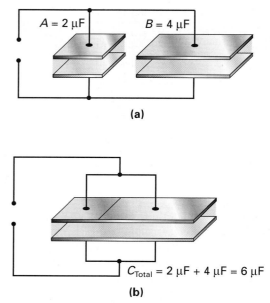

(a)

(b)

adding the plate areas, so total capacitance is equal to the sum of all the individual capacitances in parallel.

$$C_T = C_1 + C_2 + C_3 + C_4 + \cdots$$

■   **EXAMPLE:**

Determine the total capacitance of the circuit in Figure 9-15(a). What will be the voltage drop across each capacitor?

■   *Solution:*

$$C_T = C_1 + C_2 + C_3$$
$$= 1 \ \mu F + 0.5 \ \mu F + 0.75 \ \mu F$$
$$= 2.25 \ \mu F$$

As with any parallel-connected circuit, the source voltage appears across all the components. If, for example, 5 V is connected to the circuit of Figure 9-15(b), all the capacitors will charge to the same voltage of 5 V because the same voltage always exists across each section of a parallel circuit.

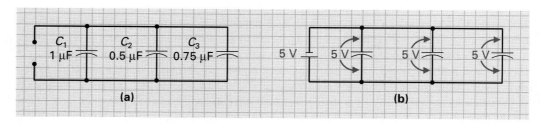

(a)   (b)

FIGURE 9-15   **Example of Parallel-Connected Capacitors.**

9-7-2   *Capacitors in Series*

In Figure 9-16(a), we have taken the two capacitors of 2 μF and 4 μF and connected them in series. Since the bottom plate of the *A* capacitor is connected to the top plate of the *B* capacitor, they can be redrawn so that they are touching, as shown in Figure 9-16(b).

The top plate of the *A* capacitor is connected to a wire into the circuit, and the bottom plate of *B* is connected to a wire into the circuit. This connection creates two center plates that are isolated from the circuit and can therefore be disregarded, as shown in Figure 9-16(c). The first thing you will notice in this illustration is that the dielectric thickness ($d\uparrow$) has increased, causing a greater separation between the plates. The effective plate area of this capacitor has decreased, as it is just the area of the top plate only. Even though the bottom plate extends out further, the electric field can only exist between the two plates, so the surplus metal of the bottom plate has no metal plate opposite for the electric field to exist.

Consequently, when capacitors are connected in series the effective plate area is decreased ($A\downarrow$) and the dielectric thickness increased ($d\uparrow$), and both of these effects result in an overall capacitance decrease ($C\downarrow\downarrow = (8.85 \times 10^{-12}) \times K \times A\downarrow/d\uparrow$).

The plate area is actually decreased to the smallest individual capacitance connected in series, which in this example is the plate area of *A*. If the plate area were the only factor, then capacitance would always equal the smallest capacitor value. However, the dielectric thickness is always equal to the sum of all the capacitor dielectrics, and this factor always causes the total capacitance (C_T) to be less than the smallest individual capacitance when capacitors are connected in series.

FIGURE 9-16   Capacitors in Series.

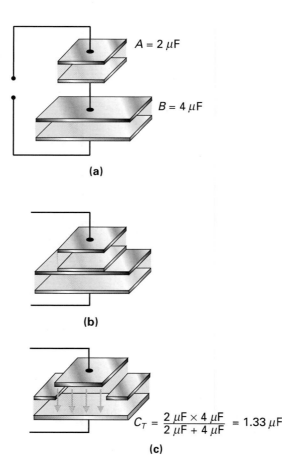

$$C_T = \frac{2\ \mu F \times 4\ \mu F}{2\ \mu F + 4\ \mu F} = 1.33\ \mu F$$

(c)

The total capacitance of two or more capacitors in series therefore is calculated by using the following formulas: For two capacitors in series,

$$C_T = \frac{C_1 \times C_2}{C_1 + C_2}$$

(product-over-sum formula)

For more than two capacitors in series,

$$C_T = \frac{1}{(1/C_1) + (1/C_2) + (1/C_3) + \cdots}$$

(reciprocal formula)

■ **EXAMPLE:**

Determine the total capacitance of the circuit in Figure 9-17.

■ *Solution:*

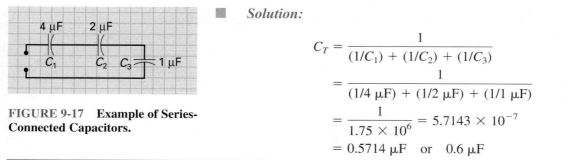

FIGURE 9-17 Example of Series-Connected Capacitors.

$$C_T = \frac{1}{(1/C_1) + (1/C_2) + (1/C_3)}$$
$$= \frac{1}{(1/4\ \mu F) + (1/2\ \mu F) + (1/1\ \mu F)}$$
$$= \frac{1}{1.75 \times 10^6} = 5.7143 \times 10^{-7}$$
$$= 0.5714\ \mu F \quad \text{or} \quad 0.6\ \mu F$$

The total capacitance for capacitors in series is calculated in the same way as total resistance when resistors are in parallel.

As with series-connected resistors, the sum of all of the voltage drops across the series-connected capacitors will equal the voltage applied (Kirchhoff's voltage law). With capacitors connected in series, the charged capacitors act as a voltage divider, and therefore the voltage-divider formula can be applied to capacitors in series.

$$V_{cx} = \frac{C_T}{C_x} \times V_T$$

where V_{cx} = voltage across desired capacitor
C_T = total capacitance
C_x = desired capacitor's value
V_T = total supplied voltage

■ **EXAMPLE:**

Using the voltage-divider formula, calculate the voltage dropped across each of the capacitors in Figure 9-17 if $V_T = 24$ V.

$$V_{C1} = \frac{C_T}{C_1} \times V_T = \frac{0.5714 \ \mu F}{4 \ \mu F} \times 24 \ V = 3.4 \ V$$

$$V_{C2} = \frac{C_T}{C_2} \times V_T = \frac{0.5714 \ \mu F}{2 \ \mu F} \times 24 \ V = 6.9 \ V$$

$$V_{C3} = \frac{C_T}{C_3} \times V_T = \frac{0.5714 \ \mu F}{1 \ \mu F} \times 24 \ V = 13.7 \ V$$

$$V_T = V_{C1} + V_{C2} + V_{C3} = 3.4 + 6.9 + 13.7 = 24 \ V$$

(Kirchhoff's voltage law)

If the capacitor values are the same, as seen in Figure 9-18(a), the voltage is divided equally across each capacitor, as each capacitor has an equal amount of charge and therefore has half of the applied voltage (in this example, 3 V across each capacitor).

When the capacitor values are different, the smaller value of capacitor will actually charge to a higher voltage than the larger capacitor. In the example in Figure 9-18(b), the smaller capacitor is actually half the size of the other capacitor, and it has charged to twice the voltage. Since Kirchhoff's voltage law has to apply to this and every series circuit, you can easily calculate that the voltage across C_1 will equal 4 V and is twice that of C_2, which is 2 V. To understand this fully, we must first understand that although the capacitance is different, both capacitors have an equal value of coulomb charge held within them, which in this example is 8 μC.

$$\begin{aligned} Q_1 &= C_1 \times V_1 \\ &= 2 \ \mu F \times 4 \ V = 8 \ \mu C \\ Q_2 &= C_2 \times V_2 \\ &= 4 \ \mu F \times 2 \ V = 8 \ \mu C \end{aligned}$$

This equal charge occurs because the same amount of current flow exists throughout a series circuit, so both capacitors are being supplied with the same number or quantity of electrons. The charge held by C_1 is large with respect to its small capacitance, whereas the same charge held by C_2 is small with respect to its larger capacitance.

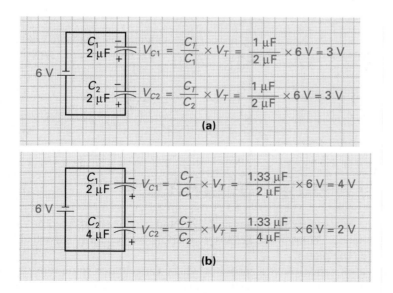

(a)

(b)

FIGURE 9-18 **Voltage Drops across Series-Connected Capacitors.**

If the charge remains the same (Q is constant) and the capacitance is small, the voltage drop across the capacitor will be large, because the charge is large with respect to the capacitance:

$$V\uparrow = \frac{Q}{C\downarrow}$$

On the other hand, for a constant charge, a large capacitance will have a small charge voltage because the charge is small with respect to the capacitance:

$$V\downarrow = \frac{Q}{C\uparrow}$$

We can apply the water analogy once more and imagine two series-connected buckets, one of which is twice the size of the other. Both are being supplied by the same series pipe, which has an equal flow of water throughout, and are consequently each holding an equal amount of water, for example, 1 gallon. The 1 gallon of water in the small bucket is large with respect to the size of the bucket, and a large amount of pressure exists within that bucket. The 1 gallon of water in the large bucket is small with respect to the size of the bucket, so a small amount of pressure exists within this bucket. The pressure within a bucket is similar to the voltage across a capacitor, and therefore a small bucket or capacitor will have a greater pressure or voltage associated with it, while a large bucket or capacitor will develop a small pressure or voltage.

To summarize capacitors in series, all the series-connected components will have the same charging current throughout the circuit, and because of this, two or more capacitors in series will always have equal amounts of coulomb charge. If the charge (Q) is equal, the voltage across the capacitor is determined by the value of the capacitor. A small capacitance will charge to a larger voltage ($V\uparrow = Q/C\downarrow$), whereas a large value of capacitance will charge to a smaller voltage ($V\downarrow = Q/C\uparrow$).

SELF-TEST EVALUATION POINT FOR SECTION 9-7

Now that you have completed this section, you should be able to:

■ **Objective 7.** *Calculate total capacitance in parallel and series capacitance circuits.*

Use the following questions to test your understanding of Section 9-7.

1. If 2 μF, 3 μF, and 5 μF capacitors are connected in series, what will be the total circuit capacitance?

2. If 7 pF, 2 pF, and 14 pF capacitors are connected in parallel, what will be the total circuit capacitance?

3. State the voltage-divider formula as it applies to capacitance.

4. True or false: With resistors, the large value of resistor will drop a larger voltage, whereas with capacitors the smaller value of capacitor will actually charge to a higher voltage.

9-8   TYPES OF CAPACITORS

Capacitors come in a variety of shapes and sizes and can be either fixed or variable in their values of capacitance. Within these groups, capacitors are generally classified by the dielectric used between the plates.

Fixed-Value Capacitor

A capacitor whose value is fixed and cannot be varied.

A **fixed-value capacitor** is a capacitor whose capacitance value remains constant and cannot be altered. Fixed capacitors normally come in a disk or a tubular package, as seen in Figure 9-19, and consist of metal foil plates separated by one of the following types of insulators (dielectric), which is the means by which we classify them.

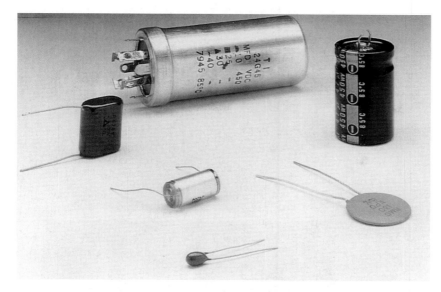

FIGURE 9-19   Fixed-Value Capacitors. (Courtesy of Sencore.)

1. Mica
2. Ceramic
3. Paper
4. Plastic
5. Electrolytic

A *variable-value capacitor* is a capacitor whose capacitance value can be changed by rotating a shaft. The variable capacitor normally consists of one electrically connected movable plate and one electrically connected stationary plate. There are basically four types of variable-value capacitors, which are also classified by the dielectric used.

1. Air
2. Mica
3. Ceramic
4. Plastic

First, let's take a closer look at the different types of fixed-value capacitors.

9-8-1   *Fixed-Value Capacitors*

Mica

Figure 9-20 illustrates the physical appearance and construction of the **mica capacitor.** In Figure 9-20(a), which illustrates the construction of the mica capacitor, you can see that thin foil plates (normally aluminum) are alternately stacked to form the plates of the capacitor, all of which are isolated from one another by a thin layer of mica dielectric. Every other plate is connected to one lead on the right, while the other metal foil plates are attached to the other connecting lead on the left. This arrangement of stacked plates provides an increase in the capacitance due to the larger overall plate area.

The complete assembly of plates and mica is then sealed inside a protective casing. Figure 9-20(b) shows some examples of dipped mica capacitors. If molding equipment is used to protect and seal the assembly the capacitor is referred to as a *molded* mica capacitor;

Mica Capacitor

Fixed capacitor that uses mica as the dielectric between its plates.

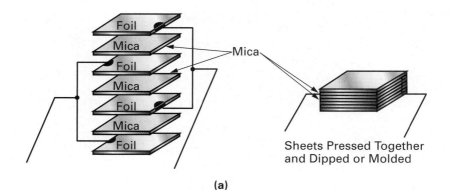

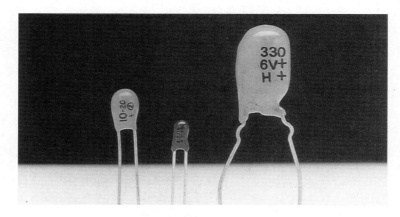

FIGURE 9-20 Mica Capacitors. (a) Construction. (b) Physical Appearance. (Courtesy of Sencore.)

however, if the plate and mica assembly is sealed by a dipping process, the capacitor is referred to as a *dipped* mica capacitor.

Ceramic

Figure 9-21 illustrates the construction and physical appearance of the molded and dipped types of **ceramic capacitor.** In their construction, you will notice the stacking of plates and isolation by ceramic, similar to the mica capacitor. Generally, the low-dielectric-constant ceramic ($K = 6.0$) is placed within the molded package, while the high-dielectric-constant ceramic (different composition means that $K = 8000$) is placed within the dipped package to obtain higher voltage ratings and high values of capacitance in small packages.

Figure 9-21(c) shows the ceramic chip surface-mount capacitors that are now available on the market for both discrete and integrated types of circuits. Values typically range from 16 to 1600 pF. Different values of capacitance can be obtained by using different types of ceramic, thereby varying the dielectric constant.

Figure 9-21(d) shows how the chip ceramic capacitors can be mounted in a SIP (single in-line package) casing.

Paper

Figure 9-22 illustrates the construction and physical appearance of tubular **paper capacitors.** Two long strips of foil are separated by a paper dielectric that has been saturated with paraffin to ensure that the paper dielectric is a good insulator. The two foil plates separated by the paper dielectric are rolled into a tubular shape and placed in either a molded or a dipped case.

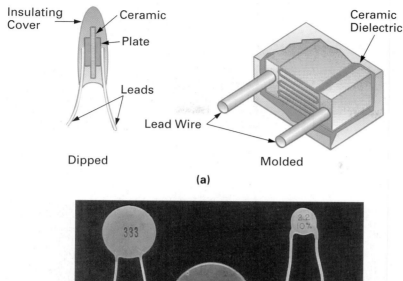

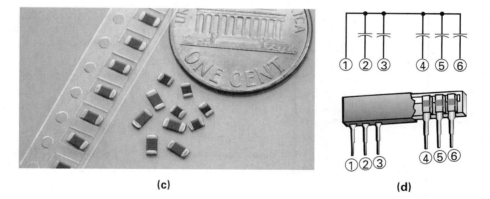

FIGURE 9-21   Ceramic Capacitors. (a) Construction. (b) Physical Appearance. (Courtesy of Sencore.) (c) Chip Ceramic Casing. (Product photo supplied as a courtesy of NIC Components Corp, Melville, N.Y.) (d) SIP Casing.

The leads of the capacitor can be either axial lead, as shown in Figure 9-22(b) (leads coming out of either end), or radial lead, as shown in Figure 9-22(c) (leads coming out of a single end).

Plastic

Plastic film capacitors have almost completely replaced the older types of paper capacitors, and their method of construction is illustrated in Figure 9-23(a). The construction is identical to the paper type; however, in this instance a plastic film dielectric strip is used as the dielectric and then rolled into a cylinder. Mylar, polycarbonate, Teflon, and polypropylene are all examples of plastic that are used as a dielectric.

The final cylindrical package can be either encased in an outer plastic layer, as shown in Figure 9-23(b), or dipped, as shown in Figure 9-23(c).

Plastic Film Capacitor

A capacitor in which alternate layers of metal aluminum foil are separated by thin films of plastic dielectric.

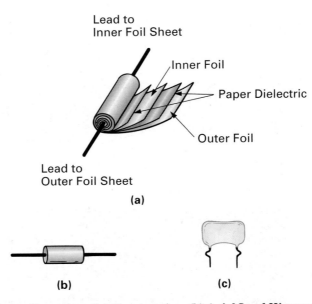

(a)

(b)   **(c)**

FIGURE 9-22   **Paper Capacitors. (a) Construction. (b) Axial Lead Wrapped. (c) Radial Lead Dipped.**

Electrolytic

    Figure 9-24 shows the construction and physical appearance of typical **electrolytic capacitors.** These types of capacitors are constructed in a similar manner as both paper and plastic types of capacitors, the difference being that the foil plates are separated by a strip of gauze that has been saturated with a conductive fluid known as an *electrolyte.* In the manufacturing process, a dc voltage causes a current flow in one direction, which causes the electrolyte to interact chemically with the aluminum foil and create a coating of aluminum oxide on the surface of the positive aluminum foil plate, as shown in Figure 9-24(a), which causes it to change chemically. This oxide, which is formed by an electrochemical reaction, becomes the dielectric for the electrolytic capacitor, and because it is extremely thin, the electrolytic capacitor can have a large capacitance for a small size. The chemical change in the positive aluminum plate during this electrochemical process makes the electrolytic capacitor **polarized.** Thus the electrolytic capacitor must always have a positive charge applied to its

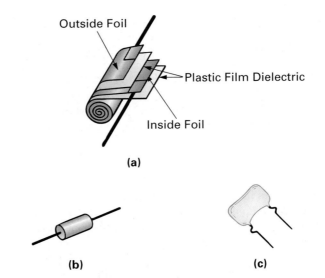

(a)

(b)   **(c)**

FIGURE 9-23   **Plastic Capacitors. (a) Construction. (b) Axial Lead. (c) Radial Lead.**

Electrolytic Capacitor

Capacitor having an electrolyte between the two plates; due to chemical action, a very thin layer of oxide is deposited on only the positive plate, which accounts for why this type of capacitor is polarized.

Polarized Electrolytic Capacitor

An electrolytic capacitor in which the dielectric is formed adjacent to one of the metal plates, creating a greater opposition to current in one direction only.

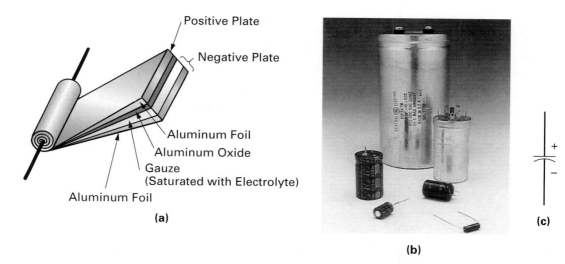

Positive Plate

Negative Plate

Aluminum Foil
Aluminum Oxide
Gauze
(Saturated with Electrolyte)
Aluminum Foil

(a)

(b)

(c)

FIGURE 9-24 Electrolytic Capacitors. (a) Construction. (b) Physical Appearance. (Courtesy of Sencore.) (c) Schematic Symbol for Electrolytic.

positive plate and a negative charge to its negative plate. If this rule is not followed, the electrolytic capacitor becomes a safety hazard, because it can in the worst-case condition explode violently.

Figure 9-24(c) shows the schematic symbols for electrolytic capacitors, which are always marked with a (+) or (−) sign to indicate polarity. In Figure 9-24(b) you can see the manufacturer's method of polarity indication.

Figure 9-24(b) shows a "can" electrolytic capacitor. With this type of electrolytic capacitor, the negative outside foil plate is connected to the metal can casing, which acts as the negative capacitor lead.

Tantalum instead of aluminum is used in some electrolytic capacitors for the plates and has advantages over the aluminum type, which are:

1. Higher capacitance per volt for a given unit volume
2. Longer life and excellent shelf life (storage)
3. Able to operate in a wider temperature range
4. Temperature stability is better
5. Construction features make them more rugged

The cost of tantalum electrolytics, however, is almost four to five times that of aluminum electrolytics, and their operating voltages are much lower.

Figure 9-25 illustrates, reviews, and includes added information on the five basic fixed capacitors discussed previously.

Tantalum Capacitor
Electrolytic capacitor having a tantalum foil anode.

9-8-2 *Variable-Value Capacitors*

Variable-value capacitors are the second basic type, and within this group, four types are commercially available. Like the fixed-value type of capacitors, they are classified by dielectric.

Variable

1. Air

The variable-value type uses a hand-rotated shaft to vary the effective plate area, and so capacitance, as shown in Figure 9-26(a).

Variable-Value Capacitor
A capacitor whose value can be varied.

Name	Construction	Approximate Range of Values and Tolerances	Characteristics	
Mica	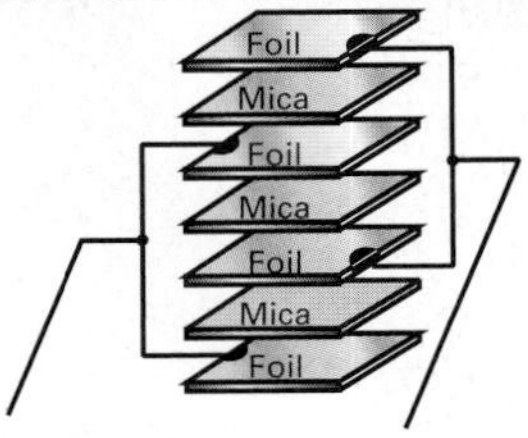 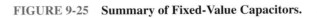	1 pF–0.1 μF ±1% to ±5%	Lower voltage rating than other capacitors of the same size	Small Capacitor Values
Ceramic	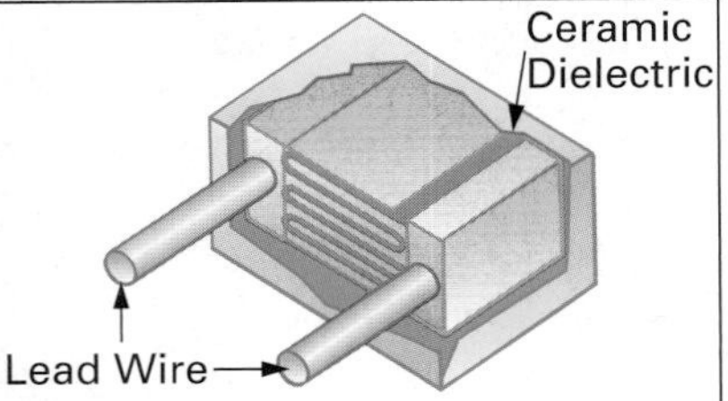	*Low Dielectric K:* 1 pF–0.01 μF ±0.5% to ±10% *High Dielectric K:* 1 pF–0.1 μF ±10% to ±80%	Most popular small value capacitor due to lower cost than mica, and its ruggedness	
Paper	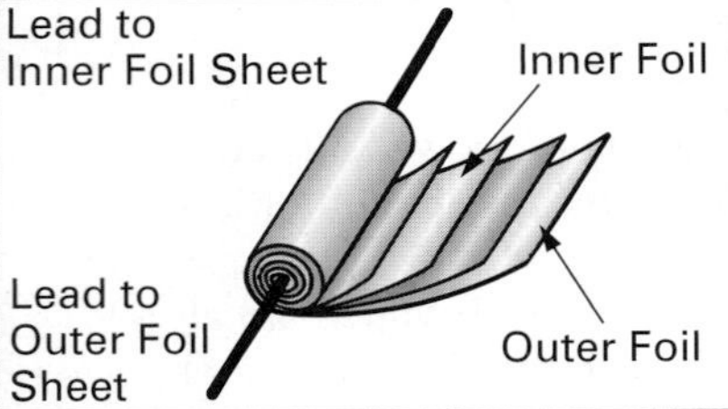	1 pF–1 μF ± 10%	Has a large plate area and therefore large capacitance for a small size	
Plastic	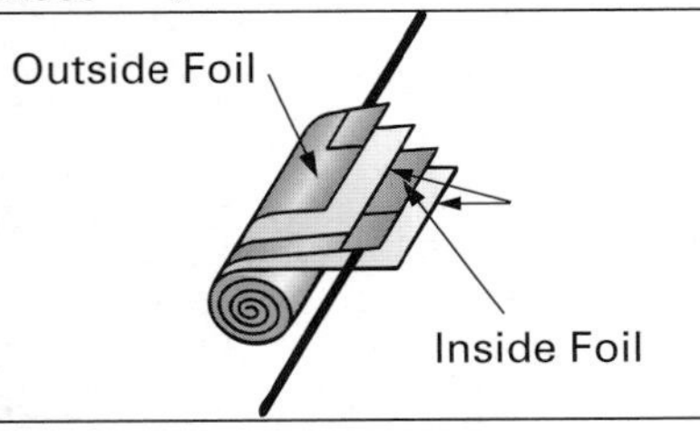	1 pF–10 μF ± 5% to ±10%	Has almost completely replaced paper capacitors; has large capacitance values for small size and high voltage ratings	Large Capacitor Values
Electrolytic (Aluminum and Tantalum)	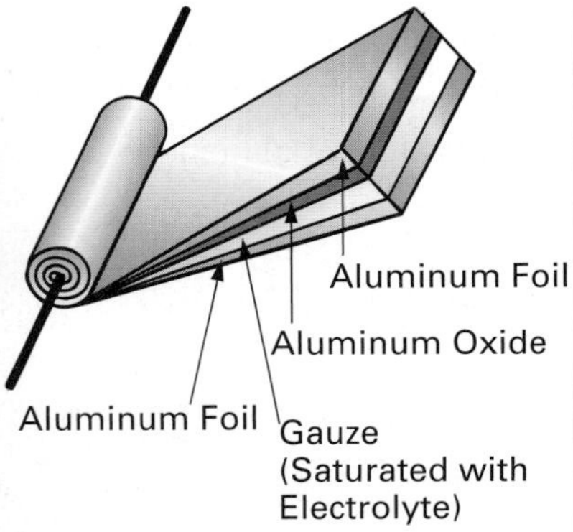	1 μF–1F ± 10% to ±50%	Most popular large value capacitor: large capacitance into small area, wide range of values. Disadvantages are: cannot be used in AC circuits as they are polarized; poor tolerances; low leakage resistance and so high leakage current. *Tantalum* advantages over aluminum include smaller size, longer life than aluminum, which has an approximate lifespan of 12 years. Disadvantages: 4 to 5 times the price.	

FIGURE 9-25 Summary of Fixed-Value Capacitors.

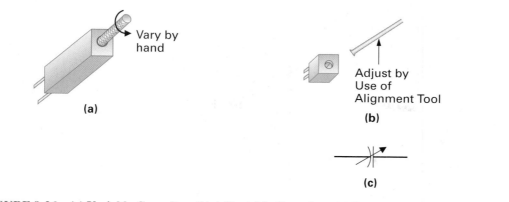

FIGURE 9-26 (a) Variable Capacitor. (b) Adjustable Capacitor. (c) Schematic Symbol for Variable or Adjustable Capacitor.

Adjustable

2. Mica

3. Ceramic

4. Plastic

The adjustable type uses screw-in or screw-out mechanical adjustment to vary the distance between the plates, and so capacitance, as shown in Figure 9-26(b).

Capacitance is dependent on effective plate area and can therefore be varied by changing either the effective plate area or distance between the plates. The dielectric constant is fixed and depends on the particular type being used. The schematic symbol for a variable capacitor is shown in Figure 9-26(c).

Air (Variable)

Figure 9-27(a) illustrates the construction of a typical air dielectric variable capacitor. With this type of capacitor, the effective plate area is adjusted to vary capacitance by causing a set of rotating plates (rotor) to mesh with a set of stationary plates (stator). When the rotor plates are fully out, the capacitance is minimum, and when the plates are fully in, the capacitance is maximum, because the maximum amount of rotor plate area is now opposite the stator plate, creating the maximum value of capacitance.

The plates are usually made of aluminum to prevent corrosion and the dielectric between the plates is air. The capacitor has to be carefully manufactured to ensure that the rotor plates do not touch the stator plates when the shaft is rotated and the plates interweave. Figure 9-27(b) illustrates a typical variable air capacitor.

In radio equipment, it is sometimes necessary to have two or more variable-value capacitors that have been constructed in such a way that a common shaft (rotor) runs through all the capacitors and varies their capacitance simultaneously, as shown in Figure 9-27(c). If you mechanically couple (gang) two or more variable capacitors so that they can all be operated from a single control, the component is known as a **ganged capacitor,** and the symbol for this arrangement is also shown in Figure 9-27(c).

Mica, Ceramic, and Plastic (Adjustable)

Figure 9-28 illustrates some of the typical packages for mica, ceramic, or plastic film types of adjustable capacitors, which are also referred to as *trimmers.* The adjustable capacitor generally has one stationary plate and one spring metal moving plate. The screw forces the spring metal plate closer or farther away from the stationary plate, varying the distance between the plates and so changing capacitance. The two plates are insulated from one another by either mica, ceramic, or plastic film, and the advantages of each are the same as for fixed-value capacitors.

Ganged Capacitor

Mechanical coupling of two or more capacitors, switches, potentiometers, or any other components so that the activation of one control will operate all sections.

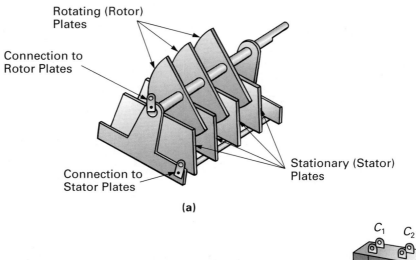

Rotating (Rotor) Plates

Connection to Rotor Plates

Connection to Stator Plates

Stationary (Stator) Plates

(a)

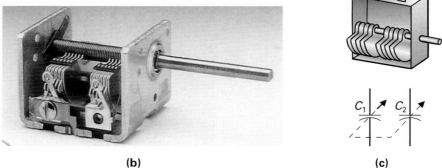

(b)

C_1 C_2

C_1 C_2

(c)

FIGURE 9-27 **Variable Air-Type Capacitors. (a) Construction. (b) Physical Appearance. (Courtesy of Sencore.) (c) Ganged.**

These types of capacitors should only be adjusted with a plastic or nonmetallic alignment tool, because a metal screwdriver may affect the capacitance of the capacitor when nearby, making it very difficult to adjust for a specific value of capacitance.

9-8-3   *The One-Farad Capacitor*

Most capacitor values are normally measured in either microfarads or picofarads. The traditional picofarad and microfarad capacitors, which have been discussed, consist of two conductor plates separated by an insulator; when voltage is applied to the capacitor, current

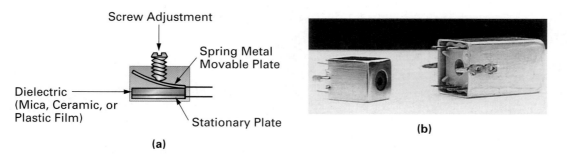

Screw Adjustment

Spring Metal Movable Plate

Dielectric (Mica, Ceramic, or Plastic Film)

Stationary Plate

(a)

(b)

FIGURE 9-28 **Adjustable Mica, Ceramic, or Plastic Capacitors. (a) Construction. (b) Physical Appearance. (Courtesy of Sencore.)**

flows and a charge builds up within the capacitor and is stored. The amount of charge stored is dependent on the plate area, plate separation, and the dielectric constant of the insulator used. All three of these factors can be varied separately or in combination to create the capacitors that we have today.

The ceramic capacitor has become the most popular small-value capacitor due to the high dielectric constants that can be obtained from ceramic (8000 or more). The electrolytic capacitor has become the most popular large-value capacitor because the extremely thin dielectric oxide layer enables us to obtain large values for a small size.

Many techniques have been tried to extend the capacitance value up toward the farad; however, the conductive surface area is the main factor that needs to be increased in order to gain high values of capacitance, as seen in Figure 9-29(a). For example, if two pieces of aluminum the size of fingernails were held together, a fingernail thickness apart, a capacitance of approximately 1 pF would be created. This 1 pF capacitor would now have to be increased by a factor of 1,000,000,000,000 to obtain a capacitor of 1 F.

To increase surface area, a process known as double etch has been used, as seen in Figure 9-29(b), where, instead of having a flat plate surface, as seen in Figure 9-29(a), large pits

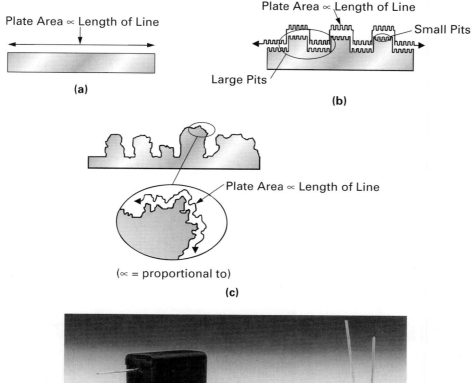

FIGURE 9-29 **The 1 F Capacitor. (a) Traditional Plate Shape. (b) Double-Etched Plate Area. (c) Activated Charcoal Surface Area. (d) Large-Farad-Value Capacitors. (Courtesy of Sencore.)**

are etched onto the surface and then smaller pits are etched onto the larger pits to create a larger surface area. This method, however, still does not get us close to 1 F.

The 1 F capacitor uses a relatively new double-layer technique, the effect of which was first noticed more than a century ago by the German scientist Hermann von Helmholtz. Helmholtz discovered that a double layer of charge will build up on the surface between a solid and a liquid.

The double-layer capacitor uses activated charcoal as the solid plates, and a liquid electrolyte between the two plates and an insulating separator allow a charge to build up within the capacitor due to the solid (charcoal) and liquid (electrolyte).

Activated charcoal is used as the plate because it has an almost infinite number of tiny particles making up its surface area, as seen in Figure 9-29(c), which yields a massive plate surface area. For example, 4 grams of activated charcoal would have an approximate surface area equivalent to a football field.

Using this technique a 1 F capacitor can be manufactured, like that seen in Figure 9-29(d). This 1 F capacitor package actually holds six 6 F capacitors connected in series, producing a total capacitance of

$$C_T \text{ (total series capacitance)} = \frac{1}{\frac{1}{6} + \frac{1}{6} + \frac{1}{6} + \frac{1}{6} + \frac{1}{6} + \frac{1}{6}}$$
$$= 1 \text{ F}$$

The reason why six 6 F capacitors are included is that the breakdown voltage of the double-layer capacitor is approximately 1 V, so by connecting six of these capacitors in series, the dielectric thickness is six times greater, increasing the breakdown voltage up to a more practical dc working voltage of 5 V. Due to the large value and the variations in activated charcoal, the tolerance of these capacitors can range from −20% to +80% of their value (0.8 to 1.8 F).

One major application for these capacitors is as a power backup for a computer's memory. When using a computer, it is important that you regularly save in a permanent memory the information you are working with in temporary memory. The reason for this is that the temporary memory will lose all of its information if there is a momentary power failure. In this situation, the 1 F capacitor could be used as a backup source of power to the computer's temporary memory during momentary power outages in order to prevent the loss of information.

SELF-TEST EVALUATION POINT FOR SECTION 9-8

Now that you have completed this section, you should be able to:

- **Objective 8.** *Describe the advantages and differences between the five basic types of fixed capacitors.*

- **Objective 9.** *Describe the advanatges and differences between the four basic types of variable capacitors.*

- **Objective 10.** *Explain the characteristics and new techniques used to create the one-farad capacitor.*

Use the following questions to test your understanding of Section 9-8.

1. List the five types of fixed-value capacitors.
2. Which fixed-value capacitor is the most popular in applications where:
   a. Large values are required?
   b. Small values are required?
3. List the four basic types of variable capacitors.
4. Which variable capacitor types are best suited for:
   a. Large value variations?
   b. Small value variations?
5. Which capacitor type is said to be polarity conscious?

9-9 CODING OF CAPACITANCE VALUES

The capacitor value, tolerance, and voltage rating need to be shown in some way on the capacitor's exterior. Presently, two methods are used: (1) alphanumeric labels and (2) color coding.

9-9-1 Alphanumeric Labels

Manufacturers today most commonly use letters of the alphabet and numbers (alphanumerics) printed on either the disk or tubular body to indicate the capacitor's specifications, as illustrated in Figure 9-30.

The tubular type, which tends to be larger in size, as seen in Figure 9-30(b), is the easier of the two since the information is basically uncoded. The value of capacitance and unit, typically the microfarad (μF or MF), tolerance figure (preceded by $\pm$ or followed by %), and voltage rating (followed by a V for voltage) are printed on all sizes of tubular cases. The remaining letters or numbers are merely manufacturers' codes for case size, series, and the like.

With disk capacitors (dipped or molded), as seen in Figure 9-30(a), certain rules have to be applied when decoding the notations. Many capacitors of this type do not define the unit of capacitance; in this situation, try to locate a decimal point. If a decimal point exists, for example 0.01 or 0.001, the value is in microfarads (10^{-6}). If no decimal point exists, for example 50 or 220, the value is in picofarads (10^{-12}) and you must analyze the number in a little more detail. Figure 9-30(b) and (c) give additional examples.

EXAMPLE:

What is the value of a capacitor if it is labeled 50, 50 V, $\pm$5?

Solution:

Since no decimal point is present, the unit is in picofarads:

$$50 \text{ pF}, \quad 50 \text{ V}, \quad \pm 5\%$$

If no decimal point is present and three digits exist and the last digit is a zero, the value is as stands and in picofarads. If the third digit is a number other than 0 (1 to 9), it is a multiplier and describes the number of zeros to be added to the picofarad value.

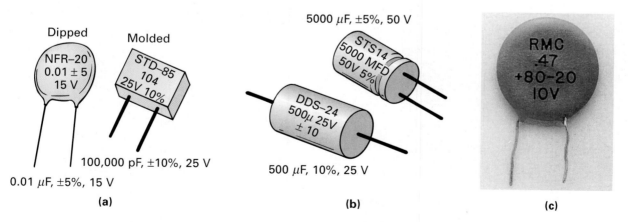

FIGURE 9-30 **Alphanumeric Coding of Capacitors. (a) Disk Type. (b) Tubular.**
(c) 0.47 μF, +80%, −20% Tolerance, 10 V. (Courtesy of Sencore.)

EXAMPLE:

If two capacitors are labeled with the following coded values, how should they be interpreted?

    a. 220

    b. 104

Solution:

Since no decimal point is present, they are both in picofarads.

    a. If the last of the three digits is a zero, the value is as it stands.

$$\underbrace{2\ 2\ 0}_{\text{three-digit value}} = 220\ \text{pF}$$

    b. If the third digit is a number from 1 to 9, it is a multiplier.

$$\underbrace{1\ 0}_{\text{two-digit value}}\underset{\text{multiplier}}{4}$$

So

$$10 \times 10^4 = 100{,}000\ \text{pF}$$

or

$$100{,}000 \times 10^{-12} = 0.1 \times 10^{-6} = 0.1\ \mu\text{F}$$

The tolerance of the capacitor is sometimes clearly indicated, for example, ± 5 or 10%; in other cases, a letter designation is used, such as

$$F = \pm 1\%$$
$$G = \pm 2\%$$
$$J = \pm 5\%$$
$$K = \pm 10\%$$
$$M = \pm 20\%$$
$$Z = -20\%, +80\%$$

TABLE 9-3

Color	1st Digit	2nd Digit	Multiplier	Tolerance
Black		0	1	$\pm 20\%$
Brown	1	1	10^{-1}	$\pm 1\%$
Red	2	2	10^{-2}	$\pm 2\%$
Orange	3	3	10^{-3}	$\pm 3\%$
Yellow	4	4	10^{-4}	$\pm 4\%$
Green	5	5	10^{-5}	$\pm 5\%$
Blue	6	6	10^{-6}	
Violet	7	7	10^{-7}	
Gray	8	8	10^{-8}	
White	9	9	10^{-9}	$\pm 10\%$

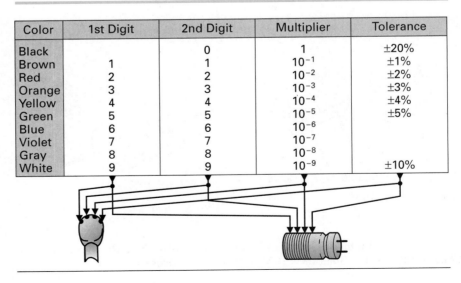

Unfortunately, there does not seem to be a standard among capacitor manufacturers, which can cause confusion when trying to determine the value of capacitance. Therefore, if you are not completely sure, you should always measure the value or consult technical data or information sheets from the manufacturer.

9-9-2   *Color Coding*

Table 9-3 illustrates the capacitor color code, which is almost identical to the resistor color code except for certain tolerances. Although the alphanumeric labels are now commonplace, the color code is still being used by some manufacturers.

SELF-TEST EVALUATION POINT FOR SECTION 9-9

Now that you have completed this section, you should be able to:

■ **Objective 11.** *Describe the coding of capacitor values on the body by use of alphanumerics or color.*

Use the following questions to test your understanding of Section 9-9.

What are the following values of capacitance?
1. 470 ± 2
2. 0.47 ± 5

9-10   CAPACITIVE TIME CONSTANT

When a capacitor is connected across a dc voltage source, it will charge to a value equal to the voltage applied. If the charged capacitor is then connected across a load, the capacitor will then discharge through the load. The time it takes a capacitor to charge or discharge can be calculated if the circuit's resistance and capacitance are known. Let us now see how we can calculate a capacitor's charge time and discharge time.

9-10-1   *DC Charging*

When a capacitor is connected across a dc voltage source, such as a battery or power supply, current will flow and the capacitor will charge up to a value equal to the dc source voltage, as shown in Figure 9-31. When the charge switch is first closed, as seen in Figure 9-31(a), there is no voltage across the capacitor at that instant and therefore a potential difference exists between the battery and capacitor. This causes current to flow and begin charging the capacitor.

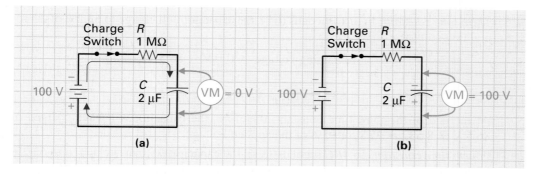

FIGURE 9-31   Capacitor Charging. (a) Switch Is Closed and Capacitor Begins to Charge. (b) Capacitor Charged.

Once the capacitor begins to charge, the voltage across the capacitor does not instantaneously rise to 100 V. It takes a certain amount of time before the capacitor voltage is equal to the battery voltage. When the capacitor is fully charged no potential difference exists between the voltage source and the capacitor. Consequently, no more current flows in the circuit as the capacitor has reached its full charge, as seen in Figure 9-31(b). The amount of time it takes for a capacitor to charge to the supplied voltage (in this example, 100 V) is dependent on the circuit's resistance and capacitance value. If the circuit's resistance is increased, the opposition to current flow will be increased, and it will take the capacitor a longer period of time to obtain the same amount of charge because the circuit current available to charge the capacitor is less.

If the value of capacitance is increased, it again takes a longer time to charge to 100 V because a greater amount of charge is required to build up the voltage across the capacitor to 100 V.

The circuit's resistance (R) and capacitance (C) are the two factors that determine the charge time (τ). Mathematically, this can be stated as

$$\tau = R \times C$$

where τ = **time constant** (s)
R = resistance (Ω)
C = capacitance (F)

In this example, we are using a resistance of 1 MΩ and a capacitance of 2 μF, which means that the time constant is equal to

$$
\begin{aligned}
\tau &= R \times C \\
&= 2\ \mu\text{F} \times 1\ \text{M}\Omega \\
&= (2 \times 10^{-6}) \times (1 \times 10^{6}) \\
&= 2\ \text{s}
\end{aligned}
$$

Two seconds is the time, so what is the constant? The constant value that should be remembered throughout this discussion is "**63.2.**"

Figure 9-32 illustrates the rise in voltage across the capacitor from 0 to a maximum of 100 V in five time constants (5 × 2 s = 10 s). So where does 63.2 come into all this?

First time constant: In 1*RC* seconds (1 × R × C = 2 s), the capacitor will charge to 63.2% of the applied voltage (63.2% × 100 V = 63.2 V).

Second time constant: In 2*RC* seconds (2 × R × C = 4 s), the capacitor will charge to 63.2% of the remaining voltage. In the example, the capacitor will be charged to 63.2 V in one time constant, and therefore the voltage remaining is equal to 100 V − 63.2 V = 36.8 V. At the end of the second time constant, therefore, the capacitor will have charged to 63.2% of the remaining voltage (63.2% × 36.8 V = 23.3 V), which means that it will have reached 86.5 V (63.2 + 23.3 = 86.5 V) or 86.5% of the applied voltage.

Third time constant: In 3*RC* seconds (6 s), the capacitor will charge to 63.2% of the remaining voltage:

$$
\begin{aligned}
\text{remaining voltage} &= 100\ \text{V} - 86.5\ \text{V} \\
&= 13.5\ \text{V} \\
63.2\%\ \text{of } 13.5\ \text{V} &= 8.532\ \text{V}
\end{aligned}
$$

At the end of the third time constant, therefore, the capacitor will have charged to 86.5 V + 8.532 = 95 V, or 95% of the applied voltage.

Fourth time constant: In 4*RC* seconds (8 s), the capacitor will have charged to 63.2% of the remaining voltage (100 V − 95 V = 5 V); therefore, 63.2% of 5 V = 3.2 V. So the capacitor will have charged to 95 V + 3.2 V = 98.2 V, or 98.2% of the applied voltage.

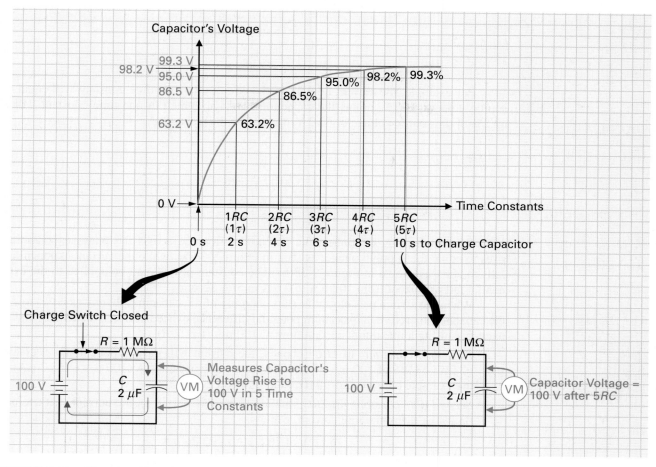

FIGURE 9-32   Charging Capacitor.

Fifth time constant: In 5*RC* seconds (10 s), the capacitor is considered to be fully charged since the capacitor will have reached 63.2% of the remaining voltage (100 V − 98.2 V = 1.8 V); therefore, 63.2% of 1.8 V = 1.1 V. So the capacitor will have charged to 98.2 V + 1.1 V = 99.3 V, or 99.3% of the applied voltage.

The voltage waveform produced by the capacitor acquiring a charge is known as an *exponential* waveform, and the voltage across the capacitor is said to rise exponentially. An exponential rise is also referred to as a *natural increase*. There are many factors that exponentially rise and fall. For example, we grow exponentially, in that there is quite a dramatic change in our height in the early years and then this increase levels off and reaches a maximum.

Before the switch is closed and even at the instant the switch is closed, the capacitor is not charged, which means that there is no capacitor voltage to oppose the supply voltage and, therefore, a maximum current of *V/R*, 100 V/1 MΩ = 100 μA flows. This current begins to charge the capacitor, and a potential difference begins to build up across the plates of the capacitor, and this voltage opposes the supply voltage, causing a decrease in charging current. As the capacitor begins to charge, less of a potential difference exists between the supply voltage and capacitor voltage and so the current begins to decrease.

To calculate the current at any time, we can use the formula

$$i = \frac{V_S - V_C}{R}$$

where i = instantaneous current
V_S = source voltage
V_C = capacitor voltage
R = resistance

For example, the current flowing in the circuit after one time constant will equal the source voltage, 100 V, minus the capacitor's voltage, which in one time constant will be 63.2% of the source voltage or 63.2 V, divided by the resistance.

$$i = \frac{V_S - V_C}{R}$$

$$= \frac{100 \text{ V} - 63.2 \text{ V}}{1 \text{ M}\Omega}$$

$$= 36.8 \text{ }\mu\text{A}$$

As the charging continues, the potential difference across the plates exponentially rises to equal the supply voltage, as seen in Figure 9-33(a), while the current exponentially falls to zero, as shown in Figure 9-33(b). The constant of 63.2 can be applied to the exponential fall of current from 100 μA to 0 μA in $5RC$ seconds.

When the switch was closed to start charging of the capacitor, there was no charge on the capacitor; therefore, a maximum potential difference existed between the battery and capacitor, causing a maximum current flow of 100 μA ($I = V/R$).

First time constant: In $1RC$ seconds, the current will have exponentially decreased 63.2% (63.2% of 100 μA = 63.2 μA) to a value of 36.8 μA (100 μA − 63.2 μA). In the example of 2 μF and 1 MΩ, this occurs in 2 s.

Second time constant: In $2RC$ seconds ($2 \times R \times C = 4$ s), the current will decrease 63.2% of the remaining current, which is

63.2% of 36.8 μA = 23.26 μA

The current will drop 23.26 μA from 36.8 μA and reach 13.5 μA or 13.5%.

Third time constant: In $3RC$ seconds (6 s), the capacitor's charge current will decrease 63.2% of the remaining current (13.5 μA) to 5 μA or 5%.

Fourth time constant: In $4RC$ seconds (8 s), the current will have decreased to 1.8 μA or 1.8%.

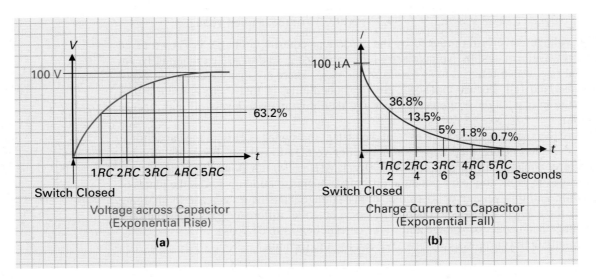

FIGURE 9-33 **Exponential Rise in Voltage and Fall in Current in a Charging Capacitive Circuit.**

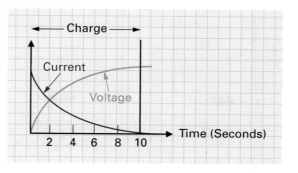

Fifth time constant: In 5RC seconds (10 s), the charge current is now 0.7 μA or 0.7%. At this time, the charge current is assumed to be zero and the capacitor is now charged to a voltage equal to the applied voltage.

Studying the exponential rise of the voltage and the exponential decay of current in a capacitive circuit, you will notice an interesting relationship. In a pure resistive circuit, the current flow through a resistor will be in step with the voltage across that same resistor, in that an increased current will cause a corresponding increase in voltage drop across the resistor. Voltage and current are consequently said to be *in step* or *in phase* with one another. With the capacitive circuit, the current flow in the circuit and voltage across the capacitor are not in step or in phase with one another. When the switch is closed to charge the capacitor, the current is maximum (100 μA), while the voltage across the capacitor is zero. After five time constants (10 s), the capacitor's voltage is now maximum (100 V) and the circuit current is zero, as seen in Figure 9-34. The circuit current flow is out of phase with the capacitor voltage, and this difference is referred to as a *phase shift*. In any circuit containing capacitance, current will lead voltage.

9-10-2   *DC Discharging*

Figure 9-35 illustrates the circuit, voltage, and current waveforms that occur when a charged capacitor is discharged from 100 V to 0 V. The 2 μF capacitor, which was charged to 100 V in 10 s (5RC), is discharged from 100 to 0 V in the same amount of time.

Looking at the voltage curve, you can see that the voltage across the capacitor decreases exponentially, dropping 63.2% to 36.8 V in 1RC seconds, another 63.2% to 13.5 V in 2RC seconds, another 63.2% to 5 V in 3RC seconds, and so on, until zero.

The current flow within the circuit is dependent on the voltage in the circuit, which is across the 2 μF capacitor. As the voltage decreases, the current will also decrease by the same amount ($I\downarrow = V\downarrow/R$).

$$\text{discharge switch closed: } I = \frac{V}{R} = \frac{100 \text{ V}}{1 \text{ M}\Omega} = 100 \text{ μA} \quad \text{maximum}$$

$$1RC \text{ (2) seconds: } I = \frac{V}{R} = \frac{36.8 \text{ V}}{1 \text{ M}\Omega} = 36.8 \text{ μA}$$

$$2RC \text{ (4) seconds: } I = \frac{V}{R} = \frac{13.5 \text{ V}}{1 \text{ M}\Omega} = 13.5 \text{ μA}$$

$$3RC \text{ (6) seconds: } I = \frac{V}{R} = \frac{5 \text{ V}}{1 \text{ M}\Omega} = 5.0 \text{ μA}$$

$$4RC \text{ (8) seconds: } I = \frac{V}{R} = \frac{1.8 \text{ V}}{1 \text{ M}\Omega} = 1.8 \text{ μA}$$

$$5RC \text{ (10) seconds: } I = \frac{V}{R} = \frac{0.7 \text{ V}}{1 \text{ M}\Omega} = 0.7 \text{ μA} \quad \text{zero}$$

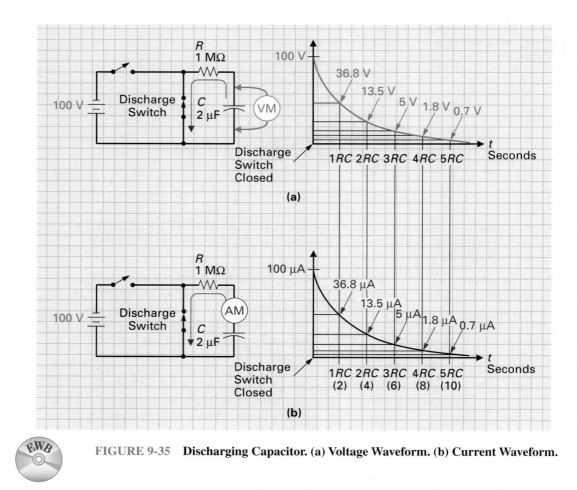

FIGURE 9-35   **Discharging Capacitor. (a) Voltage Waveform. (b) Current Waveform.**

SELF-TEST EVALUATION POINT FOR SECTION 9-10

Now that you have completed this section, you should be able to:

■ **Objective 12.**   *Explain the capacitor time constant as it relates to dc charging and discharging.*

Use the following questions to test your understanding of Section 9-10.

1. What is the capacitor time constant?
2. In one time constant, a capacitor will have charged to what percentage of the applied voltage?
3. In one time constant, a capacitor will have discharged to what percentage of its full charge?
4. True or false: The charge or discharge of a capacitor follows a linear rate of change.

9-11   AC CHARGE AND DISCHARGE

Let us begin by returning to our charged capacitor that was connected across a 100 V dc source, as shown in Figure 9-36(a). When a capacitor is connected across a dc source, it charges in five time constants and then the voltage across the capacitor will oppose the source voltage, causing current to stop. Since no current flows in this circuit, the capacitor is effectively acting as an open circuit when a constant dc voltage is applied.

If the 100 V source is reversed, as shown in Figure 9-36(b), the 100 V charge on the capacitor is now aiding the path of the battery instead of pushing or reacting against it. The discharge current will flow in the opposite direction to the charge current, until the capacitor voltage is equal to 0 V, at which time it will begin to charge in the reverse direction, as shown in Figure 9-36(c).

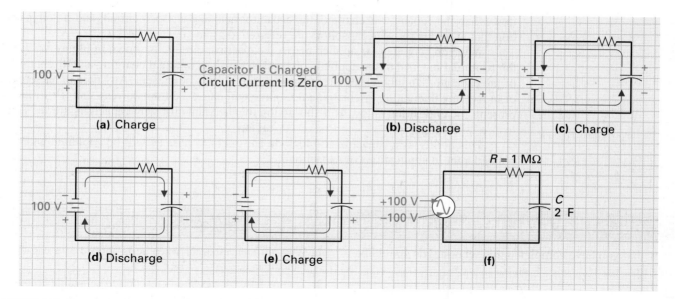

FIGURE 9-36   AC Charge and Discharge.

If the battery source is once more reversed, as shown in Figure 9-36(d), the charge current will discharge the capacitor to 0 V and then begin once more to charge it in the reverse direction, as shown in Figure 9-36(e).

The result of this switching or alternating procedure is that there is current flow in the circuit at all times (except for the instant when the battery source is removed and the polarity reversed). The switching of a dc voltage source in this way has almost the same effect as if we were applying an alternating voltage, as shown in Figure 9-36(f).

Figure 9-37 summarizes the capacitor's reaction to a dc source and an ac source. When a dc voltage is applied across a capacitive circuit, as shown in Figure 9-37(a), the capacitor will charge and then oppose the dc source, preventing current flow. Since no current can

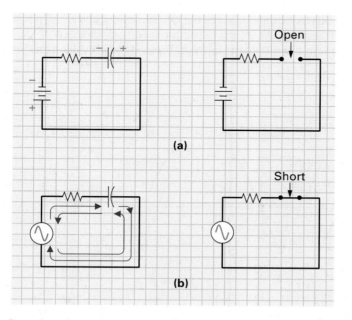

FIGURE 9-37   Capacitors' Reaction to DC and AC. (a) Capacitor Will Block DC. (b) Capacitor Will Pass AC.

flow when the capacitor is charged, the capacitor acts like an open circuit, and it is this action that accounts for why capacitors are referred to as an "open to dc."

When an ac voltage is applied across a capacitive circuit, as shown in Figure 9-37(b), the continual reversal of the applied voltage will cause a continual charging and discharging of the capacitor, and therefore circuit current will always be present. Any type of ac source or in fact any fluctuating or changing dc source will cause a circuit current, and if current is present, the opposition is low. It is this action that accounts for why capacitors are referred to as a "short to ac."

This ability of a capacitor to block dc and seem to pass ac will be exploited and explained later in applications of capacitors.

SELF-TEST EVALUATION POINT FOR SECTION 9-11

Now that you have completed this section, you should be able to:

■ **Objective 13.** *Explain how the capacitor charges and discharges when ac is applied.*

Use the following questions to test your understanding of Section 9-11.

1. What differences occur when ac is applied to a capacitor rather than dc?

2. True or false: A capacitor's reaction to ac and dc accounts for why it is known as an ac short and dc block.

9-12   PHASE RELATIONSHIP BETWEEN CAPACITOR CURRENT AND VOLTAGE

Thinking of the applied alternating voltage as a dc source that is being continually reversed sometimes makes it easier to understand how a capacitor reacts to ac. Whether the applied voltage is dc or ac, the rule holds true in that a 90° phase shift or difference exists between circuit current and capacitor voltage. The exact relationship between ac current and voltage in a capacitive circuit is illustrated in Figure 9-38.

This phase shift that exists between the circuit current and the capacitor voltage is normally expressed in degrees. At 0°, the capacitor is fully discharged (0 V), and the source is supplying a maximum circuit charge current. From 0° to 90°, the capacitor will charge to-

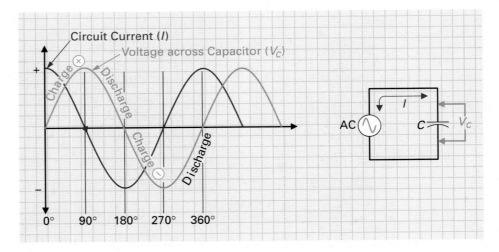

FIGURE 9-38   **AC Current Leads Voltage in a Capacitive Circuit.**

ward a maximum positive value, and this increase in capacitor voltage will oppose the source voltage, whose circuit charging current will slowly decrease to 0 A. At 90°, the capacitor is fully charged, and the circuit current is at 0 A, so the capacitor will return to the circuit the energy it consumed during the positive (+) charge cycle. This discharge circuit current is in the opposite direction to the positive charge current. At 180°, the capacitor is fully discharged (0 V), and the source is supplying a maximum circuit charge current. From 180° to 270°, the capacitor will charge toward a maximum negative value, and this negative increase in capacitor voltage will oppose the source voltage, whose circuit charging current will slowly decrease to 0 A. At 270°, the capacitor is fully charged and the circuit current is at 0 A, so the capacitor will discharge and return the energy it consumed during the negative (−) charge cycle. This discharge circuit current is in the opposite direction to the negative charge current.

Throughout this cycle, notice that the voltage across the capacitor follows the circuit current. This current, therefore, leads the voltage by 90°, and this 90° leading phase shift (current leading voltage) will occur only in a capacitive circuit.

SELF-TEST EVALUATION POINT FOR SECTION 9-12

Use the following questions to test your understanding of Section 9-12.

1. What does a voltage-current phase shift mean?

2. True or false: In a resistive circuit, the current leads the voltage by exactly 90°.

3. True or false: In a purely capacitive circuit, a 90° phase difference occurs between current and voltage.

4. True or false: Phase differences between voltage and current only occur when ac is applied.

SUMMARY

Capacitance and Capacitors (Figure 9-39)

1. Resistance will oppose the flow of current and then convert or dissipate power in the form of heat.

2. Capacitance and inductance are two circuit properties that act differently from resistance in that they will charge or store the supplied energy and then return almost all the stored energy back to the circuit, rather than lose it in wasted heat.

3. Capacitance is the ability of a circuit or device to store electrical charge. A device or component specifically designed to have this capacity or capacitance is called a *capacitor*.

4. Capacitors store electrons, and the amount of electrons stored is a measure of the capacitor's capacitance.

5. Capacitors, referred to in the past as condensers, have two leads connected to two parallel metal conductive plates, which are separated by an insulating material known as a *dielectric*. The conductive plates are normally made of metal foil, while the dielectric can be paper, air, glass, ceramic, mica, or some other form of insulator.

6. Like a secondary battery, a capacitor can be made to charge or discharge, and when discharging it will return all of the energy it consumed during charge.

7. Just as a magnetic field is produced by the flow of current, an electric field is produced by voltage.

8. A charged capacitor has an electric or electrostatic field existing between the positively charged and negatively charged plates. The strength of the electrostatic field is proportional to the charge or potential difference on the plates and inversely proportional to the distance between the plates.

9. The energy in a capacitor is actually stored in the electric or electrostatic field within the dielectric.

10. *Capacitance* is the ability of a capacitor to store an electrical charge, and the unit of capacitance is the farad (F), named in honor of Michael Faraday's work in 1831 in the field of capacitance.

11. A capacitor with the capacity of 1 farad (1 F) can store 1 coulomb of electrical charge (6.24×10^{18} electrons) if 1 volt is applied across the capacitor's plates.

12. The dielectric constant (K) is the ease with which an insulating material can establish an electrostatic (electric) field.

13. The capacitance of a capacitor is directly proportional to the dielectric constant (K) and the plates' area (A) and is inversely proportional to the dielectric thickness or distance between the plates (d).

14. The amount of charge stored by a capacitor is proportional to the capacitor's capacitance and the voltage applied across the capacitor ($Q = C \times V$).

FIGURE 9-39   **Capacitance and Capacitors.**

$$\text{Field strength (V/m)} = \frac{\text{charge difference } (V) \text{ in volts}}{\text{distance between plates } (d) \text{ in meters}}$$

(V/m = volts per meter)

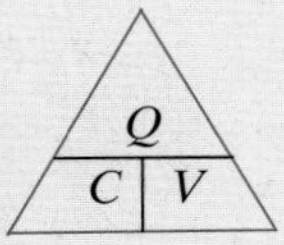

$$Q = C \times V$$

Q = charge, in coulombs
C = capacitance, in farads
V = voltage, in volts

$$= \frac{Q}{V}$$

$$V = \frac{Q}{C}$$

Dielectric Constants

MATERIAL	DIELECTRIC CONSTANT (K)[a]
Vacuum	1.0
Air	1.0006
Teflon	2.0
Wax	2.25
Paper	2.5
Amber	2.65
Rubber	3.0
Oil	4.0
Mica	5.0
Ceramic (low)	6.0
Bakelite	7.0
Glass	7.5
Water	78.0
Ceramic (high)	8000.

[a]The different material compositions can cause different values of K.

$$C = \frac{(8.85 \times 10^{-12}) \times K \times A}{d}$$

C = capacitance in farads
8.85×10^{-12} is a constant
K = dielectric constant
A = plate area, in square meters
d = distance between the plates, in meters

Capacitors in parallel:

$$C_T = C_1 + C_2 + C_3 + C_4 + \cdots$$

C_T = total capacitance, in farads
C_1 = capacitance of C_1, in farads
C_2 = capacitance of C_2, in farads
C_3 = and so on

Capacitors in series:

Dielectric Strengths

MATERIAL	DIELECTRIC STRENGTH (V/mm)
Air	787
Oil	12,764
Ceramic	39,370
Paper	49,213
Teflon	59,055
Mica	59,055
Glass	78,740

Two capacitors

$$C_T = \frac{C_1 \times C_2}{C_1 + C_2}$$

More than two capacitors

$$C_T = \frac{1}{(1/C_1) + (1/C_2) + (1/C_3) + \cdots}$$

Voltage divider formula

$$V_{cx} = \frac{C_T}{C_x} \times V_T$$

V_{cx} = desired voltage
C_T = total capacitance
C_x = selected capacitor value
V_T = supplied voltage

$$\tau = R \times C$$

τ = time constant, in seconds
R = resistance, in ohms
C = capacitance, in farads

$$i = \frac{V_S - V_C}{R}$$

i = instantaneous current, in amperes
V_S = source voltage, in volts
V_C = capacitor voltage, in volts
R = resistance, in ohms

15. If the voltage across a capacitor is increased, the charge held by the capacitor will increase until the dielectric between the two plates of the capacitor breaks down and a spark jumps or arcs between the plates.

16. The breakdown voltage of a capacitor is determined by the strength of the dielectric used.

17. The ideal or perfect insulator should have a resistance equal to infinite ohms. Insulators or the dielectric used to isolate the two plates of a capacitor are not perfect and therefore some value of resistance exists between the two plates. Although this value of resistance is very large, it will still allow a small amount of current to flow between the two plates (in most applications a few nanoamperes or picoamperes). This small current, referred to as leakage current, causes any charge in a capacitor to slowly, over a long period of time, discharge between the two plates. A capacitor should have a large leakage resistance to ensure the smallest possible leakage current.

18. If capacitors are connected in parallel, the effective plate area is increased; and since capacitance is proportional to plate area [$C\uparrow = (8.85 \times 10^{-12}) \times K \times A\uparrow/d$], the capacitance will also increase.

19. The total capacitance for capacitors connected in parallel is actually calculated by adding the plate areas, so total capacitance is equal to the sum of all the individual capacitances in parallel.

20. When capacitors are connected in series the effective plate area is decreased ($A\downarrow$) and the dielectric thickness increased ($d\uparrow$), and both of these effects result in an overall capacitance decrease [$C\downarrow\downarrow = (8.85 \times 10^{-12}) \times K \times A\downarrow/d\uparrow$]. The plate area is actually decreased to the smallest individual capacitance connected in series. If the plate area were the only factor, then capacitance would always equal the smallest capacitor value. However, the dielectric thickness is always equal to the sum of all the capacitor dielectrics, and this factor always causes the total capacitance (C_T) to be less than the smallest individual capacitance when capacitors are connected in series.

21. The total capacitance of two or more capacitors in series therefore is calculated by using the product-over-sum formula or the reciprocal formula.

22. As with series-connected resistors, the sum of all of the voltage drops across the series-connected capacitors will equal the voltage applied (Kirchhoff's voltage law). With capacitors connected in series, the charged capacitors act as a voltage divider, and therefore the voltage-divider formula can be applied to capacitors in series.

23. Capacitors in series will have the same charging current throughout the circuit, and because of this, two or more capacitors in series will always have equal amounts of coulomb charge. If the charge (Q) is equal, the voltage across the capacitor is determined by the value of the capacitor. A small capacitance will charge to a larger voltage ($V\uparrow = Q/C\downarrow$), whereas a large value of capacitance will charge to a smaller voltage ($V\downarrow = Q/C\uparrow$).

24. A fixed-value capacitor is a capacitor whose capacitance value remains constant and cannot be altered. Fixed capacitors normally come in a disk or a tubular package and consist of metal foil plates separated by one of the following types of insulators (dielectric), which is the means by which we classify them.
   a. Mica   d. Plastic
   b. Ceramic   e. Electrolytic
   c. Paper

25. A *variable-value capacitor* is a capacitor whose capacitance value can be changed by rotating a shaft. The variable capacitor normally consists of one electrically connected movable plate and one electrically connected stationary plate. There are basically four types of variable-value capacitors, which are also classified by the dielectric used.
   a. Air   c. Ceramic
   b. Mica   d. Plastic

26. Ceramic chip surface-mount capacitors are now available on the market for both discrete and integrated types of circuits.

27. The electrolytic capacitor must always have a positive charge applied to its positive plate and a negative charge to its negative plate. If this rule is not followed, the electrolytic capacitor becomes a safety hazard, because it can in the worst-case condition explode violently.

28. Tantalum instead of aluminum is used in some electrolytic capacitors for the plates and has advantages over the aluminum type, which are:
   a. Higher capacitance per volt for a given unit volume
   b. Longer life and excellent shelf life (storage)
   c. Able to operate in a wider temperature range
   d. Temperature stability is better
   e. Construction features make them more rugged

   The cost of tantulum electrolytics, however, is almost four to five times that of aluminum electrolytics, and their operating voltages are much lower.

29. The variable-value type capacitors use a hand-rotated shaft to vary the effective plate area, and so capacitance.

30. The adjustable type capacitors use screw-in or screw-out mechanical adjustment to vary the distance between the plates, and so capacitance.

31. Capacitors should only be adjusted with a plastic or nonmetallic alignment tool, because a metal screwdriver may affect the capacitance of the capacitor when nearby, making it very difficult to adjust for a specific value of capacitance.

32. The capacitor value, tolerance, and voltage rating need to be shown in some way on the capacitor's exterior. Presently, two methods are used: alphanumeric labels and color coding.

33. When a capacitor is connected across a dc voltage source, it will charge to a value equal to the voltage applied. If the charged capacitor was then connected across a load, the capacitor would then discharge through the load. The time it takes a capacitor to charge or discharge can be calculated if the circuit's resistance and capacitance are known.

34. Time constant is the time needed for either a voltage or current to rise to 63.2% of the maximum or fall to 36.8% of the initial value. The time constant of an *RC* circuit is equal to the product of *R* and *C*.

35. The voltage waveform produced by the capacitor acquiring a charge is known as an *exponential* waveform, and the voltage across the capacitor is said to rise exponentially. An exponential rise is also referred to as a *natural increase.*

36. In a pure resistive circuit, the current flow through a resistor would be in step with the voltage across that same resistor, in that an increased current would cause a corresponding increase in voltage drop across the resistor. Voltage and current are consequently said to be *in step* or *in phase* with one another.

37. With the capacitive circuit, the current flow in the circuit and voltage across the capacitor are not in step or in phase with one another. The circuit current flow is out of phase with the capacitor voltage, and this difference is referred to as a *phase shift*. In any circuit containing capacitance, current will lead voltage.

38. When a dc voltage is applied across a capacitive circuit the capacitor will charge and then oppose the dc source, preventing current flow. Since no current can flow when the capacitor is charged, the capacitor acts like an open circuit, and it is this action that accounts for why capacitors are referred to as an "open to dc."

39. When an ac voltage is applied across a capacitive circuit, the continual reversal of the applied voltage will cause a continual charging and discharging of the capacitor, and therefore circuit current will always be present. Any type of ac source or in fact any fluctuating or changing dc source will cause a circuit current, and if current is present, the opposition is low. It is this action that accounts for why capacitors are referred to as a "short to ac."

40. This ability of a capacitor to block dc and seem to pass ac will be exploited in applications of capacitors.

41. Thinking of the applied alternating voltage as a dc source that is being continually reversed sometimes makes it easier to understand how a capacitor reacts to ac. Whether the applied voltage is dc or ac, the rule holds true in that a 90° phase shift or difference exists between circuit current and capacitor voltage.

42. In a capacitive circuit, circuit current leads the capacitor voltage by 90°, and the 90° leading phase shift (current leading voltage) will occur only in a purely capacitive circuit.

REVIEW QUESTIONS

Multiple-Choice Questions

1. Capacitors were originally referred to as:
   a. Vacuum tubes   c. Inductors
   b. Condensers   d. Suppressors

2. When a capacitor charges:
   a. The voltage across the plates rises exponentially
   b. The circuit current falls exponentially
   c. The capacitor charges to the source voltage in $5RC$ seconds
   d. All of the above

3. A/an _____ field is generated by the flow of current and a/an _____ field is generated by voltage.
   a. Magnetic, electrostatic
   b. Electric, electrostatic
   c. Electric, magnetic
   d. All of the above may be considered true.

4. The strength of an electric field in a capacitor is proportional to the _____ and inversely proportional to the _____ .
   a. Plate separation, charge
   b. Plate separation, potential plate difference
   c. Plate potential difference, plate separation
   d. Both (a) and (b)

5. The plates of a capacitor are generally made out of a:
   a. Resistive material
   b. Semiconductor material
   c. Conductive material
   d. Two of the above could be true.

6. The energy of a capacitor is stored in:
   a. The magnetic field within the dielectric
   b. The magnetic field around the capacitor leads
   c. The electric field within the plates
   d. The electric field within the dielectric

7. What is the capacitance of a capacitor if it can store 24 C of charge when 6 V is applied across the plates?
   a. 2 μF   b. 3 μF   c. 4.7 μF   d. None of the above

8. The capacitance of a capacitor is directly proportional to:
   a. The plate area
   b. The distance between the plates
   c. The constant of the dielectric used
   d. Both (a) and (c)
   e. Both (a) and (b)

9. The capacitance of a capacitor is inversely proportional to:
   a. The plate area
   b. The distance between the plates
   c. The dielectric used
   d. Both (a) and (c)
   e. Both (a) and (b)

10. The breakdown voltage of a capacitor is determined by:
    a. The type of dielectric used
    b. The size of the capacitor plates
    c. The wire gauge of the connecting leads
    d. Both (a) and (b)

11. A capacitor should have a _____ leakage resistance to ensure a _____ leakage current.
    a. Large, small   c. Small, large
    b. Large, medium   d. Small, small

12. Total series capacitance of two capacitors is calculated by:
   a. Using the product-over-sum formula
   b. Using the voltage-divider formula
   c. Using the series resistance formula on the capacitors
   d. Adding all the individual values

13. Total parallel capacitance is calculated by:
   a. Using the product-over-sum formula
   b. Using the voltage-divider formula
   c. Using the parallel resistance formula on capacitors
   d. Adding all the individual values

14. The mica and ceramic fixed capacitors have:
   a. An arrangement of stacked plates
   b. An electrolyte substance between the plates
   c. An adjustable range
   d. All of the above

15. An electrolytic capacitor:
   a. Is the most popular large-value capacitor
   b. Is polarized
   c. Can have either aluminum or tantalum plates
   d. All of the above

16. Variable capacitors normally achieve a large variation in capacitance by varying _____, while adjustable trimmer capacitors only achieve a small capacitance range by varying _____ .
   a. Dielectric constant, plate area
   b. Plate area, plate separation
   c. Plate separation, dielectric constant
   d. Plate separation, plate area

17. In one time constant, a capacitor will charge to _____ of the source voltage.
   a. 86.5% **b.** 63.2% **c.** 99.3% **d.** 98.2%

18. When ac is applied across a capacitor, a _____ phase shift exists between circuit current and capacitor voltage.
   a. 45° **b.** 60° **c.** 90° **d.** 63.2°

19. A capacitor consists of:
   a. Two insulated plates separated by a conductor
   b. Two conductive plates separated by a conductor
   c. Two conductive plates separated by an insulator
   d. Two conductive plates separated by a conductive spacer

20. A 47 μF capacitor charged to 6.3 V will have a stored charge of:
   a. 296.1 μC   **d.** All of the above
   b. 2.96×10^{-4} C   **e.** Both (a) and (b)
   c. 0.296 mC

Communication Skill Questions

21. Describe the construction and main parts of a capacitor. (9-1)

22. What happens to a capacitor during:
   a. Charge (9-2-1) **b.** Discharge (9-2-2)

23. Describe how electrostatics relates to capacitance and give the formula for electric field strength. (9-3)

24. Briefly explain the relationship between capacitance, charge, and voltage. (9-4)

25. Describe the three factors affecting the capacitance of a capacitor. (9-5)

26. Briefly describe:
   a. The term *capacitor breakdown* (9-6)
   b. The term *capacitor leakage* (9-6)
   c. Why the smaller-value capacitor in a two-capacitor series circuit has the larger voltage developed across it. (9-7-2)

27. List the formula(s) used to calculate total capacitance when capacitors are connected in:
   a. Parallel (9-7-1) **b.** Series (9-7-2)

28. Describe the following types of capacitors: (9-8)
   Fixed Value
   a. Mica   **d.** Plastic
   b. Ceramic   **e.** Electrolytic
   c. Paper
   Variable Value
   a. Air **b.** Mica, ceramic, and plastic

29. Briefly explain the technique and materials used to create the 1 F capacitor. (9-8-3)

30. Explain how the constant 63.2 is used in relation to the charge and discharge of a capacitor. (9-10)

31. How is the coding of a capacitor's value, tolerance, and voltage rating indicated on a capacitor? (9-9)

32. Describe the differences between dc and ac capacitor charging and discharging. (9-11)

33. Would a mica dielectric concentrate an electrostatic field more or less than air; by how much? (9-5)

34. Calculate the total capacitance of the circuits illustrated in Figure 9-40, and describe which capacitors are in parallel and which are in series. (9-7)

35. What advantages and disadvantages do tantalum electrolytic capacitors have over the aluminum type? (9-8-1)

FIGURE 9-40

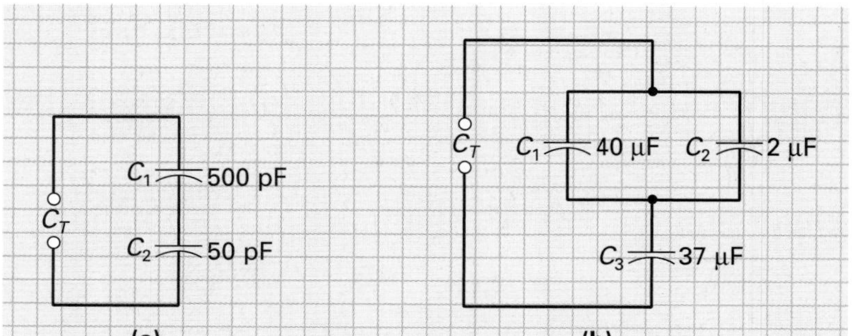

(a) $C_1 = 500$ pF $C_2 = 50$ pF C_T

(b) C_T $C_1 = 40$ μF $C_2 = 2$ μF $C_3 = 37$ μF

36. Which of the fixed-value capacitor types is polarity sensitive, and what does this mean? (9-8-1)

37. What is a ganged capacitor, and what is its application? (9-8-2)

38. Describe 1 F of capacitance. (9-4)

39. Briefly describe why series-connected capacitors are treated like parallel-connected resistors, and why parallel-connected capacitors are treated like series-connected resistors when calculating total capacitance. (9-7)

Practice Problems

40. If a 10 μF capacitor is charged to 10 V, how many coulombs of charge has it stored?

41. Calculate the electric field strength within the dielectric of a capacitor that is charged to 6 V and the dielectric thickness is 32 μm (32 millionths of a meter).

42. If a 0.006 μF capacitor has stored 125×10^{-6} C of charge, what potential difference would appear across the plates?

43. Calculate the capacitance of the capacitor that has the following parameter values: $A = 0.008$ m²; $d = 0.00095$ m; the dielectric used is paper.

44. Calculate the total capacitance if the following are connected in:
   a. Parallel: 1.7 μF, 2.6 μF, 0.03 μF, 1200 pF
   b. Series: 1.6 μF, 1.4 μF, 4 μF

45. If three capacitors of 0.025 μF, 0.04 μF, and 0.037 μF are connected in series across a 12 V source, as shown in Figure 9-41, what would be the voltage drop across each?

46. Give the value of the following alphanumeric capacitor value codes:
   a. 104   **b.** 125   **c.** 0.01   **d.** 220

47. Give the value of the following color-coded capacitor values:
   a. Yellow, violet, black   **b.** Orange, yellow, violet

48. What would be the time constant of the following *RC* circuits?
   a. $R = 6$ kΩ, $C = 14$ μF
   b. $R = 12$ MΩ, $C = 1400$ pF
   c. $R = 170$ Ω, $C = 24$ μF
   d. $R = 140$ kΩ, $C = 0.007$ μF

49. If 10 V were applied across all the *RC* circuits in Question 48, what would be the voltage across each capacitor after one time constant, and how much time would it take each capacitor to fully charge?

50. In one application, a capacitor is needed to store 25 μC of charge and will always have 125 V applied across its terminals. Calculate the capacitance value needed.

Web Site Questions

 Go to the Web site http://www.prenhall. com/cook, select the textbook *Introductory DC/AC Electronics* or *Introductory DC/AC Circuits*, this chapter, and then follow the instructions when answering the multiple-choice practice problems.

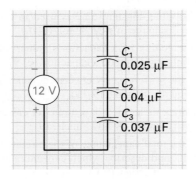

FIGURE 9-41

These tests at the end of each chapter will challenge your knowledge up to this point, and give you the practice you need for a job interview. To make this more realistic, the test will comprise both technical and personal questions. In order to take full advantage of this exercise, you may want to set up a simulation of the interview environment, have a friend read the questions to you, and record your responses for later analysis.

Company Name: **ASAP, Inc.**

Industry Branch: **Test and Measurement.**

Function: **Design and Manufacture Oscilloscopes.**

Job Title: **Field Service Technician.**

1. What do you know about our company?

2. Could you tell me what is an oscilloscope?

3. How would you feel about working alone?

4. What is a capacitor?

5. What would you do if you didn't know the answer to a question?

6. What is the difference ac and dc?

7. What are the responsibilities of a field service technician?

8. How would you handle the pressure if your service call list became overwhelming?

9. What is Ohm's law?

10. Why do you want this job?

Answers

1. Visit the company's web site before the interview to get an overall understanding of the company's ownership, product line, service, and support.
2. Chapter 8.
3. Discuss how you have worked alone on project assignments in lab and theory, and other related work history.
4. Section 9-1.
5. Report the problem to your supervisor and ask for help.
6. Chapter 8.
7. See introduction, Field Service technician.
8. Discuss how you have juggled school, studying, and a job to get to this point and other related work experience.
9. Section 2-3.
10. Describe why you were first attracted to their advertisement, why it felt compatible with your career goals, and how the company's positive attributes listed in its web site matched your expectations.

Capacitive Circuits, Testing, and Applications

Let's Toss for It!

David Packard

In 1938, Bill Hewlett and Dave Packard, close friends and engineering graduates at Stanford University, set up shop in the one-car garage behind the Packard's rented home in Palo Alto, California. In the garage, the two worked on what was to be the first product of their lifetime business together, an electronic oscillating instrument that represented a breakthrough in technology and was specifically designed to test sound equipment.

Late in the year, the oscillator (designated the 200A "because the number sounded big") was presented at a West Coast meeting of the Institute of Radio Engineers (now the Institute of Electrical and Electronics Engineers—IEEE) and the orders began to roll in. Along with the first orders for the 200A was a letter from Walt Disney Studios asking the two to build a similar oscillator covering a different frequency range. The Model 200B was born shortly thereafter, and Disney purchased eight to help develop the unique sound system for the classic film *Fantasia*. By 1940, the young company had outgrown the garage and moved into a small rented building nearby.

Over the years, the company continued a steady growth, expanding its product line to more than 10,000, including computer systems and peripheral products, test and measuring instruments, hand-held calculators, medical electronic equipment, and systems for chemical analysis. Employees have increased from 2 to almost 82,000, and the company is one of the 100 largest industrial corporations in America, with a net revenue of $8.1 billion. Whose name should go first was decided by the toss of a coin on January 1, 1939, and the outcome was Hewlett-Packard or HP.

10

Outline and Objectives

Introduction

In Chapter 9 you discovered that when a capacitor was connected to a dc circuit, it charged up until it was at a value equal to the applied dc voltage, and then from that point on current was zero. When a capacitor was connected to an ac circuit, however, the continual reversal of the applied voltage resulted in a continuous charge and discharge of the capacitor, so circuit current was always present. This ability of the capacitor makes it an extremely useful device in ac circuits.

In this chapter we will first study capacitive reactance, which is the opposition a capacitor offers to current flow. Then we will study the ac circuit characteristics of a series *RC* circuit and parallel *RC* circuit. To complete our study of capacitors, in the final two sections in this chapter we will discuss the testing of capacitors and then some of the more common applications of capacitors.

10-1 CAPACITIVE REACTANCE

Capacitive Reactance (X_C)

Measured in ohms, it is the ability of a capacitor to oppose current flow without the dissipation of energy.

Resistance (R), by definition, is the opposition to current flow with the dissipation of energy and is measured in ohms. Capacitors oppose current flow like a resistor, but a resistor dissipates energy, whereas a capacitor stores energy (when it charges) and then gives back its energy into the circuit (when it discharges). Because of this difference, a new term had to be used to describe the opposition offered by a capacitor. **Capacitive reactance (X_C),** by definition, is the opposition to current without the dissipation of energy and is also measured in ohms.

If capacitive reactance is basically opposition, it is inversely proportional to the amount of current flow. If a large current is within a circuit, the opposition or reactance must be low ($I\uparrow, X_C\downarrow$). Conversely, a small circuit current will be the result of a large opposition or reactance ($I\downarrow, X_C\uparrow$).

When a dc source is connected across a capacitor, current will flow only for a short period of time ($5RC$ seconds) to charge the capacitor. After this time, there is no further current flow. Consequently, the capacitive reactance or opposition offered by a capacitor to dc is infinite (maximum).

Alternating current is continuously reversing in polarity, resulting in the capacitor continuously charging and discharging. This means that charge and discharge currents are always flowing around the circuit, and if we have a certain value of current, we must also have a certain value of reactance or opposition.

Initially, when the capacitor's plates are uncharged, they will not oppose or react against the charging current and therefore maximum current will flow ($I\uparrow$) and the reactance will be very low ($X_C\downarrow$). As the capacitor charges, it will oppose or react against the charge current, which will decrease ($I\downarrow$), so the reactance will increase ($X_C\uparrow$). The discharge current is also highest at the start of discharge ($I\uparrow, X_C\downarrow$) as the voltage of the charged capacitor is also high; but as the capacitor discharges, its voltage decreases and the discharge current will also decrease ($I\downarrow, X_C\uparrow$).

To summarize, at the start of a capacitor charge or discharge, the current is maximum, so the reactance is low. This value of current then begins to fall to zero, so the reactance increases.

If the applied alternating current is at a high frequency, as shown in Figure 10-1(a), it is switching polarity more rapidly than a lower frequency and there is very little time between the start of charge and discharge. As the charge and discharge currents are largest at the beginning of the charge and discharge of the capacitor, the reactance has very little time

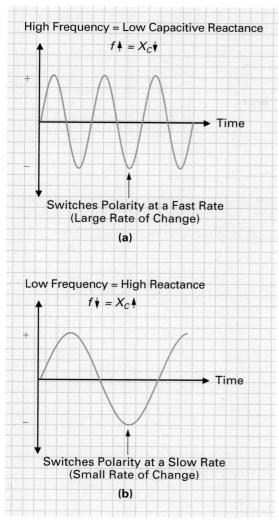

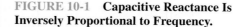
to build up and oppose the current, which is why the current is a high value and the capacitive reactance is small at higher frequencies. With lower frequencies, as shown in Figure 10-1(b), the applied alternating current is switching at a slower rate, and therefore the reactance, which is low at the beginning, has more time to build up and oppose the current.

Capacitive reactance is therefore inversely proportional to frequency:

$$\text{capacitive reactance } (X_C) \propto \frac{1}{f \text{ (frequency)}}$$

Frequency, however, is not the only factor that determines capacitive reactance. Capacitive reactance is also inversely proportional to the value of capacitance. If a larger capacitor value is used a longer time is required to charge the capacitor ($t\uparrow = C\uparrow R$), which means that current will be present for a longer period of time, so the overall current will be large ($I\uparrow$); consequently, the reactance must be small ($X_C\downarrow$). On the other hand, a small capacitance value will charge in a small amount of time ($t\downarrow = C\downarrow R$) and the current is present for only a short period of time. The overall current will therefore be small ($I\downarrow$), indicating a large reactance ($X_C\uparrow$).

$$\text{capacitive reactance } (X_C) \propto \frac{1}{C \text{ (capacitance)}}$$

Capacitive reactance (X_C) is therefore inversely proportional to both frequency and capacitance and can be calculated by using the formula

$$X_C = \frac{1}{2\pi fC}$$

where X_C = capacitive reactance, in ohms
2π = constant
f = frequency, in hertz
C = capacitance, in farads

EXAMPLE:

Calculate the reactance of a 2 μF capacitor when a 10 kHz sine wave is applied.

Solution:

CALCULATOR SEQUENCE

Step	Keypad Entry	Display Response
1.	②	2.0
2.	✕	
3.	π	3.1415927
4.	✕	6.283185
5.	① ⓪ EE ③	10E3
6.	✕	6.2831.8
7.	② EE ⑥ +/−	2.−06
8.	=	0.1256637
9.	1/x	7.9577

$$X_C = \frac{1}{2\pi fC}$$

$$= \frac{1}{2 \times \pi \times 10\ \text{kHz} \times 2\ \mu\text{F}} = 8\ \Omega$$

SELF-TEST EVALUATION POINT FOR SECTION 10-1

Now that you have completed this section, you should be able to:

■ **Objective 1.** *Define and explain capacitive reactance.*

Use the following questions to test your understanding of Section 10-1.

1. Define *capacitive reactance.*
2. State the formula for capacitive reactance.
3. Why is capacitive reactance inversely proportional to frequency and capacitance?
4. If $C = 4\ \mu\text{F}$ and $f = 4\ \text{kHz}$, calculate X_C.

10-2   SERIES *RC* CIRCUIT

In a purely resistive circuit, as shown in Figure 10-2(a), the current flowing within the circuit and the voltage across the resistor are in phase with one another. In a purely capacitive circuit, as shown in Figure 10-2(b), the current flowing in the circuit leads the voltage across the capacitor by 90°.

Purely resistive: 0° phase shift (*I* is in phase with *V*)
Purely capacitive: 90° phase shift (*I* leads *V* by 90°)

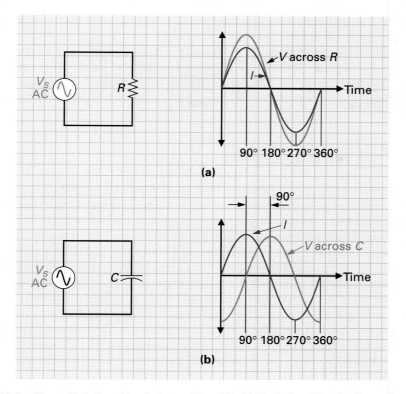

FIGURE 10-2 **Phase Relationships between *V* and *I*. (a) Resistive Circuit: Current and Voltage Are in Phase. (b) Capacitive Circuit: Current Leads Voltage by 90°.**

If we connect a resistor and capacitor in series, as shown in Figure 10-3(a), we have probably the most commonly used electronic circuit, which has many applications. The voltage across the resistor (V_R) is always in phase with the circuit current (I), as can be seen in Figure 10-3(b), because maximum points and zero crossover points occur at the same time. The voltage across the capacitor (V_C) lags the circuit current by 90°.

Since the capacitor and resistor are in series, the same current is supplied to both components; Kirchhoff's voltage law can be applied, which states that the sum of the voltage drops around a series circuit is equal to the voltage applied (V_S). The voltage drop across the resistor (V_R) and the voltage drop across the capacitor (V_C) are out of phase with one another, which means that their peaks occur at different times. The signal for the applied voltage (V_S) is therefore obtained by adding the values of V_C and V_R at each instant in time,

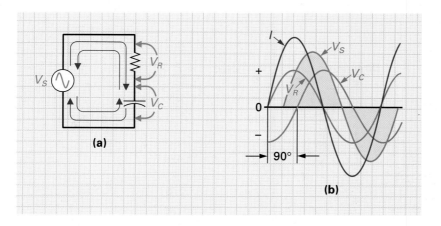

FIGURE 10-3 ***RC* Series Circuit. (a) Circuit. (b) Waveforms.**

plotting the results, and then connecting the points with a line; this is represented in Figure 10-3(b) by the shaded waveform.

Although the waveforms in Figure 10-3(b) indicate the phase relationship between I, V_S, V_R, and V_C, it seems difficult to understand clearly the relationship among all four because of the crisscrossing of waveforms. An easier method of representation is to return to the circle and vectors that were introduced in the trigonometry mini-math review in Chapter 10.

10-2-1   *Vector Diagram*

A vector (or phasor) is a quantity that has both magnitude and direction and is represented by a line terminated at the end by an arrowhead, as seen in Figure 10-4(a). Vectors are generally always used to represent a physical quantity that has two properties. For example, if you are traveling 60 miles per hour in a southeast direction, the size or magnitude of the vector would represent 60 mph, while the direction of the vector would point southeast. In an ac circuit containing a reactive component such as a capacitor, the vector is used to represent a voltage or current. The magnitude of the vector represents the value of voltage or current, while the direction of the vector represents the phase of the voltage or current.

A **vector diagram** is an arrangement of vectors to illustrate the magnitude and phase relationships between two or more quantities of the same frequency within an ac circuit. Figure 10-4(b) illustrates the basic parts of a vector diagram. As an example, the current (I) vector is at the 0° position and the size of the arrow represents the peak value of alternating current.

10-2-2   *Voltage*

Figure 10-5(a), (b), and (c) repeat our previous *RC* series circuit with waveforms and a vector diagram. In Figure 10-5(b), the current peak flowing in the series *RC* circuit occurs at 0° and will be used as a reference; therefore, the vector of current in Figure 10-5(c) is to the right in the 0° position. The voltage across the resistor (V_R) is in phase or coincident with the current (I), as shown in Figure 10-5(b), and the vector that represents the voltage across the resistor (V_R) overlaps or coincides with the I vector, at 0°.

The voltage across the capacitor (V_C) is, as shown in Figure 10-5(b), 90° out of phase (lagging) with the circuit's current, so the V_C vector in Figure 10-5(c) is drawn at −90° (minus sign indicates lag) to the current vector, and the length of this vector represents the magnitude of the voltage across the capacitor. Since the ohmic values of the resistor (R) and the capacitor (X_C) are equal, the voltage drop across both components is the same. The V_R and V_C vectors are subsequently equal in length.

The source voltage (V_S) is, by Kirchhoff's voltage law, equal to the sum of the series voltage drops (V_C and V_R). However, since these voltages are not in phase with one another,

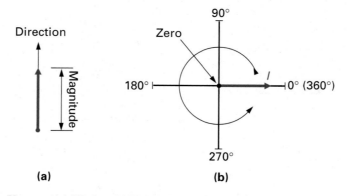

FIGURE 10-4   Vectors. (a) Vector. (b) Vector Diagram.

CHAPTER 10 / CAPACITIVE CIRCUITS, TESTING, AND APPLICATIONS

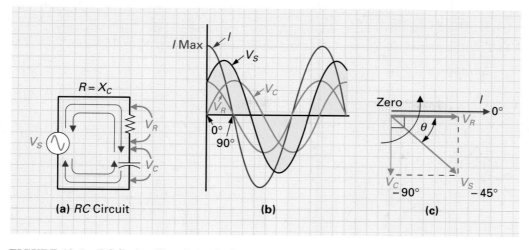

FIGURE 10-5 *RC* Series Circuit Analysis.

we cannot simply add the two together. The source voltage (V_S) will be the sum of both V_C and V_R at a particular time. Studying the waveforms in Figure 10-5(b), you will notice that peak source voltage will occur at 45°. By vectorially adding the two voltages V_R and V_C in Figure 10-5(c), we obtain a resultant V_S vector that has both magnitude and phase. The angle theta (θ) formed between circuit current (I) and source voltage (V_S) will always be less than 90°, and in this example is equal to $-45°$ because the voltage drops across R and C are equal due to R and X_C being of the same ohmic value.

If V_C and V_R are drawn to scale, the peak source voltage (V_S) can be calculated by using the same scale, and a mathematical rather than graphical method can be used to save the drafting time.

In Figure 10-6(a), we have taken the three voltages (V_R, V_C, and V_S) and formed a right-angle triangle, as shown in Figure 10-6(b). The Pythagorean theorem for right-angle triangles states that if you take the square of a (V_R) and add it to the square of b (V_C), the square root of the result will equal c (the source voltage, V_S).

$$V_S = \sqrt{V_R{}^2 + V_C{}^2}$$

By transposing the formula according to the rules of algebra we can calculate any unknown if two variables are known.

$$V_C = \sqrt{V_S{}^2 - V_R{}^2}$$
$$V_C = \sqrt{V_S{}^2 - V_R{}^2}$$

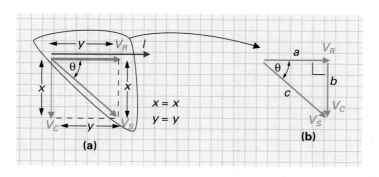

FIGURE 10-6 Voltage and Current Vector Diagram of a Series *RC* Circuit.

Calculate the source voltage applied across an *RC* series circuit if $V_R = 12$ V and $V_C = 8$ V.

☐ *Solution:*

CALCULATOR SEQUENCE

Step	Keypad Entry	Display Response
1.	☐1☐ ☐2☐	12.0
2.	☐x^2☐	144.0
3.	☐+☐	
4.	☐8☐	8.0
5.	☐x^2☐	64.0
6.	☐=☐	208.0
7.	☐$\sqrt{x}$☐	14.42220

$$V_S = \sqrt{V_R{}^2 + V_C{}^2}$$
$$= \sqrt{12\text{ V}^2 + 8\text{ V}^2}$$
$$= \sqrt{144 + 64}$$
$$= 14.42\text{ V}$$

10-2-3 *Impedance*

Impedance (Z)

Measured in ohms, it is the total opposition a circuit offers to current flow (reactive and resistive).

Since resistance is the opposition to current with the dissipation of heat, and reactance is the opposition to current without the dissipation of heat, a new term is needed to describe the total resistive and reactive opposition to current. **Impedance** (designated *Z*) is also measured in ohms and is the total circuit opposition to current flow. It is a combination of resistance (*R*) and reactance (*X*$_C$); however, in our capacitive and resistive circuit, a phase shift or difference exists, and just as V_C and V_R cannot be merely added together to obtain V_S, *R* and X_C cannot be simply summed to obtain *Z*.

If the current within a series circuit is constant (the same throughout the circuit), the resistance of a resistor (*R*) or reactance of a capacitor (*X*$_C$) will be directly proportional to the voltage across the resistor (V_R) or the capacitor (V_C).

$$V_R\updownarrow = I \times R\updownarrow, \qquad V_C\updownarrow = I \times X_C\updownarrow$$

A vector diagram can be drawn similarly to the voltage vector diagram to illustrate opposition, as shown in Figure 10-7(a). The current is used as a reference (0°); the resistance vector (*R*) is in phase with the current vector (*I*), since V_R is always in phase with *I*. The capacitive reactance (*X*$_C$) vector is at $-90°$ to the resistance vector, due to the 90° phase shift between a resistor and capacitor. The lengths of the resistance vector (*R*) and capacitive reactance vector (*X*$_C$) are equal in this example. By vectorially adding *R* and X_C, we have a resulting impedance (*Z*) vector.

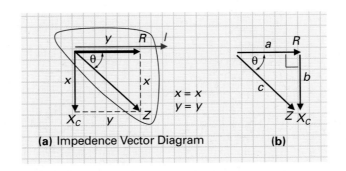

(a) Impedance Vector Diagram   (b)

FIGURE 10-7   **Resistance, Reactance, and Impedance Vector Diagram of a Series *RC* Circuit.**

By using the three variables, which have again formed a right-angle triangle [Figure 10-7(b)], we can apply the Pythagorean theorem to calculate the total opposition or impedance (Z) to current flow, taking into account both R and X_C.

$$Z = \sqrt{R^2 + X_C{}^2}$$

$$R = \sqrt{Z^2 - X_C{}^2}$$
$$X_C = \sqrt{Z^2 - R^2}$$

EXAMPLE:

Calculate the total impedance of a series RC circuit if $R = 27\ \Omega$, $C = 0.005\ \mu F$, and the source frequency = 1 kHz.

Solution:

The total opposition (Z) or impedance is equal to

$$Z = \sqrt{R^2 + X_C{}^2}$$

R is known, but X_C will need to be calculated.

$$X_C = \frac{1}{2\pi f C}$$

$$= \frac{1}{2 \times \pi \times 1\ \text{kHz} \times 0.005\ \mu F}$$
$$= 31.8\ \text{k}\Omega$$

Since $R = 27\ \Omega$ and $X_C = 31.8\ \text{k}\Omega$, then

$$Z = \sqrt{R^2 + X_C{}^2}$$
$$= \sqrt{27\ \Omega^2 + 31.8\ \text{k}\Omega^2}$$
$$= \sqrt{729 + 1 \times 10^9}$$
$$= 31.8\ \text{k}\Omega$$

As you can see in this example, the small resistance of 27 Ω has very little effect on the circuit's total opposition or impedance, due to the relatively large capacitive reactance of 31,800 Ω.

EXAMPLE:

Calculate the total impedance of a series RC circuit if $R = 45\ \text{k}\Omega$ and $X_C = 45\ \Omega$.

Solution:

$$Z = \sqrt{R^2 + X_C{}^2}$$
$$= \sqrt{45\ \text{k}\Omega^2 + 45^2}$$
$$= 45\ \text{k}\Omega$$

In this example, the relatively small value of X_C had very little effect on the circuit's opposition or impedance, due to the large circuit resistance.

Calculate the total impedance of a series *RC* circuit if $X_C = 100 \ \Omega$ and $R = 100 \ \Omega$.

▨ *Solution:*

$$
\begin{aligned}
Z &= \sqrt{R^2 + X_C{}^2} \\
&= \sqrt{100^2 + 100^2} \\
&= 141.4 \ \Omega
\end{aligned}
$$

In this example *R* was equal to X_C.

We can define the total opposition or impedance in terms of Ohm's law, in the same way as we defined resistance.

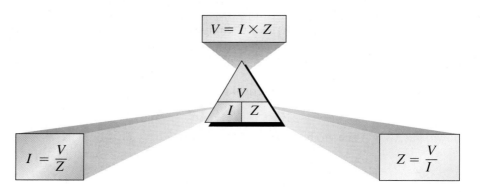

By transposition, we can arrive at the usual combinations of Ohm's law.

10-2-4 *Phase Angle or Shift (θ)*

In a purely resistive circuit, the total opposition (Z) is equal to the resistance of the resistor, so the phase shift (θ) is equal to 0° [Figure 10-8(a)].

In a purely capacitive circuit, the total opposition (Z) is equal to the capacitive reactance (X_C) of the capacitor, so the phase shift (θ) is equal to −90° [Figure 10-8(b)].

When a circuit contains both resistance and capacitive reactance, the total opposition or impedance has a phase shift that is between 0 and 90°. Referring back to the impedance vector diagram in Figure 10-7 and the preceding example, you can see that we have used a simple example where *R* has equaled X_C, so the phase shift (θ) has always been −45°. If the resistance and reactance are different from one another, as was the case in the other two examples, the phase shift (θ) will change, as shown in Figure 10-9. As can be seen in this illustration, the phase shift (θ) is dependent on the ratio of capacitive reactance to resistance

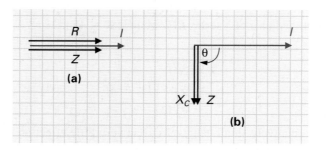

FIGURE 10-8 **Phase Angles. (a) Purely Resistive Circuit. (b) Purely Capacitive Circuit.**

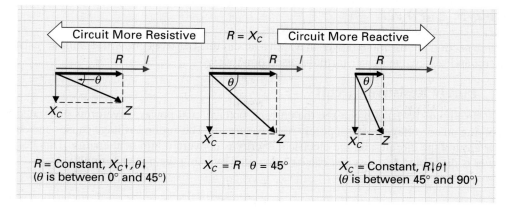

FIGURE 10-9   **Phase Angle of a Series *RC* Circuit.**

(X_C/R). A more resistive circuit will have a phase shift between 0 and 45°, while a more reactive circuit will have a phase shift between 45 and 90°. By the use of trigonometry (the science of triangles), we can derive a formula to calculate the degree of phase shift, since two quantities X_C and R are known.

The phase angle, θ, is equal to

$$\theta = \text{invtan}\, \frac{X_C}{R}$$

This formula will determine by what angle Z leads R. Since X_C/R is equal to V_C/V_R, the phase angle can also be calculated if V_R and V_C are known.

$$\theta = \text{invtan}\, \frac{V_C}{V_R}$$

This formula will determine by what angle V_R leads V_S.

EXAMPLE:

Calculate the phase shift or angle in two different series *RC* circuits if:

a. $V_R = 12$ V, $V_C = 8$ V
b. $R = 27\ \Omega$, $X_C = 31.8$ kΩ

Solution:

a. $\theta = \text{invtan}\, \dfrac{V_C}{V_R}$

$\quad = \text{invtan}\, \dfrac{8\text{ V}}{12\text{ V}}$

$\quad = 33.7°$ (V_R leads V_S by 33.7°)

b. $\theta = \text{invtan}\, \dfrac{X_C}{R}$

$\quad = \text{invtan}\, \dfrac{31.8\text{ k}\Omega}{27\ \Omega}$

$\quad = 89.95°$ (R leads Z by 89.95°)

10-2-5   *Power*

In this section we examine power in a series ac circuit. Let us begin with a simple resistive circuit and review the power formulas used previously.

Purely Resistive Circuit

In Figure 10-10 you can see the current, voltage, and power waveforms generated by applying an ac voltage across a purely resistive circuit. The applied voltage causes current to flow around the circuit, and the electrical energy is converted into heat energy. This heat or power is dissipated and lost and can be calculated by using the power formula.

$$P = V \times I$$
$$P = I^2 \times R$$
$$P = \frac{V^2}{R}$$

Voltage and current are in phase with one another in a resistive circuit, and instantaneous power is calculated by multiplying voltage by current at every instant through 360° ($P = V \times I$). The sinusoidal power waveform is totally positive, because a positive voltage multiplied by a positive current gives a positive value of power, and a negative voltage multiplied by a negative current will also produce a positive value of power. For these reasons, a resistor is said to generate a positive power waveform, which you may have noticed is twice the frequency of the voltage and current waveforms; two power cycles occur in the same time as one voltage and current cycle.

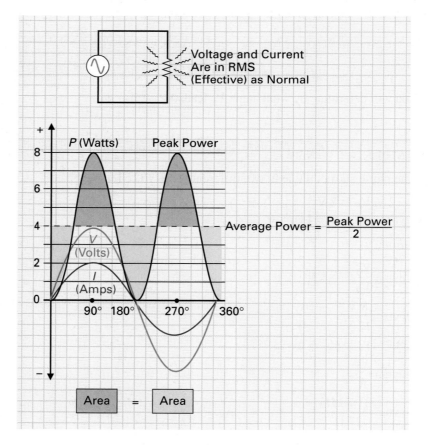

FIGURE 10-10   **Power in a Purely Resistive Circuit.**

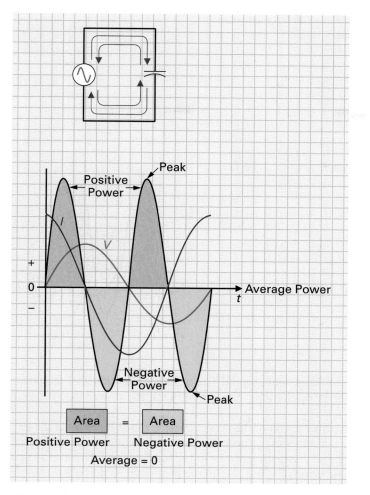

FIGURE 10-11   Power in a Purely Capacitive Circuit.

The power waveform has been split in half, and this line that exists between the maximum point (8 W) and zero point (0 W) is the average value of power (4 W) that is being dissipated by the resistor.

Purely Capacitive Circuit

In Figure 10-11, you can see the current, voltage, and power waveforms generated by applying an ac voltage source across a purely capacitive circuit. As expected, the current leads the voltage by 90°, and the power wave is calculated by multiplying voltage by current, as before, at every instant through 360°. The resulting power curve is both positive and negative. During the positive alternation of the power curve, the capacitor is taking power as the capacitor charges. When the power alternation is negative, the capacitor is giving back the power it took as it discharges back into the circuit.

The average power dissipated is once again the value that exists between the maximum positive and maximum negative points and causes the area above this line to equal the area below. This average power level calculates out to be zero, which means that no power is dissipated in a purely capacitive circuit.

Resistive and Capacitive Circuit

In Figure 10-12 you can see the current, voltage, and power waveforms generated by applying an ac voltage source across a series-connected *RC* circuit. The current leads the voltage by some phase angle less than 90°, and the power waveform is once again determined

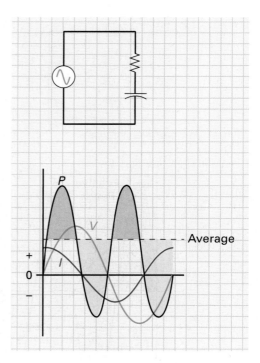

FIGURE 10-12 Power in a Resistive and Capacitive Circuit.

by the product of voltage and current. The negative alternation of the power cycle indicates that the capacitor is discharging and giving back the power that it consumed during the charge.

The positive alternation of the power cycle is much larger than the negative alternation because it is the combination of both the capacitor taking power during charge and the resistor consuming and dissipating power in the form of heat. The average power being dissipated will be some positive value, due to the heat being generated by the resistor.

Power Factor

In a purely resistive circuit, all the energy supplied to the resistor from the source is dissipated in the form of heat. This form of power is referred to as **resistive power** (P_R) or **true power,** and is calculated with the formula

$$P_R = I^2 \times R$$

In a purely capacitive circuit, all the energy supplied to the capacitor is stored from the source and then returned to the source, without energy loss. This form of power is referred to as **reactive power** (P_X) or **imaginary power.**

$$P_X = I^2 \times X_C$$

When a circuit contains both capacitance and resistance, some of the supply is stored and returned by the capacitor and some of the energy is dissipated and lost by the resistor.

Figure 10-13(a) illustrates another vector diagram. Just as the voltage across a resistor is 90° out of phase with the voltage across a capacitor, and resistance is 90° out of phase with reactance, resistive power will be 90° out of phase with reactive power.

If we take the three variables from Figure 10-13(b) to form a right-angle triangle as in Figure 10-13(c), we can vectorially add true power and imaginary power to produce a resultant **apparent power** vector. Apparent power is the power that appears to be supplied to the load and includes both the true power dissipated by the resistance and the imaginary power delivered to the capacitor.

Resistive Power or True Power

The average power consumed by a circuit during one complete cycle of alternating current.

Reactive Power or Imaginary Power

Also called wattless power, it is the power value obtained by multiplying the effective value of current by the effective value of voltage and the sine of the angular phase difference between current and voltage.

Apparent Power

The power value obtained in an ac circuit by multiplying together effective values of voltage and current, which reach their peaks at different times.

CHAPTER 10 / CAPACITIVE CIRCUITS, TESTING, AND APPLICATIONS

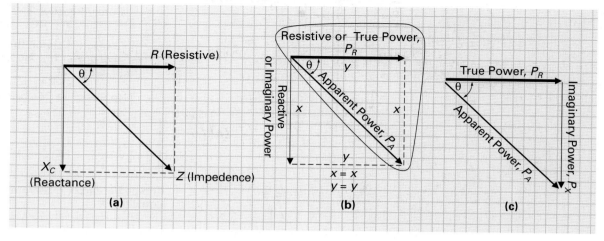

FIGURE 10-13 **Apparent Power.**

Applying the Pythagorean theorem once again, we can calculate apparent power by

$$P_A = \sqrt{P_R^2 + P_X^2}$$

where P_A = apparent power, in volt-amperes (VA)
P_R = true power, in watts (W)
P_X = reactive power, in volt-amperes reactive (VAR)

The **power factor** is a ratio of true power to apparent power and is a measure of the loss in a circuit. It can be calculated by using the formula

$$PF = \frac{\text{true power } (P_R)}{\text{apparent power } (P_A)}$$

Figure 10-14 helps explain what the power factor of a circuit actually indicates. In a purely resistive circuit, as shown in Figure 10-14(a), the apparent power will equal the true power ($P_A = \sqrt{P_R^2 + 0^2}$). The power factor will therefore equal 1 ($PF = P_R/P_A$). In a purely capacitive circuit, as shown in Figure 10-14(b), the true power will be zero, and therefore the power factor will equal 0 ($PF = 0/P_A$). A power factor of 1 therefore indicates a maximum power loss (circuit is resistive), while a power factor of 0 indicates no power loss (circuit is capacitive). With circuits that contain both resistance and reactance, the power factor will be somewhere between zero (0 = reactive) and one (1 = resistive).

Since true power is determined by resistance and apparent power is dependent on impedance, as shown in Figure 10-13(a), the power factor can also be calculated by using the formula

$$PF = \frac{R}{Z}$$

As the ratio of true power (adjacent) to apparent power (hypotenuse) determines the angle θ, the power factor can also be determined by the cosine of angle θ.

$$PF = \cos \theta$$

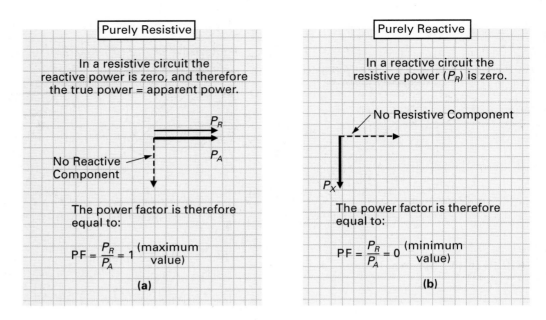

**FIGURE 10-14   Circuit's Power Factor. (a) Purely Resistive, PF = 1.
(b) Purely Reactive, PF = 0.**

EXAMPLE:

Calculate the following for a series RC circuit if R = 2.2 kΩ, X_C = 3.3 kΩ, and V_S = 5 V.

   a. Z

   b. I

   c. θ

   d. P_R

   e. P_X

   f. P_A

   g. PF

   Solution:

   a. $Z = \sqrt{R^2 + X_C^2}$
$$= \sqrt{2.2 \text{ k}\Omega^2 + 3.3 \text{ k}\Omega^2}$$
$$= 3.96 \text{ k}\Omega$$

   b. $I = \dfrac{V_S}{Z} = \dfrac{5 \text{ V}}{3.96 \text{ k}\Omega} = 1.26 \text{ mA}$

   c. $\theta = \text{invtan} \dfrac{X_C}{R} = \text{invtan} \dfrac{3.3 \text{ k}\Omega}{2.2 \text{ k}\Omega}$
$$= \text{invtan } 1.5 \ = \ 56.3°$$

   d. True power $= I^2 + R$
$$= 1.26 \text{ mA}^2 \times 2.2 \text{ k}\Omega$$
$$= 3.49 \text{ mW}$$

   e. Reactive power $= I^2 \times X_C$
$$= 1.26 \text{ mA}^2 \times 3.3 \text{ k}\Omega$$
$$= 5.24 \times 10^{-3} \text{ or } 5.24 \text{ mVAR}$$

f. Apparent power $= \sqrt{P_R{}^2 + P_X{}^2}$
$= \sqrt{3.49 \text{ mW}^2 + 5.24 \text{ mW}^2}$
$= 6.29 \times 10^{-3}$ or 6.29 mVA

g. Power factor $= \dfrac{R}{Z} = \dfrac{2.2 \text{ k}\Omega}{3.96 \text{ k}\Omega} = 0.55$

or

$= \dfrac{P_R}{P_A} = \dfrac{3.49 \text{ mW}}{6.29 \text{ mW}} = 0.55$

or

$= \cos \theta = \cos 56.3° = 0.55$

CALCULATOR SEQUENCE

Step	Keypad Entry	Display Response
1.	[3] [.] [3] [E] [3]	3.3E3
2.	[÷]	
3.	[2] [.] [2] [E] [3]	2.2E3
4.	[=]	1.5
5.	[inv] [tan]	56.309932

SELF-TEST EVALUATION POINT FOR SECTION 10-2

Now that you have completed this section, you should be able to:

■ **Objective 2.** *Describe impedance, phases angle, power, and power factor as they relate to a series and parallel RC circuit.*

Use the following questions to test your understanding of Section 10-2.

1. What is the phase relationship between current and voltage in a series *RC* circuit?

2. What is a phasor diagram?

3. Define and state the formula for impedance.

4. What is the phase angle or shift in:

  a. A purely resistive circuit?

  b. A purely capacitive circuit?

  c. A series circuit consisting of *R* and *C*?

10-3 PARALLEL *RC* CIRCUIT

Now that we have analyzed the characteristics of a series *RC* circuit, let us connect a resistor and capacitor in parallel.

10-3-1 *Voltage*

As with any parallel circuit, the voltage across all components in parallel is equal to the source voltage; therefore,

$$V_R = V_C = V_S$$

10-3-2 *Current*

In Figure 10-15(a), you will see a parallel circuit containing a resistor and a capacitor. The current through the resistor and capacitor is simply calculated by applying Ohm's law

$$\text{(resistor current) } I_R = \frac{V_S}{R}$$

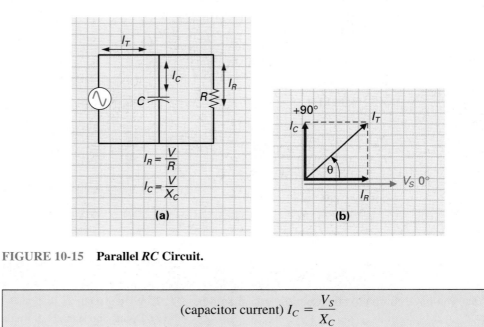

FIGURE 10-15 **Parallel *RC* Circuit.**

$$\text{(capacitor current) } I_C = \frac{V_S}{X_C}$$

Total current (I_T), however, is not as simply calculated. As expected, resistor current (I_R) is in phase with the applied voltage (V_S), as shown in the vector diagram in Figure 10-15(b). Capacitor current will always lead the applied voltage by 90°, and as the applied voltage is being used as our reference at 0° on the vector diagram, the capacitor current will have to be drawn at +90° in order to lead the applied voltage by 90°, since vector diagrams rotate in a counterclockwise direction.

Total current is therefore the vector sum of both the resistor and capacitor currents. Using the Pythagorean theorem, total current can be calculated by

$$I_T = \sqrt{I_R{}^2 + I_C{}^2}$$

10-3-3 *Phase Angle*

The angle by which the total current (I_T) leads the source voltage (V_S) can be determined with either of the following formulas:

$$\theta = \text{invtan} \frac{I_C \text{ (opposite)}}{I_R \text{ (adjacent)}} \qquad \theta = \text{invtan} \frac{R}{X_C}$$

10-3-4 *Impedance*

Since the circuit is both capacitive and resistive, the total opposition or impedance of the parallel *RC* circuit can be calculated by

$$Z = \frac{V_S}{I_T}$$

The impedance of a parallel *RC* circuit is equal to the total voltage divided by the total current. Using basic algebra, this basic formula can be rearranged to express impedance in terms of reactance and resistance.

$$Z = \frac{R \times X_C}{\sqrt{R^2 + X_C^2}}$$

10-3-5 *Power*

With respect to power, there is no difference between a series circuit and a parallel circuit. The true power or resistive power (P_R) dissipated by an RC circuit is calculated with the formula

$$P_R = I_R^2 \times R$$

The imaginary power or reactive power (P_X) of the circuit can be calculated with the formula

$$P_X = I_C^2 \times X_C$$

The apparent power is equal to the vector sum of the true power and the reactive power.

$$P_A = \sqrt{P_R^2 + P_X^2}$$

As with series RC circuits the power factor is calculated as

$$PF = \frac{P_R \ (\text{resistive power})}{P_A \ (\text{apparent power})}$$

A power factor of 1 indicates a purely resistive circuit, while a power factor of 0 indicates a purely reactive circuit.

EXAMPLE:

Calculate the following for a parallel RC circuit in which $R = 24\ \Omega$, $X_C = 14\ \Omega$, and $V_S = 10\ \text{V}$.

   a. I_R
   b. I_C
   c. I_T
   d. Z
   e. θ

Solution:

a. $I_R = \dfrac{V_S}{R} = \dfrac{10\ V}{24\ \Omega} = 416.66\ \text{mA}$

b. $I_C = \dfrac{V_S}{X_C} = \dfrac{10\ V}{14\ \Omega} = 714.28\ \text{mA}$

c. $I_T = \sqrt{I_R^2 + I_C^2}$

$\quad = \sqrt{416.66\ \text{mA}^2 + 714.28\ \text{mA}^2}$

$\quad = \sqrt{0.173 + 0.510}$

$\quad = 826.5\ \text{mA}$

$$\text{d. } Z = \frac{V_S}{I_T} = \frac{10 \text{ V}}{826.5 \text{ mA}} = 12 \text{ }\Omega$$

or

$$= \frac{R \times X_C}{\sqrt{R^2 + X_C{}^2}} = \frac{24 \times 14}{\sqrt{24^2 + 14^2}}$$

$$= 12 \text{ }\Omega$$

$$\text{e. } \theta = \arctan \frac{I_C}{I_R} = \arctan \frac{714.28 \text{ mA}}{416.66 \text{ mA}}$$

$$= \arctan 1.714 = 59.7°$$

or

$$\theta = \arctan \frac{R}{X_C} = \arctan \frac{24 \text{ }\Omega}{14 \text{ }\Omega}$$

$$= \arctan 1.714 = 59.7°$$

CALCULATOR SEQUENCE

Step	Keypad Entry	Display Response
1.	[7] [1] [4] [.] [2] [8] [E] [3] [+/−]	714.28E-3
2.	[÷]	
3.	[4] [1] [6] [.] [6] [6] [E] [3] [+/−]	416.66E-3
4.	[=]	1.7142994
5.	[inv] [tan]	59.743762

SELF-TEST EVALUATION POINT FOR SECTION 10-3

Use the following questions to test your understanding of Section 10-3.

1. What is the phase relationship between current and voltage in a parallel *RC* circuit?
2. Could a parallel *RC* circuit be called a voltage divider?
3. State the formula used to calculate:
   a. I_T
   b. Z
4. Will capacitor current lead or lag resistor current in a parallel *RC* circuit?

10-4 TESTING CAPACITORS

Now that you have a good understanding as to how a capacitor should function, let us investigate how to diagnose a capacitor malfunction.

10-4-1 *The Ohmmeter*

A faulty capacitor may have one of three basic problems:

1. A short, which is easy to detect and is caused by a contact from plate to plate.
2. An open, which is again quite easy to detect and is normally caused by one of the leads becoming disconnected from its respective plate.
3. A leaky dielectric or capacitor breakdown, which is quite difficult to detect, as it may only short at a certain voltage. This problem is usually caused by the deterioration of the dielectric, which starts displaying a much lower dielectric resistance than it was designed for. The capacitor with this type of problem is referred to as a *leaky capacitor*.

Capacitors of 0.5 μF and larger can be checked by using an analog ohmmeter or a digital multimeter with a bar graph display by using the procedure shown in Figure 10-16.

Step 1: Ensure that the capacitor is discharged by shorting the leads together.

Step 2: Set the ohmmeter to the highest ohms range scale.

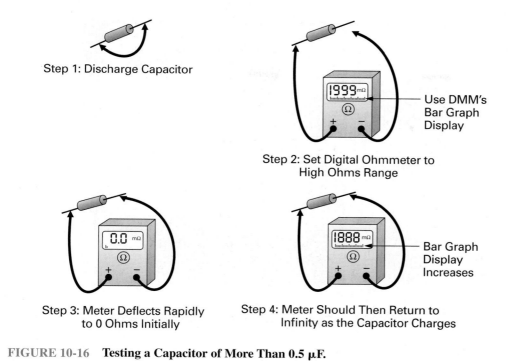

Step 1: Discharge Capacitor

Use DMM's Bar Graph Display

Step 2: Set Digital Ohmmeter to High Ohms Range

Step 3: Meter Deflects Rapidly to 0 Ohms Initially

Step 4: Meter Should Then Return to Infinity as the Capacitor Charges

Bar Graph Display Increases

FIGURE 10-16 **Testing a Capacitor of More Than 0.5 μF.**

Step 3: Connect the meter to the capacitor, observing the correct polarity if an electrolytic is being tested, and observe the bar graph display. The capacitor will initially be discharged, and therefore maximum current will flow from the meter battery to the capacitor. Maximum current means low resistance, which is why the meter's display indicates 0 Ω.

Step 4: As the capacitor charges, it will cause current flow from the meter's battery to decrease, and consequently, the bar graph display will increase.

A good capacitor will cause the meter to react as just explained. A larger capacitance will cause the meter to increase slowly to infinity (∞), as it will take a longer time to charge, while a smaller value of capacitance will charge at a much faster rate, causing the meter to increase rapidly toward ∞. For this reason, the ohmmeter cannot be reliably used to check capacitors with values of less than 0.5 μF, because the capacitor charges up too quickly and the meter does not have enough time to respond.

A shorted capacitor will cause the meter to show zero ohms and remain in that position. An open capacitor will cause a maximum reading (infinite resistance) because there is no path for current to flow.

A leaky capacitor will cause the bar graph to reduce its bars to the left, and then return almost all the way back to ∞ if only a small current is still flowing and the capacitor has a small dielectric leak. If the meter bars only come back to halfway or a large distance away from infinity, a large amount of current is still flowing and the capacitor has a large dielectric leak (defect).

When using the ohmmeter to test capacitors, there are some other points that you should be aware of:

1. Electrolytics are noted for having a small yet noticeable amount of inherent leakage, and so do not expect the meter's bar display to move all the way to the right (∞ ohms). Most electrolytic capacitors that are still functioning normally will show a resistance of 200 kΩ or more.

2. Some ohmmeters utilize internal batteries of up to 15 V, so be careful not to exceed the voltage rating of the capacitor.

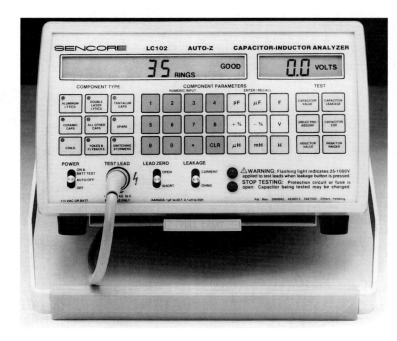

FIGURE 10-17   Capacitor Analyzer.

10-4-2   *The Capacitance Meter or Analyzer*

The ohmmeter check tests the capacitor under a low-voltage condition. This may be adequate for some capacitor malfunctions; however, a problem that often occurs with capacitors is that they short or leak at a high voltage. The ohmmeter test is also adequate for capacitors of 0.5 μF or greater; however, a smaller capacitor cannot be tested because its charge time is too fast for the meter to respond. The ohmmeter cannot check for high-voltage failure, for small-value capacitance, or if the value of capacitance has changed through age or extreme thermal exposure.

 A **capacitance meter** or analyzer, which is illustrated in Figure 10-17, can totally check all aspects of a capacitor in a range of values from approximately 1 pF to 20 F. The tests that are generally carried out include:

Capacitance Meter
Instrument used to measure the capacitance of a capacitor or circuit.

1. *Capacitor value change* [Figure 10-18(a)]: Capacitors will change their value over a period of time. Ceramic capacitors often change 10 to 15% within the first year as the ceramic material relaxes. Electrolytics change their value due to the electrolytic solution simply drying out. Some capacitors are simply labeled incorrectly by manufacturers or the technician cannot determine the correct value because of the labeling used. A value change accounts for approximately 25% of all defective capacitors.

2. *Capacitor leakage* [Figure 10-18(b)]: Leakage occurs due to an imperfection of the dielectric. Although the dielectric's resistance is very high, a small amount of leakage current will flow between the plates. This resistance, which is between the plates and therefore effectively in parallel with the capacitor, can become too low and cause the circuit that the capacitor is in to malfunction. Most capacitance meters will perform a leakage test with operating potentials up to 650 V. Leakage accounts for approximately 40% of all defective capacitors.

3. *Dielectric absorption* [Figure 10-18(c)]: This occurs mainly in electrolytics when they will take on a charge but will not fully discharge. This residual charge remains within the capacitor, similar to a small dc battery, and this changes the effective value of the capacitor once it is in circuit during operation. If this causes the value to change more

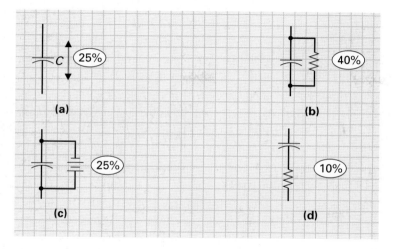

FIGURE 10-18 **Problems with Capacitors. (a) Value Change. (b) Leakage. (c) Dielectric Absorption. (d) Equivalent Series Resistance.**

than 15%, the capacitor should be rejected. Dielectric absorption accounts for approximately 25% of all defective capacitors.

4. *Equivalent series resistance* (ESR) [Figure 10-18(d)]: Series resistance is found in the capacitor leads, lead to plate connection, and electrolyte (almost always occurs in electrolytics) and causes the effective circuit capacitance value to change. Equivalent series resistance accounts for 10% of all defective capacitors.

SELF-TEST EVALUATION POINT FOR SECTION 10-4

Now that you have completed this section, you should be able to:

■ *Objective 3.* *Explain some of the more common capacitor failures and how to use an ohmmeter and capacitance analyzer to test them.*

Use the following questions to test your understanding of Section 10-4.

1. The analog ohmmeter or digital multimeter with bar graph can only be used to test capacitors that are _____ or more in value.

2. A _____ _____ tests a capacitor for value change, leakage, dielectric absorption, and equivalent series resistance.

10-5 APPLICATIONS OF CAPACITORS

There are many applications of capacitors, some of which will be discussed now; others will be presented later and in your course of electronic studies.

10-5-1 *Combining AC and DC*

Figure 10-19(a) and (b) show how the capacitor can be used to combine ac and dc. The capacitor is large in value (electrolytic typically) and can be thought of as a very large bucket that will fill or charge to a dc voltage level. The ac voltage will charge and discharge the capacitor, which is similar to pouring in and pulling out (alternating) more and less water. The resulting waveform is a combination of ac and dc that varies above and below an average dc level. In this instance, the ac is said to be superimposed on a dc level.

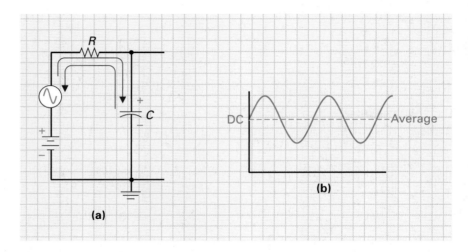

FIGURE 10-19 Superimposing AC on a DC Level.

10-5-2 *The Capacitive Voltage Divider*

In Chapter 12 we illustrated how capacitors could be connected in series across a dc voltage source to form a voltage divider. This circuit is repeated in Figure 10-20(a). Figure 10-20(b) shows how two capacitors can be used to divide up an ac voltage source of 30 V rms. Like a resistive voltage divider, voltage drop is proportional to current opposition, which with a capacitor is called *capacitive reactance* (X_C). The larger the capacitive reactance ($X_C\uparrow$), the larger the voltage drop ($V\uparrow$). Since capacitive reactance is inversely proportional to the capacitor value and the frequency of the ac source, a change in input frequency will cause a change in the capacitor's capacitive reactance, and this will change the voltage drop across the capacitor. Although a 90° phase shift is present between voltage and current in a purely capacitive circuit (*I* leads *V* by 90°), there is no phase difference between the input voltage and the output voltages.

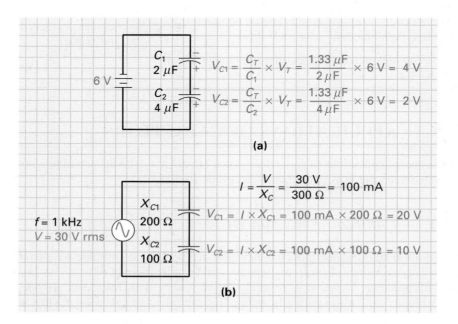

FIGURE 10-20 Capacitive Voltage Divider. (a) DC Circuit. (b) AC Circuit.

CHAPTER 10 / CAPACITIVE CIRCUITS, TESTING, AND APPLICATIONS

10-5-3   *RC Filters*

A **filter** is a circuit that allows certain frequencies to pass but blocks other frequencies. In other words, it filters out the unwanted frequencies but passes the wanted or selected ones.

There are two basic *RC* filters:

1. The **low-pass filter,** which can be seen in Figure 10-21(a); as its name implies, it passes the low frequencies but heavily **attenuates** the higher frequencies.
2. The **high-pass filter,** which can be seen in Figure 10-21(b); as its name implies, it allows the high frequencies to pass but heavily attenuates the lower frequencies.

With either the low- or high-pass filter, it is important to remember that capacitive reactance is inversely proportional to frequency ($X_C \propto 1/f$) as stated in the center of Figure 10-21.

With the low-pass filter shown in Figure 10-21(a), the output is connected across the capacitor. As the frequency of the input increases, the amplitude of the output decreases. At dc (0 Hz) and low frequencies, the capacitive reactance is very large ($X_C\uparrow = 1/f\downarrow$) with respect to the resistor. All the input will appear across the capacitor, because the capacitor and resistor form a voltage divider, as shown in Figure 10-22(a). As with any voltage divider, the

Filter

Network composed of resistor, capacitor, and inductors used to pass certain frequencies yet block others through heavy attenuation.

Low-Pass Filter

Network or circuit designed to pass any frequencies below a critical or cutoff frequency and reject or heavily attenuate all frequencies above.

Attenuate

To reduce in amplitude an action or signal.

High-Pass Filter

Network or circuit designed to pass any frequencies above a critical or cutoff frequency and reject or heavily attenuate all frequencies below.

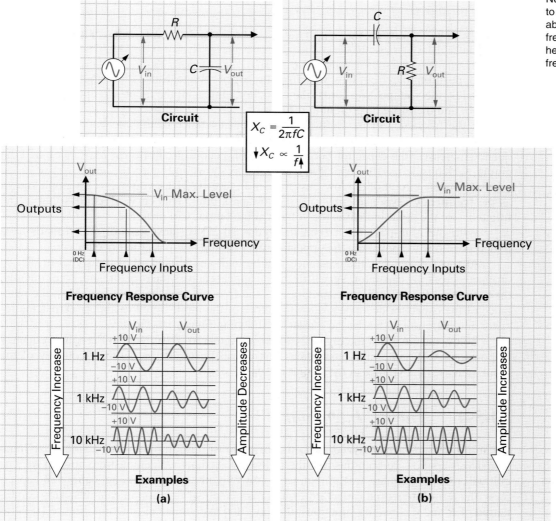

$$X_C = \frac{1}{2\pi fC}$$

$$\downarrow X_C \propto \frac{1}{f\uparrow}$$

FIGURE 10-21   *RC* Filters. (a) Low-Pass Filter. (b) High-Pass Filter.

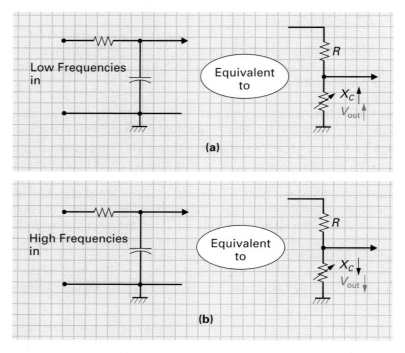

FIGURE 10-22 Low-Pass Filter. (a) Low-Pass Filter, High X_C. (b) Low-Pass Filter, Low X_C.

larger opposition to current flow will drop the larger voltage. Since the output voltage is determined by the voltage drop across the capacitor, almost all the input will appear across the capacitor and therefore be present at the output.

If the frequency of the input increases, the reactance of the capacitor will decrease $(X_C\downarrow = 1/f\uparrow)$, and a larger amount of the signal will be dropped across the resistor. As frequency increases, the capacitor becomes more of a short circuit (lower reactance), and the output, which is across the capacitor, decreases, as shown in Figure 10-22(b).

Below the circuit of the low-pass filter in Figure 10-21(a), you will see a graph known as the **frequency response curve** for the low-pass filter. This curve illustrates that as the frequency of the input increases the voltage at the output will decrease.

With the high-pass filter, seen in Figure 10-21(b), the capacitor and resistor have traded positions to show how the opposite effect for the low-pass filter occurs.

At low frequencies, the reactance will be high and almost all of the signal in will be dropped across the capacitor. Very little signal appears across the resistor, and, consequently, the output, as shown in Figure 10-23(a). As the frequency of the input increases, the reactance of the capacitor decreases, allowing more of the input signal to appear across the resistor and therefore appear at the output, as shown in Figure 10-23(b).

Below the circuit of the high-pass filter in Figure 10-21(b), you will see the frequency response curve for the high-pass filter. This curve illustrates that as the frequency of the input increases the voltage at the output increases.

10-5-4 *The RC Integrator*

Up until now we have analyzed the behavior of an RC circuit only when a sine-wave input signal was applied. In both this and the following sections we demonstrate how an RC circuit will react to a square-wave input, and show two other important applications of capacitors. The term **integrator** is derived from a mathematical function in calculus. This combination of R and C, in some situations, displays this mathematical function. Figure 10-24(a) illustrates an integrator circuit that can be recognized by the series connection of R and C, but mainly from the fact that the output is taken across the capacitor.

Frequency Response Curve

A graph indicating how effectively a circuit or device responds to the frequency spectrum.

Integrator

Device that approximates and whose output is proportional to an integral of the input signal.

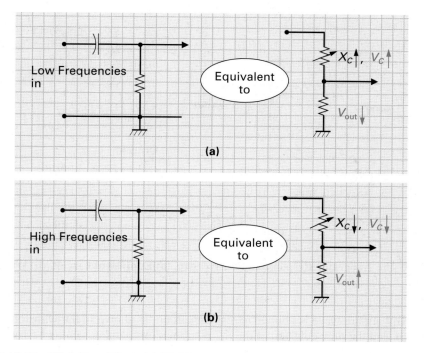

FIGURE 10-23   High-Pass Filter. (a) High X_C. (b) Low X_C.

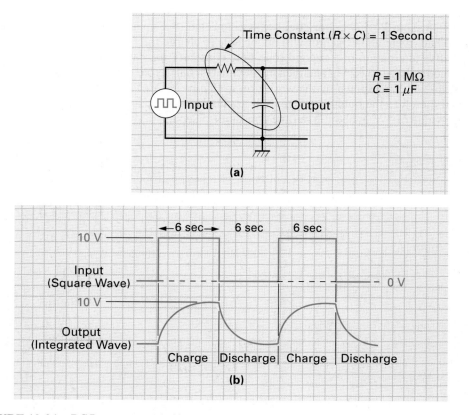

FIGURE 10-24   *RC* Integrator. (a) Circuit. (b) Waveforms.

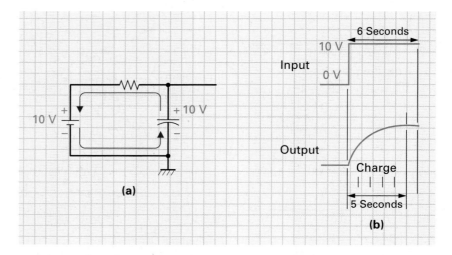

FIGURE 10-25 **Integrator Response to Positive Step. (a) Equivalent Circuit. (b) Waveforms for Charge.**

If a 10 V square wave is applied across the circuit, as shown in the waveforms in Figure 10-24(b), and the time constant of the *RC* combination calculates out to be 1 second, the capacitor will charge when the square wave input is positive toward the applied voltage (10 V) and reaches it in five time constants (5 seconds). Since the positive alternation of the square wave lasts for 6 seconds, the capacitor will be fully charged 1 second before the positive alternation ends, as shown in Figure 10-25.

When the positive half-cycle of the square-wave input ends after 6 seconds, the input falls to 0 V and the circuit is equivalent to that shown in Figure 10-26(a). The 10 V charged capacitor now has a path to discharge and in five time constants (5 seconds) is fully discharged, as shown in Figure 10-26(b).

If the same *RC* integrator circuit was connected to a square wave that has a 1-second positive half-cycle, the capacitor will not be able to charge fully toward 10 V. In fact, during the positive alternation of 1 second (one time constant), the capacitor will reach 63.2% of the applied voltage (6.32 V), and then during the 0 V half-cycle it will discharge to 63.2% of 6.32 V, to 2.33 V, as shown by the waveform in Figure 10-27. The voltage across the capacitor, and therefore the output voltage, will gradually build up and eventually level off to an average value of 5 V in about five time constants (5 seconds).

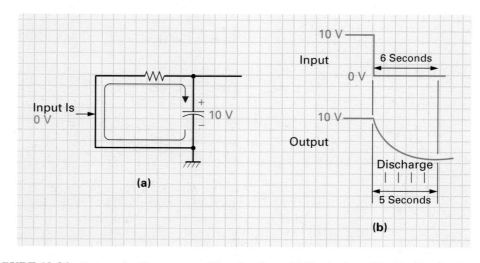

FIGURE 10-26 **Integrator Response to Negative Step. (a) Equivalent Circuit. (b) Waveforms for Discharge.**

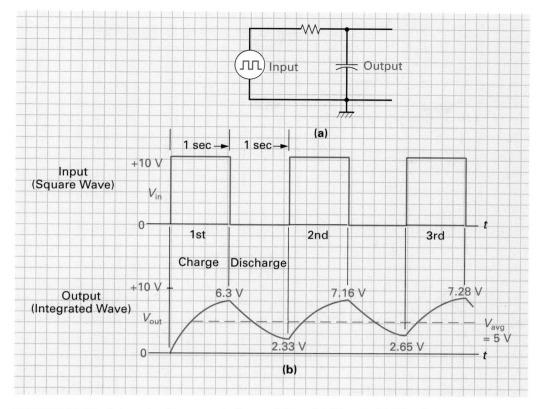

FIGURE 10-27 Integrator Response to Square Wave. (a) Circuit. (b) Waveforms.

In summary, if the period or time of the square wave is decreased, or if the time constant is increased, the same effect results. The capacitor has less time to charge and discharge and reaches an average value of the input voltage, which for a square wave is half of its amplitude.

Figure 10-28 illustrates the circuit waveform produced when the time constant is long with respect to the period of the input waveform. If the time constant is further increased so that it is even longer with respect to the input waveform, the output will have an even smaller peak-to-peak variation.

10-5-5   *The RC Differentiator*

Figure 10-29(a) illustrates the **differentiator** circuit, which is the integrator's opposite. In this case the output is taken across the resistor instead of the capacitor, and the time constant is always short with respect to the input square wave period.

The differentiator output waveform, shown in Figure 10-29(b), is taken across the resistor and is the result of the capacitor's charge and discharge. When the square wave swings positive, the equivalent circuit is that shown in Figure 10-30(a). When the 10 V is initially applied (positive step of the square wave), all the voltage is across the resistor, and therefore at the output, as the capacitor cannot charge instantly. As the capacitor begins to charge, more of the voltage is developed across the capacitor and less across the resistor. The voltage across the capacitor exponentially increases and reaches 10 V in five time constants (5×1 ms = 5 ms), while the voltage across the resistor, and therefore the output, exponentially falls from its initial 10 V to 0 V in five time constants as shown in Figure 10-30(b), at which time all the voltage is across the capacitor and no voltage will be across the resistor.

Differentiator
A circuit whose output voltage is proportional to the rate of change of the input voltage. The output waveform is then the time derivative of the input waveform, and the phase of the output waveform leads that of the input by 90°.

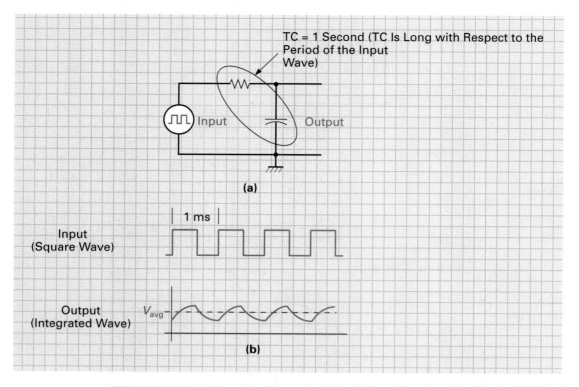

FIGURE 10-28 **Integrator Response to Long Time Constant. (a) Circuit. (b) Waveforms.**

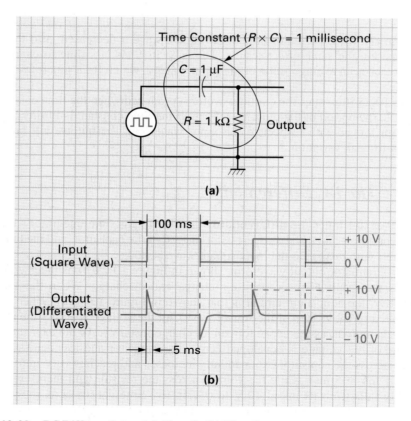

FIGURE 10-29 *RC* **Differentiator. (a) Circuit. (b) Waveforms.**

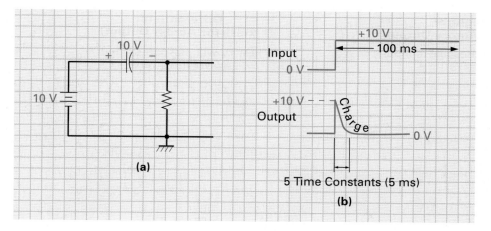

FIGURE 10-30 **Differentiator's Response to Positive Step. (a) Equivalent Circuit. (b) Waveforms.**

When the positive half-cycle of the square wave ends and the input falls to zero, the circuit is equivalent to that shown in Figure 10-31(a). The negative plate of the capacitor is now applied directly to the output. Since the capacitor cannot instantly discharge, the output drops suddenly down to -10 V as shown in Figure 10-31(b). This is the voltage across the resistor, and therefore the output. The capacitor is now in series with the resistor and therefore has a path through the resistor to discharge, which it does in five time constants to 0 V.

Figure 10-32(a) and (b) illustrates the integrator and differentiator circuits and waveforms, which are used extensively in many applications and equipment, such as computers, robots, lasers, and communications.

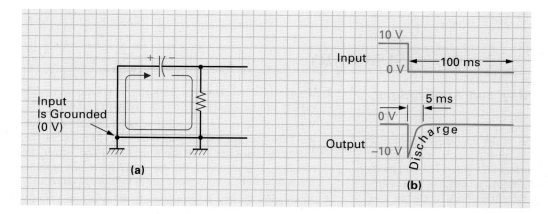

FIGURE 10-31 **Differentiator's Response to a Negative Step. (a) Equivalent Circuit. (b) Waveforms.**

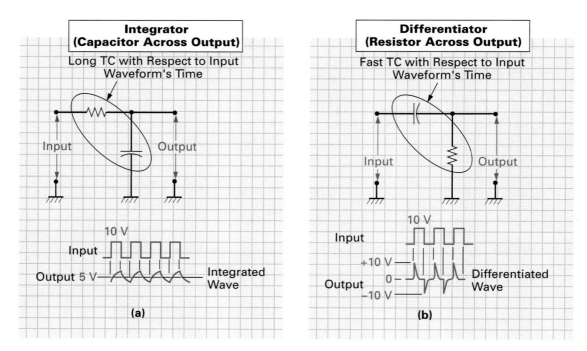

FIGURE 10-32 Summary of Integrator and Differentiator.

Now that you have completed this section, you should be able to:

■ **Objective 4.** *Describe how the capacitor and resistor in combination can be used:*
  a. To combine ac and dc
  b. To act as a voltage divider
  c. As a filter
  d. As an integrator
  e. As a differentiator

Use the following questions to test your understanding of Section 10-5.

1. Give three circuit applications for a capacitor.
2. With an *RC* low-pass filter, the _____ is connected across the output. Whereas with an *RC* high-pass filter, the _____ is connected across the output.
3. An integrator has a _____ time constant compared to the period of the input square wave.
4. A _____ circuit produces positive and negative spikes at the output when a square wave is applied at the input.

SUMMARY

Capacitive Circuits, Testing, and Applications (Figure 10-33)

1. Resistance (*R*), by definition, is the opposition to current flow with the dissipation of energy and is measured in ohms. Capacitors oppose current flow like a resistor, but a resistor dissipates energy, whereas a capacitor stores energy (when it charges) and then gives back its energy into the circuit (when it discharges). Capacitive reactance (*X_C*), by definition, is the opposition to current without the dissipation of energy and is also measured in ohms.

2. If capacitive reactance is basically opposition, it is inversely proportional to the amount of current flow.

3. When a dc source is connected across a capacitor, current will flow only for a short period of time (5*RC* seconds) to charge the capacitor. After this time, there is no further current flow. Consequently, the capacitive reactance or opposition offered by a capacitor to dc is infinite (maximum).

4. Alternating current is continuously reversing in polarity, resulting in the capacitor continuously charging and discharging. this means that charge and discharge currents are always flowing around the circuit, and if we have a certain

FIGURE 10-33 Capacitive Circuits, Testing, and Applications.

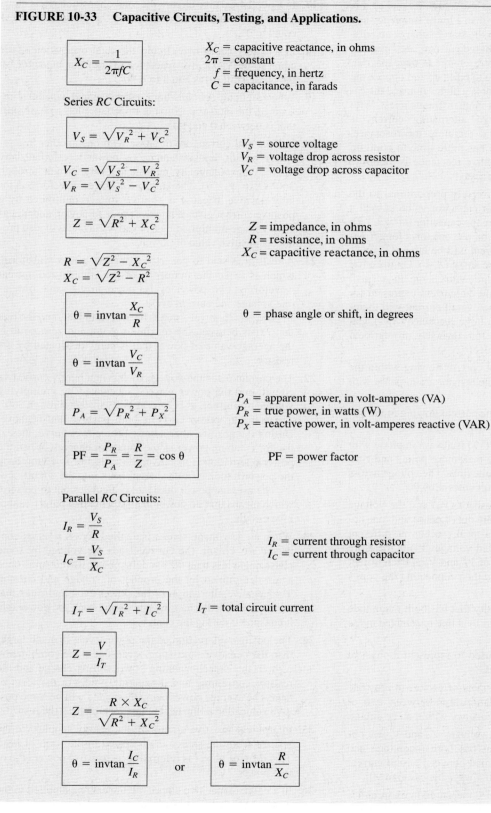

$$X_C = \frac{1}{2\pi f C}$$

X_C = capacitive reactance, in ohms
2π = constant
f = frequency, in hertz
C = capacitance, in farads

Series *RC* Circuits:

$$V_S = \sqrt{V_R{}^2 + V_C{}^2}$$

V_S = source voltage
V_R = voltage drop across resistor
V_C = voltage drop across capacitor

$$V_C = \sqrt{V_S{}^2 - V_R{}^2}$$
$$V_R = \sqrt{V_S{}^2 - V_C{}^2}$$

$$Z = \sqrt{R^2 + X_C{}^2}$$

Z = impedance, in ohms
R = resistance, in ohms
X_C = capacitive reactance, in ohms

$$R = \sqrt{Z^2 - X_C{}^2}$$
$$X_C = \sqrt{Z^2 - R^2}$$

$$\theta = \text{invtan}\,\frac{X_C}{R}$$

θ = phase angle or shift, in degrees

$$\theta = \text{invtan}\,\frac{V_C}{V_R}$$

$$P_A = \sqrt{P_R{}^2 + P_X{}^2}$$

P_A = apparent power, in volt-amperes (VA)
P_R = true power, in watts (W)
P_X = reactive power, in volt-amperes reactive (VAR)

$$PF = \frac{P_R}{P_A} = \frac{R}{Z} = \cos\theta$$

PF = power factor

Parallel *RC* Circuits:

$$I_R = \frac{V_S}{R}$$
$$I_C = \frac{V_S}{X_C}$$

I_R = current through resistor
I_C = current through capacitor

$$I_T = \sqrt{I_R{}^2 + I_C{}^2}$$

I_T = total circuit current

$$Z = \frac{V}{I_T}$$

$$Z = \frac{R \times X_C}{\sqrt{R^2 + X_C{}^2}}$$

$$\theta = \text{invtan}\,\frac{I_C}{I_R} \quad \text{or} \quad \theta = \text{invtan}\,\frac{R}{X_C}$$

value of current, we must also have a certain value of reactance or opposition.

5. If the applied alternating current is at a high frequency it is switching polarity more rapidly than a lower frequency and there is very little time between the start of charge and discharge. As the charge and discharge currents are largest at the beginning of the charge and discharge of the capacitor, the reactance has very little time to build up and oppose the current, which is why the current is a high value and the capacitive reactance is small at higher frequencies. With lower frequencies, the applied alternating current is switching at a slower rate, and therefore the reactance, which is low at the beginning, has more time to build up and oppose the current. Capacitive reactance is therefore inversely proportional to frequency.

6. Capacitive reactance is also inversely proportional to the value of capacitance. If a larger capacitor value is used a longer time is required to charge the capacitor ($t\uparrow = C\uparrow R$), which means that current will be present for a longer period of time, so the overall current will be large ($I\uparrow$); consequently, the reactance must be small ($X_C\downarrow$). On the other hand, a small capacitance value will charge in a small amount of time ($t\downarrow = C\downarrow R$) and the current is present for only a short period of time. The overall current will therefore be small ($I\downarrow$), indicating a large reactance ($X_C\uparrow$). Capacitive reactance (X_C) is therefore inversely proportional to both frequency and capacitance.

7. In a purely resistive circuit, the current flowing within the circuit and the voltage across the resistor are in phase with one another.

8. In a purely capacitive circuit, the current flowing in the circuit leads the voltage across the capacitor by 90°.

9. The voltage across the resistor (V_R) is always in phase with the circuit current (I), because maximum points and zero crossover points occur at the same time. The voltage across the capacitor (V_C) lags the circuit current by 90°.

10. The voltage drop across the resistor (V_R) and the voltage drop across the capacitor (V_C) are out of phase with one another, which means that their peaks occur at different times. The signal for the applied voltage (V_S) is therefore obtained by adding the values of V_C and V_R at each instant in time, plotting the results, and then connecting the points with a line.

11. A vector (or phasor) is a quantity that has both magnitude and direction and is represented by a line terminated at the end by an arrowhead.

12. Vectors are generally always used to represent a physical quantity that has two properties.

13. A vector diagram is an arrangement of vectors to illustrate the magnitude and phase relationships between two or more quantities of the same frequency within an ac circuit.

14. By vectorially adding the two voltages V_R and V_C we obtain a resultant V_S vector that has both magnitude and phase. The angle theta (θ) formed between circuit current (I) and source voltage (V_S) will always be less than 90°.

15. Impedance (designated Z) is also measured in ohms and is the total circuit opposition to current flow. It is a combination of resistance (R) and reactance (X_C).

16. By vectorially adding R and X_C, we have a resulting impedance (Z) vector.

17. In a purely resistive circuit, the total opposition (Z) is equal to the resistance of the resistor, so the phase shift (θ) is equal to 0°.

18. In a purely capacitive circuit, the total opposition (Z) is equal to the capacitive reactance (X_C) of the capacitor, so the phase shift (θ) is equal to −90°.

19. When a circuit contains both resistance and capacitive reactance, the total opposition or impedance has a phase shift that is between 0 and 90°.

20. Voltage and current are in phase with one another in a resistive circuit, and instantaneous power is calculated by multiplying voltage by current at every instant through 360° ($P = V \times I$). The sinusoidal power waveform is totally positive, because a positive voltage multiplied by a positive current gives a positive value of power, and a negative voltage multiplied by a negative current will also produce a positive value of power.

21. A resistor is said to generate a positive power waveform, which is twice the frequency of the voltage and current waveforms; two power cycles occur in the same time as one voltage and current cycle.

 The power waveform has been split in half, and the line that exists between the maximum point and zero point is the average value of power that is being dissipated by the resistor.

22. The current leads the voltage by 90° in a purely capacitive circuit, and the power wave is calculated by multiplying voltage by current, as before, at every instant through 360°. The resulting power curve is both positive and negative. During the positive alternation of the power curve, the capacitor is taking power as the capacitor charges. When the power alternation is negative, the capacitor is giving back the power it took as it discharges back into the circuit.

 The average power dissipated calculates out to be zero, which means that no power is dissipated in a purely capacitive circuit.

23. When an ac voltage source is applied across a series-connected RC circuit, the current leads the voltage by some phase angle less than 90°, and the power waveform is once again determined by the product of voltage and current. The negative alternation of the power cycle indicates that the capacitor is discharging and giving back the power that it consumed during the charge.

24. The positive alternation of the power cycle is much larger than the negative alternation because it is the combination of both the capacitor taking power during charge and the resistor consuming and dissipating power in the form of heat. The average power being dissipated will be some positive value, due to the heat being generated by the resistor.

25. In a purely resistive circuit, all the energy supplied to the resistor from the source is dissipated in the form of heat. This form of power is referred to as resistive power (P_R) or true power.

26. In a purely capacitive circuit, all the energy supplied to the capacitor is stored from the source and then returned to the

source, without energy loss. This form of power is referred to as reactive power (P_X) or imaginary power.

27. When a circuit contains both capacitance and resistance, some of the supply is stored and returned by the capacitor and some of the energy is dissipated and lost by the resistor.

28. Apparent power is the power that appears to be supplied to the load and includes both the true power dissipated by the resistance and the imaginary power delivered to the capacitor.

29. The power factor is a ratio of true power to apparent power and is a measure of the loss in a circuit.

30. As with any parallel circuit, the voltage across all components in parallel is equal to the source voltage.

31. The current through the resistor and capacitor is simply calculated by applying Ohm's law.

32. Capacitor current will always lead the applied voltage by 90°, and as the applied voltage is being used as our reference at 0° on the vector diagram, the capacitor current will have to be drawn at +90° in order to lead the applied voltage by 90°, since vector diagrams rotate in a counterclockwise direction.

33. Total current is therefore the vector sum of both the resistor and capacitor currents.

34. The angle by which the total current (I_T) leads the source voltage (V_S) is called the phase angle.

35. The impedance of a parallel RC circuit is equal to the total voltage divided by the total current.

36. With respect to power, there is no difference between a series circuit and a parallel circuit.

37. A faulty capacitor may have one of three basic problems:
    a. A short, which is easy to detect and is caused by a contact from plate to plate.
    b. An open, which is again quite easy to detect and is normally caused by one of the leads becoming disconnected from its respective plate.
    c. A leaky dielectric or capacitor breakdown, which is quite difficult to detect, as it may only short at a certain voltage. This problem is usually caused by the deterioration of the dielectric, which starts displaying a much lower dielectric resistance than it was designed for. The capacitor with this type of problem is referred to as a *leaky capacitor*.

38. Capacitors of 0.5 μF and larger can be checked by using a digital multimeter with a bar graph display.

39. The ohmmeter check tests the capacitor under a low-voltage condition. This may be adequate for some capacitor malfunctions; however, a problem that often occurs with capacitors is that they short or leak at a high voltage. A capacitance meter or analyzer can totally check all aspects of a capacitor in a range of values from approximately 1 pF to 20 F. The tests that are generally carried out include:
    a. Capacitor value change
    b. Capacitor leakage
    c. Dielectric absorption
    d. Equivalent series resistance

40. The capacitor can be used to combine ac and dc.

41. Capacitors could be connected in series across a dc voltage source to form a voltage divider.

42. A filter is a circuit that allows certain frequencies to pass but blocks other frequencies. In other words, it filters out the unwanted frequencies but passes the wanted or selected ones.
    There are two basic RC filters:
    a. The low-pass filter, which as its name implies, passes the low frequencies but heavily attenuates the higher frequencies.
    b. The high-pass filter, which as its name implies, allows the high frequencies to pass but heavily attenuates the lower frequencies.

43. The frequency response curve for the low-pass filter illustrates that as the frequency of the input increases the voltage at the output will decrease. The frequency response curve for the high-pass filter illustrates that as the frequency of the input increases the voltage at the output increases.

44. The term integrator is derived from a mathematical function in calculus. An integrator circuit can be recognized by the series connection of R and C, but mainly from the fact that the output is taken across the capacitor. The voltage across the capacitor of an integrator, and therefore the output voltage, will gradually build up and eventually level off to an average value in about five time constants.

45. With the differentiator circuit, the output is taken across the resistor instead of the capacitor, and the time constant is always short with respect to the input square wave period. The differentiator output waveform is taken across the resistor and is the result of the capacitor's charge and discharge.

REVIEW QUESTIONS

Multiple-Choice Questions

1. Capacitive reactance is inversely proportional to:
   a. Capacitance and resistance
   b. Frequency and capacitance
   c. Capacitance and impedance
   d. Both (a) and (c)

2. The impedance of an RC series circuit is equal to:
   a. The sum of R and X_C
   b. The square root of the sum of R^2 and X_C^2
   c. The square of the sum of R and X_C
   d. The sum of the square root of R and X_C

3. In a purely resistive circuit:
   a. The current flowing in the circuit leads the voltage across the capacitor by 90°.
   b. The circuit current and resistor voltage are in phase with one another.
   c. The current leads the voltage by 45°.
   d. The current leads the voltage by a phase angle between 0 and 90°.

4. In a purely capacitive circuit:
   a. The current flowing in the circuit leads the voltage across the capacitor by 90°.
   b. The circuit current and resistor voltage are in phase with one another.
   c. The current leads the voltage by 45°.
   d. The current leads the voltage by a phase angle between 0 and 90°.

5. In a series circuit containing both capacitance and resistance:
   a. The current flowing in the circuit leads the voltage across the capacitor by 90°.
   b. The circuit current and resistor voltage are in phase with one another.
   c. The current leads the voltage by 45°.
   d. Both (a) and (b).

6. In a series RC circuit, the source voltage is equal to:
   a. The sum of V_R and V_C
   b. The difference between V_R and V_C
   c. The vectoral sum of V_R and V_C
   d. The sum of V_R and V_C squared

7. As the source frequency is increased, the capacitive reactance will:
   a. Increase
   b. Decrease
   c. Be unaffected
   d. Increase, depending on harmonic content

8. The phase angle of a series RC circuit indicates by what angle V_S —————— V_R.
   a. Lags   c. Leads or lags
   b. Leads   d. None of the above

9. In a series RC circuit, the vector combination of R and X_C is the circuit's ——————.
   a. Phase angle   c. Source voltage
   b. Apparent power   d. Impedance

10. In a parallel RC circuit, the total current is equal to:
    a. The sum of I_R and I_C
    b. The difference between I_R and I_C
    c. The vectoral sum of I_R and I_C
    d. The sum of I_R and I_C squared

11. —————————— is the opposition offered by a capacitor to current flow without the dissipation of energy.
    a. Capacitive reactance   d. Phase angle
    b. Resistance   e. The power factor
    c. Impedance

12. —————————— is the total reactive and resistive circuit opposition to current flow.
    a. Capacitive reactance   d. Phase angle
    b. Resistance   e. The power factor
    c. Impedance

13. —————————— is the ratio of true (resistive) power to apparent power and is therefore a measure of the loss in a circuit.
    a. Capacitive reactance   d. Phase angle
    b. Resistance   e. The power factor
    c. Impedance

14. In a series RC circuit, the leading voltage will be measured across the:
    a. Resistor   c. Source
    b. Capacitor   d. Any of the choices are true.

15. In a series RC circuit, the lagging voltage will be measured across the:
    a. Resistor   c. Source
    b. Capacitor   d. Any of the choices are true.

Communication Skill Questions

16. Give the formula and define the term *capacitive reactance*. (10-1)

17. In a series RC circuit, give the formulas for calculating: (10-2)
    a. V_S, when V_R and V_C are known
    b. Z, when R and X_C are known
    c. Z, when I and V are known
    d. θ, when X_C and R are known
    e. θ, when V_C and V_R are known
    f. Power factor, when R and Z are known
    g. Power factor, when P_R and P_A are known

18. In a parallel RC circuit, give the formulas for calculating: (10-3)
    a. I_R, when V and R are known
    b. I_C, when V and X_C are known
    c. I, when I_R and I_C are known
    d. Z, when V and I are known
    e. Z, when R and X_C are known

19. What is meant by *long* or *short time constant*, and do large or small values of RC produce a long or a short time constant? (Chapter 9)

20. Describe how the inverse relationship between frequency and capacitive reactance can be used for the application of filtering. (10-5-3)

21. Illustrate the voltage and current waveforms across a resistor and capacitor in series when a dc voltage is applied during charge and when the same resistor and capacitor are connected to discharge. (Chapter 9)

22. Describe and illustrate how the capacitor can be used in the following applications:
    a. Combining ac and dc (10-5-1)
    b. A voltage divider (10-5-2)
    c. Filtering high and low frequencies (10-5-3)
    d. Integrating a square wave (10-5-4)
    e. Differentiating a square wave (10-5-5)

23. Describe the difference between reactance, resistance, and impedance. (10-2)

24. Sketch the phase relationships between: (10-2)
    a. V_R and I in a purely resistive circuit
    b. V_C and I in a purely capacitive circuit

25. What are positive power and negative power? (10-2-5)

FIGURE 10-34

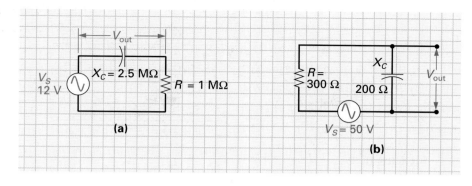

(a)

(b)

Practice Problems

26. Calculate the capacitive reactance of the capacitor circuits with the following parameters:
  a. $f = 1$ kHz, $C = 2$ μF
  b. $f = 100$ Hz, $C = 0.01$ μF
  c. $f = 17.3$ MHz, $C = 47$ μF

27. In a series RC circuit, the voltage across the capacitor is 12 V and the voltage across the resistor is 6 V. Calculate the source voltage.

28. Calculate the impedance for the following series RC circuits:
  a. 2.7 MΩ, 3.7 μF, 20 kHz
  b. 350 Ω, 0.005 μF, 3 MHz
  c. $R = 8.6$ kΩ, $X_C = 2.4$ Ω
  d. $R = 4700$ Ω, $X_C = 2$ kΩ

29. In a parallel RC circuit with parameters of $V_S = 12$ V, $R = 4$ MΩ, and $X_C = 1.3$ kΩ, calculate:
  a. I_R **c.** I_T **e.** θ
  b. I_C **d.** Z

30. Calculate the total reactance in:
  a. A series circuit where $X_{C1} = 200$ Ω, $X_{C2} = 300$ Ω, $X_{C3} = 400$ Ω
  b. A parallel circuit where $X_{C1} = 3.3$ kΩ, $X_{C2} = 2.7$ kΩ

31. Calculate the capacitance needed to produce 10 kΩ of reactance at 20 kHz.

32. At what frequency will a 4.7 μF capacitor have a reactance of 2000 Ω?

33. A series RC circuit contains a resistance of 40 Ω and a capacitive reactance of 33 Ω across a 24 V source.
  a. Sketch the schematic diagram.
  b. Calculate Z, I, V_R, V_C, I_R, I_C, and θ.

34. A parallel RC circuit contains a resistance of 10 kΩ and a capacitive reactance of 5 kΩ across a 100 V source.
  a. Sketch the schematic diagram.
  b. Calculate I_R, I_C, I_T, Z, V_R, V_C, and θ.

35. Calculate V_R and V_C for the circuits seen in Figure 10-34(a) and (b).

36. Calculate the impedance of the four circuits shown in Figure 10-35.

37. In Figure 10-36, the output voltage, since it is taken across the capacitor, will _____ the voltage across the resistor by _____ degrees.

38. If the positions of the capacitor and resistor in Figure 10-36 are reversed, the output voltage, since it is now taken across the resistor, will _____ the voltage across the capacitor by _____ degrees.

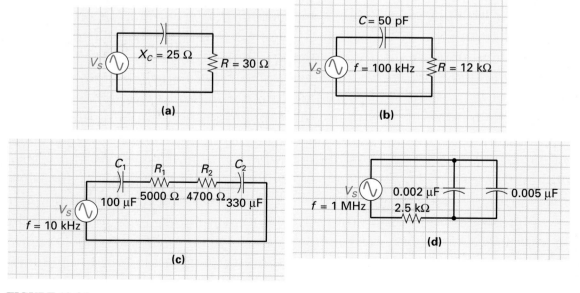

FIGURE 10-35

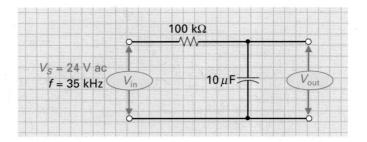

FIGURE 10-36

39. Calculate the resistive power, reactive power, apparent power, and power factor for the circuit seen in Figure 10-36; V_{in} = 24 V and f = 35 kHz.

40. Refer to Figure 10-37 and calculate the following:
   a. [Figure 10-37(a)] X_C, I, Z, I_R, θ, V_R, V_C
   b. [Figure 10-37(b)] V_R, V_C, I_R, I_C, I_T, Z, θ

Troubleshooting Problems

41. Describe the three basic problems that normally occur with faulty capacitors.

42. Describe how to use the ohmmeter to check capacitors and also explain some of its limitations.

43. Describe the four basic tests performed by a capacitor meter or analyzer.

44. Which capacitor problem accounts for the largest percent-age of defective capacitors? Explain exactly what this malfunction is.

45. If the bar graph display of a meter goes to zero ohms and remains there, the capacitor is _____. If the capacitor is _____, however, no charging will occur and the meter will indicate infinite ohms.

Web Site Questions

Go to the Web site http://www.prenhall. com/cook, select the textbook *Introductory DC/AC Electronics* or *Introductory DC/AC Circuits*, this chapter, and then follow the instructions when answering the multiple-choice practice problems.

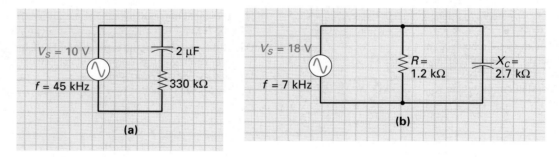

FIGURE 10-37

These tests at the end of each chapter will challenge your knowledge up to this point, and give you the practice you need for a job interview. To make this more realistic, the test will comprise both technical and personal questions. In order to take full advantage of this exercise, you may want to set up a simulation of the interview environment, have a friend read the questions to you, and record your responses for later analysis.

Company Name: S.A.F.E., Inc.

Industry Branch: Industrial

Function: Design and Manufacture Security Systems

Job Title: Engineering Technician

1. How would you rate your electronics education?

2. What do you know about surface mount technology?

3. What is reactance?

4. How do you feel about the task of constantly documenting your duties?

5. What is a differentiator circuit?

6. What is a transducer?

7. What would you say are the responsibilities of an engineering technician?

8. What would you say is the difference between electrical systems and electronic systems?

9. What are your long term professional goals?

10. What are the differences and similiarities between a rechargeable battery and a capacitor?

Answers

1. Answer must be, "very good up to this point," but go on to explain that you see your educaton as an ongoing process, and like industry, you will always be in a learning mode.
2. Chapter 3.
3. Section 10-1.
4. Documenting the engineering process is vitally important. As an engineering technician you will be called on to document design successes, and failures, in explicit detail. They are testing here to see if you have any attitude towards a process that most engineers and technicians shy away from. Discuss any lab or theory reports you have completed, and how your education has been centered around technical reading and writing.
5. Section 10-5-5.
6. Chapter 2.
7. Quote the details described in the introduction (Engineering technician) and the responsibilities listed in the job advertisement.
8. Chapter 10 introduction.
9. This is an individual choice, however, most people discuss furthering their education, and finding a job that is both stimulating and rewarding. Be sure to carefully consider your choices—you want to appear confident but not arrogant, show a desire to advance but not be overambitious.
10. Expect the interviewer to sometimes ask somewhat obscure, basic questions, that can often set you back on your heels. Take a moment before you answer, and then start. "Well, they both have plates, can be charged and discharged, but their applications are quite different. . . ."

Electromagnetism and Electromagnetic Induction

The Great Experimenter

Michael Faraday was born to James and Margaret Faraday on September 22, 1791. At twenty-two, the gifted and engaging Faraday was at the right place at the right time, and with the right talents. He impressed the brilliant professor Humphry Davvy, who made him his research assistant at Count Rumford's Royal Institution. After only two years, Faraday was given a promotion and an apartment at the Royal Institution, and in 1821 he married. He lived at the Royal Institution for the rest of his active life, working in his laboratory, giving very dynamic lectures, and publishing over 450 scientific papers. Unlike many scientific papers, Faraday's papers would never use a calculus formula to explain a concept or idea. Instead, Faraday would explain all of his findings using logic and reasoning, so that a person trying to understand science did not need to be a scientist. It was this gift of being able to make even the most complex areas of science easily accessible to the student, coupled with his motivational teaching style, that made him so popular.

In 1855 he had written three volumes of papers on electromagnetism, the first dynamo, the first transformer, and the foundations of electrochemistry, a large amount on dielectrics and even some papers on plasma. The unit of capacitance is measured in *farads,* in honor of his work in these areas of science.

Faraday began two series of lectures at the Royal Institution, which have continued to this day. Michael and Sarah Faraday were childless, but they both loved children, and in 1860 Faraday began a series of Christmas lectures expressly for children, the most popular of which was called "The Chemical History of a Candle." The other lecture series was the "Friday Evening Discourses," of which he himself delivered over a hundred. These very dynamic, enlightening, and entertaining lectures covered areas of science or technology for the lay person, and were filled with demonstrations. On one evening in 1846, an extremely nervous and shy speaker ran off just moments before he was scheduled to give the Friday Evening Discourse. Faraday had to fill in, and to this day a tradition is still enforced whereby the lecturer for the Friday Evening Discourse is locked up for half an hour before the presentation with a single glass of whiskey.

Faraday was often referred to as "the great experimenter," and it was this consistent experimentation that led to many of his findings. He was fascinated by science and technology, and was always exploring new and sometimes dangerous horizons. In fact, in one of his reports he states, "I have escaped, not quite unhurt, from four explosions." When asked to comment on experimentation, his advice was to "Let your imagination go, guiding it by judgment and principle, but holding it in and directing it by experiment. Nothing is so good as an experiment which, while it sets an error right, gives you as a reward for your humility an absolute advance in knowledge."

Outline and Objectives

Introduction

It was during a classroom lecture in 1820 that Danish physicist Hans Christian Oersted accidentally stumbled on an interesting reaction. As he laid a compass down on a bench he noticed that the compass needle pointed to an adjacent conductor that was carrying a current, instead of pointing to the earth's north pole. It was this discovery that first proved that magnetism and electricity were very closely related to one another. This phenomenon is now called *electromagnetism* since it is now known that any conductor carrying an electro or electrical current will produce a magnetic field.

In 1831, the English physicist Michael Faraday explored further Oersted's discovery of electromagnetism and found that the process could be reversed. Faraday observed that if a conductor was passed through a magnetic field, a voltage would be induced in the conductor and cause a current to flow. This phenomenon is referred to as *electromagnetic induction*.

In this chapter we examine the terms and characteristics of electromagnetism (electricity to magnetism) and electromagnetic induction (magnetism to electricity).

Electromagnetism

Relates to the magnetic field generated around a conductor when current is passed through it.

11-1   ELECTROMAGNETISM

The electron plays a very important role in magnetism. However, the real key or link between electricity and a magnetic field is motion. Anytime a charged particle moves, a magnetic field is generated. As a result, current flow, which is the movement of electrons, produces a magnetic field.

11-1-1   *Atomic Theory of Electromagnetism*

Every orbiting electron in motion around its nucleus generates a magnetic field. When electrons are forced to leave their parent atom by voltage and flow toward the positive polarity, they are all moving in the same direction and each electron's magnetic field will add to the next. The accumulation of all these electron fields will create a magnetic field around the conductor, as shown in Figure 11-1.

A simple experiment can be performed to prove that this invisible magnetic field does in fact exist around a conductor, the setup of which is illustrated in Figure 11-2. With the switch open, as shown in Figure 11-2(a), no current will flow, and therefore no mag-

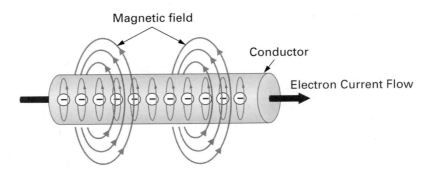

FIGURE 11-1   Electron Magnetic Fields.

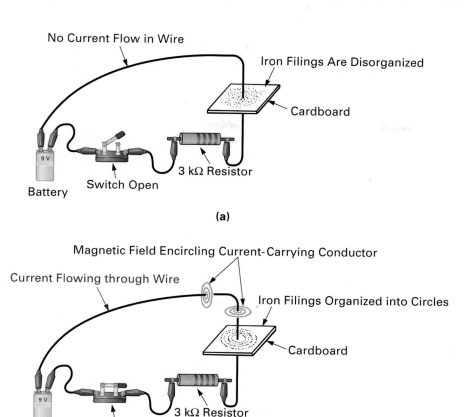

No Current Flow in Wire

Iron Filings Are Disorganized

Cardboard

9 V

3 kΩ Resistor

Switch Open

Battery

(a)

Magnetic Field Encircling Current-Carrying Conductor

Current Flowing through Wire

Iron Filings Organized into Circles

Cardboard

9 V

3 kΩ Resistor

Switch Closed

Battery

(b)

FIGURE 11-2   **Magnetic Field Experiment. (a) Switch Open, No Current Flow, No Magnetic Field. (b) Switch Closed, Current Flow, Magnetic Field.**

netic field is generated around the conductor and the iron filings on the cardboard will be disorganized.

With the switch closed, a current of 3 mA ($I = V/R = 9 \text{ V}/3 \text{ k}\Omega = 3 \text{ mA}$) will flow through the circuit and a magnetic field will be set up around the conductor. This magnetic field will cause the iron filings to become organized in circles, as shown in Figure 11-2(b).

SELF-TEST EVALUATION POINT FOR SECTION 11-1-1

Now that you have completed this section, you should be able to:

■ **Objective 1.**   *Describe what is meant by the word* electromagnetism.

■ **Objective 2.**   *Explain the atomic theory of electromagnetism.*

Use the following questions to test your understanding of Section 11-1-1.

1. True or false: An electric field always encircles a current-carrying conductor.

2. What subatomic particle is said to have its own magnetic field?

11-1-2   *DC, AC, and the Electromagnet*

A magnetic field results whenever a current flows through any piece of conductor or wire, as shown in Figure 11-3(a). If a conductor is wound to form a spiral, as illustrated in Figure 11-3(b), the conductor, which is now referred to as a **coil,** will, as a result of current flow,

Coil

Number of turns of wire wound around a core to produce magnetic flux (an electromagnet) or to react to a changing magnetic flux (an inductor).

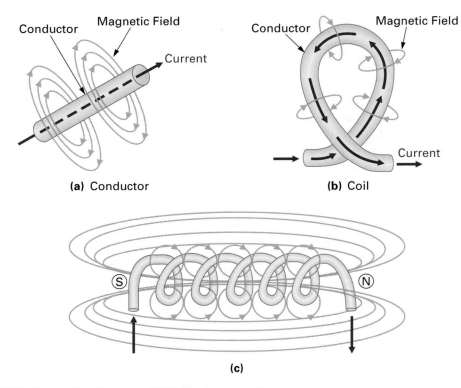

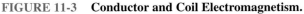

FIGURE 11-3 **Conductor and Coil Electromagnetism.**

develop a magnetic field, which will sum or intensify within the coil. If many coils are wound in the same direction, an **electromagnet** is formed, as shown in Figure 11-3(c), which will produce a concentrated magnetic field whenever a current is passed through its coils.

The **left-hand rule** can be applied to electromagnets to determine which of the poles will be the north end of the electromagnet. This rule is illustrated in Figure 11-4, which shows that if you wrap the fingers of your left hand around the coil so that your fingers are pointing in the direction of current flow, your thumb will be pointing to the north end of the electromagnet.

Up until now, we have only been discussing current flow through a coil in one direction (dc). A dc voltage produces a fixed current in one direction and therefore generates a magnetic field in a coil of fixed polarity, as shown in Figure 11-5(a).

Alternating current (ac) is continually varying, and as the polarity of the magnetic field is dependent on the direction of current flow (left-hand rule), the magnetic field will also be alternating in polarity, as shown in Figure 11-5(b). Let us look at times 1 through 4 in Figure 11-5 in more detail.

Time 1: The alternating voltage has risen to a maximum positive level and causes current to flow in the direction seen in the circuit. This will cause a magnetic field with a south pole above and a north pole below.

FIGURE 11-4 **Left-Hand Rule for Electromagnets.**

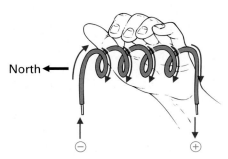

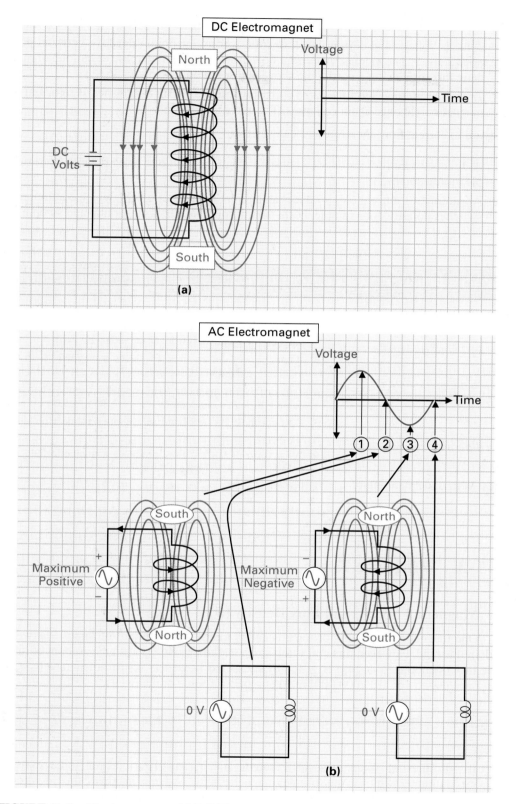

FIGURE 11-5 **Electromagnets. (a) DC Electromagnet. (b) AC Electromagnet.**

Time 2: Between positions 1 and 2, the voltage, and therefore current, will decrease from a maximum positive value to zero. This will cause a corresponding collapse of the magnetic field from maximum (time 1) to zero (time 2).

Time 3: Voltage and current increase from zero to maximum negative between positions 2 and 3. The increase in current flow causes a similar increase or buildup of magnetic flux, producing a north pole above and south pole below.

Time 4: From time 3 to time 4, the current within the circuit diminishes to zero, and the magnetic field once again collapses. The cycle then repeats.

In regard to current flow, therefore, we can say that:

1. A direct current (dc) produces a constant magnetic field of a fixed polarity, for example, north–south.
2. An alternating current (ac) produces an alternating magnetic field, which continuously switches polarity, for example, north–south, south–north, north–south, and so on.

SELF-TEST EVALUATION POINT FOR SECTION 11-1-2

Now that you have completed this section, you should be able to:

■ **Objective 3.** *Describe the left-hand rule of electromagnetism.*

■ **Objective 4.** *Describe how a magnetic field is generated by current flow through a:*
   a. Conductor
   b. Coil

■ **Objective 5.** *Describe the difference between dc and ac electromagnetism.*

Use the following questions to test your understanding of Section 11-1-2.

1. What name could be used instead of *electromagnet?*
2. True or false: The greater the number of loops in an electromagnet, the greater or stronger the magnetic field.
3. What form of current flow produces a magnetic field that maintains a fixed magnetic polarity?
4. True or false: A magnetic field is produced only when ac is passed through a conductor.
5. Describe how the left-hand rule applies to electromagnetism.
6. What form of current flow produces a magnetic field that continuously switches in polarity?

11-1-3   *Magnetic Terms*

Magnetic Flux (ϕ) and Flux Density (B)

Maxwell

One magnetic line of force or flux is called a maxwell.

One magnetic line of flux or force is called a **maxwell,** in honor of James Maxwell's work in this field. However, this unit is too small and impractical and so the amount of **magnetic flux** (symbolized ϕ, phi) is generally measured in webers instead of maxwells. One *weber* is equal to 10^8 (100,000,000) magnetic lines of force or maxwells.

Magnetic Flux

The magnetic lines of force produced by a magnet.

> magnetic flux (ϕ) = number of lines of force (or maxwells),
> in webers

If 10^8 lines of force exist in 1 square meter, and 10^8 lines of force exist in 1 square centimeter, which is the stronger magnetic field? As far as magnetic flux (ϕ) is concerned, 1 weber exists in both. So some other way is needed to specify how many lines of force exist in a given area, and this will tell us if it is a strong or weak magnetic field.

Flux Density

A measure of the strength of a wave.

Flux density (B) is equal to the number of magnetic lines of flux (ϕ) per square meter, and it is given in the unit of tesla (T).

$$\text{flux density } (B) = \frac{\text{magnetic flux } (\phi)}{\text{area } (A)}$$

where B = flux density, in teslas (T)
ϕ = Magnetic flux, in webers (Wb)
A = area, in square meters (m^2)

EXAMPLE:

If magnet A produces 10^8 (100,000,000) lines of flux (1 weber) in 1 square centimeter, and magnet B produces 10^8 lines of flux in 1 square meter, which magnet is producing the more concentrated or intense magnetic field?

Solution:

Since there are 10,000 square centimeters in 1 square meter, 1 square centimeter (1 cm^2) equals one ten-thousandth of a square meter (0.01 m^2).

$$\text{Magnet } A: \text{flux density} = \frac{1 \text{ weber}}{0.0001 \text{ (m}^2)}$$
$$= 10,000 \text{ tesla}$$
$$\text{Magnet } B: \text{flux density} = \frac{1 \text{ weber}}{1 \text{ (m}^2)}$$
$$= 1 \text{ tesla (T)}$$

Flux density determined that magnet A produced the more concentrated magnetic force.

Magnetomotive Force (MMF)

The magnetic flux produced by an electromagnet is generated by current flowing through a coil of wire. As discussed previously, electromotive force (emf) is the pressure or voltage that forces electrons to move. **Magnetomotive force** (mmf) is the magnetic pressure that produces the magnetic field. The formula for mmf is

$$\text{mmf} = I \times N \text{ (ampere-turns)}$$

where mmf = magnetomotive force, in ampere-turns (At)
I = current, in amperes
N = number of turns in the coil

The formula basically says that if you increase the current through the coil or increase the number of turns in the coil, you will increase the magnetic pressure (magnetomotive force), and by increasing magnetic pressure, you will increase the magnetic field produced, as shown in Figure 11-6.

EXAMPLE:

What is the magnetomotive force produced when 3 A of current flows through 5 turns in a coil?

Solution:

$$\text{mmf} = I \text{ (current)} \times N \text{ (number of turns)}$$
$$= 3 \text{ A} \times 5 \text{ turns}$$
$$= 15 \text{ At (ampere-turns)}$$

TIME LINE

In 1600, William Gilbert (1544-1603), an English physician, documented many years of research and experiments on magnets and magnetic bodies such as amber and lodestones. Probably his most important discovery was that when rubbed with a cloth, amber would attract lightweight objects. In 1601 he was appointed physician to Queen Elizabeth I at a salary of $150 a year. He was the first to believe that the earth was nothing but a large magnet and that its magnetic field causes a needle to align itself between north and south.

Magnetomotive Force
Force that produces a magnetic field.

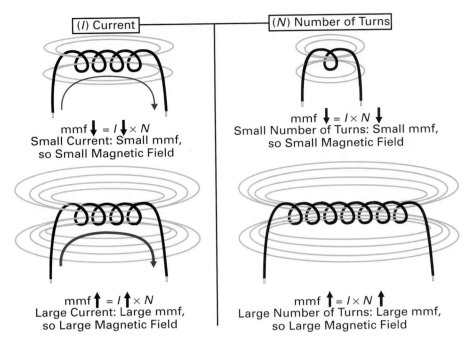

FIGURE 11-6   **Magnetomotive Force.**

Magnetizing Force (*H*)

The previous example showed that magnetomotive force (mmf) is equal to the product of current and the number of turns. But if a 20 turn coil or solenoid is stretched out to twice its length, the magnetic field or force will be half as strong, because there would be less of a reinforcing effect between the coils due to the greater distance between the coils. The length of the coil is therefore a factor that also determines the field intensity, and this new term, **magnetizing force** (*H*), is the reference we commonly use to describe the magnetic field intensity.

$$\text{magnetizing force } (H) = \frac{I \times N}{l} \quad or \quad H = \frac{\text{mmf}}{l}$$

where H = magnetizing force in ampere-turns/meter (At/m)
I = current, in amperes (A)
N = number of coil turns
l = length of coil, in meters (m)

Reluctance (ℜ)

Reluctance, in reference to magnetic energy, is equivalent to resistance in electrical energy. Reluctance is the opposition or resistance to the establishment of a magnetic field in an electromagnet. The formula for calculating reluctance is the magnetic energy equivalent of the electrical Ohm's law as shown:

$$\text{reluctance } (\Re) = \frac{\overset{\textit{Magnetic}}{\text{mmf (magnetomotive force)}}}{\phi \text{ (magnetic flux)}}$$

(measured in ampere-turns/weber)

$$\text{resistance } (R) = \frac{V \text{ (electromotive force)}}{I \text{ (current)}}$$

Electric

(measured in ohms)

Permeability (μ)

As magnetic reluctance is equivalent to electrical resistance, magnetic permeability is equivalent to electrical conductance. **Permeability** is a measure of how easily a material will allow a magnetic field to be set up within it. Permeability is symbolized by the Greek lowercase letter mu (μ) and is measured in henrys per meter (H/m). A high permeability figure ($\mu\uparrow$) indicates that a magnetic field can easily be established within a material and therefore that this material's reluctance must be low ($\Re\downarrow$). On the other hand, a low permeability figure ($\mu\downarrow$) indicates that there will be a large reluctance ($\Re\uparrow$) to establishing a magnetic field. This is mathematically stated as:

Magnetic

$$\text{permeability } (\mu) = \frac{1}{\Re \text{ (reluctance)}}$$

(Permeability is inversely proportional to reluctance and is measured in henrys per meter.)

Electric

$$\text{conductance } (G) = \frac{1}{R \text{ (resistance)}}$$

(Conductance is inversely proportional to resistance and is measured in siemens.)

The permeability figures in henrys/meter for different materials are listed in Table 11-1. **Relative** (with respect to) is a word that means that a comparison has to be made. Relative permeability (μ_r) is the measure of how well another given material will conduct magnetic lines of force with respect to, or relative to, our reference material, air, which has a

TABLE 11-1 Permeabilities of Various Materials

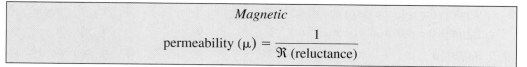

MATERIAL	RELATIVE PERMEABILITY (μ_r)	PERMEABILITY (μ)
Air or vacuum	1	1.26×10^{-6}
Nickel	50	6.28×10^{-5}
Cobalt	60	7.56×10^{-5}
Cast iron	90	1.1×10^{-4}
Machine steel	450	5.65×10^{-4}
Transformer iron	5,500	6.9×10^{-3}
Silicon iron	7,000	8.8×10^{-3}
Permalloy	100,000	0.126
Supermalloy	1,000,000	1.26

relative permeability value of 1. Referring to the relative permeability column in Table 11-1, you can see that magnetic lines of flux will pass through nickel 50 times easier than through air. The relative permeability (μ_r) of air, which is equal to 1, should not be confused with the absolute permeability (μ_0) of free space or air, which is equal to $4\pi \times 10^{-7}$ or 1.26×10^{-6}.

$$\mu = \mu_r \times \mu_0$$

Permeability = relative permeability × absolute permeability of air

Whether permeability (μ), relative permeability (μ_r), or absolute permeability (μ_0) is used as a standard makes little difference. A high permeability figure indicates a low reluctance ($\mu\uparrow$, $\Re\downarrow$), and vice versa.

■ **EXAMPLE:**

If the magnetic flux produced by a material is equal to 335 µWb and the mmf equals 15 At:

    a. What is the reluctance of the material?

    b. What is the material's permeability?

■ *Solution:*

CALCULATOR SEQUENCE

Step	Keypad Entry	Display Response
1.	[1] [5]	15
2.	[÷]	
3.	[3] [3] [5] [E] [6] [+/−]	335E-6
4.	[=]	44776.1
5.	[1/X]	22.3E-6

a. Reluctance ($\Re$) = $\dfrac{\text{mmf (magnetomotive force)}}{\phi \text{ (magnetic flux)}}$

$= \dfrac{15 \text{ At (ampere-turns)}}{335 \text{ µWb}}$

$= 44.8 \times 10^3 \text{ At/Wb}$

b. Permeability (μ) = $\dfrac{1}{44.8 \times 10^3 \text{ (reluctance)}}$

$= 22.3 \times 10^{-6}$ henrys/meter (H/m)

Summary

The magnetic field strength of an electromagnet can be increased by increasing the number of turns in its coil, increasing the current through the electromagnet, or decreasing the length of the coil ($H = I \times N/l$). The field strength can be increased further if an iron core is placed within the electromagnet, as illustrated in Figure 11-7(a). This is because an iron core has less reluctance or opposition to the magnetic lines of force than air, so the flux density (B) is increased. Another way of saying this is that the permeability or conductance of magnetic lines of force within iron is greater than that of air, and if the permeability of iron is large, its reluctance must be small. The symbol for an iron-core electromagnet is seen in Figure 11-7(b).

11-1-4   *Flux Density (B) versus Magnetizing Force (H)*

B–H Curve

Curve plotted on a graph to show successive states during magnetization of a ferromagnetic material.

The **B–H curve** in Figure 11-8 illustrates the relationship between the two most important magnetic properties: flux density (B) and magnetizing force (H). Figure 11-8(a) illustrates the B–H curve, while Figure 11-8(b) illustrates the positive rising portion of the ac current that is being applied to the iron-core electromagnetic circuit in Figure 11-8(c).

The magnetizing force is actually equal to $H = I$ (current) $\times N$ (number of turns) /l (length of coil); but since the number of turns and length of coil are fixed for the coil being used, the magnetizing force (H) is proportional to the current (I) applied, which is shown in Figure 11-8(b). The positive rise of the current from zero to maximum positive is applied

CHAPTER 11 / ELECTROMAGNETISM AND ELECTROMAGNETIC INDUCTION

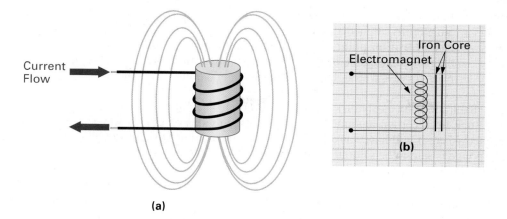

(a)

FIGURE 11-7   **Iron-Core Electromagnet. (a) Operation. (b) Symbol.**

through the electromagnet and will produce a corresponding bloom or buildup in magnetic flux at a rate indicated by the *B–H* curve shape in Figure 11-8(a).

It is important to note that as the magnetizing force (current) is increased, there are three distinct stages in the change of flux density or magnetic flux out.

Stage 1:   Up to this point, the increase in flux density is slow, as a large amount of force is required to begin alignment of the molecule magnets.

Stage 2:   Increase in flux density is now rapid and almost linear as the molecule magnets are aligning easily.

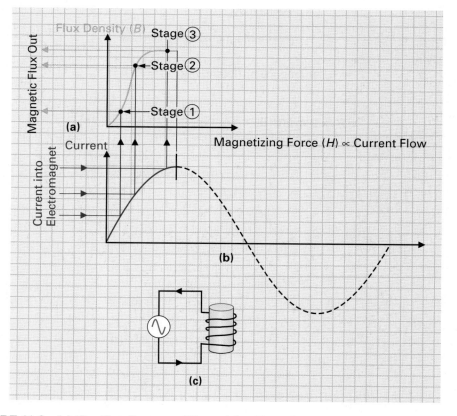

FIGURE 11-8   **(a) Flux Density versus Magnetizing Force Curve (*B–H* Curve). (b) Current Applied to Circuit. (c) Circuit.**

Saturation Point

The point beyond which an increase in one of two quantities produces no increase in the other.

Hysteresis

A lag between cause and effect. With magnetism it is the amount that the magnetization of a material lags the magnetizing force due to molecular friction.

Stage 3: In this state the molecule magnets cannot be magnetized any further because they are all fully aligned and no more flux density can be easily obtained. This is called the **saturation point,** and it is the state of magnetism beyond which an electromagnet is incapable of further magnetic strength. It is the point beyond which the *B–H* curve is a straight, horizontal line, indicating no change.

Saturation can easily be described by a simple analogy of a sponge. A dry sponge can only soak up a certain amount of water. As it continues to absorb water, a point will be reached where it will have soaked up the maximum amount of water possible. At this point, the sponge is said to be saturated with water, and no matter how much extra water you supply, it cannot hold any more.

The electromagnet is saturated at stage 3 and cannot produce any more magnetic flux, even though more magnetizing force is supplied as the sine wave continues on to its maximum positive level.

Looking at these three stages and the *B–H* curve that is produced, you can see that, in fact, the magnetization (setting up of the magnetic field, *B*) lags the magnetizing force (*H*) because of molecular friction. This lag or time difference between magnetizing force (*H*) and flux density (*B*) is known as **hysteresis.**

Figure 11-9(a) illustrates what is known as a *hysteresis loop,* which is formed when you plot magnetizing force (*H*) against flux density (*B*) through a complete cycle of alter-

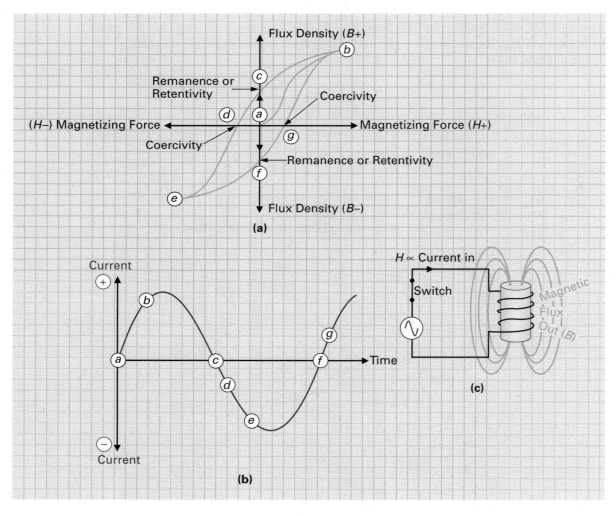

FIGURE 11-9   *B–H* Hysteresis Loop.

nating current, as seen in Figure 11-9(b). Initially, when the electric circuit switch is open, the iron core is unmagnetized. Therefore, both H and B are zero at point a. When the switch is closed, as seen in Figure 11-9(c), the current [Figure 11-9(b)] is increased and flux density [Figure 11-9(a)] increases until saturation point b is reached. This part of the waveform (from point a to b) is exactly the same as the B–H curve, discussed previously. The current continues on beyond saturation point b, but the flux density cannot increase beyond saturation.

At point c, the magnetizing force (current) is zero and B (flux density) falls to a value c that is the positive magnetic flux remaining after the removal of the magnetizing force (H). This particular value of flux density is termed **remanence or retentivity.**

The current or magnetizing force now reverses, and the amount of current in the reverse direction that causes flux density to be brought down from c to zero (point d) after the core has been saturated is termed the **coercive force.** The current, and therefore magnetizing force, continues on toward a maximum negative until saturation in the opposite magnetic polarity occurs at point e.

At point f, the magnetizing force (current) is zero and B falls to remanence, which is the negative magnetic flux remaining after the removal of the magnetizing force H. The value of current between f and g is the coercive force needed in the reverse direction to bring flux density down to zero (point g).

Remanence or Retentivity

Amount a material remains magnetized after the magnetizing force has been removed.

Coercive Force

Magnetizing force needed to reduce the residual magnetism within a material to zero.

SELF-TEST EVALUATION POINT FOR SECTION 11-1-3 AND 11-1-4

Now that you have completed this section, you should be able to:

- **Objective 6.** *Explain the following magnetic terms:*
  a. Magnetic flux
  b. Flux density
  c. Magnetizing force
  d. Magnetomotive force
  e. Reluctance
  f. Permeability (relative and absolute)
- **Objective 7.** *Explain the relationship between flux density and magnetizing force.*
- **Objective 8.** *Describe the cycle known as the hysteresis loop.*

Use the following questions to test your understanding of Sections 11-1-3 and 11-1-4.

Define the following:
1. Magnetic flux
2. Flux density
3. Magnetomotive force
4. Magnetizing force
5. Reluctance
6. Permeability
7. True or false: The hysteresis loop is formed when you plot H against B through a complete cycle of ac.
8. What is magnetic saturation?

11-1-5 *Applications of Electromagnetism*

Magnetic-Type Circuit Breaker

The magnetic-type circuit breaker shown in Figure 11-10 was first discussed in Chapter 4 and is one application of an electromagnet. This and all other circuit breakers are used for current protection. If the rated current value or below is passing through the circuit breaker, the electromagnet will not generate a strong enough magnetic field to pull the iron arm to the left and release the catch and open the contacts. However, if the current rating of the circuit breaker is exceeded, the increase in current will cause a corresponding increase in magnetic flux, attracting the iron arm, opening the contacts, and protecting the equipment from the dangerous level of current.

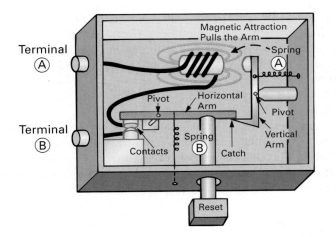

FIGURE 11-10   Magnetic-Type Circuit Breaker.

Electric Bell

The electric bell in Figure 11-11 utilizes a soft-iron core electromagnet, a striker, and a gong. When the bell push is pressed, a complete electrical circuit is made from the positive terminal to the electromagnet, the contact, the striker, and then back to the negative terminal. The electromagnet energizes (produces a magnetic field) and attracts the soft-iron striker, which strikes the gong and at the same time breaks the circuit, deenergizing the electromagnet. The striker is pulled back by the spring, and the circuit is reestablished, resulting in a continuous ringing of the bell.

Relay

Relay
Electromechanical device that opens or closes contacts when a current is passed through a coil.

A **relay** is an electromechanical device that either makes (closes) or breaks (opens) a circuit by moving contacts together or apart. Figure 11-12(a) shows the normally open (NO) relay and Figure 11-12(b) shows the normally closed (NC) relay.

Operation.   In both cases, the relay consists basically of an electromagnet connected to lines *x* and *y*, a movable iron arm known as the armature, and a set of contacts.

FIGURE 11-11   Electric Bell.

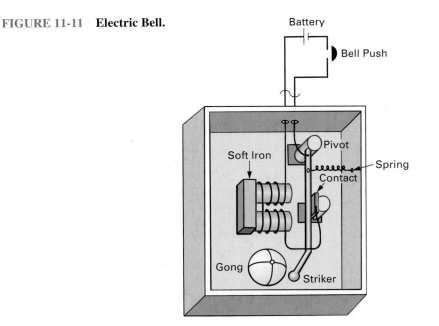

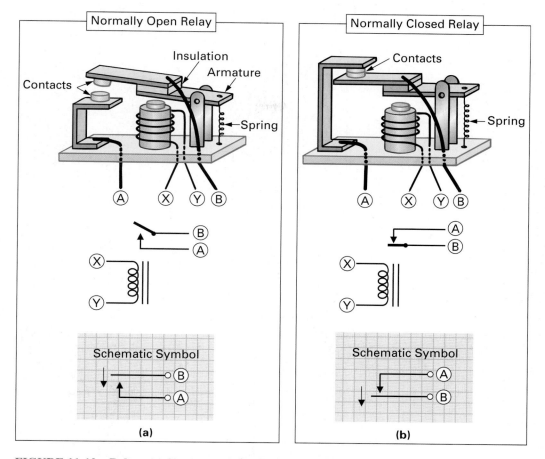

FIGURE 11-12 Relays (a) Single-Pole, Single-Throw (SPST), Normally Open (NO) Relay (Contacts Are Open until Activated). (b) Single-Pole, Single-Throw (SPST), Normally Closed (NC) Relay (Contacts Are Closed until Activated).

When current passes from x to y, the electromagnet generates a magnetic field, or is said to be **energized,** which attracts the armature toward the electromagnet. When this occurs, it closes or makes the normally open relay contacts and opens or breaks the normally closed relay contacts.

If the electromagnet is deenergized by discontinuing the current through the coil, the spring will pull back the armature to open the NO relay contacts or close the NC relay contacts between A and B.

The "normal" condition for the contacts between A and B is when the electromagnet is deenergized. In the deenergized condition, the normally open relay contacts are open and the normally closed relay contacts are closed.

The two relays discussed so far are actually single-pole, single-throw relays, as they have one movable contact (single pole) and one stationary contact that the pole can be thrown to, as shown in Figure 11-12. There are actually four basic configurations for relays and all are illustrated in Table 11-2(a). Variations on these basic four can come in all shapes and sizes, with one relay controlling sometimes several sets of contacts. Table 11-2(b) shows several different styles and packages of relays that are available.

Applications of a relay. The relay is generally used in two basic applications:

1. To enable one master switch to operate several remote or difficultly placed contact switches, as illustrated in Figure 11-13. When the master switch is closed, the relay is energized, closing all its contacts and turning on all the lights. The advantage of this is twofold in that, first, the master switch can turn on three lights at one time, which

Energized

Being electrically connected to a voltage source so that the device is activated.

TABLE 11-2 Relay Types.

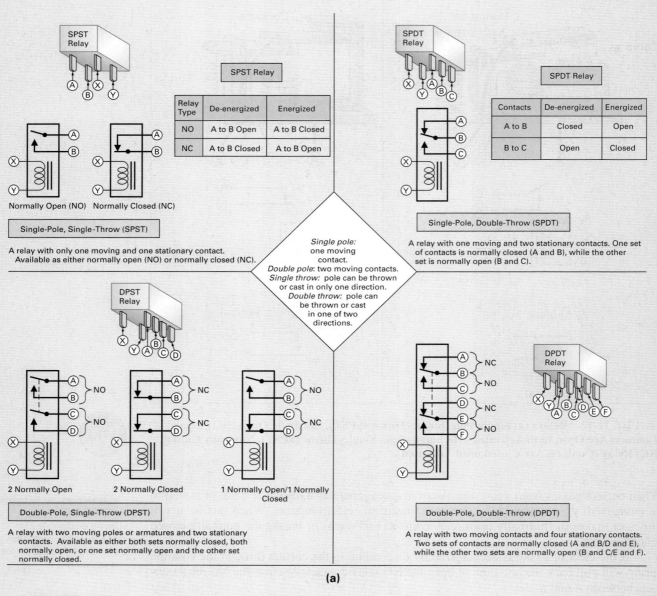

Single-Pole, Single-Throw (SPST)

SPST Relay

Relay Type	De-energized	Energized
NO	A to B Open	A to B Closed
NC	A to B Closed	A to B Open

Normally Open (NO)   Normally Closed (NC)

A relay with only one moving and one stationary contact. Available as either normally open (NO) or normally closed (NC).

Single-Pole, Double-Throw (SPDT)

SPDT Relay

Contacts	De-energized	Energized
A to B	Closed	Open
B to C	Open	Closed

A relay with one moving and two stationary contacts. One set of contacts is normally closed (A and B), while the other set is normally open (B and C).

Single pole: one moving contact.
Double pole: two moving contacts.
Single throw: pole can be thrown or cast in only one direction.
Double throw: pole can be thrown or cast in one of two directions.

Double-Pole, Single-Throw (DPST)

2 Normally Open   2 Normally Closed   1 Normally Open/1 Normally Closed

A relay with two moving poles or armatures and two stationary contacts. Available as either both sets normally closed, both normally open, or one set normally open and the other set normally closed.

Double-Pole, Double-Throw (DPDT)

A relay with two moving contacts and four stationary contacts. Two sets of contacts are normally closed (A and B/D and E), while the other two sets are normally open (B and C/E and F).

(a)

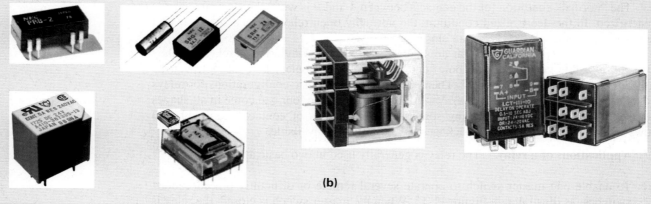

(b)

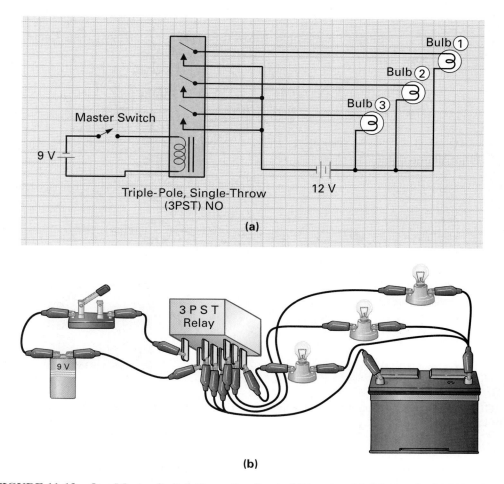

FIGURE 11-13   One Master Switch Operating Several Remotes. (a) Schematic. (b) Pictorial.

saves time for the operator, and second, only one set of wires need be taken from the master switch to the lights, rather than three sets for all three lights.

2. The second basic application of the relay is to enable a switch in a low-voltage circuit to operate relay contacts in a high-voltage circuit, as shown in Figure 11-14. The operator activates the switch in the safer low-voltage circuit, which will energize the relay, closing its contacts and connecting the more dangerous high voltage to the motor.

As an application, Figure 11-15 shows an automobile starter circuit. In this circuit a relay is being used to supply the large dc current needed to activate a starter motor in an automobile. When the ignition switch is engaged in the passenger compartment by the driver, current flows through a light-gauge wire from the negative side of the battery, through the relay's electromagnet, through the ignition switch, and back to the positive side of the battery. This current flow through the electromagnet of the relay energizes the relay and closes the relay's contacts. Closing the relay's contacts makes a path for the current to flow through the heavy-gauge cable from the negative side of the battery, through the relay contacts and starter motor, and back to the positive side of the battery. The starter motor's output shaft spins the engine, causing it to start.

This application is a perfect example of how a relay can be used to close contacts in a heavy-current (heavy-gauge cable) circuit, while the driver has only to close contacts in a small-current (light-gauge cable) circuit.

If the relay were omitted, the driver's ignition switch would have to be used to connect the 12 V and large current to the starter motor. This would mean that:

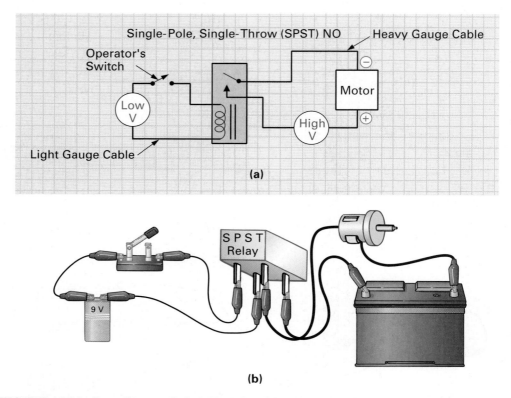

FIGURE 11-14   **Low-Current Switch Enabling a High-Current Circuit. (a) Schematic. (b) Pictorial.**

1. Heavy-gauge, expensive cable would need to be connected in a longer path between starter motor and passenger compartment.
2. The ignition switch would need to be larger to handle the heavier current.
3. The driver would be in closer proximity to a more dangerous, high-current circuit.

Reed relays and switches.   The **reed relay or switch** consists of two flat magnetic strips mounted inside a capsule, which is normally made of glass. The reed relay and reed switch are illustrated in Figure 11-16(a) and (b). The reed relay differs from the reed switch in that a reed relay has its own energizing coil, while the reed switch needs an external magnetic force to operate.

The reed relay and reed switch are both operated by a magnetic force that is provided by a coil for the reed relay and by a separate permanent magnet for the reed switch. When the magnetic force is present, opposite magnetic polarities are induced in the overlapping, high-permeability reed blades, causing them to attract one another by induced magnetism and snap together, thus closing the circuit. When the magnetic force is removed, the blades spring apart due to spring tension and open the circuit.

Referring to Figure 11-17 you will see a simple home security circuit. When the window is closed in the normal condition, the permanent magnet is directly adjacent to the reed switch, and therefore its contacts are closed, allowing current to flow from the negative side of the small 3 V battery, through the closed reed switch contacts, the relay's electromagnet, and back to the positive side of the 3 V battery. The relay is a normally closed (NC) SPST. As current is flowing through the coil of the relay, the contacts are open, preventing the large positive and negative voltage from the 12 V battery from reaching the siren.

If the window is forced and opened by an intruder, the permanent magnet will no longer be in close proximity to the reed switch and the reed switch's contacts will open. When the reed switch's contacts open, the relay's coil will no longer have current flowing through it, and therefore the relay will deenergize and its contacts will return to their normal

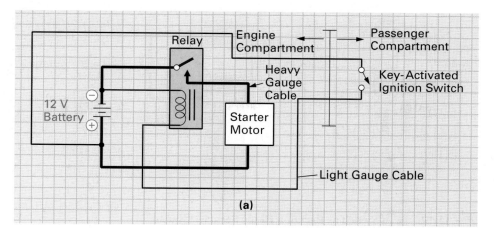

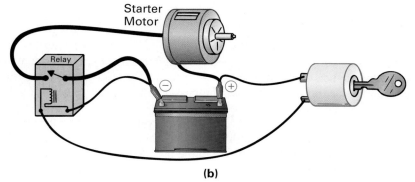

(b)

FIGURE 11-15 **Automobile Starter Circuit.**

closed condition. A large current is now permitted to flow from the negative side of the large 12 V battery, through the relay's contacts, the siren, and back to the positive side of the battery. The siren will sound, as it now has 12 V applied to it, and alert the occupant of the home.

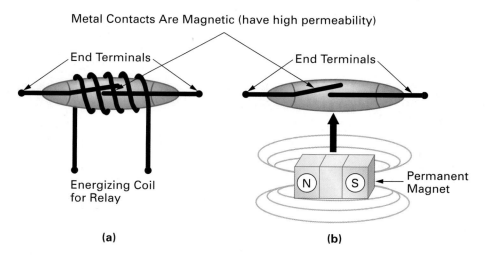

FIGURE 11-16 **Reed Relay and Switch. (a) Reed Relay: Contacts Close When Electromagnetic Coil Is Energized. (b) Reed Switch: Contacts Close When External Magnetic Field Comes in Close Proximity, Due to Induced Magnetism in the Contacts.**

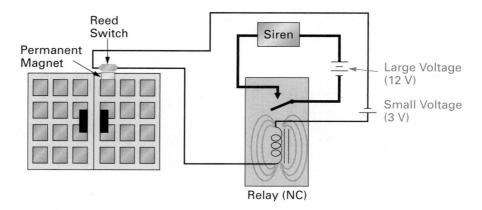

FIGURE 11-17   **Reed Switch as Part of a Home Security System.**

Solenoid-Type Electromagnet

Up to now, electromagnets have been used to close or open a set, or sets, of contacts to either make or break a current path. These electromagnets have used a stationary soft-iron core. Some electromagnets are constructed with movable iron cores, as shown in Figure 11-18, which can be used to open or block the passage of a gas or liquid through a valve. These are known as **solenoid**-type electromagnets.

When no current is flowing through the solenoid coil, no magnetic field is generated, so no magnetic force is exerted on the movable iron core, as shown in Figure 11-18(a), and therefore the compression spring maintains it in the up position, with the valve plug on the end of the core preventing the passage of either a liquid or gas through the valve (valve closed).

Solenoid

Coil and movable iron core that when energized by an alternating or direct current will pull the core into a central position.

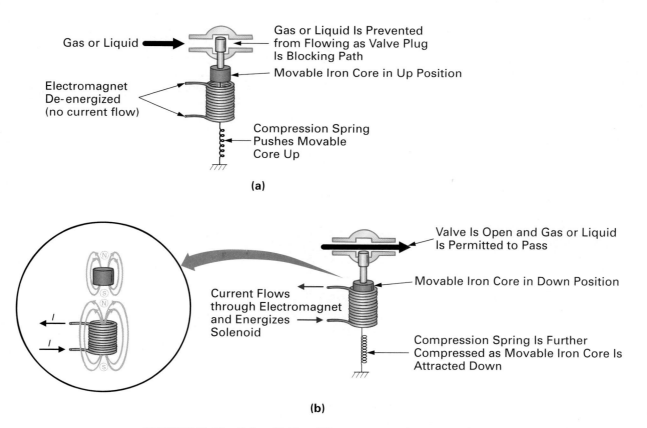

FIGURE 11-18   **Solenoid-Type Electromagnet. (a) Deenergized. (b) Energized.**

When a current flows through the electromagnet, the solenoid coil is energized, creating a magnetic field, as shown in Figure 11-18(b). Due to the influence of the coil's magnetic field, the movable soft-iron core will itself generate a magnetic field, as seen in the inset in Figure 11-18(b). This condition will create a north pole at the top of the solenoid coil and a south pole at the bottom of the movable core, and the resulting attraction will pull down the core (which is free to slide up and down), pulling with it the valve plug and opening the valve.

Solenoid-type electromagnets are actually constructed with the core partially in the coil and are used in washing machines to control water and in furnaces to control gas.

SELF-TEST EVALUATION POINT FOR SECTION 11-1-5

Now that you have completed this section, you should be able to:

■ **Objective 9.** *Describe the following applications of electromagnetism:*
  a. Magnetic-type circuit breaker
  b. Electric bell
  c. Relay
  d. Solenoid-type electromagnet

Use the following questions to test your understanding of Section 11-1-5.

1. List two applications of the electromagnet.
2. What is the difference between an NO and an NC relay?
3. What is the difference between a reed relay and a reed switch?
4. Give an application for a conventional relay and one for the reed switch.

11-2   ELECTROMAGNETIC INDUCTION

Electromagnetic induction is the name given to the action that causes electrons to flow within a conductor when that conductor is moved through a magnetic field. Stated another way, electromagnetic induction is the voltage or emf induced or produced in a coil as the magnetic lines of force link with the turns of a coil. Since this phenomenon was first discovered by Michael Faraday, let us begin by studying **Faraday's Law.**

11-2-1   *Faraday's Law*

In 1831, Michael Faraday carried out an experiment in which he used a coil, a zero center ammeter (galvanometer), and a bar permanent magnet, as shown in Figure 11-19. Faraday's discoveries, which are illustrated in Figure 11-20(a) through (f), are that:

a. When the magnet is moved into a coil the magnetic lines of flux cut the turns of the coil. This action that occurs when the magnetic lines of flux link with a conductor is known as *flux linkage*. Whenever flux linkage occurs, an emf is induced in the coil

Electromagnetic Induction

The voltage produced in a coil due to relative motion between the coil and magnetic lines of force.

Faraday's Law

1. When a magnetic field cuts a conductor, or when a conductor cuts a magnetic field, an electric current will flow in the conductor if a closed path is provided over which the current can circulate.
2. Two other laws relate to electrolytic cells.

FIGURE 11-19 Faraday's Electromagnetic Induction Experiment Components.

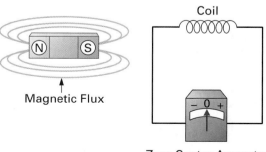

Coil

Magnetic Flux

Zero Center Ammeter

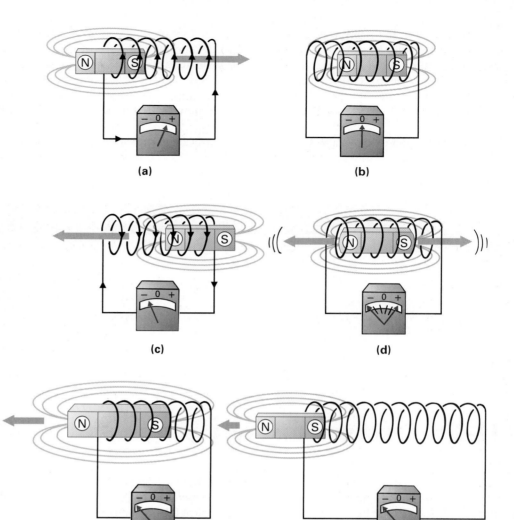

FIGURE 11-20 **Faraday's Electromagnetic Induction Discoveries.**

known as an induced voltage, which causes current to flow within the circuit and the meter to deflect in one direction, for example, to the right. Faraday discovered, in fact, that if the magnet was moved into the coil, or if the coil was moved over the magnet, an emf or voltage was induced within the coil. What actually occurs is that the electrons within the coil are pushed to one end of the coil by the magnetic field, creating an abundance of electrons at one end of the coil (negative charge) and an absence of electrons at the other end of the coil (positive charge). This potential difference across the coil will produce a current flow if a complete path for current (closed circuit) exists.

b. When the magnet is stationary within the coil, the magnetic lines are no longer cutting the turns of the coil, and so there is no induced voltage and the meter returns to zero.

c. When the magnet is pulled back out of the coil, a voltage is induced that causes current to flow in the opposite direction to that of (a) and the meter deflects in the opposite direction, for example, to the left.

d. If the magnet is moved into or out of the coil at a greater speed, the voltage induced also increases, and therefore so does current.

e. If the size of the magnet and therefore the magnetic flux strength are increased, the induced voltage also increases.

f. If the number of turns in the coil is increased, the induced voltage also increases.

In summary, whenever there is relative motion or movement between the coil of a conductor and the magnetic lines of flux, a potential difference will be induced and this action is called *electromagnetic induction*. The magnitude of the induced voltage depends on the number of turns in the coil, rate of change of flux linkage, and flux density.

11-2-2   *Lenz's Law*

About the same time a German physicist, Heinrich Lenz, performed a similar experiment along the same lines as Faraday. His law states that the current induced in a coil due to the change in the magnetic flux is such as to oppose the cause producing it.

To explain this law further, refer to Figure 11-21. When the magnet moves into the coil, a voltage is induced in the coil and the current flow resulting from this induced voltage will produce a pole at the face of the coil (left-hand rule), which opposes the entry of the magnet. In the example in Figure 11-21, we can see that as the permanent magnet moves into the coil, its magnetic lines of flux cut the turns of the coil and induce a voltage (electromagnetic induction), which causes current to flow in the coil as indicated by the meter movement. If you apply the left-hand rule to the coil, you can see that the current flow within the coil has produced a south pole on the left-hand side of the coil (electromagnetism) to oppose the entry or motion of the magnet that is producing the current.

11-2-3   *The Weber*

Consideration of Faraday's law enables us to take a closer look at the unit of flux (ϕ). The **weber** is equal to 10^8 magnetic lines of force, and from the electromagnetic induction point of view, if 1 weber of magnetic flux cuts a conductor for a period of 1 second, a voltage of 1 volt will be induced.

11-2-4   *Applications of Electromagnetic Induction*

AC Generator

The ac **generator** or alternator is an example of a device that uses electromagnetic induction to generate electricity. If you stroll around your city or town during the day or night and try to spot every piece of equipment, appliance, or device that is running from

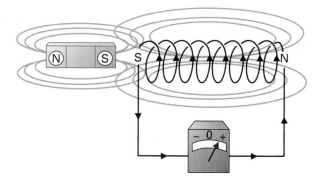

FIGURE 11-21   Lenz's Law.

the ac electricity supplied by generators from the electric power plant, it begins to make you realize how inactive, dark, and difficult our modern society might become without ac power.

When discussing Faraday's discoveries of electromagnetic induction in Figure 11-20, the coil or conductor remained stationary and the magnet was moved. It can be operated in the opposite manner so that the magnetic field remains stationary and the coil or conductor is moved. As long as the magnetic lines of force have relative motion with respect to the conductor, an emf will be induced in the coil. Figure 11-22(a) illustrates a piece of wire wound to form a coil and attached to a galvanometer; its needle rests in the center position, indicating zero current. If the conductor remains stationary within the magnetic lines of flux being generated by the permanent magnet, there is no emf or voltage induced into the wire and so no current flow through the circuit and meter.

If the conductor is moved past the permanent magnet so that it cuts the magnetic lines of flux, as seen in Figure 11-22(b), an emf is generated within the conductor, which is known as an induced voltage, and this will cause current to flow through the wire in one direction and be indicated on the meter by the deflection of the needle to the left.

If the conductor is moved in the opposite direction past the magnetic field, it will induce a voltage of the opposite polarity and cause the meter to deflect to the right, as shown in Figure 11-22(c).

The value or amount of induced voltage is indicated by how far the meter deflects and this voltage is dependent on three factors:

1. The speed at which the conductor passes through the magnetic field
2. The strength or flux density of the magnetic field
3. The number of turns in the coil

If the speed at which the conductor passes through the magnetic field or the strength of the magnetic field or the number of turns of the coil is increased, then the induced voltage will also increase. This is merely a repetition of Faraday's law, but in this case we moved the conductor instead of the magnetic field, but the results were the same.

Basic generator.   The basic generator action makes use of Faraday's discoveries of electromagnetic induction. In Chapter 8 the physical appearance of a large 700,000 kW

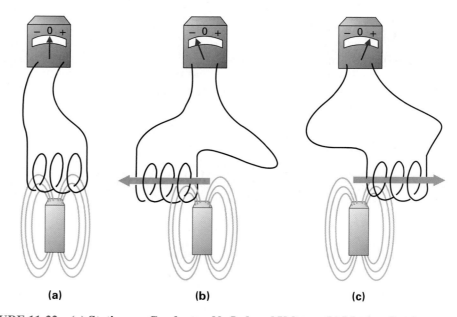

(a)	(b)	(c)

FIGURE 11-22   (a) Stationary Conductor, No Induced Voltage. (b) Moving Conductor, Induced Voltage. (c) Moving Conductor, Induced Voltage.

CHAPTER 11 / ELECTROMAGNETISM AND ELECTROMAGNETIC INDUCTION

(b)

(a)

FIGURE 11-23   AC Generator. (a) 250 kW Diesel Generator. (b) Small (2 kW) Camping Generator.

power plant generator was shown. Figure 11-23 illustrates some smaller generators that are mobile and can therefore be used in remote locations.

Figure 11-24 illustrates the basic generator's construction. The basic principle of operation is that the mechanical drive energy input will produce ac electrical energy out by means of electromagnetic induction.

A loop of conductor, known as an **armature,** is rotated continually through 360° by a mechanical drive. This armature resides within a magnetic field produced by a dc electromagnet. Voltage will be induced into the armature and will appear on slip rings, which are also being rotated. A set of stationary brushes rides on the rotating slip rings and picks off the generated voltage and applies this voltage across the load. This voltage will cause current to flow within the circuit and be indicated by the zero center ammeter.

Let's now take a closer look at the armature as it sweeps through 360°, or one complete revolution. Figure 11-25 illustrates four positions of the armature as it rotates through 360° in the clockwise direction.

Armature

Rotating or moving component of a magnetic circuit.

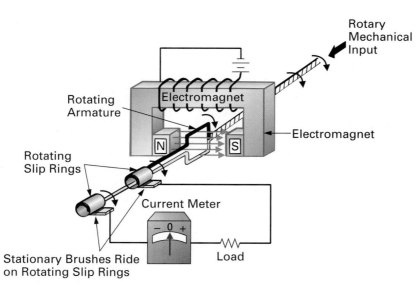

FIGURE 11-24   Basic Generator Construction.

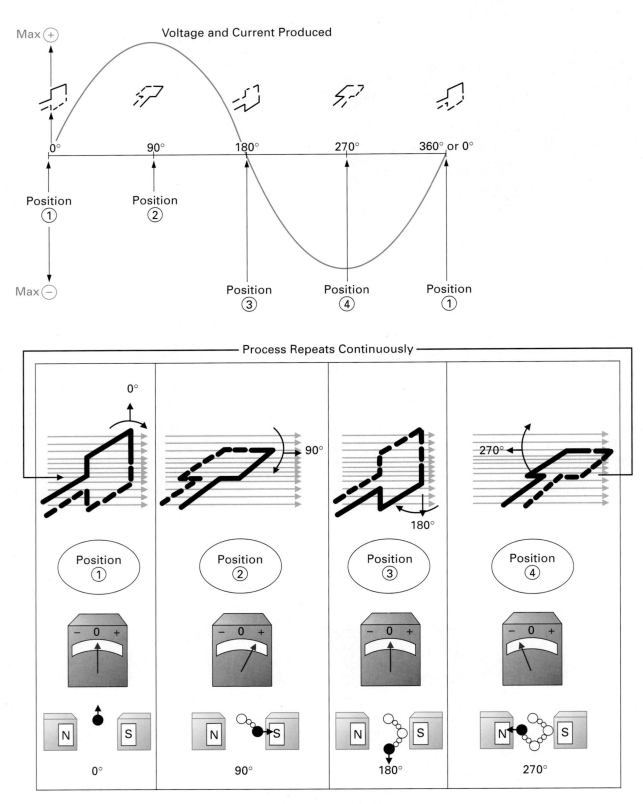

FIGURE 11-25   360° Generator Operation.

CHAPTER 11 / ELECTROMAGNETISM AND ELECTROMAGNETIC INDUCTION

Position 1: At this instant, the armature is in a position such that it does not cut any magnetic lines of force. The induced voltage in the armature conductor is equal to 0 V and there is no current flow through the circuit.

Position 2: As the conducting armature moves from position 1 to position 2, you can see that more and more magnetic lines of flux will be cut, and the induced emf in the armature (being coupled off by the brushes from the slip rings) will also increase to a maximum value. The current flow throughout the circuit will rise to a maximum as the voltage increases, and this can be seen by the zero center ammeter deflection to the right. From the ac waveform in Figure 11-25, you can see the sinusoidal increase from zero to a maximum positive as the armature is rotated from 0 to 90°.

Position 3: The armature continues its rotation from 90 to 180°, cutting through a maximum quantity and then fewer magnetic lines of force. The induced voltage decreases from the maximum positive at 90° to 0 V at 180°. At the 180° position, as with the 0° position, the armature is once again perpendicular to the magnetic field, and so no lines are cut and the induced voltage is equal to 0 V.

Position 4: From 180 to 270°, the armature is still moving in a clockwise direction, and as it travels toward 270°, it cuts more and more magnetic lines of force. The direction of the cutting action between 0 and 90° causes a positive induced voltage; and since the cutting position between 180 and 270° is the reverse, a negative induced voltage will result in the armature causing current flow in the opposite direction, as indicated by the deflection of the zero center ammeter to the left. The voltage induced when the armature is at 270° will be equal but opposite in polarity to the voltage generated when the armature was at the 90° position. The current will therefore also be equal in value but opposite in its direction of flow.

From position 4 (270°), the armature turns to the 360° or 0° position, which is equivalent to position 1, and the induced voltage decreases from maximum negative to zero. The cycle then repeats.

To summarize, in Figure 11-25, one complete revolution of the mechanical energy input causes one complete cycle of the ac electrical energy output, with a sine wave being generated as a result of circular motion.

Moving Coil Microphone

The moving coil **microphone** is an example of how electromagnetic induction is used to convert information-carrying sound waves to information-carrying electrical waves. Sound is the movement of pressure waves in the air. To create these pressure waves, a device such as a string, reed, or stretched membrane or the human vocal cords must be vibrated to compress and expand the nearby air molecules. Figure 11-26 illustrates a taut string that is vibrating back and forth and generating maximum (*A*) and minimum (*C*) pressure regions.

The frequency or pitch of the sound wave is determined by the number of complete vibrations per second (hertz or cycles/second), while the amplitude or intensity of the sound wave is determined by the amount the string shifts from left to right from its normal position (*B*).

A moving coil microphone converts mechanical sound waves into an electrical replica by use of electromagnetic induction. Figure 11-27 illustrates the physical appearance and construction of a moving coil type of microphone. A coil of wire is suspended in an air gap between magnetic poles and attached to a delicate diaphragm (flexible membrane). A strong magnetic field from a permanent magnet surrounds the coil, and a perforated protecting cover or shield is included to protect the delicate diaphragm.

Sound waves strike the diaphragm, causing it to vibrate back and forth. Since the coil is attached to the diaphragm, it will also be moved back and forth. This movement will cause the coil to cut the magnetic lines of force from the permanent magnet, and a resulting alternating voltage will be induced in the coil (electromagnetic induction). The electrical voltage produces an alternating current, which will have the same waveform shape

Microphone
Electroacoustic transducer that responds and converts a sound wave input into an equivalent electrical wave out.

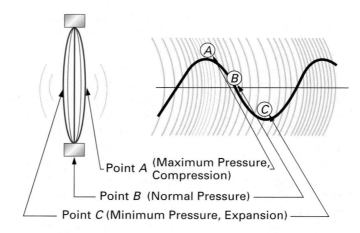

Point A (Maximum Pressure, Compression)

Point B (Normal Pressure)

Point C (Minimum Pressure, Expansion)

FIGURE 11-26   Sound Wave.

(and consequently information) as the sound wave that generated it, as seen in Figure 11-27(c). This electrical signal is often referred to as an *analog signal.* Analog is a term meaning "similar to," and as you can see in Figure 11-27(c), the electrical wave output of the microphone is an analog (similar to) of the sound wave input.

This microphone is sometimes called a dynamic or moving (dynamic is the opposite of static) coil microphone since it has a moving coil within it.

SELF-TEST EVALUATION POINT FOR SECTION 11-2

Now that you have completed this section, you should be able to:

■ **Objective 10.**   *Define* electromagnetic induction.

■ **Objective 11.**   *State Faraday's and Lenz's laws relating to electromagnetic induction.*

■ **Objective 12.**   *Describe the following applications of electromagnetic induction:*
   a. AC generator
   b. Moving coil microphone

Use the following questions to test your understanding of Section 11-2.

1. Define *electromagnetic induction.*

2. Briefly describe Faraday's and Lenz's laws in relation to electromagnetic induction.

3. What waveform shape does the ac generator produce?

4. Why do you think a microphone is called an electroacoustical transducer?

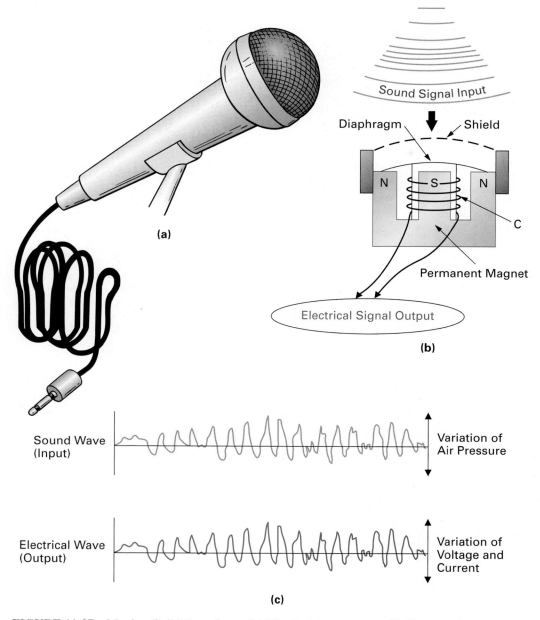

FIGURE 11-27   **Moving Coil Microphone. (a) Physical Appearance. (b) Construction. (c) Input/Output Waveforms.**

SUMMARY

Electromagnetism and Electromagnetic Induction (Figure 11-28)

1. During a classroom lecture in 1820 Danish physicist Hans Christian Oersted discovered that magnetism and electricity were very closely related to one another. This phenomenon is now called *electromagnetism* since it is now known that any conductor carrying an electrical current will produce a magnetic field.

2. In 1831, the English physicist Michael Faraday explored further Oersted's discovery of electromagnetism and found that the process could be reversed. Faraday observed that if a conductor was passed through a magnetic field, a voltage would be induced in the conductor and cause a current to flow. This phenomenon is referred to as *electromagnetic induction*.

3. Every orbiting electron in motion around its nucleus generates a magnetic field. When electrons are forced to leave

FIGURE 11-28 **Electromagnetism and Electromagnetic Induction.**

1 magnetic line of force = 1 maxwell
108 lines of force or maxwells = 1 weber

$$B = \frac{\phi}{A}$$

B = flux density, in teslas (T)
ϕ = magnetic flux, in webers (Wb)
A = area, in square meters (m^2)

$$mmf = I \times N$$

mmf = magnetomotive force, in ampere turns (At)
I = current, in amperes
N = number of turns in coil

$$H = \frac{I \times N}{l}$$

H = magnetizing force, in ampere turns per meter (At/m)
l = length of coil, in meters

$$\Re = \frac{mmf}{\phi}$$

$\Re$ = reluctance, in ampere turns per weber (At/Wb)

$$\mu = \frac{1}{\Re}$$

μ = permeability, in henrys per meter (H/m)

$$\mu = \mu_r \times \mu_0$$

μ_r = relative permeability
μ_0 = absolute permeability = $4\pi \times 10^{-7}$

Permeabilities of Various Materials

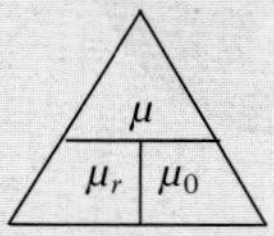

MATERIAL	RELATIVE PERMEABILITY (μ_r)	PERMEABILITY (μ)
Air or vacuum	1	1.26×10^{-6}
Nickel	50	6.28×10^{-5}
Cobalt	60	7.56×10^{-5}
Cast iron	90	1.1×10^{-4}
Transformer iron	5,500	6.9×10^{-3}
Silicon iron	7,000	8.8×10^{-3}
Permalloy	100,000	0.126
Supermalloy	1,000,000	1.26

Electromagnetic Induction

When 1 weber of magnetic flux cuts a conductor for 1 second, a voltage of 1 volt will be induced.

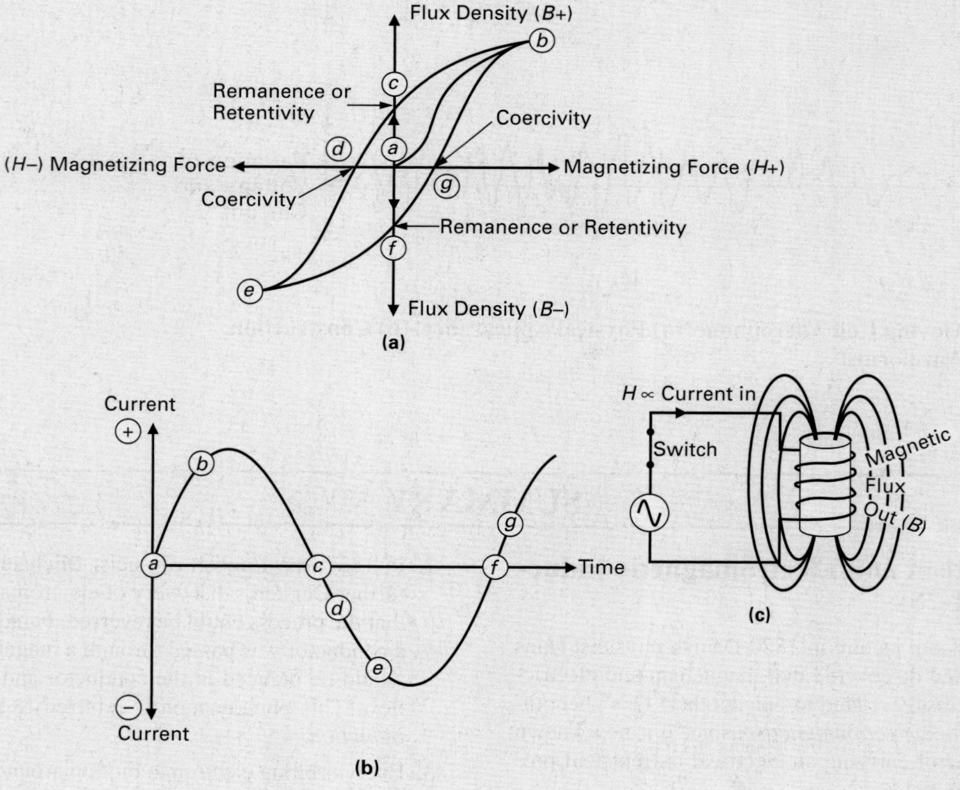

Flux Density ($B+$)

Remanence or Retentivity

Coercivity

($H-$) Magnetizing Force

Magnetizing Force ($H+$)

Coercivity

Remanence or Retentivity

Flux Density ($B-$)

(a)

Current
(+)

Time

(−)
Current

(b)

$H \propto$ Current in

Switch

Magnetic Flux Out (B)

(c)

CHAPTER 11 / ELECTROMAGNETISM AND ELECTROMAGNETIC INDUCTION

their parent atom by voltage and flow toward the positive polarity, they are all moving in the same direction and each electron's magnetic field will add to the next. The accumulation of all these electron fields will create the magnetic field around the conductor.

4. The left-hand rule can be applied to electromagnets to determine which of the poles will be the north end of the electromagnet.

5. If a conductor is wound to form a spiral, the conductor, which is now referred to as a *coil,* will, as a result of current flow, develop a magnetic field, which will sum or intensify within the coil. If many coils are wound in the same direction, an electromagnet is formed, which will produce a concentrated magnetic field whenever a current is passed through its coils.

6. A dc voltage produces a fixed current in one direction and therefore generates a magnetic field in a coil of fixed polarity, and as determined by the left-hand rule.

7. Alternating current (ac) is continually varying, and as the polarity of the magnetic field is dependent on the direction of current flow (left-hand rule), the magnetic field will also be alternating in polarity.

8. One magnetic line of flux or force is called a maxwell, in honor of James Maxwell's work in this field. One *weber* is equal to 10^8 (100,000,000) magnetic lines of force or maxwells.

9. Flux density (B) is equal to the number of magnetic lines of flux (ϕ) per square meter, and it is given in the unit of telsa (T).

10. Magnetomotive force (mmf) is the magnetic pressure that produces the magnetic field.

11. The length of the coil determines the field of intensity, and magnetizing force (H) is the reference we commonly use to describe the magnetic field intensity.

12. Reluctance is the opposition or resistance to the establishment of a magnetic field in an electromagnet.

13. Permeability is a measure of how easily a material will allow a magnetic field to be set up within it. Permeability is symbolized by the Greek lowercase letter mu (μ) and is measured in henrys per meter (H/m).

14. Relative permeability (μ_r) is the measure of how well another given material will conduct magnetic lines of force with respect to, or relative to, our reference material, air, which has a relative permeability value of 1.

15. The relative permeability (μ_r) of air, which is equal to 1, should not be confused with the absolute permeability (μ_0) of free space or air, which is equal to $4\pi \times 10^{-7}$ or 1.26×10^{-6}.

16. The magnetic field strength of an electromagnet can be increased by increasing the number of turns in its coil, increasing the current through the electromagnet, or decreasing the length of the coil ($H = I \times N/l$). The field strength can be increased further by placing an iron core within the electromagnet. An iron core has less reluctance or opposition to the magnetic lines of force than air, so the flux density (B) is increased.

17. The B–H curve illustrates the relationship between the two most important magnetic properties: flux density (B) and magnetizing force (H).

18. The saturation point is the state of magnetism beyond which an electromagnet is incapable of further magnetic strength.

19. Hysteresis is a lag between cause and effect. With magnetism it is the amount that the magnetization of a material lags the magnetizing force due to molecular friction.

20. Remanence or retentivity is the amount a material remains magnetized after the magnetizing force has been removed. Coercive force is the magnetizing force needed to reduce the residual magnetism within a material to zero.

21. The magnetic-type circuit breaker is one application of an electromagnet.

22. The electric bell is another application that makes use of a soft-iron core electromagnet.

23. A relay is an electromechanical device that either makes (closes) or breaks (opens) a circuit by moving contacts together or apart.

24. The relay consists of an electromagnet, a movable iron arm known as the armature, and some contacts. When current passes through the electromagnet, it is said to be energized. When the electromagnet is energized, it attracts the armature toward the electromagnet, and either closes or makes the normally open relay contacts, or opens or breaks the normally closed relay contacts.

   If the electromagnet is deenergized by stopping the current through the coil, the spring will pull back the armature to open the NO relay contacts or close the NC relay contacts.

25. The relay is generally used in two basic applications:
    a. To enable one master switch to operate several remote or difficultly placed contact switches.
    b. To enable a switch in a low-voltage circuit to operate relay contacts in a high-voltage circuit.

26. The reed relay or switch consists of two flat magnetic strips mounted inside a capsule, which is normally made of glass. The reed relay differs from the reed switch in that a reed relay has its own energizing coil, while the reed switch needs an external magnetic force to operate.

27. Some electromagnets are constructed with movable iron cores which can be used to open or block the passage of a gas or liquid through a valve. These are known as solenoid-type electromagnets.

28. Electromagnetic induction is the voltage or emf induced or produced in a coil as the magnetic lines of force link with the turns of a coil.

29. In 1831, Michael Faraday carried out an experiment including the use of a coil, a zero center ammeter (galvanometer), and a bar permanent magnet. Faraday discovered that whenever there is relative motion between a coil and magnetic lines of flux, a voltage will be induced.

30. Heinrich Lenz's law states that the current induced in a coil due to the change in the magnetic flux is such as to oppose the cause producing it.

31. The generator's basic principle of operation is that the mechanical drive energy input will produce ac electrical energy out by means of electromagnetic induction.

32. One complete revolution of the mechanical energy input to the generator causes one complete cycle of the ac electrical

energy output, with a sine wave being generated as a result of circular motion.

33. The moving coil microphone is an example of how electromagnetic induction is used to convert information-carrying sound waves to information-carrying electrical waves.

REVIEW QUESTIONS

Multiple-Choice Questions

1. Electromagnetism was first discovered by:
  a. Hans Christian Oersted   **c.** James Watt
  b. Heinrich Hertz   **d.** James Clark Maxwell

2. Every current-carrying conductor generates a/an:
  a. Electric field   **c.** Both (a) and (b)
  b. Magnetic field   **d.** None of the above

3. If a greater number of loops of a conductor in a coil are added, a _____ magnetic field is generated.
  a. Weaker   **b.** Stronger

4. If a smaller current is passed through a coil, a _____ magnetic field is generated.
  a. Weaker   **b.** Stronger

5. One magnetic line of flux is known as a _____.
  a. Weber   **c.** Maxwell
  b. Tesla   **d.** Oersted

6. 10^8 magnetic lines of flux are referred to as one _____.
  a. Weber   **c.** Maxwell
  b. Tesla   **d.** Oersted

7. Flux density is equal to the magnetic flux divided by area and is measured in _____.
  a. Webers   **c.** Maxwells
  b. Teslas   **d.** Oersteds

8. By increasing either current or the number of turns in a coil, the mmf, which is an abbreviation for _____, will increase.
  a. Multiple magnetic formulas   **c.** Electromotive force
  b. Magnetomotive force   **d.** None of the above

9. _____ is a term used to describe the magnetic field intensity and is equal to the mmf divided by the length of the coil.
  a. Flux density   **c.** Magnetizing force
  b. Magnetomotive force   **d.** Reluctance

10. Reluctance is equivalent to electrical:
  a. Current   **c.** Resistance
  b. Voltage   **d.** Power

11. Permeability is equivalent to electrical:
  a. Current   **c.** Resistance
  b. Conductance   **d.** Voltage

12. The absolute permeability of air is equal to:
  a. 1   **b.** 1.26×10^{-6} **c.** 4π **d.** None of the above

13. An electromagnet can be used in the application of:
  a. A relay   **c.** A circuit breaker
  b. An electric bell   **d.** All of the above

14. Relays can be used to:
  a. Allow one master switch to enable several others
  b. Allow several switches to enable one master
  c. Allow a switch in a low-current circuit to close contacts in a high-current circuit
  d. Both (a) and (b)
  e. Both (a) and (c)

15. The starter relay in an automobile is used to:
  a. Allow one master switch to enable several others
  b. Allow several switches to enable one master
  c. Allow a switch in a low-current circuit to close contacts in a high-current circuit
  d. Both (a) and (b)

16. The _____ uses an electromagnet around two flat magnetic strips mounted inside a glass capsule.
  a. Reed switch   **c.** Reed relay
  b. Magnetic circuit breaker   **d.** Starter relay

17. The reed switch could be used in a/an:
  a. Home security system
  b. Automobile starter
  c. Magnetic-type circuit breaker
  d. All of the above

18. The relative permeability of air or a vacuum is equal to:
  a. 4π **b.** 6.26×10^{-6} **c.** 1   **d.** None of the above

19. An electromagnet is also known as a:
  a. Coil   **d.** Both (a) and (c)
  b. Solenoid   **e.** Both (a) and (b)
  c. Resistor

20. A normally open relay (NO) will have:
  a. Contacts closed until activated
  b. Contacts open until activated
  c. All contacts permanently open
  d. All contacts permanently closed

21. Direct current produces a _____ magnetic field of _____ polarity.
  a. Alternating, unchanging   **c.** Constant, unchanging
  b. Constant, alternating   **d.** Both (a) and (c)

22. Magnetizing force (H) is equal to:
  a. $I \times N \times l$ **c.** $I \times N/l$
  b. $I \times N + l$ **d.** $N \times l/I$

23. Electromagnetism:
  a. Is the magnetism resulting from electrical current flow
  b. Is the electrical voltage resulting in a coil from the relative motion of a magnetic field
  c. Both (a) and (b)
  d. None of the above

24. Electromagnetic induction:
 a. Is the magnetism resulting from electrical current flow
 b. Is the electrical voltage resulting in a coil from the relative motion of a magnetic field
 c. Both (a) and (b)
 d. None of the above

25. The _____ plots magnetizing force against flux density through a complete cycle of alternating current.
 a. *B–H* curve
 c. Power curve
 b. Coercive force
 d. Hysteresis loop

26. When the magnetic flux linking a conductor is changing, an emf is induced, the magnitude of which depends on the number of coil turns, rate of change of flux linkage change, and flux density. This law was discovered by:
 a. Heinrich Lenz
 c. Michael Faraday
 b. Guglielmo Marconi
 d. Joseph Henry

27. The current induced in a coil due to the change in the magnetic flux is such as to oppose the cause producing it. This law was discovered by:
 a. Heinrich Lenz
 c. Michael Faraday
 b. Guglielmo Marconi
 d. Joseph Henry

28. An ac generator uses _____ to generate ac electricity.
 a. Electromagnetism
 c. Magnetism
 b. Electromagnetic induction
 d. None of the above

29. The generator converts _____ energy into _____ energy.
 a. Electrical, electrical
 c. Chemical, electrical
 b. Mechanical, electrical
 d. None of the above

30. The loop of conductor rotated through 360° in a generator is known as a/an:
 a. Electromagnet
 c. Armature
 b. Field coil
 d. Both (a) and (b)

31. Sound waves are a form of:
 a. Electrical energy
 c. Magnetic energy
 b. Chemical energy
 d. Mechanical energy

32. The moving coil microphone converts _____ waves into _____ waves.
 a. Sound, electrical
 c. Electromagnetic, sound
 b. Electrical, sound
 d. Sound, radio

33. There is a reciprocal relationship between permeability and _____.
 a. Flux density
 c. Remanence
 b. Magnetizing force
 d. Reluctance

34. The amount of current in the reverse direction needed to reduce the flux density (*B*) to zero is termed the:
 a. Coercive force
 c. Electromotive force
 b. Magnetizing force
 d. All of the above

Communication Skill Questions

35. Describe what is meant by the word *electromagnetism*. (11-1)

36. Briefly describe how the left-hand rule of electromagnetism is applied to: (11-1-2)
 a. Conductors
 b. Coils

37. How many maxwells make up 1 weber? (11-1-3)

38. Give the formulas for the following: (11-1-3)
 a. Flux density
 d. Reluctance
 b. Magnetomotive force
 e. Permeability
 c. Magnetizing force

39. Describe the operation of: (11-1-5)
 a. The magnetic-type circuit breaker
 b. The electric bell
 c. The NO and NC relays

40. Describe the operation of the following and the difference between them: (11-1-5)
 a. Reed relay
 b. Reed switch

41. Explain the operation and application of the solenoid-type electromagnet. (11-1-5)

42. Why is a wire looped many times to form an electromagnet more useful than a straight piece of wire? (11-2)

43. From which pole does the magnetic flux emerge and into which pole end does the flux return? (11-2)

44. What are the differences between magnetic flux and flux density? (11-1-3)

45. What effect does current have when it is passed through a coil of conductor? (11-1)

46. What effect does a magnet have when it is moved into and out of a coil? (11-2)

47. State Faraday's law. (11-2-1)

48. State Lenz's law. (11-2-2)

49. Describe the different effects when ac and dc are passed through a coil. (11-1-2)

50. Illustrate and describe all the different points on a hysteresis curve. (11-1-4)

51. Describe the meaning of the following terms:
 a. Flux density
 d. Coercive force
 b. Magnetizing force
 e. Electromagnetism
 c. Remanence
 f. Electromagnetic induction

52. Briefly describe the operation of the generator. (11-2-4)

53. Briefly describe the operation of the moving coil microphone. (11-2-4)

54. Illustrate and describe the operation of the generator through 360°. (11-2-4)

55. List some other applications of ac electromagnetism.

Practice Problems

56. If a magnet has a pole area of 6.4×10^{-3} m^2 and a 1200 μWb total flux, what flux density would the pole produce?

57. Calculate the magnetomotive force produced when 760 mA flow through 25 turns.

58. Calculate the magnetizing force (*H*) or field intensity if a 15 cm, 40 turn coil has a current of 1.2 A flowing through it.

59. Calculate the permeability (μ) of the following materials ($\mu_0 = 4\pi \times 10^{-7}$):
 a. Cast iron
 b. Nickel
 c. Machine steel

60. Calculate the reluctance of a magnetic circuit if the magnetic flux produced is equal to 2.3×10^{-4} Wb, and is produced by 3 A flowing through a solenoid of 36 turns.

61. If a coil of 50 turns is passing a current of 4 A, calculate the mmf.

62. Calculate the reluctance of an iron core when mmf = 150 At and $\phi = 360\ \mu\text{Wb}$.

63. Calculate mmf when a 9 V battery is connected across a 50 turn, 23 Ω coil.

64. Calculate the permeability of a permalloy core.

65. Calculate the magnetizing force (H) of the coil in Question 63 if it were 0.7 m long.

Web Site Questions

Go to the Web site http://www.prenhall. com/cook, select the textbook *Introductory DC/AC Electronics* or *Introductory DC/AC Circuits*, this chapter, and then follow the instructions when answering the multiple-choice practice problems.

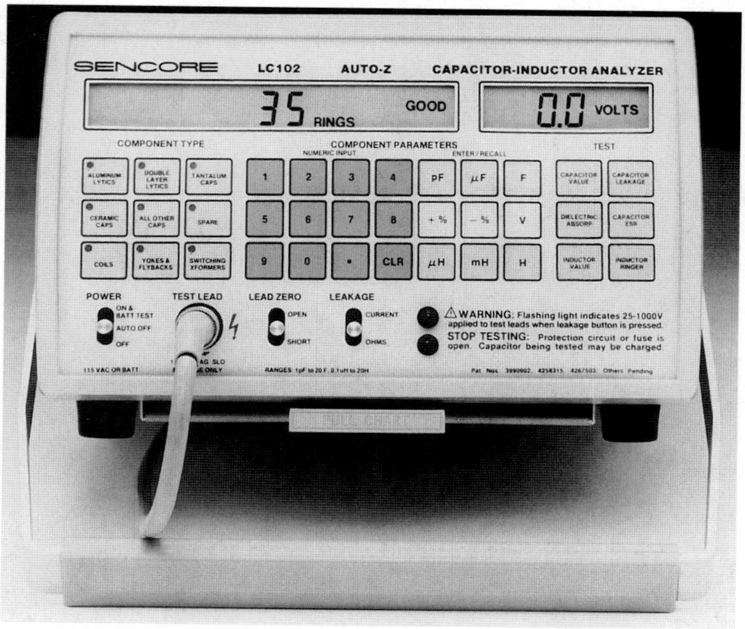

FIGURE 12-34   **Capacitor and Inductor Analyzer.**

Now that you have completed this section, you should be able to:

■   *Objective 9.*   *State the three typical malfunctions of inductors and explain how they can be recognized.*

Use the following questions to test your understanding of Section 12-10.

1. How could the following inductor malfunctions be recognized?

   a. An open

   b. A complete or section short

2. Which inductor malfunction accounts for almost 75% of all failures?

12-11   APPLICATIONS OF INDUCTORS

12-11-1   RL *Filters*

The ***RL* filter** will achieve results similar to the *RC* filter in that it will pass some frequencies and block others, as seen in Figure 12-35. The inductive reactance of the coil and the resistance of the resistor form a voltage divider. Since inductive reactance is proportional to frequency ($X_L \propto f$), the inductor will drop less voltage at lower frequencies ($f\downarrow$, $X_L\downarrow$, $V_L\downarrow$) and more voltage at higher frequencies ($f\uparrow$, $X_L\uparrow$, $V_L\uparrow$).

With the low-pass filter shown in Figure 12-35(a) the output is developed across the resistor. If the frequency of the input is low, the inductive reactance will be low, so almost all the input will be developed across the resistor and applied to the output. If the frequency of the input increases, the inductor's reactance will increase, resulting in almost all the input being dropped across the inductor and none across the resistor and therefore the output.

With the high-pass filter, shown in Figure 12-35(b), the inductor and resistor have been placed in opposite positions. If the frequency of the input is low, the inductive reactance will be low, so almost all the input will be developed across the resistor and very little will appear across the inductor and therefore at the output. If the frequency of the input is high, the

> ***RL* Filter**
>
> A selective circuit of resistors and inductors which offers little or no opposition to certain frequencies while blocking or alternating other frequencies.

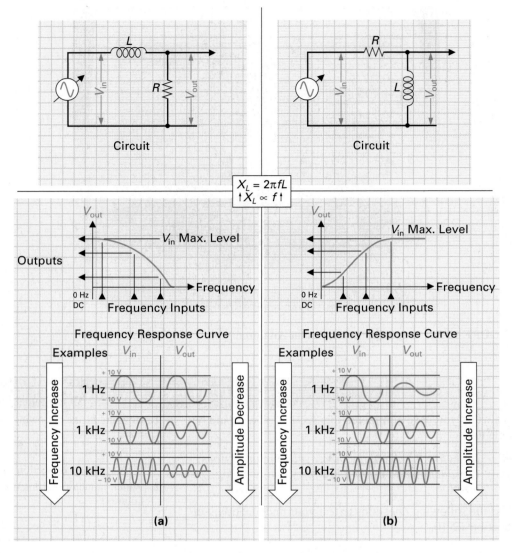

$$X_L = 2\pi fL$$
$$\uparrow X_L \propto f \uparrow$$

FIGURE 12-35   *RL* Filter. (a) Low-Pass Filter. (b) High-Pass Filter.

inductive reactance will be high, resulting in almost all of the input being developed across the inductor and therefore appearing at the output.

12-11-2   RL *Integrator*

In the preceding section on *RL* filters we saw how an *RL* circuit reacted to a sine-wave input of different frequencies. In this section and the following we will see how an *RL* circuit will react to a square-wave input, and show two other important applications of inductors. In the **RL integrator,** the output is taken across the resistor, as seen in the circuit in Figure 12-36(a). The output shown in Figure 12-36(b) is the same as the previously described *RC* integrator's output. When the input rises from 0 to 10 V at the leading positive edge of the square wave input, the situation is as seen in Figure 12-37(a).

The inductor's 10 V counter emf opposes the sudden input change from 0 to 10 V, and if 10 V is across the inductor, 0 V must be across the resistor (Kirchhoff's voltage law), and therefore appearing at the output. After five time constants ($5 \times L/R$), the inductor's current in the circuit will have built up to maximum (V_{in}/R), and the inductor will be an equivalent

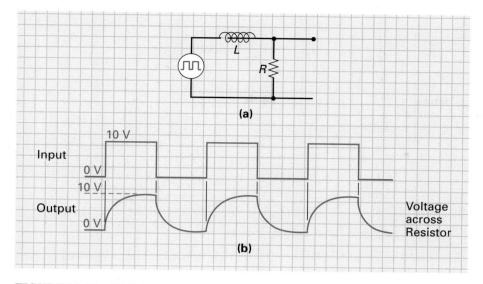

FIGURE 12-36   *RL* **Integrator.**

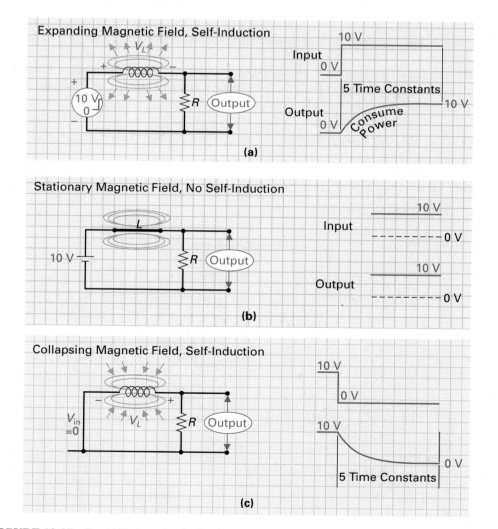

FIGURE 12-37   **Input/Output Analysis of** *RL* **Integrator.**

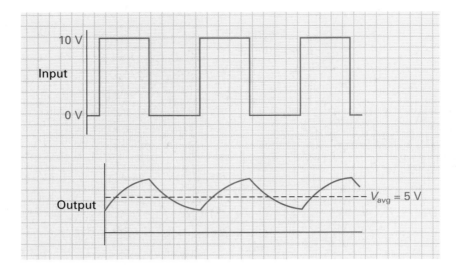

FIGURE 12-38

short circuit, because no change and consequently no back emf exist, as shown in Figure 12-37(b). All the input voltage will now be across the resistor and therefore at the output.

When the square-wave input drops to zero, the circuit is equivalent to that seen in Figure 12-37(c), and the collapsing magnetic field will cause an induced voltage within the conductor, which will cause current to flow within the circuit for five time constants, whereupon it will reach 0.

As with the *RC* integrator, if the period of the square wave is decreased or if the time constant is increased, the output will reach an average value of half the input square wave's amplitude, as shown in Figure 12-38.

12-11-3 RL *Differentiator*

RL Differentiator

An *RL* circuit whose output voltage is proportional to the rate of change of the input voltage.

With the **RL differentiator,** the output is taken across the inductor, as shown in Figure 12-39, and is the same as the *RC* differentiator's output. When the square-wave input rises from

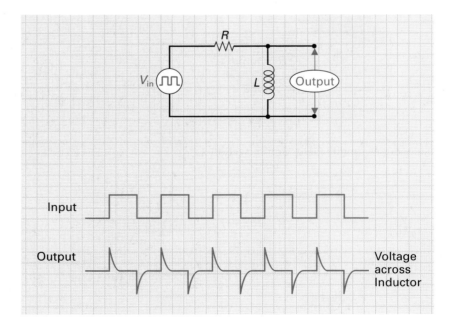

FIGURE 12-39   RL Differentiator.

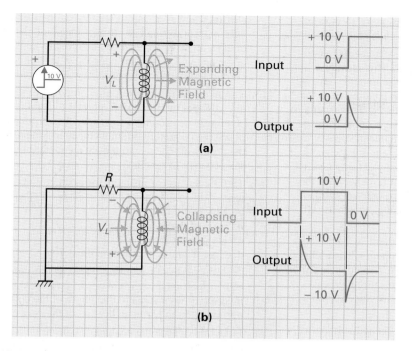

FIGURE 12-40   **Input/Output Analysis of *RL* Differentiator.**

0 to 10 V, the inductor will generate a 10 V counter emf across it, as shown in Figure 12-40(a), and this 10 V will appear across the output. As the circuit current exponentially rises, the voltage across the inductor, and therefore at the output, will fall exponentially to 0 V after five time constants.

When the square wave input falls from 10 to 0 V, the circuit is equivalent to that shown in Figure 12-40(b), and the collapsing magnetic field induces a counter emf. The sudden −10 V impulse at the output will decrease to 0 V as the circuit current decreases in five time constants.

SELF-TEST EVALUATION POINT FOR SECTION 12-11

Now that you have completed this section, you should be able to:

■ **Objective 10.**   *Describe how inductors can be used for the following applications:*
a. RL *filter*
b. RL *integrator*
c. RL *differentiator*

Use the following questions to test your understanding of Section 12-11.

1. List three circuit applications of the inductor.
2. Will an *RL* filter act as a low-pass filter or high-pass filter, or can it be made to function as either?
3. What are the waveform differences between an integrator and differentiator?

SUMMARY

Inductance and Inductors (Figure 12-41)

1. Inductance, by definition, is the ability of a device to oppose a change in current flow, and the device designed specifically to achieve this function is called the *inductor*.

2. There is, in fact, no physical difference between an inductor and an electromagnet, since they are both coils. The two devices are given different names because they are used in different applications even though their construction and principle of operation are the same. An electromagnet is

used to generate a magnetic field in response to current, while an inductor is used to oppose any change in current.

3. When an external voltage source is connected across a coil, it will force a current through the coil. This current will generate a magnetic field that will increase in field strength in a short time from zero to maximum, expanding from the center of the conductor (electromagnetism). The expanding magnetic lines of force have relative motion with respect to the stationary conductor, so an induced voltage results (electromagnetic induction). The blooming magnetic field generated by the conductor is actually causing a voltage to be induced in the conductor that is generating the magnetic field. This effect of a current-carrying coil of conductor inducing a voltage within itself is known as self-inductance.

4. The induced voltage is called a counter emf or back emf. It opposes the applied emf (or battery voltage). The ability of a coil or conductor to induce or produce a counter emf within itself as a result of a change in current is called *self-inductance,* or more commonly inductance (symbolized L).

5. The unit of inductance is the henry (H), named in honor of Joseph Henry, an American physicist, for his experimentation within this area of science.

6. The inductance of an inductor is 1 henry when a current change of 1 ampere per second causes an induced voltage of 1 volt.

7. Inductance is a measure of how much counter emf (induced voltage) can be generated by an inductor for a given amount of current change through that same inductor.

8. A larger inductance ($L\uparrow$) will create a larger induced voltage ($V_{ind}\uparrow$), and if the rate of change of current with respect to time is increased ($\Delta I/\Delta t\uparrow$), the induced voltage or counter emf will also increase ($V_{ind}\uparrow$).

9. If the current in the circuit is suddenly increased (by lowering the circuit resistance), the change in the expanding magnetic field will induce a counter emf within the inductor. This induced voltage will oppose the source voltage from the battery and attempt to hold current at its previous low level.

10. The counter emf cannot completely oppose the current increase, for if it did, the lack of current change would reduce the counter emf to zero. Current therefore incrementally increases up to a new maximum, which is determined by the applied voltage and the circuit resistance ($I = V/R$).

11. Once the new higher level of current has been reached and remains constant, there will no longer be a change. This lack of relative motion between field and conductor will no longer generate a counter emf, so the current will remain at its new higher constant value.

12. This effect also happens in the opposite respect. If current decreases (by increasing circuit resistance), the magnetic lines of force will collapse because of the reduction of current and induce a voltage in the inductor, which will produce an induced current in the same direction as the circuit current. These two combine and tend to maintain the current at the higher previous constant level. Circuit current

will fall, however, as the induced voltage and current are only present during the change. Once the new lower level of current has been reached and remains constant, the lack of change will no longer induce a voltage or current. So the current will then remain at its new lower constant value.

13. The inductor is an electronic component that will oppose any changes in circuit current, and this ability or behavior is referred to as *inductance.* Since the current change is opposed by a counter emf, inductance may also be defined as the ability of a device to induce a counter emf within itself for a change in current.

14. The inductance of an inductor is determined by four factors:
    a. Number of turns
    b. Area of the coil
    c. Length of the coil
    d. Core material used within the coil

    Inductance (L) is proportional to the number of turns (N), proportional to the cross-sectional area of the coil (A), inversely proportional to the length of the coil (l), and proportional to the permeability (μ) of the core material used.

15. Inductors oppose the change of current in a circuit and so are treated in a manner similar to resistors connected in combination.

16. Two or more inductors in series merely extend the coil length and increase inductance.

17. Inductors in parallel are treated in a manner similar to resistors, with the total inductance being less than that of the smallest inductor's value.

18. With fixed-value inductors, the inductance value remains constant and cannot be altered. They are usually made of solid copper wire with an insulating enamel around the conductor. This enamel or varnish on the conductor is needed since the turns of a coil are generally wound on top of one another. The three major types of fixed-value inductors on the market today are:
    a. Air core
    b. Iron core
    c. Ferrite core

19. The name choke is used interchangeably with the word *inductor,* because choke basically describes an inductors' behavior, which generates a counter emf to limit or choke the flow of current.

20. Ferrite is a chemical compound that is basically powdered iron oxide and ceramic. Although its permeability is less than iron, it has a higher permeability than air and can therefore obtain a higher inductance value than air for the same number of turns.

21. The toroidal (doughnut shaped) inductor has an advantage over the cylinder-shaped core in that the magnetic lines of force do not have to pass through air. With a cylindrical core, the magnetic lines of force merge out of one end of the cylinder, propagate or travel through air, and then enter in the other end of the cylinder.

22. With variable-value inductors, the inductance value can be changed by adjusting the position of the movable core with respect to the stationary coil.

QUICK REFERENCE SUMMARY SHEET

FIGURE 12-41 Inductance and Inductors.

$$V_{ind} = L \times \frac{\Delta I}{\Delta t}$$

L = inductance, in henrys
ΔI = increment of change of current
Δt = increment of change of time

$$L = \frac{N^2 \times A \times \mu}{l}$$

N = number of turns
A = cross-sectional area, in square meters (m^2)
μ = permeability
l = length of core, in meters (m)

Series *RL* Circuit:

- $L_T = L_1 + L_2 + L_3 + \ldots$ L_T = total inductance, in henrys
- $V_R = I \times R$
- $V_L = I \times X_L$
- $V_S = \sqrt{V_R^2 + V_L^2}$ $V_R = \sqrt{V_S^2 - V_L^2}$ $V_L = \sqrt{V_S^2 - V_R^2}$
- $Z = \frac{V}{I}$ $Z = \sqrt{R^2 + X_L^2}$ $R = \sqrt{Z^2 - X_L^2}$ $X_L = \sqrt{Z^2 - R^2}$
- $\theta = \arctan \frac{X_L}{R}$ $\theta = \arctan \frac{V_L}{V_R}$

Parallel *RL* Circuit:

- $I_R = \frac{V_S}{R}$
- $I_L = \sqrt{V_S / X_L}$
- $I_T = \sqrt{I_R^2 + I_L^2}$
- $Z = \frac{R \times X_L}{\sqrt{R^2 + X_L^2}}$
- $\theta = \arctan \frac{R}{X_L}$
- $L_T = \frac{1}{(1/L_1) + (1/L_2) + (1/L_3) + \ldots}$
- $L_T = \frac{L_1 \times L_2}{L_1 + L_2}$
- $P_A = \sqrt{P_R^2 + P_X^2}$ P_A = apparent power, in volt-amperes (VA)
  P_R = true power, in watts (W)
  P_X = reactive power, in volt-amperes reactive (VAR)
- $\text{PF} = \frac{\text{true power}}{\text{apparent power}}$ $\text{PF} = \frac{Z}{R}$ $\text{PF} = \cos\theta$ PF = power factor

$$\tau = \frac{L}{R}$$

τ = inductive time constant, in seconds (s)

$$X_L = 2\pi f L$$

X_L = inductive reactance, in ohms (Ω)
2π = a constant
f = frequency, in hertz (Hz)
L = inductance, in henrys (H)

$$Q = \frac{\text{energy stored}}{\text{energy dissipated}}$$

Q = quality factor of an inductor

$$Q = \frac{X_L}{R}$$

Permeabilities of Various Materials

MATERIAL	PERMEABILITY (μ)
Air or vacuum	1.26×10^{-6}
Nickel	6.28×10^{-5}
Cobalt	7.56×10^{-5}
Cast iron	1.1×10^{-4}
Machine steel	5.65×10^{-4}
Transformer iron	6.9×10^{-3}
Silicon iron	8.8×10^{-3}
Permalloy	0.126
Superalloy	1.26

23. Variable inductors should only be adjusted with a plastic or nonmetallic alignment tool since any metal object in the vicinity of the inductor will interfere with the magnetic field and change the inductance of the inductor.

24. Inductors will not have any effect on a steady value of direct current (dc) from a dc voltage source. If, however, the dc is changing (pulsating), the inductor will oppose the change whether it is an increase or decrease in direct current, because a change in current causes the magnetic field to expand or contract, and in so doing it will cut the coil of the inductor and induce a voltage that will counter the applied emf.

25. Current in an inductive circuit cannot rise instantly to its maximum value, which is determined by Ohm's law ($I = V/R$). Current will in fact take time to rise to maximum, due to the inductor's ability to oppose change.

26. It will actually take five time constants (5τ) for the current in an inductive circuit to reach maximum value.

27. The constant to remember is 63.2%. In one time constant ($1 \times L/R$) the current in the RL circuit will have reached 63.2% of its maximum value. In two time constants ($2 \times L/R$), the current will have increased 63.2% of the remaining current, and so on, through five time constants.

28. When the inductor's dc source of current is removed, the magnetic field will collapse and cut the coils of the inductor, inducing a voltage and causing a current to flow in the same direction as the original source current. This current will exponentially decay, or fall from the maximum to zero level, in five time constants.

29. If an alternating (ac) voltage is applied across an inductor, the inductor will continuously oppose the alternating current because it is always changing. The current in the circuit causes the magnetic field to expand and collapse and cut the conducting coils, resulting in an induced counter emf.

30. The voltage across the inductor (V_L) or counter emf leads the circuit current (I) by 90°.

31. Inductive reactance (X_L) is the opposition to current flow offered by an inductor without the dissipation of energy.

32. Inductive reactance is proportional to frequency ($X_L \propto f$) because a higher frequency (fast switching current) will cause a greater amount of current change, and a greater change will generate a larger counter emf, which is an opposition or reactance against current flow. When 0 Hz is applied to a coil (dc), there exists no change, so the inductive reactance of an inductor to dc is zero ($X_L = 2\pi \times 0 \times L = 0$).

33. Inductive reactance is also proportional to inductance because a larger inductance will generate a greater magnetic field and subsequent counter emf, which is the opposition to current flow.

34. Ohm's law can be applied to inductive circuits just as it can be applied to resistive and capacitive circuits. The current flow in an inductive circuit (I) is proportional to the voltage applied (V), and inversely proportional to the inductive reactance (X_L).

35. In a purely resistive circuit, the current flowing within the circuit and the voltage across the resistor are in phase with one another.

36. In a purely inductive circuit, the current will lag the applied voltage by 90°.

37. The voltage across the resistor (V_R) is in phase with the circuit current. The voltage across the inductor (V_L) leads the circuit current and the voltage across the resistor (V_R) by 90°, or, the circuit current vector lags the voltage across the inductor by 90°.

38. As with any series circuit, we have to apply Kirchhoff's voltage law when calculating the value of applied or source voltage (V_S), which, due to the phase difference between V_R and V_L, is the vector sum of all the voltage drops.

39. Impedance is the total opposition to current flow offered by a circuit with both resistance and reactance. It is measured in ohms and can be calculated by using Ohm's law. The impedance of a series RL circuit is equal to the square root of the sum of the squares of resistance and reactance, and by rearrangement, X_L and R can also be calculated if the other two values are known.

40. If a circuit is purely resistive, the phase shift (θ) between the source voltage and circuit current is zero. If a circuit is purely inductive, voltage leads current by 90°; therefore, the phase shift is $+ 90°$. If the resistance and inductive reactance are equal, the phase shift will equal $+ 45°$.

41. The phase shift in an inductive and resistive circuit is the degrees of lead between the source voltage (V_S) and current (I). The phase angle is proportional to reactance and inversely proportional to resistance.

42. As the current is the same in both the inductor and resistor in a series circuit, the voltage drops across the inductor and resistor are directly proportional to reactance and resistance.

43. When applying an ac voltage across a purely resistive circuit, voltage and current are in phase, and true power (P_R) in watts can be calculated by multiplying current by voltage ($P = V \times I$).

44. The pure inductor, like the capacitor, is a reactive component, which means that it will consume power without the dissipation of energy. The capacitor holds its energy in an electric field, while the inductor consumes and holds its energy in a magnetic field and then releases it back into the circuit. The power alternation is both positive when the inductor is consuming power and negative when the inductor is returning the power back into the circuit. As the positive and negative power alternations are equal but opposite, the average power dissipated is zero.

45. An inductor is different from a capacitor in that it has a small amount of resistance no matter how pure the inductor. For this reason, inductors will never have an average power figure of zero, because even the best inductor will have some value of inductance and resistance within it.

46. The reason an inductor has resistance is that it is simply a piece of wire, and any piece of wire has a certain value of resistance, as well as inductance.

47. The quality factor (Q) of an inductor is the ratio of the energy stored in the coil by its inductance to the energy dissipated in the coil by the resistance.

48. When a circuit contains both inductance and resistance, some of the energy is consumed and then returned by the inductor (reactive or imaginary power), and some of the energy is dissipated and lost by the resistor (resistive or true power).

49. Apparent power is the power that appears to be supplied to the load and is the vector sum of both the reactive and true power.

50. With a parallel combination of a resistor and inductor, the voltage across both components is equal because of the parallel connection, and the current through each branch is calculated by applying Ohm's law.

51. Total current (I_T) is equal to the vector combination of the resistor current and inductor current.

52. Basically, only three problems can occur with inductors:

    a. An open
    b. A complete short
    c. A section short (value change)

53. Depending on the winding's resistance, the coil should be in the range of zero to a few hundred ohms.

54. A coil with one or more shorted turns or a complete short can be checked with an ohmmeter and thought to be perfectly good because of the normally low resistance of a coil, as it is just a piece of wire.

55. An inductor analyzer can be used to check capacitance and inductance.

56. The *RL* filter will achieve results similar to the *RC* filter in that it will pass some frequencies and block others.

57. In the *RL* integrator, the output is taken across the resistor.

58. With the *RL* differentiator, the output is taken across the inductor.

REVIEW QUESTIONS

Multiple-Choice Questions

1. Self-inductance is a process by which a coil will induce a voltage within _____.
    a. Another inductor
    b. Two or more close proximity inductors
    c. Itself
    d. Both (a) and (b)

2. Inductance is a process by which a coil will induce a voltage within _____.
    a. Another inductor
    b. Two or more close proximity inductors
    c. Itself
    d. Both (a) and (b)

3. The inductor stores electrical energy in the form of a _____ field, just as a capacitor stores electrical energy in the form of a _____ field.
    a. Electric, magnetic   **b.** Magnetic, electric

4. The inductor is basically:
    a. An electromagnet
    b. A coil of wire
    c. A coil of conductor formed around a core material
    d. All of the above
    e. None of the above

5. The inductance of an inductor is proportional to _____ and inversely proportional to _____.
    a. $N, A, \mu; l$
    b. $A, \mu, l; N$
    c. $\mu, l, N; A$
    d. $N, A, l; \mu$

6. The total inductance of a series circuit is:
    a. Less than the value of the smallest inductor
    b. Equal to the sum of all the inductance values
    c. Equal to the product over sum
    d. All of the above

7. The total inductance of a parallel circuit can be calculated by:
    a. Using the product-over-sum formula
    b. Using L divided by N for equal-value inductors
    c. Using the reciprocal resistance formula
    d. All of the above

8. Air-core fixed-value inductors can use air or nonmagnetic forms, such as:
    a. Iron, cardboard
    b. Ceramic, copper
    c. Ceramic, cardboard
    d. Silicon, germanium

9. Ferrite is a chemical compound that is basically powdered:
    a. Iron oxide and ceramic   **c.** Mylar and iron
    b. Iron and steel   **d.** Gauze and electrolyte

10. The ferrite-core variable inductor varies inductance by changing:
    a. μ **b.** l **c.** N **d.** A

11. It will actually take _____ time constants for the current in an inductive circuit to reach a maximum value.
    a. 63.2 **b.** 1 **c.** 1.414 **d.** 5

12. The time constant for a series inductive/resistive circuit is equal to:
    a. $L \times R$ **c.** V/R
    b. L/R **d.** $2\pi \times f \times L$

13. Inductive reactance (X_L) is proportional to:
  a. Time or period of the ac applied
  b. Frequency of the ac applied
  c. The stray capacitance that occurs due to the air acting as a dielectric between two turns of a coil
  d. The value of inductance
  e. Two of the above are true.

14. In a series RL circuit, the source voltage (V_S) is equal to:
  a. The square root of the sum of V_R^2 and V_L^2
  b. The vector sum of V_R and V_L
  c. $I \times Z$
  d. Two of the above are partially true.
  e. Answers (a), (b), and (c) are correct.

15. In a purely resistive circuit, the phase shift is equal to _____, whereas in a purely inductive or capacitive circuit, the phase shift is _____ degrees.
  a. 45, 0   c. 45, 90
  b. 90, 0   d. None of the above

16. With an RL integrator, the output is taken across the:
  a. Inductor   c. Resistor
  b. Capacitor   d. Transformer's secondary

17. With an RL differentiator, the output is taken across the:
  a. Inductor   c. Resistor
  b. Capacitor   d. Transformer's secondary

18. An inductor or choke between the input and output forms a:
  a. High-pass filter   b. Low-pass filter

19. An inductor or choke connected to ground or in shunt forms a:
  a. Low-pass filter   b. High-pass filter

20. Lenz's law states that when current is passed through a conductor a self-induced voltage in a coil will:
  a. Aid the applied source voltage
  b. Aid the increasing current from the source
  c. Produce an opposing current
  d. Both (a) and (c)

21. When tested with an ohmmeter, an open coil would show:
  a. Zero resistance
  b. An infinite resistance
  c. A 100 Ω to 200 Ω resistance
  d. Both (b) and (c)

22. Inductive reactance:
  a. Increases with frequency
  b. Is proportional to inductance
  c. Reduces the amplitude of alternating current
  d. All of the above

23. The current through an inductor _____ the voltage across the same inductor by _____.
  a. Lags, 90°   c. Leads, 90°
  b. Lags, 45°   d. Leads, 45°

24. The phasor combination of X_L and R is the circuit's:
  a. Reactance   c. Power factor
  b. Total resistance   d. Impedance

25. In a series RL circuit, where $V_R = 200$ mV and $V_L = 0.2$ V, θ = _____
  a. 45°   c. 0°
  b. 90°   d. 1°

Communication Skill Questions

26. Briefly describe the terms: (Chapter 11, introduction)
  a. Electromagnetism   b. Electromagnetic induction

27. What is self-induction, and how does it relate to an inductor? (12-1)

28. Give the formula for inductance, and explain the four factors that determine inductance. (12-3)

29. List all the formulas for calculating total inductance when inductors are connected in: (12-4)
  a. Series   b. Parallel

30. What are a fixed- and a variable-value inductor? (12-5)

31. List the important factors of the: (12-5)
  a. Air-, iron-, and ferrite-core fixed-value inductors
  b. Ferrite-core variable-value inductor

32. Describe the current rise and fall through an inductor when: (12-6)
  a. Dc is applied   b. Ac is applied

33. Define the following terms:
  a. Inductive reactance (12-7)   d. Q factor (12-8-4)
  b. Impedance (12-8-2)   e. Power factor (12-8-4)
  c. Phase shift (12-8-3)

34. With illustrations, describe how an inductor and resistor could be used for the following applications: (12-11)
  a. Filtering high and low frequencies   c. Differentiation
  b. Integration

35. Explain briefly why an inductor acts as an open to an instantaneous change. Why does an inductor act like a short to dc? (12-1)

Practice Problems

36. Convert the following:
  a. 0.037 H to mH   c. 862 mH to H
  b. 1760 μH to mH   d. 0.256 mH to μH

37. Calculate the impedance (Z) of the following series RL combinations:
  a. 22 MΩ, 25 μH, $f = 1$ MHz
  b. 4 kΩ, 125 mH, $f = 100$ kHz
  c. 60 Ω, 0.05 H, $f = 1$ MHz

38. Calculate the voltage across a coil if: ($d = \Delta$ or delta)
  a. $d_i/d_t = 120$ mA/ms and $L = 2$ μH
  b. $d_i/d_t = 62$ μA/μs and $L = 463$ mH
  c. $d_i/d_t = 4$ A/s and $L = 25$ mH

39. Calculate the total inductance of the following series circuits:
  a. 75 μH, 61 μH, 50 mH   b. 8 mH, 4 mH, 22 mH

40. Calculate the total inductance of the following parallel circuits:
  a. 12 mH, 8 mH   b. 75 μH, 34 μH, 27 μH

41. Calculate the total inductance of the following series–parallel circuits:
  a. 12 mH in series with 4 mH, and both in parallel with 6 mH
  b. A two-branch parallel arrangement made up of 6 μH and 2 μH in series with one another, and 8 μH and 4 μH in series with one another

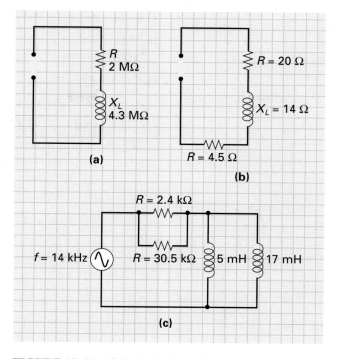

FIGURE 12-42   **Calculating Impedance.**

c. Two parallel arrangements in series with one another, made up of 1 μH and 2 μH in parallel and 4 μH and 15 μH in parallel

42. Determine the time constant for each of the examples in Question 41, and state how long it will take in each example for current to build up to maximum.

43. In a series *RL* circuit, if $V_L = 12$ V and $V_R = 6$ V, calculate:
  a. V_S **d.** Q
  b. I if $Z = 14$ kΩ   **e.** Power factor
  c. Phase angle

44. What value of inductance is needed to produce 3.3 kΩ of reactance at 15 kHz?

45. At what frequency will a 330 μH inductor have a reactance of 27 kΩ?

46. Calculate the impedance of the circuits seen in Figure 12-42.

47. Referring to Figure 12-43, calculate the voltage across the inductor for all five time constants after the switch has been closed.

48. Referring to Figure 12-44, calculate:
  a. L_T
  b. X_L
  c. Z
  d. I_{R1}, I_{L1}, and I_{L2}
  e. θ
  f. True power, reactive power, and apparent power
  g. Power factor

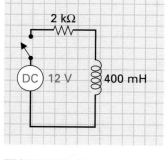

FIGURE 12-43
Inductive Time
Constant Example.

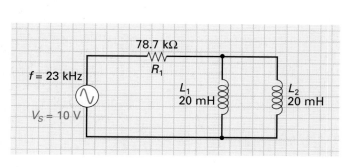

FIGURE 12-44   *RL* **Circuit.**

49. Referring to Figure 12-45, calculate:

- **a.** R_T
- **b.** L_T
- **c.** X_L
- **d.** Z
- **e.** V_R, V_L
- **f.** I_R, I_L
- **g.** I_T
- **h.** θ
- **i.** Apparent power
- **j.** PF

Troubleshooting Problem

50. What problems can occur with inductors, and how can the ohmmeter be used to determine these problems?

Web Site Questions

Go to the Web site http://www.prenhall.com/cook, select the textbook *Introductory DC/AC Electronics* or *Introductory DC/AC Circuits*, this chapter, and then follow the instructions when answering the multiple-choice practice problems.

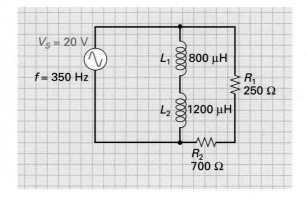

FIGURE 12-45 Parallel *RL* Circuit.

These tests at the end of each chapter will challenge your knowledge up to this point, and give you the practice you need for a job interview. To make this more realistic, the test will comprise both technical and personal questions. In order to take full advantage of this exercise, you may like to set up a simulation of the interview environment, have a friend pose the questions to you, and record your responses for later analysis.

Company Name: Konekt, Inc.

Industry Branch: Consumer Products.

Function: Cell phone repair.

Job Title: Customer Service Technician.

1. What do you know about our phones?

2. What made you choose our company?

3. What can a multimeter be used for?

4. Would you enjoy an out-of-state training course?

5. Do you mind working different shifts?

6. What is an oscilloscope?

7. Do you need constant supervision to keep you on task?

8. What is the difference between an inductor and a capacitor?

9. What do you think you will be doing on a day-in, day-out basis if we hire you as a customer service technician?

10. Do you get along well with most people?

Answers

1. Visit the company's web site before the interview to get an overall understanding of the company's ownership, product line, service, and support.

2. Describe why you were first attracted to their advertisement, why it felt compatible with your career goals, and how the company's positive attributes listed in their web site matched your expectations.

3. Chapters 1 and 2.

4. Additional training only serves to increase your skills and marketability. If you are unsure, however, don't object to anything at the time. Later, if you decide that it is something you do not want to do, you can call and decline the offer.

5. Individual choice. Once again, if you are unsure, don't object to anything at the time.

6. Chapter 8.

7. No. Discuss how you have disciplined yourself to manage school, studying and a job and how well you performed in all. Also, describe the lab environment, which was more than likely self-paced.

8. Chapter 14 introduction.

9. Quote intro. section in text and job description listed in newspaper.

10. They're asking if you can get along well with customers and colleagues. Mention any work experience in which you did direct contact with customers, and discuss how you worked with fellow students in your class.

Transformers

Woolen Mill Makes Minis

At the age of 24 in 1950, Kenneth Olsen went to work at MIT's Digital Computer Laboratory as a research assistant. For the next seven years, Olsen worked on the SAGE (Semi Automatic Ground Environment) project, which was a new computer system designed to store the constantly changing data obtained from radars tracking long-range enemy bombers capable of penetrating American air space. IBM won the contract to build the SAGE network and Olsen traveled to the plant in New York to supervise production. After completing the SAGE project, Olsen was ready to move on to bigger and better things; however, he didn't want to go back to MIT and was not happy with the rigid environment at a large corporation like IBM.

In September 1957, Olsen and fellow MIT colleague Harlan Anderson leased an old brick woolen mill in Maynard, Massachusetts, and after some cleaning and painting they began work in their new company, which they named Digital Equipment Corporation or DEC. After three years, the company came out with their first computer, the PDP 1 (programmed data processor), and its ease of use, low cost, and small size made it an almost instant success. In 1965, the desktop PDP 8 was launched, and since in that era miniskirts were popular, it was probably inevitable that the small machine was nicknamed a mini-computer. In 1969, the PDP 11 was unveiled, and like its predecessors, this number-crunching data processor went on to be accepted in many applications, such as tracking the millions of telephone calls received on the 911 emergency line, directing welding machines in automobile plants, recording experimental results in laboratories, and processing all types of data in offices, banks, and department stores. Just 20 years after introducing low-cost computing in the 1960s, DEC was second only to IBM as a manufacturer of computers in the United States.

Outline and Objectives

Introduction

When alternating current was introduced in Chapter 8, it was mentioned that an ac voltage could be stepped up to a larger voltage or stepped down to a smaller voltage by a device called a **transformer.** The transformer is an electrical device that makes use of electromagnetic induction to transfer alternating current from one circuit to another. The transformer consists of two inductors that are placed in very close proximity to one another. When an alternating current flows through the first coil or **primary winding** the inductor sets up a magnetic field. The expanding and contracting magnetic field produced by the primary cuts across the windings of the second inductor or **secondary winding** and induces a voltage in this coil.

By changing the ratio between the number of turns in the secondary winding to the number of turns in the primary winding, some characteristics of the ac signal can be changed or transformed as it passes from primary to secondary. For example, a low ac voltage could be stepped up to a higher ac voltage, or a high ac voltage could be stepped down to a lower ac voltage.

In this chapter we will examine the operation, characteristics, types, testing, and applications of transformers.

Transformer

Device consisting of two or more coils that are used to couple electric energy from one circuit to another, yet maintain electrical isolation between the two.

Primary Winding

First winding of a transformer that is connected to the source.

Secondary Winding

Output winding of a transformer that is connected across the load.

Mutual Inductance

Ability of one inductor's magnetic lines of force to link with another inductor.

13-1   MUTUAL INDUCTANCE

As was discussed in the previous chapter, self-inductance is the process by which a coil induces a voltage within itself. The principle on which a transformer is based is an inductive effect known as **mutual inductance,** which is the process by which an inductor induces a voltage in another inductor.

Figure 13-1 illustrates two inductors that are magnetically linked, yet electrically isolated from one another. As the alternating current continually rises, falls, and then rises in the opposite direction, a magnetic field will build up, collapse, and then build up in the opposite direction.

If a second inductor, or secondary coil (L_2), is in close proximity with the first inductor or primary coil (L_1), which is producing the alternating magnetic field, a voltage will be induced into the nearby inductor, which causes current to flow in the secondary circuit through the load resistor. This phenomenon is known as mutual inductance or transformer action.

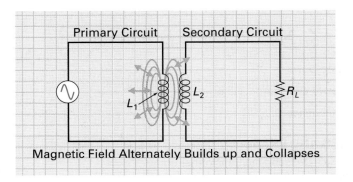

FIGURE 13-1   Mutual Inductance.

As with self-inductance, mutual inductance is dependent on change. Direct current (dc) is a constant current and produces a constant or stationary magnetic field that does not change, as reviewed in Figure 13-2(a). Alternating current, however, is continually varying, and as the polarity of the magnetic field is dependent on the direction of current flow (left-hand rule), the magnetic field will also be alternating in polarity, as seen in Figure 13-2(b); it is this continual building up and collapsing of the magnetic field that cuts the adjacent inductor's conducting coils and induces a voltage in the secondary circuit. Mutual induction is possible only with ac and cannot be achieved with dc due to the lack of change.

Self-induction is a measure of how much voltage an inductor can induce within itself. Mutual inductance is a measure of how much voltage is induced in the secondary coil due to the change in current in the primary coil.

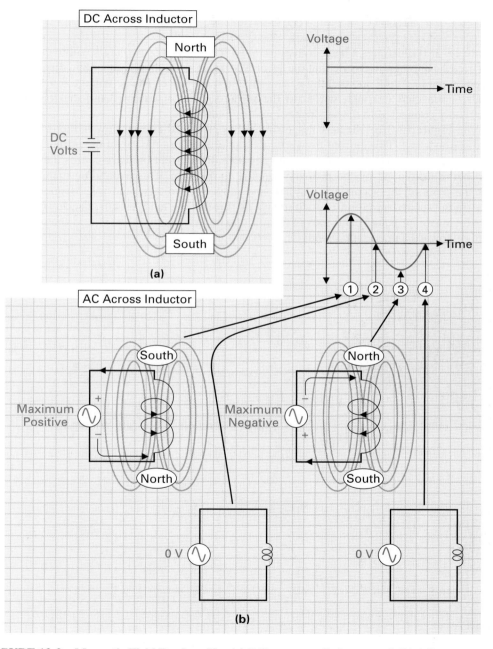

FIGURE 13-2 **Magnetic Field Produced by (a) DC across an Inductor and (b) AC across an Inductor.**

Now that you have completed this section, you should be able to:

■ **Objective 1.** *Define mutual inductance and how it relates to transformers.*

Use the following questions to test your understanding of Section 13-1.

1. Define mutual inductance and explain how it differs from self-inductance.

2. True or false: Mutual induction is possible only with direct current flow.

13-2 BASIC TRANSFORMER

Figure 13-3(a) illustrates a basic transformer, which consists of two coils within close proximity to one another, to ensure that the second coil will be cut by the magnetic flux lines produced by the first coil, and thereby ensure mutual inductance. The ac voltage source is electrically connected (through wires) to the primary coil or winding, and the load (R_L) is electrically connected to the secondary coil or winding.

In Figure 13-3(b), the ac voltage source has produced current flow in the primary circuit, as illustrated. This current flow produces a north pole at the top of the primary winding and, as the ac voltage input swings more negative, the current increase causes the magnetic field being developed by the primary winding to increase. This expanding magnetic field cuts the coils of the secondary winding and induces a voltage, and a subsequent current flows in the secondary circuit, which travels up through the load resistor. The ac voltage follows a sinusoidal pattern and moves from a maximum negative to zero and then begins to build up toward a maximum positive.

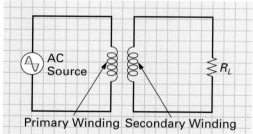

(a)

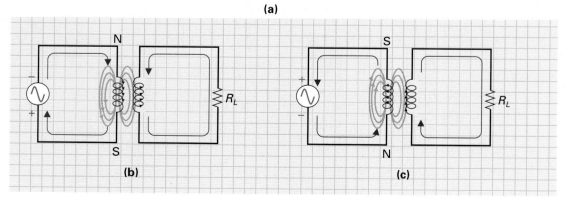

(b)   **(c)**

FIGURE 13-3   **Transformer Action.**

In Figure 13-3(c), the current flow in the primary circuit is in the opposite direction due to the ac voltage increase in the positive direction. As voltage increases, current increases, and the magnetic field expands and cuts the secondary winding, inducing a voltage and causing current to flow in the reverse direction down through the load resistor.

You may have noticed a few interesting points about the basic transformer just discussed.

1. As primary current increases, secondary current increases, and as primary current decreases, secondary current also decreases. It can therefore be said that the frequency of the alternating current in the secondary is the same as the frequency of the alternating current in the primary.

2. Although the two coils are electrically isolated from one another, energy can be transferred from primary to secondary, because the primary converts electrical energy into magnetic energy, and the secondary converts magnetic energy back into electrical energy.

13-3   TRANSFORMER LOADING

Let's now carry our discussion of the basic transformer a little further and see what occurs when the transformer is not connected to a load, as shown in Figure 13-4. Primary circuit current is determined by $I = V/Z$, where Z is the impedance of the primary coil (both its inductive reactance and resistance) and V is the applied voltage. Since no current can flow in the secondary, because an open in the circuit exists, the primary acts as a simple inductor, and the primary current is small due to the inductance of the primary winding. This small primary current lags the applied voltage due to the counter emf by approximately 90° because the coil is mainly inductive and has very little resistance.

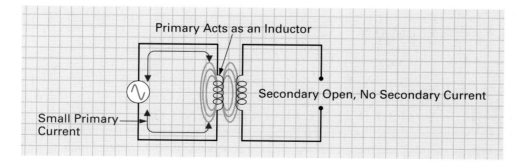

FIGURE 13-4   Unloaded Transformer.

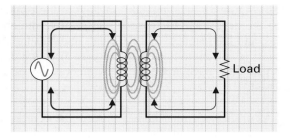

FIGURE 13-5 **Loaded Transformer.**

When a load is connected across the secondary, as shown in Figure 13-5, a change in conditions occurs and the transformer acts differently. The important point that will be observed is that as we go from a no-load to a load condition the primary current will increase due to mutual inductance. Let's follow the steps one by one.

1. The ac applied voltage sets up an alternating magnetic field in the primary winding.
2. The continually changing flux of this primary field induces and produces a counter emf into the primary to oppose the applied voltage.
3. The primary's magnetic field also induces a voltage in the secondary winding, which causes current to flow in the secondary circuit through the load.
4. The current in the secondary winding produces another magnetic field that is opposite to the field being produced by the primary.
5. This secondary magnetic field feeds back to the primary and induces a voltage that tends to cancel or weaken the counter emf that was set up in the primary by the primary current.
6. The primary's counter emf is therefore reduced, so primary current can now increase.
7. This increase in primary current is caused by the secondary's magnetic field; consequently, the greater the secondary current, the stronger the secondary magnetic field, which causes a reduction in the primary's counter emf, and therefore a primary current increase.

In summary, an increase in secondary current ($I_s \uparrow$) causes an increase in primary current ($I_p \uparrow$), and this effect in which the primary induces a voltage in the secondary (V_s) and the secondary induces a voltage into the primary (V_p) is known as *mutual inductance.*

SELF-TEST EVALUATION POINT FOR SECTION 13-3

Now that you have completed this section, you should be able to:

■ *Objective 3.* *Explain the differences between a loaded and an unloaded transformer.*

Use the following questions to test your understanding of Section 13-3.

1. True or false: An increase in secondary current causes an increase in primary current.

2. True or false: The greater the secondary current, the greater the primary's counter emf.

13-4 COEFFICIENT OF COUPLING (k)

The voltage induced into the secondary winding is dependent on the mutual inductance between the primary and secondary, which is determined by how much of the magnetic flux produced by the primary actually cuts the secondary winding.

The **coefficient of coupling** (k) is a ratio of the number of magnetic lines of force that cut the secondary compared to the total number of magnetic flux lines being produced by the primary and is a figure between 0 and 1.

Coefficient of Coupling
The degree of coupling that exists between two circuits.

$$k = \frac{\text{flux linking secondary coil}}{\text{total flux produced by primary}}$$

If, for example, all the primary flux lines cut the secondary winding, the coefficient of coupling will equal 1. If only half of the total flux lines being produced cut the secondary winding, k will equal 0.5.

EXAMPLE:

A primary coil is producing 65 μWb (microwebers) of magnetic flux. Calculate the coefficient of coupling if 52 μWb links with the secondary coil.

Solution:

$$k = \frac{52\ \mu\text{Wb}}{65\ \mu\text{Wb}} = 0.8$$

This means that 80% of the magnetic flux lines being generated by the primary coil are linking with the secondary coil.

The coefficient of coupling depends on:

1. How close together the primary and secondary are to one another
2. The type of core material used

Figure 13-6 illustrates why the primary and secondary have to be in close proximity to one another to achieve a high coefficient of coupling.

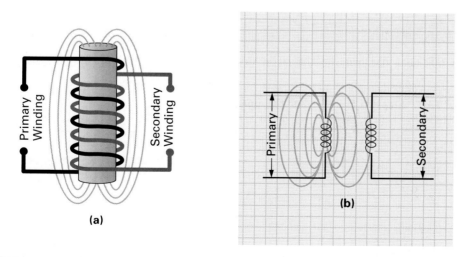

FIGURE 13-6 **Coefficient of Coupling. (a) Wound on Same Core, Small Distance between Primary and Secondary, High k. (b) Winding Far Apart, Small k.**

SELF-TEST EVALUATION POINT FOR SECTION 13-4

Now that you have completed this section, you should be able to:

■ ***Objective 4.*** *State the meaning of the coefficient of coupling.*

Use the following questions to test your understanding of Section 13-4.

1. Define and state the formula for the coefficient of coupling.
2. True or false: If $k = 0.75$, then 75% of the total flux lines produced by the primary are cutting the coils of the secondary.
3. List the two factors that determine k.

13-5 TRANSFORMER RATIOS AND APPLICATIONS

Basically, transformers are used for one of three applications:

1. To step up (increase) or step down (decrease) voltage
2. To step up (increase) or step down (decrease) current
3. To match impedances

In all three cases, any of the applications can be achieved by changing the ratio of the number of turns in the primary winding compared to the number of turns in the secondary winding. This ratio is appropriately called the turns ratio.

13-5-1 *Turns Ratio*

Turns Ratio
Ratio of the number of turns in the secondary winding to the number of turns in the primary winding of a transformer.

The **turns ratio** is the ratio between the number of turns in the secondary winding (N_s) and the number of turns in the primary winding (N_p).

$$\text{turns ratio} = \frac{N_s}{N_p}$$

Let us use a few examples to see how the turns ratio can be calculated.

EXAMPLE:

If the primary has 200 turns and the secondary has 600, what is the turns ratio (Figure 13-7)?

FIGURE 13-7 **Step-Up Transformer Example.**

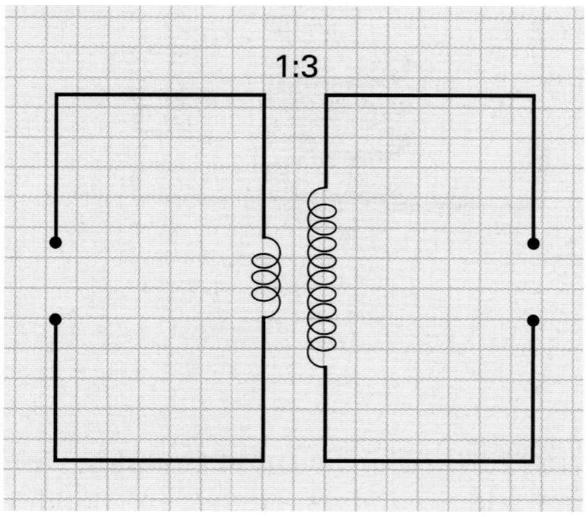

$$\text{turns ratio} = \frac{N_s}{N_p}$$

$$= \frac{600}{200}$$

$$= \frac{3 (\text{secondary})}{1 (\text{primary})}$$

$$= 3$$

This simply means that there are three windings in the secondary to every one winding in the primary. Moving from a small number (1) to a larger number (3) means that we *stepped up* in value. Stepping up always results in a turns ratio figure greater than 1, in this case, 3.

EXAMPLE:

If the primary has 120 turns and the secondary has 30 turns, what is the turns ratio (Figure 13-8)?

FIGURE 13-8   Step-Down Transformer Example.

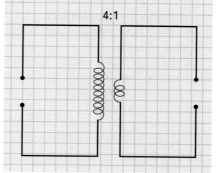

■ *Solution:*

$$\text{turns ratio} = \frac{N_s}{N_p}$$

$$= \frac{30}{120}$$

$$= \frac{1 (\text{secondary})}{4 (\text{primary})}$$

$$= 0.25$$

Said simply, there are four primary windings to every one secondary winding. Moving from a larger number (4) to a smaller number (1) means that we *stepped down* in value. Stepping down always results in a turns ratio figure of less than 1, in this case 0.25.

13-5-2   *Voltage Ratio*

Transformers are used within the power supply unit of almost every piece of electronic equipment to step up or step down the 115 V ac from the outlet. Some electronic circuits require lower-power supply voltages, while other devices may require higher-power supply

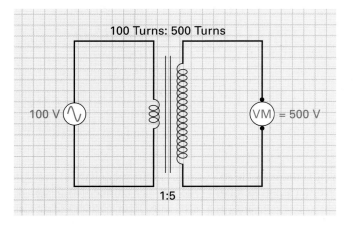

FIGURE 13-9   **Step-Up Transformer.**

voltages. The transformer is used in both instances to convert the 115 V ac to the required value of voltage.

Step Up

If the secondary voltage (V_s) is greater than the primary voltage (V_p), the transformer is called a **step-up transformer** ($V_s > V_p$), as shown in Figure 13-9. The voltage is stepped up or increased in much the same way as a generator voltage can be increased by increasing the number of turns.

If the ac primary voltage is 100 V and the turns ratio is a 1:5 step up, the secondary voltage will be five times that of the primary voltage, or 500 V, because the magnetic flux established by the primary cuts more turns in the secondary and therefore induces a larger voltage.

In this example, you can see that the ratio of the secondary voltage to the primary voltage is equal to the turns ratio; in other words,

$$\frac{V_s}{V_p} = \frac{N_s}{N_p}$$

or

$$\frac{500}{100} = \frac{500}{100}$$

To calculate V_s, therefore, we can rearrange the formula and arrive at

$$V_s = \frac{N_s}{N_p} \times V_p$$

In our example, this is

$$V_s = \frac{500}{100} \times 100 \text{ V}$$
$$= 500 \text{ V}$$

Step Down

If the secondary voltage (V_s) is smaller than the primary voltage (V_p), the transformer is called a **step-down transformer** ($V_s < V_p$), as shown in Figure 13-10. The secondary voltage will be equal to

Step-up Transformer

Transformer in which the ac voltage induced in the secondary is greater (due to more secondary windings) than the ac voltage applied to the primary.

Step-down Transformer

Transformer in which the ac voltage induced in the secondary is less (due to fewer secondary windings) than the ac voltage applied to the primary.

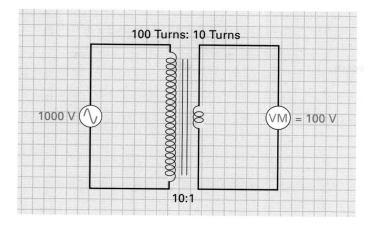

FIGURE 13-10 Step-Down Transformer.

$$V_s = \frac{N_s}{N_p} \times V_p$$

$$= \frac{10}{100} \times 1000 \text{ V} = 100 \text{ V}$$

■ **EXAMPLE:**

Calculate the secondary voltage (V_s) if a 1:6 step-up transformer has 24 V ac applied to the primary.

■ *Solution:*

$$V_s = \frac{N_s}{N_p} \times V_p$$

$$= \frac{6}{1} \times 24 \text{ V} = 144 \text{ V}$$

The coupling coefficient (k) in this formula is always assumed to be 1, which for most iron-core transformers is almost always the case. This means that all the primary magnetic flux is linking the secondary, and the secondary voltage is dependent on the number of secondary turns that are being cut by the primary magnetic flux.

The transformer can be used to transform the primary ac voltage into any other voltage, either up or down, merely by changing the transformer's turns ratio.

13-5-3 *Power and Current Ratio*

The power in the secondary of the transformer is equal to the power in the primary ($P_p = P_s$). Power, as we know, is equal to $P = V \times I$, and if voltage is stepped up or down, the current automatically is stepped down or up, respectively, in the opposite direction to voltage to maintain the power constant.

For example, if the secondary voltage is stepped up ($V_s\uparrow$), the secondary current is stepped down ($I_s\downarrow$), so the output power is the same as the input power.

$$P_s = V_s \uparrow \times I_s \downarrow$$

This is an equal but opposite change. Therefore, $P_s = P_p$; and you cannot get more power out than you put in. The current ratio is therefore inversely proportional to the voltage ratio:

$$\frac{V_s}{V_p} = \frac{I_p}{I_s}$$

If the secondary voltage is stepped up, the secondary current goes down:

$$\frac{V_s\uparrow}{V_p} = \frac{I_p}{I_s\downarrow}$$

If the secondary voltage is stepped down, the secondary current goes up:

$$\frac{V_s\downarrow}{V_p} = \frac{I_p}{I_s\uparrow}$$

If the current ratio is inversely proportional to the voltage ratio, it is also inversely proportional to the turns ratio:

$$\frac{I_p}{I_s} = \frac{V_s}{V_p} = \frac{N_s}{N_p}$$

By rearranging the current and turns ratio, we can arrive at a formula for secondary current, which is

$$I_s = \frac{N_p}{N_s} \times I_p$$

■   **EXAMPLE:**

The step-up transformer in Figure 13-11 has a turns ratio of 1 to 5. Calculate:

a. Secondary voltage (V_s)

b. Secondary current (I_s)

c. Primary power (P_p)

d. Secondary power (P_s)

FIGURE 13-11   Step-Up Transformer Example.

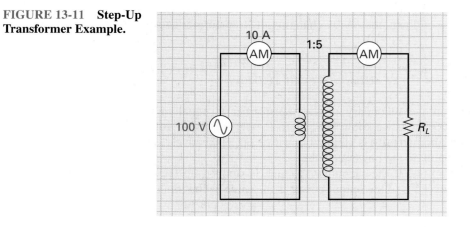

CHAPTER 13 / TRANSFORMERS

Solution:

The secondary has five times as many windings as the primary, and, consequently, the voltage will be stepped up by 5 between primary and secondary. If the secondary voltage is going to be five times that of the primary, the secondary current is going to decrease to one-fifth of the primary current.

a. $V_s = \dfrac{N_s}{N_p} \times V_p$

$\quad = \dfrac{5}{1} \times 100\ V$

$\quad = 500V$

b. $I_s = \dfrac{N_p}{N_s} \times I_p$

$\quad = \dfrac{1}{5} \times 10\ A$

$\quad = 2\ A$

c. $P_p = V_p \times I_p = 100\ V \times 10\ A = 1000\ VA$

d. $P_s = V_s \times I_s = 500\ V \times 2\ A = 1000\ VA$

Therefore, $P_p = P_s$.

13-5-4 *Impedance Ratio*

The **maximum power transfer theorem,** which has been discussed previously and is summarized in Figure 13-12, states that maximum power is transferred from source (ac generator) to load (equipment) when the impedance of the load is equal to the internal impedance of the source. If these impedances are different, a large amount of power could be wasted.

In most cases it is required to transfer maximum power from a source that has an internal impedance (Z_s) that is not equal to the load impedance (Z_L). In this situation, a transformer can be inserted between the source and the load to make the load impedance appear to equal the source's internal impedance.

For example, let's imagine that your car stereo system (source) has an internal impedance of 100 Ω and is driving a speaker (load) of 4 Ω impedance, as seen in Figure 13-13.

By choosing the correct turns ratio, the 4 Ω speaker can be made to appear as a 100 Ω load impedance, which will match the 100 Ω internal source impedance of the stereo system, resulting in maximum power transfer.

Maximum Power Transfer Theorem

The maximum power will be absorbed by the load from the source, when the impedance of the load is equal to the impedance of the source.

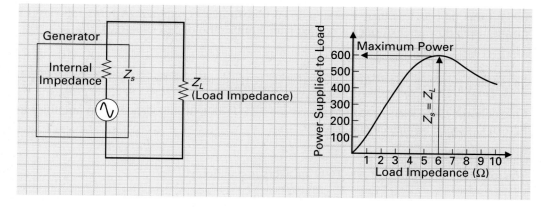

FIGURE 13-12   Maximum Power Transfer Theorem.

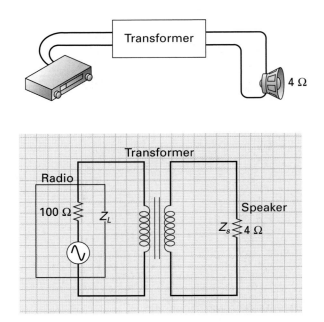

FIGURE 13-13 **Impedance Matching.**

The turns ratio can be calculated by using the formula

$$\text{turns ratio} = \sqrt{\frac{Z_L}{Z_s}}$$

Z_L = Load impedance in ohms
Z_s = Source impedance in ohms

In our example, this will calculate out to be

$$\text{turns ratio} = \sqrt{\frac{Z_L}{Z_s}}$$

$$= \sqrt{\frac{4}{100}} = \frac{\sqrt{4}}{\sqrt{100}}$$

$$= \frac{2}{10} = \frac{1}{5}$$

$$= 0.2$$

If the turns ratio is less than 1, a step-down transformer is required. A turns ratio of 0.2 means a step-down transformer is needed with a turns ratio of 5:1 (⅕ = 0.2).

▦ **EXAMPLE:**

Calculate the turns ratio needed to match the 22.2 Ω output impedance of an amplifier to two 16 Ω speakers connected in parallel.

▦ *Solution:*

The total load impedance of two 16 Ω speakers in parallel will be

$$Z_L = \frac{\text{product}}{\text{sum}} = \frac{16 \times 16}{16 + 16} = 8 \text{ Ω}$$

The turns ratio will be

$$\text{turns ratio} = \sqrt{\frac{Z_L}{Z_s}}$$

$$= \sqrt{\frac{8\ \Omega}{22.2\ \Omega}}$$

$$= \sqrt{0.36} = 0.6$$

Therefore, a step-down transformer is needed with a turns ratio of 1.67 : 1.

SELF-TEST EVALUATION POINT FOR SECTION 13-5

Now that you have completed this section, you should be able to:

■ **Objective 5.** *List the three basic applications of transformers.*

■ **Objective 6.** *Describe how a transformer's turns ratio can be used to step up or step down voltage or current, or match impedance.*

Use the following questions to test your understanding of Section 13-5.

1. What is the turns ratio of a 402 turn primary and 1608 turn secondary, and is this transformer step up or step down?
2. State the formula for calculating the secondary voltage (V_s).
3. True or false: A transformer can, by adjusting the turns ratio, be made to step up both current and voltage between primary and secondary.
4. State the formula for calculating the secondary current (I_s).
5. What turns ratio is needed to match a 25 Ω source to a 75 Ω load?
6. Calculate V_s if $N_s = 200$, $N_p = 112$, and $V_p = 115$ V.

13-6 WINDINGS AND PHASE

The way in which the primary and secondary coils are wound around the core determines the polarity of the voltage induced into the secondary relative to the polarity of the primary. In Figure 13-14, you can see that if the primary and secondary windings are both wound in a clockwise direction around the core, the voltage induced in the primary will be in phase with the voltage induced in the secondary. Both the input and output will also be in phase with one another if the primary and secondary are both wound in a counterclockwise direction.

In Figure 13-15, you will notice that in this case we have wound the primary and secondary in opposite directions, the primary being wound in a clockwise direction and the secondary in a counterclockwise direction. In this situation, the output ac sine-wave voltage is 180° out of phase with respect to the input ac voltage.

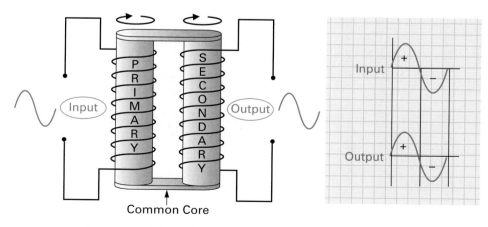

FIGURE 13-14 Primary and Secondary in Phase.

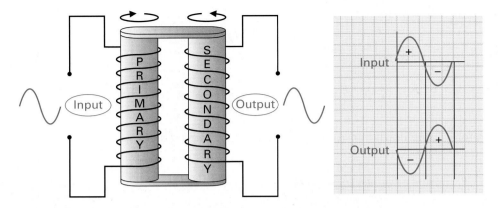

FIGURE 13-15 **Primary and Secondary out of Phase.**

In a schematic diagram, there has to be a way of indicating to the technician that the secondary voltage will be in phase or 180° out of phase with the input. The **dot convention** is a standard used with transformers and is illustrated in Figure 13-16(a) and (b).

A positive on the primary dot causes a positive on the secondary dot, and similarly, since ac is applied, a negative on the primary dot will cause a negative on the secondary dot. In Figure 13-16(a), you can see that when the top of the primary swings positive or negative, the top of the secondary will also follow suit and swing positive or negative, respectively, and you will now be able to determine that the transformer secondary voltage is in phase with the primary voltage.

In Figure 13-16(b), however, the dots are on top and bottom, which means that as the top of the primary winding swings positive, the bottom of the secondary will go positive, which means the top of the secondary will actually go negative, resulting in the secondary ac voltage being out of phase with the primary ac voltage.

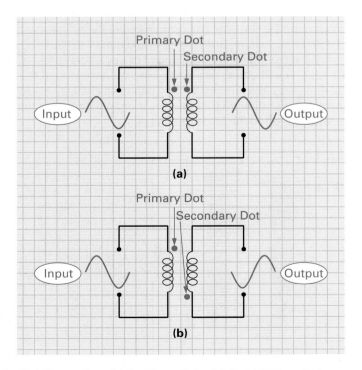

FIGURE 13-16 **Dot Convention. (a) In-Phase Dots. (b) Out-of-Phase Dots.**

Now that you have completed this section, you should be able to:

■ **Objective 7.** *Describe the transformer dot convention as it relates to windings and phase.*

Use the following questions to test your understanding of Section 13-6.

1. How does the winding of the primary and secondary affect the phase of the output with respect to the input?
2. Describe the transformer dot convention used on schematics.

13-7 TRANSFORMER TYPES

We will treat transformers the same as every other electronic component and classify them as either fixed or variable turns ratio.

13-7-1 Fixed Turns Ratio Transformers

Fixed turns ratio transformers have a turns ratio that cannot be varied. They are generally wound on a common core to ensure a high *k* and can be classified by the type of core material used. The two types of fixed transformers are:

1. Air core [Figure 13-17(a)]
2. Iron or ferrite core [Figure 13-17(b)]

The air-core transformers typically have a nonmagnetic core, such as ceramic or a cardboard hollow shell, and are used in high-frequency applications. The electronic circuit symbol just shows the primary and secondary coils. The more common iron- or ferrite-core transformers concentrate the magnetic lines of force, resulting in improved transformer performance; they are symbolized by two lines running between the primary and secondary. The iron-core transformer's lines are solid, while the ferrite-core transformer's lines are dashed.

Figure 13-18 illustrates the physical appearance of a few types of fixed transformers.

13-7-2 Variable Turns Ratio Transformers

Variable turns ratio transformers have a turns ratio that can be varied. They can be classified as follows:

1. Center-tapped secondary
2. Multiple-tapped secondary
3. Multiple windings
4. Single winding

Figure 13-19 illustrates the first two of these, the center-tapped and the multiple-tapped secondary types. Transformers can often have tapped secondaries. A tapped winding will have a lead connected to one of the loops other than the two end connections.

Center-Tapped Secondary

If the tapped lead is in the exact center of the secondary, the transformer is said to have a center-tapped secondary, as shown in Figure 13-20. With the **center-tapped transformer,**

Center-Tapped Transformer

A transformer that has a connection at the electrical center of the secondary winding.

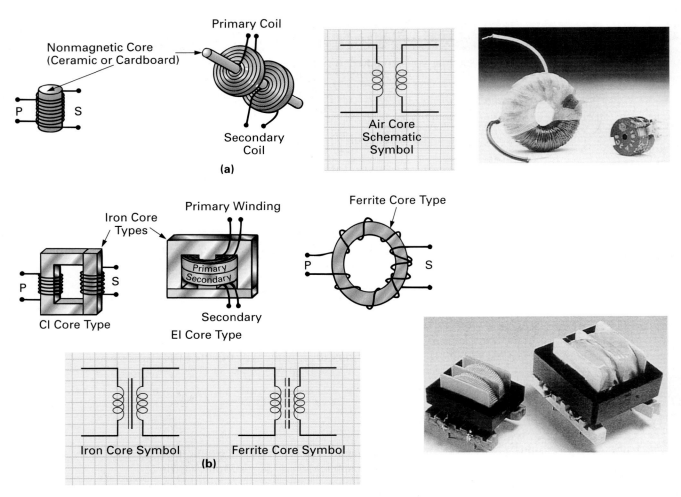

FIGURE 13-17 Fixed Turns Ratio Transformers. (a) Air-Core Transformers. (Courtesy of Sencore.) (b) Iron- or Ferrite-Core Transformers. (Courtesy of Coilcraft.)

FIGURE 13-18 Physical Appearance of Some Transformer Types. (Courtesy of Sencore.)

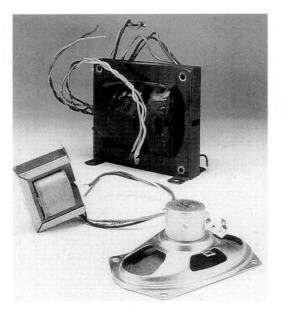

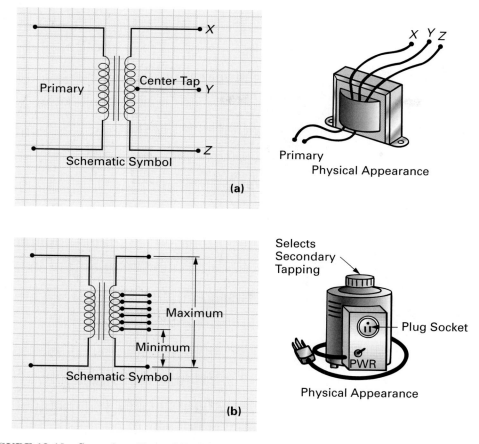

FIGURE 13-19 Secondary Tapped Transformers. (a) Center Tapped. (Courtesy of Sencore.) (b) Multiple Tapped.

the two secondary voltages are each half of the total secondary voltage. If we assume a 1:1 turns ratio (primary turns = secondary turns) and a 20 V ac primary voltage and therefore secondary voltage, each of the output voltages between either end of the secondary and the center tap will be 10 V waveforms, as shown in Figure 13-20. The two secondary outputs will be 180° out of phase with one another from the center tap. When the top of the secondary swings positive (+), the bottom of the secondary will be negative, and vice versa.

Figure 13-21 illustrates a typical power company utility pole, which as you can see makes use of a center-tapped secondary transformer. These transformers are designed to

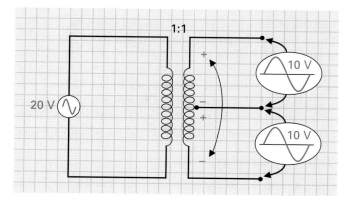

FIGURE 13-20 Center-Tapped Secondary Transformer.

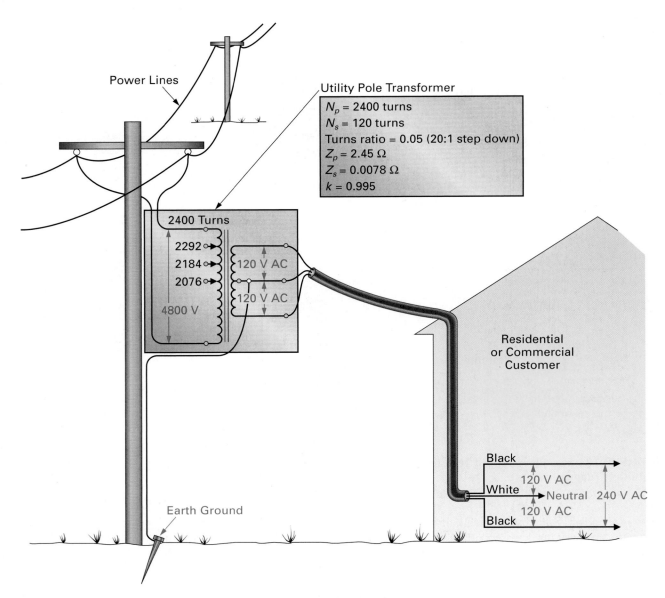

Power Lines

Utility Pole Transformer

N_p = 2400 turns
N_s = 120 turns
Turns ratio = 0.05 (20:1 step down)
Z_p = 2.45 Ω
Z_s = 0.0078 Ω
k = 0.995

2400 Turns

2292
2184
2076

4800 V

120 V AC

120 V AC

Residential
or Commercial
Customer

Earth Ground

Black
120 V AC
White ───► Neutral   240 V AC
120 V AC
Black

FIGURE 13-21   Center-Tapped Utility Pole Transformer.

step down the high 4.8 kV from the power line to 120 V/240 V for commercial and residential customers. Within the United States, the actual secondary voltage may be anywhere between 225 and 245 V, depending on the demand. The demand is actually higher during the day and in the winter months (source voltage is pulled down by smaller customer load resistance). The multiple taps on the primary are used to change the turns ratio and compensate for power line voltage differences.

The center tap of the secondary winding is connected to a copper earth grounding rod. This center tap wire is called the **neutral wire,** and within the building it is color coded with a white insulation. Residential or commercial appliances or loads that are designed to operate at 240 V are connected between the two black wires, while loads that are designed to operate at 120 V are connected between a black wire and the white wire (neutral).

Multiple-Tapped Secondary

If several leads are attached to the secondary, the transformer is said to have a multiple-tapped secondary. This multitapped transformer will tap or pick off different values of ac voltage, as seen in Figure 13-22.

Neutral Wire

The conductor of a polyphase circuit or of a single-phase three-wire circuit that is intended to have a ground potential. The potential differences between the neutral and each of the other conductors are approximately equal in magnitude and are also equally spaced in phase.

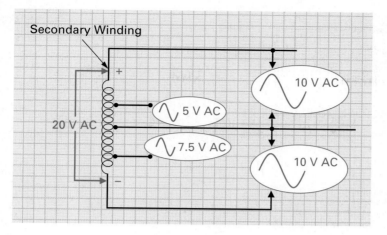

FIGURE 13-22 **Multiple-Tapped Secondary Transformer.**

Multiple Windings

The simple transformer has only one primary and one secondary. In some applications, transformers can have multiple primaries and even multiple secondaries.

Multiple primaries. The major application of this arrangement is to switch between two primary voltages and obtain the same secondary voltage. In Figure 13-23(a), the two primaries are connected in parallel, so the primary only has an equivalent 100 turns; and since the secondary has 10 turns, this 10:1 step-down ratio will result in a secondary voltage of

$$V_s = \frac{N_s}{N_p} \times V_p$$
$$= \frac{10}{100} \times 120 \text{ V}$$
$$= 12 \text{ V}$$

If the two primaries are connected in series, as shown in Figure 13-23(b), the now 200 turn primary will result in a secondary voltage of

$$V_s = \frac{N_s}{N_p} \times V_p$$
$$= \frac{10}{120} \times 240 \text{ V}$$
$$= 12 \text{ V}$$

Some portable electrical or electronic equipment, such as radios or shavers, have a switch that allows you to switch between 120 V ac (U.S. wall outlet voltage) and 240 V ac (European wall outlet voltage), as shown in Figure 13-23(c). By activating the switch, you can always obtain the correct voltage to operate the equipment, whether the wall socket is supplying 240 or 120 V.

Multiple secondaries. In some applications, more than one secondary is wound onto a common primary and core. The advantage of this arrangement can be seen in Figure 13-24, where many larger and smaller voltages can be acquired from one primary voltage.

Single Winding (Autotransformer)

The single winding transformer or autotransformer is used in the automobile ignition system (switched dc pulse system) to raise the low 12 V battery voltage up to about 20 kV

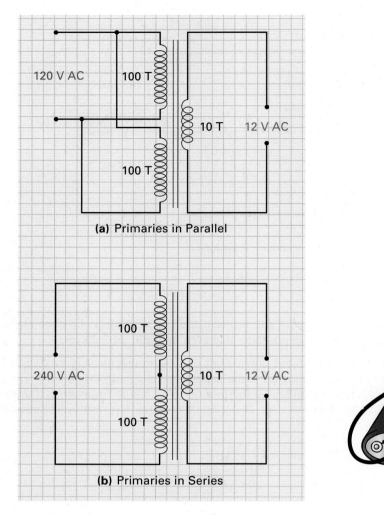

FIGURE 13-23   Multiple Primary Windings Transformer.

for the spark plugs, as shown in Figure 13-25(a). Autotransformers are constructed by winding one continuous coil onto a core, which acts as both the primary and secondary. This yields three advantages for the single winding transformer:

1. They are smaller.
2. They are cheaper.
3. They are lighter than the normal separated primary/secondary transformer types.

Its disadvantage, however, is that no electrical isolation exists between primary and secondary. It is normally used as a step-up transformer for the automobile, or in televisions to obtain 20 kV. This transformer type can also be arranged to step down voltage, as shown in Figure 13-25(b).

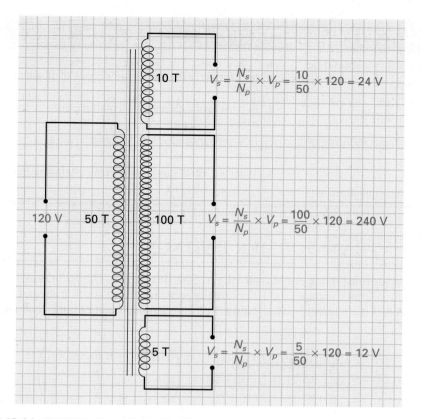

$$V_s = \frac{N_s}{N_p} \times V_p = \frac{10}{50} \times 120 = 24 \text{ V}$$

$$V_s = \frac{N_s}{N_p} \times V_p = \frac{100}{50} \times 120 = 240 \text{ V}$$

$$V_s = \frac{N_s}{N_p} \times V_p = \frac{5}{50} \times 120 = 12 \text{ V}$$

FIGURE 13-24 **Multiple Secondary Windings Transformer.**

FIGURE 13-25 **Single-Winding Transformer. (a) Step-Up. (b) Step-Down.**

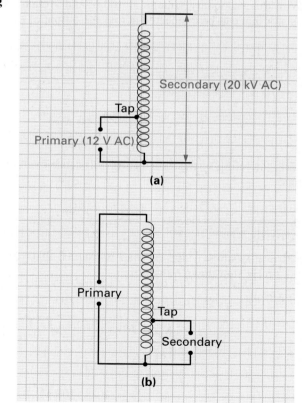

Now that you have completed this section, you should be able to:

■ **Objective 8.** *List and explain the fixed and variable transformer types.*

Use the following questions to test your understanding of Section 13-7.

1. List the two types of fixed transformers.

2. List the four basic classifications for variable transformers.

3. True or false: The two outputs of a center-tapped secondary transformer will always be 90° out of phase with one another.

4. In what application would a multiple primary transformer be used?

5. In what application would a single-winding transformer be used?

6. What are the advantages of a single-winding transformer?

13-8   TRANSFORMER RATINGS

A typical transformer rating could read 1 kVA, 500/100, 60 Hz. The 500 normally specifies the maximum primary voltage, the 100 normally specifies the maximum secondary voltage, and the 1 kVA is the apparent power rating. In this example, the maximum load current will equal

$$I_s = \frac{\text{apparent power } (P_A)}{\text{secondary voltage } (V_s)} \qquad \left(\begin{array}{c} P = V \times I; \text{ therefore,} \\ I = \dfrac{P}{V} \end{array} \right)$$

$$= \frac{1 \text{ kVA}}{100 \text{ V}} = 10 \text{ A}$$

With this secondary voltage at 100 V and a maximum current of 10 A, the smallest load resistor that can be connected across the output of the secondary is

$$R_L = \frac{V_s}{I_s}$$

$$= \frac{100 \text{ V}}{10 \text{ A}}$$

$$= 10 \text{ } \Omega$$

Exceeding the rating of the transformer will cause overheating and even burning out of the windings.

■ **EXAMPLE:**

Calculate the smallest value of load resistance that can be connected across a 3 kVA, 600/200, 60 Hz step-down transformer.

■ *Solution:*

$$I = \frac{\text{apparent power } (P_A)}{\text{secondary voltage } (V_s)}$$

$$\frac{3 \text{ kVA}}{200 \text{ V}} = 15 \text{ A}$$

$$R_L = \frac{V_s}{I_s} = \frac{200 \text{ V}}{15 \text{ A}} = 13.3 \text{ } \Omega$$

Use the following questions to test your understanding of Section 13-8.

1. What do each of the values mean when a transformer is rated as a 10 kVA, 200/100, 60 Hz?

2. If a 100 Ω resistor is connected across the secondary of a transformer that is to supply 1 kV and is rated at a maximum current of 8 A, will the transformer overheat and possibly burn out?

13-9 TESTING TRANSFORMERS

Like inductors, transformers, which are basically two or more inductors, can develop one of three problems:

1. An open winding
2. A complete short in a winding
3. A short in a section of a winding

13-9-1 *Open Primary or Secondary Winding*

An open in the primary winding will prevent any primary current, and therefore there will be no induced voltage in the secondary and therefore no voltage will be present across the load. An open secondary winding will prevent the flow of secondary current, and once again, no voltage will be present across the load.

An open in the primary or secondary winding is easily detected by disconnecting the transformer from the circuit and testing the resistance of the windings with an ohmmeter. Like an inductor, a transformer winding should have a low resistance, in the tens to hundreds range. An open will easily be recognized because of its infinite (maximum) resistance.

13-9-2 *Complete Short or Section Short in Primary or Secondary Winding*

A partial or complete short in the primary winding of a transformer will result in an excessive source current that will probably blow the circuit's fuse or trip the circuit's breaker. A partial or complete short in the secondary winding will cause an excessive secondary current, which in turn will result in an excessive primary current.

In both instances, a short in the primary or secondary will generally burn out the primary winding unless the fuse or circuit breaker opens the excessive current path. An ohmmeter can be used to test for partial or complete shorts in transformer windings; however, all coils have a naturally low resistance which can be mistaken for a short. An inductive or reactive analyzer can accurately test transformer windings, checking the inductance value of each coil.

It is virtually impossible to repair an open, partially shorted, or completely shorted transformer winding, and therefore defective transformers are always replaced with a transformer that has an identical rating.

Now that you have completed this section, you should be able to:

■ **Objective 9.** *Explain how to test the windings of a transformer for opens, partial shorts, or complete shorts.*

Use the following questions to test your understanding of Section 13-9.

1. Which test instrument could you use to test for a suspected open primary winding?

2. If a secondary winding had a resistance of 50 Ω, would this value indicate a problem?

13-10 TRANSFORMER LOSSES

Internal losses cause the power delivered from the secondary winding of a transformer in reality to be less than the power fed into the primary winding. $P_s < P_p$ in reality, due to transformer losses. There are basically three types of internal transformer loss: **copper loss,** core loss, and magnetic leakage.

13-10-1 *Copper Losses*

Due to ohmic resistance of the windings, heat ($I^2 \times R$) is dissipated in both the primary and secondary windings.

13-10-2 *Core Losses*

Hysteresis

During each ac cycle, the core is taken through a cycle of magnetization; hence energy is lost due to hysteresis and appears as heat in the core. This loss is proportional to frequency and is minimized by using a core of soft iron or an alloy such as stalloy or permalloy.

Eddy-Current Loss

The continuously changing flux induces voltages in the conducting core, which results in local **eddy currents** in the core that combine to produce a large circulating current, as

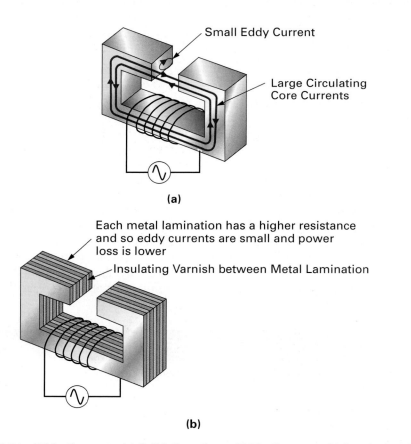

(a)

(b)

FIGURE 13-26 **Eddy Currents. (a) Solid Core, Large Eddy Currents. (b) Laminated Core, Small Eddy Currents.**

shown in Figure 13-26(a), that opposes the main flux and also generates heat. This loss, which is proportional to frequency, is minimized by using a laminated core, as shown in Figure 13-26(b).

13-10-3 *Magnetic Leakage*

Not all the flux lines produced will cut the secondary winding. The result is that for a given turns ratio, the secondary terminal voltage will be slightly lower than expected due to **magnetic leakage.**

Magnetic Leakage
The passage of magnetic flux outside the path along which it can do useful work.

SELF-TEST EVALUATION POINT FOR SECTION 13-10

Now that you have completed this section, you should be able to:

■ *Objective 10. Describe the three basic transformer power losses.*

Use the following questions to test your understanding of Section 13-10.

1. List the four types of transformer losses.
2. Why will a laminated transformer core reduce eddy-current loss?

SUMMARY

Transformers (Figure 13-27)

1. The transformer is an electrical device that makes use of electromagnetic induction to transfer alternating current from one circuit to another.

2. The transformer consists of two inductors that are placed in very close proximity to one another. When an alternating current flows through the first coil or primary winding the inductor sets up a magnetic field. The expanding and contracting magnetic field produced by the primary cuts across the windings of the second inductor or secondary winding and induces a voltage in this coil.

3. By changing the ratio between the number of turns in the secondary winding to the number of turns in the primary winding, some characteristics of the ac signal can be changed or transformed as it passes from primary to secondary.

4. The principle on which a transformer is based is an inductive effect known as mutual inductance, which is the process by which an inductor induces a voltage in another inductor.

5. As with self-inductance, mutual inductance is dependent on change. Direct current (dc) is a constant current and produces a constant or stationary magnetic field that does not change. Alternating current, however, is continually varying, and as the polarity of the magnetic field is dependent on the direction of current flow (left-hand rule), the magnetic field will also be alternating in polarity.

6. Self-induction is a measure of how much voltage an inductor can induce within itself. Mutual inductance is a measure of how much voltage is induced in the secondary coil due to the change in current in the primary coil.

7. The basic transformer consists of two coils within close proximity to one another, to ensure that the second coil will be cut by the magnetic flux lines produced by the first coil, and thereby ensure mutual inductance. The ac voltage source is electrically connected (through wires) to the primary coil or winding, and the load (R_L) is electrically connected to the secondary coil or winding.

8. As primary current increases, secondary current increases, and as primary current decreases, secondary current also decreases. It can be said that the frequency of the alternating current in the secondary is the same as the frequency of the alternating current in the primary.

9. Although the two coils are electrically isolated from one another, energy can be transferred from primary to secondary, because the primary converts electrical energy into magnetic energy, and the secondary converts magnetic energy back into electrical energy.

10. When the transformer is not connected to a load, primary circuit current is determined by $I = V/Z$, where Z is the impedance of the primary coil (both its inductive reactance and resistance) and V is the applied voltage. Primary current lags the applied voltage due to the counter emf by approximately 90° because the coil is mainly inductive and has very little resistance.

11. When a load is connected across the secondary, an increase in secondary current ($I_s \uparrow$) causes an increase in primary current ($I_p \uparrow$), and this effect in which the primary induces a voltage in the secondary (V_s) and the secondary

FIGURE 13-27   Transformers.

$$I_p = \frac{V_p}{Z_p}$$

$$\text{Coupling coefficient } (k) = \frac{\text{flux linking secondary coil}}{\text{total flux produced by primary}}$$

$$\text{Turns ratio} = \frac{N_s}{N_p}$$

$$V_s = \frac{N_s}{N_p} \times V_p$$

$$I_s = \frac{N_s}{N_p} \times I_p$$

$$\text{Turns ratio} = \sqrt{\frac{Z_L}{Z_s}}$$

$$I_s = \frac{P_A}{V_s}$$

$$R_L = \frac{V_s}{I_s}$$

Transformer Action

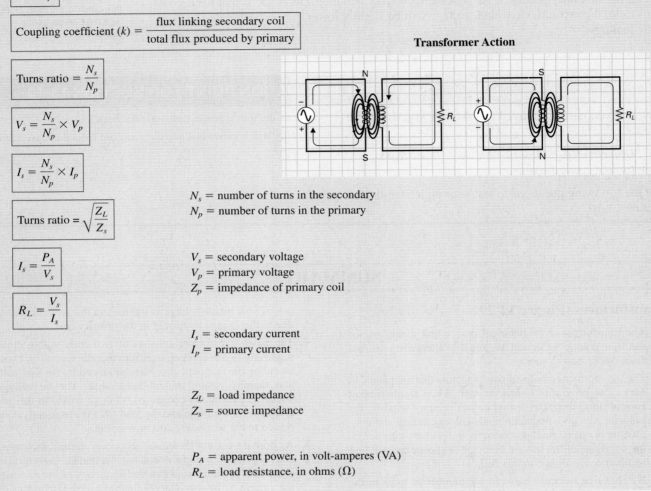

N_s = number of turns in the secondary
N_p = number of turns in the primary

V_s = secondary voltage
V_p = primary voltage
Z_p = impedance of primary coil

I_s = secondary current
I_p = primary current

Z_L = load impedance
Z_s = source impedance

P_A = apparent power, in volt-amperes (VA)
R_L = load resistance, in ohms (Ω)

induces a voltage into the primary (V_p) is known as *mutual inductance*.

12. The coefficient of coupling (k) is a ratio of the number of magnetic lines of force that cut the secondary compared to the total number of magnetic flux lines being produced by the primary and is a figure between 0 and 1.

13. Basically, transformers are used for one of three applications:
    a. To step up (increase) or step down (decrease) voltage
    b. To step up (increase) or step down (decrease) current
    c. To match impedances

    In all three cases, any of the applications can be achieved by changing the ratio of the number of turns in the primary winding compared to the number of turns in the secondary winding. This ratio is appropriately called the turns ratio.

14. The turns ratio is the ratio between the number of turns in the secondary winding (N_s) and the number of turns in the primary winding (N_p).

15. If the secondary voltage (V_s) is greater than the primary voltage (V_p), the transformer is called a step-up transformer ($V_s > V_p$).

16. If the secondary voltage (V_s) is smaller than the primary voltage (V_p), the transformer is called a step-down transformer ($V_s < V_p$).

17. The power in the secondary of the transformer is equal to the power in the primary ($P_p = P_s$).

18. The maximum power transfer theorem states that maximum power is transferred from source (ac generator) to load (equipment) when the impedance of the load is equal

to the internal impedance of the source. If these imped-ances are different, a large amount of power could be wasted.

19. In most cases a source has an internal impedance (Z_s) that is not equal to the load impedance (Z_L). In this situation, a transformer can be inserted between the source and the load to make the load impedance appear to equal the source's internal impedance.

20. The dot convention is a standard used with transformer symbols to indicate whether the secondary voltage will be in phase or out of phase with the primary voltage.

21. Fixed turns ratio transformers have a turns ratio that cannot be varied.

22. The air-core transformers typically have a nonmagnetic core, such as ceramic or a cardboard hollow shell, and are used in high-frequency applications.

23. The more common iron- or ferrite-core transformers con-centrate the magnetic lines of force, resulting in improved transformer performance.

24. Variable turns ratio transformers have a turns ratio that can be varied.

25. If the tapped lead is in the exact center of the secondary, the transformer is said to have a center-tapped secondary. With the center-tapped transformer, the two secondary voltages are each half of the total secondary voltage.

26. A typical transformer rating could read 1 kVA, 500/100, 60 Hz. The 500 normally specifies the maximum primary voltage, the 100 normally specifies the maximum second-ary voltage, and the 1 kVA is the apparent power rating.

27. An open in the primary winding will prevent any primary current, and therefore there will be no induced voltage in the secondary and no voltage will be present across the load.

28. An open secondary winding will prevent the flow of sec-ondary current, and once again, no voltage will be present across the load.

29. An open in the primary or secondary winding is easily de-tected by disconnecting the transformer from the circuit and testing the resistance of the windings with an ohmme-ter. Like an inductor, a transformer winding should have a low resistance, in the tens to hundreds range. An open will easily be recognized because of its infinite (maxi-mum) resistance.

30. A partial or complete short in the primary winding of a transformer will result in an excessive source current that will probably blow the circuit's fuse or trip the circuit's breaker.

31. A partial or complete short in the secondary winding will cause an excessive secondary current, which will result in an excessive primary current.

32. In both instances, a short in the primary or secondary will generally burn out the primary winding unless the fuse or circuit breaker opens the excessive current path.

33. An ohmmeter can be used to test for partial or complete shorts in transformer windings. All coils have a naturally low resistance which can be mistaken for a short.

34. An inductive or reactive analyzer can accurately test trans-former windings, checking the inductance value of each coil.

35. It is virtually impossible to repair an open, partially shorted, or completely shorted transformer winding, and therefore defective transformers are always replaced with a transformer that has an identical rating.

36. Internal losses cause the power delivered from the second-ary winding of a transformer in reality to be less than the power fed into the primary winding. These losses include ohmic resistance of the windings, hysteresis, local eddy currents in the core, and magnetic leakage.

REVIEW QUESTIONS

Multiple-Choice Questions

1. Transformer action is based on:
   a. Self-inductance   c. Mutual capacitance
   b. Air between the coils   d. Mutual inductance

2. An increase in transformer secondary current will cause a/an _____ in primary current.
   a. Decrease   b. Increase

3. If 50% of the magnetic lines of force produced by the pri-mary were to cut the secondary coil, the coefficient of cou-pling would be:
   a. 75   b. 50   c. 0.5   d. 0.005

4. A step-up transformer will always have a turns ratio _____, while a step-down transformer has a turns ratio _____.
   a. $< 1, > 1$ b. $> 1, > 1$ c. $> 1, < 1$ d. $< 1, < 1$

5. With an 80 V ac secondary voltage center-tapped trans-former, what would be the voltage at each output, and what would be the phase relationship between the two secondary voltages?
   a. 20 V, in phase with one another
   b. 30 V, 180° out of phase
   c. 40 V, in phase
   d. 40 V, 180° out of phase

6. One application of the autotransformer would be:
   a. To obtain the final anode high-voltage supply for the cathode-ray tube in a television
   b. To obtain two outputs, 180° out of phase with one another
   c. To tap several different voltages from the secondary
   d. To obtain the same secondary voltage for different voltages

7. Transformers can only be used with alternating current because:
   a. It produces an alternating magnetic field.
   b. It produces a fixed magnetic field.
   c. Its magnetic field is greater than that of dc.
   d. Its rms is 0.707 of the peak.

8. Eddy-current losses are reduced with laminated iron cores because:
   a. The air gap is always kept to a minimum.
   b. The resistance of iron is always low.
   c. The laminations are all insulated from one another.
   d. Current cannot flow in iron.

9. Assuming 100% efficiency, the output power, P_s, is always equal to:
   a. P_p d. Both (a) and (c)
   b. $V_s \times I_s$ e. Both (a) and (b)
   c. $0.5 \times P_p$

10. If the primary winding of a transformer were open, the result would be:
    a. No flux linkage between primary and secondary
    b. No primary or secondary current
    c. No induced voltage in the secondary
    d. All of the above

Communication Skill Questions

11. What is a transformer? (Introduction)

12. Describe mutual inductance and how it relates to transformers. (13-1)

13. Why does loading the transformer's secondary circuit affect primary transformer current? (13-3)

14. Describe what a coefficient of coupling figure of 0.9 means. (13-4)

15. List the three basic applications of transformers, and then describe how each of these applications can be achieved by merely changing the transformer's turns ratio. (13-5)

16. Briefly describe how the dot convention is used in schematics to describe phase. (13-6)

17. Illustrate the schematic symbol and main points relating to the following transformers:
    a. Fixed air core (13-7-1)
    b. Fixed iron core (13-7-1)
    c. Fixed ferrite core (13-7-1)
    d. Center-tapped secondary (13-7-2)

e. Four-output tapped secondary (13-7-2)
f. Multiple primary and multiple secondary (13-7-2)
g. Autotransformer (13-7-2)

18. Illustrate and explain the following transformer applications: (13-7)
    a. 120/240 V primary voltage to constant secondary voltage
    b. 20 kV secondary voltage using the autotransformer

19. Briefly describe what is meant by copper losses with transformers. (13-10)

20. Briefly describe the following iron and core losses and how they can be reduced: (13-10)
    a. Hysteresis   b. Eddy current

21. Briefly describe the transformer loss known as magnetic leakage. (13-10)

22. What meter could be used to check a suspected open primary coil? (13-9)

23. Could a transformer be considered a dc block? (13-1)

24. Would a step-up voltage transformer step up or step down current? What would happen to secondary power? (13-5)

25. Briefly describe how transformers are used to reduce I^2R losses in ac power distribution. (Chapter 8)

Practice Problems

26. Calculate the turns ratio of the following transformers and state whether they are step up or step down:
    a. P = 12 T, S = 24 T   c. P = 24 T, S = 5 T
    b. P = 3 T, S = 250 T   d. P = 240 T, S = 120 T

27. Calculate the secondary ac voltage for all the examples in Question 26 if the primary voltage equals 100 V.

28. Calculate the secondary ac current for all the examples in Question 26 if the primary current equals 100 mA.

29. What turns ratio would be needed to match a source impedance of 24 Ω to a load impedance of 8 Ω?

30. What turns ratio would be needed to step:
    a. 120 V to 240 V   c. 30 V to 14 V
    b. 240 V to 720 V   d. 24 V to 6 V

31. For a 24 V, 12 turn primary with a 16, 2, 1, and 4 turn multiple secondary transformer, calculate each of the secondary voltages.

32. If a 2 : 1 step-down transformer has a primary input voltage of 120 V, 60 Hz, and a 2 kVA rating, calculate the maximum secondary current and smallest load resistor that can be connected across the output.

33. Indicate the polarity of the secondary voltages in Figure 13-28(a) and (b).

34. Referring to Figure 13-29(a) and (b), sketch the outputs, showing polarity and amplitude with respect to the inputs.

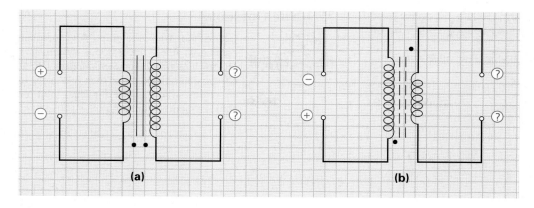

FIGURE 13-28   Dot Convention Examples.

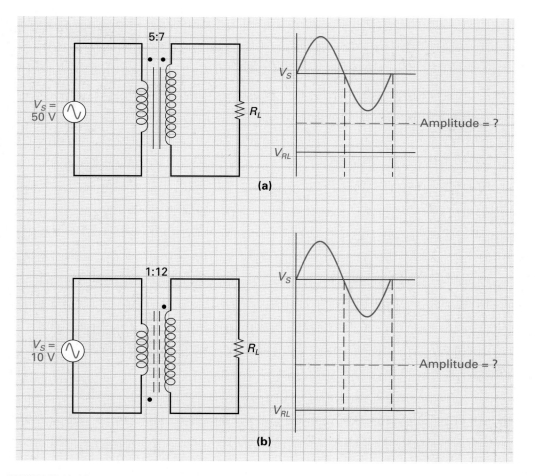

FIGURE 13-29   Input/Output Polarity Examples.

35. If a transformer is rated at 500 VA, 60 Hz, the primary voltage is 240 V ac, and the secondary voltage is 600 V ac, calculate:

a. Maximum load current   **b.** Smallest value of R_L

Web Site Questions

 Go to the Web site http://www.prenhall.com/cook, select the textbook *Introductory DC/AC Electronics* or *Introductory DC/AC Circuits*, this chapter, and then follow the instructions when answering the multiple-choice practice problems.

These tests at the end of each chapter will challenge your knowledge up to this point, and give you the practice you need for a job interview. To make this more realistic, the test will comprise both technical and personal questions. In order to take full advantage of this exercise, you may want to set up a simulation of the interview environment, have a friend read the questions to you, and record your responses for later analysis.

Company Name: SOS, Inc.

Industry Branch: Marine Electronics.

Function: Manufacture Communications Equipment.

Job Title: Calibration Technician.

1. Who was your last employer?
2. What does the term "calibration" mean?

3. What is a potentiometer?
4. Could you describe the steps involved in developing a product from start to finish?
5. Why would the load resistance connected to the secondary of a transformer determine the primary current of the same transformer?
6. What test equipment are you familiar with?
7. What is the job function of a production test technician?
8. What level of supervision would you need for this job?
9. What is the maximum power transfer theorem?
10. During a heavy production cycle we may require that you work every other Saturday for a short space of time. Would this be a problem for you?

Answers

1. Discuss your previous employer and give a copy of your letter of recommendation. If this job was not technical, refer back to how the lab work during training was structured similar to a work environment and discuss other related topics.
2. Chapter 8.
3. Chapter 3.
4. Describe the steps detailed in the Introduction section.
5. Section 13-3.
6. Up to this time, only the multimeter and oscilloscope.

7. Quote intro. section in text, and job description listed in newspaper.
8. Say that you expect that in the beginning you would have many questions, but as you became more familiar with company procedures and the product line, you would need less and less supervision.
9. Section 13-5-4.
10. If you are unsure, don't object to anything at the time. Later, if you decide that it is something you do not want to do, you can call and decline the offer.

Resistive, Inductive, and Capacitive (*RLC*) Circuits

The Fairchildren

William Shockley, John Bardeen, and Walter Brattain (left to right)

On December 23, 1947, John Bardeen, Walter Brattain, and William Shockley first demonstrated how a semiconductor device, named the *transistor,* could be made to amplify. However, the device had mysterious problems and was very unpredictable. Shockley continued his investigations, and in 1951 he presented the world with the first reliable junction transistor. In 1956, the three shared the Nobel Prize in physics for their discovery, and much later, in 1972, Bardeen would win a rare second Nobel Prize for his research at the University of Illinois in the field of superconductivity.

Shockley left Bell Labs in 1955 to start his own semiconductor company near his home in Palo Alto and began recruiting personnel. He was, however, very selective, only hiring those who were bright, young, and talented. The company was a success, although many of the employees could not tolerate Shockley's eccentricities, such as posting everyone's salary and requiring that the employees rate one another. Two years later, eight of Shockley's most talented defected. The "traitorous eight," as Shockley called them, started their own company only a dozen blocks away, named Fairchild Semiconductor.

More than 50 companies would be founded by former Fairchild employees. One of the largest was started by Robert Noyce and two other colleagues from the group of eight Shockley defectors; they named their company Intel, which was short for "intelligence."

Outline and Objectives

Introduction

In this chapter we will combine resistors (R), inductors (L), and capacitors (C) into series and parallel ac circuits. Resistors, as we have discovered, operate and react to voltage and current in a very straightforward way, in that the voltage across a resistor is in phase with the resistor current.

Inductors and capacitors operate in essentially the same way, in that they both store energy and then return it back to the circuit. However, they have completely opposite reactions to voltage and current. To help you remember the phase relationships between voltage and current for capacitors and inductors, you may wish to use the following memory phrase:

> ELI the ICE man

This phrase states that voltage (symbolized E) leads current (I) in an inductive (L) circuit (abbreviated by the word "ELI"), while current (I) leads voltage (E) in a capacitive (C) circuit (abbreviated by the word "ICE").

In this chapter we will study the relationships between voltage, current, impedance, and power in both series and parallel RLC circuits. We will also examine the important RLC circuit characteristic called resonance, and see how RLC circuits can be made to operate as filters. In the final section we will discuss how complex numbers can be used to analyze series and parallel ac circuits containing resistors, inductors, and capacitors.

14-1　SERIES *RLC* CIRCUIT

Figure 14-1 begins our analysis of series RLC circuits by illustrating the current and voltage relationships. The circuit current is always the same throughout a series circuit and can therefore be used as a reference. Studying the waveforms and vector diagrams shown alongside the components, you can see that the voltage across a resistor is always in phase with the current, while the voltage across the inductor leads the current by 90° and the voltage across the capacitor lags the current by 90°.

Now let's analyze the impedance, current, voltage, and power distribution of this circuit in a little more detail.

14-1-1　*Impedance*

Impedance is the total opposition to current flow and is a combination of both reactance (X_L, X_C) and resistance (R). An example circuit is illustrated in Figure 14-2(a).

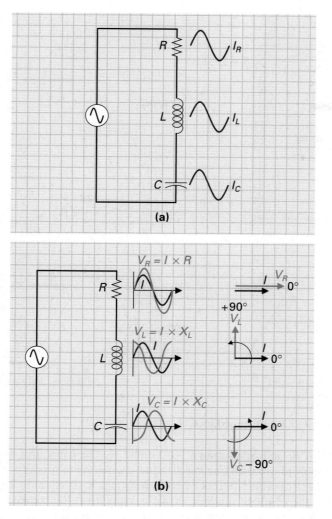

FIGURE 14-1 Series *RLC* Circuit. (a) *RLC* Series Circuit Current: Current Flow Is Always the Same in All Parts of a Series Circuit. (b) *RLC* Series Circuit Voltages: *I* Is in Phase with V_R, *I* Lags V_L by 90°, and *I* Leads V_C by 90°.

Capacitive reactance can be calculated by using the formula

$$X_C = \frac{1}{2\pi f C}$$

In the example,

$$X_C = \frac{1}{2\pi \times 60 \times 10 \ \mu F} = 265.3 \ \Omega$$

Inductive reactance is calculated by using the formula

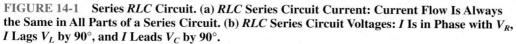

$$X_L = 2\pi f L$$

In the example,

$$X_L = 2\pi \times 60 \times 20 \ mH = 7.5 \ \Omega$$

Resistance in the example is equal to $R = 33 \ \Omega$. Figure 14-2(b) illustrates these values of resistance and reactance in a vector diagram. In this vector diagram you can see that X_L is

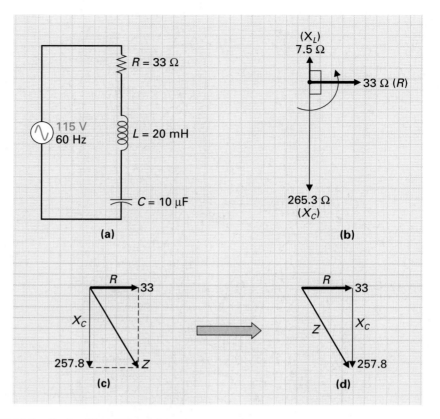

FIGURE 14-2 Series Circuit Impedance.

drawn 90° ahead of R, and X_C is drawn 90° behind R. The capacitive and inductive reactances are 180° out of phase with one another and counteract to produce the vector diagram shown in Figure 14-2(c). The difference between X_L and X_C is equal to 257.8, and since X_C is greater than X_L, the resultant reactive vector is capacitive. Reactance, however, is not in phase with resistance, and impedance is the vector sum of the reactive (X) and resistive (R) vectors. The formula, based on the Pythagorean theorem as illustrated in Figure 14-2(d), is

$$Z = \sqrt{R^2 + X^2}$$

In this example, therefore, the circuit impedance will be equal to

$$Z = \sqrt{R^2 + X^2} = \sqrt{33^2 + 257.8^2} = 260 \ \Omega$$

Since reactance (X) is equal to the difference between (symbolized $\sim$) X_L and X_C ($X_L \sim X_C$), the impedance formula can be modified slightly to incorporate the calculation to determine the difference between X_L and X_C.

$$Z = \sqrt{R^2 + (X_L \sim X_C)^2}$$

Using our example with this new formula, we arrive at the same value of impedance, and since the difference between X_L and X_C resulted in a capacitive vector, the circuit is said to act capacitively.

$$
\begin{aligned}
Z &= \sqrt{R^2 + (X_L \sim X_C)^2} \\
&= \sqrt{33^2 + (7.5 \sim 265.3)^2} \\
&= \sqrt{33^2 + 257.8^2} \\
&= 260 \ \Omega
\end{aligned}
$$

If, on the other hand, the component values were such that the difference was an inductive vector, then the circuit would be said to act inductively.

14-1-2 *Current*

The current in a series circuit is the same at all points throughout the circuit, and therefore

$$I = I_R = I_L = I_C$$

Once the total impedance of the circuit is known, Ohm's law can be applied to calculate the circuit current:

$$I = \frac{V_S}{Z}$$

In the example, circuit current is equal to

$$I = \frac{V_S}{Z}$$
$$= \frac{115 \text{ V}}{260 \text{ } \Omega}$$
$$= 0.44 \text{ A} \quad \text{or} \quad 440 \text{ mA}$$

14-1-3 *Voltage*

Now that you know the value of current flowing in the series circuit, you can calculate the voltage drops across each component, as shown in Figure 14-3(a).

$$V_R = I \times R$$

$$V_L = I \times X_L$$

$$V_C = I \times X_C$$

Since none of these voltages are in phase with one another as shown in Figure 14-3(b), they must be added vectorially to obtain the applied voltage.

The formula, based on the Pythagorean theorem and illustrated in Figure 14-3(c) and (d), can be used to calculate V_S.

$$V_S = \sqrt{V_R^2 + (V_L \sim V_C)^2}$$

In the example circuit, the applied voltage is, as we already know, 115 V.

$$V_S = \sqrt{15^2 + (3 \sim 117)^2}$$
$$= \sqrt{225 + 12{,}996}$$
$$= \sqrt{13{,}221}$$
$$= 115 \text{ V}$$

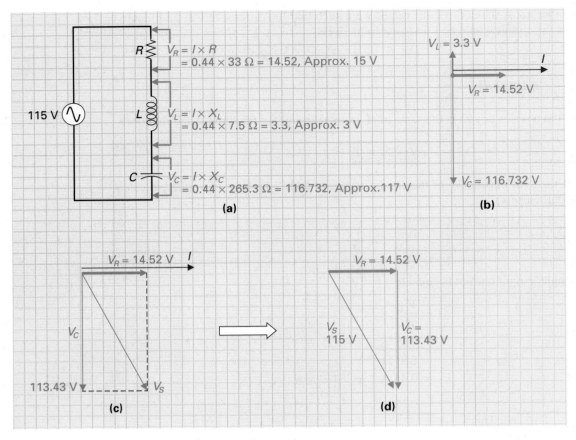

FIGURE 14-3   **Series Voltage Drops.**

14-1-4   *Phase Angle*

As can be seen in Figure 14-3(c), there is a phase difference between the source voltage (V_S) and the circuit current (I). This phase difference can be calculated with either of the following formulas.

$$\theta = \arctan \frac{V_L \sim V_C}{V_R}$$

$$\theta = \arctan \frac{X_L \sim X_C}{R}$$

In the example circuit, θ is

$$\theta = \arctan \frac{3.3 \text{ V} \sim 116.732 \text{ V}}{14.52 \text{ V}}$$

$$= \arctan 7.812$$

$$= 82.7°$$

Since the example circuit is capacitive (ICE), the phase angle will be $-82.7°$ since V_S lags I in a circuit that acts capacitively.

14-1-5 *Power*

The true power or resistive power (P_R) dissipated by a circuit can be calculated using the formula

$$P_R = I^2 \times R$$

which in our example will be

$$P_R = 0.44^2 \times 33\ \Omega = 6.4\ \text{W}$$

The apparent power (P_A) consumed by the circuit is calculated by

$$P_A = V_S \times I$$

which in our example will be

$$P_A = 115\ \text{V} \times 0.44 = 50.6\ \text{volt-amperes (VA)}$$

The true or actual power dissipated by the resistor is, as expected, smaller than the apparent power that appears to be being used.

The power factor can be calculated, as usual, by

$$\text{PF} = \cos\theta = \frac{R}{Z} = \frac{P_R}{P_A}$$

PF of 0 = reactive circuit

PF of 1 = resistive circuit

In the example circuit, PF = 0.126, indicating that the circuit is mainly reactive.

EXAMPLE:

For a series circuit where $R = 10\ \Omega$, $L = 5\ \text{mH}$, $C = 0.05\ \mu\text{F}$, and $V_S = 100\ \text{V}/2\ \text{kHz}$, calculate:

a. X_C f. Apparent power

b. X_L g. True power

c. Z h. Power factor

d. I i. Phase angle

e. V_R, V_C, and V_L

Solution:

a. $X_C = \dfrac{1}{2\pi fC} = 1.6\ \text{k}\Omega$

b. $X_L = 2\pi fL = 62.8\ \Omega$

c. $Z = \sqrt{R^2 + (X_L \sim X_C)^2}$

     $= \sqrt{(10\ \Omega)^2 + (1.6\ \text{k}\Omega \sim 62.8\ \Omega)^2}$

     $= \sqrt{10\ \Omega^2 + 1.54\ \text{k}\Omega^2}$

     $= 1.54\ \text{k}\Omega$ (capacitive circuit due to high X_C)

d. $I = \dfrac{V_S}{Z} = \dfrac{100\ \text{V}}{1.54\ \text{k}\Omega} = 64.9\ \text{mA}$

e. $V_R = I \times R = 64.9\ \text{mA} \times 10\ \Omega = 0.65\ \text{V}$

     $V_C = I \times X_C = 64.9\ \text{mA} \times 1.6\ \text{k}\Omega = 103.9\ \text{V}$

     $V_L = I \times X_L = 64.9\ \text{mA} \times 62.8\ \Omega = 4.1\ \text{V}$

f. Apparent power $= V_S \times I = 100 \text{ V} \times 64.9 \text{ mA} = 6.49 \text{ VA}$

g. True power $= I^2 \times R = 64.9^2 \times 10 \, \Omega = 42.17 \text{ mW}$

h. $\text{PF} = \dfrac{R}{Z} = \dfrac{10 \, \Omega}{1.5 \text{ k}\Omega} = 0.006$ (reactive circuit)

i. $\theta = \arctan \dfrac{V_L \sim V_C}{V_R}$

$= \arctan \dfrac{4.1 \text{ V} \sim 103.9 \text{ V}}{0.65 \text{ V}}$

$= \arctan 153.54$

$= 89.63°$

Capacitive circuit (ICE); therefore, V_S lags I by $-89.63°$.

SELF-TEST EVALUATION POINT FOR SECTION 14-1

Now that you have completed this section, you should be able to:

■ **Objective 1.** *Identify the difference between a series and a parallel RCL circuit.*

■ **Objective 2.** *Explain the following as they relate to series RLC circuits:*

a. *Impedance*
b. *Current*
c. *Voltage*
d. *Phase angle*
e. *Power*

Use the following questions to test your understanding of Section 14-1.

1. List in order the procedure that should be followed to fully analyze a series *RLC* circuit.

2. State the formulas for calculating the following in relation to a series *RLC* circuit.

a. Impedance	**f.** V_R
b. Current	**g.** V_L
c. Apparent power	**h.** V_C
d. V_S	**i.** Phase angle (θ)
e. True power	**j.** Power factor (PF)

14-2 PARALLEL *RLC* CIRCUIT

Now that the characteristics of a series circuit are understood, let us connect a resistor, inductor, and capacitor in parallel with one another. Figure 14-4(a) and (b) show the current and voltage relationships of a parallel *RLC* circuit.

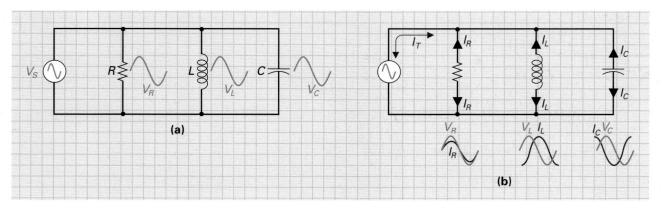

FIGURE 14-4 Parallel *RLC* Circuit. (a) *RLC* Parallel Circuit Voltage: Voltages across Each Component Are All Equal and in Phase with One Another in a Parallel Circuit. (b) *RLC* Parallel Circuit Currents: I_R Is in Phase with V_R, I_L Lags V_L by 90°, and I_C Leads V_C by 90°.

14-2-1 *Voltage*

As can be seen in Figure 14-4(a), the voltage across any parallel circuit will be equal and in phase. Therefore,

$$V_R = V_L = V_C = V_S$$

14-2-2 *Current*

With current, we must first calculate the individual branch currents (I_R, I_L, and I_C) and then calculate the total circuit current (I_T). An example circuit is illustrated in Figure 14-5(a), and the branch currents can be calculated by using the formulas

$$I_R = \frac{V}{R}$$

$$I_L = \frac{V}{X_L}$$

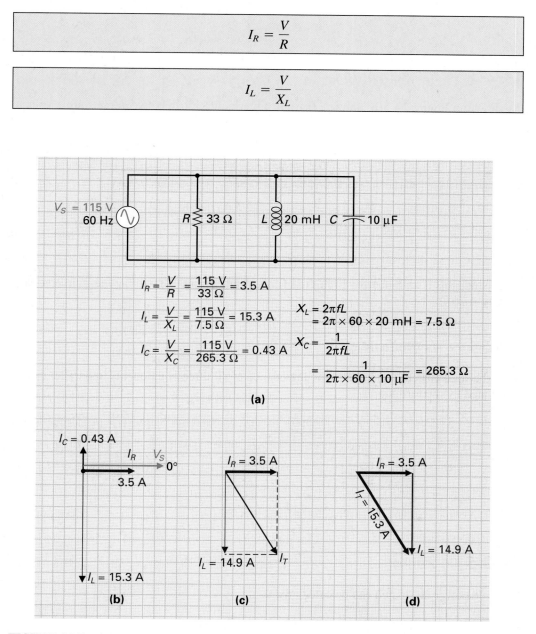

(a)

(b)

(c)

(d)

FIGURE 14-5 **Example *RLC* Parallel Circuit.**

$$I_C = \frac{V}{X_C}$$

Figure 14-5(b) illustrates these branch currents vectorially, with I_R in phase with V_S, I_L lagging by 90°, and I_C leading I_R by 90°. The 180° phase difference between I_C and I_L results in a cancellation, as shown in Figure 14-5(c).

The total current (I_T) can be calculated by using the Pythagorean theorem on the right triangle, as illustrated in Figure 14-5(d).

$$I_T = \sqrt{I_R^2 + I_X^2}$$

$$(I_X = I_L \sim I_C)$$

$$= \sqrt{3.5^2 + 14.9^2}$$
$$= 15.3 \text{ A}$$

14-2-3   *Phase Angle*

As shown in Figure 14-5(b) and (c), there is a phase difference between the source voltage (V_S) and the circuit current (I_T). This phase difference can be calculated using the formula

$$\theta = \arctan \frac{I_L \sim I_C}{I_R}$$

$$= \arctan \frac{15.3 \text{ A} - 0.43 \text{ A}}{3.5 \text{ A}}$$
$$= \arctan 4.25$$
$$= 76.7°$$

Since this is an inductive circuit (ELI), the total current (I_T) will lag the source voltage (V_S) by −76.7°.

14-2-4   *Impedance*

With the total current (I_T) known, the impedance of all three components in parallel can be calculated by the formula

$$Z = \frac{V}{I_T}$$

$$= \frac{115 \text{ V}}{15.3 \text{ A}}$$
$$= 7.5 \ \Omega$$

14-2-5   *Power*

The true power dissipated can be calculated using

$$P_R = I_R^2 \times R$$

$$= 3.5^2 \times 33 \ \Omega$$
$$= 404.3 \ \text{W}$$

The apparent power consumed by the circuit is calculated by

$$P_A = V_S \times I_T$$

$$= 115 \ \text{V} \times 15.3 \ \text{A}$$
$$= 1759.5 \ \text{volt-amperes (VA)}$$

Finally, the power factor can be calculated, as usual, with

$$PF = \cos \theta = \frac{P_R}{P_A}$$

In the example circuit, PF = 0.23.

Now that you have completed this section, you should be able to:

■ **Objective 3.** *Explain the following as they relate to parallel RCL circuits:*

   a. *Voltage*
   b. *Current*
   c. *Impedance*
   d. *Phase angle*
   e. *Power*

Use the following questions to test your understanding of Section 14-2.

1. State the formulas for calculating the following in relation to a parallel *RLC* circuit:

   a. I_R **d.** I_L
   b. I_T **e.** θ
   c. I_C **f.** Z

2. State the formulas for:

   a. P_R **c.** P_A
   b. P_X **d.** PF

14-3   RESONANCE

Resonance is a circuit condition that occurs when the inductive reactance (X_L) and the capacitive reactance (X_C) have been balanced. Figure 14-6 illustrates a parallel- and a series-connected *LC* circuit. If a dc voltage is applied to the input of either circuit, the capacitor will act as an open ($X_C =$ infinite Ω) and the inductor will act as a short ($X_L = 0 \ \Omega$).

If a low-frequency ac is now applied to the input, X_C will decrease from maximum, and X_L will increase from zero. As the ac frequency is increased further, the capacitive reactance will continue to fall ($X_C \downarrow \propto 1/f \uparrow$) and the inductive reactance to rise ($X_L \uparrow \propto f \uparrow$), as shown in Figure 14-7.

As the input ac frequency is increased further, a point will be reached where X_L will equal X_C, and this condition is known as *resonance*. The frequency at which $X_L = X_C$ in either a parallel or a series *LC* circuit is known as the *resonant frequency* (f_0) and can be calculated by the following formula, which has been derived from the capacitive and inductive reactance formulas:

$$f_0 = \frac{1}{2\pi \sqrt{LC}}$$

Resonance
Circuit condition that occurs when the inductive reactance (X_L) is equal to the capacitive reactance (X_C).

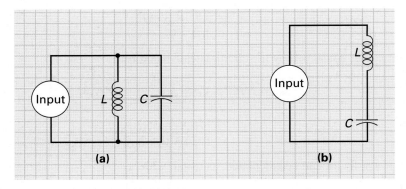

FIGURE 14-6   Resonance. (a) Parallel *LC* Circuit. (b) Series *LC* Circuit.

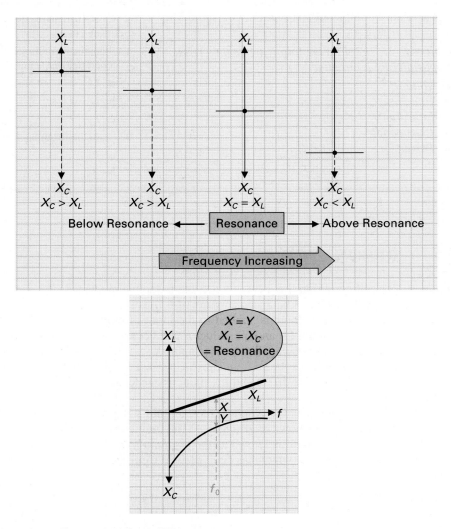

FIGURE 14-7   Frequency versus Reactance.

where f_0 = resonant frequency, in hertz (Hz)

L = inductance, in henrys (H)

C = capacitance, in farads (F)

EXAMPLE:

Calculate the resonant frequency (f_0) of a series LC circuit if L = 750 mH and C = 47 μF.

Solution:

$$f_0 = \frac{1}{2\pi\sqrt{L \times C}}$$

$$= \frac{1}{2\pi\sqrt{(750 \times 10^{-3}) \times (47 \times 10^{-6})}}$$

$$= 26.8 \text{ Hz}$$

14-3-1 Series Resonance

Figure 14-8(a) illustrates a series RLC circuit at resonance ($X_L = X_C$), or **series resonant circuit.** The ac input voltage causes current to flow around the circuit, and since all the

Series Resonant Circuit

A resonant circuit in which the capacitor and coil are in series with the applied ac voltage.

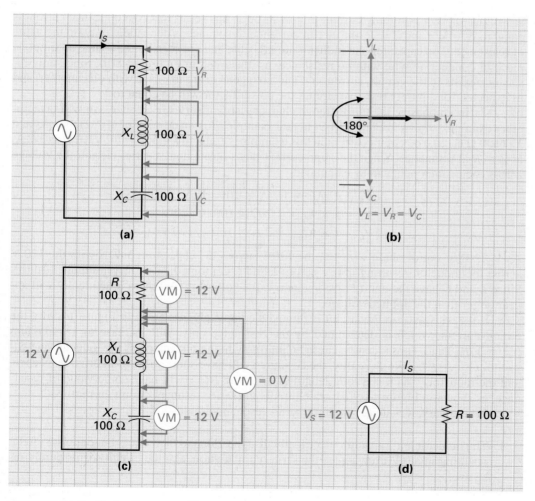

FIGURE 14-8 Series Resonant Circuit.

components are connected in series, the same value of current (I_S) will flow through all the components. Since R, X_L, and X_C are all equal to 100 Ω and the current flow is the same throughout, the voltage dropped across each component will be equal, as illustrated vectorially in Figure 14-8(b).

The voltage across the resistor is in phase with the series circuit current (I_S); however, since the voltage across the inductor (V_L) is 180° out of phase with the voltage across the capacitor (V_C), and both are equal to one another, V_L cancels V_C, when both are measured in series.

Three unusual characteristics occur when a circuit is at resonance, which do not occur at any other frequency.

(1) The first is that if V_L and V_C cancel, the voltage across L and C will measure 0 V on a voltmeter. Since there is effectively no voltage being dropped across these two components, all the voltage must be across the resistor ($V_R = 12$ V). This is true; however, since the same current flows throughout the series circuit, a voltmeter will measure 12 V across C, 12 V across L, and 12 V across R, as shown in Figure 14-8(c). It now appears that the voltage drops around the series circuit (36 V) do not equal the voltage applied (12 V). This is not true, as V_L and V_C cancel, because they are out of phase with one another, so Kirchhoff's voltage law is still valid.

(2) The second unusual characteristic of resonance is that because the total opposition or impedance (Z) is equal to

$$Z = \sqrt{R^2 + (X_L \sim X_C)^2}$$

and the difference between X_L and X_C is 0 ($Z = \sqrt{R^2 + 0}$), the impedance of a series circuit at resonance is equal to the resistance value R ($Z = \sqrt{R^2} = R$). As a result, the applied ac voltage of 12 V is forcing current to flow through this series RLC circuit. Since current is equal to $I_S = V/Z$ and $Z = R$, the circuit current at resonance is dependent only on the value of resistance. The capacitor and inductor are invisible and are seen by the source as simply a piece of conducting wire with no resistance, as illustrated in Figure 14-8(d). Since only resistance exists in the circuit, current (I_S) and voltage (V_S) are in phase with one another, and as expected for a purely resistive circuit, the power factor will be equal to 1.

(3) To emphasize the third strange characteristic of series resonance, we will take another example, shown in Figure 14-9. In this example, R is made smaller (10 Ω) than X_L and X_C (100 Ω each). The circuit current in this example is equal to $I = V/R = 12$ V/10 Ω = 1.2 A, as $Z = R$ at resonance. Since the same current flows throughout a series circuit, the voltage across each component can be calculated.

$$V_R = I \times R = 1.2 \text{ A} \times 10 = 12 \text{ V}$$
$$V_L = I \times X_L = 1.2 \text{ A} \times 100 = 120 \text{ V}$$
$$V_C = I \times X_C = 1.2 \text{ A} \times 100 = 120 \text{ V}$$

As V_L is 180° out of phase with V_C, the 120 V across the capacitor cancels with the 120 V across the inductor, resulting in 0 V across L and C combined, as shown in Figure 14-9(b). Since L and C have the ability to store energy, the voltage across them individually will appear larger than the applied voltage.

If the resistance in the circuit is removed completely, as shown in Figure 14-9(c), the circuit current, which is determined by the resistance only, will increase to a maximum ($I\uparrow = V/R\downarrow$) and, consequently, cause an infinitely high voltage across the inductor and capacitor ($V\uparrow = I\uparrow \times R$). In reality, the ac source will have some value of internal resistance, and the inductor, which is a long length of thin wire ($R\uparrow$), will have some value of resistance, as shown in Figure 14-9(d), which limits the series resonant circuit current.

In summary, we can say that in a series resonant circuit:

1. The inductor and capacitor electrically disappear due to their equal but opposite effect, resulting in a 0 V drop across the series combination, and the circuit seems purely resistive.

2. The current flow is large because the impedance of the circuit is low and equal to the series resistance (R), which has the source voltage developed across it.

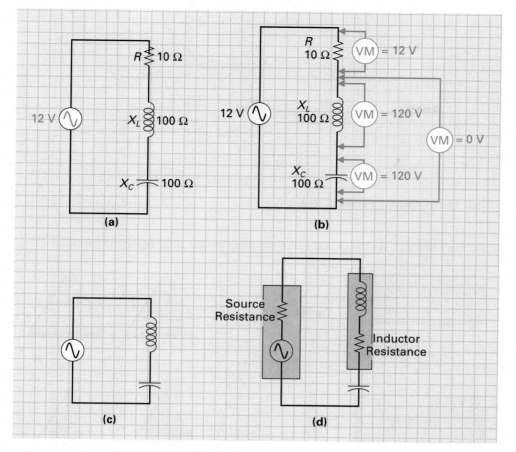

FIGURE 14-9 **Circuit Effects at Resonance.**

3. The individual voltage drops across the inductor or capacitor can be larger than the source voltage if R is smaller than X_L and X_C.

Quality Factor

As discussed previously in the inductance chapter, the Q factor is a ratio of inductive reactance to resistance and is used to express how efficiently an inductor will store rather than dissipate energy. In a series resonant circuit, the Q factor indicates the quality of the series resonant circuit, or is the ratio of the reactance to the resistance.

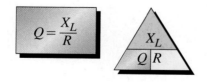

$$Q = \frac{X_L}{R}$$

or, since $X_L = X_C$,

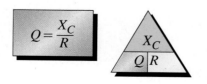

$$Q = \frac{X_C}{R}$$

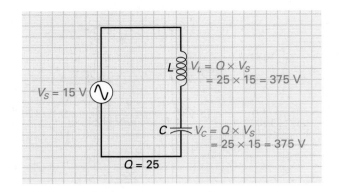

FIGURE 14-10 Quality Factor at Resonance.

Another way to calculate the Q of a series resonant circuit is by using the formula

$$Q = \frac{V_L}{V_R} = \frac{V_C}{V_R}$$

(at resonance only)

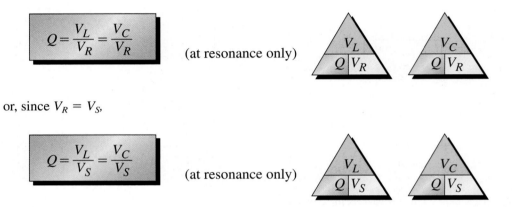

or, since $V_R = V_S$,

$$Q = \frac{V_L}{V_S} = \frac{V_C}{V_S}$$

(at resonance only)

If the Q and source voltage are known, the voltage across the inductor or capacitor can be found by transposition of the formula, as can be seen in the example in Figure 14-10.

The Q of a resonant circuit is almost entirely dependent on the inductor's coil resistance, because capacitors tend to have almost no resistance at all, only reactance, which makes them very efficient.

The inductor has a Q value of its own, and if only L and C are connected in series with one another, the Q of the series resonant circuit will be equal to the Q of the inductor, as shown in Figure 14-11. If the resistance is added in with L and C, the Q of the series resonant circuit will be less than that of the inductor's Q.

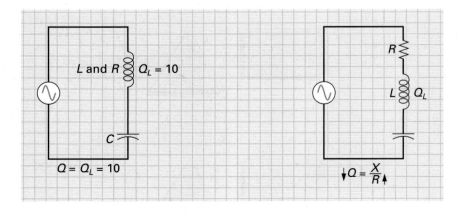

FIGURE 14-11 Resistance within Inductor.

CHAPTER 14 / RESISTIVE, INDUCTIVE, AND CAPACITIVE (*RLC*) CIRCUITS

EXAMPLE:

Calculate the resistance of the series resonant circuit illustrated in Figure 14-12.

FIGURE 14-12 Series Resonant Circuit Example.

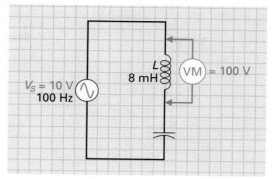

Solution:

$$Q = \frac{V_L}{V_S} = \frac{100 \text{ V}}{10 \text{ V}} = 10$$

Since $Q = X_L/R$ and $R = X_L/Q$, if the inductive reactance can be found, the R can be determined.

$$X_L = 2\pi \times f \times L$$
$$= 2\pi \times 100 \times 8 \text{ mH}$$
$$= 5 \ \Omega$$

R will therefore equal

$$R = \frac{X_L}{Q}$$
$$= \frac{5}{10}$$
$$= 0.5 \ \Omega$$

Bandwidth

A series resonant circuit is selective in that frequencies at resonance or slightly above or below will cause a larger current than frequencies well above or below the circuit's resonant frequency. The group or band of frequencies that causes the larger current is called the circuit's **bandwidth.**

Figure 14-13 illustrates a series resonant circuit and its bandwidth. The × marks on the curve illustrate where different frequencies were applied to the circuit and the resulting value of current measured in the circuit. The resulting curve produced is called a **frequency response curve,** as it illustrates the circuit's response to different frequencies. At resonance, $X_L = X_C$ and the two cancel, which is why maximum current was present in the circuit (100 mA) when the resonant frequency (100 Hz) was applied.

The bandwidth includes the group or band of frequencies that cause 70.7% or more of the maximum current to flow within the series resonant circuit. in this example, frequencies from 90 to 110 Hz cause 70.7 mA or more, which is 70.7% of maximum (100 mA), to flow. The bandwidth in this example is equal to

$$BW = 110 - 90 = 20 \text{ Hz}$$

(110 Hz and 90 Hz are known as **cutoff frequencies**).

Referring to the bandwidth curve in Figure 14-13(b), you may notice that 70.7% is also called the **half-power point,** although it does not exist halfway between 0 and

Bandwidth

Width of the group or band of frequencies between the half-power points.

Frequency Response Curve

A graph indicating a circuit's response to different frequencies.

Cutoff Frequency

Frequency at which the gain of the circuit falls below 0.707 of the maximum current or half-power (−3 dB).

Half-Power Point

A point at which power is 50%. This half-power point corresponds to 70.7% of the total current.

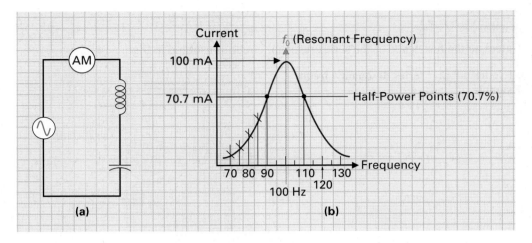

FIGURE 14-13 **Series Resonant Circuit Bandwidth. (a) Circuit. (b) Frequency Response Curve.**

maximum. This value of 70.7% is not the half-current point but the half-power point, as we can prove with a simple example.

EXAMPLE:

$R = 2$ kΩ and $I = 100$ mA; therefore, power $= I^2 \times R = 100$ mA$^2 \times 2$ k$\Omega = 20$ W. If the current is now reduced so that it is 70.7% of its original value, calculate the power dissipated.

■ *Solution:*

$$P = I^2 \times R = 70.7 \text{ mA}^2 \times 2 \text{ k}\Omega = 10 \text{ W}$$

In summary, the 70.7% current points are equal to the 50% or half-power points. A circuit's bandwidth is the band of frequencies that exists between the 70.7% current points or half-power points.

The bandwidth of a series resonant circuit can also be calculated by use of the formula

$$\text{BW} = \frac{f_0}{Q_{f_0}}$$

f_0 = Resonant frequency
Q_{f_0} = Quality factor at resonance

This formula states that the BW is proportional to the resonant frequency of the circuit and inversely proportional to the Q of the circuit.

Figure 14-14 illustrates three example response curves. In these three examples, the value of R is changed from 100 Ω to 200 Ω to 400 Ω. This does not vary the resonant frequency, but simply alters the Q and therefore the BW. The resistance value will determine the Q of the circuit, and since Q is inversely proportional to resistance, Q is proportional to current; consequently, a high value of Q will cause a high value of current.

In summary, the bandwidth of a series resonant circuit will increase as the Q of the circuit decreases (BW$\uparrow = f_0/Q\downarrow$), and vice versa.

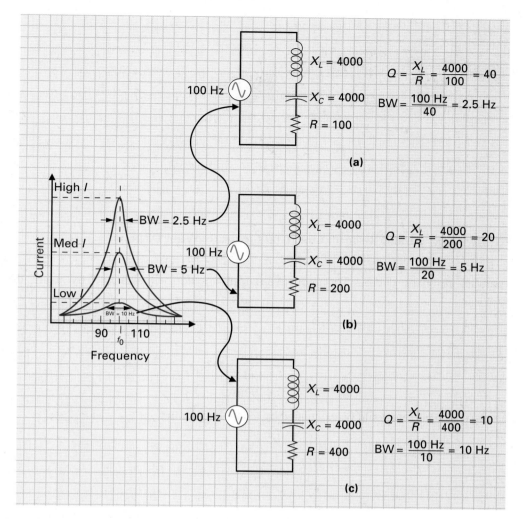

FIGURE 14-14   **Bandwidth of a Series Resonant Circuit.**

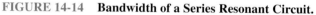
Use the following questions to test your understanding of Section 14-3-1.

1. Define *resonance*.

2. What is series resonance?

3. In a series resonant circuit, what are the three rather unusual circuit phenomena that take place?

4. How does Q relate to series resonance?

5. Define *bandwidth*.

6. Calculate BW if $f_0 = 12$ kHz and $Q = 1000$.

14-3-2   *Parallel Resonance*

The **parallel resonant circuit** acts differently from the series resonant circuit, and these different characteristics need to be analyzed and discussed. Figure 14-15 illustrates

Parallel Resonant Circuit

Circuit having an inductor and capacitor in parallel with one another, offering a high impedance at the frequency of resonance.

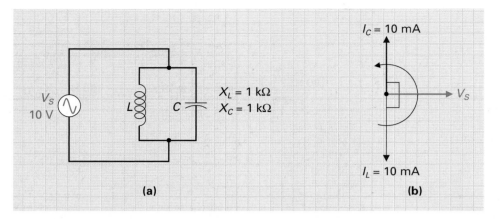

(a)   **(b)**

FIGURE 14-15   **Parallel Resonant Circuit. (a) Circuit. (b) Vector Diagram.**

a parallel resonant circuit. The inductive current could be calculated by using the formula

$$I_L = \frac{V_L}{X_L}$$
$$= \frac{10\text{ V}}{1\text{ k}\Omega}$$
$$= 10\text{ mA}$$

The capacitive current could be calculated by using the formula

$$I_C = \frac{V_C}{X_C}$$
$$= \frac{10\text{ V}}{1\text{ k}\Omega}$$
$$= 10\text{ mA}$$

Looking at the vector diagram in Figure 14-15(b), you can see that I_C leads the source voltage by 90° (ICE) and I_L lags the source voltage by 90° (ELI), creating a 180° phase difference between I_C and I_L. This means that when 10 mA of current flows up through the inductor, 10 mA of current will flow in the opposite direction down through the capacitor, as shown in Figure 14-16(a). During the opposite alternation, 10 mA will flow down through the inductor and 10 mA will travel up through the capacitor, as shown in Figure 14-16(b).

If 10 mA arrives into point X and 10 mA of current leaves point X, no current can be flowing from the source (V_S) to the parallel LC circuit; the current is simply swinging or oscillating back and forth between the capacitor and inductor.

The source voltage (V_S) is needed initially to supply power to the LC circuit and start the oscillations; but once the oscillating process is in progress (assuming the ideal case), current is only flowing back and forth between inductor and capacitor, and no current is flowing from the source. So the LC circuit appears as an infinite impedance and the source can be disconnected, as shown in Figure 14-16(c).

Flywheel Action

Sustaining effect of oscillation in an LC circuit due to the charging and discharging of the capacitor and the expansion and contraction of the magnetic field around the inductor.

Flywheel Action

Let's discuss this oscillating effect, called **flywheel action,** in a little more detail. The name is derived from the fact that it resembles a mechanical flywheel, which, once started, will keep going continually until friction reduces the magnitude of the rotations to zero.

The electronic equivalent of the mechanical flywheel is a resonant parallel-connected LC circuit. Figure 14-17(a) through (h) illustrate the continual energy transfer between capacitor and inductor, and vice versa. The direction of the circulating current reverses each

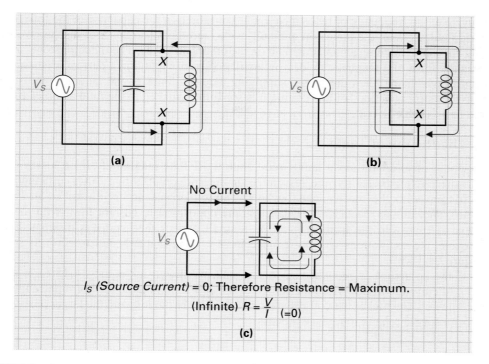

FIGURE 14-16   Current in a Parallel Resonant Circuit.

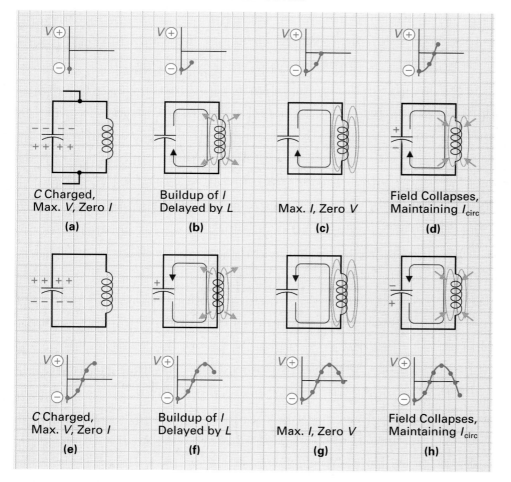

FIGURE 14-17   Energy and Current in an *LC* Parallel Circuit at Resonance.

half-cycle at the frequency of resonance. Energy is stored in the capacitor in the form of an electric field between the plates on one half-cycle, and then the capacitor discharges, supplying current to build up a magnetic field on the other half-cycle. The inductor stores its energy in the form of a magnetic field, which will collapse, supplying a current to charge the capacitor, which will then discharge, supplying a current back to the inductor, and so on. Due to the "storing action" of this circuit, it is sometimes related to the fluid analogy and referred to as a **tank circuit.**

Tank Circuit

Circuit made up of a coil and capacitor that is capable of storing electric energy.

The Reality of Tanks

Under ideal conditions a tank circuit should oscillate indefinitely if no losses occur within the circuit. In reality, the resistance of the coil reduces that 100% efficiency, as does friction with the mechanical flywheel. This coil resistance is illustrated in Figure 14-18(a), and, unlike reactance, resistance is the opposition to current flow, with the dissipation of energy in the form of heat. As a small part of the energy is dissipated with each cycle, the oscillations will be reduced in size and eventually fall to zero, as shown in Figure 14-18(b).

If the ac source is reconnected to the tank, as shown in Figure 14-18(c), a small amount of current will flow from the source to the tank to top up the tank or replace the dissipated power. The higher the coil resistance is, the higher the loss and the larger the current flow from source to tank to replace the loss.

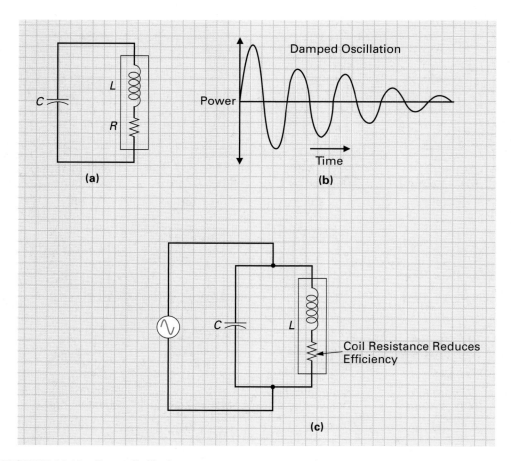

FIGURE 14-18   Losses in Tanks.

Quality Factor

In the series resonant circuit, we were concerned with voltage drops since current remains the same throughout a series circuit, so

$$Q = \frac{V_C \text{ or } V_L}{V_S} \quad \text{(at resonance only)}$$

In a parallel resonant circuit, we are concerned with circuit currents rather than voltage, so

$$Q = \frac{I_{\text{tank}}}{I_S}$$

(at resonance only)

The quality factor, Q, can also be expressed as the ratio between reactance and resistance:

$$Q = \frac{X_L}{R} \quad \text{(at any frequency)}$$

Another formula, which is the most frequently used when discussing and using parallel resonant circuits, is

$$Q = \frac{Z_{\text{tank}}}{X_L}$$

(at resonance only)

This formula states that the Q of the tank is proportional to the tank impedance. A higher tank impedance results in a smaller current flow from source to tank. This assures less power is dissipated, and that means a higher-quality tank.

Of all the three Q formulas for parallel resonant circuits, $Q = I_{\text{tank}} / I_S$, $Q = X_L / R$, and $Q = Z_{\text{tank}} / X_L$, the latter is the easiest to use as both X_L and the tank impedance can easily be determined in most cases where C, L, and R internal for the inductor are known.

Bandwidth

Figure 14-19 illustrates a parallel resonant circuit and two typical response curves. These response curves summarize what we have described previously, in that a parallel resonant circuit has maximum impedance [Figure 14-19(b)] and minimum current [Figure 14-19(c)] at resonance. The current versus frequency response curve shown in Figure 14-19(c) is the complete opposite to the series resonant response curve. At frequencies

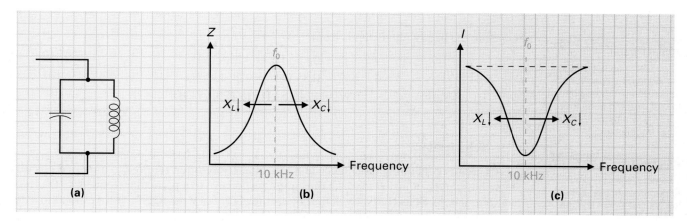

FIGURE 14-19   Parallel Resonant Circuit Bandwidth.

below resonance (< 10 kHz), X_L is low and X_C is high, and the inductor offers a low reactance, producing a high current path and low impedance. On the other hand, at frequencies above resonance (> 10 kHz), the capacitor displays a low reactance, producing a high current path and low impedance. The parallel resonant circuit is like the series resonant circuit in that it responds to a band of frequencies close to its resonant frequency.

The bandwidth (BW) can be calculated by use of the formula

$$BW = \frac{f_0}{Q_{f_0}}$$

f_0 = Resonant frequency
Q_{f_0} = Quality factor at resonance

■   **EXAMPLE:**

Calculate the bandwidth of the circuit illustrated in Figure 14-20.

FIGURE 14-20   **Bandwidth Example.**

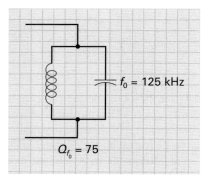

f_0 = 125 kHz

Q_{f_0} = 75

■   *Solution:*

$$BW = \frac{f_0}{Q_{f_0}}$$

$$= \frac{125 \text{ kHz}}{75}$$

$$= 1.7 \text{ kHz}$$

$$\frac{1.7 \text{ kHz}}{2} = 0.85 \text{ kHz}$$

therefore the bandwidth extends from

$$f_0 + 0.85 \text{ kHz} = 125.85 \text{ kHz}$$
$$f_0 - 0.85 \text{ kHz} = 124.15 \text{ kHz}$$
$$BW = 124.15 \text{ kHz to } 125.85 \text{ kHz}$$

Tuned Circuit

Circuit that can have its components' values varied so that the circuit responds to one selected frequency yet heavily attenuates all other frequencies.

Selectivity

Characteristic of a circuit to discriminate between the wanted signal and the unwanted signal.

Selectivity

Circuits containing inductance and capacitance are often referred to as **tuned circuits** since they can be adjusted to make the circuit responsive to a particular frequency (the resonant frequency). **Selectivity,** by definition, is the ability of a tuned circuit to respond to a desired frequency and ignore all others. Parallel resonant *LC* circuits are sometimes too selec-

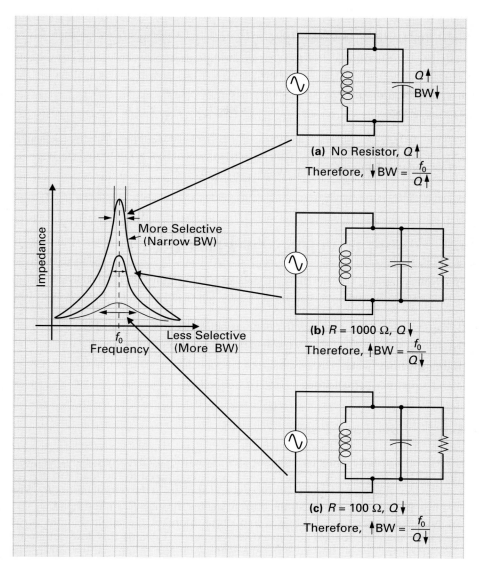

FIGURE 14-21 **Varying Bandwidth by Loading the Tank Circuit.**

tive, as the Q is too large, producing too narrow a bandwidth, as shown in Figure 14-21(a) ($\text{BW}\downarrow = f_0/Q\uparrow$).

In this situation, because of the very narrow response curve, a high resistance value can be placed in parallel with the LC circuit to provide an alternative path for source current. This process is known as *loading* or *damping* the tank and will cause an increase in source current and decrease in Q ($Q\downarrow = I_{\text{tank}}/I_{\text{source}}\uparrow$). The decrease in Q will cause a corresponding increase in BW ($\text{BW}\uparrow = f_0/Q\downarrow$), as shown by the examples in Figure 14-21, which illustrates a 1000 Ω loading resistor [Figure 14-21(b)] and a 100 Ω loading resistor [Figure 14-21(c)].

In summary, a parallel resonant circuit can be made less selective with a broader bandwidth if a resistor is added in parallel, providing an increase in current and a decrease in impedance, which widens the bandwidth.

Now that you have completed this section, you should be able to:

■ **Objective 4.** *Define resonance, and explain the characteristics of:*

   a. Series resonance
   b. Parallel resonance

Use the following questions to test your understanding of Section 14-3-2.

1. What are the differences between a series and a parallel resonant circuit?
2. Describe flywheel action.
3. Calculate the value of Q of a tank if $X_L = 50 \ \Omega$ and $R = 25 \ \Omega$.
4. When calculating bandwidth for a parallel resonant circuit, can the series resonant bandwidth formula be used?
5. What is selectivity?

14-4   APPLICATIONS OF *RLC* CIRCUITS

In the previous chapters, you saw how *RC* and *RL* filter circuits are used as low- or high-pass filters to pass some frequencies and block others. There are basically four types of filters:

1. Low-pass filter, which passes frequencies below a cutoff frequency
2. High-pass filter, which passes frequencies above a cutoff frequency
3. Bandpass filter, which passes a band of frequencies
4. Band-stop filter, which stops a band of frequencies

14-4-1   *Low-Pass Filter*

Figure 14-22(a) illustrates how an inductor and capacitor can be connected to act as a low-pass filter. At low frequencies, X_L has a small value compared to the load resistor (R_L), so nearly all the low-frequency input is developed and appears at the output across R_L. Since X_C is high at low frequencies, nearly all the current passes through R_L rather than C.

At high frequencies, X_L increases and drops more of the applied input across the inductor rather than the load. The capacitive reactance, X_C, aids this low-output-at-high-frequency effect by decreasing its reactance and providing an alternative path for current to flow.

Since the inductor basically blocks alternating current and the capacitor shunts alternating current, the net result is to prevent high-frequency signals from reaching the load. The way in which this low-pass filter responds to frequencies is graphically illustrated in Figure 14-22(b).

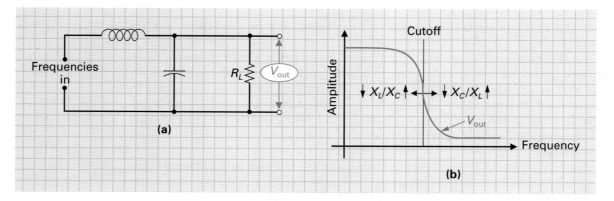

FIGURE 14-22   **Low-Pass Filter. (a) Circuit. (b) Frequency Response.**

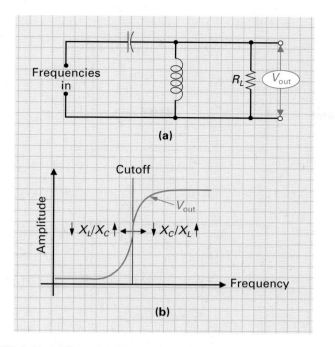

FIGURE 14-23 **High-Pass Filter. (a) Circuit. (b) Frequency Response.**

14-4-2 *High-Pass Filter*

Figure 14-23(a) illustrates how an inductor and capacitor can be connected to act as a high-pass filter. At high frequencies, the reactance of the capacitor (X_C) is low while the reactance of the inductor (X_L) is high, so all the high frequencies are easily passed by the capacitor and blocked by the inductor, so they all are routed through to the output and load.

At low frequencies, the reverse condition exists, resulting in a low X_L and a high X_C. The capacitor drops nearly all the input, and the inductor shunts the signal current away from the output load.

14-4-3 *Bandpass Filter*

Figure 14-24(a) illustrates a series resonant **bandpass filter,** and Figure 14-24(b) shows a parallel resonant bandpass filter. Figure 14-24(c) shows the frequency response curve produced by the bandpass filter. At resonance, the series resonant LC circuit has a very low impedance and will consequently pass the resonant frequency to the load with very little drop across the L and C components.

Below resonance, X_C is high, and the capacitor drops a large amount of the input signal; above resonance, X_L is high and the inductor drops most of the input frequency voltage. This circuit will therefore pass a band of frequencies centered around the resonant frequency of the series LC circuit and block all other frequencies above and below this resonant frequency.

Figure 14-24(b) illustrates how a parallel resonant LC circuit can be used to provide a bandpass response. The series resonant circuit was placed in series with the output, whereas the parallel resonant circuit will have to be placed in parallel with the output to provide the same results. At resonance, the parallel resonant circuit or tank has a high impedance, so very little current will be shunted away from the output; it will be passed on to the output, and almost all the input will appear at the output across the load.

Above resonance, X_C is small, so most of the input is shunted away from the output by the capacitor; below resonance, X_L is small, and the shunting action occurs again, but this time through the inductor.

Bandpass Filter

Filter circuit that passes a group or band of frequencies between a lower and an upper cutoff frequency, while heavily attenuating any other frequency outside this band.

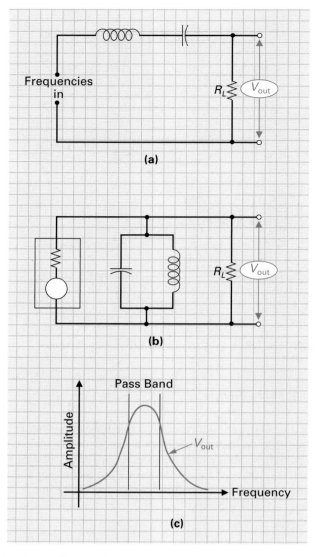

(a)

(b)

(c)

FIGURE 14-24 **Bandpass Filter. (a) Series Resonant Bandpass Filter. (b) Parallel Resonant Bandpass Filter. (c) Frequency Response.**

Figure 14-25 illustrates how a transformer can be used to replace the inductor to produce a bandpass filter. At resonance, maximum flywheel current flows within the parallel circuit made up of the capacitor and the primary of the transformer (*L*), which is known as a *tuned transformer*. With maximum flywheel current, there will be a maximum magnetic field, which means that there will be maximum power transfer between primary and second-

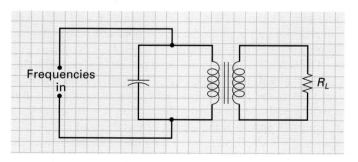

FIGURE 14-25 **Parallel Resonant Bandpass Circuit Using a Transformer.**

ary. So nearly all the input will be coupled to the output (coupling coefficient $k = 1$) and appear across the load at and around a small band of frequencies centered on resonance.

Above and below resonance, current within the parallel resonant circuit will be smaller. So the power transfer ability will be less, effectively keeping the frequencies outside the pass-band from appearing at the output.

14-4-4 *Band-Stop Filter*

Figure 14-26(a) illustrates a series resonant and Figure 14-26(b) a parallel resonant **band-stop filter.** Figure 14-26(c) shows the frequency response curve produced by a band-stop filter. The band-stop filter operates exactly the opposite to a bandpass filter in that it blocks or attenuates a band of frequencies centered on the resonant frequency of the *LC* circuit.

In the series resonant circuit in Figure 14-26(a), the *LC* impedance is very low at and around resonance, so these frequencies are rejected or shunted away from the output. Above and below resonance, the series circuit has a very high impedance, which results in almost no shunting of the signal away from the output.

In the parallel resonant circuit in Figure 14-26(b), the *LC* circuit is in series with the load and output. At resonance, the impedance of a parallel resonant circuit will be very high, and the band of frequencies centered around resonance will be blocked. Above and below resonance, the impedance of the tank is very low, so nearly all the input is developed across the output.

Filters are necessary in applications such as television or radio, where we need to tune in (select or pass) one frequency that contains the information we desire, yet block all the millions of other frequencies that are also carrying information, as shown in Figure 14-27.

Band-stop Filter

A filter that attenuates alternating currents whose frequencies are between given upper and lower cutoff values while passing frequencies above and below this band.

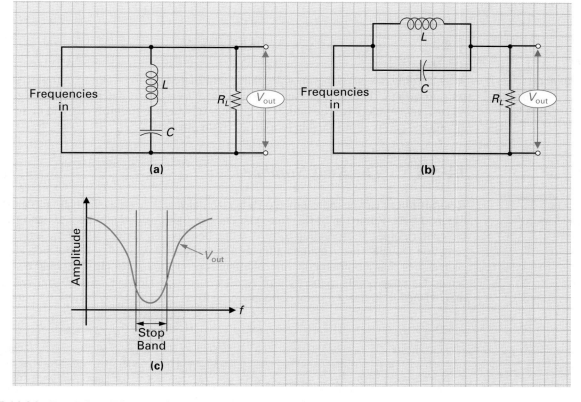

FIGURE 14-26   Band-Stop Filter. (a) Series Resonant Band-Stop Filter. (b) Parallel Resonant Band-Stop Filter. (c) Frequency Response.

FIGURE 14-27 **Tuning in of Station by Use of a Bandpass Filter.**

SELF-TEST EVALUATION POINT FOR SECTION 14-4

Now that you have completed this section, you should be able to:

■ **Objective 5.** *Identify and explain the following RLC circuit applications:*

a. Low-pass filter
b. High-pass filter
c. Bandpass filter
d. Band-stop filter

Use the following questions to test your understanding of Section 14-4.

1. Of the four types of filters, which:
   a. Would utilize the inductor as a shunt?
   b. Would utilize the capacitor as a shunt?
   c. Would use a series resonant circuit as a shunt?
   d. Would use a parallel resonant circuit as a shunt?
2. In what applications can filters be found?

14-5 COMPLEX NUMBERS

Complex Numbers

Numbers composed of a real number part and an imaginary number part.

After reading this section you will realize that there is really nothing complex about **complex numbers.** The complex number system allows us to determine the *magnitude* and *phase angle* of electrical quantities by adding, subtracting, multiplying, and dividing phasor quantities, and is an invaluable tool in ac circuit analysis.

14-5-1 *The Real Number Line*

Real Numbers

Numbers that have no imaginary parts.

Real numbers can be represented on a horizontal line, known as the real number line, as in Figure 14-28. Referring to this line, you can see that positive numbers exist to the right of the center point corresponding to zero, while negative numbers exist to the left. This representation satisfied most mathematicians for a short time, as they could indicate numbers such as 2 or 5 as a point on the line. Numbers corresponding to the $\sqrt{9}$ could also be represented, as three points to the right of zero ($\sqrt{9} = +3$). However, a problem was reached if they wished to indicate a point corresponding to $\sqrt{-9}$. The $\sqrt{-9}$ is not $+3$ [since $(+3) \times$

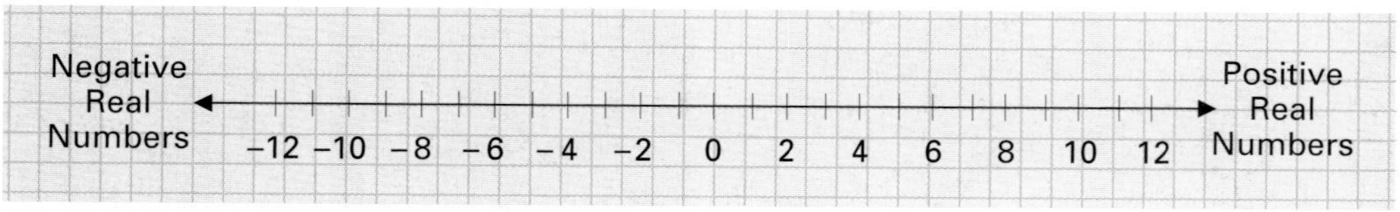

Negative Real Numbers ←——————————→ Positive Real Numbers

-12 -10 -8 -6 -4 -2 0 2 4 6 8 10 12

FIGURE 14-28 **Real Number Line.**

$(+3) = +9]$, and it is not -3 [since $(-3) \times (-3) = +9$]. So it was eventually realized that the square root of a negative number could not be indicated on the real number line, as it is not a real number.

14-5-2 *The Imaginary Number Line*

Mathematicians decided to call the square root of a negative number, such as $\sqrt{-4}$ or $\sqrt{-9}$, **imaginary numbers,** which are not fictitious or imaginary, but simply a particular type of number.

Just as real numbers can be represented on a real number line, imaginary numbers can be represented on an imaginary number line, as shown in Figure 14-29. The imaginary number line is vertical, to distinguish it from the real number line, and when working with electrical quantities a $\pm j$ prefix, known as the ***j* operator,** is used for values that appear on the imaginary number line.

14-5-3 *The Complex Plane*

A complex number is the combination of a real and imaginary number and is represented on a two-dimensional plane called the **complex plane,** shown in Figure 14-30. Generally, the real number appears first, followed by the imaginary number. Here are some examples of complex numbers.

REAL NUMBERS	IMAGINARY NUMBERS
3	$+j4$
-2	$+j4$
-3	$-j2$

Imaginary Number
A complex number whose imaginary part is not zero.

j Operator
A prefix used to indicate an imaginary number.

Complex Plane
A plane whose points are identified by means of complex numbers.

FIGURE 14-29 Imaginary Number Line.

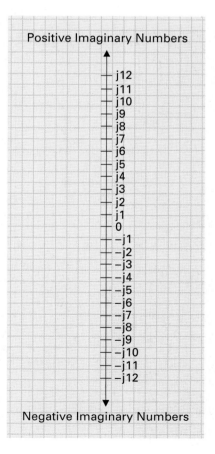

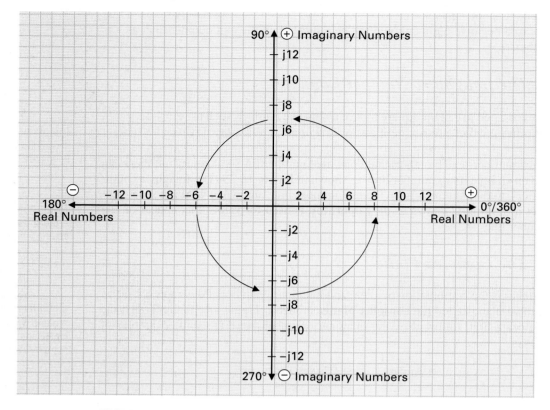

FIGURE 14-30   Complex Plane.

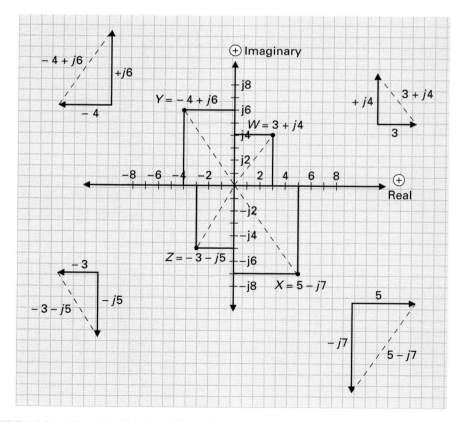

FIGURE 14-31   Complex Numbers Examples.

Complex numbers, therefore, are merely terms that need to be added as phasors, and all you have to do basically is draw a vector representing the real number and then draw another vector representing the imaginary number.

■ EXAMPLE:

Find the points in the complex plane in Figure 14-31 that correspond to the following complex numbers.

$$W = 3 + j4$$
$$X = 5 - j7$$
$$Y = -4 + j6$$
$$Z = -3 - j5$$

■ Solution:

By first locating the point corresponding to the real number on the horizontal line and then plotting it against the imaginary number on the vertical line, the points can be determined as shown in Figure 14-31.

A number like $3 + j4$ specifies two phasors in **rectangular coordinates,** so this system is the *rectangular representation of a complex number.* There are several other ways to describe a complex number, one of which is the polar representation of a complex number, using **polar coordinates,** which will be discussed next.

14-5-4   *Polar Complex Numbers*

Phasors can also be expressed in polar form, as shown in Figure 14-32, which compares rectangular and polar notation. With the rectangular notation in Figure 14-32(a), the horizontal coordinate is the real part and the vertical coordinate is the imaginary part of the complex number. With the polar notation shown in Figure 14-32(b), the magnitude of the phasor (x, or size) and the angle ($\angle \theta$, meaning "angle theta") relative to the positive real axis (measured in a counterclockwise direction) are stated.

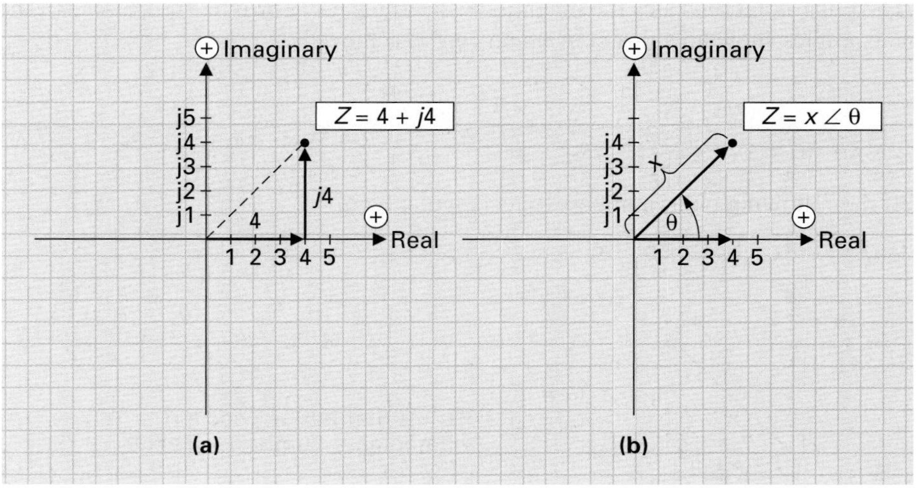

Rectangular Coordinates

A Cartesian coordinate of a Cartesian coordinate system whose straight-line axes or coordinate planes are perpendicular.

Polar Coordinates

Either of two numbers that locate a point in a plane by its distance from a fixed point on a line and the angle this line makes with a fixed line.

FIGURE 14-32   Representing Phasors. (a) Rectangular Notation. (b) Polar Notation.

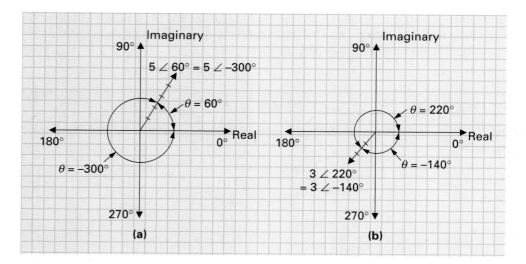

FIGURE 14-33   Polar Number Examples.

■   EXAMPLE:

Sketch the following polar numbers:

   a. $5 \angle 60°$

   b. $3 \angle 220°$

■   *Solution:*

As you can see in Figure 14-33, an equivalent negative angle, which is calculated by subtracting the given positive angle from 360°, can also be used.

14-5-5   *Rectangular/Polar Conversions*

Many scientific calculators have a feature that allows you to convert rectangular numbers to polar numbers, and vice versa. These conversions are based on the Pythagorean theorem and trigonometric functions, discussed previously in a mini-math review.

Polar-to-Rectangular Conversion

   The polar notation states the magnitude and angle, as shown in Figure 14-34. The following examples show how this conversion can be achieved.

■   EXAMPLE:

Convert the following polar numbers to rectangular form:

   a. $5 \angle 30°$ b. $18 \angle -35°$ c. $44 \angle 220°$

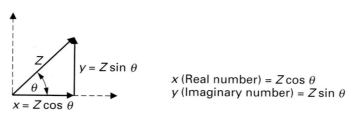

$x \text{ (Real number)} = Z \cos \theta$
$y \text{ (Imaginary number)} = Z \sin \theta$

FIGURE 14-34   Polar-to-Rectangular Conversion.

Solution:

a. Real number = 5 cos 30° = 4.33
   Imaginary number = 5 sin 30° = $j2.5$
   Polar number, 5 ∠ 30° = rectangular number, 4.33 + $j2.5$

b. Real number = 18 cos (−35°) = 14.74
   Imaginary number = 18 sin (−35°) = −$j10.32$
   Polar number, 18 ∠ −35° = rectangular number, 14.74 − $j10.32$

c. Real number = 44 cos 220° = −33.7
   Imaginary number = 44 sin 220° = −$j28.3$
   Polar number, 44 ∠ 220° = rectangular number, −33.7 − j28.3

Rectangular-to-Polar Conversion

The rectangular notation states the horizontal (real) and vertical (imaginary) sides of a triangle, as shown in Figure 14-35. The following examples show how the conversion can be achieved.

EXAMPLE:

Convert the following rectangular numbers to polar form:

   a. 4 + $j3$ b. 16 − $j14$

Solution:

a. Magnitude = $\sqrt{4^2 + 3^2}$ = 5
   Angle = arctan (3/4) = 36.9°
   Rectangular number, 4 + $j3$ = polar number, 5 ∠ 36.9°

b. Magnitude = $\sqrt{16^2 + (-14)^2}$ = 21.3
   Angle = arctan (−14/16) = − 41.2°
   Rectangular number, 16 − $j14$ = polar number, 21.3 ∠ −41.2°

14-5-6   *Complex Number Arithmetic*

Since a phase difference exists between real and imaginary (j) numbers, certain rules should be applied when adding, subtracting, multiplying, or dividing complex numbers.

FIGURE 14-35 Rectangular-to-Polar Conversion.

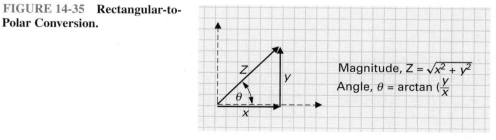

Addition

The sum of two complex numbers is equal to the sum of their real and imaginary parts.

▦ EXAMPLE:

Add the following complex numbers:

    a. $(3 + j4) + (2 + j5)$
    b. $(4 + j5) + (2 - j3)$

▦ *Solution:*

    a. $(3 + j4) + (2 + j5) = (3 + 2) + (j4 + j5) = 5 + j9$
    b. $(4 + j5) + (2 - j3) = (4 + 2) + (j5 - j3) = 6 + j2$

Subtraction

The difference of two complex numbers is equal to the difference between the separate real and imaginary parts.

▦ EXAMPLE:

Subtract the following complex numbers:

    a. $(4 + j3) - (2 + j2)$
    b. $(12 + j6) - (6 - j3)$

▦ *Solution:*

    a. $(4 + j3) - (2 + j2) = (4 - 2) + j(3 - 2) = 2 + j1$
    b. $(12 + j6) - (6 - j3) = (12 - 6) + j[6 - (-3)] = 6 + j9$

Multiplication

Multiplication of two complex numbers is achieved more easily if they are in polar form. The simple rule to remember is to multiply the magnitudes and then add the angles algebraically.

▦ EXAMPLE:

Multiply the following complex numbers:

    a. $5 \angle 35° \times 7 \angle 70°$
    b. $4 \angle 53° \times 12 \angle -44°$

▦ *Solution:*

    a. Multiply the magnitudes: $5 \times 7 = 35$.
       Algebraically add the angles: $\angle (35° + 70°) = \angle 105° = 35 \angle 105°$.
    b. Multiply the magnitudes: $4 \times 12 = 48$.
       Algebraically add the angles: $\angle [53° + (-44°)] = \angle 9° = 48 \angle 9°$.

Division

Division is also more easily carried out in polar form. The rule to remember is to divide the magnitudes, and then subtract the denominator angle from the numerator angle.

EXAMPLE:

Divide the following complex numbers:

   a. $60 \angle 30°$ by $30 \angle 15°$ b. $100 \angle 20°$ by $5 \angle -7°$

Solution:

  a. Divide the magnitudes: $60/30 = 2$.

    Subtract the denominator angle from the numerator angle: $\angle (30° - 15°) = \angle 15° = 2 \angle 15°$.

  b. Divide the magnitudes: $100/5 = 20$.

    Subtract the angles: $\angle [20° - (-7°)] = \angle 27° = 20 \angle 27°$.

14-5-7   *How Complex Numbers Apply to AC Circuits*

Complex numbers find an excellent application in ac circuits due to all the phase differences that occur between different electrical quantities, such as X_L, X_C, R, and Z as shown in Figure 14-36. The positive real number line, at an angle of $0°$, is used for resistance, which in this example is $3 \, \Omega$, as shown in Figure 14-36(a) and (b).

On the positive imaginary number line, at an angle of $90°$ $(+j)$, inductive reactance (X_L) is represented, which in this example is $j4 \, \Omega$ $(X_L = 4 \, \Omega)$. The voltage drop across an inductor (V_L) is proportional to its inductive reactance (X_L), and both are represented on the $+j$ imaginary number line, since the voltage drop across an inductor will always lead the current (which in a series circuit is always in phase with the resistance) by $90°$.

On the negative imaginary number line, at an angle of $-90°$ or $270°$ $(-j)$, capacitive reactance (X_C) is represented, which in this example is $-j2$ $(X_C = 2 \, \Omega)$. The voltage drop across a capacitor (V_C) is proportional to its capacitive reactance (X_C), and both are represented on the $-j$ imaginary number line since the voltage drop across a capacitor always lags current (charge and discharge) by $-90°$.

Series AC Circuits

Referring to Figure 14-36(a) and (b) once again, we can calculate total impedance simply by adding the phasors.

Z_T **(Rectangular).** Total series impedance is equal to the sum of all the resistances and reactances:

$$Z = R + (jX_L \sim jX_C)$$
$$= 3 + (+j4 \sim -j2)$$
$$= 3 + j2$$

Z_T **(Polar).** The total series impedance can be converted from rectangular to polar form:

$$\text{magnitude} = \sqrt{3^2 + 2^2} = 3.61 \, \Omega$$

$$\text{angle} = \arctan \frac{2}{3} = 33.7°$$

$$= 3.61 \angle 33.7°$$

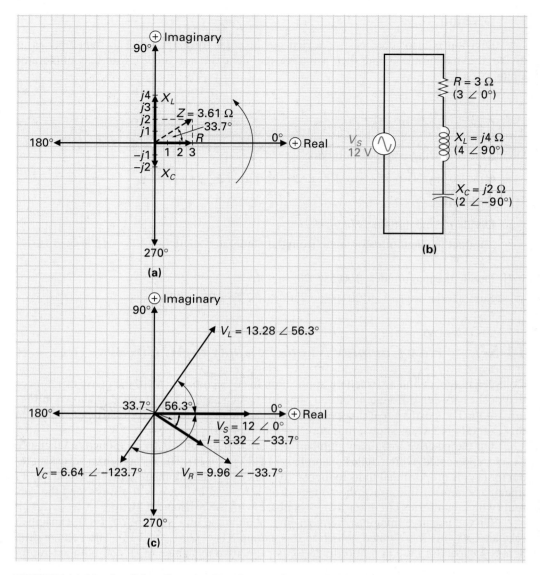

FIGURE 14-36 **Applying Complex Numbers to Series AC Circuits. (a) Impedance Phasors. (b) Series Circuit. (c) Voltage and Current Phasors.**

Current. Once the magnitude of Z_T is known (3.61 Ω), I can be calculated. The source voltage of 12 V is a real positive number (0°) and is therefore represented as $12 \angle 0°$. Current is equal to

$$I = \frac{V_S}{Z_T} = \frac{12 \angle 0°}{3.61 \angle 33.7°}$$

Polar division Divide the magnitudes: $\dfrac{12}{3.61} = 3.32 \text{ A}$

Subtract the angles: $0° - 33.7° = -33.7°$

$I = 3.32 \angle - 33.7°$

Phase angle. The circuit current has an angle of $-33.7°$, which means that it lags V_T (inductive circuit, therefore ELI). This negative phase angle is expected since this series circuit is inductive ($X_L > X_C$), in which case current should lag voltage by some phase angle. This phase angle is less than 45° because the net reactance is less than the circuit resistance, and is shown in Figure 14-36(c).

Voltage drops. The component voltage drops are calculated with the following formulas and shown in Figure 14-36(c).

$V_R = I \times R = (3.32 \angle -33.7°) \times (3 \angle 0°)$

Multiply the magnitudes: $3.32 \times 3 = 9.96$ V

Algebraically add the angles: $\angle (-33.7° + 0°) = -33.7°$

$V_R = 9.96 \angle -33.7°$

$V_L = I \times X_L = (3.32 \angle -33.7°) \times (4 \angle 90°)$

Multiply the magnitudes: $3.32 \times 4 = 13.28$ V

Algebraically add the angles: $\angle (-33.7 + 90°) = 56.3°$

$V_L = 13.28 \angle 56.3°$

$V_C = I \times X_C = (3.32 \angle -33.7°) \times (2 \angle -90°)$

Multiply the magnitudes: $3.32 \times 2 = 6.64$ V

Algebraically add the angles: $\angle (-33.7 + -90°) = -123.7°$

$V_C = 6.64 \angle -123.7°$

Phase relationships. As shown in Figure 14-36(c), the voltage across an inductor leads the circuit current by $+90°$, while the voltage across a capacitor lags the circuit current by $-90°$. The source voltage acts as the zero reference phase and leads the circuit current and the voltage across the resistor (V_R) by 33.7°.

Source voltage. Although the source voltage is known, it can be checked to verify all the previous calculations, since the sum of all the individual voltage drops should equal the source voltage.

$$\text{POLAR} \longrightarrow \text{RECTANGULAR}$$

$$
\begin{array}{ll}
V_L = 13.28 \angle 56.3 & = 7.37 + j11.05 \\
V_R = 9.96 \angle -33.7° & = 8.29 - j5.53 \\
V_C = 6.64 \angle -123.7° & = \underline{-3.68 - j5.52} \\
& 11.98 + j0
\end{array}
$$

Series–Parallel AC Circuits

Figure 14-37 illustrates a series–parallel ac circuit containing an *RL* branch, an *RC* branch, and an *RLC* branch.

Impedance of each branch. The three branches will each have a value of impedance that will be equal to:

$$\text{RECTANGULAR} \longrightarrow \text{POLAR}$$

$$
\begin{array}{l}
Z_1 = 10 + j5 = 11.2 \angle 26.6° \ \Omega \\
Z_2 = 25 - j15 = 29.2 \angle -31.0° \ \Omega
\end{array}
$$

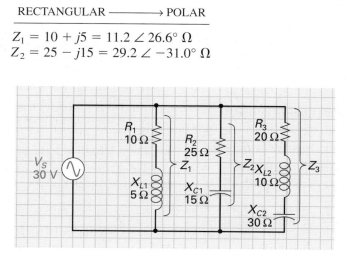

FIGURE 14-37 Applying Complex Numbers to Series–Parallel AC.

The third branch is capacitive since the difference between $-j30$ (X_{C2}) and $+ j10$ (X_{L2}) is $-j20$.

$$Z_3 = 20 - j20 = 28.3 \angle -45° \ \Omega$$

Branch currents.   The three branch currents, I_1, I_2, and I_3, are calculated by dividing the source voltage (V_S) by the individual branch impedances.

$$I_1 = \frac{V_S}{Z_1} = \frac{30 \angle 0°}{11.2 \angle 26.6°}$$

Divide magnitudes: $30 \div 11.2 = 2.68$

Subtract angles: $\angle (0° - 26.6°) = -26.6°$

$$I_1 = 2.68 \angle -26.6° = 2.4 - j1.2 \ \text{A}$$

$$I_2 = \frac{V_S}{Z_2} = \frac{30 \angle 0°}{29.2 \angle -31°} = 1.03 \angle + 31° = 0.88 + j0.5 \ \text{A}$$

$$I_3 = \frac{V_S}{Z_3} = \frac{30 \angle 0°}{28.3 \angle -45°} = 1.06 \angle + 45° = 0.75 + j0.7 \ \text{A}$$

Total current.

$$
\begin{aligned}
I_T &= I_1 + I_2 + I_3 \\
&= (2.4 - j1.2) + (0.88 + j0.5) + (0.75 + j0.7) \\
&= (2.4 + 0.88 + 0.75) + [-j1.2 + (+ j0.5) + (+ j0.7)] \\
&= 4.03 \ \text{A}
\end{aligned}
$$

In polar form, this will equal $4.03 \angle 0°$ A.

Total impedance.

$$Z_T = \frac{V_S}{I_T} = \frac{30 \angle 0°}{4.03 \angle 0°} = 7.44 \angle 0° \ \Omega$$

POLAR $\longrightarrow$ RECTANGULAR

$$7.44 \angle 0° = 7.44 + j0$$

The complex ac circuit seen in Figure 14-37 is therefore equivalent to a 7.44 Ω resistor in series with no reactance.

SELF-TEST EVALUATION POINT FOR SECTION 14-5

Now that you have completed this section, you should be able to:

■ *Objective 6.* *Describe complex numbers in both rectangular and polar form.*

■ *Objective 7.* *Perform complex number arithmetic.*

■ *Objective 8.* *Describe how complex numbers apply to ac circuits containing series–parallel RCL components.*

Use the following questions to test your understanding of Section 14-5.

1. In complex numbers, resistance is a/an _____ term and reactance is a/an _____ term. (imaginary/real)

2. Convert the following rectangular number to polar form: $5 + j6$.

3. Convert the following polar number to rectangular form: $33 \angle 25°$.

4. What is a complex number?

SUMMARY

Resistive, Inductive, and Capacitive (*RLC*) Circuits (Figure 14-38)

1. Resistors, as we have discovered, operate and react to voltage and current in a very straightforward way; the voltage across a resistor is in phase with the resistor current.

2. Inductors and capacitors operate in essentially the same way, in that they both store energy and then return it back to the circuit. However, they have completely opposite reactions to voltage and current.

3. The phrase "ELI the ICE man" states that voltage (symbolized E) leads current (I) in an inductive (L) circuit (abbreviated by the word "ELI"), while current (I) leads voltage (E) in a capacitive (C) circuit (abbreviated by the word "ICE").

4. The circuit current is always the same throughout a series circuit and can therefore be used as a reference.

5. The voltage across a resistor is always in phase with the current, while the voltage across the inductor leads the current by 90° and the voltage across the capacitor lags the current by 90°.

6. Impedance is the total opposition to current flow and is a combination of both reactance (X_L, X_C) and resistance (R).

7. The capacitive and inductive reactances are 180° out of phase with one another and counteract.

8. Impedance is the vector sum of the reactive (X) and resistance (R) vectors.

9. Since none of these voltages are in phase with one another in a series *RLC* circuit, they must be added vectorially to obtain the applied voltage.

10. The voltage across any parallel circuit will be equal and in phase.

11. With current, we must first calculate the individual branch currents (I_R, I_L, and I_C) and then calculate the total circuit current (I_T).

12. There is a phase difference between the source voltage (V_S) and the circuit current (I_T).

13. Once the total current (I_T) is known, the impedance of all three components in parallel can be calculated.

14. Resonance is a circuit condition that occurs when the inductive reactance (X_L) and the capacitive reactance (X_C) have been balanced.

15. As the input ac frequency is increased, a point will be reached where X_L will equal X_C, and this condition is known as *resonance*. The frequency at which $X_L = X_C$ in either a parallel or a series *LC* circuit is known as the *resonant frequency* (f_0).

16. A series resonant circuit is a resonant circuit in which the capacitor and coil are in series with the applied ac voltage.

17. In summary, we can say that in a series resonant circuit:
   a. The inductor and capacitor electrically disappear due to their equal but opposite effect, resulting in 0 V drops across the series combination, and the circuit consequently seems purely resistive.
   b. The current flow is large because the impedance of the circuit is low and equal to the series resistance (R), which has the source voltage developed across it.
   c. The individual voltage drops across the inductor or capacitor can be larger than the source voltage if R is smaller than X_L and X_C.

18. The Q factor is a ratio of inductive reactance to resistance and is used to express how efficiently an inductor will store rather than dissipate energy. In a series resonant circuit, the Q factor indicates the quality of the series resonant circuit, or is the ratio of the reactance to the resistance.

19. The Q of a resonant circuit is almost entirely dependent on the inductor's coil resistance, because capacitors tend to have almost no resistance figure at all, only reactive, which makes them very efficient.

20. A series resonant circuit is selective in that frequencies at resonance or slightly above or below will cause a larger current than frequencies well above or below the circuit's resonant frequency.

21. The group or band of frequencies that causes the larger current is called the circuit's bandwidth. A frequency response curve is a graph indicating a circuit's response to different frequencies.

22. The bandwidth includes the group or band of frequencies that cause 70.7% or more of the maximum current to flow within the series resonant circuit.

23. The cutoff frequency is the frequency at which the gain of the circuit falls below 0.707 of the maximum current or half-power (-3 dB).

24. The half-power point is a point at which power is 50%. This half-power point corresponds to 70.7% of the total current.

25. Bandwidth is proportional to the resonant frequency of the circuit and inversely proportional to the Q of the circuit. In summary, the bandwidth of a series resonant circuit will increase as the Q of the circuit decreases (BW $\uparrow = f_0/Q\downarrow$), and vice versa.

26. A parallel resonant circuit is a circuit having an inductor and capacitor in parallel with one another, offering a high impedance at the frequency of resonance.

27. The source voltage (V_S) is needed initially to supply power to the *LC* circuit and start the oscillations; but once the oscillating process is in progress (assuming the ideal case), current is only flowing back and forth between inductor and capacitor, and no current is flowing from the source.

28. Flywheel action is a sustaining effect of oscillation in an *LC* circuit due to the charging and discharging of the capacitor and the expansion and contraction of the magnetic field around the inductor.

29. The electronic equivalent of the mechanical flywheel is a resonant parallel-connected *LC* circuit.

30. The direction of the circulating current reverses each half-cycle at the frequency of resonance. Energy is stored in the capacitor in the form of an electric field between the plates

QUICK REFERENCE SUMMARY SHEET

FIGURE 14-38 ***RLC* Circuits.**

- **Series *RLC* Circuits**

$$X_C = \frac{1}{2\pi f C}$$

$$X_L = 2\pi f L$$

$$Z = \sqrt{R^2 + (X_L \sim X_C)^2}$$

$$I = \frac{V_S}{Z}$$

$$V_R = I \times R \quad V_L = I \times X_L \quad V_C = I \times X_C$$

$$V_S = \sqrt{V_R^{\,2} + (V_L \sim V_C)^2}$$

- **Parallel *RLC* Circuits**

$$I_R = \frac{V}{R} \quad I_L = \frac{V}{X_L} \quad I_C = \frac{V}{X_C}$$

$$I_T = \sqrt{I_R^{\,2} + I_X^{\,2}}$$

$$Z = \frac{V}{I_T}$$

- **Resonance (*XL* = *XC*)**

$$f_0 = \frac{1}{2\pi\sqrt{LC}}, f_0 = \text{resonant frequency}$$

$$Z = R$$

$$\text{BW} = \frac{f_0}{Q}, \text{BW} = \text{bandwidth}$$

- **Series Resonance**

$$Q = \frac{X_L}{R} = \frac{X_C}{R} = \frac{V_L}{V_R} = \frac{V_C}{V_R} = \frac{V_L}{V_S} = \frac{V_C}{V_S}$$

- **Parallel Resonance**

$$Q = \frac{I_{\text{tank}}}{I_S} = \frac{Z_{\text{tank}}}{X_L}$$

- **Power**

$$\text{Apparent power} = V_S \times I \text{ (volt-amperes)}$$

$$\text{True power} = I^2 \times R \text{ (watts)}$$

$$\text{PF} = \cos\theta = \frac{R}{Z} = \frac{P_R}{P_A}$$

- **Complex Numbers**

Polar-to-rectangular: Real number = magnitude × cos θ
Imaginary number = magnitude × sin θ

Rectangular-to-polar:

$$\text{Magnitude} = \sqrt{(\text{real number})^2 + (\text{imaginary number})^2}$$

$$\text{Angle} = \arctan\left(\frac{\text{imaginary number}}{\text{real number}}\right)$$

Addition = sum of the real and sum of the imaginary parts (rectangular form)
Subtraction = difference between the real and difference between the imaginary parts (rectangular form)
Multiplication = multiply the magnitudes and then add the angles algebraically (polar form)
Division = divide the magnitudes and then subtract the denominator angle from the numerator angle (polar form)

on one half-cycle, and then the capacitor discharges, supplying current to build up a magnetic field on the other half-cycle. The inductor stores its energy in the form of a magnetic field, which will collapse, supplying a current to charge the capacitor, which will then discharge, supplying a current back to the inductor, and so on. Due to the "storing action" of this circuit, it is sometimes related to the fluid analogy and referred to as a tank circuit.

31. Under ideal conditions a tank circuit should oscillate indefinitely if no losses occur within the circuit. In reality, the resistance of the coil reduces that 100% efficiency, as does friction with the mechanical flywheel.

32. As a small part of the energy is dissipated with each cycle, the oscillations will be reduced in size and eventually fall to zero.

33. If the ac source is reconnected to the tank, a small amount of current will flow from the source to the tank to top up the tank or replace the dissipated power. The higher the coil resistance is, the higher the loss and the larger the current flow from source to tank to replace the loss.

34. In the series resonant circuit, we were concerned with voltage drops since current remains the same throughout a series circuit. In a parallel resonant circuit, we are concerned with circuit currents rather than voltage.

35. Of all the three Q formulas for parallel resonant circuits, $Q = I_{\text{tank}}/I_S$, $Q = X_L/R$, or $Q = Z_{\text{tank}}/X_L$, the latter is the easiest to use as both X_L and the tank impedance can easily be determined in most cases where C, L, and R internal for the inductor are known.

36. The current versus frequency response curve in a parallel resonant circuit is the complete opposite to the series resonant response curve. At frequencies below resonance X_L is low and X_C is high, and the inductor offers a low reactance, producing a high current path and low impedance. On the other hand, at frequencies above resonance the capacitor displays a low reactance, producing a high current path and low impedance.

37. The parallel resonant circuit is like the series resonant circuit in that it responds to a band of frequencies close to its resonant frequency.

38. Circuits containing inductance and capacitance are often referred to as tuned circuits since they can be adjusted to make the circuit responsive to a particular frequency (the resonant frequency).

39. Selectivity, by definition, is the ability of a tuned circuit to respond to a desired frequency and ignore all others.

40. Parallel resonant LC circuits are sometimes too selective, as the Q is too large, producing too narrow a bandwidth. In this situation, because of the very narrow response curve, a high resistance value can be placed in parallel with the LC circuit to provide an alternative path for line current. This process is known as *loading* or *damping* the tank and will cause an increase in line current and decrease in Q ($Q\downarrow = I_{\text{tank}}/I_{\text{line}}\uparrow$). The decrease in Q will cause a corresponding increase in BW (BW $\uparrow = f_0/Q\downarrow$).

41. There are basically four types of filters:
   a. Low-pass filter, which passes frequencies below a cutoff frequency
   b. High-pass filter, which passes frequencies above a cutoff frequency
   c. Bandpass filter, which passes a band of frequencies
   d. Band-stop filter, which stops a band of frequencies

42. An inductor and capacitor can be connected to act as a low-pass filter. Since the inductor basically blocks alternating current and the capacitor shunts alternating current, the net result is to prevent high-frequency signals from reaching the load.

43. An inductor and capacitor can also be connected to act as a high-pass filter.

44. A bandpass filter circuit passes a group or band of frequencies between a lower and an upper cutoff frequency, while heavily attenuating any other frequency outside this band.

45. A band-stop filter attenuates alternating currents whose frequencies are between given upper and lower cutoff values while passing frequencies above and below this band.

46. Filters are necessary in applications such as television or radio, where we need to tune in (select or pass) one frequency that contains the information we desire, yet block all the millions of other frequencies that are also carrying information.

Complex Numbers

47. The complex number system allows us to determine the *magnitude* and *phase angle* of electrical quantities by adding, subtracting, multiplying, and dividing phasor quantities, and is an invaluable tool in ac circuit analysis.

48. Real numbers can be represented on a horizontal line, known as the real number line. Positive numbers exist to the right of the center point corresponding to zero, while negative numbers exist to the left. This representation satisfied most mathematicians for a short time, as they could indicate numbers such as 2 or 5 as a point on the line. However, a problem was reached if they wished to indicate a point corresponding to $\sqrt{-9}$. The $\sqrt{-9}$ is not $+3$ [since $(+3) \times (+3) = +9$], and it is not -3 [since $(-3) \times (-3) = +9$]. So it was eventually realized that the square root of a negative number could not be indicated on the real number line, as it is not a real number.

49. Mathematicians decided to call the square root of a negative number, such as $\sqrt{-4}$ or $\sqrt{-9}$, imaginary numbers, which are not fictitious or imaginary, but simply a particular type of number. Just as real numbers can be represented on a real number line, imaginary numbers can be represented on an imaginary number line. The imaginary number line is vertical, to distinguish it from the real number line, and when working with electrical quantities a $\pm j$ prefix, known as the *j operator,* is used for values that appear on the imaginary number line.

50. A complex number is the combination of a real and imaginary number and is represented on a two-dimensional plane called the complex plane. Generally, the real number appears first, followed by the imaginary number.

51. Complex numbers are merely terms that need to be added as phasors, and all you have to do basically is draw a vector representing the real number and then draw another vector representing the imaginary number.

52. A number like $3 + j4$ specifies two phasors in rectangular coordinates, so this system is the *rectangular representation of a complex number.*

53. There are several other ways to describe a complex number, one of which is the polar representation of a complex number, using polar coordinates.

54. With the rectangular notation the horizontal coordinate is the real part and the vertical coordinate is the imaginary part of the complex number. With the polar notation the magnitude of the phasor (x, or size) and the angle ($\angle\ \theta$, meaning "angle theta") relative to the positive real axis (measured in a counterclockwise direction) are stated.

55. Since a phase difference exists between real and imaginary (j) numbers, certain rules should be applied when adding, subtracting, multiplying, or dividing complex numbers.

56. The sum of two complex numbers is equal to the sum of their real and imaginary parts.

57. The difference of two complex numbers is equal to the difference between the separate real and imaginary parts.

58. Multiplication of two complex numbers is achieved more easily if they are in polar form. The simple rule to remember is to multiply the magnitudes and then add the angles algebraically.

59. Division is also more easily carried out in polar form. The rule to remember is to divide the magnitudes, and then subtract the denominator angle from the numerator angle.

60. Complex numbers find an excellent application in ac circuits due to all the phase differences that occur between different electrical quantities, such as X_L, X_C, R, and Z. The positive real number line, at an angle of $0°$, is used for resistance.

61. On the positive imaginary number line, at an angle of $90°$ ($+j$), inductive reactance (X_L) is represented.

62. The voltage drop across an inductor (V_L) is proportional to its inductive reactance (X_L), and both are represented on the $+j$ imaginary number line, since the voltage drop across an inductor will always lead the current (which in a series circuit is always in phase with the resistance) by $90°$.

63. On the negative imaginary number line, at an angle of $-90°$ or $270°$ ($-j$), capacitive reactance (X_C) is represented.

64. The voltage drop across a capacitor (V_C) is proportional to its capacitive reactance (X_C), and both are represented on the $-j$ imaginary number line since the voltage drop across a capacitor always lags current (charge and discharge) by $-90°$.

REVIEW QUESTIONS

Multiple-Choice Questions

1. Capacitive reactance is _____ to frequency and capacitance, while inductive reactance is _____ to frequency and inductance.
  a. Proportional, inversely proportional
  b. Inversely proportional, proportional
  c. Proportional, proportional
  d. Inversely proportional, inversely proportional

2. Resonance is a circuit condition that occurs when:
  a. V_L equals V_C **d.** Both (a) and (c)
  b. X_L equals X_C **e.** Both (a) and (b)
  c. L equals C

3. As frequency is increased, X_L will _____, while X_C will _____.
  a. Decrease, increase   **c.** Remain the same, decrease
  b. Increase, decrease   **d.** Increase, remain the same

4. In an RLC series resonant circuit, with $R = 500\ \Omega$ and $X_L = 250\ \Omega$, what would be the value of X_C?
  a. $2\ \Omega$ **c.** $250\ \Omega$
  b. $125\ \Omega$ **d.** $500\ \Omega$

5. At resonance, the voltage drop across both a series-connected inductor and capacitor will equal:
  a. 70.7 V   **c.** 10 V
  b. 50% of the source   **d.** Zero

6. In a series resonant circuit the current flow is _____, as the impedance is _____ and equal to _____.
  a. Large, small, R **c.** Large, small, X
  b. Small, large, X **d.** Small, large, R

7. A circuit's bandwidth includes a group or band of frequencies that cause _____ or more of the maximum current, or more than _____ of the maximum power to appear at the output.
  a. 110, 90   **c.** 70.7%, 50%
  b. 50%, 70.7%   **d.** Both (a) and (c)

8. The bandwidth of a circuit is proportional to the:
  a. Frequency of resonance   **c.** Tank current
  b. Q of the tank   **d.** Two of the above

9. Series or parallel resonant circuits can be used to create:
  a. Low-pass filters
  b. Low-pass and high-pass filters
  c. Bandpass and band-stop filters
  d. All of the above

10. Flywheel action occurs in:
  a. A tank circuit   **d.** Two of the above
  b. A parallel LC circuit   **e.** None of the above
  c. A series LC circuit

11. $25 \angle 39°$ is an example of a complex number in:
  a. Polar form   **c.** Algebraic form
  b. Rectangular form   **d.** None of the above

12. $3 + j10$ is an example of a complex number in:
  a. Polar form   **c.** Algebraic form
  b. Rectangular form   **d.** None of the above

13. Which complex number form is usually more convenient for addition and subtraction?
  a. Rectangular   **b.** Polar

14. Which complex number form is usually more convenient for multiplication and division?
  a. Rectangular   **b.** Polar

15. In complex numbers, *resistance* is a real term, while *reactance* is a/an _____.
  a. j term
  b. Imaginary term
  c. Value appearing on the vertical axis
  d. All of the above

Communication Skill Questions

16. Illustrate with phasors and describe the current and voltage relationships in a series *RLC* circuit. (14-1)

17. Describe the procedure for the analysis of a series *RLC* circuit. (14-1)

18. Define resonance and give the formula for calculating the frequency of resonance. (14-3)

19. Describe the three unusual characteristics of a circuit that is at resonance. (14-3)

20. Define the following: (14-3)
  a. Flywheel action   **c.** Bandwidth
  b. Quality factor   **d.** Selectivity

21. Describe a frequency response curve. (14-3)

22. Illustrate with phasors and describe the current and voltage relationships in a parallel *RLC* circuit. (14-2)

23. Describe the differences between a series and a parallel resonant circuit. (14-3)

24. Explain how loading a tank affects bandwidth and selectivity. (14-3-2)

25. Illustrate the circuit and explain the operation of the following, with their corresponding response curves. (14-4)
  a. Low-pass filter   **c.** Bandpass filter
  b. High-pass filter   **d.** Band-stop filter

26. Describe why capacitive reactance is written as $-jX_C$ and inductive reactance is written as jX_L. (14-5)

27. How are capacitive and inductive reactances written in polar form? (14-5-7)

28. List the rules used to perform complex number: (14-5-6)
  a. Addition (rectangular)   **c.** Multiplication (polar)
  b. Subtraction (rectangular)   **d.** Division (polar)

29. Describe briefly how the real number and imaginary number lines are used for ac circuit analysis and what electrical phasors are represented at $0°$, $90°$, and $-90°$. (14-5-7)

30. Referring to Figure 14-36, describe why the series circuit current (I) is not in phase with the source voltage (V_S). (14-5-7)

Practice Problems

31. Calculate the values of capacitive or inductive reactance for the following when connected across a 60 Hz source:
  a. 0.02 µF   **e.** 4 mH
  b. 18 µF   **f.** 8.18 H
  c. 360 pF   **g.** 150 mH
  d. 2700 nF   **h.** 2 H

32. If a 1.2 kΩ resistor, a 4 mH inductor, and an 8 µF capacitor are connected in series across a 120 V/60 Hz source, calculate:
  a. X_C **b.** X_L

c. Z **h.** Apparent power
d. I **i.** True power
e. V_R **j.** Resonant frequency
f. V_L **k.** Circuit quality factor
g. V_C **l.** Bandwidth

33. If a 270 Ω resistor, a 150 mH inductor, and a 20 µF capacitor are all connected in parallel with one another across a 120 V/60 Hz source, calculate:
  a. X_L **f.** I_T
  b. X_C **g.** Z
  c. I_R **h.** Resonant frequency
  d. I_L **i.** Q factor
  e. I_C **j.** Bandwidth

34. Calculate the impedance of a series circuit if $R = 750\ \Omega$, $X_L = 25\ \Omega$, and $X_C = 160\ \Omega$.

35. Calculate the impedance of a parallel circuit with the same values as those of Question 34 when a 1 V source voltage is applied.

36. State the following series circuit impedances in rectangular and polar form:
  a. $R = 33\ \Omega, X_C = 24\ \Omega$ **b.** $R = 47\ \Omega, X_L = 17\ \Omega$

37. Convert the following impedances to rectangular form:
  a. $25 \angle 37°$ **c.** $114 \angle -114°$
  b. $19 \angle -20°$ **d.** $59 \angle 99°$

38. Convert the following impedances to polar form:
  a. $-14 + j14$ **c.** $-33 - j18$
  b. $27 + j17$ **d.** $7 + j4$

39. Add the following complex numbers:
  a. $(4 + j3) + (3 + j2)$ **b.** $(100 - j50) + (12 + j9)$

40. Perform the following mathematical operations:
  a. $(35 \angle -24°) \times (13 \angle 50°)$
  b. $(100 - j25) - (25 + j5)$
  c. $(98 \angle 80°) \div (40 \angle 17°)$

41. State the impedances of the circuits seen in Figure 14-39 in rectangular and polar form. What is Z_T in ohms and its phase angle?

42. Calculate in polar form the impedances of both circuits shown in Figure 14-40. Then combine the two impedances as if the circuits were parallel connected, using the product-over-sum method. Express the combined impedance in polar form.

43. Referring to Figure 14-41, calculate:
  a. Z_T (rectangular and polar)
  b. Circuit current and phase angle
  c. Voltage drops
  d. V_C, V_L, and V_R phase relationships

44. Sketch an impedance and voltage phasor diagram for the circuits in Figure 14-39.

45. Referring to Figure 14-42, calculate:
  a. Impedance of the two branches   **c.** Total current
  b. Branch currents   **d.** Total impedance

46. Sketch an impedance and current phasor diagram for the circuit shown in Figure 14-40.

47. Referring to Figure 14-39, verify that the sum of all the individual voltage drops is equal to the total voltage.

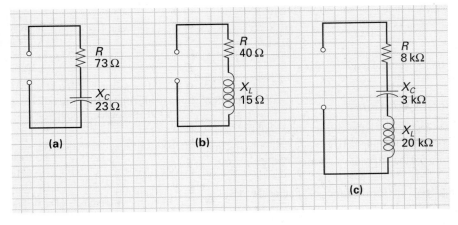

FIGURE 14-39

48. Referring to Figure 14-40, verify that the sum of all the branch currents is equal to the total current.

49. Determine whether Questions 43 and 45 would be easier to answer with or without the use of complex numbers.

50. Is the circuit in Figure 14-40 more inductive or capacitive?

Web Site Questions

Go to the Web site http://www.prenhall.com/cook, select the textbook *Introductory DC/AC Electronics* or *Introductory DC/AC Circuits*, this chapter, and then follow the instructions when answering the multiple-choice practice problems.

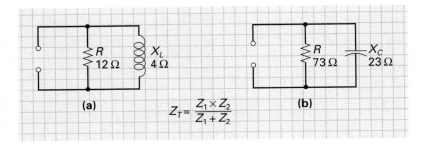

$$Z_T = \frac{Z_1 \times Z_2}{Z_1 + Z_2}$$

FIGURE 14-40

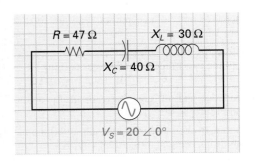

FIGURE 14-41

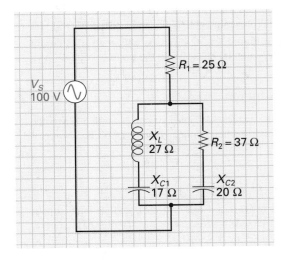

FIGURE 14-42

CHAPTER 14 / RESISTIVE, INDUCTIVE, AND CAPACITIVE (*RLC*) CIRCUITS

These tests at the end of each chapter will challenge your knowledge up to this point, and give you the practice you need for a job interview. To make this more realistic, the test will comprise both technical and personal questions. In order to take full advantage of this exercise, you may want to set up a simulation of the interview environment, have a friend read the questions to you, and record your responses for later analysis.

Company Name: PLAY, Inc.

Industry Branch: Consumer Electronics.

Function: Repair of Video Game Consoles.

Job Applying For: In-house Service Technician.

1. How did you find out about this job?
2. How do you think you would handle a job in which you are besieged with one problem after another?

3. If a resistor, capacitor, and inductor were connected in series across an ac source, what would be the phase relationship between the voltage drop across each component in comparison to the circuit current?
4. What responsibilities do you think an in-house service technician would have?
5. What is a filter?
6. Would you say that you are diplomatic?
7. What is resonance?
8. Would you work extra hours for a short period of time, if the service department were overloaded with repairs?
9. What video game systems do you own?
10. What do you know about the video game industry?

Answers

1. Describe source.
2. This has always been the job function of a technician. No matter what area of industry you are in, you will always be called upon to diagnose, isolate, and repair problems. Discuss how your education was structured to train you in testing and troubleshooting, and how you enjoy the challenge of fault finding.
3. Section 14-1.
4. Quote intro. section in text and job description listed in newspaper.
5. Section 14-4.
6. They're asking if you can get along well with customers and colleagues, and if you have the communication skills to tactfully handle daily interaction. Mention any work experience in which you had direct contact with customers and discuss how you worked with fellow students in your class.
7. Section 14-3.
8. If you are unsure, don't object to anything at the time. Later, if you decide that it is something you do not want to do, you can call and decline the offer.
9. Individual answer.
10. If you visit the company's web site before the interview, there will generally be a section describing the industry branch in which they do business—remember, the more informed you are, the better.

Semiconductor Principles

The Turing Enigma

During the Second World War, the Germans developed a cipher-generating apparatus called "Enigma." This electromechanical teleprinter would scramble messages with several randomly spinning rotors that could be set to a predetermined pattern by the sender. This key and plug pattern was changed three times a day by the Germans and cracking the secrets of Enigma became of the utmost importance to British Intelligence. With this objective in mind, every brilliant professor and eccentric researcher was gathered at a Victorian estate near London called Bletchley Park. They specialized in everything from engineering to literature and were collectively called the Backroom Boys.

By far the strangest and definitely most gifted of the group was an unconventional theoretician from Cambridge University named Alan Turing. He wore rumpled clothes and had a shrill stammer and crowing laugh that aggravated even his closest friends. He had other legendary idiosyncrasies that included setting his watch by sighting on a certain star from a specific spot and then mentally calculating the time of day. He also insisted on wearing his gas mask whenever he was out, not for fear of a gas attack, but simply because it helped his hay fever.

Turing's eccentricities may have been strange but his genius was indisputable. At the age of twenty-six he wrote a paper outlining his "universal machine" that could solve any mathematical or logical problem. The data or, in this case, the intercepted enemy messages could be entered into the machine on paper tape and then compared with known Enigma codes until a match was found.

In 1943 Turing's ideas took shape as the Backroom Boys began developing a machine that used 2,000 vacuum tubes and incorporated five photoelectric readers that could process 25,000 characters per second. It was named "Colossus," and it incorporated the stored program and other ideas from Turing's paper written seven years earlier.

Turing could have gone on to accomplish much more. However, his idiosyncrasies kept getting in his way. He became totally preoccupied with abstract questions concerning machine intelligence. His unconventional personal lifestyle led to his arrest in 1952 and, after a sentence of psychoanalysis, his suicide two years later.

Before joining the Backroom Boys at Bletchley Park, Turing's genius was clearly apparent at Cambridge. How much of a role he played in the development of Colossus is still unknown and remains a secret guarded by the British Official Secrets Act. Turing was never fully recognized for his important role in the development of this innovative machine, except by one of his Bletchley Park colleagues at his funeral who said, "I won't say what Turing did made us win the war, but I daresay we might have lost it without him."

Outline and Objectives

Introduction

Materials can be divided into three main types according to the way they react to current when a voltage is applied across them. **Insulators** (nonconductors), for example, are materials that have a very high resistance and therefore oppose current, whereas **conductors** are materials that have a very low resistance and therefore pass current easily. The third type of material is the **semiconductor** which, as its name suggests, has properties that lie between the insulator and the conductor. Semiconductor materials are not good conductors or insulators and so the next question is: What characteristic do they possess that makes them so useful in electronics? The answer is that they can be controlled to either increase their resistance and behave more like an insulator or decrease their resistance and behave more like a conductor. *It is this ability of a semiconductor material to vary its resistive properties that makes it so useful in electrical and electronic applications.*

In this chapter we will examine the characteristics of semiconductor materials so that we can better understand the operation and characteristics of semiconductor devices.

TIME LINE

In the early days of World War II, German scientist Konrad Zuse (1910-1995) who designed and built the first general-purpose computer, proposed constructing a computer that would operate 1000 times faster than anything else at that time. This proposal was rejected by Hitler, who was not interested in this long-term, two-year project, as he was sure that the war was going to be, for him, a certain, quick victory. Due to Hitler's shortsightedness, this powerful computer, which could have been used to break British communication codes, was never developed. However, unknown to both Hitler and Zuse, the British code-breaking computer project, called Ultra, had highest priority and was moving rapidly toward completion.

15-1   SEMICONDUCTOR DEVICES

Semiconductor materials such as *germanium* and *silicon* are used to construct semiconductor devices like the *diodes, transistors,* and *integrated circuits (ICs)* shown in Figure 15-1. These devices are used in electrical and electronic circuits to control current and voltage, so as to produce a desired result. For example, a diode could be used as the controlling element in a rectifier circuit that would convert ac to pulsating dc. A transistor, on the other hand, could be made to act like a variable resistance so it could amplify a radio signal. Conversely, an integrated circuit could be used to generate an oscillating signal or be made to perform arithmetic operations.

The most significant development in electronics since World War II has been a small semiconductor device called the transistor. It was first introduced in 1948 by its inventors William Schockley, Walter Bratten, and John Bardeen in the Bell Telephone Laboratories

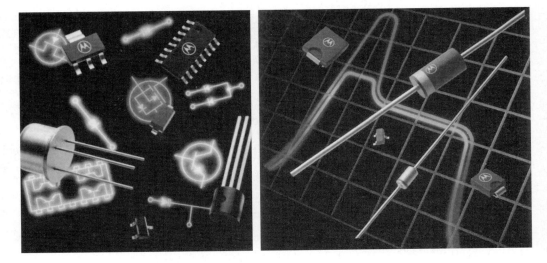

FIGURE 15-1   Semiconductor Devices. (Copyright of Motorola, Inc. Used by permission.)

and was described as a **solid state device.** This term was used because the transistor contained a solid semiconductor material between its input and output pins, unlike its predecessor the vacuum tube, which had a vacuum between its input and output pins.

The first *point-contact transistor* unveiled in 1948 was extremely unreliable, and it took its inventors another twelve years to develop the superior *bipolar junction transistor* (*BJT*) and make it available in commercial quantities.

In 1960, many electronic system manufacturers began to use the bipolar junction transistor instead of the vacuum tube in low-power and low-frequency applications. Research and development into semiconductor or solid state devices mushroomed and a variety of semiconductor devices began to appear. A different type of transistor emerged called the *field effect transistor* (*FET*), which had characteristics similar to those of the vacuum tube. Once it was discovered that semiconductor materials could also generate and sense light, a new line of optoelectronic devices became available. Later it was discovered that semiconductor materials could sense magnetism, temperature, and pressure and, as a result, a variety of sensor devices or transducers (energy converters) appeared on the market. Along with all these different types of semiconductor devices, a wide variety of semiconductor diodes emerged that could rectify, regulate, and oscillate at high frequencies. Even to this day it is clear that we have not yet seen all the potential value of semiconductors. Figure 15-2 illustrates many of these semiconductor or solid state devices.

Although semiconductor diodes and transistors are still widely used as individual or **discrete components,** in 1959 Robert Noyce discovered that more than one transistor could be constructed on a single piece of semiconductor material. Soon other components such as resistors, capacitors, and diodes were added with transistors and then interconnected to form a complete circuit on a single chip or piece of semiconductor material. This integrating of various components on a single chip of semiconductor was called an *integrated circuit* (*IC*) or *IC chip.* Today the IC is used extensively in every branch of electronics with hundreds of thousands of transistors and other components being placed on a chip of semiconduct no bigger than this ■. Figure 15-3 illustrates some of the different types of integrated circuits.

Like an evolving species, semiconductors have come to dominate the products of which they used to be only a part. For example, there used to be 400 components in a typical cellular telephone. Now there are 40, and soon only 3 or 4 IC chips will make up the entire phone circuitry. Today the semiconductor business—once regarded as a technical sideshow—occupies center stage and is key to the development of new products for all industries.

Solid State Device

Uses a solid semiconductor material, such as silicon, between the input and output whereas a vacuum tube has vacuum between input and output.

TIME LINE

In 1904 John A. Fleming, a British scientist, saw the value of an effect that was discovered by Thomas Edison but for which he saw no practical purpose. The "Edison effect" permitted Fleming to develop the "Fleming value," which passes current in only one direction. Its operation made it the first device able to convert alternating current into direct current and to detect radio waves.

Discrete Components

Separate active and passive devices that were manufactured before being used in a circuit.

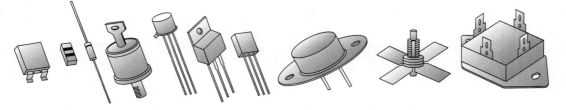

Diodes, Transistors, and Thyristor Devices

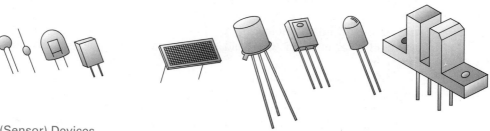

Transducer (Sensor) Devices

Optoelectronic Devices

FIGURE 15-2   Discrete Semiconductor (Solid State) Devices.

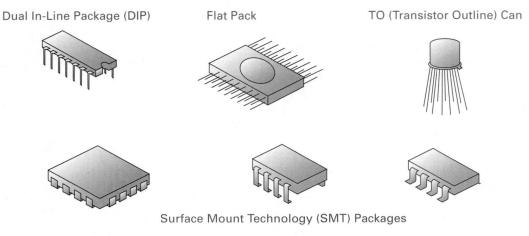

Dual In-Line Package (DIP)   Flat Pack   TO (Transistor Outline) Can

Surface Mount Technology (SMT) Packages

FIGURE 15-3   Semiconductor Integrated Circuits (ICs).

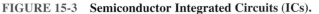

SELF-TEST EVALUATION POINT FOR SECTION 15-1

Now that you have completed this section, you should be able to:

■ **Objective 1.**   *State the relationship between the number of valence electrons in an atom and its conductivity.*

■ **Objective 2.**   *List the three semiconductor elements that are used to construct components for electrical and electronic circuit applications.*

■ **Objective 3.**   *Describe why semiconductor atoms will form a crystal lattice structure due to covalent bonding, and define the term "intrinsic."*

■ **Objective 4.**   *Explain an atom's energy gaps and energy levels along with the terms:*
   ***a.** Energy gap*
   ***b.** Conduction band*
   ***c.** Excited state*
   ***d.** Hole*

   ***e.** Electron-hole pair*
   ***f.** Recombination*
   ***g.** Lifetime*

■ **Objective 5.**   *Explain why semiconductor materials, and therefore semiconductor devices, have a negative temperature coefficient of resistance.*

■ **Objective 6.**   *Define the term "hole flow," and describe why the total current in a semiconductor is equal to the sum of the electron-flow and hole-flow currents.*

Use the following questions to test your understanding of Section 15-1.

1. Name the three most frequently used semiconductor devices in electrical and electronic equipment.

2. The main function of a semiconductor device is to control the _____ or _____ in an electrical or electronic circuit.

15-2   SEMICONDUCTOR MATERIALS

A semiconductor material is one that is neither a conductor nor a nonconductor (insulator). This means simply that it will not conduct current as well as a conductor or block current as well as an insulator. Some semiconductor materials are pure or natural elements such as carbon (C), germanium (Ge), and silicon (Si), while other semiconductor materials are compounds.

Silicon and germanium are used most frequently in the construction of semiconductor devices for electrical and electronic applications. Germanium is a brittle grayish-white element that may be recovered from the ash of certain types of coals. Silicon, the most popular semiconductor material due to its superior temperature stability, is a white element normally derived from sand. Let us now examine the silicon, germanium, and carbon semiconductor atoms in more detail.

15-2-1 Semiconductor Atoms

Figure 15-4 illustrates the silicon, germanium, and carbon atoms. The silicon atom has 14 protons in its nucleus and 14 electrons in three orbital paths distributed as 2, 8, and then 4 electrons in its valence shell. The germanium atom has 32 protons within its nucleus and 32 electrons in four orbital paths distributed as 2, 8, 18, and finally 4 electrons in the valence band or shell. The carbon atom has 6 protons in its nucleus and 6 orbiting electrons in two orbital paths distributed as 2 and 4 electrons in the valence shell. The question is: What do all these atoms have in common? The answer is that all semiconductor atoms have *four valence electrons.*

The valence shell of an atom can contain up to 8 electrons, and it is the number of electrons in this valence shell that determines the conductivity of the atom. For example, an atom with only 1 valence electron would be classed as a good conductor whereas an atom having 8 valence electrons, and therefore a complete valence shell, would be classed as an insulator.

To summarize, Figure 15-4 shows three semiconductor atoms, all of which contain four valence electrons. Since the number of valence electrons determines the conductivity of the element, semiconductor atoms are midway between conductors (which have 1 valence electron) and insulators (which have 8 valence electrons). Silicon and germanium are used to manufacture semiconductor devices, whereas carbon is combined with other elements to construct resistors.

TIME LINE

Jean Perrin discovered that cathode rays consisted of negatively charged particles, and these particles, which later became known as electrons, were measured by an English physicist, Joseph Thomson (1846-1914).

15-2-2 Crystals and Covalent Bonding

So far we have discussed only isolated atoms. When two or more similar semiconductor atoms are combined to form a solid element, they automatically arrange themselves into an orderly lattice-like structure or pattern known as a **crystal,** as shown in Figure 15-5(a). This pattern is formed because each atom shares its four valence electrons with its four neighboring atoms. Since each atom shares one electron with a neighboring atom, two atoms will share two, or a pair, of electrons between the two cores. These two atom cores are pulling the two electrons with equal but opposite force and it is this pulling action that holds the atoms together in this solid crystal-lattice structure. The joining together of two semiconductor atoms is called an **electron-pair bond** or **covalent bond.** When many atoms combine, or bond, in this way the result is a crystal (smooth, glassy, solid) lattice structure. To illustrate this bonding process, each atom in Figure 15-5(a) has been drawn as a square and

Crystal

A solid element with an orderly lattice-like structure.

Electron-Pair Bond or Covalent Bond

A pair of electrons shared by two neighboring atoms.

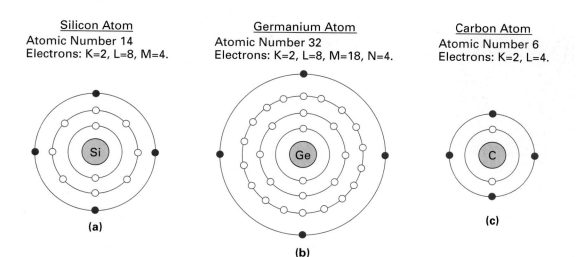

Silicon Atom
Atomic Number 14
Electrons: K=2, L=8, M=4.

Germanium Atom
Atomic Number 32
Electrons: K=2, L=8, M=18, N=4.

Carbon Atom
Atomic Number 6
Electrons: K=2, L=4.

(a)

(b)

(c)

FIGURE 15-4 **Semiconductor Atoms with Their Four Valence Shell Electrons.**

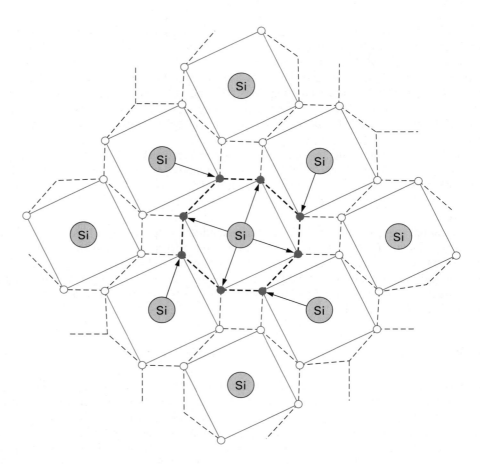

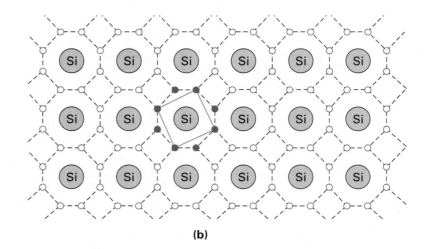

TIME LINE

Although the Fleming valve was an advance, it could not amplify or boost a signal. The "audion," developed by U.S. inventor Lee de Forest (1873-1961), sparked an era known as "vacuum-tube electronics" that brought about transcontinental telephony in 1915, radio broadcasting in 1920, radar in 1936, and television between 1927 and 1946, because of this triode vacuum tube's ability to amplify small signals.

(a)

TIME LINE

In 1947, J. Presper Eckert (right) and John Mauchly (left) unveiled ENIAC, which used over 300,000 vacuum tubes. ENIAC, which is an acronym for "electronic numerical integrator and computer," was the first large-scale electronic digital computer.

(b)

FIGURE 15-5   Covalent Bonding. (a) Silicon Atoms Sharing Valence Electrons. (b) Silicon Crystal Lattice Structure.

each valence shell has been drawn as an octagon (eight-sided figure) so that we can easily see which electrons belong to which atom. As you can see, the atom in the center of the diagram has 4 valence electrons (shown at the corners of the Si square), and shares one electron from each of its four neighbors.

Figure 15-5(b) shows a larger view of a silicon crystal structure. All of the atoms in this structure are electrically stable because all of their valence shells are complete (they all contain eight electrons). These completed valence shells cause the pure semiconductor crystal structure to act as an insulator since it will not easily give up or accept electrons. Pure semiconductor materials, which are often called **intrinsic** materials, are therefore very poor conductors. Once this pure material is available, it must then be modified by a *doping* process to give it the qualities necessary to construct semiconductor devices. Silicon is most frequently used to construct solid state or semiconductor devices such as diodes and transistors because germanium has poor temperature stability and carbon crystals (diamonds) are too expensive to use.

15-2-3 *Energy Gaps and Energy Levels*

Let us now examine the relationships between electrons and orbital shells in a little more detail so that we can better understand charge and conduction within a semiconductor material.

As mentioned previously, there are seven shells available for electrons (K, L, M, N, O, P, and Q) around the nucleus. Electrons must travel or orbit in one of these orbital paths because they cannot exist in any of the spaces between orbital shells. Each orbital shell has its own specific energy level. Therefore, electrons traveling in a specific orbital shell will contain the shell's energy level. Figure 15-6 shows an example of an atom's orbital shell energy levels. The energy levels for each shell increase as you move away from the nucleus of the atom. The valence shell and the valence electrons will always have the highest energy level for a given atom. The space between any two orbital shells is called the **energy gap.** Electrons can jump from one shell to another if they absorb enough energy to make up the difference between their initial energy level and the energy level of the shell that they are jumping to. For example, in Figure 15-6 the valence shell has an energy level of 1.0 **electron-volts (eV).** Because this atom has three orbital shells, the valence shell will be energy level 3 (e3). The second energy level or orbital shell (e2) has an energy level of 0.6 eV. Therefore, for an electron to jump from energy level 2 (shell 2) to energy level 3 (e3 or valence shell), it will have to absorb a value of energy equal to the difference between e2 and e3. This will equal:

$$1.0\ eV - 0.6\ eV = 0.4\ eV$$

In this example, when either heat, light, or electrical energy was applied, one of the electrons in shell 2 (e2) absorbed 0.4 electron-volts of energy and jumped to valence shell (e3).

If a valence (e3) electron absorbs enough energy it can jump from the valence shell into the **conduction band.** The conduction band is an energy band in which electrons can move freely or wander within a solid. When an electron jumps from the valence shell into the conduction band, it is released from the atom and no longer travels in one of its orbital paths. The electron is now free to move within the semiconductor material and is said to be in the **excited state.** An excited electron in the conduction band will eventually give up the energy it absorbed in the form of light or heat and return to its original energy level in the atom's valence shell.

When an electron jumps from the valence shell or band to the conduction band, it leaves a gap in the covalent bond called a **hole.** This action is shown in Figure 15-7(a). A hole is created every time an electron enters the conduction band. This action creates an **electron-hole pair.**

It only takes a few microseconds before a free electron in the conduction band will give up its energy and fall into one of the valence shell holes in the covalent bond. This action is called **recombination** and is shown in Figure 15-7(b). The time difference between

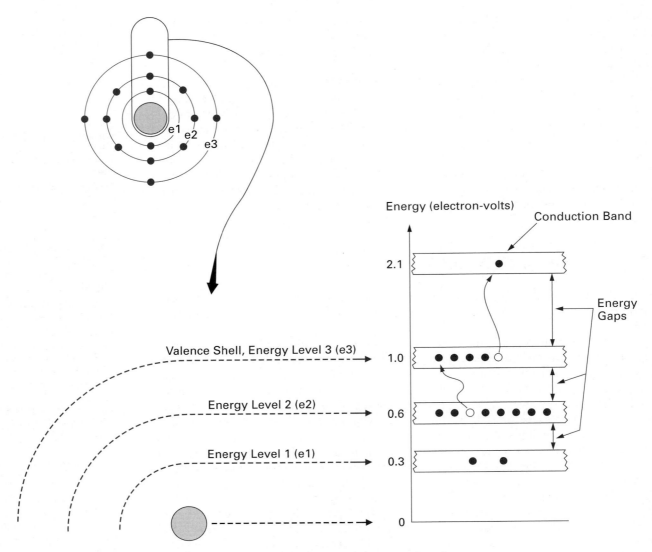

FIGURE 15-6   An Atom's Orbital Shell Energy Levels.

an electron jumping into the conduction band (becoming a free electron) and then falling back into a hole (recombination) is called the **lifetime** of the electron-hole pair.

15-2-4   *Temperature Effects on Semiconductor Materials*

At extremely low temperatures the valence electrons are tightly bound to their parent atoms, preventing valence electrons from drifting between atoms. Therefore, pure or intrinsic semiconductor materials function as insulators at temperatures close to absolute zero ($-273.16°C$ or $-459.69°F$).

At room temperature, however, the valence electrons absorb enough heat energy to break free of their covalent bonds creating electron-hole pairs, as shown in Figure 15-8. Therefore, *the conductivity of a semiconductor material is directly proportional to temperature, in that an increase in temperature will cause an increase in the semiconductor material's conductance.* This means: *an increase in temperature ($T\uparrow$) will cause an increase in a semiconductor's conductivity ($G\uparrow$) and current ($I\uparrow$).* This is why all circuits containing a semiconductor device tend to consume more current once they have warmed up.

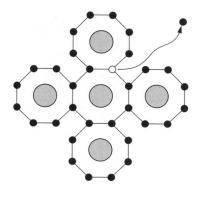

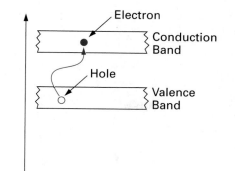

(a)

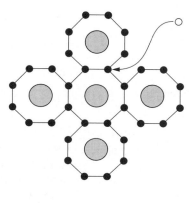

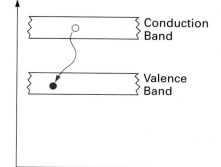

(b)

FIGURE 15-7   Valence Band and Conduction Band Actions. (a) Generating an Electron-Hole Pair. (b) Recombination.

Stated another way, *semiconductor materials, and therefore semiconductor devices, have a negative temperature coefficient of resistance, which means as temperature increases (T ↑), their resistance decreases (R ↓).*

15-2-5   *Applying a Voltage across a Semiconductor*

If a voltage was applied across a room-temperature section of intrinsic semiconductor material, **free electrons** in the conduction band would make up a small electrical current as shown in Figure 15-9. In this illustration you can see how the negatively charged free electrons are attracted to the positive terminal of the voltage source. For every free electron that leaves the semiconductor material on the right side and travels to the positive terminal of the source, another electron is generated at the negative terminal of the voltage source and is injected into the left side of the semiconductor material. These injected electrons are captured by holes in the semiconductor material (recombination). As you can see from this illustration, current in a semiconductor material is made up of both electrons and holes. The holes act like positively charged particles while the electrons act like negatively charged particles. As electrons jump between atoms in a migration to the positive terminal of the source

Free Electron

An electron that is able to move freely when an external force is applied.

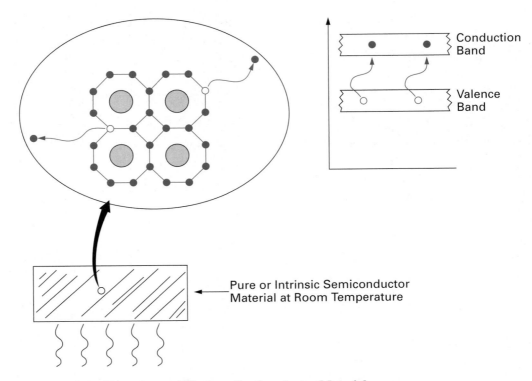

FIGURE 15-8   Temperature Effects on Semiconductor Materials.

voltage, they leave behind them holes, which are then filled by other advancing electrons. These advancing electrons leave behind them other holes, making it appear as though these holes are traveling toward the negative terminal of the source voltage. This **hole flow** is a new phenomenon to us, and it is one of the key differences between a semiconductor and a conductor. With conductors we were only interested in free-electron flow, but with semiconductors we must consider the movement of free electrons (negative charge carriers) and the apparent movement of holes (positive charge carriers).

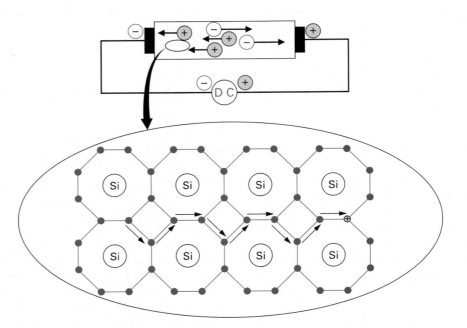

FIGURE 15-9   Electron Flow and Hole Flow in an Intrinsic Semiconductor.

In summary, therefore, *when a potential difference is applied across a semiconductor, the electrons move toward the positive potential and the holes travel toward the negative potential. The total current flow is equal to the sum of the electron flow and the hole flow currents.*

Now that you have completed this section, you should be able to:

■ **Objective 7.** *Explain why pure semiconductor materials are doped and why they remain electrically neutral.*

■ **Objective 8.** *Explain the similarities and differences between the* n-*type and* p-*type semiconductor materials, and define the terms:*
   a. Extrinsic semiconductor
   b. Majority carriers
   c. Minority carriers

Use the following questions to test your understanding of Section 15-2.

1. What do all semiconductor atoms have in common?
2. The electrical conductivity of an element is determined by the number of electrons in the valence shell. Semiconductor atoms are midway between conductors, which have _____ valence electron(s), and insulators, which have _____ valence electron(s).
3. Each of the semiconductor atoms in a crystal lattice shares its electrons with four neighboring atoms. This joining of atoms is called a _____.
4. Semiconductor materials have a _____ temperature coefficient of resistance, which means as temperature increases, resistance _____.
5. The number of electron-hole pairs within a semiconductor will increase as temperature _____.
6. When a pure or _____ semiconductor is connected across a voltage, free electrons travel toward the _____ terminal of the applied voltage, whereas holes appear to travel toward the _____ terminal of the applied voltage.

15-3 DOPING SEMICONDUCTOR MATERIALS

At room temperature pure or intrinsic semiconductors will not permit a large enough value of current. Therefore, some modification has to be applied in order to increase the semiconductor's current-carrying capability or conductivity. **Doping** is a process wherein impurities are added to the intrinsic semiconductor material either to increase the number of free electrons (negative doping) or to increase the number of holes (positive doping).

Basically, there are two types of impurities that can be added to semiconductor crystals. One type of impurity is called a *pentavalent material* because its atom has five (*penta*) valence electrons. The second type of impurity is called a *trivalent material* because its atoms have three (*tri*) valence electrons. A doped semiconductor material is referred to as an **extrinsic semiconductor** material because it is no longer pure.

15-3-1   n-*Type Semiconductor*

Figure 15-10(a) shows how a semiconductor material's atoms will appear after pentavalent atom impurities have been added. The pentavalent atoms, which are listed in Figure 15-10(b), can be added to molten silicon to create, when cooled, a crystalline structure that has an extra electron due to the pentavalent (5 valence-electron impurity) atoms. The fifth pentavalent electron is not part of the covalent bonding and requires little energy to break free and enter the conduction band, as shown in Figure 15-10(c). Because millions of pentavalent atoms are added to the pure semiconductor, there will be millions of free electrons available for flow through the material.

Doping

The process wherein impurities are added to the intrinsic semiconductor material either to increase the number of free electrons or to increase the number of holes.

Extrinsic Semiconductor

A semiconductor whose electrical properties are dependent on impurities added to the semiconductor crystal.

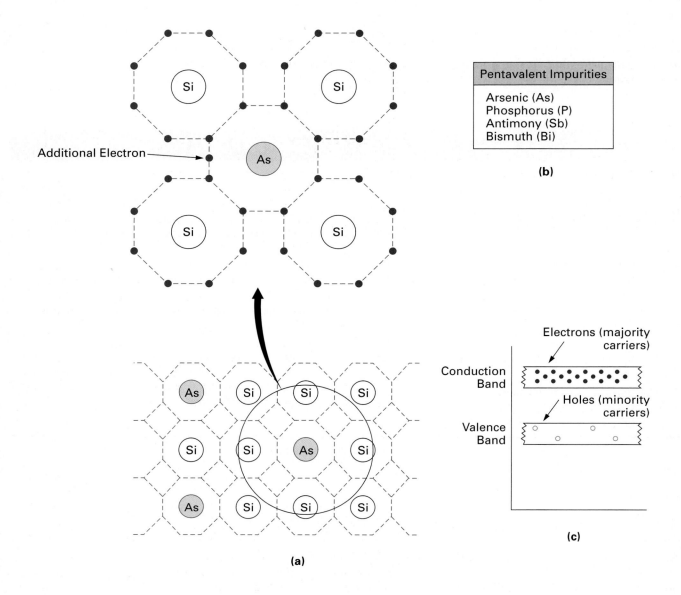

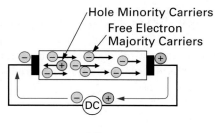

FIGURE 15-10   **Adding Pentavalent Impurities to Create an *n*-Type Semiconductor Material.**

Even though the doped semiconductor material has millions of free electrons, the material is still electrically neutral. This is because each arsenic atom has the same number of protons as electrons and so do the silicon atoms. Therefore, the overall number of protons and electrons in the semiconductor is still equal and the result is a net charge of zero. However, because we now have more electrons than valence-band holes, the material is called an **n-type semiconductor.** n-Type semiconductors have more conduction-band electrons than valence-band holes. The electrons are therefore called the **majority carriers** and the valence-band holes are called the **minority carriers.** In Figure 15-10(c) you can see the abundance of conduction-band electrons. The holes in the valence band are few and are generated by thermal energy because the semiconductor is at room temperature.

When a voltage is applied across an n-type semiconductor, as shown in Figure 15-10(d), the additional free conduction-band electrons travel toward the positive terminal of the dc source. The applied voltage will cause extra electrons to break away from their covalent bonds to create holes, resulting in an increase in current and conductivity. Although the total current flow in this n-type semiconductor is the sum of the electron and hole currents, the conduction-band electrons make up the majority of the flow.

15-3-2   p-*Type Semiconductor*

Figure 15-11(a) shows how a semiconductor material's atoms will appear after trivalent atom impurities have been added. The trivalent atoms, which are listed in Figure 15-11(b), can be added to molten silicon to create, when cooled, a crystalline structure that has a hole in the valence band of every trivalent (3 valence-electron impurity) atom. Instead of an excess of electrons, we now have an excess of holes. Because millions of trivalent atoms are added to the pure semiconductor, there will be millions of holes available for flow through the material.

Even though the doped semiconductor material has millions of holes, the material is still electrically neutral. This is because each aluminum atom has the same number of protons as electrons and so do the silicon atoms. Therefore the overall number of protons and electrons in the semiconductor is still equal and the result is a net charge of zero. However, because we now have more valence band holes than electrons the material is called a **p-type semiconductor.** p-Type semiconductors have more valence-band holes than conduction-band electrons. The holes are called the *majority carriers* and the electrons are called the *minority carriers.* In Figure 15-11(c) you can see the abundance of valence-band holes. The few electrons in the conduction band are generated by thermal energy because the semiconductor is at room temperature.

When a voltage is applied across a p-type semiconductor, as illustrated in Figure 15-11(d), the large number of holes within the material will attract electrons from the negative terminal of the dc source into the p-type semiconductor. These holes appear to move because each time an electron moves into a hole it creates a hole behind it, and the holes appear to move in the opposite direction to the electrons (toward the negative terminal of the dc source). The applied voltage will cause some electrons to break away from the covalent bond resulting in an increased current and conductivity. Although the total current flow in this p-type semiconductor is the sum of the hole and electron currents, the valence-band holes make up the majority of the flow.

n-Type Semiconductor

A material that has more conduction-band electrons than valence-band holes.

Majority Carriers

The type of carrier that makes up more than half the total number of carriers in a semiconductor device.

Minority Carriers

The type of carrier that makes up less than half of the total number of carriers in a semiconductor device.

p-Type Semiconductor

A material that has more valence-band holes than conduction-band electrons.

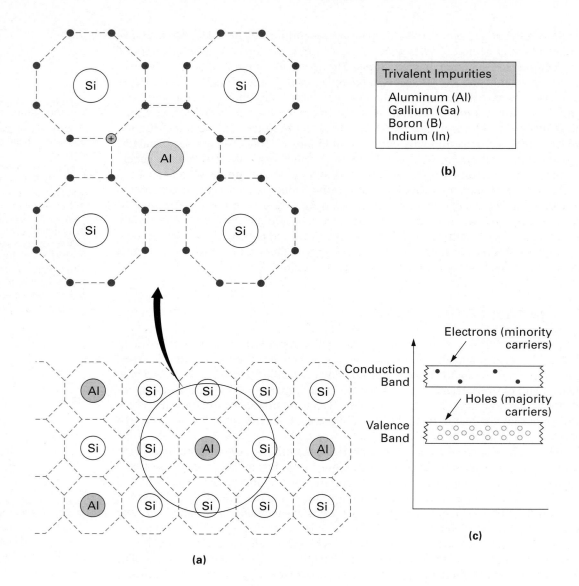

(b)

Trivalent Impurities

Aluminum (Al)
Gallium (Ga)
Boron (B)
Indium (In)

(a)

(c)

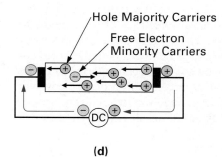

(d)

FIGURE 15-11 **Adding Trivalent Impurities to Create a *p*-Type Semiconductor Material.**

Now that you have completed this section, you should be able to:

■ **Objective 9.** *Describe the following in relation to the P-N junction:*
a. *The junction*
b. *The depletion region*
c. *The barrier voltage*

■ **Objective 10.** *Explain how P-N junctions can be:*
a. *Forward biased* b. *Reverse biased*

■ **Objective 11.** *Calculate and define the terms:*
a. *Forward voltage (V_F) drop*
b. *Forward current (I_F)*
c. *Reverse voltage (V_R) drop*

■ **Objective 12.** *Define the following terms:*
a. *Diffusion current*
b. *Leakage or reverse current*

Use the following questions to test your understanding of Section 15-3.

1. Why are impurities added to pure semiconductor materials?
2. Pentavalent atoms add _____ to semiconductor crystals, to create _____-type semiconductors.
3. Trivalent atoms add _____ to semiconductor crystals, to create _____-type semiconductors.
4. In an n-type semiconductor the majority carriers are _____, whereas in a *p*-type semiconductor the majority carriers are _____.

15-4 THE P-N JUNCTION

On their own, *n*-type semiconductor materials and *p*-type semiconductor materials are of little use. Together, however, these two form a **P-N semiconductor junction.** Semiconductor devices such as diodes and transistors are constructed using these P-N junctions, which give specific current flow characteristics. In this section we will examine the characteristics of the P-N junction in detail.

15-4-1 *The Depletion Region*

Figure 15-12(a) shows the individual *n*-type and *p*-type materials. The *n*-type material is represented as a block containing an excess of electrons (solid circles), while the *p*-type material is represented as a block containing an excess of holes (open circles). The energy diagrams below the two semiconductor sections show the differences between the two materials. Because different impurity atoms were added to the pure semiconductor material, the atomic make-up of the *n*-type and *p*-type materials is slightly different, which is why the valence bands and conduction bands are at slightly different energy levels.

Figure 15-12(b) shows the two *n*-type and *p*-type semiconductor sections joined together. A manufacturer of semiconductor devices would not join two individual pieces in this way to create a P-N junction. Instead, a single piece of pure semiconductor material would have each of its halves doped to create a *p*-type and *n*-type section.

The point at which the two oppositely doped materials come in contact with one another is called the *junction*. This junction of the two materials now permits the free electrons in the *n*-type material to combine with the holes in the *p*-type material as shown in Figure 15-12(c). As free electrons in the *n* material cross the junction and combine with holes in the *p* material, they create negative ions (atoms with more electrons than protons) in the *p* material, and leave behind positive ions (atoms with less electrons than protons) in the *n* material, as shown in Figure 15-12(d). An area or region on either side of the junction becomes emptied or depleted of free electrons and holes. This small layer containing positive and negative ions is called the **depletion region.**

As the ion layer on either side of the junction builds up, it has the effect of diminishing and eventually preventing any further recombination of free electrons and holes across the

P-N Junction
The point at which two opposite doped materials come in contact with one another.

Depletion Region
A small layer on either side of the junction that becomes empty, or depleted, of free electrons or holes.

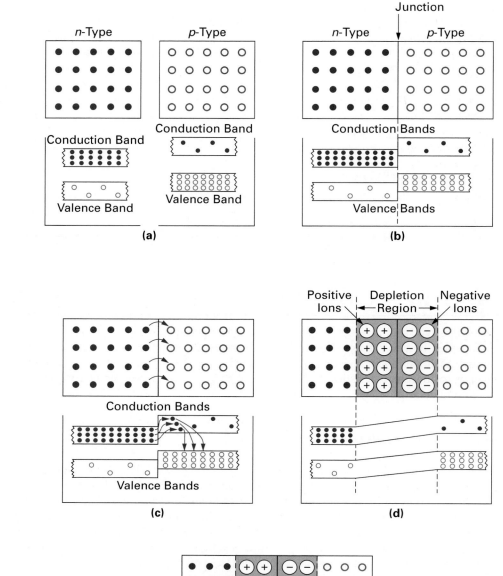

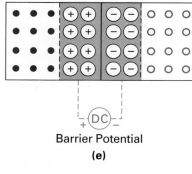

FIGURE 15-12   The Depletion Region.

junction. In other words, the negative ions in the *p* region near the junction repel and prevent free electrons in the *n* region from recombination. This action prevents the depletion region from becoming larger and larger.

These positive ions or charges and negative ions or charges accumulate a certain potential. Since these charges are opposite in polarity, a potential difference or voltage called the **barrier potential** or **barrier voltage** exists across the junction as shown in Figure 15-12(e). At room temperature, the barrier voltage of a silicon P-N junction is approximately 0.7 V, and a germanium P-N junction is approximately 0.3 V.

Barrier Potential or Barrier Voltage

The potential difference, or voltage, that exists across the junction.

15-4-2 *Biasing a P-N Junction*

Semiconductor devices are constructed using P-N junctions. These P-N junctions need voltages of a certain amplitude and polarity to control their operation. These voltages, which incline or cause the device to operate in a certain manner, are known as **bias voltages.** Bias voltages control the width of the depletion region, which in turn controls the resistance of the P-N junction and, therefore, the amount of current that can pass through the P-N junction or semiconductor device.

Bias Voltages

The dc voltages applied to control a device's operation.

To be specific, a small depletion region ($dr\downarrow$) will offer a small P-N junction resistance ($R\downarrow$) and therefore permit a large P-N junction current ($I\uparrow$). In this instance, the P-N semiconductor junction is said to be **forward biased** and acts like a conductor.

On the other hand, a large depletion region ($dr\uparrow$) will offer a large P-N junction resistance ($R\uparrow$) and therefore permit only a small P-N junction current ($I\downarrow$). In this instance the P-N semiconductor junction is said to be **reverse biased** and acts like an insulator.

Forward Biased

A small depletion region at the junction will offer a small resistance and permit a large current. Such a junction is forward biased.

Reverse Biased

A large depletion region at the junction will offer a large resistance and permit only a small current. Such a junction is reverse biased.

Forward Biasing a P-N Junction

Figure 15-13 shows in detail why a forward biased P-N junction will pass current with almost no opposition (act like a conductor). To begin, Figure 15-13(a) shows a P-N junction with wires attached. A resistor has been included to limit the amount of current passing through the P-N junction to a safe level. Energy diagrams have also been included on the right side of each part of Figure 15-13 to show the relationship between the conduction band and valence band of the *p* and *n* regions.

Let us now connect a dc voltage across the P-N junction to see how it reacts. This is shown in Figure 15-13(b). The negative potential of the dc source has been applied to the *n* region and the positive potential of the dc source has been applied to the *p* region. Referring to the energy diagram in Figure 15-13(b), you can see that the conduction band electrons in the *n* region are repelled by the negative voltage source towards the junction. On the opposite side, valence band holes in the *p* region are repelled by the positive voltage source towards the junction. A forward-conducting current will begin to flow if the external source voltage is large enough to overcome the internal barrier voltage of the P-N junction. In this example we will assume that the dc source voltage is 10 volts and therefore this will be more than enough to overcome the silicon P-N junction's barrier potential of 0.7 volts.

Conduction through the P-N junction is shown in Figure 15-13(c). When forward biased, a P-N junction will act as a conductor and have a low but finite resistance value that will cause a corresponding voltage drop across its terminals. This **forward voltage drop** (V_F) is approximately equal to the P-N junction's barrier voltage:

Forward Voltage Drop (V_F)

The forward voltage drop is equal to the junction's barrier voltage.

$$\text{Forward Voltage Drop } (V_F) \text{ for Silicon} = 0.7 \text{ V}$$
$$\text{Forward Voltage Drop } (V_F) \text{ for Germanium} = 0.3 \text{ V}$$

Figure 15-13(c) shows how a voltmeter can be used to measure the forward voltage drop of 0.7 V across a silicon P-N junction when it is forward biased.

To summarize, Figure 15-13(d) shows that when a P-N junction is forward biased ($+V \rightarrow p$ region, $-V \rightarrow n$ region) the P-N junction resistance is low ($R\downarrow$), and therefore the circuit current is high ($I\uparrow$). When forward biased, therefore, the P-N junction acts like a conductor and is equivalent to a closed switch.

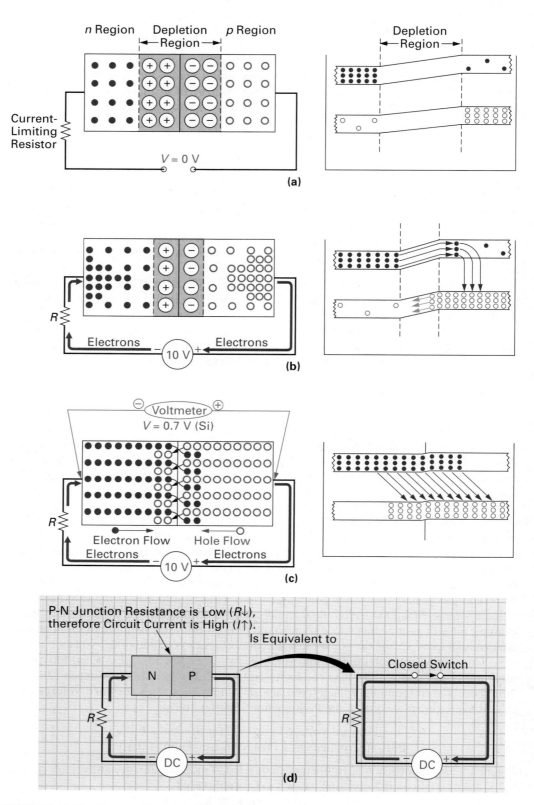

FIGURE 15-13   Forward Biasing a P-N Semiconductor Junction.

Calculate the current for the circuit in Figure 15-14.

FIGURE 15-14   **A P-N Junction Circuit.**

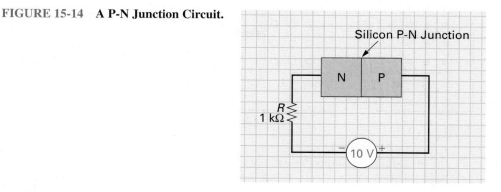

Solution:

The silicon P-N junction is forward biased ($+V \rightarrow p$ region, $-V \rightarrow n$ region). The applied voltage of 10 V will be more than enough to overcome the silicon P-N junction forward voltage drop of 0.7 volts ($V_F = 0.7$ V for silicon). Since 10 volts are applied, and the P-N junction is dropping 0.7 V, the remaining voltage of 9.3 V is being dropped across the 1 kΩ resistor. Consequently, the forward-biased current (I_F) will equal:

$$I_F = \frac{V_S - V_{P\text{-}N}}{R}$$

$$= \frac{10 \text{ V} - 0.7 \text{ V}}{1 \text{ k}\Omega}$$

$$= 9.3 \text{ mA}$$

Reverse Biasing a P-N Junction

Figure 15-15 shows in detail why a reverse biased P-N junction will reduce current to almost zero (act like an insulator). To begin, Figure 15-15(a) shows a P-N junction with wires attached and no voltage being applied. Energy diagrams have again been included to show the relationship between the conduction band and valence band of the *p* and *n* regions.

Let us now connect a dc voltage across the P-N junction to see how it reacts. This is shown in Figure 15-15(b). The positive potential of the dc source is now being applied to the *n* region, and the negative potential of the dc source has been applied to the *p* region.

A forward biased P-N junction is able to conduct current because the external bias voltage forces the majority carriers in the *n* and *p* regions to combine at the junction. In this instance, however, the dc bias voltage polarity has been reversed, causing free electrons in the *n* region to travel to the positive terminal of the voltage source leaving behind a large number of positive ions at the junction. This increases the width of the depletion region. At the same time, electrons from the negative terminal of the source are attracted to the holes in the *p* region of the P-N junction. These electrons fill the holes in the *p* region near the junction creating a large number of negative ions. This further increases the width of the depletion region. The current that is present at the time the depletion layer is expanding is called the **diffusion current.** Referring to Figure 15-15(b), you can see that the depletion region is now wider than the unbiased P-N junction shown in Figure 15-15(a).

The ions on either side of the junction build up until the P-N junction's internal-barrier voltage is equal to the external-source voltage, as shown by the voltmeter in Figure 15-15(c). When reverse biased, therefore, the **reverse voltage drop (V_R)** across a P-N junction is equal to the source or applied voltage. At this time the resistance of the junction has been increased to a point that current drops to zero.

Diffusion Current

The current that is present when the depletion layer is expanding.

Reverse Voltage Drop (V_R)

The reverse voltage drop is equal to the source voltage (applied voltage).

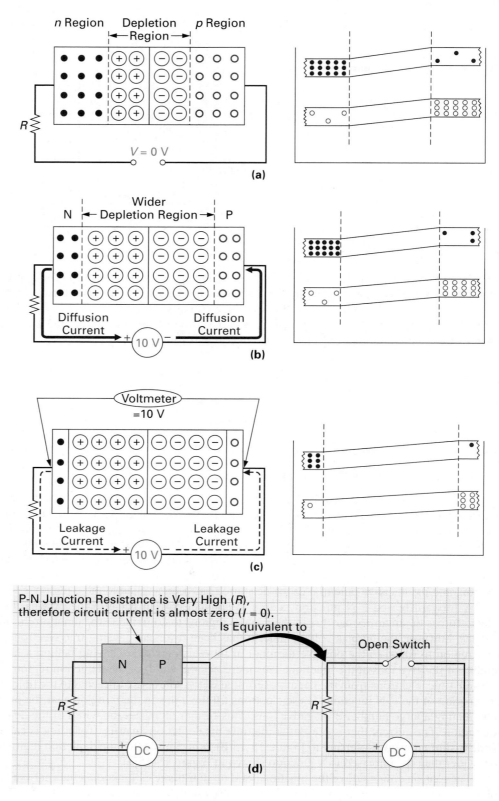

FIGURE 15-15 **Reverse Biasing a P-N Semiconductor Junction.**

Actually, an extremely small current called the **leakage current** or **reverse current** (I_R) will pass through the P-N junction, as shown in Figure 15-15(c). It is present because the minority carriers (holes in the *n* region, electrons in the *p* region) are forced toward the junction where they combine, producing a constant small current. The current in the P-N junction is still considered to be at zero because the leakage or reverse current is so small (nanoamps in silicon diodes).

Leakage Current or Reverse Current (I_R)

The extremely small current present at the junction.

To summarize, Figure 15-15(d) shows that when a P-N junction is reverse biased ($+V \rightarrow n$ region, $-V \rightarrow p$ region), the P-N junction resistance is extremely high ($R \uparrow \uparrow$), and the circuit current is effectively zero ($I = 0$ amps). When reverse biased, therefore, the P-N junction acts like an insulator and is equivalent to an open switch.

▨ EXAMPLE:

Referring to Figure 15-16(a) and (b), calculate each circuit's:

a. current value
b. P-N junction voltage drop

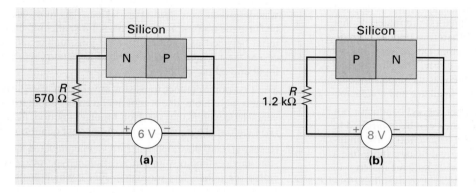

FIGURE 15-16 P-N Junction Circuit Examples.

▨ *Solution:*

The P-N junction in Figure 15-16(a) is reverse biased ($+V \rightarrow n$ region, $-V \rightarrow p$ region); therefore, the P-N junction resistance is extremely high ($R \uparrow \uparrow$), and the circuit current is effectively zero ($I = 0$). When reverse biased, the P-N junction acts like an insulator and is equivalent to an open switch, and the voltage developed across the open P-N junction will equal the source voltage applied. For Figure 15-16(a):

$$\text{Circuit Current} = 0$$
$$\text{P-N Junction Voltage Drop} = V_S = 6 \text{ V}$$

The P-N junction in Figure 15-16(b) is forward biased ($+V \rightarrow p$ region, $-V \rightarrow n$ region). Therefore, the P-N junction resistance is low ($R\downarrow$), and the circuit current is high ($I \uparrow$). When forward biased, the P-N junction acts like a conductor and is equivalent to a closed switch, and the P-N junction's voltage drop will equal the forward voltage drop (V_F) for a silicon P-N junction. For Figure 15-16(b):

$$\text{Circuit Current} = I_F = \frac{V_S - V_{P\text{-}N}}{R}$$

$$= \frac{8 \text{ V} - 0.7 \text{ V}}{1.2 \text{ k}\Omega}$$

$$= 6.08 \text{ mA}$$

$$\text{P-N Junction Voltage Drop} = V_F = 0.7 \text{ V}$$

Use the following questions to test your understanding of Section 15-4.

1. When a P-N junction is formed, a _____ region is created on either side of the junction.

2. The barrier voltage within a silicon diode is:

   a. 700 mV   **b.** 7.0 V   **c.** 0.3 V   **d.** None of the above

3. True or false: A P-N junction is forward biased when its P terminal is made positive relative to its N terminal.

4. A reverse biased P-N junction acts like a/an _____ switch, whereas a forward biased P-N junction acts like a/an _____ switch.

SUMMARY

Semiconductor Principles (Figure 15-17 and Figure 15-18)

1. Materials can be divided into three main types according to the way they react to current when a voltage is applied across them. Insulators (nonconductors) are materials that have a very high resistance and therefore oppose current, whereas conductors are materials that have a very low resistance and therefore pass current easily. The third type of material is the semiconductor which, as its name suggests, has properties that lie between the insulator and conductor.

2. Semiconductor materials are neither a good conductor nor insulator. Their advantage is that they can be controlled to either increase their resistance and behave more like an insulator, or decrease their resistance and behave more like a conductor. It is this ability of a semiconductor material to vary its resistive properties that makes it so useful in electrical and electronic applications.

3. Semiconductor materials such as germanium and silicon are used to construct semiconductor devices such as diodes, transistors, and integrated circuits (ICs). These de-

vices are used in electrical and electronic circuits to control current and voltage, so as to produce a desired result.

4. The transistor was first introduced in 1948 by its inventors William Schockley, Walter Brattain, and John Bardeen and was described as a solid state device. This term was used because the transistor contained a solid semiconductor material between its input and output pins, unlike its predecessor the vacuum tube, which had a vacuum between its input and output pins.

5. The first point-contact transistor unveiled in 1948 was extremely unreliable, and it took its inventors another twelve years to develop the superior bipolar junction transistor (BJT) and make it available in commercial quantities.

6. Research and development into semiconductor or solid state devices has resulted in a different type of transistor called the field effect transistor (FET), which has characteristics similar to those of the vacuum tube. Once it was discovered that semiconductor materials could also generate and sense light, a new line of optoelectronic devices became available. Later it was discovered that semiconductor

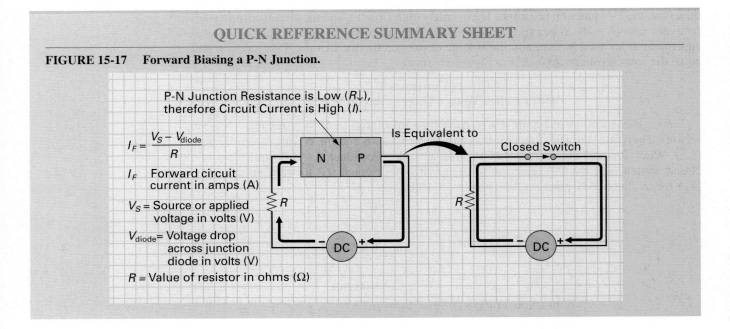

QUICK REFERENCE SUMMARY SHEET

FIGURE 15-17   Forward Biasing a P-N Junction.

FIGURE 15-18   Reverse Biasing a P-N Junction.

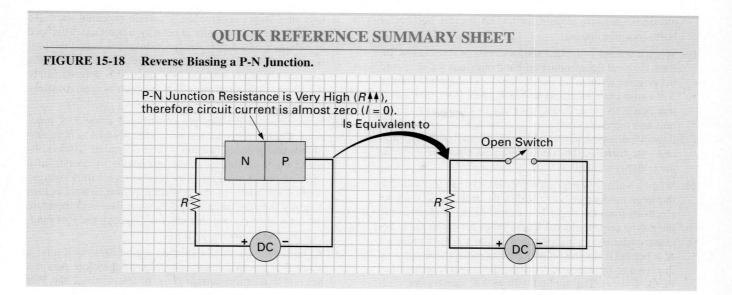

materials could sense magnetism, temperature, and pressure and, as a result, a variety of sensor devices or transducers (energy converters) appeared on the market. Along with all these different types of semiconductor devices, a wide variety of semiconductor diodes emerged that could rectify, regulate, and oscillate at high frequencies.

7. Although semiconductor diodes and transistors are still widely used as individual or discrete components, in 1959 Robert Noyce discovered that more than one transistor could be constructed on a single piece of semiconductor material. Soon other components such as resistors, capacitors, and diodes were added with transistors and then interconnected to form a complete circuit on a single chip or piece of semiconductor material. This integrating of various components on a single chip of semiconductor was called an integrated circuit (IC) or IC chip. The IC is used extensively in every branch of electronics with hundreds of thousands of transistors and other components being placed on a chip of semiconductor.

8. Some semiconductor materials are pure or natural elements such as carbon (C), germanium (Ge), and silicon (Si), while other semiconductor materials are compounds.

9. All semiconductor atoms have four valence electrons. The valence shell of an atom can contain up to eight electrons, and it is the number of electrons in this valence shell that determines the conductivity of the atom. An atom with only one valence electron would be classed as a good conductor whereas an atom having eight valence electrons, and therefore a complete valence shell, would be classed as an insulator.

10. When two or more similar semiconductor atoms are combined to form a solid element, they automatically arrange themselves into an orderly lattice-like structure or pattern known as a crystal. Since each atom shares one electron with a neighboring atom, two atoms will share two, or a pair, of electrons between the two cores. These two atom cores are pulling the two electrons with equal but opposite force and it is this pulling action that holds the atoms to-

gether in this solid crystal-lattice structure. The joining together of two semiconductor atoms is called an electron-pair bond or covalent bond.

11. All of the atoms in pure semiconductors are electrically stable because all of their valence shells are complete (they all contain eight electrons). Pure semiconductor materials, called intrinsic materials, are therefore very poor conductors.

12. A pure semiconductor must be modified by a doping process to give it the qualities necessary to construct semiconductor devices.

13. Silicon is most frequently used to construct solid state or semiconductor devices such as diodes and transistors because germanium has poor temperature stability and carbon crystals (diamonds) are too expensive to use.

14. If a valence electron absorbs enough energy it can jump from the valence shell into the conduction band.

15. When an electron jumps from the valence shell, or band, to the conduction band, it leaves a gap in the covalent bond called a hole.

16. The conductivity of a semiconductor material is directly proportional to temperature, in that an increase in temperature will cause an increase in the semiconductor material's conductance. Stated another way, semiconductor materials, and therefore devices, have a negative temperature coefficient of resistance that means as temperature increases ($T \uparrow$), their resistance decreases ($R \downarrow$).

17. When a potential difference is applied across a semiconductor, the electrons move toward the positive potential and the holes travel toward the negative potential, and the total current flow is equal to the sum of the electron flow and the hole flow currents. The holes act like positively charged particles while the electrons act like negatively charged particles.

18. At room temperature, pure or intrinsic semiconductors will not permit a large enough value of current. Doping is a process whereby impurities are added to the intrinsic semiconductor material either to increase the number of free

electrons (negative doping) or to increase the number of holes (positive doping).

19. When a voltage is applied across an n-type semiconductor, the additional free conduction-band electrons travel toward the positive terminal of the dc source. The applied voltage will cause extra electrons to break away from their covalent bonds to create holes, resulting in an increase in current and therefore conductivity. Although the total current flow in this n-type semiconductor is the sum of the electron and hole currents, the conduction-band electrons make up the majority of the flow.

20. When a voltage is applied across a p-type semiconductor, the large number of holes within the material will attract electrons from the negative terminal of the dc source into the p-type semiconductor. These holes appear to move because each time an electron moves into a hole, it creates a hole behind it and the holes appear to move in the opposite direction to the electrons (toward the negative terminal of the dc source). The applied voltage will cause some electrons to break away from the covalent bond, resulting in an increased current and, therefore, conductivity. Although the total current flow in this p-type semiconductor is the sum of the hole and electron currents, the valence-band holes make up the majority of the flow.

21. On their own, the n-type semiconductor material and p-type semiconductor material are of little use. Together, however, these two form a P-N semiconductor junction. Semiconductor devices such as diodes and transistors are constructed using these P-N junctions, which give specific current flow characteristics.

22. The point at which the two oppositely doped materials come in contact with one another is called the junction.

23. An area or region on either side of the junction becomes emptied or depleted of free electrons and holes. This small layer containing positive and negative ions is called the depletion region. These positive ions or charges and negative ions or charges accumulate a certain potential. Because these charges are opposite in polarity, a potential difference or voltage exists across the junction and is called the barrier potential or barrier voltage. At room temperature, the barrier voltage of a silicon P-N junction is approximately 0.7 V, and of a germanium P-N junction is approximately 0.3 V.

24. Semiconductor devices are constructed using P-N junctions. These P-N junctions need voltages of a certain amplitude and polarity to control their operation. These voltages, which incline or cause the device to operate in a certain manner, are known as bias voltages. Bias voltages control the width of the depletion region, which in turn controls the resistance of the P-N junction and therefore the amount of current that can pass through the P-N junction or semiconductor device.

25. To be specific, a small depletion region ($dr\downarrow$) will offer a small P-N junction resistance ($R\downarrow$) and therefore permit a large P-N junction current ($I\uparrow$). In this instance the P-N semiconductor junction is said to be forward biased and acts like a conductor.

26. On the other hand, a large depletion region ($dr\uparrow$) will offer a large P-N junction resistance ($R\uparrow$) and therefore only permit a small P-N junction current ($I\downarrow$). In this instance the P-N semiconductor junction is said to be reverse biased and acts like an insulator.

27. When a P-N junction is forward biased ($+V \rightarrow p$ region, $-V \rightarrow n$ region), the P-N junction resistance is low ($R\downarrow$); and so the circuit current is high ($I\uparrow$). When forward biased, the P-N junction acts like a conductor and is equivalent to a closed switch.

28. When a P-N junction is reverse biased ($+V \rightarrow n$ region, $-V \rightarrow p$ region), the P-N junction resistance is extremely high ($R\uparrow\uparrow$), and so the circuit current is effectively zero ($I = 0$ A). When reverse biased, the P-N junction acts like an insulator and is equivalent to an open switch.

REVIEW QUESTIONS

Multiple-Choice Questions

1. What is the atomic number of silicon?
   a. 14 **b.** 16 **c.** 10 **d.** 32

2. How many valence electrons are normally present in the valence shell of a semiconductor material?
   a. 2 **b.** 4 **c.** 6 **d.** 8

3. Adding trivalent impurities to an intrinsic semiconductor will produce a/an _____ material.
   a. Extrinsic
   b. n-type
   c. p-type
   d. Both (a) and (c) are true

4. What is the majority carrier in an n-type material?
   a. Holes **b.** Electrons **c.** Neutrons **d.** Protons

5. Adding pentavalent impurities to an intrinsic semiconductor will produce a/an _____ material.
   a. Extrinsic
   b. n-type
   c. p-type
   d. Both (a) and (b) are true

6. What are the majority carriers in a p-type semiconductor?
   a. Holes **b.** Electrons **c.** Neutrons **d.** Protons

7. A semiconductor material has a _____ temperature coefficient of resistance, which means that as temperature increases its resistance _____.
   a. Positive, increases **c.** Negative, increases
   b. Positive, decreases **d.** Negative, decreases

8. A hole is considered to be _____.
   a. Negative   c. Neutral
   b. Positive   d. Both (b) and (c) are true

9. Intrinsic semiconductors are doped to increase their _____.
   a. Resistance   c. Inductance
   b. Conductance d. Reactance

10. As temperature increases, a semiconductor acts more like a/an _____.
   a. Conductor   b. Insulator

11. A negative ion has more:
   a. Protons than electrons   c. Neutrons than protons
   b. Electrons than protons   d. Neutrons than electrons

12. A positive ion has:
   a. Lost some of its electrons   c. Lost neutrons
   b. Gained extra protons   d. Gained more electrons

13. The resistance of a semiconductor material is more than the resistance of:
   a. Glass   c. Ceramic
   b. Copper   d. Both (a) and (c) are true

14. The basic function of a semiconductor device in an electrical or electronic circuit is to:
   a. Control current
   b. Control voltage
   c. Increase the price of the equipment
   d. Both (a) and (b) are true

15. For a silicon P-N junction, $V_F = ?$
   a. The value of the applied voltage   c. 0.7 V
   b. 300 mV   d. 10 V

Communication Skill Questions

16. What two semiconductor materials are most frequently used in the manufacture of semiconductor devices? (15-2)

17. What do all semiconductor atoms have in common? (15-2-1)

18. Define the following terms:
   a. covalent bond (15-2-2)
   b. intrinsic semiconductor (15-2-2)

   c. conduction band (15-2-3)
   d. hole flow (15-2-5)
   e. doping (15-3)
   f. P-N semiconductor junction (15-4)
   g. depletion region (15-4-1)
   h. majority carriers (15-3-1)
   i. minority carriers (15-3-1)

19. Why do semiconductor materials, and therefore devices, have a negative temperature coefficient of resistance? (15-2-4)

20. Semiconductor devices are made from intrinsic or pure semiconductor materials. True or False? (15-3)

21. What is the relationship between the number of valence electrons in an atom and the conductivity of an element? (15-2-1)

22. Describe the differences between:
   a. An n-type semiconductor material (15-3-1)
   b. A p-type semiconductor material (15-3-2)

23. Why are doped semiconductor materials still electrically neutral? (15-3-1)

24. How is a P-N junction formed? (15-4)

25. What is a depletion region, and how is it formed? (15-4-1)

26. What is the barrier voltage of:
   a. A silicon P-N junction (15-4-1)
   b. A germanium P-N junction (15-4-1)

27. Describe what occurs when a P-N junction is:
   a. Forward biased (15-4-2)   b. Reverse biased (15-4-2)

28. Define the following:
   a. Forward voltage drop (15-4-2)
   b. Leakage current (15-4-2)

29. Describe why a P-N junction acts like a switch. (15-4-2)

Practice Problems

30. Which of the silicon P-N junctions in Figure 15-19 are forward biased, and which are reverse biased?

31. Determine the current for the circuits shown in Figure 15-20.

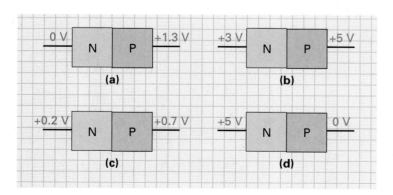

FIGURE 15-19   Biased P-N Junctions.

32. What would be the voltage drop (V_F) across each of the P-N junctions shown in Figure 15-20?

33. What would be the voltage drop across each of the resistors in Figure 15-20?

34. Which of the P-N junctions in Figure 15-20 are equivalent to open switches, and which are equivalent to closed switches?

Web Site Questions

Go to the Web site http://www.prenhall. com/cook, select the textbook *Introductory DC/AC Electronics* or *Introductory DC/AC Circuits*, this chapter, and then follow the instructions when answering the multiple-choice practice problems.

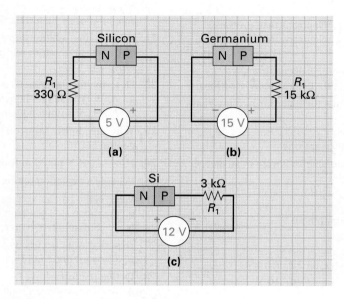

FIGURE 15-20 **P-N Junction Circuit.**

These tests at the end of each chapter will challenge your knowledge up to this point, and give you the practice you need for a job interview. To make this more realistic, the test will comprise both technical and personal questions. In order to take full advantage of this exercise, you may want to set up a simulation of the interview environment, have a friend read the questions to you, and record your responses for later analysis.

Company Name: XVP, Inc.

Industry Branch: Communications

Function: Quality testing aeronautical communications

Job Applying For: QA Technician

1. How are your written and spoken communication skills?
2. Name a semiconductor material.

3. Why are semiconductor devices called solid state components?
4. Do you work well in a team?
5. What occurs when you forward bias a P-N junction?
6. What occurs when you reverse bias a P-N junction?
7. Do you feel able to learn the emerging new technologies by yourself?
8. What is an integrated circuit?
9. What are the characteristics of a semiconductor?
10. Could you define the term "bias voltage"?

Answers

1. Your resume and cover letter demonstrated your written communication skills. Your spoken communication skills are now being tested, and it is important that all of your answers are clear and concise. Quote examples of how you practiced communication skills during your training, and any previous job related experience.
2. Section 15-2.
3. Section 15-1.
4. To achieve a high-skill, high-quality, high-performance organization at a reduced cost, industry has merged electronic system design manufacture and service. A technician, therefore, must have good teamwork skills. Emphasize your lab-team training and any previous job related experience.

5. Section 15-4-2.
6. Section 15-4-2.
7. Technology's rapid growth has meant that employees must be able to independently learn and adapt to the changes. Your math skills, logic and reasoning skills, learning skills, and relationship skills develop intelligence, which in turn develops better adaptability, and organizational and problem solving skills. Success in these areas will ensure career advancement. Discuss your training in each of these areas, and let the interviewer know that your interest in electronics is both personal and professional, so there is a strong desire on your part to understand and embrace the advances.
8. Section 15-1.
9. Introduction and section 15-2.
10. Section 15-4-2.

Diodes and Power Supply Circuits

A Problem with Early Mornings

René Descartes was born in Brittany, France, in 1596. At the age of eight he had surpassed most of his teachers at school and was sent on to the Jesuit College in La Flèche, one of the best in Europe. It was here that his genius in mathematics became apparent; however, due to his extremely delicate health, his professors allowed him to study in bed until midday.

In 1616 he had an urge to see the world and so he joined the army, which made use of Descartes' mathematical genius in military engineering. While traveling, Descartes met Dutch philosopher Isaac Beekman who convinced him to leave the army and, in his words, "turn his mind back to science and more worthier occupations."

After leaving the army Descartes traveled looking for some purpose, and then on November 10th, 1619, he found it. Descartes was in Neuberg, Germany, where he had shut himself in a well-heated room for the winter. It was on the eve of St. Martin's that a freezing blizzard forced Descartes to retire early. That night he described having an extremely vivid dream that clarified his purpose and showed him that physics and all sciences could be reduced to geometry and were therefore all interconnected like a chain.

In his time, and to this day, he is heralded as an analytical genius. In fact, Descartes' procedure can still be used as a guide to solving any problem.

Descartes' four-step procedure for solving a problem:

1. Never accept anything as true unless it is clear and distinct enough to exclude doubt from your mind.
2. Divide the problem into as many parts as necessary to reach a solution.
3. Start with the simplest things and proceed step by step toward the complex.
4. Review the solution so completely and generally that you are sure nothing was omitted.

For me, this four-step procedure has been especially helpful as a troubleshooting guide for system and circuit malfunctions.

Descartes' fame was so renowned that he was asked in 1649 to tutor Queen Christina of Sweden. The Queen demanded that her lessons begin at 5 o'clock in the morning, which conflicted with Descartes' lifetime practice of remaining in bed until midday. After several unsuccessful attempts to change her majesty's mind, and with pressure being applied by the French ambassador, Descartes agreed to the early morning lessons. A short time later on his way to the palace one cold winter morning, Descartes caught a severe chill and died within two weeks.

Outline and Objectives

Introduction

The first diode was accidentally created by Edison in 1883 when he was experimenting with his light bulb. At this time he did not place any importance on the device and its effect, as he could not see any practical application for it. The word *diode* is derived from the fact that the device has two (*di*) electrodes (*ode*).

Once the importance of diodes was realized, construction of the device began. The first diodes were vacuum-tube devices having a hot-filament negative cathode, which released free electrons that were collected by a positive plate called the anode. Today's diode is made of a P-N semiconductor junction but still operates on the same principle. The *n*-type region (cathode) is used to supply free electrons, which are then collected by the *p*-type region (anode). The operation of both the vacuum tube and semiconductor diode is identical in that the device will only pass current in one direction. That is, it will act as a conductor and pass current easily in one direction when the bias voltage across it is of one polarity, yet it will block current and imitate an insulator when the bias voltage applied is of the opposite polarity.

16-1 THE JUNCTION DIODE

The two electrodes or terminals of the diode are called the anode and cathode, as seen in Figure 16-1(a) which shows the schematic symbol of a diode. To help you remember which terminal is the anode and which is the cathode, and which terminal is positive and which is

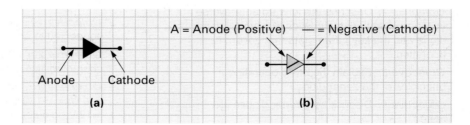

FIGURE 16-1   Schematic Symbol of a Diode.

negative, Figure 16-1(b) shows how a line drawn through the triangle section of the symbol will make the letter "A" and indicate the "anode" terminal. Similarly, if the vertical flat side of the diode symbol is aligned horizontally "—", as in Figure 16-1(b), it becomes the "negative" symbol. This memory system helps us to remember that the anode terminal of the diode is next to the triangle part of the symbol and is positive, while the cathode terminal of a diode is next to the vertical line of the symbol and is negative.

The diode is generally mounted in one of the three basic packages shown in Figure 16-2. These packages are designed to protect the diode from mechanical stresses and the environment. The difference in the size of the packages is due to the different current rating of the diode. A black band or stripe is generally placed on the package closest to the cathode terminal for identification purposes, as seen in Figure 16-2(a) and (b). Larger diode packages, like the one seen in Figure 16-2(c), usually have the diode symbol stamped on the package to indicate anode/cathode terminals.

16-1-1 *Diode Operation*

As far as operation is concerned, the diode operates like a switch. If you give the diode what it wants, that is make the anode terminal positive with respect to the cathode terminal as seen in Figure 16-3(a), the device is equivalent to a closed switch as seen in Figure 16-3(b). In this condition, the diode is said to be ON or *forward biased.*

On the other hand, if you do not give the diode what it wants, that is if you make the anode terminal negative with respect to the cathode as seen in Figure 16-3(c), the device is equivalent to an open switch as seen in Figure 16-3(d). In this condition the diode is said to be OFF or *reverse biased.*

16-1-2 *Basic Diode Application*

As an application, Figure 16-4 shows how the diode can be used as a switch within an **encoder circuit.** The pull-up resistors R_1, R_2, and R_3 ensure that lines A, B, and C are normally all at +5 V. This is the output voltage on each line when the rotary switch is in position 2, as seen in the table in Figure 16-4.

When the rotary switch is turned to position 1, D_1 is connected in-circuit and, because its anode is made positive via R_2 and its cathode is at 0 V, the diode D_1 will turn ON and be equivalent to a closed switch. The 0 V on the cathode of D_1 will be switched through to line B (all of the five volts will be dropped across R_2) producing an output voltage code of A = +5 V, B = 0 V, C = +5 V as seen in the table in Figure 16-4.

When the rotary switch is turned to position 3, D_2 and D_3 are connected in circuit and because both anodes are made positive via R_1 and R_3, and both diode cathodes are at 0 V,

Encoder Circuit

A circuit that produces different output voltage codes, depending on the position of a rotary switch.

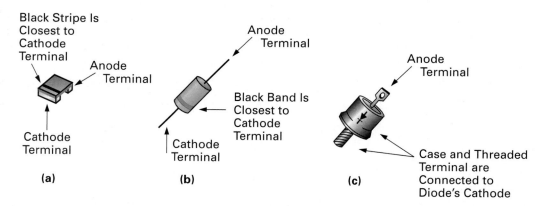

Black Stripe Is Closest to Cathode Terminal

Anode Terminal

Anode Terminal

Cathode Terminal

(a)

Anode Terminal

Black Band Is Closest to Cathode Terminal

Cathode Terminal

(b)

Anode Terminal

Case and Threaded Terminal are Connected to Diode's Cathode

(c)

FIGURE 16-2 Diode Packaging. (a) Chip Package—1/4 A. (b) Small Current Package—Less than 3 A. (c) Large Current Package—Greater than 3 A.

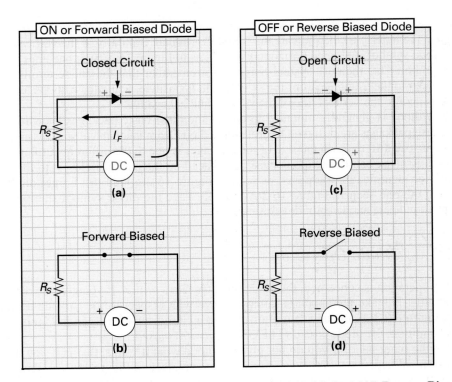

FIGURE 16-3   Diode Operation. (a)(b) Forward Biased (ON) Diode. (c)(d) Reverse Biased (OFF) Diode.

D_2 and D_3 will turn ON. These forward biased diodes will switch 0 V through to lines A and C, producing an output voltage code of $A = 0$ V, $B = +5$ V, $C = 0$ V, as seen in the table in Figure 16-4.

This *code generator* or *encoder circuit* will produce three different output voltage codes for each of the three positions of the rotary switch. These codes could then be used to initiate one of three different operations based on the operator setting of the rotary control switch.

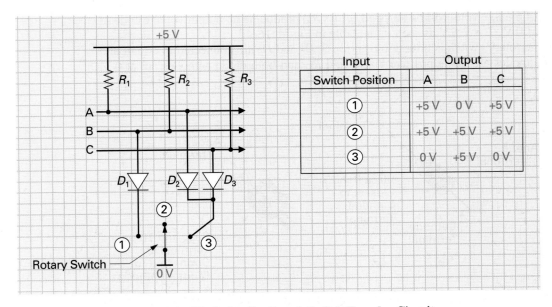

Input	Output		
Switch Position	A	B	C
①	+5 V	0 V	+5 V
②	+5 V	+5 V	+5 V
③	0 V	+5 V	0 V

FIGURE 16-4   Diode Application: A Switch Encoder Circuit.

16-1-3 *A Junction Diode's Characteristic Curve*

Semiconductor devices such as diodes and transistors are constructed using P-N junctions. *A diode, for example, has only one P-N junction and is created by doping a single piece of pure semiconductor to produce an n-type and p-type region. A bipolar junction transistor, on the other hand, has two P-N junctions and is created by doping a single piece of pure semiconductor with three alternate regions (NPN or PNP). The point at which these two opposite-doped materials come in contact with each other is called a junction, which is why these devices are called* **junction diodes** *and bipolar junction transistors.*

These P-N junctions need voltages of a certain amplitude and polarity to control their operation. These voltages, which incline or cause the diode to operate in a certain manner, are known as **bias voltages.** Bias voltages control the resistance of the junction and, therefore, the amount of current that can pass through the P-N junction diode.

The upper right quadrant of the four sections in Figure 16-5 shows what forward current will pass through the diode when a forward bias voltage is applied. As you can see from the inset, the diode is forward biased by applying a positive potential to its anode and a negative potential to its cathode. In this instance the diode is said to be ON; and is equivalent to a closed switch. Beginning at the graph origin and following the curve into the forward quadrant, you can see that the forward current through a diode is extremely small until the forward bias voltage exceeds the diode's internal barrier voltage, which for silicon is 0.7 V and for germanium is 0.3 V.

Referring to the linearly increasing current portion of the forward curve in Figure 16-5, you will notice that although there is a large change in forward current, the forward voltage drop across the diode remains almost constant between 0.7 V and 0.75 V.

Junction Diode

A semiconductor diode whose ON/OFF characteristics occur at a junction between the *n*-type and *p*-type semiconductor materials.

Bias Voltage

Voltage that inclines or causes the diode to operate in a certain manner.

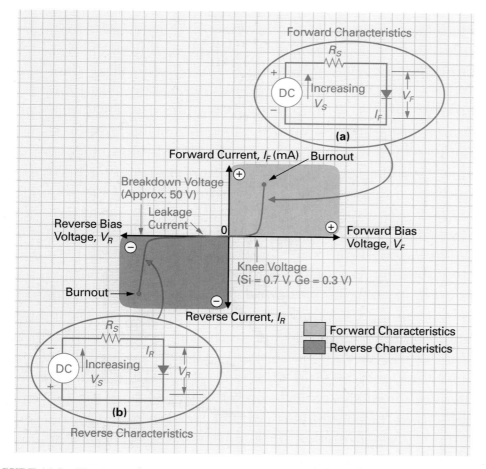

FIGURE 16-5   **The Junction Diode Voltage-Current Characteristic Curve.**

The amount of heat produced in the diode is proportional to the value of current through the diode ($P\uparrow = I^2\uparrow \times R$). For example, an IN4001 diode, which is a commonly used low-power silicon diode, has a manufacturer's maximum forward (I_F max.) rating of 1 A. If this value of current is exceeded, the diode will begin generating more heat than it can dissipate and burn out. A series current limiting resistor (R_S) is generally always included to limit the forward current, as shown in the inset in Figure 16-5. Although the series resistor will limit forward current, it cannot prevent a damaging forward current if enough pressure or forward voltage is applied ($V\uparrow = I\uparrow \times R$). The value of forward current is equal to

$$I_F = \frac{V_s - V_{\text{diode}}}{R_s}$$

EXAMPLE:

Calculate the value of current for the circuit shown in Figure 16-6.

Solution:

The diode is forward biased because the applied voltage is connected so that its positive terminal is applied to the anode and the negative terminal is applied to the cathode. Because a silicon diode is being used, the forward voltage drop will be 0.7 V. With an applied voltage of 8.5 V and a circuit resistance of 1.2 kΩ, the circuit current will equal

$$I_F = \frac{V_S - V_{\text{diode}}}{R_s}$$

$$I_F = \frac{8.5 \text{ V} - 0.7 \text{ V}}{1.2 \text{ k}\Omega}$$

$$I_F = \frac{7.8 \text{ V}}{1.2 \text{ k}\Omega}$$

$$I_F = 6.5 \text{ mA}$$

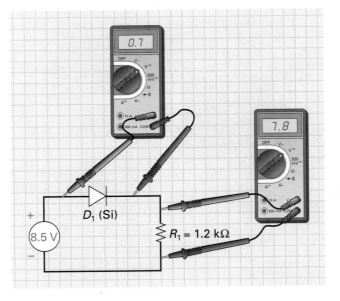

FIGURE 16-6   A P-N Junction Diode Circuit.

The lower left quadrant of the four sections in Figure 16-5 shows what reverse current will pass through the diode when a reverse bias voltage is applied. As you can see from the inset, a diode is reverse biased by applying a negative potential to its anode and a positive potential to its cathode. In this instance, current is effectively reduced to zero and the diode is said to be OFF and equivalent to an open switch.

These characteristics can be seen in the reverse curve in Figure 16-5. Beginning at the graph origin and following the curve into the reverse quadrant, you can see that the reverse current through the diode increases only slightly (approximately 100 μA). Throughout this part of the curve the diode is said to be blocking current because the leakage current is generally so small it is ignored for most practical applications. If the reverse voltage (V_R) is further increased, a point will be reached where the diode will break down, resulting in a sudden increase in current. The point on the reverse voltage scale at which the diode breaks down and there is a sudden increase in reverse current is called the **breakdown voltage.** Referring to the reverse curve in Figure 16-5, you can see that most silicon diodes break down as the reverse bias voltage approaches 50 V. For example, the IN4001 low-power silicon diode has a reverse breakdown voltage (which is sometimes referred to as the **Peak Inverse Voltage** or **PIV**) of 50 V listed on its manufacturer's data sheet. If this reverse bias voltage is exceeded, an avalanche of continuously rising current will eventually generate more heat than can be dissipated, resulting in the destruction of the diode.

Semiconductor materials, and therefore diodes, have a negative temperature coefficient of resistance. This means as temperature increases ($T\uparrow$), their resistance decreases ($R\downarrow$).

Breakdown Voltage or Peak Inverse Voltage (PIV)

The point on the reverse voltage scale at which the diode breaks down and there is a sudden increase in the reverse current.

16-1-4 *Testing Junction Diodes*

Figure 16-7 shows how an ohmmeter can be used to check whether a diode has malfunctioned or is operating correctly. A good diode should display a very low resistance when it is biased ON, and a very high resistance when it is biased OFF. Figure 16-7(a) shows how a

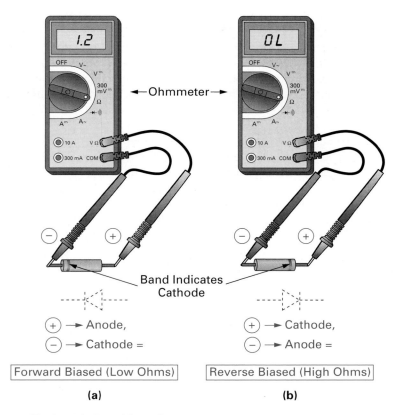

FIGURE 16-7 **Testing Diodes with an Ohmmeter.**

diode can be forward biased by an ohmmeter's internal battery (+ lead to anode, − lead to cathode), and if good, should display a low value of resistance (typically less than 10 Ω). Figure 16-7(b) shows how the diode is then flipped over and reverse biased by the ohmmeter's internal battery (+ lead to cathode, − lead to anode). If the diode is good, the ohmmeter should display a very high resistance (typically greater than 1000 MΩ). Since this value is generally off the ohmmeter's scale, you will probably have the display showing OL or OR. This is what you should expect since it means that the reverse biased diode's resistance is so high that it is over the range, or off the scale, selected.

16-2   THE ZENER DIODE

Zener Diode

Diodes constructed to operate at voltages that are equal to or greater than the reverse breakdown voltage rating.

Figure 16-8(a) shows the two schematic symbols used to represent the **zener diode.** As you can see, the zener diode symbol resembles the basic P-N junction diode symbol in appearance; however, the zener diode symbol has a zig-zag bar instead of the straight bar. This zig-zag bar at the cathode terminal is included as a memory aid since it is "**Z**" shaped and will always remind us of zener.

Figure 16-8(b) shows two typical low-power zener diode packages, and one high-power zener diode package. The surface mount low-power zener package has two metal pads for direct mounting to the surface of a circuit board, while the axial lead low-power zener package has the zener mounted in a glass or epoxy case. The high-power zener package is generally stud mounted and contained in a metal case. These packages are identical to the basic P-N junction diode low-power and high-power packages. Once again, a band or stripe is used to identify the cathode end of the zener diode in the low-power packages, whereas the threaded terminal of a high-power package is generally always the cathode.

16-2-1   *Zener Diode Voltage-Current (V-I) Characteristics*

Figure 16-9 shows the *V-I* (voltage-current) characteristic curve of a typical zener diode. This characteristic curve is almost identical to the basic P-N junction diode's characteristic curve. For example, when forward biased at or beyond 0.7 V, the zener diode will turn ON and be equivalent to a closed switch; whereas, when reverse biased, the zener diode will turn OFF and be equivalent to an open switch. The main difference, however, is that the zener diode has been specifically designed to operate in the reverse breakdown region of the curve. This is achieved, as can be seen in the inset in Figure 16-9, by making sure that the external

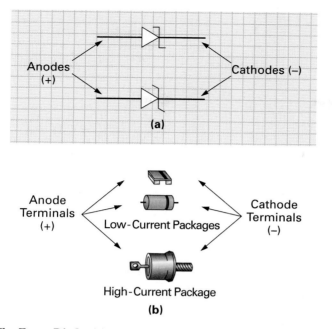

FIGURE 16-8 **The Zener Diode. (a) Zener Diode Schematic Symbols. (b) Packages.**

bias voltage applied to a zener diode will not only reverse bias the zener diode ($+\rightarrow$ cathode, $-\rightarrow$ anode) but also be large enough to drive the zener diode into its reverse breakdown region.

As the reverse voltage across the zener diode is increased from the graph origin (which represents 0 volts), the value of **reverse leakage current (I_R)** begins to increase. Comparing the voltage developed across the zener (V_Z) to the value of current through the zener (I_Z), you may have noticed that *the voltage drop across a zener diode (V_Z) remains almost constant when it is operated in the reverse zener breakdown region, even though current through the zener (I_Z) can vary considerably. This ability of the zener diode to maintain a relatively constant voltage regardless of variations in zener current is the key characteristic of the zener diode.*

Generally, manufacturers rate zener diodes based on their **zener voltage (V_Z)** rather than their breakdown voltage (V_{Br}). A wide variety of zener diode voltage ratings are available ranging from 1.8 V to several hundred volts. For example, many of the frequently used low-voltage zener diodes have ratings of 3.3 V, 4.7 V, 5.1 V, 5.6 V, 6.2 V, and 9.1 V.

**Reverse Leakage
Current (I_R)**
The undesirable flow of current through a device in the reverse direction.

Zener Voltage (V_Z)
The voltage drop across the zener when it is being operated in the reverse zener breakdown region.

16-2-2 *Testing Zener Diodes*

Because a zener diode is designed to conduct in both directions, we cannot test it with the ohmmeter as we did the basic P-N junction diode. The best way to test a zener diode is to connect the voltmeter across the zener while it is in circuit and power is applied, as seen in Figure 16-10. If the voltage across the zener is at its specified voltage, then the zener is functioning properly. If the voltage across the zener is not at the nominal value, then the following checks should be made:

1. Check the source input voltage. If this voltage (V_{in}) does not exceed the zener voltage (V_Z), the zener diode will not be at fault because the source voltage is not large enough to send the zener into its reverse breakdown region.

2. Check the series resistor (R_S) to determine that it has not opened or shorted. An open series resistor will have all of the input voltage developed across it and there will be no voltage across the zener. A shorted series resistor will not provide any current-limiting capability and the zener could possibly burn out.

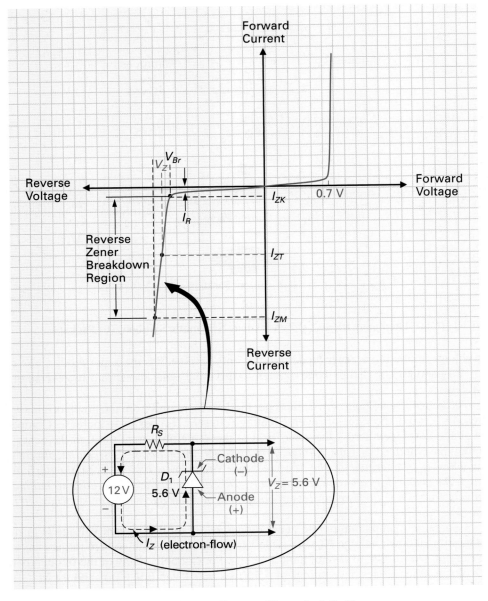

FIGURE 16-9 The Zener Diode Voltage-Current Characteristic Curve.

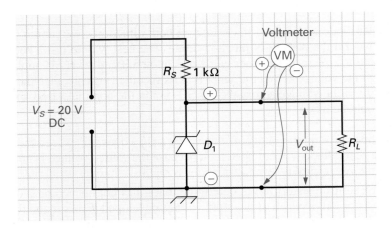

FIGURE 16-10 Zener Diode Testing.

3. Check that there is not a short across the load because this would show up as 0 V across the zener and make the zener look faulty. To isolate this problem, disconnect the load and see if the zener functions normally.

   If these three tests check out okay, the zener diode is probably at fault and should be replaced.

16-3   THE LIGHT-EMITTING DIODE

The **light-emitting diode (LED)** is a semiconductor device that produces light when an electrical current or voltage is applied to its terminals. Figure 16-11(a) shows the two schematic symbols used most frequently to represent an LED. The two arrows leaving the diode symbol represent light.

A typical LED package is shown in Figure 16-11(b). The package contains the two terminals for connection to the anode and cathode and a semi-clear case which contains the light-emitting diode and a lens. Looking at this illustration, you can see that the LED chip is

Light-Emitting Diode (LED)

A semiconductor device that produces light when an electrical current or voltage is applied to its terminals.

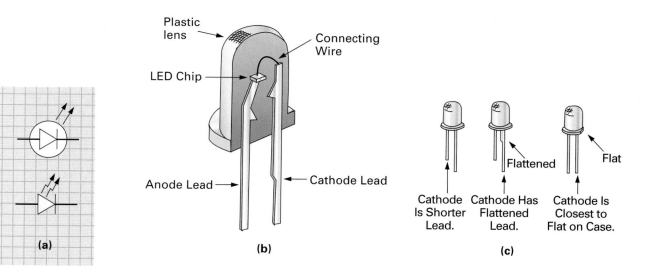

FIGURE 16-11   The Light-Emitting Diode (LED). (a) Schematic Symbol. (b) Construction. (c) Lead Identification.

directly connected to the anode lead, while the cathode lead is connected to the LED chip by a thin wire. The dome-shaped top of the plastic (epoxy) case serves as the lens and acts as a magnifier to conduct light away from the LED chip. By adjusting the lens material, lens shape, and the distance the LED chip is from the lens, manufacturers can obtain a variety of radiation patterns.

There are three methods used to identify the anode and cathode leads of the LED, and these are shown in Figure 16-11(c). In all three cases, the cathode lead is distinguished from the anode lead by having its lead shorter or its lead flattened, or being nearest to the flat side of the case.

16-3-1 *LED Characteristics*

Light-emitting diodes have *V-I* characteristic curves that are almost identical to the basic P-N junction diode as shown in Figure 16-12.

Studying the forward characteristics, you can see that LEDs have a high forward voltage rating (V_F is normally between $+1$ V and $+3$ V) and a low maximum forward current rating (I_F is normally between 20 mA and 50 mA). In most circuit applications, the LED will have a forward voltage drop of 2 V, and a forward current of 20 mA. The LED will usually always need to be protected from excessive forward current damage by a series current-limiting resistor, which will limit the forward current so that it does not exceed the LED's *maximum forward current* (I_{FM} or I_{FMAX}) rating, as seen in the inset in Figure 16-12. The circuit in the inset shows how to forward bias an LED. In this example circuit, the source voltage of 5 V is being applied to a 130 Ω series resistor (R_S) and an LED with an I_{FM}

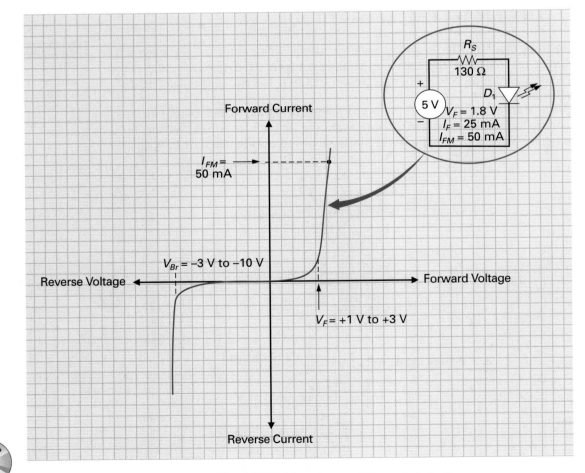

FIGURE 16-12 The LED Voltage-Current Characteristic Curve.

CHAPTER 16 / DIODES AND POWER SUPPLY CIRCUITS

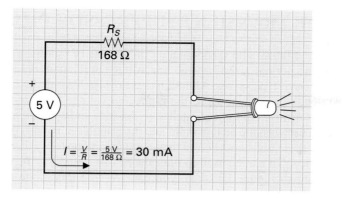

FIGURE 16-13 Testing LEDs.

rating of 50 mA and a V_F rating of 1.8 V. To calculate the value of current in this circuit, we would first deduct the 1.8 V developed across the LED (V_{LED}) from the source voltage (V_S) to determine the voltage drop across the resistor (V_{RS}). Then using Ohm's law, divide V_{RS} by R_S to determine current.

$$I_S = \frac{V_S - V_{LED}}{R_S}$$

$$= \frac{5\ V - 1.8\ V}{130\ \Omega} = \frac{3.2\ V}{130\ \Omega} = 24.6\ mA$$

Studying the reverse characteristics, you can see that LEDs have lower reverse breakdown voltage values than junction diodes (V_{Br} is typically -3 V to -10 V), which means that even a low reverse voltage will cause the LED to break down and become damaged.

16-3-2 *Testing LEDs*

Light-emitting diodes have a very long life expectancy and are much more rugged than their predecessor, the small incandescent light bulb. They do, however, break down and have to be replaced with either exactly the same device or a similar device with the same characteristics.

Figure 16-13 shows how to construct a simple LED test circuit using a dc power supply and a series current-limiting resistor. If a circuit problem is isolated to the LED, the new LED should be tested with a similar circuit to ensure it is operating correctly before it is inserted into the circuit.

SELF-TEST EVALUATION POINT FOR SECTION 16-3

Now that you have completed this section, you should be able to:

■ **Objective 10.** *Sketch the schematic symbol, and describe the different package types for the light emitting diode or LED.*

■ **Objective 11.** *Describe the forward and reverse voltage-current characteristics of a typical LED and the key parameters that govern its operation.*

■ **Objective 12.** *Explain the basic operating principles of the LED.*

■ **Objective 13.** *Discuss how LEDs are tested.*

Use the following questions to test your understanding of Section 16-3.

1. What would be the typical voltage drop across a forward biased LED?

    a. 0.7 V

    b. 2 V

    c. 0.3 V

    d. 5.6 V

2. A _____ is normally always included to limit I_F to just below its maximum.

16-4   THE TRANSIENT SUPPRESSOR DIODE

Lightning, power line faults, and the switching on and off of motors, air conditioners, and heaters can cause the normal 120 V rms ac line voltage at the wall outlet to contain under-voltage dips and over-voltage spikes. Although these *transients* only last for a few microseconds, the over-voltage spikes can cause the input line voltage to momentarily increase by 1000 V or more. In sensitive equipment, such as televisions and computers, shunt filtering devices are connected between the ac line input and the primary of the dc power supply's transformer to eliminate these transients before they get into, and possibly damage, the system.

One such device that can be used to filter the ac line voltage is the **transient suppressor diode.** Referring to Figure 16-14(a), you can see that this diode contains two zener diodes that are connected back-to-back. The schematic symbol for this diode is shown in Figure 16-14(b).

Transient suppressor diodes are also called **transorbs** because they "absorb transients." Figure 16-14(c) shows how a transorb would be connected across the ac power line input to a dc power supply. Because the zeners within the transient suppressor diode are connected back-to-back, they will operate in either direction (the device is "bi-directional") and monitor both alternations of the ac input. If a voltage surge occurs that exceeds the V_Z (zener voltage) of the diodes, they will break down and shunt the surge away from the power supply.

Most manufacturers' transorbs have a high power dissipation rating because they may have to handle momentary power line surges in the hundreds of watts. For example, the Motorola 1N5908-1N6389 series of transorbs can dissipate 1.5 kW for a period of approximately 10 ms (most surges last for a few milliseconds). The devices must also have a fast turn-on time so that they can limit or clamp any voltage spikes. For example, the Motorola P6KE6.8 series has a response time of less than 1 ns.

Transient Suppressor Diode

A device used to protect voltage-sensitive electronic devices in danger of destruction by high-energy voltage transients.

Transorb

Absorb transients. Another name for transient suppressor diode.

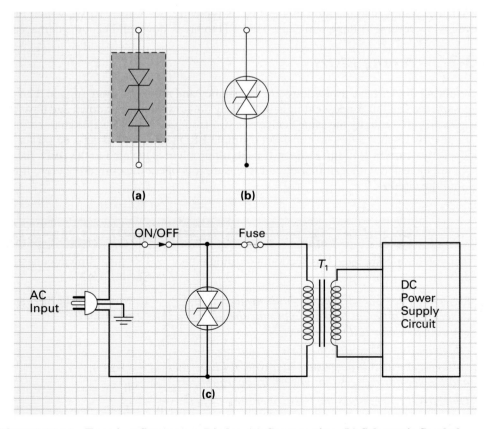

FIGURE 16-14   Transient Suppressor Diodes. (a) Construction. (b) Schematic Symbol. (c) Bidirectional Circuit Application (AC Line Voltage).

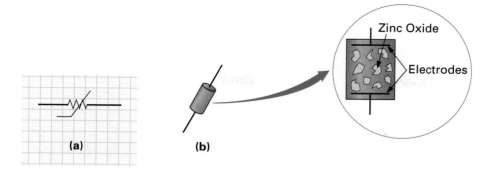

FIGURE 16-15 Metal Oxide Varistors (MOVs). (a) Schematic Symbol. (b) Physical Appearance and Construction.

In dc applications, a single unidirectional (one-direction) transient suppressor can be used instead of a bidirectional (two-direction) transient suppressor. These single transorbs have the same schematic symbol as a zener.

Metal oxide varistors (MOVs) are currently replacing zener-diode and transient-diode suppressors because they are able to shunt a much higher current surge and are cheaper. These are not semiconductor devices; in fact, they contain a zinc-oxide and bismuth-oxide compound in a ceramic body but are connected in the same way as a transient suppressor diode. They are called **varistors** because they operate as a "voltage dependent resistor" that will have a very low resistance at a certain breakdown voltage. The MOV's schematic symbol, typical appearance, and construction are shown in Figure 16-15.

SELF-TEST EVALUATION POINT FOR SECTION 16-4

Now that you have completed this section, you should be able to:

■ **Objective 14.** *Describe the characteristics and applications of the transient suppressor diode.*

Use the following questions to test your understanding of Section 16-4.

1. True or false: A bidirectional transient suppressor diode would be used to suppress ac power surges.

2. True or false: A unidirectional transient suppressor diode would be used to suppress dc power surges.

16-5 DIODE POWER SUPPLY CIRCUITS

In the previous sections we discussed the P-N junction diode, the zener diode, and the light-emitting diode. In this section we will see how we can use all of these diodes to construct a dc power supply. This section is also of great importance to our study of troubleshooting, since greater than 90% of all electronic system malfunctions are power related.

16-5-1 *Block Diagram of a DC Power Supply*

Figure 16-16(a) shows the four basic blocks of a dc power supply. Almost every piece of electronic equipment that makes use of 120 V ac as a source of power will have a built-in dc power supply, as shown in Figure 16-16(b). Stand-alone dc power supplies, such as the one shown in Figure 16-16(c), are also available for use in laboratory experimentation.

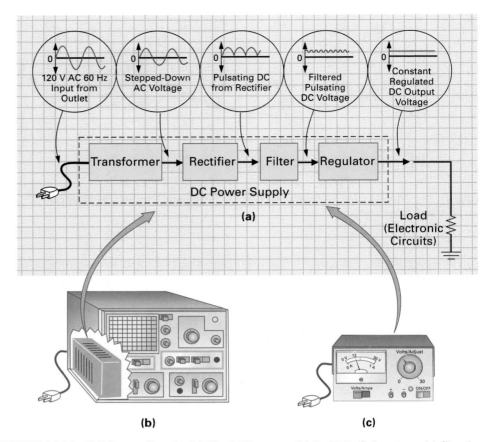

FIGURE 16-16 **DC Power Supply. (a) Block Diagram. (b) Built-In Subsystem. (c) Stand-Alone Unit.**

Let us now refer to the block diagram of the dc power supply in Figure 16-16(a) and describe the function of each block.

Since the final voltage desired is generally not 120 V, a transformer is usually included to step the ac line voltage up or down to a desired value. Electronic circuits generally require low-voltage supply values such as 12 and 5 V dc, and a step-down transformer would be used. For example, a step-down transformer with a turns ratio of 10:1 would reduce the 120 V ac input to a 12 V ac output. The output current capability of this transformer will be 1:10, which will be ideal because most electronic circuits require low supply voltages with high current capacity.

As shown by the waveforms in Figure 16-16(a), the rectifier converts the stepped-down ac input from the transformer to a pulsating dc output. This pulsating dc output could not be used to power an electronic circuit because of the continuous changes between zero volts and a peak voltage. The filter smoothes out the pulsating dc ripples into an almost constant dc level, as seen in the waveform after the filter.

The final block is called a regulator, and although there appears to be no difference between the regulator's input and output waveforms, it provides a very important function. The regulator maintains the dc output voltage from the power supply constant, or stable, despite variations in the ac input voltage or variations in the output load resistance.

Many dc power supplies have several rectifier, filter, and regulator stages, depending on how many dc output voltages are desired for the electronic system. In the following sections we will examine these electrical circuits in more detail and then combine all of them in a working dc power supply circuit.

16-5-2　*Transformers*

The turns ratio of a transformer in a dc power supply can be selected to either increase or decrease the 120 V ac input. With most electronic equipment, a supply voltage of less than 120 V is required, and therefore a step-down transformer is used. The secondary output voltage V_s from the transformer can be calculated with the following formula, which was introduced previously:

$$V_s = \frac{N_s}{N_p} \times V_p$$

To apply this formula to an example, refer to the power supply transformer shown in Figure 16-17. This transformer can be connected so that it delivers the same secondary volt-

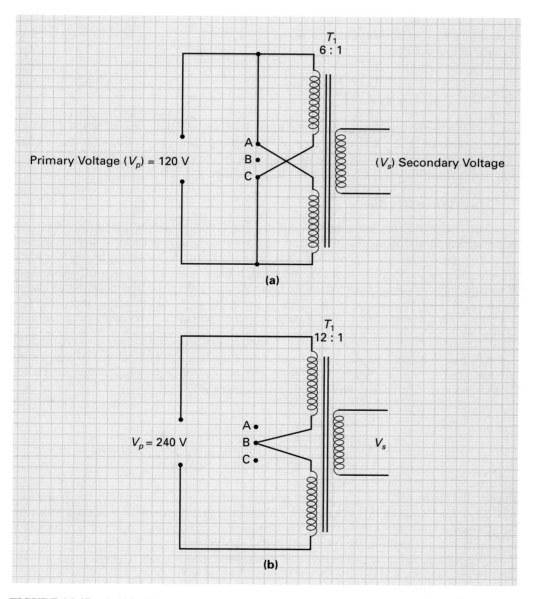

FIGURE 16-17　A 120 V/240 V Power Supply Transformer. (a) 120 V Connection. (b) 240 V Connection.

age for either a 120 V or a 240 V rms ac input. For example, when the two primary windings are connected in parallel, as shown in Figure 16-17(a), the transformer turns ratio is 6:1 step-down, and therefore the secondary voltage will be:

$$V_s = \frac{N_s}{N_p} \times V_p$$

$$= \frac{1}{6} \times 120 \text{ V rms} = 20 \text{ V rms}$$

When the two primary windings are connected in series, as shown in Figure 16-17(b), the transformer turns ratio is 12:1 step-down, and therefore the secondary voltage will be

$$V_s = \frac{N_s}{N_p} \times V_p$$

$$= \frac{1}{12} \times 240 \text{ V rms} = 20 \text{ V rms}$$

By doubling the number of primary turns, we can accept twice the input voltage and deliver the same output voltage.

16-5-3 *Rectifiers*

The junction diode's ability to switch current in only one direction makes it ideal for converting two-direction alternating current into one-direction direct current. In this section we will discuss the three basic diode rectifier circuits: the half-wave rectifier, the full-wave center-tapped rectifier, and the full-wave bridge rectifier.

Half-Wave Rectifiers

Half-Wave Rectifier Circuit

A circuit that converts ac to dc by allowing current to flow during only one-half of the ac input cycle.

The **half-wave rectifier circuit** is constructed simply by connecting a diode between the power supply transformer and the load, as shown in Figure 16-18(a). When the secondary ac voltage swings positive, as shown in Figure 16-18(b), the anode of the diode is made positive, causing the diode to turn ON and connect the positive half-cycle of the secondary ac voltage across the load (R_L). When the secondary ac voltage swings negative, as shown in Figure 16-18(c), the anode of the diode is made negative, and therefore the diode will turn OFF. This will prevent any circuit current, and no voltage will be developed across the load (R_L).

Output Voltage. Figure 16-18(d) illustrates the input and output waveforms for the half-wave rectifier circuit. The 120 V ac rms input, or 169.7 V ac peak input, is applied to the 17:1 step-down transformer, which produces an output of:

$$V_s = \frac{N_s}{N_p} \times V_p$$

$$= \frac{1}{17} \times 169.7 \text{ V peak} = 10 \text{ V peak}$$

Because the diode will only connect the positive half-cycle of this ac input across the load (R_L), the output voltage (V_{RL}) is a positive pulsating dc waveform of 10 V peak. In this final waveform in Figure 16-18(d), you can see that the circuit is called a half-wave rectifier because only half of the input wave is connected across the output.

The average value of two half-cycles is equal to 0.637 V peak. Therefore, the average value of one half-cycle is equal to 0.318 V peak (0.637/2 = 0.318):

$$V_{avg} = 0.318 \times V_{s \text{ peak}}$$

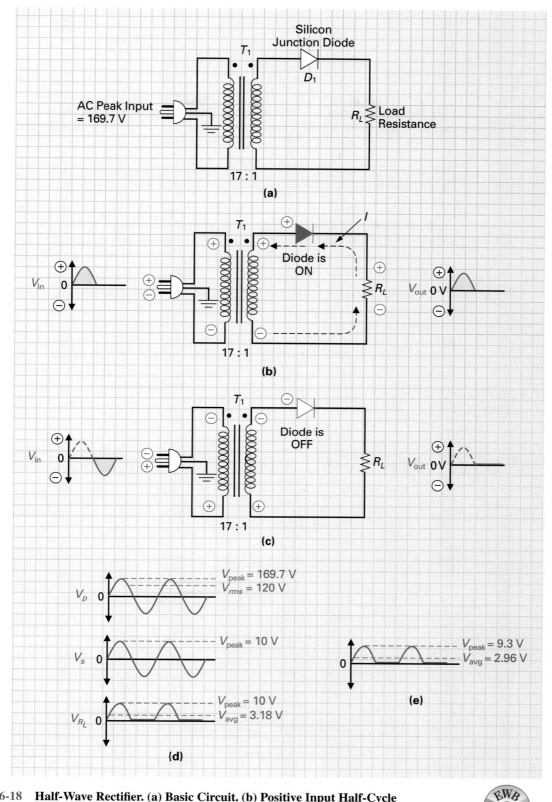

FIGURE 16-18 Half-Wave Rectifier. (a) Basic Circuit. (b) Positive Input Half-Cycle Operation. (c) Negative Input Half-Cycle Operation. (d) Input/Output Waveforms. (e) Half-Wave Output Minus Diode Barrier Voltage.

681

In the example in Figure 16-18, the average voltage of the half-wave output will be:

$$V_{avg} = 0.318 \times V_s \text{ peak}$$
$$V_{avg} = 0.318 \times 10 \text{ V}$$
$$= 3.18 \text{ V}$$

To be more accurate, there will, of course, be a small voltage drop across the diode due to its barrier voltage of 0.7 V for silicon and 0.3 V for germanium. The output from the circuit in Figure 16-18 would actually have a peak of 9.3 V (10 V − 0.7 V), and therefore an average of 2.96 V (0.318 × 9.3 V), as shown in Figure 16-18(e).

$$V_{out} = V_s - V_{diode}$$

Another point to consider is the reverse breakdown voltage of the junction or rectifier diode. When the input swings negative, as illustrated in Figure 16-18(c), the entire negative supply voltage will appear across the open or OFF diode. The maximum reverse breakdown voltage, or peak inverse voltage (PIV) rating of the rectifier diode, must therefore be larger than the peak of the ac voltage at the diode's input.

Output Polarity. The half-wave rectifier circuit can be arranged to produce either a positive pulsating dc output, as shown in Figure 16-19(a), or a negative pulsating dc output, as shown in Figure 16-19(b). Studying the difference between these circuits you can see that in Figure 16-19(a) the rectifier diode is connected to conduct the positive half-cycles of the ac input, while in Figure 16-19(b), the rectifier diode is reversed so that it will conduct the negative half-cycles of the ac input. By changing the direction of the diode in this manner, the rectifier can be made to produce either a positive or a negative dc output.

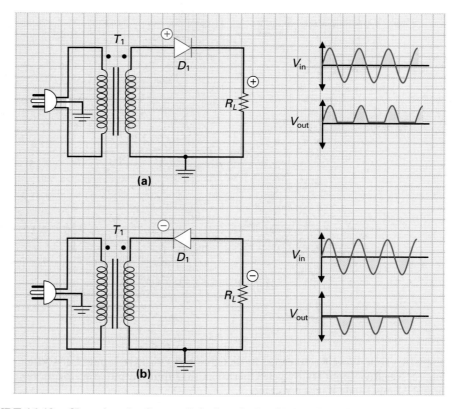

FIGURE 16-19   **Changing the Output Polarity of a Half-Wave Rectifier. (a) Positive Pulsating DC. (b) Negative Pulsating DC.**

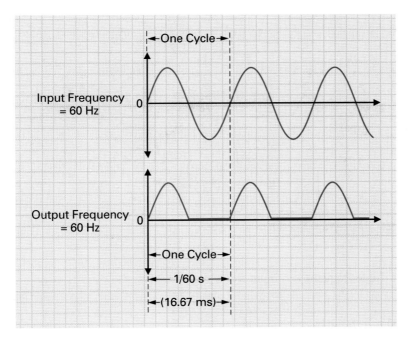

FIGURE 16-20   Ripple Frequency of a Half-Wave Rectifier.

Ripple Frequency. Referring to the input/output waveforms of the half-wave rectifier circuit shown in Figure 16-20, you can see that one output ripple is produced for every complete cycle of the ac input. Consequently, if a half-wave rectifier is driven by the 120 V ac 60 Hz line voltage, each complete cycle of the ac input and each complete cycle of the output will last for one-sixtieth or 16.67 ms (1/60 = 1 ÷ 60 = 16.67 ms). The frequency of the pulsating dc output from a rectifier is called the ripple frequency, and for half-wave rectifier circuits, the

> output pulsating dc ripple frequency = input ac frequency

EXAMPLE:

Calculate the following for the rectifier circuit shown in Figure 16-21:

a. Output polarity
b. Peak and average output voltage, taking into account the diode's barrier potential
c. Output ripple frequency

Solution:

a. Negative pulsating dc

b.
$$V_{\text{in peak}} = 169.7 \text{ V}$$
$$V_s = \frac{N_s}{N_p} \times V_p$$
$$= \frac{1}{8} \times 169.7 \text{ V peak} = -21.2 \text{ V peak}$$
$$V_{\text{out}} = V_s - V_{\text{diode}}$$
$$= 21.2 \text{ V} - 0.7 \text{ V} = -20.5 \text{ V}$$
$$V_{\text{avg}} = 0.318 \times V_{\text{out}}$$
$$= 0.318 \times 20.5 \text{ V} = -6.5 \text{ V}$$

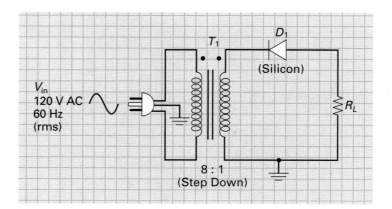

FIGURE 16-21 Half-Wave Rectifier.

c. Output ripple frequency = input frequency = 60 Hz

Full-Wave Rectifiers

The half-wave rectifier's output is difficult to filter to a smooth dc level because the output voltage and current are applied to the load for only half of each input cycle. In this section we will examine two full-wave rectifier circuits which, as their name implies, switch both half-cycles (or the full ac input wave) of the input through the load in only one direction.

**Full-Wave
Center-Tapped Rectifier**

A rectifier circuit that makes use of a center-tapped transformer to cause an output current to flow in the same direction during both half-cycles of the ac input.

Center-Tapped Rectifier. The basic **full-wave center-tapped rectifier** circuit, which is shown in Figure 16-22(a), contains a center-tapped transformer and two diodes. The center tap of the transformer secondary is grounded (0 V) to create a 180° phase difference between the top and bottom of the secondary winding.

When the ac input (V_{in}) swings positive, the circuit operates in the manner illustrated in Figure 16-22(b). A positive voltage is developed on the top of T_1 secondary turning ON D_1, while a negative voltage is developed on the bottom of T_1 secondary turning OFF D_2. This will permit a flow of electrons up through the load as indicated by the dashed current (I) line, developing a positive output half-cycle across the load (V_{out}).

When the ac input (V_{in}) swings negative, the circuit operates in the manner illustrated in Figure 16-22(c). A negative voltage is developed on the top of T_1 secondary turning OFF D_1, while a positive voltage is developed on the bottom of T_1 secondary turning ON D_2. This will permit a flow of electrons up through the load as indicated by the dashed current (I) line, again developing a positive output half-cycle across the load (V_{out}).

The two diodes and center-tapped transformer in this circuit switch the two-direction ac current developed across the secondary through the load in only one direction and develop an output voltage across the load of the same polarity. The input/output voltage details of a full-wave center-tapped rectifier are shown in Figure 16-23. A 1:1 turns ratio has been selected in this example to emphasize how the center tap in the secondary of the transformer causes the secondary voltage (V_s) to be divided into two equal halves (V_{s1} and V_{s2}). Referring to the waveforms, you can see that the ac line voltage of 120 V rms, or 169.7 V peak (V_{in}), is applied to the primary of the transformer T_1 (V_p). A 1:1 turns ratio ensures that this same voltage will be developed across the secondary winding ($V_s = V_p = 169.7$ V peak). This secondary voltage will be split in two as indicated by the waveforms V_{s1} and V_{s2}, which will each have a peak voltage of 1/2 V_s ($V_{s1\,peak} = V_{s2\,peak} = 1/2\ V_{s\,peak} = 1/2 \times 169.7$ V = 84.9 V peak). Because V_{s1} will be connected across the output during one half-cycle, and V_{s2} will be connected across the output during the other half-cycle, the output voltage (V_{out}) will also have a peak voltage equal to 1/2 V_s or 84.9 V. Compared to a half-wave rectifier, therefore, whose peak output voltage is equal to the peak secondary voltage,

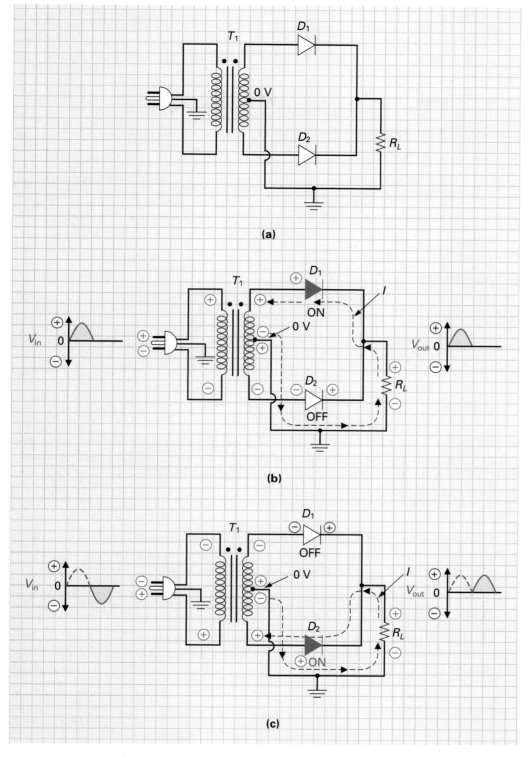

(a)

(b)

(c)

FIGURE 16-22 **Full-Wave Center-Tapped Rectifier. (a) Basic Circuit. (b) Positive Input Half-Cycle Operation. (c) Negative Input Half-Cycle Operation.**

the full-wave center-tapped rectifier seems to do worse, delivering a peak output of 1/2 V_s. However, the full-wave center-tapped rectifier compensates for the halving of the peak output voltage by doubling the number of half-cycles at the output, compared to a half-wave rectifier. The average output voltage (V_{avg}) can be calculated for a full-wave rectifier with the following formula:

$$V_{avg} = 0.636 \times \frac{1}{2} V_{s\,peak}$$

■ EXAMPLE:

Calculate the average output voltage from a half-wave and full-wave center-tapped rectifier if $V_{s\,peak} = 169.7$ V.

■ *Solution:*

Half-wave rectifier:

$$V_{avg} = 0.318 \times V_{s\,peak}$$
$$= 0.318 \times 169.7\ V = 54\ V$$

Full-wave center-tapped rectifier:

$$\text{If } V_{s\,peak} = 169.7\ V, \text{ then } V_{s1\,peak} \text{ and } V_{s2\,peak} = 84.9\ V$$
$$V_{avg} = 0.636 \times \frac{1}{2} V_{s\,peak}$$
$$= 0.636 \times 84.9\ V = 54\ V$$

As you can see from this example, even though the peak output of the full-wave center-tapped rectifier was half that of the half-wave rectifier, the average output was the same because the full-wave center-tapped rectifier doubles the number of half-cycles at the output, compared to a half-wave rectifier.

Because two half-cycles appear at the output for every one cycle at the input, as shown in Figure 16-23, the ripple frequency will be twice that of the input frequency, and this higher frequency will be easier to filter or smooth of fluctuations.

$$\text{output pulsating dc ripple frequency} = 2 \times \text{input ac frequency}$$

To be completely accurate, we should take into account the barrier voltage drop across the diodes. Since only one diode is on for each half-cycle, the peak output voltage will only be less 0.7 V (silicon diode) as shown by the last waveform in Figure 16-23. Therefore

$$V_{out} = \frac{1}{2} V_{s\,peak} - 0.7\ V$$

With regard to the peak inverse voltage, if you refer back to Figure 16-22(b), you will see that the full secondary peak voltage ($V_{s\,peak}$) appears across the OFF diode D_2 for one half-cycle of the input. Similarly, the full $V_{s\,peak}$ voltage appears across the OFF D_1 during the other half-cycle of the input, as shown in Figure 16-22(c). Both diodes must therefore have a peak inverse voltage (PIV) rating that is larger than the peak secondary voltage ($V_{s\,peak}$). For the circuit in Figure 16-23, the diode's maximum reverse voltage rating (PIV) must be greater than 169.7 V.

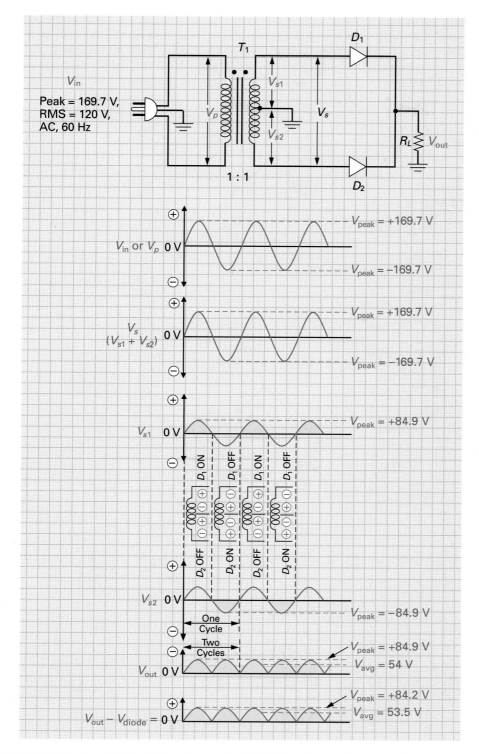

FIGURE 16-23 Input/Output Waveforms of a Full-Wave Center-Tapped Rectifier.

Bridge Rectifier. By center tapping the secondary of the transformer and having two diodes instead of one, we were able to double the ripple frequency and ease the filtering process. However, that seems to be the only advantage the full-wave center-tapped rectifier has over the half-wave rectifier, and the price we pay is having a more expensive center-tapped transformer and an extra diode.

With the **bridge rectifier circuit** shown in Figure 16-24(a), we can have the peak secondary output voltage of the half-wave circuit and the full-wave ripple frequency of the center-tapped circuit. This circuit was originally called a "bridge" rectifier because its shape resembled the framework of a suspension bridge. Figures 16-24(b) and (c) illustrate how the bridge rectifier circuit will behave when an ac input cycle is applied.

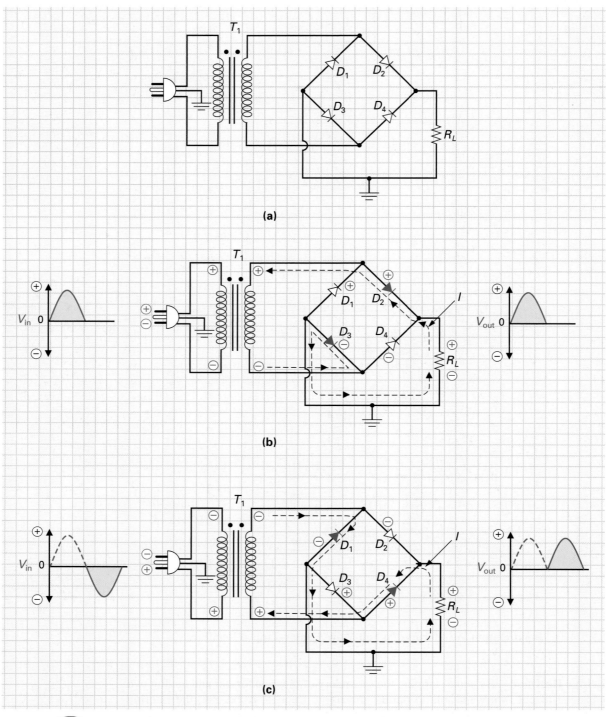

FIGURE 16-24   Full-Wave Bridge Rectifier. (a) Basic Circuit. (b) Positive Half-Cycle Operation. (c) Negative Half-Cycle Operation.

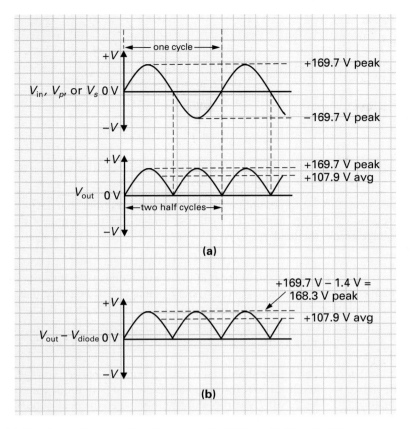

FIGURE 16-25   **Input/Output Waveforms of a Full-Wave Bridge Rectifier.**

When the ac input (V_{in}) swings positive, the circuit operates in the manner illustrated in Figure 16-24(b). A positive potential is applied to the top of the bridge, causing D_2 to turn ON, while a negative potential is applied to the bottom of the bridge, causing D_3 to turn ON. With D_2 and D_3 ON, and D_1 and D_4 OFF, electrons will flow up through the load as indicated by the dashed current line (I), developing a positive output half-cycle across the load (V_{out}).

When the ac input (V_{in}) swings negative, the circuit operates in the manner illustrated in Figure 16-24(c). A negative potential is applied to the top of the bridge, causing D_1 to turn ON, while a positive potential is applied to the bottom of the bridge, causing D_4 to turn ON. With D_1 and D_4 ON and D_2 and D_3 OFF, electrons will flow up through the load as indicated by the dashed current line (I), again developing a positive output half-cycle across the load (V_{out}).

Like the center-tapped rectifier, the bridge rectifier switches two half-cycles of the same polarity through to the load. However, unlike the center-tapped rectifier, the bridge rectifier connects the total peak secondary voltage across the load. A secondary peak voltage of 169.7 V ac would produce a pulsating dc peak across the load of 169.7 V, or the same voltage, as shown in Figure 16-25(a). The average voltage in this case would be larger than that of a center-tapped rectifier, and equal to

$$V_{avg} = 0.636 \times V_{s\,peak}$$

$$= 0.636 \times 169.7\text{ V} = 107.9\text{ V}$$

EXAMPLE:

Calculate the average output voltage from a center-tapped rectifier and a bridge rectifier if $V_{s\,peak} = 169.7$ V.

Solution:

Center-tapped:

$$V_{\text{avg}} = 0.636 \times 1/2\ V_s$$
$$= 0.636 \times 84.9\ \text{V}$$
$$= 54\ \text{V}$$

Bridge:

$$V_{\text{avg}} = 0.636 \times V_s$$
$$= 0.636 \times 169.7\ \text{V}$$
$$= 107.9\ \text{V}$$

As you can see from this example, unlike the center-tapped rectifier that only connects half of the peak secondary voltage across the load, the bridge rectifier connects the total peak secondary voltage across the load.

Because two half-cycles appear at the output for every one cycle at the input, as shown in Figure 16-25(a) on the previous page, the ripple frequency will be twice that of the input frequency, and this higher frequency will be easier to filter or smooth of fluctuations.

$$\boxed{\text{output pulsating dc ripple frequency} = 2 \times \text{input ac frequency}}$$

To be completely accurate, we should take into account the barrier voltage drop across the diodes. Because two diodes are ON for each half-cycle and both diodes are connected in series with the load, a 1.4 V (2 × 0.7 V) drop will occur between the peak secondary voltage (V_s) and the peak output voltage, as shown in Figure 16-25(b).

$$\boxed{V_{\text{out}} = V_{s\ \text{peak}} - 1.4\ \text{V}}$$

With regard to the peak inverse, if you refer to Figure 16-26, you can see that the full $V_{s\ \text{peak}}$ voltage appears across the two OFF diodes. Therefore, each diode must have a peak inverse voltage (PIV) rating that is greater than the peak secondary voltage (V_s). For the circuit and values given in Figures 16-24 and 16-25, the diode's maximum reverse voltage rating (PIV) must be greater than 169.7 V.

16-5-4 *Filters*

The filter in a dc power supply converts the pulsating dc output from the half-wave or full-wave rectifier into an unvarying dc voltage, as was shown in the waveforms in the basic

FIGURE 16-26   Peak Inverse Voltage.

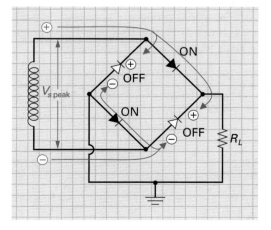

block diagram in Figure 16-16. In this section, we will examine the three basic types of filters: the **capacitive filter,** the *RC* **filter,** and the *LC* **filter.**

The Capacitive Filter

Figure 16-27(a) shows a positive output half-wave rectifier and capacitive filter circuit. The waveforms in Figure 16-27(b), (c), and (d) show how the capacitive filter will respond to the half-wave pulsating dc output from the rectifier. Let us examine how the filter operates by referring to the dashed lines in the circuit in Figure 16-27(a), the output waveform from

Capacitive Filter

A capacitor used in a power supply filter system to suppress ripple currents while not affecting direct currents.

RC Filter

A selective circuit which makes use of a resistance-capacitance network.

LC Filter

A selective circuit which makes use of an inductance-capacitance network.

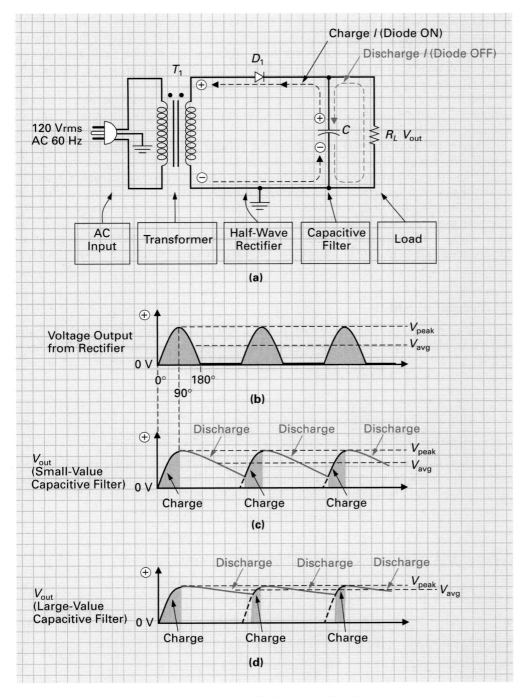

FIGURE 16-27   Capacitive Filtering of a Half-Wave Rectifier Output.

the rectifier shown in Figure 16-27(b), and the small-value capacitive filter waveform shown in Figure 16-27(c). When the ac input swings positive, the diode is turned ON and the capacitor charges as indicated by the black dashed charge current line in Figure 16-27(a). The charge time constant will be small because no resistance exists in the charge path except for that of the resistance of the connecting wires (charge $\tau\downarrow = R\downarrow \times C$). This charge time is indicated by the gray shaded section between 0 and 90° in Figure 16-27(c). When the ac input begins to fall from its positive peak (at 90°), the diode is turned OFF by the large positive potential on the diode's cathode being supplied by the charged capacitor, and the decreasing positive potential on the diode's anode being supplied by the input. With the diode OFF, the capacitor begins to discharge as indicated by the colored dashed discharge current line in Figure 16-27(a). The discharge time constant is a lot longer than the charge time because of the load resistance (discharge $\tau\uparrow = R\uparrow \times C$). The decreasing slope in Figure 16-27(c) illustrates the decreasing voltage across the capacitor, and therefore across the load at the output (V_{out}), as the capacitor discharges. As the ac input and the rectifier's pulsating dc cycle repeat, the output (V_{out}) is, as shown in Figure 16-27(c), an almost constant dc output with a slight variation or ripple above and below the average value.

Figure 16-27(d) shows how a larger-value capacitor will make the charge and discharge time constants longer ($\tau\uparrow = R \times C\uparrow$), therefore decreasing the amount of ripple and increasing the average output voltage.

Figure 16-28(a) shows a positive output voltage full-wave bridge rectifier and capacitive filter circuit. Figures 16-28(b), (c), and (d) show the output waveforms from the rectifier and the filtered output from a small-value and a large-value capacitor filter. The ripple frequency from a full-wave rectifier is twice that of a half-wave rectifier, and therefore the capacitor does not have too much time to discharge before another positive half-cycle reoccurs. This results in a higher average voltage than in a half-wave circuit, and if a larger-value capacitor is used, the ripple will be even less because of the greater time constant ($\tau\uparrow = R \times C\uparrow$), causing an increased average voltage, as shown in Figure 16-28(d).

Percent of Ripple

In the previous two capacitive filter circuits, you could see that even though the output from a filter should be a constant dc level, there is a slight fluctuation or ripple. This fluctuation is called the percent ripple and its value is used to rate the action of the filter. It can be calculated with the following formula:

$$\text{Percent ripple} = \frac{V_{rms} \text{ of ripple}}{V_{avg} \text{ of ripple}} \times 100$$

To help show how this is calculated, let us apply this formula to the example in Figure 16-29. In this example, the filter's output is fluctuating between $+20$ and $+30$ V, and therefore the ripple's peak-to-peak value is

$$\text{Peak-to-peak of ripple} = 20 \text{ to } 30 \text{ V}$$
$$= 10 \text{ V pk-pk}$$

A peak-to-peak value of 10 V means that the ripple has a peak value of

$$\text{Peak of ripple} = 1/2 \text{ of pk-pk value}$$
$$= 1/2 \text{ of } 10 \text{ V}$$
$$= 5 \text{ V peak}$$

Once the peak of the ripple is known, the ripple rms value can be calculated:

$$\text{RMS of ripple} = 0.707 \text{ of peak}$$
$$= 0.707 \text{ of } 5 \text{ V}$$
$$= 3.54 \text{ V}$$

CHAPTER 16 / DIODES AND POWER SUPPLY CIRCUITS

FIGURE 16-28   Capacitive
Filtering of a Full-Wave
Rectifier Output.

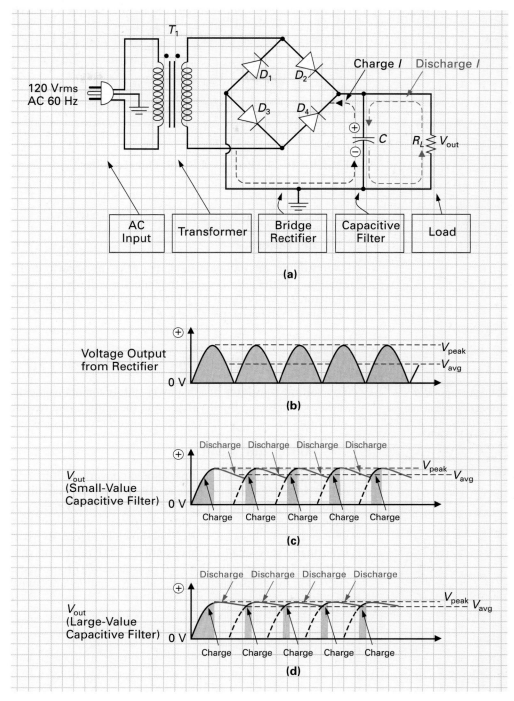

(a)

(b)

(c)

(d)

The average of the ripple is approximately midway between peaks, which in the example in Figure 16-29 is +25 V. Inserting these values in the formula, we can calculate the percent of ripple:

$$\% \text{ ripple} = \frac{V_{rms} \text{ of ripple}}{V_{avg} \text{ of ripple}} \times 100$$

$$= \frac{3.54 \text{ V}}{25 \text{ V}} \times 100$$

$$= 14\%$$

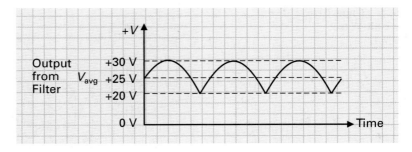

FIGURE 16-29   Percent of Ripple Out of a Filter.

This means that the average voltage out of the rectifier (+25 V) will fluctuate 14%.

LC Filters

Adding an inductor in series (L_1), as shown in Figure 16-30, will increase the efficiency of the filter because the inductor will oppose current without generating heat. The inductor, L_1, will offer a very low opposition or reactance to dc ($X_L\downarrow \propto f\downarrow$) and a very high reactance to the ac ripple ($X_L\uparrow \propto f\uparrow$).

16-5-5   *Regulators*

The **voltage regulator** in a dc power supply maintains the dc output voltage constant despite variations in the ac input voltage and the output load resistance. To help explain why these variations occur, let us first examine the relationship between a source and its load.

Source and Load

Ideally, a dc power supply should convert all of its *ac electrical energy input* into a *dc electrical energy output*. However, like all devices, circuits, and systems, a dc power supply is not 100% efficient and, along with the electrical energy output, the dc power supply generates wasted heat. This is why a dc power supply can be represented as a voltage source with an internal resistance (R_{int}), as shown in Figure 16-31(a). The internal resistance, which is normally very small (in this example, 1 Ω), represents the heat energy loss or inefficiency of the dc power supply. The load resistance (R_L) represents the resistance of the electronic circuits in the electronic system. Since R_{int} and R_L are connected in series with one another,

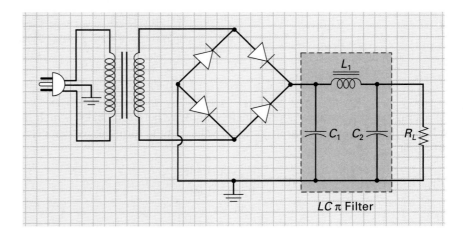

FIGURE 16-30   *LC* π Filter.

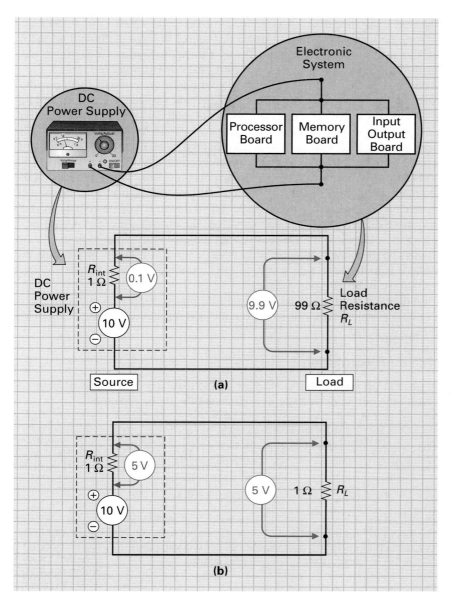

FIGURE 16-31   Source and Load.

the 10 V supply in this example will be developed proportionally across these resistors, resulting in

$$V_{R_{int}} = \quad 0.1 \text{ V}$$
$$V_{R_L} = \frac{9.9 \text{ V}}{10 \text{ V total}}$$

If the load resistance (R_L) were to decrease dramatically to 1 Ω, as shown in Figure 16-31(b), the decrease in load resistance ($R_L\downarrow$) would cause an increase in load current ($I_L\uparrow$). This current increase would cause an increase in the heat generated by the power supply ($P_{R_{int}}\uparrow = I^2\uparrow \times R$). This loss is seen at the output of the dc power supply, which is now delivering only a 5 V output to the load because the 10 V source voltage is being divided equally across R_{int} and R_L.

$$V_{R_{int}} = \quad 5 \text{ V}$$
$$V_{R_L} = \frac{5 \text{ V}}{10 \text{ V total}}$$

If the load resistance decreases, the load current will increase, and if too much current is drawn from the source, it will pull down the source voltage. Although a load resistance will generally not change as dramatically as shown in Figure 16-31(b), it will change slightly as different control settings are selected, and this will *load the source*. For example, the load resistance of the electronic circuits in your music system will change as the volume, bass, treble, and other controls are adjusted. This load resistance change would affect the output of a dc power supply if a regulator were not included to maintain the output voltage constant despite variations in load resistance.

To the electric company, each user appliance is a small part of its load, and it is the source. The ac voltage from the electric company can be anywhere between 105 V and 125 V ac rms (148 to 177 V ac peak), depending on consumer use, which depends on the time of day. When many appliances are in use, the load current will be high and the overall load resistance low, causing the source voltage to be pulled down. If a dc power supply did not have a regulator, these different input ac voltages from the electric company would produce different dc output voltages, when we really want the dc supply voltage to the electronic circuits to always be the same value no matter what ac input is present. This is the other reason that a regulator is included in a dc power supply; it maintains the dc output voltage constant despite variations in the ac input voltage.

In summary, a regulator is included in a dc power supply to maintain the dc output voltage constant despite variations in the output load resistance and the ac input voltage, as shown in Figure 16-32.

Percent of Regulation

The percent of regulation is a measure of the regulator's ability to regulate—or maintain constant—the output dc voltage. It is calculated with the formula

$$\text{percent regulation} = \frac{V_{nl} - V_{fl}}{V_{nl}} \times 100$$

V_{nl} = no-load voltage

V_{fl} = full-load voltage

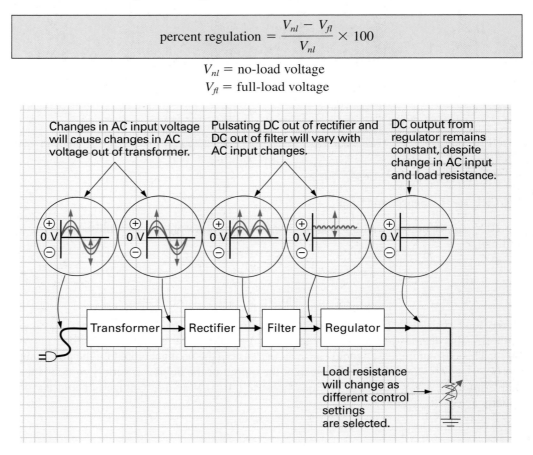

FIGURE 16-32   Function of a Regulator.

If we apply this formula to the example in Figure 16-31, we get

$$\% \text{ regulation} = \frac{V_{nl} - V_{fl}}{V_{nl}} \times 100$$

$$= \frac{10\text{ V} - 5\text{ V}}{10\text{ V}} \times 100 = 50\%$$

In the ideal situation, a regulator would be included and would maintain the output voltage constant between no-load and full-load, resulting in a percent regulation figure of

$$\% \text{ regulation} = \frac{V_{nl} - V_{fl}}{V_{nl}} \times 100$$

$$= \frac{10\text{ V} - 10\text{ V}}{10\text{ V}} \times 100$$

$$= 0\%$$

Most regulators achieve a percent regulation figure that is not perfect (0%) but is generally in single digits (2% to 8%).

Zener Regulator

Figure 16-33 shows how a zener diode (D_5) and a series resistor (R_1) would be connected in a dc power supply circuit to provide regulation. An increase in the ac input voltage would cause an increase in the dc output from the filter, which would cause an increase in the current through the series resistor (I_S), the zener diode (I_Z), and the load (I_L). The reverse biased zener diode will, however, maintain a constant voltage across its anode and cathode (in this example, 12 V) despite these input voltage and current variations, with the additional voltage being dropped across the series resistor. For example, if the output from the filter could be between +15 and +20 V, the zener diode would always drop 12 V, while the series resistor would drop between 3 V (when input is 15 V) and 8 V (when input is 20 V). Consequently, the zener regulator will maintain a constant output voltage despite variation in the ac input voltage.

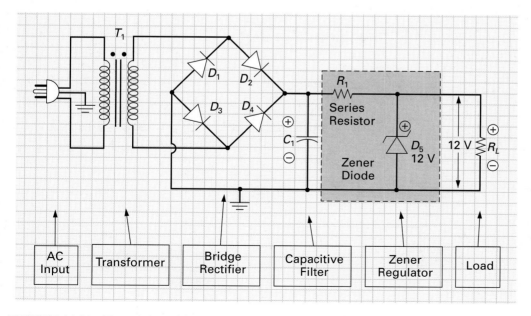

FIGURE 16-33   Zener Diode Regulator.

From the other standpoint, the zener regulator will also maintain a constant output voltage despite variations in load resistance. The zener diode achieves this by increasing and decreasing its current (I_Z) in response to load resistance changes. Despite these changes in zener and load current however, the zener voltage (V_Z), and therefore the output voltage (V_{out} or V_{R_L}), always remains constant. The disadvantage with this regulator is that the series-connected resistor will limit load current and, in addition, generate unwanted heat.

The IC Regulator

Most dc power supply circuits today make use of integrated circuit (IC) regulators, such as the one shown in Figure 16-34. These IC regulators contain about 50 individual or discrete components all integrated on one silicon semiconductor chip and then encapsulated in a three-pin package. The inset in Figure 16-34 shows how all of the IC regulator's internal components form circuits which are shown as blocks. For example, a short-circuit protection and thermal shutdown circuit will protect the IC regulator by turning it OFF if the

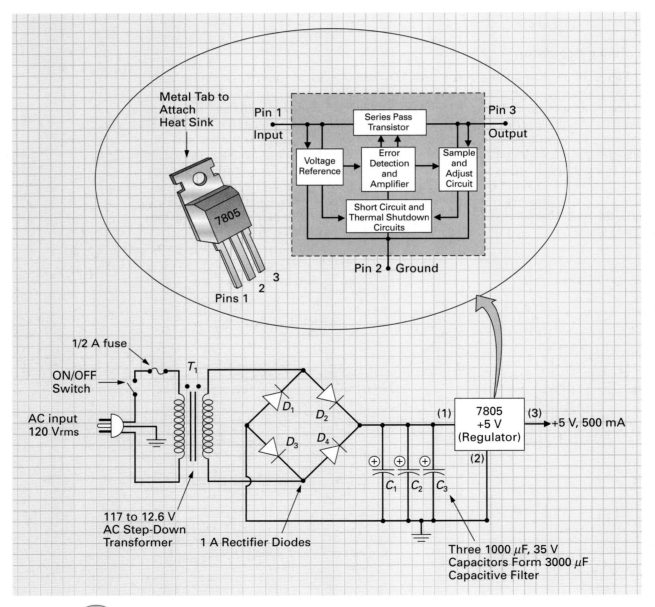

FIGURE 16-34   Integrated Circuit Regulator within a Power Supply Circuit.

load current drawn exceeds the regulator's rated current, or if the heat sink is too small and the IC regulator is generating heat faster than it can dissipate it.

16-5-6   *Troubleshooting a DC Power Supply*

In this section we will discuss how to troubleshoot a typical dc power supply circuit. The three-step troubleshooting procedure for fixing a failure can be broken down into three basic steps:

Step 1: DIAGNOSE

The first step is to determine whether a problem really exists. To carry out this step, a technician must collect as much information as possible about the system, circuit, and components used, and then diagnose the problem.

Step 2: ISOLATE

The second step is to apply a logical and sequential reasoning process to isolate the problem. In this step, a technician will operate, observe, test, and apply troubleshooting techniques in order to isolate the malfunction.

Step 3: REPAIR

The third and final step is to make the actual repair and then test the circuit.

Let us examine this three-step troubleshooting process in more detail, and apply it to a typical dc power supply circuit.

Typical Power Supply Circuit

As an example, Figure 16-35 shows the schematic diagram for a dc power supply circuit that can generate a $+5$ V, $+12$ V, and -12 V dc output from a 120 V or 240 V ac input. You should recognize most of the pieces of this picture because this circuit contains nearly all of the devices discussed in the previous sections of this chapter.

The power supply ON/OFF switch (S_1) switches a 120 V ac input to the 120 V primary-winding connection, or a 240 V ac input to the 240 V primary-winding connection of the transformer (T_1). By doubling the number of primary turns, we can accept twice the input voltage and deliver the same output voltage. The upper secondary winding of T_1 supplies a bridge rectifier module (BR_1), that generates a $+12$ V peak full-wave pulsating dc output, which is filtered by C_1 and regulated by the 7805 (U1) to produce a $+5$ V dc 750 mA output. The $+5$ V output will turn on a "power ON" LED (D_3), which is mounted on the power supply printed circuit board (PCB) along with the filter and regulator. R_1 is a current-limiting resistor for D_3.

The lower secondary winding of T_1 supplies the two half-wave rectifier diodes D_1 and D_2, which generate -50 V and $+50$ V peak outputs, respectively. These inputs are filtered by C_2 and C_3 and then regulated by a 7912 (U_2), which generates a -12 V dc 500 mA output, and by a 7812 (U_3), which generates a $+12$ V dc 500 mA output. The devices C_2, C_3, U_2, and U_3 are also all mounted on the power supply printed circuit board (PCB).

Step 1: Diagnose

It is extremely important that you first understand how a system, circuit, and all of its components are supposed to work so that you can determine whether or not a problem exists. If you were preparing to troubleshoot the dc power supply circuit in Figure 16-35, your first step should be to read through the circuit description and review the operation of each device used in the circuit until you feel completely confident with the correct operation of the circuit.

For other circuits and systems, technicians usually refer to service or technical manuals which generally contain circuit descriptions and troubleshooting guides. As far as each of the devices is concerned, technicians refer to manufacturer data books that contain a full description of the device.

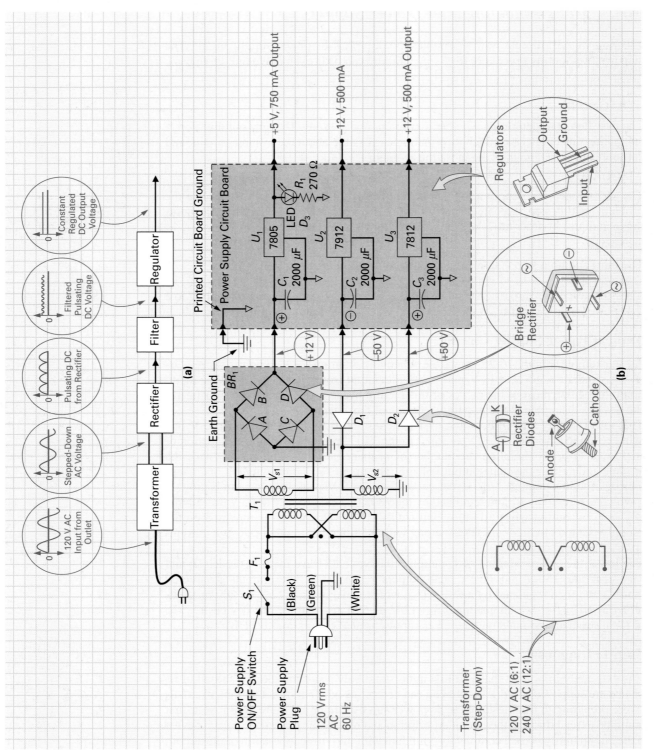

(a)

(b)

FIGURE 16-35   DC Power Supply Circuit. (a) Block Diagram. (b) Circuit Diagram.

Referring to all of this documentation before you begin troubleshooting will generally speed up and simplify the isolation process. Once you are fully familiar with the operation of the circuit, you will be ready to *diagnose the problem* as either an *operator error* or as a *circuit malfunction.*

Here are some examples of operator errors that can be mistaken for circuit malfunctions if a technician is not fully familiar with the operation of the dc power supply circuit in Figure 16-35.

SYMPTOMS	DIAGNOSIS
1. No +5 V, −12 V, or +12 V output.	Check first that the power supply is plugged in and turned ON because either of these operator errors will give the appearance of a circuit malfunction.
2. The +5 V LED is ON, but the −12 V and +12 V LEDs are not lit.	There are no LED indicators for the −12 V and +12 V outputs, therefore this is a normal condition.

Once you have determined that the problem is not an operator error but, in fact, a circuit malfunction, proceed to Step 2 and isolate the circuit problem.

Step 2: Isolate

A technician will generally spend most of the time in this phase of the process isolating a circuit problem. Steps 1 (diagnose) and 3 (repair) may only take a few minutes to complete compared to Step 2, which could take a few hours to complete. However, with practice and a good logical and sequential reasoning process you can quickly isolate even the most obscure of problems.

The directions you take as you troubleshoot a circuit malfunction may be different for each problem. However, certain troubleshooting techniques apply to all problems. Let us now review some of these techniques and apply them to our dc power supply circuit in Figure 16-35.

1. Check first for **obvious errors:**
   a. Power fuse blown
   b. Wiring errors if circuit is newly constructed
   c. Devices are incorrectly oriented. For example, if the rectifier diodes, bridge rectifier module, IC regulators, or the LED were placed in the circuit in the opposite direction to what is shown in Figure 16-35, that device would not operate as it should and therefore neither would the circuit.
2. Use your **senses** to check for broken wires, loose connections, overheating or smoking components, leads or pins not making contact, and so on.
3. and 4. Use a **cause-and-effect troubleshooting process** which involves studying the effects you are getting from the faulty circuit and then trying to reason out what could be the cause. Also, apply the **half-split method** of troubleshooting first to the entire system, then to a circuit within the system, and then to a section within the circuit to help speed up the isolation process. To explain how to use these two methods with the circuit in Figure 16-35, let us examine a few examples.

EXAMPLE:

The dc power supply in Figure 16-35 was being used to supply power to several circuits in an electronic system. If a power problem were to occur, how would we apply the half-split method?

FIGURE 16-36   Using the Half-Split
Method to Isolate a System Problem.

FIGURE 16-36   Using the Half-Split Method to Isolate a System Problem.

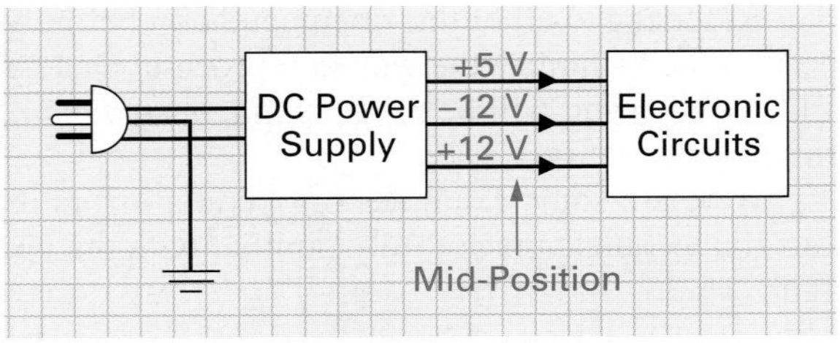

■ *Solution:*

The half-split process involves first choosing a point roughly in the middle of the problem area. By making a test at this midpoint you can determine whether the problem exists either before or after this point. For example, Figure 16-36 shows how this half-split method can be applied to a power supply problem. By testing the output of the dc power supply (+5 V, −12 V, +12 V) you can determine whether the power problem is in the power supply or within the circuits that are being supplied power.

    a. If the dc supply voltages are present and of the correct value, then the dc power supply is functioning normally.

    b. If these dc supply voltages are not present or are not of the correct value, then we will need to isolate the problem.

    In some instances you will have to be careful not to assume the obvious when applying the half-split method. For example, let us imagine that we have tested the output of the dc power supply and its output voltage is incorrect. Your first instinct would be to assume that the problem lies within the dc power supply. However, as we discovered in this chapter, *a short in the load can pull down a source, making it appear to be supplying a faulty output.* Therefore, most problems need to be further isolated because the problem could be in the source or it could be in the load. Let us look at another example in which the source appears to be the problem when, in fact, the fault is in the load.

□ **EXAMPLE:**

The +5 V output from a dc power supply in Figure 16-37 measures +2 V on the multimeter. How would we proceed to troubleshoot this problem?

■ *Solution:*

Figure 16-37 shows how a +5 V output from a dc power supply could be pulled down to +2 V due to a short or low resistance path in one of the electronic circuits that are being supplied power. These problems can cause the power supply input fuse to blow or the IC regulator to disconnect its output due to the excessive current being drawn by the short. Because the output of a power supply generally goes to several boards, and within each board it is connected across many devices, it becomes very difficult to isolate a short. If the fuse does not blow and the regulator IC does not switch OFF, a short will definitely load or pull down the dc supply voltage. At first, it appears that the power supply itself is malfunctioning because of the decreased output voltage, but once you disconnect the output of the dc power supply (at point *A*) from the external circuit boards, the +5 V should go back up to its correct voltage indicating that the problem lies in the load and not in the source. To isolate the faulty circuit board, you will have to reconnect the supply voltage (at point *A*) and then disconnect each circuit board one at a time in the following way to isolate the faulty board:

    a. First disconnect board 1 at point *B* and then check the +5 V output from the dc power supply to see if it has returned to its normal voltage. If the dc supply voltage at point *A*

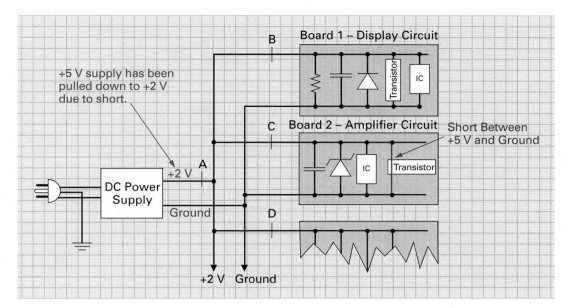

FIGURE 16-37   Locating Power Line Shorts.

remains LOW, then the short is not within board 1 and you should reconnect this board.

b. Disconnect board 2 at point *C* and again check the +5 V output from the dc power supply at point *A* to see if it has returned to its normal voltage. If its output voltage remains LOW, then the short is not within board 2, and you should reconnect this board. In the example in Figure 16-37, however, the short does exist within board 2 and, once this board is disconnected at point *C,* the output from the dc power supply at point *A* will return to +5 V.

c. If the fault was not in board 2, you would have continued to disconnect each board until you isolated the short.

Once the faulty board is located, you will have to troubleshoot that circuit until the short is isolated. For example, if this board were a seven-segment encoder and display circuit, we would follow the troubleshooting procedure for that board. A dramatic approach to localizing a difficult-to-find short on a complex printed circuit board is to freeze spray the entire board using an aerosol freeze can. Once the board is reconnected to the power supply, the short will become quickly visible since its path of excess current will generate heat and therefore defrost faster than the rest of the board.

What would happen if an open developed in one of the lines connecting power to a circuit? Let us answer this question by again looking at another example.

EXAMPLE:

How would you isolate the open shown in Figure 16-38?

Solution:

An open in any of the dc supply lines connecting power to a circuit will cause that circuit not to operate. These opens can be easily traced with a voltmeter to determine the point at which voltage is no longer present.

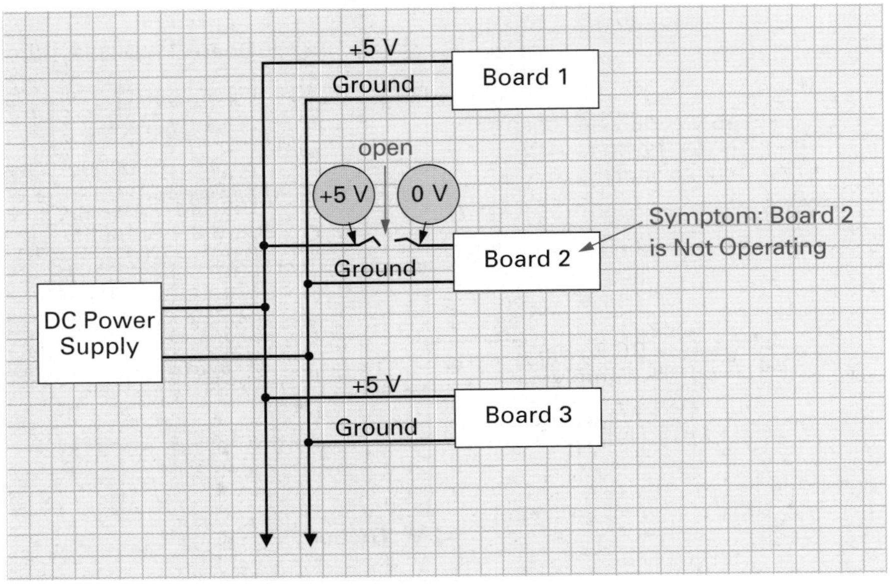

FIGURE 16-38 Locating Power Line Opens.

In summary, once you have disconnected the circuit load from a dc power supply's output, you can isolate the problem as being either a source fault or load fault, because, *if the power supply's output voltage goes back to its rated voltage after it has been disconnected from the circuit load, the problem is with the load. Whereas, if the output voltage remains LOW or at zero after the source has been disconnected from the load, the problem is with the source.*

Assuming that we have isolated the problem to the power supply, let us now examine several types of common failures. Once again, we will apply these problems to the example circuit in Figure 16-35.

▦ **EXAMPLE:**

What would happen to the circuit in Figure 16-35 if the power supply's fuse were to blow?

▦ *Solution:*

A short in the power supply or the load will cause an excessive current to be drawn from the secondary of the transformer. The primary of the transformer will try to respond and deliver the power needed by increasing its current. Since the power supply's fuse (F_1) is connected in series with the transformer's primary, as shown in Figure 16-35, the increase in primary current will blow or open the fuse. When a fuse needs to be replaced, always make sure that the replacement has the same voltage and current rating. In most cases, power supplies use a **slow-blow fuse,** which allows a surge of current to pass through the power supply when it is first turned on. This surge current is momentary, usually lasting for less than a second, while all of the system circuits are warming up. If the surge or high current persists for longer than a second, however, the fuse will blow and protect the power supply. Under no circumstances should you ever use a fuse that has a higher current rating because this will permit a higher value of current to pass through to the power supply and load circuits. Never be tempted to "defeat" or bypass the fuse if you do not have a replacement by placing a piece of wire across the fuse terminals. This could introduce other faults, start a fire, or possibly electrocute somebody, since there is now nothing to prevent a damaging value of current.

Slow-Blow Fuse

Allows a surge of current to pass through the power supply when it is first turned on.

EXAMPLE:

What would happen to the circuit in Figure 16-35 if the power supply transformer (T_1) developed the following problems:

a. An open primary or secondary winding

b. A shorted primary or secondary winding

c. A short between the primary winding and its grounded case, or between the secondary and its grounded case

Solution:

a. An open in the primary or secondary winding will cause the power supply's output voltage to drop to zero. Making a few checks with the voltmeter, you will find that you have primary ac voltage but no secondary ac voltage. If you unplug the power supply and disconnect the transformer and check the resistance of its windings, you will find the open winding using the ohmmeter (it will have an infinite resistance).

b. A shorted primary or secondary will provide a low resistance path, resulting in an excessive primary current and a blown fuse. If the fuse does not blow because the current is not quite high enough, the shorted transformer will produce almost no output voltage and be generating a lot of heat. Once again, by unplugging and disconnecting the transformer and then checking its primary and secondary winding resistance, you should find the short (the transformer in Figure 16-35 should typically have a primary resistance of 50 Ω, and a secondary resistance of about 10 Ω).

c. If either the primary winding or secondary winding were to short to the transformer's casing (which is generally grounded), the result will be a low-resistance path to ground, resulting in an excessive primary current and a blown fuse. Once again, by unplugging and disconnecting the transformer and then checking the winding-to-case resistance for both the primary and secondary, you should be able to find out if a winding has shorted to the case. This winding-to-case resistance should normally be infinite ohms.

Once you have determined that the transformer has developed an internal problem, you will have to get a replacement because these problems cannot be repaired. Be sure that the replacement has the same transformer rating.

EXAMPLE:

What would happen to the circuit in Figure 16-35 if one of the rectifier diodes were to open in the

a. Half-wave rectifiers

b. Full-wave rectifier

Solution:

a. An open rectifier diode is quite a common power supply problem. If a half-wave rectifier diode such as D_1 or D_2 in Figure 16-35 were to open, the symptoms would be easy to diagnose because no pulsating dc would be applied to the filters and regulators, and there would not be a -12 or $+12$ V output from the power supply. Like all opens, all of the supply voltage (V_{s2}) will appear across the defective device's terminals when measured with the multimeter.

b. If one of the diodes in the bridge rectifier in Figure 16-35 were to open, it would prevent current flow for one-half of the input cycle. This would result in a half-wave output that would be difficult to filter, and if the average voltage into U_1 is less than $+5$ V, the regulator would not function, producing no $+5$ V output.

EXAMPLE:

EXAMPLE:

What would happen to the circuit in Figure 16-35 if one of the rectifier diodes were to short in the

    a. Half-wave rectifiers

    b. Full-wave rectifier

Solution:

    a. A shorted half-wave rectifier diode, such as D_1 or D_2 in Figure 16-35, would connect ac to the filter capacitor, causing the filter capacitor to charge and discharge. Because a constant dc voltage is not being applied to the regulator, there would be no dc output voltage.

    b. If a diode, such as *B,* in the bridge rectifier module BR_1 in Figure 16-35 were to short, it would act like a forward biased diode when diodes *B* and *C* are turned ON. However, when diodes *A* and *D* are forward biased, the ON *D* and shorted *B* will short the secondary winding, causing an excessive current. It is likely that this high current will burn open the shorted diode *B* and the ON diode *D.*

EXAMPLE:

What would happen to the circuit in Figure 16-35 if a filter capacitor were leaky or were to short?

Solution:

A shorted or leaky filter capacitor will generally cause an excessive current that will usually burn open the rectifier diodes, producing no dc output. If the leakage is not too large, the problem will be an increase in the ripple voltage into the regulator.

EXAMPLE:

What would happen to the circuit in Figure 16-35 if a regulator's output current rating was exceeded?

Solution:

Referring to Figure 16-35, you can see that each output of the regulator has a rated voltage and maximum load current value. If this current is exceeded due to a short in the load, the regulator will shut down to protect itself. These built-in short-circuit protection and thermal shutdown circuits normally prevent the regulator from damage. In most cases, if the regulator's input is fine but there is no output voltage, the regulator has probably shut itself down, due to a short in the load or an insufficient heat sink. By disconnecting the output of the regulator from the load, you should be able to determine if the problem is the source or the load.

Step 3: Repair

The final step is to repair the circuit, which could involve simply removing a wire clipping or some excess solder, resoldering a broken connection, reconnecting a connector, or some other easy repair. In most instances, however, the repair will involve the replacement of a faulty component. For a circuit that has been constructed on a breadboard or prototyping board, the removal and replacement of the component is simple. However, when a printed circuit board is involved you should make a note of the component's orientation (in the case of rectifier diodes, rectifier modules, electrolytic capacitors, IC regulators, and other similar devices), and observe good desoldering and soldering practices.

When the circuit has been repaired, always perform a final test to see that the circuit is now fully operational.

Now that you have completed this section, you should be able to:

■ **Objective 15.** *Define the function of the transformer, rectifier, filter, and regulator in a dc power supply.*

■ **Objective 16.** *Describe the characteristics of and differences between the half-wave rectifier, the full-wave center-tapped rectifier, and the full-wave bridge rectifier.*

■ **Objective 17.** *Describe the characteristics of the capacitive filter, RC filter, and LC filter.*

■ **Objective 18.** *Describe the relationship between source and load.*

■ **Objective 19.** *Explain how a zener diode can be connected to act as a power supply regulator.*

■ **Objective 20.** *Describe the operation and characteristics of an IC regulator.*

■ **Objective 21.** *Explain the function of each device in a typical 120 V/240 V ac input +5 V, +12 V, and −12 V dc output power supply.*

■ **Objective 22.** *Describe the three-step troubleshooting process and apply it to a typical dc power supply circuit.*

■ **Objective 23.** *Examine some of the more common dc power supply circuit malfunctions such as transformer, rectifier, filter, and regulator problems.*

Use the following questions to test your understanding of Section 16-5.

1. What are the four blocks that make up a dc power supply?
2. What is the function of
   a. A Rectifier
   b. A Regulator
3. Why is a filter needed in a dc power supply?
4. What is the three-step troubleshooting procedure?

SUMMARY

1. The first diode was accidentally created by Edison in 1883 when he was experimenting with his light bulb.
2. The word *diode* is derived from the fact that the device has two (*di*) electrodes (*ode*).
3. A diode will act as a conductor and pass current easily in one direction when the bias voltage across it is of one polarity, yet it will block current and imitate an insulator when the bias voltage applied is of the opposite polarity.

The Junction Diode

4. The two electrodes or terminals of the diode are called the anode and cathode.
5. A black band or stripe is generally placed on the package closest to the cathode terminal for identification purposes.
6. If you make the anode terminal of a diode positive with respect to the cathode terminal, the device is equivalent to a closed switch and is said to be ON, or forward biased.
7. If you make the anode terminal of a diode negative with respect to the cathode, the device is equivalent to an open switch and is said to be OFF, or reverse biased.

8. Semiconductor devices such as diodes and transistors are constructed using P-N junctions. A diode for example, has only one P-N junction and is created by doping a single piece of pure semiconductor to produce an *n*-type and *p*-type region. A bipolar junction transistor, on the other hand, has two P-N junctions and is created by doping a single piece of pure semiconductor with three alternate regions (NPN or PNP).
9. The *n*-type region is called the cathode, while the *p*-type region is called the anode.

Junction Diode Characteristics (Figure 16-39)

10. The voltages, which incline or cause the diode to operate in a certain manner, are known as bias voltages. Bias voltages control the width of the depletion region, which in turn controls the resistance of the junction and therefore the amount of current that can pass through the P-N junction diode.
11. Like the P-N junction, the junction diode's operation is determined by the polarity of the applied voltage.

FIGURE 16-39   Junction Diode Characteristics.

Voltage/Current Characteristics

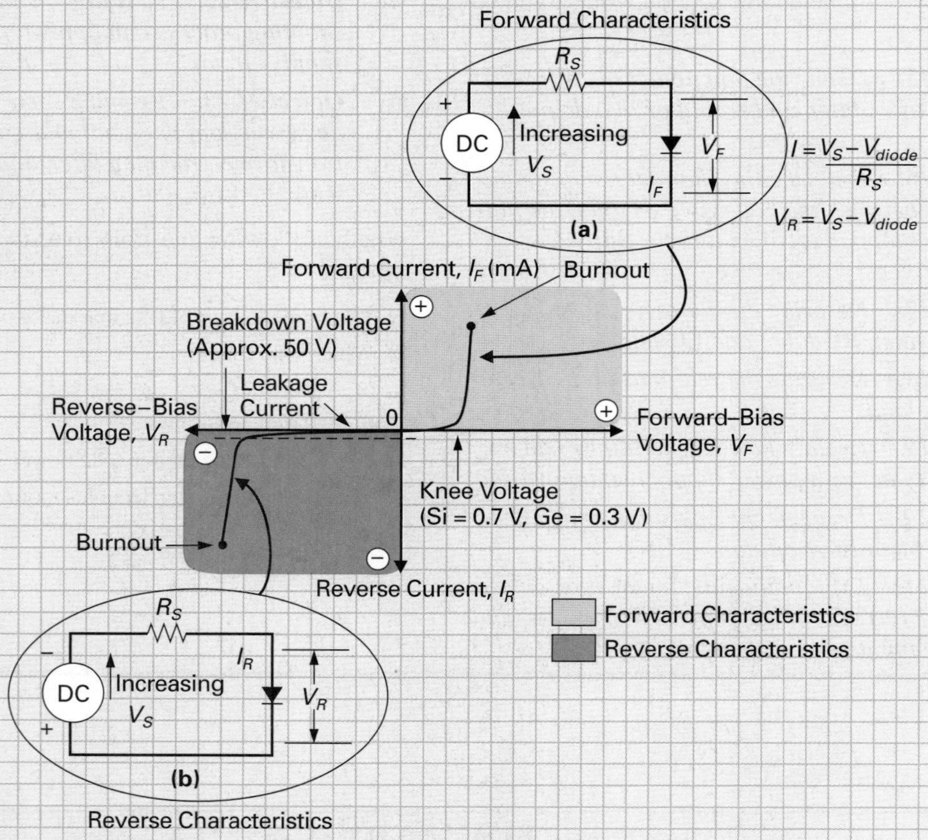

Forward Characteristics

R_S

DC

$+$

$-$

Increasing
V_S

V_F

I_F

$I = \dfrac{V_S - V_{diode}}{R_S}$

$V_R = V_S - V_{diode}$

(a)

Forward Current, I_F (mA)   Burnout

Breakdown Voltage
(Approx. 50 V)

Leakage
Current

Reverse–Bias
Voltage, V_R

$\ominus$

0

$\oplus$

Forward–Bias
Voltage, V_F

$\oplus$

Knee Voltage
(Si = 0.7 V, Ge = 0.3 V)

Burnout

$\ominus$

Reverse Current, I_R

Forward Characteristics

Reverse Characteristics

R_S

DC

$-$

$+$

Increasing
V_S

I_R

V_R

(b)

Reverse Characteristics

Temperature Characteristics

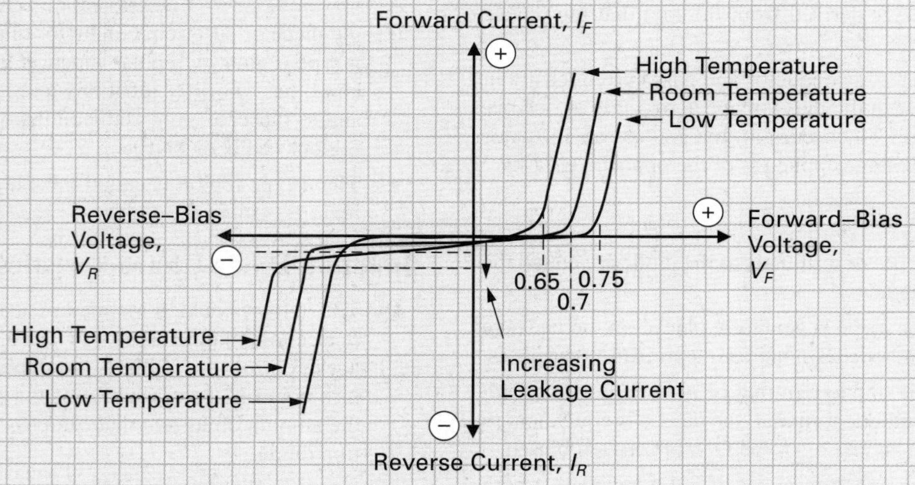

Forward Current, I_F

$\oplus$

High Temperature
Room Temperature
Low Temperature

Reverse–Bias
Voltage,
V_R

$\ominus$

$\oplus$

Forward–Bias
Voltage,
V_F

0.65   0.75
0.7

High Temperature
Room Temperature
Low Temperature

Increasing
Leakage Current

$\ominus$

Reverse Current, I_R

12. A diode is forward biased when the negative terminal of the applied voltage is connected to the *n* region of the diode, and the positive terminal of the applied voltage is connected to the *p* region of the diode ($+V \rightarrow p$ region, $-V \rightarrow n$ region). When forward biased, free electrons are repelled from the *n* region by the negative source and attracted to the positive terminal of the voltage source. This forward-conducting electron flow will only occur if the external source voltage is large enough to overcome the internal barrier voltage of the junction diode. For a silicon diode, the external source voltage must be equal to or greater than 0.7 V, whereas for a germanium diode the applied voltage must be equal to or greater than 0.3 V.

13. The diode symbol actually points in the direction of forward hole flow and reflects conventional flow.

14. A diode is reverse biased when the positive terminal of the applied voltage is connected to the *n* region of the diode, and the negative terminal of the applied voltage is connected to the *p* region of the diode ($+V \rightarrow n$ region, $-V \rightarrow p$ region). This applied voltage polarity will reverse bias the junction diode because free electrons in the *n* region are attracted and therefore travel to the positive terminal of the applied source voltage. At the same time, holes feel the attraction of the negative terminal of the applied voltage. As a result, the depletion region will increase in width until its internal voltage is equal but opposite to the external applied voltage. At this time, the diode will stop conducting, cutting off all current flow.

15. A very small leakage current leaks through the diode when it is reverse biased. This current is extremely small (microamps to nanoamps in value) and in most cases can be ignored. It is caused by minority carriers, which are holes in the *n* region and electrons in the *p* region, that are forced towards the junction by the applied source voltage.

16. The voltage-current characteristic curve is a graph that shows how much current will pass through a typical junction diode when it is forward biased or reverse biased.

17. The point on the forward voltage scale of the *V-I* characteristic curve at which the curve suddenly rises and resembles the shape of a human knee is called the knee voltage. This knee voltage is just another name for the diode's internal barrier voltage, which for a silicon diode is about 0.7 V.

18. The forward voltage drop across the diode remains almost constant between 0.7 V to 0.75 V.

19. The amount of heat produced in the diode is proportional to the value of current through the diode ($P\uparrow = I^2\uparrow \times R$).

20. To prevent current rising to a damaging level, a series current limiting resistor (R_S) is always included to limit the forward current.

21. If the reverse voltage (V_R) is increased, a point will be reached where the diode will break down, resulting in a sudden increase in current. This excessive current is due to the large external reverse bias voltage that is now strong enough to pull valence electrons away from their parent atoms, resulting in a large increase in minority carriers. The point on the reverse voltage *V-I* characteristic scale at which the diode breaks down and there is a sudden increase in reverse current is called the breakdown voltage, or peak inverse voltage (PIV).

22. Semiconductor materials, and therefore diodes, have a negative temperature coefficient of resistance. This means as temperature increases ($T\uparrow$), their resistance decreases ($R\downarrow$).

23. The forward voltage drop across a conducting diode is inversely proportional to temperature, which means that the voltage drop across a diode will be less at higher temperatures.

24. At higher temperatures, the reverse leakage current increases to a point that the reverse biased diode is no longer equivalent to an open switch.

25. A manufacturer's specification sheet or data sheet details the characteristics and maximum and minimum values of operation for a given device.

26. As a technician, you should be familiar with some of the basic operating limits of a device so that you can isolate component malfunctions within a circuit by determining whether the device is operating to specifications.

Junction Diode Testing

27. Diodes that are operated beyond their maximum forward and reverse ratings will, more than likely, malfunction. These malfunctions can result in one of two types of failure. The diode may burn out and then act as a permanently open switch (less common), or effectively melt the semiconductor material and act as a permanently closed switch (more common). A defective diode will no longer be able to be switched ON or OFF. Instead, it will either remain permanently OFF or permanently ON.

28. A good diode should display a very low resistance when it is biased ON and a very high resistance when it is biased OFF. If good, when a diode is forward biased by an ohmmeter it should display a low value of resistance (typically less than 10 Ω). When a diode is then flipped over and reverse biased by the ohmmeter's internal battery (+ lead to cathode, − lead to anode), the ohmmeter should display a very high resistance (typically greater than 1000 MΩ). Since this value is generally off the ohmmeter's scale, you will probably have the display showing OL or OR.

29. A diode that has been damaged and is permanently shorted between its anode and cathode will display a LOW resistance on the ohmmeter when it is both forward and reverse biased by the ohmmeter.

30. A diode that has been damaged and is permanently opened between its anode and cathode will display a HIGH resistance on the ohmmeter when it is both forward and reverse biased by the ohmmeter.

31. After the basic P-N junction diode, the two most widely used diodes are the zener diode and the light-emitting diode.

32. The basic P-N junction diode, zener diode, and light-emitting diode can be used in a variety of circuit applications such as rectifying, regulating, detecting, clipping, clamping, multiplying, switching, displaying, and oscillating.

The Zener Diode (Figure 16-40)

33. It was American inventor Clarence Zener who first began investigating in detail the reverse breakdown of a diode. He discovered that the basic P-N junction diode would conduct a high reverse current when a sufficiently high reverse bias voltage was applied across its terminals. This large external reverse voltage would cause an increase in the number of minority carriers in the *n* and *p* regions as valence electrons are pulled away from their parent atoms. Clarence Zener was fascinated by the reverse breakdown characteristics of a basic P-N junction diode and, after many years of study, developed a special diode that could handle high values of reverse current and still safely dissipate any heat that was generated. These special diodes, constructed to operate at voltages that are equal to or greater than the reverse breakdown voltage rating, were named zener diodes after their inventor.

34. The zener diode symbol resembles the basic P-N junction diode symbol in appearance. However, the zener diode symbol has a zig-zag bar instead of the straight bar.

35. The zener diode packages are identical to those of the basic P-N junction diode in that a band or stripe is used to identify the cathode end of the zener diode in the low-power packages, whereas the threaded terminal of a high-power package is generally always the cathode.

36. The characteristic curve of a zener diode is almost identical to the basic P-N junction diode's characteristic curve. For example, when forward biased at or beyond 0.7 V, the zener diode will turn ON and be equivalent to a closed switch. When reverse biased, the zener diode will turn OFF and be equivalent to an open switch. The main difference, however, is that the zener diode has been specifically designed to operate in the reverse breakdown region of the curve. This is achieved by making sure that the external bias voltage applied to a zener diode will not only reverse bias the zener diode ($+ \rightarrow$ cathode, $- \rightarrow$ anode) but also be large enough to drive the zener diode into its reverse breakdown region.

37. As the reverse voltage across the zener diode is increased from the graph origin (which represents 0 volts), the value of reverse leakage current (I_R) begins to increase. The reverse zener breakdown region of the curve begins when the current through the zener suddenly increases. This point on the reverse bias voltage scale is called the zener breakdown voltage (V_{Br}). At this knee-shaped point, a minimum value of reverse current, known as the zener knee current (I_{ZK}), will pass through the zener. If the applied voltage is increased beyond V_{Br}, the value of zener current (I_Z) will increase. The voltage drop developed across the zener when it is being operated in the reverse zener breakdown region is called the zener voltage (V_Z).

38. The voltage drop across a zener diode (V_Z) remains almost constant when it is operated in the reverse zener breakdown region, even though current through the zener (I_Z) can vary considerably. This ability of the zener diode to maintain a relatively constant voltage regardless of variations in zener current is the key characteristic of the zener diode.

39. Manufacturers rate zener diodes based on their zener voltage (V_Z). A wide variety of zener diode voltage ratings are available ranging from 1.8 V to several hundred volts.

40. If the maximum zener current (I_{ZM}) of a zener diode is exceeded, the diode will more than likely be destroyed because this value of current will generate more heat than the zener can safely dissipate (P_D).

41. The voltage developed across a zener diode will be equal to the zener diode's voltage rating, and the voltage developed across the circuit's series resistor will be equal to the difference between the diode's zener voltage and the input voltage.

42. Zener impedance (Z_Z) is the opposition offered by a zener diode to current.

Zener Diode Testing

43. The best way to test a zener diode is to connect the volt meter across the zener while it is in circuit and power is applied. If the voltage across the zener is at its specified voltage, then the zener is functioning properly. If the voltage across the zener is not at the nominal value, you should check the input voltage, the series resistor, the load, and then the zener.

The Light Emitting Diode (Figure 16-41)

44. The light-emitting diode (LED) is a semiconductor device that produces light when an electrical current or voltage is applied to its terminals. In other words, it converts electrical energy into light energy. It is classified as an optoelectronic device because it combines optics and electronics.

45. The LED is probably the most widely used light source in electronic equipment. It has replaced the incandescent lamp due to its longer life expectancy and lower operating power.

46. When forward biased, the LED will emit energy in response to a forward current. This emission of energy may be in the form of heat energy, light energy, or both heat and light energy depending on the type of semiconductor material used. It is the type of material used to construct the LED that determines the color, and therefore frequency, of the light emitted. For example, different compounds are available that will cause the LED to emit red, yellow, green, blue, white, orange, or infrared light when it is forward biased.

47. The LED schematic symbol has two arrows leaving the standard diode symbol that represent light.

48. The package contains the two terminals for connection to the anode and cathode, and a semi-clear case which contains the light emitting diode and a lens. The dome-shaped top of the plastic (epoxy) case serves as the lens and acts as a magnifier to conduct light away from the LED chip. By adjusting the lens material, lens shape, and the distance of the LED chip from the lens, manufacturers can obtain a variety of radiation patterns.

49. There are three methods used to identify the anode and cathode leads of the LED. In all three cases the cathode lead is distinguished from the anode lead by either having

FIGURE 16-40   The Zener Diode.

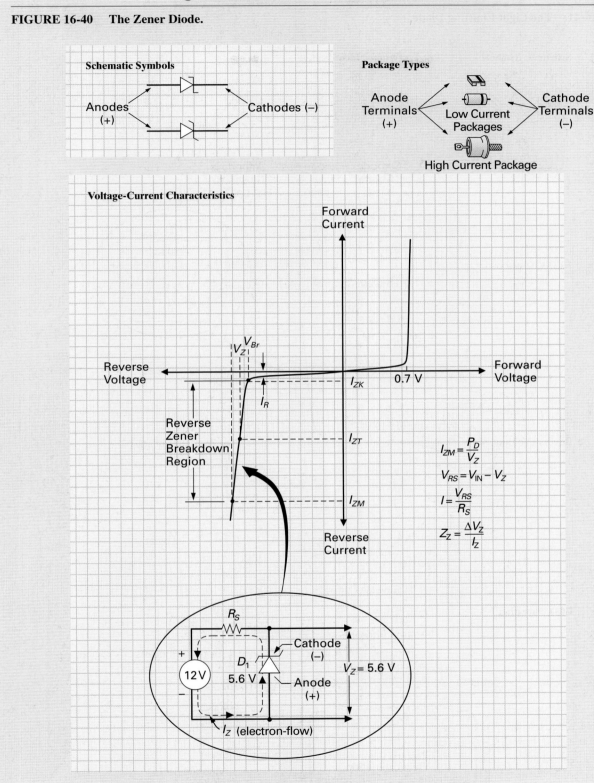

Schematic Symbols

Anodes (+)   Cathodes (−)

Package Types

Anode Terminals (+)   Low Current Packages   Cathode Terminals (−)

High Current Package

Voltage-Current Characteristics

Forward Current

Reverse Voltage

V_Z V_{Br}

Reverse Zener Breakdown Region

I_{ZK}

I_R

I_{ZT}

I_{ZM}

0.7 V

Forward Voltage

Reverse Current

$$I_{ZM} = \frac{P_D}{V_Z}$$

$$V_{RS} = V_{IN} - V_Z$$

$$I = \frac{V_{RS}}{R_S}$$

$$Z_Z = \frac{\Delta V_Z}{I_Z}$$

R_S

D_1

12 V

5.6 V

Cathode (−)

Anode (+)

$V_Z = 5.6$ V

I_Z (electron-flow)

711

FIGURE 16-41   The Light Emitting Diode.

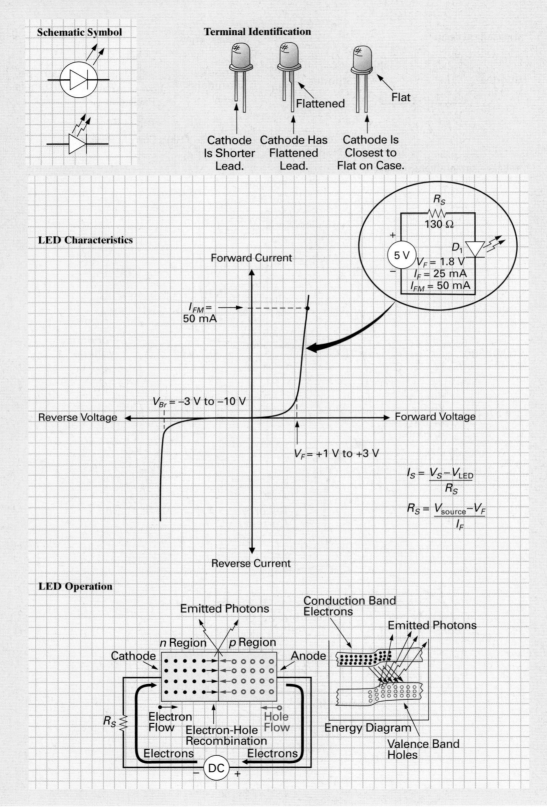

its lead shorter, its lead flattened, or the side of the case nearest to the cathode lead.

50. LEDs have a high forward voltage rating (V_F is typically $+1$ V to $+3$ V) and a low maximum forward current rating (I_F is typically 20 mA to 100 mA). These ratings mean that the LED will develop more voltage across its terminals when it is forward biased. The low forward current ratings means that the LED will usually always need to be protected from excessive current damage by a series current-limiting resistor.

51. LEDs have low reverse breakdown voltage values (V_{Br} is typically -3 V to -10 V), which means that only a low reverse voltage will cause the LED to break down and become damaged.

52. When forward biased, the negative terminal of the dc source injects electrons into the n-type region, or cathode, of the LED. These electrons travel towards the P-N junction. At the same time, electrons are attracted out of the p-type region, or anode, and travel towards the positive terminal of the source voltage making it appear as though holes are moving towards the P-N junction. The electrons from the n-type region and the holes from the p-type region combine at the P-N junction. Because the electrons are at a higher energy level (conduction band), they release energy as they drop into a valence hole. The energy is a small packet of light called a photon. Most forward biased LEDs release billions of photons per second as billions of electrons combine with holes at the P-N junction. The semi-transparent semiconductor material used to construct LEDs allows the light energy to escape where it is captured and focused by the dome shaped lens of the LED package.

53. The light-emitting diode's key specifications are maximum forward current, forward voltage drop, and maximum reverse voltage.

54. The brightness of an LED is determined by the value of current that passes through the LED. Therefore, more current will mean more light, and less current will mean less light.

Testing LEDs

55. Infrared-emitting diodes are used in security and fiber optic communication applications.

56. Light-emitting diodes have a very long life expectancy and are much more rugged than their predecessor, the small incandescent light bulb.

57. A problem with an LED circuit is normally easily noticed because the LED fails to light up. As with all circuit troubleshooting, you can isolate the circuit fault by systematically checking all of the logical possibilities.

Power Supply Transformers

58. The turns ratio of a transformer in a dc power supply can be selected to either increase or decrease the 120 V ac input. With most electronic equipment, a supply voltage of less than 120 V is required, and therefore a step-down transformer is used.

Half-Wave Rectifier Circuits

59. The P-N junction diode's ability to switch current in only one direction makes it ideal for converting two-direction alternating current into one-direction direct current.

60. The half-wave rectifier circuit is constructed simply by connecting a diode between the power supply transformer and the load.

61. When the secondary ac voltage swings positive, the anode of the diode is made positive, causing the diode to turn ON and connect the positive half-cycle of the secondary ac voltage across the load (R_L). When the secondary ac voltage swings negative, the anode of the diode is made negative, and therefore the diode will turn OFF. This will prevent any circuit current, and therefore no voltage will be developed across the load (R_L).

62. The circuit is called a half-wave rectifier because only half of the input wave is connected across the output.

63. To be more accurate, there will, of course, be a small voltage drop across the diode due to its barrier voltage of 0.7 V for silicon and 0.3 V for germanium.

64. When the input swings negative, the entire negative supply voltage will appear across the open or OFF diode. The maximum reverse breakdown voltage, or peak inverse voltage (PIV) rating of the rectifier diode, must therefore be larger than the peak of the ac voltage at the diode's input.

65. By changing the direction of the rectifier diode, the rectifier can be made to produce either a positive or a negative dc output.

66. The frequency of the pulsating dc output from a rectifier is called the ripple frequency, and for half-wave rectifier circuits, the output pulsating dc ripple frequency = input ac frequency.

Full-Wave Center-Tapped Rectifier Circuits

67. The half-wave rectifier's output is difficult to filter to a smooth dc level because the output voltage and current are applied to the load for only half of each input cycle.

68. The full-wave rectifier circuit switches both half-cycles (or the full ac input wave) of the input through the load in only one direction.

69. The basic full-wave center-tapped rectifier circuit contains a center-tapped transformer and two diodes. The center tap of the transformer secondary is grounded (0 V) to create a 180° phase difference between the top and bottom of the secondary winding.

70. The two diodes and center-tapped transformer in this circuit switch the two-direction ac current developed across the secondary through the load in only one direction and therefore develop an output voltage across the load of the same polarity.

71. Even though the peak output of the full-wave center-tapped rectifier was half that of the half-wave rectifier, the average output was the same because the full-wave center-tapped rectifier doubles the number of half-cycles at the output, compared to a half-wave rectifier.

72. Since two half-cycles appear at the output for every one cycle at the input, the ripple frequency will be twice that of the input frequency, and this higher frequency will be easier to filter or smooth of fluctuations.

73. Since only one diode is on for each half-cycle, the peak output voltage will only be less 0.7 V (silicon diode).

74. The full $V_{S\,peak}$ voltage appears across the OFF D_1 during the other half cycle of the input, and therefore both diodes must have a peak inverse voltage (PIV) rating that is larger than the peak secondary voltage ($V_{S\,peak}$).

Full-Wave Bridge Rectifier Circuits

75. By center tapping the secondary of the transformer and having two diodes instead of one, we were able to double the ripple frequency and therefore ease the filtering process. However, that seems to be the only advantage the full-wave center-tapped rectifier has over the half-wave rectifier, and the price we pay is having to have a more expensive center-tapped transformer and an extra diode.

76. With the bridge rectifier circuit, we can have the peak secondary output voltage of the half-wave circuit and the full-wave ripple frequency of the center-tapped circuit. This circuit was originally called a "bridge" rectifier because its shape resembled the framework of a suspension bridge.

77. Like the center-tapped rectifier, the bridge rectifier switches two half-cycles of the same polarity through to the load; however, unlike the center-tapped rectifier, the bridge rectifier connects the total peak secondary voltage across the load.

78. Since two half-cycles appear at the output for every one cycle at the input, the ripple frequency will be twice that of the input frequency, and this higher frequency will be easier to filter or smooth of fluctuations.

79. To be completely accurate, we should take into account the barrier voltage drop across the diodes. Since two diodes are ON for each half-cycle and both diodes are connected in series with the load, a 1.4 V (2×0.7 V) drop will occur between the peak secondary voltage (V_S) and the peak output voltage.

80. Each diode must have a peak inverse voltage (PIV) rating that is greater than the peak secondary voltage (V_S).

Capacitive Filtering of a Rectifier Output

81. The filter in a dc power supply converts the pulsating dc output from the half-wave or full-wave rectifier into an unvarying dc voltage.

82. The three basic types of filters are the capacitive filter, the RC filter, and the LC filter.

83. Even though the output from a filter should be a constant dc level, there is a slight fluctuation or ripple. This fluctuation is called the percent ripple and its value is used to rate the action of the filter.

84. The low-pass RC pi (π) filter could be used to further reduce ripple, due to its ability to pass dc but block any fluctuations. It is called a π filter because it contains two

vertical parts and one horizontal part and therefore resembles π.

85. The LC filter replaces the series resistor in the RC π filter with an inductor L_1 to increase the efficiency of the filter since the inductor will oppose current without generating heat.

Voltage Regulator Circuits

86. The voltage regulator in a dc power supply maintains the dc output voltage constant despite variations in the ac input voltage and the output load resistance.

87. Ideally, a dc power supply should convert all of its ac electrical energy input into a dc electrical energy output. However, like all devices, circuits, and systems, a dc power supply is not 100% efficient and along with the electrical energy output, the dc power supply generates wasted heat.

88. Although a load resistance will generate not change dramatically, it will change slightly as different control settings are selected, and this will load the source. This load resistance change would affect the output of a dc power supply if a regulator were not included to maintain the output voltage constant despite variations in load resistance.

89. To the electric company, each user appliance is a small part of its load, and it is the source. The ac voltage from the electric company can be anywhere between 105 V and 125 V ac rms (148 to 177 V ac peak), depending on consumer use, which depends on the time of day. When many appliances are in use, the load current will be high and the overall load resistance low, causing the source voltage to be pulled down. If a dc power supply did not have a regulator, these different input ac voltages from the electric company would produce different dc output voltages, when we really want the dc supply voltage to the electronic circuits to always be the same value no matter what ac input is present.

90. The percent of regulation is a measure of the regulator's ability to regulate, or maintain constant, the output dc voltage. Most regulators achieve a percent regulation figure that is not perfect (0%) but is generally in single digits (2% to 8%, typically).

91. A zener diode and a series resistor could be connected in a dc power supply circuit to provide regulation. An increase in the ac input voltage would cause an increase in the dc output from the filter, which would cause an increase in the current through the series resistor, the zener diode, and the load. The reverse-biased zener diode will, however, maintain a constant voltage across its anode and cathode despite these input voltage and current variations, with the additional voltage being dropped across the series resistor.

92. The zener regulator will also maintain a constant output voltage despite variations in load resistance. The zener diode achieves this by increasing and decreasing its current (I_Z) in response to load resistance changes. Despite these changes in zener and load current however, the zener voltage (V_Z), and therefore the output voltage (V_{out} or V_{RL}), always remains constant.

93. The zener regulator's disadvantage is that the series connected resistor will limit load current, and, in addition, generate unwanted heat.

94. Most dc power supply circuits today make use of integrated circuit (IC) regulators. These IC regulators contain about fifty individual or discrete components all integrated together on one silicon semiconductor chip and then encapsulated in a three-pin package.

95. The three terminals of the IC regulator are labeled "input" (pin 1), "output" (pin 2), and "ground" (pin 3). The package is generally given the identification code of either 78XX or 79XX. The 78XX (seventy-eight hundred) series of regulators are used to supply a positive output voltage, with the last two digits specifying the output voltage. For example, 7805 = +5 V, 7812 = +12 V, and so on. The 79XX (seventy-nine hundred) series of regulators are used to supply a negative output voltage, with the last two digits, once again, specifying the output voltage. These regulators will deliver a constant regulated output voltage as long as the input voltage is greater than the regulator's rated output voltage, and can deliver a maximum output current of up to 1.5 A if properly heat sunk.

Troubleshooting a DC Power Supply

96. The procedure for fixing a failure can be broken down into three basic steps, and these steps are:

Step 1: DIAGNOSE

The first step is to determine whether a problem really exists. To carry out this step, a technician must collect as much information as possible about the system, circuit and components used, and then diagnose the problem.

Step 2: ISOLATE

The second step is to apply a logical and sequential reasoning process to isolate the problem. In this step, a technician will operate, observe, test and apply troubleshooting techniques in order to isolate the malfunction.

Step 3: REPAIR

The third and final step is to make the actual repair, and then final test the circuit.

97. It is extremely important that you first understand how a system, circuit, and all of its components are supposed to work so that you can determine whether or not a problem exists.

98. Technicians usually refer to service or technical manuals which generally contain circuit descriptions and troubleshooting guides. As far as each of the devices are concerned, technicians refer to manufacturer data books which contain a full description of the device.

99. Referring to all of this documentation before you begin troubleshooting will generally speed up and simplify the isolation process. Once you are fully familiar with the operation of the circuit, you will be ready to diagnose the problem as either an operator error or as a circuit malfunction.

100. Once you have determined that the problem is not an operator error but, in fact, a circuit malfunction, proceed to Step 2 and isolate the circuit problem.

101. A technician will generally spend most of his or her time isolating a circuit problem. Steps 1 (diagnose) and 3 (repair) may only take a few minutes to complete compared to Step 2 which could take a few hours to complete. However, with practice and a good logical and sequential reasoning process, you can quickly isolate even the most obscure of problems.

102. The directions you take as you troubleshoot a circuit malfunction may be different for each problem. However, certain troubleshooting techniques apply to all problems.
    a. Check first for obvious errors.
       1. Power fuse
       2. Wiring errors if circuit is newly constructed
       3. Devices are correctly oriented.
    b. Use your senses to check for broken wires, loose connections, overheating or smoking components, leads or pins not making contact, and so on.
    c. Use a cause and effect troubleshooting process, which means study the effects you are getting from the faulty circuit and then try to reason out what could be the cause.
    d. Apply the half-split method of troubleshooting first to the entire system, then to a circuit within the system, and then to a section within the circuit to help speed up the isolation process. Once you have disconnected the circuit load from a dc power supply's output, you can isolate the problem as being either a source fault or load fault. If the power supply's output voltage goes back to its rated voltage after it has been disconnected from the circuit load, the problem is with the load. Whereas if the output voltage remains LOW or at zero after the source has been disconnected from the load, the problem is with the source.

103. The final step is to repair the circuit, which could involve simply removing a wire clipping or some excess solder, resoldering a broken connection, reconnecting a connector, or some other easy repair. In most instances, however, the repair will involve the replacement of a faulty component. For a circuit that has been constructed on a breadboard or prototyping board, the removal and replacement of the component is simple. However, when a printed circuit board is involved, you should make a note of the component's orientation, and observe good desoldering and soldering practices.

104. When the circuit has been repaired, always perform a final test to see that the circuit is now fully operational.

Multiple-Choice Questions

1. What is the barrier voltage for a silicon junction diode?
   a. 0.3 V   **b.** 0.4 V   **c.** 0.7 V   **d.** 2.0 V

2. Which of the following junction diodes are forward biased?
   a. Anode = +7 V, cathode = +10 V
   b. Anode = +5 V, cathode = +3 V
   c. Anode = +0.3 V, cathode = +5 V
   d. Anode = −9.6 V, cathode = −10 V

3. The junction diode _____ current when it is forward biased, and _____ current when it is reverse biased.
   a. Blocks, conducts   **c.** Blocks, prevents
   b. Conducts, passes   **d.** Conducts, blocks

4. The *n*-type region of a junction diode is connected to the _____ terminal and the *p*-type region is connected to the _____.
   a. Cathode, anode   **b.** Anode, cathode

5. Semiconductor devices need voltages of a certain amplitude and polarity to control their operation. These voltages are called:
   a. Barrier potentials   **c.** Knee voltages
   b. Depletion voltages   **d.** Bias voltages

6. When reverse biased, a junction diode has a leakage current passing through it which is typically measured in:
   a. Amps   **b.** Milliamps   **c.** Microamps   **d.** Kiloamps

7. What happens to the forward voltage drop across the diode (V_F) if temperature increases? V_F will:
   a. Decrease   **c.** Remain the same
   b. Increase   **d.** Be unpredictable

8. When forward biased, a junction diode is equivalent to a/an _____ switch, whereas when it is reverse biased it is equivalent to a/an _____ switch.
   a. Open, closed   **c.** Open, open
   b. Closed, closed   **d.** Closed, open

9. The black band on a diode's package is always closest to the _____.
   a. Anode   **c.** *p*-type material
   b. Cathode   **d.** Both (a) and (c) are true

10. When current dramatically increases, the voltage point on the diode's forward *V-I* characteristic curve is called the:
    a. Breakdown voltage   **c.** Barrier voltage
    b. Knee voltage   **d.** Both (b) and (c) are true

11. When current dramatically increases, the voltage point on the diode's reverse *V-I* characteristic curve is called the:
    a. Breakdown voltage   **c.** Barrier voltage
    b. Knee voltage   **d.** Both (b) and (c) are true

12. A logic gate is:
    a. A circuit that converts ac to dc
    b. An analog circuit
    c. A two-state decision making circuit
    d. A circuit that converts dc to ac

13. A rectifier is:
    a. A circuit that converts ac to dc
    b. An analog circuit
    c. A two-state decision making circuit
    d. A circuit that converts dc to ac

14. What resistance should a good diode have when it is reverse biased?
    a. Less than 10 Ω
    b. More than 1000 MΩ
    c. Between 120 Ω and 1.2 kΩ
    d. Both (b) and (c) are true

15. What resistance should a good diode have when it is forward biased?
    a. Less than 10 Ω
    b. More than 1000 MΩ
    c. Between 120 W and 1.2 kΩ
    d. Both (b) and (c) are true

16. The _____ diode is designed to withstand high reverse currents that result when the diode is operated in the reverse breakdown region.
    a. Basic P-N junction   **c.** Zener
    b. Light emitting   **d.** Both (a) and (c) are true

17. When a zener diode's breakdown voltage is exceeded, the reverse current through the diode increases from a small leakage value to a high reverse current value.
    a. True   **b.** False

18. When operating in the reverse breakdown region, the _____ the zener will vary over a wide range, while the _____ the zener will vary by only a small amount.
    a. Voltage drop across, forward current through
    b. Forward current through, voltage drop across
    c. Reverse current through, forward current through
    d. Reverse current through, forward drop across

19. The zener diode's symbol is different from all other diode symbols due to its:
    a. Z shaped cathode bar   **c.** Straight bar cathode
    b. Two exiting arrows   **d.** None of the above

20. The ability of a zener diode to maintain a relatively constant _____ regardless of variations in zener _____ is the key characteristic of a zener diode.
    a. Current, voltage   **c.** Current, impedance
    b. Impedance, voltage   **d.** Voltage, current

21. A 12 V ± 5% zener diode will have a voltage drop in the _____ range.
    a. 10.8 V to 13.2 V   **c.** 5.04 V to 18.96 V
    b. 11.4 V to 12.6 V   **d.** 11.88 V to 12.12 V

22. A 9.1 V zener diode has a power rating of 10 W. What is the diode's value of maximum zener current?
    a. 1.1 mA   **b.** 91 mA   **c.** 1.1 A   **d.** None of the above

23. A/an _____ circuit maintains the output voltage of a voltage source constant despite variations in the input voltage and the load resistance.
a. Encoder   **c.** Comparator
b. Logic gate   **d.** Voltage regulator

24. The zener diode is able to maintain the voltage drop across its terminals constant by continually changing its _____ in response to a change in input voltage.
a. Impedance   **c.** Power rating
b. Voltage   **d.** Both (a) and (c) are true

25. In a voltage regulator circuit, the voltage developed across the zener diode remains constant and therefore any changes in the input voltage must appear across the:
a. Load   **c.** Source terminals
b. Series resistor   **d.** Zener diode

26. The light emitting diode is a semiconductor device that converts _____ energy into _____ energy.
a. Chemical, electrical   **c.** Electrical, light
b. Light, electrical   **d.** Heat, electrical

27. The _____ lead of an LED is distinguished from the other terminal by its longer lead, flattened lead, or its close proximity to the flat side of the case.
a. Anode   **b.** Cathode

28. The forward voltage drop across a typical LED is usually:
a. 0.7 V   **b.** 0.3 V   **c.** 5 V   **d.** 2.0 V

29. Since the forward voltage rating of an LED remains almost constant, the output power of an LED is directly proportional to its:
a. Impedance   **c.** Reverse voltage
b. Forward current   **d.** Both (a) and (c) are true

30. If all seven cathodes in a seven-segment display are connected to 0 V, the display is referred to as a common _____ configuration and requires a _____ voltage input to turn on an LED segment.
a. Anode, LOW   **c.** Anode, HIGH
b. Cathode, LOW   **d.** Cathode, HIGH

31. Electrical systems manage the flow of _____, while electronic systems manage the flow of _____.

a. Information, power   **b.** Power, information

32. The four main circuit blocks of a dc power supply listed in order from input to output are:
a. Transformer, rectifier, filter, regulator
b. Filter, regulator, rectifier, transformer
c. Transformer, rectifier, regulator, filter
d. Rectifier, filter, regulator, transformer

33. Which of the four circuit blocks of a dc power supply converts an ac input into a pulsating dc output?
a. Transformer   **c.** Filter
b. Regulator   **d.** Rectifier

34. Which of the four circuit blocks of a dc power supply converts a high ac voltage into a low ac voltage?
a. Transformer   **c.** Filter
b. Regulator   **d.** Rectifier

35. Which of the four circuit blocks of a dc power supply maintains the output voltage constant despite variations in the ac input and output load?
a. Transformer   **c.** Filter
b. Regulator   **d.** Rectifier

36. Which of the four circuit blocks of a dc power supply converts a pulsating dc input into a steady dc output?
a. Transformer   **c.** Filter
b. Regulator   **d.** Rectifier

37. Which rectifier circuit can be used to generate a negative pulsating dc output?
a. Half-wave   **c.** Bridge
b. Center-tapped   **d.** All of the above

38. Which rectifier uses four diodes?
a. Half-wave   **c.** Bridge
b. Center-tapped   **d.** All of the above

39. A typical dc power supply will supply a _____-voltage _____-current output.
a. Low, high   **c.** Low, low
b. High, low   **d.** High, high

40. Which is typically the most commonly used filter for a dc power supply?
a. Pi   **b.** *RC*   **c.** Capacitive   **d.** *LC*

41. What rating is used to measure the action of a filter?
a. Percent rectification   **c.** Percent regulation
b. Percent ripple   **d.** Percent filtration

42. A small load resistance will cause a _____ load current and possibly pull _____ the source voltage.
a. Small, down   **c.** Small, up
b. Large, up   **d.** Large, down

43. What would be the output voltage of a 7915 IC regulator?
a. +5 V   **c.** −5 V
b. +15 V   **d.** −15 V

44. A _____ diode would typically be used in a power supply as a power indicator.
a. Rectifier   **c.** Zener
b. Light-emitting   **d.** Junction

45. A _____ diode would typically be used as a regulator.
a. Rectifier   **c.** Zener
b. Light-emitting   **d.** Junction

Communication Skill Questions

46. What is a junction diode? (16-1)

47. Sketch the schematic symbol for a diode and label its terminals. (16-1)

48. Describe the basic operation of a diode. (16-1)

49. What is the relationship between a P-N junction and a diode? (16-1)

50. Explain in detail the differences between a forward biased diode and a reverse biased diode. (16-1)

51. Referring to the junction diode's *V-I* characteristic curve, describe the: (16-1)
a. Forward bias curve   **b.** Reverse bias curve

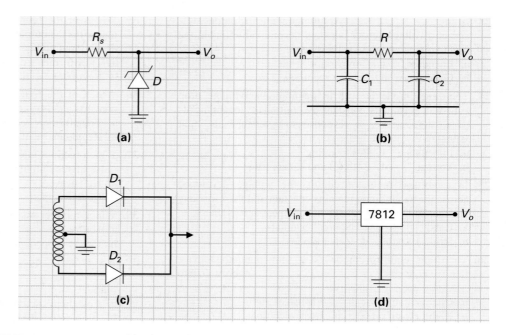

FIGURE 16-42 **Power Supply Circuits.**

52. Describe the temperature characteristics of a junction diode. (16-1)

53. What type of diode malfunction is most common? (16-1)

54. How can an ohmmeter be used to test for an open or short within a diode? (16-1)

55. Who invented the zener diode? (16-2)

56. Sketch the zener diode schematic symbol. (16-2)

57. What is the zener diode's key characteristic? (16-2)

58. Define the following terms: (16-2)
    a. Reverse leakage current
    b. Zener breakdown voltage
    c. Zener knee current

    d. Zener current
    e. Zener voltage
    f. Zener impedance
    g. Maximum zener power dissipation
    h. Zener test current
    i. Maximum zener current

59. LED is an abbreviation for _____. (16-3)

60. Sketch the schematic symbol used to represent an LED. (16-3)

61. What elements are contained in a typical LED package? (16-3)

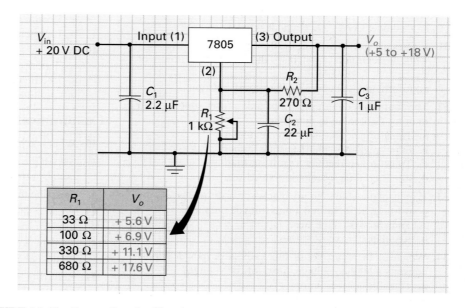

FIGURE 16-43 **Power Supply Circuit.**

R_1	V_o
33 Ω	+ 5.6 V
100 Ω	+ 6.9 V
330 Ω	+ 11.1 V
680 Ω	+ 17.6 V

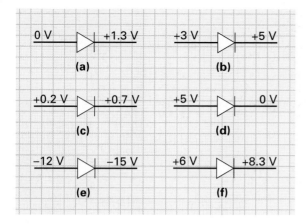

FIGURE 16-44 Biased Junction Diodes.

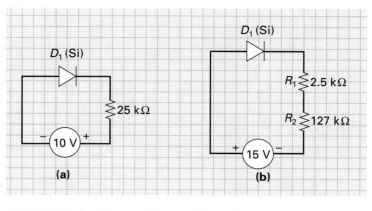

FIGURE 16-45 Forward Current Examples.

62. What method(s) is used to distinguish an LED's cathode from its anode? (16-3)

63. Why is a dc power supply considered an electrical system? (Introduction)

64. Define the function of the four main blocks in a dc power supply. (16-5)

65. Sketch the input and output waveforms of each of the four blocks in a dc power supply. (16-5)

66. Sketch a half-wave rectifier circuit, then describe the circuit's operation and characteristics. (16-5)

67. Sketch a full-wave center-tapped rectifier circuit, and describe the circuit's operation and characteristics. (16-5)

68. Sketch a full-wave bridge rectifier circuit, and describe the circuit's operation and characteristics. (16-5)

69. Define ripple frequency and what it will be for a half-wave rectifier and a full-wave rectifier. (16-5)

70. With a sketch, describe the action of the capacitive filter. (16-5)

71. Define percent of ripple. (16-5)

72. How can a load resistance pull down the source voltage? (16-5)

73. Define percent of regulation. (16-5)

74. Identify circuits (a) to (d) in Figure 16-42. (16-5)

75. What are the advantages/disadvantages of *RC* and *LC* filters? (16-5)

76. Describe what you think the circuit in Figure 16-43 will do.

77. Briefly describe how the 7805 can supply an output that is greater than +5 V.

Practice Problems

78. Which of the silicon diodes in Figure 16-44 are forward biased and which are reverse biased?

79. Calculate I_F for the circuits in Figure 16-45.

80. What would be the voltage drop across each of the diodes in Figure 16-45?

81. What would be the voltage drop across each of the resistors in Figure 16-45?

82. In reference to the polarity of the applied voltage, which of the circuits in Figure 16-46 are correctly biased for normal zener operation?

83. In reference to the magnitude of the applied voltage, which of the circuits in Figure 16-46 are correctly biased for normal zener operation?

84. Calculate the value of circuit current for each of the zener diode circuits in Figure 16-46.

85. Would a 1 watt zener diode have a suitable power dissipation rating for the circuit in Figure 16-46(a)?

86. What wattage or maximum power dissipation rating would you choose for the zener diode in Figure 16-46(e)?

87. What would be the regulated output voltage and polarity at points *X* and *Y* for the circuits shown in Figure 16-47 (a) and (b)?

88. Knowing that the zener diode must always be reverse biased, how could we change the circuit connection in Figure 16-47(a) and (b) to obtain opposite polarity supply voltages?

89. Calculate the value of circuit current for both the circuits shown in Figure 16-48.

90. Calculate the value of I_{RS}, I_Z, I_{RL}, and V_{RL} for the two extreme low and high voltages shown in Figure 16-48. Assume R_L remains constant at 500 Ω.

91. Calculate the values of I_{RS}, I_{RL}, and I_Z for all values of maximum and minimum input voltage and load resistance for the circuit in Figure 16-48.

92. Which of the light-emitting diodes in Figure 16-49 are biased correctly?

93. Calculate the value of circuit current for each of the LEDs in Figure 16-49.

94. Would a maximum forward current rating of 18 mA be adequate for the LED in Figure 16-49(a)?

95. Would a maximum reverse voltage rating of 5 V be adequate for the LEDs in Figure 16-49(c)?

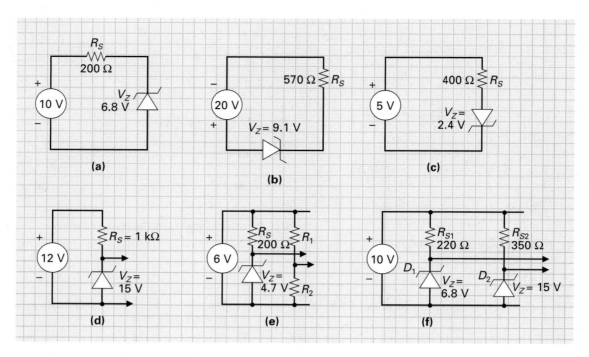

(a) **(b)** **(c)**

(d) **(e)** **(f)**

FIGURE 16-46   Biasing Voltage Polarity and Magnitude.

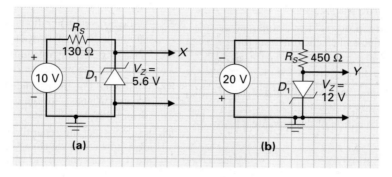

(a) **(b)**

FIGURE 16-47   Voltage Regulator Circuits (Unloaded).

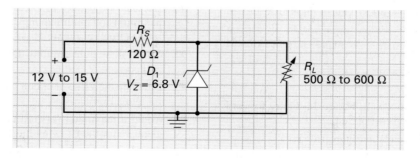

FIGURE 16-48   A Voltage Regulator Circuit (Loaded).

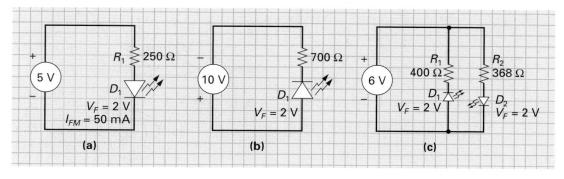

FIGURE 16-49   Biasing Light Emitting Diodes.

96. Figure 16-50 shows how two basic P-N junction diodes and a resistor can be used to construct an OR gate circuit. In this circuit, an LED has been connected to the output so that any HIGH or positive 5 V output will turn ON the LED. Considering the A and B input combinations shown in the table, indicate whether the LED will be ON or OFF for each of these input conditions.

97. Assuming a 0.7 V voltage drop across the P-N junction diode and a 2 V voltage drop across the LED, what would be the value of current through the LED if one of the inputs in Figure 16-50 was HIGH (+5 V)?

98. Which of the bi-color LEDs will be ON in Figure 16-51 when the input voltage is +10 V, and which will be ON when the input voltage is −10 V?

99. In Figure 16-51, a basic P-N junction diode (D_1) has been included across R_1 so that when the input voltage goes negative this diode will bypass the additional current-limiting resistor R_1. When the input voltage is positive, D_1 is reversed biased and therefore both R_1 and R_2 will limit the value of series current. Which of the colors in the bi-color LED will be brighter?

100. Calculate the value of green and red LED current for the circuit in Figure 16-51.

101. In Figure 16-52 a comparator is used to compare 0 V (at the negative input) to a sine wave which varies between

+5 V and −5 V (at the positive input). When the comparator's inputs are "true" (positive input is positive with respect to negative input), the comparator's output will be switched HIGH (+12 V). When the comparator's inputs are "not true" (positive input is negative with respect to the negative input), the comparator's output will be switched LOW (−12 V). What value of series current-limiting resistor should be used to ensure that the LEDs are bright, but do not burn out, for the +12 V and −12 V source voltages? Aim for a forward current that is about 90% of the maximum rating.

102. Would the 16-LED display in Figure 16-53 (D_{11}-D_{14}) be classified as a common-anode or common-cathode display?

103. In regard to Figure 16-53, construct a table to show the HIGH and LOW encoder outputs on A, B, C, and D for each of the rotary switch positions. Also show in the table which of the LEDs are ON or OFF for each of these codes.

104. In Figure 16-54, seven switches are used to turn ON and OFF the seven LEDs in a seven-segment display. Construct a table to show which switches will be CLOSED and which will be OPEN to display the digits 0 through 9 on the seven-segment display.

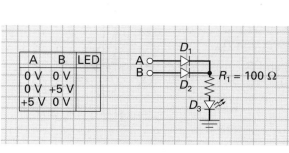

FIGURE 16-50   An OR Gate Circuit with an LED Output.

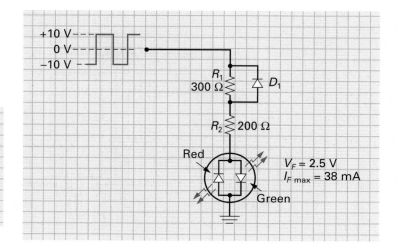

FIGURE 16-51   A Bi-Color Output Display Circuit.

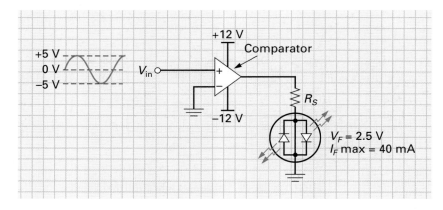

FIGURE 16-52   A Comparator Circuit with a Bi-Color Output Display.

105. Are the seven-segment displays in Figure 16-54 and Figure 16-55 common-anode or common-cathode types?

106. Figure 16-55(a) shows how a 5.1 V zener voltage regulator circuit can be used to supply power to a seven-segment encoder and display circuit. Figure 16-55(b) shows how the LOW and HIGH codes generated by the encoder circuit will turn ON and OFF the necessary LEDs in the seven-segment display to produce the digits 0 through 9. For example, when the rotary switch is in position 0, a single P-N junction diode makes line *G* LOW. This will turn segment "g" of the seven-segment display OFF, while the pull up resistors R_1 through R_6 will make lines *A* through *F* HIGH and therefore turn ON segments "a" through "f." This condition is shown in the first row of the table in Figure 16-55(b).

Calculate the following encoder circuit details:

a. The amount of current that will be drawn by each of the basic P-N junction diodes in the encoder circuit.

b. What rotary switch position will cause the maximum value of current, and what will this current value be?

Calculate the following display circuit details:

c. The amount of current that will be drawn by each of the LEDs in the seven-segment display.

d. What digit on the seven-segment display will cause the maximum value of current, and what will this current value be?

Calculate the following encoder-display circuit details:

e. Which path draws more current: an encoder diode or a display diode?

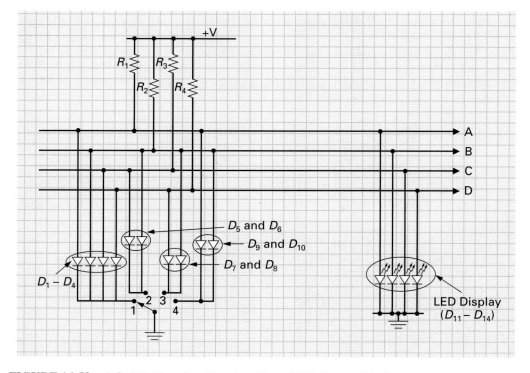

FIGURE 16-53   A Switch Encoder Circuit with an LED Output Display.

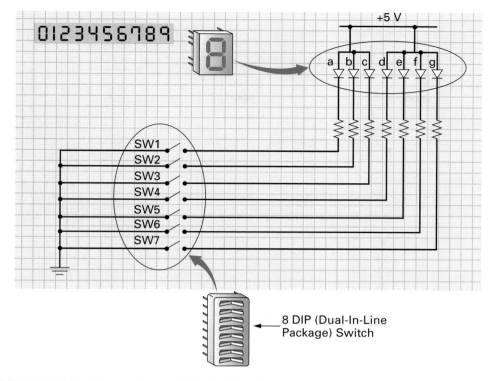

FIGURE 16-54   A Seven-Segment Display Circuit.

f. When the digit 8 is being displayed, a total of 99.2 mA of current is being drawn from the 5.1 V source. When the digit 1 is being displayed, a total of 112.8 mA is being drawn from the 5.1 V source. Because the source only sees the encoder-display circuit as a single load resistance, what will this resistance be equal to?

107. If a 240 V rms ac 60 Hz input is applied to the 19:1 step-down transformer shown in Figure 16-56, what would be the peak secondary output voltage?

108. What would be the average voltage out of a positive half-wave rectifier if the transformer secondary in question 130 were applied? (Take into account the barrier voltage drop.)

109. What would be the ripple frequency at the output for the circuits described in questions 130 and 131?

110. Calculate the peak output voltage (taking into account V_{diode}) for the circuit shown in Figure 16-57.

111. What would be the output ripple frequency from the circuit in Figure 16-57?

112. Calculate the average output voltage of the bridge rectifier circuit shown in Figure 16-58, taking into account the diode voltage drop.

113. A capacitive filter produces the output shown in Figure 16-59. What is the filter's percent of ripple?

114. If a regulator delivers +12 V when no load is connected, and +10.6 V when a full load is connected, what would be the regulator's percent of regulation?

Troubleshooting Questions

115. Referring to the switch encoder circuit in Figure 16-60, first determine what digital codes will be generated at A, B, and C for each of the switch positions 1, 2 and 3. Next, determine what problems would occur for each of the following circuit malfunctions or conditions:
  a. What would happen if D_1 were to open permanently?
  b. Would a 200 mA fuse protect the +5 V supply voltage?

116. Describe briefly how you would test a zener diode.

117. Describe briefly how you would test a light-emitting diode.

118. Define troubleshooting.

119. Describe the three steps used to locate and repair a circuit malfunction.

120. In regard to Figure 16-61 on p. 726, describe what symptoms would occur from the following circuit malfunctions:
  a. The zener diode shorts
  b. The zener diode opens
  c. R_s opens
  d. R_S shorts
  e. The dc supply voltage decreases to 10 V
  f. The dc supply voltage increases to 15 V
  g. One of the encoder diodes opens
  h. One of the encoder diodes shorts
  i. There is an open in one of the control lines A through G between the encoder and the seven-segment display
  j. One of the seven-segment LEDs opens
  k. One of the seven-segment LEDs shorts
  l. One of the resistors R_1 through R_7 shorts
  m. One of the resistors R_1 through R_7 opens

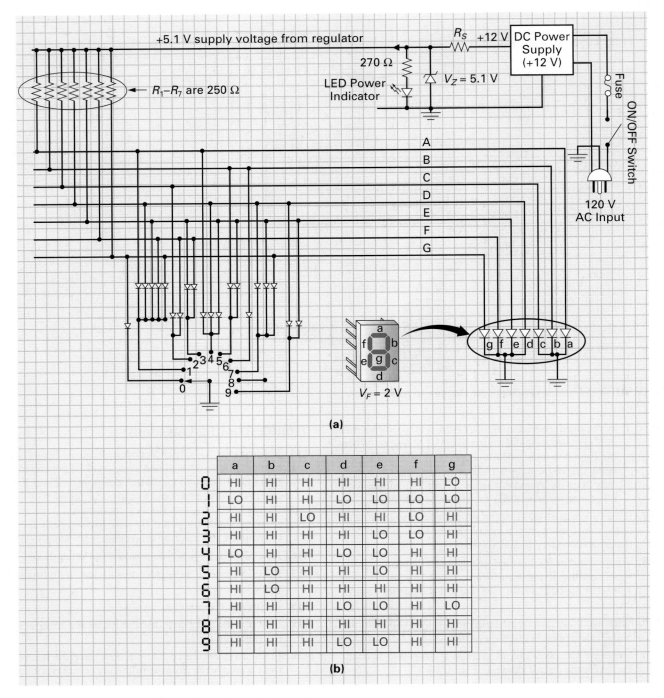

	a	b	c	d	e	f	g
0	HI	HI	HI	HI	HI	HI	LO
1	LO	HI	HI	LO	LO	LO	LO
2	HI	HI	LO	HI	HI	LO	HI
3	HI	HI	HI	HI	LO	LO	HI
4	LO	HI	HI	LO	LO	HI	HI
5	HI	LO	HI	HI	LO	HI	HI
6	HI	LO	HI	HI	HI	HI	HI
7	HI	HI	HI	LO	LO	HI	LO
8	HI	HI	HI	HI	HI	HI	HI
9	HI	HI	HI	LO	LO	HI	HI

(b)

FIGURE 16-55   A Seven-Segment Encoder and Display Circuit.

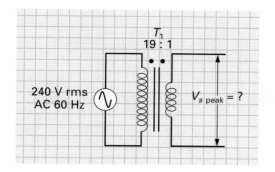

FIGURE 16-56

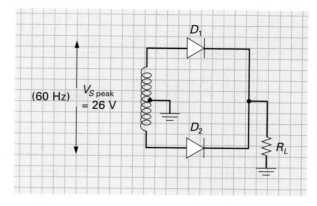

FIGURE 16-57

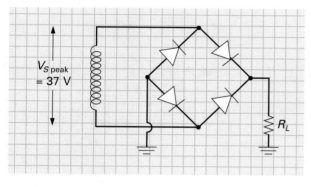

FIGURE 16-58

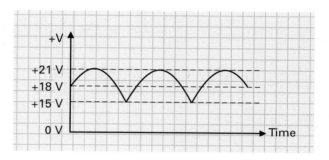

FIGURE 16-59

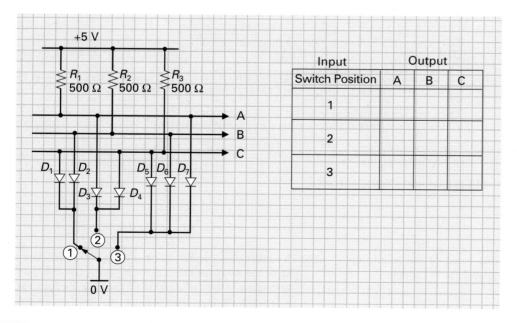

FIGURE 16-60   A Switch Encoder Circuit.

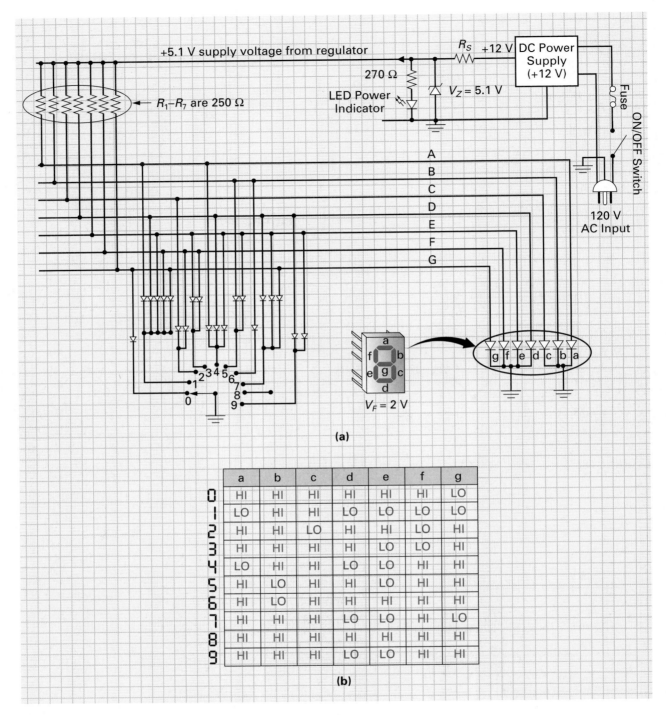

FIGURE 16-61 **A Seven-Segment Encoder and Display Circuit.**

121. Briefly describe the effect that some of the following problems would have on a typical dc power supply:
  a. An open rectifier diode
  b. A shorted rectifier diode
  c. A shorted filter capacitor
  d. A shutdown regulator

122. As a technician troubleshoots a circuit it is important that he or she study the effect given by the circuit and then try to determine the cause. Most of the time an open circuit will not cause any circuit damage, and these problems can be introduced without concern; however, shorts between two points should be contemplated carefully because of possible hazard to the technician and damage to the circuit. Here are a few examples of circuit problems. Describe what circuit problems they will produce:

a. Remove F_1 (Caution: Like all fuses, the 120 V ac will be present across the fuse terminals when it is removed.)

b. Reverse the orientation of D_2 or D_1

c. Open one of the inputs to the bridge rectifier

d. Disconnect one of the filtering capacitors

e. Disconnect the ground input to the 7805

f. Short the 7912 output to ground

g. Reverse the orientation of the LED

Web Site Questions

Go to the Web site http://www.prenhall. com/cook, select the textbook *Introductory DC/AC Electronics* or *Introductory DC/AC Circuits*, this chapter, and then follow the instructions when answering the multiple-choice practice problems.

JOB INTERVIEW TEST

These tests at the end of each chapter will challenge your knowledge up to this point, and give you the practice you need for a job interview. To make this more realistic, the test will comprise both technical and personal questions. In order to take full advantage of this exercise, you may want to set up a simulation of the interview environment, have a friend read the questions to you, and record your responses for later analysis.

Company Name: DCPS, Inc.

Industry Branch: ALL

Function: Design and Manufacture DC Power Supplies

Job Title: Engineering Technician

1. What have you heard about us?
2. What is the difference between dc and ac?

Answers

1. Visit the company's web site before the interview to get an overall understanding of the company's ownership, product line, service, and support.
2. Chapter 8.
3. Discuss how you have worked alone on project assignments in lab and theory, and other related work history.
4. Introduction.
5. Report the problem to my supervisor and ask for help.

3. In many instances an engineer will assign you a task and their time will be limited, so you will be on your own. How do you feel about that?
4. What is a dc power supply?
5. What would you do if you didn't know the answer to a question?
6. What is the function of a regulator?
7. What are the responsibilities of an engineering technician?
8. How would you handle the pressure if you became overloaded with assignments?
9. How would you troubleshoot a dc power supply?
10. Why do you want this particular job?

6. Section 16-5-5.
7. See introduction, engineering technician, and job listing.
8. Discuss how you have juggled school, studying, and a job to get to this point, and other related work experience.
9. Section 16-5-6.
10. Discuss why you were first attracted to their advertisement, why it felt compatible with your career goals, and how the company's positive attributes listed in its web site matched your expectations.

Bipolar Junction Transistors (BJTs)

There's No Sleeping When He's Around!

Carl Friedrich Gauss was born April 30, 1777, to poor, uneducated parents in Brunswick, Germany. He was a child of precocious abilities, particularly in mental computation. In elementary school he soon impressed his teachers, who said that mathematical ability came easier to Gauss than speech.

In secondary school he rapidly distinguished himself in ancient languages and mathematics. At 14, Gauss was presented to the court of the duke of Brunswick, where he displayed his computing skill. Until his death in 1806, the duke generously supported Gauss and his family, encouraging the boy with textbooks and a laboratory.

In the early years of the nineteenth century, Gauss's interest was in astronomy, and his accumulated work on celestial mechanics was published in 1809. In 1828, at a conference in Berlin, Gauss met physicist Wilhelm Weber, who would eventually become famous for his work on electricity. They worked together for many years and became close friends, investigating electromagnetism and the use of a magnetic needle for current measurement. In 1833 they constructed an electric telegraph system that could communicate across Göttingen from Gauss's observatory to Weber's physics laboratory. (This telegraph system of communication was later developed independently by U.S. inventor Samuel Morse.)

Gauss conceived almost all of his fundamental mathematical discoveries between the ages of 14 and 17. There are many stories of his genius in his early years, one of which involved a sarcastic teacher who liked giving his students long-winded problems and then resting, or on some occasions sleeping, in class. On his first day with Gauss, who was 8 years old, the teacher began, as usual, by telling the students to find the sum of all the numbers from 1 to 100. The teacher barely had a chance to sit down before Gauss raised his hand and said "5050." The dumbfounded teacher, who believed Gauss must have heard the problem before and memorized the answer, asked Gauss to explain how he had solved the problem. He replied: "The numbers 1, 2, 3, 4, 5, and so on to 100 can be paired as 1 and 100, 2 and 99, 3 and 98, and so on. Since each pair has a sum of 101, and there are 50 pairs, the total is 5050."

Outline and Objectives

Introduction

In 1948, a component known as a transistor sparked a whole new era in electronics, the effects of which have not been fully realized even to this day. A transistor is a three-element device made of semiconductor materials used to control electron flow, the amount of which can be controlled by varying the voltages applied to its three elements. Having the ability to control the amount of current through the transistor allows us to achieve two very important applications: switching and amplification.

Like the diode, transistors are formed by *p* and *n* regions and, as we are already aware, the point at which a *p* and an *n* region join is known as a junction. Transistors in general are classified as being either the *bipolar* or *unipolar* type. The bipolar type has two P-N junctions, while unipolar transistors have only one P-N junction. In this chapter we will study all of the details relating to the *bipolar* transistor, or as it is also known, the *bipolar junction transistor* or *BJT*.

17-1   FIRST APPROXIMATION DESCRIPTION OF A BIPOLAR TRANSISTOR

In most cases it is easier to build a jigsaw puzzle when you can refer to the completed picture on the box. The same is true whenever anyone is trying to learn anything new, especially a science that contains many small pieces. These first approximation descriptions are a means for you to quickly see the complete picture without having to wait until you connect all of the pieces. Like the diode's first approximation description, this general overview will cover the transistor's basic construction, schematic symbol, physical appearance, basic operation, and main applications.

NPN Transistor

A thin, lightly doped *p*-type region (base) is sandwiched between two *n*-type regions (emitter and collector).

Base

The region that lies between an emitter and a collector of a transistor and into which minority carriers are injected.

17-1-1   *Transistor Types (NPN and PNP)*

Like the diode, a bipolar transistor is constructed from a semiconductor material. However, unlike the diode, which has two oppositely doped regions and one P-N junction, the transistor has three alternately doped semiconductor regions and two P-N junctions. These three alternately doped regions are arranged in one of two different ways, as shown in Figure 17-1.

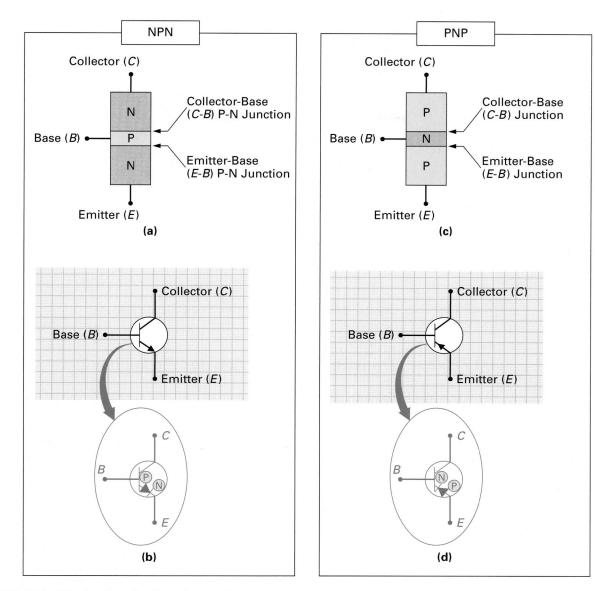

FIGURE 17-1   Bipolar Junction Transistor (BJT) Types.

With the **NPN transistor** shown in Figure 17-1(a), a thin, lightly doped p-type region known as the **base** (symbolized B) is sandwiched between two n-type regions called the **emitter** (symbolized E) and the **collector** (symbolized C). Looking at the NPN transistor's schematic symbol in Figure 17-1(b), you can see that an arrow is used to indicate the emitter lead. As a memory aid for the NPN transistor's schematic symbol, you may want to remember that when the emitter arrow is "**N**ot **P**ointing i**N**" to the base, the transistor is an "**NPN**." An easier method is to think of the arrow as a diode, with the tip of the arrow or cathode pointing to an n terminal and the back of the arrow or anode pointing to a p terminal, as seen in the inset in Figure 17-1(b).

The **PNP transistor** can be seen in Figure 17-1(c). With this transistor type, a thin, lightly doped n-type region (base) is placed between two p-type regions (emitter and collector). Figure 17-1(d) illustrates the PNP transistor's schematic symbol. Once again, if you think of the emitter arrow as a diode, as shown in the inset in Figure 17-1(d), the tip of the arrow or cathode is pointing to an n terminal and the back of the arrow or anode is pointing to a p terminal.

Emitter

A transistor region from which charge carriers are injected into the base.

Collector

A semiconductor region through which a flow of charge carriers leaves the base of the transistor.

PNP Transistor

A thin, lightly doped n-type region (base) is placed between two p-type regions (emitter and collector).

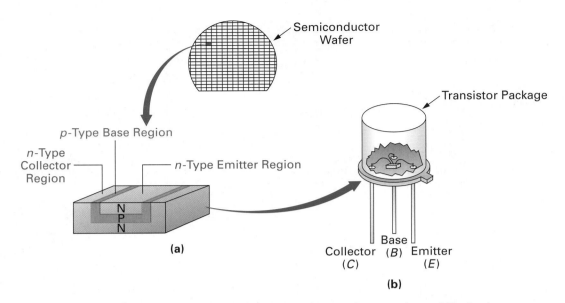

FIGURE 17-2   Bipolar Junction Transistor Construction and Packaging.

17-1-2   *Transistor Construction and Packaging*

Like the diode, the three layers of an NPN or PNP transistor are not formed by joining three alternately doped regions. These three layers are formed by a "diffusion process," which first melts the base region into the collector region, and then melts the emitter region into the base region. For example, with the NPN transistor shown in Figure 17-2(a), the construction process would begin by diffusing or melting a *p*-type base region into the *n*-type collector region. Once this *p*-type base region is formed, an *n*-type emitter region is diffused or melted into the newly diffused *p*-type base region to form an NPN transistor. Keep in mind that manufacturers will generally construct thousands of these transistors simultaneously on a thin semiconductor wafer or disc, as shown in Figure 17-2(a). Once tested, these discs, which are about 3 inches in diameter, are cut to separate the individual transistors. Each transistor is placed in a package, as shown in Figure 17-2(b). The package will protect the transistor from humidity and dust, provide a means for electrical connection between the three semiconductor regions and the three transistor terminals, and serve as a heat sink to conduct away any heat generated by the transistor.

Figure 17-3 illustrates some of the typical low-power and high-power transistor packages. Most low-power, small-signal transistors are hermetically sealed in a metal, plastic, or epoxy package. Four of the low-power packages shown in Figure 17-3(a) have their three leads protruding from the bottom of the package because these package types are usually inserted and soldered into holes in printed circuit boards (PCBs). The surface mount technology (SMT) low-power transistor package, on the other hand, has flat metal legs that mount directly onto the surface of the PCB. These transistor packages are generally used in high component density PCBs because they use less space than a "through-hole" package. To explain this in more detail, a through-hole transistor package needs a hole through the PCB and a connecting pad around the hole to make a connection to the circuit. With an SMT package, however, no holes are needed, only a small connecting pad. Without the need for holes, pads on printed circuit boards can be smaller and placed closer together, resulting in considerable space saving.

The high-power packages, shown in Figure 17-3(b), are designed to be mounted onto the equipment's metal frame or chassis so that the additional metal will act as a heat sink and conduct the heat away from the transistor. With these high-power transistor packages, two or three leads may protrude from the package. If only two leads are present, the metal case will serve as a collector connection, and the two pins will be the base and emitter.

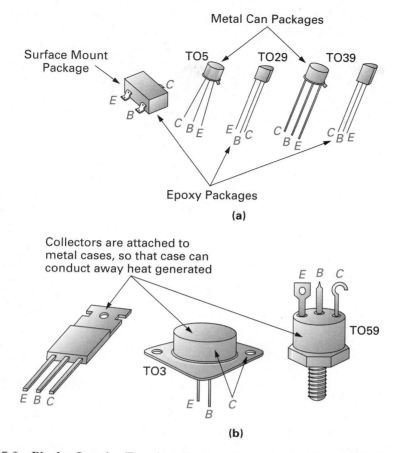

FIGURE 17-3   **Bipolar Junction Transistor Package Types. (a) Low Power. (b) High Power.**

Transistor package types are normally given a reference number. These designations begin with the letters "TO," which stands for transistor outline, and are followed by a number. Figure 17-3 includes some examples of TO reference designators.

17-1-3   *Transistor Operation*

Figure 17-4 shows an NPN bipolar transistor, and the inset shows how a transistor can be thought of as containing two diodes: a *base-to-collector diode* and a *base-to-emitter diode*. With an NPN transistor, both diodes will be back-to-back and "**N**ot be **P**ointing i**N**" (NPN) to the base, as shown in the inset in Figure 17-4. For a PNP transistor, the base-collector and base-emitter diodes will both be pointing into the base.

Transistors are basically controlled to operate as a switch, or they are controlled to operate as a variable resistor. Let us now examine each of these operating modes.

The Transistor's ON/OFF Switching Action

Figure 17-5 illustrates how the transistor can be made to operate as a switch. This ON/OFF switching action of the transistor is controlled by the transistor's base-to-emitter (*B-E*) diode. If the *B-E* diode of the transistor is forward biased, the transistor will turn ON; if the *B-E* diode of the transistor is reverse biased, the transistor will turn OFF.

To begin with, let us see how the transistor can be switched ON. In Figure 17-5(a), the *B-E* diode of the transistor is forward biased (anode at base is +5 V, cathode at emitter is 0 V), and the transistor will turn ON. Its collector and emitter output terminals will be equivalent to a closed switch, as shown in Figure 17-5(b). This low resistance between the tran-

FIGURE 17-4 The Base-Collector and Base-Emitter Diodes Within a Bipolar Transistor.

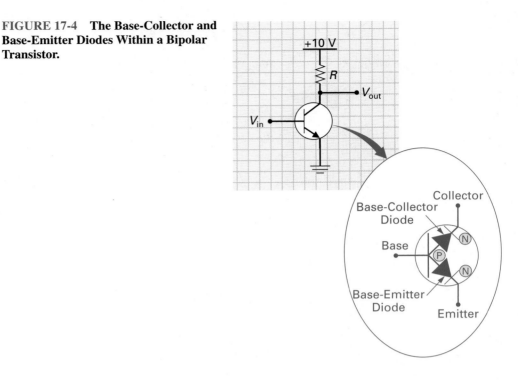

sistor's collector and emitter will cause a current (I), as shown in Figure 17-5(b). The output voltage in this condition will be zero volts because all of the $+10$ V supply voltage will be dropped across R. Another way to describe this would be to say that the low resistance path between the transistor's emitter and collector connects the zero volt emitter potential through to the output.

Now let us see how the transistor can be switched OFF. In Figure 17-5(c), the transistor has 0 V being applied to its base input. In this condition, the B-E diode of the transistor is reverse biased (anode at base is 0 V, cathode at emitter is 0 V), and so the transistor will turn OFF and its collector and emitter output terminals will be equivalent to an open switch, as shown in Figure 17-5(d). This high resistance between the transistor's collector and emit-

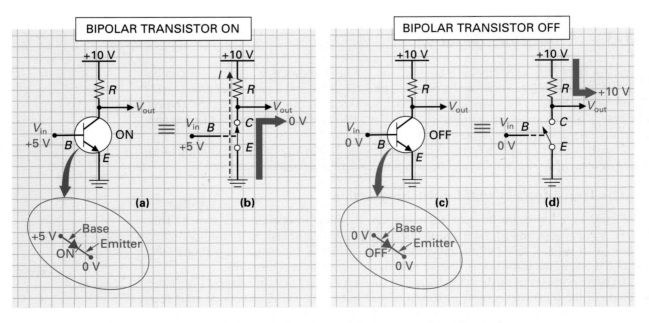

FIGURE 17-5 The Bipolar Transistor's ON/OFF Switching Action.

CHAPTER 17 / BIPOLAR JUNCTION TRANSISTORS (BJTs)

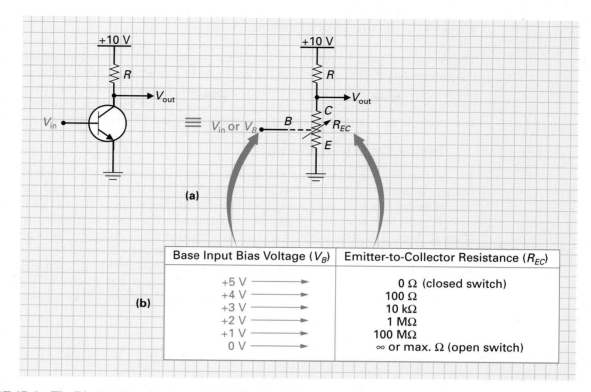

(a)

Base Input Bias Voltage (V_B)	Emitter-to-Collector Resistance (R_{EC})
+5 V ⟶	0 Ω (closed switch)
+4 V ⟶	100 Ω
+3 V ⟶	10 kΩ
+2 V ⟶	1 MΩ
+1 V ⟶	100 MΩ
0 V ⟶	∞ or max. Ω (open switch)

(b)

FIGURE 17-6   The Bipolar Transistor's Variable-Resistor Action.

ter will prevent any current and any voltage drop, resulting in the full +10 V supply voltage being applied to the output, as shown in Figure 17-5(d).

The Transistor's Variable-Resistor Action

In the previous section we saw how the transistor can be biased to operate in one of two states: ON or OFF. When operated in this two-state way, the transistor is being switched ON and OFF in almost the same way as a junction diode. The transistor, however, has another ability that the diode does not have—it can also function as a variable resistor, as shown in the equivalent circuit in Figure 17-6(a). In Figure 17-5 we saw how +5 V base input bias voltage would result in a low resistance between emitter and collector (closed switch) and how a 0 V base input bias would result in a high resistance between emitter and collector (open switch). The table in Figure 17-6(b) shows an example of the relationship between base input bias voltage (V_B) and emitter-to-collector resistance (R_{CE}). In this table, you can see that the transistor is not only going to be driven between the two extremes of fully ON and fully OFF. When the base input voltage is at some voltage level between +5 V and 0 V, the transistor is partially ON; therefore, the transistor's emitter-to-collector resistance is somewhere between 0 Ω and maximum Ω. For example, when $V_B = +4$ V, the transistor is not fully ON, and its emitter-to-collector resistance will be slightly higher, at 100 Ω. If the base input bias voltage is further reduced to +3 V, for example, you can see in the table that the emitter-to-collector resistance will further increase to 10 kΩ. Further decreases in base input voltage ($V_B\downarrow$) will cause further increases in emitter-to-collector resistance ($R_{CE}\uparrow$) until $V_B = 0$ V and $R_{CE} =$ maximum Ω.

As a matter of interest, the name transistor was derived from the fact that through base control we can "transfer" different values of "resistance" between the emitter and collector. This effect of "transferring resistance" is known as **transistance** and the component that functions in this manner is called the transistor.

Now that we have seen how the transistor can be made to operate as either a switch or a variable resistor, let us see how these characteristics can be made use of in circuit applications.

Transistance

The effect of transferring resistance.

17-1-4   *Transistor Applications*

The transistor's impact on electronics has been phenomenal. It initiated the multibillion dollar semiconductor industry and was the key element behind many other inventions, such as integrated circuits (ICs), optoelectronic devices, and digital computer electronics. In all of these applications, however, the transistor is basically made to operate in one of two ways: as a switch or as a variable resistor. Let us now briefly examine an example of each.

Digital Logic Gate Circuit

A digital logic gate circuit makes use of the transistor's ON/OFF switching action. Digital circuits are often referred to as "switching" or "two-state" circuits because their main control device (the transistor) is switched between the two states of ON and OFF. The transistor is at the very heart of all digital electronic circuits. For example, transistors are used to construct logic gate circuits, gates are used to construct flip-flop circuits, flip-flops are used to construct register and counter circuits, and these circuits are used to construct microprocessor, memory, and input/output circuits—the three basic blocks of a digital computer.

Figure 17-7(a) shows how the transistor can be used to construct a NOT gate or INVERTER gate. The basic NOT gate circuit is constructed using one NPN transistor and two

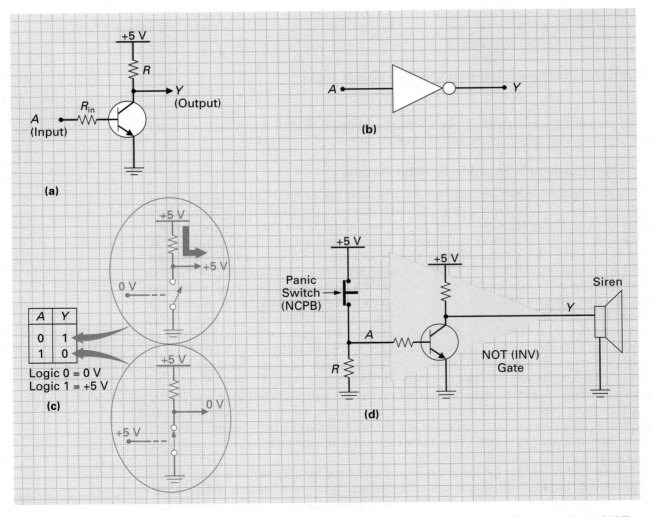

FIGURE 17-7   The Transistor Being Used in a Digital Electronic Circuit. (a) Basic NOT or INVERTER Gate Circuit. (b) NOT Gate Schematic Symbol. (c) NOT Gate Function Table. (d) NOT Gate Security System Application.

resistors. This logic gate has only one input (A) and one output (Y), and its schematic symbol is shown in Figure 17-7(b). Figure 17-7(c) shows how this logic gate will react to the two different input possibilities. When the input is 0 V (logic 0), the transistor's base-emitter P-N diode will be reverse biased and so the transistor will turn OFF. Referring to the inset for this circuit condition in Figure 17-7(c), you can see that the OFF transistor is equivalent to an open switch between emitter and collector, and therefore the +5 V supply voltage will be connected to the output. In summary, a logic 0 input (0 V) will be converted to a logic 1 output (+5 V). On the other hand, when the input is +5 V (logic 1), the transistor's base-emitter P-N diode will be forward biased and so the transistor will turn ON. Referring to the inset for this circuit condition in Figure 17-7(c), you can see that the ON transistor is equivalent to a closed switch between emitter and collector, and therefore 0 V will be connected to the output. In summary, a logic 1 input (+5 V) will be converted to a logic 0 output (0 V).

Referring to the function table in Figure 17-7(c), you can see that the output logic level is "NOT" the same as the input logic level—hence the name NOT gate.

As an application, Figure 17-7(d) shows how a NOT or INV gate can be used to invert an input control signal. In this circuit, you can see that a normally closed push button (NCPB) switch is used as a panic switch to activate a siren in a security system. Because the push button is normally closed, it will produce +5 V at A when it is not in alarm. If this voltage were connected directly to the siren, the siren would be activated incorrectly. By including the NOT gate between the switch circuit and the siren, the normally HIGH output of the NCPB will be inverted to a LOW, and not activate the siren when we are not in alarm. When the panic switch is pressed however, the NCPB contacts will open producing a LOW input voltage to the NOT gate. This LOW input will be inverted to a HIGH output and activate the siren.

Analog Amplifier Circuit

When used as a variable resistor, the transistor is the controlling element in many analog or linear circuit applications such as amplifiers, oscillators, modulators, detectors, regulators, and so on. The most important of these applications is **amplification,** which is the boosting in strength or increasing in amplitude of electronic signals.

Figure 17-8(a) shows a simplified transistor amplifier circuit, while Figure 17-8(b) shows the voltage waveforms present at different points in the circuit. As you can see, the transistor is labeled Q_1 because the letter "Q" is the standard letter designation used for transistors.

Before applying an ac sine-wave input signal, let us determine the dc voltage levels at the transistor's base and collector. The 6.3 kΩ/3.7 kΩ resistance ratio of the voltage divider R_1 and R_2 causes the +10 V supply voltage to be proportionally divided, producing +3.7 V dc across R_2. This +3.7 V dc will be applied to the base of the transistor, causing the base-emitter junction of Q_1 to be forward biased and Q_1 to turn ON. With transistor Q_1 ON, a certain value of resistance will exist between the transistor's collector and emitter (R_{CE} or R_{EC}), and this resistance will form a voltage divider with R_E and R_C, as seen in the inset in Figure 17-8(a). The dc voltage at the collector of Q_1 relative to ground (V_C) will be equal to the voltage developed across Q_1's collector-emitter resistance (R_{CE}) and R_E. In this example circuit, with no ac signal applied, a V_B of +3.7 V dc will cause R_{CE} and R_E to cumulatively develop +8 V at the collector of Q_1. The transistor has a base bias voltage (V_B) that is +3.7 V dc relative to ground and a collector voltage (V_C) that is +8 V dc relative to ground. Capacitors C_1 and C_2 are included to act as dc blocks, with C_1 preventing the +3.7 V dc base bias voltage (V_B) from being applied back to the input (V_{in}) and C_2 preventing the +8 V dc collector reference voltage (V_C) from being applied across the output (V_{R_L} or V_{out}).

Let us now apply an input signal and see how it is amplified by the amplifier circuit in Figure 17-8(a). The alternating input sine-wave signal (V_{in}) is applied to the base of Q_1 via C_1, which, like most capacitors, offers no opposition to this ac signal. This input signal, which has a peak-to-peak voltage change of 200 mV, is shown in the first waveform in Figure 17-8(b). The alternating input signal will be superimposed on the +3.7 V dc base bias voltage and cause the +3.7 V dc at the base of Q_1 to increase by 100 mv (3.7 V +

Amplification
Boosting in strength, or increasing amplitude, of electronic signals.

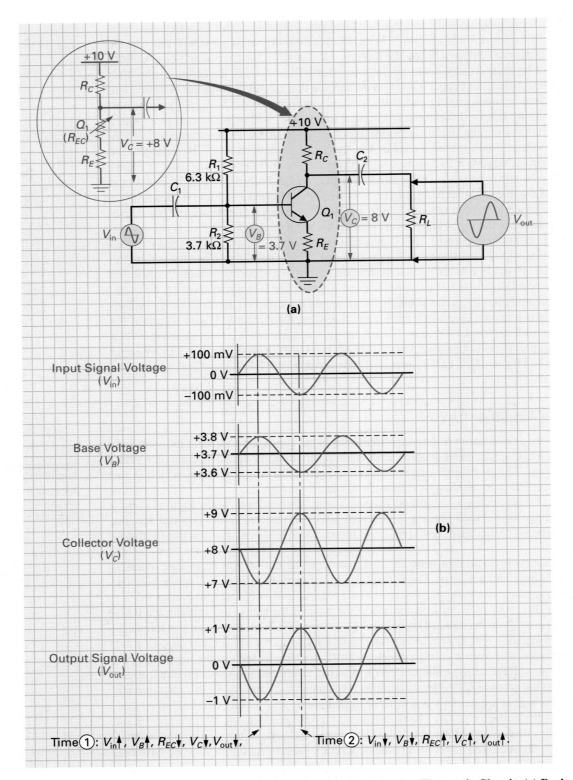

FIGURE 17-8 The Transistor Being Used in an Analog Electronic Circuit. (a) Basic Amplifier Circuit. (b) Input/Output Voltage Waveforms.

100 mV = 3.8 V), and decrease by 100 mv (3.7 V − 100 mV = 3.6 V), as seen in the second wave-form in Figure 17-8(b). An increase in the input signal ($V_{in}\uparrow$), and therefore the base voltage ($V_B\uparrow$), will cause an increase in the emitter diode's forward bias, causing Q_1 to turn more ON and the emitter-to-collector resistance of Q_1 to decrease ($R_{EC}\downarrow$). Because voltage drop is always proportional to resistance, a decrease in $R_{EC}\downarrow$ will cause a decrease in the voltage drop across R_{CE} and R_E ($V_C\downarrow$), and this decrease in V_C will be coupled to the output via C_2, causing a decrease in the output voltage developed across the load ($V_{out}\downarrow$).

Now let us examine what will happen when the sine-wave input signal decreases. A decrease in the input signal ($V_{in}\downarrow$), and therefore the base voltage ($V_B\downarrow$), will cause a decrease in the emitter diode's forward bias, causing Q_1 to turn less ON and the emitter-to-collector resistance of Q_1 to increase ($R_{EC}\uparrow$). Because voltage drop is always proportional to resistance, an increase in $R_{EC}\uparrow$ will cause an increase in the voltage drop across R_{EC} and R_E ($V_C\uparrow$). This increase in V_C will be coupled to the output via C_2, causing an increase in the output voltage developed across the load ($V_{out}\uparrow$).

Comparing the input signal voltage (V_{in}) to the output signal voltage (V_{out}) in Figure 17-8(b), you can see that a change in the input signal voltage produces a corresponding greater change in the output signal voltage. The ratio (comparison) of output signal voltage change to input signal voltage change is a measure of this circuit's **voltage gain (A_V).** In this example, the **output signal voltage change (ΔV_{out})** is between +1 V and −1 V, and the **input signal voltage change (ΔV_{in})** is between +100 mV and −100 mV. The circuit's voltage gain between input and output will therefore be:

$$\text{Voltage Gain } (A_V) = \frac{\text{Output Voltage Change } (\Delta V_{out})}{\text{Input Voltage Change } (\Delta V_{in})}$$

$$A_V = \frac{+1 \text{ to} -1 \text{ V}}{+100 \text{ mV to} -100 \text{ mV}} = \frac{2 \text{ V}}{200 \text{ mV}} = 10$$

A voltage gain of 10 means that the output voltage is ten times larger than the input voltage. The transistor does not produce this gain magically within its NPN semiconductor structure. The gain or amplification is achieved by the input signal controlling the conduction of the transistor, which takes energy from the collector supply voltage and develops this energy across the load resistor. Amplification is achieved by having a small input voltage control a transistor and its large collector supply voltage, so that a small input voltage change results in a similar but larger output voltage change.

Comparing the input signal to the output signal at time 1 and time 2 in Figure 17-8(b), you can see that this circuit will invert the input signal voltage in the same way that the NOT gate inverts its input voltage (positive input voltage swing produces a negative output voltage swing, and vice versa). This inversion always occurs with this particular transistor circuit arrangement; however, it is not a problem since the shape of the input signal is still preserved at the output (both input and output signals are sinusoidal).

A Switching Regulator Circuit

In the previous chapter, it was shown how a series resistor and zener diode could be used in a dc power supply to function as a voltage regulator. These regulator types maintain a constant output voltage because variations in input voltage or load current are dissipated as heat. These **series dissipative regulators** generally have a low "conversion efficiency" of typically 60% to 70% and should be used only in low- to medium-load current applications.

Series switching regulators, on the other hand, have a conversion efficiency of typically 90%. To explain the operation of these regulator types, refer to the simplified circuit in Figure 17-9(a). To improve efficiency, a series-pass transistor (Q_1) is operated as a switch, rather than as a variable resistor. This means that Q_1 is switched ON and OFF, and therefore either switches the +12 V input at its collector through to its emitter, or blocks the +12 V from passing through to the emitter. These +12 V pulses at the emitter of Q_1

Voltage Gain (A_V)
The ratio of the output signal voltage change to input signal voltage change.

Output Signal Voltage Change
Change in output signal voltage in response to a change in the input signal voltage.

Input Signal Voltage Change
The input voltage change that causes a corresponding change in the output voltage.

Series Dissipative Regulators
Voltage regulators that maintain a constant output voltage by causing variations in input voltage or load current to be dissipated as heat.

Series Switching Regulators
A regulator circuit containing a power transistor in series with the load, that is switched ON and OFF to regulate the dc output voltage delivered to the load.

charge capacitor C_1 to an average voltage (which in this example is $+5$ V) and this voltage is applied to the load (R_L). To explain this in more detail, when Q_1 is turned ON by a HIGH base voltage from the switching regulator IC, the unregulated $+12$ V at Q_1's collector is switched through to Q_1's emitter, where it reverse biases D_1, and is applied to the series-connected inductor L_1 and parallel capacitor C_1. Inductor L_1 and capacitor C_1 act as a low-pass filter because series-connected L_1 opposes the ON/OFF changes in current and passes a relatively constant current to the load: shunt- or parallel-connected C_1 opposes the ON/OFF changes in voltage and holds the output voltage relatively constant at $+5$ V. When Q_1 is turned OFF by a LOW base voltage from the switching regulator IC, the unregulated $+12$ V input is disconnected from the LC filter, the inductor's magnetic field will collapse and produce a currrent through the load, and the $+5$ V charge held by C_1 will still be applied across the load. Inductor L_1, therefore, smoothes out the current changes, while capacitor C_1 smoothes out the voltage changes caused by the ON/OFF switching of transistor Q_1. The next, and most important, question is: how does this circuit regulate, or maintain constant, the output voltage? The answer is: throught a closed-loop "sense and adjust" system controlled by a switching regulator IC. The switching regulator IC operates by comparing an internal fixed reference voltage to a sense input, which is taken from the $+5$ V output, as is shown in Figure 17-9(a). Referring to the waveforms shown in Figure 17-9(b), you can see that whenever the output voltage falls below $+5$ V (from time t_1 to

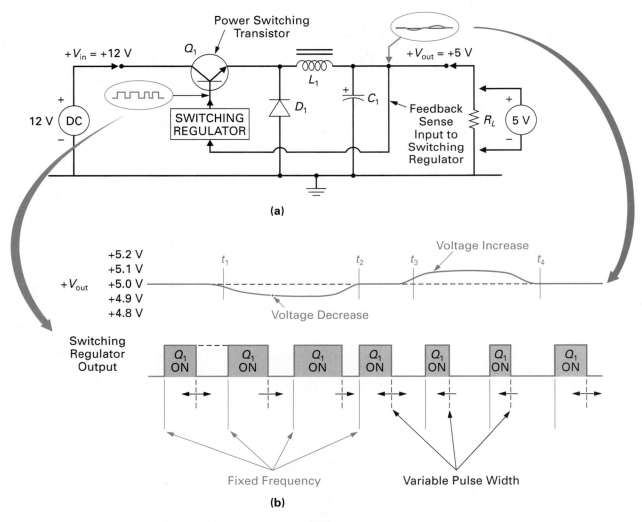

FIGURE 17-9   Basic Switching Regulator Action.

t_2), the switching regulator responds by increasing the width of the positive output pulse applied to the base of Q_1. This increases the ON time of Q_1, which raises the average output voltage, bringing the output back up to $+5$ V. On the other hand, whenever the output voltage rises above $+5$ V (between time t_3 and t_4), the switching regulator responds by decreasing the width of the positive output pulse applied to the base of Q_1. This decreases the ON time of Q_1, which lowers the average output voltage, bringing the output back down to $+5$ V. The net result is that the output voltage will remain locked at $+5$ V despite variations in the input voltage and variations in the load.

A switching regulator, therefore, is a voltage regulator that chops up, or switches ON and OFF (at typically a 20 kHz rate), a dc input voltage to efficiently produce a regulated dc output voltage. A **switching power supply** uses switching regulators and is generally small in size and very efficient. The only disadvantage is that the circuitry is generally a little more complex, and therefore a little more costly.

Switching Power Supply

A dc power supply that makes use of a series switching regulator controlled by a pulse-width-modulator to regulate the output voltage.

SELF-TEST EVALUATION POINT FOR SECTION 17-1

Now that you have completed this section, you should be able to:

- **Objective 1.** Name the three terminals of the bipolar junction transistor.

- **Objective 2.** Describe the difference between the construction and schematic symbol of the NPN and PNP bipolar transistor.

- **Objective 3.** Identify the base, collector, and emitter terminals of typical low-power and high-power transistor packages.

- **Objective 4.** Describe the two basic actions of a bipolar transistor:
  a. ON/OFF switching action
  b. Variable-resistor action

- **Objective 5.** Define the terms transistor and transistance.

- **Objective 6.** Explain how the transistor can be used in the following basic applications:
  a. A digital (two-state) circuit, such as a logic gate
  b. An analog (linear) circuit, such as an amplifier

Use the following questions to test your understanding of Section 17-1.

1. What are the two basic types of bipolar transistor?
2. Name the three terminals of a bipolar transistor.
3. What are the two basic ways in which a transistor is made to operate?
4. Which of the modes of operation mentioned in question 3 is made use of in digital circuits and which is made use of in analog circuits?

17-2 SECOND APPROXIMATION DESCRIPTION OF A BIPOLAR TRANSISTOR

Now that we have a good understanding of the bipolar junction transistor's (BJT's) general characteristics, operation, and applications, let us examine all of these aspects in a little more detail.

17-2-1 Basic Bipolar Transistor Action

When describing diodes previously, we saw how the P-N junction of a diode could be either forward or reverse biased to either permit or block the flow of current through the device. The transistor must also be biased correctly; however, in this case, two P-N junctions rather than one must have the correct external supply voltages applied.

A Correctly Biased NPN Transistor Circuit

Figure 17-10(a) shows how an NPN transistor should be biased for normal operation. In this circuit, a $+10$ V supply voltage is connected to the transistor's collector (C) via a

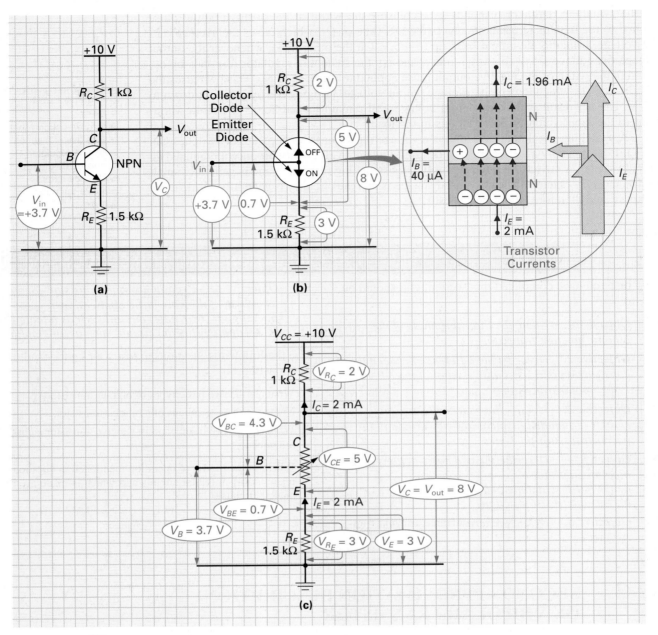

FIGURE 17-10   A Correctly Biased NPN Transistor Circuit.

1 kΩ collector resistor (R_C). The emitter (E) of the transistor is connected to ground via a 1.5 kΩ emitter resistor (R_E), and, as an example, an input voltage of +3.7 V is being applied to the base (B). The output voltage (V_{out}) is taken from the collector, and this collector voltage (V_C) will be equal to the voltage developed across the transistor's collector-to-emitter and the emitter resistor R_E.

As previously mentioned in the first approximation description of the transistor, the transistor can be thought of as containing two diodes, as shown in Figure 17-9(b). In normal operation, *the transistor's emitter diode or junction is forward biased, while the transistor's collector diode or junction is reverse biased.* To explain how these junctions are biased ON and OFF simultaneously, let us see how the input voltage of +3.7 V will affect this transistor circuit. An input voltage of +3.7 V is large enough to overcome the barrier voltage of the emitter diode (base-emitter junction), and so it will turn ON (base or anode is +, emitter or cathode is connected to ground or 0 V). Like any forward biased silicon diode, the emitter diode will drop 0.7 V between base and emitter, and so the +3.7 V at the base will produce +3.0 V at the emitter. Knowing the voltage drop across the emitter resistor ($V_{R_E} = 3$ V) and the resistance of the emitter resistor ($R_E = 1.5$ kΩ), we can calculate the value of current through the emitter resistor.

$$I_{R_E} = \frac{V_{R_E}}{R_E} = \frac{3 \text{ V}}{1.5 \text{ k}\Omega} = 2 \text{ mA}$$

This emitter resistor current of 2 mA will leave ground, travel through R_E, and then enter the transistor's n-type emitter region. This current at the transistor's emitter terminal is called the **emitter current (I_E).** The forward biased emitter diode will cause the steady stream of electrons entering the emitter to head toward the base region, as shown in the inset in Figure 17-10(b). The base is a very thin, lightly doped region with very few holes in relation to the number of electrons entering the transistor from the emitter. Consequently, only a few electrons combine with the holes in the base region and flow out of the base region. This relatively small current at the transistor's base terminal is called the **base current (I_B).** Because only a few electrons combine with holes in the base region, there is an accumulation of electrons in the base's p layer. These free electrons, feeling the attraction of the large positive collector supply voltage (+10 V), will travel through the n-type collector junction and out of the transistor to the positive external collector supply voltage. The current emerging out of the transistor's collector is called the **collector current (I_C).** Because both the collector current and base current are derived from the emitter current, we can state that:

$$I_E = I_B + I_C$$

In the example in the inset in Figure 17-10(b), you can see that this is true because

$$I_E = I_B + I_C$$
$$I_E = 40 \text{ }\mu\text{A} + 1.96 \text{ mA} = 2 \text{ mA} \quad (40 \text{ }\mu\text{A} = 0.04 \text{ mA})$$

Stated another way, we can say that the collector current is equal to the emitter current minus the current that is lost out of the base.

$$I_C = I_E - I_B$$
$$I_C = 2 \text{ mA} - 40 \text{ }\mu\text{A} = 1.96 \text{ mA}$$

Approximately 98% of the electrons entering the emitter of a transistor will arrive at the collector. Because of the very small percentage of current flowing out of the base (I_B equals about 2% of I_E), we can approximate and assume that I_C is equal to I_E.

$$I_C \cong I_E$$

(I_C approximately equals I_E)

The Current-Controlled Transistor

In the previous section, we discovered that because the collector and base currents (I_C and I_B) are derived from the emitter current (I_E), an increase in the emitter current ($I_E\uparrow$), for example, will cause a corresponding increase in collector and base current ($I_C\uparrow$, $I_B\uparrow$). Looking at this from a different angle, an increase in the applied base voltage (base input increases to +3.8 V) will increase the forward bias applied to the emitter diode of the transistor, which will draw more electrons up from the emitter and cause an increase in I_E, I_B, and I_C. Similarly, a decrease in the applied base voltage (base input decreases to +3.6 V) will decrease the forward bias applied to the emitter diode of the transistor, which will decrease the number of electrons being drawn up from the emitter and cause a decrease in I_E, I_B, and I_C. The applied input base voltage will control the amount of base current, which will in turn control the amount of emitter and collector current, and therefore the conduction of the transistor. This is why *the bipolar transistor is known as a current-controlled device.*

Emitter Current (I_E)
The current at the transistor's emitter terminal.

Base Current (I_B)
The relatively small current at the transistor's base terminal.

Collector Current (I_C)
The current emerging out of the transistor's collector.

Continuing our calculations for the example circuit in Figure 17-10(b), let us apply this current relationship and assume that I_C is equal to I_E, which, as we previously calculated, is equal to 2 mA. Knowing the value of current for the collector resistor ($I_{R_C} = 2$ mA) and the resistance of the collector resistor ($R_C = 1$ kΩ), we can calculate the voltage drop across the collector resistor.

$$V_{R_C} = I_{R_C} \times R_C = 2 \text{ mA} \times 1 \text{ k}\Omega = 2 \text{ V}$$

With 2 V being dropped across R_C, the voltage at the transistor's collector (V_C) will be:

$$V_C = +10 \text{ V} - V_{R_C} = 10 \text{ V} - 2 \text{ V} = 8 \text{ V}$$

Because the voltage at the transistor's collector relative to ground is applied to the output, the output voltage will also be equal to 8 V.

$$V_C = V_{\text{out}} = 8 \text{ V}$$

At this stage, we can determine a very important point about any correctly biased NPN transistor circuit. *A properly biased transistor will have a forward biased base-emitter junction (emitter diode is ON), and a reverse biased base-collector junction (collector diode is OFF).* We can confirm this with our example circuit in Figure 17-10(b), because we now know the voltages at each of the transistor's terminals.

Emitter diode (base-emitter junction) is forward biased (ON) because
Anode (base) is connected to +3.7 V (V_{in})
Cathode (emitter) is connected to 0 V via R_E

Collector diode (base-collector junction) is reverse biased (OFF) because
Anode (base) is connected to +3.7 V (V_{in})
Cathode (collector) is at +8 V (due to 2 V drop across R_C)

Keep in mind that even though the collector diode (base-collector junction) is reverse biased, current will still flow through the collector region. This is because most of the electrons traveling from emitter-to-base (through the forward biased emitter diode) do not find many holes in the thin, lightly doped base region, and therefore the base current is always very small. Almost 98% of the electrons accumulating in the base region feel the strong attraction of the positive collector supply voltage and flow up into the collector region and then out of the collector as collector current.

With the example circuit in Figure 17-10(a) and (b), the emitter diode is ON and the collector diode is OFF, and the transistor is said to be operating in its normal, or *active, region.*

Operating a Transistor in the Active Region

Active Operation or in the Active Region

When the base-emitter junction is forward biased and the base-collector junction is reverse biased. In this mode, the transistor is equivalent to a variable resistor between collector and emitter.

A transistor is said to be in **active operation,** or in the **active region,** when its base-emitter junction is forward biased (emitter diode is ON), and the base-collector junction is reverse biased (collector diode is OFF). In this mode, the transistor is equivalent to a variable resistor between collector and emitter.

In Figure 17-10(c), our transistor circuit example has been redrawn with the transistor this time being shown as a variable resistor between collector and emitter and with all of our calculated voltage and current values inserted. Before we go any further with this circuit, let us discuss some of the letter abbreviations used in transistor circuits. To begin with, the term V_{CC} is used to denote the "stable collector voltage" and this dc supply voltage will typically be positive for an NPN transistor. Two Cs are used in this abbreviation ($+V_{CC}$) because V_C (*V* sub single *C*) is used to describe the voltage at the transistor's collector relative to ground. The doubling up of letters such as V_{CC}, V_{EE}, or V_{BB} is used to denote a constant dc bias voltage for the collector (V_{CC}), emitter (V_{EE}), and base (V_{BB}). A single sub letter abbreviation such as V_C, V_E, or V_B is used to denote a transistor terminal voltage relative to ground. The other voltage abbreviations, V_{CE}, V_{BE}, and V_{CB}, are used for the voltage difference between two terminals of the transistor. For example, V_{CE} is used to denote the potential difference between the transistor's collector and emitter terminals. Finally, I_E, I_B, and I_C are, as previously stated, used to denote the transistor's emitter current (I_E), base current (I_B), and collector current (I_C).

Because the transistor's resistance between emitter and collector in Figure 17-10(c) is in series with R_C and R_E, we can calculate the voltage drop between collector and emitter (V_{CE}) because V_{RC} and V_{RE} are known.

$$V_{CE} = V_{CC} - (V_{R_E} + V_{R_C})$$
$$V_{CE} = 10 \text{ V} - (3 \text{ V} + 2 \text{ V}) = 10 \text{ V} - 5 \text{ V} = 5 \text{ V}$$

Now that we know the voltage drop between the transistor's collector and emitter (V_{CE}), we can calculate the transistor's equivalent resistance between collector and emitter (R_{CE}) because we know that the current through the transistor is 2 mA.

$$R_{CE} = \frac{V_{CE}}{I_C} = \frac{5 \text{ V}}{2 \text{ mA}} = 2.5 \text{ k}\Omega$$

Operating the Transistor in Cutoff and Saturation

Figure 17-11 shows the three basic ways in which a transistor can be operated. As we have already discovered, the bias voltages applied to a transistor control the transistor's operation by controlling the two P-N junctions (or diodes) in a bipolar transistor. For example, the center column reviews how a transistor will operate in the active region. As you can see, our previous circuit example with all of its values has been used. To summarize: *when a transistor is operated in the active region, its emitter diode is biased ON, its collector diode is biased OFF, and the transistor is equivalent to a variable resistor between the collector and the emitter.*

The left column in Figure 17-11 shows how the same transistor circuit can be driven into **cutoff.** A transistor is in cutoff when the bias voltage is reduced to a point that it stops current in the transistor. In this example circuit, you can see that when the base input bias voltage (V_B) is reduced to 0 V, the transistor is cut off. *In cutoff, both the emitter and the collector diode of the transistor will be biased OFF, the transistor is equivalent to an open switch between the collector and the emitter, and the transistor current is zero.*

The right column in Figure 17-11(c) shows how the same transistor circuit can be driven into **saturation.** A transistor is in saturation when the bias voltage is increased to such a point that any further increase in bias voltage will not cause any further increase in current through the transistor. In the equivalent circuit in Figure 17-11(c), you can see that when the base input bias voltage (V_B) is increased to +6.7 V, the emitter diode of the transistor will be heavily forward biased and the emitter current will be large.

$$I_E = \frac{V_B - V_{BE}}{R_E} = \frac{6.7 \text{ V} - 0.7 \text{ V}}{1.5 \text{ k}\Omega} = 4.\text{mA}$$

Because I_B and I_C are both derived from I_E, an increase in I_E will cause a corresponding increase in both I_B and I_C. These high values of current through the transistor account for why a transistor operating in saturation is said to be equivalent to a closed switch (high conductance, low resistance). Although the transistor's resistance between the collector and emitter (R_{CE}) is assumed to be 0 Ω, there is still some small value of R_{CE}. Typically, a saturated transistor will have a 0.3 V drop between the collector and emitter ($V_{CE} = 0.3$ V), as shown in the equivalent circuit in Figure 17-11(c). If $V_{BE} = 0.7$ V and $V_{CE} = 0.3$ V, then the voltage drop across the collector diode of a saturated transistor (V_{BC}) will be 0.4 V. This means that the base of the transistor (anode) is now +0.4 V relative to the collector (cathode), and there is not enough reverse bias voltage to turn OFF the collector diode. *In saturation, both the emitter and collector diodes are said to be forward biased, the transistor is equivalent to a closed switch, and any further increase in bias voltage will not cause any further increase in current through the transistor.*

Cutoff

A transistor is in cutoff when the bias voltage is reduced to a point that it stops current in the transistor.

Saturation

A transistor is in saturation when the bias voltage is increased to such a point that further increase will not cause any increase in current through the transistor.

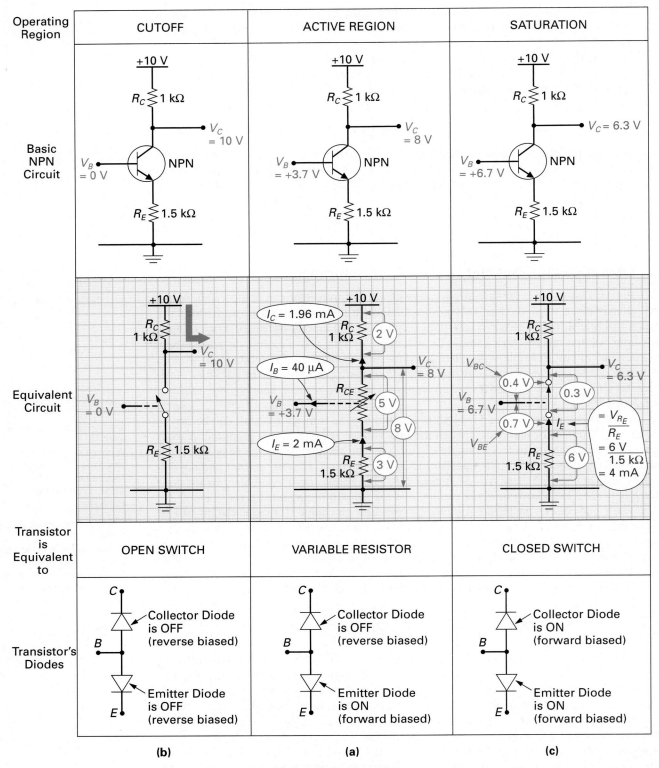

FIGURE 17-11 **The Three Bipolar Transistor Operating Regions.**

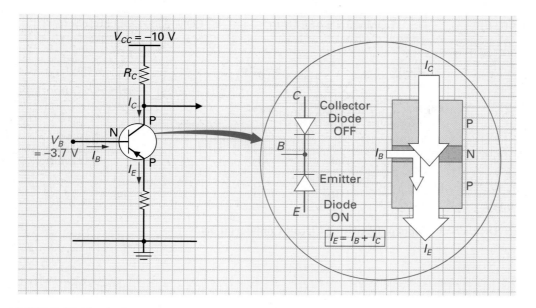

FIGURE 17-12   A Correctly Biased PNP Transistor Circuit.

Biasing PNP Bipolar Transistors

Generally the PNP transistor is not employed as much as the NPN transistor in most circuit applications. The only difference that occurs with PNP transistor circuits is that the polarity of V_{CC} and the base bias voltage (V_B) need to be reversed to a negative voltage, as shown in Figure 17-12. The PNP transistor has the same basic operating characteristics as the NPN transistor, and all of the previously discussed equations still apply. Referring to the inset in Figure 17-12, you will see that the -3.7 V base bias voltage will forward bias the emitter diode, and the -10 V V_{CC} will reverse bias the collector diode, so that the transistor is operating in the active region. Also, the electron transistor currents are in the opposite direction. This, however, makes no difference because the sum of the collector current entering the collector and base current entering the base is equal to the value of emitter current leaving the emitter, so $I_E = I_B + I_C$ still applies.

17-2-2   *Bipolar Transistor Circuit Configurations and Characteristics*

In the previous sections, we have seen how the bipolar junction transistor can be used in digital two-state switching circuits and analog or linear circuits such as the amplifier. In all of these different circuit interconnections or **configurations,** the bipolar transistor was used as the main controlling element, with one of its three leads being used as a common reference and the other two leads being used as an input and an output. Although there are many thousands of different bipolar transistor circuit applications, all of these circuits can be classified in one of three groups based on which of the transistor's leads is used as the **common** reference. These three different circuit configurations are shown in Figure 17-13. With the *common-emitter (C-E)* bipolar transistor circuit configuration, shown in Figure 17-13(a), the input signal is applied between the base and emitter, while the output signal appears between the transistor's collector and emitter. With this circuit arrangement, the input signal controls the transistor's base current, which in turn controls the transistor's output collector current, and the emitter lead is common to both the input and output. Similarly, with the *common-base (C-B)* circuit configuration shown in Figure 17-13(b), the input signal is applied between the transistor's emitter and base, the output signal is developed across the transistor's collector and base, and the base is common to both input and output. Finally,

Configurations

Different circuit interconnections.

Common

Shared by two or more services, circuits, or devices. Although the term "common ground" is frequently used to describe two or more connections sharing a common ground, the term "common" alone does not indicate a ground connection, only a shared connection.

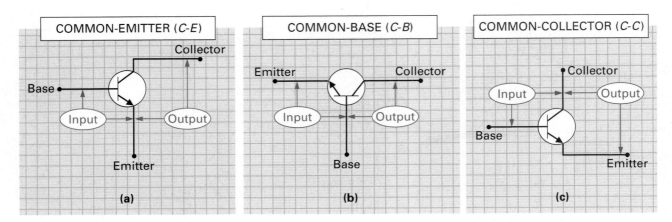

FIGURE 17-13 Bipolar Transistor Circuit Configurations.

with the *common-collector (C-C)* circuit configuration shown in Figure 17-13(c), the input is applied between the base and collector, the output is developed across the emitter and collector, and the collector is common to both the input and output.

To begin with, we will discuss the common-emitter circuit configuration characteristics because it has been this circuit arrangement that we have been using in all of the circuit examples in this chapter.

Common-Emitter Circuits

With the **common-emitter circuit,** the transistor's emitter lead is common to both the input and output signals. In this circuit configuration, the base serves as the input lead, and the collector serves as the output lead. Figure 17-14 contains a basic common-emitter (*C-E*) circuit, its associated input/output voltage and current waveforms, characteristic curves, and table of typical characteristics. Using this illustration, we will examine the operation and characteristics of the *C-E* circuit configuration.

DC current gain. Referring to the *C-E* circuit in Figure 17-14(a), and its associated waveforms in Figure 17-14(b), let us now examine this circuit's basic operation.

Before applying the ac sine-wave input signal (V_{in}), let us assume that $V_{in} = 0$ V and examine the transistor's dc operating characteristics, or the "no input signal" condition. The voltage divider R_1 and R_2 will divide the V_{CC} supply voltage, producing a positive dc base bias voltage across R_2. This base bias voltage will be applied to the base of Q_1, and because it is generally greater than 0.7 V, it will forward bias the transistor's base-emitter junction, turning Q_1 ON (in the example in Figure 17-14, $V_B = 3$ V dc). Capacitor C_1 is included to act as a dc block, preventing the base bias voltage (V_B) from being applied back to the input (V_{in}). The value of dc base bias voltage will determine the value of base current (I_B) flowing out of the transistor's base, and this value of I_B will in turn determine the value of collector current (I_C) flowing out of the transistor's collector and through R_C. Because the transistor's output current (I_C) is so much larger than the transistor's very small input current (I_B), the circuit produces an increase in current, or a **current gain.** The current gain in a common-emitter configuration is called the transistor's **beta** (symbolized β). A transistor's **dc beta** (β_{DC}) indicates a common-emitter transistor's "dc current gain," and it is the ratio of its output current (I_C) to its input current (I_B). This ratio can be expressed mathematically as:

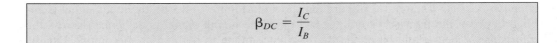

$$\beta_{DC} = \frac{I_C}{I_B}$$

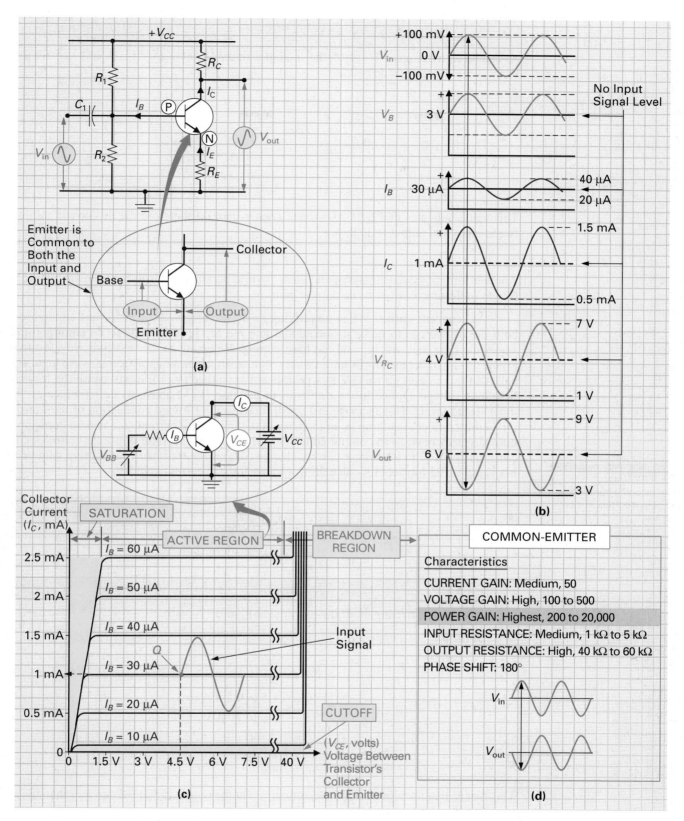

FIGURE 17-14 **Common-Emitter** *(C-E)* **Circuit Configuration Characteristics.**

EXAMPLE:

As can be seen in Figure 17-14(b), the "no input signal" level of $I_C = 1$ mA, and $I_B = 30$ μA. What is the transistor's dc beta?

Solution:

$$\beta_{DC} = \frac{I_C}{I_B}$$
$$= \frac{1\text{mA}}{30 \text{ μA}}$$
$$= 33.3$$

This value indicates that I_C is 33.3 times greater than I_B, therefore the dc current gain between input and output is 33.3.

Another form of system analysis, known as "hybrid parameters," uses the term *hfe* instead of β_{DC} to indicate a transistor's dc current gain.

AC current gain. When amplifying an ac waveform, a transistor has applied to it both the dc voltage to make it operational and the ac signal voltage to be amplified that varies the base bias voltage and the base current. Referring to the first four waveforms in Figure 17-14(b), let us now examine what will happen when we apply a sine-wave input signal. As V_{in} increases above 0 V to a peak positive voltage ($V_{in}\uparrow$), it will cause the forward base bias voltage applied to the transistor to increase ($V_B\uparrow$) above its dc reference or "no input signal level." This increase in V_B will increase the forward conduction of the transistor's emitter diode, resulting in an increase in the input base current ($I_B\uparrow$) and a corresponding larger increase in the output collector current ($I_C\uparrow$). The ratio of output collector current change (ΔI_C) to input base current change (ΔI_B) is the transistor's ac current gain or **ac beta (β_{AC}).** The formula for calculating a transistor's ac current gain is:

AC Beta (β_{AC})
The ratio of a transistor's ac output current to input current.

$$\beta_{AC} = \frac{\Delta I_C}{\Delta I_B}$$

EXAMPLE:

Calculate the ac current gain of the example in Figure 17-14(b).

Solution:

$$\beta_{AC} = \frac{\Delta I_C}{\Delta I_B}$$
$$= \frac{1.5 \text{ to } 0.5 \text{ mA}}{40 \text{ to } 20 \text{ μA}}$$
$$= \frac{1 \text{ mA}}{20 \text{ μA}}$$
$$= 50$$

This means that the alternating collector current at the output is 50 times greater than the alternating base current at the input.

Common-emitter transistors will typically have beta, or current gain values, of 50.

Voltage gain. The common-emitter circuit is not only used to increase the level of current between input and output. It can also be used to increase the amplitude of the input signal voltage, or produce a voltage gain. This action can be seen by examining the *C-E* circuit in Figure 17-14(a) and by following the changes in the associated waveforms in Figure 17-14(b). An input signal voltage increase from 0 V to a positive peak ($V_{in}\uparrow$) causes an increase in the dc base bias voltage ($V_B\uparrow$), causing the emitter diode of Q_1 to turn more ON and result in an increase in both I_B and I_C. Because the I_C flows through R_C, and because voltage drop is proportional to current, an increase in I_C will cause an increase in the voltage drop across R_C ($V_{R_C}\uparrow$). The output voltage is equal to the voltage developed across Q_1's collector-to-emitter resistance (R_{CE}) and R_E. Because R_C is in series with Q_1's collector-to-emitter resistance (R_{CE}) and R_E, an increase in $V_{R_C}\uparrow$ will cause a decrease in the voltage developed across R_{CE} and R_E, which is $V_{out}\downarrow$. This action is summarized with Kirchhoff's voltage law, which states that the sum of the voltages in a series circuit is equal to the voltage applied ($V_{R_C} + V_{out} = V_{CC}$). Using the example in Figure 17-14(b), you can see that when:

$V_{R_C}\uparrow$ to 7 V, $V_{out}\downarrow$ to 3 V.   ($V_{R_C} + V_{out} = V_{CC}$, 7 V + 3 V = 10 V)

$V_{R_C}\downarrow$ to 1 V, $V_{out}\uparrow$ to 9 V.   ($V_{R_C} + V_{out} = V_{CC}$, 1 V + 9 V = 10 V)

Although the input voltage (V_{in}) and output voltage (V_{out}) are out of phase with one another, you can see from the example values in Figure 17-14(b) that there is an increase in the signal voltage between input and output. This voltage gain between input and output is possible because the output current (I_C) is so much larger than the input current (I_B). The amount of voltage gain (which is symbolized A_V) can be calculated by comparing the output voltage change (ΔV_{out}) to the input voltage change (ΔV_{in}).

$$A_V = \frac{\Delta V_{out}}{\Delta V_{in}}$$

EXAMPLE:

Calculate the voltage gain of the circuit and its associated waveforms in Figure 17-14(a) and (b).

Solution:

$$A_V = \frac{\Delta V_{out}}{\Delta V_{in}} = \frac{+9 \text{ V to } +3 \text{ V}}{+100 \text{ mV to } -100 \text{ mV}} = \frac{6 \text{ V}}{200 \text{ mV}} = 30$$

This value indicates that the output ac signal voltage is 30 times larger than the input ac signal voltage.

Most common-emitter transistor circuits have high voltage gains between 100 to 500.

Power gain. As we have seen so far, the common-emitter circuit provides both current gain and voltage gain. Because power is equal to the product of current and voltage ($P = V \times I$), it is not surprising that the *C-E* circuit configuration also provides **power gain** (A_P). The power gain of a circuit can be calculated by dividing the output signal power (P_{out}) by the input signal power (P_{in}).

Power Gain (A_P)
The ratio of the output signal power to the input signal power.

$$A_P = \frac{P_{out}}{P_{in}}$$

To calculate the amount of input power (P_{in}) applied to the C-E circuit, we will have to multiply the change in input signal voltage (ΔV_{in}) by the accompanying change in input signal current (ΔI_{in} or ΔI_B).

$$P_{in} = \Delta V_{in} \times \Delta I_{in}$$

To calculate the amount of output power (P_{out}) delivered by the C-E circuit, we will have to multiply the change in output signal voltage (ΔV_{out}) produced by the change in output signal current (ΔI_{out} or ΔI_C).

$$P_{out} = \Delta V_{out} \times \Delta I_{out}$$

The power gain of a common-emitter circuit is therefore calculated with the formula:

$$A_P = \frac{P_{out}}{P_{in}} = \frac{\Delta V_{out} \times \Delta I_{out}}{\Delta V_{in} \times \Delta I_{in}}$$

◼ EXAMPLE:

Calculate the power gain of the example circuit in Figure 17-14.

◼ *Solution:*

$$
\begin{aligned}
A_P &= \frac{P_{out}}{P_{in}} = \frac{\Delta V_{out} \times \Delta I_{out}}{\Delta V_{in} \times \Delta I_{in}} \\
&= \frac{(9\text{ V to }3\text{ V}) \times (1.5\text{ mA to }0.5\text{ mA})}{(+100\text{ mV to }-100\text{ mV}) \times (40\ \mu\text{A to }20\ \mu\text{A})} \\
&= \frac{6\text{ V} \times 1\text{ mA}}{200\text{ mV} \times 20\ \mu\text{A}} = \frac{6\text{ mW}}{4\ \mu\text{W}} = 1500
\end{aligned}
$$

In this example, the common-emitter circuit has increased the input signal power from $4\ \mu\text{W}$ to 6 mW—a power gain of 1500.

The power gain of the circuit in Figure 17-14 can also be calculated by multiplying the previously calculated C-E circuit voltage gain (A_V) by the previously calculated C-E circuit current gain (β_{AC}).

$$\text{Since } A_V = \frac{\Delta V_{out}}{\Delta V_{in}}, \text{ and } \beta_{AC} = \frac{\Delta I_{out}\ (\text{or } \Delta I_C)}{\Delta I_{in}\ (\text{or } \Delta I_B)}$$

$$A_P = V \times I = \frac{\Delta V_{out}}{\Delta V_{in}} \times \frac{\Delta I_{out}\ (\text{or } \Delta I_C)}{\Delta I_{in}\ (\text{or } \Delta I_B)} \text{ or } A_P = A_V \times \beta_{AC}$$

$$A_P = A_V \times \beta_{AC}$$

For the example circuit in Figure 17-14, this will be:

$$A_P = A_V \times \beta_{AC} = 30 \times 50 = 1500$$

indicating that the common-emitter circuit's output power in Figure 17-14 is 1500 times larger than the input power.

The power gain of common-emitter transistor circuits can be as high as 20,000, making this characteristic the circuit's key advantage.

Collector characteristic curves.   One of the easiest ways to compare several variables is to combine all of the values in a graph. Figure 17-14(c) shows a special graph, called the **collector characteristic curves,** for a typical common-emitter transistor circuit. The

Collector Characteristic Curves

Graph for a typical common-emitter transistor circuit.

CHAPTER 17 / BIPOLAR JUNCTION TRANSISTORS (BJTs)

data for this graph are obtained by using the transistor test circuit, shown in the inset in Figure 17-14(c), which will apply different values of base bias voltage (V_{BB}) and collector bias voltage (V_{CC}) to an NPN transistor. The two ammeters and one voltmeter in this test circuit are used to measure the circuit's I_B, I_C, and V_{CE} response to each different circuit condition. The values obtained from this test circuit are then used to plot the transistor's collector current (I_C in mA) on the vertical axis against the transistor's collector-emitter voltage (V_{CE}) drop on the horizontal axis for various values of base current (I_B in μA). This graph shows the relationship between a transistor's input base current, output collector current, and collector-to-emitter voltage drop.

Let us now examine the typical set of collector characteristic curves shown in Figure 17-14(c). When V_{CE} is increased from zero, by increasing V_{CC}, the collector current rises very rapidly, as indicated by the rapid vertical rise in any of the curves. When the collector diode of the transistor is reverse biased by the voltage V_{CE}, the collector current levels off. At this point, any one of the curves can be followed based on the amount of base current, which is determined by the value of base bias voltage (V_{BB}) applied. This flat part of the curve is known as the transistor's **active region.** The transistor is normally operated in this region, where it is equivalent to a variable resistor between the collector and emitter.

Active Region

Flat part of the collector characteristic curve. A transistor is normally operated in this region, where it is equivalent to a variable resistor between the collector and emitter.

As an example, let us use these curves in Figure 17-14(c) to calculate the value of output current (I_C) for a given value of input current (I_B) and collector-to-emitter voltage (V_{CE}). If V_{BB} is adjusted to produce a base current of 30 μA, and V_{CC} is adjusted until the voltage between the transistor's collector and emitter (V_{CE}) is 4.5 V, the output collector current (I_C) will be equal to approximately 1 mA. This is determined by first locating 4.5 V on the horizontal axis (V_{CE} = 4.5 V), following this point directly up to the I_B = 30 μA curve, and then moving directly to the left to determine the value of output current on the vertical axis (I_C = 1 mA). When these values of V_{CE} and I_B are present, the transistor is said to be operating at point "Q." This dc operating point is often referred to as a **quiescent operating point (Q point),** which means a dc steady-state or no input signal operating point. The Q point of a transistor is set by the circuit's dc bias components and supply voltages. For instance, in the circuit in Figure 17-14(a), R_1 and R_2 were used to set the dc base bias voltage (V_B, and therefore I_B), and the values of V_{CC}, R_C, and R_E were chosen to set the transistor's dc collector-emitter voltage (V_{CE}). At this dc operating point, we can calculate the transistor's dc current gain (β_{DC}), since both I_B and I_C are known:

Quiescent Operating Point (Q Point)

The voltage or current value that sets up the no input signal or operating point bias voltage.

$$\beta_{DC} = \frac{I_C}{I_B} = \frac{1 \text{ mA}}{30 \text{ μA}} = 33.3$$

If a sine-wave signal was applied to the circuit, as shown in the waveforms in Figure 17-14(b) and the characteristic curves in Figure 17-14(c), it would cause the transistor's input base current, and therefore output collector current, to alternate above and below the transistor's Q point (dc operating point). This ac input signal voltage will cause the input base current (I_B) to increase between 20 μA and 40 μA, and this input base current change will generate an output collector current change of 0.5 mA to 1.5 mA. The transistor's ac current gain (β_{AC}) will be:

$$\beta_{AC} = \frac{\Delta I_C}{\Delta I_B}$$
$$= \frac{1.5 \text{ to } 0.5 \text{ mA}}{40 \text{ to } 20 \text{ μA}}$$
$$= \frac{1 \text{ mA}}{20 \text{ μA}}$$
$$= 50$$

Returning to the collector characteristic curves in Figure 17-14(c), you can see that if the collector supply voltage (V_{CC}) is increased to an extreme, a point will be reached where the V_{CE} voltage across the transistor will cause the transistor to break down, as indicated by the rapid rise in I_C. This section of the curve is called the **breakdown region** of the graph, and the damaging value of current through the transistor will generally burn out and destroy

Breakdown Region

The point at which the collector supply voltage will cause a damaging value of current through the transistor.

the device. As an example, for the 2N3904 bipolar transistor, breakdown will occur at a V_{CE} voltage of 40 V.

There are two shaded sections shown in the set of collector characteristic curves in Figure 17-14(c). These two shaded sections represent the other two operating regions of the transistor. To begin with, let us examine the vertically shaded **saturation region.** If the base bias voltage (V_{BB}) is increased to a large positive value, the emitter diode of the transistor will turn ON heavily, I_B will be a large value, and the transistor will be operating in saturation. In this operating region, the transistor is equivalent to a closed switch between its collector and emitter (both the emitter and collector diode are forward biased), and therefore the voltage drop between collector and emitter will be almost zero (V_{CE} = typically 0.3 V, when transistor is saturated), and I_C will be a large value that is limited only by the externally connected components. The horizontally shaded section represents the **cutoff region** of the transistor. If the base bias voltage (V_{BB}) is decreased to zero, the emitter diode of the transistor will turn OFF, I_B will be zero, and the transistor will be operating in cutoff. In this operating region, the transistor is equivalent to an open switch between its collector and emitter (both the emitter and collector diode are reverse biased), and the voltage drop between collector and emitter will be equal to V_{CC} and I_C will be zero.

A set of collector characteristic curves are therefore generally included in a manufacturer's device data sheet, and can be used to determine the values of I_B, I_C, and V_{CE} at any operating point.

Input resistance. The **input resistance (R_{in})** of a common-emitter transistor is the amount of opposition offered to an input signal by the input base-emitter junction (emitter diode). Because the base-emitter junction is normally forward biased when the transistor is operating in the active region, the opposition to input current is relatively small. However, the extremely small base region will only support a very small input base current. On average, if no additional components are connected in series with the transistor's base-emitter junction, the input resistance of a *C-E* transistor circuit is typically a medium value between 1 kΩ and 5 kΩ. This typical value is an average because the transistor's input resistance is a "dynamic or changing quantity" that will vary slightly as the input signal changes the conduction of the *C-E* transistor's emitter diode, and this changes I_B ($R \updownarrow = V/I \updownarrow$).

Because the transistor has a small value of input P-N junction capacitance and input terminal inductance, the opposition to the input signal is not only resistive but, to a small extent, reactive. For this reason, the total opposition offered by the transistor to an input signal is often referred to as the **input impedance (Z_{in})** because impedance is the total combined resistive and reactive input opposition.

Output resistance. The **output resistance (R_{out})** of a common-emitter transistor is the amount of opposition offered to an output signal by the output base-collector junction (collector diode). This junction is normally reverse biased when the transistor is operating in the active region, and therefore the *C-E* transistor's output resistance is relatively high. However, because a unique action occurs within the transistor and allows current to flow through this reverse biased junction (electron accumulation at the base and then conduction through collector diode due to attraction of $+V_{CC}$), the output current (I_C) is normally large, and so the output resistance is not an extremely large value. On average, if no load resistor is connected in series with the transistor's collector diode, the output resistance of a *C-E* transistor circuit is typically a high value between 40 kΩ and 60 kΩ.

Because the transistor has a small value of output P-N junction capacitance and output terminal inductance, the opposition to the output signal is not only resistive but also reactive. For this reason, the total opposition offered by the transistor to the output signal is often referred to as the **output impedance (Z_{out}).**

Common-Base Circuits

With the **common-base circuit,** the transistor's base lead is common to both the input and output signal. In this circuit configuration, the emitter serves as the input lead, and the collector serves as the output lead. Figure 17-15 contains a basic common-base (*C-B*) cir-

Saturation Region

The point at which the collector supply voltage has the transistor operating at saturation.

Cutoff Region

The point at which the collector supply voltage has the transistor operating in cutoff.

Input Resistance (R_{in})

The amount of opposition offered to an input signal by the input base-emitter junction (emitter diode).

Input Impedance (Z_{in})

The total opposition offered by the transistor to an input signal.

Output Resistance (R_{out})

The amount of opposition offered to an output signal by the output base-collector junction (collector diode).

Output Impedance (Z_{out})

The total opposition offered by the transistor to the output signal.

Common-Base (C-B) Circuit

Configuration in which the input signal is applied between the transistor's emitter and base, while the output is developed across the transistor's collector and base.

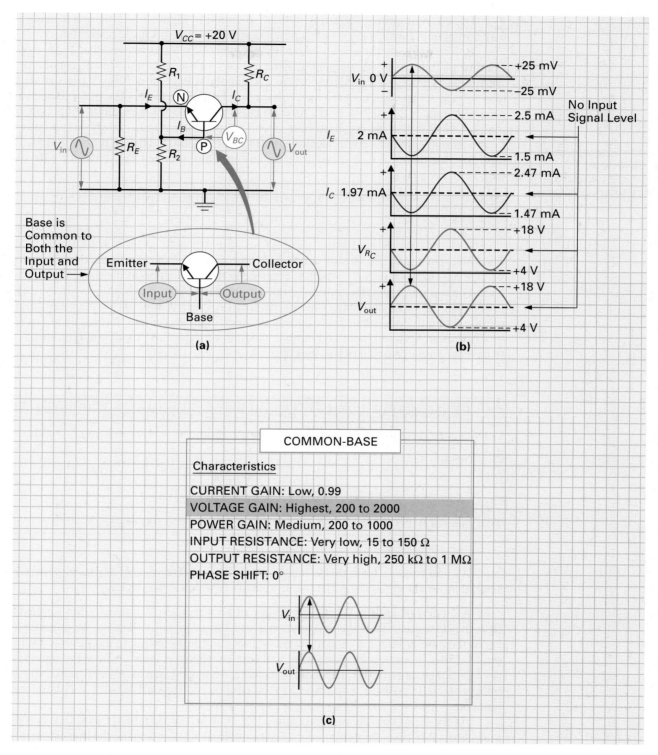

FIGURE 17-15   **Common-Base (*C-B*) Circuit Configuration Characteristics.**

cuit, its associated input/output voltage and current waveforms, and table of typical characteristics. Using this illustration, we will examine the operation and characteristics of the *C-B* circuit configuration.

DC current gain. Referring to the *C-B* circuit in Figure 17-15(a), and its associated waveforms in Figure 17-15(b), let us now examine this circuit's basic operation.

Before applying the ac sine-wave input signal (V_{in}), let us assume that $V_{in} = 0$ V and examine the transistor's dc operating characteristics, or the "no input signal" condition. The voltage divider R_1 and R_2 will divide the V_{CC} supply voltage, producing a positive dc base bias voltage across R_2. This base bias voltage will be applied to the base of Q_1. Because it is generally greater than 0.7 V, it will forward bias the transistor's base-emitter junction, turning Q_1 ON. Because the common-base circuit's input current is I_E and its output current is I_C, the current gain between input and output will be determined by the ratio of I_C to I_E. This ratio for calculating a *C-B* transistor's dc current gain is called the transistor's **dc alpha** (α_{DC}) and is equal to

$$\alpha_{DC} = \frac{I_C}{I_E}$$

The "no input signal" or "steady state" dc levels of I_E and I_C are determined by the value of voltage developed across R_2, which is controlling the conduction of the transistor's forward biased base-emitter junction (emitter diode). Because the output current I_C is always slightly lower than the input current I_E (due to the small I_B current flow out of the base), the *C-B* transistor circuit does not increase current between input and output. In fact, there is a slight loss in current between input and output, which is why the *C-B* circuit configuration is said to have a current gain that is less than 1.

■ **EXAMPLE:**

Calculate the dc alpha of the circuit in Figure 17-15(a) if $I_C = 1.97$ mA and $I_E = 2$ mA.

■ *Solution:*

$$\alpha_{DC} = \frac{I_C}{I_E} = \frac{1.97 \text{ mA}}{2 \text{ mA}} = 0.985$$

This value of 0.985 indicates that I_C is 98.5% of I_E (0.985 × 100 = 98.5).

As you can see from this example, the difference between I_C and I_E is generally so small that we always assume that the dc alpha is 1, which means that $I_C = I_E$.

AC current gain. When amplifying an ac waveform, a transistor has applied to it both the dc voltage to make it operational and the ac signal voltage to be amplified that varies the base-emitter bias and the input emitter current. Referring to the waveforms in Figure 17-15(b), let us now examine what will happen when we apply a sine-wave input signal. As mentioned previously, the positive voltage developed across R_2 will make the NPN transistor's base positive with respect to the emitter and forward bias the P-N base-emitter junction.

As the input voltage swings positive ($V_{in}\uparrow$), it will reduce the forward bias across the transistor's P-N base-emitter junction. For example, if $V_B = +5$ V and the transistor's *n*-type emitter is made positive, the P-N base-emitter diode will be turned more OFF. Turning the transistor's emitter diode less ON will cause a decrease in emitter current ($I_E\downarrow$), a decrease in collector current ($I_C\downarrow$), and a decrease in the voltage developed across R_C ($V_{R_C}\downarrow$). Because V_{R_C} and V_{out} are connected in series across V_{CC}, a decrease in $V_{R_C}\downarrow$ must be accompanied by an increase in $V_{out}\uparrow$. To explain this another way, the decrease in I_E and the subsequent decrease in both I_C and I_B means that the conduction of the transistor has decreased. This decrease in conduction means that the normally forward biased base-emitter junction has turned less ON, and the normally reverse biased base-collector junction has turned more OFF. Because the transistor's base-collector junction is in series with R_C and R_2 across V_{CC}, an increase in the transistor's base-collector resistance ($R_{BC}\uparrow$) will cause an increase in the voltage developed across the transistor's base-collector junction ($V_{BC}\uparrow$), which will cause an increase in $V_{out}\uparrow$.

Similarly, as the input voltage swings negative ($V_{in}\downarrow$), it will increase the forward bias across the transistor's P-N base-emitter junction. For example, if $V_B = -5$ V and the transistor's n-type emitter is made negative, the P-N base-emitter diode will be turned more ON. Turning the transistor's emitter diode more ON will cause an increase in emitter current ($I_E\uparrow$), an increase in collector current ($I_C\uparrow$), and an increase in the voltage developed across R_C ($V_{R_C}\uparrow$). Because V_{R_C} and V_{out} are connected in series across V_{CC}, an increase in $V_{R_C}\uparrow$ must be accompanied by a decrease in $V_{out}\downarrow$.

Now that we have seen how the input voltage causes a change in input current (I_E) and output current (I_C), let us examine the C-B circuit's ac current gain. The ratio of input emitter current change (ΔI_E) to output collector current change (ΔI_C) is the C-B transistor's ac current gain or **AC alpha (α_{AC})**. The formula for calculating a C-B transistor's ac current gain is:

$$\alpha_{AC} = \frac{\Delta I_C}{\Delta I_E}$$

EXAMPLE:

Calculate the ac current gain of the example in Figure 17-15(b).

Solution:

$$\begin{aligned}
\alpha_{AC} &= \frac{\Delta I_C}{\Delta I_E} \\
&= \frac{2.47 \text{ to } 1.47 \text{ mA}}{2.5 \text{ to } 1.5 \text{ mA}} \\
&= \frac{1 \text{ mA}}{1 \text{ mA}} \\
&= 1
\end{aligned}$$

This means that the change in output collector current is equal to the change in input emitter current, and therefore the ac current gain is 1. (The output is 1 times larger than the input, $1 \text{ mA} \times 1 = 1 \text{ mA}$.)

Common-base transistors will typically have an ac alpha, or ac current gain, of 0.99.

Voltage gain. Although the common-base circuit does not achieve any current gain, it does make up for this disadvantage by achieving a very large voltage gain between input and output. Returning to the C-B circuit in Figure 17-15(a) and its waveforms in Figure 17-15(b), let us see how this very high voltage gain is obtained. Only a small input voltage (V_{in}) is needed to control the conduction of the transistor's emitter diode, and therefore the input emitter current (I_E) and output collector current (I_C). Even though I_C is slightly lower than I_E, it is still a relatively large value of current and will develop a large voltage change across R_C for a very small change in V_{in}. Because V_{R_C} and V_{out} are in series and connected across V_{CC}, a large change in voltage across R_C will cause a large change in the voltage developed across the transistor output (V_{BC}) and V_{out}. As before, the amount of voltage gain (which is symbolized A_V) can be calculated by comparing the output voltage change (ΔV_{out}) to the input voltage change (ΔV_{in}).

$$A_V = \frac{\Delta V_{out}}{\Delta V_{in}}$$

EXAMPLE:

Calculate the voltage gain of the circuit and its associated waveforms in Figure 17-15(a) and (b).

Solution:

$$A_V = \frac{\Delta V_{out}}{\Delta V_{in}} = \frac{+18 \text{ V to } +4 \text{ V}}{+25 \text{ mV to } -25 \text{ mV}} = \frac{14 \text{ V}}{50 \text{ mV}} = 280$$

This value indicates that the output ac signal voltage is 280 times larger than the ac input signal voltage.

Most common-base transistor circuits have very high voltage gains between 200 and 2000. While on the topic of comparing the input voltage to output voltage, you can see by looking at Figure 17-15(b) that, unlike the *C-E* circuit, the common-base circuit has no phase shift between input and output (V_{out} is in phase with V_{in}).

Power gain. Although the common-base circuit achieves no current gain, it does have a very high voltage gain and therefore can provide a medium amount of power gain ($P\uparrow = V\uparrow \times I$). The power gain of a circuit can be calculated by dividing the output signal power (P_{out}) by the input signal power (P_{in}).

$$A_P = \frac{P_{out}}{P_{in}}$$

To calculate the amount of input power (P_{in}) applied to the *C-B* circuit, we will have to multiply the change in input signal voltage (ΔV_{in}) by the accompanying change in input signal current (ΔI_{in} or ΔI_E).

$$P_{in} = \Delta V_{in} \times \Delta I_{in}$$
$$P_{in} = 50 \text{ mV} \times 1 \text{ mA} = 50 \text{ } \mu\text{W}$$

To calculate the amount of output power (P_{out}) delivered by the *C-B* circuit, we will have to multiply the change in output signal voltage (ΔV_{out}) produced by the change in output signal current (ΔI_{out} or ΔI_C).

$$P_{out} = \Delta V_{out} \times \Delta I_{out}$$
$$P_{out} = 14 \text{ V} \times 1 \text{ mA} = 14 \text{ mW}$$

The power gain of a common-base circuit is calculated with the formula:

$$A_P = \frac{P_{out}}{P_{in}} = \frac{\Delta V_{out} \times \Delta I_{out}}{\Delta V_{in} \times \Delta I_{in}}$$

EXAMPLE:

Calculate the power gain of the example circuit in Figure 17-15.

$$A_P = \frac{P_{\text{out}}}{P_{\text{in}}} = \frac{\Delta V_{\text{out}} \times \Delta I_{\text{out}}}{\Delta V_{\text{in}} \times \Delta I_{\text{in}}}$$

$$= \frac{14 \text{ V} \times 1 \text{ mA}}{50 \text{ mV} \times 1 \text{ mA}} = \frac{14 \text{ mW}}{50 \text{ }\mu\text{W}} = 280$$

In this example, the common-base circuit has increased the input signal power from 50 μW to 14 mW—a power gain of 280.

The power gain of the circuit in Figure 17-15 can also be calculated by multiplying the previously calculated *C-B* circuit voltage gain (A_V) by the previously calculated *C-B* circuit current gain (α_{AC}).

$$A_P = A_V \times \alpha_{AC}$$

For the example circuit in Figure 17-15, this will be

$$A_P = A_V \times \alpha_{AC} = 280 \times 1 = 280$$

indicating that the common-base circuit's output power in Figure 17-15 is 280 times larger than the input power.

Typical common-base circuits will have power gains from 200 to 1000.

Input resistance. The input resistance (R_{in}) of a common-base transistor is the amount of opposition offered to an input signal by the input base-emitter junction (emitter diode). Because the base-emitter junction is normally forward biased and the input emitter current (I_E) is relatively large, the input signal sees a very low input resistance. On average, if no additional components are connected in series with the transistor's base-emitter junction, the input resistance of a *C-B* transistor circuit is typically a low value between 15 Ω and 150 Ω. This typical value is an average because the transistor's input resistance is a dynamic or changing quantity that will vary slightly as the input signal changes the conduction of the *C-B* transistor's emitter diode, and this changes I_E ($R\updownarrow = V/I\updownarrow$).

Output resistance. The output resistance (R_{out}) of a common-base transistor is the amount of opposition offered to an output signal by the output base-collector junction (collector diode). This junction is normally reverse biased when the transistor is operating in the active region, and therefore the *C-B* transistor's output resistance is relatively high. On average, if no load resistor is connected in series with the transistor's collector diode, the output resistance of a *C-B* transistor circuit is typically a very high value between 250 kΩ and 1 MΩ.

Common-Collector Circuits

With the **common-collector circuit,** the transistor's collector lead is common to both the input and output signal. In this circuit configuration therefore, the base serves as the input lead, and the emitter serves as the output lead. Figure 17-15 contains a basic common-collector (*C-C*) circuit, its associated input/output voltage and current waveforms, and table of typical characteristics. Using this illustration, we will examine the operation and characteristics of the *C-C* circuit configuration.

DC current gain. Referring to the *C-C* circuit in Figure 17-16(a) and its associated waveforms in Figure 17-16(b), let us now examine this circuit's basic operation.

Before applying the ac sine-wave input signal (V_{in}), let us assume that $V_{\text{in}} = 0$ V and examine the transistor's dc operating characteristics, or the "no input signal" condition. The voltage divider R_1 and R_2 will divide the V_{CC} supply voltage, producing a positive dc base bias voltage across R_2. This base bias voltage will be applied to the base of Q_1. Because it is generally greater than 0.7 V, it will forward bias the transistor's base-emitter junction,

Common-Collector (C-C) Circuit

Configuration in which the input signal is applied between the transistor's base and collector, while the output is developed across the transistor's collector and emitter.

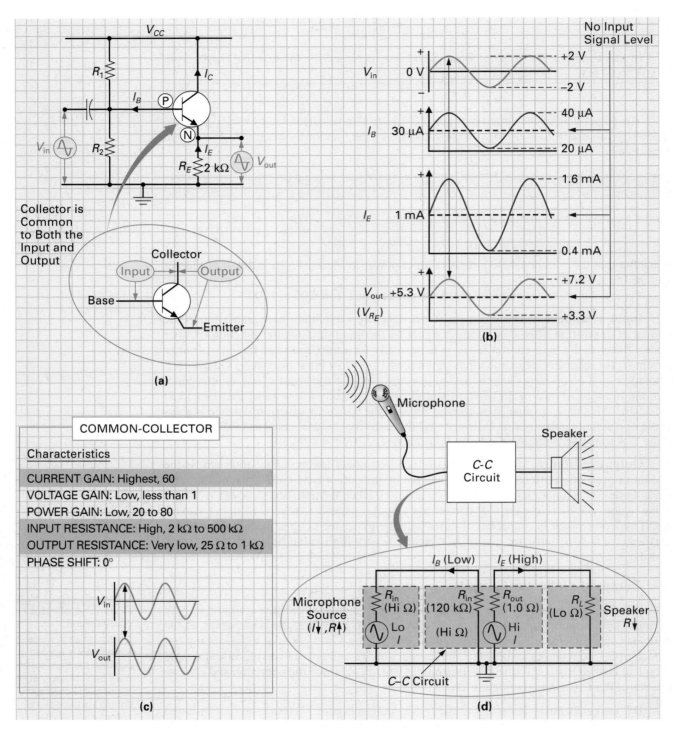

FIGURE 17-16 Common-Collector *(C-C)* Circuit Configuration Characteristics.

turning Q_1 ON. Because the common-collector (C-C) circuit's input current I_B is much smaller than the output current I_E, the circuit provides a high current gain. In fact, the common-collector circuit provides a slightly higher gain than the C-E circuit because the common-collector's output current (I_E) is slightly higher than the C-E's output current (I_C).

Like any circuit configuration, the dc current gain is equal to the ratio of output current to input current. For the C-C circuit, this is equal to the ratio of I_E to I_B:

$$\text{DC Current Gain} = \frac{I_E}{I_B}$$

Transistor manufacturers will generally not provide specifications for all three circuit configurations. In most cases, because the common-emitter (C-E) circuit configuration is most frequently used, manufacturers will give the transistor's characteristics for only the C-E circuit configuration. In these instances, we will have to convert this C-E circuit data to equivalent specifications for other configurations. For example, in most data sheets the transistor's dc current gain will be listed as β_{DC}. As we know, dc beta is the measure of a C-E circuit's current gain because it compares input current I_B to output current I_C. How, then, can we convert this value so that it indicates the dc current gain of a common-collector circuit? The answer is as follows:

$$\text{Common-Collector DC Current Gain} = \frac{\text{Output Current}}{\text{Input Current}} = \frac{I_E}{I_B}$$

Since $I_E = I_B + I_C$,

$$\text{DC Current Gain} = \frac{I_E}{I_B} = \frac{(I_B + I_C)}{I_B}$$

Since $I_B \div I_B = 1$,

$$\text{DC Current Gain} = 1 + \frac{I_C}{I_B}$$

Since $\dfrac{I_C}{I_B} = \beta_{DC}$,

$$\text{DC Current Gain} = 1 + \frac{I_C}{I_B} = 1 + \beta_{DC}$$

$$\text{DC Current Gain} = 1 + \beta_{DC}$$

■ EXAMPLE:

Calculate the dc current gain of the C-C circuit in Figure 17-16 if the transistor's $\beta_{DC} = 32.33$.

■ Solution:

$$\text{DC Current Gain} = 1 + \frac{I_C}{I_B} = 1 + \frac{(I_E - I_B)}{I_B} = 1 + \frac{1\ \text{mA} - 30\ \mu\text{A}}{30\ \mu\text{A}}$$

$$= 1 + \frac{970\ \mu\text{A}}{30\ \mu\text{A}} = 1 + 32.33 = 33.33$$

or,

$$\text{DC Current Gain} = 1 + \beta_{DC} = 1 + 32.33 = 33.33$$

As you can see in this example, the dc current gain of a common-collector circuit ($\beta_{DC} + 1$) is slightly higher than the dc current gain of a *C-E* circuit (β_{DC}). In most instances, the extra 1 makes so little difference when the transistor's dc current gain is a large value of about 30, as in this example, that we assume that the current gain of a common-collector circuit is equal to the current gain of a *C-E* circuit.

C-C DC Current Gain $\cong$ *C-E* DC Current Gain (β_{DC})

AC current gain.   When amplifying an ac waveform, a transistor has applied to it both the dc voltage to make it operational and the ac signal voltage that varies the base-emitter bias and the input base current. Referring to the waveforms in Figure 17-15(b), let us now examine what will happen when we apply a sine-wave input signal. As mentioned previously, the positive voltage developed across R_2 will make the NPN transistor's base positive with respect to the emitter, and therefore forward bias the P-N base-emitter junction.

As the input voltage swings positive ($V_{in}\uparrow$), it will add to the forward bias applied across the transistor's P-N base-emitter junction. This means that the transistor's emitter diode will turn more ON, cause an increase in the $I_B\uparrow$, and therefore a proportional but much larger increase in the output current $I_E\uparrow$.

Similarly, as the input voltage swings negative ($V_{in}\downarrow$), it will subtract from the forward bias applied across the transistor's P-N base-emitter junction. This means that the transistor's emitter diode will turn less ON, cause a decrease in the $I_B\downarrow$, and therefore a proportional but larger decrease in the output current $I_E\downarrow$.

The ac current gain of a common-collector transistor is calculated using the same formula as dc current gain. However, with an ac current, we will compare the output current change (ΔI_E) to the input current change (ΔI_B).

$$\text{AC Current Gain} = \frac{\Delta I_E}{\Delta I_B}$$

EXAMPLE:

Calculate the ac current gain of the circuit in Figure 17-16(a), using the values in Figure 17-16(b).

Solution:

$$\text{AC Current Gain} = \frac{\Delta I_E}{\Delta I_B} = \frac{1.6 \text{ mA} - 0.4 \text{ mA}}{40 \text{ } \mu\text{A} - 20 \text{ } \mu\text{A}} = 60$$

Like the common-collector's dc current gain, because there is so little difference between I_E and I_C, we can assume that the ac current gain is equivalent to β_{AC}.

C-C AC Current Gain $\cong$ *C-E* AC Current Gain (β_{AC})

Common-collector transistor circuit configurations can have current gains as high as 60, indicating that I_E is sixty times larger than I_B.

Voltage gain.   Although the common-collector circuit has a very high current gain rating, it cannot increase voltage between input and output. Returning to the *C-C* circuit in Figure 17-16(a) and its waveforms in Figure 17-16(b), let us see why this circuit has a very low voltage gain.

As the input voltage swings positive ($V_{in}\uparrow$), it will add to the forward bias applied across the transistor's P-N base-emitter junction ($V_{BE}\uparrow$). As the transistor's emitter diode turns more ON, it will cause an increase in $I_B\uparrow$, a proportional but larger increase in $I_E\uparrow$, and therefore an increase in the voltage developed across R_E (V_{R_E}, V_{out}, or $V_E\uparrow$). This increase in the voltage developed across R_E has a **degenerative effect** because an increase in the emitter voltage ($V_E\uparrow$) will counter the initial increase in base voltage ($V_B\uparrow$), and therefore the voltage difference between the transistor's base and emitter will remain almost constant (V_{BE} is almost constant). In other words, if the base goes positive and then the emitter goes positive, there is almost no increase in the potential difference between the base and the emitter and so the change in forward bias is almost zero. There is, in fact, a very small change in forward bias between base and emitter, and this will cause a small change in I_B and I_E, and therefore a small output voltage will be developed across R_E. Comparing the input and output voltage signals in Figure 17-15(b), you can see that both are about 4 V pk-pk, and both are in phase with one another. The common-collector circuit is often referred to as an **emitter-follower** or **voltage-follower** because the emitter output voltage seems to track or follow the phase and amplitude of the input voltage.

As with all circuit configurations, the amount of voltage gain (A_V) can be calculated by comparing the output voltage change (ΔV_{out}) to the input voltage change (ΔV_{in}).

$$A_V = \frac{\Delta V_{out}}{\Delta V_{in}}$$

Degenerative Effect

An effect that causes a reduction in amplification due to negative feedback.

Emitter-Follower or Voltage-Follower

The common-collector circuit in which the emitter output voltage seems to track or follow the phase and amplitude of the input voltage.

■ EXAMPLE:

Calculate the voltage gain of the circuit and its associated waveforms in Figure 17-16(a) and (b).

■ *Solution:*

$$A_V = \frac{\Delta V_{out}}{\Delta V_{in}} = \frac{+7.2 \text{ V to } +3.3 \text{ V}}{+2 \text{ V to } -2 \text{ V}} = \frac{3.9 \text{ V}}{4 \text{ V}} = 0.975$$

This value indicates that the output ac signal voltage is 0.975 or 97.5% of the ac input signal voltage ($0.975 \times 4 \text{ V} = 3.9 \text{ V}$).

Most common-collector transistor circuits have a voltage gain that is less than 1. However, in most circuit examples it is assumed that output voltage change equals input voltage change.

Power gain. Although the common-collector circuit achieves no voltage gain, it does have a very high current gain, and therefore can provide a small amount of power gain ($P\uparrow = V \times I\uparrow$). As before, the power gain of a circuit can be calculated by dividing the output signal power (P_{out}) by the input signal power (P_{in}).

$$A_P = \frac{P_{out}}{P_{in}}$$

To calculate the amount of input power (P_{in}) applied to the C-C circuit, we will have to multiply the change in input signal voltage (ΔV_{in}) by the accompanying change in input signal current (ΔI_{in} or ΔI_B).

$$P_{in} = \Delta V_{in} \times \Delta I_{in}$$
$$P_{in} = 4 \text{ V} \times 20 \text{ }\mu\text{A} = 80 \text{ }\mu\text{W}$$

To calculate the amount of output power (P_{out}) delivered by the C-C circuit, we will have to multiply the change in output signal voltage (ΔV_{out}) produced by the change in output signal current (ΔI_{out} or ΔI_E).

$$P_{out} = \Delta V_{out} \times \Delta I_{out}$$
$$P_{out} = 3.9 \text{ V} \times 1.2 \text{ mA} = 4.68 \text{ mW}$$

The power gain of a common-collector circuit is therefore calculated with the formula:

$$A_P = \frac{P_{out}}{P_{in}} = \frac{\Delta V_{out} \times \Delta I_{out}}{\Delta V_{in} \times \Delta I_{in}}$$

EXAMPLE:

Calculate the power gain of the example circuit in Figure 17-16.

Solution:

$$A_P = \frac{P_{out}}{P_{in}} = \frac{\Delta V_{out} \times \Delta I_{out}}{\Delta V_{in} \times \Delta I_{in}}$$
$$= \frac{3.9 \text{ V} \times 1.2 \text{ mA}}{4 \text{ V} \times 20 \text{ } \mu\text{A}} = \frac{4.68 \text{ mW}}{80 \text{ } \mu\text{W}} = 58.5$$

In this example, the common-collector circuit has increased the input signal power from 80 μW to 4.68 mW—a power gain of 58.5.

The power gain of the circuit in Figure 17-16 can also be calculated by multiplying the previously calculated *C-C* circuit voltage gain (A_V) by the previously calculated *C-C* circuit current gain.

$$A_P = A_V \times \text{AC Current Gain}$$

For the example circuit in Figure 17-15, this will be

$$A_P = A_V \times \text{AC Current Gain} = 0.975 \times 60 = 58.5$$

indicating that the common-collector circuit's output power in Figure 17-16 is 58.5 times larger than the input power.

Typical common-collector circuits will have power gains from 20 to 80.

Input resistance. An input signal voltage will see a very large input resistance when it is applied to a common-collector circuit. This is because the input signal sees the very large emitter-connected resistor ($R_E \uparrow\uparrow$) and, to a smaller extent, the resistance of the forward biased base-emitter junction ($R_{in} \uparrow = V_{in}/I_B \downarrow$: R_{in} is large because I_{in} or I_B is small). Using these two elements, we can derive a formula for calculating the input resistance of a *C-C* transistor circuit.

$$R_{in} = R_E \times \text{AC Current Gain}$$

Since

$$C\text{-}C \text{ AC Current Gain} \cong C\text{-}E \text{ AC Current Gain } (\beta_{AC})$$

the input resistance can also be calculated with the formula:

$$R_{in} = R_E \times \beta_{AC}$$

EXAMPLE:

Calculate the input resistance of the circuit in Figure 17-16, assuming $\beta_{AC} = 60$.

$$R_{in} = R_E \times \beta_{AC} = 2 \text{ k}\Omega \times 60 = 120 \text{ k}\Omega$$

This means that an input voltage signal will see this *C-C* circuit as a resistance of 120 kΩ.

The input resistance of a *C-C* transistor circuit is typically a very large value between 2 kΩ and 500 kΩ.

Output resistance. The output signal from a common-collector circuit sees a very low output resistance, as proved by this circuit's very high output current gain. Like this circuit's input resistance, the output resistance is largely dependent on the value of the emitter resistor R_E.

The output resistance of a typical *C-C* transistor circuit is a very low value between 25 Ω and 1 kΩ.

Impedance or resistance matching. Do not be misled into thinking that the very high input resistance and low output resistance of the common-collector transistor circuit are disadvantages. On the contrary, the very high input resistance and low output resistance of this configuration are made use of in many circuit applications, along with the *C-C* circuit's other advantage of high current gain.

To explain why a high input resistance and a low output resistance are good circuit characteristics, refer to the application circuit in Figure 17-16(d). In this example, a microphone is connected to the input of a *C-C* amplifier, and the output of this circuit is applied to a speaker. As we know, the sound wave input to the microphone will physically move a magnet within the microphone, which will in turn interact to induce a signal voltage into a stationary coil. This voice signal voltage from the microphone, which is our source, is then applied across the input resistance of our example *C-C* circuit, which is our load.

In the inset in Figure 17-16(d), you can see that the microphone has been represented as a low-current ac source with a high internal resistance, and the input resistance of the *C-C* circuit is shown as a high value (in the previous example, 120 kΩ) resistor. Remembering our previous discussion on sources and loads, we know that a small load resistance will cause a large current to be drawn from the source, and this large current will drain or pull down the source voltage. Many signal sources, such as microphones, can only generate a small signal source voltage because they have a high internal resistance. If this small signal source voltage is applied across an amplifier with a small input resistance, a large current will be drawn from the source. This heavy load will pull the signal voltage down to such a small value that it will not be large enough to control the amplifier. A large amplifier input resistance ($R_{in}\uparrow$), on the other hand, will not load the source. Therefore the input voltage applied to the amplifier will be large enough to control the amplifier circuit, to vary its transistor currents, and achieve the gain between amplifier input and output. In summary, the high input resistance of the *C-C* circuit can be connected to a high resistance source because it will not draw an excessive current and pull down the source voltage.

Referring again to the inset in Figure 17-16(d), you can see that at the output end, the *C-C*'s output circuit has been represented as a high current source with a low value internal output resistor, and the speaker has been represented as a low resistance load. The low output resistance of the *C-C* circuit means that this circuit can deliver the high current output that is needed to drive the low resistance load.

As you will see later in application circuits, most *C-C* circuits are used as a resistance or **impedance matching circuit** that can match, or isolate, a high resistance (low current) source, such as a microphone, to a low resistance (high current) load, such as the speaker. By acting as a **buffer current amplifier,** the *C-C* circuit can ensure that power is efficiently transferred from source to load.

Impedance Matching Circuit

A circuit that can match, or isolate, a high resistance (low current) source.

Buffer Current Amplifier

The *C-C* circuit that can ensure that power is efficiently transferred from source to load.

17-2-3 Bipolar Junction Transistor Data Sheet

Like the diode's data sheets, manufacturer's bipolar transistor data sheets list the typical dc and ac operating characteristics of the device. Figure 17-17 shows the data sheet of a typical *general purpose switching or amplifying bipolar transistor,* with annotations describing the key characteristics.

As before, notes are included in these data sheets to call out important characteristics and to explain some of the terms that have not been previously used.

17-2-4 Testing Bipolar Junction Transistors

Although transistors are exceptionally more reliable than their counterpart, the vacuum tube, they still will malfunction. These failures are normally the result of excessive temperature, current, or mechanical abuse and generally result in one of three problems:

1. An open between two or three of the transistor's leads
2. A short between two or three of the transistor's leads
3. A change in the transistor's characteristics

Transistor Tester

The **transistor tester** shown in Figure 17-18 is a special test instrument that can be used to test both NPN and PNP bipolar transistors. This special meter can be used to determine whether an open or short exists between any of the transistor's three terminals, the transistor's dc current gain (β_{DC}), and whether an undesirable value of leakage current is present through one of the transistor's junctions.

Ohmmeter Transistor Test

If the transistor tester is not available, the ohmmeter can be used to detect open and shorted junctions, which are the most common transistor failures. Figure 17-19 shows the step-by-step procedure for testing an NPN transistor. Following through this test procedure, we begin by reverse biasing the collector diode and then the emitter diode, and then forward biasing the collector diode and then the emitter diode. The table in Figure 17-19 shows the order and action to be performed for each step and the reading that should result if the NPN transistor junction is operating correctly.

17-2-5 Bipolar Transistor Biasing Circuits

As we discovered in the previous discussion on transistor circuit configurations, the ac operation of a transistor is determined by the "dc bias level," or "no input signal level." This steady-state or dc operating level is set by the value of the circuit's dc supply voltage (V_{CC}) and the value of the circuit's biasing resistors. This single supply voltage and the one or more biasing resistors set up the initial dc values of transistor current (I_B, I_E and I_C) and transistor voltage (V_{BE}, V_{CE} and V_{BC}).

In this section, we will examine some of the more commonly used methods for setting the "initial dc operating point" of a bipolar transistor circuit. As you encounter different circuit applications, you will see that many of these circuits include combinations of these basic biasing techniques and additional special-purpose components for specific functions. Because the common-emitter (*C-E*) circuit configuration is used more extensively than the *C-B* and the *C-C,* we will use this configuration in all of the following basic biasing circuit examples.

DEVICE: 2N3903 and 2N3904—NPN Silicon Switching and Amplifier Transistors

Maximum continuous collector current (I_c) = 200 mA.

MAXIMUM RATINGS

Rating	Symbol	Value	Unit
Collector-Emitter Voltage	V_{CEO}	40	Vdc
Collector-Base Voltge	V_{CBO}	60	Vdc
Emitter-Base Voltage	V_{EBO}	6.0	Vdc
Collector Current — Continuous	I_C	200	mAdc
Total Device Dissipation @ T_A = 25°C Derate above 25°C	P_D	625 5.0	mW mW/°C
*Total Device Dissipation @ T_C = 25°C Derate above 25°C	P_D	1.5 12	Watts mW/°C
Operating and Storage Junction Temperature Range	T_J, T_{stg}	−55 to +150	°C

*THERMAL CHARACTERISTICS

Characteristic	Symbol	Max	Unit
Thermal Resistance, Junction to Ambient	$R_{\theta JA}$	200	°C/W
Thermal Resistance, Junction to Case	$R_{\theta JC}$	83.3	°C/W

2N3903
2N3904★

CASE 29-04, STYLE 1
TO-92 (TO-226AA)

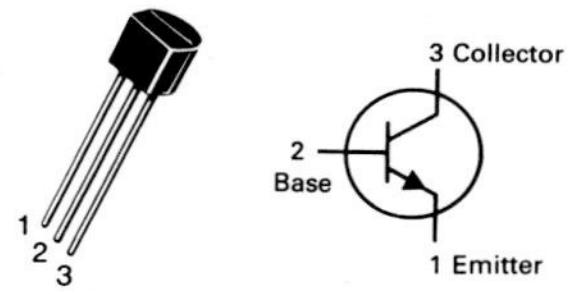

3 Collector

2 Base

1 Emitter

GENERAL PURPOSE TRANSISTORS

NPN SILICON

★This is a Motorola
designated preferred device.

ELECTRICAL CHARACTERISTICS (T_A = 25°C unless otherwise noted.)

OFF Characteristic (operated in cutoff)		Symbol	Min	Max	Unit
Collector-Emitter Breakdown Voltage(1) (I_C = 1.0 mAdc, I_B = 0)		$V_{(BR)CEO}$	40	—	Vdc
Collector-Base Breakdown Voltage (I_C = 10 μAdc, I_E = 0)		$V_{(BR)CBO}$	60	—	Vdc
Emitter-Base Breakdown Voltage (I_E = 10 μAdc, I_C = 0)		$V_{(BR)EBO}$	6.0	—	Vdc
Base Cutoff Current (V_{CE} = 30 Vdc, V_{EB} = 3.0 Vdc)		I_{BL}	—	50	nAdc
Collector Cutoff Current (V_{CE} = 30 Vdc, V_{EB} = 3.0 Vdc)		I_{CEX}	—	50	nAdc
ON Characteristic (operated in active and saturation region)					
DC Current Gain(1) (I_C = 0.1 mAdc, V_{CE} = 1.0 Vdc)	2N3903 2N3904	h_{FE}	20 40	— —	—
(I_C = 1.0 mAdc, V_{CE} = 1.0 Vdc)	2N3903 2N3904		35 70	— —	
(I_C = 10 mAdc, V_{CE} = 1.0 Vdc)	2N3903 2N3904		50 100	150 300	
(I_C = 50 mAdc, V_{CE} = 1.0 Vdc)	2N3903 2N3904		30 60	— —	
(I_C = 100 mAdc, V_{CE} = 1.0 Vdc)	2N3903 2N3904		15 30	— —	
Collector-Emitter Saturation Voltage(1) (I_C = 10 mAdc, I_B = 1.0 mAdc) (I_C = 50 mAdc, I_B = 5.0 mAdc)		$V_{CE(sat)}$	— —	0.2 0.3	Vdc
Base-Emitter Saturation Voltage(1) (I_C = 10 mAdc, I_B = 1.0 mAdc) (I_C = 50 mAdc, I_B = 5.0 mAdc)		$V_{BE(sat)}$	0.65 —	0.85 0.95	Vdc

NOTE: The "O" following CBO, CEO, EBO indicates the third terminal is "open." For example, $V_{(BR)\,CEO}$ means the breakdown voltage between collector and emitter with the base open.

h_{FE} = β_{DC}, dc current gain is measured at different values of I_c.

Maximum base-emitter voltage (V_{BE}) when transistor is saturated

Maximum value of voltage between collector and emitter (V_{CE}) when transistor is in saturation.

FIGURE 17-17 A General Purpose NPN Silicon Transistor. (Copyright of Motorola. Used by permission.)

Step	Action	Result if OK
1	Select Low Resistance Range	
2	Connect $\ominus$ of Ohmmeter to Base	
3	With $\oplus$ of Ohmmeter, Probe Collector —————— High Ω Emitter —————— High Ω (If a Low Ω Reading Results from Step 3, Respective Collector or Emitter Diode Is Shorted)	High Ω High Ω
4	Connect $\oplus$ of Ohmmeter to Base	
5	With $\ominus$ of Ohmmeter, Probe Collector —————— Low Ω Emitter —————— Low Ω (If a High Ω Reading Results from Step 5, Respective Collector or Emitter Diode Is Open)	Low Ω Low Ω

(a)

(b)

FIGURE 17-19   NPN Ohmmeter Test Procedure. (a) Reverse-Biasing Collector, Then Emitter,
Diode. (b) Forward-Biasing Collector, Then Emitter, Diode.

Base Biasing

Figure 17-20(a) shows how a common-emitter transistor circuit could be base biased. With **base biasing,** the emitter diode of the transistor is forward biased by applying a positive base bias voltage ($+V_{BB}$) via a current-limiting resistor (R_B) to the base of Q_1. In Figure 17-20(b), the transistor circuit from Figure 17-20(a) has been redrawn so as to simplify the analysis of the circuit. The transistor is now represented as a diode between base and emitter (emitter diode), and the transistor's emitter to collector has been represented as a variable resistor. Assuming Q_1 is a silicon bipolar transistor, the forward biased emitter diode will have a standard base-emitter voltage drop of 0.7 V (emitter diode drop = 0.7 V).

$$V_{BE} = 0.7 \text{ V}$$

The base bias resistor (R_B) and the transistor's emitter diode form a series circuit across V_{BB}, as seen in Figure 17-19(b). Therefore, the voltage drop across R_B (V_{RB}) will be equal to the difference between V_{BB} and V_{BE}.

$$V_{R_B} = V_{BB} - V_{BE}$$
$$= V_{BB} - 0.7 \text{ V}$$

$$V_{R_B} = V_{BB} - V_{BE} = 10 \text{ V} - 0.7 \text{ V} = 9.3 \text{ V}$$

Now that the resistance and voltage drop across R_B are known, we can calculate the current through R_B (I_{R_B}). Because a series circuit is involved, the current through R_B (I_{R_B}) will also be equal to the transistor base current I_B.

$$I_B = \frac{V_{R_B}}{R_B}$$

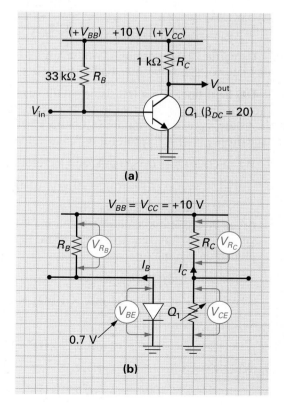

FIGURE 17-20   A Base Biased Common Emitter Circuit. (a) Basic Circuit. (b) Simplified Equivalent Circuit.

$$I_B = \frac{V_{R_B}}{R_B} = \frac{9.3 \text{ V}}{33 \text{ k}\Omega} = 282 \text{ }\mu\text{A}$$

Because the transistor's dc current gain (β_{DC}) is given in Figure 17-20(a), we can calculate I_C because β_{DC} tells us how much greater the output current I_C is compared to the input current I_B.

$$\boxed{I_C = I_B \times \beta_{DC}}$$

$$I_C = I_B \times \beta_{DC} = 282 \text{ }\mu\text{A} \times 20 = 5.6 \text{ mA}$$

Because the current through R_C is I_C, we can now calculate the voltage drop across R_C (V_{R_C}).

$$\boxed{V_{R_C} = I_C \times R_C}$$

$$V_{R_C} = I_C \times R_C = 5.6 \text{ mA} \times 1 \text{ k}\Omega = 5.6 \text{ V}$$

Now that V_{R_C} is known, we can calculate the voltage drop across the transistor's collector-to-emitter because V_{CE} and V_{R_C} are in series and will be equal to the applied voltage V_{CC}.

$$\boxed{V_{CE} = V_{CC} - V_{R_C}}$$

$$V_{CE} = V_{CC} - V_{R_C} = 10 \text{ V} - 5.6 \text{ V} = 4.4 \text{ V}$$

Combining the previous two equations, we can obtain the following V_{CE} formula:

$$V_{CE} = V_{CC} - V_{R_C}$$

Since

$$V_{R_C} = I_C \times R_C$$

$$\boxed{V_{CE} = V_{CC} - (I_C \times R_C)}$$

$$V_{CE} = V_{CC} - (I_C \times R_C) = 10 \text{ V} - (5.6 \text{ mA} \times 1 \text{ k}\Omega) = 4.4 \text{ V}$$

Using the above formulas, which are all basically Ohm's law, you can calculate the current and voltage values in a base biased circuit.

DC load line. In a transistor circuit, such as the example in Figure 17-20, V_{CC} and V_{R_C} are constants. On the other hand, the input current I_B and the output current I_C are variables. Using the example circuit in Figure 17-20, let us calculate what collector-to-emitter voltage drops (V_{CE}) will result for different values of I_C.

a. When Q_1 is OFF, $I_C = 0$ mA, and therefore V_{CE} equals:

$$V_{CE} = V_{CC} - (I_C \times R_C) = 10 \text{ V} - (0 \text{ mA} \times 1 \text{ k}\Omega) = 10 \text{ V} - 0 \text{ V} = 10 \text{ V}$$

This would make sense because Q_1 would be equivalent to an open switch between collector and emitter when it is OFF, and therefore all of the 10 V V_{CC} supply voltage would appear across the open. Figure 17-21 shows how this point would be plotted on a graph (point A).

b. When $I_C = 1$ mA,

$$V_{CE} = 10 \text{ V} - (1 \text{ mA} \times 1 \text{ k}\Omega) = 10 \text{ V} - 1 \text{ V} = 9 \text{ V} \text{ (point } B)$$

c. When $I_C = 2$ mA,

$$V_{CE} = 10 \text{ V} - (2 \text{ mA} \times 1 \text{ k}\Omega) = 10 \text{ V} - 2 \text{ V} = 8 \text{ V} \text{ (point } C)$$

d. When $I_C = 3$ mA,

$$V_{CE} = 10 \text{ V} - (3 \text{ mA} \times 1 \text{ k}\Omega) = 10 \text{ V} - 3 \text{ V} = 7 \text{ V} \text{ (point } D)$$

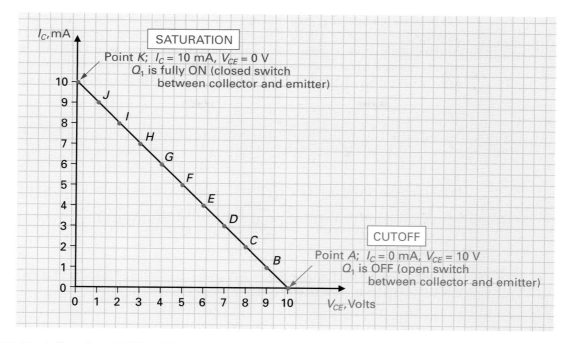

FIGURE 17-21   **A Transistor DC Load Line with Cutoff and Saturation Points.**

e. When $I_C = 4$ mA, $V_{CE} = 6$ V (point E)

f. When $I_C = 5$ mA, $V_{CE} = 5$ V (point F)

g. When $I_C = 6$ mA, $V_{CE} = 4$ V (point G)

h. When $I_C = 7$ mA, $V_{CE} = 3$ V (point H)

i. When $I_C = 8$ mA, $V_{CE} = 2$ V (point I)

j. When $I_C = 9$ mA, $V_{CE} = 1$ V (point J)

k. When $I_C = 10$ mA, the only resistance is that of R_C because Q_1 is fully ON and is equivalent to a closed switch between collector and emitter. It is not a surprise that the voltage drop across Q_1's collector-to-emitter is almost 0 V.

$$V_{CE} = 10 \text{ V} - (10 \text{ mA} \times 1 \text{ k}\Omega) = 10 \text{ V} - 10 \text{ V} = 0 \text{ V (point } K)$$

The line drawn in the graph in Figure 17-21 is called the **dc load line** because it is a line representing all the dc operating points of the transistor for a given load resistance. In this example, the transistor's load was the 1 kΩ collector-connected resistor R_C in Figure 17-20.

Cutoff and saturation points.   Let us now examine the two extreme points in a transistor's dc load line, which in the example in Figure 17-21 were points A and K. If a transistor's base input bias voltage is reduced to zero, its input current I_B will be zero, Q_1 will turn OFF and be equivalent to an open switch between the collector and emitter, the output current I_C will be 0 mA, and a V_{CE} will be 10 V. This point in the transistor dc load line is called *cutoff* (point A in Figure 17-20) because the output collector current is reduced to zero, or cut off. In summary, at cutoff:

$$I_{C(Cutoff)} = 0 \text{ mA}$$
$$V_{CE(Cutoff)} = V_{CC}$$

In the example circuit in Figure 17-20 and its dc load line in Figure 17-21, with Q_1 cut OFF:

$$I_{C(Cutoff)} = 0 \text{ mA}, \ V_{CE(Cutoff)} = V_{CC} = 10 \text{ V}$$

If the base input bias voltage is increased to a large positive value, the transistor's collector diode (which is normally reverse biased) will be forward biased. In this condition, I_B

DC Load Line

A line representing all the dc operating points of the transistor for a given load resistance.

will be at its maximum, Q_1 will be fully ON and equivalent to a closed switch between the collector and emitter, I_C will be at its maximum of 10 mA, and V_{CE} will be 0 V. This point in the transistor's dc load line is called *saturation* (point K in Figure 17-21) because, just as a point is reached where a wet sponge is saturated and cannot hold any more water, the transistor at saturation cannot increase I_C beyond this point. In summary, at saturation:

$$I_{C(Sat.)} = \frac{V_{CC}}{R_C}$$

$$V_{CE(Sat.)} = 0.V$$

In the example circuit in Figure 17-20 and its dc load line in Figure 17-21, with Q_1 saturated:

$$I_{C(Sat.)} = \frac{V_{CC}}{R_C} = \frac{10\text{ V}}{1\text{ k}\Omega} = 10\text{ mA}$$

$$V_{CE(Sat.)} = 0\text{ V}$$

Rearranging the formula $\beta_{DC} = I_C/I_B$, we can calculate the value of input base current that causes the output saturation current:

$$\beta_{DC} = \frac{I_C}{I_B}\text{, therefore,}$$

$$I_{B(Sat.)} = \frac{I_{C(Sat.)}}{\beta_{DC}}$$

In the example circuit in Figure 17-20 and its dc load line in Figure 17-21, the input current that will cause saturation will be

$$I_{B(Sat.)} = \frac{I_{C(Sat.)}}{\beta_{DC}} = \frac{10\text{ mA}}{20} = 500\text{ }\mu\text{A}$$

Figure 17-22 summarizes all of our base bias circuit calculations so far by including the dc load line from Figure 17-21 in a set of collector characteristic curves for the transis-

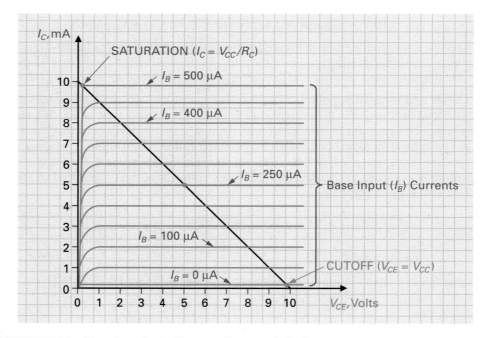

FIGURE 17-22   **Transistor Input/Output Characteristic Graph.**

tor circuit example in Figure 17-20. As you can see in the graph in Figure 17-22, at cutoff: $I_B = 0$ μA, $I_C = 0$ mA, and $V_{CE} = V_{CC}$ which is 10 V. On the other hand, at saturation: $I_B = 500$ μA, $I_C = 10$ mA, and $V_{CE} = 0$ V.

Quiescent point. Generally, the value of the base bias resistor (R_B) is chosen so that the value of base current (I_B) is near the middle of the dc load line. For example, if a base bias resistance of 37.2 kΩ was used in the example circuit in Figure 17-20 ($R_B = 37.2$ kΩ), it would produce a base current of 250 mA ($I_B = 9.3$ V/37.2 kΩ = 250 μA). Referring to the dc load line in Figure 17-22, you can see that this value of base current is half way between cutoff at 0 μA, and saturation at 500 μA. This point is called the *quiescent* (at rest) or *Q point* and is defined as *the dc bias point at which the circuit rests when no ac input signal is applied.* An ac input signal voltage will vary I_B above and below this *Q* point, resulting in a corresponding but larger change in I_C.

EXAMPLE:

Complete the following for the circuit shown in Figure 17-23.

a. Calculate I_B.

b. Calculate I_C.

c. Calculate V_{CE}.

d. Sketch the circuit's dc load line with saturation and cutoff points.

e. Indicate where the Q point is on the circuit's dc load line.

Solution:

a. Since $V_{BE} = 0.7$ V and $V_{BB} = 12$ V,

$$V_{R_B} = 12 \text{ V} - 0.7 \text{ V} = 11.3 \text{ V}$$

$$I_B = \frac{V_{R_B}}{R_B} = \frac{11.3 \text{ V}}{220 \text{ kΩ}} = 51.4 \text{ μA}$$

b.
$$I_C = I_B \times \beta_{DC} = 51.4 \text{ μA} \times 80 = 4.1 \text{ mA}$$

c.
$$V_{R_C} = I_C \times R_C = 4.1 \text{ mA} \times 1.2 \text{ kΩ} = 4.92 \text{ V}$$
$$V_{CE} = V_{CC} - V_{R_C} = 12 \text{ V} - 4.92 \text{ V} = 7.08 \text{ V}$$

d. At cutoff, the transistor is OFF and therefore equivalent to an open switch between collector and emitter. All of the V_{CC} supply voltage will therefore be across Q_1.

At cutoff, $V_{CE} = V_{CC} = 12$ V (see cutoff in the dc load line in Figure 17-24).

At saturation, the transistor is fully ON and therefore equivalent to a closed switch between the collector and emitter. The only resistance is that of R_C, and so:

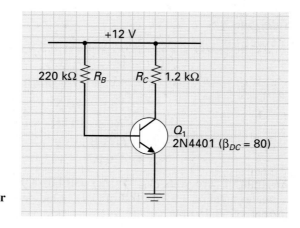

FIGURE 17-23 Bipolar Transistor Example.

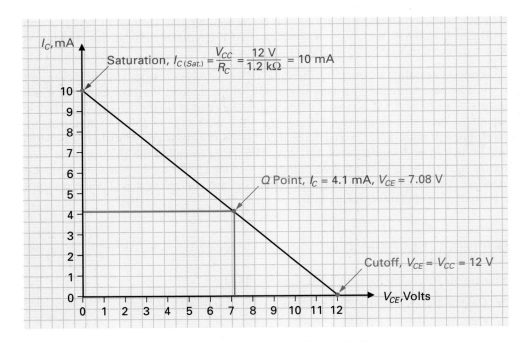

FIGURE 17-24   The DC Load Line for the Circuit in Figure 17-23.

At saturation,

$$I_{C(Sat.)} = \frac{V_{CC}}{R_C} = \frac{12\ V}{1.2\ k\Omega} = 10\ mA$$

(see saturation in the dc load line in Figure 17-23).

e. The operating point or Q point of this circuit is set by the base bias resistor R_B. This Q point will be at

$$I_C = 4.1\ mA$$

which produces a

$$V_{CE} = 7.08\ V$$

This quiescent (Q) point is also shown on Figure 17-24.

EXAMPLE:

Calculate the current through the lamp in Figure 17-25.

Solution:

$$V_{BE} = 0.7\ V,\ V_{in} = +5\ V,\ \text{therefore,}$$
$$V_{R_B} = V_{in} - 0.7\ V$$
$$= 5\ V - 0.7\ V = 4.3\ V$$
$$I_B = \frac{V_{R_B}}{R_B} = \frac{4.3\ V}{10\ k\Omega} = 430\ \mu A$$
$$I_C = I_B \times \beta_{DC}$$
$$= 430\ \mu A \times 125 = 53.75\ mA$$

An input of zero volts ($V_{in} = 0\ V$) will turn OFF Q_1, and therefore lamp L_1. On the other hand, an input of $+5\ V$ will turn ON Q_1 and permit a collector current, and therefore lamp current, of 53.75 mA.

FIGURE 17-25   Two-State Lamp
Circuit.

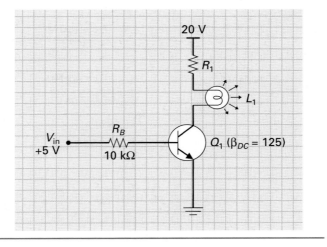

Base biasing applications.

Base bias circuits are used in switching circuit applications like the two-state ON/OFF lamp circuit discussed in the previous example. In these circuits, the bipolar transistor is equivalent to a switch and is controlled by a HIGH/LOW input voltage that drives the transistor between the two extremes of cutoff and saturation.

The advantage of this biasing technique is circuit simplicity because only one resistor is needed to set the base bias voltage. The disadvantage of the base biased circuit is that it cannot compensate for changes in its dc bias current due to changes in temperature. To explain this in more detail, a change in temperature will result in a change in the internal resistance of the transistor (all semiconductor devices have a negative temperature coefficient of resistance—temperature $\uparrow$ causes internal resistance $\downarrow$). This change in the transistor's internal resistance will change the transistor's dc bias currents (I_B and I_C), which will change or shift the transistor's dc operating point or Q point away from the desired midpoint.

Voltage-Divider Biasing

Figure 17-26(a) shows how a common-emitter transistor circuit could be **voltage-divider biased.** The name of this biasing method comes from the two- resistor series voltage divider (R_1 and R_2) connected to the transistor's base. In this most widely used biasing method, the emitter diode of Q_1 is forward biased by the voltage developed across R_2 (V_{R_2}), as seen in the simplified equivalent circuit in Figure 17-26(b). To calculate the voltage developed across R_2, and therefore the voltage applied to Q_1's base, we can use the voltage-divider formula.

$$V_{R_2} \text{ or } V_B = \frac{R_2}{R_1 + R_2} \times V_{CC}$$

$$V_{R_2} \text{ or } V_B = \frac{R_2}{R_1 + R_2} \times V_{CC} = \frac{10 \text{ k}\Omega}{20 \text{ k}\Omega + 10 \text{ k}\Omega} \times 20 \text{ V} = 0.333 \times 20 \text{ V} = 6.7 \text{ V}$$

Because the current through R_1 and R_2 (from ground to $+V_{CC}$) is generally more than 10 times greater than the base current of Q_1 (I_B), it is normally assumed that I_B will have no effect on the voltage-divider current through R_1 and R_2. The R_1 and R_2 voltage divider can be assumed to be independent of Q_1, and the previous voltage-divider formula can be used to calculate V_{R_2} or V_B.

Because $V_B = 6.7$ V, the emitter diode of Q_1 will be forward biased. Assuming a 0.7 V drop across the transistor's base-emitter junction ($V_{BE} = 0.7$ V), the voltage at the emitter terminal of Q_1 (V_E) will be:

$$V_{R_E} \text{ or } V_E = V_B - 0.7 \text{ V}$$

Voltage-Divider Biasing

A biasing method used with amplifiers in which a series arrangement of two fixed-value resistors is connected across the voltage source. The result is that a desired fraction of the total voltage is obtained at the center of the two resistors and is used to bias the amplifier.

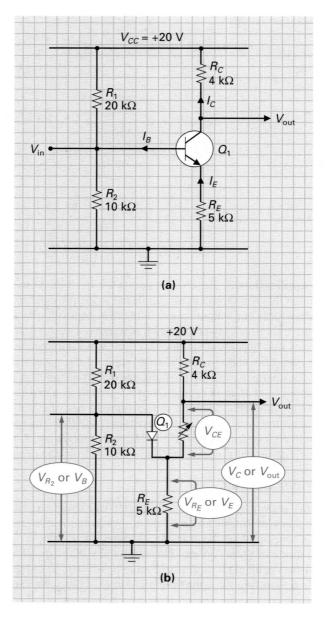

FIGURE 17-26 A Voltage-Divider Biased Common-Emitter Circuit. (a) Basic Circuit. (b) Simplified Equivalent Circuit.

$$V_{R_E} \text{ or } V_E = V_B - 0.7 \text{ V} = 6.7 \text{ V} - 0.7 \text{ V} = 6 \text{ V}$$

Now that the voltage drop across R_E (V_{R_E}) is known, along with its resistance, we can calculate the current through R_E and the value of current being injected into the transistor's emitter.

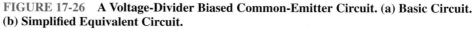

$$I_{R_E} = I_E = \frac{V_{R_E}}{R_E}$$

$$I_{R_E} = I_E = \frac{V_{R_E}}{R_E} = \frac{6 \text{ V}}{5 \text{ k}\Omega} = 1.2 \text{ mA}$$

Because we know that a transistor collector current (I_C) is approximately equal to the emitter current (I_E), we can state that

$$I_E \cong I_C$$

$$I_E \cong I_C = 1.2 \text{ mA}$$

Now that I_C is known, we can calculate the voltage drop across R_C (V_{R_C}) because both its resistance and current are known.

$$V_{R_C} = I_C \times R_C$$

$$V_{R_C} = I_C \times R_C = 1.2 \text{ mA} \times 4 \text{ k}\Omega = 4.8 \text{ V}$$

The dc quiescent voltage at the collector of Q_1 with respect to ground (V_C), which is also V_{out}, will be equal to the dc supply voltage (V_{CC}) minus the voltage drop across R_C.

$$V_C \text{ or } V_{out} = V_{CC} - V_{R_C}$$

$$V_C \text{ or } V_{out} = V_{CC} - V_{R_C} = 20 \text{ V} - 4.8 \text{ V} = 15.2 \text{ V}$$

Because V_{CC} is connected across the series voltage divider formed by R_C, Q_1's collector-to-emitter resistance (R_{CE}), and R_E, we can calculate V_{CE} if both V_{R_C} and V_E are known:

$$V_{CE} = V_{CC} - (V_{R_C} + V_E)$$

$$V_{CE} = V_{CC} - (V_{R_C} + V_E) = 20 \text{ V} - (4.8 \text{ V} + 6 \text{ V}) = 20 \text{ V} - 10.8 \text{ V} = 9.2 \text{ V}$$

DC load line. Figure 17-27 shows the dc load line for the example circuit in Figure 17-26. Referring to the dc load line's two end points, let us examine this circuit's saturation and cutoff points.

When transistor Q_1 is fully ON or saturated, it will have approximately 0 Ω of resistance between its collector and emitter. As a result, R_C and R_E determine the value of I_C when Q_1 is saturated.

$$I_{C(Sat.)} = \frac{V_{CC}}{R_C + R_E}$$

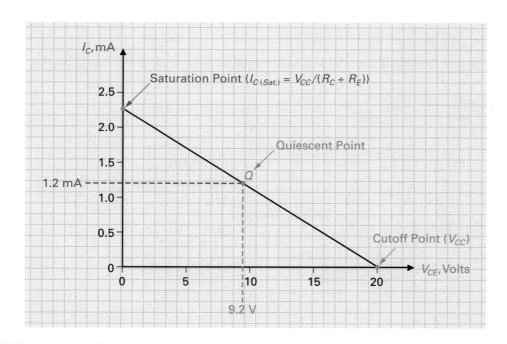

FIGURE 17-27 **The DC Load Line for the Circuit in Figure 17-25.**

$$I_{C(Sat.)} = \frac{V_{CC}}{R_C + R_E} = \frac{20\text{ V}}{4\text{ k}\Omega + 5\text{ k}\Omega} = \frac{20\text{ V}}{9\text{ k}\Omega} = 2.2\text{ mA}$$

As you can see in Figure 17-27, at saturation, I_C is maximum at 2.2 mA, and V_{CE} is 0 V because Q_1 is equivalent to a closed switch (0 Ω) between Q_1's collector and emitter.

$$V_{CE(Sat.)} = 0\text{ V}$$

At the other end of the dc load line in Figure 17-27, we can see how the transistor's characteristics are plotted when it is cut off. When Q_1 is cut OFF, it is equivalent to an open switch between collector-to-emitter. Therefore all of the V_{CC} supply voltage will appear across the series circuit open.

$$V_{CE(Cutoff)} = V_{CC}$$

$$V_{CE(Cutoff)} = V_{CC} = 20\text{ V}$$

As you can see in Figure 17-27, when Q_1 is cut OFF, all of the V_{CC} supply voltage will appear across Q_1's collector-to-emitter terminals, and I_C will be blocked and equal to zero.

$$I_{C(Cutoff)} = 0\text{ mA}$$

Generally, the value of the voltage-divider resistors R_1 and R_2 are chosen so that the value of base current (I_B) is near the middle of the dc load line. Referring to Figure 17-27, you can see that by plotting our previously calculated values of I_C (which at rest was 1.2 mA) and V_{CE} ($V_{CE} = 9.2$ V), we obtain a Q point that is near the middle of the dc load line.

EXAMPLE:

Calculate the following for the circuit shown in Figure 17-28.

    a. V_B and V_E
    b. Determine whether C_E will have any effect on the dc operating voltages
    c. I_C
    d. V_C and V_{CE}
    e. Sketch the circuit's dc load line and include the saturation, cutoff, and Q points

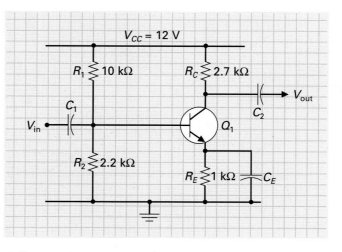

FIGURE 17-28   **A Common-Emitter Amplifier Circuit Example.**

Solution:

a. $$V_B = \frac{R_2}{R_1 + R_2} \times V_{CC} = \frac{2.2\ k\Omega}{10\ k\Omega + 2.2\ k\Omega} \times 12\ V = 2.16\ V$$

$$V_E = V_B - 0.7\ V = 2.16\ V - 0.7\ V = 1.46\ V$$

b. Since all capacitors can be thought of as a dc block, C_E will have no effect on the circuit's dc operating voltages.

c. $$I_E = \frac{V_E}{R_E} = \frac{1.46\ V}{1\ k\Omega} = 1.46\ mA$$

$$I_C \cong I_E = 1.46\ mA$$

d. $V_{R_C} = I_C \times R_C = 1.46\ mA \times 2.7\ k\Omega = 3.9\ V$

V_{out} or $V_C = V_{CC} - V_{R_C} = 12\ V - 3.9\ V = 8.1\ V$

$V_{CE} = V_{CC} - (V_{R_C} + V_E) = 12\ V - (3.9\ V + 1.46\ V) = 12\ V - 5.36\ V = 6.64\ V$

e. $$I_{C(Sat.)} = \frac{V_{CC}}{R_C + R_E} = \frac{12\ V}{2.7\ k\Omega + 1\ k\Omega} = \frac{12\ V}{3.7\ k\Omega} = 3.24\ mA$$

$V_{CE(Cutoff)} = V_{CC} = 12\ V$

Q point, $I_C = 1.46\ mA$ and $V_{CE} = 6.64\ V$

(This information is plotted on the graph in Figure 17-29.)

Voltage-divider bias applications. Voltage-divider biased circuits are used in analog or linear circuit applications such as the amplifier circuit discussed in the previous example. In these circuits, the bipolar transistor is equivalent to a variable resistor and is controlled by an alternating input signal voltage.

Unlike the base biased circuit, the voltage-divider biased circuit has very good temperature stability due to the emitter resistor R_E. To explain this in more detail, let us assume that there is an increase in the temperature surrounding a voltage-divider circuit, such as the example circuit in Figure 17-28. As temperature increases, it causes an increase in the transistor's internal currents ($I_B\uparrow$, $I_E\uparrow$, $I_C\uparrow$) because all semiconductor devices have a negative temperature coefficient of resistance (temperature $\uparrow$, $R\downarrow$, $I\uparrow$). An increase in $I_E\uparrow$ will cause

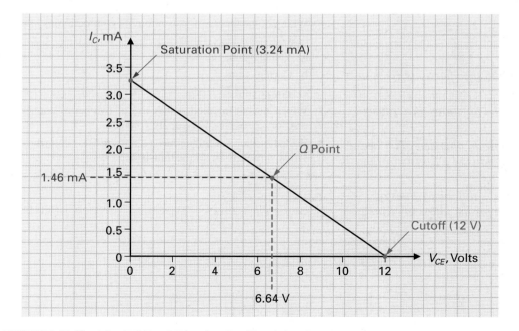

FIGURE 17-29 **The DC Load Line for the Circuit in Figure 17-28.**

an increase in the voltage drop across $R_E\uparrow$, which will decrease the voltage difference between the transistor's base and emitter ($V_{BE}\downarrow$). Decreasing the forward bias applied to the transistor's emitter diode will decrease all of the transistor's internal currents ($I_B\downarrow$, $I_E\downarrow$, $I_C\downarrow$) and return them to their original values. Therefore, a change in output current (I_C) due to temperature will effectively be fed back to the input and change the input current (I_B), which is why a circuit containing an emitter resistor is said to have **emitter feedback** for temperature stability.

Emitter Feedback
The coupling from the emitter output to the base input in a transistor amplifier.

SELF-TEST EVALUATION POINT FOR SECTION 17-2

Now that you have completed this section, you should be able to:

■ **Objective 7.** Describe in more detail the bipolar junction transistor action.

■ **Objective 8.** Explain in more detail how the transistor can be used as an amplifier.

■ **Objective 9.** Identify and list the characteristics of the three transistor circuit combinations, namely:
   a. Common base
   b. Common emitter
   c. Common collector

■ **Objective 10.** Explain the meaning of the following:
   a. Transistor voltage and current abbreviations
   b. DC alpha
   c. DC beta
   d. Collector characteristic curve
   e. AC beta
   f. Input resistance or impedance
   g. Output resistance or impedance

■ **Objective 11.** Calculate a transistor circuit's:
   a. DC current gain
   b. AC current gain
   c. Voltage gain
   d. Power gain

■ **Objective 12.** Interpret the specifications given in a manufacturer's data sheet.

■ **Objective 13.** Describe the functions of a transistor tester.

■ **Objective 14.** Explain how an ohmmeter can be used to check the P-N junctions of a transistor for opens and shorts.

■ **Objective 15.** Calculate the different values of circuit voltage and current for the following transistor biasing methods:

   a. Base biasing
   b. Voltage-divider biasing

■ **Objective 16.** Define the following terms:
   a. DC load line
   b. Cutoff point
   c. Saturation point
   d. Quiescent point

Use the following questions to test your understanding of Section 17-2.

1. The bipolar transistor is a _____ (voltage/current) controlled device.

2. When a bipolar transistor is being operated in the active region, its emitter diode is _____ biased and its collector diode is _____ biased.

3. Which of the following is correct:
   a. $I_E = I_C - I_B$
   b. $I_C = I_E - I_B$
   c. $I_B = I_C - I_E$

4. When a transistor is in cutoff, it is equivalent to a/an _____ between its collector and emitter.

5. When a transistor is in saturation, it is equivalent to a/an _____ between its collector and emitter.

6. Which of the bipolar transistor circuit configurations has the best
   a. Voltage gain
   b. Current gain
   c. Power gain

7. Which biasing method makes use of two series-connected resistors across the V_{CC} supply voltage?

8. Which biasing technique has a single resistor connected in series with the base of the transistor?

SUMMARY

Bipolar Junction Transistors (Figure 17-30 and Figure 17-31)

1. The bipolar transistor has three alternately doped semiconductor regions called the emitter (E), base (B), and collector (C).

2. A transistor can be thought of as containing two diodes: a base-to-collector diode and a base-to-emitter diode.

3. The name transistor was derived from the fact that through base control we can "transfer" different values of "resistance" between the emitter and collector. This effect of "transferring resistance" is known as transistance, and the component that functions in this manner is called the transistor.

4. A digital logic gate circuit makes use of the transistor's ON/OFF switching action.

5. When used as a variable resistor, the transistor is the controlling element in many analog or linear circuit applications such as amplifiers, oscillators, modulators, detectors, regulators, and so on. The most important of these applications is amplification, which is the boosting in strength or increasing in amplitude of electronic signals. The gain or amplification is achieved by the input signal controlling the conduction of the transistor, which takes energy from the collector supply voltage and develops this energy across the load resistor. Amplification is achieved by having a small input voltage control a transistor and its large collector supply voltage so that a small input voltage change results in a similar but larger output voltage change.

6. Since both the collector current and base current are derived from the emitter current, we can state that the sum of the base current and the collector current will equal the emitter current.

7. An increase in the emitter current ($I_E\uparrow$) will cause a corresponding increase in collector and base current ($I_C\uparrow$, $I_B\uparrow$). The applied input base voltage will control the amount of base current, which will in turn control the amount of emitter and collector current and the conduction of the transistor. This is why the bipolar transistor is known as a current-controlled device.

8. A transistor is said to be in the active region when its base-emitter junction is forward biased (emitter diode is ON), and the base-collector junction is reverse biased (collector diode is OFF). In this mode, the transistor is equivalent to a variable resistor between the collector and emitter.

9. In cutoff, both the emitter and collector diode of the transistor will be biased OFF, the transistor is equivalent to an open switch between the collector and emitter, and transistor current is zero.

10. A transistor is in saturation when the bias voltage is increased to such a point that any further increase in bias voltage will not cause any further increase in current through the transistor. In saturation, both the emitter and collector diodes are said to be forward biased, and the transistor is equivalent to a closed switch.

11. Although there are many thousands of different bipolar transistor circuits, all of these circuits can be classified into one of three groups based on which of the transistor's leads is used as the "common reference." These three different circuit configurations are the:
    a. Common-emitter (C-E), in which the input signal is applied between the base and emitter, while the output signal appears between the transistor's collector and emitter. With this circuit arrangement, the input signal controls the transistor's base current, which in turn controls the transistor's output collector current, and the emitter lead is common to both the input and output.
    b. Common-base (C-B), in which the input signal is applied between the transistor's emitter and base, the output signal is developed across the transistor's collector and base, and the base is common to both input and output.
    c. Common-collector (C-C), in which the input is applied between the base and the collector, the output is developed across the emitter and the collector, and the collector is common to both the input and output.

12. The current gain in a common-emitter configuration is called the transistor's beta (symbolized β) and is the ratio of output collector current to input base current. The term hfe is often used instead of dc current gain (β_{DC}) in manufacturer data sheets.

13. The common-emitter circuit provides both current gain and voltage gain. Since power is equal to the product of current and voltage ($P = V \times I$), it is not surprising that the C-E circuit configuration also provides a high value of power gain (A_P).

14. The set of collector characteristic curves graph shows the relationship between a transistor's input base current, output collector current, and collector-to-emitter voltage drop.

15. The input resistance (R_{in}), or input impedance (Z_{in}), of a transistor is the amount of opposition offered to an input signal by the transistor's input junction.

16. The output resistance (R_{out}), or output impedance (Z_{out}), of a transistor is the amount of opposition offered to an output signal by the transistor's output junction.

17. The current gain in a common-base configuration is called the transistor's alpha (symbolized α) and is the ratio of output collector current to input emitter current.

18. Although the common-base circuit configuration does not achieve any current gain, it does have a very high voltage gain.

19. The common-collector circuit provides a slightly higher gain than the C-E circuit because the common-collector's output current (I_E) is slightly higher than the C-E's output current (I_C).

FIGURE 17-30 BJT Circuit Configuration Characteristics.

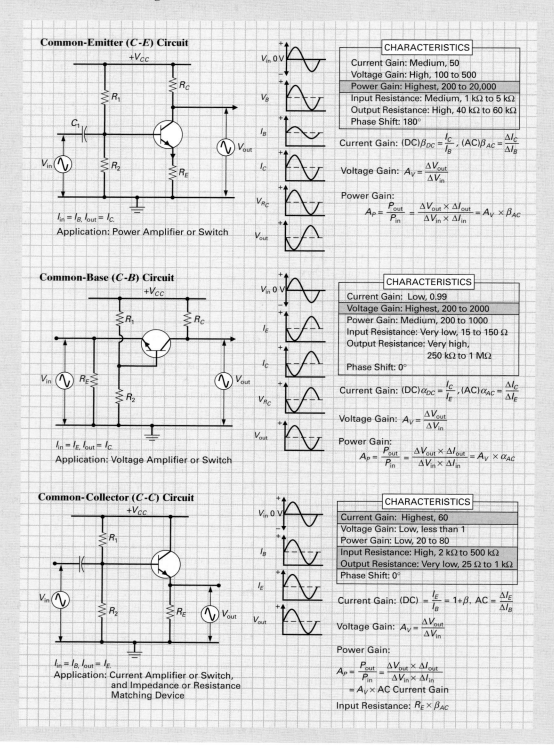

Common-Emitter (*C-E*) Circuit

$I_{in} = I_B$, $I_{out} = I_C$.

Application: Power Amplifier or Switch

CHARACTERISTICS

Current Gain: Medium, 50
Voltage Gain: High, 100 to 500
Power Gain: Highest, 200 to 20,000
Input Resistance: Medium, 1 kΩ to 5 kΩ
Output Resistance: High, 40 kΩ to 60 kΩ
Phase Shift: 180°

Current Gain: (DC)$\beta_{DC} = \dfrac{I_C}{I_B}$, (AC)$\beta_{AC} = \dfrac{\Delta I_C}{\Delta I_B}$

Voltage Gain: $A_V = \dfrac{\Delta V_{out}}{\Delta V_{in}}$

Power Gain:

$$A_P = \dfrac{P_{out}}{P_{in}} = \dfrac{\Delta V_{out} \times \Delta I_{out}}{\Delta V_{in} \times \Delta I_{in}} = A_V \times \beta_{AC}$$

Common-Base (*C-B*) Circuit

$I_{in} = I_E$, $I_{out} = I_C$.

Application: Voltage Amplifier or Switch

CHARACTERISTICS

Current Gain: Low, 0.99
Voltage Gain: Highest, 200 to 2000
Power Gain: Medium, 200 to 1000
Input Resistance: Very low, 15 to 150 Ω
Output Resistance: Very high,
                      250 kΩ to 1 MΩ
Phase Shift: 0°

Current Gain: (DC)$\alpha_{DC} = \dfrac{I_C}{I_E}$, (AC)$\alpha_{AC} = \dfrac{\Delta I_C}{\Delta I_E}$

Voltage Gain: $A_V = \dfrac{\Delta V_{out}}{\Delta V_{in}}$

Power Gain:

$$A_P = \dfrac{P_{out}}{P_{in}} = \dfrac{\Delta V_{out} \times \Delta I_{out}}{\Delta V_{in} \times \Delta I_{in}} = A_V \times \alpha_{AC}$$

Common-Collector (*C-C*) Circuit

$I_{in} = I_B$, $I_{out} = I_E$.

Application: Current Amplifier or Switch,
      and Impedance or Resistance
      Matching Device

CHARACTERISTICS

Current Gain: Highest, 60
Voltage Gain: Low, less than 1
Power Gain: Low, 20 to 80
Input Resistance: High, 2 kΩ to 500 kΩ
Output Resistance: Very low, 25 Ω to 1 kΩ
Phase Shift: 0°

Current Gain: (DC) $= \dfrac{I_E}{I_B} = 1+\beta$, AC $= \dfrac{\Delta I_E}{\Delta I_B}$

Voltage Gain: $A_V = \dfrac{\Delta V_{out}}{\Delta V_{in}}$

Power Gain:

$$A_P = \dfrac{P_{out}}{P_{in}} = \dfrac{\Delta V_{out} \times \Delta I_{out}}{\Delta V_{in} \times \Delta I_{in}}$$
$$= A_V \times \text{AC Current Gain}$$

Input Resistance: $R_E \times \beta_{AC}$

FIGURE 17-31   BJT Biasing.

Base-Biasing

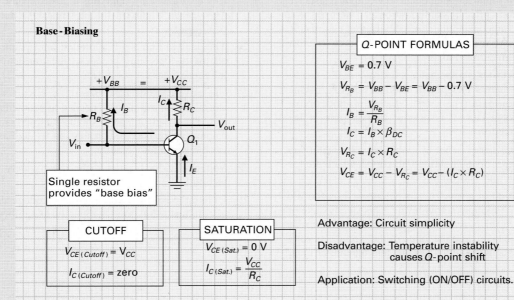

Single resistor provides "base bias"

Q-POINT FORMULAS

$$V_{BE} = 0.7 \text{ V}$$

$$V_{R_B} = V_{BB} - V_{BE} = V_{BB} - 0.7 \text{ V}$$

$$I_B = \frac{V_{R_B}}{R_B}$$

$$I_C = I_B \times \beta_{DC}$$

$$V_{R_C} = I_C \times R_C$$

$$V_{CE} = V_{CC} - V_{R_C} = V_{CC} - (I_C \times R_C)$$

CUTOFF
$V_{CE(Cutoff)} = V_{CC}$
$I_{C(Cutoff)} = $ zero

SATURATION
$V_{CE(Sat.)} = 0$ V
$I_{C(Sat.)} = \dfrac{V_{CC}}{R_C}$

Advantage: Circuit simplicity

Disadvantage: Temperature instability causes Q-point shift

Application: Switching (ON/OFF) circuits.

Voltage-Divider Biasing

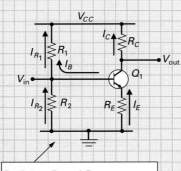

Resistors R_1 and R_2 provide "voltage-divider bias."

Q-POINT FORMULAS

$$V_{R_2} \text{ or } V_B = \frac{R_2}{R_1 + R_2} \times V_{CC}$$

$$V_{R_E} \text{ or } V_E = V_B - 0.7 \text{ V}$$

$$I_E = \frac{V_E}{R_E}$$

$$I_C \cong I_E$$

$$V_{R_C} = I_C \times R_C$$

$$V_{out} \text{ or } V_C = V_{CC} - V_{R_C}$$

$$V_{CE} = V_{CC} - (V_{R_C} + V_{CE})$$

CUTOFF
$V_{CE(Cutoff)} = V_{CC}$
$I_{C(Cutoff)} = $ zero

SATURATION
$V_{CE(Sat.)} = 0$ V
$I_{C(Sat.)} = \dfrac{V_{CC}}{R_C + R_E}$

Advantage: Good temperature stability

Disadvantage: Needs additional bias resistor

Application: Analog or linear circuits, such as the amplifier

783

20. The common-collector circuit is often referred to as an emitter-follower or voltage-follower because the emitter output voltage seems to track or follow the phase and amplitude of the input voltage.

21. The very high input resistance and low output resistance of the common-collector configuration can be used as a resistance or impedance matching device that can match, or isolate, a high resistance (low current) source to a low resistance (high current) load.

22. The transistor tester is a special test instrument that can be used to test whether an open or short exists, the transistor's dc current gain (β_{DC}), and whether an undesirable value of leakage current is present through one of the transistor's junctions.

23. If the transistor tester is not available, the ohmmeter can be used to detect open or shorted junctions, which are the most common transistor failures.

24. The ac operation of a transistor is determined by the "dc bias level," or "no input signal level." This steady-state or dc operating level is set by the value of the circuit's dc supply voltage (V_{CC}) and the value of the circuit's biasing resistors. This single supply voltage, and the one or more biasing resistors, set up the initial dc values of transistor current (I_B, I_E, and I_C) and transistor voltage (V_{BE}, V_{CE}, and V_{BC}).

25. With base biasing, the emitter diode of the transistor is forward biased by applying a positive base bias voltage ($+V_{BB}$) via a current-limiting resistor (R_B) to the base of Q_1.

26. The dc load line is a line plotted on a graph representing all the dc operating points of the transistor for a given load resistance.

27. The advantage of base biasing is circuit simplicity because only one resistor is needed to set the base bias voltage. The disadvantage of the base biased circuit is that it cannot compensate for changes in its dc bias current due to changes in temperature.

28. Voltage-divider biasing is the most widely used biasing technique.

29. Unlike the base biased circuit, the voltage-divider biased circuit has very good temperature stability due to the emitter resistor R_E which provides emitter feedback.

REVIEW QUESTIONS

Multiple-Choice Questions

1. The bipolar junction transistor has three terminals called the:
   a. Drain, source, gate
   b. Anode, cathode, gate
   c. Main terminal 1, main terminal 2, gate
   d. Emitter, base, collector

2. The term bipolar junction transistor was given to the device because it has:
   a. Two P-N junctions
   b. Two magnetic poles
   c. One p region and one n region
   d. Two magnetic junctions

3. An NPN transistor is normally biased so that its base is _____.
   a. Positive   **b.** Negative

4. Which is considered the most common bipolar junction transistor configuration?
   a. Common-base   **c.** Common-emitter
   b. Common-collector   **d.** None of the above

5. A common-collector circuit is often called a/an _____.
   a. Base-follower   **c.** Collector-follower
   b. Emitter-follower   **d.** None of the above

6. With the NPN transistor schematic symbol, the emitter arrow will point _____ the base, whereas with the PNP transistor schematic symbol, the emitter arrow will point _____ the base.
   a. Toward, away from   **b.** Away from, toward

7. The transistor's ON/OFF switching action is made use of in _____ circuits.
   a. Analog   **c.** Linear
   b. Digital   **d.** Both (a) and (c) are true

8. The transistor's variable resistor action is made use of in _____ circuits.
   a. Analog   **c.** Linear
   b. Digital   **d.** Both (a) and (c) are true

9. Approximately 98 percent of the electrons entering the _____ of a bipolar transistor will arrive at the _____, and the remainder will flow out of the _____.
   a. Emitter, collector, base   **c.** Collector, emitter, base
   b. Base, collector, emitter   **d.** Emitter, base, collector

10. The common-base circuit configuration achieves the highest _____ gain, the common-emitter achieves the highest _____ gain, and the common-collector achieves the highest _____ gain.
   a. Voltage, current, power   **c.** Voltage, power, current
   b. Current, power, voltage   **d.** Power, voltage, current

11. Which of the following abbreviations is used to denote the voltage drop between a transistor's base and emitter?
   a. I_{BE} **b.** V_{CC} **c.** V_{CE} **d.** V_{BE}

12. Which of the following abbreviations is used to denote the voltage drop between a transistor's collector and emitter?
   a. V_C **b.** V_{CE} **c.** V_E **d.** V_{CC}

13. A transistor's _____ specification indicates the gain in dc current between the input and output of a common-emitter circuit.
a. α_{AC} **b.** α_{DC} **c.** β_{AC} **d.** β_{DC}

14. Consider the following for a base biased bipolar transistor circuit: $R_B = 33$ kΩ, $R_C = 560$ Ω, Q_1 (β_{DC}) = 25, V_{CC} = +10 V. What is V_{BE}?
a. 1.43 mV
b. 25×33 kΩ
c. 0.7 V
d. Not enough information given to calculate

15. Which point on the dc load line results in an $I_C = V_{CC}/R_C$ and a $V_{CE} = 0$ V?
a. Saturation point **c.** Q point
b. Cutoff point **d.** None of the above

16. Which point on the dc load line results in a $V_{CE} = V_{CC}$, and an $I_C = 0$?
a. Saturation point **c.** Q point
b. Cutoff point **d.** None of the above

17. The midway point on the dc load line at which a transistor is biased with dc voltages when no input signal is applied is called the:
a. Saturation point **c.** Q point
b. Cutoff point **d.** None of the above

18. Which transistor biasing method makes use of one current limiting resistor in the base circuit?
a. Base bias **c.** Emitter-follower bias
b. Voltage-divider bias **d.** Current-divider bias

19. A forward biased transistor emitter or collector diode should have a _____ resistance, while a reverse biased emitter or collector diode should have a _____ resistance.
a. Low, low **c.** High, high
b. High, low **d.** Low, high

20. A transistor tester will check a transistor's:
a. Opens or shorts between any of the terminals
b. Gain
c. Reverse leakage current value
d. All of the above

Communication Skill Questions

21. Briefly describe how the BJT is used in: (17-1-4)

a. Digital circuit applications
b. Analog circuit applications

22. Sketch and briefly describe the operation of: (17-1-4)
a. A transistor logic gate **b.** A transistor amplifier

23. Briefly describe how a transistor achieves a gain in voltage between input and output. (17-1-4)

24. What is the relationship between a bipolar transistor's emitter current, base current, and collector current? (17-2-1)

25. Why is the bipolar transistor known as a current-controlled device? (17-2-1)

26. In normal operation, which of the bipolar transistor's P-N junctions or diodes is forward biased, and which is reverse biased? (17-2-1)

27. What is the bipolar transistor equivalent to when it is operated in: (17-2-1)
a. Saturation **b.** The active region **c.** Cutoff

28. Briefly describe the following terms: (17-2-2)
a. DC beta **c.** DC alpha
b. AC beta **d.** AC alpha

29. What are the collector characteristic curves? (17-2-2)

30. Why is the common-emitter circuit configuration the most widely used? (17-2-2)

31. Why is the common-collector circuit configuration well suited as an impedance matching device? (17-2-2)

Practice Problems

32. Identify the type and terminals of the transistors shown in Figure 17-32.

33. A bipolar transistor is correctly biased for operation in the active region when its emitter diode is forward biased and its collector diode is reverse biased. Referring to Figure 17-33, which of the bipolar transistor circuits is correctly biased?

34. Calculate the value of the missing current in the following examples:
a. $I_E = 25$ mA, $I_C = 24.6$ mA, $I_B = $?
b. $I_B = 600$ μA, $I_C = 14$ mA, $I_E = $?
c. $I_E = 4.1$ mA, $I_B = 56.7$ μA, $I_C = $?

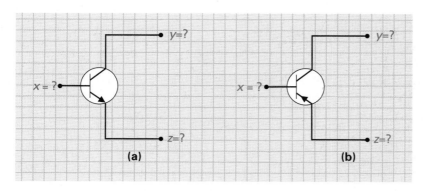

FIGURE 17-32 Identify the Transistor Type and Terminals.

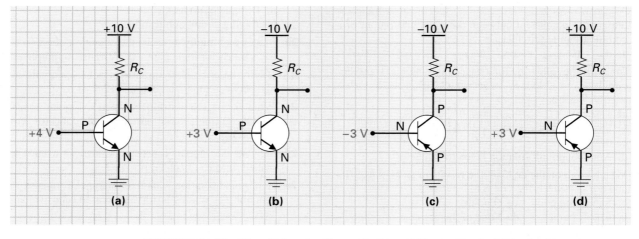

FIGURE 17-33 Identifying the Correctly Biased (Active Region) Bipolar Transistors.

35. Calculate the voltage gain (A_V) of the transistor amplifier whose input/output waveforms are shown in Figure 17-34.

36. Identify the configuration of the actual bipolar transistor electronic system circuits shown in Figure 17-35.

37. Identify the bipolar transistor type and the biasing technique used in each of the circuits in Figure 17-35.

38. Calculate the following for the base biased transistor circuit shown in Figure 17-36:
a. I_B **b.** I_C **c.** V_{CE}

39. Sketch the dc load line for the circuit in Figure 17-36, showing the saturation, cutoff, and Q points.

40. Calculate the following for the voltage-divider biased transistor circuit shown in Figure 17-37:
a. V_B and V_E **b.** I_C **c.** V_C **d.** V_{CE}

41. Sketch the dc load line for the circuit in Figure 17-37, showing the saturation, cutoff, and Q points.

42. In Figure 17-37, does Kirchhoff's voltage law apply to the voltage divider made up of R_C, R_{CE}, and R_E?

Troubleshooting Questions

43. Briefly describe what characteristics a transistor tester can check.

44. How can an ohmmeter be used to test transistors?

45. Which of the transistors being tested in Figure 17-38 are good or bad? If bad, state the suspected problem.

Web Site Questions

 Go to the Web site http://www.prenhall. com/cook, select the textbook *Introductory DC/AC Electronics* or *Introductory DC/AC Circuits*, this chapter, and then follow the instructions when answering the multiple-choice practice problems.

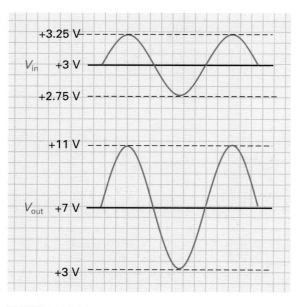

FIGURE 17-34 Transistor Amplifier Input/Output Waveforms.

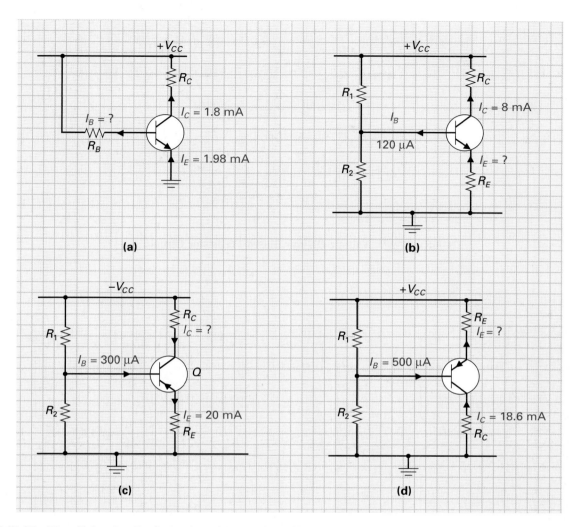

FIGURE 17-35   Identifying the Configuration of Actual BJT Circuits.

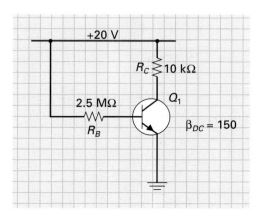

FIGURE 17-36   Base Biased Transistor Circuit.

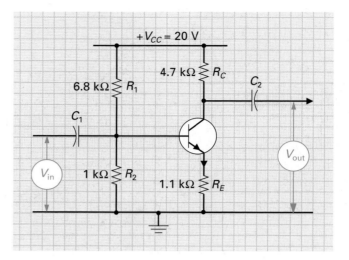

FIGURE 17-37 **Voltage-Divider Biased Transistor Circuit.**

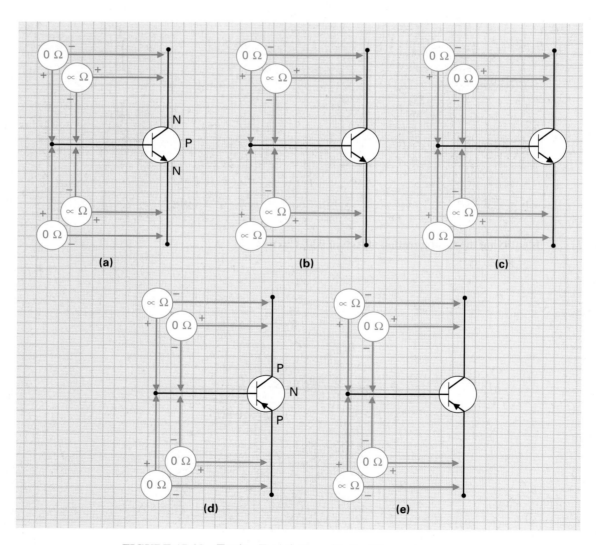

FIGURE 17-38 **Testing Transistors with the Ohmmeter.**

These tests at the end of each chapter will challenge your knowledge up to this point, and give you the practice you need for a job interview. To make this more realistic, the test will comprise both technical and personal questions. In order to take full advantage of this exercise, you may want to set up a simulation of the interview environment, have a friend read the questions to you, and record your responses for later analysis.

Company Name: PST, Inc.

Industry Branch: Computer/Communications

Function: Design and Manufacture wireless handhelds

Job Title: QA Technician

1. Tell me what you know about us.
2. What steps would you follow to test a discrete BJT?

Answers

1. Be prepared and do some homework beforehand. visit the company's web site before the interview to get an overall understanding of the company;s ownership, product line, service, and support.
2. Section 17-2-4.
3. Discuss how you have worked with a partner in the lab on experiments, and any other related work history.
4. Section 17-1-3.

3. What experience have you had working in a team environment?
4. In what applications would you use the BJT's switching action, and in what applications would you use the BJT's variable-resistor action?
5. What responsibilities do you think a QA technician would have?
6. What are the key characteristics of the three BJT circuit configurations?
7. Could you use the multimeter to test a BJT?
8. Would you say that you have faults?
9. If you discovered an error in the cirucit design, would you search out the engineer responsible, mention it to the CEO, or report it to your supervisor?
10. Why did you choose our company?

5. See introduction, QA technician, and refer to job advertisement.
6. Section 17-2-2.
7. Yes, section 17-2-4.
8. Yes, but make sure your faults are positive. For example, you always like to finsih what you start, and so on.
9. Your supervisor.
10. Refer back to the company's positive attributes listed in its web site.

Field Effect Transistors (FETs)

Spitting Lightning Bolts

Nikola Tesla

Nikola Tesla was born in Yugoslavia in 1856. He studied mathematics and physics in Prague, and in 1884 he emigrated to the United States.

In New York he met Thomas Edison, the self-educated inventor who is best known for his development of the phonograph and the incandescent light bulb. Both men were gifted and eccentric and, due to their common interest in "invention," they got along famously. Because Tesla was unemployed, Edison offered him a job. In his lifetime, Edison would go on to take out 1,033 patents and become one of the most prolific inventors of all time. Tesla would go on to invent many different types of motors, generators, and transformers, one of which is named the "Tesla coil" and produces five-foot lightning bolts. With this coil, Tesla investigated "wireless power transmission," the only one of his theories that has not come into being.

Both Tesla and Edison had very strong views on different aspects of electricity and, as time passed, the two men began to engage in very long, loud, and angry arguments. One such discussion concerned whether power should be distributed as alternating current or direct current. Eventually, the world would side with Tesla and choose ac. At the time, this topic, like many others, would cause a hatred to develop between the two men. Eventually Tesla left Edison and started his own company. However, the anger remained, and on one occasion when they were both asked to attend a party for their friend Mark Twain, both refused to come because the other had been invited.

In 1912, Tesla and Edison were both nominated for the Nobel prize in physics, but because neither one would have anything to do with the other, the prize went to a third party—proving that bitterness really will cause a person to cut off his nose to spite his face.

When angry count up to four; when very angry, swear.

Mark Twain

Outline and Objectives

Introduction

In the previous chapters, we have concentrated on all aspects of the bipolar junction transistor or BJT. We have examined its construction, operation, characteristics, testing, and basic circuit applications. In this chapter we will examine another type of transistor called the *field effect transistor,* which is more commonly called an FET (pronounced "eff-ee-tee"). Like the BJT, the FET has three terminals and can operate as a switch and can be used in digital circuit applications. It can also operate as a variable resistor and be used in analog or linear circuit applications. In fact, as we step through this chapter, you will see many similarities between the BJT and FET. You will also notice a few distinct differences between these two transistor types, and these differences are what make the BJT ideal in some applications and the FET ideal in other applications.

There are two types of field effect transistors or FETs. One type is the *junction field effect transistor,* which is more typically called a JFET (pronounced "jay-fet"). The other type is the *metal oxide semiconductor field effect transistor,* which is more commonly called a MOSFET (pronounced "moss-fet"). In this chapter we will examine the operation, characteristics, applications, and testing of these two types of field effect transistors.

18-1 JUNCTION FIELD EFFECT TRANSISTOR (JFET)

Like the bipolar junction transistor, the **junction field effect transistor** or **JFET** is constructed from *n*-type and *p*-type semiconductor materials. However, the JFET's construction is very different from the BJT's construction, and therefore we will need to first see how the JFET device is built before we can understand how it operates.

18-1-1 *JFET Construction*

Just as the bipolar junction transistor has two basic types (NPN BJT or PNP BJT), there are two types of junction field effect transistor called the **n-channel JFET** and **p-channel JFET.** The construction and schematic symbols for these two JFET types are shown in Figure 18-1.

To begin with, let us examine the construction of the more frequently used *n*-channel JFET, shown in Figure 18-1(a). This type of JFET basically consists of an *n*-type block of semiconductor material on top of a *p*-type substrate, with a "U" shaped *p*-type section attached to the surface of a *p*-type substrate. Like the BJT, the JFET has three terminals called

---

TIME LINE

In 1961, Steven Hofstein devised the field-effect transistor used in MOS (metal oxide semiconductor) integrated circuits.

Junction Field Effect Transistor (JFET)

A field-effect transistor made up of a gate region diffused into a channel region. When a control voltage is applied to the gate, the channel is depleted or enhanced, and the current between source and drain is thereby controlled.

---

n-Channel JFET

A junction field effect transistor having an *n*-type channel between source and drain.

---

p-Channel JFET

A junction field effect transistor having a *p*-type channel between source and drain.

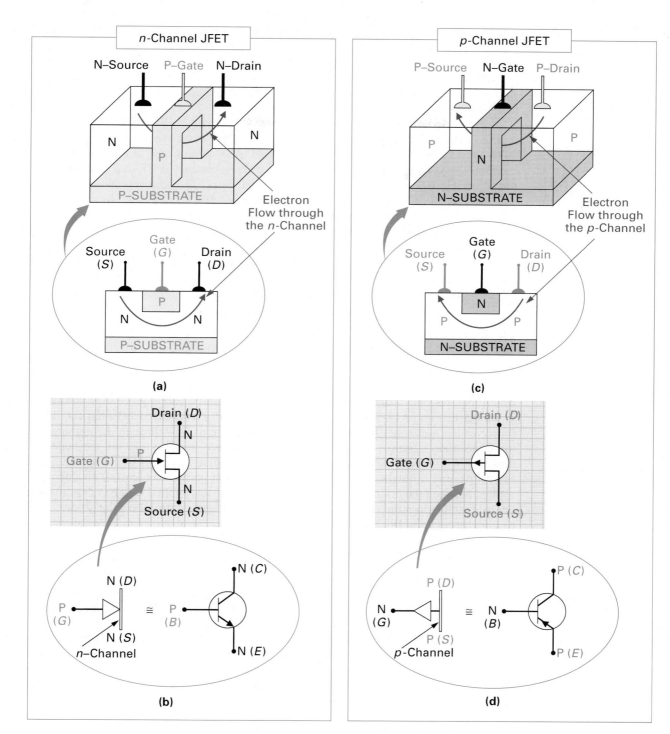

FIGURE 18-1 **The Junction Field Effect Transistor (JFET) Types.**

the **gate, source,** and **drain.** The gate lead is attached to the *p*-type substrate, and the source and drain leads are attached to either end of an *n*-type channel that runs through the middle of the "U" shaped *p*-type section. In the simplified two-dimensional view in the inset in Figure 18-1(a), you can see the *n*-type **channel** that exists between the *n*-channel JFET's source and drain. The schematic symbol for the *n*-channel JFET is shown in Figure 18-1(b), and, as you can see, the gate lead's arrowhead points into the device. To aid your memory, you can imagine this arrowhead as a P-N junction diode as shown in the inset in Figure 18-1(b). The gate lead is connected to the diode's anode, which is a *p*-type material, and the source and drain leads are connected to either end of the diode's cathode, which is an *n*-type material. Because the source-to-drain channel is made from an *n*-type material, this is an *n*-channel JFET. The *p*-type gate and *n*-type source and drain makes this *n*-channel JFET equivalent to an NPN BJT, which has a *p*-type base and *n*-type emitter and collector, as shown in the inset in Figure 18-1(b).

Figure 18-1(c) shows the construction of the *p*-channel JFET. This JFET type is constructed in exactly the same way as the *n*-channel JFET except that the gate lead is attached to an *n*-type substrate, and the source and drain leads are attached to either end of a *p*-type channel. Looking at the schematic symbol for the *p*-channel JFET in Figure 18-1(d), you can see that the gate's arrowhead points out of the device. To help you distinguish the symbols used for the *n*-channel JFET from the *p*-channel JFET, once again imagine the arrowhead as a P-N junction diode as shown in the inset in Figure 18-1(d). Because the gate lead is connected to the diode's cathode, the gate must be an *n*-type material. Because the source and drain leads are connected to either end of the diode's anode, the source-to-drain channel is therefore made from a *p*-type material, and this is a *p*-channel JFET. The *n*-type gate and *p*-type source and drain makes this *p*-channel JFET equivalent to a PNP BJT, which has an *n*-type base and *p*-type emitter and collector, as shown in the inset in Figure 18-1(d).

18-1-2   *JFET Operation*

As you know, an NPN bipolar transistor needs both a collector supply voltage ($+V_{CC}$) and a base-emitter bias voltage (V_{BE}) in order to operate correctly. The same is true for the JFET, which requires both a **drain supply voltage ($+V_{DD}$)** and a **gate-source bias voltage (V_{GS})**, as shown in Figure 18-2(a). The $+V_{DD}$ bias voltage is connected between the drain and source of the *n*-channel JFET and will cause a current to flow through the *n*-channel. This source-to-drain current—which is made up of electrons because they are the majority carriers within an *n*-type material—is called the JFET's **drain current (I_D)**. The value of drain current passing through a JFET's channel is dependent on two elements: the value of $+V_{DD}$ applied between the drain and source, and the value of V_{GS} applied between gate and source. Let us examine in more detail why these applied voltages control the value of drain current passing through the JFET's channel.

The Relationship between $+V_{DD}$ and I_D

The value of $+V_{DD}$ controls the amount of drain current between source and drain because it is this supply voltage that controls the potential difference applied across the channel, as seen in Figure 18-2(a). Therefore, an increase in the voltage applied across the JFET's drain and source ($+V_{DD}\uparrow$) will increase the amount of drain current ($I_D\uparrow$) passing through the channel. Similarly, a decrease in the voltage applied across the JFET's drain and source ($+V_{DD}\downarrow$) will decrease the amount of drain current ($I_D\downarrow$) passing through the channel. In most cases, a schematic diagram will show the V_{DD} supply voltage connection to the JFET as a source connection to ground and a drain connection up to $+V_{DD}$, as shown in Figure 18-2(b).

The Relationship between V_{GS} and I_D

The value of V_{GS} controls the amount of drain current between source and drain because it is this voltage that controls the resistance of the channel. Figure 18-3(a) shows that when the V_{GS} bias voltage is 0 volts (which is the same as connecting the gate to ground), there is no potential difference between the gate and source. With no bias voltage applied, a small depletion layer will form and spread into the channel. Although it appears as though two depletion regions exist, in fact they are both part of the same depletion region that extends around the wall of the *n*-channel. This extremely small depletion region will offer very little opposition to I_D, and so drain current will be large. In Figure 18-3(b), V_{GS} is increased to -2 V (made more negative), and therefore the gate-to-source P-N junction will

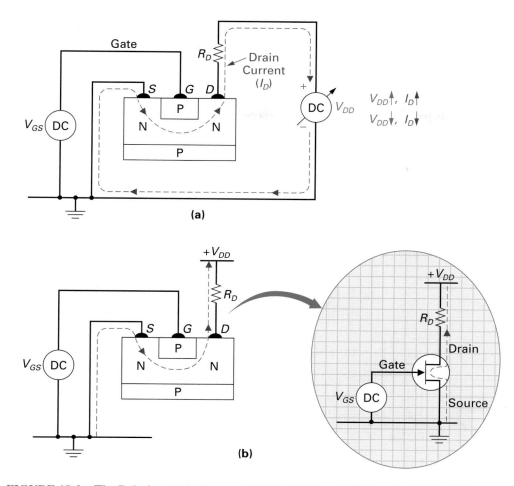

FIGURE 18-2 The Relationship between a JFET's DC Supply Voltage ($+V_{DD}$) and Output Current (I_D).

be further reverse biased. This causes an increase in the depletion region, decrease in the channel's width, and a decrease in drain current. In Figure 18-3(c), V_{GS} is increased to -4 V, and therefore the gate-to-source P-N junction will be further reverse biased, causing an increase in the depletion region, decrease in the channel's width, and a further decrease in drain current.

In most circuit applications, the $+V_{DD}$ supply voltage is maintained constant and the V_{GS} input voltage is used to control the resistance of the channel and the value of the output current, I_D. This can be seen more clearly in the insets in Figure 18-3. Because the output voltage (V_D or V_{out}) is dependent on the resistance between the JFET's source and drain, by controlling the value of I_D, we can control the output voltage. For instance, if the input voltage V_{GS} is made more negative, the resistance of the channel will be increased, causing the output current I_D to decrease, and therefore the voltage developed between drain and ground (V_D or V_{out}) to increase. The gate-to-source junction of an FET is normally always reverse biased by the input voltage (V_{GS}), and it is this input voltage that controls the output current (I_D) and output voltage (V_{out}). Because the gate-to-source junction of an FET is normally always reverse biased by the input voltage (V_{GS}), there will be no input current. This characteristic accounts for the FET's naturally high input impedance. The operation of an FET is very different from the BJT, which normally uses an input voltage to forward bias the base-to-emitter junction and vary the input current (which varies the output current), and therefore the output voltage. This is the distinct difference between an FET and a BJT. An FET's input junction is normally reverse biased, and therefore the input voltage controls the output current. A BJT's input junction is normally forward biased and therefore the input current

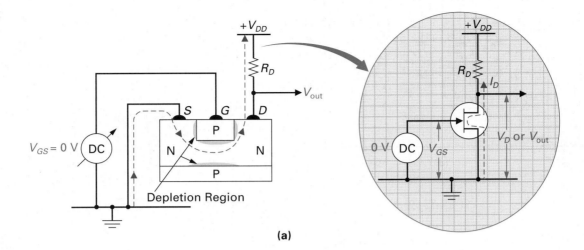

(a)

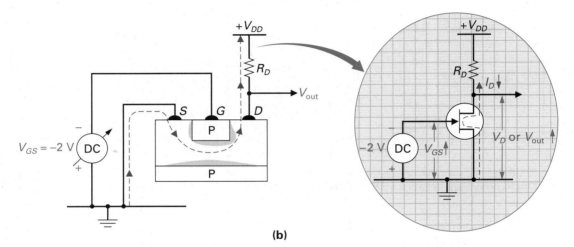

(b)

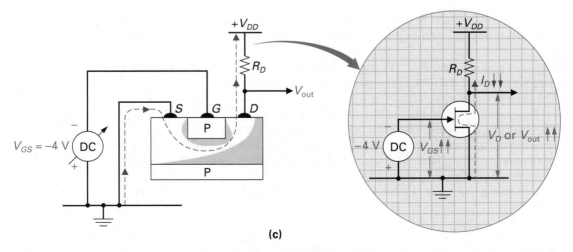

(c)

FIGURE 18-3 **The Relationship between a JFET's Input Voltage (V_{GS}) and Output Current (I_D).**

controls the output current. This difference is why BJTs are known as **current-controlled devices** and FETs are known as **voltage-controlled devices.** In fact, the name "field effect transistor" is derived from this voltage-control action because the applied input voltage will generate an electric field. It is this electric field that varies the size of the depletion region, and therefore the resistance of the channel between the FET's drain and source output terminals. In other words, the "effect" of the electric "field" causes "transistance," which is the transferring of different values of resistance between the output terminals.

The term "junction" is attached to this type of FET because of the single P-N junction formed between the gate and the source-to-drain channel. Therefore, an *n*-channel JFET has a single P-N junction between gate to channel, and a *p*-channel JFET has a single N-P junction between gate to channel.

The field effect transistor is also often referred to as a **unipolar device** because only one type of semiconductor material exists between the output terminals (*n*-type or *p*-type channel between source and drain), and therefore the charge carriers have only one polarity (unipolar). Compare this to a BJT, which is a **bipolar device** because there is a change in semiconductor material between the output terminals (NPN or PNP between emitter and collector), and the charge carriers can be one of two polarities (bipolar, because both majority and minority carriers are used).

18-1-3 *JFET Characteristics*

Like the BJT, the JFET's response to certain variables is best described by using a graph. Figure 18-4(a) shows a graph plotting drain current (I_D) against drain-to-source voltage (V_{DS}). As you have probably already observed, this **drain characteristic curve** is very similar to a bipolar transistor's collector characteristic curve. Starting at 0 V and moving right along the horizontal axis, you can see that an increase in the drain supply voltage ($+V_{DD}$), and therefore an increase in V_{DS}, will result in a continual increase in I_D. At a certain V_{DS} voltage (in this example 5 V), further increases in V_{DS} will cause no further increase in I_D. This value of V_{DS} is called the **pinch-off voltage (V_P)** because it is the point at which the bias voltage has caused the depletion region to pinch off or restrict drain current. From this point on, further increases in V_{DS} are counteracted by increases in the resistance of the channel, and therefore I_D remains constant. This is shown by the flat portion of the graph in Figure 18-4(a) and is called the **constant-current region** because I_D remains constant despite changes in V_{DS}. If V_{DS} is further increased (by increasing $+V_{DD}$), the JFET will eventually reach its **breakdown voltage (V_{BR}),** at which time a damaging value of I_D will pass through the JFET.

In the example graph in Figure 18-4(a), we plotted what would happen to I_D as V_{DS} increased with V_{GS} at 0 V. In Figure 18-4(b) we will examine what will happen to an *n*-channel JFET when the gate-source junction is reverse biased by several negative voltages. As previously described in the JFET operation section, a negative voltage is normally applied to reverse bias the gate and set up a depletion region. As V_{GS} is made more negative, the gate will be further reverse biased, and the corresponding I_D value will be smaller. Therefore, when V_{GS} is at 0 V, a maximum value of drain current is passing through the JFET's channel. This maximum value of drain current is called the **drain-to-source current with shorted gate (I_{DSS}).** This name is derived from the fact that when $V_{GS} = 0$ V, as shown in the inset in Figure 18-4(a), the gate and source terminals of the JFET are at the same potential of zero volts, and therefore the gate is effectively shorted to the source as shown by the dashed line. The drain-to-source current with shorted gate (I_{DSS}) rating is therefore the maximum current that can pass through the channel of a given JFET. When given on a specification sheet, this rating is equivalent to a bipolar transistor's $I_{C(Sat.)}$ rating.

Returning to Figure 18-4(b), you can see that if V_{GS} is made more negative, the depletion regions within the JFET will get closer and closer and eventually touch, cutting off drain current. This negative V_{GS} bias voltage that causes I_D to drop to approximately zero is called the **gate-to-source cutoff voltage or $V_{GS(OFF)}$.** In the example in Figure 18-4(b), when $V_{GS} = -5$ V, I_D is almost zero and therefore $V_{GS(OFF)} = -5$ V. When cut OFF, the

Current-Controlled Device

A device in which the input junction is normally forward biased and the input current controls the output current.

Voltage-Controlled Device

A device in which the input junction is normally reverse biased and the input voltage controls the output current.

Unipolar Device

A device in which only one type of semiconductor material exists between the output terminals and therefore the charge carriers have only one polarity (unipolar).

Bipolar Device

A device in which there is a change in semiconductor material between the output terminals (NPN or PNP between emitter and collector), so the charge carriers can be one of two polarities (bipolar).

Drain Characteristic Curve

A plot of the drain current (I_D) versus the drain-to-source voltage (V_{DS}).

Pinch-Off Voltage (V_P)

The value of V_{DS} at which further increases in V_{DS} will cause no further increase in I_D.

Constant-Current Region

The flat portion of the drain characteristic curve. In this region I_D remains constant despite changes in V_{DS}.

Breakdown Voltage (V_{BR})

The voltage at which a damaging value of I_D will pass through the JFET.

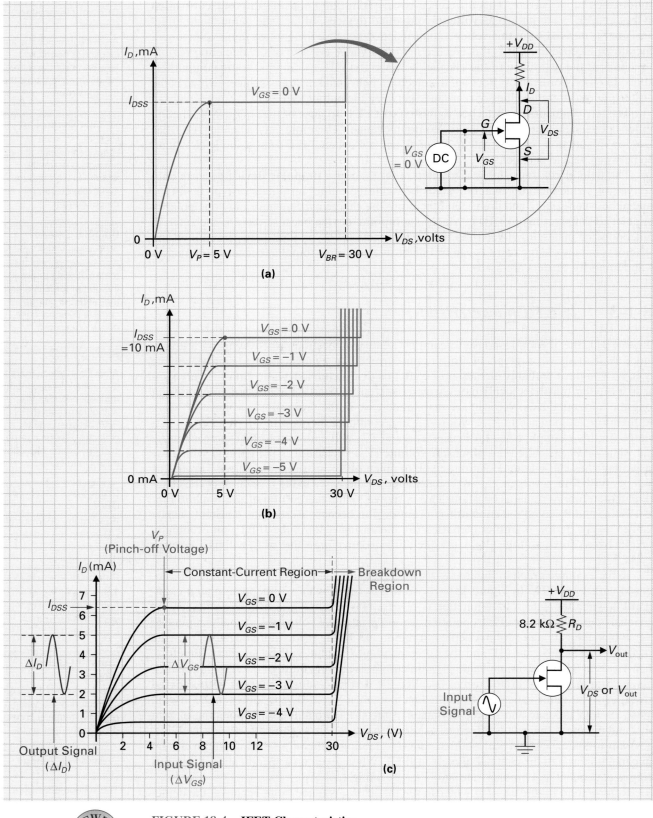

(a)

(b)

(c)

 FIGURE 18-4   JFET Characteristics.

JFET will be equivalent to an open circuit between drain and source, and subsequently all of the drain supply voltage (V_{DD}) will appear across the open JFET ($V_{DS} = V_{DD}$).

To summarize the specifications in Figure 18-4(b),

When $V_{GS} = 0$ V, $I_D = I_{DSS} = 10$ mA

$V_P = 5$ V

$V_{BR} = 30$ V

$V_{GS(OFF)} = -5$ V

Constant-Current Region $= V_P$ to $V_{BR} = 5$ V to 30 V

Drain-to-Source Current with Shorted Gate (I_{DSS})
The maximum value of drain current, achieved by holding V_{GS} at 0 V.

Gate-to-Source Cutoff Voltage or $V_{GS(OFF)}$
The negative V_{GS} bias voltage that causes I_D to drop to approximately zero.

18-1-4 *Transconductance*

Figure 18-4(c) illustrates a JFET test circuit and its associated characteristic graph. Before we see how a JFET can be made to amplify, let's first summarize the details given in this graph. If we first consider the curve when $V_{GS} = 0$ V, you can see that up to V_P, I_D increases in almost direct proportion to V_{DS}. This is because the depletion region is not sufficiently large enough to affect I_D, so the channel is simply behaving as a semiconductor with a fixed resistance value between source and drain.

When V_{DS} is equal to V_P, the drain current (I_D) will be pinched into an extremely narrow channel between the wedge-shaped depletion region. Any further increase in V_{DS} will have two effects:

1. increase the pinching effect on the channel, which will resist current flow, and
2. increase the potential between the drain and source, which will encourage current flow.

The net result is that channel resistance increases in direct proportion with V_{DS} and consequently I_D remains constant, as shown by the flat portion of the characteristic curve.

Assuming a fixed value of V_{DD}, any increase in the negative voltage of V_{GS} will cause a corresponding decrease in I_D. Therefore, beyond V_P, I_D is controlled by small-signal changes (such as the input signal) in V_{GS} and is independent of changes in V_{DS}. This section of the curve between V_P and V_{BR} is called the constant-current region.

Like the bipolar transistor, an FET can be used to amplify a signal, as shown in Figure 18-4(c). As before, the amount of amplification achieved is a ratio between output and input. For a bipolar transistor, the amount of gain is equal to the ratio of input current to output current (beta). For an FET, there is no input current, and therefore an FET's gain is equal to the ratio of output current change (ΔI_D) to input voltage change (ΔV_{GS}). This ratio is called the FET's **transconductance** (symbolized δ_m).

$$\delta_m = \frac{\Delta I_D}{\Delta V_{GS}}$$

Transconductance
Also called mutual conductance, it is the ratio of a change in output current to the initiating change in input voltage.

$\delta_m =$ transconductance in siemens (S)

$\Delta I_D =$ change in drain current

$\Delta V_{GS} =$ change in gate-source voltage

EXAMPLE:

Calculate the transconductance of the FET for the example shown in Figure 18-4(c).

Solution:

$$\delta_m = \frac{\Delta I_D}{\Delta V_{GS}} = \frac{5 \text{ mA} - 2 \text{ mA}}{-1 \text{ V} - (-3 \text{ V})} = \frac{3 \text{ mA}}{2 \text{ V}} = 1.5 \text{ millisiemens}$$

A high-gain FET will produce a large change in I_D for a small change in V_{GS}, resulting in a high transconductance figure ($\delta_m \uparrow$).

18-1-5 *Voltage Gain*

Because transconductance is the ratio of output current change (ΔI_D) to input voltage change (ΔV_{GS}), it is no surprise that this ratio is used to determine a JFET's voltage gain. The voltage gain formula is as follows

$$A_V = \delta_m \times R_D$$

▧ EXAMPLE:

Calculate the voltage gain for the circuit example shown in Figure 18-4(c).

▧ *Solution:*

$$A_V = \delta_m \times R_D = 1.5\ \text{mS} \times 8.2\ \text{k}\Omega = 12.3$$

This means that the output voltage will be 12.3 times greater than the input voltage.

18-1-6 *JFET Data Sheet*

Throughout this section, we have used certain JFET specifications in our calculations, such as $V_{GS(OFF)}$ and I_{DSS}. As an example, Figure 18-5 shows the data sheet for a typical *n*-channel JFET. As before, notes have been inserted within the data sheet to describe any confusing ratings; however, most of these ratings are self-explanatory.

18-1-7 *JFET Biasing*

The biasing methods used in FET circuits are very similar to those employed in BJT circuits. In this section we will examine the circuit calculations for the three most frequently used JFET biasing methods: gate biasing, self biasing and voltage-divider biasing.

Gate Biasing

Figure 18-6(a) shows a gate biased JFET circuit. The gate supply voltage ($-V_{GG}$) is used to reverse bias the gate-source junction of the JFET. With no gate current, there can be no voltage drop across R_G and therefore the voltage at the gate of the JFET will equal the dc gate supply voltage.

$$V_{GS} = V_{GG}$$

In the example in Figure 18-6,

$$V_{GS} = V_{GG} = -1.5\ \text{V}$$

Knowing V_{GS}, we can calculate I_D if the JFET's current (I_{DSS}) and voltage ($V_{GS(OFF)}$) specification limits are known by using the following formula.

$$I_D = I_{DSS}\left(1 - \frac{V_{GS}}{V_{GS(OFF)}}\right)^2$$

DEVICE: 2N5484 Through 2N5486—N-Channel JFET

Most of these specifications are self-explanatory. $V_{GS\,(OFF)}$ which has been given throughout is listed in the OFF characteristics, while I_{DSS} is listed in the ON characteristics. The only confusing maximum rating is "Forward Gate Current" because the gate is never normally forward biased (V_{GS} = a negative voltage or zero volts). This rating indicates that if the gate accidentally becomes forward biased, gate current must not exceed 10 mA dc or the JFET will be destroyed.

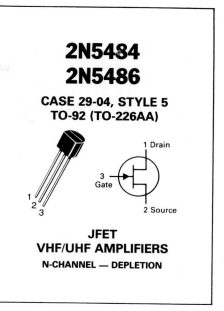

2N5484
2N5486

CASE 29-04, STYLE 5
TO-92 (TO-226AA)

JFET
VHF/UHF AMPLIFIERS
N-CHANNEL — DEPLETION

MAXIMUM RATINGS

Rating	Symbol	Value	Unit
Drain-Gate Voltage	V_{DG}	25	Vdc
Reverse Gate-Source Voltage	V_{GSR}	25	Vdc
Drain Current	I_D	30	mAdc
Forward Gate Current	$I_{G(f)}$		mAdc
Total Device Dissipation @ T_C = 25°C Derate above 25°C	P_D	310 2.82	mW mW/°C
Operating and Storage Junction Temperature Range	T_J, T_{stg}	−65 to +150	°C

ELECTRICAL CHARACTERISTICS (T_A = 25°C unless otherwise noted.)

Characteristic		Symbol	Min	Typ	Max	Unit
OFF CHARACTERISTICS						
Gate-Source Breakdown Voltage (I_G = −1.0 μAdc, V_{DS} = 0)		$V_{(BR)GSS}$	−25	—	—	Vdc

Gate Reverse Current
  (V_{GS} = −20 Vdc, V_{DS} = 0)
  (V_{GS} = −20 Vdc, V_{DS} = 0, T_A = 100°C)
		I_{GSS}	
			—
			—
				
				—
				—
					
					−1.0
					−0.2
						
						nAdc
						μAdc

Gate Source Cutoff Voltage (V_{DS} = 15 Vdc, I_D = 10 nAdc)	2N5484 2N5485 2N5486	$V_{GS(off)}$	 −0.3 −0.5 −2.0	 — — —	 −3.0 −4.0 −6.0	Vdc
ON CHARACTERISTICS						
Zero-Gate-Voltage Drain Current (V_{DS} = 15 Vdc, V_{GS} = 0)	2N5484 2N5485 2N5486	I_{DSS}	 1.0 4.0 8.0	 — — —	 5.0 10 20	mAdc

$V_{GS(OFF)}$

I_{DSS}

FIGURE 18-5 **Data Sheet for an *n*-Channel JFET. (Copyright of Motorola. Used by permission.)**

In the example in Figure 18-6,

$$I_D = 20\text{ mA}\left(1 - \frac{-1.5\text{ V}}{-4\text{ V}}\right)^2$$
$$= 20\text{ mA }(1 - 0.375)^2$$
$$= 20\text{ mA} \times 0.625^2$$
$$= 20\text{ mA} \times 0.39 = 7.8\text{ mA}$$

Now that I_D is known, we can calculate the voltage drop across R_D using Ohm's law.

$$V_{R_D} = I_D \times R_D$$

In the example in Figure 18-6,

$$V_{R_D} = I_D \times R_D = 7.8\text{ mA} \times 1\text{ k}\Omega = 7.8\text{ V}$$

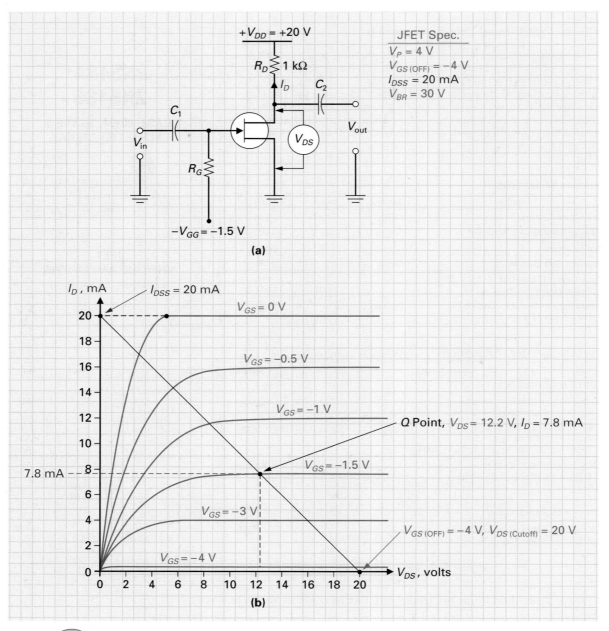

FIGURE 18-6 **A Gate-Biased JFET Circuit. (a) Basic Circuit. (b) Drain Characteristic Curves and DC Load Line.**

Because V_{R_D} plus V_{DS} will equal V_{DD}, we calculate V_{DS} once V_{R_D} is known with the following formula.

$$V_{DS} = V_{DD} - V_{R_D}$$

In the example in Figure 18-6,

$$V_{DS} = V_{DD} - V_{R_D} = 20\,\text{V} - 7.8\,\text{V} = 12.2\,\text{V}$$

Figure 18-6(b) shows the drain characteristic curves and dc load line for the example JFET circuit in Figure 18-6(a). Like the bipolar transistor, the JFET's dc load line extends between the maximum output current point, or saturation point (when the JFET is fully ON,

I_{DSS} = 20 mA), to the maximum output voltage point (when the JFET is cut OFF, V_{DS} = 20 V). The dc operating point, or Q point, which was determined with the previous calculations, is also plotted on the dc load line in Figure 18-6(b).

EXAMPLE:

Calculate the following for the circuit shown in Figure 18-7.

a. V_{GS}

b. I_D

c. V_{DS}

d. Maximum value of I_D

e. V_{DS} when $V_{GS} = V_{GS(OFF)}$

f. Q point

Solution:

a. $V_{GS} = V_{GG} = -3$ V

b. $I_D = I_{DSS}\left(1 - \dfrac{V_{GS}}{V_{GS(OFF)}}\right)^2 = 15 \text{ mA}\left(1 - \dfrac{-3 \text{ V}}{-6 \text{ V}}\right)^2 = 15 \text{ mA } (1 - 0.5)^2 = 3.75 \text{ mA}$

c. $V_{DS} = V_{DD} - V_{R_D}$ (since $V_{R_D} = I_D \times R_D$, we can substitute)

$V_{DS} = V_{DD} - (I_D \times R_D) = 15 \text{ V} - (3.75 \text{ mA} \times 1.2 \text{ k}\Omega) = 15 \text{ V} - 4.5 \text{ V} = 10.5 \text{ V}$

d. Maximum value of $I_D = I_{DSS} = 15$ mA

e. When $V_{GS} = V_{GS(OFF)}$, the JFET is cut off and equivalent to an open switch between drain and source. In this condition all of the drain supply voltage will appear across the open JFET.

$$V_{DS(Cutoff)} = V_{DD} = 15 \text{ V}$$

f. The dc operating or Q point is

$$V_{GS} = -3 \text{ V}$$
$$I_D = 3.75 \text{ mA}$$
$$V_{DS} = 10.5 \text{ V}$$

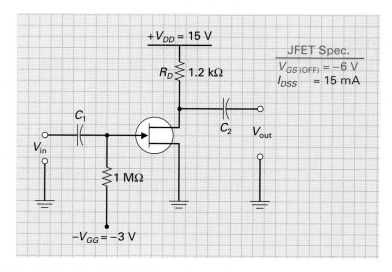

FIGURE 18-7   A Gate Biased JFET Circuit Example.

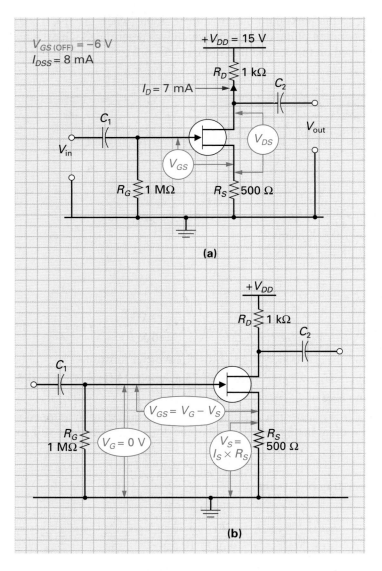

FIGURE 18-8 **A Self Biased JFET Circuit. (a) Basic Circuit. (b) How R_S Develops a $-V_{GS}$.**

Self Biasing

Figure 18-8(a) shows how to self bias a JFET circuit. One advantage of this biasing method over gate biasing is that only a single drain supply voltage is needed (V_{DD}) instead of both V_{DD} and a negative gate supply voltage ($-V_{GG}$). The other difference you may have noticed is that a source resistor (R_S) has been included, and R_G has been connected to ground. Although this arrangement seems completely different to the gate bias circuit, the inclusion of R_S and the grounding of R_G will achieve the same result, which is to reverse bias the JFET's gate-source junction. Figure 18-8(b) illustrates how this is achieved. Since there is no gate current in a JFET circuit ($I_G = 0$), all of the current flowing into the source will travel through the channel and flow out of the drain. Therefore

$$I_S = I_D$$

For the example in Figure 18-8,

$$I_S = I_D = 7 \text{ mA}$$

Now that I_S is known, we can calculate the voltage drop across the source resistor (V_{R_S}), and therefore the voltage at the JFET's source (V_S).

$$V_{R_S} = V_S = I_S \times R_S$$

For the example in Figure 18-8,

$$V_{RS} = V_S = I_S \times R_S = 7\text{ mA} \times 500\ \Omega = +3.5\text{ V}$$

Because $I_G = 0$ A, there will be no voltage drop across R_G, and so the voltage at the gate of the JFET will be 0 V.

$$V_G = 0\text{ V}$$

Now that we know that $V_S = +3.5$ V and $V_G = 0$ V, we can see how the JFET's gate-source junction is reverse biased. To reverse bias a gate biased JFET, we simply made the gate voltage negative with respect to the source that is at 0 V. With a self biased JFET, we achieve the same result by making the source voltage positive with respect to the gate that is at 0 V. This makes the gate of the JFET negative with respect to the source. This potential difference from gate-to-source (V_{GS}) is therefore equal to

$$V_{GS} = V_G - V_S$$

Because $V_S = I_S \times R_S$ and $V_G = 0$ V, we can substitute the previous formula to obtain

$$V_{GS} = 0\text{ V} - (I_S \times R_S)$$

or

$$V_{GS} = -(I_S \times R_S)$$

or because $I_S = I_D$

$$V_{GS} = -(I_D \times R_S)$$

In the example in Figure 18-8,

$$\begin{aligned} V_{GS} &= -(I_S \text{ or } I_D \times R_S) \\ &= -(7\text{ mA} \times 500\ \Omega) \\ &= -3.5\text{ V} \end{aligned}$$

If $-V_{GS}$ and R_S are known, we could transpose the above equation to calculate I_D.

$$I_D = \frac{V_{GS}}{R_S}$$

In the example in Figure 18-8,

$$I_D = \frac{V_{GS}}{R_S} = \frac{3.5\text{ V}}{500\ \Omega} = 7\text{ mA}$$

The final calculation is to determine the voltage at the JFET's drain with respect to ground (V_D) and the drain-to-source voltage drop across the JFET.

$$V_D = V_{DD} - V_{R_D}$$

Because $V_{R_D} = I_D \times R_D$,

$$V_D = V_{DD} - (I_D \times R_D)$$

In the example in Figure 18-8,

$$V_D = V_{DD} - (I_D \times R_D) = 15\,\text{V} - (7\,\text{mA} \times 1\,\text{k}\Omega) = 15\,\text{V} - 7\,\text{V} = 8\,\text{V}$$

Now that the voltage drops across R_D (V_{R_D}) and R_S (V_{R_S}) are known, we can calculate the voltage drop across the JFET's drain to source (V_{DS}).

$$V_{DS} = V_{D_D} - (V_{R_D} + V_{R_S})$$

In the example in Figure 18-8,

$$V_{DS} = V_{DD} - (V_{R_D} + V_{R_S}) = 15\,\text{V} - (7\,\text{V} + 3.5\,\text{V}) = 15\,\text{V} - 10.5\,\text{V} = 4.5\,\text{V}$$

■ EXAMPLE:

Calculate the following for the circuit shown in Figure 18-9.

a. V_S
b. V_{GS}
c. V_{DS}
d. I_D maximum
e. V_{DS} when the JFET is OFF
f. V_D

■ Solution:

a. Because $I_S = I_D$, $V_S = I_S \times R_S = 4\,\text{mA} \times 500\,\Omega = 2\,\text{V}$
b. $V_{GS} = V_G - V_S = 0\,\text{V} - 2\,\text{V} = -2\,\text{V}$
c. $V_{DS} = V_{DD} - (V_{R_D} + V_{R_S}) = 10\,\text{V} - [(I_D R_D) + 2\,\text{V}]$
   $= 10\,\text{V} - [(4\,\text{mA} \times 1.2\,\text{k}\Omega) + 2\,\text{V}] = 10\,\text{V} - (4.8\,\text{V} + 2\,\text{V})$
   $= 10\,\text{V} - 6.8\,\text{V} = 3.2\,\text{V}$

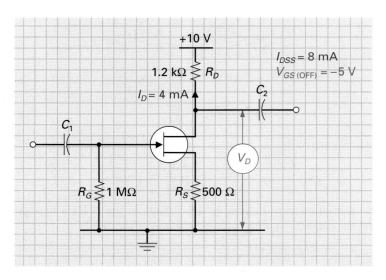

FIGURE 18-9 **A Self Biased JFET Circuit Example.**

CHAPTER 18 / FIELD EFFECT TRANSISTORS (FETs)

d. I_D maximum $= I_{DSS} = 8$ mA

e. $V_{DS(Cutoff)} = V_{DD} = 10$ V

f. $V_D = V_{DS} + V_{R_S} = 3.2$ V $+ 2$ V $= 5.2$ V

As previously mentioned, one advantage of this self biased JFET method is that only a drain supply voltage is needed (V_{DD}). The gate supply voltage (V_{GG}) is not needed due to the inclusion of a source resistor that reverse biases the JFET's gate-source junction by applying a positive voltage to the source with respect to the 0 V on the gate. This method of effectively sending back a negative voltage from the source to the gate is known as "negative feedback." It not only enables us to bias a JFET with one supply voltage, it also provides temperature stability. Any change in the ambient temperature will cause a change in the semiconductor JFET's conduction, which would move the JFET's Q point away from its desired setting. The inclusion of R_S will prevent the Q point from shifting due to temperature in the same way as a BJT's emitter resistor. If temperature were to increase, for instance ($T\uparrow$), the resistance of the semiconductor would decrease ($R\downarrow$) because all semiconductor materials have a negative temperature coefficient of resistance ($T\uparrow, R\downarrow$), and this will cause the channel current to increase. If the drain current increases ($I_D\uparrow$), the voltage drop across R_S will increase (V_{R_S} or $V_S\uparrow = I_D\uparrow \times R_S$). This increase in V_S will increase the gate-source reverse voltage ($-V_{GS}\uparrow$), causing the JFET's channel to get narrower and the drain current to decrease ($I_D\downarrow$) and counteract the original increase. Similarly, a decrease in temperature will cause a decrease in I_D, which will decrease the gate-source reverse bias, resulting in an increase in I_D. The Q point will remain relatively stable despite changes in temperature when a JFET circuit has a source resistor included.

Voltage-Divider Biasing

Referring to the voltage-divider biased JFET circuit shown in Figure 18-10, you will probably notice that it is very similar to the voltage-divider biased BJT circuit discussed previously. Like the self biased circuit, the inclusion of a source resistor stabilizes the Q point despite ambient temperature changes. In addition, using a voltage divider to determine the gate-source bias voltage ensures that V_{GS}, and therefore the circuit, has increased stability.

The gate voltage (V_G) is calculated using the following voltage-divider formula:

$$V_{R_2} \text{ or } V_G = \frac{R_2}{R_1 + R_2} \times V_{DD}$$

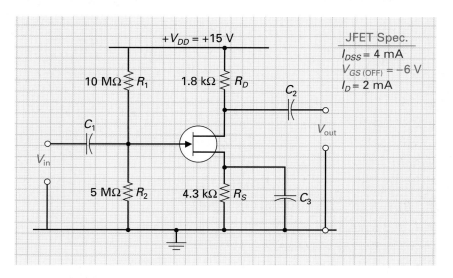

FIGURE 18-10 A Voltage-Divider Biased JFET Circuit.

For the example in Figure 18-10,

$$V_{R_2} \text{ or } V_G = \frac{R_2}{R_1 + R_2} \times V_{DD} = \frac{5 \text{ M}\Omega}{10 \text{ M}\Omega + 5 \text{ M}\Omega} \times 15 \text{ V} = 5 \text{V}$$

Because $I_D = I_S$ ($I_G = 0$) and the drain resistance and current are known, we can next calculate the voltage drop across the source resistor (V_{R_S}), drain resistor (V_{R_D}), and the JFET's source-drain junction (V_{DS}).

$$I_S = I_D$$

$$I_S = I_D = 2 \text{ mA}$$

$$V_{R_S} = I_S \times R_S$$

$$V_{R_S} = 2 \text{ mA} \times 4.3 \text{ k}\Omega = 8.6 \text{ V}$$

$$V_{R_D} = I_D \times R_D$$

$$V_{R_D} = 2 \text{ mA} \times 1.8 \text{ k}\Omega = 3.6 \text{ V}$$

$$V_{DS} = V_{DD} - (V_{R_S} + V_{R_D})$$

$$V_{DS} = 15 \text{ V} - (8.6 \text{ V} + 3.6 \text{ V}) = 2.8 \text{ V}$$

Now that the JFET's gate and source voltages are known (V_G and V_S), we can calculate the value of gate-source reverse bias ($-V_{GS}$).

$$V_{GS} = V_G - V_S$$

$$(V_G = V_{R_2}, V_S = V_{R_S})$$

For the example in Figure 18-10,

$$V_{GS} = 5 \text{ V} - 8.6 \text{ V} = -3.6 \text{ V}$$

■ EXAMPLE:

Calculate the following for the voltage-divider biased JFET circuit shown in Figure 18-11:

a. V_G

b. I_S

c. V_S

d. V_{DS}

e. V_{GS}

f. V_D, when $V_{GS} = V_{GS(OFF)}$

g. I_D, when $V_{GS} = 0$ V

■ Solution:

a. $V_G = \dfrac{R_2}{R_1 + R_2} \times V_{DD} = \dfrac{10 \text{ M}\Omega}{100 \text{ M}\Omega + 10 \text{ M}\Omega} \times 30 \text{ V} = 2.7 \text{ V}$

b. $I_S = I_D = 3.6 \text{ mA}$

c. $V_S = V_{R_S} = I_S \times R_S = 3.6 \text{ mA} \times 2.7 \text{ k}\Omega = 9.7 \text{ V}$

d. $V_{DS} = V_{DD} - (V_{R_S} + V_{R_D}) = 30 \text{ V} - [9.7 \text{ V} + (I_D \times R_D)]$

$= 30 \text{ V} - [9.7 \text{ V} + (3.6 \text{ mA} \times 5 \text{ k}\Omega)] = 30 \text{ V} - (9.7 \text{ V} + 18 \text{ V}) = 2.3 \text{ V}$

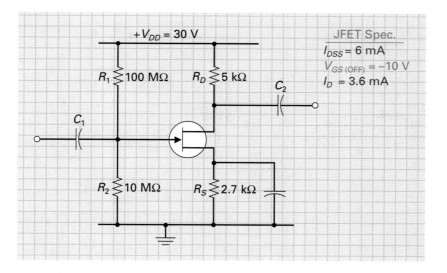

FIGURE 18-11   A Voltage-Divider Biased Circuit Example.

e. $V_{GS} = V_G - V_S = 2.7\,\text{V} - 9.7\,\text{V} = -7\,\text{V}$

f. When $V_{GS} = V_{GS(OFF)}$, JFET is OFF and $V_D = V_{DD} = 30\,\text{V}$

g. When $V_{GS} = 0\,\text{V}$, $I_D = \text{maximum} = I_{DSS} = 6\,\text{mA}$

18-1-8   *JFET Circuit Configurations*

The three JFET circuit configurations are illustrated in Figure 18-12 along with their typical circuit characteristics. Like the bipolar transistor configurations, the term "common" is used to indicate which of the JFET's leads is common to both the input and output. In this section we will examine the characteristics of these three configurations: common-source, common-gate, and common-drain.

Common-Source (*C-S*) Circuits

Similar to its bipolar counterpart, the common-emitter configuration, the **common-source configuration** is the most widely used JFET circuit and is detailed in Figure 18-12(a). The input is applied between the gate and source and the output is taken between the drain and source, with the source being common to both input and output. The ac input will pass through the coupling capacitor C_1 and be superimposed on the dc gate-source bias voltage provided by resistor R_1, which sets up the dc operating or Q point. As the signal input changes, it will cause a change in gate voltage, which will cause a corresponding change in the output drain current. The output voltage developed between the FET's drain and ground is 180° out of phase with the input because an increase in $V_{\text{in}}\uparrow$, and therefore $V_{GS}\uparrow$, will cause an increase in $I_D\uparrow$, a decrease in the voltage drop across the FET ($V_{DS}\downarrow$), and a decrease in the output voltage $V_{\text{out}}\downarrow$. Resistor R_S is included to provide temperature stability and, as with the bipolar transistor, the source decoupling capacitor C_2 is included to prevent degenerative feedback.

When a small ac input signal is applied to the gate of a common-source amplifier, the variations in voltage at the gate control the JFET, which effectively acts as a variable resistor, varying the output drain current. These changes in drain current will vary the voltage drop across R_D and the drain-to-source voltage drop, which, with R_S, determines the output voltage. Referring to the characteristics listed in Figure 18-12(a), you can see that the output voltage (V_{out}) of the common-source JFET configuration can be five to ten times larger than the gate control input voltage (V_{in}). If a high amount of voltage gain is desired, R_D is made

Common-Source Configuration

An FET configuration in which the source is grounded and common to the input and output signal.

	Voltage Gain	Input Impedance	Output Impedance	Circuit Appearance and Application	Waveforms
Common-Source (a)	5–10 (Voltage Amp)	Very High 1–15 MΩ	Low 2–10 kΩ	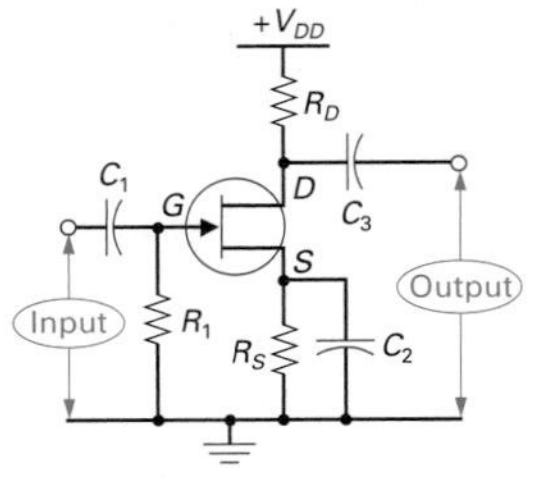 Most widely used FET configuration. It is mainly used as a voltage amplifier, however it is also used as an impedance matching device and can handle the high radio frequency signals.	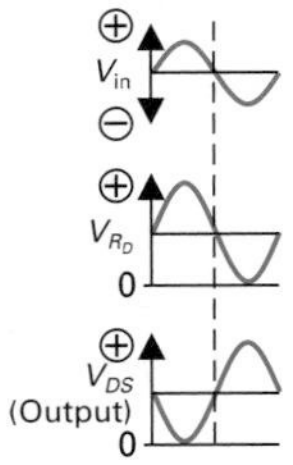 V_{in} and V_{out} are out of phase (180° phase shift)
Common-Gate (b)	2–5	Very Low 200–1500 Ω	Medium 5–15 kΩ	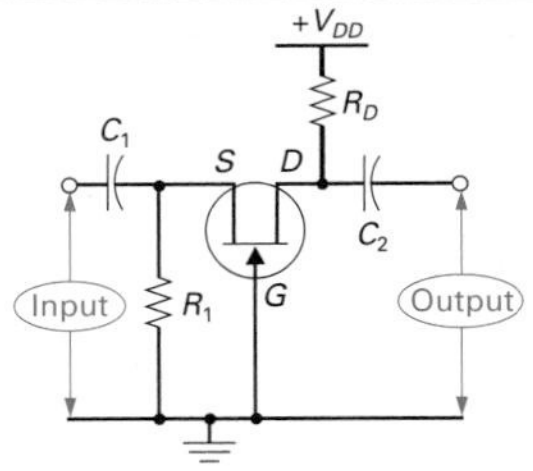 This configuration is used to amplify radio frequency signals due to its very stable nature at high frequencies. It is also used as a buffer to match a low impedance source to a high impedance load.	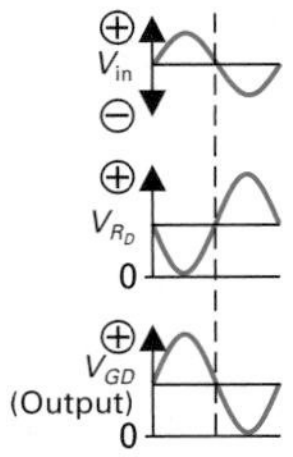 V_{in} and V_{out} are in phase (0° phase shift)
Common-Drain (c)	0.98	Very High 1500 MΩ	Low 10 kΩ	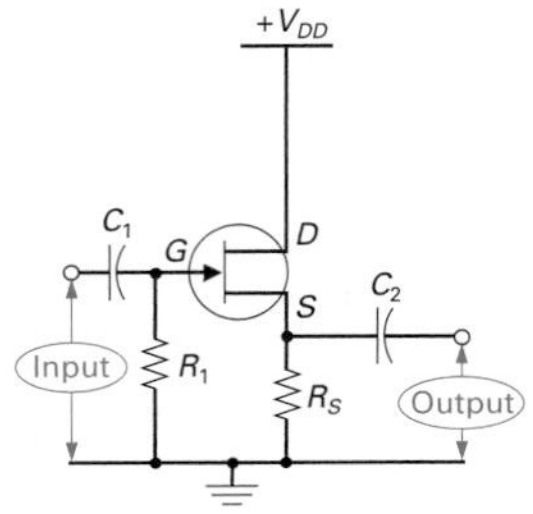 This amplifier is commonly called a source-follower as the source follows whatever is applied to the gate. Its very high input impedance will not load down (and therefore not distort) signals from high-impedance signal sources, such as a microphone, and its low output impedance is ideal to drive a low-impedance load such as an audio amplifier.	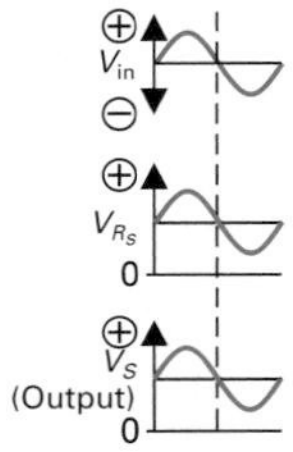 V_{in} and V_{out} are in phase (0° phase shift)

FIGURE 18-12   JFET Circuit Configurations.

relatively large (typically greater than 20 kΩ) and the JFET is biased so that its drain-to-source resistance is also high. A larger resistance will develop a larger voltage.

Also listed in the common-source characteristics in Figure 18-12(a) is the very high input resistance and the relatively low output resistance of this circuit. The high input resistance is due to the JFET's reverse biased gate-source junction, which permits no gate input current, and therefore has a very large resistance. This key characteristic means that the common-source JFET circuit is ideal in applications where we need to provide voltage amplification but do not want to load down a source that can only generate a small input signal. Such applications include the following:

1. Digital circuits in which the outputs of many circuits are connected to one another, and therefore the output resistances of all the circuits load one another. As a result, the signals generated by these circuits are small, and a circuit is needed that will not load the signal source but will still provide voltage gain.

2. Analog circuits in which it can amplify both dc and low- and high-frequency ac input signal voltages. The *C-S* circuit's high input impedance makes it ideal at the front end of systems such as the first RF amplifier stage following the antenna and the first stage in a voltmeter, in which it will not load the source yet will amplify a wide range of input signal voltages.

Common-Gate (*C-G*) Circuits

The **common-gate circuit configuration** shown in Figure 18-12(b) is very similar to its bipolar counterpart, the common-base circuit. The input is applied between the source and gate, while the output appears across the drain and gate. Self bias resistor R_1 sets up the static Q point, and the input is applied through the coupling capacitor C_1 and will cause a change in the JFET's source voltage. An increase in source voltage will cause a decrease in the V_{GS} forward bias (*n*-type source is driven positive), a decrease in I_D, a decrease in the voltage drop across R_D, and therefore an increase in the voltage dropped between the FET's drain and gate. Because the voltage developed across the JFET's drain and gate is applied to the output, an increase in the input produces an increase in the output, and so the input and output voltage are in phase with one another. Similarly, as the input voltage decreases, the gate-source forward bias will increase. Therefore, I_D will increase, and there will be more voltage developed across R_D and less voltage developed at the output.

Referring to the common-gate characteristics listed in Figure 18-12(b), you can see that this circuit can be used to provide a small voltage gain. Because the input is applied to the JFET's high-current source terminal, the input resistance is very low. This low input resistance and relatively high output resistance makes the circuit ideal in applications where we need to efficiently transfer power between a low-resistance source and a high-resistance load.

Common-Drain (*C-D*) Circuits

Comparable to the bipolar transistor's common-collector or emitter-follower, the **common-drain configuration** shown in Figure 18-12(c) is sometimes called a **source-follower** because the source output voltage follows in polarity and amplitude the input voltage at the gate. Once again, self bias resistor R_1 sets up the quiescent operating point, and an ac gate input voltage will cause a variation in I_D. When the input voltage at the gate swings positive, the FET will conduct more current, less voltage will be developed across the FET drain to source, and therefore more voltage will be developed across R_S and the output. Similarly, a decrease in the input voltage will cause the resistance of the JFET's drain-source junction to increase. Therefore, V_{DS} will increase and V_{R_L}, or V_{out}, will decrease.

Referring to the common-drain characteristics listed in Figure 18-12(c), you can see that the output voltage is slightly less than the input voltage (circuit does not provide any voltage gain). The input resistance of the common-drain circuit configuration is extremely high due to the JFET's reverse biased gate and R_S connection, and the output resistance is relatively very low. Inserting a common-drain circuit between a high-resistance source and a low-resistance load will ensure that the two opposite resistances are matched and power is efficiently transferred.

18-1-9   *JFET Applications*

It is the high input impedance of the JFET, and therefore its ability not to load a source, and the voltage amplification ability that are mainly made use of in circuit application. Like the bipolar junction transistor, the JFET can be made to function as a switch or as a variable resistor. Let us begin by examining how the JFET's switching ability can be made use of in digital or two-state circuits.

Digital (Two-State) JFET Circuits

As a switch, the JFET makes use of only two points on the load line: saturation (in which it is equivalent to a closed switch between source and drain) and cutoff (in which it is

---

Common-Gate Configuration

An FET configuration in which the gate is grounded and common to the input and output signal.

Common-Drain Configuration

An FET configuration in which the drain is grounded and common to the input and output signal.

Source-Follower

Another name used for a common-drain circuit configuration.

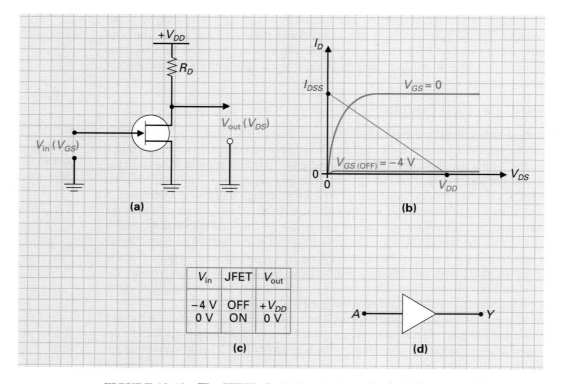

FIGURE 18-13 **The JFET's Switching Action—Digital Circuit Applications. (a) Basic Switching Circuit. (b) Load Line. (c) Input/Output Voltages. (d) Digital Buffer-Driver Schematic Symbol.**

equivalent to an open switch between source and drain). Figure 18-13(a) shows an ON/OFF JFET switch circuit and its associated load line in Figure 18-13(b). Figure 18-13(c) shows the input/output voltages for each of the circuit's two operating states. When $V_{GS} = V_{GS(OFF)}$ (-4 V), the JFET is cut OFF (lower end of the load line) and is equivalent to an open switch between source and drain. With the JFET's drain-source open, $I_D = 0$ mA, and the drain supply voltage will be applied to the output ($V_{DS} = V_{out} = +V_{DD}$). On the other hand, when $V_{GS} = 0$ V, the JFET is saturated (upper end of the load line) and is equivalent to a closed switch between source and drain. With the JFET's drain-source closed, $I_D = $ max. $= I_{DSS}$, and the 0 V at the source will be applied to the output ($V_{DS} = V_{out} = 0$ V).

A typical FET application in digital circuits would be a buffer circuit, which is used to isolate one device from another. The high input impedance of the FET does not load the input circuit or circuits, while the low output impedance of the FET provides a high output current to the output circuit. The high output current and buffering or isolating characteristics of these circuits account for why they are also called buffer-drivers. The schematic symbol of the buffer-driver is shown in Figure 18-13(d).

The high input impedance ($Z_{in}\uparrow$) of the FET is also made use of in other FET integrated circuits (ICs). When Z_{in} is high, circuit current is low ($I\downarrow$), and therefore power dissipation is low ($P_D\downarrow$). This condition is ideal for digital integrated circuits (ICs), which contain thousands of transistors all formed onto one small piece of silicon. The low power dissipation of the JFET enables us to densely pack many more components into a very small area.

Analog (Linear) JFET Circuits

In two-state applications, the JFET is made to operate between the two extreme points of saturation (0 Ω) and cutoff (max. Ω). By controlling the gate-source bias voltage (V_{GS}), the resistance between the JFET's drain and source (R_{DS}) can be changed to be any value between 0 Ω and maximum Ω. The JFET can therefore be made to act as a variable resistor, with an increase in negative V_{GS} causing a larger R_{DS}. In contrast, a decrease in negative V_{GS} causes a smaller R_{DS}. This is illustrated in Figure 18-14(a).

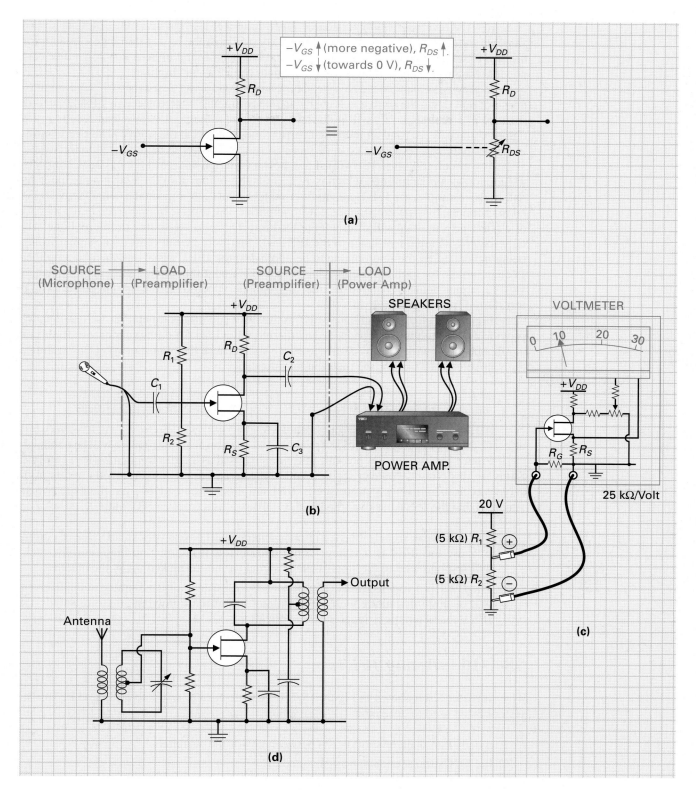

FIGURE 18-14 The JFET's Variable-Resistor Action—Analog Circuit Applications.
(a) Equivalent Circuit. (b) Application 1: An Audio Preamplifier Circuit. (c) Application 2:
A Voltmeter High Input Impedance Circuit. (d) Application 3: An RF Amplifier Circuit.

It is the reverse biased gate-source junction of a JFET that gives the JFET its key advantage: *an extremely high input impedance* (typically in the high-MΩ range). In addition, the JFET can provide a small voltage gain and has been found to be a very low noise component. All these characteristics make it an ideal choice as an amplifier.

In previous chapters, we have seen how a light load (large resistance $R_L\uparrow$ and small $I\downarrow$) does not pull down the source voltage by any large amount, whereas a heavy load (small resistance $R_L\downarrow$ and large $I\uparrow$) will pull down the source voltage. A heavy load results in less output voltage ($V_{R_L}\downarrow$) and an increased current and heat loss at the source. The overall effect is that a small load resistance or impedance causes less power to be delivered to the load.

The circuit in Figure 18-14(b) shows how a common-source JFET has been connected to function as a preamplifier, which is a circuit that provides gain for a very weak input signal. In this example, the JFET preamplifier matches the high-impedance (small signal) crystal microphone to a low-impedance power amplifier. The reverse biased gate-source junction of a JFET preamplifier will offer a large load impedance to the source or microphone. This light load ($R_L\uparrow$) input resistance of the JFET will therefore permit most of the signal voltage being generated by the microphone to be applied to the JFET's gate and then be amplified. In other words, the high input impedance of the JFET amplifier circuit will not pull down the voltage signal being generated by the microphone. Therefore, maximum power will be transferred from source to load.

Figure 18-14(c) shows how a JFET at the front end of a voltmeter or oscilloscope will provide a very high input impedance and therefore not load the circuit under test. In this example, the meter will measure 10 V across R_2, and because the ohms per volt (Ω/V) rating of the voltmeter is 125 kΩ/V, the meter input impedance is

$$Z_{\text{in}} = \Omega/\text{V} \times V_{\text{measured}}$$
$$= 125 \text{ k}\Omega/\text{V} \times 10 \text{ V} = 1.25 \text{ M}\Omega$$

A 1.25 MΩ meter resistance in parallel with the 5 kΩ resistance of R_2 will have very little effect (1.25 MΩ in parallel with 5 kΩ = about 5 kΩ), so an accurate reading will be obtained.

Figure 18-14(d) shows how the JFET can be used as a radio frequency (RF) amplifier. Studying this circuit, you can see that both the gate and drain contain tuned circuits in the same way as the previously discussed BJT RF amplifier circuits. However, there are two advantages that the JFET has over the bipolar transistor as a front end RF amp.

1. The very weak signals injected into the antenna will have a very small value of current. Because the JFET is a voltage-controlled device, it requires no input current, and it will respond well to the small voltage signal variations picked up by the antenna.
2. The JFET is a very low noise component. Because any noise generated at the front end will be amplified along with the signal at each of the following amplifier stages, this JFET characteristic is ideal in this application.

18-1-10 *Testing JFETs*

The transistor tester shown in Figure 18-15(a) can be used to test both BJTs and FETs. This tester can be used to determine

1. whether an open or short exists between any of the terminals,
2. the FET's transconductance/gain, and
3. the FET's value of I_{DSS} and leakage current.

If a transistor tester is not available, the ohmmeter can be used to detect the most common failures: opens and shorts. Figure 18-15(b) indicates what resistance values should be obtained between the terminals of a good n-channel and p-channel JFET. Looking at these ohmmeter readings, you can see that because the JFET has only one P-N junction (gate-to-channel), it is relatively simple to test with an ohmmeter for an open or shorted junction.

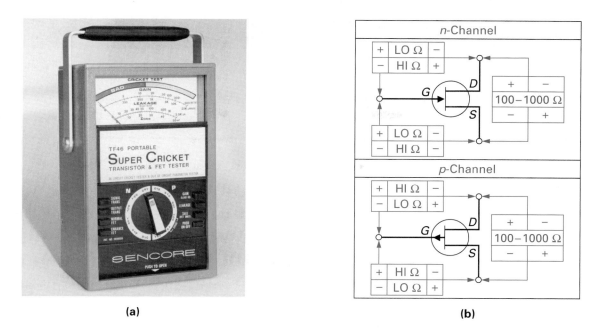

(a)

(b)

FIGURE 18-15   **Testing JFETs.**

Now that you have completed this section, you should be able to:

■ **Objective 1.**   *Name the two different types of field effect transistors.*

■ **Objective 2.**   *Describe the physical construction and operation of the function FET (JFET), and name and identify its three terminals.*

■ **Objective 3.**   *Describe why the term "field effect" is used in the name of an FET and why the term "junction" is used in JFET.*

■ **Objective 4.**   *Explain the JFET operation, and the following characteristics:*
a. V_P, V_{BR}, I_{DSS}, $V_{GS(OFF)}$
b. Transconductance
c. High input impedance

■ **Objective 5.**   *Interpret the JFET specificiations given in a typical manufacturer's data sheet.*

■ **Objective 6.**   *Calculate the different values of circuit voltage and current for the following transistor biasing methods:*
a. Gate biasing
b. Self biasing
c. Voltage-divider biasing

■ **Objective 7.**   *Identify and list the characteristics of the following three FET circuit configurations:*
a. Common source
b. Common gate
c. Common drain

■ **Objective 8.**   *Explain how the JFET's characteristics can be used in digital and analog circuit applications.*

■ **Objective 9.**   *Describe how to test a JFET.*

Use the following questions to test your understanding of Section 18-1.

1. The BJT is a _____ controlled device while the FET is a _____ controlled device.

2. The gate-source junction of a JFET is always _____ biased.

3. True or false: When $V_{GS} = 0$ V, $I_D = I_{DSS}$.

4. True or false: When $V_{GS} = V_{GS(OFF)}$, I_D = max.

5. _____ is a ratio of an FET's output current change to input voltage change.

6. A JFET has a _____ input impedance due to its _____ biased gate-source junction.

7. What component in a JFET circuit provides temperature stability?

8. Like self bias, _____ bias has negative feedback and therefore maintains the Q point stable.

9. Which FET circuit configuration is most widely used like its BJT common-emitter counterpart?

10. Which JFET circuit configuration could provide a high input impedance and a good value of voltage gain?

11. Which JFET circuit configuration is best suited for providing a very high input impedance and low output impedance?

Continued

12. Which JFET characteristic is made use of in most circuit applications?

13. Why is the JFET ideal as an RF preamplifier?

14. Using a transistor tester to test a 2N5484 JFET, what typical readings should we obtain for the following:

    a. $V_{GS(OFF)}$

    b. I_{DSS}

15. Will the gate-to-source and gate-to-drain of a JFET test with an ohmmeter like any other P-N junction?

18-2   THE METAL OXIDE SEMICONDUCTOR FIELD EFFECT TRANSISTOR (MOSFET)

With the JFET, an input voltage of zero volts would reverse bias the P-N junction, resulting in a maximum channel size and a maximum value of source-to-drain current. To decrease the size of the channel, the input voltage was made negative to further reverse bias the gate-source junction. This action would deplete the channel of free carriers, reducing the size of the channel and therefore the source-to-drain current. This type of action is actually called "depletion-mode operation," since an input voltage is used to deplete the channel and, therefore, reduce the channel's size and current. The MOSFET does not have a P-N gate-channel junction like the JFET. It has a "metal gate" that is insulated from the "semiconductor channel" by a layer of "silicon dioxide," hence the name "metal oxide semiconductor." Like all "field effect transistors" (FETs), the input voltage will generate an "electric field" which will have the "effect" of changing the channel's size.

The key difference between the JFET and MOSFET is that the JFET's input voltage would always have to be zero or a negative voltage in order to reverse bias the gate-source junction. With the MOSFET, the input voltage can be either a positive or negative voltage since gate current will always be zero because the gate is insulated from the channel. To examine each of these input voltage possibilities:

    a. If the input voltage is negative, the resulting electric field depletes the channel, reducing its size, and the MOSFET is said to be operating in the depletion mode.

    b. If, on the other hand, the input voltage is positive, the resulting electric field enhances the channel, increasing its size, and the MOSFET is said to be operating in the enhancement mode.

The MOSFET can therefore be operated in either the depletion or enhancement mode due to its insulated gate. The two different types of MOSFETs are given names based on their normal mode of operation. For instance, the *depletion-type MOSFET (D-type MOSFET or D-MOSFET),* should actually be called a DE-MOSFET because it can be operated in both the depletion mode and the enhancement mode, whereas the *enhancement-type MOSFET (E-type MOSFET or E-MOSFET)* is correctly named since it can only be operated in the enhancement mode. In this chapter, we will examine the construction, operation, characteristics, circuit biasing, applications, and testing of these two MOSFET types.

Depletion-Type MOSFET

A field effect transistor with an insulated gate (MOSFET) that can be operated in either the depletion or enhancement mode.

18-2-1   *The Depletion-Type (D-Type) MOSFET*

The **depletion-type MOSFET** construction is slightly different from the JFET, and therefore we will need to first see how the D-MOSFET device is built before we can understand how it operates.

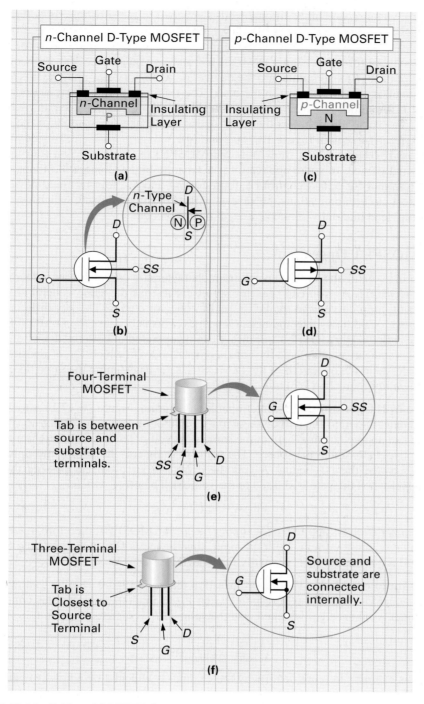

n-Channel D-Type MOSFET

Source Gate Drain

n-Channel | Insulating
P | Layer

Substrate

(a)

n-Type
Channel

(b)

p-Channel D-Type MOSFET

Source Gate Drain

Insulating | *p*-Channel
Layer | N

Substrate

(c)

(d)

Four-Terminal
MOSFET

Tab is between
source and
substrate
terminals.

SS S G D

(e)

Three-Terminal
MOSFET

Tab is
Closest to
Source
Terminal

S G D

Source and
substrate are
connected
internally.

(f)

FIGURE 18-16   D-Type MOSFET Construction and Types.

D-MOSFET Construction

Like the JFET and BJT, the D-MOSFET has two basic transistor types called the **n-channel D-type MOSFET** and **p-channel D-type MOSFET.** The construction and schematic symbols for these two D-MOSFET types are shown in Figure 18-16.

To begin with, let us examine the construction of the more frequently used *n*-channel D-type MOSFET, shown in Figure 18-16(a). This type of MOSFET basically consists of an *n*-type channel formed on a *p*-type substrate. A source and drain lead are connected to either

n-Channel D-Type MOSFET

A depletion type MOSFET having an *n*-type channel between its source and drain terminals.

p-Channel D-Type MOSFET

A depletion type MOSFET having a *p*-type channel between its source and drain terminal.

end of the *n*-channel, and an additional lead is attached to the substrate. In addition, a thin insulating (silicon dioxide) layer is placed on top of the *n*-channel, and a metal plated area with a gate lead attached is formed on top of this insulating layer. Figure 18-16(b) shows the schematic symbol for an *n*-channel D-MOSFET, and, as you can see, the arrow on the substrate (*SS*) or base (*B*) lead points into the device. As a memory aid, imagine this arrowhead as a P-N junction diode as shown in the inset. The source and drain leads are connected to either end of the diode's *n*-type cathode. Therefore, this device must be an *n*-channel D-MOSFET. The basic difference in the construction and schematic symbol of the *p*-channel D-MOSFET can be seen in Figure 18-16(c) and (d).

Figure 18-16(e) and (f) show how the MOSFET is available as a four-terminal or three-terminal device. In some applications, a separate bias voltage will be applied to the substrate terminal for added control of drain current and the four-terminal device will be used. In most circuit applications, however, the three-terminal device, which has its source and substrate lead internally connected, is all that is needed.

D-MOSFET Operation

Figure 18-17(a) shows the typical drain characteristic curves for an *n*-channel depletion-type MOSFET. As you can see, this set of curves has the same general shape as the JFET's set of drain curves and the BJT's set of collector curves. The key difference is that V_{GS} is plotted for both positive and negative values. This is because the D-MOSFET should actually be called a DE-MOSFET because it can be operated in both the depletion mode (in which V_{GS} is a negative value) and the enhancement mode (in which V_{GS} is a positive value). To best understand the operation of the D-MOSFET, let us examine the three operation diagrams shown in Figure 18-17(b), (c), and (d).

Zero-Volt Operation: The center operation diagram, Figure 18-17(b), shows how the *n*-channel D-MOSFET will respond to a V_{GS} input of zero volts. When $V_{GS} = 0$ V, the gate and source terminals are at the same zero volt potential, and therefore the gate is effectively shorted to the source. The value of drain current passing through the channel is called the I_{DSS} value (I_{DSS} is the drain-to-source current passing through the channel when the gate is shorted to the source). Therefore, when

$$V_{GS} = 0 \text{ V}, I_D = I_{DSS}$$

When zero volts is applied to the input of a D-type MOSFET, therefore, it will conduct a value of drain current. With no input, this device is ON, which is why the D-type MOSFET is known as a "normally ON" device.

Enhancement Mode: The upper operation diagram, Figure 18-17(c), shows how the *n*-channel D-MOSFET will respond when V_{GS} is made positive. In this condition, the channel is enhanced or widened, and the value of I_D is increased above I_{DSS}. Therefore, when

$$V_{GS} = +V, I_D > I_{DSS}$$

Let us examine in more detail why the channel is widened by a positive gate voltage. Because the valence-band holes in the *p*-type material (majority carriers) will be repelled by a positive gate voltage, and the conduction band electrons in the *p*-type material (minority carriers) will be attracted to the channel by the positive gate voltage, there will be a build-up of electrons in the *p*-type material near the channel. This build-up of electrons in the *p*-type material below the channel will effectively widen the size of the channel, reducing its resistance, and therefore increasing I_D to a value greater than I_{DSS}.

Depletion Mode: The lower operation diagram, Figure 18-17(d), shows how the *n*-channel D-MOSFET will respond when V_{GS} is made negative. In this condition, the channel is depleted of free carriers, and therefore the value of I_D is decreased below I_{DSS}. Therefore, when

$$V_{GS} = -V, I_D < I_{DSS}$$

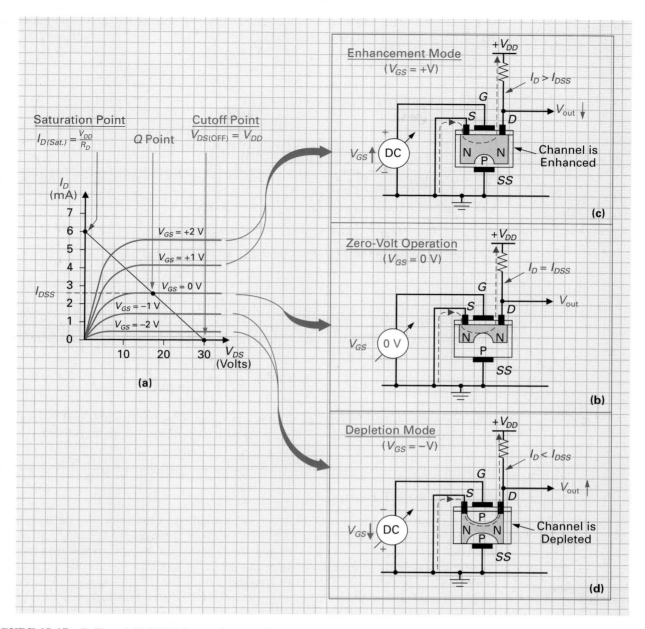

FIGURE 18-17 **D-Type MOSFET Operation and Characteristics.**

To summarize the *n*-channel MOSFET's operation, when V_{GS} was either zero volts or a negative voltage, the *n*-channel D-MOSFET acted in almost exactly the same way as an *n*-channel JFET. However, unlike the JFET, the D-MOSFET can have a forward biased gate-to-source P-N junction because the silicon dioxide insulating layer prevents any current from passing through the gate and will still maintain a high input resistance. This dual operating ability is why the depletion-type MOSFET or D-MOSFET should actually be called a depletion-enhancement or DE-MOSFET.

The drain characteristic curves of the D-MOSFET can be used to plot the device's dc load line, as shown in Figure 18-17(a), with

$$I_{D(Sat.)} = \frac{V_{DD}}{R_D} \text{ at saturation, and}$$

$$V_{DS(OFF)} = V_{DD} \text{ at cutoff}$$

As with the JFET, the D-MOSFET's transconductance is equal to the ratio of output current change (ΔI_D) to input voltage change (ΔV_{GS}),

$$\delta_m = \frac{\Delta I_D}{\Delta V_{GS}}$$

and the D-MOSFET's voltage gain is equal to

$$A_V = \delta_m \times R_D$$

▢ EXAMPLE:

A D-MOSFET circuit has the following specifications:

$$I_{DSS} = 2 \text{ mA}, V_{GS(OFF)} = -6 \text{ V}, R_D = 3 \text{ k}\Omega, V_{DD} = 12 \text{ V}$$

Calculate the following two extremes on the D-MOSFET's load line:

a. $I_{D(Sat.)}$

b. $V_{DS(OFF)}$

▢ *Solution:*

a. When the D-MOSFET is saturated, it is equivalent to a closed switch and therefore the only resistance is that of R_D.

$$I_{D(Sat.)} = \frac{V_{DD}}{R_D} = \frac{12 \text{ V}}{3 \text{ k}\Omega} = 4 \text{ mA}$$

b. When the D-MOSFET is cut off, it is equivalent to an open switch, and therefore the full drain supply voltage will appear across the open between drain and source.

$$V_{DS(OFF)} = V_{DD} = 12 \text{ V}$$

D-MOSFET Biasing

Like the JFET, the D-MOSFET can be configured in the same way as a common-drain, common-gate, or common-source circuit, with all of the dc and ac configuration characteristics being the same. As far as biasing, the D-MOSFET is easier to bias than the JFET because of its ability to operate in either the depletion mode ($-V_{GS}$) or the enhancement mode ($+V_{GS}$). In fact, one of the most frequently used D-MOSFET biasing methods is to simply have no biasing at all. This biasing method is called **zero biasing** because the Q point is set at zero volts ($V_{GS} = 0$ V), as seen in Figure 18-18. This makes biasing the D-MOSFET very simple because no gate or source bias voltages are needed. The ac input signal developed across R_G is therefore applied to the extremely high input impedance of the D-MOSFET, causing an increase and decrease in the conduction of the MOSFET above and below the $V_{GS} = 0$ V, Q point.

Zero Biasing

A configuration in which no bias voltage is applied at all.

D-MOSFET Applications

The D-MOSFET is most frequently used in analog or linear circuit applications. This is because the D-MOSFET can be very simply biased at a midpoint in the load line and then have its output current varied above and below this natural Q point in a linear fashion. This, coupled with the D-MOSFET's almost infinite input impedance and low noise properties, makes it ideal as a preamplifier at the front end of a system. Figure 18-19 shows how two D-MOSFETs can be used to construct a typical front end **cascode amplifier circuit.** This cascode amplifier circuit consists of a self biased common-source amplifier (Q_1) in series with a voltage-divider biased common-gate amplifier (Q_2). The input signal (V_{in}) is applied to Q_1's gate, and the amplified output at Q_1's drain is then passed to Q_2's source, where it is further amplified by Q_2 before appearing at Q_2's drain, and therefore at the output (V_{out}).

Cascode Amplifier Circuit

An amplifier circuit consisting of a self-biased common-source amplifier in series with a voltage-divider biased common-gate amplifier.

CHAPTER 18 / FIELD EFFECT TRANSISTORS (FETs)

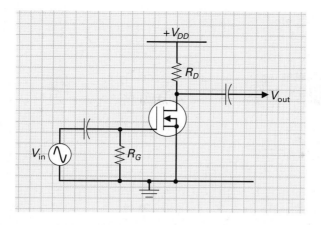

FIGURE 18-18   Zero Biasing a D-MOSFET.

The FET's only limiting factor is that its high input impedance starts to decrease as the input signal's frequency increases. Refer to the inset in Figure 18-19, which shows how the gate, insulator, and channel of a D-MOSFET form a capacitor. This input capacitance of typically 5 pF has very little effect at low input signal frequencies ($X_C\uparrow = 1/2\pi f\downarrow C$) because the input impedance is high ($X_C\uparrow$ therefore $Z_{in}\uparrow$) and the loading effect is negligible. At higher radio frequency ($X_C\downarrow = 1/2\pi f\uparrow C$), however, the input impedance is lowered ($X_C\downarrow$ therefore $Z_{in}\downarrow$) and the D-MOSFET loses its high input impedance advantage. To compensate for this disadvantage, FETs are often connected in series, as in Figure 18-19, so that their input capacitances are also in series. Recall that series-connected capacitors have a lower total capacitance than any of the individual capacitance values. Therefore, the overall input capacitance of two series-connected D-MOSFETs will be less than that of a single D-MOSFET, making this cascode amplifier ideal as a high radio frequency (RF) amplifier: a low input capacitance ($C_{in}\downarrow$) means a high input reactance ($X_{C(in)}\uparrow$), and therefore a high input impedance ($Z_{in}\uparrow$) at high frequencies.

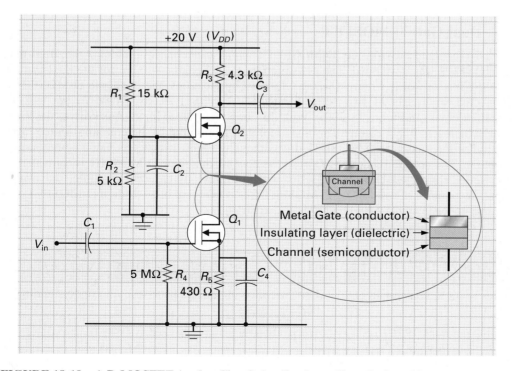

FIGURE 18-19   A D-MOSFET Analog Circuit Application—Cascode Amplifier.

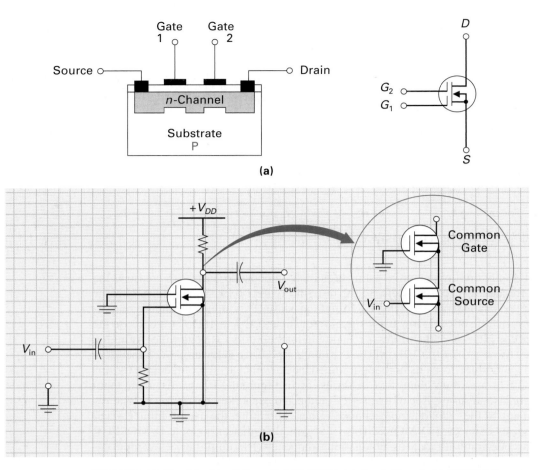

FIGURE 18-20 The Dual-Gate D-MOSFET. (a) Dual-Gate D-MOSFET Construction and Schematic Symbol. (b) Dual-Gate D-MOSFET Application—Cascode Amplifier.

Dual-Gate D-MOSFET

Dual-Gate D-MOSFET

A metal oxide semiconductor FET having two separate gate electrodes.

To compensate for the D-MOSFET's input capacitance problem, the **dual-gate D-MOSFET** was developed. The construction and schematic symbol for the dual-gate D-MOSFET is shown in Figure 18-20(a). In most applications, the dual-gate D-MOSFET is connected so it acts as two series-connected D-MOSFETs, as shown in the cascode amplifier circuit in Figure 18-20(b). With this amplifier, the ac input signal drives the lower gate, which acts like a common-source amplifier. The output of the common-source lower section of the dual-gate D-MOSFET drives the upper half, which acts like a common-gate amplifier. The inset in Figure 18-20(b) shows how the dual-gate D-MOSFET is equivalent to two series-connected D-MOSFETs. As with the previous cascode amplifier, the overall input capacitance of a dual-gate D-MOSFET is less than that of a standard D-MOSFET, and if capacitance is low, X_C, and therefore Z_{in}, are high.

SELF-TEST EVALUATION POINT FOR SECTION 18-2-1

Now that you have completed this section, you should be able to:

■ *Objective 12.* *Describe how the D-MOSFET is constructed, how it operates, and its characteristics.*

Use the following questions to test your understanding of Section 18-2-1.

1. The two different types of MOSFETs are called the _____ type MOSFET and _____ type MOSFET.

Continued

2. True or False: The D-type MOSFET can be operated in both the depletion and enhancement mode.

3. The D-type MOSFET is a normally _____ (ON/OFF) device.

4. Which FET has a higher input impedance: JFET or MOSFET?

5. Why is the D-MOSFET ideal as a preamplifier?

   a. It can be mid-load-line biased when 0 V is applied.

b. It has a high input impedance.

c. It has low noise properties.

d. All of the above

6. The _____ MOSFET was developed to lower input capacitance so that it can handle high-frequency signals.

18-2-2   The Enhancement-Type (E-Type) MOSFET

With an input of zero volts, a D-MOSFET will be ON, and a certain value of current will pass through the channel between source and drain. If the input to the D-MOSFET is made positive, the channel is enhanced, causing the source-to-drain current to increase. If the input is made negative, the channel is depleted, causing the source-to-drain current to decrease. The D-MOSFET can therefore operate in either the enhancement or depletion mode and is called a "normally ON" device because it is ON when nothing (0 V) is applied.

The **enhancement-type MOSFET** or **E-MOSFET** can only operate in the enhancement mode. In other words, when the input is either zero volts or a negative voltage, the transistor is OFF and there is no source-to-drain current. However, when the input is made positive, the E-MOSFET will turn ON, resulting in a source-to-drain channel current. The E-MOSFET is therefore a "normally OFF" device because it is OFF when nothing (0 V) is applied.

E-MOSFET Construction

As with all of the other transistor types, it is easier to understand the operation and characteristics of a device once we have seen how the component is constructed. The E-MOSFET has two basic transistor types called the **n-channel E-type MOSFET** and **p-channel E-type MOSFET**. The construction and schematic symbols for these two E-MOSFET types are shown in Figure 18-21.

To begin with, let us examine the construction of the more frequently used n-channel E-type MOSFET, shown in Figure 18-21(a). Studying the construction of this E-MOSFET, notice that no channel exists between the source and drain. Consequently, with no gate bias voltage, the device will be OFF. When the gate is made positive, however, electrons will be attracted from the substrate, causing a channel to be induced between the source and drain. This enhanced channel will permit drain current to flow, and any further increase in gate voltage will cause a corresponding increase in the size of the channel and therefore the value of I_D.

The schematic symbol for the n-channel E-type MOSFET can be seen in Figure 18-21(b). The construction and schematic symbol for the p-channel E-MOSFET can be seen in Figure 18-21(c) and (d). As with the D-MOSFET, both four-terminal and three-terminal devices are available, with the three-terminal device having a common connection between source and substrate as shown in Figure 18-21(e).

The only difference between the E-type and D-type MOSFET symbols is the three dashed lines representing the drain, substrate, and source regions. The dashed line is used instead of the solid line to indicate that an E-MOSFET has a normally broken path, or channel, between drain, substrate, and source. To aid memory, the three dashed lines used in the "E"-MOSFET symbol could be thought of as the three horizontal prongs in the capital letter "E," as shown in Figure 18-21(f).

Enhancement-Type MOSFET or E-MOSFET

A field effect transistor with an insulated gate (MOSFET) that can only be turned on if the channel is enhanced.

n-Channel E-Type MOSFET

An enhancement type MOSFET having an n-type channel between its source and drain terminals.

p-Channel E-Type MOSFET

An enhancement type MOSFET having a p-type channel between its source and drain terminals.

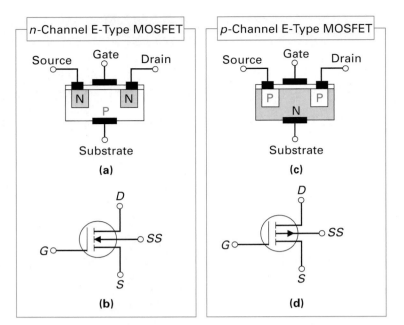

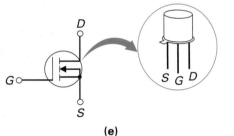

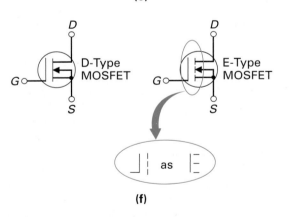

FIGURE 18-21   E-Type MOSFET Construction and Type.

E-MOSFET Operation

Figure 18-22(a) shows the typical drain characteristic curves for an *n*-channel enhancement-type MOSFET. As you can see, this set of drain curves is very similar to the D-MOSFET's set of drain curves; however, in this case only positive values of V_{GS} are plotted. Looking at the relationship between V_{GS} and I_D, you may have noticed that any increase in V_{GS} will cause a corresponding increase in I_D.

To best understand the operation of the E-MOSFET, let us examine the three operation diagrams shown in Figure 18-22(b), (c), and (d).

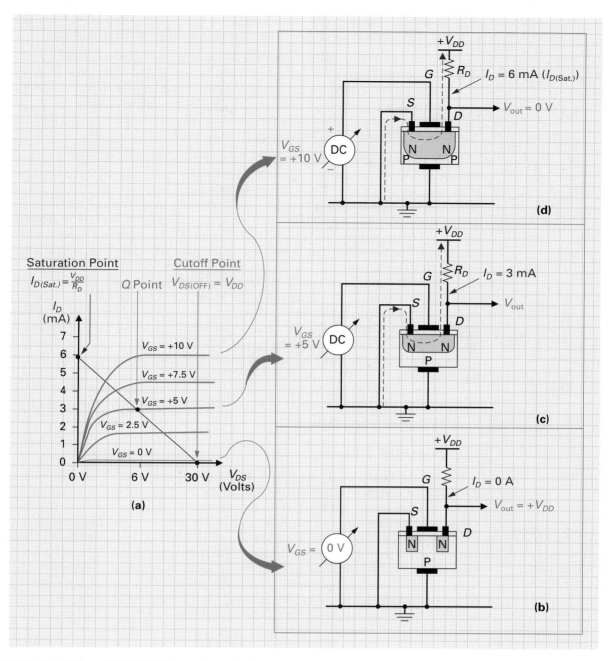

FIGURE 18-22   E-Type MOSFET Operation and Characteristics.

$V_{GS} = 0$ **V Curve:** The lower operation diagram, Figure 18-22(b), shows how the n-channel E-MOSFET will respond to a V_{GS} input of zero volts. When $V_{GS} = 0$ V, there is no channel connecting the source and drain, and therefore the drain current will be zero. As a result, $+V_{DD}$ will be present at the output because the E-MOSFET is equivalent to an open switch between drain and source.

$V_{GS} = +5$ **V Curve:** The center operation diagram, Figure 18-22(c), shows how the n-channel E-MOSFET will respond to a V_{GS} input of +5 volts. When $V_{GS} = +5$ V, the E-MOSFET will act in almost exactly the same way as an enhanced D-MOSFET. The positive gate voltage will repel the p-type material's majority carriers (holes) away from the gate, while attracting the p-type material's minority carriers (electrons) toward the gate. This action will form an *n*-type bridge between the source and drain, and therefore a value of I_D will flow between source and drain, as shown.

$V_{GS} = +10$ V Curve: The upper operation diagram, Figure 18-22(d), shows how the *n*-channel E-MOSFET will respond if the V_{GS} input is further increased to +10 volts. When $V_{GS} = +10$ V, the attraction of electrons and repulsion of holes within the *p*-type material is increased, causing the channel's width to increase and I_D to also increase. As a result, 0 V will be present at the output because the E-MOSFET is equivalent to a closed switch between drain and source.

In summary, when V_{GS} is zero volts, drain current is also zero. As the value of V_{GS} is increased (made more positive), the channel becomes wider, causing I_D to increase. On the other hand, as the value of V_{GS} is decreased (made less positive), the channel becomes narrower, causing I_D to decrease. In other words, when the input is either zero volts or a negative voltage, the transistor is OFF, and there is no source-to-drain current. However, when the input is made positive, the E-MOSFET will turn ON, resulting in a source-to-drain channel current. The E-MOSFET is therefore a "normally OFF" device because it is OFF when nothing (0 V) is applied.

Although it cannot be seen in the set of drain curves in Figure 18-22(a), V_{GS} will have to increase to a positive threshold voltage of about +1 V before a channel will be induced and a small value of drain current will flow. This "threshold level" is a highly desirable characteristic because it prevents noise or any low-level input signal voltage from turning ON the device. This advantage makes the E-type MOSFET ideally suited as a switch because it can be turned ON by an input voltage and turned OFF once the input voltage falls below the threshold level.

The *p*-channel E-type MOSFET operates in much the same way as the *n*-channel device except that holes are attracted from the substrate to form a *p*-channel and the V_{GS} and V_{DS} bias voltages are reversed.

E-MOSFET Biasing

Like the D-MOSFET, the E-MOSFET can be configured as a common-drain, common-gate, or common-source circuit. Unlike the D-MOSFET and JFET, the E-MOSFET cannot be biased using self bias or zero bias because V_{GS} must be a positive voltage. As a result, gate bias and voltage-divider bias can be used; however, more frequently E-MOSFETs are **drain-feedback biased,** as shown in Figure 18-23(a). In this example, R_D equals 8 kΩ, and R_G (which feeds back a positive voltage from the drain, hence the name drain-feedback bias) equals 100 MΩ. Because an E-MOSFET has an extremely high input impedance (due to the insulated gate), no current will flow in the gate circuit. With no gate current, there will be no voltage drop across the gate resistor ($V_{R_G} = 0$ V), and therefore the voltage at the gate will be at the same potential as the voltage at the drain.

Drain-Feedback Biased

A configuration in which the gate receives a bias voltage fed back from the drain.

$$V_{GS} = V_{DS}$$

To help set up the Q point, most manufacturer's data sheets specify a load line midpoint drain current $I_{D(ON)}$ and drain voltage $V_{DS(ON)}$. In the example in Figure 18-23(a), when the E-MOSFET is ON, or conducting, and $I_D = 3$ mA, the E-MOSFET's drain-to-source voltage drop ($V_{DS(ON)}$) is 6 V. The value of R_D in Figure 18-23(a) has been chosen so that this E-MOSFET circuit will be biased at its specified Q point. To check, we can use the formula

$$V_{DS} = V_{DD} - (I_{D(ON)} \times R_D)$$

In the example in Figure 18-23(a)

$$V_{DS} = V_{DD} - (I_{D(ON)} \times R_D) = 30\,\text{V} - (3\,\text{mA} \times 8\,\text{k}\Omega)$$
$$= 30\,\text{V} - 24\,\text{V} = 6\,\text{V}$$

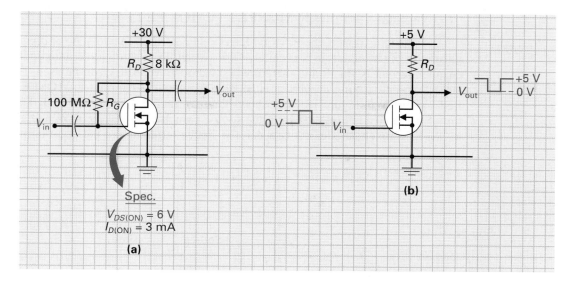

FIGURE 18-23 Biasing E-Type MOSFETs. (a) Drain-Feedback Biasing. (b) Voltage Controlled Switching.

▨ EXAMPLE:

A drain-feedback biased E-MOSFET circuit has the following specifications: $R_D = 1\ k\Omega$, $I_{D(ON)} = 10\ mA$, $V_{DD} = 20\ V$. Calculate V_{DS} and V_{R_D}.

▨ *Solution:*

$$V_{DS} = V_{DD} - (I_{D(ON)} \times R_D)$$
$$= 20\ V - (10\ mA \times 1\ k\Omega)$$
$$= 20\ V - 10\ V = 10\ V$$

The constant drain supply voltage (V_{DD}) will be evenly divided across R_D and the E-MOSFET's drain-to-source ($V_{DS} = 10\ V$, $V_{R_D} = 10\ V$).

In most instances, the E-MOSFET will be used in digital two-state switching circuit applications. In this case, there will be no need for a gate resistor (R_G) because the input voltage (V_{in}) will either turn ON or OFF the E-MOSFET, as shown in Figure 18-23(b). For example, when $V_{in} = 0\ V$, the E-MOSFET is OFF, therefore $V_{out} = +V_{DD} = +5\ V$, whereas when $V_{in} = +5\ V$, the E-MOSFET is ON and therefore $V_{out} = 0\ V$.

E-MOSFET Applications

The E-MOSFET is more frequently used in digital or two-state circuit applications. One reason is that it naturally operates as a "normally OFF voltage-controlled switch" because it can be turned ON when the gate voltage is positive and turned OFF when the gate voltage falls below a threshold level. This threshold level is a highly desirable characteristic because it prevents noise from false triggering, or accidentally turning ON, the device. The other E-MOSFET advantage is its extremely high input impedance, which means that the device's circuit current, and therefore power dissipation, are low. This enables us to densely pack or integrate many thousands of E-MOSFETs onto one small piece of silicon, forming a high component density integrated circuit (IC). These low-power and high-density advantages make the E-MOSFET ideal in battery-powered small-size (portable) applications such as calculators, wristwatches, notebook computers, hand-held video games, digital cellular phones, and so on.

Vertical-Channel E-MOSFET (VMOS FET)

Vertical-Channel E-MOSFET

An enhancement type MOSFET that, when turned on, forms a vertical channel between source and drain.

As just mentioned, the E-MOSFET is generally used in two-state or digital circuit applications, where it acts as a normally OFF switch. In most digital circuit applications, the channel current is small, and therefore a standard E-MOSFET can be used. However, if a larger current-carrying capability is needed, the **vertical-channel E-MOSFET** or VMOS FET can be used. Figure 18-24(a) shows the construction of a VMOS FET. The gate at the top of the device is insulated from the source (which is also at the top), and as with all E-MOSFETs, no channel exists between the source terminal and the drain terminal with no

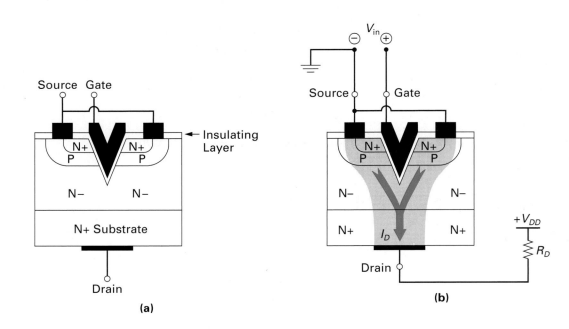

(a)

(b)

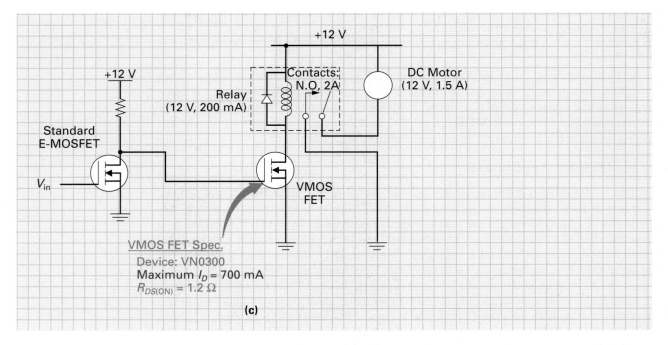

(c)

FIGURE 18-24   The Vertical Channel E-MOSFET (VMOS FET). (a) VMOS FET OFF. (b) VMOS FET ON. (c) VMOS FET Application Circuit—Low-Power Signal to High-Power Load Circuit.

bias voltage applied. The VMOS FET's semiconductor materials are labeled P, N+, and N− and indicate different levels of doping.

When this *n*-channel VMOS FET is biased ON (gate is made positive with respect to source), as seen in Figure 18-24(b), a vertical *n*-type channel is formed between source and drain. This channel is much wider than a standard E-MOSFET's horizontal channel, which is why VMOS FETs can handle a much higher drain current.

Figure 18-24(c) shows how the high-current capability of a VMOS FET can be made use of in an interfacing circuit application. In this circuit, a VMOS FET is used to interface a low-power source input signal to a high-power load. Referring to the current specifications of the VMOS FET, relay, and motor, you can see that each device is used to step up current. The standard E-MOSFET supplies a low-power input signal to the VMOS FET, which can handle enough current to actuate a relay whose contacts can handle enough current to switch power to the dc motor. To be more specific, if V_{in} is LOW, the standard E-MOSFET will turn OFF, producing a HIGH output to the gate of the VMOS FET, turning it ON. When the VMOS FET turns ON, it effectively switches ground through to the lower end of the relay coil, energizing the relay and closing its normally open contacts. The closed relay contacts switch ground through to the lower end of the motor, which turns ON because it now has the full +12 V supply across its terminals. The motor will stay ON as long as V_{in} stays LOW. If V_{in} were to go HIGH, the standard E-MOSFET inverter would produce a LOW output, which would turn OFF the VMOS FET, relay, and motor.

Other than its high-current capability, the VMOS FET also has a positive temperature coefficient of resistance, which means an increase in temperature will cause a decrease in drain current ($T\uparrow$, $R\uparrow$, $I\downarrow$), and this will prevent thermal runaway. This gives the VMOS FET a distinct advantage over the BJT power amplifier, which has a negative temperature coefficient of resistance ($T\uparrow$, $R\downarrow$, $I\uparrow$). This means that if maximum ratings are exceeded, an increase in temperature will cause an increase in current, which will generate a further increase in heat (temperature) and current, and so on.

MOSFET Data Sheets

Throughout this chapter, we have used certain MOSFET specifications in our calculations, such as I_{DSS}, $V_{DS(ON)}$, $I_{D(ON)}$, and so on. As an example, Figure 18-25 shows the data sheet for a typical E-type MOSFET, and, as you can see, these specifications are listed.

MOSFET Handling Precautions

Certain precautions must be taken when handling any MOSFET devices. The very thin insulating layer between the gate and the substrate of a MOSFET can easily be punctured if an excessive voltage is applied. Your body can build up extremely large electrostatic charges due to friction. If this charge came in contact with the pins of a MOSFET device, an electrostatic-discharge (ESD) would occur, resulting in a possible arc across the thin insulating layer causing permanent damage. Most MOSFETs presently manufactured have zeners internally connected between gate and source to bypass high-voltage static or in-circuit potentials and protect the MOSFET. However, it is important to remember the following.

1. All MOS devices are shipped and stored in a "conductive foam" or "protective foil" so that all of the IC pins are kept at the same potential, and therefore electrostatic voltages cannot build up between terminals.

2. When MOS devices are removed from the conductive foam, be sure not to touch the pins because your body may have built up an electrostatic charge.

3. When MOS devices are removed from the conductive foam, always place them on a grounded surface such as a metal tray.

4. When continually working with MOS devices, use a "wrist grounding strap," which is a length of cable with a 1 MΩ resistor in series. This prevents electrical shock if you come in contact with a voltage source.

5. All test equipment, soldering irons, and work benches should be properly grounded.

DEVICE: 2N7002LT1—N-Channel Silicon E-MOSFET

MAXIMUM RATINGS

Rating	Symbol	Value	Unit
Drain-Source Voltage	V_{DSS}	60	Vdc
Drain-Gate Voltage (R_{GS} = 1 MΩ)	V_{DGR}	60	Vdc
Drain Current — Continuous TC = 25°C(1) TC = 100°C(1) — Pulsed (2)	I_D I_D I_{DM}	±115 ±75 ±800	mA
Gate-Source Voltage — Continuous — Non-repetitive (tp ≤50 μs)	V_{GS} V_{GSM}	±20 ±40	Vdc Vpk
Total Power Dissipation TC = 25°C TC = 100°C Derate above 25°C ambient	P_D	200 80 1.6	mW mW/°C

2N7002LT1★

CASE 318-07 STYLE 21
SOT-23 (TO-236AB)

TMOS FET
TRANSISTOR

N-CHANNEL

ELECTRICAL CHARACTERISTICS (T_A = 25°C unless otherwise noted.)

Characteristic	Symbol	Min	Typ	Max	Unit
OFF CHARACTERISTICS					
Drain-Source Breakdown Voltage (V_{GS} = 0, I_D = 10 μA)	$V_{(BR)DSS}$	60	—	—	Vdc
Zero Gate Voltage Drain Current (V_{GS} = 0, V_{DS} = 60 V) T_J = 25°C T_J = 125°C	I_{DSS}	— —	— —	1.0 500	μAdc
Gate-Body Leakage Current Forward (V_{GS} = 20 Vdc)	I_{GSSF}	—	—	100	nAdc
Gate-Body Leakage Current Reverse (V_{GS} = −20 Vdc)	I_{GSSR}	—	—	−100	nAdc

(1) The Power Dissipation of the package may result in a lower continuous drain current.
(2) Pulse Width ≤ 300 μs, Duty Cycle ≤ 2.0%.

Characteristic	Symbol	Min	Typ	Max	Unit
ON CHARACTERISTICS*					
Gate Threshold Voltage (V_{DS} = V_{GS}, I_D = 250 μA)	$V_{GS(th)}$	1.0	—	2.5	Vdc
On-State Drain Current (V_{DS} ≥ 2.0 $V_{DS(on)}$, V_{GS} = 10 V)	$I_{D(on)}$	500	—	—	mA
Static Drain-Source On-State Voltage (V_{GS} = 10 V, I_D = 500 mA) (V_{GS} = 5.0 V, I_D = 50 mA)	$V_{DS(on)}$	 — —	 — —	 3.75 .375	Vdc
Static Drain-Source On-State Resistance (V_{GS} = 10 V, I_D = 500 mA) T_C = 25°C T_C = 125°C (V_{GS} = 5.0 V, I_D = 50 mA) T_C = 25°C T_C = 125°C	$r_{DS(on)}$	 — — — —	 — — — —	 7.5 13.5 7.5 13.5	Ohms
Forward Transconductance (V_{DS} ≥ 2.0 $V_{DS(on)}$, I_D = 200 mA)	g_{FS}	80	—	—	mmhos
DYNAMIC CHARACTERISTICS					
Input Capacitance (V_{DS} = 25 V, V_{GS} = 0, f = 1.0 MHz)	C_{iss}	—	—	50	pF
Output Capacitance (V_{DS} = 25 V, V_{GS} = 0, f = 1.0 MHz)	C_{oss}	—	—	25	pF
Reverse Transfer Capacitance (V_{DS} = 25 V, V_{GS} = 0, f = 1.0 MHz)	C_{rss}	—	—	5.0	pF
SWITCHING CHARACTERISTICS*					
Turn-On Delay Time (V_{DD} = 25 V, I_D ≅ 500 mA,	$t_{d(on)}$	—	—	30	ns
Turn-Off Delay Time R_G = 25 Ω, R_L = 50 Ω)	$t_{d(off)}$	—	—	40	ns
BODY-DRAIN DIODE RATINGS					
Diode Forward On-Voltage (I_S = 11.5 mA, V_{GS} = 0 V)	V_{SD}	—	—	−1.5	V
Source Current Continuous (Body Diode)	I_S	—	—	−115	mA
Source Current Pulsed	I_{SM}	—	—	−800	mA

$I_{D (ON)}$

$V_{DS (ON)}$

*Pulse Test: Pulse Width ≤ 300 μs, Duty Cycle ≤ 2.0%.

FIGURE 18-25   An E-Type MOSFET Data Sheet. (Courtesy of Motorola. Used by permission.)

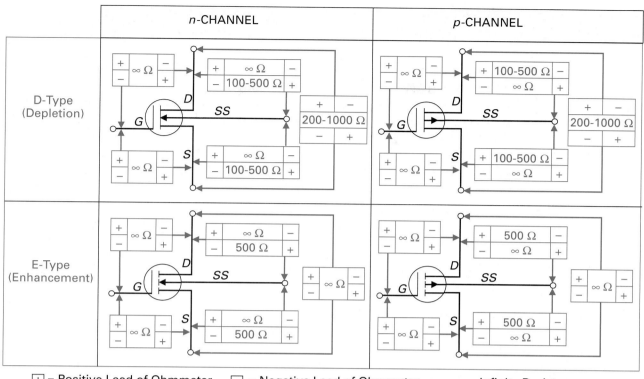

+ = Positive Lead of Ohmmeter   − = Negative Lead of Ohmmeter   ∞ = Infinite Resistance

FIGURE 18-26   Testing MOSFETs with the Ohmmeter.

6. All power in equipment should be off before MOS devices are removed or inserted into printed circuit boards.

7. Any unused MOSFET terminals must be connected because an unused input left open can build up an electrostatic charge and float to high voltage levels.

8. Any boards containing MOS devices should be shipped or stored with the connection side of the board in conductive foam.

MOSFET Testing

Like the BJT and JFET, MOSFETs can be tested with the transistor tester to determine: first, whether an open or short exists between any of the terminals; second, the transistor's transconductance/gain; and third, the value of I_{DSS} and leakage current.

If a transistor tester is not available, the ohmmeter can be used to determine the most common failures: opens and shorts. Figure 18-26 shows what resistance values should be obtained between the terminals of a good *n*-channel or *p*-channel D-MOSFET or E-MOSFET when testing with an ohmmeter. Remember that the gate-to-channel resistance of a MOSFET is always an open due to the insulated gate.

SELF-TEST EVALUATION POINT FOR SECTION 18-2-2

Now that you have completed this section, you should be able to:

■ **Objective 10.** *Explain the meaning of the term "metal oxide semiconductor field effect transistor."*

■ **Objective 11.** *Define the terms "depletion-mode operation" and "enchancement-mode operation."*

■ **Objective 12.** *Describe how the D-MOSFET is constructed, how it operates, and its characteristics.*

Continued

■ **Objective 13.** *Explain D-MOSFET biasing and typical applications.*

■ **Objective 14.** *Discuss the operation, characteristics, and application of the dual-gate D-MOSFET.*

■ **Objective 15.** *Describe how the E-MOSFET is constructed, how it operates, its characteristics, and its biasing.*

■ **Objective 16.** *List the typical applications of the E-MOSFET.*

Use the following questions to test your understanding of Section 18-2-2.

1. True or false: The E-type MOSFET can be operated in both the depletion and enhancement mode.

2. The E-type MOSFET is a normally _____ (ON/OFF) device.

3. The E-type MOSFET is generally
   a. Zero biased   c. Drain-feedback biased
   b. Base biased   d. None of the above

4. True or false: The E-MOSFET naturally operates as a voltage-controlled switch.

5. What are the $I_{D(ON)}$ and $V_{DS(ON)}$ values for a 2N4351?

6. If the 2N4351 E-MOSFET were used as a voltage-controlled switch, how long would it take
   a. To switch from OFF to ON
   b. To switch from ON to OFF

7. What general precautions should be observed when handling MOSFETs?

8. The gate-to-source or gate-to-drain resistance of a good E-MOSFET should always measure _____ ohms.

SUMMARY

Field Effect Transistors (Figure 18-27)

1. The field effect transistor, more commonly called an FET (pronounced "eff-ee-tee"), has three terminals and can operate as a switch—and therefore be used in digital circuit applications—or as a variable resistor—and therefore be used in analog or linear circuit applications.

2. There are two types of field effect transistors or FETs. One type is the junction field effect transistor, which is more typically called a JFET (pronounced "jay-fet"). The other type is the metal oxide semiconductor field effect transistor, which is more commonly called a MOSFET (pronounced "moss-fet").

3. Just as the bipolar junction transistor has two basic types (NPN BJT or PNP BJT), there are two types of junction field effect transistors called the *n*-channel JFET and *p*-channel JFET.

4. Like the BJT, the JFET has three terminals, which are called the gate, source, and drain.

5. The JFET requires both a drain supply voltage ($+V_{DD}$) and a gate-source bias voltage (V_{GS}) in order to operate.

6. In most circuit applications, the $+V_{DD}$ supply voltage is maintained constant and the V_{GS} input voltage is used to control the resistance of the channel and the value of the output current, I_D.

7. The difference between an FET and a BJT is that an FET's input junction is normally reverse biased, and therefore the input voltage controls the output current. A BJT's input junction is normally forward biased, and therefore the input current controls the output current. This is why BJTs are known as current-controlled devices, and FETs are known as voltage-controlled devices.

8. The name "field effect transistor" is derived from the device's voltage control action because the applied input voltage will generate an electric field. It is this electric field that varies the size of the depletion region, and therefore the resistance of the channel between the FET's drain and source output terminals.

9. The term "junction" is attached to this type of FET because of the single P-N junction formed between the gate and the source-to-drain channel.

10. The drain characteristic curve is a graph plotting drain current (I_D) against drain-to-source voltage (V_{DS}).

11. Like the bipolar transistor, an FET can be used to amplify a signal. As before, the amount of amplification achieved is a ratio between output and input. For a bipolar transistor, the amount of gain is equal to the ratio of input current to output current (beta). For an FET, there is no input current, and therefore an FET's gain is equal to the ratio of output current change (ΔI_D) to input voltage change (ΔV_{GS}). This ratio is called the FET's transconductance (symbolized δ_m).

12. The three most frequently used JFET biasing methods are gate biasing, self biasing, and voltage-divider biasing.

13. Gate biasing is the most simple of the three biasing methods.

14. With self biasing, only a single drain supply voltage is needed (V_{DD}), whereas gate biasing requires both V_{DD} and a negative gate supply voltage ($-V_{GG}$).

15. The Q point will remain relatively stable despite changes in temperature when a JFET circuit has a source resistor included.

16. Like the self biased circuit, the voltage-divider biased circuit includes a source resistor to stabilize the Q point de-

FIGURE 18-27   Field Effect Transistor Circuit Configurations.

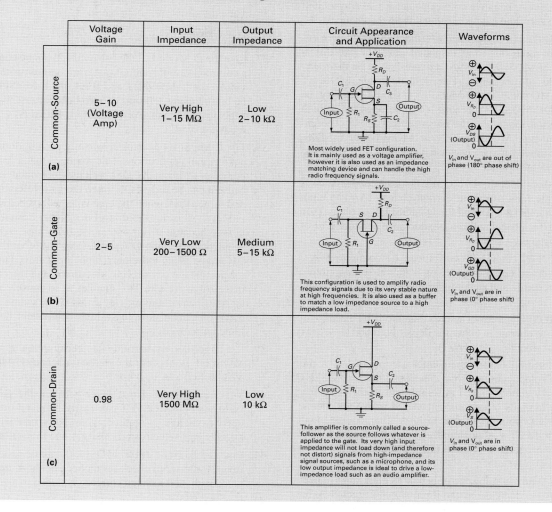

	Voltage Gain	Input Impedance	Output Impedance	Circuit Appearance and Application	Waveforms
Common-Source (a)	5–10 (Voltage Amp)	Very High 1–15 MΩ	Low 2–10 kΩ	Most widely used FET configuration. It is mainly used as a voltage amplifier, however it is also used as an impedance matching device and can handle the high radio frequency signals.	V_{in} and V_{out} are out of phase (180° phase shift)
Common-Gate (b)	2–5	Very Low 200–1500 Ω	Medium 5–15 kΩ	This configuration is used to amplify radio frequency signals due to its very stable nature at high frequencies. It is also used as a buffer to match a low impedance source to a high impedance load.	V_{in} and V_{out} are in phase (0° phase shift)
Common-Drain (c)	0.98	Very High 1500 MΩ	Low 10 kΩ	This amplifier is commonly called a source-follower as the source follows whatever is applied to the gate. Its very high input impedance will not load down (and therefore not distort) signals from high-impedance signal sources, such as a microphone, and its low output impedance is ideal to drive a low-impedance load such as an audio amplifier.	V_{in} and V_{out} are in phase (0° phase shift)

spite ambient temperature changes. In addition, using a voltage divider to determine the gate bias voltage ensures that V_{GS}, and therefore the circuit, has increased stability.

17. The three JFET circuit configurations are called common-source, common-gate, and common-drain.

18. Similar to its bipolar counterpart, the common-emitter configuration, the common-source configuration is the most widely used JFET circuit.

19. The output voltage (V_{out}) of the common-source JFET configuration can be five to ten times larger than the gate-control input voltage (V_{in}).

20. The common-source circuit has a very high input resistance and relatively low output resistance.

21. The common-gate circuit can be used to provide a small voltage gain.

22. Comparable to the bipolar transistor's common-collector, or emitter-follower, the common-drain configuration is

sometimes called a source-follower because the source output voltage follows in polarity and amplitude the input voltage at the gate.

23. It is the high input impedance of the JFET, and therefore its ability not to load a source, and the voltage amplification ability that are mainly made use of in circuit applications.

24. The transistor tester can be used to test both BJTs and FETs. This tester can be used to determine
  a. whether an open or short exists between any of the terminals,
  b. the FET's transconductance/gain, and
  c. the FET's value of I_{DSS} and leakage current.

25. The ohmmeter can be used to detect the most common JFET failures: opens and shorts.

26. The MOSFET does not have a P-N gate-channel junction like the JFET. It has a "metal gate" that is insulated from the "semiconductor channel" by a layer of "silicon dioxide," hence the name "metal oxide semiconductor."

27. The key difference between the JFET and MOSFET is that the JFET's input voltage would always have to be zero or a negative voltage in order to reverse bias the gate-source junction. With the MOSFET, the input voltage can be either a positive or negative voltage because gate current will always be zero (the gate is insulated from the channel).

28. The MOSFET can be operated in either the depletion or enhancement mode due to its insulated gate. The two different types of MOSFET are given names based on their normal mode of operation. The depletion-type MOSFET (D-type MOSFET or D-MOSFET) should actually be called a DE-MOSFET because it can be operated in both the depletion mode and the enhancement mode. The enhancement-type MOSFET (E-type MOSFET or E-MOSFET) is correctly named because it can only be operated in the enhancement mode.

29. As with the JFET, the D-MOSFET's transconductance is equal to the ratio of output current change (ΔI_D) to input voltage change (ΔV_{GS}), and the D-MOSFET's voltage gain is equal to the product of transconductance and the value of drain resistance.

30. Like the JFET, the D-MOSFET can be configured as a common-drain, common-gate, or common-source circuit, with all of the dc and ac configuration characteristics being the same.

31. The D-MOSFET is easier to bias than the JFET because of its ability to operate in either the depletion mode ($-V_{GS}$) or the enhancement mode ($+V_{GS}$).

32. The D-MOSFET is most frequently used in analog or linear circuit applications. This is because the D-MOSFET can be very simply biased at a midpoint in the load line, and then have its output current varied above and below this natural Q point in a linear fashion. This, coupled with the D-MOSFET's almost infinite input impedance and low-noise properties, makes it ideal as a preamplifier at the front end of a system.

33. The FET's only limiting factor is that its high input impedance starts to decrease as the input signal's frequency increases due to the insulated gate's input capacitance of typically 5 pF. To compensate for this disadvantage, FETs are often connected in series because the overall input capacitance of two series-connected D-MOSFETs will be less than that of a single D-MOSFET.

34. To compensate for the D-MOSFET's input capacitance problem, the dual-gate D-MOSFET was developed.

35. The enhancement-type MOSFET or E-MOSFET can only operate in the enhancement mode.

36. The only difference between the E-type and D-type MOSFET symbols is the three dashed lines representing the drain, substrate, and source regions.

37. Like the D-MOSFET, the E-MOSFET can be configured as a common-drain, common-gate, or common-source circuit. Unlike the D-MOSFET and JFET, the E-MOSFET cannot be biased using self bias or zero bias because V_{GS} must be a positive voltage. As a result, E-MOSFETs are drain-feedback biased.

38. The E-MOSFET is more frequently used in digital or two-state circuit applications. One reason is that it naturally operates as a "normally OFF voltage-controlled switch" because it can be turned ON when the gate voltage is positive and turned OFF when the gate voltage falls below a threshold level. This threshold level is a highly desirable characteristic because it prevents noise from false triggering, or accidentally turning ON, the device. The other E-MOSFET advantage is its extremely high input impedance, which means that the device's circuit current, and therefore power dissipation, are low. This enables us to densely pack or integrate many thousands of E-MOSFETs onto one small piece of silicon, forming a high component density integrated circuit (IC).

39. If a larger current-carrying capability than the standard E-MOSFET is needed, the vertical-channel E-MOSFET or VMOS FET can be used.

40. The very thin insulating layer between the gate and the substrate of a MOSFET can easily be punctured if an excessive voltage is applied. Your body can build up extremely large electrostatic charges due to friction.

41. Like the BJT and JFET, MOSFETs can be tested with the transistor tester to check for an open or short, transconductance/gain, and leakage current.

42. If a transistor tester is not available, the ohmmeter can be used to check for opens and shorts.

REVIEW QUESTIONS

Multiple-Choice Questions

1. The BJT is a _____ controlled device whereas the FET is a _____ controlled device.
 a. voltage, current **b.** current, voltage

2. The gate-source junction of a JFET is always _____ biased since the input voltage is normally _____ or some _____ voltage.
 a. reverse, 0 V, negative **c.** reverse, −4 V, positive
 b. forward, 0 V, negative **d.** forward, −4 V, negative

3. When $V_{GS} = V_{GS(OFF)}$, $I_D = $?
 a. I_{DSS} **c.** Maximum
 b. Zero **d.** V_P

4. Which JFET circuit configuration is also known as a source-follower?
  a. Common-drain   **c.** Common-gate
  b. Common-source **d.** Both (a) and (b) are true

5. Which JFET circuit configuration provides a high input impedance and a good voltage gain?
  a. Common-drain   **c.** Common-gate
  b. Common-source **d.** Both (a) and (c) are true

6. Which biasing method makes use of a $-V_{GG}$ supply voltage?
  a. Self biasing   **c.** Base biasing
  b. Gate biasing   **d.** Voltage-divider biasing

7. Which biasing method uses a source resistor?
  a. Self-biasing   **c.** Voltage-divider biasing
  b. Gate biasing **d.** Both (a) and (c)

8. Transconductance is a ratio of
  a. ΔI_D to ΔV_{DS} **c.** ΔI_D to ΔV_{GS}
  b. ΔV_{GD} to ΔI_D **d.** ΔV_{GS} to ΔV_{DS}

9. Which JFET circuit configuration has phase inversion between input and output?
  a. Common-drain   **c.** Common-gate
  b. Common-source **d.** Both (b) and (c)

10. The current between _____ leads of a JFET is controlled by varying the reverse bias voltage applied to the _____ leads.
  a. gate and source, source and drain
  b. source and drain, drain and gate
  c. source and drain, gate and source
  d. drain and gate, gate and source

11. The _____ has an insulated gate that allows the use of either a positive or negative gate input voltage.
  a. JFET   **c.** BJT
  b. MOSFET **d.** VJT

12. The _____ has drain current when the gate voltage is zero and is therefore known as a normally _____ device.
  a. D-MOSFET, OFF **c.** E-MOSFET, OFF
  b. D-MOSFET, ON   **d.** E-MOSFET, ON

13. Which type of MOSFET can use zero biasing to bias it to a mid-load-line point?
  a. D-MOSFET **c.** VMOS FET
  b. E-MOSFET **d.** Both (b) and (c)

14. With the _____, the gate voltage must be greater than the device's threshold voltage to produce drain current.
  a. JFET   **c.** E-MOSFET
  b. D-MOSFET **d.** All of the above

15. The vertical channel MOSFET is ideal for interfacing low-voltage devices to high-power loads.
  a. True **b.** False

16. Which of the following transistors has the highest input impedance?
  a. BJT   **c.** MOSFET
  b. JFET **d.** Both (a) and (c) are true

17. D-MOSFETs are more frequently used in _____ circuit applications whereas E-MOSFETs are more often used in _____ circuit applications.
  a. analog, digital **c.** digital, linear
  b. linear, analog   **d.** two-state, digital

18. A digital code consists of a series of HIGH and LOW voltages called _____.
  a. NOT gates **c.** Logic levels
  b. CMOS   **d.** Both (a) and (b)

19. The _____ has made its mark in two-state circuits in which it uses only two points on the load line and acts like a voltage-controlled _____.
  a. E-MOSFET, variable resistor
  b. E-MOSFET, switch
  c. D-MOSFET, variable resistor
  d. D-MOSFET, switch

20. To avoid possible damage to MOS devices while handling, testing, or in operation, the following precaution should be taken.
  a. All leads should be connected except when being tested or in actual operation.
  b. Pick up devices by plastic case instead of leads.
  c. Do not insert or remove devices when power is applied.
  d. All of the above

Communication Skill Questions

21. Identify the schematic symbols shown in Figure 18-28. (18-1-1)

22. Name the two different types of JFETs. (18-1-1)

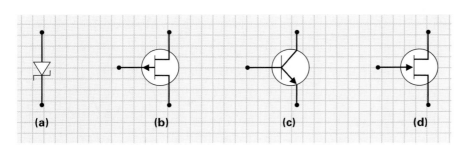

FIGURE 18-28 Schematic Symbols.

FIGURE 18-29 **MOSFET Schematic Symbols.**

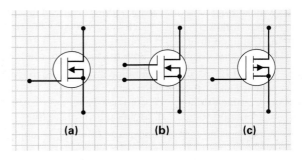

23. Briefly describe the operation of a JFET. (18-1-2)
24. Why is an FET referred to as a voltage-controlled device, and the BJT referred to as a current-contolled device? (18-1-2)
25. What is transconductance? (18-1-4)
26. What key JFET characteristic is made use of in application circuits? (18-1-9)
27. List the three different JFET circuit configurations, and briefly describe their characteristics. (18-1-8)
28. Briefly describe the following JFET biasing methods: (18-1-7)
    a. Gate biasing
    b. Self biasing
    c. Voltage-divider biasing
29. Why is the FET better suited to miniaturization than the BJT? (18-1-9)
30. Describe some of the typical digital and analog circuit applications of the JFET. (18-1-9)

31. How does a MOSFET differ from a JFET? (18-2)
32. Why should a D-MOSFET really be called a DE-MOSFET? (18-2-1)
33. Identify the device schematic symbols shown in Figure 18-29. (18-2-1 and 18-2-2)
34. Briefly describe the operation and application of the dual-gate MOSFET. (18-2-1)
35. Why are MOSFETs ideally suited for digital circuit applications? (18-2-2)

Practice Problems

36. Calculate the following for the amplifier circuit in Figure 18-30:
    a. Transconductance **b.** Voltage gain
37. Calculate the following for the circuit in Figure 18-31:
    a. V_{GS} **d.** $I_{D(maximum)}$
    b. I_D **e.** Q point
    c. V_{DS}

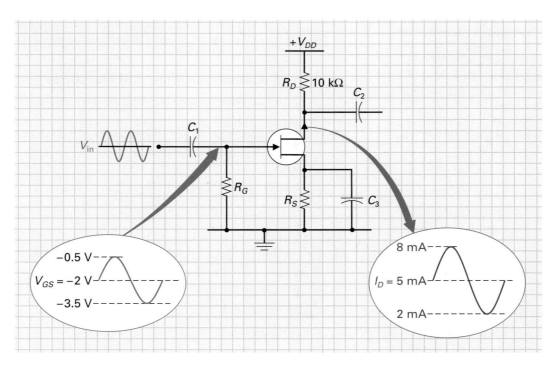

FIGURE 18-30 A Common-Source Amplifier.

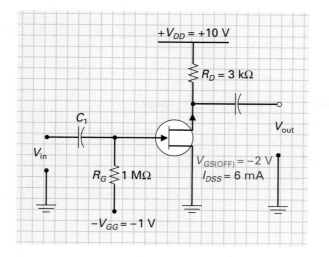

FIGURE 18-31   A Gate-Biased JFET Circuit.

38. Calculate the following for the circuit in Figure 18-32:
    a. V_S **b.** V_{GS} **c.** V_{DS}

39. Calculate the following for the circuit in Figure 18-33:
    a. V_G **b.** V_S **c.** V_{GS} **d.** V_{DS}

40. Calculate the following for the circuit shown in Figure 18-34:
    a. $I_{D(Sat.)}$ **b.** $V_{DS(OFF)}$

41. Calculate the following for the circuit shown in Figure 18-35:
    a. I_D **b.** V_{DS}

42. What biasing method was used in Figures 18-34 and 18-35?

Troubleshooting Questions

43. How can an ohmmeter test JFETs?

44. Which of the JFETs in Figure 18-36 are good, and which are bad? If bad, state the suspected problem.

45. Are the MOSFETs in Figure 18-37 good or bad?

Web Site Questions

Go to the Web site http://www.prenhall.com/cook, select the textbook *Introductory DC/AC Electronics* or *Introductory DC/AC Circuits*, this chapter, and then follow the instructions when answering the multiple-choice practice problems.

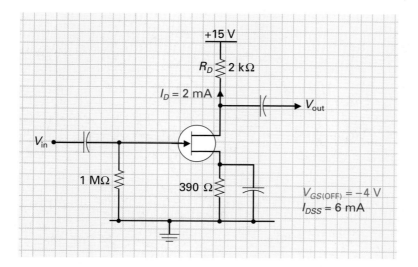

FIGURE 18-32   A Self Biased JFET Circuit.

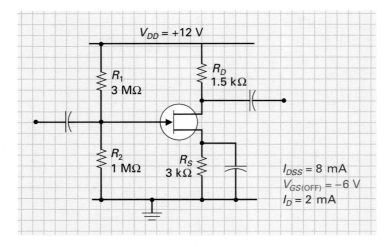

FIGURE 18-33   A Voltage-Divider Biased JFET Circuit.

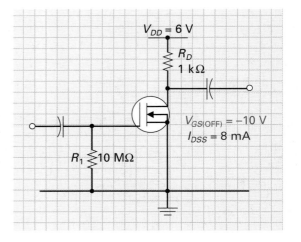

FIGURE 18-34   A D-MOSFET Amplifier Circuit.

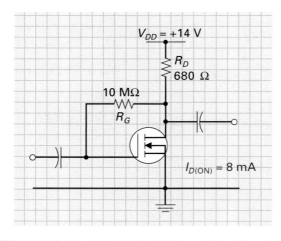

FIGURE 18-35   An E-MOSFET Amplifier Circuit.

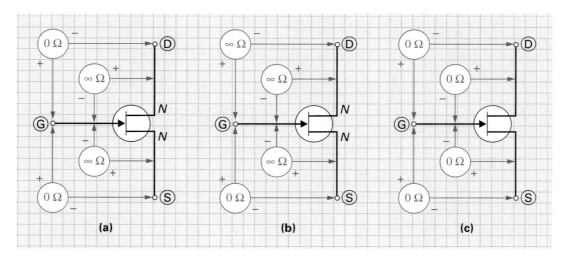

FIGURE 18-36   Testing JFETs with the Ohmmeter.

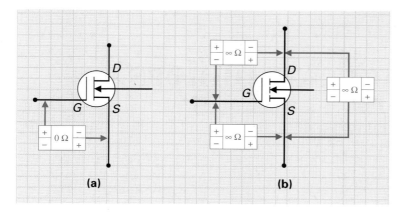

FIGURE 18-37   Testing MOSFETs with the Ohmmeter.

JOB INTERVIEW TEST

These tests at the end of each chapter will challenge your knowledge up to this point, and give you the practice you need for a job interview. To make this more realistic, the test will comprise both technical and personal questions. In order to take full advantage of this exercise, you may want to set up a simulation of the interview environment, have a friend read the questions to you, and record your responses for later analysis.

Company Name: Vital, Inc.

Industry Branch: Medical

Function: Maintain and Service ICU Equipment

Job Applying For: Field Service Technician

1. Where else have you applied for a job?
2. What is the difference between a BJT and an FET?

3. How would you feel about working on your own?
4. What is transconductance?
5. Do you think that there will be occasions in which you don't know the answer to a problem, and if so what would you do?
6. How would you test a JFET?
7. What would you say is the job function of a field service technician?
8. Without supervision in the field, could you discipline yourself to get through a heavy work load?
9. What is the difference between a D-MOSFET and an E-MOSFET?
10. Why do you feel we should hire you for this job?

Answers

1. The question is inapproporiate, but it would not be advisable to say so. Discuss the other areas of electronics you are interested in, but refrain from discussing the companies you have submitted resumes to.
2. Introduction.
3. Discuss how you have worked alone on project assignments and other related work history.
4. Section 18-1-4.
5. Report the problem to my supervisor and ask for help.
6. Section 18-1-10.

7. See introduction, Field Service technician, and company's job listing.
8. Discuss how you have juggled school, studying and a job to get to the position you are, and other related work experience.
9. Section 18-2-2.
10. Describe why you were first attracted to their advertisement, why it felt compatible with your career goals, and how the company's positive attributes listed in its web site matched your expectations.

Operational Amplifiers

I See

Standing six feet six inches tall, Jack St. Clair Kilby was a quiet, introverted man from Kansas. Excited about the prospect of joining the very well-respected Massachusetts Institute of Technology (MIT) to further his education, he was thoroughly disappointed when he failed the mathematics entrance exam by three points. For the next ten years he worked for a manufacturer of radio and television parts, paying particularly close attention to the new component on the block, the transistor.

In May of 1958, Texas Instruments, a new, fast-growing company who developed the first commercially available silicon transistor just four years earlier, offered him a job in their development lab. A project was under way to print electronic components on ceramic wafers and then wire them and stack them together to make a circuit. The more Kilby became involved in the project, the more he realized how complicated and ridiculous the method was. The idea to miniaturize was a good one, but a different solution was needed.

Two months later in July, the company shut down for summer vacation, but Kilby was forced to work because he had not accrued vacation time. This proved to be a blessing in disguise for Kilby who found himself in the lab with a lot of time and resources available to him to develop his idea. His idea was to build resistors and capacitors from the same semiconductor material that was being used to manufacture transistors. This would mean that all of the components that make up a circuit could be manufactured simultaneously on a single slice of semiconductor material.

A few months later Kilby presented his prototype to a very skeptical boss. It contained five components all connected by tiny wires with the complete assembly held together by large blobs of wax. Kilby suddenly found himself the owner of a patent and the richly deserved acclaim that always goes along with being first. The first integrated circuit or IC was born on a thin wafer of germanium just two-fifths of an inch long. Texas Instruments demonstrated its miniaturization advantage by building a computer for the Air Force using their newly developed technique in a 587 IC. Its rewards were immediately apparent when a 78 cubic foot monster computer was replaced with a more powerful unit measuring only 6.5 cubic inches.

Outline and Objectives

Introduction

The operational amplifier was initially a vacuum tube circuit used in the early 1940s in analog computers. The name "operational amplifier" or "op-amp" was chosen because the circuit was used as a high-gain dc "amplifier" performing mathematical "operations." These early circuits were expensive and bulky, and they found very little application until the semiconductor integrated circuit was developed in 1958 by Jack Kilby at Texas Instruments. Circuits that once needed hundreds of discrete or individual components can now be integrated into a single IC, making equipment smaller, more energy efficient, cheaper, and easier to design and troubleshoot.

Today's IC op-amp is a very high-gain dc amplifier that can have its operating characteristics changed by connecting different external components. This makes the op-amp very versatile, and it is this versatility that has made the op-amp the most widely used linear IC.

In this chapter, we will be examining the op-amp's operation, characteristics, typical circuit applications, and troubleshooting.

19-1   OPERATIONAL AMPLIFIER BASICS

Operational Amplifier (Op-Amp)

Special type of high-gain amplifier.

To begin with, Figure 19-1 introduces the **operational amplifier,** or **op-amp,** by showing its schematic symbol in Figure 19-1(a) and internal circuit in Figure 19-1(b). It would be safe to say that you will not really be learning anything that has not already been covered because the op-amp's internal circuit is simply a combination of three previously covered amplifier circuits. These three circuits are all interconnected and contained within a single IC, and together they function as a "high-gain, high input impedance, low output impedance amplifier."

19-1-1   Op-Amp Symbol

Referring again to Figure 19-1(a), you can see that the triangle-shaped amplifier symbol is used to represent the op-amp in an electronic schematic diagram. Comparing the two symbols, you may have noticed that in some cases the two power supply connections are not shown, even though power is obviously applied.

Inverting Input

The inverting or negative input of an op-amp.

Noninverting Input

The noninverting or positive input of an op-amp.

Let us now examine the op-amp's input and output terminals shown in Figure 19-1. The two op-amp inputs are labeled "−" and "+." The "−" or negative input is called the **inverting input** because any signal applied to this input will be amplified and inverted between input and output (output is 180° out of phase with input). On the other hand, the "+" or positive input is called the **noninverting input** because any signal applied to this input will be amplified but not inverted between input and output (output is in phase with input). An input signal will normally be applied to only one of these inputs, while the other input is used to control the op-amp's operating characteristics.

The two power supply connections to the op-amp are labeled "$+V$" and "$-V$." Figure 19-1(c) shows how power to the op-amp can be supplied by dual supply voltages or by a single supply voltage. When two supply voltages are used (dual supply voltages), the voltage values are of the same value but of opposite polarity (for example, $+12$ V and -12 V). On the other hand, when only one supply voltage is used (single supply voltage), a positive or negative voltage is applied to its respective terminal while the other terminal is grounded (for example, $+5$ V and ground or -5 V and ground). Having both a positive and negative power supply voltage will allow the output signal to swing positive and negative, above and below zero. As with all high gain amplifiers, however, the output voltage can never exceed the value of the $+V$ and $-V$ supply voltages.

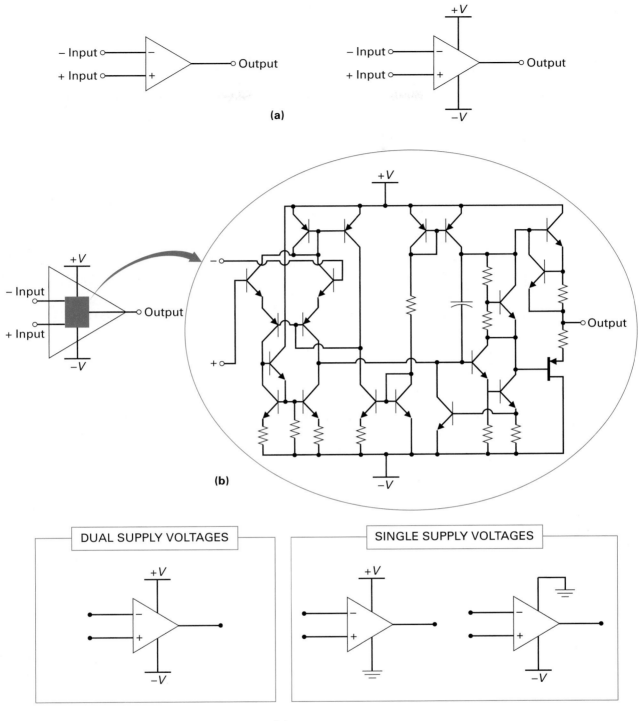

FIGURE 19-1   The Operational Amplifier. (a) Schematic Symbols. (b) Internal Circuit.
(c) Power Supply Connections.

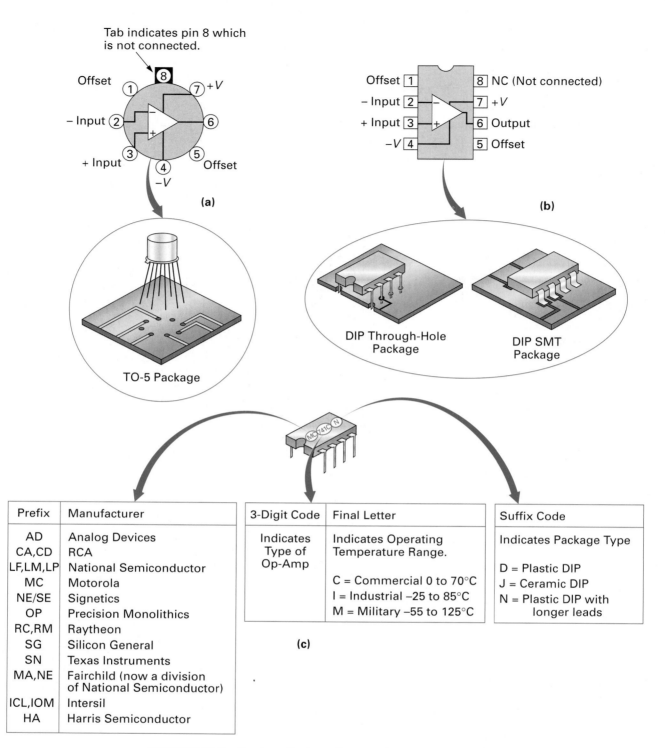

Tab indicates pin 8 which is not connected.

Offset ① ⑦ +V
– Input ② ⑥
+ Input ③ ⑤ Offset
④
–V

(a)

TO-5 Package

Offset 1   8 NC (Not connected)
– Input 2   7 +V
+ Input 3   6 Output
–V 4   5 Offset

(b)

DIP Through-Hole Package   DIP SMT Package

Prefix	Manufacturer
AD	Analog Devices
CA,CD	RCA
LF,LM,LP	National Semiconductor
MC	Motorola
NE/SE	Signetics
OP	Precision Monolithics
RC,RM	Raytheon
SG	Silicon General
SN	Texas Instruments
MA,NE	Fairchild (now a division of National Semiconductor)
ICL,IOM	Intersil
HA	Harris Semiconductor

3-Digit Code	Final Letter
Indicates Type of Op-Amp	Indicates Operating Temperature Range. C = Commercial 0 to 70°C I = Industrial –25 to 85°C M = Military –55 to 125°C

Suffix Code
Indicates Package Type D = Plastic DIP J = Ceramic DIP N = Plastic DIP with longer leads

(c)

FIGURE 19-2 **Op-Amp Package Types and Identification Codes.**

19-1-2 *Op-Amp Packages*

The entire op-amp circuit is placed within one of two basic packages, shown in Figure 19-2(a) and (b). The TO-5 metal can package is available with 8, 10, or 12 leads, while the dual in-line through-hole and surface-mount packages typically have 8 or 14 pins.

Like all ICs, an identification code is used to indicate the device manufacturer, device type, and key characteristics. Figure 19-2(c) lists some of the more common manufacturer prefix codes, operating temperature codes, and package codes. In this example, the "MC 741C N" code indicates that the 741 op-amp is made by Motorola, it is designed for commercial application in which the temperature range is between 0 to 70°C, and the package is a through-hole DIP with longer leads.

Referring back to the IC packages in Figure 19-2(a) and (b), you can see that in addition to the two inputs, single output, and two power supply terminals, there are two additional leads labeled offset. These two inputs will normally be connected to a potentiometer that can be adjusted to set the output at zero volts when both inverting and noninverting inputs are at zero volts. **Balancing** the op-amp in this way is generally needed due to imbalances within the op-amp's internal circuit.

SELF-TEST EVALUATION POINT FOR SECTION 19-1

Now that you have completed this section, you should be able to:

■ **Objective 1.** *Describe the symbol, package types, and internal block diagram of the operational amplifier or op-amp.*

Use the following questions to test your understanding of Section 19-1.

1. Is the operational amplifier a discrete component or an integrated circuit?
2. List the names of the op-amp's five terminals.
3. Name the two basic op-amp package types.
4. Briefly describe the meaning of each of the three parts in an op-amp's identification code.

19-2 OP-AMP OPERATION AND CHARACTERISTICS

As mentioned previously, the operational amplifier contains three amplifier circuits, and these three circuits are all interconnected and contained within a single IC. Referring to the block diagram of the op-amp in Figure 19-3(a) you can see that these three circuits are *a differential amplifier, a voltage amplifier, and an output amplifier.* Combined, these three circuits give the op-amp its key characteristics, which are *high gain, high input impedance,* and *low output impedance.* We will briefly review their characteristics because combined they determine the characteristics of the op-amp.

19-2-1 *The Differential Amplifier within the Op-Amp*

The differential amplifier within the op-amp is connected to operate in its "differential-input, single-output mode." The operation of the differential amplifier in this mode is reviewed in Figure 19-3(b). When both input signals are equal in amplitude and in phase with one another, they are referred to as **common-mode input signals,** as seen in the waveforms in Figure 19-3(b). On the other hand, if the input signals are out of phase with one another, they are referred to as **differential-mode input signals,** as seen in the other set of waveforms in Figure 19-3(b). The differential amplifier will amplify differential input signals while rejecting common-mode input signals. The questions you may be asking at this stage are what are common-mode input signals and why should we want to reject them? The answers are as follows: temperature changes and noise are common-mode input signals, and they are unwanted signals. Let us examine these common-mode signals in more detail.

1. Temperature variations within electronic equipment affect the operation of semiconductor materials and, therefore, the operation of semiconductor devices. These temperature variations can cause the dc output voltage of the first stage to drift away from its

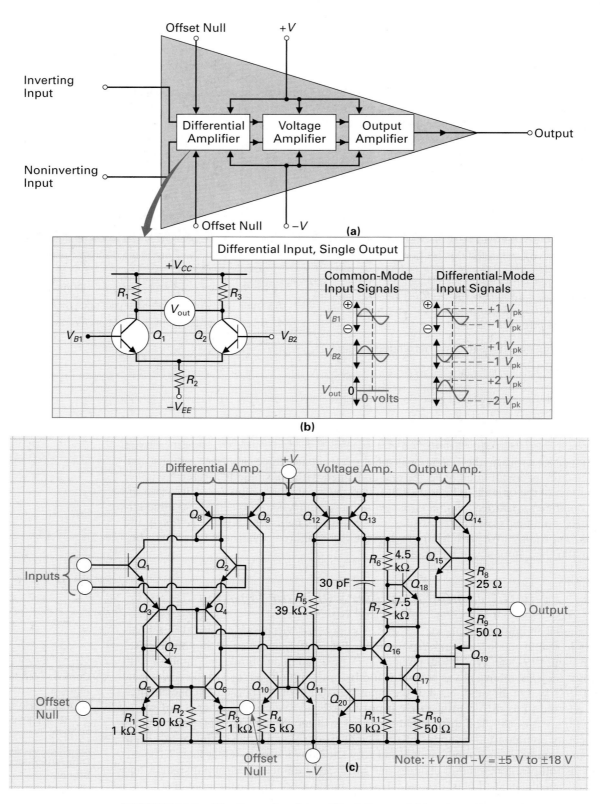

FIGURE 19-3 The Operational Amplifier's Internal Circuit. (a) Block Diagram.
(b) The Differential Amplifier's Operation. (c) Circuit Diagram.

normal Q point. The second-stage amplifier will amplify this voltage change in the same way as it would amplify any dc input signal and so will all of the following amplifier stages. The increase or decrease in the normal Q-point bias for all of the amplifier stages will get progressively worse due to this thermal instability, and the final stage may have a Q point that is so far off of its mid position that an input signal may drive it into saturation or cutoff, causing signal distortion. With the differential amplifier, any change that occurs due to temperature changes will affect both stages and so will not appear at the output of the differential amplifier, due to its **common-mode rejection.**

Common-Mode Rejection

The ability of a device to reject a voltage signal applied simultaneously to both input terminals.

2. The second common-mode input signal that the differential amplifier removes is noise. It is often necessary to amplify low-level signals from low-sensitivity sources such as microphones, light detectors, and other transducers. High-gain amplifiers are used to increase the amplitude of these small input signals up to a more usable level that is large enough to drive or control a load, such as a loudspeaker. The 60 Hz ac power line, or any other electrical variation, can induce a noise signal along with the input signal at the input of this high-gain amplifier. Since these noise signals will be induced at all points in the circuit, and be identical in amplitude and phase, the differential amplifier will block these unwanted signals because they will be present at both inputs of the differential amplifier (noise will be a common-mode input). A true input signal, on the other hand, will appear at the two inputs of the differential amplifier as a differential input signal, and therefore be amplified.

19-2-2 *Common-Mode Rejection Ratio*

To summarize, a differential input will be amplified by the op-amp's differential amplifier and passed to the output, while unwanted signals caused by temperature variations or noise will appear as common-mode input signals and therefore be rejected. An op-amp's ability to provide a high **differential gain (A_{VD})** and a low **common-mode gain (A_{CM})** is directly dependent on its internal differential amplifier and is a measure of an op-amp's performance. This ratio is called the **common-mode rejection ratio (CMRR)** and is calculated with the following formula:

Differential Gain (A_{VD})

The amplification of differential-mode input.

$$CMMR = \frac{A_{VD}}{A_{CM}}$$

Looking at this formula, you can see that the higher the A_{VD} (differential gain), or the smaller the A_{CM} (common-mode gain), the higher the CMRR value, and therefore the better the operational amplifier. This ratio can also be expressed in dBs by using the following formula:

Common-Mode Gain (A_{CM})

The amplification of common-mode input.

Common-Mode Rejection Ratio (CMRR)

The ratio of an operational amplifier's differential gain to the common-mode gain.

$$CMMR = 20 \times \log \frac{A_{VD}}{A_{CM}}$$

EXAMPLE:

If an op-amp's differential amplifier has a differential gain of 5000 and a common-mode gain of 0.5, what is the operational amplifier's CMRR? Express the answer in standard gain and dBs.

Solution:

$$CMMR = \frac{A_{VD}}{A_{CM}} = \frac{5000}{0.5} = 10{,}000$$

$$CMMR = 20 \times \log \frac{A_{VD}}{A_{CM}} = 20 \times \log 10{,}000 = 20 \times 4 = 80 \text{ dB}$$

A CMRR of 10,000 or 80 dB means that the op-amp's desired input signals will be amplified 10,000 times more than the unwanted common-mode input signals.

19-2-3   *The Op-Amp Block Diagram*

Now that we have reviewed the differential amplifier's circuit characteristics, let us return to the op-amp block diagram in Figure 19-3(a). It is the op-amp's differential-amplifier stage that provides the good common-mode rejection and high differential gain. Because the op-amp's "−" and "+" inputs are applied to either base of the diff-amp, we know that input current will be very small. It is this circuit characteristic that provides the op-amp with another key feature, which is a high input impedance ($I\downarrow, Z\uparrow$). The voltage-amplifier stage following the diff-amp usually consists of several darlington-pair stages that provide an overall op-amp voltage gain of typically 50,000 to 200,000. The final-output stage consists of a complementary emitter-follower stage to provide a low output impedance and high current gain, so that the op-amp can deliver up to several milliamps, depending on the value of the load.

19-2-4   *The Op-Amp Circuit Diagram*

The complete internal circuit of a typical op-amp can be seen in Figure 19-3(c). With integrated circuits, it is better to have transistors function as resistors wherever possible because they occupy less chip space than actual resistors. This accounts for why the circuit seems to contain many transistors that have their base and collector leads connected. You may also have noticed that no coupling capacitors have been used so that the op-amp can amplify both ac and dc input signals. As with most schematics, the inputs are shown on the left, output on the right, and power is above and below. As discussed previously, the two balancing, or **offset null, inputs** will normally be connected to an external potentiometer that can be adjusted to set the output at zero volts when both the inverting and noninverting inputs are at zero volts. Balancing the op-amp to find the zero-volt output point, or null, in this way is generally needed due to slight imbalances within the op-amp's internal circuit.

An important point to realize at this time is that the op-amp is a single component, and up until this time we have concentrated on an understanding of the op-amp's internal circuitry because this helps us to better understand the circuit's normal input/output relationships and characteristics. These operational characteristics are important if we are going to be able to isolate whether a circuit malfunction is internal or external to the op-amp. However, because it is impossible to repair any internal op-amp failures, we will not concentrate on every detail of the op-amp's internal circuit.

19-2-5   *Basic Op-Amp Circuit Applications*

Now that we have an understanding of the op-amp's characteristics, let us now put it to use in some basic circuit applications. To begin with, we will examine the comparator circuit.

The Open-Loop Comparator Circuit

Figure 19-4 shows how the op-amp can be used to function as a **comparator,** which is a circuit that is used to detect changes in voltage level. In Figure 19-4(a), the inverting input (−) of the op-amp is grounded and the input signal is applied to the op-amp's noninverting input (+). Referring to the associated waveforms, you can see that when the input swings positive relative to the negative input (which is 0), the output of the amplifier goes into immediate saturation due to the very large gain of the op-amp. For example, if the op-amp had a voltage gain of 25,000 ($A_V = 25,000$), even a small input of $+25$ mV would cause the op-amp to try and drive its output to 625 V ($V_{out} = V_{in} \times A_V = 25$ mV $\times$ 25,000 $= 625$ V).

Offset Null Inputs

The two balancing inputs used to balance an op-amp.

Comparator

An op-amp used without feedback to detect changes in voltage level.

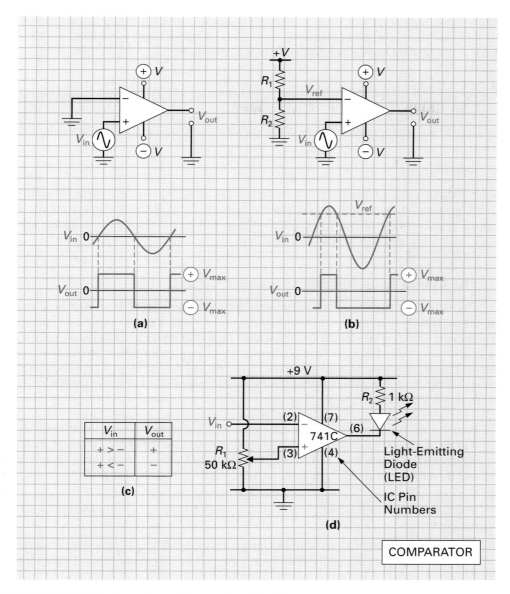

FIGURE 19-4   The Open-Loop Comparator Circuit.

Since the maximum possible positive output voltage cannot exceed the positive supply voltage ($+V$), the output goes to its maximum positive limit, which is equal to the $+V$ supply voltage. When the input swings negative, the amplifier is driven immediately into its opposite state (cutoff), and the output goes to its maximum negative limit, which is equal to the $-V$ supply voltage.

Figure 19-4(b) shows how a voltage divider made up of R_1 and R_2 can be used to supply the inverting input ($-$) of the op-amp with a reference voltage (V_{ref}) that can be determined by using the voltage-divider formula.

$$V_{ref} = \frac{R_2}{R_1 + R_2} \times (+V)$$

Referring to the associated waveforms in Figure 19-4(b), you can see that whenever the ac input signal is more positive than the reference voltage, the output is positive. On the other hand, whenever the ac input signal is less than the reference voltage, the output is negative.

The table in Figure 19-4(c) summarizes the operation of the comparator circuit. In the first line of the table you can see that when the negative input is negative relative to the positive input, the output will go to its positive limit ($V_{out} = +V$). In the second line of the table you can see that when the opposite occurs (the negative input is positive, or the positive input is negative), the output will go to its negative limit ($V_{out} = -V$).

■ EXAMPLE:

Briefly describe the operation of the 741C op-amp comparator circuit shown in Figure 19-4(d).

■ *Solution:*

Potentiometer R_1 sets up the reference voltage (V_{ref}), which can be anywhere between 0 V and +9 V. When the input voltage (V_{in}) to the negative input of the op-amp is greater than the positive reference voltage, the op-amp's output will be equal to the $-V$ voltage supply (which is ground), and so the LED will turn ON.

The Closed-Loop Inverting Amplifier Circuit

The op-amp is usually operated in either the **open-loop mode** or **closed-loop mode.** With the previously discussed comparator circuit, the op-amp was operating in its open-loop mode because there was no signal feedback from output to input. In most instances, the op-amp is operated in the closed-loop mode, in which there is signal feedback from output back to input. This feedback signal will always be out of phase with the input signal and therefore oppose the original signal, which is why it is called "degenerative or negative feedback." Negative feedback, however, is necessary in nearly all op-amp circuits for the following reasons:

1. Because the op-amp has such an extremely high gain, even a very small input signal will be amplified to a very large signal, which will drive the op-amp out of its linear region and into saturation and cutoff. Negative feedback will lower the op-amp's gain, and therefore control the op-amp to prevent output waveform distortion.

2. Having such a high gain can cause the amplifier to go into oscillation due to positive feedback. Negative feedback prevents an amplifier from going into oscillation by reducing the op-amp's gain.

3. The open-loop gain of an op-amp can have a very large range of value for the same device. For example, the 741's open-loop gain can be anywhere from a minimum of 25,000 to 200,000. Including negative feedback in the op-amp circuit will reduce the gain to a consistent value so that the same part can be relied on to provide the same response.

Figure 19-5(a) shows how an op-amp can be connected as an **inverting amplifier circuit,** which produces an amplified output signal that is 180° out of phase with the input signal. Looking at the output voltage label ($-V_{out}$), notice that the negative symbol preceding V_{out} is being used to indicate the 180° phase inversion between input and output. In this circuit arrangement, the input signal (V_{in}) is applied through an input resistor (R_{in}) to the inverting input ($-$) of the op-amp, while the noninverting input ($+$) is connected to ground. A feed-back loop is connected from the output back to the inverting input via the feed-back resistor R_F.

Let us now take a closer look at the closed-loop feedback system that occurs within this amplifier circuit. If the applied input voltage was zero volts ($V_{in} = 0$ V), the differential input signal (which is the difference between the op-amp's "+" and "−" inputs) will be 0 V, because both the inverting and noninverting inputs will now be at 0 V. A differential input of zero volts therefore will generate an output of zero volts. If the input signal were now to swing positive toward +5 V, the output (V_{out}) would swing negative due to the internal op-amp circuit phase inversions. This negative output voltage swing would be applied back to the inverting input via R_F to counteract the original positive input change. The feedback path

---

Open-Loop Mode
A control system that has no means of comparing the output with the input for control purposes.

Closed-Loop Mode
A control system containing one or more feedback control loops in which functions of the controlled signals are combined with functions of the commands that tend to maintain prescribed relationships between the commands and the controlled signals.

Inverting Amplifier Circuit
An op-amp circuit that produces an amplified output signal that is 180° out of phase with the input signal.

CHAPTER 19 / OPERATIONAL AMPLIFIERS

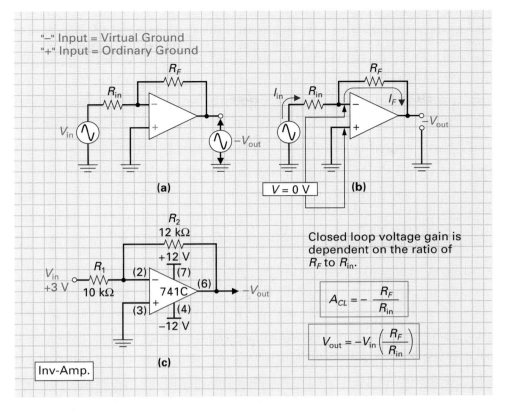

FIGURE 19-5　The Closed-Loop Inverting Amplifier Circuit.

is designed so that it cannot completely cancel the input signal, for if it did, there would be no input, and therefore no output or feedback. In most instances, the feedback voltage (V_F) will greatly restrain the input voltage change to the point that a $+5$ V input change at V_{in} will only be felt as a $+5$ microvolt change at the op-amp's inverting input. Therefore, even though V_{in} seems to change in values measured in volts, the inverting input of the op-amp will only change in values measured in microvolts. In fact, if the voltage at the inverting input of the op-amp is measured with a voltmeter as shown in Figure 19-5(b), the "$-$" input appears to remain at 0 V due to the very minute change at the "$-$" input. The inverting input of the op-amp in this case would be defined as a **virtual ground,** which is different from an **ordinary ground.** A virtual ground is a voltage ground because this point is at zero volts; however, it is not a current ground because it cannot sink or conduct away any current. An ordinary ground, on the other hand, is at zero volts and can sink any amount of current.

　　Returning to the inverting amplifier circuit in Figure 19-5(b), we can now analyze this circuit in a little more detail now that we know its basic operation. To begin with, if the "$-$" input of the op-amp is at 0 V, then all of the input voltage (V_{in}) will be dropped across the input resistor (R_{in}). Therefore, the input current can be calculated if we know the value of V_{in} and R_{in}, with the following formula:

$$I_{in} = \frac{V_{in}}{R_{in}}$$

Knowing that the left side of R_F is at 0 V means that all of the output voltage will be developed across R_F, and therefore the value of feedback current (I_F) can be calculated with the formula:

$$I_F = \frac{-V_{out}}{R_F}$$

Virtual Ground

A ground for voltage but not for current.

Ordinary Ground

A connection in the circuit that is said to be at ground potential or zero volts. Because of its connection to earth it has the ability to conduct electrical current to and from earth.

The extremely high input impedance of the op-amp means that only a very small fraction of the input current will enter the inverting input. In fact, nearly all of the current flowing through R_{in} and reaching the "−" op-amp input will leave this virtual ground point via the easiest path, which is through R_F. Therefore, it can be said that the feedback current is equal to the input current, or

$$I_{in} = I_F$$

If I_{in} (which equals V_{in}/R_{in}) and I_F (which equals $-V_{out}/R_F$) are equal, then

$$\frac{V_{in}}{R_{in}} = \frac{-V_{out}}{R_F}$$

If this is rearranged, we arrive at the following:

$$\frac{V_{out}}{V_{in}} = -\frac{R_F}{R_{in}}$$

Because the voltage gain of an amplifier is equal to

$$A_V = \frac{V_{out}}{V_{in}}$$

Closed-Loop Voltage Gain (A_{CL})

The voltage gain of an amplifier when it is operated in the closed-loop mode.

and because $V_{out}/V_{in} = -R_F/R_{in}$, the **closed-loop voltage gain (A_{CL})** of the inverting operational amplifier is equal to the ratio of R_F to R_{in}.

$$A_{CL} = -\frac{R_F}{R_{in}}$$

(Negative symbol preceding R_F/R_{in} indicates signal inversion)

By rearranging the previous equation $V_{out}/V_{in} = -R_F/R_{in}$, we can arrive at the following formula for calculating the output voltage of this circuit.

Since

$$\frac{V_{out}}{V_{in}} = -\frac{R_F}{R_{in}}$$

$$V_{out} = -V_{in}\left(\frac{R_F}{R_{in}}\right)$$

The input impedance of this inverting op-amp is equal to the value of R_{in} because the input voltage (V_{in}) is developed across the input resistor (R_{in}).

EXAMPLE:

Calculate the output voltage of the inverting amplifier shown in Figure 19-5(c).

Solution:

Noninverting Amplifier Circuit

An op-amp circuit that produces an amplified output signal that is in phase with the input signal.

$$V_{out} = -V_{in}\left(\frac{R_F}{R_{in}}\right) = -3\text{ V} \times \frac{12\text{ k}\Omega}{10\text{ k}\Omega} = -3\text{ V} \times 1.2 = -3.6\text{ V}$$

The Closed-Loop Noninverting Amplifier Circuit

Figure 19-6(a) shows how an op-amp can be configured to operate as a **noninverting amplifier circuit.** The input voltage (V_{in}) is applied to the op-amp's noninverting input (+),

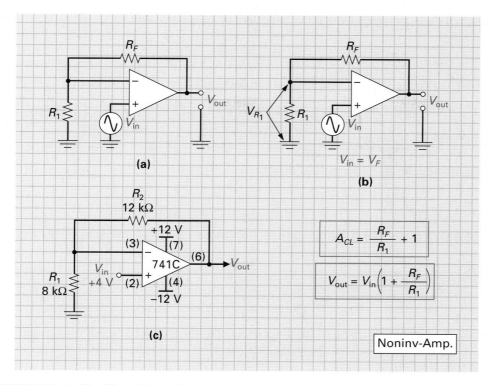

FIGURE 19-6   The Closed-Loop Noninverting Amplifier Circuit.

and therefore the output voltage (V_{out}) will be in phase with the input. To achieve negative feedback, the output is applied back to the inverting input ($-$) of the op-amp via the feedback network formed by R_F and R_1.

Let us now take a closer look at the closed-loop negative feedback system that occurs within this amplifier circuit. The output voltage (V_{out}) is proportionally divided across R_F and R_1, with the feedback voltage (V_F) developed across R_1 being applied to the inverting input ($-$) of the op-amp, as shown in Figure 19-6(b). Because V_{out} is in phase with V_{in}, the feedback voltage (V_F) will also be in phase with V_{in}, and therefore these two in-phase inputs to the op-amp will be common-mode input signals. As a result, feedback will be degenerative. However, because V_{out} is slightly larger than V_{in}, there will be a small difference between V_{in} and V_F, and this differential input will be amplified. To summarize, the noninverting op-amp provides negative feedback by feeding back an in-phase common-mode signal, and this degenerative feedback will lower the op-amp's gain to

1. prevent output waveform distortion,
2. prevent the amplifier from going into oscillation, and
3. reduce the gain of the op-amp to a consistent value.

The very small microvolt difference between V_{in} and V_F will be amplified; however, since there is such a very small difference between these two input signals, it can be said that

$$V_{\text{in}} = V_F$$

Because the closed-loop voltage gain of any op-amp is equal to

$$A_{CL} = \frac{V_{\text{out}}}{V_{\text{in}}}$$

the gain of the noninverting amplifier could also be calculated with

$$A_{CL} = \frac{V_{out}}{V_F}$$

By using the voltage-divider formula, we can develop a formula for calculating V_F.

$$V_F = \frac{R_1}{R_1 + R_2} \times V_{out}$$

By rearranging the above formula as follows:

$$\frac{V_{out}}{V_F} = \frac{R_1 + R_F}{R_1} \quad \text{or} \quad \frac{V_{out}}{V_F} = \frac{R_1}{R_1} + \frac{R_F}{R_1} = \frac{R_F}{R_1} + 1$$

and because $V_{out}/V_F = A_{CL}$, the noninverting op-amp's gain can also be calculated with the formula

$$A_{CL} = \frac{R_F}{R_1} + 1$$

Because the output voltage of an amplifier is equal to the product of input voltage and gain ($V_{out} = V_{in} \times A_{CL}$), we can add V_{in} to the previous closed-loop gain formula in order to calculate output voltage.

$$V_{out} = V_{in}\left(1 + \frac{R_F}{R_1}\right)$$

With the inverting op-amp, the input impedance is determined by the input resistor (R_{in}). With the noninverting amplifier, there is no resistor connected because V_{in} is applied directly into the very high input impedance of the op-amp. As a result, the noninverting amplifier circuit has an extremely high input impedance.

EXAMPLE:

Calculate the output voltage from the noninverting amplifier shown in Figure 19-6(c).

Solution:

$$V_{out} = V_{in}\left(1 + \frac{R_F}{R_1}\right) = 4\,\text{V}\left(1 + \frac{12\,\text{k}\Omega}{8\,\text{k}\Omega}\right) = 4\,\text{V} \times (1 + 1.5) = 4\,\text{V} \times 2.5 = +10\,\text{V}$$

The open-loop comparator and closed-loop inverting and noninverting amplifier circuits are just three of many op-amp application circuits. In the final section of this chapter, we will examine many other typical op-amp circuit applications, but for now let us review our understanding of the op-amp by examining a typical manufacturer's data sheet.

19-2-6   *An Op-Amp Data Sheet*

To better understand the characteristics of the op-amp, Figure 19-7 shows the data sheet for a "747," which is a 14-pin IC containing two 741 op-amps. Like most data sheets, this one contains a general description of the device, a pin-configuration diagram, a listing of

DEVICE: µA 747—Dual Linear Op-Amp IC

DESCRIPTION
The 747 is a pair of high-performance monolithic operational amplifiers constructed on a single silicon chip. High common-mode voltage range and absence of "latch-up" make the 747 ideal for use as a voltage-follower. The high gain and wide range of operating voltage provides superior performance in integrator, summing amplifier, and general feedback applications. The 747 is short-circuit protected and requires no external components for frequency compensation. The internal 6dB/octave roll-off insures stability in closed-loop applications. For single amplifier performance, see µA741 data sheet.

FEATURES
- No frequency compensation required
- Short-circuit protection
- Offset voltage null capability
- Large common-mode and differential voltage ranges
- Low power consumption
- No latch-up

PIN CONFIGURATION

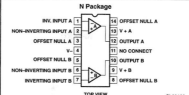

ABSOLUTE MAXIMUM RATINGS

SYMBOL	PARAMETER	RATING	UNIT
V_S	Supply voltage	±18	V
$P_{D\,MAX}$	Maximum power dissipation T_A=25°C (still air)[1]	1500	mW
V_{IN}	Differential input voltage	±30	V
V_{IN}	Input voltage[2]	±15	V
	Voltage between offset null and V-	±0.5	V
T_{STG}	Storage temperature range	-65 to +150	°C
T_A	Operating temperature range	0 to +70	°C
T_{SOLD}	Lead temperature (soldering, 10sec)	300	°C
I_{SC}	Output short-circuit duration	Indefinite	

Explanation of Key Maximum Ratings

Supply Voltage: This is the maximum voltage that can be used to power the op-amp.

Maximum Power Dissipation: This is the maximum power the op-amp can dissipate.

Differential Input Voltage: This is the maximum voltage that can be applied across the + and – inputs.

Input Voltage: This is the maximum voltage that can be applied between an input and ground.

Operating Temperature Range: The temperature range in which the op-amp will operate within the manufacturer's specifications.

Output Short Circuit Duration: This is the amount of time that the op-amp's output can be short circuited to ground or to a supply voltage. This op-amp has an "indefinite" rating since it has an internal circuit that will turn OFF the op-amp's output and protect the internal circuitry if an output short occurs.

FIGURE 19-7   An Op-Amp Data Sheet. (Courtesy of Philips Semiconductors.)   *Continued*

maximum ratings, an internal-circuit diagram, and a listing of input/output characteristics. Notes have been included so that you will be able to understand the meaning of most key terms.

Figure 19-8 compares the characteristics and cost of several different op-amp types to the 741, which is very popular due to its good performance and low price. Referring to the cost-factor column, you can see, for example, that the LF351 is twice the price of the 741. Studying the key characteristics in this figure, you can see that the 741, for example, will typically have an input impedance of 2 MΩ, an output impedance of 75Ω, an open-loop gain of 25,000, and a common-mode rejection ratio of 70 dB for any input signal from 0 Hz up to 1 MHz. The gain bandwidth column lists only the upper frequency limit because the op-amp has no internal coupling capacitors, and therefore the lower frequency limit extends down to dc signals, or 0 Hz.

DEVICE: µA 747—Dual Linear Op-Amp IC

EQUIVALENT SCHEMATIC

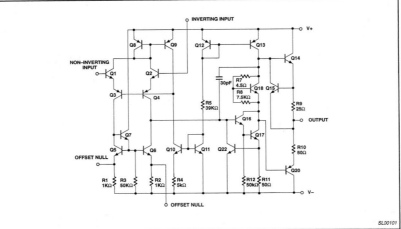

SL00101

DC ELECTRICAL CHARACTERISTICS

T_A=25°C, V_{CC} = ±15V unless otherwise specified.

SYMBOL	PARAMETER	TEST CONDITIONS	µA747C Min	µA747C Typ	µA747C Max	UNIT
V_{OS}	Offset voltage	R_S≤10kΩ		2.0	6.0	mV
		R_S≤10kΩ, over temp.		3.0	7.5	mV
$\Delta V_{OS}/\Delta T$				10		µV/°C
I_{OS}	Offset current			20	200	nA
		Over temperature		7.0	300	nA
$\Delta I_{OS}/\Delta T$				200		pA/°C
I_{BIAS}	Input current			80	500	nA
		Over temperature		30	800	nA
$\Delta I_B/\Delta T$				1		nA/°C
V_{OUT}	Output voltage swing	R_L≥2kΩ, over temp.	±10	±13		V
		R_L≥10kΩ, over temp.	±12	±14		V
I_{CC}	Supply current each side			1.7	2.8	mA
		Over temperature		2.0	3.3	mA
P_d	Power consumption			50	85	mW
		Over temperature		60	100	mW
C_{IN}	Input capacitance			1.4		pF
	Offset voltage adjustment range			±15		mV
R_{OUT}	Output resistance			75		Ω
	Channel separation			120		dB
PSRR	Supply voltage rejection ratio	R_S≤10kΩ, over temp.		30	150	µV/V
A_{VOL}	Large-signal voltage gain (DC)	R_L≥2kΩ, V_{OUT}=±10V	25,000			V/V
		Over temperature	15,000			V/V
CMRR	Common-mode rejection ratio	R_S≤10kΩ, V_{CM}=±12V	70			dB
		Over temperature				

Explanation of Key Ratings

Input Offset Voltage: The voltage that must be applied to one input for the output voltage to be zero.

Input Offset Current: The difference of the two input bias currents when the output voltage is zero.

Input Bias Current: The average of the currents flowing into both inputs (ideally input bias currents are equal).

Input Resistance: This is the resistance of either input when the other input is grounded.

Output Resistance: This is the resistance of the op-amp's output.

Output Short-Circuit Current: Maximum output current that the op-amp can deliver to a load.

Supply Current: The current that the op-amp circuit will draw from the power supply.

Slew rate: The maximum rate of change of output voltage under large signal conditions.

Channel Separation: When two op-amps are within one package, there will be a certain amount of interference or "crosstalk" between op-amps.

FIGURE 19-7   (continued)

OP-AMP TYPE	INPUT IMPEDANCE	OUTPUT IMPEDANCE	OPEN-LOOP GAIN (Min.)	CMRR (dB)	COST FACTOR	SLEW RATE (V/μs)	GAIN BANDWIDTH	FEATURES
741 C	2 MΩ	75Ω	25,000	70	1	0.5	1 MHz	Low cost
101	800 kΩ	Low	25,000	70	1	0.5	1 MHz	Low cost
108 A	70 MΩ	Low	80,000	96	—	0.3	1 MHz	Precision low drift
351	High	Low	25,000	70	2	13	4 MHz	Low bias current
318	High	Low	25,000	70	5	70	15 MHz	High slew rate
357	High	Low	50,000	80	4	30	20 MHz	High CMRR
363	High	Low	1,000,000	94	45	—	2 MHz	Low noise; high rejection
356	High	Low	25,000	80	3	10	5 MHz	Improved 741

FIGURE 19-8 **Comparing the Characteristics of Several Op-Amps.**

SELF-TEST EVALUATION POINT FOR SECTION 19-2

Now that you have completed this section, you should be able to:

- ***Objective 2.*** *Explain how the op-amp's first differential amplifier stage gives the op-amp its high input impedance, high differential gain, and low common-mode gain characteristics.*
- ***Objective 3.*** *Define common-mode rejection ratio.*
- ***Objective 4.*** *Describe how the internal second stage voltage amplifier gives the op-amp its very high gain characteristic.*
- ***Objective 5.*** *Describe how the internal third stage emitter-follower amplifier gives the op-amp its low output impedance and current gain characteristics.*
- ***Objective 6.*** *Explain how the op-amp operates in the following three basic circuit applications:*
  a. *Open-loop comparator circuit*
  b. *Closed-loop inverting amplifier circuit*
  c. *Closed-loop noninverting amplifier circuit*
- ***Objective 7.*** *Interpret a typical op-amp data sheet.*
- ***Objective 8.*** *Compare the characteristics of several different op-amp types.*

Use the following questions to test your understanding of Section 19-2.

1. Can the op-amp be used to amplify dc as well as ac signal inputs?
2. What is the difference between an open-loop and closed-loop op-amp circuit?
3. List the key characteristics of an op-amp.
4. What are the three basic amplifier types within an op-amp?
5. Which of the following circuits is connected in an open-loop mode?
   a. Comparator
   b. Inverting amplifier
   c. Noninverting amplifier
6. Why is it important for an op-amp circuit to have negative feedback?

19-3 ADDITIONAL OP-AMP CIRCUIT APPLICATIONS

The operational amplifier's flexibility and characteristics make it the ideal choice for a wide variety of circuit applications. In fact, because the op-amp is the most frequently used linear IC, it is safe to say that you will find several op-amps in almost every electronic system. Although it is impossible to cover all of these applications, many circuits predominate, and others are merely variations on the same basic theme. In this section we will concentrate on the operation and characteristics of all of the most frequently used op-amp circuit applications.

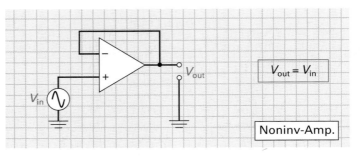

FIGURE 19-9   Voltage-Follower Circuit.

19-3-1   *The Voltage-Follower Circuit*

Figure 19-9 shows how an op-amp can be connected to form a noninverting **voltage-follower circuit.** Using the noninverting closed-loop gain formula discussed previously, we can calculate the voltage gain of this circuit.

$$A_{CL} = \frac{R_F}{R_1} + 1 = \frac{0\ \Omega}{0\ \Omega} + 1 = 0 + 1 = 1$$

With a gain of 1, the output voltage will be equal to the input voltage—so what is the advantage of this circuit? The answer is the op-amp characteristics of a high input impedance and a low output impedance. Similar to the BJT's emitter-follower and the FET's source-follower, the op-amp voltage-follower circuit derives its name from the fact that the output voltage follows the input voltage in both polarity and amplitude. This circuit is therefore ideal as a buffer, interfacing a high-impedance source to a low-impedance load.

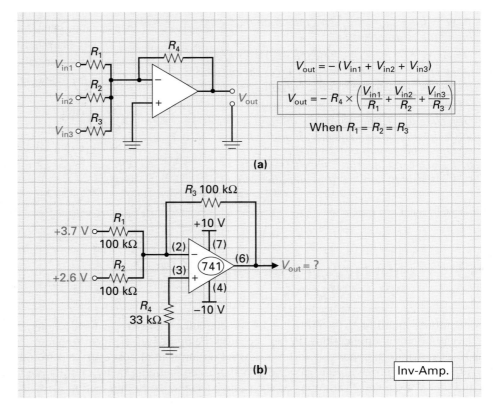

FIGURE 19-10   Summing Amplifier Circuit.

19-3-2 *The Summing Amplifier Circuit*

The **summing amplifier circuit,** or adder amplifier, consists of two or more input resistors connected to the inverting input of an op-amp as shown in Figure 19-10(a). This circuit will sum or add all of the input voltages, and therefore the output voltage will be

$$V_{out} = -(V_{in1} + V_{in2} + V_{in3})$$

Using Ohm's law ($V = R \times I$), we can also calculate the output voltage with the formula

$$V_{out} = -R_4 \times \left(\frac{V_{in1}}{R_1} + \frac{V_{in2}}{R_2} + \frac{V_{in3}}{R_3} \right)$$

Once again, the negative sign preceding the formula indicates that the output signal will be opposite in polarity to the two or three input signals.

> **Summing Amplifier Circuit (or Adder Circuit)**
> An op-amp circuit that will sum or add all of the input voltages.

■ **EXAMPLE:**

Calculate the output voltage of the summing amplifier circuit shown in Figure 19-10(b).

■ *Solution:*

$$V_{out} = -(V_{in1} + V_{in2}) = -(3.7\,V + 2.6\,V) = -6.3\,V$$

19-3-3 *The Difference Amplifier Circuit*

Figure 19-11(a) shows how the op-amp can be connected to operate as a difference or **differential amplifier circuit.** In this circuit application, the op-amp will simply be making use of its first internal amplifier stage, which (as mentioned earlier) is a diff-amp. In this circuit, all four resistors are normally of the same value, and the output voltage is equal to the difference between the two input voltages.

$$V_{out} = V_{in2} - V_{in1}$$

> **Differential Amplifier Circuit**
> An op-amp circuit in which the output voltage is equal to the difference between the two input voltages.

■ **EXAMPLE:**

Calculate the output voltage of the difference amplifier circuit shown in Figure 19-11(b).

■ *Solution:*

$$V_{out} = V_{in2} - V_{in1} = 6\,V - 3.2\,V = 2.8\,V$$

19-3-4 *The Differentiator Circuit*

The op-amp **differentiator circuit** (not to be confused with the previous differential circuit) is similar to the basic inverting amplifier except that R_{in} is replaced by a capacitor, as shown in Figure 19-12(a). Including a capacitor in any circuit means that we will develop problems as the frequency of the input signal increases because capacitive reactance is inversely proportional to frequency. This means that the reactance of the input capacitor will decrease for input signals that are higher in frequency, and therefore the input voltage applied to the

> **Differentiator Circuit**
> A circuit whose output voltage is proportional to the rate of change of the input voltage. The output waveform is the time derivative of the input waveform.

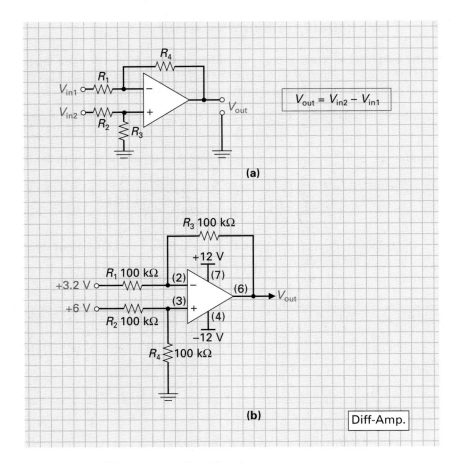

(a)

$$V_{out} = V_{in2} - V_{in1}$$

R_4

R_1 — V_{in1}

V_{in2} — R_2

R_3

V_{out}

R_3 100 kΩ

R_1 100 kΩ (2) +12 V (7)

+3.2 V

(3)

+6 V — R_2 100 kΩ

(6) V_{out}

(4)

−12 V

R_4 100 kΩ

(b)

Diff-Amp.

FIGURE 19-11   The Difference Amplifier Circuit.

op-amp and output voltage from the op-amp will increase with frequency. Including an additional resistor (R_S) in series with the input capacitor, as shown in Figure 19-12(b), will decrease the high-frequency gain because gain will now be a ratio of R_F/R_S.

Figure 19-12(c) shows the differentiator's input/output waveforms, with the peak of the output square wave being equal to

$$V_{out(pk)} = 2\,\pi f \times R_1 \times C_1 \times V_{in(pk)}$$

◻ **EXAMPLE:**

Calculate the peak output voltage of the differentiator circuit shown in Figure 19-12(d).

◻ *Solution:*

$$V_{out(pk)} = 2\,\pi f \times R_1 \times C_1 \times V_{in(pk)}$$
$$= (2 \times \pi \times 500\ \text{Hz}) \times 100\ \text{k}\Omega \times 0.01\ \mu\text{F} \times (1.5\ \text{V pk}) = 4.7\ \text{V}$$

19-3-5   *The Integrator Circuit*

Integrator Circuit

A circuit with an output
which is the integral of its
input with respect to time.

Figure 19-13(a) shows how the position of the resistor and capacitor in the differentiator circuit can be reversed to construct an op-amp **integrator circuit.** As in the differentiator circuit, the capacitor will alter the gain of the op-amp because its capacitive reactance changes with frequency. To compensate for this effect, a parallel resistor (R_P) is included in

CHAPTER 19 / OPERATIONAL AMPLIFIERS

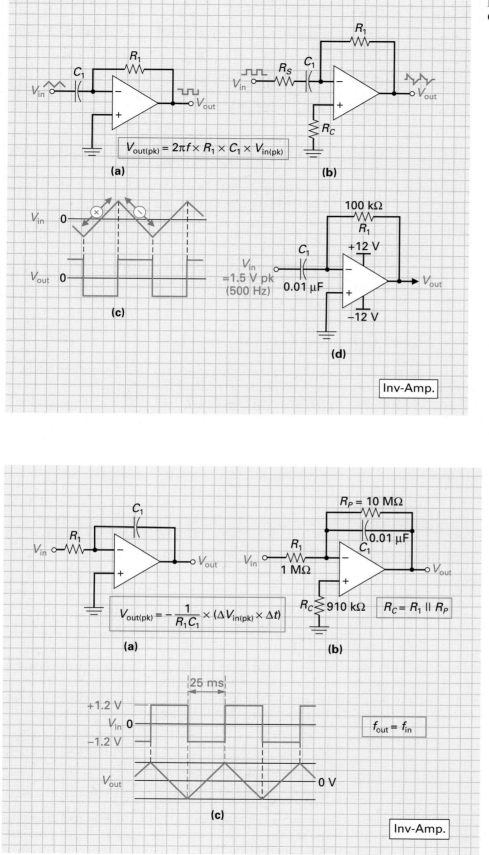

$$V_{out(pk)} = 2\pi f \times R_1 \times C_1 \times V_{in(pk)}$$

(a)

(b)

(c)

V_{in}
$= 1.5 \text{ V pk}$
(500 Hz)

C_1
0.01 μF

100 kΩ
R_1
$+12 \text{ V}$
-12 V

V_{out}

(d)

Inv-Amp.

FIGURE 19-12   **Differentiator Circuit.**

$$V_{out(pk)} = -\frac{1}{R_1 C_1} \times (\Delta V_{in(pk)} \times \Delta t)$$

(a)

C_1

R_1

V_{in}

V_{out}

$R_P = 10 \text{ MΩ}$

0.01 μF

C_1

R_1
1 MΩ

V_{in}

V_{out}

$R_C \lessgtr 910 \text{ kΩ}$

$R_C = R_1 \parallel R_P$

(b)

25 ms

$+1.2 \text{ V}$
$V_{in} \ 0$
-1.2 V

$f_{out} = f_{in}$

V_{out}

0 V

(c)

Inv-Amp.

FIGURE 19-13   **Integrator Circuit.**

shunt with the capacitor, as shown in Figure 19-13(b), to decrease the low-frequency gain: gain will now be a ratio of R_P/R_1.

Figure 19-13(c) shows the integrator's input/output waveforms, with the peak of the output triangular wave being equal to

$$V_{\text{out(pk)}} = \frac{1}{R_1 C_1} \times (\Delta V_{\text{in(pk)}} \times \Delta t)$$

■ EXAMPLE:

Calculate the peak output voltage of the integrator circuit shown in Figure 19-13(b) considering the input/output waveforms given in Figure 19-13(c).

■ *Solution:*

$$V_{\text{out(pk)}} = \frac{1}{R_1 C_1} \times (\Delta V_{\text{in(pk)}} \times \Delta t) = \frac{1}{1\ \text{M}\Omega \times 0.01\ \mu\text{F}} \times (0\ \text{V to } 1.2\ \text{V pk} \times 25\ \text{ms})$$

$$= 100 \times (1.2\ \text{V pk} \times 25\ \text{ms}) = 100 \times 0.03 = 3\ \text{V pk}$$

19-3-6 *Signal Generator Circuits*

Figure 19-14 shows how the op-amp can be connected to act as a signal generator, which is a circuit that will convert a dc supply voltage into a repeating output signal.

In Figure 19-14(a), the **twin-T sine-wave oscillator,** which has two T-shaped feedback networks, will generate a repeating sine-wave output at a frequency equal to

$$f_0 = \frac{1}{2\pi RC}$$

■ EXAMPLE:

Calculate the frequency of the oscillator shown in Figure 19-14(a).

■ *Solution:*

$$f_0 = \frac{1}{2\pi RC} = \frac{1}{2 \times \pi \times 6.8\ \text{k}\Omega \times 0.033\ \mu\text{F}} = 709\ \text{Hz}$$

In Figure 19-14(b), the op-amp has been connected to act as a **square-wave generator,** or as it is more frequently called a **relaxation oscillator.** The output signal is fed back to both the inverting and noninverting inputs of the op-amp. The capacitor will charge and discharge through R controlling the frequency of the output square wave, which is equal to

$$f_0 = \frac{1}{2\ RC \log\left(\dfrac{2R_1}{R_2} + 1\right)}$$

Connecting the output of the square-wave generator in Figure 19-14(b) into the inputs of the circuits in Figure 19-14(c) and (d), we can generate a triangular or staircase output waveform. The **triangular-wave generator** in Figure 19-14(c) is simply the integrator

---

Twin-T Sine-Wave Oscillator

An oscillator circuit that makes use of two T-shaped feedback networks.

Square-Wave Generator

A circuit that generates a continuously repeating square wave.

Relaxation Oscillator

An oscillator circuit whose frequency is determined by an *RL* or *RC* network, producing a rectangular or sawtooth output waveform.

Triangular-Wave Generator

A signal generator circuit that produces a continuously repeating triangular wave output.

CHAPTER 19 / OPERATIONAL AMPLIFIERS

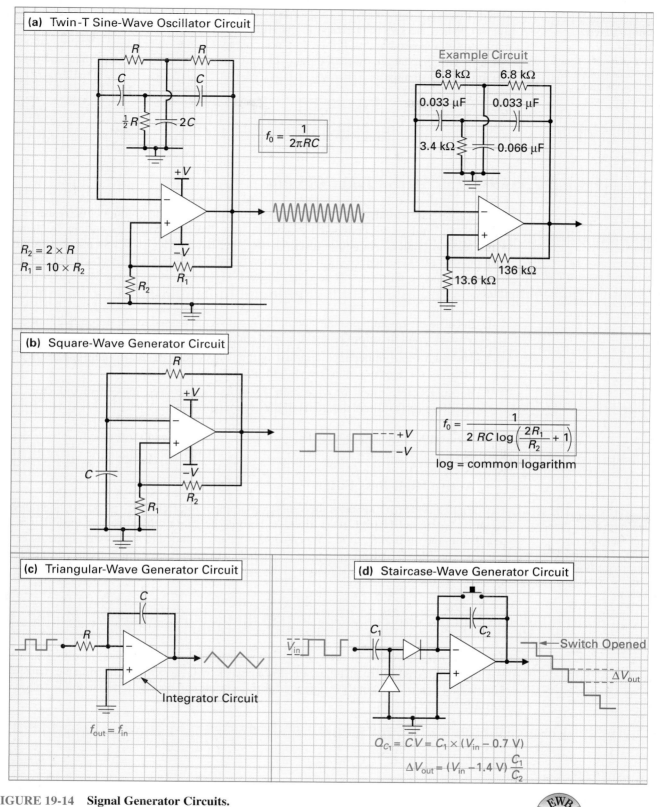

(a) Twin-T Sine-Wave Oscillator Circuit

Example Circuit

$$f_0 = \frac{1}{2\pi RC}$$

$R_2 = 2 \times R$
$R_1 = 10 \times R_2$

(b) Square-Wave Generator Circuit

$$f_0 = \frac{1}{2\,RC\log\left(\dfrac{2R_1}{R_2} + 1\right)}$$

log = common logarithm

(c) Triangular-Wave Generator Circuit

Integrator Circuit

$f_{out} = f_{in}$

(d) Staircase-Wave Generator Circuit

Switch Opened

ΔV_{out}

$Q_{C_1} = CV = C_1 \times (V_{in} - 0.7\ V)$

$\Delta V_{out} = (V_{in} - 1.4\ V)\dfrac{C_1}{C_2}$

FIGURE 19-14   Signal Generator Circuits.

EWB

Staircase-Wave Generator

A signal generator circuit that produces an output signal voltage that increases in steps.

circuit discussed previously, with its output frequency equal to the square-wave input frequency. When the switch is closed in the **staircase-wave generator** in Figure 19-14(d) the capacitor is bypassed and will therefore not charge. On the other hand, when the switch is open, C_2 will be charged by each input cycle, producing equal output steps that have the following voltage change

$$\Delta V_{out} = (V_{in} - 1.4 \text{ V}) \frac{C_1}{C_2}$$

19-3-7 *Active Filter Circuits*

Active Filter

A circuit that uses an amplifier with passive filter elements to provide frequency paths with rejection characteristics.

Passive filters are circuits that contain passive or nonamplifying components (resistors, capacitors, and inductors) connected in such a way that they will pass certain frequencies while rejecting others. An **active filter,** on the other hand, is a circuit that uses an amplifier with passive filter elements to provide frequency paths with rejection characteristics. Active filters, like the op-amp circuits seen in Figure 19-15, have several advantages over passive filters.

1. Because the op-amp provides gain, the input signal passed to the output will not be attenuated, and therefore better response curves can be obtained.
2. The high input impedance and low output impedance of the op-amp means that the filter circuit does not interfere with the signal source or load.
3. Because active filters provide gain, resistors can be used instead of inductors, and therefore active filters are generally less expensive.

Figure 19-15 illustrates how the op-amp can be connected to form the four basic active filter types.

Active High-Pass Filter

Active High-Pass Filter

A circuit that uses an amplifier with passive filter elements to pass all frequencies above a cutoff frequency.

Figure 19-15(a) illustrates the simple op-amp circuit, frequency response, and relevant formulas for an **active high-pass filter.** As before, the gain of this inverting amplifier is dependent on the ratio of R_F to R_{in}. When capacitors are included in any circuit, impedance (Z) must be considered instead of simply resistance, and gain is now equal to the ratio of feedback impedance to input impedance.

$$A_{CL} = -\frac{Z_F}{Z_{in}}$$

The input RC network will offer a high impedance to low frequencies, resulting in a low voltage gain. At high frequencies, the RC network will have a low impedance, causing a high voltage gain. The cutoff frequency for this circuit can be calculated with the following formula when $C_1 = C_2$.

$$f_C = \frac{1}{2\pi RC}$$

Active Low-Pass Filter

Active Low-Pass Filter

Amplifier circuit with passive filter elements to pass all frequencies below a cutoff frequency.

Figure 19-15(b) illustrates the op-amp circuit, frequency response curve, and relevant formulas for an **active low-pass filter.** At low frequencies, the capacitor's reactance is high, and low-frequency signals will be passed to the op-amp's input to be amplified and passed to

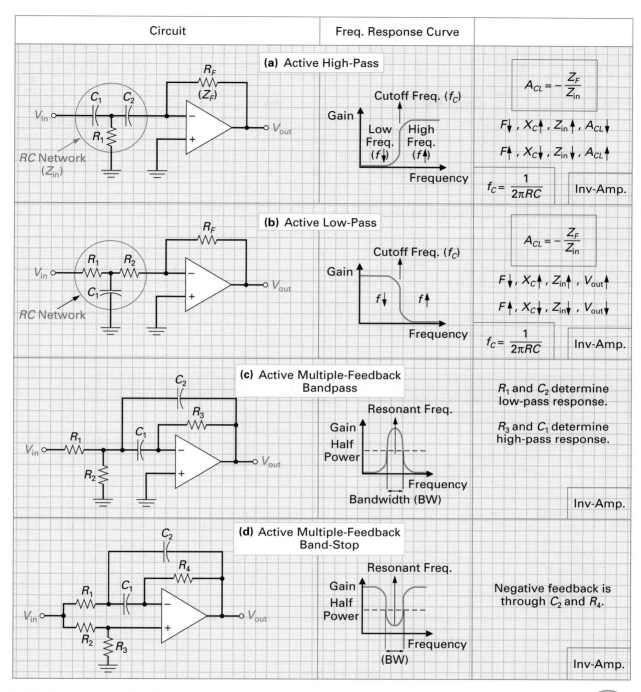

Circuit	Freq. Response Curve	

(a) Active High-Pass

$$A_{CL} = -\frac{Z_F}{Z_{in}}$$

$F\downarrow, X_C\uparrow, Z_{in}\uparrow, A_{CL}\downarrow$

$F\uparrow, X_C\downarrow, Z_{in}\downarrow, A_{CL}\uparrow$

$$f_C = \frac{1}{2\pi RC}$$ Inv-Amp.

(b) Active Low-Pass

$$A_{CL} = -\frac{Z_F}{Z_{in}}$$

$F\downarrow, X_C\uparrow, Z_{in}\uparrow, V_{out}\uparrow$

$F\uparrow, X_C\downarrow, Z_{in}\downarrow, V_{out}\downarrow$

$$f_C = \frac{1}{2\pi RC}$$ Inv-Amp.

(c) Active Multiple-Feedback Bandpass

R_1 and C_2 determine low-pass response.

R_3 and C_1 determine high-pass response.

Inv-Amp.

(d) Active Multiple-Feedback Band-Stop

Negative feedback is through C_2 and R_4.

Inv-Amp.

FIGURE 19-15 **Active Filter Circuits.**

the output. As frequency increases, the capacitive reactance of C_1 will decrease; more of the signal will be shunted away from the op-amp and will not appear at the output. The cutoff frequency for this circuit can be calculated with the following formula when $R_1 = R_2$.

$$f_C = \frac{1}{2\pi RC}$$

Active Bandpass Filter

Figure 19-15(c) illustrates how the op-amp can be connected to form an **active bandpass filter.** At frequencies outside of the band, V_{out} is fed back to the input without being attenuated, and therefore the input signal amplitude is almost equal to the feedback signal amplitude. This results in almost complete cancellation of the signal and therefore a very small output voltage. On the other hand, for the narrow band of frequencies within the band, the feedback network will increase its amount of attenuation. This increase of attenuation means that a very small feedback signal will appear back at the negative input of the op-amp and will have a very small degenerative effect. As a result, the change at the input of the op-amp will be larger when the input signal frequencies are within this band, and the voltage out will also be larger.

Active Band-Stop Filter

Figure 19-15(d) illustrates how the op-amp can be connected to form an **active band-stop filter,** also known as a band-reject or notch filter. The basic operation of this circuit is opposite to that of the previously discussed bandpass filter. At frequencies outside of the band, the feedback signal will be heavily attenuated, and therefore the degenerative effect will be small and the output voltage large. On the other hand, at frequencies within the band, the feedback signal will not be heavily attenuated, and therefore the degenerative effect will be large and the output voltage small.

SELF-TEST EVALUATION POINT FOR SECTION 19-3

Now that you have completed this section, you should be able to:

■ *Objective 9.* *Identify and describe the operation of the following op-amp circuit applications:*
  a. Voltage-follower circuit
  b. Summing amplifier circuit
  c. Difference amplifier circuit
  d. Differentiator circuit
  e. Integrator circuit
  f. Signal generator circuits
  g. Active filter circuits

Use the following questions to test your understanding of Section 19-3.

1. Which op-amp circuit provides a voltage gain of 1 and is used as a buffer?

2. Which op-amp circuit will sum all of the input voltages?

3. What is the basic circuit difference and input/output waveform difference between the integrator and differentiator circuit?

4. Which op-amp circuit will generate an output that is equal to the difference between the two inputs?

5. Sketch a circuit showing how the op-amp can be connected to generate a repeating square-wave output.

6. What is the difference between an active filter and a passive filter?

SUMMARY

The Operational Amplifier (Figure 19-16)

1. The name "operational amplifier" or "op-amp" was chosen because the circuit was used as a high-gain dc "amplifier" performing mathematical "operations."

2. Today's IC op-amp is a very high-gain dc amplifier that can have its operating characteristics changed by connecting different external components.

3. The two op-amp inputs are labeled "−" and "+." The "−" or negative input is called the inverting input because any signal applied to this input will be amplified and inverted between input and output. On the other hand, the "+" or

positive input is called the noninverting input because any signal applied to this input will be amplified but not inverted between input and output.

4. The two power supply connections to the op-amp are labeled "+V" and "−V." Having both a positive and negative power supply voltage will allow the output signal to swing positive and negative, above and below zero. As with all high-gain amplifiers, however, the output voltage can never exceed the value of the +V and −V supply voltages.

5. The two offset or balancing inputs will normally be connected to a potentiometer that can be adjusted to set the

FIGURE 19-16   The Operational Amplifier (Op-Amp).

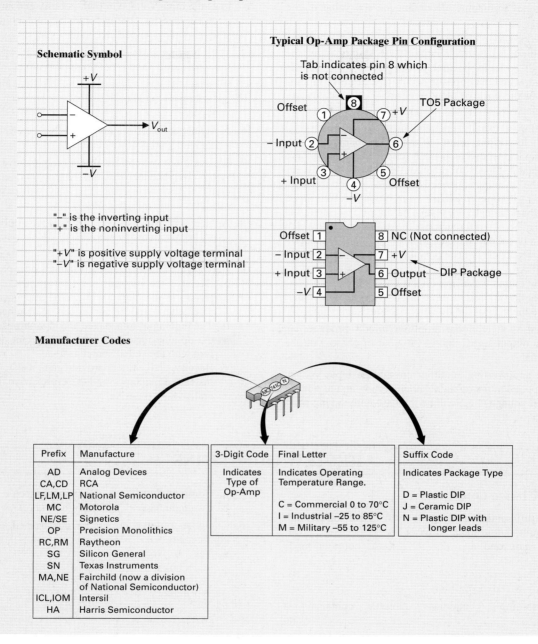

Schematic Symbol

$+V$

$-$

$+$

V_{out}

$-V$

"$-$" is the inverting input
"$+$" is the noninverting input

"$+V$" is positive supply voltage terminal
"$-V$" is negative supply voltage terminal

Typical Op-Amp Package Pin Configuration

Tab indicates pin 8 which
is not connected

TO5 Package

Offset   1   7 $+V$

$-$ Input 2   6

$+$ Input 3   5

4   Offset

$-V$

Offset   1   8 NC (Not connected)

$-$ Input 2   7 $+V$

$+$ Input 3   6 Output   DIP Package

$-V$ 4   5 Offset

Manufacturer Codes

Prefix	Manufacture	3-Digit Code	Final Letter	Suffix Code
AD	Analog Devices	Indicates Type of Op-Amp	Indicates Operating Temperature Range.	Indicates Package Type
CA,CD	RCA			
LF,LM,LP	National Semiconductor		C = Commercial 0 to 70°C	D = Plastic DIP
MC	Motorola		I = Industrial −25 to 85°C	J = Ceramic DIP
NE/SE	Signetics		M = Military −55 to 125°C	N = Plastic DIP with longer leads
OP	Precision Monolithics			
RC,RM	Raytheon			
SG	Silicon General			
SN	Texas Instruments			
MA,NE	Fairchild (now a division of National Semiconductor)			
ICL,IOM	Intersil			
HA	Harris Semiconductor			

output at zero volts when both the inverting and noninverting inputs are at zero volts.

6. The three circuits all interconnected and contained within the single op-amp IC are a differential amplifier, a voltage amplifier, and an output amplifier. Combined, these three circuits give the op-amp its key characteristics, which are high gain, high input impedance, and low output impedance.

7. A differential input will be amplified by the op-amp's first-stage differential amplifier and passed to the output, while

unwanted signals caused by temperature variations or noise will appear as common-mode input signals and therefore be rejected. An op-amp's ability to provide a high differential gain (A_{VD}) and a low common-mode gain (A_{CM}) is directly dependent on its internal differential amplifier and is a measure of an op-amp's performance. This ratio is called the common-mode rejection ratio (CMRR).

8. The voltage-amplifier stage following the diff-amp usually consists of several darlington-pair stages that provide an overall op-amp voltage gain of typically 50,000 to 200,000. The final-output stage consists of a complementary emitter-follower stage to provide a low output impedance and high current gain so that the op-amp can deliver up to several milliamps, depending on the value of the load.

9. With the open-loop op-amp comparator circuit, when the negative input is negative relative to the positive input, the output will go to its positive limit ($V_{out} = +V$). On the other hand, when the opposite occurs (the negative input is positive, or the positive input is negative), the output will go to its negative limit ($V_{out} = -V$).

10. The op-amp is usually operated in either the open-loop mode or closed-loop mode. With the comparator circuit, the op-amp is operated in its open-loop mode because there is no signal feedback from output to input. In most instances, the op-amp is operated in the closed-loop mode, in which there is signal feedback from output back to input. This feedback signal will always be out of phase with the input signal and therefore oppose the original signal, which is why it is called "degenerative or negative feedback." Negative feedback is necessary in nearly all op-amp circuits to
   a. prevent output waveform distortion,
   b. prevent the amplifier from going into oscillation, and
   c. reduce the gain of the op-amp to a consistent value.

11. The inverting operational amplifier circuit produces an amplified output signal that is 180° out of phase with the input signal.

12. With the noninverting amplifier circuit, the input voltage (V_{in}) is applied to the op-amp's noninverting input (+), and the output voltage (V_{out}) will be in phase with the input.

13. The noninverting voltage-follower circuit has a voltage gain of 1. Similar to the BJT's emitter-follower and the FET's source-follower, the op-amp voltage-follower circuit derives its name from the fact that the output voltage follows the input voltage in both polarity and amplitude. This circuit is ideal as a buffer for interfacing a high-impedance source to a low-impedance load.

14. The inverting summing amplifier circuit, or adder amplifier, will sum or add all of the input voltages.

15. With the difference or differential amplifier circuit, the output voltage is equal to the difference between the two input voltages.

16. A signal generator is a circuit that will convert a dc supply voltage into a repeating output signal.

17. Passive filters are circuits that contain passive or nonamplifying components (resistors, capacitors, and inductors) connected in such a way that they will pass certain frequencies while rejecting others.

18. An active filter is a circuit that uses an amplifier with passive filter elements to provide frequency paths with rejection characteristics.

19. Active filters have several advantages over passive filters.
   a. Because the op-amp provides gain, the input signal passed to the output will not be attenuated, and therefore better response curves can be obtained.
   b. The high input impedance and low output impedance of the op-amp means that the filter circuit does not interfere with the signal source or load.
   c. Because active filters provide gain, resistors can be used instead of inductors, and therefore active filters are generally less expensive.

REVIEW QUESTIONS

Multiple-Choice Questions

1. When a differential amplifier is used in the differential-input, single-output mode, it has a _____ differential gain and a _____ common-mode gain.
   a. High, high   c. Low, low
   b. High, low   d. Low, high

2. The op-amp's internal circuit contains a _____, _____, and _____ amplifier stage.
   a. Differentiator, current, power
   b. Integrator, voltage, output
   c. Darlington-pair, emitter-follower, summing
   d. A differential, darlington-pair, emitter-follower

3. The op-amp's differential-amplifier stage provides the op-amp with a
   a. Low common-mode gain   c. High input impedance
   b. High differential gain   d. All of the above

4. Which transistor circuit is used in the op-amp's final-output stage to provide a low output impedance and high current gain?
   a. Common emitter   c. Common collector
   b. Common base   d. Both (a) and (c)

5. Could an op-amp circuit be constructed using discrete components?
   a. Yes   b. No

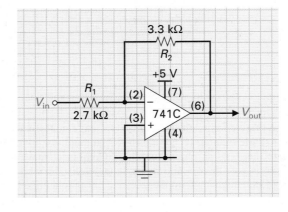

FIGURE 19-17   A 741C Op-Amp Circuit.

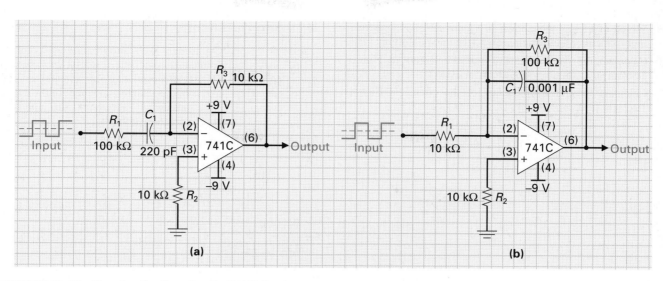

FIGURE 19-18   Two Applications for the 741C Op-Amp.

6. The comparator is considered a/an _____ loop op-amp circuit.
   a. Common   **c.** Differential
   b. Open   **d.** Closed

7. What is the lower frequency limit of an op-amp?
   a. 20 Hz   **b.** 6 Hz   **c.** DC   **d.** 7.34 Hz

8. A virtual ground is a ground to _____ but not to _____.
   a. Current, voltage   **b.** Voltage, current

9. The feedback loop in a closed-loop op-amp circuit provides
   a. Positive feedback
   b. Negative feedback
   c. Degenerative feedback
   d. Both (a) and (b)
   e. Both (b) and (c)

10. The _____ input(s) of an op-amp is used to compensate for slight differences in the transistors in the differential-amplifier stage.
   a. Inverting   **c.** +V
   b. DC offset   **d.** V_{out}

Communication Skill Questions

11. Why is the term "operational" used to describe the op-amp? (Intro.)

12. Sketch the op-amp schematic symbol. (19-1-1)

13. What are the three basic amplifier blocks within an op-amp? (19-2)

14. In what mode is the differential amplifier used within the op-amp? (19-2-1)

15. What is the difference between common-mode input signals and differential-mode input signals? (19-2-1)

16. Define "common-mode rejection ratio" in relation to the op-amp. (19-2-2)

17. Sketch the basic block diagram of an op-amp's internal circuit, and list the characteristics of each block. (19-2-3)

18. What is the difference between an open-loop and closed-loop op-amp circuit? (19-2-5)

19. Why is it necessary for an op-amp to have negative feedback? (19-2-5)

20. Sketch a simple inverting operational amplifier circuit, and briefly describe its operation. (19-2-5)

21. How is the gain, and therefore output voltage, of an inverting op-amp circuit determined? (19-2-5)

22. How is the gain, and therefore output voltage, of a noninverting op-amp circuit determined? (19-2-5)

Practice Problems

23. Identify the circuit shown in Figure 19-17. Why must the input to this circuit always be a negative voltage?

24. What would be the output voltage from the circuit in Figure 19-17 if the input voltage were −1.6 V?

25. Identify the circuits shown in Figure 19-18(a) and (b), and then sketch the shape of the output waveform if a square wave were applied to the inputs.

26. Referring to the circuit shown in Figure 19-19, what would be the voltage out if a +7.3 V input were applied?

27. Identify the circuits shown in Figure 19-20(a) and (b), and then calculate the output voltages for the given input voltages.

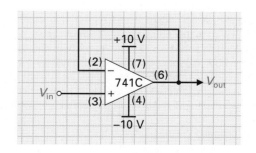

FIGURE 19-19   An Op-Amp Circuit.

28. Which of the circuits shown in Figure 19-21 is an active high-pass filter and which is an active low-pass filter?

29. Calculate the cutoff frequency of the circuit in Figure 19-21(a) if $C_1 = C_2 = 0.1$ μF, $R_1 = 10$ kΩ.

30. Calculate the cutoff frequency of the circuit in Figure 19-21(b) if $R_1 = R_2 = 33$ kΩ, $C_1 = 0.33$ μF.

Web Site Questions

Go to the Web site http://www.prenhall. com/cook, select the textbook *Introductory DC/AC Electronics* or *Introductory DC/AC Circuits*, this chapter, and then follow the instructions when answering the multiple-choice practice problems.

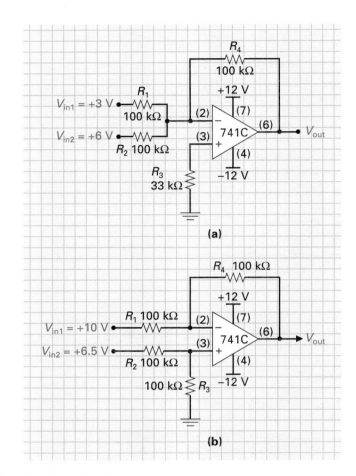

FIGURE 19-20   **Op-Amp Circuit Examples.**

JOB INTERVIEW TEST

These tests at the end of each chapter will challenge your knowledge up to this point, and give you the practice you need for a job interview. To make this more realistic, the test will comprise both technical and personal questions. In order to take full advantage of this exercise, you may want to set up a simulation of the interview environment, have a friend read the questions to you, and record your responses for later analysis.

Company Name: Automate, Inc.

Industry Branch: Industrial

Function: Design and Manufacture Industrial Automation

Job Title: Engineering Technician

1. How would you rate your electronics training?

2. What is an integrated circuit?

3. Are you familiar with interpreting data sheets?

4. You will probably spend a part of your day involved with documentation. How do you feel about that?

continued

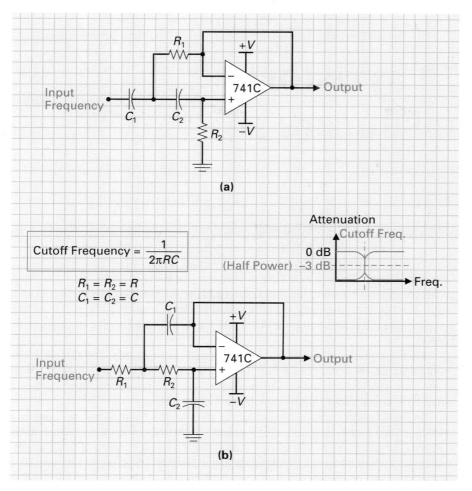

FIGURE 19-21   **Op-Amp Active Filter Circuits.**

5. What is an op-amp?

6. What is the difference between an active filter and a passive filter?

7. What would you say are the responsibilities of an engineering technician?

8. What is the difference between standard logic devices and programmable logic devices?

9. How could an op-amp be used as a comparator?

10. What is common-mode rejection?

Answers

1. Very good. Go on to explain that you see your education as an ongoing process, and like industry, you will always be in a learning mode.

2. Introduction.

3. Yes, cite examples explained in text and devices used in lab.

4. Documenting the engineering process is vitally important. As an engineering technician you will be called on to document design successes, and failures, in explicit detail. They are testing here to see if you have a talent for technical writing, since this is a must in this position.

5. Introduction.

6. Section 19-3-7.

7. See introduction, Engineering technician, and job listing.

8. Always be prepared for the interviewer to test your limits. Be honest and say "I haven't heard of those terms" if that's the case.

9. Section 19-2-5.

10. Section 19-2.

Timers, Thyristors, and Transducers

Leibniz's Language of Logic

Gottfried Wilhelm Leibniz was born in Leipzig, Germany, in 1646. His father was a professor of moral philosophy and spent much of his time discussing thoughts and ideas with his son. Tragically, his father died when he was only six, and from that time on Leibniz spent hour upon hour in his late father's library reading through all of his books.

By the age of twelve he had taught himself history, Latin, and Greek. At fifteen he entered the University of Leipzig, and it was here that he came across the works of scholars such as Johannes Kepler and Galileo. The new frontiers of science fascinated him, so he added mathematics to his curriculum.

In 1666, while finishing his university studies, the twenty-year-old Leibniz wrote what he called modestly a schoolboy essay, "De Arte Combinatoria," which means "On the Art of Combination." In this work, he described how all thinking of any sort on any subject could be reduced to exact mathematical statements. Logic or, as he called it, the laws of thought, could be converted from the verbal realm—which is full of ambiguities—into precise mathematical statements. In order to achieve this, however, Leibniz stated that a "universal language" would be needed. Most of his professors found the paper either baffling or outrageous and this caused Leibniz not to pursue the idea any further.

After graduating from university, Leibniz was offered a professorship at the University of Nuremberg, which he turned down for a position as an international diplomat. This career proved not to be as glamorous as he imagined because most of his time was spent in uncomfortable horse-drawn coaches traveling between the European capitals.

In 1672, his duties took him to Paris where he met Dutch mathematician and astronomer Christian Huygens. After seeing the hours that Huygens spent on endless computations, Leibniz set out to develop a mechanical calculator. A year later Leibniz unveiled the first machine that could add, subtract, multiply, and divide decimal numbers.

In 1676 Leibniz began to concentrate more on mathematics. It was at this time that he invented calculus, which was also independently discovered by Isaac Newton in England. Leibniz's focus, however, was on the binary number system, which occupied him for years. He worked tirelessly to document the long combinations of ones and zeros that make up the modern binary number system and to perfect binary arithmetic. What is ironic is that for all his genius, Leibniz failed to make the connection between his 1666 essay and binary, which was the universal language of logic that he was seeking. It would be a century and a quarter after Leibniz's death (in 1716) when another self-taught mathematician named George Boole would discover it.

Outline and Objectives

Introduction

In this chapter, we will examine timers, thyristors, and transducers. As their name states, timer circuits generate timing signals that synchronize operations within electronic systems. Thyristors are generally used as dc and ac power control devices, while transducers are generally used as sensing, displaying, and actuating devices.

In the first section of this chapter, you will be introduced to the astable, monostable, and bistable multivibrator circuits, and the 555 timer integrated circuit.

The semiconductor thyristor acts as an electronically controlled switch, switching power ON and OFF to adjust the average amount of power delivered to a load or connecting or disconnecting power from a load. Some thyristors are "unidirectional," which means that they will only conduct current in one direction (dc), while others are "bi-directional," which means that they can conduct current in either direction (ac). Thyristors are generally used as electronically controlled switches instead of the transistors because they have better power handling capabilities and are more efficient.

A semiconductor transducer is an electronic device that converts one form of energy to another. Electronic transducers can be classified as either input transducers (such as the photodiode) that generate input control signals, or output transducers (such as the LED) that convert output electrical signals to some other energy form. Input transducers are sensors that convert thermal, optical, mechanical, and magnetic energy variations into equivalent voltage and current variations. Output transducers, on the other hand, perform the exact opposite, converting voltage and current variations into optical or mechanical energy variations.

Clock Signal

Generally a square wave used for the synchronization and timing of several circuits.

Clock Oscillator

A device for generating a clock signal.

Astable Multivibrator

A device commonly used as a clock oscillator.

Monostable Multivibrator

A device that when triggered will generate a rectangular pulse of fixed duration.

20-1   DIGITIAL TIMER AND CONTROL CIRCUITS

Timing is everything in digital logic circuits. To control the timing of digital circuits, a clock signal is distributed throughout the digital system. This square wave **clock signal** is generated by a **clock oscillator,** and its sharp positive (leading) and negative (trailing) edges are used to control the sequence of operations in a digital circuit. In this section we will discuss the **astable multivibrator,** which is commonly used as a clock oscillator, and the **monostable multivibrator,** which when triggered will generate a rectangular pulse of a fixed duration. To complete our discussion on multivibrator circuits, we will also discuss the **bistable multivibrator,** which is a digital control device that can be either set or reset.

20-1-1 *The Astable Multivibrator Circuit*

The astable multivibrator circuit, seen in Figure 20-1(a), is used to produce an alternating two-state square or rectangular output waveform. This circuit is often called a **free-running multivibrator** because the circuit requires no input signal to start its operation. It will simply begin oscillating the moment the dc supply voltage is applied.

The circuit consists of two *cross-coupled bipolar transistors,* which means that there is a cross connection between the base and the collector of the two transistors Q_1 and Q_2. This circuit also contains two RC timing networks: R_1/C_1 and R_2/C_2.

Let us now examine the operation of this astable multivibrator circuit. When no dc supply voltage is present ($V_{CC} = 0$ V), both transistors are OFF, and therefore there is no output. When a V_{CC} supply voltage is applied to the circuit (for example, $+5$ V), both transistors will receive a positive bias base voltage via R_1 and R_2. Although both Q_1 and Q_2 are matched bipolar transistors, which means that their manufacturer ratings are identical, no two transistors are ever the same. This difference, and the differences in R_1 and R_2 due to resistor tolerances, means that one transistor will turn ON faster than the other. Let us assume that Q_1 turns on first, as seen in Figure 20-1(b). As Q_1 conducts, its collector voltage decreases because it is like a closed switch between the collector and emitter. This decrease in collector voltage is coupled through C_1 to the base of Q_2, causing it to conduct less and eventually turn OFF. With Q_2 OFF, its collector voltage will be high ($+5$ V) because Q_2 is equivalent to an open switch between the collector and emitter. This increase in collector voltage is coupled through C_2 to the base of Q_1 causing it to conduct more and eventually turn fully ON. The cross coupling between these two bipolar transistors will reinforce this condition with the LOW Q_1 collector voltage keeping Q_2 OFF and the HIGH Q_2 collector voltage keeping Q_1 ON. With Q_1 equivalent to a closed switch, a current path now exists for C_1 to charge as seen in Figure 20-1(b). As soon as the charge on C_1 reaches about 0.7 V, Q_2 will conduct because its base-emitter junction will be forward biased. This condition is shown in Figure 20-1(c). When Q_2 conducts, its collector voltage will drop, cutting OFF Q_1 and creating a charge path for C_2. As soon as the charge on C_2 reaches 0.7 V, Q_1 will conduct again and the cycle will repeat.

The output waveforms switch between the supply voltage ($+V_{CC}$) when a transistor is cut off, and zero volts when a transistor is saturated (ON). The result is two square-wave outputs that are out of phase with one another, as seen in the waveforms in Figure 20-1(d). Referring to the output waveforms in Figure 20-1(d), you can see that the time constant of R_1 and C_1, and R_2 and C_2 determine the complete cycle time. If the R_1/C_1 time constant is equal to the R_2/C_2 time constant, both halves of the cycle will be equal (50% duty cycle) and the result will be a square wave. Referring to Figure 20-1(d), you can see that the R_1/C_1 time constant will determine the time of one half-cycle, while the R_2/C_2 time constant will determine the time of the other half-cycle. The formula for calculating the time of one half-cycle is equal to:

$$t = 0.7 \times (R_1 \times C_1) \quad \text{or} \quad T = 0.7 \times (R_2 \times C_2)$$

The frequency of this square wave can be calculated by taking the reciprocal of both half-cycles, which will be

$$f = \frac{1}{1.4 \times RC}$$

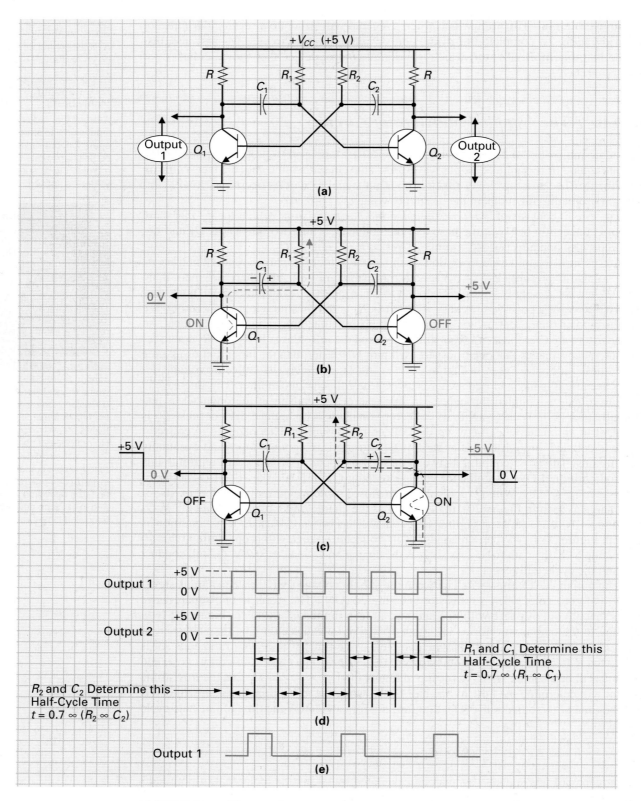

FIGURE 20-1 The Astable Multivibrator. (a) Basic Circuit. (b) Q_1 ON Condition. (c) Q_2 ON Condition. (d) Square Wave Mode. (e) Rectangular Wave Mode.

EXAMPLE:

Calculate the positive and negative cycle time and circuit frequency of the astable multivibrator circuit in Figure 20-1, if

$$R_1 \text{ and } R_2 = 100 \text{ k}\Omega \qquad C_1 \text{ and } C_2 = 1 \text{ μF}$$

Solution:

Because the time constant of R_1/C_1 and R_2/C_2 are the same, each half-cycle time will be the same and equal to

$$t = 0.7 \times (R \times C)$$
$$= 0.7 \times (100 \text{ k}\Omega \times 1 \text{ μF}) = 0.07 \text{ s or } 70 \text{ ms}$$

The frequency of the astable circuit will be equal to the reciprocal of the complete cycle, or the reciprocal of twice the half-alternation time.

$$f = \frac{1}{1.4 \times RC} = \frac{1}{1.4 \times (100 \text{ k}\Omega \times 1 \text{ μF})} = 7.14 \text{ Hz}$$
$$\text{or } f = \frac{1}{2 \times t} = \frac{1}{2 \times 70 \text{ ms}} = \frac{1}{0.14} = 7.14 \text{ Hz}$$

If the time constants of the two RC timing networks in the astable circuit are different, however, the result will be a rectangular or pulse waveform, as seen in Figure 20-1(e). In this instance, the same formula can be used to calculate the time for each alternation, and the frequency will be equal to the reciprocal of the time for both alternations.

20-1-2   *The Monostable Multivibrator Circuit*

The astable multivibrator is often referred to as an "unstable multivibrator" because it is continually alternating or switching back and forth, and therefore it has no stable condition or state. The monostable multivibrator has, as its name implies, one (mono) stable state. The circuit will remain in this stable state indefinitely until a trigger is applied and forces the monostable multivibrator into its unstable state. It will remain in its unstable state for a small period of time and then switch back to its stable state and await another trigger. The monostable multivibrator is often compared to a gun and is called a **one-shot multivibrator** because it will produce one output pulse or shot for each input trigger.

Referring to the monostable multivibrator circuit in Figure 20-2(a), you can see that the monostable is similar to the astable except for the trigger input circuit and for the fact that it has only one RC timing network. To begin with, let us consider the stable state of the monostable. Components R_2, D_1, and R_5 form a voltage divider, the values of which are chosen to produce a large positive Q_2 base voltage. This large positive base bias voltage will cause Q_2 to saturate (turn heavily ON), which in turn will produce a LOW Q_2 collector voltage, which will be coupled via R_4 to the base of Q_1, cutting it OFF. The circuit remains in this stable state (Q_2 ON, Q_1 OFF) until a **trigger input** is received.

Referring to the timing waveforms in Figure 20-2(b), you can see how the circuit reacts when a positive input trigger is applied. The pulse is first applied to a differentiator circuit (C_2 and R_5) that converts the pulse into a positive and a negative spike. These spikes are then applied to the positive clipper diode D_1, which only allows the negative spike to pass to the base of Q_2. This negative spike will reverse bias Q_2's base-emitter junction, turning Q_2 OFF and causing its collector voltage to rise to $+V_{CC}$, as seen in the waveforms. This increased Q_2 collector voltage will be coupled to the base of Q_1, turning it ON. The monostable multivibrator is now in its unstable state, which is indicated in the second color in Figure 20-2(a). In this condition, C_1 will charge as shown by the dashed current line. However, as soon as the voltage across C_1 reaches 0.7 V (which is dependent on the R_2/C_2 time

During his years as a student at the University of Utah, Nolan Bushnell (1943–) enjoyed playing the ancient Chinese game called GO. In this game, a word is used frequently throughout as a warning to your opponents that they are in jeopardy of losing with your next move.

With an initial investment of $500, Bushnell launched the video game industry in 1972, developing the first coin-operated Ping-Pong game, Pong. After almost an overnight success, Bushnell sold his company four years later for $15 million; the company was named after the polite Chinese term he learned in college, and as a warning to the competition, ATARI.

Following ATARI, Busnell stepped up his ambition, taking high-tech to the masses through his Chuck E. Cheese's Pizza Time Theatres.

One-Shot Multivibrator
Produces one output pulse or shot for each input trigger.

Trigger Input
Pulse used to initiate a circuit action.

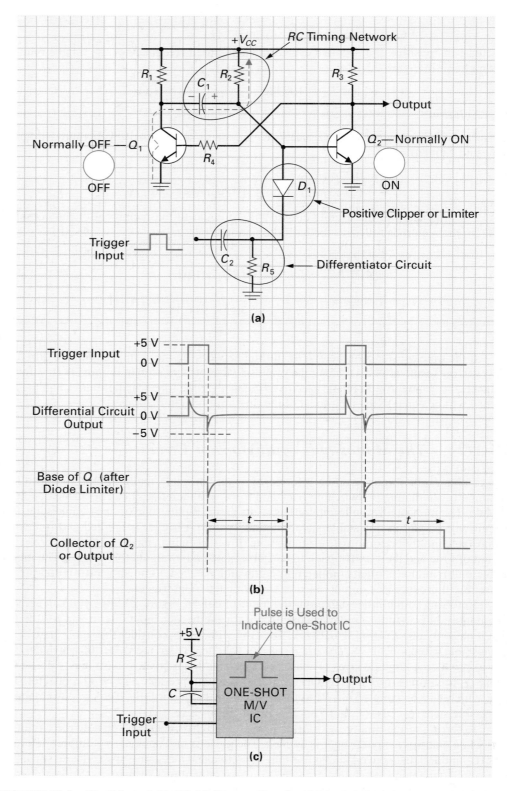

FIGURE 20-2   The Monostable Multivibrator Circuit. (a) Bipolar Transistor Circuit.
(b) Input/Output Timing Waveforms. (c) One-Shot IC.

constant), it will force Q_2 to conduct, which in turn will cause Q_1 to cut OFF and the monostable to return to its stable state. The output pulse width or pulse time (t) seen in Figure 20-2(b), can be calculated with the same formula used for the astable multivibrator:

$$t = 0.7 \times (R_2 \times C_1)$$

The one-shot multivibrator is sometimes used in *pulse-stretching* applications. For example, referring to the waveforms in Figure 20-2(b), imagine the input positive trigger pulse is $1 \mu s$ in width, and the RC time constant of R_2 and C_1 is such that the output pulse width (t) is 500 μs. In this example, the input pulse would be effectively stretched from 1 μs to 500 μs. The monostable multivibrator, or one-shot timer circuit, is also used to introduce a *time delay*. Referring again to the waveforms in Figure 20-2(b), imagine a differentiator circuit connected to the output of the monostable circuit. If the output pulse width was again set to 500 μs, there would be a 500 μs delay between the differentiated negative edge of the input pulse and the differentiated negative edge of the output pulse.

Figure 20-2(c) shows the logic symbol for a monostable (one-shot) multivibrator. Nearly all one-shot circuits in use today are in integrated form. These IC one-shots operate in exactly the same way as their discrete component counterparts. For example, the 74LS123 IC contains two fully independent monostable multivibrators. Like the symbol in Figure 20-2(c), the 74LS123 has pins for connecting external timing resistors and capacitors.

20-1-3   *The Bistable Multivibrator Circuit*

The bistable multivibrator has two (bi) stable states, and its bipolar transistor circuit is illustrated in Figure 20-3(a). The circuit has two inputs called the "SET" and "RESET" inputs, and these inputs drive the base of Q_1 and Q_2. The two outputs from this circuit are taken from the collectors of Q_1 and Q_2 and are called "Q" and "$\overline{Q}$" (pronounced "Q not"). The $\overline{Q}$ output derives its name from the fact that its voltage level is always the opposite of the Q output. For example, if Q is HIGH, $\overline{Q}$ will be LOW, and if Q is LOW, $\overline{Q}$ will be HIGH. The bistable multivibrator circuit is often called an **S-R (set-rest) flip-flop** because

A pulse on the SET input will "flip" the circuit into the set state (Q output is set HIGH), while

A pulse on the RESET input will "flop" the circuit into its reset state (Q output is reset LOW).

Figure 20-3(b) shows the logic symbol for a *S-R* or *R-S* flip-flop, or bistable multivibrator circuit.

To fully understand the operation of the circuit, refer to the waveforms in Figure 20-3(c). When power is first applied, one of the transistors will turn "ON" first and, because of the cross coupling, turn the other transistor "OFF." Let us assume that the circuit in Figure 20-3(a) starts with Q_1 ON and Q_2 OFF. The low voltage (approximately 0.3 V) on Q_1's collector will be coupled to Q_2's base, thus keeping it OFF. The high voltage on Q_2's collector (approximately $+5$ V) will be coupled to the base of Q_1, keeping it ON. This condition is called the **reset state** because the primary output (output 1, or Q) has been reset to binary 0, or 0 V. The cross-coupling action between the transistors will keep the transistors in the reset state (output 1 or $Q = 0$ V, and output 2 or $\overline{Q} = 5$ V) until an input appears.

Following the waveforms in Figure 20-3(c), you can see that the first input to go active is the SET input at time "t_1." This positive pulse will be applied to the base of Q_2 and forward bias its base-emitter junction. As a result, Q_2 will go ON and its LOW collector voltage will be applied to output 2 ($\overline{Q}$). This LOW on Q_2's collector will also be cross-coupled to the base of Q_1, turning it OFF and therefore making output 1 (Q) go HIGH. When a pulse appears on the SET input, the circuit will be put in its **set state** which means that the primary output (output 1 or Q) will be set HIGH. Studying the waveforms in Figure 20-3(c) once again, you will notice that after the positive SET input pulse has ended, the bistable will still remain **latched** or held in its last state, due to the cross-coupling between the transistors.

TIME LINE

In 1971, Ted Hoff of Intel Corporation designed a microprocessor, the 4004, that had all the basic parts of a central processor. Intel improved on the 4-bit 4004 microprocessor and unveiled an 8-bit microprocessor in 1974 that could add two numbers in 3.2 billionths of a second.

S-R (Set-Reset) Flip-Flop

A multivibrator circuit in which a pulse on the SET input will "flip" the circuit into the set state while a pulse on the RESET input will "flop" the circuit into its reset state.

Reset State

Primary output set LOW.

Set State

Primary output set HIGH.

Latched

Held in the last state.

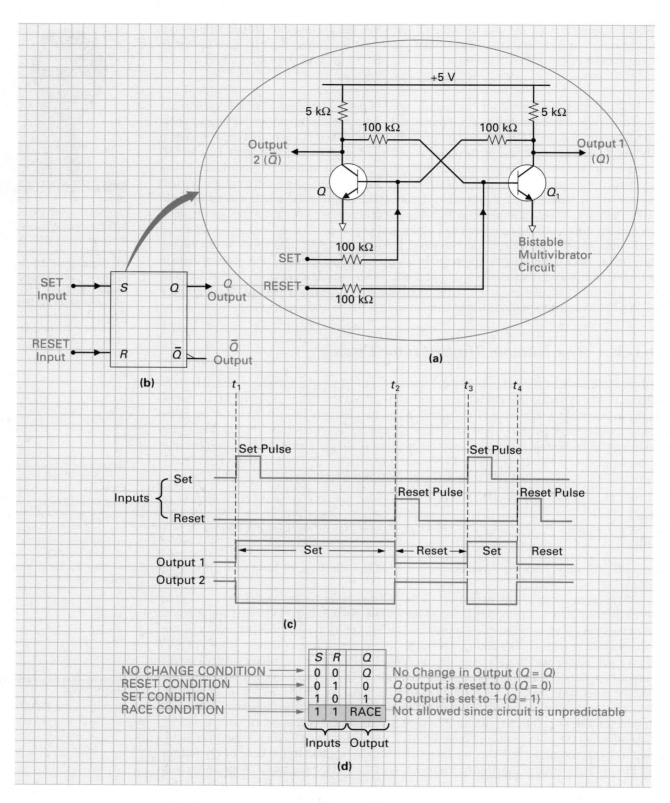

FIGURE 20-3 The Bistable Multivibrator or Set-Rest ($S = R$ or $R = S$) Flip-Flop. (a) Bipolar Transistor Circuit. (b) SR Symbol. (c) Timing Waveforms. (d) Truth Table.

This ability of the bistable multivibrator to remain in its last condition or state explains why the *S-R* flip-flop is also called an **S-R latch.**

Following the waveforms you can see that a RESET pulse occurs at time t_2, and resets the primary output (output 1 or Q) LOW. The flip-flop then remains latched in its reset state, until a SET pulse is applied to the set input at time t_3, setting Q HIGH. Finally, a positive RESET pulse is applied to the reset input at time t_4, and Q is reset LOW.

The operation of the *S-R* flip-flop is summarized in the truth table or function table shown in Figure 20-3(d). When only the R input is pulsed HIGH (reset condition), the Q output is reset to a binary 0 or reset LOW ($\overline{Q}$ will be the opposite, or HIGH). On the other hand, when only the S input is pulsed HIGH (set condition), the Q output is set to a binary 1 or set HIGH ($\overline{Q}$ will be the opposite, or LOW). When both the S and R inputs are LOW, the *S-R* flip-flop is said to be in the **no-change** or latch condition because there will be no change in the output Q. For example, if the output Q is SET, and then the S and R inputs are made LOW, the Q output will remain SET, or HIGH. On the other hand, if the output Q is RESET, and then the S and R inputs are made LOW, the Q output will remain RESET, or LOW.

The external circuits driving the S and R inputs will be designed so that these inputs are never both HIGH, as shown in the last condition in the table in Figure 20-3(d). This is called the *race condition* because both bipolar transistors will have their bases made positive, and therefore they will race to turn ON, and then shut the other transistor OFF via the cross-coupling. This input condition is not normally applied because the output condition is unpredictable.

The schmitt trigger circuit discussed previously is a bistable multivibrator circuit because its output voltage can be either one of two states. That is, its output can be either SET HIGH or RESET LOW, based on the voltage level of the input control voltage.

The bistable multivibrator *S-R* flip-flop or *S-R* latch has become one of the most important circuits in digital electronics. It is used in a variety of applications ranging from data storage to counting and frequency division. These applications will be covered later in the digital circuits chapter of this textbook. You will, however, see how this *S-R* flip-flop circuit is made use of in the following 555 timer circuit.

20-1-4 *The 555 Timer Circuit*

One of the most frequently used low-cost integrated circuit timers is the *555 timer*. Its IC package consists of 8 pins, as seen in Figure 20-4(a), and derives its number identification from the distinctive voltage divider circuit, seen in Figure 20-4(b), consisting of three 5 kΩ resistors. It is a highly versatile timer that can be made to function as an astable multivibrator, monostable multivibrator, frequency divider, or modulator depending on the connection of external components.

Nearly all the IC manufacturers produce a version of the 555 timer, which can be labeled in different ways, for example: SN72 555, MC14 555, SE 555, and so on. Two 555 timers are also available in a 16 pin dual IC package that is labeled with the numbers 556.

Basic 555 Timer Circuit Action

Referring to the basic block diagram in Figure 20-4(b), let us examine the basic action of all the devices in a 555 timer circuit. The three 5 kΩ resistors develop reference voltages at the inputs of the two comparators A and B. As previously mentioned, a comparator is a circuit that compares an input signal voltage to an input reference voltage and then produces a YES/NO or HIGH/LOW decision output. The negative input of comparator A will have a reference that is 2/3 of V_{CC}, and therefore the positive input (pin 6, threshold) will have to be more positive than 2/3 of V_{CC} for the output of comparator A to go HIGH. With comparator B, the positive input has a reference that is 1/3 of V_{CC}, and therefore the negative input (pin 2, trigger) will have to be more negative, or fall below, 1/3 of V_{CC} for the output of comparator B to go HIGH. If the output of comparator A were to go HIGH, the set/reset flip-flop output would be reset LOW to 0 V. This LOW output would be inverted by the INVERTER to a HIGH, and then inverted and buffered (boosted in current) by the final INVERTER to

S-R Latch

Another name for *S-R* flip-flop, so called because the output remains latched in the set or reset state even though the input is removed.

No-Change or Latch Condition

When both inputs are LOW, the *S-R* flip-flop is said to be in the no-change or latch condition because there will be no change in the output.

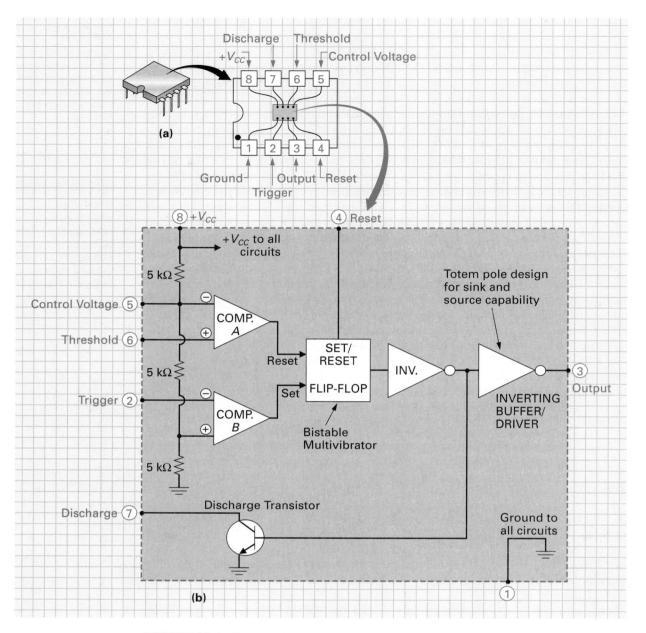

FIGURE 20-4 The 555 Timer. (a) IC Pin Layout. (b) Basic Block Diagram.

appear as a LOW at the output pin 3. If the output of comparator B were to go high, the set/reset flip-flop would be set HIGH to +5 V. This HIGH output would be inverted by the INVERTER to a LOW, and then inverted and buffered by the final INVERTER to appear as a HIGH at the output pin 3. When the output of the set/reset flip-flop is LOW (reset), the input at the base of the discharge transistor will be HIGH. The transistor will therefore turn ON, and its low emitter-to-collector resistance will ground pin 7. When the output of the set/reset flip-flop is HIGH (set), the input at the base of the discharge transistor will be LOW. The transistor will therefore turn OFF and its high emitter-to-collector resistance will cause pin 7 to ground.

The 555 Timer as an Astable Multivibrator

Figure 20-5(a) shows how the 555 timer can be connected to operate as an astable or free-running multivibrator. The waveforms in Figure 20-5(b) show how the externally

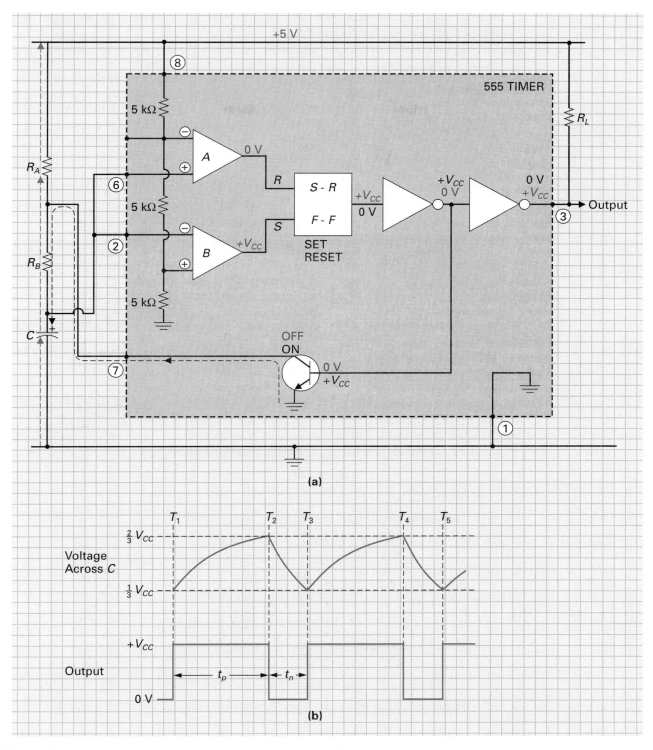

FIGURE 20-5   The 555 Timer as an Astable Multivibrator. (a) Circuit. (b) Waveforms.

connected capacitor C will charge and discharge and how the output will continually switch between its positive ($+V_{CC}$) and negative (0 V) peaks.

To begin with, let us assume that the output of the S-R flip-flop is set HIGH, and therefore the output will be HIGH (blue condition in Figure 20-5(a), time T_1 in Figure 20-5(b)). The HIGH output of the S-R flip-flop will be inverted to a LOW and turn OFF the 555's internal discharge transistor. With this transistor OFF, the external capacitor C can begin to charge towards $+V_{CC}$ via R_A and R_B.

At time T_2, the capacitor's charge has increased beyond 2/3 of V_{CC}, and therefore the output of comparator A will go HIGH and RESET the S-R flip-flop's output LOW. This will cause the output (pin 3) of the 555 to go LOW. However, the discharge transistor's base will be HIGH, and so it will turn ON. With the discharge transistor ON, the capacitor can begin to discharge (black condition in Figure 20-5(a), time T_2 in Figure 20-5(b)). At time T_3, the capacitor's charge has fallen below 1/3 of V_{CC}, or the trigger level of comparator B. As a result, comparator B's output will go HIGH and SET the output of the S-R flip-flop HIGH, or back to its original state. The discharge transistor will once again be cut OFF, allowing the capacitor to charge and the cycle to repeat.

As you can see in Figure 20-5(a), the capacitor charges through R_A and R_B to 2/3 of V_{CC}, and then discharges through R_B to 1/3 of V_{CC}. As a result, the positive half-cycle time (t_p) can be calculated with the formula:

$$t_p = 0.7 \times C \times (R_A + R_B)$$

The negative half-cycle time (t_n) can be calculated with the formula:

$$t_n = 0.7 \times C \times R_B$$

The total cycle time will equal the sum of both half-cycles ($t = t_p + t_n$), and the frequency will equal the reciprocal of time ($f = 1/t$).

■ **EXAMPLE:**

Calculate the positive half-cycle time, negative half-cycle time, complete cycle time, and frequency of the 555 astable multivibrator circuit in Figure 20-5, if

$$R_A = 1 \text{ k}\Omega \quad R_B = 2 \text{ k}\Omega \quad C = 1 \text{ }\mu\text{F}$$

■ *Solution:*

The positive half-cycle will last for

$$t_p = 0.7 \times C \times (R_A + R_B)$$
$$= 0.7 \times 1 \text{ }\mu\text{F} \times (1 \text{ k}\Omega + 2 \text{ k}\Omega) = 2.1 \text{ ms}$$

The negative half-cycle will last for

$$t_n = 0.7 \times C \times R_B$$
$$= 0.7 \times 1 \text{ }\mu\text{F} \times 2 \text{ k}\Omega = 1.4 \text{ ms}$$

The complete cycle time will be

$$t = t_p + t_n$$
$$= 2.1 \text{ ms} + 1.4 \text{ ms} = 3.5 \text{ ms}$$

The frequency of this 555 astable multivibrator will be

$$f = \frac{1}{t}$$
$$= \frac{1}{3.5 \text{ ms}} = 285.7 \text{ Hz}$$

The 555 Timer as a Monostable Multivibrator

Figure 20-6(a) shows how the 555 timer can be connected to operate as a monostable or one-shot multivibrator. The waveforms in Figure 20-6(b) show the time relationships

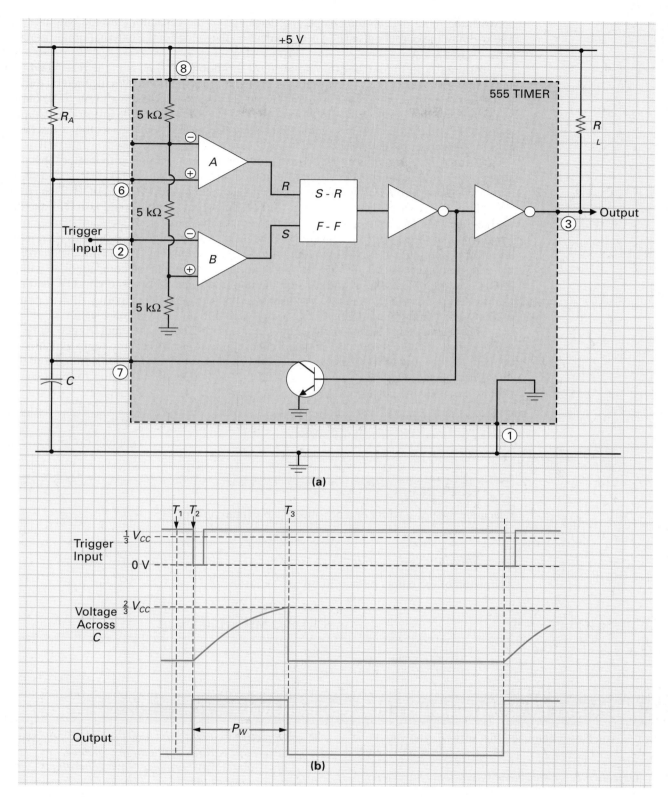

(a)

(b)

FIGURE 20-6   The 555 Timer as a Monostable Multivibrator. (a) Circuit. (b) Waveforms.

between the input trigger, the charge and discharge of the capacitor, and the output pulse. The width of the output pulse (P_W) is dependent on the values of the external timing components R_A and C.

At time T_1 in Figure 20-6(b), the set-reset flip-flop (*S-R F-F*) is in the reset condition and is therefore producing a LOW output. This LOW from the *S-R F-F* is inverted by the INVERTER, and then inverted and buffered by the final stage to produce a LOW (0 V) output from the 555 timer at pin 3. The LOW output from the *S-R F-F* will be inverted and appear as a HIGH at the base of the discharge transistor, turning it ON and providing a discharge path for the capacitor to ground.

At time T_2, a trigger is applied to pin 2 of the 555 monostable multivibrator. This negative trigger will cause negative input of comparator B to fall below 1/3 of V_{CC}, and so the output of comparator B will go HIGH and SET the output of the *S-R F-F* HIGH. This HIGH from the *S-R F-F* will send the output of the 555 timer (pin 3) HIGH, and turn OFF the discharge transistor. Once the path to ground through the discharge transistor has been removed from across the capacitor, the capacitor can begin to charge via R_A to $+ V_{CC}$, as seen in the waveforms in Figure 20-6(b). The output of the 555 timer remains HIGH until the charge on the capacitor exceeds 2/3 of V_{CC}. At this time (T_3), the output of comparator A will go HIGH, resetting the *S-R F-F* and causing the output of the 555 timer to go LOW and also turning ON the discharge transistor to discharge the capacitor. The circuit will then remain in this stable condition until a new trigger arrives to initiate the cycle once again.

The leading edge of the positive output pulse is initiated by the input trigger, while the trailing edge of the output pulse is determined by the R_A and C charge time, which is dependent on their values. Because the capacitor can charge to 2/3 of V_{CC} in a little more than one time constant (1 time constant = 0.632, 2/3 = 0.633), the following formula can be used to calculate the pulse width (P_W):

$$P_W = 1.1 \times (R_A \times C)$$

EXAMPLE:

Calculate the pulse width of a 555 monostable multivibrator, if

$$R_A = 2 \text{ M}\Omega, \text{ and } C = 1 \text{ } \mu\text{F.}$$

Solution:

The width of the output pulse will be

$$P_W = 1.1 \times (R_A \times C)$$
$$= 1.1 \times (2 \text{ M}\Omega \times 1 \text{ } \mu\text{F}) = 2.2$$

SELF-TEST EVALUATION POINT FOR SECTION 20-1

Now that you have completed this section, you should be able to:

■ **Objective 1.** *Explain the operation, characteristics, and application of the following bipolar transistor digital timer and control circuits:*
a. The astable multivibrator circuit
b. The monostable multivibrator circuit
c. The bistable multivibrator circuit
d. The 555 8-pin IC timer circuit

■ **Objective 2.** *Describe how the 555 IC timer can be used in various applications, such as:*
a. A square wave or pulse wave generator
b. A one-shot timer

Use the following questions to test your understanding of Section 20-1.

1. The astable multivibrator is also called the _____ multivibrator.

continued

2. Which of the multivibrator circuits has only one stable state?

3. Which multivibrator is also called a set-reset flip-flop?

4. Why is the set-reset flip-flop also called a latch?

5. The 555 timer derived its number identification from _____.

6. List two applications of the 555 timer.

20-2 THYRISTORS

The most frequently used thyristors are the "silicon controlled rectifier (SCR)," "triode ac semiconductor switch (TRIAC)," "diode ac semiconductor switch (DIAC)," "unijunction transistor (UJT)," and "programmable unijunction transistor (PUT)." In this section, we will examine the operation, characteristics, applications, and testing of all these electronic devices.

20-2-1 The Silicon Controlled Rectifier (SCR)

The **silicon controlled rectifier** or **SCR** is the most frequently used of the thyristor family. Figure 20-7(a) shows how this device is a four-layered, alternately-doped component, with three terminals labeled anode (*A*), cathode (*K*) and gate (*G*).

SCR Operation

To simplify the operation of the SCR, Figure 20-7(b) shows how the four layers of the SCR can be thought of, when split, as being a PNP and NPN transistor, and Figure 20-7(c) shows how this interconnection forms a **complementary latch circuit.** To operate as an ON/OFF switch, the SCR must be biased like a diode, with the anode of the SCR made positive relative to the cathode. The gate of the SCR is an active-HIGH input and must be triggered by a voltage that is positive relative to the cathode. In the example in Figure 20-7(c), you can see that the anode of the SCR is connected to +50 V via the load (R_L), the cathode is connected to 0 V, and the gate has a control input that is either 0 V or +5 V. When the gate input is at 0 V, the NPN and PNP transistors are OFF, and the SCR is equivalent to an open switch, as shown in the inset in Figure 20-7(c). On the other hand, when the gate input is +5 V, the NPN transistor's base is made positive with respect to the emitter and so the NPN transistor will turn ON. Turning the NPN transistor ON will connect the 0 V at the emitter of the NPN transistor through to the PNP's base, causing it to also turn ON. Turning the PNP transistor ON will connect the large positive voltage on the PNP's emitter through to the collector and the base of the NPN transistor, keeping it ON even after the +5 V input trigger is removed. As a result, both transistors will be latched ON and held ON by one another, allowing a continuous flow of current from cathode to anode. A momentary positive input trigger, therefore, will cause the SCR to be latched ON and be equivalent to a closed switch, as seen in the inset in Figure 20-7(c).

SCR Characteristics

Figure 20-7(d) shows a correctly biased SCR: the left inset shows a typical low-power and high-power package, and the right inset shows a typical SCR characteristic curve. This graph plots the forward and reverse voltage applied across the SCR's anode-to-cathode against the current through the SCR between cathode and anode. Looking at the forward conduction quadrant, you can see that the voltage needed to turn ON an SCR is called the **forward breakover voltage.** An SCR's forward breakover voltage, or turn ON voltage, is inversely proportional to the value of gate current. For example, a larger gate current will cause the SCR to turn ON when only a small forward voltage is applied between the SCR's anode-to-cathode. If no gate current is applied, the forward voltage applied across the SCR's anode to cathode will have to be very large to make the SCR turn ON and conduct current between

Silicon Controlled Rectifier or SCR

A three-junction, three-terminal, unidirectional P-N-P-N thyristor that is normally an open circuit. When triggered with the proper gate signal it switches to a conducting state and allows current to flow in one direction.

Complementary Latch Circuit

A circuit containing an NPN and PNP transistor that once triggered ON will remain latched ON.

Forward Breakover Voltage

Voltage needed to turn ON an SCR.

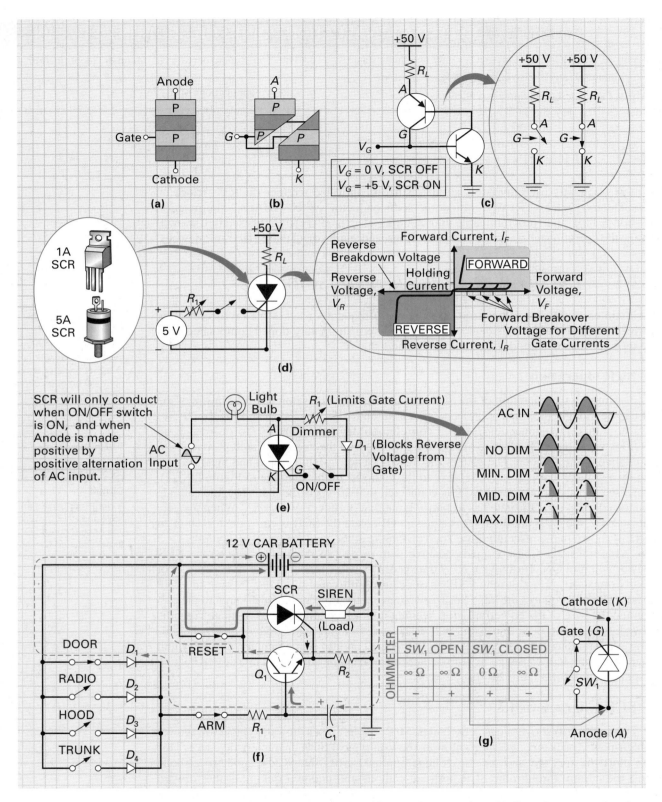

FIGURE 20-7 Silicon-Controlled Rectifier. (a) Construction. (b) Complementary Latch. (c) Closed, Open Latch Action. (d) Correctly Biased. (e) SCR Application—Basic Light Dimmer. (f) Application—Alarm System. (g) Testing SCRs.

cathode and anode. Once the SCR is turned ON, a holding current latches the SCR's complementary latch ON, independent of the gate current. If the forward current between the SCR's cathode and anode falls below this minimum holding current value, the SCR will turn OFF because the NPN/PNP latch will not have enough current to keep each other ON. A gate trigger is therefore used to turn ON the SCR; however, once the SCR is ON, it can only be turned OFF by decreasing the anode-to-cathode voltage so that the cathode-to-anode current passing through the SCR drops below the holding current value.

In the reverse direction with the gate switch open, the SCR acts in almost the same way as a diode because a large reverse voltage is needed between anode and cathode to cause the SCR to break down.

SCR Applications

The SCR is used in a variety of electrical power applications, such as light dimmer circuits, motor-speed control circuits, battery charger circuits, temperature control systems, and power regulator circuits. From the characteristic curve in Figure 20-7(d), we discovered that the SCR will only conduct current in the forward direction, which is why it is classed as a **unidirectional device.** This means that if an ac signal is applied across the SCR, it will only respond to a gate trigger during the time that the ac alternation makes the anode positive with respect to the cathode. For example, Figure 20-7(e) shows how the SCR could be connected to form a simple light dimmer circuit. When the ON/OFF switch is open, the SCR will be OFF because its gate current is zero, and the light will be OFF. When the ON/OFF switch is closed, diode D_1 will connect a positive voltage to the gate of the SCR whenever the ac input is positive. The value of gate current applied to the SCR is controlled by the variable dimmer resistor (R_1), and therefore this resistor value will determine the SCR's forward breakover (turn ON) voltage. Referring to the waveforms in the inset in Figure 20-7(e), you can see that when the resistance of R_1 is zero (no dim), gate current will be maximum, the SCR will turn ON for the full positive half cycle of the ac input, and the average power delivered to the light bulb will be HIGH. As the resistance of R_1 is increased, the SCR gate current is decreased, causing the SCR to turn ON for less of the positive alternation, and therefore the average power delivered to the light bulb to decrease. You may ask: why don't we simply connect a variable resistor in series with the light bulb to vary the light bulb's current and therefore brightness? The problem with this arrangement is identical to the disadvantages of the previously discussed "series dissipative regulator." The wattage rating, size, and cost of the variable resistor would be very large, and the circuit would be very inefficient because power is taken away from the light bulb by dissipating it away as heat from the resistor. Varying the SCR's ON/OFF time is a much more efficient system because we will only switch through to the light bulb the power that is desired, and therefore vary the average power applied to the load in almost exactly the same way as the previously discussed "switching regulator circuit."

Now that we have seen the SCR in an ac circuit application, let us now see how it could be used in a dc circuit application. Figure 20-7(f) shows a basic car-alarm system. When both the ARM switch and RESET switch are closed, the alarm system is active, and any of the four sensor switches (door, radio, hood, or trunk) will activate the alarm. For example, Figure 20-7(f) shows what will happen if the car door is opened when the alarm system is armed. Opening the door will cause capacitor C_1 to charge via diode D_1 and resistor R_1. After a short delay, the charge on C_1 is large enough to turn ON Q_1, which will switch the positive potential on its collector from the battery through to its emitter and to the gate of the SCR. This positive gate trigger will turn ON the SCR and activate the siren because the SCR is equivalent to a closed switch when ON and when triggered will connect the full 12-V battery across the siren. Once the SCR is turned ON, it will remain latched ON independent of the ARM switch and the sensor switches. Only by opening the RESET switch, which is hidden within the vehicle, can the siren be shut OFF. The values of R_1 and C_1 are chosen so that a small delay occurs before Q_1 and the SCR are triggered. This delay is included so that the vehicle owner has enough time to enter the car and disarm the alarm system by opening the ARM switch.

Unidirectional Device
A device that will conduct current in only one direction.

SCR Testing

Using the oscilloscope, you can monitor the gate trigger input and ON/OFF switching of an SCR while it is operating in circuit. If you suspect that the SCR is the cause of a circuit malfunction, you should remove the SCR and use the ohmmeter test circuit shown in Figure 20-7(g) to check for terminal-to-terminal opens and shorts. With this test, we will be using the ohmmeter's internal battery to apply different polarities to the different terminals of the SCR and SW_1 to either apply or disconnect gate current. To explain the ohmmeter response table in this illustration, you can see that if SW_1 is open (no gate current), the resistance between anode and cathode should be almost infinite ohms (actually about 250 kΩ), no matter what polarity is applied between anode and cathode. On the other hand, if SW_1 is closed and the anode is made positive with respect to the cathode, the gate will also be made positive due to SW_1, and so the SCR should turn ON and have a very low resistance between anode and cathode. If SW_1 is closed, and a reverse polarity is applied across the SCR (anode is made negative, cathode is made positive), the ohmmeter should once again read infinite ohms.

20-1-2   *The Triode AC Semiconductor Switch (TRIAC)*

The disadvantage with the SCR is that it is unidirectional, which means that it can only be activated when the applied anode-to-cathode voltage makes the anode positive with respect to the cathode, and it will conduct current in one direction. As a result, the SCR can only control a dc supply voltage, or one-half cycle of the ac supply voltage. To gain control of the complete ac input cycle, we would need to connect two SCRs in parallel, facing in opposite directions, as shown in Figure 20-8(a). This is exactly what was done to construct the **triode ac semiconductor switch** or **TRIAC,** which has three terminals called main terminal 1 (MT_1), main terminal 2 (MT_2), and gate (G). The *P-N* doping for this **bidirectional device** is shown in Figure 20-8(b), and its schematic symbol is shown in Figure 20-8(c).

TRIAC Operation and Characteristics

Since the TRIAC is basically two SCRs connected in parallel, back-to-back, it comes as no surprise that its operation and characteristics are very similar to the SCR. The characteristic curve for a typical TRIAC is shown in Figure 20-8(d). Looking at the identical forward and reverse curves, you can see that the key difference with the TRIAC is that it can be triggered or activated by either a positive or negative input gate trigger. This means that the TRIAC can be used to control both the positive and negative alternation of an ac supply voltage. To explain this in more detail, Figure 20-8(e) shows how a TRIAC could be connected across an ac input. When the ON/OFF gate switch is open, the TRIAC will not receive a gate trigger and so it will remain OFF. When the ON/OFF gate switch is closed, the TRIAC will be triggered by the positive and negative cycles of the ac input, via R_1. By adjusting the resistance of R_1, we can control the TRIAC's value of gate current. By controlling gate current, we can control the TRIAC's turn-ON voltage (positive and negative breakover voltage), so that the ac input voltage can be chopped up to adjust the average value of voltage applied to the load.

TRIAC Applications

Figure 20-8(f) shows how a TRIAC could be connected as an automatic night light for home or business security and safety. Later in this chapter, we will discuss the photocell in more detail. However, for now just think of it as a variable resistor that changes its resistance based on the amount of light present. During the day when the photocell is exposed to light, its resistance is less than a few ohms, and since it is in parallel with the energizing coil of a reed relay, most of the current will pass through the photocell keeping the reed relay de-energized. As the sun goes down and the photocell is deprived of light, its resistance increases to a few mega ohms, and this high resistance will cause the current through the reed relay coil to increase, and therefore the reed relay to energize. With the reed-relay switch

---

Triode AC Semiconductor Switch or TRIAC

A bidirectional gate-controlled thyristor that provides full-wave control of ac power.

Bidirectional Device

A device that will conduct current in either direction.

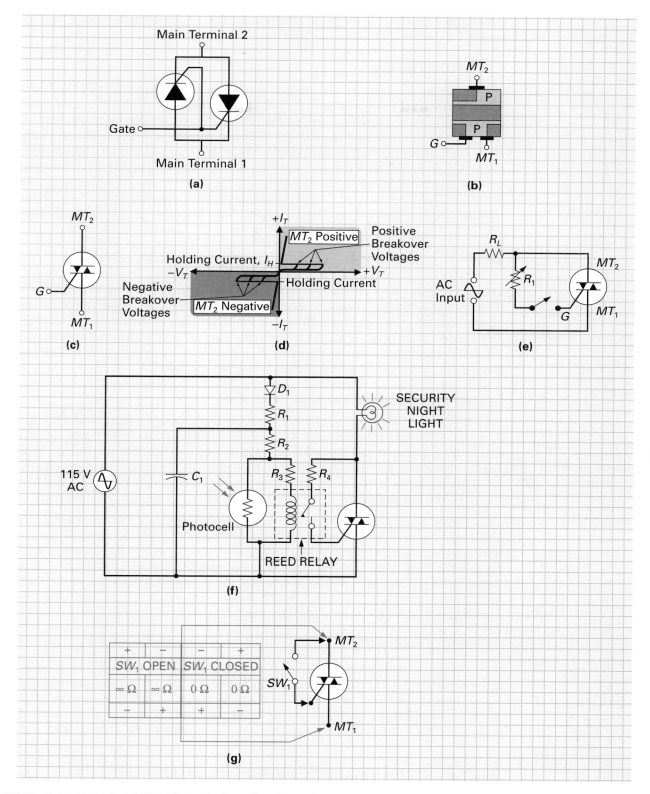

FIGURE 20-8 TRIAC. (a) TRIAC Equivalent Circuit. (b) Construction. (c) Schematic Symbol. (d) *V-I* Characteristics. (e) Application—AC TRIAC Switch. (f) Application— Automatic Night Light. (g) Testing TRIACs.

closed, the TRIAC will be triggered by each half cycle of the ac supply voltage, causing it to turn ON and connect power to the night light.

TRIAC Testing

Using the oscilloscope, you can monitor the gate trigger input, and ON/OFF switching of a TRIAC, while it is operating in-circuit. If you suspect that the TRIAC is the cause of a circuit malfunction, you should remove the TRIAC and use the ohmmeter test circuit shown in Figure 20-8(g), to check for terminal-to-terminal opens and shorts. The ohmmeter response table in this illustration shows that if SW_1 is open (no gate current), the resistance between MT_2 and MT_1 should be almost infinite ohms (actually about 250 kΩ), no matter what polarity is applied. When switch 1 is closed, the gate of the TRIAC will receive a trigger. Since the TRIAC operates on either a positive or negative trigger, it should turn ON no matter what polarity is applied, and therefore have a very low resistance between MT_2 and MT_1.

20-2-3 *The Diode AC Semiconductor Switch (DIAC)*

Diode AC Semiconductor Switch or DIAC
A bidirectional diode that has a symmetrical switching mode.

One disadvantage with the TRIAC is that its positive breakover voltage is usually slightly different from its negative breakover voltage. This nonsymmetrical trigger characteristic can be compensated for by using a **diode ac semiconductor switch** or **DIAC** to trigger a TRIAC. The DIAC's construction is shown in Figure 20-9(a), its schematic symbol in Figure 20-9(b) and its equivalent circuit in Figure 20-9(c). Equivalent to two back-to-back, series-connected junction diodes, the DIAC has two terminals.

DIAC Operation and Characteristics

Symmetrical Bidirectional Switch
A device that has the same value of breakover voltage in both the forward and reverse direction.

Since the PNP regions of a DIAC are all equally doped, the DIAC will have the same forward and reverse characteristics, as shown in Figure 20-9(d). As a result, the DIAC is classed as a **symmetrical bidirectional switch,** which means that it will have the same value of breakover voltage in both the forward and reverse direction.

DIAC Applications

Figure 20-9(e) shows how a DIAC could be connected as a pulse-triggering device in a TRIAC ac power control circuit. The DIAC will turn ON when the capacitor has charged to either the positive or negative breakover voltage ($+V_{BO}$ or $-V_{BO}$). Once this voltage is reached the DIAC turns ON, and the capacitor discharges through the DIAC, triggering the TRIAC into conduction, which then connects the ac supply voltage across the load. The variable resistor R_1 is used to adjust the RC charge time constant, so that the DIAC turn-ON time, and therefore TRIAC turn-ON time, can be changed.

DIAC Testing

Using the oscilloscope, you can monitor the ON/OFF switching of a DIAC while it is operating in-circuit. If you suspect that the DIAC is the cause of a circuit malfunction, you should remove the DIAC and check it with the ohmmeter, as seen in Figure 20-9(f). Since the DIAC is basically two diodes connected back-to-back in series, the ohmmeter should show a low resistance reading between its terminals no matter what polarity is applied.

20-2-4 *The Unijunction Transistor (UJT)*

Unijunction Transistor or UJT
A P-N device that has an emitter connected to the P-N junction on one side of the bar and two bases at either end of the bar. Used primarily as a switching device.

The **unijunction transistor** or **UJT** operates in a very different way to the SCR, TRIAC, and DIAC. Although it is given the name transistor, it is never used as an amplifying device like the BJT and FET: it is only ever used as a voltage-controlled switch. Figure 20-10(a) shows the construction of the UJT and illustrates how the uni (one) junction transistor derives its name from the fact that it has only one P-N junction. Looking at this illustration you

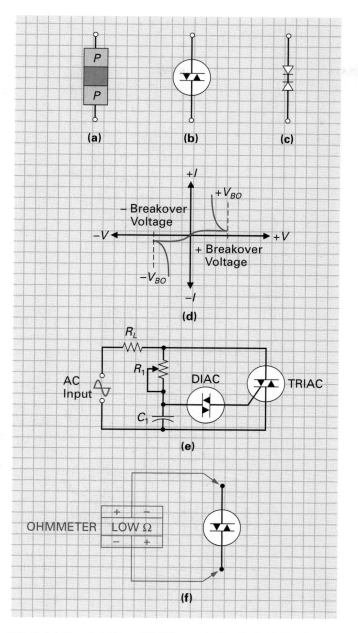

FIGURE 20-9 DIAC. (a) Construction. (b) Schematic Symbol. (c) Equivalent Circuit. (d) *V-I* Characteristics. (e) Application—TRIAC Control. (f) Testing DIACs.

can see that the UJT is a three-terminal device with an emitter lead (E) attached to a small p-type pellet, that is fused into a bar of n-type silicon with contacts at either end [labeled base 1 (B_1) and base 2 (B_2)].

The schematic symbol for the UJT is shown in Figure 20-10(b). To remember whether the junction is a P-N or N-P type, think of the arrow as a junction diode: the emitter (diode-anode) is positive and the bar (diode-cathode) is negative.

UJT Operation and Characteristics

Figure 20-10(c) illustrates the UJT's equivalent circuit. The emitter-to-bar P-N junction is equivalent to a junction diode, and the bar is equivalent to a two-resistor voltage divider (R_{B1} and R_{B2}). Referring once again to the UJT construction in Figure 20-10(a), you can see that the emitter pellet is closer to terminal B_2 than B_1, and this is why the resistance

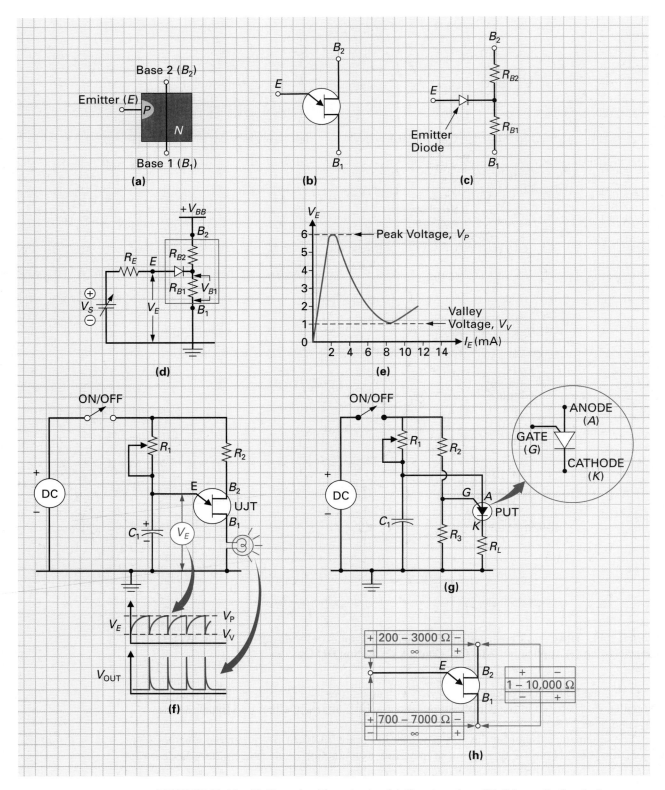

FIGURE 20-10 **Unijunction Transistors. (a) Construction. (b) Schematic Symbol. (c) Equivalent Circuit. (d) Correctly Biased. (e) V-I Characteristic Curve. (f) Application— Relaxation Oscillator. (g) Programmable Unijunction Transistor (PUT). (h) Ohmmeter Resistances.**

of R_{B2} is smaller than the resistance of R_{B1}. Figure 20-10(d) shows how a UJT should be correctly biased. The voltage source V_{BB} is connected to make B_2 positive relative to B_1 and the input voltage V_S is connected to make the emitter positive with respect to B_1. The resistor R_E is used to limit emitter current. When V_S is zero, the UJT's emitter diode is OFF, and the resistance between B_2 and B_1 allows only a very small amount of current between ground and $+V_{BB}$. If the emitter supply voltage is increased so that V_E exceeds the voltage at B_1, the emitter diode will turn ON and inject holes into the p region. Flooding the lower half of the UJT with holes increases the amount of current flow through the UJT, dramatically reducing the resistance of R_{B1}. A lower R_{B1} resistance results in a lower V_{B1} voltage drop, and this further increases the E to B_1 P-N forward bias permitting more holes to be injected into the n-type bar between E and B_1. Figure 20-10(e) graphically illustrates this action by plotting the UJT's emitter voltage (between E and B_1) against the UJT's emitter current. An increase in V_S, and therefore V_E, produces very little emitter current until the **peak voltage (V_P)** is reached. Beyond V_P, V_E has exceeded V_{B1} and the emitter diode is forward biased, causing IE to increase and V_E to decrease due to the lower resistance of R_{B1}. This negative resistance region reaches a low point known as the **valley voltage (V_V),** which is a point at which V_E begins to increase and the UJT no longer exhibits a negative resistance. The point at which a UJT turns ON and increases the current between B_1 and B_2 can be controlled, and used in switching applications.

Peak Voltage (V_P)
The maximum value of voltage.

Valley Voltage (V_V)
The voltage at the dip or valley in the characteristic curve.

UJT Applications

The UJT's negative resistance characteristic is useful in switching and timing applications. Figure 20-10(f) shows how a UJT could be connected to form a relaxation oscillator in an emergency flasher circuit. When the ON/OFF switch is closed, capacitor C_1 charges by resistor R_1. When the voltage across C_1 reaches the UJT's VP value, the UJT will turn ON and its resistance between E and B_1 will drop LOW. This low resistance will allow C_1 to discharge through the UJT's E-to-B_1 junction and into the flasher light bulb, causing it to momentarily flash. As C_1 discharges, its voltage decreases and this causes the UJT to turn OFF. The cycle then repeats since the off UJT will allow capacitor C_1 to begin charging towards V_P, at which time it will trigger the UJT and repeat the process. The circuit's repetition rate, or frequency, is determined by the UJT's V_P rating, the supply voltage, and the RC time constant. To change the flashing rate, the value of R_1 can be changed to vary the rate at which C_1 is charged and therefore how soon the UJT is triggered.

The Programmable UJT (PUT)

The **programmable unijunction transistor** or **PUT** is a variation on the basic UJT thyristor. This four-layer thyristor has three terminals labeled cathode (K), anode (A), and gate (G). The key difference between the basic UJT and the PUT is that the PUT's peak voltage (V_P) can be controlled. Figure 20-10(g) shows how a PUT could also be connected to form a relaxation oscillator circuit. To differentiate the PUT's schematic symbol from the SCR, the gate input is connected into the anode side of the diode symbol instead of the cathode side. This circuit will produce exactly the same output waveform as the circuit shown in Figure 20-10(f). The gate-to-cathode voltage is derived from R_3, which is connected with R_2 to form a voltage divider. This circuit will operate in exactly the same way as the previous relaxation oscillator, in that C_1 will charge via R_1 until the charge across C reaches the V_P value. In this circuit, however, the V_P trigger voltage is set by R_3. When the PUT's anode-to-cathode voltage exceeds the gate voltage by 0.7 V (single diode voltage drop), the PUT will turn ON, and C_1 will discharge through the PUT and develop an output pulse across R_L. To vary the frequency of this circuit, we can change the resistance of R_1 as before, or change the ratio of R_2 to R_3, which controls the V_P value of the PUT. For example, if R_3 is made larger than R_2, the gate voltage and therefore V_P voltage will be larger. A high V_P value will mean that C_1 will have to charge to a larger voltage before the PUT will turn ON. Increasing the time needed for C_1 to charge will decrease the triggering rate of the PUT and therefore decrease the circuit's frequency of operation.

Programmable Unijunction Transistor or PUT
A unijunction transistor that can have its peak voltage controlled.

UJT Testing

Using the oscilloscope, you can monitor the ON/OFF switching of a UJT while it is operating in-circuit. If you suspect that the UJT is the cause of a circuit malfunction, you should remove it and check it with the ohmmeter, as seen in Figure 20-10(h).

SELF-TEST EVALUATION POINT FOR SECTION 20-2

Now that you have completed this section, you should be able to:

■ **Objective 3.** *Define the term* thyristor.

■ **Objective 4.** *Describe the operation, characteristics, applications, and testing of the following thyristors:*
   a. *Silicon controlled rectifier or SCR*
   b. *Triode ac semiconductor switch or TRIAC*
   c. *Diode ac semiconductor switch or TRIAC*
   d. *Unijunction transistor or UJT*
   e. *Programmable unijunction transistor or PUT*

■ **Objective 5.** *Identify the different types of thyristor from their schematic symbols.*

Use the following questions to test your understanding of Section 20-2.

1. The SCR's forward breakover voltage will _____ as gate current is increased.
2. The SCR can only be used to control dc power. (True/False)
3. The TRIAC is a _____ directional device that can be either positive or negative triggered.
4. Which device can be used to trigger the TRIAC to compensate for the TRIAC's non-symmetrical triggering characteristic?
5. TRIACs should always be used instead of SCRs when we want to control the complete ac power cycle. (True/False)
6. The unijunction transistor has _____ P-N semiconductor junction(s).
7. The UJT can, like the bipolar and field effect transistors, function as either a switch or an amplifier. (True/False)
8. The PUT's peak voltage can be varied by changing the voltage between _____ and cathode.

20-3 TRANSDUCERS

A **semiconductor transducer** is an electronic device that converts one form of energy to another. Electronic transducers can be classified as either **input transducers** (such as the photodiode) that generate input control signals or **output transducers** (such as the LED) that convert output electrical signals to some other energy form. Input transducers are sensors that convert thermal, optical, mechanical, and magnetic energy variations into equivalent voltage and current variations. Output transducers, on the other hand, perform the exact opposite, converting voltage and current variations into optical or mechanical energy variations.

20-3-1 *Optoelectronic Transducers*

The LED and photodiode are the most frequently used optoelectronic devices; however, they are not the only semiconductor devices in the optoelectronic family. In this section, we will review the operation of the LED and photodiode, along with the operation of other types of semiconductor light transducers.

Light-Sensitive Devices

Figure 20-11 illustrates some of the different types of **light-sensitive semiconductor devices.**

The **photoconductive cell** or **light-dependent resistor (LDR)** shown in Figure 20-11(a) is a two-terminal device that changes its resistance (conductance) when light (photo) is applied. The photoconductive cell is normally mounted in a metal or plastic case with a glass window that allows the sensed light to strike the S-shaped light-sensitive material (typically

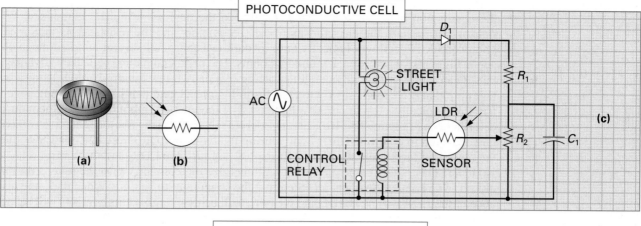

PHOTOCONDUCTIVE CELL

(a)

(b)

(c)

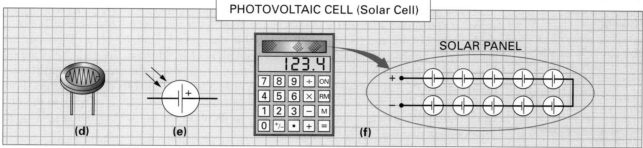

PHOTOVOLTAIC CELL (Solar Cell)

(d)

(e)

123.4

(f)

SOLAR PANEL

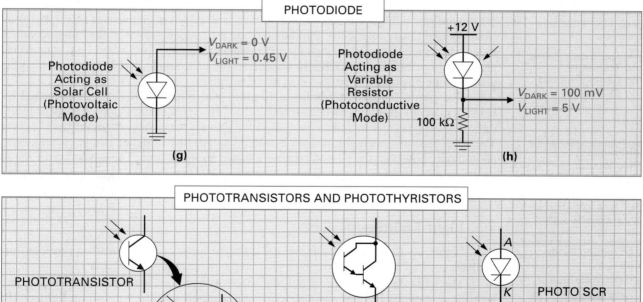

PHOTODIODE

Photodiode
Acting as
Solar Cell
(Photovoltaic
Mode)

$V_{DARK} = 0$ V
$V_{LIGHT} = 0.45$ V

(g)

Photodiode
Acting as
Variable
Resistor
(Photoconductive
Mode)

+12 V

100 kΩ

$V_{DARK} = 100$ mV
$V_{LIGHT} = 5$ V

(h)

PHOTOTRANSISTORS AND PHOTOTHYRISTORS

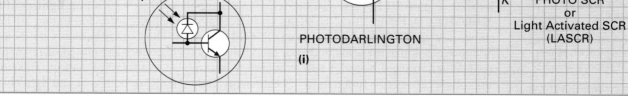

PHOTOTRANSISTOR

PHOTODARLINGTON

PHOTO SCR
or
Light Activated SCR
(LASCR)

(i)

FIGURE 20-11   Light-Sensitive Devices.

897

cadmium sulfide). When light strikes the photoconductive atoms, electrons are released into the conduction band and the resistance between the device's terminals is reduced. When light is not present, the electrons and holes recombine, and the resistance is increased. A photoconductive cell will typically have a "dark resistance" of several hundred mega ohms and a "light resistance" of a few hundred ohms. The photoconductive cell's key advantage is that it can withstand a high operating voltage (typically a few hundred volts). Its disadvantages are that it responds slowly to changes in light level, and that its power rating is generally low (typically a few hundred milliwatts). The schematic symbol for the photoconductive cell is shown in Figure 20-11(b). Figure 20-11(c) shows how the photoconductive cell could be connected to control a street light. During the day, the LDR's resistance is LOW due to the high light levels, and the current through D_1, R_1, R_2, the LDR, and the control relay coil is HIGH. This HIGH value of current will energize the coil and cause the relay's normally closed (*NC*) contacts to open, and therefore the street light to be OFF. When dark, the resistance of the LDR will be HIGH, and the relay will be de-energized, its contacts will return to their normal condition, which is closed, and the street light will be ON.

The **photovoltaic cell** or **solar cell,** shown in Figure 20-11(d), generates a voltage across its terminals that will increase as the light level increases. The solar cell is usually made from silicon, and its schematic symbol is shown in Figure 20-11(e). The solar cell is available as either a discrete device or as a solar panel in which many solar cells are interconnected to form a series-aiding power source, as shown in the calculator application in Figure 20-11(f). The output of a solar cell is normally rated in volts and milliamps. For example, a typical photovoltaic cell could generate 0.5 V and 40 mA. To increase the output voltage simply connect solar cells in series; to increase the output current simply connect solar cells in parallel.

A **photodiode** is a photo-detecting or light-receiving device that contains a semiconductor P-N junction. When used in the "photovoltaic mode" as shown in Figure 20-11(g), the photodiode will generate an output voltage (voltaic) in response to a light (photo) input (will operate like a solar cell). Photodiodes are most widely used in the "photoconductive mode," in which they will change their conductance (conductive) when light (photo) is applied (will operate like an LDR). In this mode, the photodiode is reverse biased (*n*-type region is made positive, *p*-type region is made negative).

The **photo-transistor, photo-darlington,** and **photo-SCR** (or light-activated SCR, LASCR), are all examples of light-reactive devices. As seen in the phototransistor inset, all three devices basically include a photodiode to activate the device whenever light is present. The advantage of the phototransistor is that it can produce a higher output current than the photodiode; however, the photodiode has a faster response time. The advantage of the photodarlington is its high current gain, and therefore it is ideal in low-light applications. When larger current switching is needed (typically a few amps), the LASCR can be used.

Light-Emitting Devices

Figure 20-12 illustrates some of the different types of **light-emitting semiconductor devices.**

The LED was covered in detail in Chapter 4; however, let us review its basic characteristics. Figure 20-12(a) shows the schematic symbols used to represent a **light-emitting diode** or **LED,** Figure 20-12(b) shows its typical physical appearance, and Figure 20-12(c) shows how the LED can be used as an ON/OFF indicator. The LED is basically a P-N junction diode and like all semiconductor diodes it can be either forward biased or reverse biased. When forward biased it will emit energy in response to a forward current. This emission of energy may be in the form of heat energy, light energy, or both heat and light energy depending on the type of semiconductor material used. The type of material also determines the color, and therefore frequency, of the light emitted. For example, different compounds are available that will cause the LED to emit red, yellow, green, blue, white, orange, or infrared light when it is forward biased.

Figure 20-12(d) shows the schematic symbol for the **injection laser diode** or **ILD,** Figure 20-12(e) shows its typical physical appearance, and Figure 20-12(f) shows how it is

Photovoltaic Cell or Solar Cell

A device that generates a voltage across its terminal that will increase as the light level increases.

Photodiode

A photo-detecting or light-receiving device that contains a semiconductor P-N junction.

Photo-Transistor, Photo-Darlington, and Photo-SCR

Examples of light-reactive devices.

Light-Emitting Semiconductor Devices

Semiconductor devices that will emit light when an electrical signal is applied.

Light-Emitting Diode or LED

A semiconductor diode that converts electric energy into electromagnetic radiation at visible and near infrared frequencies when its P-N junction is forward biased.

Injection Laser Diode or ILD

A semiconductor P-N junction diode that uses a lasing action to increase and concentrate the light output.

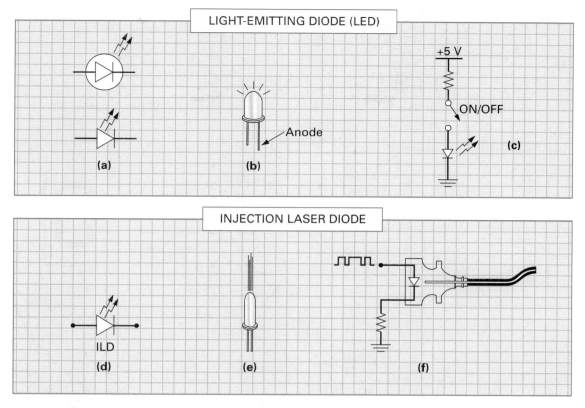

FIGURE 20-12   Light-Emitting Devices.

usually used in fiber optic communication. The ILD differs from the LED in that it generates monochromatic (one color, or one frequency) light. Although it appears as though the LED is only generating one color light, it is in fact emitting several wavelengths or different frequencies that combined make up a color. Generating only one frequency is ideal in applications such as fiber optics, where we want to keep the light beam tightly focused so that the beam can travel long distances down a very thin piece of glass or plastic fiber. To explain this point in more detail, let us compare the light from a light bulb to the light produced by an ILD. An ILD generates light of only one frequency (monochromatic), and since all of the small packets of light being generated are of the same frequency, they act like an organized army, marching together in the same direction as a tight concentrated beam. The light bulb, on the other hand, generates white light that is composed of every color or frequency (panchromatic), and since these frequencies are all different, they act completely disorganized, and therefore radiate in all directions.

Optoelectronic Application—Character Displays

One of the biggest applications of LEDs is in the multisegment display. Figure 20-13(a) reviews the variety of LED display types available, which were discussed in detail in Chapter 4. Figure 20-13(b) shows the construction of a common-anode seven segment display. The need to display letters, in addition to numerals, resulted in a display with more segments such as the 14-segment display. For greater character definition, dot matrix displays are used, the most popular of which contains 35 small LEDs arranged in a grid of five vertical columns and seven horizontal rows, forming a 5×7 matrix. The bar display is rapidly replacing analog meter movements for displaying a quantity. For example, a quantity of fuel can be represented by the number of activated segments: no bars lit being zero and all bars lit being maximum.

The variety of character displays shown in Figure 20-13(a) are also available as **liquid crystal displays (LCDs),** as shown in Figure 20-13(c). The two key differences between an

Liquid Crystal Displays (LCDs)

A digital display having two sheets of glass separated by a sealed quantity of liquid crystal material. When a voltage is applied across the front and back electrodes, the liquid crystal's molecules become disorganized, causing the liquid to darken.

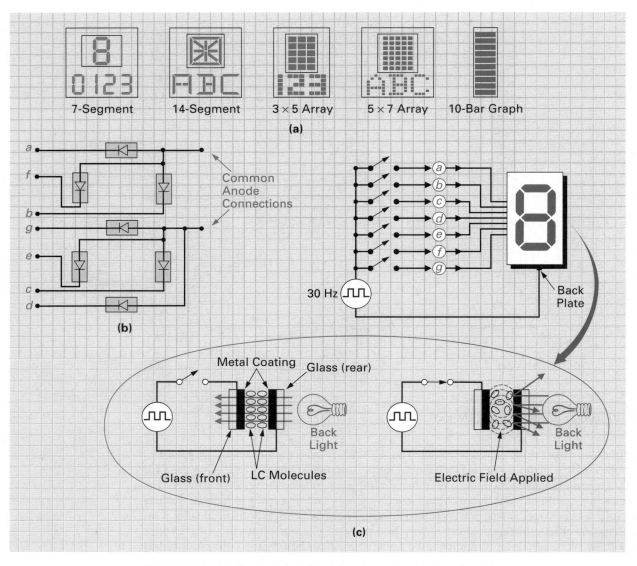

FIGURE 20-13 Optoelectronic Application—Character Displays.

LED and LCD display are that an LED display generates light while the LCD display controls light, and the LCD display consumes a lot less power than an LED display. The low power consumption feature of the LCD display makes it ideal in portable battery-operated systems such as wristwatches, calculators, video games, portable test equipment, and so on. The LCD's only disadvantage is that the display is hard to see, but this can be compensated for by including back-lighting to highlight the characters. The liquid crystal display contains two pieces of glass that act as a sandwich for a "nematic liquid" or liquid crystal material, as seen in the inset in Figure 20-13(c). The rear piece of glass is completely coated with a very thin layer of transparent metal, while the front piece of glass is coated with the same transparent metal segments in the shape of the desired display. The operation of the liquid crystal display is explained in the two illustrations in the inset in Figure 20-13(c). When a segment switch is open, no electric field is generated between the two LCD metal plates, the nematic liquid molecules remain in their normal state which is parallel to the plane of the glass, and so all of the back-lighting passes through to the front display making the segment invisible. When a segment switch is closed, an ac voltage is applied between the two metal plates, generating an electric field between the two LCD metal plates. This electric field will cause the nematic liquid molecules to turn by 90°, and so all of the back-lighting for that segment is blocked making the segment visible.

Optoelectronic Application—Optically Coupled Isolators

Up until this point, we have considered the optoelectronic emitter and detector as discrete or individual devices. There are, however, some devices available that include both a light-emitting and light-sensing device in one package. These devices are called **optically coupled isolators,** a name that describes their basic function: they are used to *optically couple two electrically isolated points.* To examine these devices in more detail, refer to the three basic types shown in Figure 20-14.

Figure 20-14(a) shows an **optically coupled isolator DIP module.** This optocoupler contains an infrared-emitting diode (IRED) and a silicon phototransistor and is generally used to transfer switching information between two electrically isolated points. Figure 20-14(b) shows how DIP optically coupled isolator ICs can be used in a stepper-motor control circuit (discussed previously in Chapter 10). Since both circuits are identical, let us explain only the upper circuit's operation. Digital logic circuits generate the $+5$ V peak multiphase ON/OFF switching signal that is applied to IRED in the optocoupler. A HIGH input, for example, will turn ON the IRED, which will emit light to the phototransistor, turning it ON and switching the $+12$ V on its collector through to the emitter and the base of Q_1. This $+12$ V input to the base of Q_1 will cause it to turn ON and act as a closed switch, connecting ground to the top of the A winding. Since $+12$ V is connected to the center of the A/B winding, the A winding will be energized and the motor will move a step. The optocoupler, therefore, couples the ON/OFF switching information from the digital logic circuits and at the same time provides the necessary electrical isolation between the $+5$ V digital logic circuits and the $+12$ V motor supply circuits. In the past, isolation was provided by relays or isolation transformers that were larger in size, consumed more power, and were more expensive.

Figure 20-14(c) shows a different type of optocoupler or optoisolator called the **optically coupled isolator interrupter module.** This device consists of a matched and aligned emitter and detector and is used to detect opaque or nontransparent targets. Figure 20-14(d) shows how the interrupter optocoupler module could be used as an optical tachometer. The IRED is permanently ON, as seen in the inset in Figure 20-14(d), and emits a constant light beam towards the phototransistor. A transparent disc mounted to a shaft has opaque targets evenly spaced around the disc. As these opaque targets pass through the infrared beam, they will cause the phototransistor to momentarily turn OFF. On the other hand, a transparent section of the disc will not block the infrared light, and so the phototransistor will remain ON. As a result the phototransistor will turn ON and OFF, generating pulses, the number of which is an indication of the shaft's speed of rotation.

Figure 20-14(e) shows another type of optocoupler called the **optically coupled isolator reflector module.** Like the interrupter module, this device consists of a matched and aligned emitter and detector, and it is also used to detect targets. Figure 20-14(f) shows how the reflector module could also be used as an optical tachometer. The disadvantage with the interrupter module is that the disc has to be exactly aligned between the emitter or detector sections. The reflector module does not have to be positioned so close to the target disc, and therefore the alignment is not so crucial. The target disc used for reflector modules is different in that it is composed of reflective and nonreflective target areas. As the disc rotates, almost no light will be reflected by the dark areas when they are present at the focal point of the IRED and phototransistor. However, when a reflective target is present at the focal point, light is reflected directly back to the phototransistor, turning it ON and therefore generating an output pulse.

20-3-2 *Temperature Transducers*

A **thermistor** is a semiconductor device that acts as a temperature sensitive resistor. Figure 20-15(a) shows a few different types of thermistors. There are basically two different types: **positive temperature coefficient (PTC) thermistors** and the more frequently used **negative temperature coefficient (NTC) thermistors.** To explain the difference, Figure

---

Optically Coupled Isolators

Devices that contain a light-emitting and light-sensing device in one package. They are used to optically couple two electrically isolated points.

Optically Coupled Isolator DIP Module

An optocoupler that contains an infrared-emitting diode and a silicon photo-transistor and is generally used to transfer switching information between two electrically isolated points.

Optically Coupled Isolator Interrupter Module

A device that consists of a matched and aligned emitter and detector and that is used to detect opaque or nontransparent targets.

Optically Coupled Isolator Reflector Module

A device that consists of a matched and aligned emitter and detector and that is used to detect targets.

Thermistor

A semiconductor device that acts as a temperature sensitive resistor.

Positive Temperature Coefficient (PTC) Thermistors

A thermistor in which a temperature increase causes the resistance to increase.

Negative Temperature Coefficient (NTC) Thermistor

A thermistor in which a temperature increase causes the resistance to decrease.

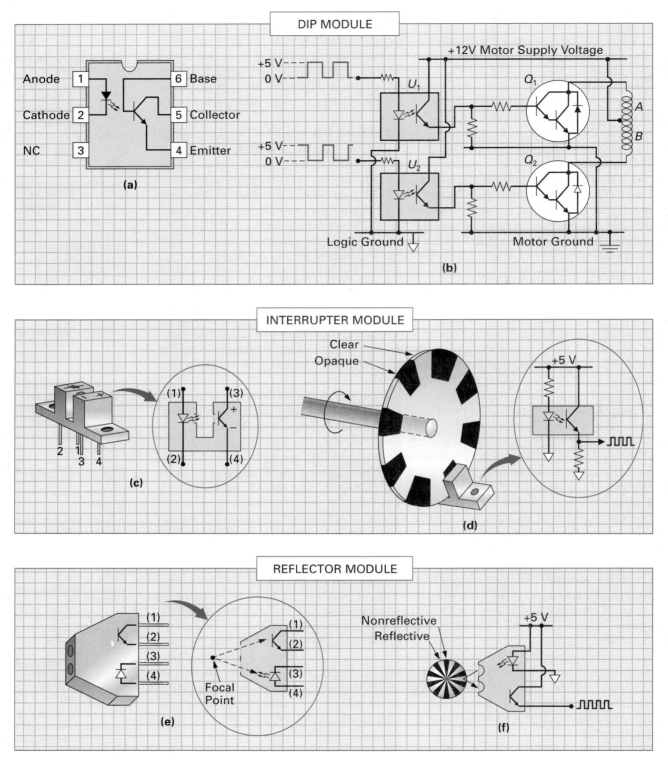

FIGURE 20-14   Optoelectronic Application—Optically Coupled Isolators.

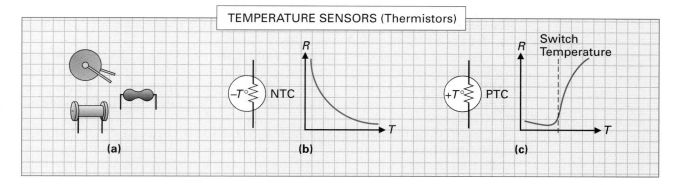

(a) (b) (c)

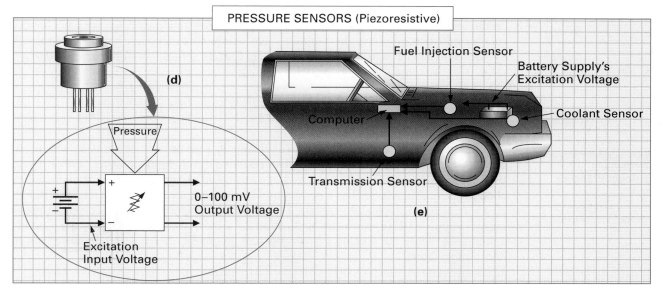

(d) (e)

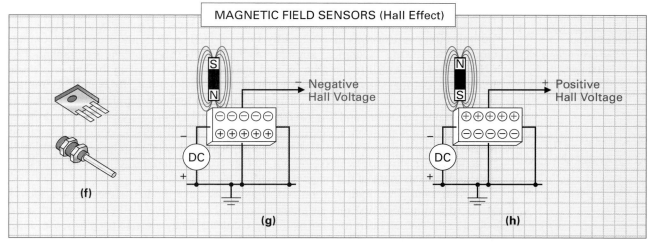

(f) (g) (h)

FIGURE 20-15 **Semiconductor Transducers.**

20-15(b) shows how a temperature increase causes the resistance of an NTC thermistor to decrease. On the other hand, Figure 20-15(c) shows how a temperature increase causes the resistance of a PTC thermistor to increase.

Thermistors are used in a variety of applications. For example, an NTC could be used in a fire alarm circuit. When the ambient temperatures are LOW, the resistance of the NTC is HIGH, and therefore current cannot energize the alarm. If the ambient temperature increases to a HIGH level, however, the resistance of the thermistor drops LOW, and the alarm circuit is energized.

The PTC, on the other hand, could be used as a sort of circuit protection device (similar to a circuit breaker). With the PTC, the "switch temperature" (which is the temperature at which the resistance rapidly increases) can be varied by different construction techniques from below 0°C to above 160°C. The PTC therefore could be used as a current-limiting circuit protection device because currents lower than a limiting value will not generate enough heat to cause the PTC thermistor to switch to its high resistance state. On the other hand, circuit currents that go above the limiting value will generate enough heat to cause the PTC thermistor to switch to its high-resistance state. Current surges therefore are limited to a safe value until the surge is over.

20-3-3 *Pressure Transducers*

A semiconductor pressure transducer will change its resistance in accordance with changes in pressure. Figure 20-15(d) shows a **piezoresistive diaphragm pressure sensor.** Piezoresistance of a semiconductor is described as a change in resistance due to a change in the applied pressure. This device has a dc excitation voltage applied, as seen in the inset in Figure 20-15(d), and will typically generate a 0 to 100 mV output voltage based on the pressure sensed. Figure 20-15(e) shows an application for these devices in which they are used in automobiles to sense the cooling-system pressure, hydraulic-transmission pressure, and fuel-injection pressure.

20-3-4 *Magnetic Transducers*

The **hall effect sensor** was discovered by Edward Hall in 1879 and is used in computers, automobiles, sewing machines, aircraft, machine tools, and medical equipment. Figure 20-15(f) shows a few different types of hall effect sensors. To explain the hall effect principle, Figure 20-15(g) shows that when a magnetic field whose polarity is north is applied to the sensor, it causes a separation of charges, generating in this example a negative hall voltage output. On the other hand, Figure 20-15(h) shows that if the magnetic field polarity is reversed so that a south pole is applied to the sensor, it will cause a polarity separation of charges within the sensor that will result in a positive hall voltage output. The amplitude of the generated positive or negative output voltage from the hall effect sensor is directly dependent on the strength of the magnetic field.

Their small size, light weight, and ruggedness make the hall effect sensors ideal in a variety of commercial and industrial applications. For example, hall effect sensors are embedded in the human heart to serve as timing elements. They are also used to sense shaft rotation, camera shutter positioning, rotary position, flow rate, and so on.

Now that you have completed this section, you should be able to:

■ **Objective 6.** *Define the term* transducer.

■ **Objective 7.** *Describe the application, characteristics, applications, and testing of the following light-sensitive optoelectronic devices:*
   a. *Photoconductive cell or light-dependent resistor*
   b. *Photovoltaic cell or solar cell*
   c. *Photodiode*
   d. *Phototransistors and photothyristors*

■ **Objective 8.** *Describe the operations, characteristics, applications and testing of the following light-reactive optoelectronic devices:*
   a. *Light-emitting diode or LED*
   b. *Injection laser diode or ILD*

■ **Objective 9.** *Explain the operation and characteristics of light-emitting diode displays and liquid crystal diode displays.*

■ **Objective 10.** *Describe the operation, characteristics, and applications of the following optically coupled isolators:*
   a. *DIP module*
   b. *Interruptor module*
   c. *Reflector module*

■ **Objective 11.** *Describe the operation, characteristics, and applications of the following semiconductor sensors:*

   a. *NTC thermistor*
   b. *PTC thermistor*
   c. *Piezoresistive pressure transducer*
   d. *Hall effect magnetic transducer*

Use the following questions to test your understanding of Section 20-3.

1. The photoconductive cell basically operates as a light-sensitive _____.

2. The photovoltaic cell converts light energy into _____ energy, and is often referred to as a _____ cell.

3. What are the two modes of operation for the photodiode?

4. With the photoconductive cell, photodiode, phototransistor, photo-darlington, photothyristor, and other light-sensitive devices, a light increase always causes a _____ in conductance.

5. LEDs convert _____ energy into _____ energy, whereas an ILD will convert _____ energy into _____ energy.

6. LED displays emit light while LCD displays control light. (True/False)

7. An optically coupled isolator DIP module will electrically connect two points, while isolating information from the two points. (True/False)

8. Would a PTC thermistor or NTC thermistor be ideal as a series current limiter?

SUMMARY

Basic Digital Timer and Control Circuits

1. To control the timing of digital circuits, a clock signal is distributed throughout the digital system. This square wave clock signal is generated by a clock oscillator, and its sharp positive (leading) and negative (trailing) edges are used to control the sequence of operations in a digital circuit.

2. The astable multivibrator circuit is used to produce an alternating two-state square or rectangular output waveform. This circuit is often called a free-running multivibrator because the circuit requires no input signal to start its operation. It will simply begin oscillating the moment the dc supply voltage is applied. If the time constants of the two *RC* timing networks in the astable circuit are different, the result will be a rectangular or pulse waveform.

3. The astable multivibrator is often referred to as an "unstable multivibrator" because it is continually alternating or switching back and forth, and therefore it has no stable condition or state. The monostable multivibrator has, as its name implies, one (mono) stable state. The circuit will remain in this stable state indefinitely until a trigger is ap-

plied and forces the monostable multivibrator into its unstable state. It will then remain in its unstable state for a small period of time, and then switch back to its stable state and await another trigger. The monostable multivibrator is often compared to a gun and called a one-shot multivibrator because it will produce one output pulse or shot for each input trigger. This one-shot timer circuit is often used in pulse-stretching and time delay applications.

4. The bistable multivibrator has two (bi) stable states, two inputs called the "SET" and "RESET," and two outputs called "*Q*" and "*Q̄*." The bistable multivibrator circuit is often called an *S-R* (set-reset) flip-flop because a pulse on the SET input will "flip" the circuit into the set state (*Q* output is set HIGH), while a pulse on the RESET input will "flop" the circuit into its reset state (*Q* output is reset LOW). The ability of the bistable multivibrator to remain in its last condition or state accounts for why the *S-R* flip-flop is also called an *S-R* latch. The bistable multivibrator *S-R* flip-flop or *S-R* latch has become one of the most important circuits in digital electronics. It is used in a variety of applications ranging from data storage to counting and frequency division.

The 555 Timer Circuit (Figure 20-16)

5. One of the most frequently used low-cost integrated circuit timers is the 555 timer. Its IC package consists of 8 pins, and it derives its number identification from the distinctive input voltage divider circuit consisting of three 5 kΩ resistors. It is a highly versatile timer that can be made to function as an astable multivibrator, monostable multivibrator, frequency divider, or modulator depending on the connection of external components.

6. Thyristors are generally used as dc and ac power control devices, while transducers are generally used as sensing, displaying, and actuating devices.

Thyristors

7. The semiconductor thyristor acts as an electronically controlled switch, connecting or disconnecting power from a load, or switching power ON and OFF to adjust the average amount of power delivered to a load.

8. Some thyristors are "unidirectional," which means that they will only conduct current in one direction (dc), while others are "bidirectional," which means that they can conduct current in either direction (ac).

9. Thyristors are generally used as electronically controlled switches instead of the transistors, because they have better power handling capabilities and are more efficient.

10. The most frequently used thyristors are the "silicon controlled rectifier (SCR)," "triode ac semiconductor switch (TRIAC)," "diode ac semiconductor switch (DIAC)", "unijunction transistor (UJT)," and "programmable unijunction transistor (PUT)."

11. The silicon controlled rectifier, or SCR, is a four-layered alternately-doped component with three terminals labeled anode (A), cathode (K) and gate (G).

12. An SCR's forward breakover voltage, or turn-ON voltage, is inversely proportional to the value of gate current.

13. Once the SCR is turned ON, a holding-current latches the SCR ON, independent of the gate current. If the forward current between the SCR's cathode and anode falls below this minimum holding-current value, the SCR will turn OFF.

14. A gate trigger is used to turn ON the SCR; however, once the SCR is ON, it can only be turned OFF by decreasing the anode-to-cathode voltage so that the cathode-to-anode current passing through the SCR drops below the holding-current value.

15. The SCR is used in a variety of electrical power applications, such as light dimmer circuits, motor speed control circuits, battery charger circuits, temperature control systems, and power regulator circuits.

16. The SCR will only conduct current in the forward direction, which is why it is classed as a unidirectional device.

17. If an ac signal is applied across the SCR, it will only respond to a gate trigger during the time that the ac alternation makes the anode positive with respect to the cathode.

18. Using the oscilloscope, you can monitor the gate-trigger input and ON/OFF switching of an SCR while it is operating in circuit. If you suspect that the SCR is the cause of a circuit malfunction, you should remove the SCR and use the ohmmeter to check for terminal-to-terminal opens and shorts.

19. The disadvantage with the SCR is that it is unidirectional, which means that it can only be activated when the applied anode-to-cathode voltage makes the anode positive with respect to the cathode and will therefore only conduct current in one direction. As a result, the SCR can only control a dc supply voltage, or one-half cycle of the ac supply voltage.

20. To gain control of the complete ac input cycle, we would need to connect two SCRs in parallel, facing in opposite directions. This is exactly what was done to construct the bidirectional triode ac semiconductor switch or TRIAC, which has three terminals called main terminal 1 (MT_1), main terminal 2 (MT_2), and gate (G).

21. The TRIAC can be triggered or activated by either a positive or negative input gate trigger. This means that the TRIAC can be used to control both the positive and negative alternation of an ac supply voltage.

QUICK REFERENCE SUMMARY SHEET

FIGURE 20-16   The 555 Timer Circuit.

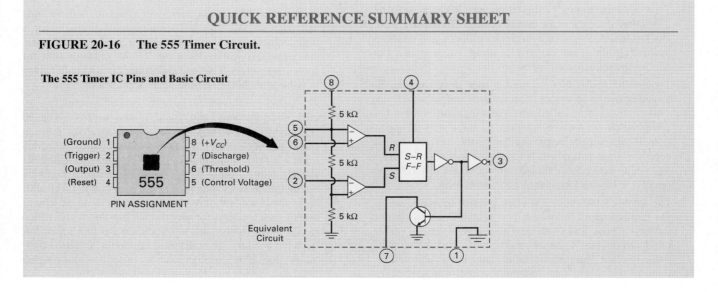

The 555 Timer IC Pins and Basic Circuit

22. By controlling gate current, we can control the TRIAC's turn-ON voltage (positive and negative breakover voltage), so that the ac input voltage can be chopped up to adjust the average value of voltage applied to the load.

23. Using the oscilloscope, you can monitor the gate-trigger input and ON/OFF switching of a TRIAC while it is operating in-circuit. If you suspect that the TRIAC is the cause of a circuit malfunction, you should remove the TRIAC and use the ohmmeter to check for terminal-to-terminal opens and shorts.

24. One disadvantage with the TRIAC is that its positive breakover voltage is usually slightly different from its negative breakover voltage. This nonsymmetrical trigger characteristic can be compensated for by using a diode ac semiconductor switch or DIAC to trigger a TRIAC. The DIAC is equivalent to two back-to-back, series connected junction diodes and has two terminals.

25. Since the PNP regions of a DIAC are all equally doped the DIAC will have the same forward and reverse characteristics. As a result, the DIAC is classed as a symmetrical bidirectional switch, which means that it will have the same value of breakover voltage in both the forward and reverse direction.

26. Using the oscilloscope, you can monitor the ON/OFF switching of a DIAC while it is operating in-circuit. If you suspect that the DIAC is the cause of a circuit malfunction, you should remove the DIAC and check it with the ohmmeter.

27. The unijunction transistor or UJT is never used as an amplifying device like the BJT and FET; it is only ever used as a voltage-controlled switch.

28. The UJT has only one P-N junction and is a three-terminal device with an emitter lead (E) attached to a small p-type pellet that is fused into a bar of n-type silicon with contacts at either end labeled base 1 (B_1) and base 2 (B_2).

29. The emitter-to-bar P-N junction is equivalent to a junction diode, and the bar is equivalent to a two-resistor voltage divider (R_{B1} and R_{B1}).

30. An increase in V_S, and therefore V_E, produces very little emitter current until the peak voltage (V_P) is reached. Beyond V_P, V_E has exceeded V_{B1} and the emitter diode is forward biased, causing I_E to increase and V_E to decrease, due to the lower resistance of R_{B1}. This negative resistance region reaches a low point known as the valley voltage (V_V), which is a point at which V_E begins to increase and the UJT no longer exhibits a negative resistance. The point at which a UJT turns ON and increases the current between B_1 and B_2 can be controlled, and therefore used in switching applications.

31. The programmable unijunction transistor or PUT is a variation on the basic UJT thyristor. This four-layer thyristor has three terminals labeled cathode (K), anode (A), and gate (G). The key difference between the basic UJT and the PUT is that the PUT's peak voltage (V_P) can be controlled.

32. Using the oscilloscope, you can monitor the ON/OFF switching of a UJT while it is operating in-circuit. If you suspect that the UJT is the cause of a circuit malfunction, you should remove it and check it with the ohmmeter.

Transducers

33. A semiconductor transducer is an electronic device that converts one form of energy to another.

34. Electronic transducers can be classified as either input transducers (such as the photodiode) that generate input control signals, or output transducers (such as the LED) that convert output electrical signals to some other energy form.

35. Input transducers are sensors that convert thermal, optical, mechanical and magnetic energy variations into equivalent voltage and current variations.

36. Output transducers on the other hand, perform the exact opposite, converting voltage and current variations into optical or mechanical energy variations.

Optoelectronic Transducers

37. The photoconductive cell, or light-dependent resistor (LDR), is a two-terminal device that changes its resistance (conductance) when light (photo) is applied.

38. When light strikes the photoconductive atoms, electrons are released into the conduction band, and therefore the resistance between the device's terminals is reduced. When light is not present, the electrons and holes recombine, and therefore the resistance is increased.

39. A photoconductive cell will typically have a "dark resistance" of several hundred mega ohms, and a "light resistance" of a few hundred ohms.

40. The photoconductive cell's key advantage is that it can withstand a high operating voltage (typically a few hundred volts). Its disadvantages are that it responds slowly to changes in light level, and that its power rating is generally low (typically a few hundred milliwatts).

41. The photovoltaic cell, or solar cell, generates a voltage across its terminals that will increase as the light level increases.

42. The solar cell is available as either a discrete device, or as a solar panel in which many solar cells are interconnected to form a series-aiding power source.

43. The output of a solar cell is normally rated in volts and milliamps. To increase the output voltage simply connect solar cells in series, and to increase the output current simply connect solar cells in parallel.

44. The phototransistor, photodarlington, and photo SCR (or light-activated SCR, LASCR), are all examples of light-reactive devices. All three devices basically include a photodiode to activate the device whenever light is present.

45. The advantage of the phototransistor is that it can produce a higher output current than the photodiode; however, the photodiode has a faster response time. The advantage of the photodarlington is its high current gain, and therefore it is ideal in low-light level applications. When larger current switching is needed (typically a few amps), the LASCR can be used.

46. The LED is basically a P-N junction diode, and like all semiconductor diodes it can be either forward biased or reverse biased. When forward biased it will emit energy in response to a forward current. This emission of energy may be in the form of heat energy, light energy, or both heat and light energy depending on the type of semiconductor material used.

47. The type of material used to construct the LED determines the color, and therefore frequency, of the light emitted. For example, different compounds are available that will cause the LED to emit red, yellow, green, blue, white, orange, or infrared light when it is forward biased.

48. The injection laser diode, or ILD, generates light of only one frequency (monochromatic), and since all of the small packets of light being generated are of the same frequency, they act like an organized army, marching together in the same direction as a tight concentrated beam. The light bulb on the other hand, generates white light that is composed of every color or frequency (panchromatic), and since these frequencies are all different, they act completely disorganized, and therefore radiate in all directions.

49. Generating only one frequency is ideal in applications such as fiber optics, where we want to keep the light beam tightly focused so that the information beam can travel long distances down a very thin piece of glass or plastic fiber.

50. Although it appears as though the LED is only generating one color light, it is in fact emitting several wavelengths or different frequencies that combined make up a color.

Optoelectronic Character Displays

51. One of the biggest applications of LEDs is in the multisegment display.

52. The seven-segment display is an example of a multisegment display. The need to display letters, in addition to numerals, resulted in a display with more segments such as the 14-segment display. For greater character definition, dot matrix displays are used, the most popular of which contains 35 small LEDs arranged in a grid of five vertical columns and seven horizontal rows to form a 5×7 matrix. The bar display is rapidly replacing analog meter movements for displaying the magnitude of a quantity.

53. A variety of character displays are available as liquid crystal displays (LCDs).

54. The two key differences between an LED and LCD display are that an LED display generates light while the LCD display controls light, and the LCD display consumes a lot less power than an LED display.

55. The low power consumption feature of the LCD display makes it ideal in portable battery-operated systems such as wristwatches, calculators, video games, portable test equipment, and so on.

56. The LCD's only disadvantage is that the display is hard to see; however, this can be compensated for by including back-lighting to highlight the characters.

57. When an LCD segment switch is open, no electric field is generated between the two LCD metal plates, the nematic liquid molecules remain in their normal state, which is parallel to the plane of the glass, and all of the back-lighting passes through to the front display making the segment invisible. On the other hand, when a segment switch is closed, an ac voltage is applied between the two metal plates and an electric field is generated between the two LCD metal plates. This electric field will cause the nematic liquid molecules to turn by 90°, so all of the back-lighting for that segment is blocked, making the segment visible.

Optically Coupled Isolators

58. There are devices available that include both a light-emitting and light-sensing device in one package. These devices are called optically coupled isolators, a name that describes their basic function because they are used to optically couple two electrically isolated points.

59. An optically coupled isolator DIP module contains an infrared emitting diode (IRED) and a silicon phototransistor. It is generally used to transfer switching information between two electrically isolated points.

60. The optically coupled isolator interrupter module is a device that contains a matched and aligned emitter and detector and is used to detect opaque or nontransparent targets.

61. The optically coupled isolator reflector module consists of a matched and aligned emitter and detector, and it is also used to detect targets.

62. The disadvantage with the interrupter module is that the disc has to be exactly aligned between the emitter or detector sections. The reflector module does not have to be positioned so close to the target disc, and therefore the alignment is not so crucial.

Temperature, Pressure and Magnetic Semiconductor Transducers

63. A thermistor is a semiconductor device that acts as a temperature sensitive resistor.

64. There are basically two different types of thermistors: positive temperature coefficient (PTC) thermistors and the more frequently used negative temperature coefficient (NTC) thermistors.

65. A temperature increase causes the resistance of an NTC thermistor to decrease. On the other hand, a temperature increase causes the resistance of a PTC thermistor to increase.

66. A semiconductor pressure transducer will change its resistance in accordance with changes in pressure.

67. Piezoresistance of a semiconductor is described as a change in resistance due to a change in the applied pressure. This device has a dc excitation voltage applied and will typically generate a 0 to 100 mV output voltage based on the pressure sensed.

68. The hall effect sensor was discovered by Edward Hall in 1879 and is used in computers, automobiles, sewing machines, aircraft, machine tools, and medical equipment.

69. When a magnetic field whose polarity is south is applied to a hall effect sensor, it causes a separation of charges, generating a negative hall voltage output. On the other hand, if the magnetic field polarity is reversed so that a north pole is applied to the sensor, it will cause an opposite polarity separation of charges within the sensor, resulting in a positive hall voltage output.

70. Their small size, light weight, and ruggedness make the hall effect sensors ideal in a variety of commercial and industrial applications such as sensing shaft rotation, camera shutter positioning, rotary position, flow rate, and so on.

Multiple-Choice Questions

1. The _____ multivibrator will produce a continuously alternating square wave or pulse wave output.
   a. Astable   b. Monostable   c. Bistable   d. Schmitt

2. The _____ multivibrator is also called a one-shot.
   a. Astable   b. Monostable   c. Bistable   d. Schmitt

3. Which multivibrator is also called a set-reset flip-flop?
   a. Astable   b. Monostable   c. Bistable   d. Schmitt

4. The output pulse width of a 555 timer is determined by the externally connected _____ and _____.
   a. Power supply, resistor
   b. Load resistance, capacitor
   c. Capacitor, load resistor
   d. Input resistor, capacitor

5. The 555 timer consists of a _____ resistor voltage divider, _____ comparator(s), _____ R-S flip-flop, an INVERTER, output stage and discharge transistor on a single IC.
   a. 2, 2, 2   b. 3, 2, 1   c. 1, 2, 3   d. 3, 1, 2

6. A monostable multivibrator will generally make use of a _____ and _____ circuit on the trigger input.
   a. Integrator, clipper   c. Schmitt, clipper
   b. Differentiator, Schmitt   d. Differentiator, clipper

7. Which of the following thyristors would normally be used to control dc power?
   a. TRIAC   d. DIAC
   b. SCR   e. Both (b) and (c)
   c. UJT

8. Which of the following thyristors is a bidirectional device?
   a. TRIAC   c. UJT
   b. SCR   d. ENIAC

9. _____ an SCR's gate current will _____ its forward breakover voltage.
   a. Increasing, decrease   c. Decreasing, increase
   b. Increasing, increase   d. Decreasing, decrease

10. Which of the following thyristors has a symmetrical switching characteristic?
    a. TRIAC   d. DIAC
    b. SCR   e. Both (b) and (c)
    c. UJT

11. Which of the following light sensitive devices will generate a voltage when light is applied?
    a. Photoconductive cell   d. Phototransistor
    b. Photodiode   e. Both (b) and (c)
    c. Photovoltaic cell

12. Which of the following is a monochromatic light source?
    a. LED   d. LCD
    b. Light bulb   e. Both (b) and (c)
    c. ILD

13. Which of the following devices would be best suited to detect whenever a coin has been inserted in a slot?
    a. Piezoresistive sensor   d. Interrupter optoisolator
    b. Hall Effect sensor   e. Both (a) and (c)
    c. PTC thermistor

14. What advantage does a liquid crystal display have over an LED display?
    a. Consumes less power   d. Both (a) and (c)
    b. Is easier to read   e. Both (a) and (b)
    c. Is ideal in portable applications

15. What device could be used as a series connected current limiter?
    a. Hall effect sensor   c. PTC thermistor
    b. Piezoresistive sensor   d. NTC thermistor

16. Which of the following sensors could be used to control the switching of current through a motor's stator windings by sensing the position of the motor's permanent magnet rotor?
    a. Hall effect sensor   c. PTC thermistor
    b. Piezoresistive sensor   d. NTC thermistor

Communication Skill Questions

17. Sketch an astable multivibrator circuit, and briefly describe its operation. (20-1-1)

18. What is the purpose of the differentiator and clipper on the trigger input of a monostable multivibrator? (20-1-2)

19. How can the monostable multivibrator be used to stretch a pulse? (20-1-2)

20. Sketch a monostable multivibrator circuit, and briefly describe its operation. (20-1-2)

21. What is the similarity between the bistable multivibrator and the schmitt trigger? (20-1-3)

22. Sketch the bistable multivibrator circuit, and briefly describe its operation. (20-1-3)

23. Give the full name for the following abbreviations:
    a. SCR   d. UJT   g. PTC
    b. TRIAC   e. PUT   h. NTC
    c. DIAC   f. ILD   i. LCD

24. Explain how the SCR is turned ON, its latching action, and how it is turned OFF. (20-2-1)

25. Why is the SCR classed as a unidirectional device? (20-2-1)

26. Can SCRs be used in ac and dc power control applications? (20-2-1)

27. What is the basic difference between an SCR and a TRIAC? (20-2-2)

28. Why is the TRIAC classed as a bidirectional device? (20-2-2)

29. Which of the thyristors is a symmetrical bidirectional switch? (20-2-3)

30. Briefly describe the operation of the unijunction transistor. (20-2-4)

31. What advantage does a PUT have over the basic UJT? (20-2-4)

32. Define the term "transducer." (20-3)

33. What is the difference between a light-dependent resistor and a solar cell? (20-3-1)

34. Which of the light-sensitive devices would be best suited to detect a very high-speed data transmission from a fiber optic cable? (20-3-1)

35. What advantage does the phototransistor have over the photodiode? (20-3-1)

36. In what application would the LASCR be used? (20-3-1)

37. What is the basic difference between an LED and an ILD? (20-3-1)

38. Briefly describe the operation of a liquid crystal display. (20-3-1)

39. What is an optically coupled isolator? (20-3-1)

40. Describe an application for the following optically coupled isolators: (20-3-1)
    a. DIP module   **c.** Reflector module
    b. Interrupter module

41. What is the difference between a PTC thermistor and an NTC thermistor? (20-3-2)

42. Briefly describe the operation of the piezoresistive pressure sensor and the hall effect sensor. (20-3-3 and 20-3-4)

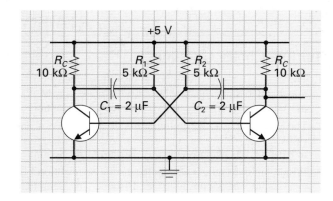

FIGURE 20-17   A Two-State Switching Circuit.

Practice Problems

43. Identify the circuit shown in Figure 20-17.

44. Calculate the frequency of the output for the circuit shown in Figure 20-17.

45. Identify the circuit shown in Figure 20-18.

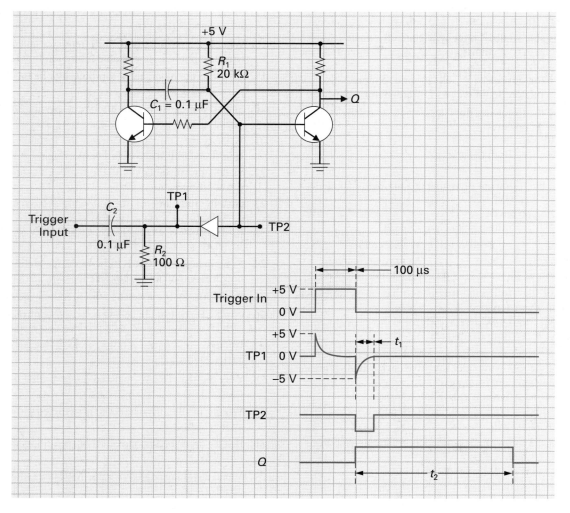

FIGURE 20-18   A Pulse Stretching Circuit.

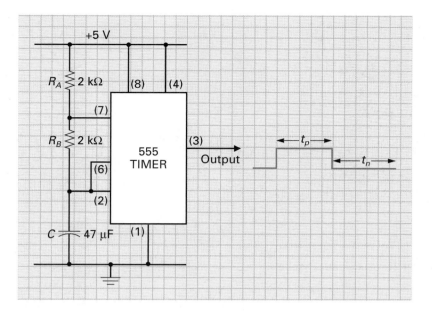

FIGURE 20-19   A 555 Timer Circuit.

46. Calculate the width of the output pulse (t_2) from the circuit shown in Figure 20-18.

47. Comparing the input pulse to the output pulse in Figure 20-18, how much pulse stretching has actually occured?

48. What would be the approximate width of the trigger pulse at *TP*2 in Figure 20-21?

49. Identify the circuit shown in Figure 20-21.

50. Calculate the following in relation to the circuit shown in Figure 20-19:
   a. The positive half-cycle time
   b. The negative half-cycle time
   c. The cycle's period
   d. The cycle's frequency

51. What would be the duty cycle of the output waveform generated by the 555 timer circuit in Figure 20-19?

52. Referring to Figure 20-20, what is the waveform's:
   a. Pulse width   **c.** Frequency
   b. Period   **d.** Duty cycle

53. Identify the schematic symbols shown in Figure 20-21.

To practice your circuit recognition and operation ability, refer to the circuit in Figure 20-22, and answer the following questions.

54. Describe the operation of the circuit when the input is normally HIGH.

55. Describe the operation of the circuit when the input is taken LOW.

56. How is the SCR turned ON, and how is the SCR turned OFF?

To practice your circuit recognition and operation ability, refer to the circuit in Figure 20-23, and answer the following questions.

57. Is the 555 timer connected to function as an astable, monostable, or bistable multivibrator?

58. Describe the basic operation of this circuit.

59. Why do you think a TRIAC is being used in this circuit instead of an SCR?

To practice your circuit recognition and operation ability, refer to the circuit in Figure 20-24(a) and (b), and answer the following questions.

60. Identify the light-emitting and light-sensitive devices used in these circuits.

61. Why was a different light-reactive device used for each of these circuits?

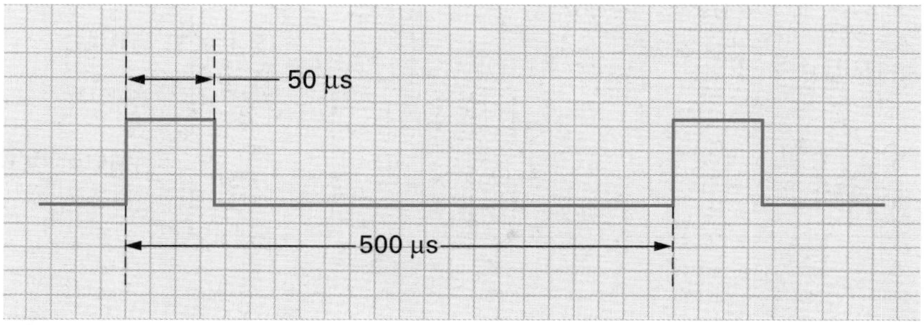

FIGURE 20-20   A Rectangular or Pulse Waveform.

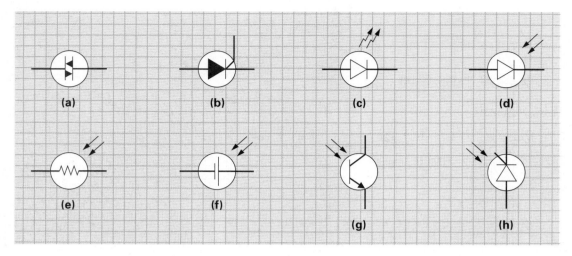

FIGURE 20-21   Schematic Symbols.

FIGURE 20-22   Low-Power to
High-Power ON/OFF Switching.

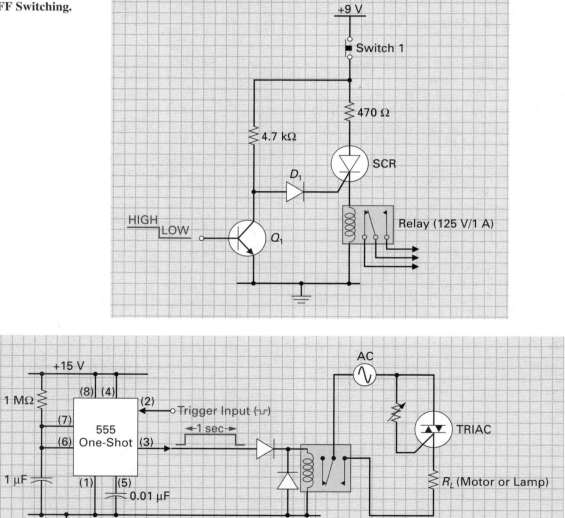

FIGURE 20-23   One-Shot Timer Control of TRIAC.

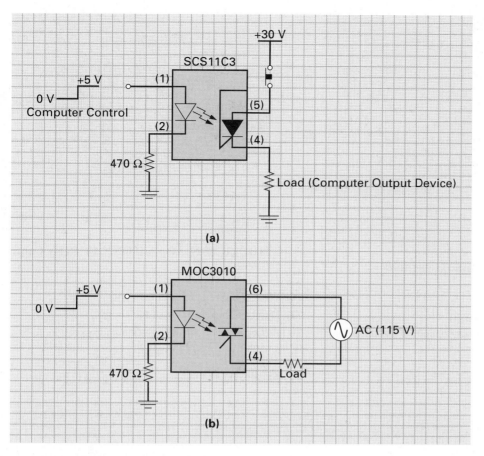

FIGURE 20-24   Computer Output Device Control.

62. Why are these optically coupled isolators needed for computer output device control?

To practice your circuit recognition and operation ability, refer to the circuit in Figure 20-25, and answer the following questions.

63. Identify the light-sensitive device used in this circuit.

64. What is the function of this circuit?

65. Describe the operation of the circuit.

To practice your circuit recognition and operation ability, refer to the circuit in Figure 20-26, and answer the following questions.

66. Identify the light-sensitive device used in this circuit.

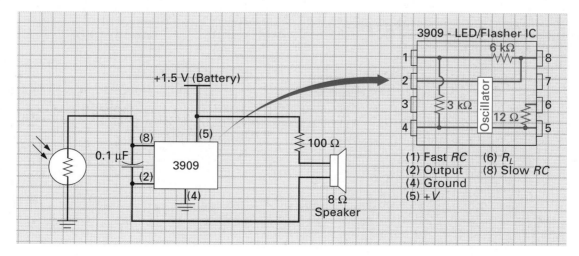

FIGURE 20-25   Light-Controlled Tone Generator Circuit.

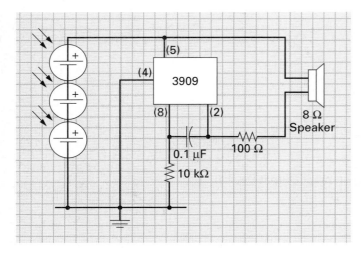

FIGURE 20-26 Solar-Powered Tone Generator.

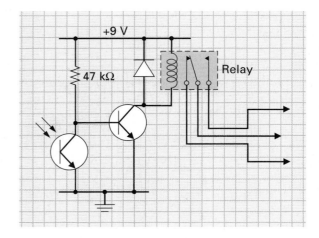

FIGURE 20-27 Optical ON/OFF Relay Control Circuit.

67. What is the function of this circuit?

68. Describe the operation of the circuit.

To practice your circuit recognition and operation ability, refer to the circuit in Figure 20-27, and answer the following questions.

69. Identify the light-sensitive device used in this circuit.

70. What is the function of this circuit?

71. Describe the operation of the circuit.

To practice your circuit recognition and operation ability, refer to the circuit in Figure 20-28, and answer the following questions.

72. Identify the light-sensitive device used in this circuit.

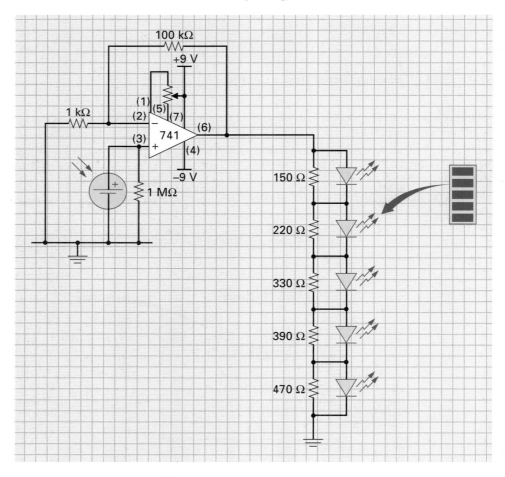

FIGURE 20-28 Lightmeter with Bargraph Display.

73. What is the function of this circuit?

74. Briefly describe the function of the 741 op-amp in this circuit.

75. Why is a variable resistor connected to the 741 op-amp?

76. What is the normal voltage drop across an LED?

77. Describe how the bargraph display in this circuit operates.

Web Site Questions

Go to the Web site http://www.prenhall.com/cook, select the textbook *Introductory DC/AC Electronics* or *Introductory DC/AC Circuits*, this chapter, and then follow the instructions when answering the multiple-choice practice problems.

JOB INTERVIEW TEST

These tests at the end of each chapter will challenge your knowledge up to this point, and give you the practice you need for a job interview. To make this more realistic, the test will comprise both technical and personal questions. In order to take full advantage of this exercise, you may want to set up a simulation of the interview environment, have a friend pose the questions to you, and record your responses for later analysis.

Company Name: ASI, Inc.

Industry Branch: Computers

Function: Design, manufacture and service printers

Job Title: Production Test Technician

1. Do you have a genuine interest in computers?

2. In what applications could you use a 555 timer?

3. What is a transducer?

4. What application software are you familiar with?

5. What do you know about thyristors?

6. Would you call yourself trustworthy?

7. Would you say that you are personable?

8. Do you normally arrive early, late, or right on time for work?

9. How long have you been interested in electronics?

10. Would you still say that you still have a lot to learn in electronics?

Answers

1. The answer must be yes. Go on to explain what computer systems you have at home and school, and in what capacity you have used them personally and professionally.

2. Section 20-1.

3. Section 20-3.

4. Discuss EWB's MultiSym and/or Circuit Maker, and all other application software you are familiar with.

5. Sections 20-2.

6. Yes.

7. They're asking if you can get along well with colleagues. Mention any work experience in which you have direct contact with people, and discuss how you worked with fellow students in your class.

8. Early.

9. Explain.

10. Yes. From World War II onward, no branch of science has contributed more to the development of the modern world as electronics. It has stimulated dramatic advances in the fields of communication, computing, consumer products, industrial automation, test and measurement, and health care. It has now become the largest single industry in the world, exceeding the automobile and oil industries, with annual sales of electronic systems exceeding $2 trillion. Keeping up with the advances will in itself be a full time job.

Electronics Dictionary

Absorption Loss or dissipation of energy as it travels through a medium. For example, radio waves lose some of their electromagnetic energy as they travel through the atmosphere.

AC Abbreviation for "alternating current."

AC alpha (α_{AC}) The ratio of input emitter current change to output collector current change.

AC beta (β_{AC}) The ratio of a transistor's ac output current to input current.

Accelerate To go faster.

AC coupling Circuit or component that couples or passes the ac signal yet blocks any dc level.

AC/DC If indicated on a piece of equipment, it means that the equipment will operate from either an ac or dc supply.

AC generator Device that transforms or converts a mechanical input into an ac electrical power output.

Acoustic Relating to sound or the science of sound.

AC power supply Power supply that delivers one or more sources of ac voltage.

Activate To put ac voltage to work or make active. The application of an enabling signal.

Active bandpass filter A circuit that uses an amplifier with passive filter elements to pass only a band of input frequencies.

Active band-stop filter A circuit that uses an amplifier with passive filter elements to block a band of input frequencies.

Active component Component that amplifies a signal or achieves some sort of gain between input and output.

Active equipment Equipment that will transmit and receive, as opposed to passive equipment, which only receives.

Active filter A filter that uses an amplifier with passive filter elements to provide pass or rejection characteristics.

Active high-pass filter A circuit that uses an amplifier with passive filter elements to pass all frequencies above a cut-off frequency.

Active low-pass filter A circuit that uses an amplifier with passive filter elements to pass all frequencies below a cut-off frequency.

Active operation or in the active region When the base-emitter junction is forward biased and the base-collector junction is reverse biased. In this mode, the transistor is equivalent to a variable resistor between collector and emitter.

Active region Flat part of the collector characteristic curve. A transistor is normally operated in this region, where it is equivalent to a variable resistor between the collector and emitter.

AC voltage Alternating voltage.

ADC Abbreviation for "analog-to-digital converter."

Adjacent The side of a right-angled triangle that has a common endpoint, in that it extends between the hypotenuse and the vertical side.

Adjustable resistor Resistor whose value can be changed.

Admittance (symbolized Y) Measure of how easily ac will flow through a circuit. It is equal to the reciprocal of impedance and is measured in siemens.

Aerial Another term used for antenna.

AF Abbreviation for "audio frequency."

AFC Abbreviation for "automatic frequency control."

AGC Abbreviation for "automatic gain control."

Air-core inductor An inductor that has no metal core.

Algebra A generalization of arithmetic in which letters representing numbers are combined according to the rules of arithmetic.

Aligning tool Small, nonconductive, nonmagnetic screwdriver used to adjust receiver or transmitter tuned circuits.

Alkaline cell Also called an alkaline manganese cell, it is a primary cell that delivers more current than the carbon–zinc cell.

Alligator clip Spring clip normally found on the end of test leads and used to make temporary connections.

Alphanumeric Having numerals, symbols, and letters of the alphabet.

Alpha wave Also called alpha rhythm, it is a brain wave between 9 and 14 Hz.

Alternating current Electric current that rises from zero to a maximum in one direction, falls to zero, and then rises to a maximum in the opposite direction, and then repeats another cycle, the positive and negative alternations being equal.

Alternation One ac cycle consists of a positive and negative alternation.

Alternator Another term used to describe an ac generator.

AM Abbreviation for "amplitude modulation."

Ambient temperature Temperature of the air surrounding the components.

American wire gauge American wire gauge (AWG) is a system of numerical designations of wire sizes, with the first being 0000 (the largest size) and then going to 000, 00, 0, 1, 2, 3, and so on up to the smallest sizes of 40 and above.

Ammeter Meter placed in the circuit path to measure the amount of current flow.

Ampere (A) Unit of electric current.

Ampere-hours If you multiply the amount of current, in amperes, that can be supplied for a given time frame in hours, a value of ampere-hours is obtained.

Ampere-hour meter Meter that measures the amount of current drawn in a unit of time (hours).

Ampere-turn Unit of magnetomotive force.

Ampere-turn per meter Base unit of magnetic field strength.

Amplification Process of making bigger or increasing the voltage, current, thus increasing the power of a signal.

Amplifier Circuit or device that achieves amplification.

Amplitude Magnitude or size an alternation varies from zero.

Amplitude distortion Changing of a wave shape so that it does not match its original form.

Analog Relating to devices or circuits in which the output varies in direct proportion to the input.

Analog data Information that is represented in a continuously varying form, as opposed to digital data, which have two distinct and discrete values.

Analog display Display in which a moving pointer or some other analog device is used to display the data.

Analog electrical circuits A power circuit designed to control power in a continuously varying form.

Analog electronic circuits Electronic circuits that represent information in a continuously varying form.

Analog multimeter Electronic test instrument that can perform multiple tasks in that it can be used to measure voltage, current, or resistance.

Anode Positive electrode or terminal.

Antenna Device that converts an electrical wave into an electromagnetic wave that radiates away from the antenna.

Antenna transmission line System of conductors connecting the transmitter or receiver to the antenna.

Apparent power The power value obtained in an ac circuit by multiplying together effective values of voltage and current, which reach their peaks at different times.

Arc Discharge of electricity through a gas. For example, lightning is a discharge of static electricity buildup through air.

Armature Rotating or moving component of a magnetic circuit.

Armstrong oscillator A tuned transformer oscillator developed by E. H. Armstrong.

Astable multivibrator A device commonly used as a clock oscillator.

Atom Smallest particle of an element.

Atomic number Number of positive charges or protons in the nucleus of an atom.

Atomic weight The relative weight of a neutral atom of an element, based on a neutral oxygen atom having an atomic weight of 16.

Attenuate To reduce in amplitude an action or signal.

Attenuation Loss or decrease in energy of a signal.

Audio Relating to all the frequencies that can be heard by the human ear; from the Latin word meaning "I hear."

Audio frequency Frequency that can be detected by a human ear. From approximately 20 Hz to 20 kHz.

Audio-frequency amplifier An amplifier that has one or more transistor stages designed to amplify audio-frequency signals.

Audio-frequency generator A signal generator that can be set to generate a sinusoidal AF signal voltage at any desired frequency in the audio spectrum.

Autotransformer Single-winding tapped transformer.

Average A single value that summarizes or represents the general significance of a set of unequal values.

Average value Mean value found when the area of a wave above a line is equal to the area of the wave below the line.

Avionics Field of aviation electronics.

AWG Abbreviation for "American wire gauge."

Axial lead Component that has its connecting leads extending from either end of its body.

Balanced bridge Condition that occurs when a bridge is adjusted to produce a zero output.

Balancing Setting the output of an op-amp to zero volts when both inverting and noninverting inputs are at zero volts.

Ballast resistor A resistor that increases in resistance when current increases. It can therefore maintain a constant current despite variations in line voltage.

Bandpass filter Filter circuit that passes a group or band of frequencies between a lower and an upper cutoff frequency, while heavily attenuating any other frequency outside this band.

Band-stop filter A filter that attenuates alternating currents whose frequencies are between given upper and lower cutoff values while passing frequencies above and below this band.

Bandwidth Width of the group or band of frequencies between the half-power points.

Bar graph or bar chart A graphic means of quantitative comparison by rectangles with lengths proportional to the measure of the data being compared.

Bar graph display A left-to-right or low-to-high set of bars that are turned on in successive order based on the magnitude they are meant to represent.

Barometer Meter used to measure atmospheric pressure.

Barretter Temperature-sensitive device having a positive temperature coefficient; that is, as temperature increases, resistance increases.

Barrier potential or voltage The potential difference, or voltage, that exists across the junction.

Base The region that lies between an emitter and a collector of a transistor and into which minority carriers are injected.

Base biasing A transistor biasing method in which the dc supply voltage is applied to the base of the transistor via a base bias resistor.

Base current (I_B) The relatively small current at the transistor's base terminal.

Bass Low audio frequencies normally handled by a woofer in a sound system.

Battery DC voltage source containing two or more cells that converts chemical energy into electrical energy.

Battery charger Piece of electrical equipment that converts ac input power to a pulsating dc output, which is then used to charge a battery.

Baud Unit of signaling speed describing the number of elements per second.

Beta β The transistor's current gain in a common-emitter configuration.

B-H curve Curve plotted on a graph to show successive states during magnetization of a ferromagnetic material.

Bias voltage The dc voltage applied to a semiconductor device to control its operation.

Bidirectional device A device that will conduct current in either direction.

Binary Number system having only two levels and used in digital electronics.

Binary word A numerical value expressed as a group of binary digits.

Bipolar device A device in which there is a change in semiconductor material between the output terminals (NPN or PNP between emitter and collector), so the charge carriers can be one of two polarities (bipolar).

Bipolar family A group of digital logic circuits that make use of the bipolar junction transistor.

Bipolar transistor A bipolar junction transistor in which excess minority carriers are injected from an emitter region into a base region and from there pass into a collector region.

Bistable multivibrator A digital control device that can be either set or reset.

Bits An abbreviation for "binary digits."

Bleeder current Current drawn continuously from a voltage source. A bleeder resistor is generally added to lessen the effect of load changes or provide a voltage drop across a resistor.

Body resistance The resistance of the human body.

Bolometer Device whose resistance changes when heated.

Branch current A portion of the total current that is present in one path of a parallel circuit.

Breakdown region The point at which the collector supply voltage will cause a damaging value of current through the transistor.

Breakdown voltage The voltage at which breakdown of a dielectric or an insulator occurs. Also, the voltage (V_{BR}) at which a damaging value of I will pass through the JFET.

Bridge rectifier circuit A full-wave rectifier circuit using four diodes that will convert an alternating voltage input into a direct voltage output.

Buffer current amplifier The C-C circuit that can ensure that power is efficiently transferred from source to load.

BW Abbreviation for "bandwidth."

Bypass capacitor Capacitor that is connected to provide a low-impedance path for ac.

Byte Group of 8 binary digits or bits.

Cable Group of two or more insulated wires.

CAD Abbreviation for "computer-aided design."

Calculator Device that can come in a pocket (battery-operated) or desktop (ac power) size and is used to achieve arithmetic operations.

Calibrate A procedure in which a meter or other device is adjusted to ensure its accuracy. The unit being calibrated is normally placed in a certain condition and then adjusted until a measured value is equal to an expected standard value.

Calibration To determine by measurement or comparison with a standard the correct value of each scale reading.

Capacitance (C) Measured in farads, it is the ability of a capacitor to store an electrical charge.

Capacitance meter Instrument used to measure the capacitance of a capacitor or circuit.

Capacitive filter A capacitor used in a power supply filter system to supress ripple currents while not affecting direct currents.

Capacitive reactance (X_C) Measured in ohms, it is the ability of a capacitor to oppose current flow without the dissipation of energy.

Capacitor Device that stores electric energy in the form of an electric field that exists within a dielectric (insulator) between two conducting plates, each connected to a connecting lead. This device was originally called a condenser.

Capacitor-input filter Filter in which a capacitor is connected to shunt away low-frequency signals such as ripple from a power supply.

Capacitor microphone Microphone that contains a stationary and movable plate separated by air. The arriving sound waves cause movements in the movable plate, which affects the distance between the plates and therefore the capacitance, generating a varying AF electrical wave. Also called an electrostatic microphone.

Capacitor speaker Also called the electrostatic speaker, its operation depends on mechanical forces generated by an electrostatic field.

Capacity Output current capabilities of a device or circuit.

Capture The act of gaining control of a signal.

Carbon composition resistor Fixed resistor consisting of carbon particles mixed with a binder, which is molded and then baked. Also called a composition resistor.

Carbon film resistor Thin carbon film deposited on a ceramic form to create resistance.

Carbon microphone Microphone whose operation depends on the pressure variation in the carbon granules changing their resistance.

Cardiac pacemaker Device used to control the frequency or rhythm of the heart by stimulating it through electrodes.

Carrier Wave that has one of its characteristics varied by an information signal.

The carrier will then transport this information between two points.

Cascoded amplifier An amplifier that contains two or more stages arranged in a series manner.

Cascode amplifier circuit An amplifier circuit consisting of a self biased common-source amplifier in series with a voltage-divider biased common-gate amplifier.

Cassette Thin, flat, rectangular device containing a length of magnetic tape that can be recorded onto or played back.

Cathode Term used to describe a negative electrode or terminal.

Cathode ray tube A vacuum tube in which electrons emitted by a hot cathode are focused into a narrow beam by an electron gun and then applied to a fluorescent screen. The beam can be varied in intensity and position to produce a pattern or picture on the screen.

Cell Single unit having only two plates that convert chemical energy into a dc electrical voltage.

Celsius-temperature scale Scale that defines the freezing point of water as $0°C$ and the boiling point as $100°C$ (named after Anders Celsius).

Center tap Midway connection between the two ends of a winding or resistor.

Center-tapped rectifier Rectifier that makes use of a center-tapped transformer and two diodes to achieve full-wave rectification.

Center-tapped transformer A transformer that has a connection at the electrical center of the secondary winding.

Ceramic capacitor Capacitor in which the dielectric material used between the plates is ceramic.

Cermet Ceramic-metal mixture used in film resistors.

Channel A path for a signal.

Charge Quantity of electrical energy stored in a battery or capacitor.

Charging current Current that flows to charge a capacitor or battery when a voltage is applied.

Chassis Metal box or frame in which components, boards, and units are mounted.

Chassis ground Connection to the metal box or frame that houses the components and associated circuitry.

Choke Inductor used to impede the flow of alternating or pulsating current.

Circuit Interconnection of many components to provide an electrical path between two or more points.

Circuit breaker Reusable fuse. This device will open a current-carrying path without damaging itself once the current value exceeds its maximum current rating.

Circuit failure A malfunction or breakdown of a component in the circuit.

Circular mil (cmil) A unit of area equal to the area of a circle whose diameter is 1 mil.

Circumference The perimeter of a circle.

Clapp oscillator A series tuned colpitts oscillator circuit.

Class A amplifier An amplifier in which the transistor is in its active region for the entire signal cycle.

Class AB amplifier An amplifier in which the transistor is in its active region for slightly more than half the signal cycle.

Class B amplifier An amplifier in which the transistor is in its active region for approximately half the signal cycle.

Class C amplifier An amplifier in which the transistor is in its active region for less than half the signal cycle.

Clock Generally, a square waveform used for the synchronization and timing of several circuits.

Clock input (C) An accurate timing signal used for synchronizing the operation of digital circuits.

Clock oscillator A device for generating a clock signal.

Clock signal Generally a square wave used for the synchronization and timing of several circuits.

Closed circuit Circuit having a complete path for current to flow.

Closed-loop mode A control system containing one or more feedback control loops, in which functions of the controlled signals are combined with functions of the commands that tend to maintain prescribed relationships between the commands and the controlled signals.

Closed-looped voltage gain (A_{CL}) The voltage gain of an amplifier when it is operated in the closed-loop mode.

Coaxial cable Transmission line in which a center signal-carrying conductor is completely covered by a solid dielectric, another conductor, and then an outer insulating sleeve.

Coefficient This is related to the ratio of change under certain conditions.

Coefficient of coupling The degree of coupling that exists between two circuits.

Coercive force Magnetizing force needed to reduce the residual magnetism within a material to zero.

Coil Number of turns of wire wound around a core to produce magnetic flux (an electromagnet) or to react to a changing magnetic flux (an inductor).

Cold resistance The resistance of a device when cold.

Collector A semiconductor region through which a flow of charge carriers leaves the base of the transistor.

Collector characteristic curve A set of characteristic curves of collector voltage versus collector current for a fixed value of transistor base current.

Collector current (I_C) The current emerging out of the transistor's collector.

Color code Set of colors used to indicate a component's value, tolerance, and rating.

Colpitts oscillator An LC tuned oscillator circuit in which two tank capacitors are used instead of a tapped coil.

Common Shared by two or more services, circuits, or devices. Although the term "common ground" is frequently used to describe two or more connections sharing a common ground, the term "common" alone does not indicate a ground connection, only a shared connection.

Common-base (C-B) circuit Configuration in which the input signal is applied between the transistor's emitter and base, while the output is developed across the transistor's collector and base.

Common-collector (C-C) circuit Configuration in which the input signal is applied between the transistor's base and collector, while the output is developed across the transistor's collector and emitter.

Common-drain configuration An FET configuration in which the drain is grounded and common to the input and output signal.

Common-emitter (C-E) circuit Configuration in which the input signal is applied between the base and the emitter, while the output signal appears between the transistor's collector and emitter.

Common-gate configuration An FET configuration in which the gate is grounded and common to the input and output signal.

Common-mode gain (A_{CM}) The amplification of common-mode input.

Common-mode input An input signal applied equally to both ungrounded inputs of a balanced amplifier. Also called in-phase input.

Common-mode rejection The ability of a device to reject a voltage signal applied simultaneously to both input terminals.

Common-mode rejection ratio Abbreviated CMRR, it is the ratio of an operational amplifier's differential gain to common-mode gain.

Common-source configuration An FET configuration in which the source is grounded and common to the input and output signal.

Communication Transmission of information between two points.

Comparator An operational amplifier used without feedback to detect changes in voltage level.

Complementary dc amplifier An amplifier in which NPN and PNP transistors are used in an alternating sequence.

Complementary latch circuit A circuit containing a NPN and PNP transistor that once triggered ON will remain latched ON.

Complex numbers Numbers composed of a real number part and an imaginary number part.

Complex plane A plane whose points are identified by means of complex numbers.

Component (1) Device or part in a circuit or piece of equipment. (2) In vector diagrams, it can mean a part of a wave, voltage, or current.

Compound A material composed of united separate elements.

Computer Piece of equipment used to process information or data.

Conductance (G) Measure of how well a circuit or path conducts or passes current. It is equal to the reciprocal of resistance.

Conduction Use of a material to pass electricity or heat.

Conduction band An energy band in which electrons can move freely within a solid.

Conductivity Reciprocal of resistivity; it is the ability of a material to conduct current.

Conductors Materials that have a very low resistance and pass current easily.

Configurations Different circuit interconnections.

Connection When two or more devices are attached together so that a path exists between them, there is said to be a connection between the two.

Connector Conductive device that makes a connection between two points.

Constant Fixed value.

Constant current Current that remains at a fixed, unvarying level despite variations in load resistance.

Constant-current region The flat portion of the drain characteristic curve. In this region I_D remains constant despite changes in V_{DS}.

Contact Current-carrying part of a switch, relay, or connector.

Continuity Occurs when a complete path for electric current exists.

Continuity test Resistance test to determine whether a path for electric current exists.

Contrast The difference between the light and dark areas in a video picture. High contrast pictures have dark blacks and brilliant whites. Low contrast pictures have an overall gray appearance.

Control voltage A voltage signal that starts, stops, or adjusts the operation of a device, circuit, or system.

Conventional current flow A current produced by the movement of positive charges toward a negative terminal.

Copper loss Also called I^2R loss, it is the power lost in transformers, generators, connecting wires, and other parts of a circuit because of the current flow (I) through the resistance (R) of the conductors.

Core Magnetic material within a coil used to intensify or concentrate the magnetic field.

Cosine The trigonometric function that for an acute angle is the ratio between the leg adjacent to the angle when it is considered part of a right triangle, and the hypotenuse.

Coulomb Unit of electric charge. One coulomb equals 6.24×10^{18} electrons.

Counter electromotive force Abbreviated "counter emf," or sometimes called back emf, it is the voltage generated in an inductor due to an alternating or pulsating current and is always of opposite polarity to that of the applied voltage.

Covalent bond A pair of electrons shared by two neighboring atoms.

Crossover distortion Distortion of a signal at the zero crossover point.

CRT Abbreviation for "cathode-ray tube."

Crystal A natural or synthetic piezoelectric or semiconductor material whose atoms are arranged with some degree of geometric regularity.

Cube root A number that when multiplied by itself three times gives the number.

Current (I) Measured in amperes or amps, it is the flow of electrons through a conductor.

Current amplifier An amplifier designed to build up the current of a signal.

Current clamp A device used in conjunction with an ac ammeter, containing a magnetic core in the form of hinged jaws that can be snapped around the current-carrying wire.

Current-controlled device A device in which the input junction is normally forward biased and the input current controls the output current.

Current divider A parallel network designed to proportionally divide the circuit's total current.

Current gain The increase in current produced by the transistor circuit.

Current-limiting resistor Resistor that is inserted into the path of current flow to limit the amount of current to some desired level.

Current sink A circuit or device that absorbs current from a load circuit.

Current source A circuit or device that supplies current to a load circuit.

Current-to-voltage converter An amplifier circuit used for converting a low dc current into a proportional low dc voltage.

Cutoff The minimum value of bias voltage that stops output current in a transistor.

Cutoff frequency Frequency at which the gain of the circuit falls below 0.707 of the maximum current or half-power (-3 dB).

Cutoff region The point at which the collector supply voltage has the transistor operating in cutoff.

Cycle When a repeating wave rises from zero to some positive value, back to zero, and then to a maximum negative value before returning back to zero, it is said to have completed one cycle. The number of cycles occurring in one second is the frequency measured in hertz (cycles per second).

DAC Abbreviation for "digital-to-analog converter."

Damping Reduction in the magnitude of oscillation due to absorption.

Darlington pair A current amplifier consisting of two separate transistors with their collectors connected together and the emitter of one connected to the base of the other.

D'Arsonval movement When a direct current is passed through a small, lightweight, movable coil, the magnetic field produced interacts with a fixed permanent magnetic field that rotates while moving an attached pointer positioned over a scale.

DC Abbreviation for "direct current."

DC alpha Term for the dc current amplification factor of a transistor.

DC beta (β_{DC}) The ratio of a transistor's dc output current to its input current.

DC block Component used to prevent the passage of direct current, normally a capacitor.

DC generator Device used to convert a mechanical energy input into a dc electrical output.

DC load line A line representing all the dc operating points of the transistor for a given load resistance.

DC offsets The change in input voltage that is required to produce an output voltage of zero when no input signal is present.

DC power supply A power line, generator, battery, power pack, or other dc source of power for electrical equipment.

Dead short Short circuit having almost no resistance.

Decibel (dB) One tenth of a bel (1×10^{-1} bel). The logarithmic unit for the difference in power at the output of an amplifier compared to the input.

Degerative effect An effect that causes a reduction in amplification due to negative feedback.

Degenerative (negative) feedback Feedback in which a portion of the output signal is fed back 180° out of phase with the input signal. Also called degenerative feedback, inverse feedback, or stabilized feedback.

Degree A unit of measure for angles equal to an angle with its vertex at the center of a circle and its sides cutting off $\frac{1}{360}$ of the circumference.

Depletion region A small layer on either side of the junction that becomes empty, or depleted, of free electrons or holes.

Depletion-type MOSFET A field effect transistor with an insulated gate (MOSFET) that can be operated in either the depletion or enhancement mode.

Design engineer Engineer responsible for the design of a product for a specific application.

Desoldering The process of removing solder from a connection.

Device Component or part.

DIAC A bidirectional diode that has a symmetrical switching mode.

Diagnostic Related to the detection and isolation of a problem or malfunction.

Diameter The length of a straight line passing through the center of an object.

Dielectric Insulating material between two (*di*) plates in which the *electric* field exists.

Dielectric constant (K) The property of a material that determines how much electrostatic energy can be stored per unit volume when unit voltage is applied. Also called permittivity.

Dielectric strength The maximum potential a material can withstand without rupture.

Difference amplifier An amplifier whose output is proportional to the difference between the voltages applied to the two inputs.

Differential amplifier An amplifier whose output is proportional to the difference between the voltages applied to its two inputs.

Differential gain (A_{VD}) The amplification of differential-mode input.

Differential-mode input signals Input signals to an op-amp that are out of phase with one another.

Differentiator A circuit whose output voltage is proportional to the rate of change of the input voltage. The output waveform is then the time derivative of the input waveform, and the phase of the output waveform leads that of the input by 90°.

Differentiator amplifier A circuit whose output voltage is proportional to the rate of change of the input voltage. The output waveform is the time derivative of the input waveform.

Diffusion A method of producing a junction by diffusing an impurity metal into a semiconductor at a high temperature.

Diffusion current The current that is present when the depletion layer is expanding.

Digital Relating to devices or circuits that have outputs of only two distinct levels or steps, for example, on–off, 0–1, open–closed, and so on.

Digital computer Piece of electronic equipment used to process digital data.

Digital data Data represented in digital form.

Digital display Display in which either light-emitting diodes (LEDs) or liquid crystal diodes (LCDs) are used and each of the segments can be either turned on or off.

Digital electrical (power) circuit An electrical circuit designed to turn power either ON or OFF (two-state).

Digital electronic circuits Electronic circuits that encode information into a group of pulses consisting of HIGH or LOW voltages.

Digital multimeter Multimeter used to measure amperes, volts, and ohms and indicate the results on a digital readout display.

Diode ac semiconductor switch or DIAC A bidirectional diode that has a symmetrical switching mode.

DIP Abbreviation for "dual in-line package."

Direct coupling The coupling of two circuits by means of a wire.

Direct current (dc) Current flow in only one direction.

Direct-current amplifier An amplifier capable of amplifying dc voltages and slowly varying voltages.

Direct voltage (dc voltage) Voltage that causes electrons to flow in only one direction.

Directly proportional The relation of one part to one or more other parts in which a change in one causes a similar change in the other.

Discharge Release of energy from either a battery or capacitor.

Disconnect Breaking or opening of an electric circuit.

Discrete components Separate active and passive devices that were manufactured before being used in a circuit.

Display Visual presentation of information or data.

Display unit Unit designed to display digital data.

Dissipation Release of electrical energy in the form of heat.

Distortion An undesired change in the waveform of a signal.

Distributed capacitance Also known as self-capacitance, it is any capacitance other than that within a capacitor, for example, the capacitance between the coil of an inductor or between two conductors (lead capacitance).

Distributed inductance Any inductance other than that within an inductor, for example, the inductance of a length of wire (line inductance).

Domain Also known as a magnetic domain, it is a moveable magnetized area in a magnetic material.

Doping The process wherein impurities are added to the intrinsic semiconductor material either to increase the number of free electrons or to increase the number of holes.

Dot convention A standard used with transformer symbols to indicate whether the secondary voltage will be in phase or out of phase with the primary voltage.

Double-pole, double-throw switch (DPDT) Switch having two movable contacts (double pole) that can be thrown or positioned in one of two positions (double throw).

Double-pole, single-throw switch (DPST) Switch having two movable contacts (double pole) that can be thrown in only one position (single throw) to either make a dual connection, or opened to disconnect both poles.

Drain One of the field effect transistor's electrodes.

Drain characteristic curve A plot of the drain current (I_D) versus the drain-to-source voltage (V_{DS}).

Drain current (I_D) A JFET's source-to-drain current.

Drain-feedback biased A configuration in which the gate receives a bias voltage fed back from the drain.

Drain supply voltage ($+V_{DD}$) The bias voltage connected between the drain and source of the JFET, which causes current to flow.

Drain-to-source current with shorted gate (I_{DSS}) The maximum value of drain current, achieved by holding V_{GS} at 0 V.

Dropping resistor Resistor whose value has been chosen to drop or develop a given voltage across it.

Dry cell DC voltage-generating chemical cell using a nonliquid (paste) type of electrolyte.

Dual-gate D-MOSFET A metal oxide semiconductor FET having two separate gate electrodes.

Dual in-line package (DIP) Package that has two (dual) sets or lines of connecting pins.

Dual-trace oscilloscope Oscilloscope that can simultaneously display two signals.

Duty cycle A term used to describe the amount of ON time versus OFF time. The

ON time is usually expressed as a percentage.

Dynamic Related to conditions or parameters that change, i.e. motion.

E-core Laminated form, in the shape of the letter E, onto which inductors and transformers are wound.

Eddy currents Small currents induced in a conducting core due to the variations in alternating magnetic flux.

Efficiency Ratio of output power to input power, normally expressed in percent.

Electrical components Components, circuits, and systems that manage the flow of power.

Electrical equipment Equipment designed to manage or control the flow of power.

Electrical wave Traveling wave propagated in a conductive medium that is a variation in voltage or current and travels at slightly less than the speed of light.

Electric charge Electric energy stored on the surface of a material.

Electric current (I) Electron movement or motion.

Electric field Also called a voltage field, it is a field or force that exists in the space between two different potentials or voltages.

Electrician Person involved with the design, repair, and assembly of electrical equipment.

Electricity Science that states certain particles possess a force field, which with electrons is negative and with protons is positive. Electricity can be divided into two groups: static and dynamic. Static electricity deals with charges at rest, while dynamic electricity deals with charges in motion.

Electric motor Device that converts an electrical energy input into a mechanical energy output.

Electric polarization A displacement of bound charges in a dielectric when placed in an electric field.

Electroacoustic transducer Device that achieves an energy transfer from electric to acoustic (sound), and vice versa. Examples include a microphone and a loudspeaker.

Electroluminescence Transmission or conversion of electrical energy into light energy.

Electrolyte Electrically conducting liquid (wet) or paste (dry).

Electrolytic capacitor Capacitor having an electrolyte between the two plates; due to chemical action, a very thin layer of oxide is deposited on only the positive plate, which accounts for why this type of capacitor is polarized.

Electromagnet A magnet consisting of a coil wound on a soft iron or steel core. When current is passed through the coil

a magnetic field is generated and the core is strongly magnetized to concentrate the magnetic field.

Electromagnetic communication Use of an electromagnetic wave to pass information between two points. Also called wireless communication.

Electromagnetic energy Radiant electromagnetic energy, such as radio and light waves.

Electromagnetic field Field having both an electric (voltage) and magnetic (current) field.

Electromagnetic induction The voltage produced in a coil due to relative motion between the coil and magnetic lines of force.

Electromagnetic spectrum List or diagram showing the entire range of electromagnetic radiation.

Electromagnetic wave Wave that consists of both an electric and magnetic variation.

Electromagnetism Relates to the magnetic field generated around a conductor when current is passed through it.

Electromechanical transducer Device that transforms electrical energy into mechanical energy (motor), and vice versa (generator).

Electromotive force (emf) Force that causes the motion of electrons due to a potential difference between two points.

Electron Smallest subatomic particle of negative charge that orbits the nucleus of the atom.

Electron flow A current produced by the movement of free electrons toward a positive terminal.

Electron-hole pair When an electron jumps from the valence shell or band to the conduction band, it leaves a gap in the covalent bond called a hole. This action creates an electron-hole pair.

Electronic components Components, circuits, and systems that manage the flow of information.

Electronic equipment Equipment designed to control the flow of information.

Electronics Science related to the behavior of electrons in devices.

Electron-pair bond or covalent bond A pair of electrons shared by two neighboring atoms.

Electron-volt (eV) A unit of energy equal to the energy acquired by an electron when it passes through a potential difference of 1 V in a vacuum.

Electrostatic Related to static electric charge.

Electrostatic field Force field produced by static electrical charges.

Element There are 107 different natural chemical substances, or elements, that exist on earth. These can be categorized as gas, solid, or liquid.

Emitter A transistor region from which charge carriers are injected into the base.

Emitter current (I_E) The current at the transistor's emitter terminal.

Emitter feedback The coupling from the emitter output to the base input in a transistor amplifier.

Emitter follower or voltage-follower The common-collector circuit in which the emitter output voltage seems to track or follow the phase and amplitude of the input voltage.

Emitter modulator A modulating circuit in which the modulating signal is applied to the emitter of a bipolar transistor.

Encoder circuit A circuit that produces different output voltage codes, depending on the position of a rotary switch.

Energized Being electrically connected to a voltage source so that the device is activated.

Energy Capacity to do work.

Energy gap The space between two orbital shells.

Engineer Person who designs and develops materials to achieve desired results.

Engineering notation A floating-point system in which numbers are expressed as products consisting of a number that is greater than one multiplied by an appropriate power of ten that is some multiple of three.

Enhancement-mode MOSFET A field effect transistor in which there are no charge carriers in the channel when the gate-source voltage is zero.

Equation A formal statement of the equality or equivalence of mathematical or logical expressions.

Equipment Term used to describe electrical or electronic units.

Equivalent resistance (R_{eq}) Total resistance of all the individual resistances in a circuit.

Error voltage A voltage that is proportional to the error that exists between input and output.

Excited state An energy level in which an electron may exist if given sufficient energy to reach this state from a lower state.

Exponent A symbol written above and to the right of a mathematical expression to indicate the operation of raising to a power.

Extrinsic semiconductor A semiconductor whose electrical properties are dependent on impurities added to the semiconductor crystal.

Facsimile Electronic process whereby pictures or images are scanned and the graphical information is converted into electrical signals that can be reproduced locally or transmitted to a remote point, where a likeness or facsimile of the original can be produced.

Fahrenheit temperature scale Temperature scale that indicates the freezing point of water at 32°F and boiling point at 212°F.

Fall time Time it takes a negative edge of a pulse to fall from 90% to 10% of its peak value.

Farad (F) Unit of capacitance.

Faraday's law 1. When a magnetic field cuts a conductor, or when a conductor cuts a magnetic field, an electric current will flow in the conductor if a closed path is provided over which the current can circulate. 2. Two other laws related to electrolytic cells.

FCC Abbreviation for "Federal Communications Commission."

Ferrite A powdered, compressed, and sintered magnetic material having high resistivity. The high resistance makes eddy-current losses low at high frequencies.

Ferrite bead Ferrite composition in the form of a bead.

Ferrite core Ferrite core normally shaped like a doughnut.

Ferrite-core inductor An inductor containing a ferrite core.

Ferrites Compound composed of iron oxide, a metallic oxide, and ceramic. The metal oxides include zinc, nickel, cobalt, or iron.

Ferrous Composed of and/or containing iron. A ferrous metal exhibits magnetic characteristics, as opposed to nonferrous metals.

Fiber optics Laser's light output carries information that is conveyed between two points by thin glass optical fibers.

Field effect transistor Abbreviated FET, it is a transistor in which the resistance of the source to drain current path is changed by applying a transverse electric field to the gate.

Field strength The strength of an electric, magnetic, or electromagnetic field at a given point.

Filament Thin thread of, for example, carbon or tungsten, which when heated by the passage of electric current will emit light.

Filament resistor The resistor in a light bulb or electron tube.

Filter Network composed of resistor, capacitor, and inductors used to pass certain frequencies yet block others through heavy attenuation.

Fission Atomic or nuclear fission is the process of splitting the nucleus of heavy elements such as uranium and plutonium into two parts, which results in large releases of radioactivity and heat (fission or division of a nuclei).

Fixed biasing A constant value of bias for an FET in which the voltage is independent of the input signal strength.

Fixed component Component whose value or characteristics cannot be varied or changed.

Fixed-value capacitor A capacitor whose value is fixed and cannot be varied.

Fixed-value resistor A resistor whose value cannot be changed.

Floating ground Ground potential that is not tied or in reference to earth.

Flow-soldering Flow or wave soldering technique is used in large-scale electronic assembly to solder all the connections on a printed circuit board by moving the board over a wave or flowing bath of molten solder.

Fluorescent lamp Gas-filled glass tube that when bombarded by the flow of electric current causes the gas to ionize and then release light.

Flux A material used to remove oxide films from the surfaces of metals in preparation for soldering.

Flux density A measure of the strength of a wave.

Flywheel action Sustaining effect of oscillation in an *LC* circuit due to the charging and discharging of the capacitor and the expansion and contraction of the magnetic field around the inductor.

Force Physical action capable of moving a body or modifying its movement.

Formula A general fact, rule, or principle expressed usually in mathematical symbols.

Forward biased A small depletion region at the junction will offer a small resistance and permit a large current. Such a junction is forward biased.

Forward breakover voltage Voltage needed to turn ON an SCR.

Forward voltage drop (V_F) The forward voltage drop is equal to the junction's barrier voltage.

Free electrons Electrons that are not in any orbit around a nucleus.

Free-running Operating without any external control.

Free-running multivibrator A circuit that requires no input signal to start its operation, but simply begins to oscillate the moment the dc supply voltage is applied.

Frequency Rate or recurrences of a periodic wave normally within a unit of one second, measured in hertz (cycles/second).

Frequency counter Meter used to measure the frequency or cycles per second of a periodic wave.

Frequency-division multiplex (FDM) Transmission of two or more signals over a common path by using a different frequency band for each signal.

Frequency-domain analysis A method of representing a waveform by plotting its amplitude versus frequency.

Frequency drift   A slow change in the frequency of a circuit due to temperature or frequency determining component value changes.

Frequency multiplier   A harmonic conversion circuit in which the frequency of the output signal is an exact multiple of the input signal.

Frequency response   Indication of how well a device or circuit responds to the different frequencies applied to it.

Frequency response curve   A graph indicating a circuit's response to different frequencies.

Friction   The rubbing or resistance to relative motion between two bodies in contact.

Frictional electricity   Generation of electric charges by rubbing one material against another.

Full-scale deflection (FSD)   Deflection of a meter's pointer to the farthest position on the scale.

Full-wave center-tapped rectifier   A rectifier circuit that makes use of a center-tapped transformer to cause an output current to flow in the same direction during both half-cycles of the ac input.

Full-wave rectifier   Rectifier that makes use of the full ac wave (in both the positive and negative cycle) when converting ac to dc.

Full-wave voltage-doubler circuit   A rectifier circuit that doubles the output voltage by charging capacitors during both alternations of the ac input—making use of the full ac input wave.

Function generator   Signal generator that can function as a sine, square, rectangular, triangular, or sawtooth waveform generator.

Fundamental frequency   This sine wave is always the lowest frequency and largest amplitude component of any waveform shape and is used as a reference.

Fuse   This circuit- or equipment-protecting device consists of a short, thin piece of wire that melts and breaks the current path if the current exceeds a rated damaging level.

Fuse holder   Housing used to support a fuse with two connections.

Fusion   Atomic or nuclear fusion is the process of melting the nuclei of two light atoms together to create a heavier nucleus, which results in a large release of energy (fusion or combining of nuclei).

Gain   Increase in power from one point to another. Normally expressed in decibels.

Gamma rays   High-frequency electromagnetic radiation from radioactive particles.

Ganged   Mechanical coupling of two or more capacitors, switches, potentiome-ters, or any other components so that the activation of one control will operate all.

Gas   Any aeriform or completely elastic fluid, which is not a solid or liquid. All gases are produced by the heating of a liquid beyond its boiling point.

Gate   One of the field effect transistor's electrodes (also used for thyristor devices).

Gate-source bias voltage (V_{GS})   The bias voltage applied between the gate and source of a field effect transistor.

Gate-to-source cutoff voltage or $V_{GS(OFF)}$ The negative V_{GS} bias voltage that causes I_D to drop to approximately zero.

Geiger counter   Device used to detect nuclear particles.

Generator   Device used to convert a mechanical energy input into an electrical energy output.

Giga   Prefix for 1 billion (10^9).

Graph   A diagram (as a series of one or more points, lines, line segments, curves, or areas) that represents the variation of a variable in comparison with that of one or more other variables.

Graph origin   Center of the graph where the horizontal axis and vertical axis cross.

Greenwich mean time (GMT)   Also known as universal time, it is a standard based on the earth's rotation with respect to the sun's position. The solar time at the meridian of Greenwich, England, which is at zero longitude.

Ground   An intentional or accidental conducting path between an electrical circuit or system and the earth, or some conducting body acting in the place of the earth.

Gunn diode   A semiconductor diode that utilizes the Gunn effect to produce microwave frequency oscillation or to amplify a microwave-frequency signal.

Half-power point   A point at which power is 50%. This corresponds to 70.7% of the total current.

Half-split method   A troubleshooting technique used to isolate a faulty block in a circuit or system. In this method a mid-point is chosen and tested to determine which half is malfunctioning.

Half-wave rectifier circuit   A circuit that converts ac to dc by only allowing current to flow during one-half of the ac input cycle.

Hall effect sensor   A sensor that generates a voltage in response to a magnetic field.

Hardware   Electrical, electronic, mechanical, or magnetic devices or components. The physical equipment.

Harmonic frequency   Sine wave that is smaller in amplitude and is some multiple of the fundamental frequency.

Hartley oscillator   An *LC* tuned oscillator circuit in which the tank coil has an intermediate tap.

Henry (H)   Unit of inductance.

Hertz (Hz)   Unit of frequency. One hertz is equal to one cycle per second.

Hexadecimal number system   A base 16 number system, using the digits 0 through 9 and the six additional digits represented by A, B, C, D, E, and F.

High fidelity (hi-fi)   Sound reproduction equipment that is as near to the original sound as possible.

High-pass filter   Network or circuit designed to pass any frequencies above a critical or cutoff frequency and reject or heavily attenuate all frequencies below.

High *Q*   Abbreviation for quality and generally related to inductors that have a high value of inductance and very little coil resistance.

High tension   Lethal voltage in the kilovolt range and above.

High-voltage probe   Accessory to the voltmeter that has added multiplier resistors within the probe to divide up the large potential being measured by the probe.

***H*–lines**   Invisible lines of magnetic flux.

Hole   A mobile vacancy in the valance structure of a semiconductor. This hole exists when an atom has less than its normal number of electrons, and is equivalent to a positive charge.

Hole flow   Conduction in a semiconductor when electrons move into holes when a voltage is applied.

Hologram   Three-dimensional picture created with a laser.

Holography   Science dealing with three-dimensional optical recording.

Horizontal   Parallel to the horizon or perpendicular to the force of gravity.

Horizontally polarized wave   Electromagnetic wave that has the electric field lying in the horizontal plane.

Hot resistance   The resistance of a device when hot due to the generation of heat by electric current.

Hybrid circuit   Circuit that combines two technologies (passive and active or discrete and integrated components) onto one microelectronic circuit. Passive components are generally made by thin-film techniques, while active components are made utilizing semiconductor techniques. Integrated circuits can be mounted on the microelectronic circuit and connected to discrete components also on the small postage-stamp-size boards.

Hydroelectric   Generation of electric power by the use of water in motion.

Hypotenuse   The side of a right-angled triangle that is opposite the right angle.

Hysteresis Amount that the magnetization of a material lags the magnetizing force due to molecular friction.

IC Abbreviation for "integrated circuit."

Imaginary number A complex number whose imaginary part is not zero.

IMPATT diode A semiconductor diode that has a negative resistance characteristic produced by a combination of impact avalanche breakdown and charge carrier transmit time effects in a thin semiconductor chip.

Impedance (Z) Measured in ohms, it is the total opposition a circuit offers to current flow (reactive and resistive).

Impedance coupling The coupling of two signal amplifier circuits through the use of an impedance, such as a choke.

Impedance matching circuit A circuit that can match, or isolate, a high resistance (low current) source. Matching of the source impedance to the load impedance causes maximum power to be transferred.

Incandescence State of a material when it is heated to such a high temperature that it emits light.

Incandescent lamp An electric lamp that generates light when an electric current is passed through its filament of resistance, causing it to heat to incandescence.

Induced current Current that flows due to an induced voltage.

Induced voltage Voltage generated in a conductor when it is moved through a magnetic field.

Inductance Property of a circuit or component to oppose any change in current as the magnetic field produced by the change in current causes an induced countercurrent to oppose the original change.

Inductive circuit Circuit that has a greater inductive reactance figure than capacitive reactance figure.

Inductive reactance (X_L) Measured in ohms, it is the opposition to alternating or pulsating current flow without the dissipation of energy.

Inductor Length of conductor used to introduce inductance into a circuit.

Inductor analyzer A test instrument designed to test inductors.

Infinite Having no limits.

Infinity Amount larger than any number can indicate.

Information Data or meaningful signals.

Infrared Electromagnetic heat radiation whose frequencies are above the microwave frequency band and below red in the visible band.

Ingot A mass of metal cast into a convenient shape.

Inhibit To stop an action or block data from passing.

Injection laser diode or ILD A semiconductor P-N junction diode that uses a lasing action to increase and concentrate the light output.

In phase Two or more waves of the same frequency whose maximum positive and negative peaks occur at the same time.

Input impedance (Z_{in}) The total opposition offered by the transistor to an input signal.

Input resistance (R_{in}) The amount of opposition offered to an input signal by the input base-emitter junction (emitter diode).

Input signal voltage change The input voltage change that causes a corresponding change in the output voltage.

Input transducers A transducer that generates input control signals.

Insulated When a nonconductive material is used to isolate conducting materials from one another.

Insulating material Material that will in nearly all cases prevent the flow of current due to its chemical composition.

Insulation resistance Resistance of the insulating material. The greater the insulation resistance, the better the insulator.

Insulator Material that has few electrons per atom and those electrons are close to the nucleus and cannot be easily removed.

Integrated When two or more components are combined (into a circuit) and then incorporated in one package or housing.

Integrator Device that approximates and whose output is proportional to an integral of the input signal.

Integrator amplifier An amplifier whose output is the integral of its input with respect to time.

Integrator circuit A circuit with an output which is the integral of its input with respect to time.

Intermediate-frequency amplifier An amplifier that has one or more transistor stages designed to amplify intermediate frequency signals.

Intermittent Occurring at random intervals of time. An intermittent component, circuit, or equipment problem is undesirable and difficult to troubleshoot, as the problem needs to occur before isolation of the fault can begin.

Internal resistance No voltage source is 100% efficient in that not all the energy in is converted to electrical energy out; some is wasted in the form of heat dissipation. The internal series resistance of a voltage source represents this inefficiency.

Intrinsic semiconductor materials Pure semiconductor materials.

Inversely proportional The relation of one part to one or more other parts in which a change in one causes an opposite change in the other.

Inverting amplifier circuit An op-amp circuit that produces an amplified output signal that is 180° out of phase with the input signal.

Inverting input The inverting or negative input of an operational amplifier.

Ion Atom that has an equal number of protons (+charge) and electrons (−charge) is considered a neutral atom. If more electrons exist in orbit around the atom it is considered a negative ion, whereas if less electrons are in orbit around the atom it is referred to as a positive ion.

Jack Socket or connector into which a plug may be inserted.

JFET A field effect transistor made up of a gate region diffused into a channel region. When a control voltage is applied to the gate, the channel is depleted or enhanced, and the current between source and drain is thereby controlled.

j operator A prefix used to indicate an imaginary number.

Joule The unit of work and energy.

Junction Contact or connection between two or more wires or cables.

Junction diode A semiconductor diode in which the rectifying characteristics occur at a junction between the n-type and p-type semiconductor materials.

Kilovolt-ampere 1000 volts at 1 ampere.

Kilowatt-hour 1000 watts for 1 hour.

Kilowatt-hour meter A meter used by electric companies to measure a customer's electric power use in kilowatt-hours.

Kinetic energy Energy associated with motion.

Kirchhoff's current law The sum of the currents flowing into a point in a circuit is equal to the sum of the currents flowing out of that same point.

Kirchhoff's voltage law The algebraic sum of the voltage drops in a closed-path circuit is equal to the algebraic sum of the source voltage applied.

Knee voltage The voltage at which a curve joins two relatively straight portions of a characteristic curve.

Lag Difference in time between two waves of the same frequency expressed in degrees, i.e., one waveform lags another by a certain number of degrees.

Laminated core Core made up of sheets of magnetic material insulated from one another by an oxide or varnish.

Lamp Device that produces light.

Laser Device that produces a very narrow, intense beam of light. The name is an acronym for "light amplification by stimulated emission of radiation."

Latched A bistable circuit action that causes the circuit to remain held or locked in its last activated state.

LC filter A selective circuit which makes use of an inductance-capacitance network.

Lead The angle by which one alternating signal leads another in time, or a wire that connects two points in a circuit.

Lead–acid cell Cell made up of lead plates immersed in a sulfuric acid electrolyte.

Leakage current or reverse current (I_R) The extremely small current present at the junction.

Least significant bit (LSB) The rightmost binary digit.

LED Abbreviation for "light-emitting diode."

Left-hand rule If the fingers of the left hand are placed around the wire so that the thumb points in the direction of the electron flow, the fingers will be pointing in the direction of the magnetic field being produced by the conductor.

Lenz's law The current induced in a circuit due to a change in the magnetic field is so directed as to oppose the change in flux, or to exert a mechanical force opposing the motion.

Lie detector Piece of electronic equipment, also called a polygraph, that determines whether a person is telling the truth by looking for dramatic changes in blood pressure, body temperature, breathing rate, heart rate, and skin moisture in response to certain questions.

Lifetime The time difference between an electron jumping into the conduction band and then falling back into a hole.

Light Electromagnetic radiation in a band of frequencies that can be received by the human eye.

Light-emitting diode A semiconductor diode that converts electric energy into electromagnetic radiation at visible and near infrared frequencies when its P-N junction is forward biased.

Light-emitting semiconductor devices Semiconductor devices that will emit light when an electrical signal is applied.

Light-sensitive semiconductor devices Semiconductor devices that change their characteristics in response to light.

Limiter Circuit or device that prevents some portion of its input from reaching the output.

Linear Relationship between input and output in which the output varies in direct proportion to the input.

Linear scale A scale whose divisions are uniformly spaced.

Line graph A graph in which points representing values of a variable are connected by a broken line.

Liquid crystal displays (LCDs) A digital display having two sheets of glass separated by a sealed quantity of liquid crystal material. When a voltage is applied across the front and back elec-

trodes, the liquid crystals' molecules become disorganized, causing the liquid to darken.

Literal number A number expressed as a letter.

Live Term used to describe a circuit or piece of equipment that is on and has current flow within it.

Load Source drives a load, and whatever component, circuit, or piece of equipment is connected to the source can be called a load and will have a certain load resistance, which will consequently determine the load current.

Load current The current that is present in the load.

Load impedance Total reactive and resistive opposition of a load.

Loading The adding of a load to a source.

Loading effect Large load resistance will cause a small load current to flow, and so the loading down of the source or loading effect will be small (light load), whereas a small load resistance will cause a large load current to flow from the source, which will load down the source (heavy load).

Load resistance The resistance of the load.

Locked To automatically follow a signal.

Lodestone A magnetite stone possessing magnetic polarity.

Logarithms The exponent that indicates the power to which a number is raised to produce a given number.

Logic Science dealing with the principles and applications of gates, relays, and switches.

Logic gate circuits Two-state (ON/OFF) circuits used for decision-making functions in digital logic circuits.

Loss Term used to describe a decrease in power.

Lower sideband A group of frequencies that are equal to the differences between the carrier and modulation frequencies.

Low-pass filter Network or circuit designed to pass any frequencies below a critical or cutoff frequency and reject or heavily attenuate all frequencies above.

Magnet Body that can be used to attract or repel magnetic materials.

Magnetic circuit breaker Circuit breaker that is tripped or activated by use of an electromagnet.

Magnetic coil Spiral of a conductor, which is called an electromagnet.

Magnetic core Material that exists in the center of the magnetic coil to either support the windings (nonmagnetic-material) or intensify the magnetic flux (magnetic-material).

Magnetic field Magnetic lines of force traveling from the north to the south pole of a magnet.

Magnetic flux The magnetic lines of force produced by a magnet.

Magnetic leakage The passage of magnetic flux outside the path along which it can do useful work.

Magnetic poles Points of a magnet from which magnetic lines of force leave (north pole) and arrive (south pole).

Magnetism Property of some materials to attract and repel others.

Magnetizing force Also called magnetic field strength, it is the magnetomotive force per unit length at any given point in a magnetic circuit.

Magnetomotive force Force that produces a magnetic field.

Mainframe Large computers that initially were only affordable to medium-sized and large businesses who had the space for them. Mini-computers came after mainframes and were affordable to any business, and now we have micro-computers which are easily affordable to anyone.

Majority carriers The type of carrier that constitutes more than half the total number of carriers in a semiconductor material. In n-type materials electrons are the majority carriers, whereas in p-type materials holes are the majority carriers.

Matched impedance Condition that occurs when the source impedance is equal to the load impedance, resulting in maximum power being transferred.

Matching Connection of two components or circuits so that maximum energy is transferred or coupled between the two.

Maximum power transfer theorem A theorem that states maximum power will be transferred from source to load when the source resistance is equal to the load resistance.

Maxwell One magnetic line of force or flux is called a maxwell.

Measurement Determining the presence and magnitude of variables.

Medical electronics Branch of electronics involved with therapeutic or diagnostic practices in medicine.

Mercury cell Primary cell that has mercuric oxide cathode, a zinc anode, and a potassium hydroxide electrolyte.

Metal film resistor A resistor in which a film of metal, metal oxide, or alloy is deposited on an insulating substrate.

Metal oxide resistor A metal film resistor in which an oxide of a metal (such as tin) is deposited as a film onto the substrate.

Metal oxide varistors (MOVs) Devices that are replacing zener-diode and transient-diode suppressors because they are able to shunt a much higher current surge and are cheaper.

Meter (1) Any electrical or electronic measuring device. (2) In the metric system, it is a unit of length equal to 39.37 inches or 3.28 feet.

Meter FSD current The value of current needed to cause the meter movement to deflect the needle to its full-scale deflection (FSD) position.

Meter resistance The resistance of a meter's armature coil.

Metric system A decimal system of weights and measures based on the meter and the kilogram.

Mica capacitor Fixed capacitor that uses mica as the dielectric between its plates.

Microphone Electroacoustic transducer that responds and converts a sound wave input into an equivalent electrical wave out.

Microwave Term used to describe a band of very small wavelength radio waves within the UHF, SHF, and EHF bands.

Mil One thousandth of an inch (0.001 in.).

Miller-effect capacitance (C_M) An undesirable inherent capacitance that exists between the junctions of transistors.

Minority carriers The type of carrier that constitutes less than half the total number of carriers in a semiconductor material. In n-type materials holes are the minority carriers, whereas in p-type materials electrons are the minority carriers.

Mismatch Term used to describe a difference between the source impedance and load impedance, which will prevent maximum power transfer.

Modulation Process whereby an information signal is used to modify some characteristic of another higher-frequency wave known as a carrier.

Molecule Smallest particle of a compound that still retains its chemical characteristics.

Monostable multivibrator A device that when triggered will generate a rectangular pulse of fixed duration.

MOS family A group of digital logic circuits that make use of metal oxide semiconductor field effect transistors (MOSFETs).

MOSFET A field effect transistor in which the insulating layer between the gate electrode and the channel is a metal-oxide layer. Either a p or n substrate.

Most significant bit (MSB) The leftmost binary digit.

Moving-coil microphone Microphone that makes use of a moving coil between a fixed magnetic field. Also called a dynamic microphone.

Moving-coil pickup Dynamic phonograph pickup that uses a coil between a fixed magnetic field, which is moved back and forth by the needle or stylus.

Moving-coil speaker Dynamic speaker that uses a coil placed between a fixed magnetic field and converts the electrical wave input into sound waves.

Multimeter Piece of electronic test equipment that can perform multiple tasks in that it can be used to measure voltage, current, or resistance.

Multiple-emitter input transistor A transistor specially constructed to have more than one emitter.

Multiplier resistor A resistor connected in series with the meter movement of a voltmeter.

Mutual inductance Ability of one inductor's magnetic lines of force to link with another inductor.

Navigation equipment Electronic equipment designed to aid in the direction of aircraft and ships to their destination.

n-Channel D-type MOSFET A depletion-type MOSFET having an n-type channel between its source and drain terminals.

n-Channel E-type MOSFET An enhancement-type MOSFET having an n-type channel between its source and drain terminals.

n-Channel JFET A junction field effect transistor having an n-type channel between source and drain.

Negative (neg.) (1) Some value less than zero. (2) Terminal that has an excess of electrons.

Negative charge An electric charge that has more electrons than protons.

Negative feedback Feedback in which a portion of the output signal is fed back 180° out of phase with the input signal.

Negative ground A system whereby the negative terminal of the voltage source is connected to the system's conducting chassis or body.

Negative ion Atom that has more than the normal neutral amount of electrons.

Negative resistance A resistance such that when the current through it increases, the voltage drop across the resistance decreases.

Negative temperature coefficient Effect that describes that if temperature increases, resistance or capacitance will decrease.

Negative temperature coefficient (NTC) thermistor A thermistor in which a temperature increase causes the resistance to decrease.

Neon bulb Glass envelope filled with neon gas, which when ionized by an applied voltage will glow red.

Network Combination and interconnection of components, circuits, or systems.

Neutral When an object is neither positive nor negative.

Neutral atom An atom in which the number of positive charges in the nucleus (protons) is equal to the number of negative charges (electrons) that surround the nucleus.

Neutral wire The conductor of a polyphase circuit or a single-phase three-

wire circuit that is intended to have a ground potential. The potential differences between the neutral and each of the other conductors are approximately equal in magnitude and are also equally spaced in phase.

Neutron Subatomic particle residing within the nucleus and having no electrical charge.

Nickel-cadmium cell Most popular secondary cell; it uses a nickel oxide positive electrode and cadmium negative electrode.

No-change or latch condition When both inputs are LOW, the S-R flip-flop is said to be in the no-change or latch condition because there will be no change in the output.

Node Junction or branch point.

Noise Unwanted electromagnetic radiation within an electrical or mechanical system.

Noninverting amplifier An operational amplifier in which the input signal is applied to the ungrounded positive input terminal to give a gain that is greater than one, and make the output change in phase with the input voltage.

Noninverting input The noninverting or positive input of an operational amplifier.

Nonlinear scale A scale whose divisions are not uniformly spaced.

Normally closed (N.C.) Designation which states that the contacts of a switch or relay are connected normally; however, when activated, these contacts will open.

Normally open (N.O.) Designation which states that the contacts of a switch or relay are normally not connected; however, when activated, these contacts will close.

North pole Pole of a magnet out of which magnetic lines of force are assumed to originate.

Norton's theorem Any network of voltage sources and resistors can be replaced by a single equivalent current source (I_N) in parallel with a single equivalent resistance (R_N).

NPN transistor Negative-positive-negative transistor in which a layer of p-type conductive semiconductor is located between two n-type regions.

n-Type semiconductor A material that has more conduction-band electrons than valence-band holes.

Nuclear energy Atomic energy or power released in a nuclear reaction when either a neutron is used to split an atom into smaller atoms (fission) or when two smaller nuclei are joined together (fusion).

Nuclear reactor Unit that maintains a continuous self-supporting nuclear reaction (fission).

Nucleus Core of an atom; it contains both positive (protons) and neutral (neutrons) subatomic particles.

Octave Interval between two sounds whose fundamental frequencies differ by a ratio of 2 to 1.

Offset null inputs The two balancing inputs used to balance an op-amp.

Ohm Unit of resistance, symbolized by the Greek capital letter omega (Ω).

Ohmmeter Measurement device used to measure electric resistance.

Ohm's law Relationship between the three electrical properties of voltage, current, and resistance. Ohm's law states that the current flow within a circuit is directly proportional to the voltage applied across the circuit and inversely proportional to its resistance.

Ohms per volt Value that indicates the sensitivity of a voltmeter. The higher the ohms per volt rating, the more sensitive the meter.

One-shot multivibrator Produces one output pulse or shot for each input trigger.

Open circuit Break in the path of current flow.

Open collector output A type of output structure found in certain bipolar logic families. Resistive pull-ups are generally added to provide the high level output voltage.

Open-loop mode A control system that has no means of comparing the output with the input for control purposes.

Operational amplifier Special type of high-gain amplifier; also called an op-amp.

Operator error An incorrect use of the controls of a circuit or system.

Opposite The side of a right-angle triangle that is opposite the angle theta.

Optically coupled isolator DIP module An optocoupler that contains an infrared-emitting diode and a silicon photo-transistor and is generally used to transfer switching information between two electrically isolated points.

Optically coupled isolator interrupter module A device that consists of a matched and aligned emitter and detector and that is used to detect opaque or nontransparent targets.

Optically coupled isolator reflector module A device that consists of a matched and aligned emitter and detector and that is used to detect targets.

Optically coupled isolators Devices that contain a light-emitting and light-sensing device in one package. They are used to optically couple two electrically isolated points.

Optimum power transfer Since the ideal maximum power transfer conditions cannot always be achieved, most designers try to achieve maximum

power transfer and have the source resistance and load resistance as close in value as possible.

Ordinary ground A connection in the circuit that is said to be at ground potential or zero volts and, because of its connection to earth, has the ability to conduct electrical current to and from the earth.

OR gate When either input *A* OR *B* is HIGH, the output will be HIGH.

Oscillate Continual repetition of or passing through a cycle.

Oscillator Electronic circuit that converts dc to a continuous alternating current out.

Oscilloscope Instrument used to view signal amplitude, period, and shape at different points throughout a circuit.

Out of phase When the maximum and minimum points of two or more waveforms do not occur at the same time.

Output Terminals at which a component, circuit, or piece of equipment delivers current, voltage, or power.

Output impedance (Z_{out}) The total opposition offered by the transistor to the output signal.

Output power Amount of power a component, circuit, or system can deliver to its load.

Output resistance (R_{out}) The amount of opposition offered to an output signal by the output base-collector junction (collector diode).

Output signal voltage change Change in output signal voltage in response to a change in the input signal voltage.

Output transducers Transducers that convert output electrical signals to some other energy form.

Overload Situation that occurs when the load is greater than the component, circuit, or system was designated to handle (load resistance too small, load current too high), resulting in waveform distortion and/or overheating.

Overload protection Protective device such as a fuse or circuit breaker that automatically disconnects or opens a current path when it exceeds an excessive value.

Paper capacitor Fixed capacitor using oiled or waxed paper as a dielectric.

Parallax error The apparent displacement of an object's position caused by a shift in the point of observation of the object.

Parallel Also called shunt; circuit having two or more paths for current flow.

Parallel circuit Also called shunt; circuit having two or more paths for current flow.

Parallel data transmission The transfer of information simultaneously over a set of parallel paths or channels.

Parallel resonant circuit Circuit having an inductor and capacitor in parallel with one another, offering a high impedance at the frequency of resonance.

Parallel-to-serial converter An IC that converts the parallel word applied to its inputs into a serial data stream.

Parasitic oscillations An unwanted self-sustaining oscillation or self generated random pulse.

Passband Band or range of frequencies that will be passed by a filter circuit.

Passive component Component that does not amplify a signal, such as a resistor or capacitor.

Passive filter A filter that contains only passive (nonamplifying) components and provides pass or rejection characteristics.

Passive system System that emits no energy; in other words it only receives; it does not transmit and consequently reveal its position.

p-Channel D-type MOSFET A depletion-type MOSFET having a *p*-type channel between its source and drain terminals.

p-Channel E-type MOSFET An enhancement-type MOSFET having a *p*-type channel between its source and drain terminals.

p-channel JFET A junction field effect transistor having a *p*-type channel between source and drain.

Peak Maximum or highest-amplitude level.

Peak inverse voltage Abbreviated PIV, it is the maximum rated value of an ac voltage acting in the direction opposite to that in which a device is designed to pass current.

Peak to peak Difference between the maximum positive and maximum negative values.

Peak voltage (V_P) The maximum value of voltage.

Pentode A five-electrode electron tube that has an anode, cathode, control grid, and two additional electrodes.

Percent In the hundred; of each hundred.

Percent of regulation The change in output voltage that occurs between no-load and full-load in a dc voltage source. Dividing this change by the rated full-load value and multiplying the result by 100 gives percent regulation.

Percent of ripple The ratio of the effective or rms value of ripple voltage to the average value of the total voltage. Expressed as a percentage.

Period Time taken to complete one complete cycle of a periodic or repeating waveform.

Permanence Magnetic equivalent of electrical conductance and consequently equal to the reciprocal of reluctance, just as conductance is equal to the reciprocal of resistance.

Permanent magnet Magnet, normally made of hardened steel, that retains its magnetism indefinitely.

Permeability Measure of how much better a material is as a path for magnetic lines of force with respect to air, which has a permeability of 1 (symbolized by the Greek lowercase letter mu, μ).

Phase Angular relationship between two waves, normally between current and voltage in an ac circuit.

Phase angle Phase difference between two waves, normally expressed in degrees.

Phase-locked loop (PLL) circuit A circuit consisting of a phase comparator that compares the output frequency of a voltage controlled oscillator with an input frequency. The error voltage out of the phase comparator is then coupled via an amplifier and low-pass filter to the control input of the voltage controlled oscillator to keep it in phase, and therefore at exactly the same frequency as the input.

Phase shift Change in phase of a waveform between two points, given in degrees of lead or lag.

Phase-shift oscillator An *RC* oscillator circuit in which the 180° phase shift is achieved with several *RC* networks.

Phase splitter A circuit that takes a single input signal and produces two output signals that are 180° apart in phase.

Phonograph Piece of equipment used to reproduce sound.

Phosphor Luminescent material applied to the inner surface of a cathode ray tube that when bombarded with electrons will emit light.

Photoconduction A process by which the conductance of a material is changed by incident electromagnetic radiation in the light spectrum.

Photoconductive cell Material whose resistance decreases or conductance increases when light strikes it.

Photoconductive cell or light-dependent resistor (LDR) A two-terminal device that changes its resistance when light is applied.

Photodetector Component used to detect or sense light.

Photodiode A semiconductor diode that changes its electrical characteristics in response to illumination.

Photometer Meter used to measure light intensity.

Photon Discrete portion of electromagnetic energy. A small packet of light.

Photoresistor Also known as a photoconductive cell or light-dependent resistor, it is a device whose resistance varies with the illumination of the cell.

Photo-transistor, photo-darlington, and photo-SCR Examples of light-reactive devices.

Photovoltaic action A process by which a device generates a voltage as a result of exposure to radiation.

Photovoltaic cell Component, commonly called a solar cell, used to convert light energy into electric energy (voltage).

Photovoltaic cell or solar cell A device that generates a voltage across its terminal that will increase as the light level increases.

Pi Value representing the ratio between the circumference and diameter of a circle and equal to approximately 3.142 (symbolized by the lowercase Greek letter π).

Pie chart or circle graph A circular chart cut by radii into segments illustrating relative magnitudes.

Pierce crystal oscillator An oscillator circuit in which a piezoelectric crystal is connected in a tank between output and input.

Piezoelectric crystal Crystal material that will generate a voltage when mechanical pressure is applied and conversely will undergo mechanical stress when subjected to a voltage.

Piezoelectric effect The operation of a voltage between the opposite sides of a piezoelectric crystal as a result of pressure or twisting. Also, the reverse effect in which the application of a voltage to opposite sides causes deformation to occur at the frequency of the applied voltage.

Piezoresistive diaphragm pressure sensor A sensor that changes its resistance in response to pressure.

Pinch-off region A region on the characteristic curve of an FET in which the gate bias causes the depletion region to extend completely across the channel.

Pinch-off voltage (V_P) The value of V_{DS} at which further increases in V_{DS} will cause no further increase in I_D.

PIN diode A semiconductor diode that has a high-resistance intrinsic region between its low-resistance *p*-type and *n*-type regions.

Pitch Term used to describe the inflection or frequency scale of sounds. When the pitch is increased by one octave, twice the original frequency will be the result.

Plastic film capacitor A capacitor in which alternate layers of metal aluminum foil are separated by thin films of plastic dielectric.

Plate Conductive electrode in either a capacitor or battery.

Plug Movable connector that is normally inserted into a socket.

P-N junction The point at which two opposite doped materials come in contact with one another.

PNP transistor Positive-negative-positive transistor in which a layer of *n*-type conductive semiconductor is located between two *p*-type regions.

Polar coordinates Either of two numbers that locate a point in a plane by its distance from a fixed point on a line and the angle this line makes with a fixed line.

Polarity Term used to describe positive and negative charges.

Polarized electrolytic capacitor An electrolytic capacitor in which the dielectric is formed adjacent to one of the metal plates, creating a greater opposition to current in one direction only.

Positive Point that attracts electrons, as opposed to negative, which supplies electrons.

Positive charge The charge that exists in a body which has fewer electrons than normal.

Positive ground A system whereby the positive terminal of the voltage source is connected to the system's conducting chassis or body.

Positive ion Atom that has lost one or more of its electrons and therefore has more protons than electrons, resulting in a net positive charge.

Positive shunt clipper A circuit that has a diode connected in shunt with the lead or output, and its orientation is such that it will clip off the positive alternation of the ac input.

Positive temperature coefficient of resistance The rate of increase in resistance relative to an increase in temperature.

Positive temperature coefficient (PTC) thermistors A thermistor in which a temperature increase causes the resistance to increase.

Potential difference (PD) Voltage difference between two points, which will cause current to flow in a closed circuit.

Potential energy Energy that has the potential to do work because of its position relative to others.

Potentiometer Three-lead variable resistor that through mechanical turning of a shaft can be used to produce a variable voltage or potential.

Power Amount of energy converted by a component or circuit in a unit of time, normally seconds. It is measured in units of watts (joules/second).

Power amplifier An amplifier designed to deliver maximum output power to a load.

Power dissipation Amount of heat energy generated by a device in one second when current flows through it.

Power factor Ratio of actual power to apparent power. A pure resistor has a power factor of 1 or 100%, while a capacitor has a power factor of 0 or 0%.

Power gain (A_P) The ratio of the output signal power to the input signal power.

Power loss Ratio of power absorbed to power delivered.

Power supply Piece of electrical equipment used to deliver either ac or dc voltage.

Prefix Name used to designate a factor or multiplier.

Pressure The application of force to something by something else in direct contact with it.

Primary First winding of a transformer that is connected to the source, as opposed to the secondary that is connected to the load.

Primary cell Cell that produces electrical energy through an internal electrochemical action; once discharged, it cannot be reused.

Printed circuit board (PCB) Insulating board that has conductive tracks printed onto the board to make the circuit.

Programmable unijunction transistor or PUT A unijunction transistor that can have its peak voltage controlled.

Propagation Traveling of electromagnetic, electrical, or sound waves through a medium.

Propagation time Time it takes for a wave to travel between two points.

Proportional A term used to describe the relationship between two quantities that have the same ratio.

Protoboard An experimental arrangement of a circuit on a board. Also called a breadboard.

Proton Subatomic particle within the nucleus that has a positive charge.

p-Type semiconductor A material that has more valence-band holes than conduction-band electrons.

Pull-up resistor A resistor connected between a signal output and positive V_{CC} to pull the signal line HIGH when it is not being driven LOW.

Pulse Rise and fall of some quantity for a period of time.

Pulse fall time Time it takes for a pulse to decrease from 90% to 10% of its maximum value.

Pulse repetition frequency The number of times per second that a pulse is transmitted.

Pulse repetition time The time interval between the start of two consecutive pulses.

Pulse rise time Time it takes for a pulse to increase from 10% to 90% of its maximum value.

Pulse width, pulse length, or pulse duration The time interval between the leading edge and trailing edge of a pulse at which the amplitude reaches 50% of the peak pulse amplitude.

Push-pull amplifier A balanced amplifier that uses two similar equivalent amplifying transistors working in phase opposition.

Pythagorean theorem A theorem in geometry: The square of the length of the hypotenuse of a right triangle equals the sum of the squares of the lengths of the other two sides.

Q Quality factor of an inductor or capacitor; it is the ratio of a component's reactance (energy stored) to its effective series resistance (energy dissipated).

Quiescent point The voltage or current value that sets up the no signal input or operating point bias voltage.

Race condition An unpredictable bistable circuit condition.

Radar Acronym for "radio detection and ranging"; it is a system that measures the distance and direction of objects.

Radioastronomy Branch of astronomy that studies the radio waves generated by celestial bodies and uses these emissions to obtain more information about them.

Radio broadcast Transmission of music, voice, and other information on radio carrier waves that can be received by the general public.

Radio communication Term used to describe the transfer of information between two or more points by use of radio or electromagnetic waves.

Radio-frequency (RF) amplifier An amplifier that has one or more transistor stages designed to amplify radio-frequency signals.

Radio-frequency generator A generator capable of supplying RF energy at any desired frequency in the radio spectrum.

Radio-frequency probe A probe used in conjunction with an ac meter to measure high-frequency RF signals.

Radius A line segment extending from the center of a circle or sphere to the circumference.

Ratio The relationship in quantity, amount, or size between two or more things.

RC Abbreviation for "resistance–capacitance."

RC circuit Circuit containing both a resistor and capacitor.

RC coupling A coupling method in which resistors are used as the input and output impedances of the two stages. A coupling capacitor is generally used between the stages to couple the ac signal and block the dc supply bias voltages.

RC filter A selective circuit which makes use of a resistance-capacitance network.

RC time constant In one time constant, which is equal to the product of resistance and capacitance in seconds, a capacitor will have charged or discharged 63.2% of the maximum applied voltage.

Reactance (X) Opposition to current flow without the dissipation of energy.

Reactive power Also called imaginary power or wattless power, it is the power value obtained by multiplying the effective value of current by the effective value of voltage and the sine of the angular phase difference between current and voltage.

Real number A number that has no imaginary part.

Receiver Unit or piece of equipment used for the reception of information.

Recombination The combination and resultant neutralization of particles or objects having unlike charges. For example, a hole and an electron, or a positive ion and negative ion.

Rectangular coordinates A Cartesian coordinate of a Cartesian coordinate system whose straight-line axes or coordinate planes are perpendicular.

Rectangular wave Also known as a pulse wave; it is a repeating wave that only alternates between two levels or values and remains at one of these values for a small amount of time relative to the other.

Rectification Process that converts alternating current (ac) into direct current (dc).

Rectifier A device that converts alternating current into a unidirectional or dc current.

Rectifier circuit Achieves rectification.

Rectifier diodes or rectifiers Junction diodes that achieve rectification.

Reed relay Relay that consists of two thin magnetic strips within a glass envelope with a coil wrapped around the envelope so that when it is energized the relay's contacts or strips will snap together, making a connection between the leads attached to each of the reed strips.

Regulator A device that maintains a desired quantity at a predetermined voltage.

Relative Not independent; compared with or with respect to some other value of a measured quantity.

Relaxation oscillator An oscillator circuit whose frequency is determined by an RL or RC network, producing a rectangular or sawtooth output waveform.

Relay Electromechanical device that opens or closes contacts when a current is passed through a coil.

Reluctance Resistance to the flow of magnetic lines of force.

Remanence Amount a material remains magnetized after the magnetizing force has been removed.

Reset and carry An action that occurs in any column that reaches its maximum count.

Reset state A circuit condition in which the output is reset to binary 0.

Residual magnetism Magnetism remaining in the core of an electromagnet after the coil current has been removed.

Resistance Symbolized R and measured in ohms (Ω), it is the opposition to current flow with the dissipation of energy in the form of heat.

Resistive power (true power) The average power consumed by a circuit during one complete cycle of alternating current.

Resistive temperature detector (RTD) A temperature detector consisting of a fine coil of conducting wire (such as platinum) that will produce a relatively linear increase in resistance as temperature increases.

Resistivity Measure of a material's resistance to current flow.

Resistor Component made of a material that opposes the flow of current and therefore has some value of resistance.

Resistor color code Coding system of colored stripes on a resistor that indicates the resistor's value and tolerance.

Resonance Circuit condition that occurs when the inductive reactance (X_L) is equal to the capacitive reactance (X_C).

Resonant circuit Circuit containing an inductor and capacitor tuned to resonate at a certain frequency.

Resonant frequency Frequency at which a circuit or object will produce a maximum amplitude output.

Resultant vector A vector derived from or resulting from two or more other vectors.

Reverse biased A large depletion region at the junction will offer a large resistance and permit only a small current. Such a junction is reverse biased.

Reverse leakage current (I_R) The undesirable flow of current through a device in the reverse direction.

Reverse voltage drop (V_R) The reverse voltage drop is equal to the source voltage (applied voltage).

RF Abbreviation for "radio frequency."

RF amplifier/frequency tripler circuit A radio frequency amplifier circuit in which the frequency of the output is three times that of the input frequency.

RF amplifier/multiplier circuit A radio frequency amplifier circuit in which the frequency of the output is an exact multiple of the input frequency.

RF local oscillator circuit The radio frequency oscillator in a superheterodyne receiver.

Rheostat Two-terminal variable resistor used to control current.

Right-angle triangle A triangle having a 90° or square corner.

Ripple frequency The frequency of the ripple present in the output of a dc source.

Rise time Time it takes a positive edge of a pulse to rise from 10% to 90% of its high value.

RL differentiator An RL circuit whose output voltage is proportional to the rate of change of the input voltage.

RL filter A selective circuit of resistors and inductors that offers little or no opposition to certain frequencies while blocking or attenuating other frequencies.

RL integrator An RL circuit with an output proportionate to the integral of the input signal.

RMS Abbreviation for "root mean square."

RMS value Rms value of an ac voltage, current, or power waveform is equal to 0.707 times the peak value. The rms value is the effective or dc value equivalent of the ac wave.

Rotary switch Electromechanical device that has a rotating shaft connected to one terminal that is capable of making or breaking a connection.

R–2R ladder circuit A network or circuit composed of a sequence of L networks connected in tandem. This R–$2R$ circuit is used in digital-to-analog converters.

Rounding off An operation in which a value is abbreviated by applying the following rule: When the first digit to be dropped is a 6 or more, or a 5 followed by a digit that is more than zero, increase the previous digit by 1. When the first digit to be dropped is a 4 or less, or a 5 followed by a zero, do not change the previous digit.

Saturation The condition in which a further increase in one variable produces no further increase in the resultant effect.

Saturation point The point beyond which an increase in one of two quantities produces no increase in the other.

Saturation region The point at which the collector supply voltage has the transistor operating at saturation.

Sawtooth wave Repeating waveform that rises from zero to a maximum value linearly and then falls to zero and repeats.

Scale Set of markings used for measurement.

Schematic diagram Illustration of the electrical or electronic scheme of a circuit, with all the components represented by their respective symbols.

Schottky diode A semiconductor diode formed by contact between a semiconductor layer and a metal coating. Hot carriers (electrons for n-type material and holes for p-type material) are emitted from the Schottky barrier of the semiconductor and move to the metal coating. Since majority carriers predominate, there is essentially no injection or storage of minority carriers to limit switching speed.

Scientific notation Numbers are entered and displayed in terms of a power of 10.

For example:

Number	Scientific Notation
7642	7.642×10^3
0.000096	96×10^{-6}
0.0012	1.2×10^{-3}
64,000,000	64×10^6

Scopemeter A hand-held, battery-operated instrument that combines a multimeter, oscilloscope, frequency counter, and signal generator in one.

Secondary Output winding of a transformer that is connected across the load.

Secondary cells Electrolytic cells for generating electricity. Once discharged the cell may be restored or recharged by sending an electric current through the cell in the opposite direction to that of the discharge current.

Selectivity Characteristic of a circuit to discriminate between the wanted signal and the unwanted signal.

Self biasing Gate bias for an FET in which a resistor is used to drop the supply voltage and provide gate bias.

Self-inductance The property that causes a counter electromotive force to be produced in a conductor when the magnetic field expands or collapses with a change in current.

Semiconductor transducer An electronic device that converts one form of energy to another.

Semiconductors Materials that have properties that lie between insulators and conductors.

Serial data transmission The transfer of information sequentially through a single path or channel.

Serial-to-parallel converter An IC that converts a serial data stream applied to its input into a parallel output word.

Series circuit Circuit in which the components are connected end to end so that current has only one path to follow throughout the circuit.

Series clipper A circuit that will clip off part of the input signal. Also known as a limitor since the circuit will limit the ac input. A series clipper circuit has a clipping or limiting device in series with the load.

Series–parallel circuit Network or circuit that contains components that are connected in both series and parallel.

Series resonance Condition that occurs when the inductive and capacitive reactances are equal and both components are connected in series with one another, and therefore the impedance is minimum.

Series resonant circuit A resonant circuit in which the capacitor and coil are in series with the applied ac voltage.

Series switching regulators A regulator circuit containing a power transistor in series with the load, that is switched ON and OFF to regulate the dc output voltage delivered to the load.

Set-reset flip-flop A bistable multivibrator circuit that has two inputs that are used to either SET or RESET the output.

Set state A circuit condition in which the output is set to binary 1.

Seven-segment display Component that normally has eight LEDs, seven of which are mounted into segments or bars that make up the number 8, and the eighth LED is used as a decimal point.

Shells or bands An orbital path containing a group of electrons that have a common energy level.

Shield Metal grounded cover that is used to protect a wire, component, or piece of equipment from stray magnetic and/or electric fields.

Shock The sudden pain, convulsion, unconsciousness, or death produced by the passage of electric current through the body.

Short circuit Also called a short; it is a low-resistance connection between two points in a circuit, typically causing a large amount of current flow.

Shorted out Term used to describe a component that has either internally malfunctioned, resulting in a low-resistance path through the component, or a component that has been bypassed by a low-resistance path.

Shunt clipper A circuit that will clip off part of the input signal. Also known as a limitor since the circuit will limit the ac input. A shunt clipper circuit has a clipping or limiting device in shunt with the load.

Shunt resistor A resistor connected in parallel or shunt with the meter movement of an ammeter.

Side bands A band of frequencies on both sides of the carrier frequency of a modulated signal produced by modulation.

Signal Conveyor of information.

Signal-to-noise ratio Ratio of the magnitude of the signal to the magnitude of the noise, normally expressed in decibels.

Signal voltage RMS or effective voltage value of a signal.

Silicon (Si) Nonmetallic element (atomic number 14) used in pure form as a semiconductor.

Silicon controlled rectifier Abbreviated SCR, it is a three-junction, three-terminal unidirectional P-N-P-N thyristor that is normally an open circuit. When triggered with the proper gate signal it switches to a conducting state and allows current to flow in one direction.

Silicon transistor Transistor using silicon as the semiconducting material.

Silver (Ag) Precious metal that does not easily corrode and is more conductive than copper.

Silvered mica capacitor Mica capacitor with silver deposited directly onto the mica sheets, instead of using conducting metal foil.

Silver solder Solder composed of silver, copper, and zinc with a melting point lower than silver but higher than the standard lead-tin solder.

Simplex Communication in only one direction at a time, for example, facsimile and television.

Simulcast Broadcasting of a program simultaneously in two different forms, for example, a program on both AM and FM.

Sine Sine of an angle of a right-angle triangle is equal to the opposite side divided by the hypotenuse.

Sine wave Wave whose amplitude is the sine of a linear function of time. It is drawn on a graph that plots amplitude against time or radial degrees relative to the angular rotation of an alternator.

Single in-line package (SIP) Package containing several electronic components (generally resistors) with a single row of external connecting pins.

Single-pole, double-throw (SPDT) Three-terminal switch or relay in which one terminal can be thrown in one of two positions.

Single-pole, single-throw (SPST) Two-terminal switch or relay that can either open or close one circuit.

Single sideband (SSB) AM radio communication technique in which the transmitter suppresses one sideband and therefore only transmits a single sideband.

Single-throw switch Switch containing only one set of contacts, which can be either opened or closed.

Sink Device, such as a load, that consumes power, or conducts away heat.

Sintering Process of bonding either a metal or powder by cold-pressing it into a desired shape and then heating to form a strong, cohesive body.

Sinusoidal Varying in proportion to the sine of an angle or time function; for example, alternating current (ac) is sinusoidal.

SIP Abbreviation for "single in-line package."

Skin effect Tendency of high-frequency (RF) currents to flow near the surface layer of a conductor.

Slide switch Switch having a sliding bar, button, or knob.

Slow-acting relay Slow-operating relay that when energized may not pull up the armature for several seconds.

Slow-blow fuse Fuse that can withstand a heavy current (up to ten times its rated value) for a small period of time without opening.

Snap switch Switch containing a spring under tension or compression that causes the contacts to come together suddenly when activated.

SNR Abbreviation for "signal-to-noise ratio."

Soft magnetic material Ferromagnetic material that is easily demagnetized.

Software Program of instructions that directs the operation of a computer.

Solar cell Photovoltaic cell that converts light into electric energy. They are especially useful as a power source for space vehicles.

Solder Metallic alloy that is used to join two metal surfaces.

Soldering Process of joining two metallic surfaces to make an electrical contact by melting solder (usually tin and lead) across them.

Soldering gun Soldering tool having a trigger switch and pistol shape that at its tip has a fast-heating resistive element for soldering.

Soldering iron Soldering tool having an internal heating element that is used for heating a connection to melt solder.

Solenoid Coil and movable iron core that when energized by an alternating or direct current will pull the core into a central position.

Solid conductor Conductor having a single solid wire, as opposed to strands.

Solid state Pertaining to circuits and devices that use solid semiconductors such as silicon. Solid state electronic devices have a solid material between their input and output pins (transistors, diodes), whereas vacuum tube electronics uses tubes, which have a vacuum between input and output.

Sonar Acronym for "sound navigation and ranging." A system using sound waves to determine a target's direction and distance.

Sonic Pertaining to the speed of sound waves.

Sound wave Traveling wave propagated in an elastic medium that travels at a speed of approximately 1133 ft/s.

Source Device that supplies the signal power or electric energy to a load. Also, one of the field effect transistor's electrodes.

Source-follower An FET amplifier in which the signal is applied between the gate and drain and the output is taken between the source and drain. Used to handle large-input signals and applications requiring a low-input capacitance.

Source impedance Impedance a source presents to a load.

South pole Pole of a magnet into which magnetic lines of force are assumed to enter.

Spark Momentary discharge of electric energy due to the breakdown of air or some other dielectric material separating two terminals.

SPDT Abbreviation for "single-pole, double-throw."

Speaker Also called a loudspeaker; it is an electroacoustic transducer that converts an electrical wave input into a mechanical sound wave (acoustic) output into the air.

Specification sheet or data sheet Details the characteristics and maximum and minimum values of operation of a device.

Spectrum The frequency spectrum displays all the frequencies and their applications.

Spectrum analyzer Instrument that can display the frequency domain of a waveform, plotting amplitude against frequency of the signals present.

Speed of light Physical constant—the speed at which light travels through a vacuum—equal to 186,282.397 miles/s, 2.997925×10^8 m/s, 161,870 nautical miles/s, or 328 yards/μs.

Speed of sound Speed at which a sound wave travels through a medium. In air it is equal to about 1133 ft/s or 334 m/s, while in water it is equal to approximately 4800 ft/s or 1463 m/s. Also known as sonic speed.

Speedup capacitor Capacitor connected in a circuit to speed up an action due to its inherent behavior.

SPST Abbreviation for "single-pole, single-throw."

Square The product of a number multiplied by itself.

Square root A factor of a number that when squared gives the number.

Square wave Wave that alternates between two fixed values for an equal amount of time.

Square-wave generator A circuit that generates a continuously repeating square wave.

S-R latch Another name for *S-R* flip-flop, so called because the output remains latched in the set or reset state even though the input is removed.

S-R (set-reset) flip-flop A multivibrator circuit in which a pulse on the SET input will "flip" the circuit into the set state while a pulse on the RESET input will "flop" the circuit into its reset state.

Staircase-wave generator A signal generator circuit that produces an output signal voltage that increases in steps.

Standard Exact value used as a basis for comparison or calibration.

Static Crackling noise heard on radio receivers caused by electric storms or electric devices in the vicinity.

Static electricity Electricity at rest or stationary.

Statistics A branch of mathematics dealing with the collection, analysis, interpretation, and presentation of masses of numerical data.

Stator Stationary part of some rotating device.

Statute mile Distance unit equal to 5280 ft or 1.61 km.

Step-down transformer Transformer in which the ac voltage induced in the secondary is less (due to fewer secondary windings) than the ac voltage applied to the primary.

Stepper motor A motor that rotates in small angular steps.

Step-up transformer Transformer in which the ac voltage induced in the secondary is greater (due to more secondary windings) than the ac voltage applied to the primary.

Stereo sound Sound system in which the sound is delivered through at least two channels and loudspeakers arranged to give the listener a replica of the original performance.

Stranded conductor Conductor composed by a group of twisted wires.

Stray capacitance Undesirable capacitance that exists between two conductors such as two leads or a lead and a metal chassis.

Subassembly Components contained in a unit for convenience in assembling or servicing the equipment.

Subatomic Particles such as electrons, protons, and neutrons that are smaller than atoms.

Substrate The mechanical insulating support on which a device is fabricated.

Summing amplifier circuit or adder circuit An op-amp circuit that will sum or add all of the input voltages.

Superconductor Metal such as lead or niobium that, when cooled to within a few degrees of absolute zero, can conduct current with no electrical resistance.

Superheterodyne receiver Radio-frequency receiver that converts all RF inputs to a common intermediate frequency (IF) before demodulation.

Superhigh frequency (SHF) Frequency band between 3 and 30 GHz, so designated by the Federal Communications Commission (FCC).

Superposition theorem Theorem designed to simplify networks containing two or more sources. It states: In a network containing more than one source, the current at any point is equal to the algebraic sum of the currents produced by each source acting separately.

Supersonic Faster than the speed or velocity of sound (Mach 1).

Supply voltage Voltage produced by a power source or supply.

Surface mount technology A method of installing tiny electronic components on the same side of a circuit board as the printed wiring pattern that interconnects them.

SW Abbreviation for "shortwave."

Sweep generator Test instrument designed to generate a radio-frequency voltage that continually and automatically varies in frequency within a selected frequency range.

Swing Amount a frequency or amplitude varies.

Switch Manual, mechanical, or electrical device used for making or breaking an electric circuit.

Switching power supply A dc power supply that makes use of a series switching regulator controlled by a pulse-width-modulator to regulate the output voltage.

Switching transistor Transistor designed to switch either on or off.

Symmetrical bidirectional switch A device that has the same value of breakover voltage in both the forward and reverse direction.

Synchronization Also called sync; it is the precise matching or keeping in step of two waves or functions.

Synchronous Two or more circuits or devices in step or in phase.

Sync pulse or signal Pulse waveform generated to synchronize two processes.

System Combination or linking of several parts or pieces of equipment to perform a particular function.

Tachometer Instrument that produces an output voltage that indicates the angular speed of the input in revolutions per minute.

Tangent The trigonometric function that for an acute angle is the ratio between the leg opposite the angle when it is considered part of a right triangle, and the leg adjacent.

Tank circuit Circuit made up of a coil and capacitor that is capable of storing electric energy.

Tantalum capacitor Electrolytic capacitor having a tantalum foil anode.

Tap Electrical connection to some point, other than the ends, on the element of a resistor or coil.

Tapered Nonuniform distribution of resistance per unit length throughout the element.

Technician Expert in troubleshooting circuit and system malfunctions. Along with a thorough knowledge of all test equipment and how to use it to diagnose problems, the technician is also familiar with how to repair or replace faulty components. Technicians basically translate theory into action.

Telegraphy Communication between two points by sending a series of coded current pulses either through wires or by radio.

Telemetry Transmission of instrument reading to a remote location either through wires or by radio waves.

Telephone Apparatus designed to convert sound waves into electrical waves, which are then sent to and reproduced at a distant point.

Telephone line Wires existing between subscribers and central stations.

Telephony Telecommunications system involving the transmission of speech information, therefore allowing two or more persons to converse verbally.

Teletypewriter Electric typewriter that like a teleprinter can produce coded signals corresponding to the keys pressed or print characters corresponding to the coded signals received.

Television (TV) System that converts both audio and visual information into corresponding electric signals that are then transmitted through wires or by radio to a receiver, which reproduces the original information.

Telex Teletypewriter exchange service.

Temperature coefficient of frequency Rate frequency changes with temperature.

Temperature coefficient of resistance Rate resistance changes with temperature.

Temperature stability The ability of a resistor to maintain its value of resistance despite changes in temperature.

Tera (T) Prefix that represents 10^{12}.

Terminal Connecting point for making electric connections.

Tesla (T) SI unit of magnetic flux density (1 tesla = 1 Wb/m^2).

Test Sequence of operations designed to verify the correct operation or malfunction of a system.

Tetrode A four-electrode electron tube that has an anode, cathode, control grid, and an additional electrode.

Thermal relay Relay activated by a heating element.

Thermal stability The ability of a circuit to maintain stable characteristics despite changes in the ambient temperature.

Thermistor Temperature-sensitive semiconductor that has a negative temperature coefficient of resistance (as temperature increases, resistance decreases).

Thermocouple Temperature transducer consisting of two dissimilar metals welded together at one end to form a junction that when heated will generate a voltage.

Thermometry Relating to the measurement of temperature.

Thermostat Temperature-sensitive device that opens or closes a circuit.

Theta The eighth letter of the Greek alphabet, used to represent an angle.

Thévenin's theorem Theorem that replaces any complex network with a single voltage source in series with a single resistance. It states: Any network of resistors can be replaced with an equivalent voltage source (V_{TH}) and an equivalent series resistance (R_{TH}).

Thick-film capacitor Capacitor consisting of two thick-film layers of conductive film separated by a deposited thick-layer dielectric film.

Thick-film resistor Fixed-value resistor consisting of a thick-film resistive element made from metal particles and glass powder.

Thin-film capacitor Capacitor in which both the electrodes and the dielectric are deposited in layers on a substrate.

Thin-film detector (TFD) A temperature detector containing a thin layer of platinum, and used for very precise temperature readings.

Three-phase supply AC supply that consists of three ac voltages that are 120° out of phase with one another.

Threshold Minimum point at which an effect is produced or indicated.

Thyristor A semiconductor switching device in which bistable action depends on P-N-P-N regenerative feedback. A thyristor can be bidirectional or unidirectional and have from two to four terminals.

Time constant Time needed for either a voltage or current to rise to 63.2% of the maximum or fall to 36.8% of the initial value. The time constant of an *RC* circuit is equal to the product of *R* and *C*, while the time constant of an *RL* circuit is equal to the inductance divided by the resistance.

Time-division multiplex (TDM) Transmission of two or more signals on the same path but at different times.

Time-domain analysis A method of representing a waveform by plotting its amplitude versus time.

Tinned Coated with a layer of tin or solder to prevent corrosion and simplify the soldering of connections.

Toggle switch Spring-loaded switch that is put in one of two positions, either on or off.

Tolerance Permissible deviation from a specified value, normally expressed as a percentage.

Tone A term describing both the bass and treble of a sound signal.

TO package Cylindrical, metal can type of package for some semiconductor components.

Toroidal coil Coil wound on a doughnut-shaped core.

Torque Moving force.

Totem pole circuit A transistor circuit containing two transistors connected one on top of the other with two inputs and one output.

Transconductance Also called mutual conductance, it is the ratio of a change in output current to the initiating change in input voltage.

Transducer Any device that converts energy from one form to another.

Transformer Device consisting of two or more coils that are used to couple electric energy from one circuit to another, yet maintain electrical isolation between the two.

Transformer coupling Also called inductive coupling, it is the coupling of two circuits by means of mutual inductance provided by a transformer.

Transient suppressor diode A device used to protect voltage sensitive electronic devices in danger of destruction by high energy voltage transients.

Transistance The characteristic achieved by a transistor that makes possible the control of voltages and currents so as to achieve gain or switching action.

Transistor (TRANSfer resISTOR) Semiconductor device having three main electrodes called the emitter, base, and collector that can be made to either amplify or rectify.

Transistor tester Special test instrument that can be used to test both NPN and PNP bipolar transistors.

Transmission Sending of information.

Transmission line Conducting line used to couple signal energy between two points.

Transmitter Equipment used to achieve transmission.

Transorb Absorb transients. Another name for transient suppressor diode.

Transposition The transfer of any term of an equation from one side over the the other side with a corresponding change of the sign.

Treble High audio frequencies normally handled by a tweeter in a sound system.

TRIAC A bidirectional gate-controlled thyristor that provides full-wave control of ac power.

Triangular wave A repeating wave that has equal positive and negative ramps that have linear rates of change with time.

Triangular-wave generator A signal generator circuit that produces a continuously repeating triangular wave output.

Trigger Pulse used to initiate a circuit action.

Triggering Initiation of an action in a circuit which then functions for a predetermined time, for example, the duration of one sweep in a cathode-ray tube.

Trigger input Pulse used to initiate a circuit action.

Trigonometry The study of the properties of triangles and trigonometric functions and of their applications.

Trimmer Small-value variable resistor, capacitor, or inductor.

Triode A three-electrode vacuum tube that has an anode, cathode, and control grid.

Troubleshooting The process of locating and diagnosing malfunctions or breakdowns in equipment by means of systematic checking or analysis.

Tune To adjust the resonance of a circuit so that it will select the desired frequency.

Tuned circuit Circuit that can have its components' values varied so that the circuit responds to one selected frequency yet heavily attenuates all other frequencies.

Tunnel diode A heavily doped junction diode that has negative resistance in the forward direction of its operating range due to quantum mechanical tunnelling.

Turns ratio Ratio of the number of turns in the secondary winding to the number of turns in the primary winding of a transformer.

Twin-T sine-wave oscillator An oscillator circuit that makes use of two T-shaped feedback networks.

Two phase Two repeating waveforms having a phase difference of 90°.

UHF Abbreviation for "ultrahigh frequency."

Ultrasonic Signals that are just above the range of human hearing of approximately 20 kHz.

Uncharged Having a normal number of electrons and therefore no electrical charge.

Unidirectional device A device that will conduct current in only one direction.

Unijunction transistor Abbreviated UJT, it is a P-N device that has an emitter connected to the P-N junction on one side of the bar, and two bases at either end of the bar. Used primarily as a switching device.

Unipolar device A device in which only one type of semiconductor material exists between the output terminals and therefore the charge carriers have only one polarity (unipolar).

Unit A determinate quantity adopted as a standard of measurement.

Upper sideband A group of frequencies that are equal to the sums of the carrier and modulation frequencies.

VA Abbreviation for "volt-ampere."

Vacuum tube Electron tube evacuated to such a degree that its electrical characteristics are essentially unaffected by the presence of residual gas or vapor.

Eventually replaced by the transistor for amplification and rectification.

Valence shell or ring Outermost shell formed by electrons.

Valley voltage (V_V) The voltage at the dip or valley in the characteristic curve.

Varactor diode A P-N semiconductor diode that is reverse biased to increase or decrease its depletion region width and vary device capacitance.

Variable Quantity that can be altered or controlled to assume a number of distinct values.

Variable capacitor Capacitor whose capacitance can be changed by varying the effective area of the plates or the distance between the plates.

Variable resistor A resistor whose value can be changed. *See also* rheostat *and* potentiometer.

Variable-tuned RF amplifier A tuned radio frequency amplifier in which the tuned circuit(s) can be adjusted to select the desired station carrier frequency.

Variable-value capacitor A capacitor whose value can be varied.

Variable-value inductor An inductor whose value can be varied.

Varistor Voltage dependent resistor.

VCR Abbreviation for "videocassette recorder."

Vector or phasor Quantity that has both magnitude and direction. They are normally represented as a line, the length of which indicates magnitude and the orientation of which, due to the arrowhead on one end, indicates direction.

Vector addition Determination of the sum of two out-of-phase vectors using the Pythagorean theorem.

Vector diagram Arrangement of vectors showing the phase relationships between two or more ac quantities of the same frequency.

Vertical-channel E-MOSFET An enhancement type MOSFET that, when turned ON, forms a vertical channel between source and drain.

Very high frequency (VHF) Electromagnetic frequency band from 30 to 300 MHz as set by the FCC.

Very low frequency (VLF) Frequency band from 3 to 30 kHz as set by the FCC.

Video Relating to any picture or visual information, from the Latin word meaning "I see."

Video amplifier An amplifier that has one or more transistor stages designed to amplify video signals.

Video signals A signal that contains visual information for television or radar systems.

Virtual ground A ground for voltage but not for current.

Voice coil Coil attached to the diaphragm of a moving coil speaker, which is moved through an air gap between the pole pieces of a permanent magnet.

Voice synthesizer Synthesizer that can simulate speech in any language by stringing together phonemes.

Volt (V) Unit of voltage, potential difference, or electromotive force. One volt is the force needed to produce 1 ampere of current in a circuit containing 1 ohm of resistance.

Voltage (V or E) Term used to designate electrical pressure or the force that causes current to flow.

Voltage amplifier An amplifier designed to build up the voltage of a signal.

Voltage-controlled device A device in which the input junction is normally reverse biased and the input voltage controls the output current.

Voltage divider Fixed or variable series resistor network that is connected across a voltage to obtain a desired fraction of the total voltage.

Voltage-divider biasing A biasing method used with amplifiers in which a series arrangement of two fixed-value resistors is connected across the voltage source. The result is that a desired fraction of the total voltage is obtained at the center of the two resistors and is used to bias the amplifier.

Voltage drop Voltage or difference in potential developed across a component or conductor due to the loss of electric pressure as a result of current flow.

Voltage-follower An operational amplifier that has a direct feedback to give unity gain so that the output voltage follows the input voltage. Used in applications where a very high input impedance and very low output impedance are desired.

Voltage gain Also called voltage amplification, it is equal to the difference between the output voltage level and the input signal voltage level. This value is normally expressed in decibels, which are equal to 20 times the logarithm of the ratio of the output voltage to the input voltage.

Voltage rating Maximum voltage a component can safely withstand without breaking down.

Voltage regulator A device that maintains the output voltage of a voltage source within required limits despite variations in input voltage and load resistance.

Voltage source A circuit or device that supplies voltage to a load circuit.

Voltaic cell A battery cell having two unlike metal electrodes immersed in a solution that chemically interacts with the plates to produce an emf.

Volt-ampere (VA) Unit of apparent power in an ac circuit containing reactance. Apparent power is equal to the product of voltage and current.

Voltmeter Instrument designed to measure the voltage or potential difference. Its scale can be graduated in kilovolts, volts, or millivolts.

Volume Magnitude or power level of a complex audio frequency (AF) wave, expressed in volume units (VU).

VOM meter Abbreviation for volt-ohm-milliamp meter.

VRMS Abbreviation for "volts root-mean-square."

Wall outlet Spring-contact outlet mounted on the wall to which a portable appliance is connected to obtain electric power.

Watt (W) Unit of electric power required to do work at a rate of 1 joule/second. One watt of power is expended when 1 ampere direct current flows through a resistance of 1 ohm. In an ac circuit, the true power is effective volts multiplied by effective amperes, multiplied by the power factor.

Wattage rating Maximum power a device can safely handle continuously.

Watt-hour (Wh) Unit of electrical work, equal to a power of 1 watt being absorbed continuously for 1 hour.

Wattmeter A meter used to measure electric power in watts.

Wave Electric, electromagnetic, acoustic, mechanical, or other form whose physical activity rises and falls or advances and retreats periodically as it travels through some medium.

Waveform Shape of a wave.

Waveguide Rectangular or circular metal pipe used to guide electromagnetic waves at microwave frequencies.

Wavelength (λ) Distance between two points of corresponding phase and is equal to waveform velocity or speed divided by frequency.

Weber (Wb) Unit of magnetic flux. One weber is the amount of flux that when linked with a single turn of wire for an interval of 1 second, will induce an electromotive force of 1 V.

Wet cell Cell using a liquid electrolyte.

Wetting The coating of a contact surface.

Wheatstone bridge A four-arm, generally resistive bridge that is used to measure resistance.

Wideband amplifier Also called broadband amplifier, it is an amplifier that has a flat response over a wide range of frequencies.

Wien-Bridge oscillator An *RC* oscillator circuit in which a Wien Bridge determines frequency.

Winding One or more turns of a conductor wound to form a coil.

Wire Single solid or stranded group of conductors having a low resistance to current flow.

Wire gauge American wire gauge (AWG) is a system of numerical designations of wire sizes, with the first being 0000 (the largest size) and then going to 000, 00, 0, 1, 2, 3, and so on up to the smallest sizes of 40 and above.

Wireless Term describing radio communication that requires no wires between the two communicating points.

Wirewound resistor Resistor in which the resistive element is a length of high-resistance wire or ribbon, usually nichrome, wound onto an insulating form.

Wire-wrapping Method of prototyping in which solderless connections are made by wrapping wire around a rectangular terminal.

Woofer Large loudspeaker designed primarily to reproduce low audio-frequency signals at large power levels.

Work Work is done anytime energy is transformed from one type to another, and the amount of work done is dependent on the amount of energy transformed.

X Symbol for reactance.

X axis Horizontal axis.

Y Symbol for admittance.

Y axis Vertical axis.

Z axis Axis perpendicular to both the *X* and *Y* axes.

Zener diode A semiconductor diode in which a reverse breakdown voltage current causes the diode to develop a constant voltage.

Zener voltage (V_Z) The voltage drop across thee zener when it is being operated in the reverse zener breakdown region.

Zero biasing A configuration in which no bias voltage is applied at all.

Zeroing Calibrating a meter so that it shows a value of zero when zero is being measured.

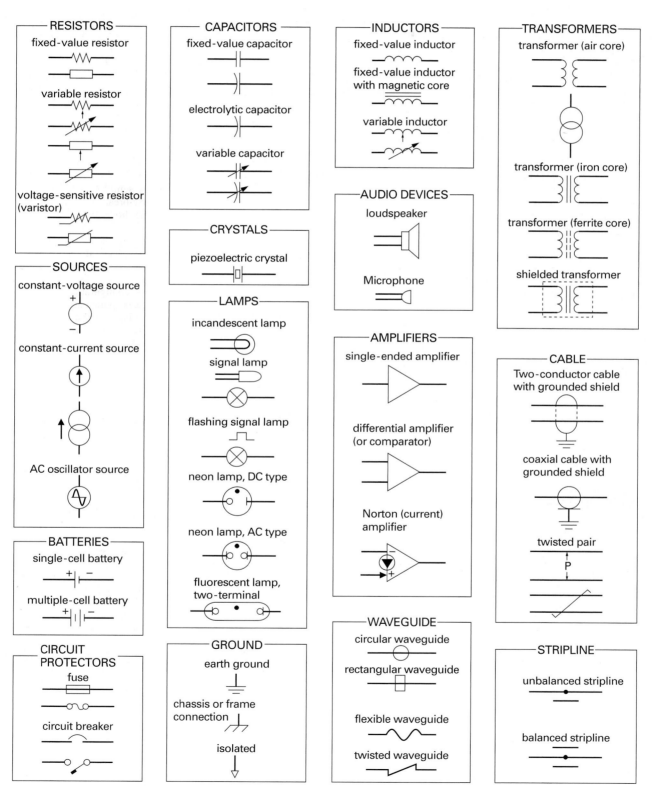

Electronic Schematic Symbols

RESISTORS
fixed-value resistor

variable resistor

voltage-sensitive resistor
(varistor)

SOURCES
constant-voltage source

constant-current source

AC oscillator source

BATTERIES
single-cell battery

multiple-cell battery

CIRCUIT PROTECTORS
fuse

circuit breaker

CAPACITORS
fixed-value capacitor

electrolytic capacitor

variable capacitor

CRYSTALS
piezoelectric crystal

LAMPS
incandescent lamp

signal lamp

flashing signal lamp

neon lamp, DC type

neon lamp, AC type

fluorescent lamp,
two-terminal

GROUND
earth ground

chassis or frame
connection

isolated

INDUCTORS
fixed-value inductor

fixed-value inductor
with magnetic core

variable inductor

AUDIO DEVICES
loudspeaker

Microphone

AMPLIFIERS
single-ended amplifier

differential amplifier
(or comparator)

Norton (current)
amplifier

WAVEGUIDE
circular waveguide

rectangular waveguide

flexible waveguide

twisted waveguide

TRANSFORMERS
transformer (air core)

transformer (iron core)

transformer (ferrite core)

shielded transformer

CABLE
Two-conductor cable
with grounded shield

coaxial cable with
grounded shield

twisted pair

P

STRIPLINE
unbalanced stripline

balanced stripline

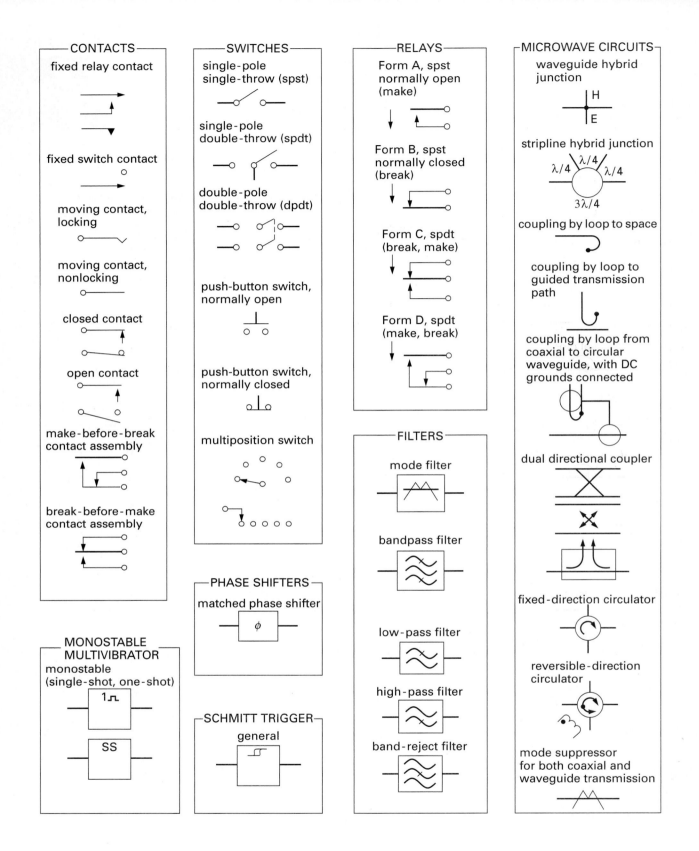

CONTACTS

fixed relay contact

fixed switch contact

moving contact, locking

moving contact, nonlocking

closed contact

open contact

make-before-break contact assembly

break-before-make contact assembly

MONOSTABLE MULTIVIBRATOR

monostable (single-shot, one-shot)

1

SS

SWITCHES

single-pole single-throw (spst)

single-pole double-throw (spdt)

double-pole double-throw (dpdt)

push-button switch, normally open

push-button switch, normally closed

multiposition switch

PHASE SHIFTERS

matched phase shifter

φ

SCHMITT TRIGGER

general

RELAYS

Form A, spst normally open (make)

Form B, spst normally closed (break)

Form C, spdt (break, make)

Form D, spdt (make, break)

FILTERS

mode filter

bandpass filter

low-pass filter

high-pass filter

band-reject filter

MICROWAVE CIRCUITS

waveguide hybrid junction

H
E

stripline hybrid junction

λ/4 λ/4 λ/4

3λ/4

coupling by loop to space

coupling by loop to guided transmission path

coupling by loop from coaxial to circular waveguide, with DC grounds connected

dual directional coupler

fixed-direction circulator

reversible-direction circulator

mode suppressor for both coaxial and waveguide transmission

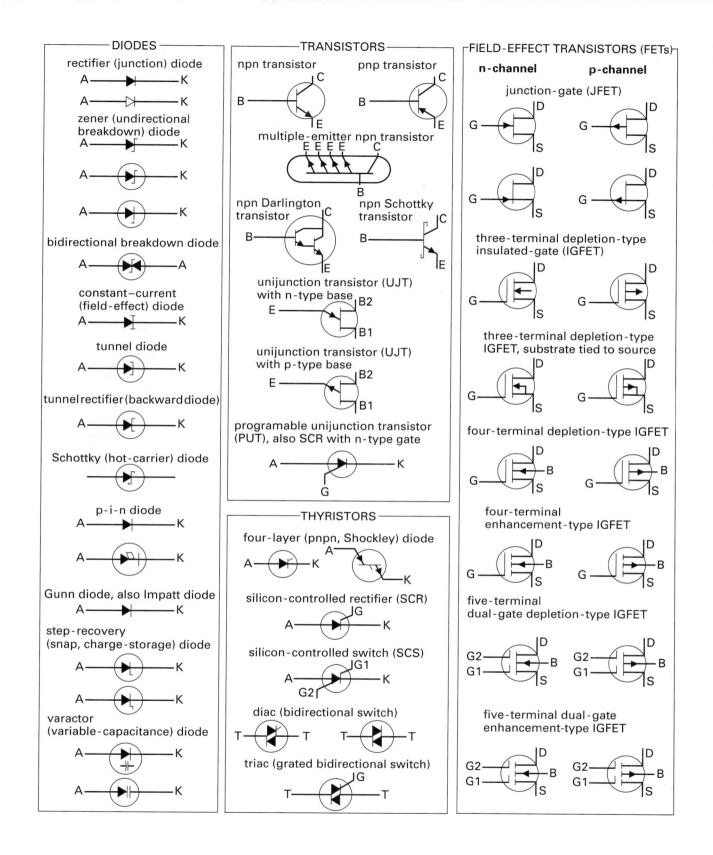

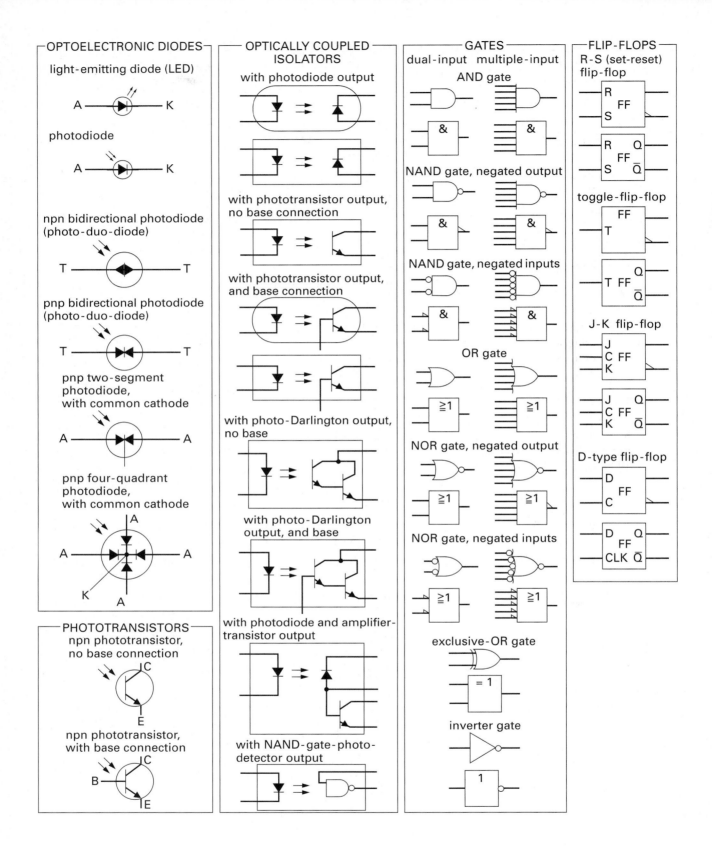

OPTOELECTRONIC DIODES

light-emitting diode (LED)

A ———▷|— K

photodiode

A ———▷|— K

npn bidirectional photodiode (photo-duo-diode)

T ——————— T

pnp bidirectional photodiode (photo-duo-diode)

T ——————— T

pnp two-segment photodiode, with common cathode

A ——————— A

pnp four-quadrant photodiode, with common cathode

A | A
A ——————— A
K A

PHOTOTRANSISTORS

npn phototransistor, no base connection

C
E

npn phototransistor, with base connection

C
B
E

OPTICALLY COUPLED ISOLATORS

with photodiode output

with phototransistor output, no base connection

with phototransistor output, and base connection

with photo-Darlington output, no base

with photo-Darlington output, and base

with photodiode and amplifier-transistor output

with NAND-gate-photo-detector output

GATES

dual-input   multiple-input

AND gate

&   &

NAND gate, negated output

&   &

NAND gate, negated inputs

&   &

OR gate

≥1   ≥1

NOR gate, negated output

≥1   ≥1

NOR gate, negated inputs

≥1   ≥1

exclusive-OR gate

= 1

inverter gate

1

FLIP-FLOPS

R-S (set-reset) flip-flop

R
FF
S

R   Q
FF
S   Q̄

toggle-flip-flop

FF
T

Q
T FF
Q̄

J-K flip-flop

J
C FF
K

J   Q
C FF
K   Q̄

D-type flip-flop

D
FF
C

D   Q
FF
CLK Q̄

Soldering Tools and Techniques

Electronic components such as resistors, capacitors, diodes, transistors, and integrated circuits are combined to form circuits, which in turn are combined to form electronic systems. These components have their leads physically interconnected and then bonded with solder. Like the expression "the straw that broke the camel's back," one sloppy solder connection can cause an entire electronic system to fail.

C-1   *Soldering Tools*

Soldering is a skill that can be developed with practice and knowledge. A solder connection is the joining together of two metal parts by applying both heat and solder. The heat provided by the soldering iron is at a high enough temperature to melt the solder, making it a liquid that flows onto and slightly penetrates the two metal surfaces that need to be connected. Once the soldering iron and therefore the heat are removed, the solder cools. In so doing, it solidifies and bonds the two metal surfaces, producing a good *electrical and mechanical connection*. This is the purpose of soldering: to bond two metal surfaces together so that they are both electrically and mechanically connected.

> **Soldering**
> Process of joining two metallic surfaces to make an electrical contact by melting solder (usually tin and lead) across them.

You should at this time have a set of tools and toolbox with all the components needed for the experiments listed in the lab manual. Figure C-1 illustrates the set of basic electronic tools that will be needed for soldering and experimentation. These tools include:

1. Solder wick: Used to remove solder from terminals.
2. Heat sink: During the soldering process, this device is clamped onto the lead of especially heat-sensitive components to conduct the heat generated by the soldering iron away from the component.
3. Soldering brush: Used to clean off flux after soldering.
4. Pliers
5. Long-nose pliers: Used for gripping and bending; they are also known as needle-nose pliers.
6. Cutters: These are available in many different sizes and are used to cut wires or cables.
7. Wire strippers: The strippers are designed to be adjustable and they are used for stripping the insulating sleeve off a wire.
8. Solder: 60/40 rosin-core solder is most commonly used in electronics.
9. Nut driver
10. Blade screwdriver
11. Phillips screwdriver
12. Soldering iron

C-2   *Wetting*

Every time you make a good electrical and mechanical connection, the solder will flow, when heated to its melting temperature, over the lead and terminal to be connected, as shown in Figure C-2. The solder actually penetrates the metals, and this embedding of the solder into the metal is called **wetting.** If the solder feathers out to a thin edge, good wetting

> **Wetting**
> The coating of a contact surface.

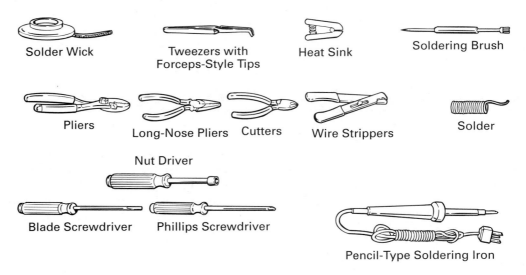

FIGURE C-1   Basic Electronic Tools.

is said to have occurred, and it is this that gives the connection its physical strength and electrical connection.

C-3   *Solder and Flux*

Solder is a mixture of tin and lead, both of which have a low melting temperature with respect to other metals. This is necessary so that the soldering iron can melt the solder and not the terminals or leads.

Different proportions of tin and lead are available to produce solder with different characteristics. For example, Figure C-3(a) shows a 60/40 solder, which is a mixture of 60% tin and 40% lead, whereas Figure C-3(b) shows a 40/60 solder that consists of 40% tin and 60% lead. The proportions of tin to lead determine the melting temperature of the solder; for example, 60/40 solder has a lower melting temperature than 40/60 solder. In electronics, 60/40 is most commonly used because of its low melting temperature, which means that a component lead or terminal will not have to be heated to a high temperature in order to make a connection. A 63/37 solder has an even lower melting temperature and is therefore even safer than 60/40 solder. It is, however, more expensive.

Any lead or terminal is always exposed to the air, which forms an invisible insulating layer on the surface of leads, pins, terminals, and any other surface. A chemical substance is needed to remove this layer, otherwise the solder would not be able to flow and stick to the metal contacts. **Flux** removes this invisible insulating oxide layer, and nearly all solder used in electronics contains the flux inside a type of solder tube, as shown in Figure C-4. As the solder is applied to a heated connection, the flux will automatically remove any oxide. The two most common types of flux are rosin and acid. Acid-core solder is used

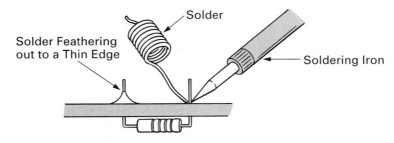

FIGURE C-2   Wetting.

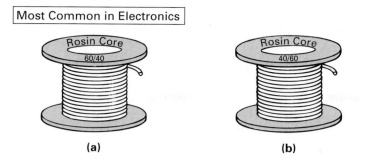

FIGURE C-3 **Tin/Lead Solder Ratios. (a) 60/40. (b) 40/60.**

only in sheet metal work and should never be used in electronics since it is highly corrosive. In electronics, you should only use rosin-core solder, and this is normally indicated as shown in Figure C-3.

A variety of solder diameters are available. The larger-diameter solder is used for terminals and large component leads and connections, whereas the smaller-diameter solder is used for soldering terminals that are very close to one another on a printed circuit board, so the amount of solder being applied to the connection needs to be carefully controlled.

C-4   *Soldering Irons*

Figure C-5 illustrates the two basic types of **soldering irons.** They are rated in terms of wattage, which indicates the amount of power consumed. More important, the wattage rating of a soldering iron indicates the amount of heat it produces. When an iron is applied to a connection, heat transfer takes place and heat is drained away from the iron to the metals to be connected, so a larger connection will need a larger-wattage soldering iron. A 25 to 60 W pencil iron is ideal for most electronic work.

For your safety and to protect voltage- and current-sensitive components, you should always use a soldering iron with a three-pin plug and three-wire ac power cord. The third ground wire will ground the iron's exposed metal areas and the tip to prevent electrical shock. This grounded tip will protect delicate MOS integrated circuits from leakage electricity and static charges. On the subject of safety, always turn off the equipment before soldering, as the grounded tip may cause a short circuit in the equipment and possible damage.

The tip of a soldering iron can sometimes be changed, and a different temperature tip will allow your iron to be used for different applications. The temperature of the tip selected should be governed by the size of the connection and the temperature sensitivity of the components. In general, a 700°F tip is ideal for most electronic applications; however, for delicate printed circuit boards, use a 600°F tip.

The tip of a soldering iron comes in a variety of shapes and sizes, as shown in Figure C-5(a). The heat produced by the soldering iron heats the connection and melts the solder so that it flows over the connection. A tip shape should be chosen that will conduct the heat to the connection as quickly as possible, so as not to damage the component or PCB terminal.

Soldering Iron
Tool having an internal heating element that is used for heating a connection to melt solder.

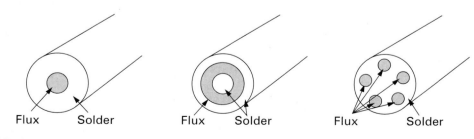

Flux   Solder   Flux   Solder   Flux   Solder

FIGURE C-4 **Rosin-Core Solder.**

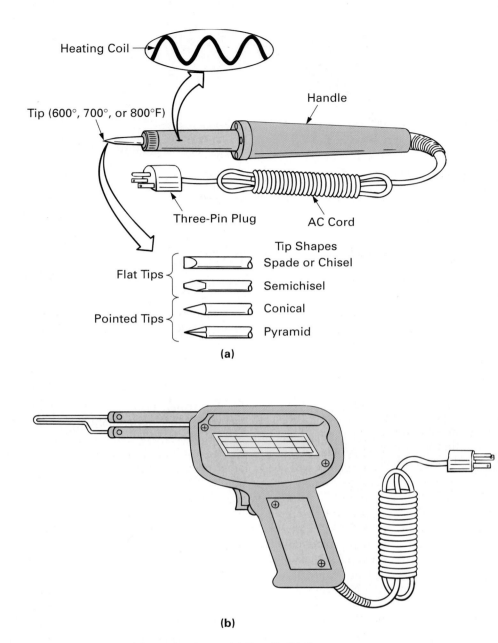

Heating Coil

Tip (600°, 700°, or 800°F)

Handle

Three-Pin Plug

AC Cord

Tip Shapes

Flat Tips {
Spade or Chisel
Semichisel

Pointed Tips {
Conical
Pyramid

(a)

(b)

FIGURE C-5   **Types of Soldering Irons. (a) Pencil. (b) Gun.**

The tip shape that makes the best contact with the connection, and will therefore conduct the most heat, should be used.

As you regularly use your soldering iron, the tip will naturally accumulate dirt and oxide, and this contamination will reduce its effectiveness. The tip should be regularly cleaned by wiping it across a damp sponge, as shown in Figure C-6(a). Soldering iron tips are generally made of copper, plated with either iron or nickel, so you should never clean a tip by filing the end, as this will probably remove the plating.

The tip of the soldering iron should always be **tinned** after it has been cleaned. Tinning the tip can be seen in Figure C-6(b) and is achieved by applying a small layer of rosin-core solder to the tip. This will protect the tip from oxidation and increase the amount of heat transfer to the connection.

Tinned

Coated with a layer of tin or solder to prevent corrosion and simplify the soldering of connections.

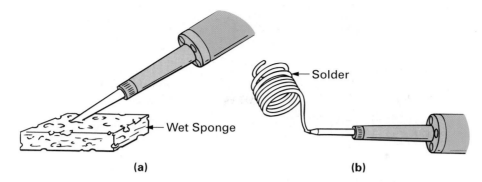

FIGURE C-6 Cleaning a Soldering Iron. (a) Cleaning the Tip. (b) Tinning the Tip.

In summary, it is important to remember that *a soldering iron should not be used to melt the solder. Its purpose is to heat the connection so that the solder will melt when it makes contact with the connection.*

C-5   *Soldering Techniques*

Before carrying out the six basic soldering steps, wires and components should be prepared. Wires should be wrapped around a terminal to give them a more solid physical support, as shown in Figure C-7(a). Their insulation should be stripped back so as to leave a small gap. Too large a gap will expose the bare wire and possibly cause a short to another terminal, whereas too small a gap will cause the insulation to burn during soldering. Stranded wire should be tinned, as shown in Figure C-7(b), and components should be mounted flat on the board, as shown in Figure C-7(c).

Six-Step Soldering Procedure

To begin, always remember to wear safety goggles and a protective apron, and then proceed.

Step 1: Clean the tip on a damp sponge.

Step 2: Tin the tip if necessary.

Step 3: Heat the connection.

Step 4: Apply the solder to the opposite side of the connection.

Step 5: Leave the tip of the soldering iron on the connection only long enough to melt the solder.

Step 6: Remove the solder and then the iron; then let the melted solder cool and solidify undisturbed.

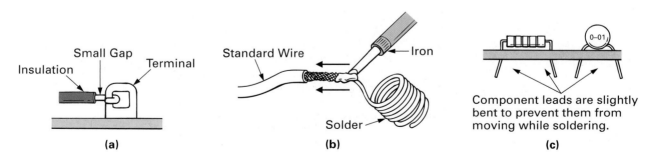

FIGURE C-7 Wire and Component Preparation.

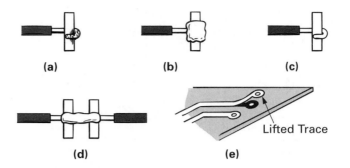

FIGURE C-8   Poor Solder Connections.

Once this procedure is complete, you should remove the excess leads with the cutters and remove the flux residue, because it collects dust and dirt that could produce electrical leakage paths. When inspecting your work, you should notice the following:

1. The connection is *smooth and shiny.*
2. The solder should *feather out to a thin edge.*

Poor Solder Connections

1. *Cold solder joint:* This connection has a dull gray appearance and it makes a poor mechanical and electrical connection. It is normally caused due to connector lead movement while the solder was cooling, or to insufficient heat [Figure C-8(a)]. The cure is to resolder the connection.
2. *Excessive solder joint:* This is caused by too much solder being applied to the connection or by using solder of too large diameter [Figure C-8(b)]. The cure is to desolder the excess with a wick.
3. *Insufficient solder joint:* Not enough solder was applied, so a poor bond exists between the lead and the terminal [Figure C-8(c)]. The cure is to resolder as if it were a new joint.
4. *Solder bridge:* This occurs when adjacent terminals or traces on printed circuit boards are connected accidentally as a result of using too much solder [Figure C-8(d)]. Most can be removed by desoldering the excess with a wick.
5. *Excessive heat:* On printed circuit boards, too much heat can lift the traces or track and ruin the entire board [Figure C-8(e)].

C-6   *Desoldering*

If a faulty component has to be replaced or if a component or wire was soldered in an incorrect place, it will have to be desoldered. Figure C-9 illustrates some of the tools used to des-

FIGURE C-9   Desoldering Tools.

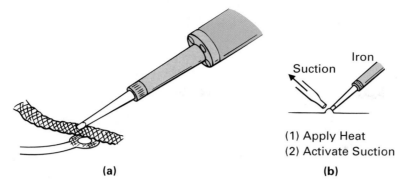

(a)

Suction   Iron

(1) Apply Heat
(2) Activate Suction

(b)

FIGURE C-10

older. The easiest way to remove a component is to remove all the solder and then disconnect the leads. All three tools shown in Figure C-9 are designed to remove solder.

Use of the braided wick to remove solder can be seen in Figure C-10(a). The braid contains many small wire strands, and when it is placed between the solder and the iron, the melted solder flows into the braid by capillary action. The solder-filled braid is then cut off and discarded.

The **desoldering** bulb or spring-loaded plunger relies on suction rather than capillary action to remove the solder, as shown in Figure C-10(b). The steps to follow are:

Step 1:   Squeeze the desoldering bulb or cock the spring-loaded plunger.
Step 2:   Melt the solder with the soldering iron.
Step 3:   Remove the iron's tip.
Step 4:   Quickly insert the tip of the bulb or plunger into the molten solder and then activate the suction.

When as much of the solder as possible has been removed, use the long-nose pliers to hold the lead and remove the component. It may be necessary to use the iron to heat the lead slightly to loosen up the residual solder so that the component lead can be removed with the pliers. Always be careful not to apply an excessive amount of heat or stress to either the component or the board.

Desoldering
The process of removing solder from a connection.

C-7   *How to Solder and Desolder Surface Mount Components (SMCs)*

In this section, we will discuss the soldering and desoldering of surface mount components (SMCs). One of the key differences between SMC soldering and through-hole component soldering is that through-hole components, like the resistor shown in Figure C-11(a), have wire leads that hold the device in place. Surface mount components, on the other hand, have nothing to hold them in place while you apply solder to their terminals.

One of the easiest ways to hold the SMC in place is to apply a small spot of liquid flux to the pretinned circuit board footprints, as shown in the upper diagram in Figure C-11(b), and then place the SMC on the sticky footprints. To solder an SMC device onto a printed circuit board (PCB) you should use a 10 watt to 15 watt soldering iron and 15 mil solder. When soldering a chip SMC, like the one shown in Figure C-11(b), follow this procedure:

Step 1:   Place the chip SMC on its flux-coated footprint using tweezers.
Step 2:   Touch the soldering iron tip to the footprint (not the SMC).
Step 3:   Apply solder to the junction until a neat fillet, about half the height of the end terminal, is formed.
Step 4:   Remove solder and iron.

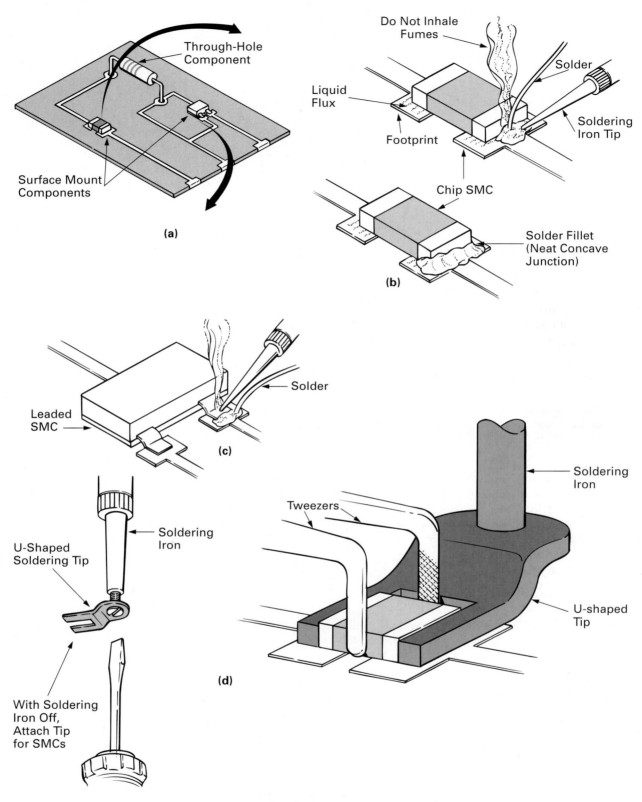

(a)

(b)

(c)

(d)

FIGURE C-11   **Soldering Surface Mount Components (SMCs).**

When soldering a leaded SMC, like the one shown in Figure C-11(c), follow this procedure:

Step 1:   Place the leaded SMC on its flux-coated footprint using tweezers.
Step 2:   Touch the soldering iron tip to the SMC pin.
Step 3:   Apply solder to the junction, and allow solder to flow around the footprint and pin.
Step 4:   Remove solder and iron.

If you have a U-shaped tip available, like the one shown in Figure C-11(d), the soldering of SMCs will be a lot easier. In this instance the procedure is as follows:

Step 1:   With iron off, install U-shaped tip to the end of iron.
Step 2:   Melt some solder over the tip, and then clean off tip by wiping it across damp sponge.
Step 3:   Pre-tin the footprint on the printed circuit board with a thin layer of solder.
Step 4:   Holding SMC in place with tweezers, place U-shaped soldering iron tip over the SMC so that both forks touch the pretinned footprint.
Step 5:   As pretinned solder on footprint melts, use tweezers to gently press SMC on its footprint pads.
Step 6:   Remove U-shaped soldering tip, wait a few seconds until solder has solidified and bonded SMC to its footprint, and then remove tweezers.
Step 7:   Apply additional solder to footprint as shown in Figure C-11(b).

Several U-shaped soldering iron tips are available for all sizes of SMCs. These specialized tips are designed specifically for soldering and desoldering. To desolder and remove an SMC using a U-shaped tip, follow this procedure:

Step 1:   Place the hot U-shaped iron tip over the SMC so that both forks touch the footprint of the SMC.
Step 2:   When the solder liquefies, twist the tip of the iron to free the SMC from its footprint.
Step 3:   Use tweezers to remove the SMC.

To remove an SMC using a wedge-shaped soldering iron tip, follow the standard desoldering procedure described in Section C-6.

C-8   *Safety*

Molten solder, like any other liquid, can splash or spill, and soldering irons are even hotter than the solder. Consequently, you should always wear protective safety glasses and an apron to protect yourself from burns. NEVER fling hot solder off an iron; always use the sponge.

Safety When Troubleshooting

The following procedures and precautions have been acquired from experienced technicians and should be applied whenever possible.

1. Except when absolutely necessary, do not work on electrical or electronic circuits or equipment when power is on.

2. When troubleshooting inside equipment, people tend to lean one hand on the chassis (the metal framework of the equipment) and hold the test lead or probe in the other hand. If the probing hand comes in contact with a high voltage, current will flow from one hand, through your chest (heart and lungs), and finally through the other hand to the chassis ground. To avoid this dangerous situation, always place the free hand in a pocket or behind your back while testing a piece of equipment with power on.

3. When troubleshooting equipment with power on, try to insulate yourself by wearing rubber-soled electrical safety shoes or standing on a rubber mat.

4. Electrolytic and other large-value capacitors (which will be covered in Chapters 13 and 14) can hold a dangerous voltage charge even after the equipment has been turned off.

5. All tools should be well insulated. If not, it is not only dangerous to you, but probes that are not well insulated right down to the tip can cause short circuits between two points, resulting in additional circuit and equipment problems.

6. Always switch off the equipment and disconnect the power (since some equipment has power present even when off) before replacing any components. When removing components that get hot, such as resistors, allow enough time for cooling after the equipment has been turned off.

7. Necklaces, rings, and wristwatches have a low resistance and should be removed when working on equipment.

8. Inspect the equipment before working on it, and if it is in poor condition (frayed, cracked, or burned power cords, chipped plugs), turn off the equipment and replace these hazards.

9. Make sure that someone is present who can render assistance in the event of an emergency.

10. Make it a point to know the location of the power-off switch.

11. Cathode-ray tubes, which are the picture tubes within televisions and computer monitors, are highly evacuated and should therefore be handled with extreme care. If broken, the relatively high external pressure will cause an implosion (burst inward), which will result in the inner metal parts and glass fragments being expelled violently outward.

D-1   *First Aid, Treatment, Resuscitation*

Shock.   Electric shock is the effect produced on the body and in particular on the nerves by an electric current passing through it. Its magnitude depends on the strength of the current, which, in turn, depends on the voltage value present. Its effect varies according to the ohmic resistance of the body, which varies in different persons, and also according to the parts of the body between which the current flows (contact in the cardiac region can be particularly dangerous). It also depends on the current flow and on the surface resistance of the skin, which is much reduced when the skin is wet and is reduced to zero if the skin is penetrated.

Shock can be felt from voltages as low as 15 V. At 20 to 25 V most people will experience pain, and the victim may be unable to let go of the conductor. It is believed that under certain conditions, death can be caused by voltages as low as 70 V, but generally the danger below 120 V ac is believed to be small (although not entirely negligible). Most serious and fatal accidents occur at the industrial 200 to 240 V ac when current is greater than 25 mA.

Injury can also be caused by a minor shock, not serious in itself but which has the effect of contracting the muscles sufficiently to result in a fall or other reaction.

Burns. Burns can be caused by the passage of a heavy current through the body or by direct contact with an electrically heated surface. Burns can also be caused by the intense heat generated by the arcing from a short circuit. All cases of burns require immediate medical attention.

Explosion. An explosion can be caused by the ignition of flammable gases by a spark from an electric contact.

Eye injuries. These can be caused by exposure to the strong ultraviolet rays of an electric arc. In these cases, the eyes may become inflamed and painful after a lapse of several hours, and there may be temporary loss of sight. Although very painful, the condition usually passes off within 24 hours.

Lasers are also dangerous to the eyes due to their intensely concentrated beam, and therefore specially filtered protective glasses should be worn, as seen in Figure D-1.

Precautions to protect the eyes must always be taken by wearing protective goggles when clipping leads or soldering.

Body injuries from microwave and radio-frequency equipment. The energy in microwave and radio-frequency equipment can damage the body, especially those parts with a low blood supply. The eyes are particularly vulnerable. The highest energy level to which operators should be subject is 1.0 mW/cm^2, and intensities exceeding 10 mW/cm^2 should always be avoided.

Resuscitation. You should familiarize yourself with the various methods of artificial respiration by contacting your local Red Cross for complete instruction. The mouth-to-mouth method of artificial respiration is the most effective of the resuscitation techniques. It is comparatively simple and produces the best and quickest results when done correctly.

FIGURE D-1 Protective Glasses for Laser Experimentation. (Photo courtesy of Hewlett-Packard Company.)

It is essential to begin artificial respiration without delay. *Do not touch the victim with your bare hands until the circuit is broken.* If this is not possible, *protect yourself* with dry insulating material and pull the victim clear of the conductor.

Step 1:   Lay the patient on his or her back and, if on a slope, have the stomach slightly lower than the chest.

Step 2:   Make a brief inspection of the mouth and throat to ensure that they are clear of obvious obstructions.

Step 3:   Give the patient's head the maximum backward tilt so that the chin is prominent and the neck stretched to give a clear airway, as shown in Figure D-2(a).

Step 4:   Sealing off the patient's nose with your thumb and finger, open your mouth wide and make an airtight seal over the patient's open mouth and then blow, as shown in Figure D-2(b). (New steps indicate using a plastic bag airway).

Step 5:   After exhaling, turn your head to watch for chest movement, while inhaling deeply in readiness for blowing again, as shown in Figure D-2(c).

Step 6:   If the chest does not rise, check that the patient's mouth and throat are free of obstruction and that the head is tilted back as far as possible, and then blow again.

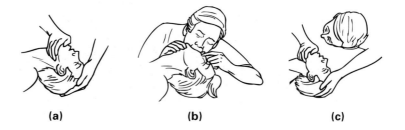

(a)   **(b)**   **(c)**

FIGURE D-2   **Mouth-toMouth Resuscitation.**

Frequency Spectrum

The term "frequency" describes the number of alternations occurring in one second. Direct current (dc) is a steady or constant current that does not alternate and is therefore listed as zero cycles per second or 0 Hz. This appendix illustrates in detail the entire range of frequencies from the lowest subaudible frequency to the highest cosmic rays, along with their applications.

The following page is an overall summary of the complete range of frequencies or spectrum, as it is normally called. The subsequent pages cover each band or section of frequencies in a lot more detail and list the different frequency applications.

This and all of the other appendixes will be useful as references throughout your course of electronic study.

OVERVIEW OF FREQUENCY SPECTRUM

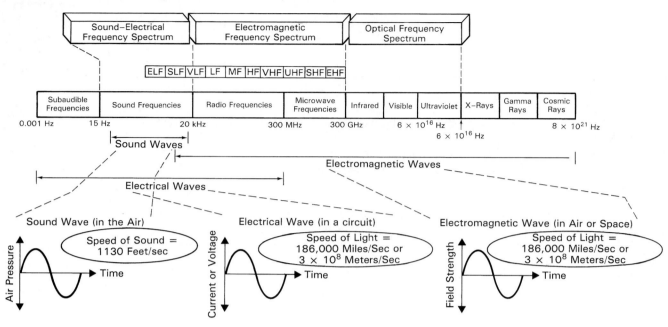

0 Hz	16 ⟶ 16 kHz	16 ⟶ 30 kHz	10 kHz ⟶ 300 GHz
Direct Current (DC) Motors, Relays, Supply Voltages	Audio Frequencies (AC) Motors, Amplifiers, Music Equipment, Speakers, Microphones, Oscillators	Sound (Ultrasonic) Frequencies Sonar, Music, Speech	Electromagnetic (Radio) Frequencies Voice Communications Television Navigation Medical, Scientific, and Military

300 GHz ⟶ 4 × 10¹⁴	4 × 10¹⁴ ⟶ 7.69 × 10¹⁴	4 × 10¹⁴ ⟶ 6 × 10¹⁶	9.375 × 10¹⁵ ⟶ 3 × 10¹⁹
Infrared (R) Heating, Photography, Sensing, Military	Visible Color, Photography, Movies, TV	Ultraviolet (UV) Sterilizing, Medical	X–Rays Medical, Gauge Thickness, Inspection

3 × 10¹⁹ ⟶ 5 × 10²⁰	5 × 10²⁰ ⟶
Gamma Rays Deeper Penetrating Than X–Rays, Detection of Radiation	Cosmic Rays Present in Outer Space

THE SOUND SPECTRUM

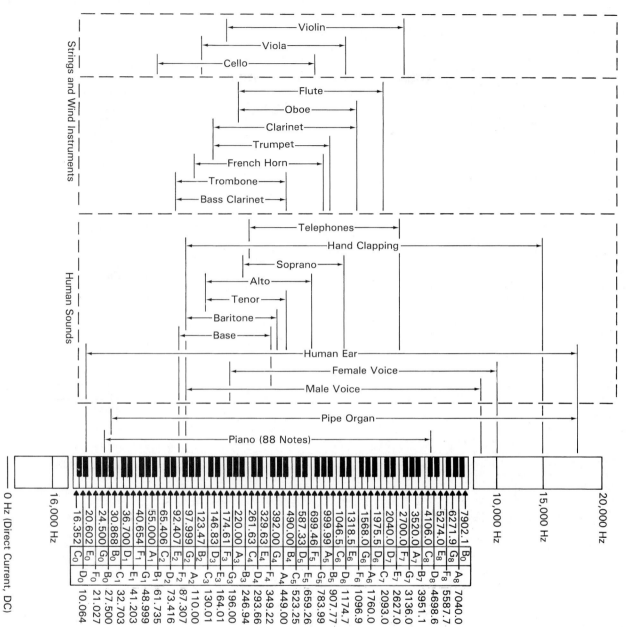

VLF (Very Low–Frequency Band)

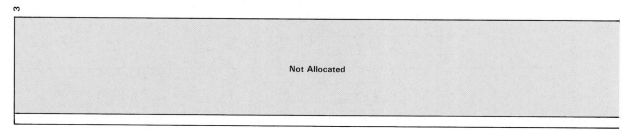

3 kHz

Not Allocated

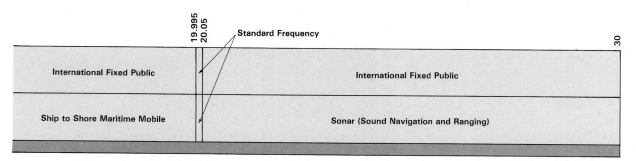

30 kHz

LF (Low–Frequency Band)

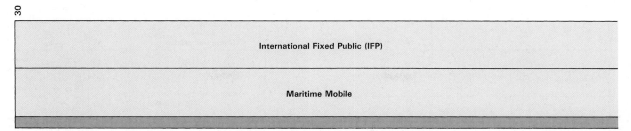

30 kHz

| 30 | | | | | | 59 | 61 | | 70 | | 90 |

International Fixed Public (IFP)					
Maritime Mobile					

International Fixed Public (IFP)	Standard Frequency	(IFP)	(IFP)	Loran C. Radionavigation
			Decca Maritime Mobile Navigation (British)	
Maritime Mobile		**MARITIME MOBILE**	Radiolocation (Radar)	

| 110 | | 130 | | 160 |

Loran C. Radionavigation	(IFP)	(IFP)	(IFP)
	Maritime Mobile	Maritime Mobile (MM)	Maritime Mobile (MM)
	Radar (Radio Detection and Ranging)		

| 200 | | 285 | 300 |

(IFP)	Aeronautical Mobile Communications	Maritime Radio–navigation
Maritime Mobile (MM)	Maritime Radionavigation	Aeronautical Radio–navigation

300 kHz

MF (Medium–Frequency Band)

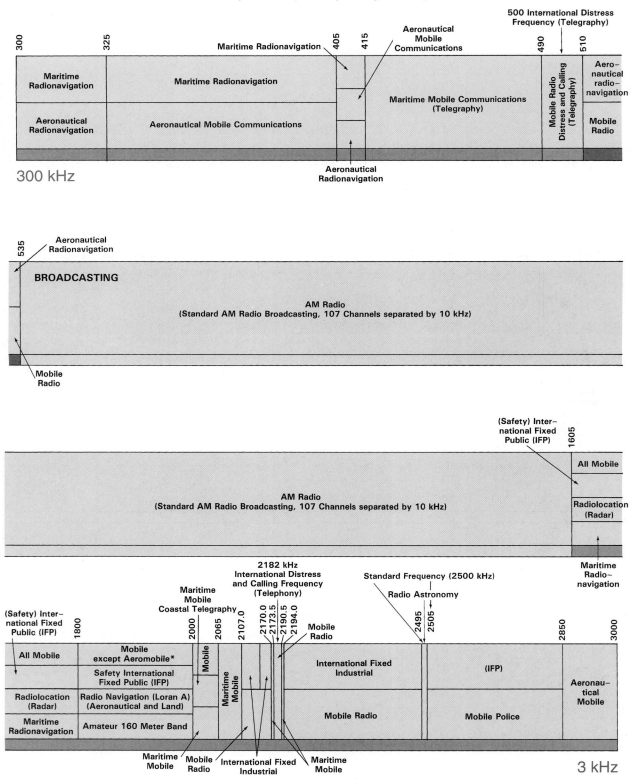

300 kHz

3 kHz

APPENDIX E / FREQUENCY SPECTRUM

HF (High–Frequency Band)

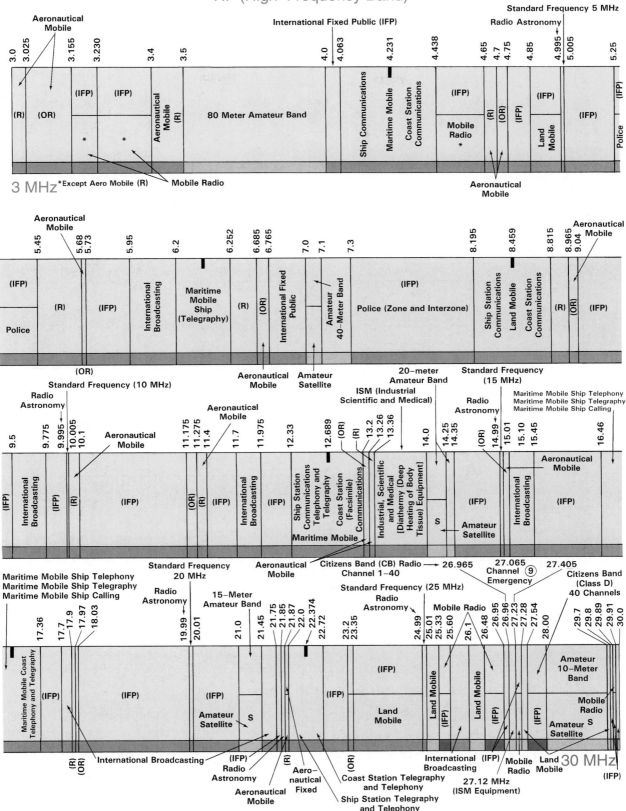

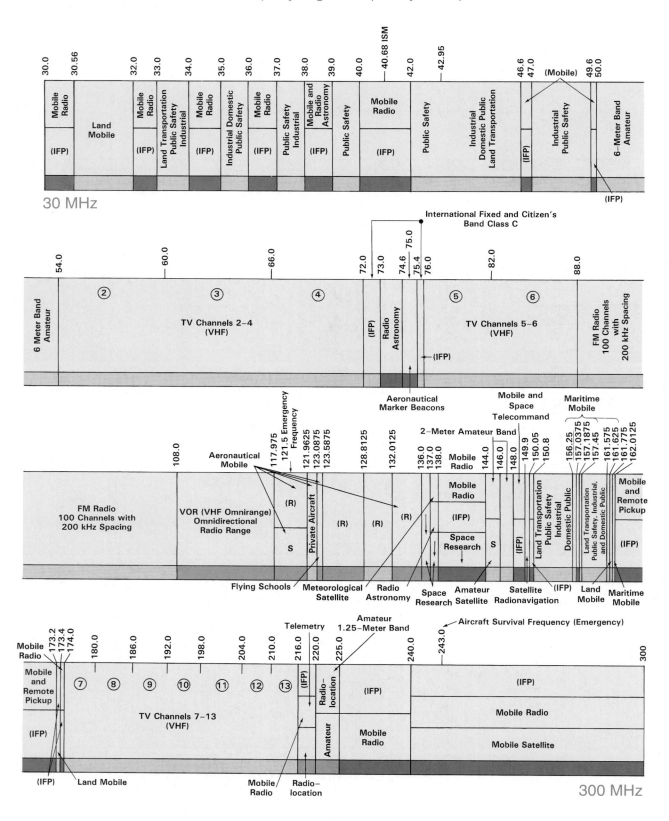

VHF (Very High–Frequency Band)

30 MHz

300 MHz

APPENDIX E / FREQUENCY SPECTRUM

UHF (Ultra High–Frequency Band)

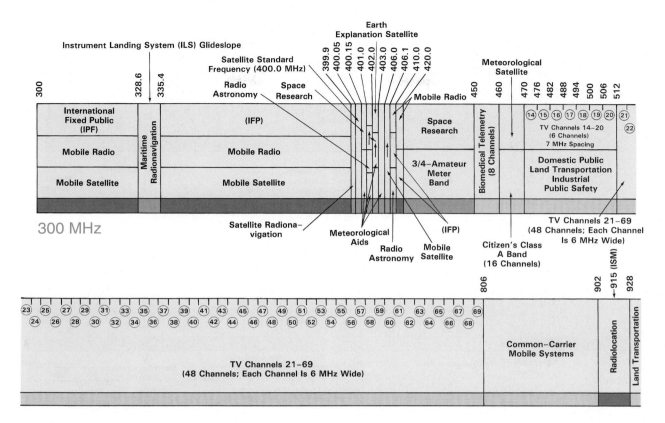

300 MHz

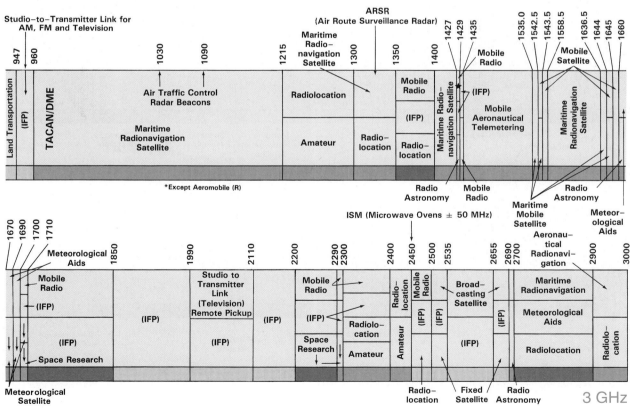

3 GHz

SHF (Super High–Frequency Band)

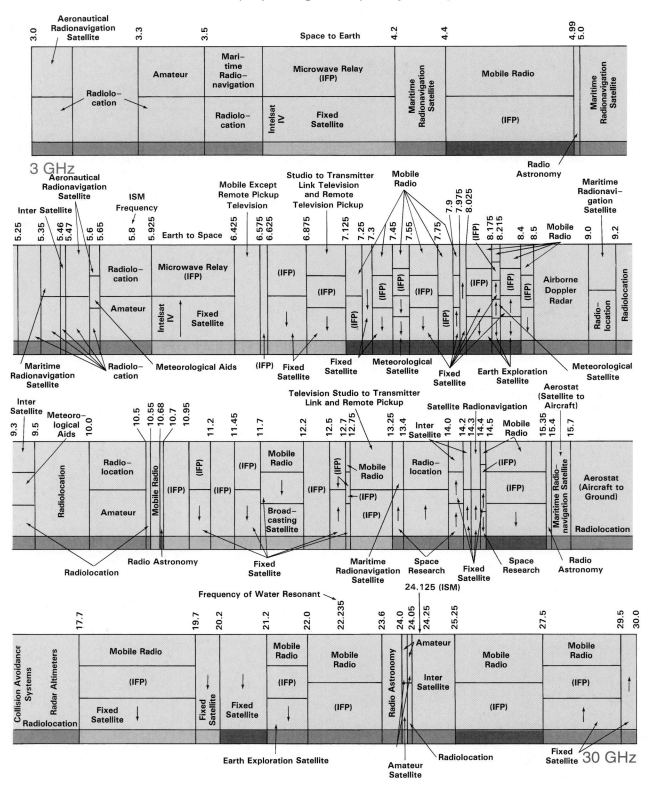

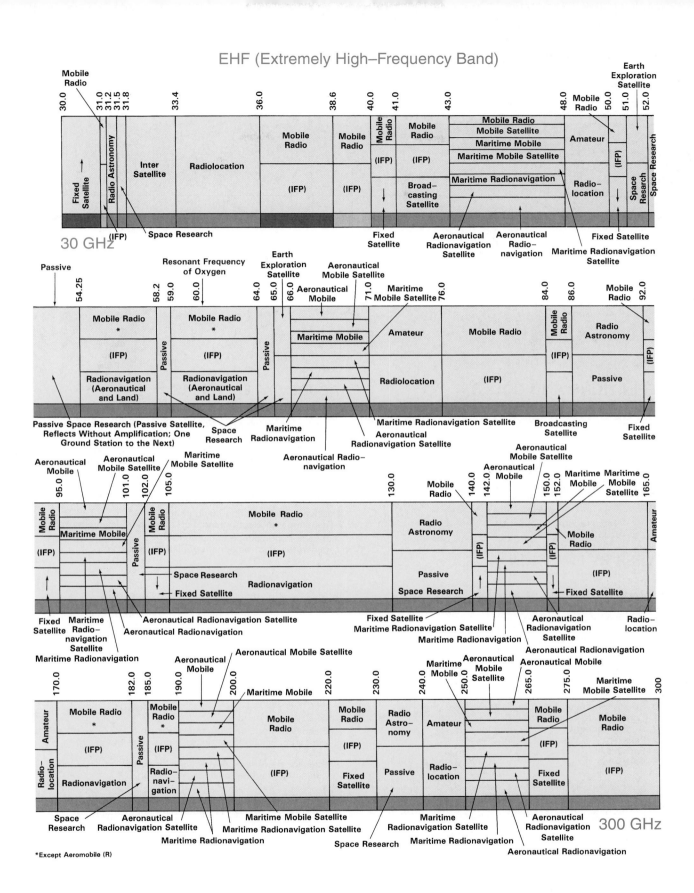

EHF (Extremely High–Frequency Band)

30 GHz

300 GHz

*Except Aeromobile (R)

Optical Spectrum

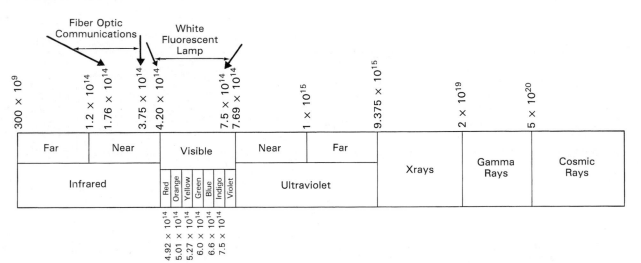

DEFINITIONS AND CODES

USAGE

 Exclusively used by government, state, and federal agencies

 Shared by government and nongovernment agencies

Nongovernment; publicly used frequencies

TYPES OF EMISSION

Amplitude (AM; Amplitude Modulation)
A0—Steady Unmodulated Pure Carrier
A1—Telegraphy on Pure Continuous Waves
A2—Amplitude Tone—modulated Telegraphy
A3—AM Telephony Including Single and Double Sideband
 with full, Reduced, or Suppressed Carrier
A4—Facsimile (Pictures)
A5—Television

Frequency FM; Frequency Modulation)
F0—Steady Unmodulated Pure Carrier
F1—Carrier—Shift Telegraphy
F2—Audio—frequency—shift Telegraphy
F3—Frequency or Phase—modulated Telephony
F4—Facsimile
F5—Television

Pulse (PM; Pulse Modulation)
P0—Pulsed Radar
P1—Telegraphy: On/Off Keying of Pulsed Carrier
P2—Telegraphy: Pulse Modulation of Pulse Carrier,
 Pulse Width, Phase or Position Tone Modulated
P3—Telephony: Amplitude (PAM), Width (PWM),
 Phase or Position (PPM) Modulated Pulses

ABBREVIATED TERMS

(R), International major air route (air traffic control)
(OR), Off route military (air traffic control)
 S Assigned satellite frequency
 ↑ Earth to space communication (uplink)
 ↓ Space to earth communication (downlink)

ISM: industrial, scientific, medical
Passive communication equipment:
 generates no electromagnetic radiation (receive only)
Active communication equipment:
 generates electromagnetic radiation (transmits)

APPLICATIONS

- International Fixed: public radiocommunication
service. Transmitters designed for one frequency.

- Aeronautical Fixed: a service intended for the
transmission of air navigation and preparation for
safety of flight.

- Fixed Satellite: satellites that maintain a
geostationary orbit (same position above the earth)

- Radiolocation: radio waves used to detect an
object's direction, position, or motion.

- Radio Astronomy: radio waves emitted by celestial
bodies that are used to obtain data about them.

- Space Operations

- Space Research

APPLICATIONS (continued)

- Mobile: Radio service between a fixed location and one or more mobile stations or between mobile stations.

- Aeronautical Mobile: As above, except mobile stations are aircraft.

- Maritime Mobile: As above, except mobile stations are marine.

- Land Mobile: As above, except mobile stations are automobiles.

- Mobile Satellite

- Aeronautical Mobile Satellite

- Maritime Mobile Satellite

- Radionavigation (Aeronautical and Land): Navigational use of radiolocation equipment such as direction finders, radio compass, radio homing beacons, etc.

- Aeronautical Radionavigation

- Maritime Radionavigation

- Satellite Radionavigation

- Aeronautical Radionavigation Satellite

- Maritime Radionavigation Satellite

- Inter Satellite

APPLICATIONS (continued)

- Broadcasting: Transmission of speech, music, or visual programs for commercial or public service purposes.

- Broadcasting Satellite

- Amateur: a frequency used by persons licensed to operate radio transmitters as a hobby. Person is also called a radio ham.

- Amateur Satellite: Communication by radio hams via satellite.

- Citizen's: a radiocommunications service of fixed, land, and mobile stations intended for short-distance personal or business purposes.

- Standard Frequency: Highly accurate signal broadcasted by the national bureau of standards (NBS) radio station (WWV) to provide frequency, time, solar flare, and other standards.

- Standard Frequency Satellite: NBS broadcast via satellite.

- Meteorological Aids: a radio service in which the emission consists of signals used solely for meteorological use.

- Meteorological Satellite: Meteorological broadcast via satellite.

- Earth Exploration Satellite: Radio frequencies used for earth exploration.

STEP 1-1
1. **a.** $16^4 = 16 \times 16 \times 16 \times 16 = 65,536$
   b. $32^3 = 32 \times 32 \times 32 = 32,768$
   c. $112^2 = 112 \times 112 = 12,544$
   d. $15^6 = 15 \times 15 \times 15 \times 15 \times 15 \times 15 = 11,390,625$
   e. $2^3 = 2 \times 2 \times 2 = 8$
   f. $3^{12} = 3 \times 3 \times 3 \times 3 \times 3 \times 3 \times 3 \times 3 \times 3 \times 3 \times 3 \times 3 = 531,441$
2. **a.** $\sqrt[2]{144} = 12$
   b. $\sqrt[3]{3375} = 15$
   c. $\sqrt[2]{20} = 4.47$
   d. $\sqrt[3]{9} = 2.08$
3. **a.** 10^2
   b. 10^0
   c. 10^1
   d. 10^6
   e. 10^{-3}
   f. 10^{-6}
4. **a.** $6.3 \times 10^3 = 6.3\,0\,0. = 6300.0$ or 6300
   b. $114,000 \times 10^{-3} = 114.0\,0\,0. = 114.0$ or 114
   c. $7,114,632 \times 10^{-6} = 7.1\,1\,4\,6\,3\,2. = 7.114632$
   d. $6624 \times 10^6 = 6624.0\,0\,0\,0\,0\,0. = 6,624,000,000.0$
5. **a.** $\sqrt{3 \times 10^6} = \sqrt{3,000,000} = 1732.05$
   b. $(2.6 \times 10^{-6}) - (9.7 \times 10^{-9}) = 0.0000025$ or 2.5×10^{-6}
   c. $\dfrac{(4.7 \times 10^3)^2}{3.6 \times 10^6} = (4.7 \times 10^3)^2 \div (3.6 \times 10^6) = 6.14$
6. **a.** $47,000 = 47\,0\,0\,0. = 47 \times 10^3$
   b. $0.00000025 = 0.0\,0\,0\,0\,0\,0\,2\,5\,0. = 250 \times 10^{-9}$
   c. $250,000,000 = 250.0\,0\,0\,0\,0\,0. = 250 \times 10^6$
   d. $0.0042 = 0.0\,0\,4.2 = 4.2 \times 10^{-3}$
7. **a.** 10^3
   b. 10^{-2}
   c. 10^{-3}
   d. 10^6
   e. 10^{-6}

STEP 1-2
1. Elements are made up of similar atoms; compounds are made up of similar molecules.
2. Protons, neutrons, electrons
3. Copper
4. Like charges repel; unlike charges attract.

STEP 1-3
1. Amp
2. Current = Q/t (number of coulombs divided by time in seconds).
3. The direction of flow is different: negative to positive is termed electron current flow; positive to negative is known as conventional current flow.
4. The ammeter

STEP 1-4
1. Volts
2. 3000 kV

3. The voltmeter
4. They are directly proportional.

STEP 1-5
1. False
2. **a.** Electrons are far away from nucleus.
   b. Many electrons per atom.
   c. Incomplete valence shell.
3. Copper
4. 28.6 millisiemens

STEP 1-6
1. True
2. Mica
3. The voltage needed to cause current to flow through a material
4. Small

STEP 1-7
1. No
2. Yes
3. A short circuit provides an unintentional path for current to flow. A closed circuit has a complete path for current.
4. An open switch breaks a current path, while a closed switch makes a current path.

STEP 2–1
1. Yes
2. $\dfrac{144}{12} \times \square = \dfrac{36}{6} \times 2 \times \square = 60$
   $12 \times \square = 6 \times 2 \times \square = 60$
   $12 \times 5 = 12 \times 5 = 60$
   $\square = 5$
3. Yes
4. **a.** $x + 14 = 30$
   $x + 14 - 14 = 30 - 14 \qquad (-14 \text{ from both sides})$
   $x = 30 - 14$
   $x = 16$
   b. $8 \times x = \dfrac{80 - 40}{10} \times 12$
   $8 \times x = 4 \times 12$
   $8 \times x = 48$
   $\dfrac{8 \times x}{8} = \dfrac{48}{8} \qquad (\div 8)$
   $x = \dfrac{48}{8}$
   $x = 6$
   c. $y - 4 = 8$
   $y - 4 + 4 = 8 + 4 \qquad (+4)$
   $y = 8 + 4$
   $y = 12$

d.
$$(x \times 3) - 2 = \frac{26}{2}$$
$$(x \times 3) - 2 + 2 = \frac{26}{2} + 2 \quad (+2)$$
$$x \times 3 = \frac{26}{2} + 2$$
$$x \times 3 = 15$$
$$\frac{x \times 3}{3} = \frac{15}{3} \quad (\div 3)$$
$$x = \frac{15}{3}$$
$$x = 5$$

e.
$$x^2 + 5 = 14$$
$$x^2 + 5 - 5 = 14 - 5 \quad (-5)$$
$$x^2 = 14 - 5$$
$$\sqrt{x^2} = \sqrt{14 - 5} \quad (\sqrt{\ })$$
$$x = \sqrt{14 - 5}$$
$$x = \sqrt{9}$$
$$x = 3$$

f.
$$2(3 + 4x) = 2(x + 13)$$
$$6 + 8x = 2x + 26 \quad \text{(remove parentheses)}$$
$$6 + 8x - 2x = 2x + 26 - 2x \quad (-2x)$$
$$6 + (8x - 2x) = 26$$
$$6 + 6x = 26$$
$$6 + 6x - 6 = 26 - 6 \quad (-6)$$
$$6x = 26 - 6$$
$$6x = 20$$
$$\frac{6 \times x}{6} = \frac{20}{6} \quad (\div 6)$$
$$x = \frac{20}{6}$$
$$x = 3.3333$$

5. a. $x + y = z, y = ?$
$$x + y - x = z - x \quad (-x)$$
$$y = z - x$$

b. $Q = C \times V, C = ?$
$$\frac{Q}{V} = \frac{C \times V}{V} \quad (\div V)$$
$$\frac{Q}{V} = C$$
$$C = \frac{Q}{V}$$

c. $X_L = 2 \times \pi \times f \times L, L = ?$
$$\frac{X_L}{2 \times \pi \times f} = \frac{2 \times \pi \times f \times L}{2 \times \pi \times f} \quad (\div 2 \times \pi \times f)$$
$$\frac{X_L}{2 \times \pi \times f} = L$$
$$L = \frac{X_L}{2 \times \pi \times f}$$

d. $V = I \times R, R = ?$
$$\frac{V}{I} = \frac{I \times R}{I} \quad (\div I)$$
$$\frac{V}{I} = R$$
$$R = \frac{V}{I}$$

6. a. $I^2 = 9$
$$\sqrt{I^2} = \sqrt{9} \quad (\sqrt{\ })$$
$$I = \sqrt{9}$$
$$I = 3$$

b.
$$\sqrt{Z} = 8$$
$$\sqrt{Z}^2 = 8^2$$
$$Z = 8^2$$
$$Z = 64$$

7. $x = y \times z$ and $a = x \times y, y = 14, z = 5, a = ?$
$$a = x \times y$$
$$a = y \times z \times y \quad \text{(substitute } x \text{ for } y \times z\text{)}$$
$$a = y^2 \times z \quad (y \times y = y^2)$$
$$a = 14^2 \times 5$$
$$a = 980$$

STEP 2-2
1. It is the opposition to current flow.
2. The difference is in how much or how little current will flow.
3. They are inversely proportional.
4. Small

STEP 2-3
1. A circuit is said to have a resistance of 1 ohm when 1 volt produces a current of 1 ampere.
2. $I = V/R = 24\ \text{V}/6\ \Omega = 4\ \text{A}$.
3. A memory aid to help remember Ohm's law:

4. Current is proportional to voltage and inversely proportional to resistance.
5. $V = I \times R = 25\ \text{mA} \times 1\ \text{k}\Omega = (25 \times 10^{-3}) \times (1 \times 10^3) = 25\ \text{V}$.
6. $R = V/I = 12\ \text{V}/100\ \mu\text{A} = 12\ \text{V}/100 \times 10^{-6} = 120\ \text{k}\Omega$.

STEP 2-4
1. Type of conducting material used, cross-sectional area, length, temperature
2. The resistance of a conductor is proportional to its length and resistivity, and inversely proportional to its cross-sectional area.
3. True
4. True

STEP 2-5
1. Light, heat, magnetic, chemical, electrical, mechanical
2. When energy is transformed, work is done. Power is the rate at which work is done or energy is transformed.
3. $W = Q \times V, P = V \times I$
4. When 1 kW of power is used in 1 hour

STEP 3-1
1. $\dfrac{176 \div 8}{8 \div 8} = \dfrac{22}{1}$ or $22:1$ (ratio of twenty-two to one)
2. a. 86.44
   b. 12,263,415.01
   c. 0.18
3. 86.43760   **a.** 7
              b. 6
   12,263,415.00510   **a.** 13
                      b. 12
   0.176600   **a.** 6
              b. 4
4. a. 0.075   **c.** 0.235
   b. 220   **d.** 19.2
5. a. 35   **b.** 4004

6.

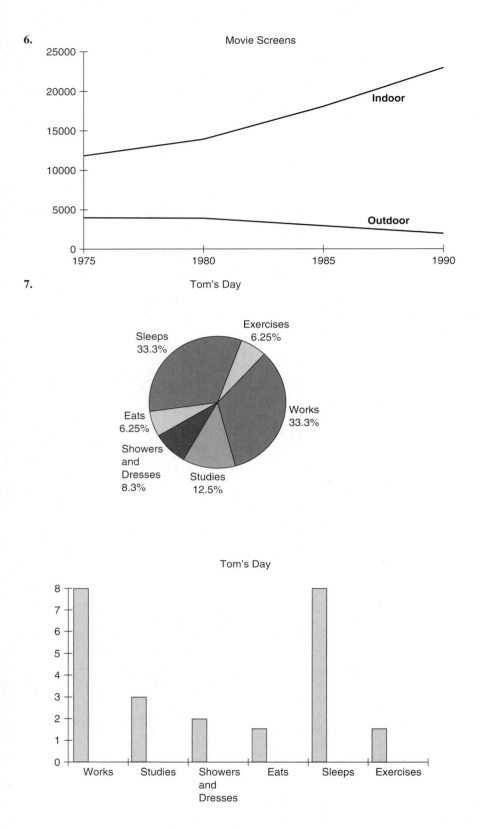

Movie Screens

7. Tom's Day

Tom's Day

STEP 3-2

1. Carbon composition, carbon film, metal film, wirewound, metal oxide, thick film
2. SIPs have one row of connecting pins, while DIPs have two rows of connecting pins.
3. Rheostat has two terminals; potentiometer has three terminals.
4. Linear means that the resistance changes in direct proportion to the amount of change of the input, while a tapered potentiometer varies nonuniformly.

5. True

6. Photoresistor

STEP 3-3

1. Ohmmeter

2. See Figure 3-20.

STEP 3-4

1. General purpose are ±5% or greater; precision are ±2% or less.

2. 222×0.1, ±0.25% = 22.2 Ω, ±0.25%

3. Yellow, violet, red, silver

4. True (General purpose, ±20% tolerance)

STEP 3-5 and 3-6

1. $P = I^2 \times R$

2. Small, large

STEP 3-7

1. Opens

2. Ohmmeter

STEP 4-1

1. Load resistance is the resistance of a device, circuit, or system. Load current is the amount of current drawn from the source by the device, circuit, or system. A dc voltage source is one that will supply a constant polarity output voltage, while a dc current is a unidirectional (one direction) current.

2. False

3. They are inversely proportional.

4. A load that will draw a large amount of current.

STEP 4-2

1. A piezoresistive device changes its resistance in response to pressure, whereas a piezoelectric device generates a voltage output in response to a pressure input.

2. A device that generates a voltage output when heat is applied to its two-metal junction.

3. By using a photovoltaic cell.

4. Generator

5. Negative plate, positive plate, electrolyte.

6. False.

STEP 4-3 and 4-4

1. Indicate the maximum current that can flow without blowing the fuse and the maximum voltage value that will not cause an arc.

2. Fast-blow fuses open almost instantly after an excessive current occurs, whereas a slow-blow fuse will have a certain time delay until it opens.

3. Thermal, magnetic, thermomagnetic

4. SPST, SPDT, DPST, DPDT, NOPB, NCPB, rotary, and DIP

STEP 4-5

1. It is designed to accommodate and interconnect components to form experimental circuits.

2. A power supply voltage is generally connected across these two points so that voltages can be tapped off to supply the circuit.

STEP 5-1 and 5-2

1. A circuit in which current has only one path

2. 8 A

STEP 5-3

1. $R_T = R_1 + R_2 + R_3 + \cdots$

2. $R_T = R_1 + R_2 + R_3 = 2 \text{ k}\Omega + 3 \text{ k}\Omega + 4700 = 9.7 \text{ k}\Omega$

STEP 5-4

1. True

2. True

3. $R_T = R_1 + R_2 = 6 + 12 = 18 \ \Omega$; $I_T = V_S/R_T = 18/18 = 1 \text{ A}$
$V_{R_1} = 1 \text{ A} \times 6 \ \Omega = 6 \text{ V}$
$V_{R_2} = 1 \text{ A} \times 12 \ \Omega = 12 \text{ V}$

4. $V_X = (R_X/R_T) = V_S$

5. Potentiometer

6. No

STEP 5-5

1. $P = I \times V$ or $P = V^2/R$ or $P = I^2 \times R$

2. $P = V^2/R = 12^2/12 = 144/12 = 12 \text{ W}$

3. Wirewound, at least 12 W, ideally a 15 W

4. $P_T = P_1 + P_2 = 25 \text{ W} + 3800 \text{ mW} = 28.8 \text{ W}$

STEP 5-6

1. Component will open, component's value will change, and component will short.

2. No current will flow; source voltage dropped across it.

3. False

4. No voltage drop across component, yet current still flows in circuit; resistance equals zero for component.

STEP 6-1 and 6-2

1. When two or more components are connected across the same voltage source so that current can branch out over two or more paths

2. False

3. $V_{R_1} = V_S = 12 \text{ V}$

4. No

STEP 6-3

1. The sum of all currents entering a junction is equal to sum of all currents leaving that junction.

2. $I_2 = I_T - I_1 = 4 \text{ A} - 2.7 \text{ A} = 1.3 \text{ A}$

3. $I_X = (R_T/R_X) \times I_T$

4. $I_T = V_T/R_T = 12/1 \text{ k}\Omega = 12 \text{ mA}$; $I_1 = 1 \text{ k}\Omega/2 \text{ k}\Omega \times 12 \text{ mA} = 6 \text{ mA}$

STEP 6-4

1. $R_T = \dfrac{R_1 \times R_2}{R_1 + R_2}$

2. $R_T = \dfrac{1}{(1/R_1) + (1/R_2) + (1/R_3)} + \cdots$

3. $R_T = \dfrac{\text{common value of resistors } (R)}{\text{number of parallel resistors } (n)}$

4. $R_T = \dfrac{1}{(1/2.7 \text{ k}\Omega) + (1/24 \text{ k}\Omega) + (1/1 \text{ M}\Omega)} = 2.421 \text{ k}\Omega$

STEP 6-5

1. True

2. $P_1 = I_1 \times V = 2 \text{ mA} \times 24 \text{ V} = 48 \text{ mW}$

3. $P_T = P_1 + P_2 = 22 \text{ mW} + 6400 \ \mu\text{W} = 28.4 \text{ mW}$

4. Yes

STEP 6-6

1. No current will flow in the open branch; total current will decrease.

2. Maximum current is through shorted branch; total current will increase.

3. Will cause a corresponding opposite change in branch current and total current

4. False

STEP 7-1 and 7-2

1. By tracing current to see if it has one path (series connection) or more than one path (parallel connection)

2. $R_{1,2} = R_1 + R_2 = 12\text{ k}\Omega + 12\text{ k}\Omega = 24\text{ k}\Omega$

$$= R_{1,2,3} = \frac{R_{1,2} \times R_3}{R_{1,2} + R_3} = \frac{24\text{ k}\Omega \times 6\text{ k}\Omega}{24\text{ k}\Omega + 6\text{ k}\Omega}$$

$$= \frac{144\text{ k}\Omega}{30\text{ k}\Omega} = 4.8\text{ k}\Omega$$

3. STEP A: Find equivalent resistances of series-connected resistors. STEP B: Find equivalent resistances of parallel-connected combinations. STEP C: Find equivalent resistances of remaining series-connected resistances.

4. $R_{1,2} = R_1 + R_2 = 470 + 330 = 800\ \Omega$

$$R_{1,2,3} = \frac{R_{1,2} \times R_3}{R_{1,2} + R_3} = \frac{800 \times 270}{800 + 270} = \frac{216\text{ k}\Omega}{1.07\text{ k}\Omega} = 201.9\ \Omega$$

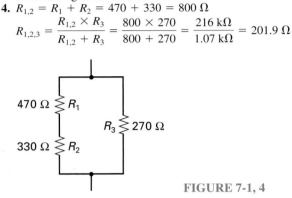

FIGURE 7-1, 4

STEP 7-3

1. STEP 1: Find total resistance (STEPs A, B and C). STEP 2: Find total current. STEP 3: Find voltage drop with $I_T \times R_X$.

2. The voltage drops previously calculated would not change since the ratio of the series resistor to the series equivalent resistors remains the same, and therefore the voltage division will remain the same.

STEP 7-4, 7-5, and 7-6

1. Find total resistance; find total current; find the voltage across each series and parallel combination resistors; find the current through each branch of parallel resistors; find the total and individual power dissipated.

2. This is a do-it-yourself question; each answer will vary.

STEP 7-7

1. When a load resistance changes the circuit and lowers output voltage

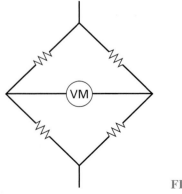

FIGURE 7-7, 2

2. Used to check for an unknown resistor's resistance
3. R
4. Current divider for a digital-to-analog converter

STEP 7-8

1. If component is in series with circuit, no current will flow and source voltage will be across bad component. If component is parallel, no current will flow in that branch.

2. Total resistance will decrease and bad component will have 0 V dropped across it.

3. Will cause the circuit's behavior to vary

STEP 7-9

1. Small, large
2. Voltage source
3. Voltage, resistor
4. Current

STEP 8–1

1. $C = \sqrt{A^2 + B^2}$

   $= \sqrt{15^2 + 22^2}$

   $= \sqrt{225 + 484}$

   $= \sqrt{709}$

   $= 26.63$ meters

2. **a.** $A = \sqrt{C^2 - B^2}$

   $= \sqrt{3^2 - 2^2}$

   $= \sqrt{9 - 4}$

   $= \sqrt{5}$

   $= 2.24$ feet

   b. $B = \sqrt{C^2 - A^2}$

   $= \sqrt{160^2 - 80^2}$

   $= \sqrt{25,000 - 6400}$

   $= \sqrt{19,200}$

   $= 138.56$ km

   c. $C = \sqrt{A^2 + B^2}$

   $= \sqrt{112^2 + 25^2}$

   $= \sqrt{12,544 + 625}$

   $= \sqrt{13,169}$

   $= 114.76$ mm

3. **a.** $O = 35$ mm $\qquad \theta = 36°$

   $H = \ ?$

   $SOH \qquad$ or $\quad \sin \theta = \dfrac{O}{H}$

   $\sin 36° = \dfrac{35\text{ mm}}{H}$

   $0.59 = \dfrac{35}{H}$

   $0.59 \times H = \dfrac{35}{\cancel{H}} \times \cancel{H} \qquad (\times H)$

   $\dfrac{0.59 \times H}{0.59} = \dfrac{35}{0.59} \qquad (\div 0.59)$

   $H = \dfrac{35}{0.59} = 59.32$ mm

   b. $H = 160$ km $\qquad \theta = 38°$

   $A = \ ?$

   $CAH \qquad$ or $\quad \cos \theta = \dfrac{A}{H}$

   $\cos 38° = \dfrac{A}{160\text{ km}}$

   $0.79 = \dfrac{A}{160}$

$$0.79 \times 160 = \frac{A}{160} \times 160 \quad (\times 160)$$
$$0.79 \times 160 = A$$
$$A = 126.4 \text{ km}$$

c. $O = 163$ cm $\theta = 72°$
$A = \quad ?$

$$TOA \quad \text{or} \quad \tan \theta = \frac{O}{A}$$
$$\tan 72° = \frac{163 \text{ cm}}{A}$$
$$3.08 = \frac{163}{A}$$
$$3.08 \times A = \frac{163 \times A}{A} \quad (\times A)$$
$$\frac{3.08 \times A}{3.08} = \frac{163}{3.08} \quad (\div 3.08)$$
$$A = \frac{163}{3.08} = 52.92 \text{ cm}$$

4. a. $H = 120$ miles $\theta = ?$
$O = 38$ miles

$$SOH \quad \text{or} \quad \sin \theta = \frac{O}{H}$$
$$\sin \theta = \frac{38 \text{ miles}}{120 \text{ miles}}$$
$$\sin \theta = 0.32$$
$$\text{invsin} \times \sin \theta = \text{invsin } 0.32 \quad (\times \text{ invsin})$$
$$\theta = \text{invsin } 0.32$$
$$\theta = 18.66°$$

b. $H = 25$ feet $\theta = ?$
$A = 17$ feet

$$CAH \quad \text{or} \quad \cos \theta = \frac{A}{H}$$
$$\cos \theta = \frac{17 \text{ feet}}{25 \text{ feet}}$$
$$\cos \theta = 0.68$$
$$\text{invcos} \times \cos \theta = \text{invcos } 0.68$$
$$\theta = \text{invcos } 0.68$$
$$\theta = 47.16°$$

c. $A = 69$ cm $\theta = ?$
$O = 51$ cm

$$TOA \quad \text{or} \quad \tan \theta = \frac{O}{A}$$
$$\tan \theta = \frac{51 \text{ cm}}{69 \text{ cm}}$$
$$\tan \theta = 0.74$$
$$\text{invtan} \times \tan \theta = \text{invtan } 0.74$$
$$\theta = \text{invtan } 0.74$$
$$\theta = 36.5°$$

5. a. $x = \sqrt{A^2 + B^2}$
$= \sqrt{40^2 + 30^2}$
$= \sqrt{1600 + 900}$
$= \sqrt{2500}$
$= 50$ volts

b. $x = \sqrt{A^2 + B^2}$
$= \sqrt{75^2 + 26^2}$
$= \sqrt{5625 + 676}$
$= \sqrt{6301}$
$= 79.38$ watts

c. $x = \sqrt{A^2 + B^2}$
$= \sqrt{93^2 + 36^2}$
$= \sqrt{8649 + 1296}$
$= \sqrt{9945}$
$= 99.72$ mm
$y = \sqrt{A^2 + B^2}$
$= \sqrt{48^2 + 96^2}$
$= \sqrt{2304 + 9216}$
$= \sqrt{11,520}$
$= 107.33$ mm

STEP 8-2
1. **a.** Alternating current
   b. Direct current
2. DC; current only flows in one direction
3. AC
4. DC
5. Power transfer, information transfer
6. DC flows in one direction, whereas ac first flows in one direction and then in the opposite direction.

STEP 8-3-1
1. AC generators can be larger, less complex, and cheaper to run; transformers can be used with ac to step up/down, so low-current power lines can be used; easy to change ac to dc, but hard the other way around.
2. False
3. $P = I^2 \times R$
4. A device that can step up or down ac voltages
5. 120 V ac
6. AC, DC

STEP 8-3-2
1. The property of a signal or message that conveys something meaningful to the recipient; the transfer of information between two points
2. Sound, electromagnetic, electrical
3. **a.** Sound wave
   b. Electromagnetic wave
   c. Electrical wave
4. 1133 feet per second; 186,000 miles per second
5. Sound, electromagnetic, electrical
6. **a.** Microphone   **c.** Antenna
   b. Speaker   **d.** Human ear
7. Electronic
8. Electrical

STEP 8-4
1.

(a)   (b)   (c)   (d)   (e)

2. Time, frequency
3. (Sound wave); $\lambda(\text{mm}) = \dfrac{344.4 \text{ m/s}}{f(\text{Hz})}$;

   (Electromagnetic wave): $\lambda(\text{m}) = \dfrac{3 \times 10^8 \text{ m/s}}{f(\text{Hz})}$;

different because sound waves travel at a different speed than do electromagnetic waves.
4. Odd

STEP 8-5-1
1. True; no
2. RF probe is used to measure high-frequency electrical waves and uses a special high-frequency rectifier. HV probe is used to measure high voltages and uses multiplier resistors.

STEP 8-5-2
1. Oscilloscope
2. Cathode ray tube (CRT)
3. $t = 80\ \mu s$
   $f = 12.5$ kHz
4. $V_p = 4$ V
   $V_{p\text{-}p} = 8$ V
5. Waveforms can be compared.

STEP 8-5-3
1. Sine, square, triangular
2. Measure the frequency of a periodic wave

STEP 8-5-4
1. A hand-held test instrument that combined an oscilloscope and multimeter
2. Portability, ease of set-up, and ease of use

STEP 9-1
1. Two plates, dielectric
2. Dielectric

STEP 9-2
1. True
2. None

STEP 9-3
1. (b)
2. Field strength (V/cm) = charge difference (V), volts/distance between plates (d), meters
3. The displacement of an atom's electrons within the dielectric toward the positive plate
4. "Di" because it is between two plates; "electric" because an electric field exists within it

STEP 9-4
1. The farad
2. Capacitance, C (farads) = charge, Q (coulombs)/voltage, V (volts)
3. $30,000 \times 10^{-6} = 0.03 \times 10^0 = 0.03$ F
4. $C = Q/V = 17.5/9 = 1.94$ F

STEP 9-5
1. Plate area, distance between plates, type of dielectric used
2. $C = \dfrac{(8.85 \times 10^{-12}) \times K \times A}{d}$
3. Double
4. Double

STEP 9-6
1. Glass
2. False
3. Large, small
4. Nanoamperes (b)

STEP 9-7
1. $C_T = \dfrac{1}{(1/C_1) + (1/C_2) + (1/C_3)}$
   $\quad\ = \dfrac{1}{(1/2\ \mu F) + (1/3\ \mu F) + (1/5\ \mu F)}$
   $\quad\ = 0.968\ \mu F$ or $1\ \mu F$
2. $C_T = C_1 + C_2 + C_3 = 7$ pF $+ 2$ pF $+ 14$ pF $= 23$ pF
3. $V_{CX} = (C_T/C_X) \times V_T$
4. True

STEP 9-8
1. Mica, ceramic, paper, plastic, electrolytic
2. **a.** Electrolytic; **b.** ceramic
3. Air, mica, ceramic, plastic
4. **a.** Air; **b.** mica, ceramic, plastic
5. Electrolytic

STEP 9-9
1. 470 pF, 2% tolerance
2. 0.47 μF or 470 nF, 5% tolerance

STEP 9-10
1. The time it takes a capacitor to charge to 63.2%
2. 63.2%
3. 36.8%
4. False

STEP 9-11
1. Capacitive opposition is low for ac (short) and high for dc (open).
2. True

STEP 9-12
1. It means that the voltage and current maximums and minimums occur at different times.
2. False
3. True
4. False

STEP 10-1
1. Opposition to current flow without the dissipation of energy
2. $X_C = 1/2\pi f C$
3. When frequency or capacitance goes up, there is more charge and discharge current; so X_C is lower.
4. $X_C = 1/2\pi f C = 1/2\pi \times 4$ kHz $\times 4\ \mu F = 9.95\ \Omega$

STEP 10-2
1. Current leads voltage by some phase angle less than 90%.
2. An arrangement of vectors to illustrate the magnitude and phase relationships between two or more quantities of the same frequency
3. Z = total opposition to current flow; $Z = \sqrt{R^2 + X_C^2}$
4. **a.** 0°
   b. 90°
   c. Between 0 and 90°

STEP 10-3
1. Resistor current is in phase with voltage; capacitor current is 90° out of phase (leading) with voltage.
2. No
3. **a.** $I_T = \sqrt{I_R^2 + I_C^2}$; **b.** $Z = V_S/I_T$;
   also $Z = (R \times X_C)/\sqrt{R^2 + X_C^2}$
4. Lead

STEP 10-4

1. 0.5 μF
2. Capacitance meter or analyzer

STEP 10-5

1. **a.** Filter
   b. Voltage divider
   c. Differentiator
2. Capacitor, resistor
3. Long
4. Differentiator

STEP 11-1-1

1. False
2. Electron

STEP 11-1-2

1. Coil
2. True
3. DC
4. False
5. Used to determine the magnetic polarity; placing your left-hand fingers in the direction of the current, your thumb points in the direction of the north magnetic pole.
6. AC

STEP 11-1-3 and 11-1-4

1. Number of lines of force (or maxwells) in webers
2. Number of magnetic lines of flux per square meter
3. The magnetic pressure that produces the magnetic field
4. Magnetomotive force divided by length of coil
5. Opposition or resistance to the establishment of a magnetic field
6. A measure of how easily a material will allow a magnetic field to be set up within it
7. True
8. The state beyond which an electromagnet is incapable of further magnetic strength

STEP 11-1-5

1. Magnetic-type circuit breaker, relays
2. An NO relay is open when off; NC is closed when off.
3. A reed relay has an electromagnet around it, while a reed switch needs an external magnet to operate.
4. Home security circuits

STEP 11-2

1. The voltage induced or produced in a coil as the magnetic lines of force link with the turns of a coil
2. Faraday's: When the magnetic flux linking a coil is changing, an emf is induced.
   Lenz's: The current induced in a coil due to the change in the magnetic flux is such as to oppose the cause producing it.
3. Sine wave
4. Because it converts sound waves of air pressure (acoustical) to electrical waves (electro)

STEP 12-1

1. When the current-carrying coil of a conductor induces a voltage within itself
2. The induced voltage, which opposes the applied emf; $V_{ind} = L \times (\Delta i/\Delta t)$
3. $V_{ind} = L \times (\Delta i/\Delta t) = 2 \text{ mH} \times 4\text{kA/s} = 8 \text{ V}$

STEP 12-2

1. Different applications: An electromagnet is used to generate a magnetic field; an inductor is used to oppose any changes of circuit current.
2. True

STEP 12-3

1. Number of turns, area of coil, length of coil, core material used
2. $L = \dfrac{N^2 \times A \times \mu}{l}$

STEP 12-4

1. False
2. **a.** $L_T = L_1 + L_2 + L_3 + \cdots$
   b. $L_T = 1/(1/L_1) + (1/L_2) + (1/L_3) + \cdots$
3. **a.** $L_T = L_1 + L_2 = 4 \text{ mH} + 2 \text{ mH} = 6 \text{ mH}$
   b. $L_T = \dfrac{L_1 \times L_2}{L_1 + L_2}$ (using product over sum)
   $= \dfrac{4 \text{ mH} \times 2 \text{ mH}}{4 \text{ mH} + 2 \text{ mH}} = \dfrac{8 \text{ mH}}{6 \text{ mH}} = 1.33 \text{ mH}$

STEP 12-5

1. Air core, iron core, ferrite core
2. Air core
3. Chemical compound, basically powdered iron oxide and ceramic
4. Toroidal-type inductors have a greater inductance.
5. Ferrite-core variable inductor
6. Core permeability

STEP 12-6

1. Current in an inductive circuit builds up in the same way that voltage does in a capacitive circuit, but the capacitive time constant is proportional to resistance, where the inductive time constant is inversely proportional to resistance.
2. True
3. False
4. It will continuously oppose the alternating current.

STEP 12-7

1. The opposition to current flow offered by an inductor without the dissipation of energy: $X_L = 2 \times \pi \times f \times L$
2. The higher the frequency (and therefore the faster the change in current) or the larger the inductance of the inductor, the larger the magnetic field created, the larger the counter emf will be to oppose applied emf; so the inductive reactance (opposition) will be large also.
3. Inductive reactance is an opposition, so it can be used in Ohm's law in place of resistance.
4. False

STEP 12-8

1. False
2. $V_S = \sqrt{V_R^2 + V_L^2} = \sqrt{4^2 + 2^2} = \sqrt{16 + 4} = \sqrt{20} = 4.47 \text{ V}$
3. The total opposition to current flow offered by a circuit with both resistance and reactance: $Z = \sqrt{R^2 + X_L^2}$
4. $+45°$
5. Power consumption above the zero line, caused by positive current and voltage or negative current and voltage
6. Quality factor of an inductor that is the ratio of the energy stored in the coil by its inductance to the energy dissipated in the coil by the resistance: $Q = X_L/R$

7. True power is energy dissipated and lost by resistance; reactive power is energy consumed and then returned by a reactive device.

8. PF = true power (P_R)/apparent power (P_A) or PF = R/Z or PF = $\cos \theta$

STEP 12-9
1. False

2. $I_T = \sqrt{I_R{}^2 + I_L{}^2}$

STEP 12-10
1. a. With an ohmmeter; resistance will be infinite instead of low.

b. With an inductor analyzer to check its inductance value

2. An open

STEP 12-11
1. *RL* integrator, *RL* differentiator, *RL* filter

2. Can be made to function as either a high- or a low-pass filter

3. The integrator will approximate a dc level; the differentiator will output spikes of current.

STEP 13-1
1. Mutual inductance is the process by which an inductor induces a voltage in another inductor, whereas self-inductance is the process by which a coil induces a voltage within itself.

2. False

STEP 13-2
1. True

2. True

3. Primary and secondary

STEP 13-3
1. True

2. False

STEP 13-4
1. It is the ratio of the number of magnetic lines of force that cut the secondary compared to the total number of magnetic flux lines being produced by the primary: *k* = flux linking secondary coil/total flux produced by primary.

2. True

3. How close together the primary and secondary are to one another and the type of core material used

STEP 13-5
1. Turns ratio = N_s/N_p = 1608/402 = 4; step up

2. $V_s = N_s/N_p \times V_p$

3. False

4. $I_s = N_p/N_s \times I_p$

5. Turns ratio = $\sqrt{Z_L/Z_S} = \sqrt{75\ \Omega/25\ \Omega} = \sqrt{3} = 1.732$

6. $V_s = (N_s/N_p) \times V_p = (200/112) \times 115 = 205.4$ V

STEP 13-6
1. If both primary and secondary are wound in the same direction, output is in phase with input; but if primary and secondary are wound in different directions, the output will be 180° out of phase with the input.

2. When there is a positive on the primary dot, there will be a positive on the secondary dot. If there is a negative on the primary dot, there will be a negative on the secondary dot.

STEP 13-7
1. Air core, iron or ferrite core

2. Center-tapped secondary, multiple-tapped secondary, multiple winding, single winding

3. False

4. To be able to switch between two primary voltages and obtain the same secondary voltage

5. Automobile ignition system

6. Smaller, cheaper, lighter than normal separated primary/secondary transformer types

STEP 13-8
1. 10 kVA is the apparent power rating, 200 V the maximum primary voltage, 100 V the maximum secondary voltage, at 60 cycles per second (Hz)

2. $R_L = V_s/I_s$ = 1 kV/8 A = 125 Ω; 125 > 100, so the transformer will overheat and possibly burn out

STEP 13-9
1. An ohmmeter

2. Probably not since 50 Ω is a normal coil resistance

STEP 13-10
1. Copper losses, hysteresis, eddy-current loss, magnetic leakage

2. Eddy currents cannot cross the laminated parts of the core, so reduced eddy currents mean reduced opposition to main flux and reduced heat, resulting in a reduced loss.

STEP 14-1
1. Calculate the inductive and capacitive reactance (X_L and X_C), the circuit impedance (Z), the circuit current (I), the component voltage drops (V_R, V_L, and V_C), and the power distribution and power factor (PF).

2. a. $Z = \sqrt{R^2 + (X_L \sim X_C)^2}$

b. $I = V_s/Z$

c. Apparent power = $V_s \times I$ (volt-amperes)

d. $V_S = \sqrt{V_R{}^2 + (V_L \sim V_C)^2}$

e. True power = $I^2 \times R$ (watts)

f. $V_R = I \times R$

g. $V_L = I \times X_L$

h. $V_C = I \times X_C$

i. $\theta = \arctan \dfrac{V_L \sim V_C}{V_R}$

j. PF = $\cos \theta$

STEP 14-2
1. a. $I_R = V/R$ **c.** $I_C = V/X_C$

b. $I_T = \sqrt{I_R{}^2 + I_X{}^2}$ **d.** $I_L = V/X_L$

2. a. $P_R = I^2 \times R$ **c.** $P_A = V_s \times I_T$

b. $P_X = I^2 \times X_L$ **d.** PF = $\cos \theta$

STEP 14-3-1
1. A circuit condition that occurs when the inductive reactance (X_L) and the capacitive reactance (X_C) have been balanced

2. A series *RLC* circuit that at resonance X_L equals X_C, so V_L and V_C will cancel, and $Z = R$

3. Voltage across *L* and *C* will measure 0; impedance only equals *R*; voltage drops across inductor or capacitor can be higher than source voltage

4. *Q* factor indicates the quality of the series resonant circuit, or is the ratio of the reactance to the resistance

5. Group or band of frequencies that causes the larger current flow

6. BW = f_0/Q = 12 kHz/1000 = 12 Hz

STEP 14-3-2

1. In a series resonant *RLC* circuit, source current is maximum; in a parallel resonant circuit, source current is minimum at resonance
2. Oscillating effect with continual energy transfer between capacitor and inductor
3. $Q = X_L/R = 50\ \Omega/25\ \Omega = 2$
4. Yes
5. Ability of a tuned circuit to respond to a desired frequency and ignore all others

STEP 14-4

1. **a.** High pass **c.** Band stop
   b. Low pass **d.** Band pass
2. Television, radio, and other communications equipment

STEP 14-5

1. Real, imaginary
2. Magnitude $= \sqrt{5^2 + 6^2} = 7.81$
   Angle $= \arctan(6/5) = 50.2°$
   Rectangular number $65 + j6 =$ polar number $7.81 \angle 50.2°$
3. Real number $= 33 \cos 25° = 29.9$
   Imaginary number $= 33 \sin 25° = 13.9$
   Polar number $33 \angle 25° =$ rectangular number $29.9 + j13.9$
4. Combination of a real and imaginary number

STEP 15-1

1. Diodes, transistors and integrated circuits (ICs)
2. Current or voltage

STEP 15-2

1. They all have 4 valence electrons.
2. 1, 4
3. Covalent bond
4. Negative, decreases
5. Increases
6. Intrinsic, positive, negative

STEP 15-3

1. To increase their conductivity
2. Electrons, N
3. Holes, P
4. Electrons, holes

STEP 15-4

1. Depletion region
2. (a) 700 mV
3. True
4. Open, closed

STEP 16-1

1. 0.7 V
2. 0.7 V
3. One

STEP 16-2

1. True
2. Reverse
3. The cathode bar is shaped like a "z"

STEP 16-3

1. (a)
2. Series current limiting resistor

STEP 16-4

1. True
2. True

STEP 16-5

1. Transformer, rectifier, filter, regulator
2. **a.** Section 16-5-3
   b. Section 16-5-5
3. Section 16-5-4
4. Diagnose, Isolate, Repair

STEP 17-1

1. NPN and PNP
2. Emitter, base, and collector
3. As a switch, and as a variable-resistor
4. The two-state switching action is used in digital circuits, while the variable-resistor action is used in analog circuits.

STEP 17-2

1. Current
2. Forward, reverse
3. (b)
4. Open switch
5. Closed switch
6. **a.** Common-base
   b. Common-collector
   c. Common-emitter
7. Voltage-divider bias
8. Base biasing

STEP 18-1

1. Current, voltage
2. Reverse
3. True
4. False
5. Transconductance
6. Very high, reverse
7. R_s
8. Voltage divider
9. Common-source
10. Common-source
11. Common-drain
12. Its high input impedance
13. Because it responds well to the small signal inputs from the antenna, and because it is a low noise component
14. **a.** 0.3 V dc (min), 3.0 V dc (max)
    b. 5 mA dc
15. Yes P+, N−, = LO Ω
    P−, N+, = HI Ω

STEP 18-2-1

1. Depletion, enhancement
2. True
3. ON
4. MOSFET
5. (d)
6. Dual-gate D-MOSFET

STEP 18-2-2

1. False
2. OFF
3. (c) Drain feedback biased
4. True
5. $I_{D(ON)} = 3$ mA, and $V_{DS(ON)} = 1.0$ V
6. **a.** Turn ON delay $= 45$ ns
   b. Turn OFF delay $= 60$ ns
7. **a.** Ensure all MOSFET device pins are kept at same voltage level.
   b. Use a wrist grounding strap.
   c. All test equipment, soldering irons, and work benches should be properly grounded.

d. All power in equipment should be off before MOS devices are removed or inserted into printed circuit boards.

e. Any unused MOSFET terminals must be connected.

8. Infinite ohms

STEP 19-1

1. Integrated circuits

2. $-$, $+$, $+V$, $-V$, output

3. TO5 can, DIP package

4. The prefix indicates manufacturer, the following digit code indicates the op-amp type, the letter after the part number indicates the operating temperature range, and the final suffix code indicates the package type.

STEP 19-2

1. Yes

2. An open-loop op-amp circuit has no feedback, whereas a closed-loop op-amp circuit does have a feedback path.

3. High gain, very high input impedance, and very low output impedance

4. Differential amplifier, darlington-pair voltage amplifier, emitter-follower output amplifier

5. (a) Comparator

6. Negative or degenerative feedback will lower the op-amp's gain to

    a. prevent output waveform distortion,

    b. prevent the amplifier from going into oscillation, and

    c. reduce the gain of the op-amp to a consistent value.

STEP 19-3

1. Voltage follower

2. Summing amplifier

3. The differential circuit has an input capacitor and a feedback resistor, while the integrator circuit has an input resistor and a feedback capacitor.

4. Differential or differential amplifier

5. See Figure 19-15(b).

6. An active filter circuit will filter and amplify the signal input, while a passive filter will only filter the signal input.

STEP 20-1

1. Free-running

2. Mono-stable

3. Bistable

4. Because it can be latched in the set or reset condition

5. The three $5k\Omega$ voltage divider

6. Astable, monostable

STEP 20-2

1. Decrease

2. False

3. Bidirectional

4. DIAC

5. True

6. One

7. False

8. Gate

STEP 20-3

1. Resistor

2. Electrical, solar

3. Photoconductive and photovoltaic

4. Decrease

5. Electrical to light, electrical to light

6. True

7. False

8. PTC

Answers to Odd-Numbered Problems

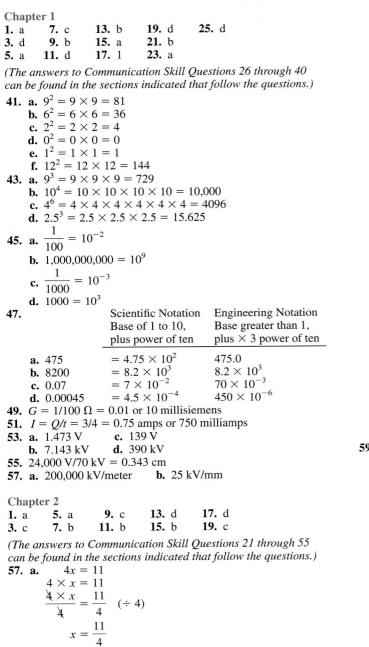

Chapter 1

1. a   **7.** c   **13.** b   **19.** d   **25.** d
3. d   **9.** b   **15.** a   **21.** b
5. a   **11.** d   **17.** 1   **23.** a

(The answers to Communication Skill Questions 26 through 40 can be found in the sections indicated that follow the questions.)

41. a. $9^2 = 9 \times 9 = 81$
   b. $6^2 = 6 \times 6 = 36$
   c. $2^2 = 2 \times 2 = 4$
   d. $0^2 = 0 \times 0 = 0$
   e. $1^2 = 1 \times 1 = 1$
   f. $12^2 = 12 \times 12 = 144$

43. a. $9^3 = 9 \times 9 \times 9 = 729$
   b. $10^4 = 10 \times 10 \times 10 \times 10 = 10{,}000$
   c. $4^6 = 4 \times 4 \times 4 \times 4 \times 4 \times 4 = 4096$
   d. $2.5^3 = 2.5 \times 2.5 \times 2.5 = 15.625$

45. a. $\dfrac{1}{100} = 10^{-2}$
   b. $1{,}000{,}000{,}000 = 10^9$
   c. $\dfrac{1}{1000} = 10^{-3}$
   d. $1000 = 10^3$

47.

	Scientific Notation Base of 1 to 10, plus power of ten	Engineering Notation Base greater than 1, plus $\times$ 3 power of ten
a. 475	$= 4.75 \times 10^2$	475.0
b. 8200	$= 8.2 \times 10^3$	8.2×10^3
c. 0.07	$= 7 \times 10^{-2}$	70×10^{-3}
d. 0.00045	$= 4.5 \times 10^{-4}$	450×10^{-6}

49. $G = 1/100\ \Omega = 0.01$ or 10 millisiemens
51. $I = Q/t = 3/4 = 0.75$ amps or 750 milliamps
53. a. 1.473 V   **c.** 139 V
   b. 7.143 kV   **d.** 390 kV
55. 24,000 V/70 kV = 0.343 cm
57. a. 200,000 kV/meter   **b.** 25 kV/mm

Chapter 2

1. a   **5.** a   **9.** c   **13.** d   **17.** d
3. c   **7.** b   **11.** b   **15.** b   **19.** c

(The answers to Communication Skill Questions 21 through 55 can be found in the sections indicated that follow the questions.)

57. a.
$$4x = 11$$
$$4 \times x = 11$$
$$\frac{\cancel{4} \times x}{\cancel{4}} = \frac{11}{4} \quad (\div 4)$$
$$x = \frac{11}{4}$$
$$x = 2.75$$

b.
$$6a + 4a = 70$$
$$10a = 70$$
$$10 \times a = 70$$
$$\frac{\cancel{10} \times a}{\cancel{10}} = \frac{70}{10} \quad (\div 10)$$
$$a = \frac{70}{10}$$
$$a = 7$$

c.
$$5b - 4b = \frac{7.5}{1.25}$$
$$1b = \frac{7.5}{1.25} \quad (1b = 1 \times b = b)$$
$$b = \frac{7.5}{1.25}$$
$$b = 6$$

d.
$$\frac{2z \times 3z}{4.5} = 2z \qquad \begin{bmatrix} 2z \times 3z = 2 \times z \times 3 \times z \\ = (2 \times 3) \times (z \times z) \\ = 6 \times z^2 \\ = 6z^2 \end{bmatrix}$$
$$\frac{6z^2}{4.5} = 2z$$
$$\frac{6z^2}{\cancel{4.5}} \times \cancel{4.5} = 2z \times 4.5 \quad (\times 4.5)$$
$$6z^2 = 2z \times 4.5$$
$$\frac{6z \times \cancel{z} \times z}{\cancel{2} \times \cancel{z}} = \frac{(2 \times z) \times 4.5}{2 \times z} \quad (\div 2z)$$
$$3z = 4.5$$
$$\frac{\cancel{3} \times z}{\cancel{3}} = \frac{4.5}{3} \quad (\div 3)$$
$$z = \frac{4.5}{3}$$
$$z = 1.5$$

59. a. Power (P) = voltage (V) $\times$ current (I)
    1500 watts = 120 volts $\times$?
    Transpose formula to solve for I.
$$P = V \times I$$
$$\frac{P}{V} = \frac{\cancel{V} \times I}{\cancel{V}} \quad (\div V)$$
$$\frac{P}{V} = I$$
$$I = \frac{P}{V}$$

Current (I) in amperes $= \dfrac{\text{power }(P)\text{ in watts}}{\text{voltage }(V)\text{ in volts}}$
$$I = \frac{1500\ \text{W}}{120\ \text{V}}$$
Current $(I) = 12.5$ amperes

b. Voltage (V) in volts = current (I) in amperes ×
resistance (R) in ohms

120 volts = 12.5 amperes × ?

Transpose formula to solve for R.

$$V = I \times R$$

$$\frac{V}{I} = \frac{\cancel{I} \times R}{\cancel{I}}$$

$$\frac{V}{I} = R$$

$$R = \frac{V}{I}$$

Resistance (R) in ohms = $\dfrac{\text{voltage } (V) \text{ in volts}}{\text{current } (I) \text{ in amperes}}$

$$R = \frac{120 \text{ V}}{12.5 \text{ A}}$$

Resistance (R) = 9.6 ohms

61. $V = I \times R = 8$ mA × 16 kΩ = 128 V

63. $P = I \times V$, 2.4 kW, 1.024 W, 1.2 kW

65. Total cost = power in (kW) × time × cost per hour
  a. 0.3 × 10 × 9 = 27 cents   **c.** 0.06 × 10 × 9 = 5 cents
  b. 0.1 × 10 × 9 = 9 cents   **d.** 0.025 × 10 × 9 = 2 cents

67. $R = \rho \times l/a = 10.7 \times 200/80^2 = 334$ mΩ

69. $V = I \times R = 7.5$ A × 0.2485 Ω = 1.86375 V

71. $W = Q \times V$, $V = W/Q = 1000$ J/40 C = 25 V

73. $R = V/I$
  a. 120 V/20 mA = 6 kΩ
  b. 12 V/2 A = 6 Ω
  c. 9 V/100 μA = 90 kΩ
  d. 1.5 V/4 mA = 375 Ω; so **c.** has the largest, **b.** the smallest

75. $P = W/t = 5000$ J/25 s = 200 W

77. $P = I^2 \times R$, $R = P/I^2 = 100$ W/4 A² = 6.25 Ω

79. $P = V \times I = 12$ V × 300 mA = 3.6 W

Chapter 3

1. b   **7.** 1   **13.** a   **17.** c   **21.** c
3. b   **9.** c   **15.** c   **19.** a   **23.** b
5. d   **11.** b; $P = I^2R$ **25.** d
            $= 20$ mA² × 2 kΩ
            $= 0.8$ W

*(The answers to Communication Skill Questions 26 through 45
can be found in the sections indicated that follow the questions.)*

47. a. 48   **c.** 0.93
  b. 156.4

49. a. 11.5
  b. 7.8
  c. 0.045
  d. 660
  e. 500

51. a. Rock
  b. Rap
  c. Country
  d. Rap = 7%
     Pop = 10%
     Urban Contemporary = 12%
     Country = 17%
     Rock = 34%

53. 5.6 kΩ × 0.1 = 560 Ω, 5.6 kΩ ± 560 kΩ = 5.04 kΩ to
6.16 kΩ

55. $P = V^2/R = 12$ V²/12 Ω = 144/12 = 12 W

57.

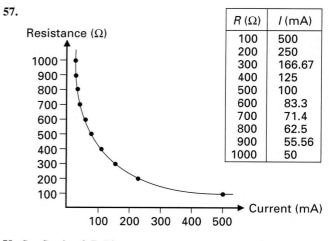

R (Ω)	I (mA)
100	500
200	250
300	166.67
400	125
500	100
600	83.3
700	71.4
800	62.5
900	55.56
1000	50

59. See Section 3-7, #4.

61. It will not burn up.

Chapter 4

1. b   **7.** d   **13.** b
3. d   **9.** a   **15.** b
5. c   **11.** c

*(The answers to Communication Skill Questions 16 through 25
can be found in the sections indicated that follow the questions.)*

27.

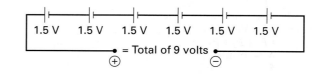

(a)

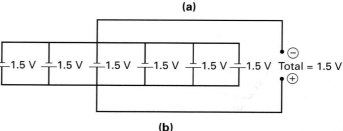

(b)

FIGURE Ch. 5.27

29.

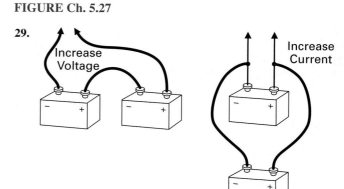

FIGURE Ch. 5.29

Chapter 5

1. d   **5.** a   **9.** a   **13.** d
3. c   **7.** d   **11.** c   **15.** d

(The answers to Communication Skill Questions 16 through 25 can be found in the sections indicated that follow the questions.)

27. $I = \dfrac{V_S}{R_T}$, $R_T = R_1 + R_2 = 40 + 35 = 75 \ \Omega$

$I = 24/75 = 320 \ \text{mA}$; $150 \ \Omega$ (double $75 \ \Omega$) needed to halve current

29. $40 \ \Omega$, $20 \ \Omega$, $60 \ \Omega$. (Any values can be used, as long as the ratio remains the same.)

31. $I_{R_1} = I_T = 6.5 \ \text{mA}$

33. $P_T = P_1 + P_2 + P_3 = 120 + 60 + 200 = 380 \ \text{W}$; $I_T = P_T/V_S = 380/120 = 3.17 \ \text{A}$

$V_1 = P_1/I_T = 120 \ \text{W}/3.17 \ \text{A} = 38 \ \text{V}$; $V_2 = P_2/I_T = 60 \ \text{W}/3.17 \ \text{A} = 18.9 \ \text{V}$

$V_3 = P_3/I_T = 200 \ \text{W}/3.17 \ \text{A} = 63.1 \ \text{V}$

35. a. $R_T = R_1 + R_2 + R_3 = 22 \ \text{k}\Omega + 3.7 \ \text{k}\Omega + 18 \ \text{k}\Omega = 43.7 \ \text{k}\Omega$

     $I = V/R = 12 \ \text{V}/43.7 \ \text{k}\Omega = 274.6 \ \mu\text{A}$

   b. $R_T = V/I = 12 \ \text{V}/10 \ \text{mA} = 1.2 \ \text{k}\Omega$

     $P_T = V \times I = 12 \ \text{V} \times 10 \ \text{mA} = 120 \ \text{mW}$

   c. $R_T = R_1 + R_2 + R_3 + R_4 = 5 + 10 + 6 + 4 = 25 \ \Omega$; $V_S = I \times R_T = 100 \ \text{mA} \times 25 \ \Omega = 2.5 \ \text{V}$

     $V_{R_1} = I \times R_1 = 100 \ \text{mA} \times 5 = 500 \ \text{mV}$, $V_{R_2} = I \times R_2 = 100 \ \text{mA} \times 10 = 1 \ \text{V}$, $V_{R_3} = I \times R_3 = 100 \ \text{mA} \times 6 = 600 \ \text{mV}$

     $V_{R_4} = I \times R_4 = 100 \ \text{mA} \times 4 = 400 \ \text{mV}$, $P_1 = I \times V_1 = 100 \ \text{mA} \times 500 \ \text{mV} = 50 \ \text{mW}$

     $P_2 = I \times V_2 = 100 \ \text{mA} \times 1 \ \text{V} = 100 \ \text{mW}$, $P_3 = I \times V_3 = 100 \ \text{mA} \times 600 \ \text{mV} = 60 \ \text{mW}$

     $P_4 = I \times V_4 = 100 \ \text{mA} \times 400 \ \text{mV} = 40 \ \text{mW}$

   d. $P_T = P_1 + P_2 + P_3 + P_4 = 12 \ \text{mW} + 7 \ \text{mW} + 16 \ \text{mW} + 3 \ \text{mW} = 38 \ \text{mW}$

     $I = P_T/V_S = 38 \ \text{mW}/12.5 \ \text{V} = 3.04 \ \text{mA}$

     $R_1 = P_1/I^2 = 12 \ \text{mW}/(3.04 \ \text{mA})^2 = 1.3 \ \text{k}\Omega$

     $R_2 = P_2/I^2 = 7 \ \text{mW}/(3.04 \ \text{mA})^2 = 757.4 \ \Omega$

     $R_3 = P_3/I^2 = 16 \ \text{mW}/(3.04 \ \text{mA})^2 = 1.73 \ \text{k}\Omega$

     $R_4 = P_4/I^2 = 3 \ \text{mW}/(3.04 \ \text{mA})^2 = 324.6 \ \Omega$

37. Zero voltage drop across the shorted component, while there is also an increase in voltage across the others.

39. a. No current at all (zero)

   b. Go to infinity (∞)

   c. Measure source voltage

   d. No voltage across any other component

Chapter 6

1. b   **5.** c   **9.** a
3. d   **7.** b

(The answers to Communication Skill Questions 11 through 20 can be found in the sections indicated that follow the questions.)

21. $R_T = R/\text{no. of } R\text{'s} = 30 \ \text{k}\Omega/4 = 7.5 \ \text{k}\Omega$

23. $R_T = R/\text{no. of } R\text{'s} = 25/3 = 8.33 \ \Omega$, $I_T = V_S R_T = 10/8.33 = 1.2 \ \text{A}$

$I_1 = I_2 = I_3 = R_T/R_x \times I_T = 8.33/25 \times 1.2 = I_T/\text{no. of } R\text{'s} = 1.2 \ \text{A}/3 = 400 \ \text{mA}$

25. $I_T = V_S/R_T = 14/700 = 20 \ \text{mA}$; $I_X = I_T/\text{no. of } R = 20 \ \text{mA}/3 = 6.67 \ \text{mA}$

27. a. $R_T = \dfrac{R_1 \times R_2}{R_1 + R_2} = \dfrac{33 \times 22}{33 + 22} = \dfrac{726 \ \text{k}\Omega}{55 \ \text{k}\Omega} = 13.2 \ \text{k}\Omega$

   b. $I_T = V_S/R_T = 20/13.2 \ \text{k}\Omega = 1.5 \ \text{mA}$

   c. $I_1 = R_T/R_1 \times I_T = (13.2 \ \text{k}\Omega/33 \ \text{k}\Omega) \times 1.5 \ \text{mA} = 600 \ \mu\text{A}$

     $I_2 = R_T/R_2 \times I_T = 13.2 \ \text{k}\Omega/22 \ \text{k}\Omega \times 1.5 \ \text{mA} = 900 \ \mu\text{A}$

   d. $P_T = I_T \times V_S = 1.5 \ \text{mA} \times 20 = 30 \ \text{mW}$

   e. $P_1 = I_1 \times V_1$, $(V_1 = V_S = 20 \ \text{V})$, $600 \ \mu\text{A} \times 20 = 12 \ \text{mW}$

     $P_2 = I_2 \times V_2$, $(V_2 = V_S = 20 \ \text{V})$, $900 \ \mu\text{A} \times 20 = 18 \ \text{mW}$

29. a. $R_T = \dfrac{R_1 \times R_2}{R_1 + R_2} = \dfrac{22 \ \text{k}\Omega \times 33 \ \text{k}\Omega}{22 \ \text{k}\Omega + 33 \ \text{k}\Omega} = \dfrac{726 \ \text{k}\Omega}{55 \ \text{k}\Omega} = 13.2 \ \text{k}\Omega$

     $I_T = \dfrac{V_S}{R_T} = \dfrac{10}{13.2 \ \text{k}\Omega} = 757.6 \ \mu\text{A}$

     $I_1 = \dfrac{R_T}{R_1} \times I_T = \dfrac{13.2 \ \text{k}\Omega}{22 \ \text{k}\Omega} \times 757.6 \ \mu\text{A} = 454.56 \ \mu\text{A}$

     $I_2 = \dfrac{R_T}{R_1} \times I_T = \dfrac{13.2 \ \text{k}\Omega}{33 \ \text{k}\Omega} \times 757.6 \ \mu\text{A} = 303.04 \ \mu\text{A}$

   b. $R_T = \dfrac{1}{(1/R_1) + (1/R_2) + (1/R_3)}$

     $= \dfrac{1}{(1/220 \ \Omega) + (1/330 \ \Omega) + (1/470 \ \Omega)} = 103 \ \Omega$

     $I_T = V_S/R_T = 10/103 = 97 \ \text{mA}$, $I_1 = R_T/R_1 \times I_T = 103/220 \times 97 \ \text{mA} = 45.4 \ \text{mA}$

     $I_2 = (R_T/R_2) \times I_T = (103/330) \times 97 \ \text{mA} = 30.3 \ \text{mA}$

     $I_3 = (R_T/R_2) \times I_T = (103/470) \times 97 \ \text{mA} = 21.3 \ \text{mA}$

31. a. $G_T = \dfrac{1}{R_1} + \dfrac{1}{R_2} + \dfrac{1}{R_3} = \dfrac{1}{5} + \dfrac{1}{5} + \dfrac{1}{5} = 0.6 \ \text{S}$,

     $R_T = \dfrac{1}{G} = \dfrac{1}{0.6} = 1.67 \ \Omega$

   b. $G_T = \dfrac{1}{R_1} + \dfrac{1}{R_2} = \dfrac{1}{200} + \dfrac{1}{200} = 10 \ \text{mS}$,

     $R_T = \dfrac{1}{G} = \dfrac{1}{10 \ \text{mS}} = 100 \ \Omega$

   c. $G_T = \dfrac{1}{R_1} + \dfrac{1}{R_2} + \dfrac{1}{R_3} = \dfrac{1}{1 \ \text{M}\Omega} + \dfrac{1}{500 \ \text{M}\Omega} + \dfrac{1}{3.3 \ \text{M}\Omega}$

     $= 1.305 \ \mu\text{S}$, $R_T = \dfrac{1}{G} = \dfrac{1}{1.305 \ \mu\text{S}} = 766.3 \ \text{k}\Omega$

   d. $G_T = \dfrac{1}{R_1} + \dfrac{1}{R_2} + \dfrac{1}{R_3} = \dfrac{1}{5} + \dfrac{1}{3} + \dfrac{1}{2} = 1.033 \ \text{S}$,

     $R_T = \dfrac{1}{G} = \dfrac{1}{1.033} = 967.7 \ \text{m}\Omega$

33. a. $R_T = \dfrac{R_1 \times R_2}{R_1 + R_2} = \dfrac{15 \times 7}{15 + 7} = \dfrac{105}{22} = 4.77 \ \Omega$

   b. $R_T = \dfrac{1}{(1/R_1) + (1/R_2) + (1/R_3)}$

     $= \dfrac{1}{(1/26 \ \Omega) + (1/15 \ \Omega) + (1/30 \ \Omega)} = 7.22 \ \Omega$

   c. $R_T = \dfrac{R_1 \times R_2}{R_1 + R_2} = \dfrac{5.6 \ \text{k}\Omega \times 2.2 \ \text{k}\Omega}{5.6 \ \text{k}\Omega + 2.2 \ \text{k}\Omega} = \dfrac{12.32 \ \text{M}\Omega}{7.8 \ \text{k}\Omega}$

     $= 1.58 \ \Omega$

   d. $R_T = \dfrac{1}{(1/R_1) + (1/R_2) + (1/R_3) + (1/R_4) + (1/R_5)} =$

$\dfrac{1}{(1/1 \ \text{M}\Omega) + (1/3 \ \text{M}\Omega) + (1/4.7 \ \text{M}\Omega) + (1/10 \ \text{M}\Omega) + (1/33 \ \text{M}\Omega)}$

     $= 596.5 \ \text{k}\Omega$

35. a. $I_2 = I_T - I_1 - I_3 = 6 \ \text{mA} - 2 \ \text{mA} - 3.7 \ \text{mA} = 300 \ \mu\text{A}$

   b. $I_T = I_1 + I_2 + I_3 = 6 \ \text{A} + 4 \ \text{A} + 3 \ \text{A} = 13 \ \text{A}$

   c. $R_T = \dfrac{R_1 \times R_2}{R_1 + R_2} = \dfrac{5.6 \ \text{M} \times 3.3 \ \text{M}}{5.6 \ \text{M} + 3.3 \ \text{M}} = \dfrac{18.48 \ \text{M}\Omega}{8.9 \ \text{M}\Omega}$

     $= 2.08 \ \text{M}\Omega$

     $V_S = I_T \times R_T = 100 \ \text{mA} \times 2.08 \ \text{M}\Omega = 208 \ \text{kV}$

     $I_1 = \dfrac{R_T}{R_1} \times I_T = \dfrac{2.08 \ \text{M}\Omega}{5.6 \ \text{M}\Omega} \times 100 \ \text{mA} = 37 \ \text{mA}$

     $I_2 = \dfrac{R_T}{R_2} \times I_T = \dfrac{2.08 \ \text{m}\Omega}{3.3 \ \text{m}\Omega} \times 100 \text{mA} = 63 \ \text{mA}$

   d. $I_1 = V_{R_1}/R_1 = 2/200 \ \text{k}\Omega = 10 \ \mu\text{A}$, $I_2 = I_T - I_1 = 100 \ \text{mA} - 10 \ \mu\text{A} = 99.99 \ \text{mA}$

     $R_2 = \dfrac{V_{R_2}}{I_2} = \dfrac{2}{99.99 \ \text{mA}} = 20.002 \ \Omega$,

     $P_T = I_T \times V_S = 100 \ \text{mA} \times 2 \ \text{V} = 200 \ \text{mW}$

37. a

39. Total current would increase, and the branch current with the shorted resistor would increase to 20 V/1 Ω = 20 A.

Chapter 7

1. c   **7.** a   **13.** a   **17.** c
3. b   **9.** c   **15.** a   **19.** c
5. c   **11.** b

(The answers to Communication Skill Questions 21 through 35 can be found in the sections indicated that follow the questions.)

37. $V_A = V_S = 100$ V, $V_B = V_S - V_R = 100 - 11.125 = 88.875$ V
$V_C = V_S - V_{R_1} - V_{R_2} = 44.3$, $V_D = I_{R_4} \times R_4 = 4.43$ mA $\times$ 2.5 kΩ = 11.075 V
$V_E = 0$ V

39. a. $V_{RL} = \dfrac{R_L}{R_T} \times V_S = \dfrac{R_L}{R_L + R_{int}} \times V_S = \dfrac{25}{25 + 15} \times 15$
$= 9.375$

   b. $V_{RL} = \dfrac{R_L}{R_T} \times V_S = \dfrac{R_L}{R_L + R_{int}} \times V_S = \dfrac{2.5 \text{ k}\Omega}{2.5 \text{ k}\Omega + 15} \times 15$
$= 14.91$ V

   c. $V_{RL} = \dfrac{R_L}{R_T} \times V_S = \dfrac{R_L}{R_L + R_{int}} \times V_S = \dfrac{2.5 \text{ M}\Omega}{2.5 \text{ M}\Omega + 15} \times 15$
$= 14.99991$ V

41. a. For $V_1: R_T = R_1 + R_{2,3}$, $R_{2,3} = \dfrac{2 \text{ k}\Omega \times 6 \text{ k}\Omega}{2 \text{k}\Omega + 6 \text{ k}\Omega} = \dfrac{12 \text{M}\Omega}{8 \text{ k}\Omega}$
$= 1.5$ kΩ, $R_T = 8$ kΩ $+ 1.5$ kΩ $= 9.5$ kΩ, $I_T = V_S/R_T$
$= \dfrac{28}{9.5 \text{ k}\Omega} = 2.95$ mA, $I_{R_2} = R_{2,3}/R_2 \times I_T$
$= \dfrac{1.5 \text{ k}\Omega}{2 \text{ k}\Omega} \times 2.95$ mA $= 2.21$ mA

For $V_2: R_T = R_3 + R_{1,2}$, $R_{1,2} = \dfrac{8 \text{ k}\Omega \times 2 \text{ k}\Omega}{8 \text{ k}\Omega + 2 \text{ k}\Omega} = \dfrac{16 \text{ M}\Omega}{10 \text{ k}\Omega}$
$= 1.6$ kΩ, $R_T = 6$ kΩ $+ 1.6$ kΩ $= 7.6$ kΩ, $I_T = V_S/R_T$
$= 20/7.6$ kΩ $= 2.63$ mA, $I_{R_2} = R_{1,2}/R_2 \times I_T$
$= 1.6$ kΩ/2 kΩ $\times 2.63$ mA $= 2.1$ mA, I_{R_2} total
$= 2.21$ mA $+ 2.1$ mA $= 4.31$ mA

   b. For $V_1: R_T = R_1 + R_{2,3} + R_4$, $R_{2,3} = \dfrac{R_2 \times R_3}{R_2 + R_3} = \dfrac{15 \times 75}{15 + 75}$
$= \dfrac{1125}{90} = 12.5$ Ω
$= R_T = 10 + 12.5 + 5 = 27.5$, $I_T = V_S/R_T = 3.5/27.5$
$= 127$ mA, $I_{R_3} = (R_{2,3}/R_3) \times I_T = (12.5/75) \times 127$ mA
$= 21$ mA
For $V_2: R_T = R_2 + \dfrac{1}{(1/R_{1,4}) + (1/R_3)}$, $R_{1,4} = R_1 + R_4 =$
$10 \, \Omega + 5 \, \Omega = 15 \, \Omega$, $R_T = 15 \, \Omega + \dfrac{1}{(1/15 \, \Omega) + (1/75 \, \Omega)} =$
$27.5 \, \Omega$, $I_T = V_S/R_T = 1.5$ V/27.5 Ω $= 54.5$ mA, $I_{R_3} =$
$(R_{1,3,4}/R_3) \times I_T = (12.5 \, \Omega/75 \, \Omega) \times 54.5$ mA $= 9$ mA, I_{R_3}
Total $= 21$ mA $- 9$ mA $= 12$ mA

43. $V_{TH} = 5$ V, $R_{TH} = 3$ kΩ
$I_{RL} = V_{TH}/R_T = 5/(3 \text{ k}\Omega + 1 \text{ k}\Omega) = 5/4$ kΩ $= 1.25$ mA
$I_N = I_{R_2} = V_S/R_2 = 5/3$ kΩ
$= 1.67$ mA, $R_N = 3$ kΩ

45. a. $V = I \times R = 5$ mA $\times 5$ MΩ $= 25$ kV
   b. $V = I \times R = 10$ A $\times 10$ kΩ $= 100$ kV
   c. $V = I \times R = 0.0001$ A $\times 2.5$ kΩ $= 250$ mV

47. This answer will vary with each person.

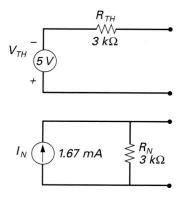

FIGURE Ch. 7, 43

49. $V_{TH} = \dfrac{R_4}{R_T} \times V_S$, $R_T = R_{1,2} + R_4$, $R_{1,2} = \dfrac{R_1 \times R_2}{R_1 + R_2} = \dfrac{2 \times 3}{2 + 3}$
$= \dfrac{6}{5} = 1.2 \, \Omega$

$R_T = 1.2 + 7 = 8.2 \, \Omega$, $V_{TH} = \dfrac{7}{8.2} \times 20 = 17.1$ V

$R_{TH} = \dfrac{1}{(1/R_1) + (1/R_2) + (1/R_4)} = \dfrac{1}{(1/2) + (1/3) + (1/7)}$
$= 1.024 \, \Omega$

   a. If R_2 shorts, $V_{TH} = 20$ V, $R_{TH} = 0 \, \Omega$

   b. If R_2 opens, $V_{TH} = \dfrac{R_4}{R_T} \times V_S$, $R_T = R_1 + R_4 = 2 + 7 = 9 \, \Omega$,

$V_{TH} = \dfrac{7}{9} \times 20 = 15.56$ V, $R_{TH} = \dfrac{R_1 \times R_4}{R_1 + R_4} = \dfrac{2 \times 7}{2 + 7} = \dfrac{14}{9}$
$= 1.56 \, \Omega$

Chapter 8

1. d   **7.** d   **13.** d   **19.** a   **25.** b
3. b   **9.** a   **15.** b   **21.** d   **27.** c
5. c   **11.** c   **17.** b   **23.** b   **29.** d

(The answers to Communication Skill Questions 31 through 50 can be found in the sections indicated that follow the questions.)

51. a. $C = \sqrt{A^2 + B^2}$
$= \sqrt{20^2 + 53^2}$
$= \sqrt{400 + 2809}$
$= \sqrt{3209}$
$= 56.65$ mi

   b. $C = \sqrt{2^2 + 3^2}$
$= 3.6$ km

   c. $C = \sqrt{4^2 + 3^2}$
$= 5$ inches

   d. $C = \sqrt{12^2 + 12^2}$
$= 16.97$ mm

53. a. $\sin 0° = 0$ **i.** $\cos 60° = 0.5$
   b. $\sin 30° = 0.5$ **j.** $\cos 90° = 0$
   c. $\sin 45° = 0.707$ **k.** $\tan 0° = 0$
   d. $\sin 60° = 0.866$ **l.** $\tan 30° = 0.577$
   e. $\sin 90° = 1.0$ **m.** $\tan 45° = 1.0$
   f. $\cos 0° = 1.0$ **n.** $\tan 60° = 1.73$
   h. $\cos 45° = 0.707$ **o.** $\tan 90° = \infty$ (infinity)

55. a.

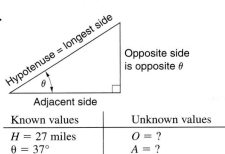

Known values	Unknown values
$H = 27$ miles	$O = ?$
$\theta = 37°$	$A = ?$

We must calculate the length of the opposite and adjacent sides. To achieve this we can use either H or θ to calculate O (*SOH*), or H and θ to calculate A (*CAH*).

$$\sin \theta = \frac{O}{H}$$

$$\sin 37° = \frac{O}{27}$$

$$0.6 = \frac{O}{27}$$

$$0.6 \times 27 = \frac{O}{27} \times 27$$

$$O = 0.6 \times 27$$

Opposite $= 16.2$ mi

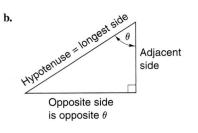

The next step is to use the Pythagorean theorem to calculate the length of the unknown side now that the length of two sides are known.

$$B = \sqrt{C^2 - A^2}$$
$$= \sqrt{27^2 - 16.2^2}$$
$$= \sqrt{729 - 262.44}$$
$$= \sqrt{466.56}$$

Adjacent or $B = 21.6$ mi

b.

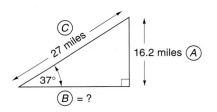

Known values	Unknown values
$O = 29$ cm	$H = ?$
$\theta = 21°$	$A = ?$

We can use either O and θ to calculate H (*SOH*), or O and θ to calculate A (*TOA*).

$$\tan \theta = \frac{O}{A}$$

$$\tan 21° = \frac{29 \text{ cm}}{A}$$

$$0.384 = \frac{29}{A}$$

$$0.384 \times A = \frac{29}{A} \times A \ (\times A)$$

$$\frac{0.384 \times A}{0.384} = \frac{29}{0.384} \ (\div 0.384)$$

$$A = \frac{29}{0.384}$$

$$A = 75.5 \text{ cm}$$

Now that O and A are known, we can calculate H.

$$C \text{ (or } H) = \sqrt{A^2 + B^2}$$
$$= \sqrt{75.5^2 - 29^2}$$
$$= \sqrt{5700.25 + 841}$$
$$= \sqrt{6541.25}$$
$$= 80.88 \text{ cm}$$

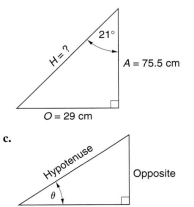

c.

Known values	Unknown values
$H = 34$ volts	$\theta = ?$
$A = 18.5$ volts	$O = ?$

Because both H and A are known, we can use cosine to calculate θ (*CAH*).

$$\cos \theta = \frac{A}{H}$$

$$\cos \theta = \frac{18.5 \text{ V}}{34 \text{ V}}$$

$$\cos \theta = 0.544$$

$$\cos \times \text{invcos } \theta = \text{invcos } 0.544 \ (\times \text{invcos})$$

$$\theta = \text{invcos } 0.544$$

$$\theta = 57°$$

To calculate the length of the unknown side, we can use the Pythagorean theorem.

$$B = \sqrt{C^2 - A^2}$$
$$= \sqrt{34^2 - 18.5^2}$$
$$= \sqrt{813.75}$$
$$= 28.5 \text{ V}$$

57. Frequency = 1/time: **a.** 1/16 ms = 62.5 Hz; **b.** 1/1 s = 1 Hz; **c.** 1/15 μs = 66.67 kHz; **d.** 1/0.05s = 20 Hz; **e.** 1/200 μs = 5 kHz; **f.** 1/350 ms = 2.86 Hz

59. **a.** Peak = 1.414 × rms = 1.414 × 40 mA = 56.56 mA
b. Peak to peak = 2 × peak = 2 × 56.56 mA = 113.12 mA
c. Average = 0.637 × peak = 0.637 × 56.56 mA = 36 mA

61. Duty cycle % = $(P_w/t) \times 100$, $t = 1/f = 1/10$ kHz = 100 μs; duty cycle = (10 μs/100 μs) × 100 = 10%

63. I_{avg} = baseline + (duty cycle × I_p), duty cycle = $P_w/t \times 100$, $t = 1/f = 1/10$ kHz = 100 μs, duty cycle = 10 μs/100 μs × 100 = 10%, $I_{avg} = 0 + (0.1 \times 15$ A) = 1.5 A

65. **a.** Third harmonic = 3 × fundamental = 3 × 1 kHz = 3 kHz
b. Second harmonic = 2 × fundamental = 2 × 1 kHz = 2 kHz
c. Seventh harmonic = 7 × fundamental = 7 × 1 kHz = 7 kHz

67. $V_{pk-pk} = 2 \times V_{peak}$, $V_{peak} = V_{rms} \times 1.414 = 6 \times 1.414 = 8.484$; $V_{pk-pk} = 2 \times 8.484 = 16.968$, 2 V/cm × 8 would only show 16 V, so 5 V/cm is lowest setting. $t = 1/f = 1/350$ kHz = 2.857 μs; 0.2 μs/cm × 10 would only show 2 μs, so 0.5 μs/cm is lowest setting.

69. $t = 5.5$ cm × 1 μs/cm = 5.5 μs; frequency = $1/t = 1/5.5$ μs = 181.818 kHz

Chapter 9

1. b	**7.** d	**13.** d	**19.** c
3. a	**9.** b	**15.** d	
5. c	**11.** a	**17.** b	

(The answers to Communication Skill Questions 21 through 39 can be found in the sections indicated that follow the questions.)

41. Field strength = V/d = 6 V/32 μm = 187.5 kV/m

43. $C = \dfrac{(8.85 \times 10^{-12}) \times k \times A}{d} =$

$\dfrac{(8.85 \times 10^{-12}) \times 2.5 \times 0.008 \text{ m}^2}{0.00095 \text{ m}} = 186.3$ pF

45. $C_T = \dfrac{1}{(1/C_1) + (1/C_2) + (1/C_3)} =$

$\dfrac{1}{(1/0.025 \text{ μF}) + (1/0.04 \text{ μF}) + (1/0.037 \text{ μF})} = 0.0109$ μF

$V_{C_1} = \dfrac{C_T}{C_1} \times V_T = \dfrac{0.0109 \text{ μF}}{0.025 \text{ μF}} \times 12$ V = 5.2 V

$V_{C_2} = \dfrac{C_T}{C_2} \times V_T = \dfrac{0.0109 \text{ μF}}{0.04 \text{ μF}} \times 12$ V = 3.27 V

$V_{C_3} = \dfrac{C_T}{C_3} \times V_T = \dfrac{0.0109 \text{ μF}}{0.037 \text{ μF}} \times 12$ V = 3.53 V

47. **a.** 47 F
b. 34×10^{-7} F or 3.4 μF

49. **a.** V_{1TC} = 63.2% of V_S = 0.632 × 10 V = 6.32 V; 5TC = 5 × 84 ms = 420 ms
b. V_{1TC} = 63.2% of V_S = 0.632 × 10 V = 6.32 V; 5TC = 5 × 16.8 ms = 84 ms
c. V_{1TC} = 63.2% of V_S = 0.632 × 10 V = 6.32 V; 5TC = 5 × 4.08 ms = 20.4 ms
d. V_{1TC} = 63.2% of V_S = 0.632 × 10 V = 6.32 V; 5TC = 5 × 980 μs = 4.9 ms

Chapter 10

1. b	**5.** d	**9.** d	**13.** e
3. b	**7.** b	**11.** a	**15.** b

(The answers to Communication Skill Questions 16 through 25 can be found in the sections indicated that follow the questions.)

27. $V_S = \sqrt{V_R^2 + V_C^2} = \sqrt{6^2 + 12^2} = \sqrt{36 + 144} = \sqrt{180} = 13.4$ V

29. **a.** $I_R = V/R$ = 12 V/4 MΩ = 3 μA
b. $I_C = V/X_C$ = 12 V/1.3 kΩ = 9.23 mA
c. $I_T = \sqrt{I_R^2 + I_C^2} = \sqrt{(3 \text{ μA})^2 + (9.23 \text{ mA})^2}$ = 9.23 mA
d. $Z = V/I_T$ = 12/9.23 mA = 1.3 kΩ
e. θ = arctan (R/X_C) = arctan (4 MΩ/1.3 kΩ) = 89.98°

31. $C = \dfrac{1}{2\pi f X_C} = \dfrac{1}{2\pi(20 \text{ kHz}) 10 \text{ kΩ}} = 795.8$ pF

33.

FIGURE Ch. 10.33(a)

b. $Z = \sqrt{R^2 + X_C^2} = \sqrt{40^2 + 33^2} = 51.9$ Ω
$I = V/Z$ = 24 V/51.9 Ω = 462.4 mA
$V_R = I \times R$ = 462.4 mA × 40 Ω = 18.496 V
$V_C = I \times X_C$ = 462.4 mA × 33 Ω = 15.2592 V
$I_R = I_C = I_T$ = 462.4 mA
θ = arctan (X_C/R) = arctan (33 Ω/40 Ω) 39.5°

35. **a.** $Z = \sqrt{R^2 + X_C^2} = \sqrt{1 \text{ M}^2 + 2.5 \text{ M}^2} = 2.69$ MΩ
$I = V/Z$ = 12/2.69 MΩ = 4.5 μA
$V_R = I \times R$ = 4.5 μA × 1 MΩ = 4.5 V
$V_C = I \times X_C$ = 4.5 μA × 2.5 MΩ = 11.25 V
b. $Z = \sqrt{R^2 + X_C^2} = \sqrt{300^2 + 200^2} = 360.6$ Ω
$I = V/Z$ = 50 V/360.6 Ω = 138.7 mA
$V_R = I \times R$ = 138.7 mA × 300 Ω = 41.61 V
$V_C = I \times R$ = 138.7 mA × 200 Ω = 27.74 V

37. Lag, 90

39. $X_C = \dfrac{1}{2\pi fc} = \dfrac{1}{2\pi(35 \text{ kHz}) 10 \text{ μF}} = 455$ mΩ.
$Z = \sqrt{R^2 + X_C^2} = \sqrt{100 \text{ kΩ}^2 + 455 \text{ mΩ}^2} = 100$ kΩ.
$I = V/Z$ = 24 V/100 kΩ = 240 μA
True power = $I^2 \times R$ = 240 μA^2 × 100 kΩ = 5.76 mW
Reactive power = $I^2 \times X_C$ = 240 μA^2 × 455 mΩ = 26 nVAR
Apparent power = $\sqrt{P_R^2 + P_X^2}$ = 5.76 mVA
Power factor = R/Z = 100 kΩ/100 kΩ = 1

41. Short: contact from plate to plate
Open: lead disconnected from its plate
Breakdown: deterioration of dielectric

43. Capacitor value change, capacitor leakage: leakage current will flow between plates.
Dielectric absorption: will not fully discharge, leaving residual charge.
Equivalent series resistance: resistance of leads, lead to plate connection, electrolyte causing effective circuit capacitance value to change.

45. Shorted, Open

Chapter 11

1. a	**11.** b	**19.** a	**27.** 1
3. b	**13.** d	**21.** c	**29.** b
5. c	**15.** c	**23.** 1	**31.** d
7. b	**17.** a	**25.** d	**33.** d
9. c			

(The answers to Communication Skill Questions 35 through 55 can be found in the sections indicated that follow the questions.)

57. mmf = $I \times N$; 760 mA × 25 = 19 A · t

59. μ = relative $\mu \times \mu^0$

   a. $90 \times 4\pi \times 10^{-7} = 1.13 \times 10^{-4}$ H/m

   b. $50 \times 4\pi \times 10^{-7} = 6.28 \times 10^{-5}$ H/m

   c. $450 \times 4\pi \times 10^{-7} = 5.65 \times 10^{-4}$ H/m

61. mmf = $I \times N = 4 \times 50 = 200$ ampere-turns

63. mmf = $I \times N$; $I = V/R = 9/23 = 391$ mA; 391 mA $\times 50$
    = 19.55 ampere-turns

65. H = mmf/l = 19.55/0.7 = 27.9 ampere-turns/meter

Chapter 12

1. c   **9.** a   **17.** a   **25.** a

3. b   **11.** d   **19.** b

5. a   **13.** e   **21.** b

7. d   **15.** d   **23.** a

(The answers to Communication Skill Questions 26 through 35 can be found in the sections indicated that follow the questions.)

37. a. 22 MΩ

   b. 78.6 kΩ

   c. 314.2 kΩ

39. a. $L_T = L_1 + L_2 + L_3 = 75\ \mu\text{H} + 61\ \mu\text{H} + 50\ \text{mH}$
     $= 50.136$ mH

   b. $L_T = L_1 + L_2 + L_3 = 8\ \text{mH} + 4\ \text{mH} + 22\ \text{mH} = 34\ \text{mH}$

41. a. 4.36 mH

   b. 4.7 μH

   c. 3.82 μH

43. a. $V_S = \sqrt{V_R^2 + V_L^2} = \sqrt{12^2 + 6^2} = \sqrt{144 + 36}$
     $= \sqrt{180} = 13.4$ V

   b. $I = V_S/Z = 13.4$ V/14 kΩ = 957.1 μA

   c. $\angle = \arctan V_L/V_R = \arctan 2 = 63.4°$.

   d. $Q = V_L/V_R = 12/6 = 2$.

   e. PF = $\cos\theta = 0.448$.

45. $f = X_L/2\pi L = 27$ kΩ/$2\pi 330$ μH = 13.02 MHz

47. $\tau = \dfrac{L}{R} = \dfrac{400\ \text{mH}}{2\ \text{k}\Omega} = 200\ \mu\text{s}$

V_L will start at 12 V and then exponentially drop to 0 V

Time	Factor	V_S	V_L
0	1.0	12	12
1 TC	0.365	12	4.416
2 TC	0.135	12	1.62
3 TC	0.05	12	0.6
4 TC	0.018	12	0.216
5 TC	0.007	12	0.084

49. a. $R_T = R_1 + R_2 = 250 + 700 = 950\ \Omega$

   b. $L_T = L_1 + L_2 = 800\ \mu\text{H} + 1200\ \mu\text{H} = 2\ \text{mH}$

   c. $X_L = 2\pi f L = 2\pi(350\ \text{Hz})\ 2\ \text{mH} = 4.4\ \Omega$

   d. $Z = \dfrac{R \times X_L}{\sqrt{R^2 + X_L^2}} = \dfrac{950 \times 4.4}{\sqrt{950^2 + 4.42^2}} = 4.39$

   e. $V_{R_T} = V_{L_T} = V_S = 20$ V

   f. $I_{R_T} = \dfrac{V_S}{R_T} = \dfrac{20\ \text{V}}{950\ \Omega} = 21$ mA,

     $I_{L_T} = \dfrac{V_{S_1}}{X_{L_T}} = \dfrac{20\ \text{V}}{4.4\ \Omega} = 4.5$ A

   g. $I_T = \sqrt{I_R^2 + I_L^2} = \sqrt{21\ \text{mA}^2 + 4.5\ \text{A}^2} = 4.5$ A

   h. $\theta = \arctan(R/X_L) = \arctan 950/4.4 = 89.7°$

   i. $P_R = I^2 \times R = 21\ \text{mA}^2 \times 950 = 418.95$ mW
     $P_X = I^2 \times X_L = 4.5\ \text{A}^2 \times 4.4 = 89.1$ VAR
     $P_A = \sqrt{P_R^2 + P_X^2} = \sqrt{418.95\ \text{mW}^2 + 89.1\ \text{A}^2}$
     $= 89.1$ VA

   j. PF = $P_R/P_A = 418.95$ mW/89.1 W = 0.0047

Chapter 13

1. d   **5.** d   **9.** a

3. c   **7.** a

(The answers to Communication Skill Questions 11 through 25 can be found in the sections indicated that follow the questions.)

27. a. $V_s = \dfrac{N_s}{N_P} \times V_P = 24/12 \times 100$ V = 200 V

   b. $V_s = \dfrac{N_s}{N_P} \times V_P = 250/3 \times 100$ V = 8.33 kV

   c. $V_s = \dfrac{N_s}{N_P} \times V_P = 5/24 \times 100$ V = 20.83 V

   d. $V_s = \dfrac{N_s}{N_P} \times V_P = 120/240 \times 100$ V = 50 V

29. Turns ratio = $\sqrt{Z_L/Z_S} = \sqrt{8\ \Omega/24\ \Omega} = \sqrt{1/3} = 0.58$

31. For 16 turns: $V_s = (N_s/N_p) \times V_p = (16/12) \times 24$ V = 32 V
    For 2 turns: $V_s = (N_s/N_p) \times V_p = (2/12) \times 24$ V = 4 V
    For 1 turn: $V_s = (N_s/N_p) \times V_p = (1/12) \times 24$ V = 2 V
    For 4 turns: $V_s = (N_s/N_p) \times V_p = (4/12) \times 24$ V = 8 V

33. Follow polarity dots.

35. a. I_s = apparent power/V_s = 500 VA/600 V = 833.3 mA

   b. $R_L = V_s/I_s = 600$ V/833.3 mA = 720 Ω

Chapter 14

1. b   **5.** d   **9.** c   **13.** a

3. b   **7.** c   **11.** a   **15.** d

(The answers to Communication Skill Questions 16 through 30 can be found in the sections indicated that follow the questions.)

31. a. $X_C = \dfrac{1}{2\pi f C} = \dfrac{1}{2\pi(60)0.02\mu\text{F}} = 132.6$ kΩ

   b. $X_C = \dfrac{1}{2\pi f C} = \dfrac{1}{2\pi(60)18\mu\text{F}} = 147.4$ Ω

   c. $X_C = \dfrac{1}{2\pi f C} = \dfrac{1}{2\pi(60)360\ \text{pF}} = 7.37$ MΩ

   d. $X_C = \dfrac{1}{2\pi f C} = \dfrac{1}{2\pi(60)2700\ \text{nF}} = 982.4$ Ω

   e. $X_L = 2\pi f L = 2\pi(60)4\ \text{mH} = 1.5$ Ω

   f. $X_L = 2\pi f L = 2\pi(60)8.18\ \text{H} = 3.08$ kΩ

   g. $X_L = 2\pi f L = 2\pi(60)150\ \text{mH} = 56.5$ Ω

   h. $X_L = 2\pi f L = 2\pi(60)2\ \text{H} = 753.98$ Ω

33. a. $X_L = 2\pi f L = 2\pi(60\ \text{Hz})150\ \text{mH} = 56.5$ Ω

   b. $X_C = \dfrac{1}{2\pi f c} = \dfrac{1}{2\pi(60\ \text{Hz})20\ \mu\text{F}} = 132.6$ Ω

   c. $I_R = V/R = 120$ V/270 Ω = 444.4 mA

   d. $I_L = V/X_L = 120$ V/56.5 Ω = 2.12 A

   e. $I_C = V/X_C = 120$ V/132.6 Ω = 905 mA

   f. $I_T = \sqrt{I_R^2 + I_X^2} = \sqrt{(444.4\ \text{mA})^2 + (1.215)^2} = 1.29$ A

   g. $Z = V/I_T = 120$ V/1.29 A = 93.02 Ω

   h. Resonant frequency = $\dfrac{1}{2\pi\sqrt{LC}}$

     $= \dfrac{1}{2\pi\sqrt{150\ \text{mH} \times 20\ \mu\text{F}}}$
     $= 91.89$ Hz

   i. $X_L = 2\pi f L$
     $= 6.28 \times 91.89\ \text{Hz} \times 150\ \text{mH}$
     $= 86.54$ Ω
     $Q = \dfrac{X_L}{R} = \dfrac{86.54\ \Omega}{270\ \Omega} = 0.5769$

   j. BW = $\dfrac{f_0}{Q} = \dfrac{91.89\ \text{Hz}}{0.5769} = 159.28$ Hz

35. Using a source voltage of 1 volt:

$Z = V/I_T$, $I_T = \sqrt{I_R^2 + I_X^2}$, $I_R = V/R = 1/750 = 1.33$ mA

$I_L = V/X_L = 1/25 = 40$ mA, $I_C = V/X_C = 1/160 = 6.25$ mA

$I_X = I_L - I_C = 40$ mA $- 6.25$ mA $= 33.75$ mA

$I_T = \sqrt{(1.33\text{ mA})^2 + (33.75\text{ mA})^2} = 33.78$ mA

$Z = 1/33.78$ mA $= 29.6\ \Omega$

37. a. Real number $= 25 \cos 37° = 19.97$; imaginary number $= 25 \sin 37° = 15$, $19.97 + j15$

  b. Real number $= 19 \cos -20° = 17.9$; imaginary number $= 19 \sin -20° = -6.5$, $17.9 - j6.5$

  c. Real number $= 114 \cos -114° = -46.4$; imaginary number $= 114 \sin -114° = -104.1$, $-46.4 - j104.1$*

  d. Real number $= 59 \cos 99° = +9.2$; imaginary number $= 59 \sin 99° = 58.3$, $+9.2 + j58.3$

39. a. $(4 + j3) + (3 + j2) = (4 + 3) + (j3 + j2) = 7 + j5$

  b. $(100 - j50) + (12 + j9) = (100 + 12) + (-j50 + j9) = 112 - j41$

41. a. $Z_T = 73 - j23$, $\sqrt{73^2 + 23^2} = 76.5$, $\angle = \arctan(23/73) = -17.5°$, $76.5 \angle -17.5°$; $Z_T = 76.5\ \Omega$ at $-17.5°$ phase angle

  b. $Z_T = 40 + j15$, $\sqrt{40^2 + 15^2} = 42.7$, $\angle = \arctan(15/40) = 20.6°$, $42.7 \angle 20.6°$; $Z_T = 42.7\ \Omega$ at $20.6°$ phase angle

  c. $Z_T = 8\text{ k}\Omega - j3\text{ k}\Omega + j20\text{ k}\Omega = 8\text{ k}\Omega + 17\text{ k}\Omega$, $\sqrt{8\text{ k}\Omega^2 + 17\text{ k}\Omega^2} = 18.8\text{ k}\Omega$, $\angle = \arctan(17\text{ k}\Omega/8\text{ k}\Omega) = 64.8°$, $18.8\text{ k}\Omega \angle 64.8°$; $Z_T = 18.8\text{ k}\Omega$ at $64.8°$ phase angle

43. a. $Z_T = 47 - j40 + j30 = 47 - j10$, $\sqrt{47^2 + 10^2} = 48.05$, $\angle = \arctan(-10/47) = -12°$, $Z_T = 48.05 \angle -12°$

  b. $I = \dfrac{V_S}{Z_T} = \dfrac{20 \angle 0°}{48.05 \angle -12} = \dfrac{20}{48.05 \angle 0 - (-12)°} = 416.2$ mA $\angle 12°$

  c. $V_R = I \times R = 416.2$ mA $\angle 12° \times 47 \angle 0 = 416.2$ mA $\times 47 \angle (12 + 0) = 19.56$ V $\angle 12°$

  $V_C = I \times X_C = 416.2$ mA $\angle 12° \times 40 \angle -90° = 416.2$ mA $\times 40 \angle (12° - 90°) = 16.65 \angle -78°$

  $V_L = I \times X_L = 416.2$ mA $\angle 12° \times 30 \angle 90° = 416.2$ mA $\times 30 \angle (12° + 90°) = 12.49$ V $\angle 102°$

  d. V_C lags I_T by 90°, V_L leads I by 90°, V_R is in phase with I.

45. a. $Z_1 = 0 + j27 - j17 = j10$, $\sqrt{0^2 + 10^2} = 10 \angle = 90°$, $Z_1 = 10 \angle 90°$; $Z_2 = 37 - j20 = 42.06 \angle -28.39°$; Z combined $=$ Product/Sum; Product $= 420.59 \angle 61.61°$; Sum $= 37 - j10 = 38.33 \angle -15.12°$; Z combined $= 10.97 \angle 76.73°$

  b. $I_1 = 3.72 \angle -34.48$ A; $I_2 = 884 \angle 83.91$ mA

  c. $I_T = 3.39 \angle -21.21°$ A

  d. $Z_T = 29.52 \angle 21.21°\ \Omega$

47. $V_R = 19.56 \angle 12° = 19.56 \cos 12° + j19.56 \sin 12° = 19.13 + j4.07$

  $V_C = 16.65 \angle -78° = 16.65 \cos -78° + j16.65 \sin -78° = 3.46 - j16.29$

  $V_L = 12.49 \angle 102° = 12.49 \cos 102° + j12.49 \sin 102° = -2.6 + j12.22$

49. Easier with use of complex numbers

Chapter 15

1. a
3. d
5. d
7. d
9. b
11. b
13. b
15. c

*These examples were for practice purposes only; real numbers are always positive when they represent impedances.

(The answers to Communication Skill Questions 16 through 29 can be found in the sections indicated that follow the questions.)

31. a. P-N junction is forward biased. $I = V_S - V_{P\text{-}N}/R = 5$ V $- 0.7$ V$/330\ \Omega = 13$ mA.

  b. P-N junction is forward biased. $I = V_S - V_{P\text{-}N}/R = 15$ V $- 0.3$ V$/15\ \Omega = 980$ µA.

  c. P-N junction is reverse biased. $I = 0$ A.

33. a. $V_{R_1} = V_S - V_{P\text{-}N} = 5$ V $- 0.7$ V $= 4.3$ V

  b. $V_{R_1} = V_S - V_{P\text{-}N} = 15$ V $- 0.3$ V $= 14.7$ V

  c. Since the P-N junction is open, or reverse biased, all of the applied voltage will appear across this series open. Therefore, the voltage across $R_1 = 0$ V.

Chapter 16

1. c	**11.** a	**21.** b	**31.** b	**41.** b
3. d	**13.** a	**23.** d	**33.** d	**43.** d
5. d	**15.** a	**25.** b	**35.** b	**45.** c
7. a	**17.** a	**27.** a	**37.** d	
9. b	**19.** a	**29.** b	**39.** a	

(The answers to Communication Skill Questions 46 through 77 can be found in the sections indicated that follow the questions.)

79. a. Since D_1 is reverse biased, I_F will be 0.

  b. $I = V_S - V_{\text{diode}}/R$
  $I = 15$ V $- 0.7$ V$/2.5$ K$\Omega + 127$ KΩ
  $I = 14.3$ V$/129.5$ KΩ
  $I = 110.4$ µA

81. a. Since the diode is reverse biased and therefore equivalent to an open switch, all of the applied voltage will appear across the open diode ($V_{\text{diode}} = V_S = 10$ V). The voltage drop across the resistor will therefore be 0 V ($V_R = 0$ V).

  b. With 15 V being applied, and a 0.7 V drop across the diode, 14.3 V is being proportionally developed across R_1 and R_2. Since circuit current is known, we can calculate the individual voltage drops by using Ohms law:
  $V_{R_1} = I \times R_1$
  $V_{R_1} = 110.4$ µA $\times 2.5$ k$\Omega = 0.276$ V
  $V_{R_2} = I \times R_2$
  $V_{R_2} = 110.4$ µA $\times 127$ k$\Omega = 14.021$ V

83. a. Polarity correct, magnitude correct (10 V is larger than 6.8 V)

  b. Polarity incorrect, magnitude correct

  c. Polarity incorrect, magnitude correct

  d. Polarity correct, magnitude incorrect

  e. Polarity correct, magnitude correct

  f. $D_1 = $ Polarity correct, magnitude correct

  g. $D_2 = $ Polarity correct, magnitude incorrect

85. In question 43(a) we determined that $I = 160$ mA for the circuit in Figure 16.18(a). Since $V_Z = 6.8$ V, the zener will be dissipating;
  $I_{ZM} = P_D/V_Z$, therefore
  $P_D = I_{ZM} \times V_Z = 160$ mA $\times 6.8$ V $- 1.088$ *Watts*
  The 1 W maximum power dissipation rating is therefore being exceeded and could result in the zener diode burning out. A 1.25 W zener would be better in this application.

87. a. The voltage at point $x = +5.6$ V

  b. The voltage at point $y = -12$ V

89. a. $I_S = V_{in} - V_Z/R_S = 10$ V $- 5.6$ V$/130\ \Omega = 33.8$ mA

  b. $I_S = V_{in} - V_Z/R_S = 20$ V $- 12$ V$/450\ \Omega = 17.8$ mA

91. $V_{in} = 12$ V, $V_Z = 6.8$ V,
  $V_{RS} = V_{in} - V_Z = 12$ V $- 6.8$ V $= 5.2$ V
  $I_{RS} = V_{RS}/R_S = 5.2$ V$/120\ \Omega = 43.3$ mA
  $R_L = 500\ \Omega$
  $I_L = V_L/R_L = 6.8$ V$/500\ \Omega = 13.6$ mA
  $I_Z = I_S - I_L = 43.3$ mA $- 13.6$ mA $= 29.7$ mA
  $R_L = 600\ \Omega$

$I_L = V_L/R_L = 6.8 \text{ V}/600 \ \Omega = 11.3 \text{ mA}$
$I_Z = I_S - I_L = 43.3 \text{ mA} - 11.3 \text{ mA} = 32 \text{ mA}$
$V_{in} = 15 \text{ V}, V_Z = 6.8 \text{ V},$
$V_{RS} = V_{in} - V_Z = 15 \text{ V} - 6.8 \text{ V} = 8.2 \text{ V}$
$I_{RS} = V_{RS}/R_S = 8.2 \text{ V}/120 \ \Omega = 68.3 \text{ mA}$
$R_L = 500 \ \Omega$
$I_L = V_L/R_L = 6.8 \text{ V}/500 \ \Omega = 13.6 \text{ mA}$
$I_Z = I_S - I_L = 68.3 \text{ mA} - 13.6 \text{ mA} = 54.7 \text{ mA}$
$R_L = 600 \ \Omega$
$I_L = V_L/R_L = 6.8 \text{ V}/600 \ \Omega = 11.3 \text{ mA}$
$I_Z = I_S - I_L = 68.3 \text{ mA} - 11.3 \text{ mA} = 57 \text{ mA}$

93. a. $I_S = V_S - V_{LED}/R_S = 5 \text{ V} - 2 \text{ V}/250 \ \Omega = 12 \text{ mA}$
  b. $I_S = V_S - V_{LED}/R_S = 5 \text{ V} - 2 \text{ V}/700 \ \Omega = 4.3 \text{ mA}$
  c. Since D_1 is reverse biased, $I =$ zero.
   For D_2, $I_S = V_S - V_{LED}/R_S = 6 \text{ V} - 2 \text{ V}/368 \ \Omega$
   $= 10.9 \text{ mA}$

95. No
97. $V_{R1} = V_{in} - V_{D1} - V_{D3} = 5 \text{ V} - 0.7 \text{ V} - 2 \text{ V} = 2.3 \text{ V}$
  $I_{R1} = I_{LED} = V_{R1}/R_1 = 2.3 \text{ V}/100 \ \Omega = 23 \text{ mA}$
99. The red LED will be more bright because the -10 V input will forward bias D_1 and therefore bypass the additional current limiting provided by R_1.
101. 90% of $I_{FMAX} = 90\% \times 40 \text{ mA} = 36 \text{ mA}$
  $R_S = V_{Source} - V_F/I_F = 12 \text{ V} - 2.5 \text{ V}/36 \text{ mA} = 263.9 \ \Omega$
  Since the closest standard-resistor (Appendix E Table 1) value is 262 V ($\pm 5\%$), current will be equal to
  $I_S = V_S - V_{LED}/R_S = 12 \text{ V} - 2.5 \text{ V}/262 \ \Omega = 36.26 \text{ mA}$
103. HI output from decoder will turn ON LED.

ROTARY SWITCH	A	B	C	D	D_1	D_2	D_3	D_4
1	LO	LO	LO	LO	OFF	OFF	OFF	OFF
2	HI	LO	LO	HI	ON	OFF	OFF	ON
3	HI	LO	HI	LO	ON	OFF	ON	OFF
4	LO	LO	HI	HI	OFF	OFF	ON	ON

105. Figure 16–27 = common-anode
  Figure 16–28 = common-cathode
107. $V_p = V_{rms} \times 1.414 = 240 \text{ V} \times 1.414 = 339.4 \ V_{peak}$
  $V_S = N_S/N_P \times V_P = 1/19 \times 339.4 \text{ V} = 17.86 \ V_{peak}$

109. DC ripple frequency out = AC ripple frequency in = 60 Hz
111. Full wave output ripple frequency = $2 \times$ Input Freq. = $2 \times 60 \text{ Hz} = 120 \text{ Hz}$
113. Peak to Peak of ripple = 15 V to 21 V = 6 V pk-pk
  Peak of ripple = 1/2 of pk-pk value = 1/2 of 6 V = 3 V
  rms of ripple = $0.707 \times 3 \text{ V} = 2.12 \text{ V}$
  % of Ripple = V_{rms} of ripple/V_{avg} of ripple $\times 100$
  = $2.12 \text{ V}/18 \text{ V} \times 100 = 11.78\%$
115.

SWITCH POSITION	A	B	C
1.	Logic 1	Logic 0	Logic 1
2.	Logic 0	Logic 1	Logic 0
3.	Logic 0	Logic 0	Logic 0

  Logic 0 = 0 V, Logic 1 = +5 V

  a. With D_1 open, the code generated in position 1 will change since the output line C cannot be pulled low. Therefore, the code generated when the switch is in position 1 with D_1 permanently open will be:

SWITCH POSITION	A	B	C
1.	Logic 1	Logic 0	Logic 1
2.	Logic 0	Logic 1	Logic 0
3.	Logic 0	Logic 0	Logic 0

  Output line C will give a high in switch position 1.

b. Yes, since the maximum operation current will be 25.8 mA when the switch is in position 3 and D_5 D_6 and D_7 are on. Each parallel branch in this case will have the following value of current;
  $I = V_S - V_{Diode}/R$
  $I = 5 \text{ V} - 0.7 \text{ V}/500 \ \Omega$
  $I = 8.6 \text{ mA}$
  Since there are three branches, the total current will be $3 \times 8.6 \text{ mA} = 25.8 \text{ mA}$.
  A 200 mA fuse will therefore blow whenever the switch is put in position 3, because a current of 25.8 mA will be drawn.
117. Section 16-3-2
119. Section 16-5

Chapter 17

1. d	**7.** b	**13.** d	**19.** d	
3. a	**9.** a	**15.** a		
5. b	**11.** d	**17.** c		

(The answers to Communication Skill Questions 21 through 31 can be found in the sections indicated that follow the questions.)

33. a. NPN transistor is correctly biased. Base is positive relative to emitter (emitter diode ON), collector diode is positive relative to base (collector diode OFF).
  b. NPN transistor is incorrectly biased. Base is positive relative to emitter (emitter diode ON), collector diode is negative relative to base (collector diode ON).
  c. PNP transistor is correctly biased. Base is negative relative to emitter (emitter diode ON), collector diode is negative relative to base (collector diode OFF).
  d. PNP transistor is incorrectly biased. Base is positive relative to emitter (emitter diode OFF), collector diode is positive relative to base (collector diode ON).
35. $A_V = \Delta V_{out}/\Delta V_{in} = 11 \text{ V} - 3 \text{ V}/3.25 \text{ V} - 2.75 \text{ V} = 8 \text{ V}/0.5 \text{ V} = 16$
  The output voltage is 16 times greater than the input voltage.
37. a. NPN, base biasing.
  b. NPN, base biasing.
  c. NPN, voltage-divider biasing.
  d. PNP, base biasing.
39. Saturation Point, $I_{C(Sat.)} = V_{CC}/R_C = 20 \text{ V}/10 \text{ k}\Omega = 2 \text{ mA}$
  Q Point, $I_C = 1.16 \text{ mA}, V_{CE} = 8.4 \text{ V}$
  Cutoff Point, $V_{CE(Cutoff)} = V_{CC} = 20 \text{ V}$

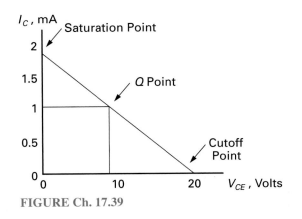

FIGURE Ch. 17.39

41. Saturation Point, $I_{C(Sat.)} = V_{CC}/R_C + R_E = 20$ V/4.7 kΩ + 1.1 kΩ = 3.45 mA

Q Point, $I_C = 1.7$ mA, $V_{CE} = 10.14$ V

Cutoff Point, $V_{CE(Cutoff)} = V_{CC} = 20$ V

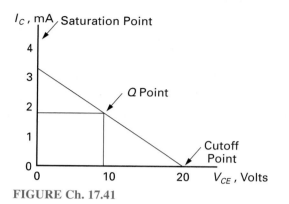

FIGURE Ch. 17.41

43. 21-2-4

45. a. Transistor's diodes both test OK.
  b. Collector diode is open, collector diode is OK—replace transistor.
  c. Collector diode is shorted, emitter diode is OK—replace transistor.
  d. Emitter diode is shorted, collector diode is OK—replace transistor.
  e. Transistor's diodes both test OK.

Chapter 18

1. b	**7.** d	**13.** a	**19.** b
3. b	**9.** b	**15.** a	
5. b	**11.** b	**17.** a	

(The answers to Communication Skill Questions 21 through 35 can be found in the indicated sections that follow the questions.)

37. a. $V_{GS} = V_{CC} = -1$ V
  b. $I_D = I_{DSS}(1 - V_{GS}/V_{GS(OFF)})^2 = 6$ mA$(1 - -1$ V/-2 V$)^2 = 1.5$ mA
  c. $V_{DS} = V_{DD} - (I_D \times R_D) = 10$ V $- (1.5$ mA $\times 3$ k$\Omega) = 10$ V $- 4.5$ V $= 5.5$ V
  d. $I_{D(maximum)} = I_{DSS} = 6$ mA
  e. Q point, $V_{GS} = -1$ V, $I_D = 1.5$ mA, $V_{DS} = 5.5$ V

39. a. $V_G = R_2/R_1 + R_2 \times V_{DD} = 1$ MΩ/3 MΩ + 1 MΩ $\times$ 12 V $= 3$ V
  b. $V_S = V_{RS} = I_S \times R_S = 2$ mA $\times 3$ k$\Omega = 6$ V
  c. $V_{GS} = V_G - V_S = 3$ V $- 6$ V $= -3$ V
  d. $V_{DS} = V_{DD} - (V_{RS} + V_{RD}) = 12$ V $- [6$ V $+ (I_D \times R_D)] = 12$ V $- [6$ V $+ (2$ mA $\times 1.5$ k$\Omega)] = 12$ V $- (6$ V $+ 3$ V$) = 3$ V

41. a. $I_D = I_{D(ON)} = 8$ mA
  b. $V_{DS} = V_{DD} - (I_{D(ON)} \times R_D)$
    $= 14$ V $- (8$ mA $\times 680$ $\Omega)$
    $= 14$ V $- 5.44$ V $= 8.56$ V

43. Section 22-1-10

45. a. Bad; gate-source short, should be infinite Ω.
  b. Good; gate-source and gate-drain should be infinite Ω, and since unbiased E-MOSFETs do not have a channel between source and drain this should also measure infinite Ω.

Chapter 19

1. b	**7.** c
3. d	**9.** e
5. a	

(The answers to Communication Skill Questions 11 through 22 can be found in the indicated sections that follow the questions.)

23. A single polarity supply (+5 V) inverting op-amp. The inverting amplifier would gave to have a negative input because the output can only be between 0 V (−V supply voltage) and +5 V (+V supply voltage).

25. a. Differentiator (see waveforms in Figure 19-13)
  b. Integrator (see waveforms in Figure 19-14)

27. a. Summing amplifier
    $V_{out} = -(V_{in1} + V_{in2}) = -(3$ V $+ 6$ V$) = -9$ V
  b. Difference amplifier
    $V_{out} = V_{in2} - V_{in1} = 10$ V $- 6.5$ V $= 3.5$ V

29. $f_C = 1/2\pi RC = 1/2\pi \times 10$ k$\Omega \times 0.1$ μF $= 159.2$ Hz

Chapter 20

1. a	**5.** b	**9.** a	**13.** d
3. c	**7.** e	**11.** e	**15.** c

(The answers to Communication Skill Questions 17 through 42 can be found in the indicated sections that follow the questions.)

43. A square-wave astable multivibrator circuit.

45. A monostable multivibrator circuit

47. Ratio = Output/Input = 1.4 ms or 1400 μs/100 μs = 14
The output is 14 times longer than the input.

49. A 555 timer IC connected as an astable multivibrator circuit

51. Duty cycle = $P_W t \times 100\%$ = 131.6 ms./197.4 ms $\times 100\%$
    $= 66.67\%$

53. a. DIAC
  b. SCR
  c. LED or ILD
  d. Photodiode
  e. Photoresistor
  f. Solar cell
  g. Phototransistor
  h. LASCR

55. A LOW input will turn OFF Q_1, causing its collector output to go HIGH, which will trigger the SCR ON, and connect power to the relay's energizing coil

57. Monostable

59. Because the TRIAC can be used to control the full ac cycle.

61. The SCR was used to connect dc power to a load, while the TRIAC was used to connect ac power to a load.

63. Photoresistor

65. As the light level changes, the resistance of the photoresistor changes, changing the oscillator's RC frequency-determining value, and therefore changing the frequency at the output of the 3909 being applied to the speaker.

67. To power a constant-frequency tone generator circuit using solar cells.

69. Phototransistor

71. When light is present, phototransistor is ON, NPN transistor is OFF, relay is de-energized, normally closed contacts are closed, normally open contacts are open. When light is not present, phototransistor is OFF, NPN transistor is ON, relay is energized, normally closed contacts open, normally open contacts closed.

73. To have a bargraph display represent the level of ambient light present.

75. To control the dc output offset of the op-amp

77. When a voltage of 2 V is dropped across a resistor, its associated LED will turn ON. The larger value resistors in the bargraph voltage divider will develop a larger voltage drop. This means that the resistor at the bottom of the rung will turn ON first, when only a small voltage is applied from the 741. As the light level increases, the voltage out of the 741 increases, the voltage drop across all of the resistors increases proportionally, and the LEDs turn ON in order from the bottom up.

Index